AF750621

Birkhäuser

Advances in Mathematical Fluid Mechanics

Series Editors

Giovanni P. Galdi, Pittsburgh, USA
John G. Heywood, Vancouver, Canada
Rolf Rannacher, Heidelberg, Germany

Advances in Mathematical Fluid Mechanics is a forum for the publication of high quality monographs, or collections of works, on the mathematical theory of fluid mechanics, with special regards to the Navier-Stokes equations. Its mathematical aims and scope are similar to those of the *Journal of Mathematical Fluid Mechanics*. In particular, mathematical aspects of computational methods and of applications to science and engineering are welcome as an important part of the theory. So also are works in related areas of mathematics that have a direct bearing on fluid mechanics.

The monographs and collections of works published here may be written in a more expository style than is usual for research journals, with the intention of reaching a wide audience. Collections of review articles will also be sought from time to time.

More information about this series at http://www.springer.com/series/5032

Šárka Nečasová • Milan Pokorný •
Vladimír Šverák
Editors

Selected Works of Jindřich Nečas

PDEs, Continuum Mechanics and Regularity

Birkhäuser

Editors
Šárka Nečasová
Mathematical Institute
Academy of Sciences
Prague
Czech Republic

Milan Pokorný
Faculty of Mathematics and Physics
Charles University in Prague
Prague
Czech Republic

Vladimír Šverák
School of Mathematics
University of Minnesota
Minneapolis
Minnesota
USA

Advances in Mathematical Fluid Mechanics
ISBN 978-3-0348-0230-7 ISBN 978-3-0348-0231-4 (eBook)
DOI 10.1007/978-3-0348-0231-4

Library of Congress Control Number: 2013954387

Mathematics Subject Classification (2010): 35J, 58J10, 8J20, 35Q, 35P, 35J, 35F, 35K, 35Q35, 76H, 74B, 46T, 74C, 76D, 76N

Springer Basel Heidelberg New York Dordrecht London
© Springer Basel 2015
This work is subject to copyright. All rights are reserved by the Publisher, whether the whole or part of the material is concerned, specifically the rights of translation, reprinting, reuse of illustrations, recitation, broadcasting, reproduction on microfilms or in any other physical way, and transmission or information storage and retrieval, electronic adaptation, computer software, or by similar or dissimilar methodology now known or hereafter developed.
The use of general descriptive names, registered names, trademarks, service marks, etc. in this publication does not imply, even in the absence of a specific statement, that such names are exempt from the relevant protective laws and regulations and therefore free for general use.
The publisher, the authors and the editors are safe to assume that the advice and information in this book are believed to be true and accurate at the date of publication. Neither the publisher nor the authors or the editors give a warranty, express or implied, with respect to the material contained herein or for any errors or omissions that may have been made.

Cover design: deblik, Berlin

Printed on acid-free paper

Springer Basel AG is part of Springer Science+Business Media (www.birkhauser-science.com)

Preface

This book contains the most influential works of Jindřich Nečas, one of the founders of the Prague School of Partial Differential Equations and one of its most prominent members in the period 1955–2002. Professor Nečas's work had an impact on several branches of the Theory of PDEs, and the selected papers are organized according to the following areas: linear elliptic operators, regularity of solutions of nonlinear elliptic and parabolic systems, nonlinear functional analysis, continuum mechanics, and fluid mechanics.

Each group of papers is complemented by comments of experts in the corresponding fields, with the aim to put J. Nečas's work into the context of the latest developments in the area. In addition, personal recollections of some of J. Nečas's collaborators are included, as well as a few pieces of J. Nečas's correspondence and some photographs.

J. Nečas's work continues to play an important role in the Theory of PDEs and many of his papers are just as illuminating today as they were at the time they first appeared. We believe it is important to collect these works in one place, and this book is the result of our efforts in this direction.

We are very grateful for the kind support of Professor G.P. Galdi which was essential for the success of this project. We would also like to thank Birkhäuser representatives B. Hellriegel and S. Gasser for their support. The project would not have been possible without help and support from the contributors: I. Babuška, P. Ciarlet, P. Drábek, M. Feistauer, I. Hlaváček, J. Jarušek, O. John, J. Kristensen, A. Kufner, J. Málek, G. Mingione, P. Quittner, T. Roubíček, G. Seregin, and J. Stará. We would like to express our gratitude to all of them. Last but not least, we thank Mrs. Bauerová and Mrs. Švecová whose help in technical and secretarial matters was invaluable.

Prague, Czech Republic
Minneapolis, MN
2013

Š. Nečasová and M. Pokorný
V. Šverák

Contents

Preface ... v

Contents ... vii

List of Contributors ... xiii

Part I Jindřich Nečas

1. **Short Biography of Jindřich Nečas** ... 3
   Šárka Nečasová and Milan Pokorný
2. **Complete Bibliography of Jindřich Nečas** ... 7
3. **Jindřich Nečas in the Early 1960s** ... 21
   Ivo Babuška
4. **Jindřich Nečas and the Prague School of PDEs** ... 23
   Alois Kufner

Part II Linear Operators

5. **Selected Papers on Linear Elliptic Operators** ... 27
   - 5.1 JN-C17 On domaines of type N ... 27
   - 5.2 JN-C28 Sur les normes équivalentes dans $W_p^{(k)}(\Omega)$ et sur la coercitivité des formes formellement positives ... 41
   - 5.3 JN-C23 Sur la coercivité des formes sesquilinéaires, elliptiques ... 68
   - 5.4 JN-C24 L'application de l'égalité de Rellich sur les systèmes elliptiques du deuxième ordre ... 91

5.5 JN-C43 On inequalities of Korn's type. I. Boundary-value problems for elliptic systems of partial differential equations ... 106
Ivan Hlaváček and Jindřich Nečas

5.6 JN-C47 On inequalities of Korn's Type. II. Applications to linear elasticity ... 113
Ivan Hlaváček and Jindřich Nečas

6. Commentary ... 137

6.1 Linear Elliptic Operators ... 137
Alois Kufner

Part III Regularity Theory

7. Selected Papers on Regularity Theory ... 143

7.1 JN-C30 Sur l'appartenance dans la classe $C^{(k),\mu}$ des solutions variationnelles des équations elliptiques non-linéaires de l'ordre 2k en deux dimensions ... 143

7.2 JN-C96 On the regularity of weak solutions to nonlinear elliptic systems via Liouville's type property ... 152
Marianno Giaquinta and Jindřich Nečas

7.3 JN-C100 On the regularity of weak solutions to non linear elliptic systems of partial differential equations ... 163
Marianno Giaquinta and Jindřich Nečas

7.4 JN-C104 Counterexample to the regularity of weak solution of elliptic systems ... 183
Jindřich Nečas, Oldřich John and Jana Stará

7.5 JN-C113 A Liouville theorem for nonlinear elliptic systems with isotropic nonlinearities ... 193
Pierre Louis Lions, Jindřich Nečas and Ivan Netuka

7.6 JN-C114 On the regularity up to the boundary for second order nonlinear elliptic systems ... 204
Marianno Giaquinta, Jindřich Nečas, Oldřich John and Jana Stará

7.7 JN-C156 On regularity of solutions of nonlinear parabolic systems ... 221
Jindřich Nečas and Vladimír Šverák

7.8 JN-C86 Example of an irregular solution to a nonlinear elliptic system with analytic coefficients and conditions for regularity, Theory of Nonlinear Operators 232

8. Commentaries 243

8.1. Regularity for Nonlinear Equations and Systems 243
Jana Stará and Oldřich John

8.2 Sketches of Regularity Theory from the Twentieth Century and the work of Jindřich Nečas 247
Jan Kristensen and Giuseppe Mingione

Part IV Nonlinear Functional Analysis, Calculus of Variations

9. Selected Papers on Nonlinear Functional Analysis and Calculus of Variations 287

9.1 JN-C40 Sur l'alternative de Fredholm pour les opérateurs non-linéaires avec applications aux problèmes aux limites 287

9.2 JN-C53 Fredholm alternative for nonlinear operators and applications to partial differential equations and integral equations 302

9.3 JN-C73 Application of Rothe's method to abstract parabolic equations 309

9.4 JN-C94 On a class of nonlinear initial value problems in Hilbert spaces 314
Konrad Gröger and Jindřich Nečas

9.5 JN-C130 On the construction of Lusternik-Schnirelmann critical values with application to bifurcation problems 325
Jindřich Nečas, Ari Lehtonen and Pekka Neittaanmäki

10. Commentary 339

10.1 Nonlinear Functional Analysis, Calculus of Variations 339
Pavel Drábek

Part V Solid Mechanics, Elasticity, Thermoelasticity, and Plasticity

11. Selected Papers on Solid Mechanics, Elasticity, Thermoelasticity, and Plasticity 345

11.1 JN-C95 Dynamic deformation processes of elastic-plastic systems 345
Konrad Gröger, Jindřich Nečas and Luděk Trávníček

11.2 JN-C101 On the solution of the variational inequality to the signorini problem with small friction 351
Jindřich Nečas, Jiří Jarušek and Jaroslav Haslinger

11.3 JN-C125 Unilateral problems in nonlinear, three-dimensional elasticity 367
Philippe G. Ciarlet and Jindřich Nečas

11.4 JN-C131 Injectivity and self-contact in nonlinear elasticity 387
Philippe G. Ciarlet and Jindřich Nečas

11.5 JN-C164 Existence of global weak solutions to the dynamical problem for a three-dimensional elastic body with singular memory 405
Hamid Bellout, Frederick Bloom and Jindřich Nečas

11.6 JN-C169 Existence of global weak solutions for a class of quasilinear hyperbolic integro-differential equations describing visco-elastic materials 415
Hamid Bellout and Jindřich Nečas

12. Commentaries 433

12.1 Mechanics of Solids 433
Ivan Hlaváček

12.2 The Contribution of Jindřich Nečas to the Development of the Theory of Solvability of Contact Problems 436
Jiří Jarušek

Part VI Fluid Mechanics

13. Selected Papers on Fluid Mechanics 441

13.1 JN-C136 Remarks on the solvability of transonic flow problems 441
Miloslav Feistauer and Jindřich Nečas

13.2 JN-C151 Global solution to the compressible isothermal multipolar fluid 453
Jindřich Nečas, Antonín Novotný and Miroslav Šilhavý

13.3 JN-C153 Some qualitative properties of the viscous compressible heat conductive multipolar fluid 477
Jindřich Nečas and Antonín Novotný

13.4 JN-C155 Multipolar viscous fluids 501
Jindřich Nečas and Miroslav Šilhavý

13.5 JN-C167 Young measure-valued solutions for non-Newtonian incompressible fluids 520
Hamid Bellout, Frederick Bloom and Jindřich Nečas

13.6 JN-C176 On Leray's self-similar solutions of the Navier-Stokes equations 561
Jindřich Nečas, Michael Růžička and Vladimír Šverák

13.7 JN-C184 On weak solutions to a class of non-Newtonian incompressible fluids in bounded three-dimensional domains: the case $p \geq 2$ 573
Josef Málek, Jindřich Nečas and Michael Růžička

13.8 JN-C178 A Finite-dimensional attractor for three-dimensional flow of incompressible fluids 619
Josef Málek and Jindřich Nečas

13.9 JN-C187 Global analysis of the flows of fluids with pressure-dependent viscosities 640
Josef Málek, Jindřich Nečas and Kumbakonan R. Rajagopal

13.10 JN-C165 On the non-Newtonian incompressible fluids 667
Josef Málek, Jindřich Nečas and Michael Růžička

14. Commentary 697

14.1 Comments on Papers on Fluid Mechanics 697
Josef Málek, Gregory Seregin and Vladimír Šverák

Part VII Recollections, Correspondence and Photographs

15. Comments and Recollections by Collaborators and Students 709

15.1 My Memories of My Friend Jindřich Nečas 709
Philippe G. Ciarlet

15.2 My Memories of My Supervisor Professor Jindřich Nečas 711
Pavol Quittner

15.3 My Memories of Professor Jindřich Nečas 713
Miloslav Feistauer

15.4 Jindřich Nečas and Industrial Mathematics 716
Tomáš Roubíček

16. Some Notable Personal Correspondence 719

17. Photographs 757

17.1 Childhood... 757
17.2 Jindřich Nečas in the 1960s and 1970s.. 760
17.3 Jindřich Nečas in the 1980s and 1990s.. 767
17.4 Italy 2000 and USA 2001 783

List of Contributors

Ivo Babuška Department of Aerospace Engineering and Engineering Mechanics, University of Texas, Austin, TX, USA

Philippe G. Ciarlet City University of Hong Kong, Kowloon Tong, Hong Kong

Pavel Drábek Department of Mathematics, University of West Bohemia, Plze, Czech Republic

Miloslav Feistauer Faculty of Mathematics and Physics, Charles University in Prague, Prague, Czech Republic

Ivan Hlaváček Institute of Mathematics, Academy of Sciences, Prague, Czech Republic

Jiří Jarušek Institute of Mathematics, Academy of Sciences, Prague, Czech Republic

Oldřich John Faculty of Mathematics and Physics, Charles University in Prague, Prague, Czech Republic

Jan Kristensen Mathematical Institute, University of Oxford, Oxford, UK

Alois Kufner Institute of Mathematics, Academy of Sciences, Prague, Czech Republic

Josef Málek Faculty of Mathematics and Physics, Charles University in Prague, Prague, Czech Republic

Giuseppe Mingione Department of Mathematics and Computer Science, University of Parma, Parma, Italy

Šárka Nečasová Institute of Mathematics, Academy of Sciences, Prague, Czech Republic

Milan Pokorný Faculty of Mathematics and Physics, Charles University in Prague, Prague, Czech Republic

Pavol Quittner Comenius University, Bratislava, Slovakia

Tomáš Roubíček Faculty of Mathematics and Physics, Charles University in Prague, Prague, Czech Republic

Gregory Seregin Mathematical Institute, University of Oxford, Oxford, UK

Jana Stará Faculty of Mathematics and Physics, Charles University in Prague, Prague, Czech Republic

Vladimír Šverák School of Mathematics, University of Minnesota, Minneapolis, MN, USA

Part I
Jindřich Nečas

Chapter 1
Short Biography of Jindřich Nečas

Šárka Nečasová and Milan Pokorný

Jindřich Nečas, one of the most renowned Czech mathematicians of the twentieth century, was born in Prague on December 14, 1929. At the age of 4 years, his family moved to the town of Mělník (approximately 30 km north of Prague). He spent his youth here and always delighted in return visits.

His father was born in Moravia and little Jindřich spent every holiday with his grandparents in Veselice near Brno. His mother, Paula Nardelli, was originally from Italy, her parents having moved from Genova to Prague. He inherited from her the Italian temperament. Little Jindřich loved days when all mother's relatives came to their house. They made lively conversation, told jokes and sang, all of them being very talented in music, a characteristic he inherited. Jindřich started to play violin when he was 6 years old. He retained the love of music for his whole life. He loved to play Mozart concerts and to sing. During weekends with his family he played on violin accompanied by his wife or daughter Jindřiška on piano.

In secondary school he was excellent in mathematics and physics and in particular loved the French language, in which he even wrote a poem. Young Jindřich was very much impressed by construction of new things. He wanted to construct a perpetual motion machine which led him to the decision to study physics. He started his studies at Faculty of Natural Sciences of the Charles University in Prague in 1948 and successfully finished his studies of Mathematical Analysis in 1952.

After a short stay at the Czech Technical University (CTU), he started his graduate studies at the Mathematical Institute of the Czechoslovak Academy of Sciences (MU CSAV) under the supervision of Professor Ivo Babuška. Already at this time, Jindřich Nečas met the wonderful world of the cross section between mathematics and mechanics; as the student of Ivo Babuška he cooperated in preparation of the monograph *Mathematical Methods of the Theory of Plane Elasticity* by Babuška, Rektorys and Vyčichlo (published in 1955 see [BRV1, BRV2]), and participated in the computations required for construction of the Orlík Dam. In 1957 he defended his thesis *Solution of the biharmonic problem for convex polygons* and received the degree of Candidate of Science (CSc.). More details about this period can be found in the contribution written by Ivo Babuška. In this period he married Zdeňka

© Springer Basel 2015

Š. Nečasová et al. (eds.), *Selected Works of Jindřich Nečas*, Advances in Mathematical Fluid Mechanics, DOI 10.1007/978-3-0348-0231-4_1

Staňková, who became his lifelong love, wife and mother of their two daughters, Jindřiška and Šárka.

The next period of his scientific life was connected with functional-analytic methods in the theory of partial differential equations, i.e. the theory of weak solutions. His supervisor introduced him to S.L. Sobolev and organized his trip to Italy. Here and later on in France he met the most significant mathematicians working in this area: M. Picone, G. Fischera, E. Magenes, J.-L. Lions and many others. Jindřich Nečas also became independent from the organizational point of view; in 1960 he became head of the newly created Department of the Theory of Partial Differential Equations. In this period, he also introduced the theory of weak solutions for partial differential equations in regular courses for the students at the Faculty of Mathematics and Physics, where he was promoted to the academic rank of Docent in 1964. In the period of 1965–1971, even though working at the Academy of Science, he was also head of the Department of Mathematical Analysis at the Faculty of Mathematics and Physics. In 1966 he received the highest scientific degree that time in the Czech Republic, Doctor of Science (DrSc.). Unfortunately, promotion to the academic rank of Professor remained unfinished due to changes after the year 1968. In 1962, he established in the Mathematical Institute the seminar on Partial Differential Equations which even now takes place regularly on Tuesday mornings. His fundamental results of this period can be found in the monograph *Les Méthodes Directes en Théorie des Equations Elliptiques* which appeared in 1967 and recently has been translated into English. More details about this period can be found in the contribution of A. Kufner.

Together with his family, Jindřich Nečas spent the years 1969–1970 in the United States. He was invited by F. Browder to Illinois State University in Chicago. He traveled throughout the United States and among other opportunities he received an offer from Berkeley for a permanent position. He refused it, as he felt it a moral duty to come back even in that so difficult time (after the Prague Spring, so-called normalization started in Czechoslovakia in 1969) and to continue his work in the country which he loved.

At the beginning of the 1970s, his new, nonlinear period started. It is connected with many new important results from nonlinear functional analysis (as e.g. the infinite dimensional version of the Sard Theorem), collected in the monograph *Spectral Analysis of Nonlinear Operators* (together with S. Fučík, V. Souček and J. Souček) as well as with regularity results in the theory of nonlinear partial differential equations (collected in the monograph *Introduction to the Theory of Nonlinear Elliptic Equations* (1983 and 1986). During this period, he also left the Academy of Science and moved permanently to the Faculty of Mathematics and Physics. More details about this period and results can be found in the contributions of P. Drábek, of R. Mingione and J. Kristensen and of O. John and J. Stará.

The next period in his scientific life is connected to rigorous mathematical formulation and well posedness of problems in continuum mechanics. The return of interest for these kind of problems can be traced back to 1967, when he (together with I. Hlaváček, under the name of Seminar of Theoretical Mechanics on Variational Principles) started the seminar on mathematical problems in mechanics.

This seminar, under the name of Nečas's Seminar on Continuum Mechanics, continues to exist today. The first part of this period was devoted mostly to problems in solid mechanics. He worked together with outstanding scientists both in theory and numerical methods (we mention e.g. Ph. Ciarlet, see also his contribution in this book). The results of this period can be found in monographs *Mathematical Theory of Elastic and Elasto-Plastic Bodies* (together with I. Hlaváček 1981 and 1983) and *Solutions of Variational Inequalities in Mechanics* (together with I. Hlaváček, J. Haslinger and J. Lovíšek, 1982, 1986 and 1988). See also the contribution of J. Jarušek for more details about this period.

Jindřich Nečas became the head of the Division of Mathematical Modelling at the Mathematical Institute of Charles University and together with I. Marek and J. Kratochvíl started a new, in fact the first multidisciplinary study branch at the Faculty of Mathematics and Physics, called Mathematical Modelling in Physics and Science. This study branch reflects perfectly the ideology of Jindřich Nečas: the connection of physics, numerical mathematics and mathematical analysis. After a number of changes in the political situation in the Czech Republic, Jindřich Nečas became Professor at the Charles University in 1990. In that period he also was recruited for a position at the University of Northern Illinois, DeKalb. Here, in 1998, he was promoted to Presidential Research Professor and after 4 years, he attained the position of Distinguished Research Professor. In 1999, Jindřich Nečas became Professor Emeritus at the Charles University and most of his activities have been since then connected with the University of Northern Illinois.

By the end of the 1980s, his attention turned to problems in fluid mechanics. Discussions with J. Poláček in Prague and R. Glowinski and O. Pironneau in Paris played an important role in motivating this decision. The first topic concerned models of transonic flow. Results of these investigations are reported in the monograph *Ecoulements de fluide. Compacité par entropie*. He also obtained many fundamental results both in the theory of non-Newtonian fluids (see the monograph *Weak and Measure-Valued Solutions to Evolutionary PDEs*, 1996, together with J. Málek, M. Rokyta and M. Růžička) and in the theory of the Navier–Stokes equations (e.g. the joint papers with R. Růžička,V. Šverák and with J. Málek, M. Pokorný, M. Schonbek). More details about this period of scientific life can be found in the contributions by V. Šverák, G. Seregin and J. Málek.

After a long serious illness, Professor Jindřich Nečas passed away on December 5, 2002, thinking about a problem which had become his favorite one in the last decade: regularity or singularity for the Navier–Stokes equations.

During his life, he received many honors, including the Order of Merit of the Czech Republic, which he was awarded by President Václav Havel on October 28, 1998. The Technical University of Dresden awarded him the Doctorate Honoris Causa. Preparation of his Doctor Honoris Causa nomination at the University of Bayreuth was interrupted by his death.

Professor Jindřich Nečas was in influential force to the mathematical community all over the world. According to Mathscinet, he is the author of more than 200 publications with more than 2300 citations. His results cover many aspects of partial differential equations, calculus of variations, nonlinear functional analysis, and mathematical theory in continuum mechanics. But he was also an excellent teacher; he created the Prague school of partial differential equations, his former master degree and Ph.D. students are professors at Universities in the Czech Republic, Slovakia and many other countries not only in Europe but also in the United States. We will remember him not only for his mathematics, but also for his friendly, engaging personality and the energy and enthusiasm which characterized his activities.

Let us quote the words of C. Simader in his condolence letter to Š. Nečasová: "The scientific community of the whole world, not only that from Czech Republic or from Charles University lost an outstanding mathematician who influenced decisively with his fundamental contributions for more than half a century central fields of real analysis like partial differential equations, functional analysis and function spaces. His skill in teaching and his friendly attitude towards students made him an extremely successful academic teacher too. I liked him so much. Let me mention another thing: I often talked with your father on recent history and I estimated highly that I never felt resentments against me as a German. Although he was a representative of a generation who very much suffered from the war and the occupation by Germans and as a consequence of this war, by 40 years of communism."

We finish with words of Professor Willi Jäger: "Last year, Olga Oleinik died, who we considered as a friend. Now Jindřich Nečas! Two bright names in the scientific world, especially important for European science. Your father has given important contributions and impulses and was one of the main people to rebuild science in Europe which suffered a lot under the hot and cold war. I had always hoped that he would return to Europe and to Prague. I am very sad that he will not return alive."

References

[BRV1] Babuška, Ivo; Rektorys, Karel; Vyčichlo, František, Matematická theorie rovinné pružnosti. (Czech) [Mathematical theory of plane elasticity.] Nakladatelství Československé Akademie Věd, Praha, 1955.

[BRV2] Babuška, Ivo; Rektorys, Karel; Vyčichlo, František, Mathematische Elastizitätstheorie der ebenen Probleme. (German) In deutscher Sprache herausgegeben von Winfried Heinrich, Akademie-Verlag, Berlin 1960

Šárka Nečasová
Institute of Mathemaics, Academy of Sciences, Žitná 25, 11567 Prague 1, Czech Republic
email: matus@math.cas.cz

Milan Pokorný
Faculty of Mathematics and Physics, Charles University in Prague, Sokolovská 83, 18675 Prague 8, Czech Republic
email: pokorny@karlin.mff.cuni.cz

Chapter 2
Complete Bibliography of Jindřich Nečas

Preliminary Note

All contributions which are reprinted in this book are labelld by * in this bibliography. In the text they are referred with the prefix "JN-", e.g. [JN-C17] instead of [C17] as in the list below.

Monographs

[A1] J. Nečas, *Les méthodes directes en théorie des équations elliptiques*, Academia, Prague, and Masson et Cie, Éditeurs, Paris, 1967. English translation see [A14].

[A2] S. Fučík, J. Nečas, J. Souček, and V. Souček, *Spectral analysis of nonlinear operators*, Lecture Notes in Mathematics, vol. 346, Springer-Verlag, Berlin, 1973.

[A3] S. Fučík, J. Nečas, and V. Souček, *Einführung in die Variationsrechnung*, Teubner-Texte zur Mathematik, BSB B. G. Teubner Verlagsgesellschaft, Leipzig, 1977.

[A4] J. Nečas and I. Hlaváček, *Mathematical theory of elastic and elasto-plastic bodies: an introduction*, Studies in Applied Mechanics, vol. 3, Elsevier Scientific Publishing Co., Amsterdam, 1980. Czech edition see [A7].

[A5] I. Hlaváček, J. Haslinger, J. Nečas, and J. Lovíšek, *Riešenie variačných nerovností v mechanike*, Alfa—Vydavatel' stvo Technickej a Ekonomickej Literatúry, Bratislava, 1982. In Slovac. Russian edition see [A8] and English edition [A10].

[A6] J. Nečas, *Introduction to the theory of nonlinear elliptic equations*, Teubner-Texte zur Mathematik, BSB B. G. Teubner Verlagsgesellschaft, Leipzig, 1983. See also [A9].

[A7] J. Nečas and I. Hlaváček, *Úvod do matematické teorie pružných a pružně plastických těles*, SNTL Praha, 1983. Czech edition of [A4].

[A8] I. Hlaváček, J. Haslinger, J. Nečas, and J. Lovíšek, *Reshenie variatsionnykh neravenstv v mekhanike*, "Mir", Moscow, 1986. Russian edition of [A5].

[A9] J. Nečas, *Introduction to the theory of nonlinear elliptic equations*, A Wiley-Interscience Publication, John Wiley & Sons Ltd., Chichester, 1986. Reprint of [A6].

[A10] I. Hlaváček, J. Haslinger, J. Nečas, and J. Lovíšek, *Solution of variational inequalities in mechanics*, Applied Mathematical Sciences, vol. 66, Springer-Verlag, New York, 1988. English translation of [A5].

[A11] J. Nečas, *Écoulements de fluide: compacité par entropie*, RMA: Research Notes in Applied Mathematics, vol. 10, Masson, Paris, 1989.

[A12] J. Málek, J. Nečas, M. Rokyta, and M. Ružička, *Weak and measure-valued solutions to evolutionary PDEs*, Applied Mathematics and Mathematical Computation, vol. 13, Chapman & Hall, London, 1996.

[A13] J. Haslinger, I. Hlaváček, and J. Nečas, *Numerical methods for unilateral problems in solid mechanics*, in.: *Handbook of numerical analysis*, 4, 313–485, 1996.

© Springer Basel 2015

Š. Nečasová et al. (eds.), *Selected Works of Jindřich Nečas*, Advances in Mathematical Fluid Mechanics, DOI 10.1007/978-3-0348-0231-4_2

[A14] J. Nečas, *Direct methods in the theory of elliptic equations*, Springer Monographs in Mathematics, Springer, Heidelberg, 2012. English translation of [A1].

Textbooks

[B1] J. Nečas, *Variational methods for solutions of linear elliptic equations*, ÚTAM ČSAV, 1960. In Czech.
[B2] J. Nečas and A. Kufner, *Functional analytic methods of solutions to elliptic partial differential equations*, MÚ ČSAV, 1963. In Czech.
[B3] K. Rektorys et al., *A survey of applied mathematics*, Tech. Lit., Prague, 1963. In Czech.
[B4] J. Nečas, S. Fučík, and V. Souček, *Introduction to the calculus of variations*, SPN Praha, 1972. In Czech.
[B5] J. Nečas and O. John, *Partial differential equations of mathematical physics*, SPN Praha, 1972. In Czech.
[B6] I. Hlaváček and J. Nečas, *Introduction to the mathematical theory of elastic and elasto-plastic bodies*, SPN Praha, 1976. In Czech.
[B7] J. Nečas, *Regularité des solutions faibles d'equations elliptiques non-linéaires; applications à l'élasticité*, Université Pierre et Marie Curie, Laboratoire d'analyse numérique, No.81046, 1960.
[B8] J. Nečas, *On the regularity of weak solutions to nonlinear elliptic systems of partial differential equations*, Scuola Normale Superiore Pisa, 1979.

Published Papers

1956

[C1] J. Nečas, *Influence of outside temperature on the strain of dams and other concrete masses*, Apl. Mat., 1956, vol. 1, pp.103–118. In Czech.
[C2] J. Nečas, *Influence of outside temperature on the strain of dams and other concrete masses (second part)*, Apl. Mat., 1956, vol. 1, pp. 186–199. In Czech.
[C3] D. Mayer and J. Nečas, *Das Addieren unendlicher Reihen unter Benützung von Integraltransformationen*, Apl. Mat., 1956, vol. 1, pp. 165–185.

1958

[C4] J. Nečas, *Une note sur la propriété caractéristique de la transformée de Laplace d'une fonction et sur certains espace de Hilbert* $\bar{H}_{kl}$ *dont la somme* $\sum_{k=0,l=0}^{\infty} \bar{H}_{kl}$ *est l'ensemble des transformées de Laplace de distributions*, Časopis Pěst. Mat., 1958, vol. 83, pp. 160–170.
[C5] J. Nečas, *Solution du problème biharmonique pour le coin infini I*, Časopis Pěst. Mat., 1958, vol. 83, pp. 257–286. In Czech.
[C6] J. Nečas, *Solution du problème biharmonique pour le coin infini II*, Časopis Pěst. Mat., 1958, vol. 83, pp. 399–424. In Czech.

1959

[C7] J. Nečas, *Résolution du problème de Dirichlet pour les équations elliptiques aux dérivées partielles du second ordre*, Czechoslovak Math. J., 1959, vol. 9 (84), pp. 467–469. In Russian.
[C8] J. Nečas, *Solution du problème biharmonique pour le coin infini pas convex*, Časopis Pěst. Mat., 1959, vol. 84, pp. 90–98. In Czech.

[C9] J. Nečas, *L'extension de l'espace des conditions aux limites du problème biharmonique pour les domaines à points anguleux*, Czechoslovak Math. J., 1959, vol. 9 (84), pp. 339–371.

1960

[C10] J. Nečas, *Über Grenzwerte von Funktionen, welche ein endliches Dirichlet'sches Integral haben*, Apl. Mat., 1960, vol. 5, pp. 202–209.
[C11] J. Nečas, *Sur les solutions des équations elliptiques aux dérivées partielles du second ordre avec intégrale de Dirichlet non-bornée*, Czechoslovak Math. J., 1960, vol. 10 (85), pp. 283–289. In Russian.

1961

[C12] J. Nečas, *Sur une méthode pour résoudre les équations aux dérivées partielles du type elliptique, voisine de la variationnelle*, Czechoslovak Math. J., 1961, vol. 11 (86), pp. 662–683.
[C13] J. Nečas, *Sur le problème de Dirichlet pour l'équation aux dérivées partielles du quatrième ordre du type elliptique*, Rend. Sem. Mat. Univ. Padova, 1961, vol. 31, pp. 198–231.

1962

[C14] J. Nečas, *Application de l'egalité de Rellich aux problémes aux limites*, Collége de France, 1962, Sem. Eq. Der. Art., pp. 143–167.
[C15] J. Dvořák and J. Nečas, *Determination of stresses in a rectangular wedge by superposition of stressed states of halfplanes*, Rev. Math. Pures Appl. (Bucarest), 1962, vol. 7, pp. 467–480. In Russian.
[C16] J. Nečas, *On the regularity of solutions of second-order elliptic partial differential equations with an unbounded Dirichlet integral*, Arch. Rational Mech. Anal., 1962 vol. 9, pp. 134–144.
[C17]* J. Nečas, *On domains of type $\mathcal{N}$*, Czechoslovak Math. J., 1962, vol. 12 (87), pp. 274–287. In Russian.
[C18] J. Nečas, *Sur une méthode pour résoudre les équations aux dérivées partielles du type elliptique, voisine de la variationnelle*, Ann. Scuola Norm. Sup. Pisa (3), 1962, vol. 16, pp. 305–326.
[C19] J. Nečas, *Sur l'existence de la solution classique du problème de Poisson pour les domaines plans*, Ann. Scuola Norm. Sup. Pisa (3), 1962, vol. 16, pp. 285–296.

1963

[C20] J. Nečas, *On the solution of elliptic partial differential equations with an unbounded Dirichlet integral*, in: *Differential Equations and Their Applications (Proc. Conf., Prague, 1962)*, pp. 93–104. Publ. House Czechoslovak Acad. Sci., Prague, 1963.
[C21] J. Nečas, *Sur la coercivité des formes bilinéaires pour les equations elliptiques*, in: *Atti del VII Congresso UMI* Genova 1–2, 1963.

1964

[C22] J. Nečas, *Sur les équations différentielles aux dérivées partielles du type elliptique du deuxième ordre*, Czechoslovak Math. J., 1964, vol. 14 (89) pp. 125–146.
[C23]* J. Nečas, *Sur la coercivité des formes sesquilinéaires, elliptiques*, Rev. Roumaine Math. Pures Appl., 1964, vol. 9, pp. 47–69.

1965

[C24]* J. Nečas, *L'application de l'égalité de Rellich sur les systèmes elliptiques du deuxième ordre*, J. Math. Pures Appl. (9), 1965, vol. 44, pp. 133–147.

1966

[C25] J. Nečas, *Sur une méthode générale pour la solution des problèmes aux limites non-linéaires*, Ann. Scuola Norm. Sup. Pisa (3), 1966, vol. 20, pp. 655–674.

[C26] J. Nečas and Z. Poracká, *On extrema of functionals*, Comment. Math. Univ. Carolinae, 1966, vol. 7, pp. 509–520.

[C27] J. Nečas, *Sur la méthode variationnelle pour les équations aux dérivées partielles non linéaires du type elliptique; l'existence et la régularité des solutions*, Comment. Math. Univ. Carolinae, 1966, vol. 7, pp. 301–317.

[C28]* J. Nečas, *Sur les normes equivalentes dans* W_p^k *et sur la coercivité des formes formellement positives*, Sem. math. sup., Université Montréal, 1966, pp. 102–128.

1967

[C29] J. Kadlec and J. Nečas, *Sulla regolarità delle soluzioni di equazioni ellittiche negli spazi* $H^{k,\lambda}$, Ann. Scuola Norm. Sup. Pisa (3), 1967, vol. 21, pp. 527–545.

[C30]* J. Nečas, *Sur l'appartenance dans la classe* $C^{(k),\mu}$ *des solutions variationnelles des équations elliptiques non-linéaires de l'ordre* $2k$ *en deux dimensions*, Comment. Math. Univ. Carolinae, 1967, vol. 8, pp. 209–217.

[C31] J. Nečas, *Sur la régularité des solutions faibles des equations elliptiques non-lineaires*, Collége de France, 1967–1968, pp. 20–57.

[C32] J. Nečas, *Sur l'existence de la solution régulière pour le problème de Dirichlet de l'équation elliptique non linéaire d'ordre* $2k$, Atti Accad. Naz. Lincei Rend. Cl. Sci. Fis. Mat. Natur. (8), 1967, vol. 42, pp. 347–354.

[C33] J. Nečas, *Sur la régularité des solutions variationelles des équations elliptiques non-linéaires d'ordre* $2k$ *en deux dimensions*, Ann. Scuola Norm. Sup. Pisa (3), 1967, vol. 21, pp. 427–457.

[C34] J. Nečas, *On the existence and regularity of solutions of nonlinear elliptic equations*, in: *Diff. Equations and their Applications, Equadiff II (Proc. Conf., Bratislava, 1966)*, pp. 101–119, Slov. Ped. Nakladatel'., Bratislava, 1967.

[C35] A. Kufner and J. Nečas, *In memoriam Jana Kadlece*, Časopis Pěst. Mat., 1967, vol. 92, pp. 490–492. In Czech.

1968

[C36] A. Kufner and J. Nečas, *In memoriam of Jan Kadlec*, Czechoslovak Math. J., 1968, vol. 18, pp. 190–192.

[C37] J. Nečas, *Sur la régularité des solutions faibles des équations elliptiques non linéaires*, Comment. Math. Univ. Carolinae, 1968, vol. 9, pp. 365–413.

[C38] J. Kačur, J. Nečas, J. Polák, and J. Souček, *Convergence of a method for solving the magnetostatic field in nonlinear media*, Apl. Mat., 1968, vol. 13, pp. 456–465.

1969

[C39] J. Nečas, *Les équations elliptiques non linéaires*, Czechoslovak Math. J., 1969, vol. 19 (94), pp. 252–274.

[C40]* J. Nečas, *Sur l'alternative de Fredholm pour les opérateurs non-linéaires avec applications aux problèmes aux limites*, Ann. Scuola Norm. Sup. Pisa (3), 1969, vol. 23, pp. 331–345.

[C41] J. Nečas, *Integral transforms (Operational calculus)*, in: *Survey of applicable mathematics*, pp. 1125–1136, Iliffe books LTD, London, 1st Ed. 1969.

1970

[C42] J. Nečas, Z. Poracká, and R. Kodnár, *Remarks on a nonlinear theory of thin elastic plates*, Mat. Časopis S. A. V., 1970, vol. 20, pp. 62–71.

[C43]* I. Hlaváček and J. Nečas, *On inequalities of Korn's type. I. Boundary-value problems for elliptic system of partial differential equations*, Arch. Rational Mech. Anal., 1970, vol. 36, pp. 305–311.

[C44]* I. Hlaváček and J. Nečas, *On inequalities of Korn's type. II. Applications to linear elasticity.*, Arch. Rational Mech. Anal., 1970, vol. 36, pp. 312–334.

1971

[C45] J. Nečas, *The discreteness of the spectrum of a nonlinear Sturm-Liouville equation*, Dokl. Akad. Nauk SSSR, 1971, vol. 201, pp. 1045–1048. In Russian.

[C46] A. Kratochvíl and J. Nečas, *The discreteness of the spectrum of a nonlinear Sturm-Liouville equation of fourth order*, Comment. Math. Univ. Carolinae, 1971, vol. 12, pp. 639–653. In Russian.

[C47] J. Nečas, *On the demiregularity of weak solutions of nonlinear elliptic equations*, Bull. Amer. Math. Soc., 1971, vol. 77, pp. 151–156.

1972

[C48] J. Nečas, *Fredholm alternatives and application to boundary value problems*, in: *Trudy III-ovo sovetsko-cechoslovackogo sovescanija*, pp. 162–171, Novosibirsk, SOAN SSSR, 1972. In Russian.

[C49] S. Fučík, J. Nečas, J. Souček, and V. Souček, *Strengthening upper bound for the number of critical levels of nonlinear functionals*, Comment. Math. Univ. Carolinae, 1972, vol. 13, pp. 297–310.

[C50] J. Nečas and J. Stará, *Principio di massimo per i sistemi ellittici quasi-lineari non diagonali*, Boll. Un. Mat. Ital. (4), 1972, vol. 6, pp. 1–10.

[C51] S. Fučík and J. Nečas, *Ljusternik-Schnirelmann theorem and nonlinear eigenvalue problems*, Math. Nachr., 1972, vol. 53, pp. 277–289.

[C52] J. Nečas, *Remark on the Fredholm alternative for nonlinear operators with application to nonlinear integral equations of generalized Hammerstein type*, Comment. Math. Univ. Carolinae, 1972, vol. 13, pp. 109–120.

[C53]* J. Nečas, *Fredholm alternative for nonlinear operators and applications to partial differential equations and integral equations.* Časopis Pěst. Mat., 1972, vol. 97, pp. 65–71.

[C54] S. Fučík, J. Nečas, J. Souček, and V. Souček, *New infinite dimensional versions of Morse-Sard theorem*, Boll. Un. Mat. Ital. (4), 1972, vol. 6, pp. 317–322.

[C55] S. Fučík, O. John, and J. Nečas, *On the existence of Schauder bases in Sobolev spaces*, Comment. Math. Univ. Carolinae, 1972, vol. 13, pp. 163–175.

[C56] S. Fučík, J. Nečas, J. Souček, and V. Souček, *Upper bound for the number of critical levels for nonlinear operators in Banach spaces of the type of second order nonlinear partial differential operators*, J. Functional Analysis, 1972, vol. 11, pp. 314–333.

[C57] S. Fučík, J. Nečas, J. Souček, and V. Souček, *Upper bound for the number of eigenvalues for nonlinear operators*, Comment. Math. Univ. Carolinae, 1972, vol. 13, pp. 191–195.

[C58] J. Nečas, *On the discreteness to the spectrum of a non-linear Sturm-Liouville equation of the fourth order*, in: *Trudy III-ovo sovetsko-cechoslovackogo sovescanija*, pp. 107–121, Novosibirsk, SOAN SSSR, 1972. In Russian.

1973

[C59] S. Fučík and J. Nečas, *Spectral theory of nonlinear operators*, in: *Proceedings of Equadiff 3*, pp. 163–174. Folia Fac. Sci. Natur. Univ. Purkynianae Brunensis, Ser. Monograph., Tomus 1, 1973.

[C60] J. Nečas, *Variational methods in nonlinear elasticity*, Acta Polytechnica, 1973, pp. 129–133. In Czech.

[C61] S. Fučík, J. Nečas, J. Souček, and V. Souček, *Note to nonlinear spectral theory: application to the nonlinear integral equations of the Lichtenstein type*, Math. Nachr., 1973, vol. 58, pp. 257–267.

[C62] J. Nečas and J. Kratochvíl, *On the existence of solutions of boundary-value problems for elastic-inelastic solids*, Comment. Math. Univ. Carolinae, 1973, vol. 14, pp. 755–760.

[C63] J. Nečas, *On the formulation of the traction problem for the flow theory of plasticity*, Apl. Mat., 1973, vol. 18, pp. 119–127.

[C64] S. Fučík, A. Kratochvíl, and J. Nečas, *Kačanov–Galerkin method*, Comment. Math. Univ. Carolinae, 1973, vol. 14, pp. 651–659.

[C65] J. Nečas, *Range of nonlinear operators and application to boundary value problems*, Mathematica Balcanica, 1973, vol. 3, pp. 383–387. Proc. Internat. Conf. Integral, Differential and Functional Equations (Bled, 1973).

[C66] J. Nečas, *On the range of nonlinear operators with linear asymptotes which are not invertible*, Comment. Math. Univ. Carolinae, 1973, vol. 14, pp. 63–72.

[C67] J. Nečas, *Fredholm theory of boundary value problems for nonlinear ordinary differential operators*, in: *Proceedings of Summer School at Babylon in 1971*, pp. 85–119, Academia Prague, 1973.

[C68] S. Fučík, J. Nečas, J. Souček, and V. Souček, *Upper bound for the number of eigenvalues for nonlinear operators*, Ann. Scuola Norm. Sup. Pisa (3), 1973, vol. 27, pp. 53–71.

1974

[C69] J. Nečas, *Mathematical models of elastic-inelastic materials*, in: *Schriftenreihe Zentralinst. Math. Mech. Akad. Wiss. DDR*, No. 20, pp. 227–232, 1974.

[C70] S. Fučík, J. Nečas, J. Souček, and V. Souček, *Krasnoselskii's main bifurcation theorem*, Arch. Rational Mech. Anal., 1974, vol. 54, pp. 328–339.

[C71] J. Nečas and J. Naumann, *On a boundary value problem in nonlinear theory of thin elastic plates*, Apl. Mat., 1974, vol. 19, pp. 7–16.

[C72] S. Fučík, M. Kučera, J. Nečas, J. Souček, and V. Souček, *Morse-Sard theorem in infinite dimensional Banach spaces and investigation of the set of all critical levels*, Časopis Pěst. Mat., 1974, vol. 99, pp. 217–243.

[C73]* J. Nečas, *Application of Rothe's method to abstract parabolic equations*, Czechoslovak Math. J., 1974, vol. 24 (99), pp. 496–500.

[C74] S. Fučík, A. Kratochvíl, and J. Nečas, *Kačanov-Galerkin method and its application*, Acta Univ. Carolinae—Math. et Phys., 1974, vol. 15, pp. 31–33.

[C75] O. John and J. Nečas, *Regularity results for non linear integro-differential operators in two dimensions*, in: *Schriftenreihe Zentralinst. Math. Mech. Akad. Wiss. DDR*, No. 20, pp. 117–123, 1974.

1975

[C76] J. Nečas, *On regularity of solutions to nonlinear variational inequalities for second-order elliptic systems*, Rend. Mat. (6), 1975, vol. 8, pp. 481–498.

[C77] O. John and J. Nečas, *On the solvability of von Kármán equations*, Apl. Mat., 1975, vol. 20, pp. 48–62.

[C78] S. Fučík, A. Kratochvíl, and J. Nečas, *Kačanov's method and its application*, Rev. Roumaine Math. Pures Appl., 1975, vol. 20, pp. 907–916.

[C79] S. Fučík, M. Kučera, and J. Nečas, *Ranges of nonlinear asymptotically linear operators*, J. Differential Equations, 1975, vol. 17, pp. 375–394.

[C80] J. Nečas, *An approximate method for finding the critical points of even functionals*, Trudy Mat. Inst. Steklov., 1975, vol. 134, pp. 235–239. In Russian.

[C81] M. Müller and J. Nečas. *Über die Regularität der schwachen Lösungen von Randwertaufgaben für quasilineare elliptische Differentialgleichungen höherer Ordnung*, Czechoslovak Math. J., 1975, vol. 25 (100), pp. 227–239.

1976

[C82] J. Nečas, *Theory of locally monotone operators modeled on the finite displacement theory for hyperelasticity*, Beiträge Anal., 1976, Heft 8, pp. 103–114.

[C83] J. Nečas and M. Štípl, *A paradox in the theory of linear elasticity*, Apl. Mat., 1976, vol. 21, pp.431–433.

[C84] A. Doktor, J. Nečas, and R. Švarc, *A remark on applications of the Laplace transform to abstract differential equations of parabolic type*, Časopis Pěst. Mat., 1976, vol. 101, pp. 7–19.

[C85] J. Nečas, *Introduction to variational methods of solution of elliptic equations with applications to theory of elasticity*, in: *Proceedings of summer school on numerical solution of elliptic equations by finite elements method*, Charles University in Prague, 1976.

1977

[C86]* J. Nečas, *Example of an irregular solution to a nonlinear elliptic system with analytic coefficients and conditions for regularity*, Abh. Akad. Wiss. DDR Abt. Math.–Natur.–Tech., Jahrgang 1977, 1977, vol. 1, pp. 197–206.

[C87] M. Kučera and J. Nečas, *Interior regularity of solutions to systems of variational inequalities*, Časopis Pěst. Mat., 1977, vol. 102, pp.73–82.

[C88] J. Jarušek and J. Nečas, *Sur les domaines des valeurs des opérateurs non-linéaires*, Časopis Pěst. Mat., 1977, vol. 102, pp. 61–72.

1978

[C89] J. Nečas, J. Stará, and R. Švarc, *Classical solution to a second order nonlinear elliptic system in* $\mathbf{R}^3$, Ann. Scuola Norm. Sup. Pisa Cl. Sci. (4), 1978, vol. 5, pp. 605–631.

[C90] M. Kučera, J. Nečas, and J. Souček, *The eigenvalue problem for variational inequalities and a new version of the Ljusternik-Schnirelmann theory*, in: *Nonlinear Analysis*, pp. 125–143, Academic Press, New York, 1978.

[C91] J. Nečas and L. Trávníček, *Variational inequalities of elastoplasticity with internal state variables*, in: *Theory of nonlinear operators (Proc. Fifth Internat. Summer School, Central Inst. Math. Mech. Acad. Sci. GDR, Berlin, 1977)*, Abh. Akad. Wiss. DDR, Abt. Math. Naturwiss. Tech., 1978, vol. 6, pp. 195–204, Akademie-Verlag, Berlin, 1978.

[C92] J. Nečas, *An example of a nonsmooth solution of a nonlinear elliptic system with analytic coefficients and a smoothness condition*, in: *Proceedings of an All-Union Conference on Partial Differential Equations (Moscow State Univ., Moscow, 1976)*, pp. 174–177, Moskov. Gos. Univ., 1978.

1979

[C93] A. Kratochvíl and J. Nečas, *Secant modulus method for the construction of a solution of nonlinear eigenvalue problems*, Boll. Un. Mat. Ital. B (5), 1979, vol. 16, pp. 694–710.

[C94]* K. Gröger and J. Nečas, *On a class of nonlinear initial value problems in Hilbert spaces*, Math. Nachr., 1979, vol. 93, pp. 21–31.

[C95]* K. Gröger, J. Nečas, and L. Trávníček, *Dynamic deformation processes of elastic-plastic systems*, Z. Angew. Math. Mech., 1979, vol. 59, pp. 567–572.

[C96]* M. Giaquinta and J. Nečas, *On the regularity of weak solutions to nonlinear elliptic systems via Liouville's type property*, Comment. Math. Univ. Carolin., 1979, vol. 20, pp. 111–121.

[C97] J. Nečas, *On the regularity of weak solutions to variational equations and inequalities for nonlinear second order elliptic systems*, in: *Equadiff IV (Proc. Czechoslovak Conf. Differential Equations and their Applications, Prague, 1977)*, Lecture Notes in Math., vol. 703, pp. 286–299, Springer, Berlin, 1979.

[C98] J. Nečas, *Functional analytic method in the theory of partial differential inequalities and mathematical models of continuum mechanics*, in: *Mathematics, Sci. Conf. Fac. Math. Phys., Charles Univ., Prague 1978, Part 1*, pp. 117–120, Praha, 1979. In Czech.

1980

[C99] J. Nečas and O. A. Oleĭnik, *Liouville theorems for elliptic systems*, Dokl. Akad. Nauk SSSR, 1980, vol. 252, pp. 1312–1316. In Russian.

[C100]* M. Giaquinta and J. Nečas, *On the regularity of weak solutions to nonlinear elliptic systems of partial differential equations*, J. Reine Angew. Math., 1980, vol. 316, pp. 140–159.

[C101]* J. Nečas, J. Jarušek, and J. Haslinger, *On the solution of the variational inequality to the Signorini problem with small friction*, Boll. Un. Mat. Ital. B (5), 1980, vol. 17, pp. 796–811.

[C102] J. Nečas and L. Trávníček, *Evolutionary variational inequalities and applications in plasticity*, Apl. Mat., 1980, vol. 25, pp. 241–256.

[C103] J. Mawhin, J. Nečas, and B. Novák, *Zemřel docent Svatopluk Fučík*, Časopis Pěst. Mat., 1980, vol. 105, pp. 91–101. In Czech.

[C104]* J. Nečas, O. John, and J. Stará, *Counterexample to the regularity of weak solution of elliptic systems*, Comment. Math. Univ. Carolin., 1980, vol. 21, pp. 145–154.

[C105] J. Nečas, *Variational inequalities in elasticity and plasticity with application to Signorini's problems and to flow theory of plasticity*, Z. Angew. Math. Mech., 1980, vol. 60, pp. 20–26.

[C106] A. Kratochvíl and J. Nečas, *Gradient methods for the construction of Ljusternik-Schnirelmann critical values*, RAIRO Anal. Numér., 1980, vol. 14, pp. 43–54.

[C107] J. Mawhin, J. Nečas, and B. Novák, *In memoriam professor Svatopluk Fučík*, Czechoslovak Math. J., 1980, vol. 30 (105), pp. 153–162.

1981

[C108] J. Nečas, *A necessary and sufficient condition for the regularity of weak solutions to nonlinear elliptic systems of partial differential equations*, in: *Nonlinear analysis (Berlin, 1979)*, Abh. Akad. Wiss. DDR, Abt. Math. Naturwiss. Tech., 1981, vol. 2, pp. 201–209, Akademie-Verlag, Berlin, 1981.

[C109] J. Nečas, *Elliptic differential equations*, in: *Mathem. Kongress DDR 1981, Math. Gesellschaft DDR*, Leipzig, 1981.

[C110] I. Hlaváček and J. Nečas, *Optimization of the domain in elliptic unilateral boundary value problems by finite element methods*, *Equadiff 5, Proceedings 1981*, pp. 131–135, Teubner Texte B 47, Leipzig, 1981.

[C111] O. John, J. Nečas, and J. Stará, *On the regularity for 2nd order nonlinear elliptic systems*, *Equadiff 5, Proceedings 1981*, pp. 165–168, Teubner Texte B 47, Leipzig, 1981.

1982

[C112] I. Hlaváček and J. Nečas, *Optimization of the domain in elliptic unilateral boundary value problems by finite element method*, RAIRO Anal. Numér., 1982, vol. 16, pp. 351–373.

[C113]* P.-L. Lions, J. Nečas, and I. Netuka, *A Liouville theorem for nonlinear elliptic systems with isotropic nonlinearities*, Comment. Math. Univ. Carolin., 1982, vol. 23, pp. 645–655.

[C114]* M. Giaquinta, J. Nečas, O. John, and J. Stará, *On the regularity up to the boundary for second order nonlinear elliptic systems*, Pacific J. Math., 1982, vol. 99, pp. 1–17.

[C115] J. Nečas, *On the solution of the 19th Hilbert's problem*, in: *Recent trends in mathematics. Reinhardsbrun 1982*, pp. 214–223, Teubner Texte, vol. 50, BSB B. G. Teubner Verlagsgesellschaft, Leipzig, 1982.

[C116] I. Hlaváček and J. Nečas, *Optimization of the domain in elliptic unilateral boundary value problems by FEM*, Wiss. Z. Hochsch. Archit. Bauwes. Weimar, 1982, vol. 28, pp. 145–147.

1983

[C117] I. Marek and J. Nečas, *60'th Anniversary of Birthday of Professor Karel Rektorys*, Czechoslovak Math. J., 1983, vol. 33 (108), pp. 320–323.

[C118] I. Marek and J. Nečas, *Šedesátiny Prof. RNDr. Karla Rektoryse, DrSc.*, Časopis Pěst. Mat., 1983, vol. 108, pp. 104–109.
[C119] J. Nečas and I. Hlaváček, *Solution of Signorini's contact problem in the deformation theory of plasticity by secant modules method*, Apl. Mat., 1983, vol. 28, pp. 199–214.
[C120] J. Nečas, *On regular solutions to the displacement boundary value problem in finite elasticity*, in: *Trends in applications of pure mathematics to mechanics, Vol. IV (Bratislava, 1981)*, Monographs Stud. Math., vol. 20, pp. 176–185, Pitman, Boston, Mass., 1983.

1984

[C121] M. Feistauer, J. Mandel, and J. Nečas, *Entropy regularization of the transonic potential flow problem*, Comment. Math. Univ. Carolin., 1984, vol. 25, pp. 431–443.
[C122] Ph. G. Ciarlet and J. Nečas, *Problèmes unilatéraux en élasticité non linéaire tridimensionnelle*, C. R. Acad. Sci. Paris Sér. I Math., 1984, vol. 298, pp. 189–192.

1985

[C123] M. Feistauer and J. Nečas, *On the solvability of transonic potential flow problems*, Z. Anal. Anwendungen, 1985, vol. 4, pp. 305–329.
[C124] Ph. G. Ciarlet and J. Nečas, *Injectivité presque partout, auto-contact, et non-interpénétrabilité en élasticité non-linéaire tridimensionnelle*, C. R. Acad. Sci. Paris Sér. I Math., 1985, vol. 301, pp. 621–624.
[C125]* Ph. G. Ciarlet and J. Nečas, *Unilateral problems in nonlinear, three-dimensional elasticity*, Arch. Rational Mech. Anal., 1985, vol. 87, pp. 319–338.

1986

[C126] J. Nečas, *Entropy compactification of the transonic flow*, in: *Equadiff 6 (Brno, 1985), Lecture Notes in Math.*, vol. 1192, pp. 399–408, Springer, Berlin, 1986.
[C127] J. Nečas, *On singularities of solutions to nonlinear elliptic systems of partial differential equations*, in: *Nonlinear functional analysis and its applications, Part 2 (Berkeley, Calif., 1983)*, Proc. Sympos. Pure Math., vol. 45, pp. 219–228, Amer. Math. Soc., Providence, R.I., 1986.
[C128] M. Feistauer and J. Nečas, *On the solution of transonic flows with weak shocks*, Comment. Math. Univ. Carolin., 1986, vol. 27, pp. 791–804.
[C129] J. Nečas, *Entropy compactification of the transonic flow*, *Mathematical methods in engineering*, Karlovy Vary I, pp. 97–101, 1986.

1987

[C130]* J. Nečas, A. Lehtonen, and P. Neittaanmäki, *On the construction of Lusternik-Schnirelmann critical values with application to bifurcation problems*, Appl. Anal., 1987, vol. 25, pp. 253–268.
[C131]* Ph. G. Ciarlet and J. Nečas, *Injectivity and self-contact in nonlinear elasticity*, Arch. Rational Mech. Anal., 1987, vol. 97, pp. 171–188.
[C132] J. Mandel and J. Nečas, *Convergence of finite elements for transonic potential flows*, SIAM J. Numer. Anal., 1987, vol. 24, pp. 985–996.
[C133] J. Nečas, *Strongly dissipative nonlinear hyperbolic systems*, in: *Proceeding of a seminar at Sous*, Fyzika a matematika, JČMF Praha, pp. 7 –13, 1987. In Czech.

1988

[C134] J. Nečas, *Finite element approach to the transonic flow problem*, in: *Proceedings of the Second International Symposium on Numerical Analysis (Prague, 1987)*, pp. 70–74, Teubner-Texte Math., vol. 107, BSB B. G. Teubner Verlagsgesellschaft, Leipzig, 1988.

[C135] M. Feistauer and J. Nečas, *Viscosity method in a transonic flow*, Comm. Partial Differential Equations, 1988, vol. 13, pp. 775–812.

[C136]* M. Feistauer and J. Nečas, *Remarks on the solvability of transonic flow problems*, Manuscripta Math., 1988, vol. 61, pp. 417–428.

[C137] J. Nečas, *A viscosity solution method for transonic flow*, in: *Functional and numerical methods in mathematical physics*, pp. 155–161, Naukova Dumka, Kiev, 1988. In Russian.

[C138] A. Friedman and J. Nečas, *Systems of nonlinear wave equations with nonlinear viscosity*, Pacific J. Math., 1988, vol. 135, pp. 29–55.

[C139] O. John, V. A. Kondrat'ev, D. M. Lekveishvili, J. Nečas, and O. A. Oleĭnik, *Solvability of the system of von Kármán equations with nonhomogeneous boundary conditions in nonsmooth domains*, Trudy Sem. Petrovsk., 1988, vol. 13, pp. 197–205. In Russian.

1989

[C140] J. Nečas, *Dynamics in the nonlinear thermo-visco-elasticity*, in: *Symposium Partial Differential Equations (Holzhau, 1988)*, Teubner-Texte Math. vol. 112, pp. 197–203, BSB B. G. Teubner Verlagsgesellschaft, Leipzig, 1989.

[C141] J. Nečas, A. Novotný, and M. Šilhavý, *Global solution to the ideal compressible heat conductive multipolar fluid*, Comment. Math. Univ. Carolinae, 1989, vol. 30, pp. 551–564.

[C142] M. Feistauer, J. Nečas, and V. Šverák, *On the weak compactness of solutions to the equations of compressible flow*, Appl. Anal., 1989, vol. 34, pp. 35–52.

1990

[C143] J. Nečas, *The current state and future of nonlinear analysis in Czechoslovakia*, Pokroky Mat. Fyz. Astronom., 1990, vol. 35, pp. 250–255. In Czech.

[C144] J. Nečas, A. Novotný, and V. Šverák, *Uniqueness of solutions to the systems for thermoelastic bodies with strong viscosity*, Math. Nachr., 1990, vol. 149, pp. 319–324.

[C145] P. Klouček and J. Nečas, *The solution of transonic flow problems by the method of stabilization*, Appl. Anal., 1990, vol. 37, pp. 143–167.

[C146] J. Milota, J. Nečas, and V. Šverák, *On weak solutions to a viscoelasticity model*, Comment. Math. Univ. Carolinae, 1990, vol. 31, pp. 557–565.

[C147] J. Nečas and T. Roubíček, *Approximation of a nonlinear thermoelastic problem with a moving boundary via a fixed-domain method.* Apl. Mat., 1990, vol. 35, pp. 361–372.

[C148] O. John, V. A. Kondrat'ev, D. M. Lekveishvili, J. Nečas, and O. A. Oleĭnik, *Solvability of the von Kármán equations with nonhomogeneous boundary conditions in nonsmooth domains*, J. Sov. Math., 1990, vol. 50, pp. 1491–1497, 1990. English translation of the Russian original.

1991

[C149] J. Nečas, *Theory of multipolar viscous fluids*, in: *The mathematics of finite elements and applications, VII (Uxbridge, 1990)*, pp. 233–244, Academic Press, London, 1991.

[C150] H. Bellout, F. Bloom, and J. Nečas, *Global existence of weak solutions to the nonlinear transmission line problem*, Nonlinear Anal., 1991, vol. 17, pp. 903–921.

[C151]* J. Nečas, A. Novotný, and M. Šilhavý, *Global solution to the compressible isothermal multipolar fluid*, J. Math. Anal. Appl., 1991, vol. 162, pp. 223–241.

[C152] J. Nečas and M. Růžička, *A dynamic problem of thermoelasticity*, Z. Anal. Anwendungen, 1991, vol. 10, pp. 357–368.

[C153]* J. Nečas and A. Novotný, *Some qualitative properties of the viscous compressible heat conductive multipolar fluid*, Comm. Partial Differential Equations, 1991, vol. 16, pp. 197–220.

[C154] Ch. P. Gupta, Y. C. Kwong, and J. Nečas, *Nonresonance conditions for the strong solvability of a general elliptic partial differential operator*, Nonlinear Anal., 1991, vol. 17, pp. 613–625.

[C155]* J. Nečas and M. Šilhavý, *Multipolar viscous fluids*, Quart. Appl. Math., 1991, vol. 49, pp. 247–265.

[C156]* J. Nečas and V. Šverák, *On regularity of solutions of nonlinear parabolic systems*, Ann. Scuola Norm. Sup. Pisa Cl. Sci. (4), 1991, vol. 18, pp. 1–11.

1992

[C157] H. Bellout, F. Bloom, and J. Nečas, *A model of wave propagation in a nonlinear superconducting dielectric*, Differential Integral Equations, 1992, vol. 5, pp. 1185–1199.

[C158] J. Málek, J. Nečas, and A. Novotný, *Measure-valued solutions and asymptotic behavior of a multipolar model of a boundary layer*, Czechoslovak Math. J., 1992, vol. 42(117), pp. 549–576.

[C159] J. Nečas, A. Novotný, and M. Šilhavý, *Global solution to the viscous compressible barotropic multipolar gas*, Theoret. Comput. Fluid Dynamics, 1992, vol. 4, pp. 1–11.

[C160] H. Bellout, F. Bloom, and J. Nečas, *Phenomenological behavior of multipolar viscous fluids*, Quart. Appl. Math., 1992, vol. 50, pp. 559–583.

[C161] J. Nečas and M. Růžička, *Global solution to the incompressible viscous-multipolar material problem*, J. Elasticity, 1992, vol. 29, pp. 175–202.

1993

[C162] J. Jarušek, J. Málek, J. Nečas, and V. Šverák, *Variational inequality for a viscous drum vibrating in the presence of an obstacle*, Rend. Mat. Appl. (7), 1993, vol. 12, pp. 943–958.

[C163] H. Bellout, F. Bloom, and J. Nečas, *Solutions for incompressible non-Newtonian fluids*, C. R. Acad. Sci. Paris Sér. I Math., 1993, vol. 317, pp. 795–800.

[C164]* H. Bellout, F. Bloom, and J. Nečas, *Existence of global weak solutions to the dynamical problem for a three-dimensional elastic body with singular memory*, SIAM J. Math. Anal., 1993, vol. 24, pp. 36–45.

[C165]* J. Málek, J. Nečas, and M. Růžička, *On the non-Newtonian incompressible fluids*, Math. Models Methods Appl. Sci., 1993, vol. 3, pp. 35–63.

1994

[C166] J. Nečas, *Theory of multipolar fluids*, in: *Problems and methods in mathematical physics (Chemnitz, 1993)*, Teubner-Texte Math., vol. 134, pp. 111–119, Teubner, Stuttgart, 1994.

[C167]* H. Bellout, F. Bloom, and J. Nečas, *Young measure-valued solutions for non-Newtonian incompressible fluids*, Comm. Partial Differential Equations, 1994, vol. 19, pp. 1763–1803.

[C168] Ch. P. Gupta, Y. C. Kwong, and J. Nečas, *Landesman-Lazer condition for properly elliptic operators*, Boll. Un. Mat. Ital. A (7), 1994, vol. 8, pp. 65–74.

[C169]* H. Bellout and J. Nečas, *Existence of global weak solutions for a class of quasilinear hyperbolic integro-differential equations describing viscoelastic materials*, Math. Ann., 1994, vol. 299, pp. 275–291.

1995

[C170] H. Bellout, F. Bloom, and J. Nečas, *Bounds for the dimensions of the attractors of non-linear bipolar viscous fluids*, Asymptotic Anal., 1995, vol. 11, pp. 131–167.

[C171] H. Bellout, F. Bloom, and J. Nečas, *Existence, uniqueness, and stability of solutions to the initial-boundary value problem for bipolar viscous fluids*, Differential Integral Equations, 1995, vol. 8, pp. 453–464.

[C172] H. Bellout, F. Bloom, and J. Nečas, *Bounds for the dimensions of the attractors of non-linear bipolar viscous fluids*, Asymp. Anal., 1995, vol. 11, pp. 131–167.

1996

[C173] J. Nečas, *Theory of multipolar fluids*, in: *World Congress of Nonlinear Analysts '92, Vol. I–IV (Tampa, FL, 1992)*, pp. 1073–1081, de Gruyter, Berlin, 1996.

[C174] H. Bellout and J. Nečas, *The exterior problem in the plane for a non-Newtonian incompressible bipolar viscous fluid*, Rocky Mountain J. Math., 1996, vol. 26, pp. 1245–1260.

[C175] J. Nečas, M. Růžička, and V. Šverák, *Sur une remarque de J. Leray concernant la construction de solutions singulières des équations de Navier-Stokes*, C. R. Acad. Sci. Paris Sér. I Math., 1996, vol. 323, pp. 245–249.

[C176]* J. Nečas, M. Růžička, and V. Šverák, *On Leray's self-similar solutions of the Navier-Stokes equations*, Acta Math., 1996, vol. 176, pp. 283–294.

[C177] W. Hao, S. Leonardi, and J. Nečas, *An example of irregular solution to a nonlinear Euler-Lagrange elliptic system with real analytic coefficients*, Ann. Scuola Norm. Sup. Pisa Cl. Sci. (4), 1996, vol. 23, pp. 57–67.

[C178]* J. Málek and J. Nečas, *A finite-dimensional attractor for three-dimensional flow of incompressible fluids*, J. Differential Equations, 1996, vol. 127, pp. 498–518.

1997

[C179] J. Nečas, *Regularity results for the solution of the Cauchy problem to Navier-Stokes equations*, in: *Computational science for the 21st century. Dedicated to Prof. Roland Glowinski on the occasion of his 60th birthday. Symposium, Tours, France, May 5–7, 1997.*, pp. 23–30, Chichester, John Wiley & Sons., 1997.

1999

[C180] S. Leonardi, J. Málek, J. Nečas, and M. Pokorný, *On Axially Symmetric Flows in R^3*, Z. Anal. Anwendungen, 1999, vol. 18, pp. 639–649.

[C181] J. Málek, J. Nečas, M. Pokorný, and M.E. Schonbek, *On possible singular solutions to the Navier-Stokes equations*, Math. Nachr., 1999, vol. 199, pp. 97–114.

[C182] H. Bellout, J. Nečas, and K.R. Rajagopal, *On the existence and uniqueness of flows of multipolar fluids of grade 3 and their stability*, Int. Journ. Eng. Sciences, 1999, vol. 37, pp. 75–96.

[C183] F. Mošna and J. Nečas, *Nonlinear Hyperbolic equations with Dissipative Temporal and spatial Non-Local Memory*, Z. Anal. Anwendungen, 1999, vol. 18, pp. 939–951.

2001

[C184]* J. Málek, J. Nečas, and M. Růžička, *On the weak solutions to a class of non-Newtonian incompressible fluids in bounded three-dimensional domains: the case $p \geq 2$*, Adv. Differential Equations, 2001, vol. 6, pp. 257–302.

[C185] J. Nečas and T. Roubíček, *Buoyancy-driven viscous flow with L^1-data*, Nonlinear Anal. Ser. A: Theory Methods and Appl., 2001, vol. 46, pp. 737–755.

[C186] E. Behr, J. Nečas, and H. Wu, *On blow up of solution for Euler equations*, M2AN, Math. Model. Numer. Anal. 2001, vol. 35, pp. 229–238.

2002

[C187]* J. Málek, J. Nečas, and K.R. Rajagopal, *Global analysis of the flows of fluids with pressure-dependent viscosities*, Arch. Rational. Mech. Anal., 2002, vol. 165, pp. 243–269.

[C188] J. Nečas and J. Neustupa, *New conditions for local regularity of a suitable weak solution to the Navier-Stokes equation*, J. Math. Fluid Mech., 2002 vol. 4, pp. 237–256.

[C189] J. Málek, J. Nečas, and K.R. Rajagopal, *Global existence of solutions for flows of fluids with pressure and shear dependent viscosities*, Appl. Math. Lett., 2002, vol. 15, 961–967.

[C190] H. Bellout, E. Cornea, and J. Nečas, *On the concept of very weak L^2 solutions to Euler's equations*, SIAM J. Math. Anal., 2002, vol. 33, pp. 995–1006.

2003

[C191] J. Hron, J. Málek, J. Nečas, and K.R. Rajagopal, *Numerical simulations and global existence of solutions of two-dimensional flows of fluids with pressure- and shear-dependent viscosities*, in: *Proceedings: Modelling 2001 (Pilsen)*, Math. Comput. Simulation, 2003, vol. 61, pp. 297–315.

Editorial Work

[D1] J. Nečas, J. Kolomý, R. Kluge, and A. Langenbach (eds.), *Theory of nonlinear operators. Proceedings of a summer-school held in October 1972 at Neuendorf (Hiddensee), GDR*, Schriftenreihe des Zentralinstituts für Mathematik und Mechanik bei der Akademie der Wissenschaften der DDR, Heft 20, Akademie-Verlag, Berlin, 1974.

[D2] G.P. Galdi and J. Nečas (eds.), *Recent developments in theoretical fluid mechanics. Papers presented at the 2nd winter school in fluid dynamics, held in Paseky, Czech Republic, in November 29-December 4, 1992*, Pitman Research Notes in Mathematics Series. 291. Harlow: Longman Scientific & Technical, New York, NY: Wiley, 1993.

[D3] G.P. Galdi, J. Málek, and J. Nečas (eds.), *Progress in theoretical and computational fluid mechanics. 3rd winter school, Paseky, Czech Republic, December 12–18, 1993*, Pitman Research Notes in Mathematics Series. 308. Harlow, Essex: Longman Scientific & Technical. New York, NY: Wiley, 1994.

[D4] G.P. Galdi, J. Málek, and J. Nečas (eds.), *Mathematical theory in fluid mechanics. Lectures of the 4th winter school, Paseky, Czech Republic, December 3–9, 1995*, Pitman Research Notes in Mathematics Series. 354. Harlow, Essex: Longman, 1996.

[D5] J. Málek, J. Nečas, and M. Rokyta (eds.), *Advanced topics in theoretical fluid mechanics. Lectures of the 5th Winter School on mathematical theory in fluid mechanics, Paseky nad Jizerou, Czech Republic, December 6–14, 1997*, Pitman Research Notes in Mathematics Series. 392. Harlow: Longman, 1998.

[D6] W. Jäger, J. Nečas, O. John, K. Najzar, and J. Stará (eds.), *Partial differential equations: theory and numerical solution. Proceedings of the ICM'98 satellite conference, Prague, Czech Republic, August 10–16, 1998*, Chapman & Hall/CRC Research Notes in Mathematics, 406, Boca Raton, FL: Chapman & Hall/CRC, 2000.

[D7] J. Málek, J. Nečas, and M. Rokyta (eds.) *Advances in mathematical fluid mechanics. Lecture notes of the 6th international school on mathematical theory in fluid mechanics, Paseky, Czech Republic, September 19–26, 1999*, Springer, Berlin, 2000.

Chapter 3
Jindřich Nečas in the Early 1960s

Ivo Babuška

Abstract The first period of J. Nečas scientific career is described.

Jindřich Nečas studied in mid-1950s. The 1950s were under the influence of World War II. In 1938 Hitler occupied Czechoslovakia and closed all universities. A gap in higher education and in scientific generation occurred. In 1949–1951 a higher education program was created. A group of about ten young promising people was selected; they were awarded fellowships and studied under the leading Czech mathematicians, E. Čech, V. Jarník, V. Kořínek, V. Knichal. I was one of them. Let me mention J. Kurzweil, M. Fiedler and V. Pták as a few of the others in the group. The study was very intense. The above mentioned teachers were completely devoted to the education of this group of young promising men. These people then later became advisers of the younger students. J. Nečas studied then under my supervision. Before my mathematical studies I studied civil engineering at the Czech Technical University and became Dipl. Engineer. I was interested and working in the field of differential equations and their numerical treatment as well as applications in solid mechanics. During the 1950s in Czechoslovakia a large gravitational concrete dam was built at Orlík on the Vltava river about 50 miles from Prague. One of the problems with building such a dam is that the concrete creates heat when it is hardening. This creates stresses which could create cracks when the dam is cooling down. (For example the Hoover Dam in the USA had cooling pipes to prevent high temperatures.) One of the problems of Orlík was whether the cooling pipes could be avoided by using a special cement. No cooling pipes were used. This problem was mathematically and numerically addressed by a group under my leadership and J. Nečas became a member of this group. For more about it, see [1]. These years very much influenced Jindřich's scientific life. Mathematical problems of partial differential equations stemming from the problems of thermoelasticity became his major interest. He achieved outstanding results in this field. His interests in these specific applications in mechanics are very visible from the monographs he wrote. Let me now mention some of his work during the Orlík project. There were no electronic computers and all numerical computations were done on mechanical

© Springer Basel 2015

Š. Nečasová et al. (eds.), *Selected Works of Jindřich Nečas*, Advances in Mathematical Fluid Mechanics, DOI 10.1007/978-3-0348-0231-4_3

calculators. Various methodologies adequate to this environment had to be used. One of them was the method of the integral transformations as Laplace and Mellin transformation. The dam is nearly a convex two-dimensional cone. Typical for Jindřich was that he concentrated on the general mathematical aspects. So he addressed among other things the biharmonic problem on an infinite cone and the most general boundary conditions leading to the existence and uniqueness of the solution of the problem. This problem is directly related to the two-dimensional elasticity. Jindřich addressed the problem and the properties of its solution by the Mellin transformations, see [C5, C6]. The major problem was characterization of the solution properties in the neighborhood of corners and infinity. Problems on cones later became very important problems and were addressed by world-leading mathematicians (for example in papers by the mathematicians Kondratjev, Grisvard, Oleinik, Mazya and others). Analysis of the singular behavior of solutions in the neighborhood of corner points is very important for the numerical treatment of these problems. The years of Nečas's studies during the 1950s were very formative and influenced his future work in the theory of PDEs. He always concentrated on theoretical mathematical problems which were clearly, directly or indirectly, related to applications in mechanics.

Nečas' Works Quoted Above

The references [Cxx] refer to the respective papers listed in Chapter 2 "Complete Bibliography of Jindřich Nečas".

[C5] J. Nečas, *Solution du problème biharmonique pour le coin infini I*, Časopis Pěst. Mat., 1958, vol. 83, pp. 257–286. In Czech.

[C6] J. Nečas, *Solution du problème biharmonique pour le coin infini II*, Časopis Pěst. Mat., 1958, vol. 83, pp. 399–424. In Czech.

References

[1] I. Babuška, Development of Computational Mathematics in Czechoslovakia and in the USSR. In History of Scientific Computing, S.G.Nash editor, ACM Press, 1990, 314–328.

Ivo Babuška
Department of Aerospace Engineering and Engineering Mechanics,
University of Texas, 105 W. Dean Keeton Street, SHC 328, Austin, TX 78712, USA
email: babuska@ices.utexas.edu

Chapter 4
Jindřich Nečas and the Prague School of PDEs

Alois Kufner

Abstract The organization of research in Partial Differential Equations at the Academy of Sciences and at the Charles University in connection with the activities of J. Nečas is described.

I do not know exactly at what time the Department of PDEs of the Mathematical Institute was founded, but it was in connection with some practical problems (the building of the Orlík Dam), probably soon after foundation of the Academy of Sciences in 1952.

In any case, in 1958, as I started working at the Mathematical Institute, this Department existed and was headed by Ivo BABUŠKA. At the same time, there was also a Department of ODEs, and in this department there was a group headed (informally) by Otto VEJVODA, focusing on evolution equations. At that time a great deal of attention in the PDE department was given to the book of Sobolev *Some Applications of Functional Analysis to Mathematical Physics* (maybe also due to close contacts of Babuška with Sobolev—he visited Sobolev in the newly founded research center Akademgorodok near Novosibirsk) and, in connection with this topic, also to the theory of weak solutions (called mainly variational solutions at that time). This theory, together with modern tools such as Sobolev spaces, was among the main mathematical interests of Jindřich NEČAS, who had also close contacts with people like Lions, Magenes etc.

Babuška, due his connection with applied problems, started to work more on practical and numerical problems (such as problems of the stability of solutions with respect to variations of the domain and problems of numerical stability), and consequently some divergence of research interests with the more theoretical part of a department (presented by J. Nečas) became transparent. Eventually the "theoretical" group became autonomous and moved to a new place, still as a part of the Mathematical Institute. The new department was finally named a Department of the Theory of PDEs. It was headed by J. Nečas and the main topic was the modern theory of elliptic equations. Over time new young people arrived (originally, only J. Fuka and A. Kufner moved with J. Nečas to the new place): J. Kadlec,

© Springer Basel 2015

Š. Nečasová et al. (eds.), *Selected Works of Jindřich Nečas*, Advances in Mathematical Fluid Mechanics, DOI 10.1007/978-3-0348-0231-4_4

J. Souček, A. Kratochvíl, M. Kučera, Z. Poracká, together with mathematicians from the university, J. Kopáček and J. Král. The main attention was devoted to the investigation of regularity of weak solutions. Other topics have been investigated (e.g. potential theory—J. Král, and the theory of function spaces). There was a close collaboration with the Faculty of Mathematics and Physics of the Charles University, and finally J. Nečas started to work simultaneously also as the head of the Department of Mathematical Analysis at the University. He has taken new students and co-workers (O. John, J. Stará, and his most gifted student, S. Fučík) and started what was a new direction of research for the Prague group, namely the NONLINEAR theory of PDEs. Besides Nečas's department, a Department of Evolution Differential Equations was also created in the Mathematical Institute, originally headed by O. Vejvoda. It still exists under the name of Department of Evolution Differential Equations.

The former Department of the Theory of PDEs existed in the Mathematical Institute until recently, although it was after some time renamed as the Department of Qualitative Methods of Mathematical Analysis. Finally, in 2010 it was merged with the Department of Evolution Differential Equations and the Department of Topology and Functional Analysis.

Alois Kufner
Institute of Mathematics, Academy of Sciences, Žitná 25, 11567 Prague 1, Czech Republic
email: kufner@math.cas.cz

Part II
Linear Operators

Chapter 5
Selected Papers on Linear Elliptic Operators

Чехословацкий математический журнал, т. 12 (87) 1962, Прага

ОБ ОБЛАСТЯХ ТИПА $\mathfrak{N}$

ИНДРЖИХ НЕЧАС (Jindřich Nečas), Прага

(Поступило в редакцию 28/VI 1960 г.)

Области с границей, представленной функциями, удовлетворяющими условию Липшица — т. е. области типа $\mathfrak{N}$ — имеют важное значение в теории пространств Беппо Леви (Beppo Levi), которая является фундаментальной для решения эллиптических дифференциальных уравнений по вариационному методу. В этой статье приводится доказательство теоремы об аппроксимации такой области при помощи областей с бесконечно-дифференцируемой границей.

1. ПОСТАНОВКА ПРОБЛЕМЫ И ОКОНЧАТЕЛЬНОЕ УТВЕРЖДЕНИЕ

Пусть E_n — n-мерное пространство. Точку в E_n обозначим через $[x_1, x_2, \ldots, x_n]$ или короче X.

Ограниченная область, расположенная в E_n, является областью типа $\mathfrak{N}$, если существует:

1. m систем координат в E_n и m функций a_r, $r = 1, 2, \ldots, m$ так, что каждая точка границы области Ω представлена по крайней мере в одной из этих систем в виде: $[x_{r1}, x_{r2}, \ldots, x_{rn-1}, a_r(x_{r1}, \ldots, x_{rn-1})]$, короче $[X_r, a_r(X_r)]$. Функции a_r определены в сферической окрестности

$$|X_r| = \left(\sum_{i=1}^{n-1} x_{ri}^2\right)^{1/2} < \alpha$$

и удовлетворяют там условию Липшица:

$$|a_r(X_r) - a_r(Y_r)| \leqq K|X_r - Y_r| ;$$

2. число $0 < \beta \leqq 1$ так, что точки $|X_r| < \alpha$, $a_r(X_r) - \beta < x_{rn} < a_r(X_r)$ в Ω и точки $|X_r| < \alpha$, $a_r(X_r) < x_{rn} < a_r(X_r) + \beta$ вне $\bar{\Omega}$.

В третьем разделе приведем доказательство следующей теоремы об аппроксимации:

Теорема. *Пусть Ω типа $\mathfrak{N}$. Пусть $[X_r, x_{rn}]$ или же a_r, $r = 1, 2, \ldots, m$ — системы координат или же функции, упомянутые при определении области типа $\mathfrak{N}$. Пусть $p \geqq 1$.*

© Institute of Mathematics, Academy of Sciences of the Czech Republic 1962. Reprinted with kind permission from: Nečas, J., On domaines of type N (Russian), Czechoslovak Mathematical Journal 12, 274-287, http://dml.cz.

Тогда существует совокупность областей Ω_h, $0 < h < h_0$, $h_1 < h_2 \Rightarrow \bar{\Omega}_{h_2} \subset \subset \Omega_{h_1}$ *так, что* $\lim_{h\to 0} \Omega_h = \Omega$ *и границы* Ω_h *представлены в системах координат* $[X_r, x_{rn}]$ *функциями* $a_r(h)$. *Функции* $a_r(h)$ *бесконечно дифференцируемы в* $|X_r| < \alpha$, *удовлетворяют там условию Липшица*

$$(1.1) \qquad |a_r(h, X_r) - a_r(h, Y_r)| \leqq K|X_r - Y_r| \ ,$$

причем постоянная K *не зависит от* h. *Далее,*

$$(1.2) \qquad \lim_{h\to 0} \big(\sup_{|Xr|<\alpha} |a_r(h, X_r) - a_r(X_r)| \big) = 0 \, ,$$

$$(1.3) \quad \lim_{h\to 0} \int_{|X_r|<\alpha} \left| \frac{\partial a_r}{\partial x_{ri}} - \frac{\partial a_r(h)}{\partial x_{ri}} \right|^p \mathrm{d}X_r = 0 \, , \quad r = 1, 2, \ldots, m, \quad i = 1, 2, \ldots, n-1 \, .$$

(*Первые производные функции* $a_r(X_r)$ *существуют почти всюду.*)

Мы докажем теорему посредством лемм об аппроксимации функций и лемм о замене систем координат. Важно заметить, что значение лемм об аппроксимации функций выходит за пределы этой статьи, и можно ими с успехом воспользоваться после модификации, пригодной для изучения свойств пространств Беппо Леви.

2. ЛЕММЫ ОБ АППРОКСИМАЦИИ ФУНКЦИЙ

Через Δ_α обозначаем сферическую окрестность $|X| < \alpha$ в E_n. Множество функций f, суммируемых с p-ой степенью в множестве M, мы будем обозначать через $L_p(M)$. Введем в множестве $L_p(M)$ норму

$$\|f\| = \left[\int_M |f|^p \, \mathrm{d}\Omega \right]^{1/p} .$$

Лемма 2.1. *Пусть* Ω — *ограниченная область. Пусть* $f \in L_p(\Omega)$, *где* $p \geqq 1$. *Пусть* h — *непрерывная функция вместе со своими первыми производными в* Ω, *ограниченными здесь постоянной* K, *и пусть* $-1 \leqq h(X) \leqq 1$ *для* $X \in \Omega$. *Пусть* $f = 0$ *вне* Ω.

Тогда для каждого $\varepsilon > 0$ *найдется такое число* $\delta > 0$, *что* $|Z| < \delta \Rightarrow$

$$(2.1) \qquad \left[\int_\Omega |f(X + Zh(X)) - f(X)|^p \, \mathrm{d}X \right]^{1/p} < \varepsilon \, .$$

Доказательство. Пусть $\varepsilon > 0$. Теперь найдется число v_1, $0 < v_1 < \mu(\Omega)$, такое, что

$$\left[\int_M |f(X)|^p \, \mathrm{d}X \right]^{1/p} < \frac{\varepsilon}{6} \quad \text{если} \quad \mu(M) < v_1 \, .$$

Здесь $\mu(M)$ означает меру множества M. По теореме Лузина найдется замкнутое множество $F \subset \Omega$ так, что $\mu(F) > \mu(\Omega) - v_1/4$ и функция f непрерывна в F. Пусть теперь $\delta_1 > 0$ такое, что

$$|Z| < \delta_1 \Rightarrow H_Z = \mathop{\mathscr{E}}_Y \big(Y = X + Zh(X) \, , \quad X \in F \big) \subset \Omega \, .$$

Теперь найдется $\delta_2 > 0$ так, что

$$|Z| < \delta_2 \Rightarrow \frac{\mu(\Omega) - \nu_1/4}{\mu(F)} < D(X) < 2 \quad \text{для} \quad X \in \Omega\,,$$

и отображение $Y(X) = X + Zh(X)$ — простое. Здесь $D(X)$ — определитель Якоби отображения $Y(X)$.

Отсюда следует, что

$$\mu(H_z) = \int_F D(X)\,\mathrm{d}X > \mu(\Omega) - \frac{\nu_1}{4}\,.$$

Теперь существует $\delta_3 > 0$ так, что

$$|Z| < \delta_3 \Rightarrow |f(X + Zh(X)) - f(X)| < \frac{\varepsilon}{3\mu(\Omega)^{1/p}}\,,$$

если X и $X + Zh(X)$ — элементы F. Пусть $L_Z = FH_Z$. Имеем $L_Z = F - (F - H_Z)$ и поэтому

$$\mu(L_Z) > \mu(F) - \mu(\Omega) + \mu(H_Z) > \mu(\Omega) - \nu_1/2\,.$$

Пусть $\delta = \min(\delta_1, \delta_2, \delta_3)$ и пусть $|Z| < \delta$. Тогда получаем

$$(2.2)\quad \left[\int_\Omega |f(X + Zh(X)) - f(X)|^p\,\mathrm{d}X\right]^{1/p} \leqq \left[\int_{L_Z} |f(X + Zh(X)) - f(X)|^p\,\mathrm{d}X\right]^{1/p} + \\ + \left[\int_{\Omega - L_Z} |f(X + Zh(X))|^p\,\mathrm{d}X\right]^{1/p} + \left[\int_{\Omega - L_Z} |f(X)|^p\,\mathrm{d}X\right]^{1/p}.$$

Первое слагаемое в правой части (2.2) $< \varepsilon/3$. Не трудно видеть, что

$$\left[\int_{\Omega - L_Z} |f(X + Zh(X))|^p\,\mathrm{d}X\right]^{1/p} \leqq \left(\frac{\mu(\Omega)}{\mu(\Omega) - \nu_1/2}\right)^{1/p} \left[\int_{Y(\Omega - L_Z)} |f(Y)|^p\,\mathrm{d}Y\right]^{1/p}.$$

Теперь $\mu(Y(\Omega - L_z)) \leqq \mu(\Omega - L_z)\,.\,2 < \nu_1$. Так как

$$\left(\frac{\mu(\Omega)}{\mu(\Omega) - \nu_1/2}\right)^{1/p} \leqq 2\,,$$

то второе слагаемое в правой части (2.2) $< \varepsilon/3$. Но третий интеграл в правой части (2.2) меньше чем $\varepsilon/6$, чем завершается доказательство нашей леммы.

Пусть $f \in L_p(\Delta_\alpha)$. Следует определить среднюю функцию для f с изменяющимся параметром усреднения следующим образом:

$$(2.3)\qquad f(h, X) = \frac{1}{\kappa h^n(X)} \int_{|X - Y| < h(X)} \exp \frac{|X - Y|^2}{|X - Y|^2 - h^2(X)} f(Y)\,\mathrm{d}Y\,,$$

где

$$X \in \Delta_\alpha\,, \quad h(X) = h \exp \frac{|X|^2}{|X|^2 - \alpha^2}\,,$$

$$\kappa = \int_{|X| < 1} \exp \frac{|X|^2}{|X|^2 - 1}\,\mathrm{d}X\,.$$

Прежде всего, очевидно, существует постоянная H так, что для $h \leqq H\alpha$ множество $\underset{Y}{\mathscr{E}}(|Y - X| < h(X),\ X \in \Delta_\alpha) \subset \Delta_\alpha$. Всегда мы будем рассматривать только такие h. Можно сказать, что остают в силе те свойства средней функции (2.3), которые были доказаны для средних функций, введеных С. Л. Соболевым в [1]. Но кроме упомянутых выше свойств мы докажем еще другие свойства, касающиеся поведения средней функции вблизи границы Δ_α.

Лемма 2.2. *Если $f \in L_p(\Delta_\alpha)$, $p \geqq 1$, то $f(h) \in L_p(\Delta_\alpha)$, $f(h)$ бесконечно дифференцируема в Δ_α и $\lim_{h \to 0} f(h) = f$ в $L_p(\Delta_\alpha)$.*

Доказательство. Мы можем записать интеграл (2.3) в виде

$$f(h, X) = \frac{1}{\kappa h^n(X)} \int_{\Delta_\alpha} \omega(X, Y, h) f(Y)\,\mathrm{d}Y, \tag{2.4}$$

полагая

$$\omega(X, Y, h) \begin{cases} \exp \dfrac{|X - Y|^2}{|X - Y|^2 - h^2(X)} & \text{для} \quad |X - Y| < h(X), \\ 0 \quad \text{для} \quad |X - Y| \geqq h(X). \end{cases}$$

Так как функция $\omega(X, Y, h)$ бесконечно непрерывно дифференцируема для $X \in \Delta_\alpha$, $Y \in E_n$, то мы получим второе утверждение леммы 2.2 из хорошо известных теорем о дифференцировании за знаком интеграла.

В результате замены переменной мы получим из (2.3)

$$f(h, X) = \frac{1}{\kappa} \int_{|Z| < 1} \exp \frac{|Z|^2}{|Z|^2 - 1} [f(X + Zh(X))]\,\mathrm{d}Z, \tag{2.5}$$

одкуда следует

$$\|f(h) - f\|^p_{Lp(\Delta\alpha)} = \int_{\Delta_\alpha} \left| \frac{1}{\kappa} \int_{|Z| < 1} \exp \frac{|Z|^2}{|Z|^2 - 1} [f(X + Zh(X)) - f(X)]\,\mathrm{d}Z \right|^p \mathrm{d}X \leqq$$

$$\leqq \mathrm{const} \int_{\Delta_\alpha} \mathrm{d}X \int_{|Z| < 1} |f(X + Zh(X)) - f(X)|^p\,\mathrm{d}Z =$$

$$= \mathrm{const} \int_{|Z| < 1} \mathrm{d}Z \int_{\Delta_\alpha} |f(X) + Zh(X)) - f(X)|^p\,\mathrm{d}X.$$

Но теперь по лемме 2.1 ко всякому $\varepsilon > 0$ можно подобрать такое $\delta > 0$, что из $h < \delta$ следует

$$\int_{\Delta_\alpha} |f(X + Zh(X)) - f(X)|^p\,\mathrm{d}X < \varepsilon$$

для $Z \in \Delta_1$, чем лемма 2.2 полностью доказана.

Множество функций, непрерывных в $\overline{\Delta}_\alpha$, мы будем обозначать через $C(\Delta_\alpha)$, а норму в этом пространстве определять как обычно

$$\|f\|_{C(\Delta_\alpha)} = \max_{X \in \overline{\Delta}_\alpha} |f(X)| .$$

Мы докажем теперь следуюшую лемму:

Лемма 2.3. *Если* $f \in C(\Delta_\alpha)$, *то* $\lim_{h \to 0} f(h) = f$ *в* $C(\Delta_\alpha)$ *и* $f(h, X) = f(X)$ *для* $|X| = \alpha$.

Доказательство. Начнем с доказательством второй части леммы 2.3. В самом деле, пусть $|X| = \alpha$ и пусть $X_k \in \Delta_\alpha$, $X_k \to X$. Тогда

$$f(h, X_k) - f(X_k) = \tag{2.6}$$
$$= \frac{1}{\kappa h^n(X_k)} \int_{|X_k - Y| < h(X_k)} \exp \frac{|X_k - Y|^2}{|X_k - Y|^2 - h^2(X_k)} [f(Y) - f(X_k)] \, \mathrm{d}Y .$$

Так как $h(X_k) \to 0$, то из равномерной непрерывности в $\overline{\Delta}_\alpha$ и из (2.6) следует, что

$$\lim_{k \to \infty} (f(h, X_k) - f(X_k)) = 0 , \quad \text{откуда} \quad \lim_{k \to \infty} f(h, X_k) = f(X) .$$

Функция $f(h, X)$ лежит в $C(\Delta_\alpha)$. Второе утверждение леммы 2.3 является почти очевидным следствием равномерной непрерывности f в $\overline{\Delta}_\alpha$ и неравенства

$$|f(h, X) - f(X)| \leqq \frac{1}{\kappa h^n(X)} \int_{|X - Y| < h(X)} \exp \frac{|X - Y|^2}{|X - Y|^2 - h^2(X)} |f(Y) - f(X)| \, \mathrm{d}Y .$$

Таким образом лемма 2.3 полностью доказана.

Пусть $f \in C(\Delta_\alpha)$ и пусть она удовлетворяет в $\overline{\Delta}_\alpha$ условию Липшица. Тогда, определим

$$K(f) = \sup_{X, Y \in \overline{\Delta}_\alpha} \frac{|f(X) - f(Y)|}{|X - Y|} .$$

Мы будем в дальнейшем пользоваться тем, что f имеет в Δ_α почти всюду полный дифференциал, и что первые производные этой функции совпадают с первыми обобщенными производными. Доказательства этих утверждений приведены в [2] и в работе [3], стр. 315.

Лемма 2.4. *Если* $f \in C(\Delta_\alpha)$ *и удовлетворяет там условию Липшица, то*

$$\overline{\lim_{h \to 0}}\, K(f(h)) = K(f) , \quad \lim_{h \to 0} \frac{\partial f(h)}{\partial x_i} = \frac{\partial f}{\partial x_i} \quad \text{в} \quad L_p(\Delta_\alpha),\ p \geqq 1,\ i = 1, 2, \ldots, n .$$

Доказательство. Лишь для целей доказательства леммы 2.4 вводим обозначение $X_s = [x_1, x_2, \ldots, x_i + s, x_{i+1}, \ldots, x_n]$.

Из (2.5) имеем

$$\frac{f(h, X_s) - f(h, X)}{s} = \frac{1}{\kappa} \int_{|Z| < 1} \exp \frac{|Z|^2}{|Z|^2 - 1} \frac{f(X_s + Zh(X_s)) - f(X + Zh(X))}{s} \, \mathrm{d}Z . \tag{2.7}$$

Имея в виду, что у f полный дифференциал почти всюду в Δ_α получаем для почти всех Z из Δ_1

$$\lim_{s\to 0}\frac{f(X_s + Zh(X_s)) - f(X + Zh(X))}{s} =$$

$$= \sum_{j=1}^{n}\frac{\partial f}{\partial x_j}(X + Zh(X))\, z_j\,\frac{\partial h}{\partial x_i}(X) + \frac{\partial f}{\partial x_i}(X + Zh(X))\,.$$

С другой стороны,

$$\frac{|f(X_s + Zh(X_s)) - f(X + Zh(X))|}{|X_s + Zh(X_s) - X - Zh(X)|}\cdot\frac{|X_s + Zh(X_s) - X - Zh(X)|}{s} \leqq \text{const}\,,$$

и потому можно сделать предельный переход под знаком интеграла в (2.7).

Итак,

$$(2.8)\qquad \frac{\partial f(h, X)}{\partial x_i} =$$

$$= \frac{1}{\kappa}\int_{|Z|<1}\exp\frac{|Z|^2}{|Z|^2 - 1}\left[\sum_{j=1}^{n}\frac{\partial f}{\partial x_j}(X + Zh(X))\, z_j\,\frac{\partial h}{\partial x_i}(X)\right]dZ +$$

$$+ \frac{1}{\kappa}\int_{|Z|<1}\exp\frac{|Z|^2}{|Z|^2 - 1}\,\frac{\partial f}{\partial x_i}(X + Zh(X))\,dZ\,.$$

Но $\lim\limits_{h\to 0}\partial h/\partial x_i(X) = 0$ в $C(\Delta_\alpha)$; итак, из (2.8), пользуясь леммой 2.2, получим вторую часть леммы 2.4.

С другой стороны, из (2.8) следует

$$(2.9)\qquad \overline{\lim_{h\to 0}}\left|\frac{\partial f(h, X)}{\partial x_i}\right| \leqq K(f)\,.$$

Оценка (2.9) не зависит от выбора системы координат, из чего следует ее независимость от направления, в котором происходит дифференцирование. Итак, $\overline{\lim\limits_{h\to 0}}\, K(f(h)) \leqq K(f)$. Но знак $<$ не возможен, потому что, во-первых, для каждого $\varepsilon > 0$ найдется направление $\vec{v}$ и точка $X \in \Delta_\alpha$ таким образом, что

$$\frac{\partial f}{\partial \vec{v}}(X) + \varepsilon > K(f)\,,$$

и, во-вторых, существует последовательность $h_k \to 0$ для $k \to \infty$, для которой

$$\frac{\partial f}{\partial \vec{v}}(h_k, X) \to \frac{\partial f}{\partial \vec{v}}(X)$$

по смыслу точечной сходимости. Лемма 2.4 полностью доказана.

Теперь положим

$$(2.10)\qquad \tilde{f}(h, X) = f(h, X) - [nK(f) + 1]\, h(X)\,.$$

Тогда имеет место следующая

Лемма 2.5. *Если f удовлетворяет условию Липшица, то $\tilde{f}(h)$ имеет те же свойства, что и функция f, приведенные в лемме* 2.4, *и кроме того*

$$\tilde{f}(h, X) < f(X) \quad \text{и} \quad \frac{\partial \tilde{f}}{\partial h}(h, X) < 0 \quad \text{для} \quad X \in \Delta_\alpha .$$

В самом деле,

$$f(X) - \tilde{f}(h, X) = \frac{1}{\kappa h^n(X)} \int_{|X-Y|<h(X)} \exp \frac{|X - Y|^2}{|X - Y|^2 - h^2(X)} \,.$$
$$. \left[f(X) - f(Y) + nK(f)\, h(X) + h(X)\right] \mathrm{d}Y > 0 ,$$

так как $f(X) - f(Y) + nK(f)\, h(X) + h(X) > 0$ для $X \in \Delta_\alpha$.

Для доказательства последнего утверждения леммы 2.5 можно поступить так как же при доказательстве леммы 2.4. Таким образом получим

$$\frac{\partial \tilde{f}}{\partial h}(h, X) = \frac{1}{\kappa} \int_{|Z|<1} \exp \frac{|Z|^2}{|Z|^2 - 1} \left[\sum_{j=1}^{n} \frac{\partial f}{\partial x_j}(X + Zh(X))\, z_j \,.\, a(x) - \right.$$
$$\left. - nK(f)\, a(X) - a(X)\right] \mathrm{d}Z \leqq - a(X) < 0 . \quad \text{Здесь} \quad a(X) = \exp \frac{|X|^2}{|X|^2 - \alpha^2} .$$

Этим лемма 2.5 доказана.

Пусть $\alpha_1 > \alpha$ и функция f непрерывна в Δ_{α_1}; мы определим $f(h, X)$ в Δ_{α_1} следующим образом:

$$(2.10) \quad f(h, X) \begin{cases} \dfrac{1}{\kappa h^n(X)} \displaystyle\int_{|X-Y|<h(X)} \exp \frac{|X - Y|^2}{|X - Y|^2 - h^2(X)} f(Y)\, \mathrm{d}Y & \text{для} \quad X \in \Delta_\alpha , \\ f(X) \quad \text{для} \quad X \in \Delta_{\alpha_1} - \Delta_\alpha . \end{cases}$$

Из леммы 2.3 вытекает, что $f(h, X)$ непрерывна в Δ_{α_1}.

Что касается поведения производных, на этот вопрос дает частичный ответ лемма 2.6.

Лемма 2.6. *Пусть $f \in C(\Delta_{\alpha_1})$ (это предположение не существенно) и пусть f бесконечно непрерывно дифференцируема в некотором открытом множестве G, расположенном целиком в Δ_{α_1}. Тогда $f(h)$, определенная в* (2.10), *имеет то же свойство.*

Доказательство. Обозначим

$$\Gamma_\alpha = \mathop{\mathscr{E}}_{X} (|X| = \alpha) \quad (\alpha < \alpha_1) .$$

Если бы $\Gamma_\alpha \,.\, G = 0$, то нечего доказывать.

Пусть $\Gamma_\alpha G \neq 0$. Если i — целое, неотрицательное число и $i_1, i_2, \ldots, i_n$ —

целые неотрицательные числа, сумма которых равна i, и $X_0 \in \Gamma_\alpha G$, то для доказательства нашей леммы достаточно доказать, что

$$\lim_{k\to\infty} \frac{\partial^i f(h, X_k)}{\partial x_1^{i_1} \ldots \partial x_n^{i_n}} = \frac{\partial^i f(X_0)}{\partial x_1^{i_1} \ldots \partial x_n^{i_n}}$$

при условии, что $X_k \in \Delta_\alpha$ и $\lim\limits_{k\to\infty} X_k = X_0$.

В самом деле, исследуя формулу (2.5), мы получаем, начиная с k, достаточно большим

$$\frac{\partial^i f(h, X_k)}{\partial x_1^{i_1} \ldots \partial x_n^{i_n}} = \frac{1}{\kappa}\int_{|Z|<1} \exp\frac{|Z|^2}{|Z|^2-1}\,\frac{\partial^i f(X_k + Zh(X_k))}{\partial x_1^{i_1}\ldots\partial x_n^{i_n}}\,dZ + \tag{2.11}$$

$$+\frac{1}{\kappa}\int_{|Z|<1} \exp\frac{|Z|^2}{|Z|^2-1}\, g_{i_1 \ldots i_n}(X_k, Z)\,dZ\,,$$

причем $\lim\limits_{k\to\infty} g_{i_1 \ldots i_n}(X_k, Z) = 0$ равномерно для $Z \in \Delta_1$.

Повторяя теперь первую часть доказательства леммы 2.3, мы получим таким образом наше утверждение, а, следовательно, и доказательство леммы 2.6.

3. ЛЕММЫ ОБ ИЗМЕНЕНИИ СИСТЕМЫ КООРДИНАТ И ДОКАЗАТЕЛЬСТВО ОСНОВНОЙ ТЕОРЕМЫ

Леммы этого раздела тесно примыкаются к доказательству основной теоремы, и поэтому мы будем пользоваться в этом разделе обозначением, принятым для этого доказательства. U_r обозначает область, состоящую из точек

$$\mathop{\mathscr{E}}_{X}\left(|X_r| < \alpha,\ a_r(X_r) - \beta < x_{rn} < a_r(X_r) + \beta\right).$$

Через $\dot{\Omega}$ мы обозначаем границу области Ω типа $\mathfrak{N}$. Пусть $\Gamma_1, \Gamma_2, \ldots, \Gamma_p$ — составляющие $\dot{\Omega}$. Очевидно, что $\dot{\Omega} \subset \sum\limits_{r=1}^{m} U_r$. Мы будем предполагать, что из $U_{r_1}\Gamma_{s_1} \neq 0$, $U_{r_2}\Gamma_{s_2} \neq 0$, $s_1 \neq s_2$ следует $U_{r_1}U_{r_2} = 0$, не нарушая применения в общих случаях. Мы предполагаем еще, что каждая из окрестностей U_r необходима для покрытия Ω.

Очевидно, что существуют убывающие последовательности $\alpha = \alpha_0 > \alpha_1 > \alpha_2, \ldots, \beta = \beta_0 > \beta_1 > \beta_2 > \ldots$ таким образом, что области

$$U_r^k = \mathop{\mathscr{E}}_{X}\left(|X_r| < \alpha_k\,,\quad a_r(X_r) - \beta_k < x_{rn} < a_r(X_r) + \beta_k\right.$$

имеют те же свойства, что и U_r.

Пусть теперь для $r = 1, 2, \ldots, m$

(3.1) $$a_r(h, X_r) =$$
$$= \frac{1}{\kappa h^{n-1}(X_r)} \int_{|X_r - Y_r| < h(X_r)} \exp \frac{|X_r - Y_r|^2}{|X_r - Y_r|^2 - h^2(X_r)} a_r(Y_r) \, dY_r - [nK(a_r) + 1] \, h(X_r)$$

для $X_r \in \Delta_{\alpha_1}$, где

$$h(X_r) = h \exp \frac{|X_r|^2}{|X_r|^2 - \alpha_1^2} .$$

По лемме 2.5 можно найти $h_1 \leqq \alpha - \alpha_1$ так, что

$$0 < h \leqq h_1 \Rightarrow a_r(X_r) - \beta_1 < a_r(h, X_r) < a_r(X_r) .$$

Пусть теперь $r = 1$. Можно предполагать, что $U_1^1 U_1^2 \dot{\Omega} = \Gamma^1 \neq \emptyset$. Обозначим через $\Lambda(X)$ прямую, проходящую через точку X на Γ^1 и параллельную оси x_{2n}, и через

$$\Gamma_h = \mathop{\mathscr{E}}_X \left(X = [X_1, a_1(h, X_1)] , \quad |X_1| < \alpha_1\right) .$$

Покажем следующую лемму:

Лемма 3.1. *Имеется такое* $0 < l \leqq h_1$, *что для* $0 < h \leqq l$ *пересечение* $\Gamma_h \Lambda(X)$ *состоит лишь из одной точки.*

Доказательство. Обозначая через $\Gamma = U_1 U_2 \dot{\Omega}$, под G будем понимать проекцию множества Γ на плоскость $x_{1n} = 0$ и под G_h или G_0 проекцию множества $\Gamma_h U_1^1 U_2^1$ или $\Gamma_0 U_1^1 U_2^1$ на ту же плоскость. Из непрерывности функции a_r и леммы 2.3 следует, что существует $0 < h_2 \leqq h_1$ так, что $h \leqq h_2 \Rightarrow \bar{G}_h \subset G$ и расстояние $\bar{G}_h$ до границы множества G больше чем $c > 0$.

Для доказательства леммы 3.1 мы можем предполагать, что первая система координат связана со второй соотношениями

(3.2) $$x_{1i} = x_{2i} + \lambda_i , \quad i = 1, 2, \ldots, n-2 , \quad x_{1n-1} = x_{2n-1} \cos \varphi - x_{2n} \sin \varphi + \lambda_{n-1} , \quad x_{1n} = x_{2n-1} \sin \varphi + x_{2n} \cos \varphi + \lambda_n ,$$

где $-\pi < \varphi < \pi$.

Если $\varphi = 0$, то нечего доказывать. Итак, пусть $\varphi \neq 0$. Так как граница $\dot{\Omega}$ в первой и второй системе координат представлена в виде функции, удовлетворяющей условию Липшица, то имеем:

(3.3) $$0 < \varphi < \pi \Rightarrow \inf_{X_1 \in G} \frac{\partial a_1}{\partial x_{1n-1}} (X_1) + \operatorname{cotg} \varphi > d > 0 ,$$
$$-\pi < \varphi < 0 \Rightarrow \sup_{X_1 \in G} \frac{\partial a_1}{\partial x_{1n-1}} (X_1) + \operatorname{cotg} \varphi < -d < 0 .$$

Если $h \leqq h_3 \leqq h_2$, где h_3 достаточно мало (нужно иметь $h_3 \leqq c$), то из (2.8) следует, что для

$$(3.4) \qquad 0 < \varphi < \pi \Rightarrow \frac{\partial a_1(h, X_1)}{\partial x_{1n-1}} + \operatorname{cotg} \varphi > d\,,$$

$$-\pi < \varphi < 0 \Rightarrow \frac{\partial a_1(h, X_1)}{\partial x_{1n-1}} + \operatorname{cotg} \varphi < -d \quad \text{для} \quad X_1 \in G_h\,.$$

Рассмотрим ближе случай $0 < \varphi < \pi$. Итак, пусть

$$X = [X_2, x_{2n}] \in \Gamma_h U_2^1\,.$$

Из уравнения (3.5) имеем

$$a_1(h, x_{21} + \lambda_1, \ldots, x_{2n-2} + \lambda_{n-2}\,,$$
$$x_{2n-1} \cos \varphi - x_{2n} \sin \varphi + \lambda_{n-1}) - x_{2n-1} \sin \varphi - x_{2n} \cos \varphi - \lambda_n = 0\,.$$

Имея в виду теорему о неявных функциях и неравенство (3.4), можно выразить x_{2n} как функцию X_2, $x_{2n} = b(h, X_2)$. Обозначив теперь через K область определения этой функции, мы сразу видим, что $b(h)$ в K бесконечно непрерывно дифференцируема. Но из этого следует, что можно найти $0 < h_4 \leqq h_3$ таким образом, что для $h \leqq h_4$ множество K не зависит от h и, точнее говоря, оно является проекцией Γ^1 на плоскость $x_{2n} = 0$. Полагая $h_4 = l$, мы завершаем доказательство леммы 3.1. Сейчас же видно, что из доказательства леммы 3.1 вытекают и другие следствия.

Почти очевидна

Лемма 3.2. *Если* $X = [X_2, x_{2n}]$, $X_2 \in \Delta_{\alpha_1}$ *и* $X \notin \Gamma^1$, *то* $\Lambda(X)\,\Gamma_h = 0$. *Далее,*

$$\lim_{h \to 0} \left(\sup_{X_2 \in K} |b(h, X_2) - a_2(X_2)|\right) = 0\,.$$

Весьма важное значение имеет следующая

Лемма 3.3. *Функция* $b(h, X_2)$ *в* K *бесконечно непрерывно дифференцируема, первые производные ее ограничены постоянной, независящей от* $h \leqq l$, *так что она удовлетворяет в* K *условию Липшица и, кроме того,*

$$\lim_{h \to 0} \int_K \left| \frac{\partial b(h, X_2)}{\partial x_{2i}} - \frac{\partial a_2(X_2)}{\partial x_{2i}} \right|^p \mathrm{d}X_2 = 0\,, \quad p \geqq 1 \quad \text{для} \quad i = 1, 2, \ldots, n-1\,.$$

Доказательство. Так как в правильности первого утверждения леммы 3.3 мы уже убедились, то теперь приступим к доказательству второго.

В самом деле,

$$(3.6) \qquad \frac{\partial b(h, X_2)}{\partial x_{2i}} = \frac{[\partial a_1(h\; x_{21} + \lambda_1, \ldots)/\partial x_{1i}]}{[\partial a_1(h, x_{21} + \lambda_1, \ldots)/\partial x_{1n-1}] \sin \varphi + \cos \varphi}$$

для $i = 1, 2, \ldots, n-2$ и

$$\frac{\partial b(h, X_2)}{\partial x_{2n-1}} = \frac{[\partial a_1(h, x_{21} + \lambda_1, \ldots)/\partial x_{1n-1}] \cos \varphi - \sin \varphi}{[\partial a_1(h, x_{21} + \lambda_1, \ldots/\partial x_{1n-1}] \sin \varphi + \cos \varphi}\,.$$

Теперь из леммы 2.4 и неравенства (3.4) следует второе утверждение.

Посредством функции $a_1(h, X_1)$ мы определим теперь отображение множества G_h в K следующим образом:

$$(3.7) \qquad x_{2i} = x_{1i} - \lambda_i\,, \quad i = 1, 2, \ldots, n-2\,,$$
$$x_{2n-1} = x_{1n-1}\cos\varphi + a_1(h, X_1)\sin\varphi - \lambda_{n-1}\cos\varphi - \lambda_n \sin\varphi$$

и обозначим его через Z_h. Это отображение непрерывно и взаимно однозначно, первые производные этого отображения непрерывны и его определитель Якоби

$$D_h(X_1) = \sin\varphi\left[\frac{\partial a_1(h, X_1)}{\partial x_{1n-1}} + \operatorname{cotg}\varphi\right] \neq 0\,.$$

Посредством функции $a_1(X_1)$ мы определим отображение множества G_0 на K при помощи (3.7), если заменим $a_1(h, X_1)$ через $a_1(X_1)$. Обозначим это отображение через Z. Это обтоtражение непрерывно и взаимно однозначно и обладает важным свойством

$$(3.8) \qquad |Z(X_1) - Z(Y_1)| \leqq C|X_1 - Y_1|\,.$$

Также и отображение Z^{-1} непрерывно и обладает свойством

$$(3.9) \qquad |Z^{-1}(X_2) - Z^{-1}(Y_2)| \leqq C|X_2 - Y_2|\,.$$

Из (3.8), (3.9) следует: если $m \subset G_0$ и $\mu(m) = 0$, то и $\mu(Z(M)) = 0$, и наоборот.

Итак, имеем для почти всех X_2 из K (полный дифференциал от $a_1(X_1)$ существует почти всюду в G_0)

$$(3.10) \qquad \frac{\partial a_2(X_2)}{\partial x_{2i}} = \frac{\partial a_1(Z^{-1}(X_1))/\partial x_{1i}}{[\partial a_1(Z^{-1}(X_1))/\partial x_{1n-1}]\sin\varphi + \cos\varphi}\,, \quad i = 1, 2, \ldots n-2\,,$$
$$\frac{\partial a_2(X_2)}{\partial x_{2n-1}} = \frac{[\partial a_1(Z^{-1}(X_1))/\partial x_{1n-1}]\cos\varphi - \sin\varphi}{[\partial a_1(Z^{-1}(X_1))/\partial x_{1n-1}]\sin\varphi + \cos\varphi}\,.$$

Из (3.6) и (3.10) следует, что для доказательства третьего утверждения леммы 3.3 достаточно показать, что

$$\lim_{h\to 0}\left[\int_K\left|\frac{\partial a_1}{\partial x_{1i}}(h, Z_h^{-1}(X_2)) - \frac{\partial a_1}{\partial x_{1i}}(Z^{-1}(X_2))\right|^p \mathrm{d}X_2\right]^{1/p} = 0$$

для $i = 1, 2, \ldots, h-1$. Имеем

$$\left\|\frac{\partial a_1(h, Z_h^{-1})}{\partial x_{1i}} - \frac{\partial a_1(Z^{-1})}{\partial x_{1i}}\right\|_{L_p(K)} \leqq$$
$$\leqq \left\|\frac{\partial a_1(h, Z_h^{-1})}{\partial x_{1i}} - \frac{\partial a_1(Z_h^{-1})}{\partial x_{1i}}\right\|_{L_p(K)} + \left\|\frac{\partial a_1}{\partial x_{1i}}(Z_h^{-1}) - \frac{\partial a_1}{\partial x_{1i}}(Z^{-1})\right\|_{L_p(K)}.$$

Что касается первого слагаемого в правой части то оно стремится к нулю, вследствие возможности применения теоремы о замене переменных.

Теперь существует последовательность функций f_l, непрерывных в $\bar{\Delta}_\alpha$, равномерно ограниченных постоянной C и таких ,что

$$\lim_{l\to\infty}\left\|f_l - \frac{\partial a_1}{\partial x_{1i}}\right\|_{L_p(\Delta_{\alpha 1})} = 0\,, \quad \lim_{l\to\infty} f_l(X_1) = \frac{\partial a_1(X_1)}{\partial x_{1i}}$$

почти всюду в Δ_{α_1}. Пусть $\varepsilon > 0$. Имеем

$$(3.11)\qquad \left[\int_K \left|\frac{\partial a_1}{\partial x_{1i}}(Z_h^{-1}(X_2)) - \frac{\partial a_1}{\partial x_{1i}}(Z^{-1}(X_2))\right| \mathrm{d}X_2\right]^{1/p} \leqq$$

$$\leqq \left[\int_K \left|\frac{\partial a_1}{\partial x_{1i}}(Z_h^{-1}(X_2)) - f_l(Z_h^{-1}(X_2))\right|^p \mathrm{d}X_2\right]^{1/p} +$$

$$+ \left[\int_K |f_l(Z_h^{-1}(X_2)) - f_l(Z^{-1}(X_2))|^p \,\mathrm{d}X_2\right]^{1/p} +$$

$$+ \left[\int_K \left|f_l(Z^{-1}(X_2)) - \frac{\partial a_1}{\partial x_{1i}}(Z^{-1}(X_2))\right|^p \mathrm{d}X_2\right]^{1/p}.$$

Ввиду того, что якобиян D_h ограничен и независим от h, можно найти такое l_0, что

$$l \geqq l_0 \Rightarrow \left[\int_K \left|\frac{\partial a_1}{\partial x_{1i}}(Z_h^{-1}(X_2)) - f_l(Z_h^{-1}(X_2))\right|^p \mathrm{d}X_2\right]^{1/p} < \frac{\varepsilon}{3}.$$

Что касается третьего интеграла в правой части (3.11), то он стремится к нулю ввиду того, что

$$f_l(Z^{-1}(X_2)) \to \frac{\partial a_1}{\partial x_{1i}}(Z^{-1}(X_2))$$

почти всюду в K и ввиду того, что существует суммируемая мажоранта. Итак, можно считать l_0 столь большим, что

$$\left[\int_K \left|f_l(Z^{-1}(X_2)) - \frac{\partial a_1}{\partial x_{1i}}(Z^{-1}(X_2))\right|^p \mathrm{d}X_2\right]^{1/p} < \frac{\varepsilon}{3}.$$

Но из леммы 3.2 следует, что $\lim_{h\to 0} Z_h^{-1} = Z^{-1}$ равномерно относительно X_2 из K. Так как f – непрерывная функция, то существует h_0 таким образом, что для $h \leqq h_0$ имеем

$$\left[\int_K |f_{l_0}(Z^{-1}_h(X_p)) - f_{l_0}(Z^{-1}(X_2))|^p \,\mathrm{d}X_2\right]^{1/p} < \frac{\varepsilon}{3},$$

в результате чего лемма 3.3 полностью доказана.

Обозначим

$$M_h^1 = \mathop{\mathscr{E}}_X(X = [X_1, x_{1n}]\,, \quad X_1 \in \Delta_{\alpha_1}, \quad a_1(h, X_1) \leqq x_{1n} < a_1(X_1))$$

и положим $\Omega_h^1 = \Omega - M_h^1$. Тогда из леммы 3.1 – 3.3 вытекает следующая

Лемма 3.4. *Можно найти* $0 < h_0$ *таким образом, что для* $h \leqq h_0$ Ω_h^1 *будут области, расположенные в* Ω *и удовлетворять следующим условиям:*

$$\Omega_h^1 \subset \Omega\,, \quad \lim_{h\to 0} \Omega_h^1 = \Omega\,,$$

границы областей Ω_h^1 *можно представить в системах координат* $[X_r, x_{rn}]$ *в виде функций* $a_r^1(h, X_r)$ *удовлетворяющих в* $|X_r| < \alpha_1$ *условию Липшица, причем* $K(a_r^1(h)) \leqq \text{const}$. *Далее,*

$$\lim_{h\to 0}\big(\sup_{X_r \in \Delta_{\alpha_1}} |a_r^1(h, X_r) - a_r^1(X_r)|\big) = 0\,,$$

$$\lim_{h\to 0} \int_{\Delta_{\alpha_1}} \left| \frac{\partial a_r^1(h, X_r)}{\partial x_{ri}} - \frac{\partial a_r^1(X_r)}{\partial x_{ri}} \right|^p \mathrm{d}X_r = 0\,, \quad r = 1, 2, \ldots, m,\ i = 1, 2, \ldots, n-1\,.$$

При этом $a_1^1(h, X_1)$ *— бесконечно непрерывно дифференцируемые функции.*

Во избежание излишнего усложнения этого доказательства мы не будем останавливаться на некоторых деталях, которые можно рассмотреть при помощи рассуждений, сделанных в предыдущих леммах; для ориентировки мы укажем надлежащую лемму.

В самом деле, первой стадией построения совокупности Ω_h является лемма 3.4. Возмем теперь α_2, β_2, построим области U_r^2, $r = 1, 2, \ldots, m$ и заменим функции $a_r^1(h, X_r)$ функциями $a_r^2(h, X_r)$. (h в первой и второй стадии то же самое). Это значит, что исходной областью теперь будет Ω_h^2, причем граница изменяется в области U_2^2. Таким образом получим область Ω_h^2, для которой 3.4 остается в силе при замене Ω_h^1 посредством Ω_h^2, α_1 посредством α_2. (Если нужно, уменьшаем h_0.) Третья стадия: Исходной областью является Ω_h^2, граница изменяется в U_3^3, и получаем Ω_h^3 и. т. д. Таким образом получаем для $h \leqq h_0$ область Ω_h так, что

$$\overline{\Omega}_h \subset \Omega\,, \quad h_1 < h_2 \Rightarrow \overline{\Omega}_{h_2} \subset \Omega_{h_1}$$

и что границы Ω_h представлены в виде функций $a_r(h, X_r)$ в $|X_r| < \alpha_m$ и удовлетворяют там условию Липшица

$$|a_r(h, X_r) - a_r(h, Y_r)| \leqq K|X_r - Y_r|\,,$$

где K — постоянная и

$$\lim_{h\to 0}\big(\sup_{X_r \in |X_r| < \alpha_m} |a_r(h, X_r) - a_r(X_r)|\big) = 0\,,$$

$$\lim_{h\to 0} \int_{\Delta_{\alpha_m}} \left| \frac{\partial a_r(h, X_r)}{\partial x_{ri}} - \frac{\partial a_r(X_r)}{\partial x_{ri}} \right|^p \mathrm{d}X_r = 0\,, \quad i = 1, 2, \ldots, n-1,\ r = 1, 2, \ldots, m\,.$$

Надо еще доказать, что $a_r(h, X_r)$ в Δ_{α_m} бесконечно непрерывно дифференцируемы. Рассмотрим функцию $a_r(h, X_r)$. Из построения вытекает, что бесконечно непрерывно дифференцируемая функция — $a_r^r(h, X_r)$. Но в дальнейших стадиях это свойство не может быть потеряно. В самом деле, пусть $X_r^0 \in \Delta_{\alpha_m}$.

Пусть $s > r$ — первый индекс такой, что $U_s^s \,.\, U_r^m \neq \emptyset$. Если точка

$$X_r^0 = [X, a_r^r(h, X^0)] \quad \text{лежит в} \quad U_r^m U_s^m ,$$

то она может быть представлена в виде

$$X = [X_s^0, a_s^{s-1}(h, X_s^0)] .$$

В s-той стадии функция $a_s^{s-1}(h, X_s)$ изменяется и приобретает вид $a_s^s(h, X_s)$, которая бесконечно непрерывно дифференцируема — и по лемме 3.3 функция $a_r^s(h, X_r)$ бесконечно непрерывно дифференцируема в окрестности точки X_r^0.

Но может теперь случиться, что точка $[X_r^0, a_r^r(h, X_r^0)]$ будет на границе U_s^s, т. е. на границе цилиндра $|X_s| < \alpha_s$. Но опять по лемме 3.3 $a_s^{s-1}(h, X_s)$ бесконечно непрерывно дифференцируема в окрестности точки X_s^0.

Но из леммы 2.6 следует, что функция $a_s^s(h, X_s)$ также бесконечно непрерывно дифференцируема в этой окрестности в случае, если ее определить посредством $a_s^{s-1}(h, X_s)$ для $X_s \notin \Delta_{\alpha_s}$. (Таким образом была определена граница области Ω_h^s.) Но опять по лемме 3.3 функция $a_r^s(h, X_r)$ бесконечно непрерывно дифференцируема в окрестности точки X_r^0. Итак, доказательство основной теоремы завершено.

Литература

[1] *С. Л. Соболев:* Некоторые применения функционального анализа в математической физике. Ленинград 1950.

[2] *H. Rademacher:* Über partielle und totale Differenzierbarkeit von Funktionen mehrer Variabeln. Mathematische Annalen, *79* Band, 1919, 340—359.

[3] *J. Deny, J. L. Lions:* Les espaces du type de Beppo Levi. Annales de l'Institut Fourier, tome *5*, 1953—54, 305—370.

Résumé

SUR LES DOMAINES DU TYPE $\mathfrak{N}$

JINDŘICH NEČAS, Praha

Les domaines dont la frontière est représentable par les fonctions satisfaisants à la condition de Lipschitz semblent avoir une grande importance dans la théorie des espaces de Beppo Levi. On montre dans cet article un théorème sur l'approximation d'un tel domaine par les domaines aux frontières indéfiniment différentiables. Ce théorème est énoncé dans la première partie de l'article.

SUR LES NORMES ÉQUIVALENTES DANS $W_p^{(k)}(\Omega)$ ET SUR LA COERCITIVITÉ DES FORMES FORMELLEMENT POSITIVES

Jindřich Nečas

© Séminaire de Mathématiques Supérieures Université Montréal 1966. Reprinted with kind permission from his daughter Šárka Nečasová from: Nečas, J., Sur les normes équivalentes dans $W_p^k(\Omega)$ et sur la coercitivité des formes formellement positives, Séminaire Equations aux Dérivées Partielles, 102-128.

1. INTRODUCTION

L'essentiel du travail est l'étude de la coercitivité des formes formellement positives. On se donne $\varkappa_1, \varkappa_2, \ldots, \varkappa_m$, des entiers positifs, et $P_1(D), \ldots P_h(D)$, des opérateurs différentiels linéaires appliquant $W_2^{(\varkappa_1)}(\Omega) \times W_2^{(\varkappa_2)}(\Omega) \times \ldots \times W_2^{(\varkappa_m)}(\Omega)$ dans $L_2(\Omega)$.

Une forme sesquilinéaire, formellement positive

(1.1) $A(\vec{u}, \vec{u}) = \int_\Omega \sum_{\ell=1}^{h} |P_\ell(D)\vec{u}|^2 \, dx$ est $\prod_{r=1}^{m} W_2^{(\varkappa_r)}(\Omega)$

coercitive si pour $\vec{u} \in \prod_{r=1}^{m} W_2^{(\varkappa_r)}(\Omega)$ on a

(1.2) $$\sum_{r=1}^{m} |u_r|^2_{W_2(\varkappa_r)} \leq c \int_\Omega \sum_{\ell=1}^{h} |P_\ell(D)\vec{u}|^2 dx + c \sum_{r=1}^{m} |u_r|^2_{L_2}, \quad c > 0.$$

Par $W_2^{(\varkappa_r)}(\Omega)$, l'espace de Sobolev, des fonctions de carré sommable avec leurs dérivées jusqu'à l'ordre $\varkappa_r$ est désigné.

Soit $$P_\ell(D)\vec{u} = \sum_{r=1}^{m} \sum_{|\alpha| = \varkappa_r} p_{r\ell\alpha} D^\alpha u_r \equiv$$

$$\equiv \sum_{r=1}^{m} P_{r\ell}(D) u_r \quad \text{avec } p_{r\ell\alpha} \text{ constantes.}$$

Il est démontré que la condition nécessaire est suffisante pour la $\prod_{r=1}^{m} W_2^{(\varkappa_r)}(\Omega)$ coercitivité de la forme (1.1) est:

(1.3) rang ($P_{rl}(\xi)$) = m pour chaque $\xi \in C_N$, $\xi \neq 0$,

C_N étant l'espace euclidien complexe à N dimensions. La frontière $\partial\Omega$ est supposée lipschitzienne.

La généralisation de (1.2) pour les espaces $W_p^{(\varkappa)}(\Omega)$ avec $1 < p < \infty$ est démontrée.

(1.2) découle d'une inégalité, sur les normes équivalentes dont la généralisation pour $1 < p < \infty$ est aussi prouvée, à savoir pour l'entier, ν entier positif, on a:

(1.4) $$|u|_{W_p^{(\ell)}(\Omega)} \leq c\left(\sum_{|\alpha|=\nu} |D^\alpha u|_{W_p^{(l-\nu)}(\Omega)} + |u|_{W_p^{(l-\nu)}(\Omega)}\right).$$

Ici le cas $\ell < \nu$ est intéressant.

L'étude de la $W_2^{(\varkappa)}$ coercitivité des formes formellement positives au cas d'une fonction, a été commencée par N. Aronszajn [1], sa méthode a été généralisée par M. Schechter dans [2] pour V-coercitivité, où V est un certain sousespace fermé de $W_2^{(\varkappa)}$. Les formes sesquilinéaires ont été considérées par S.Agmon [3] et sa méthode a été utilisée au travail de D.G. de Figueiredo [4] pour les systèmes. Dans tous ces travaux cités, il était au fond des choses l'hypothèse sur la régularité de la frontière.

Les formes d'élasticité ont été étudiées par S. Campanato [5], K.O. Fridrichs [6], S.G. Michlin [7].

La condition (1.3) pour $m = 1$ se trouve au travail cité de N. Aronszajn, sous une autre forme. Au cas $m = 1$, mais sous l'hypothèse faible sur la frontière de Ω, à savoir sous l'hypothèse que $\partial\Omega$ est lipschitzienne, suffisance de la condition (1.3) a été démontrée par K.T. Smith [8].

2. LES NORMES ÉQUIVALENTES

On désigne par E_N l'espace euclidien réel à N dimensions et par $x = (x_1, x_2 \dots x_n)$ son point générique $+ |x| = \left(\sum_{i=1}^{N} x_i^2\right)^{1/2}$.

Un domaine borné Ω est à frontière lipschitzienne et on l'écrira $\Omega \in \mathcal{H}^{(o),1}$, s'il existe des cartes locales (x_r', x_{rn}), $r = 1, 2, \dots M$, $o <$ ~ et les fonctions lipschitziennes dans

$$\bar{\Delta}_r = \{(x' \in E_{N-1},\ |x_{ri}| \leq \alpha,\ i = 1,2,\ldots N-1\},$$ soient

a_r, telles que chaque point de la frontière de Ω qu'on désignera $\partial\Omega$, s'écrit au moins dans un des systèmes mentionnés sous la forme $\left(x_r',\ a_r(x_r')\right),\ x_r' \in \Delta_r$. De plus, on suppose le domaine d'un seul côté de la frontière.

Soit $1 \leq p < \infty$, k un entier positif, Ω un domaine borné ou non. On désigne par $W_p^{(k)}(\Omega)$ l'espace des fonctions de p-ième puissance sommable sur Ω avec leurs dérivées jusqu'à l'ordre k; les dérivées sont prises au sens des distributions et le symbole usuel

$$D^\alpha u = \frac{\partial^{|\alpha|} u}{\partial x_1^{\alpha_1} \ldots \partial x_N^{\alpha_N}}$$ est utilisé.

On munit $W_p^{(k)}(\Omega)$ de la norme

(2.1) $$|u|_{W_p^{(k)}} = \left(\int_\Omega \sum_{|\alpha| \leq k} |D^\alpha u|^p \, dx \right)^{1/p} .$$

On désigne encore par $\mathcal{D}(\Omega)$ l'espace des fonctions indéfiniment différentiables à support compact dans Ω et par $\overset{\circ}{W}{}_p^{(k)}(\Omega)$ sa fermeture dans la topologie de $W_p^{(k)}(\Omega)$. Si k est un entier négatif, $1 < p < \infty$,

$\frac{1}{p} + \frac{1}{\mathcal{L}} = 1$, on désigne par $W_p^{(k)}(\Omega) = \left(\overset{\circ}{W}{}_{\mathcal{L}}^{(-k)}(\Omega)\right)$, le dual du $\overset{\circ}{W}{}_{\mathcal{L}}^{(-k)}(\Omega)$.

Si $f \in Lp(E_N)$, $1 \leq p < \infty$, on désigne par $\hat{f} = \mathcal{F}f$ sa transformée de Fourier. Nous avons le théorème suivant sur les multiplicateurs (cf. P.I.Lizorkin [9]):

Lemme 1. Soit $\Phi(\xi)$ une fonction continue avec toutes les dérivées au sens usuel

$$\frac{\partial^k \Phi}{\partial \xi_1^{k_1} \partial \xi_2^{k_2} \cdots \partial \xi_N^{k_N}}, \text{ où } k_j \text{ est 1 ou 0, } 0 \leq k \leq N$$

pour $|\xi_j| > 0$, $j = 1,2,\ldots N$

Soit $1 < p \leq q < \infty$, $\beta = \frac{1}{p} - \frac{1}{\lambda}$ et

$$\left| \xi_1^{k_1} \cdots \xi_N^{k_N} \frac{\partial^k \phi}{\partial \xi_1^{k_1} \cdots \partial \xi_N^{k_N}} \right| \leq M < \infty$$

pour toutes les dérivées en question. Alors la transformation $F^{-1}\left(\phi(\xi)\hat{f}(\xi)\right)$ est linéaire et bornée de $L_p(E_N)$ dans $L_p(E_N)$.

Nous pouvons maintenant démontrer:

Théorème 1. Soit $\Omega \in \mathcal{H}^{(o),1}$, $f \in W_p^{(\ell)}(\Omega)$, $1 < p < \infty$, ℓ un entier positif ou non. Alors

$$(2.2) \quad |f|_{W_p^{(\ell)}(\Omega)} \leq c\left(\sum_{i=1}^{N}\left|\frac{\partial f}{\partial x_i}\right|_{W_p^{(\ell-1)}(\Omega)} + |f|_{W_p^{(\ell-1)}(\Omega)}\right).$$

Démonstration. Si $\ell \geq 1$, le théorème est trivial. Soit alors $\ell \leq 0$. $\mathcal{D}(\Omega)$ étant dense au $W_p^{(\ell)}(\Omega)$, il suffit de considérer $f \in \mathcal{D}(\Omega)$. Soit K un ensemble fermé de Ω. Démontrons d'abord la validité de (2.2) pour les

f à support dans K avec une constante ne dépendant que de K. On a évidemment

$$(2.3)\quad \sum_{i=1}^{N} \left| \frac{\partial f}{\partial x_i} \right|_{W_p^{(\ell-1)}(E_N)} + |f|_{W_p^{(\ell-1)}(E_N)} \leq$$

$$\leq c_1(K) \left(\sum_{i=1}^{N} \left| \frac{\partial f}{\partial x_i} \right|_{W_p^{(\ell-1)}(\Omega)} + |f|_{W_p^{(\ell-1)}(\Omega)} \right).$$

D'autre part, en vertu de l'égalité de Parseval et du lemme 1, pour $1 < p < \infty$ et k un entier, $W_p^{(k)}(E_N)$ est l'ensemble de telles distributions à croissance polynomiale, pour lesquelles $F^{-1}\left((1+|\xi|^2)^{\frac{k}{2}} \hat{f}\right) \in L_p(E_N)$. On en tire aisément l'inégalité

$$(2.4)\quad |f|_{W_p^{(\ell)}(E_N)} \leq c_2 \left(\sum_{i=1}^{N} \left| \frac{\partial f}{\partial x_i} \right|_{W_p^{(\ell-1)}(E_N)} + |f|_{W_p^{(\ell-1)}(E_N)} \right),$$

d'où en vertu du (2.3) et de l'inégalité triviale

$|f|_{W_p^{(\ell)}(\Omega)} \leq |f|_{W_p^{(\ell)}(E_N)}$ suit finalement (2.2) pour f à support dans K avec une constante $C_3(K)$.

Soit $\delta > 0$, si petit que l'ensemble des points (x_r', x_{rN}), $x_r' \in \Delta_r$,

$a_r(x_r') < x_{rN} < a_r(x_r') + \delta$ appartient dans Ω ; pour fixer les idées, on suppose l'axe x_{rN} pénétrant dans Ω. Désignons cet ensemble par

A_r et soit $\Lambda_r = \partial\Omega \cap \partial A_r$.

Considérons un ensemble fermé $L \subset A_r \cup \Lambda_r$ et seulement telles f dont support $f \subset L$. Pour simplifier l'écriture, on omettra l'indice r.

Posons
$$a(x';h) = \frac{1}{\chi h^{N-1}} \int_{\Delta} \exp \frac{|x'-y'|^2}{|x'-y'|^2 - h^2}\, a(y')\, dy'$$

pour $x' \in \Delta' = E(x' \in E_{N-1}, |x_i| < \alpha, i = 1,2...N-1)$,

où $0 < \alpha' < \alpha$, $0 < \ldots$ et $\ldots$ sont choisis de sorte que $L \subset A' + \Lambda'$, $0 < h < \alpha - \alpha'$; $\displaystyle\frac{1}{\chi h^{N-1}} \int_{\Delta} \exp \frac{|x'|^2}{|x'|^2 - h^2}\, dx' = 1.$

La fonction $a(x';h)$ est continue sur $\bar{\Delta} \times (0, \alpha - \alpha')$, indéfiniment différentiable dans $\Delta' \times (0, \alpha - \alpha')$ et telle que pour $|i| \geq 1$

(2.5.) $$\frac{\partial^{|i|} a}{\partial x_1^{i_1} \ldots \partial x_{N-1}^{i_{N-1}}\, \partial h^{i_N}} \leq C_i\, h^{1-|i|}.$$

Soit maintenant ε un nombre positif tel que $\varepsilon \leq \frac{\kappa . \lambda}{\delta'}$ et tel que pour $0 < y_N < \delta'$, $x' \in \Delta'$

$$(2.6) \quad \left| \varepsilon \frac{\partial a}{\partial h}(x', \varepsilon y_N) \right| \leq \frac{1}{2}\left(\frac{\kappa}{\delta'} - 1 \right) = c_5 .$$

On désigne $K' = \Delta'_X (0, \delta')$ et définit par $Ty = x$ l'application de K' dans A de la manière suivante

$$(2.7) \quad \begin{cases} y' = x' \\ a(y', \varepsilon y_N) + (1 + c_5)\, y_N = x_N \end{cases}$$

Désignons $T(K') = B'$. On obtient $A' \subset B'$, T est un homeomorphisme de $\overline{K'}$ sur $\overline{B'}$, indéfiniment différentiable dans K' au Jacobien

$$\frac{dT}{dy} = \varepsilon \frac{\partial a}{\partial h} + 1 + c_5 .$$

Soit maintenant $g(y) = f(T(y))$.

Nous obtenons

$$(2.8) \quad \sum_{i=1}^{N} \left| \frac{\partial g}{\partial y_i} \right|_{W_p^{(\ell-1)}(K')} + |g|_{W_p^{(\ell-1)}(K')} \leq$$

$$c_6 \left(\sum_{i=1}^{N} \left| \frac{\partial f}{\partial x_i} \right|_{W_p^{(\ell-1)}(A)} + |f|_{W_p^{(\ell-1)}(A)} \right).$$

En effet, soit $\psi \in \mathcal{B}(K')$; il suit de l'inégalité de Hardy, l'inégalité

$$\int_{K'} \left| D_y^{\alpha} \psi(y) \right|^p y_N^{\,p(|\alpha| + \ell - 1)} \, dy \leq$$

$$\leq c_7 \, |\psi|^p_{W_p^{(1-\ell)}(K')}$$

pour $|\alpha| \leq 1-\ell$. Il s'en suit

$$|\psi(T^{-1}(x))|_{W_p^{(1-\ell)}(A)} \leq c_8 \, |\psi|_{W_p^{(1-\ell)}(K')}$$

d'où
$$\left| \frac{\partial x_j}{\partial y_i} \psi(T^{-1}(x)) \left(\frac{dT}{dy} \right)^{-1} \right|_{W_p^{(1-\ell)}(A)} \leq c_9 \, |\psi|_{W_p^{(1-\ell)}(K')},$$

On en tire finalement (2.8)

Désignons par $C = E(y \in E_N,\ y = (y', y_N),\ y' \in \Lambda',\ 2-\ell < y_N < \beta)$ et soient λ_i, $i = 1, 2, \ldots, 2-\ell$ les nombres déterminés du système

(2.9) $$\sum_{i=1}^{2-\ell} \lambda_i \, (-i)^h = 1, \quad h = 0, 1, 2, \ldots, \ell + 1 \; .$$

Posons pour $2-\ell < y_N < 0$; $g(y', y_N) =$

$$= \sum_{i=1}^{2-\ell} \lambda_i \, g\left(y', -\frac{y_N}{i}\right).$$ Nous démontrerons:

(2.9) $$\sum_{i=1}^{N} \left|\frac{\partial g}{\partial y_i}\right|_{W_p^{(\ell-1)}(C)} + |g|_{W_p^{(\ell-1)}(C)} \leq$$

$$\leq c_{10} \left(\sum_{i=1}^{N} \left|\frac{\partial g}{\partial y_i}\right|_{W_p^{(\ell-1)}(K')} + |g|_{W_p^{(\ell-1)}(K')} \right).$$

En effet, soit $v \in \overset{\circ}{W}{}_p^{(1-\ell)}(C)$, $1 \leq i \leq N-1$. On a

$$\int_C \frac{\partial g}{\partial y_i} v \, dy = \int_{K'} \frac{\partial g}{\partial y_i}(y', y_N) \left[v(y', y_N) + \right.$$

$$\left. + \sum_{j=1}^{2-\ell} \lambda_j \, j \, v(y', -j\, y_N) \right] dy = \int_{K'} \frac{\partial g}{\partial y_i}(y', y_N).$$

$$\cdot \, w(y) \, dy.$$

Il suit de (2.9) que $w \in \overset{\circ}{W}{}_p^{(1-\ell)}(K')$

avec $|w|_{W^{(1-\frac{\ell}{p})}(K')} \leq c_{11} |v|_{W^{(1-\frac{\ell}{p})}(C)}$.

On raisonne ainsi pour $\int_C g\, v\, dy$. Puis nous avons

$$\int_C \frac{\partial g}{\partial y_N} v\, dy = \int_{K'} \frac{\partial g}{\partial y_N} (y', y_N) \,.$$

$$\left[v(y', y_N) - \sum_{j=1}^{2-\ell} \lambda_j\, v(y', -j\, y_N) \right] dy =$$

$$= \int_{K'} \frac{\partial g}{\partial y_N} (y', y_N)\, \omega (y)\ dy.$$ Il suit de nouveau

de (2.9) que $\omega \in \overset{o}{W}{}^{(1-\frac{\ell}{p})}(K')$ et

$$|\omega|_{W^{(1-\frac{\ell}{p})}(K')} \leq c_{12} |v|_{W^{(1-\frac{\ell}{p})}(C)}.$$

(2.9) est alors complètement démontrée. On démontre comme au-dessus:

$$(2.10) \quad |g|_{W^{(\frac{\ell}{p})}(C)} \leq c_{13} \left(\sum_{i=1}^{N} \left| \frac{\partial g}{\partial y_i} \right|_{W^{(\frac{\ell}{p}-1)}(C)} + |g|_{W^{(\frac{\ell}{p}-1)}(C)} \right).$$

Mais on a évidemment

$$(2.11) \quad |g|_{W^{(\frac{\ell}{p})}(K')} \leq |g|_{W^{(\frac{\ell}{p})}(C)}$$

et on démontre comme (2.8):

$$|f|_{W_p^{(\ell)}(A')} \leq c_{14} \; |g|_{W_p^{(\ell)}(K')} \quad ,$$

d'où et de (2.11), (2.10), (2.9), (2.8) suit finalement

$$|f|_{W_p^{(\ell)}(A)} \leq c_{15}\left(\sum_{i=1}^{N} \left| \frac{\partial f}{\partial x_i} \right|_{W_p^{(\ell-1)}(A)} + |f|_{W_p^{(\ell-1)}(A)} \right).$$

La partition de l'unité habituelle nous donne le résultat, c.q.f.d.

Rappelons un lemme bien connu, cf. p. ex. E. Magenes, G. Stampacchia [10]:

<u>Lemme 2</u> Soit $\Omega \in \mathcal{H}^{(o,1)}$, $1 \leq p < \infty$. Alors à chaque $\varepsilon > 0$, il existe $\lambda(\varepsilon) > 0$ de sorte que pour $u \in W_p^{(k)}(\Omega)$, où k est un entier positif,

$$\|u\|_{W_p^{(k-1)}(\Omega)} \leq \varepsilon \|u\|_{W_p^{(k)}(\Omega)} + \lambda(\varepsilon) \|u\|_{Lp(\Omega)} .$$

Cela étant, on démontre par récurrence, compte tenu du théorème 1:

<u>Théorème</u> 2. Soit $\Omega \in \{ \}^{(o),1}$, $1 < p < \infty$, ℓ un entier (aussi négatif), ν un entier positif. Alors

$$(2.12)\quad |f|_{W_p^{(\ell)}(\Omega)} \leq c \Big(\sum_{|\alpha| = \nu} |D^\alpha f|_{W_p^{(\ell-\nu)}(\Omega)} + |f|_{W_p^{(\ell-\nu)}(\Omega)} \Big).$$

3. <u>Coercitivité des formes formellement positives, d'autres normes équivalentes</u>

Soient $\varkappa_1, \varkappa_2, \ldots\ldots \varkappa_m$ des entiers positifs, $1 < p < \infty$, et $P_\ell(D)$, $\ell = 1,2,\ldots\ldots h$, des opérateurs différentiels de $\prod_{r=1}^{m} W_p^{(\varkappa_r)}(\Omega)$ dans $L_p(\Omega)$ définis par $P_\ell(D)\,\vec{u} = \sum_{r=1}^{m} \sum_{|\alpha| \leq \varkappa_r} p_{r\ell\alpha}\, D^\alpha u_r$, avec $p_{r\ell\alpha} \in L_\infty(\Omega)$; désignons $P_{r\ell}(D)\,u_r = \sum_{|\alpha| = \varkappa_r} p_{r\ell\alpha} D^\alpha u_r$.

Si p = 2, on définit une forme formellement positive

$$B(\vec{v}, \vec{u}) = \int_\Omega \left(\sum_{\ell=1}^{h} P_\ell(D)\,\vec{v}\cdot \overline{P_\ell(D)\,\vec{u}} \right) dx.$$

Soit encore V un sousespace fermé de $\prod_{r=1}^{m} W_p^{(\varkappa r)}(\Omega)$

tel que $\prod_{r=1}^{m} \overset{o}{W}{}_p^{(\varkappa r)}(\Omega) \subset V \subset \prod_{r=1}^{m} W_p^{(\varkappa r)}(\Omega)$.

Le système $P_\ell(D)$ (et pour $p=2$ aussi la forme B) est dit V-coercitive, si pour $\vec{u} \in V$:

$$(3.1) \quad \sum_{r=1}^{m} |u_r|^p_{W_p^{(\varkappa r)}} \le c \sum_{\ell=1}^{h} |P_\ell(D)\,\vec{u}|^p_{L_p} + c \sum_{r=1}^{m} |u_r|^p_{L_p}.$$

Considérons d'abord le cas

$$V = \prod_{r=1}^{m} \overset{o}{W}{}_p^{(\varkappa r)}(\Omega), \text{ le cas facile.}$$

<u>Théorème</u> 3. Soit Ω un domaine quelconque, $1 < p < \infty$,

$$P_\ell(D) = \sum_{r=1}^{m} \sum_{|\alpha| = \varkappa_r} p_{r\ell\alpha} D^\alpha u_r, \text{ avec } p_{r\ell\alpha} \text{ cons-}$$

tantes. Alors la condition nécessaire et suffisante pour $\prod_{r=1}^{m} \overset{o}{W}{}_p^{(\chi i)}(\Omega)$ coercitivité du système $P_l(D)$ est:

(3.2) rang $\left(P_{rl}(\xi)\right) = m$ pour $\xi \in E_N$, $\xi \neq 0$.

Démonstration: La condition est nécessaire. Supposons le contraire. On a alors (3.1) avec $C_1 > 0$ et soit $\xi \neq 0$ en question tel que rang $\left(P_{rl}(\xi)\right) < m$. Il existe des nombres $b_1, \dots, b_m$ tels que $\sum_{r=1}^{m} P_{rl}(\xi)\, b_r = 0$, $l = 1,2\dots h$,

$$\sum_{r=1}^{m} |b_r|^2 \neq 0.$$

Pour $\lambda > 0$ et $\varphi \in \mathcal{D}(\Omega)$, $\varphi \neq 0$, posons

(3.3) $u_{r\lambda}(x) = \varphi(x)(i\lambda)^{-\chi_r} b_r\, e^{i\lambda(\xi,x)}$. Si nous substituons $u_{r\lambda}$ dans (3.1), nous obtenons $\lim_{\lambda\to\infty} \sum_{r=1}^{m} |u_{r\lambda}|^p_{W_p^{(\chi r)}} =$

$$= c_2 > 0, \quad \lim_{\lambda\to\infty} \sum_{l=1}^{h} |P_l(D)\, \overline{u}_\lambda|_{Lp} + \lim_{\lambda\to\infty} \sum_{r=1}^{m} |u_{r\lambda}|^p_{Lp} = 0$$

ce qui n'est pas possible.

La condition est suffisante. Désignons par Δi, $i=1,2,\ldots \tau$, les déterminants $m \times m$ de la matrice $P_{r\ell}(\xi)$. Soit $\vec{u} \in [\mathcal{D}(\Omega)]^m$. La transformation de Fourier nous donne

$$\widehat{P_\ell(D)\vec{u}} = \sum_{r=1}^{m} P_{r\ell}(\xi)\, i^{\alpha r}\, \hat{u}_r.$$ Si nous désignons par $\Delta_{i,r\ell}$ les mineurs algébriques $m-1 \times m-1$, correspondants, où $\Delta i, r\ell = 0$ si $P_{r\ell}(\xi)$ n'appartient pas dans la matrice engendrée par Δi, nous obtenons pour $r=1, 2, \ldots .m$:

$$i^{\alpha r}\hat{u}_r = \frac{\sum_{j=1}^{\tau}\sum_{\ell=1}^{h} \overline{\Delta j}\, \Delta j_{,r\ell}\, \widehat{P_\ell(D)\vec{u}}}{\sum_{j=1}^{\tau} |\Delta_j|^2}$$

Si $|\alpha| = \alpha_r$, on obtient que

$$\frac{\bar{\Delta} j\, \Delta j_{,r\ell}\, \xi^{\alpha}}{\sum_{j=1}^{\tau} |\Delta_j|^2}$$ est un multiplicateur au sens

du lemme 1, d'où l'assertion.

Nous obtenons maintenant immédiatement par la partition de l'unité standarte:

<u>Théorème</u> 4. Soit Ω un domaine borné, $1 < p < \infty$,

$$P_l(D) = \sum_{r=1}^{m}{}' \sum_{|\alpha| \leq \varkappa_r} p_{r l_\alpha} D^\alpha u_r \quad \text{avec } p_{r l_\alpha} \in L_\infty(\Omega),$$

$p_{r l_\alpha}$ continues dans $\bar{\Omega}$ pour $|\alpha| = \varkappa_r$. Alors la condition nécessaire et suffisante pour la $\prod_{r=1}^{m} \overset{o}{W}{}_p^{(\varkappa_r)}(\Omega)$ coercitivité est que pour $x \in \bar{\Omega}$, $\xi \in E_N$, $\xi \neq 0$;

rang $\left(P_{r l}(x, \xi)\right) = m$.

En effet, la nécessité se démontre de la même manière: si rang $\left(P_{r l}(x, \xi)\right) < m$, $x \in \bar{\Omega}$, on obtient la contradiction avec (3.3) en prenant support φ assez petit et assez proche au point x.

La condition suffisante se démontre par la partition de l'unité, cf. L. Garding [12] et par l'application du lemme 2.

Considérons maintenant le cas de la

$$\prod_{i=1}^{m} W_2^{(\chi i)}(\Omega)$$ - coercitivité.

Nous avons (cf. E. Gagliardo [11]):

Lemme 3. Soit $\Omega \in \mathcal{H}^{(o),1}$. Alors l'ensemble des restrictions des fonctions de $\mathcal{D}(E_N)$ sur Ω est dense dans $W_p^{(k)}(\Omega)$, $1 \leq p < \infty$.

Cela étant, nous avons

Théorème 5. Soit $\Omega \in \mathcal{H}^{(o),1}$, $1 < p < \infty$,

$$P_\ell(D)\,\vec{u} = \sum_{r=1}^{m} \sum_{|\alpha| = \chi_r} p_{r\ell\alpha} D^\alpha u_r, \quad p_{r\ell\alpha} \text{ constantes.}$$

Alors la condition nécessaire et suffisante pour

$\prod_{r=1}^{m} W_p^{(\alpha i)}(\Omega)$ coercitivité du système $P_\ell(D)$ est:

(3.5) $\quad \operatorname{rang}\left(P_{r\ell}(\xi)\right) = m$ pour $\xi \in C_N$, $\xi \neq 0$.

(C_N est l'espace euclidien complexe à N dimensions.)

Démonstration. La condition est nécessaire. Supposons le contraire. Soit $\xi \neq 0$ tel que $\operatorname{rang}\left(P_{r\ell}(\xi)\right) < m$. Il existe des nombres $b_1 \ldots b_m$ comme dans la démonstration du théorème 3. Pour $\lambda \neq 0$, complexe, posons

$u_{r\lambda}(x) = \lambda^{-\chi r} b_r e^{\lambda(\xi, x)}$. Si nous substituons $u_{r\lambda}$ dans (3.1), nous obtenons:

$$\sum_{r=1}^{m} |u_{r\lambda}|^{p}_{W_p^{(\chi r)}(\Omega)} \leq c \sum_{r=1}^{m} |u_{r\lambda}|^{p}_{Lp(\Omega)}.$$

Mais l'application identique de $W_p^{(\chi r)}(\Omega)$ dans $Lp(\Omega)$ étant complètement continue, les boules fermées de l'enveloppe linéaire de l'ensemble des $\vec{u}\lambda$, λ complexe, sont compactes, ce qui n'est pas vrai, chaque boule ayant la dimension infinie. La condition est suffisante. Soient $\Delta_j, \Delta_{j,\ell}$ les déterminants du théorème 3. La seule racine commune et $\xi = 0$. En vertu du théorème de Hilbert, du Nullstellensatz, cf. V.d. Waerden [12], il existe les polynômes B_j tels que $\sum_{j=1}^{\tau} B_j \Delta_j = \xi_1^{\rho}$ 1), ρ un entier positif.

Posons $C_\ell = \sum_{j=1}^{\tau} \Delta_{j,1\ell} B_j$. Nous obtenons

1) L'idée de cette démonstration est à Z. Frolik Université de Prague.

$$(3.6)\quad \sum_{l=1}^{h} P_{rl}\, c_l = \begin{cases} \xi^{\rho}_{1} & r = 1 \\ 0 & r > 1 \end{cases} .$$

Evidemment, après le raisonnement évident, on peut trouver ν un entier positif tel que pour chaque multiindice α^r, $|\alpha^r| = \chi_r + \nu$, $r = 1,2,\dots m$, il existe des polynômes homogènes du degré ν, soient c_l tels que

$$(3.7)\quad \sum_{l=1}^{h} P_{sl}\, c_l = \begin{cases} \xi^{\alpha^r} & \text{pour } s = r \\ 0 & \text{pour } s \neq r \end{cases} .$$

Soit maintenant $\vec{u} \in [\mathcal{D}(E_N)]^m$, $\varphi \in \mathcal{D}(\Omega)$, $|\delta| = \nu$, $\alpha^r = \rho + \delta$. On a

$$(3.8)\quad \int_{\Omega} D^{\rho} u_r\, D^{\delta} \bar{\varphi}\, dx = \frac{1}{(2\pi)^N} \int_{E_N} (i)^{\chi_r - \nu} \xi^{\rho+\delta} \hat{u}_r \bar{\hat{\varphi}}\, d\xi =$$

$$= \frac{1}{(2\pi)^N} \int_{E_N} \sum_{s=1}^{m} (i)^{\chi_s - \nu} \Big(\sum_{l=1}^{h} P_{sl}\, c_l \Big) \hat{u}_s \bar{\hat{\varphi}}\, d\xi =$$

$$= \frac{1}{(2\pi)^N} \int_{E_N} \sum_{l=1}^{h} \Big(\sum_{s=1}^{m} (i)^{\chi_s - \nu} P_{sl}\, c_l\, \hat{u}_s \Big) \bar{\hat{\varphi}}\, d\xi =$$

$$= \int_{\Omega} \sum_{\ell=1}^{h} P_\ell(D)\vec{u}\; C_\ell(D)\bar{\vec{\varphi}}\; dx$$ d'où

$$(3.9)\qquad \sum_{|\delta|=\nu} |D^{\gamma+\delta} u_r|_{W_p^{(-\nu)}(\Omega)} \leq c_1 \sum_{\ell=1}^{h} |P_\ell(D)\vec{u}|_{Lp(\Omega)}$$

d'où en vertu du théorème 2 suit:

$$(3.10)\qquad \sum_{r=1}^{m} \sum_{|\gamma|=\varkappa_r} |D^\gamma u_r|_{Lp(\Omega)} \leq$$

$$\leq c_2 \sum_{\ell=1}^{h} |P_\ell(D)\vec{u}|_{Lp(\Omega)} + c_2 \sum_{r=1}^{m} \sum_{|\gamma|=\varkappa_r} |D^\gamma u_r|_{W_p^{(-\nu)}(\Omega)} \leq$$

$$\leq c_3 \sum_{\ell=1}^{h} |P_\ell(D)\vec{u}|_{Lp(\Omega)} + c_3 \sum_{r=1}^{m} |u_r|_{W_p^{(\varkappa_r-\nu)}(\Omega)}.$$

Il suit maintenant des lemmes 2,3 et de (3.10) l'inégalité (3.1), c.q.f.d.

Nous obtenons maintenant par la partition de l'unité :

<u>Théorème 6</u> Soit $\Omega \in \mathcal{H}^{(o),1}$, $1 < p < \infty$,

$$P_\ell(D) = \sum_{r=1}^{m} \sum_{|\alpha| \le \ell_\alpha} p_{r\ell_\alpha} D^\alpha u_r \quad \text{avec} \quad p_{r\ell_\alpha} \in L_\infty(\Omega),$$

$p_{r\ell_\alpha} \in C(\bar{\Omega})$ pour $|\alpha| = \varkappa_r$. Alors si pour $x \in \Omega$,

$\xi \in E_n, \xi \neq 0$

rang $\left(P_{r\ell}(x, \xi)\right) = m$ et si pour $x \in \partial\Omega$,

$\xi \in C_N, \xi \neq 0$, rang $\left(P_{r\ell}(x, \xi)\right) = m$,

le système $\Gamma_\ell(D)$ est $\prod_{r=1}^{m} W_p^{(\varkappa r)}(\Omega)$ coercitif.

Remarque. Il faut remarquer que les théorèmes 3-6 nous engendrent dans $\prod_{r=1}^{m} \overset{o}{W}{}_p^{(\varkappa r)}(\Omega)$, $\prod_{r=1}^{m} W_p^{(\varkappa r)}(\Omega)$ respectivement, une vaste classe des normes équivalentes.

Pour les applications des résultats de ce § aux systèmes elliptiques, cf. S.G.Michlin [7], S.Campanato [5].

Exemple. Soit $\varkappa_1 = \varkappa_2 = \dots = \varkappa_m = 1$, $m = N$, $P_\ell(D)$ définis par:

$$\frac{\partial u_1}{\partial x_1},\ \frac{\partial u_1}{\partial x_2}+\frac{\partial u_2}{\partial x_1},\ \frac{\partial u_1}{\partial x_3}+\frac{\partial u_3}{\partial x_1},\ldots\ \frac{\partial u_1}{\partial x_N}+\frac{\partial u_N}{\partial x_1},\ \frac{\partial u_2}{\partial x_2},\ \frac{\partial u_2}{\partial x_3}+\frac{\partial u_3}{\partial x_2},\ldots$$

$$\ldots\ \frac{\partial u_2}{\partial x_N}+\frac{\partial u_N}{\partial x_2},\ \frac{\partial u_3}{\partial x_3},\ldots\ \frac{\partial u_N}{\partial x_N}$$

. La matrice correspondante:

$$\begin{pmatrix}
\xi_1, & \xi_2, & \xi_3, \cdots & \xi_N, & 0, & 0, & 0, & 0, & 0, & 0, & 0, & 0, \\
0, & \xi_1, & 0, \ldots & 0, & \xi_2 & \xi_3, & \xi_N, & 0, & 0, & 0, & 0, & 0, \\
0, & 0, & \xi_1, \ldots & 0, & 0, & \xi_2, 0 & 0, & \xi_3 & \xi_4, & \xi_N, & 0, & 0, \\
 & & & & & & & 0, & \xi_3, 0, & 0, & \xi_4, & 0, \\
0, & 0, & \ldots\ldots & \xi_1 & 0, & \ldots 0, & \xi_2, & 0, & 0, \ldots & \xi_3, & 0, \ldots 0, & \xi_N
\end{pmatrix}$$

On a évidemment rang $\left(P_{r\ell}(\xi)\right) = N$.

Le système mentionné est celui de l'élasticité.

BIBLIOGRAPHIE

1. N.Aronszajn : On coercive integro-differential quadratic forms,Conf. Part. Dif. Eq., Univ. Kansas 14 (1954), 94-106

2. M.Schechter : Integral inequalities for partial differential operators and functions satisfying general boundary conditions. Comm. Appl. Math. 12 (1959), 37-66

3. S. Agmon : The coercivness problem for integro-differential forms. J. Anal. Math. 6 (1958), 183-223

4. D.G. de Figueiredo : The coercivness problem for forms over vector valued functions. Comm. Pure Appl. Math. 16 (1963), 63-94

5. S. Campanato: Sui problemi al contorno per i sistemi di equazioni differenziali lineari del tipo dell'elasticita I, I, Ann. Sc. Norm. Sup. Pisa 13 (1959), 223-258, 275-302.

6. K.O.Fridrichs: On the boundary-value problems of the theory of elasticity and Korn's inequality. Ann. Math. 48 (1947),

7. S.G. Michlin: Le problème de minimum pour la fonctionnelle quadratique (en russe), Moscou 1952.

8. K.T. Smith : Inequalities formally positive integro-differential forms. Bull. Amer. Math. Soc.67 (1961), 368-370.

9. P.I. Lizorkin : (Lp, L$_\varphi$) multiplicateurs des intégrales de Fourier (en russe). Dokl. Akad Nauk SSSR, 152 (1963) 808-811.

10. E. Magenes, G. Stampacchia : I problemi al contorno per le equazioni differenziali di tipo ellitico. Ann. Scuola Norm. Sup. Pisa 12, (1958) 247-354.

11. E. Gagliardo : Proprieta di alcune classi di funzioni in piu variabili. Ricerche Mat. 7 (1958), 102-137,

12. V.d Waerden : Moderne Algebra, 2 t.

SUR LA COERCIVITÉ DES FORMES SESQUILINÉAIRES, ELLIPTIQUES

PAR

JINDŘICH NEČAS *)

(Prague)

1. INTRODUCTION

Étant donné Ω, un domaine de E_N, désignons par $W_p^{(k)}(\Omega)$ l'espace
s fonctions de $p^{\text{ième}}$ puissance sommable avec toutes leurs dérivées,
ises au sens des distributions, jusqu'à l'ordre k (k un entier non négatif).
nsidérons un opérateur elliptique

.1) $$A = \sum_{|i|,|j|\leqslant k} (-1)^{|i|} D^i(a_{ij} D^j),$$

. $i = (i_1, i_2, \ldots, i_N)$, $j = (j_1, j_2, \ldots, j_N)$, i_r, j_r des entiers $\geqslant 0$, $|i| = \sum_{r=1}^{N} i_r$, $D^i = \dfrac{\partial^{|i|}}{\partial x_1^{i_1} \partial x_2^{i_2}, \ldots, \partial x_N^{i_N}}$. On coordonne à l'opérateur A une
me sesquilinéaire, définie sur $W_2^{(k)}(\Omega) \times W_2^{(k)}(\Omega)$:

2) $$A(v, u) = \int_\Omega \sum_{|i|,|j|\leqslant k} \bar{a}_{ij} D^i v D^j \bar{u} dx.$$

Soit V un sous-espace fermé de $W_2^{(k)}(\Omega)$. On étudie le problème de la
ercivité de la forme (1.1) pour l'espace V : on cherche $\lambda \geqslant 0$ de sorte
e pour $v \in V \Rightarrow |A(v, v)| \geqslant c |v|^2_{W_2^{(k)}} - \lambda c |v|^2_{L_2}$, $c > 0$. Il est utile
ajouter à la forme (1.1) une autre forme, soit $a(v, u)$, qui ne dépend que
s valeurs frontières de v, u, et d'étudier la coercivité de la forme $((v, u)) = A(v, u) + a(v, u)$. L'inégalité (1.2) est une condition suffisante pour
e les problèmes aux limites soient résolubles. L'étude des conditions
;ébriques entraînantes (1.2) a été faite par plusieurs auteurs (cf. L. Gar-

*) Matematický ústav ČSAV, Žitná 25, Praha 1, Tchécoslovaquie.

© Publishing House of the Romanian Academy 1964. Reprinted with kind permission from: Nečas, J., Sur la coercivité des formes sesquilinéaires, elliptiques, Revue Roumaine de Mathematiques Pures et Appliquées 9, 47-69.

ding [7], M. I. Višik [16], N. Aronszajn [3], S. Agmon [1]). La solution du problème aux limites à conditions aux limites homogènes se trouve dans l'espace mentionné V.

Pour étudier le problème sur la régularité de la solution ou le problème aux limites « sans supposer l'intégrale de Dirichlet finie », on peut introduire la notion de la coercivité d'une forme sesquilinéaire, soit $B(v, u)$, sur $H_1 \times H_2$, H_1, H_2 étant deux espaces de Hilbert, en exigeant :

$$(1.3)\quad (a) \sup_{|v|_{H_1} \leqslant 1} |B(v, u)| \geqslant c_1 |u|_{H_2}, \quad (b) \sup_{|u|_{H_2} \leqslant 1} |B(v, u)| \geqslant c_2 |v|_{H_1}.$$

L'auteur a employé ce point de vue dans [12], en posant pour H_1, H_2 certains espaces avec le poids.

Dans cet article, on étudiera le problème (1.3) pour $H_1 = \mathring{W}_2^{(k-\theta)}(\Omega)$, $H_2 = \mathring{W}_2^{(k+\theta)}(\Omega)$, avec $|\theta| < 1/2$, où $\mathring{W}_2^{(k-\theta)}(\Omega)$, resp. $\mathring{W}_2^{(k+\theta)'}(\Omega)$, est l'adhérence de $\mathfrak{D}(\Omega)$ (de l'espace des fonctions indéfiniment différentiables à support compact dans Ω) dans la topologie induite par $W_2^{(k-\theta)}(\Omega)$, resp. $W_2^{(k+\theta)}(\Omega)$. On définit $W_2^{(m)}(\Omega)$ pour $0 < m < 1$ comme l'espace des restrictions des fonctions de $W_2^{(m)}(E_N)$ sur Ω, en définissant $W_2^{(m)}(E_N)$ comme sous-espace de $L_2(E_N)$ de telles fonctions f pour lesquelles $\int_{E_N} |\hat{f}(\xi)|^2 (1 + |\xi|^2)^m d\xi < \infty$, $\hat{f}(\xi)$ étant la transformée de Fourier de f. Pour $m > 0$, non entier, on définit $W_2^{(m)}(\Omega)$ comme le sous-espace de $W_2^{([m])}(\Omega)$ des fonctions f, telles que $D^i f \in W_2^{(m-[m])}(\Omega)$ pour $|i| = [m]$. Pour Ω on prendra soit un demi-espace « déformé » : $E(x;\ x = (x', x_N),\ x_N > a(x'))$, où $a(x')$ est une fonction lipschitzienne, définie dans E_{N-1}, nulle au voisinage de l'infini, soit un domaine borné à frontière lipschitzienne. Au premier cas, on démontre au travail présenté la validité de (1.3) pour $B(v, u) = \int_\Omega \sum_{|i|, |j| = k} \bar{a}_{ij} D^i v D^j \bar{u} dx + \lambda \int_\Omega v\bar{u} dx$, où les coefficients a_{ij} sont constants, $\lambda > 0$, $|\theta| < 1/2$ et assez petit, l'opérateur A fortement elliptique ; si a_{ij} sont réels, on a le résultat pour tous $|\theta| < 1/2$. Au deuxième cas, on considère la forme sesquilinéaire $B(v, u) = \int_\Omega \sum_{|i|, |j| \leqslant k} \bar{a}_{ij} D^i v D^j \bar{u} dx + \lambda \int_\Omega v\bar{u} dx$ à coefficients assez réguliers, l'opérateur A est supposé uniformément fortement elliptique. On a (1.3) pour $(|\theta| < 1/2$, assez petit, λ assez grand ; si les coefficients a_{ij}, $|i| = |j| = k$ sont réels, on a (1.3) pour tous $|\theta| < 1/2$ et λ assez grand. Puis, sous les hypothèses mentionnées, on déduit de (1.3) :

(1.4) $A + \lambda$ est un isomorphisme de $\mathring{W}_2^{(k+\theta)}(\Omega)$ sur $(\mathring{W}_2^{(k-\theta)}(\Omega))'$

(où $(\mathring{W}_2^{(k-\theta)}(\Omega))'$ est le dual du $\mathring{W}_2^{(k-\theta)}(\Omega)$). D'un coup, on résout le problème non homogène.

Signalons que le dernier résultat (d'où on peut obtenir (1.3)) est pour Ω régulier une conséquence des résultats de J. L. Lions, E. Magenes [9]; on peut étudier ainsi les autres problèmes comme celui de Neumann, etc. Ω doit être supposé assez régulier, le procédé étant basé sur les estimations à priori (cf. S. Agmon, A. Douglis, L. Nirenberg [2]) qui n'ont pas lieu pour notre type de domaines. On précisera plus loin que, dans un certain sens, le résultat (1.4) sous nos hypothèses sur le domaine ne peut pas être élargi pour $\theta > 1/2$.

2. LES ESPACES $W_2^{(m)}(\Omega)$

Soit Ω un domaine de E_N, k un entier $\geqslant 0$, $p \geqslant 1$. On désigne par $W_p^{(k)}(\Omega)$ l'espace des fonctions de $p^{\text{ième}}$ puissance sommable dans Ω avec toutes leurs dérivées (prises au sens des distributions)*) jusqu'à l'ordre k. On le munit de la norme $|u|_{W_p^{(k)}(\Omega)} = \left(\int_\Omega \sum_{|i| \leqslant k} |D^i u|^p \, dx\right)^{1/p}$ et pour $p=2$ du produit scalaire $(v, u)_{W_2^{(k)}(\Omega)} = \int_\Omega \sum_{|i| \leqslant m} D^i v D^i \bar{u} dx$.

On désigne par $\mathring{W}_p^{(k)}(\Omega)$ l'adhérence de $\mathfrak{D}(\Omega)$ moyennant la norme de $W_p^{(k)}(\Omega)$. $W_p^{(k)}(\Omega)$ est un espace de Banach, pour $p = 2$ de Hilbert.

Soit $v \in W_2^{(k)}(E_N)$ et désignons par $\hat{v}$ sa transformée de Fourier-Plancherel**). On tire de l'égalité de Parceval pour v, u $W_2^{(k)}(E_N)$:

$$(v, u)_{W_2^{(k)}(E_N)} = (1/2\,\pi)^N \int_{E_N} (\hat{v} \hat{\bar{u}} \sum_{|i| \leqslant m} \xi^{2i}) \, d\xi, \quad \text{où} \quad \xi^j = \xi_1^{j_1} \xi_2^{j_2} \dots \xi_N^{j_N} N.$$

Désignons $|\xi|^2 = \sum_{i=1}^N \xi_i^2$. On a évidemment pour k un entier

$$c_1 |v|_{W_2^{(k)}(E_N)} \leqslant \left(\int_{E_N} |\hat{v}|^2 (1 + |\xi|^2)^k d\xi\right)^{1/2} \leqslant c_2 |v|_{W_2^{(k)}(E_N)}. \tag{2.1}$$

Soit maintenant $m \geqslant 0$. On désigne par $W_2^{(m)}(E_N)$ le sous-espace de $L_2(E_N)$ des fonctions, pour lesquelles

$$|v|_{W_2^{(m)}(E_N)} = \left(\int_{E_N} |\hat{v}|^2 (1 + |\xi|^2)^m \, d\xi\right)^{1/2} < \infty. \tag{2.2}$$

On le munit de la norme (2.2) et on obtient ainsi un espace de Hilbert à produit scalaire $(v, u)_{W_2^{(m)}(E_N)} = \int_{E_N} \hat{v} \hat{\bar{u}} (1 + |\xi|^2)^m \, d\xi$,

Sans difficulté, on démontre que $\mathfrak{D}(E_N)$ est dense dans $W_2^{(m)}(E_N)$.

Soit maintenant $m < 0$. On désigne par $W_2^{(m)}(E_N)$ le dual du $W_2^{(-m)}(E_N)$.

*) $v \in W_p^{(k)}(\Omega)$, k un entier, si $v \in L_p(\Omega)$ et il existe des fonctions de $L_p(\Omega)$, soient ω_i, $|i| \leqslant m$, de sorte que l'on ait pour $\varphi \in \mathfrak{D}(\Omega)$: $\int_\Omega \varphi\, \omega_i \, dx = (-1)^{|i|} \int_\Omega D^i\varphi \, v dx$.

**) Si $v \in W_2^{(k)}(E_N) \cap L_1(E_N)$, on a $\hat{v}(\xi) = \int_{E_N} v(x)\, e^{-i(\xi,\, x)}\, dx$.

Soit $\Omega \subset E_N$, $0 \leqslant m < 1$. On désigne par $W_2^{(m)}(\Omega)$ l'espace des restrictions sur Ω des fonctions de $W_2^{(m)}(E_N)$. Désignons par Q le sous-espace fermé du $W_2^{(m)}(E_N)$, des fonctions, nulles sur Ω. On munit $W_2^{(m)}(\Omega)$ de la norme $|u|_{W_2^{(m)}(\Omega)} = |\tilde{u}|_{W_2^{(m)}(E_N)/Q}$. Soit $m \geqslant 0$, m non entier. On désigne par $W_2^{(m)}(\Omega) \subset W^{([m])}(\Omega)$ le sous-espace des fonctions, pour lesquelles $D^i u \in W_2^{(m-[m])}(\Omega)$, $|i| = m$. On pose :

$$(2.3) \qquad |u|_{W_2^{(m)}(\Omega)} = \Big(|u|^2_{W_2^{([m])}(\Omega)} + \sum_{|i|=[m]} |D^i u|^2_{W^{(m-[m])}(\Omega)}\Big)^{1/2}.$$

On désigne pour $m \geqslant 0$ par $\mathring{W}_2^{(m)}(\Omega)$ l'adhérence de $\mathfrak{D}(\Omega)$ moyennant de la norme de $W_2^{(m)}(\Omega)$; si $m > 0$, on désigne par $W_2^{(-m)}(\Omega) = (\mathring{W}_2^{(m)}(\Omega))'$.

La norme (2.3) n'est pas trop pratique. C'est pourquoi, on introduit pour les espaces $W_2^{(m)}(\Omega)$ une autre norme (cf. p. ex. N. Aronszajn, K. T. Smith [4]) qui sous les hypothèses mentionnées plus loin équivaut à (2.3)*). Soit $0 < m < 1$. On introduit l'espace ${}_*W_2^{(m)}(\Omega)$ des fonctions de $L_2(\Omega)$ pour lesquelles

$$(2.4) \qquad |u|_{{}_*W_2^{(m)}(\Omega)} = \left(|u|^2_{L_2(\Omega)} + \int_\Omega\int_\Omega \frac{|u(x) - u(y)|^2}{|x-y|^{N+2m}}\, dx dy\right)^{1/2} < \infty.$$

Si $m \geqslant 0$, m non entier, on définit $W_2^{(m)}(\Omega)$ comme l'espace de telles $f \in W_2^{([m])}(\Omega)$, pour lesquelles $D^i f \in {}_*W_2^{(m-[m])}(\Omega)$, $|i| = k$. On pose

$$(2.5) \qquad |u|_{{}_*W_2^{(m)}(\Omega)} = \Big(|u|^2_{W_2^{([m])}(\Omega)} + \sum_{|i|=[m]} |D^i u|^2_{{}_*W_2^{(m-[m])}(\Omega)}\Big)^{1/2}.$$

Introduisons ${}_{**}W_2^{(m)}(E_N)$, $0 < m < 1$, comme l'espace des fonctions de $L_2(E_N)$, pour lesquelles

$$(2.6) \qquad |u|_{{}_{**}W_2^{(m)}(E_N)} = \Bigg(|u|^2_{L_2(E_N)} + \sum_{i=1}^N \int_0^\infty \int_{E_N} \frac{|u(x_1, x_2, \ldots, x_{i-1}, x_i + t, x_{i+1}, \ldots, x_N) - u(x_1, \ldots, x_N)|^2}{t^{1+2m}} \times dt\, dx\Bigg)^{1/2} < \infty.$$

Si $m \geqslant 0$, non entier, on définit ${}_{**}W_2^{(m)}(E_N)$ par

$$(2.7) \qquad |u|_{{}_{**}W_2^{(m)}(E_N)} = \Big(|u|^2_{W_2^{([m])}(E_N)} + \sum_{|i|=[m]} |D^i u|^2_{{}_{**}W_2^{(m-[m])}(E_N)}\Big)^{1/2} < \infty.$$

On démontre, sans difficulté, en utilisant la transformation de Fourier au cas de $\Omega = E_N$, l'équivalence des normes $|\;|_{W_2^{(m)}(E_N)}$ et $|\;|_{{}_{**}W_2^{(m)}(E_N)}$, d'où, par un calcul élémentaire, il s'ensuit l'équivalence des normes $|\;|_*$, $|\;|_{**}$.

*) On peut ainsi introduire les espaces $W_p^{(m)}(\Omega)$, $p > 1$, $m \geqslant 0$; cf. p. ex. E. Gagliardo [5], L. N. Slobodětskij [14]. Une autre méthode d'introduire ces espaces est d'utiliser l'interpolation. Cf. J. L. Lions, E. Magenes [9], [10], [11].

Résumons :

LEMME 2.1. *Soit* $\Omega = E_N$. *Alors les normes* (2.2), (2.5) *et* (2.7) *sont équivalentes.*

Considérons maintenant le cas de $\Omega = E_N^+$, où E_N^+ est le demi-espace $x_N > 0$. On définit ${}_{**}W_2^{(m)}(E_N^+)$ par (2.6) et (2.7) et on obtient l'équivalence des normes $|\ |_*$, $|\ |_{**}$; pour démontrer l'équivalence des normes $|\ |_{**}$ et $|\ |$, il suffit d'étudier le cas $0 < m < 1$. Soit $|u|_{{}_{**}W_2^{(m)}(E_N)} < \infty$. On prolonge u sur E_N en posant pour $x_N < 0$ $(x = (x', x_N))$:

(2.8) $$u(x', x_N) = u(x', -x_N)$$

et on obtient $u \in W_2^{(m)}(E_N)$ avec $|u|_{W_2^{(m)}(E_N)} \leqslant c\,|u|_{W_2^{(m)}(E_N^+)}$.

Résumons :

LEMME 2.2. *Soit* $\Omega = E_N^+$. *Alors les normes* (2.2), (2.5) *et* (2.7) *sont équivalentes.*

Voilà un cas particulier de deux résultats de J. L. Lions, E. Magenes [11] :

LEMME 2.3. *Soit* $\Omega = E_N^+$, $0 \leqslant m < 1/2$. *Soit* P *le prolongement par zéro des fonctions de* $W_2^{(m)}(E_N^+)$ *sur* E_N. *Alors* P *est une application linéaire et continue de* $W_2^{(m)}(E_N^+)$ *dans* $W_2^{(m)}(E_N)$. *On a* $\overset{\circ}{W}{}_2^{(m)}(E_N^+) = W_2^{(m)}(E_N^+)$.

LEMME 2.4. *Soit* $\Omega = E_N^+$, $(1/2) < m \leqslant 1$. *Soit* P *le prolongement par zéro des fonctions de* $\overset{\circ}{W}{}_2^{(m)}(E_N^+)$ *sur* E_N. *Alors* P *est une application linéaire et continue de* $\overset{\circ}{W}{}_2^{(m)}(E_N^+)$ *dans* $W_2^{(m)}(E_N)$.

Conséquence du lemme 2.3 : P est une application linéaire et continue de $W_2^{(m)}(E_N^+)$ dans $W_2^{(m)}(E_N)$ pour $0 \leqslant m \leqslant 1$, $m \neq 1/2$. Pour le cas $m = 1/2$, ce résultat n'est pas valable (cf. [11]).

On considère maintenant le cas du demi-espace «déformé» et les normes (2.3), (2.5); on pose $\Omega = E(x;\ x_N > a\ (x'))$. On a d'après le lemme 2.1 pour $u \in W_2^{(m)}(\Omega)$: $|u|_{{}_*W_2^{(m)}(\Omega)} \leqslant c_1 |u|_{W_2^{(m)}(\Omega)}$. Pour démontrer l'inégalité contraire, il suffit de construire un opérateur-prolongement, linéaire et borné de ${}_*W_2^{(m)}(\Omega)$ dans $W_2^{(m)}(E_N)$, $0 < m < 1$. Pour ce but, soit $x = T(y)$ la transformation de E_N^+ sur Ω définie par

(2.9) $$(x = x', x_N),\ y = (y', y_N),\ \ x' = y',\ x_N = y_N + a\,(y').$$

Cette transformation étant lipschitzienne, avec déterminant 1, on obtient, en posant pour $u \in {}_*W_2^{(m)}(\Omega)$, $v\,(y) = u\,(T\,(y))$:

(2.10) $$|v|_{{}_{**}W_2^{(m)}(E_N^+)} \leqslant c_2\,|v|_{{}_*W_2^{(m)}(E_N^+)} \leqslant c_3\,|u|_{{}_*W_2^{(m)}(\Omega)}.$$

En prolongeant v de E_N^+ sur E_N par (2.8), on obtient, tenant compte du lemme 2.2 :

(2.11) $$|v|_{{}_{**}W_2^{(m)}(E_N)} \leqslant c_4\,|v|_{{}_{**}W_2^{(m)}(E_N^+)}.$$

Il suffit maintenant de revenir aux cartes x et on obtient de (2.10) et (2.11) :

$$|u|_{W_2^{(m)}(E_N)} \leqslant c_5\,|u|_{{}_*W_2^{(m)}(\Omega)},$$

d'où le résultat. Résumons :

LEMME 2.5. *Soit $a(x')$ une fonction lipschitzienne, définie dans E_{N-1}, nulle au voisinage de l'infini. Soit $\Omega = E(x;\ x = (x', x_N),\ x_N > a(x'))$. Alors les normes* (2.3) *et* (2.5) *équivalent.*

En utilisant la transformation (2.9), on voit, tenant compte du lemme précédent que l'on a

LEMME 2.6. *Soit Ω du lemme précédent. Alors le lemme 2.3 a lieu.*

Pour démontrer au cas présent du demi-espace « déformé » la modification du lemme 2.4, il suffit de nouveau d'utiliser la transformation (2.9) et de s'appuyer sur le fait que cette transformation fait correspondre d'une manière linéaire et bornée les éléments de $\mathring{W}_2^{(1)}(\Omega)$ sur $\mathring{W}_2^{(1)}(E_N^+)$ et au contraire. On le voit immédiatement, en utilisant la technique usuelle de la régularisation (cf. p. ex. S. L. Sobolev [15]). On a alors :

LEMME 2.7. *Soit Ω du lemme précédent. Alors le lemme 2.4 reste en vigueur.*

On démontre sans difficulté :

LEMME 2.8. *Soit $0 \leqslant m < n$. Alors à chaque $\varepsilon > 0$ on peut coordonner $\lambda(\varepsilon) > 0$ de sorte que pour $u \in W_2^{(n)}(E_N)$:*

$$(2.12) \qquad |u|_{W_2^{(m)}(E_N)} \leqslant \varepsilon\, |u|_{W_2^{(n)}(E_N)} + \lambda(\varepsilon)\, |u|_{L_2(E_N)}.$$

Démonstration. Soit $\lambda(\varepsilon)$ choisie de sorte que $(1 + |\xi|^2)^m \leqslant \varepsilon^2 (1 + |\xi|^2)^n + \lambda^2$ pour $\xi \in E_N$; on en tire pour $u \in W_2^{(n)}(E_N)$:

$$\int_{E_N} (1 + |\xi|^2)^m |\hat{u}(\xi)|^2\, d\xi \leqslant \varepsilon^2 \int_{E_N} (1 + |\xi|^2)^n |\hat{u}(\xi)|^2\, d\xi + \lambda^2 \int_{E_N} |\hat{u}(\xi)|^2\, d\xi,$$

d'où l'assertion.

On dit qu'un domaine Ω, borné de E_N, est à la frontière lipschitzienne s'il existe M systèmes des cartes $[x_{r1}, x_{r2}, \ldots, x_{rN}]$, en bref, $[x_r', x_{rN}]$ et M fonctions lipschitziennes :

$$(2.13) \qquad a_r(x_r'), \text{ dans la fermeture du cube } |x_{ri}| < \alpha,\ i = 1, 2 \ldots N-1,$$

telles que chaque point de la frontière de Ω (qu'on désigne par $\dot{\Omega}$) se peut écrire au moins dans un de ces systèmes sous la forme $[x_r', a_r(x_r')]$. On suppose encore l'existence d'un nombre $\beta > 0$, tel que les points $|x_{ri}| < \alpha$, $i = 1, 2, \ldots, N-1$, $a_r(x_r') < x_{rN} < a_r(x_r') + \beta$ soient dans Ω, tandis que les points $|x_{ri}| < \alpha$, $i = 1, 2, \ldots, N-1$, $a_r(x_r') - \beta < x_{rN} < a_r(x_r')$ ne soient pas dans $\bar{\Omega}$ (la fermeture de Ω). On désigne encore

$$(2.14) \qquad U_r = E(x;\ |x_{ri}| < \alpha,\ i=1, 2, \ldots, N-1,\ a_r(x_r') - \beta < x_{rN} < a_r(x_r') + \beta),$$

$$V_r = E(x;\ |x_{ri}| < \alpha,\ i = 1, 2, \ldots, N-1,\ a_r(x_r') < x_{rN} < a_r(x_r') + \beta)\,;$$

on peut trouver des fonctions dans $\mathfrak{D}(U_r)$, soient :

$$(2.15) \qquad \varphi_r,\ r = 1, 2, \ldots, M, \text{ de sorte que } x \in \dot{\Omega} \Rightarrow \sum_{r=1}^{M} \varphi_r(x) = 1.$$

On prend encore une autre fonction dans $\mathfrak{D}(\Omega)$, soit φ_{M+1}, de sorte que $x \in \Omega \Rightarrow \sum_{r=1}^{M+1} \varphi_r(x) = 1$.

On a :

LEMME 2.9. *Soit Ω un domaine à frontière lipschitzienne. Alors les normes* (2.3) *et* (2.5) *équivalent.*

Démonstration. On a évidemment pour $u \in W_2^{(m)}(\Omega)$: $|u|_{{}_*W_2^{(m)}(\Omega)} \leqslant \leqslant c_1 |u|_{W_2^{(m)}(\Omega)}$. Soit $|u|_{{}_*W_2^{(m)}(\Omega)} < \infty$. On désigne par $u_r = u\,\varphi_r$. On a

$$|u_r|_{{}_*W_2^{(m)}(\Omega)} \leqslant c_2 |u|_{{}_*W_2^{(m)}(\Omega)}. \tag{2.16}$$

Pour le voir, il suffit de considérer le cas où $0 < m < 1$. On a

(2.17)

$$\int_\Omega \int_\Omega \frac{|\varphi_r(x)\,u(x) - \varphi_r(y)\,u(y)|^2}{|x-y|^{N+2m}}\,dxdy \leqslant 2\int_\Omega \int_\Omega \frac{|\varphi_r(x)\,(u(x) - u(y))|^2}{|x-y|^{N+2m}}\,dxdy + $$
$$+ 2\int_\Omega \int_\Omega \frac{|u(y)|^2 |\,\varphi_r(x) - \varphi_r(y)|^2}{|x-y|^{N+2m}}\,dxdy\,;$$

la première intégrale du second membre se peut évidemment estimer par $|u|^2_{{}_*W_2^{(m)}(\Omega)}$ tandis que pour la deuxième on a

(2.18)

$$\int_\Omega \int_\Omega \frac{|u(y)|^2 |\,\varphi_r(x) - \varphi_r(y)|^2}{|x-y|^{N+2m}}\,dxdy \leqslant c_3 \int_\Omega |u(y)|^2\,dy \int_\Omega \frac{dx}{|x-y|^{N-2+2m}} \leqslant$$
$$\leqslant c_3 \int_\Omega |u(y)|^2\,dy \int_B \frac{dx}{|x|^{N-2+2m}} \leqslant c_4 \int_\Omega |u(y)|^2\,dy,$$

où B est une boule centrée à l'origine, du rayon assez grand. Alors il s'ensuit de (2.18) et (2.17) l'inégalité (2.16). Prolongeons u_{M+1} par 0 hors de Ω. On obtient

$$\int_{E_N} \int_{E_N} \frac{|u_{M+1}(x) - u_{M+1}(y)|^2}{|x-y|^{N+2m}}\,dxdy = \int_\Omega \int_\Omega + 2\int_{E_N - \Omega} \int_\Omega . \tag{2.19}$$

Désignons $\Omega^* \subset \overline{\Omega^*} \subset \Omega$ un domaine contenant le support de φ_{M+1}. On a

$$\int_{E_N-\Omega} \int_\Omega \frac{|u_{M+1}(y)|^2}{|x-y|^{N+2m}}\,dxdy = \int_{E_N-\Omega} \int_{\Omega^*} \frac{|u_{M+1}(y)|^2}{|x-y|^{N+2m}}\,dxdy \leqslant c_5 \int_{\Omega^*} |u_{M+1}(y)|^2\,dy,$$

d'où

$$|u_{M+1}|_{W_2^{(m)}(E_N)} \leqslant c_6 |u_{M+1}|_{{}_*W_2^{(m)}(\Omega)}. \tag{2.20}$$

Considérons maintenant $r \leqslant M$. Soit P_r la projection du support de φ_r (cf. 2.15) sur l'hyperplan $x_{rN} = 0$ et ψ_r une fonction réelle, indéfiniment différentiable dans E_{N-1}, à support dans le cube $|x_{ri}| < \alpha$, $i = 1, 2, \ldots, N-1$, et égale à 1 au voisinage de P_r. Posons $a(x_r') = \psi_r(x_r')a_r(x_r')$ dans $|x_{ri}| < \alpha$ et $= 0$ ailleurs. Désignons par $\sigma = E(x;\, x = (x_r', x_{rN}), x_{rN} > a(x_r'))$. En raisonnant comme pour démontrer (2.19) on obtient $|u_r|_{{}_*W_2^{(m)}(\sigma)} \leqslant c_7 |u_r|_{{}_*W_2^{(m)}(\Omega)}$. Utilisant (2.8) et (2.9), on prolonge u_r sur E_N de sorte que $|u_r|_{W_2^{(m)}(E_N)} \leqslant c_8 |u_r|_{W_2^{(m)}(\sigma)}$, d'où

$$|u_r|_{W_2^{(m)}(\Omega)} \leqslant c_7 |u_r|_{{}_*W_2^{(m)}(\Omega)}. \tag{2.21}$$

Alors (2.20) et (2.21) nous donnent l'assertion du lemme.

En définissant le prolongement cité dans la démonstration précédente hors σ par zéro, on obtient, tenant compte des lemmes 2.6 et 2.7 :

LEMME 2.10. *Soit Ω un domaine à frontière lipschitzienne. Alors les lemmes 2.3 et 2.4 ont lieu.*

On obtient ainsi :

LEMME 2.11. *Soit Ω un domaine à frontière lipschitzienne ou un demi-espace « déformé ». Soit $m \geqslant 0$, $m - [m] \neq 1/2$. Soit P l'opérateur du prolongement par zéro des fonctions de $\mathring{W}_2^{(m)}(\Omega)$ sur E_N. Alors P est une application linéaire et continue de $\mathring{W}_2^{(m)}(\Omega)$ dans $W_2^{(m)}(E_N)$.*

Il s'ensuit du lemme 2.9 :

LEMME 2.12. *Soit $0 \leqslant m \leqslant n$, Ω un domaine à frontière lipschitzienne. Alors $W_2^{(m)}(\Omega) \supset W_2^{(n)}(\Omega)$ algébriquement et topologiquement*).*

Il s'ensuit des lemmes 2.8 et 2.11 :

LEMME 2.13. *Sous les hypothèses du lemme 2.11, avec $0 \leqslant m < n$, $\varepsilon > 0$ arbitrairement petit, il existe $\lambda(\varepsilon) > 0$ de sorte que pour $u \in \mathring{W}_2^{(n)}(\Omega)$*

$$|u|_{W_2^{(m)}(\Omega)} \leqslant \varepsilon |u|_{W_2^{(n)}(\Omega)} + \lambda(\varepsilon) |u|_{L_2(\Omega)}. \tag{2.22}$$

Remarquons encore que pour les domaines à frontières lipschitziennes, l'espace $W_p^{(1)}(\Omega)$ peut être défini comme l'espace des restrictions. On obtient ainsi une topologie équivalente.

Finissons la 2e partie de notre article par :

LEMME 2.14. *Soit Ω un domaine à frontière lipschitzienne, $m \geqslant 0$. Alors la restriction de $\mathfrak{D}(E_N)$ sur Ω est dense dans $W_2^{(m)}(\Omega)$**).*

Démonstration. Utilisant la technique usuelle de la partition de l'unité (on utilise les fonctions (2.15)), tout revient à voir si l'assertion est vraie pour $u \in W_2^{(m)}(\Omega)$ à support dans $\overline{\Omega} \cap U_r$ (cf. 2.14). Soit $\varepsilon > 0$ et u_h définie comme suit : $u_h(x) = u_h(x', x_{rN}) = u(x', x_{rN} - h)$. On a pour $h \to 0$: $u_h \to u$ dans $W_2^{(m)}(\Omega)$, alors on peut trouver h_0 de sorte que $|u_{h_0} - u|_{W_2^{(m)}(\Omega)} < \varepsilon/2$. Mais $u_{h_0} \in W_2^{(m)}(\Omega^*)$ avec $\overline{\Omega} \subset \Omega^*$; on peut trouver

*) Évidemment, la condition $0 \leqslant m$ est inutile.

**) Il suffit que la frontière de Ω soit continue.

$\psi \in \mathfrak{D}(\Omega^*)$, $\psi(x) = 1$ pour $x \in \overline{\Omega}$, et l'on obtient $u_{h_0}\psi \in W_2^{(m)}(E_N)$. Cela étant, il existe $\varphi \in \mathfrak{D}(E_N)$ de sorte que $|\varphi - u_{h_0}\psi|_{W_2^{(m)}(E_N)} < \varepsilon/2 \Rightarrow |\varphi - u|_{W_2^{(m)}(\Omega)} < \varepsilon$, c.q.f.d.

3. LA COERCIVITÉ DES FORMES DANS LE DEMI-ESPACE « DÉFORMÉ »

Soit Ω un demi-espace « déformé » $\Omega = E(x;\ x = (x', x_N),\ x_N > a(x'))$, k un entier > 0 et λ un nombre. On désigne par

$$B_\lambda(v, u) = \int_\Omega \sum_{|i|,|j| = k} \overline{a}_{ij}\, D^i v D^j \overline{u} dx + \lambda \int_\Omega v\overline{u} dx \tag{3.1}$$

une forme sesquilinéaire dans $W_2^{(k)}(\Omega) \times W_2^{(k)}(\Omega)$, à coefficients constants, et fortement elliptiques :

$$(\xi_1, \xi_2, \xi_3, \ldots, \xi_N) \in E_N \Rightarrow \mathcal{Re} \sum_{|i|,|j| = k} \overline{a}_{ij}\, \xi^{i+j} \geqslant \varepsilon \left(\sum_{i=1}^N \xi_i^2\right)^k, \quad \varepsilon > 0. \tag{3.2}$$

On désigne :

$$|\mathcal{Im} \sum_{|i|,|j| = k} \overline{a}_{ij}\, \xi^{i+j}| \leqslant \delta\, |\xi|^{2k}. \tag{3.2'}$$

LEMME 3.1. *La forme* (3.1), *définie sur* $\mathfrak{D}(\Omega) \times \mathfrak{D}(\Omega)$, *peut être prolongée pour* $|\theta| \leqslant k$ *par continuité en une forme sesquilinéaire sur* $\mathring{W}_2^{(k-\theta)}(\Omega) \times \mathring{W}_2^{(k+\theta)}(\Omega)$.

La démonstration du 3.1 se fait immédiatement, par la transformation de Fourier, tenant compte des lemmes 2.6 et 2.7.

On désigne par R_α une transformation de $\mathfrak{D}(E_N)$ dans $L_2(E_N)$, définie par :

$$R_\alpha(\hat{\varphi})(\xi) = \Gamma(\alpha+1)\, \hat{\varphi}(\xi) \int_{\dot{K}} (1 + I(\xi, \sigma))^{-\alpha} \nu(\sigma) dS, \tag{3.3}$$

où K est la boule unité, centrée à l'origine, $\nu(\sigma)$ une fonction définie sur la surface $\dot{K}$ de la boule, $\nu(\sigma) \geqslant 0$, indéfiniment différentiable avec $\int_{\dot{K}} \nu(\sigma)\, dS = 1$. (ξ, σ) désigne le produit scalaire dans E_N.

LEMME 3.2. *Soit* $m \geqslant 0$, $\alpha + m \geqslant 0$, $|\alpha| < 1$. *Alors la transformation* R_α *peut être prolongée par continuité en une transformation linéaire et bornée de* $W_2^{(m)}(E_N)$ *sur* $W_2^{(m+\alpha)}(E_N)$ *et on a*

$$\left|\int_{\dot{K}} (1+I(\xi, \sigma))^{-\alpha}\, \nu(\sigma)\, dS\right| \leqslant [c_1(1-\alpha)]\,(1 + |\xi|^2)^{-\alpha/2}, \tag{3.4}$$

$$\mathcal{Re} \int_{\dot{K}} (1 + I(\xi, \sigma))^{-\alpha}\, \nu(\sigma)\, dS \geqslant c_2\,(1 - |\alpha|)\,(1 + |\xi|^2)^{-\alpha/2}, \tag{3.5}$$

$$\left|\mathcal{Im} \int_{\dot{K}} (1+I(\xi, \sigma))^{-\alpha}\, \nu(\sigma)\, dS\right| \leqslant c_3\,[|\alpha|/(1-\alpha)]\,(1+|\xi|^2)^{-\alpha/2}. \tag{3.6}$$

Démonstration. On a pour $|\alpha|<1$

$$(3.7)\qquad \left|\int_{\dot K}\left(1+I\,|\xi|\left(\frac{\xi}{|\xi|},\sigma\right)\right)^{-\alpha}\nu(\sigma)\,dS\right|\leqslant\int_{\dot K}\left(1+|\xi|^2\left(\frac{\xi}{|\xi|},\sigma\right)^2\right)\nu(\sigma)\,dS.$$

Si $\alpha\leqslant 0$, on tire de (3.7)

$$(3.8)\qquad \int_{\dot K}\left(1+|\xi|^2\left(\frac{\xi}{|\xi|},\sigma\right)^2\right)^{-\alpha/2}\nu(\sigma)\,dS\leqslant(1+|\xi|^2)^{-\alpha/2},$$

si $\alpha>0$, on a pour $|\xi|\leqslant 1$

$$(3.9)\qquad \int_{\dot K}\left(1+|\xi|^2\left(\frac{\xi}{|\xi|},\sigma\right)^2\right)^{-\alpha/2}\nu(\sigma)\,dS\leqslant 1,$$

et pour $|\xi|>1$ on obtient

$$(3.10)\qquad \int_{\dot K}\left(1+|\xi|^2\left(\frac{\xi}{|\xi|},\sigma\right)^2\right)^{-\alpha/2}\nu(\sigma)\,dS=|\xi|^{-\alpha}\int_{\dot K}\left(|\xi|^{-2}+\right.$$

$$\left.+\left(\frac{\xi}{|\xi|},\sigma\right)^2\right)^{-\alpha/2}\nu(\sigma)dS\leqslant|\xi|^{-\alpha}\int_{\dot K}\left|\left(\frac{\xi}{|\xi|},\sigma\right)\right|^{-\alpha}\nu(\sigma)dS\leqslant[c_4/(1-\alpha)]\,|\xi|^{-\alpha}.$$

On obtient alors de (3.8) — (3.10) l'inégalité (3.4). Pour démontrer (3.5) considérons d'abord de nouveau $\alpha\leqslant 0$. On a $\mathcal{Re}\,(1+I\,(\xi,\sigma))^{-\alpha}=$ $=\left(1+|\xi|^2\left(\frac{\xi}{|\xi|},\sigma\right)^2\right)^{-\alpha/2}\cdot\cos\alpha\vartheta$, où $\operatorname{arctg}\vartheta=(\xi,\sigma)$; étant $|\vartheta|<\pi/2$, on a $\cos\alpha\vartheta\geqslant 1-|\alpha|$, d'où $\mathcal{Re}\int_{\dot K}(1+I(\xi,\sigma))^{-\alpha}\nu(\sigma)\,dS\geqslant(1-|\alpha|)\int_{\dot K}$ $\left(1+|\xi|^2\left(\frac{\xi}{|\xi|},\sigma\right)^2\right)^{-\alpha/2}\nu(\sigma)\,dS$. Si $|\xi|\leqslant 1$, on a $\int_{\dot K}(1+|\xi|^2(\xi,\sigma)^2)^{-\alpha/2}\nu(\sigma)$ $dS\geqslant 1$, si $|\xi|>1$, on obtient $\int_{\dot K}(1+|\xi|^2(\xi,\sigma)^2)^{-\alpha/2}\nu(\sigma)\,dS=|\xi|^{-\alpha}$ $\int_{\dot K}\left(|\xi|^{-2}+\left(\frac{\xi}{|\xi|},\sigma\right)^2\right)^{-\alpha/2}\nu(\sigma)\,dS\geqslant|\xi|^{-\alpha}\int_{\dot K}\left|\left(\frac{\xi}{|\xi|},\sigma\right)\right|^{-\alpha}\nu(\sigma)\,dS\geqslant c_5\,|\xi|^{-\alpha}.$

On a alors pour $\alpha\leqslant 0$

$$(3.11)\qquad \mathcal{Re}\int_{\dot K}(1+I\,(\xi,\sigma))^{-\alpha}\,\nu(\sigma)\,dS\geqslant c_6\,(1-|\alpha|)\,(1+|\xi|^2)^{-\alpha/2}.$$

Considérons $\alpha>0$. On obtient pour $|\xi|\leqslant 1:\mathcal{Re}\int_{\dot K}(1+I\,(\xi,\sigma))^{-\alpha}$ $\nu(\sigma)\,dS\geqslant(1-|\alpha|)\,2^{-\alpha/2}\int_{\dot K}\nu(\sigma)\,dS$, pour $|\xi|>1:\mathcal{Re}\int_{\dot K}(1+I\,(\xi,\sigma))^{-\alpha}$ $\nu(\sigma)\,dS\geqslant(1-|\alpha|)\,|\xi|^{-\alpha}2^{-\alpha/2}\int_{\dot K}\nu(\sigma)\,dS$, d'où pour $\alpha>0$

$$(3.12)\qquad \mathcal{Re}\int_{\dot K}(1+I\,(\xi,\sigma))^{-\alpha}\,\nu(\sigma)\,dS\geqslant c_7\,(1-|\alpha|)\,(1+|\xi|^2)^{-\alpha/2}.$$

(3.11) et (3.12) nous donnent (3.5). Démontrons finalement (3.6). On a $|\mathcal{I}m\,(1+I(\xi,\sigma))^{-\alpha}| = \left(1+|\xi|^2\left(\frac{\xi}{|\xi|},\sigma\right)\right)^{-\alpha/2} |\sin\alpha\vartheta| \leqslant (\pi/2)|\alpha|(1+$ $+|\xi|^2\left(\frac{\xi}{|\xi|},\sigma\right)\right)^{-\alpha/2}$.

Si $\alpha \leqslant 0$, on tire de (3.8)

$$\left|\mathcal{I}m\int_{\dot K}(1+I(\xi,\sigma))^{-\alpha}\,\nu(\sigma)\,dS\right| \leqslant (\pi/2)|\alpha|\,(1+|\xi|^2)^{-\alpha/2}, \tag{3.13}$$

pour $\alpha > 0$ on obtient de (3.9) et (3.10)

$$\left|\mathcal{I}m\int_{\dot K}(1+I(\xi,\sigma))^{-\alpha}\,\nu(\sigma)\,dS\right| \leqslant c_8[|\alpha|/(1-\alpha)]\,(1+|\xi|^2)^{-\alpha/2}; \tag{3.14}$$

(3.13) avec (3.14) entraînent (3.6), d'où le lemme.

THÉORÈME 1. *Soient* Ω *un demi-espace «déformé», $B_\lambda(v,u)$ la forme* (3.1) *jouissant de* (3.2) *et* (3.3). *Alors pour* $|\theta| < 1/2$ *et* θ *assez petit, on a pour* $\lambda \geqslant 2$:

$$\mathcal{R}e\,B_\lambda(v,u) \geqslant c_1(\theta)|u|^2_{W_2^{(k+\theta)}(\Omega)} + c_2(\lambda-1)^{1/2}|u|^2_{L_2(\Omega)}, \tag{3.15}$$

pour $\lambda \geqslant 1$:

$$\mathcal{R}e\,B_\lambda(v,u) \geqslant c_3(\theta)\,|u|^2_{W_2^{(k+\theta)}(\Omega)}, \tag{3.16}$$

où $u \in \overset{\circ}{W}{}_2^{(k+\theta)}(\Omega)$, $v \in \overset{\circ}{W}{}_2^{(k-\theta)}(\Omega)$, $v = R_{-2\theta}(u)$, R *est une transformation linéaire et continue de* $\overset{\circ}{W}{}_2^{(k+\theta)}(\Omega)$ *dans* $\overset{\circ}{W}{}_2^{(k-\theta)}(\Omega)$; *si les coefficients* a_{ij} *sont réels, on a l'assertion pour tous* $|\theta| < 1/2$.

Démonstration. Soient $\varphi \in \mathfrak{D}(\Omega)$, K la boule, $\nu(\sigma)$ la fonction du lemme 3.2. On modifie l'identité de S. L. Sobolev (cf. S. L. Sobolev [15]), et on obtient : $\int_0^\infty \frac{d}{dt}(e^{-t}\varphi(x+t\sigma))\,dt = -\varphi(x)$; ayant $\frac{d}{dt}\varphi(x+t\sigma) = $ $= \sum_{i=1}^N \frac{\partial\varphi}{\partial x_i}(x+t\sigma)\sigma_i$, on obtient $\varphi(x) = \int_0^\infty\left(e^{-t}\varphi(x+t\sigma) - e^{-t}\sum_{i=1}^N\frac{\partial\varphi}{\partial x_i}(x+t\sigma)\sigma_i\right)dt$, d'où $\varphi(x) = \int_{\dot K}\nu(-\sigma)\,dS\int_0^\infty e^{-t}\left(\varphi(x+t\sigma) - \sum_{i=1}^N\frac{\partial\varphi}{\partial x_i}(x+t\sigma)\,\sigma_i\right)dt$. En posant $x + t\sigma = y$, on obtient

$$\varphi(x) = \int_{E_N}\left(\sum_{i=1}^N\frac{\partial\varphi}{\partial x_i}(y)\frac{x_i-y_i}{|x-y|^N}e^{-|x-y|} + \varphi(y)\cdot\frac{1}{|x-y|^{N-1}}e^{-|x-y|}\right)\nu\left(\frac{x-y}{|x-y|}\right)dy. \tag{3.17}$$

Définissons pour $\varphi \in \mathfrak{D}(\Omega)$ l'opérateur $R_\alpha(\varphi)$, en posant $(-1 < \alpha < \infty)$

$$R_\alpha(\varphi) = \int_{E_N}\left(\sum_{i=1}^N\frac{\partial\varphi}{\partial x_i}(y)\frac{x_i-y_i}{|x-y|^{N-\alpha}} + \frac{\varphi(y)}{|x-y|^{N-1-\alpha}}\right)e^{-|x-y|}\,\nu\left(\frac{x-y}{|x-y|}\right)dy\cdot \tag{3.18}$$

On a $R_0(\varphi) = \varphi$. Ayant $\int_{E_N} \frac{e^{-z}}{|z|^{N-1-\alpha}} dz < \infty$ pour $-1 < \alpha < \infty$, R_α est une transformation linéaire et bornée de $W_2^{(1)}(E_N)$ dans $L_2(E_N)$. Après calculs élémentaires on obtient que la transformée de Fourier de $R_\alpha(\varphi)$ coïncide avec (3.3).

On peut maintenant trouver un cône fixe, à l'axe parallèle à l'axe x_N, à savoir $\sum_{i=1}^{N-1} x_i^2 < \varkappa x_N^2$, $x_N < 0$, $\varkappa > 0$, de la manière que pour $y \in \dot{\Omega}$, les points x pour lesquels $x_N < y_N$, $\sum_{i=1}^{N-1} (x_i - y_i)^2 < \varkappa (y_N - x_N)^2$ n'appartiennent plus à Ω.

Soit maintenant $\nu(\sigma) \neq 0$ seulement pour tels σ, pour lesquels $\sigma_N < 0$, $\sum_{i=1}^{N-1} \sigma_i^2 < \varkappa \sigma_N^2$. Soit $u \in \mathfrak{D}(\Omega)$, $v = R_{-2\theta}(u)$. On a $v \in W_2^{(m)}(E_N)$ pour chaque $m \geqslant 0$, en vertu de (3.4) ; d'autre part, le support de v est contenu dans Ω, d'où il s'ensuit que $v \in \overset{\circ}{W}{}_2^{(m)}(\Omega)$, à plus forte raison dans $\overset{\circ}{W}{}_2^{(k-\theta)}(\Omega)$. On a maintenant pour $\lambda \geqslant 1$ (c_4 resp. c_5 c'est la constante c_2 resp. c_3 du lemme précédent) :

$$(3.19) \quad \mathcal{R}e\, B_\lambda(v,u) = (1/2\pi)^N \mathcal{R}e \int_{E_N} \sum_{|i|,|j|=k} \bar{a}_{ij} \xi^{i+j} \widehat{v(\xi)}\, \overline{\widehat{u(\xi)}}\, d\xi +$$

$$+ \mathcal{R}e \lambda \int_{E_N} \widehat{v(\xi)}\, \overline{\widehat{u(\xi)}}\, d\xi = \Gamma(1-2\theta)\, \mathcal{R}e \int_{E_N} \sum_{|i|,|j|=k} \bar{a}_{ij} \xi^{i+j} \left(\int_{\dot{K}} (1 + \right.$$

$$\left. + I(\xi, \sigma))^{2\theta} \nu(\sigma)\, dS \right) |\hat{u}(\xi)|^2 d\xi + \Gamma(1-2\theta)\, \mathcal{R}e \lambda \int_{E_N} \left(\int_{\dot{K}} (1 + \right.$$

$$\left. + I(\xi, \sigma))^{2\theta} \nu(\sigma)\, dS \right) |\hat{u}(\xi)|^2 d\xi \geqslant \Gamma(1-2\theta)\, \varepsilon\, c_4 (1 -$$

$$- |2\theta|) \int_{E_N} |\xi|^{2k} (1 + |\xi|^2)^\theta |\hat{u}(\xi)|^2 d\xi + \Gamma(1-2\theta)\, c_4 (1 -$$

$$- |2\theta|) \lambda \int_{E_N} (1 + |\xi|^2)^\theta |\hat{u}(\xi)|^2 d\xi - \Gamma(1-2\theta)\, \delta c_5 (|2\theta|/(1 +$$

$$+ 2\theta)) \int_{E_N} |\xi|^{2k} (1 + |\xi|^2)^\theta |\hat{u}(\xi)|^2 d\xi \geqslant \Gamma(1-2\theta) (\varepsilon c_4 (1 - |2\theta|) -$$

$$- \delta c_5 (|2\theta|/(1 + 2\theta))) \int_{E_N} |\xi|^{2k} (1 + |\xi|^2)^\theta |\hat{u}(\xi)|^2 d\xi +$$

$$+ \Gamma(1-2\theta)\, c_4 (1 - |2\theta|) \lambda \int_{E_N} (1 + |\xi|^2)^\theta |\hat{u}(\xi)|^2 d\xi.$$

Si l'on prend θ de sorte que $\varepsilon c_4(1-|2\theta|) - \delta c_5(|2\theta|/(1+2\theta)) = = c_5(2\theta) > 0$, on obtient, en désignant par

$c_6(2\theta) = \min(\Gamma(1-2\theta)c_5(2\theta), \Gamma(1-2\theta)\, c_4(1-|2\theta|))$, de (3.19) :

$$(3.20)\qquad \mathcal{R}e B(v, u) \geqslant c_6(2\theta)\int_{E_N} (\lambda + |\xi|^{2k})(1+|\xi|^2)^\theta\, |\hat{u}(\xi)|^2\, d\xi.$$

Par un calcul élémentaire, on obtient pour $\lambda \geqslant 2$:

$$(3.21)\qquad (\lambda + |\xi|^{2k})(1+|\xi|^2)^\theta \geqslant c_7(1+|\xi|^2)^k + c_8(\lambda-1)^{1/2},$$

pour $\lambda \geqslant 1$:

$$(3.22)\qquad (\lambda + |\xi|^{2k})(1+|\xi|^2)^\theta \geqslant c_9(1+|\xi|^2)^{k+\theta};$$

en tenant compte que $\delta = 0$, si les coefficients a_{ij} sont réels, on obtient de (3.20) — (3.22) et des lemmes 2.6 et 2.7 l'assertion du théorème.

Remarquons que l'inégalité (3.16) entraîne sans difficulté la première inégalité de (1.3) pour $B_\lambda(v, u)$, $\lambda \geqslant 1$ (et, naturellement, en changeant les rôles de v et u, la deuxième). En effet, ayant $|R_{-2\theta}(u)|_{W_2^{(k-\theta)}(\Omega)} \leqslant$

$\leqslant c_1 |u|_{W_2^{(k+\theta)}(\Omega)}$, $|B_\lambda(R_{-2\theta}(u), u)| \geqslant c_2 |u|^2_{W_2^{(k+\theta)}(\Omega)}$, d'où $B_\lambda\left(\dfrac{R_{-2\theta}(u)}{|R_{-2\theta}(u)|_{W_2^{(k-\theta)}(\Omega)}}, u\right) \geqslant$

$\geqslant (c_2/c_1)\, |u|_{W_2^{(k+\theta)}(\Omega)}$.

On peut démontrer sans difficulté sous les hypothèses du théorème 1 l'inégalité (3.16) pour $B(v, u) = \int_\Omega \sum_{|i|,\, |j| \leqslant k} \bar{a}_{ij} D^i v D^j \bar{u}\, dx + \lambda \int_\Omega v\bar{u}\, dx$ avec a_{ij} constantes, λ assez grand. On peut aussi étudier le cas des coefficients variables ; on s'en occupera pour les domaines bornés dans la partie suivante.

4. LA COERCIVITÉ DES FORMES AUX DOMAINES À FRONTIÈRES LIPSCHITZIENNES

Soit Ω un domaine borné à frontière lipschitzienne, k un nombre entier > 0 et

$$(4.1)\qquad B_\lambda(v, u) = \int_\Omega \sum_{|i|,\, |j| \leqslant k} \bar{a}_{ij}\, D^i v D^j \bar{u}\, dx + \lambda \int_\Omega v\bar{u}\, dx$$

une forme sesquilinéaire dans $W_2^{(k)}(\Omega) \times W_2^{(k)}(\Omega)$, à coefficients mesurables, bornés, et uniformément fortement elliptique :

$(4.2)\qquad (\xi_1, \xi_2, \ldots \xi_N) \in E_N \Rightarrow \mathcal{R}e \sum_{|i|=|j|=k} \bar{a}_{ij}\, \xi^{i+j} \geqslant \varepsilon |\xi|^{2k}$, $\varepsilon > 0$, presque partout.

On désigne par $\mathfrak{D}_\Omega(E_N)$ l'espace des restrictions des fonctions de $\mathfrak{D}(E_N)$ sur Ω.

Considérons sur $\mathfrak{D}_\Omega(E_N) \times \mathfrak{D}_\Omega(E_N)$ une forme sesquilinéaire

$$(4.3)\qquad \int_\Omega \bar{a}\, \frac{\partial f}{\partial x_i}\, \bar{g}\, dx.$$

On a

LEMME 4.1. *Soit* $0 \leqslant \theta < 1/2$. *Alors la forme* (4.3) *peut être prolongée par continuité en une forme sesquilinéaire sur* $W_2^{(1-\theta)}(\Omega) \times W_2^{(\theta)}(\Omega)$ *si a est une fonction* $\varkappa$-*höldérienne dans* $\overline{\Omega}$ *avec* $\varkappa > \theta$.

Démonstration. Soit $f \in \mathfrak{D}(E_N)$, $g \in \mathfrak{D}(E_N)$. On tire de (2.17) que $ag \in W_2^{(\theta)}(\Omega)$ et l'inégalité

(4.4) $$|ag|_{W_2^{(\theta)}(\Omega)} \leqslant c_1 |a|_{C^{(0),\varkappa}\overline{\Omega}} |g|_{W_2^{(\theta)}(\Omega)}.$$

Désignons $ag = h$ et posons $h = 0$ pour $x \notin \Omega$. On a $\int_{E_N} \frac{\partial f}{\partial x_i} \bar{h}\, dx =$

$= (2\pi)^{-N} \int_{E_N} I \xi_i \hat{f}(\xi) \overline{\hat{h}(\xi)}\, d\xi$, d'où

(4.5) $$\left|\int_\Omega \bar{a} \frac{\partial f}{\partial x_i} \bar{g}\, dx\right| \leqslant \left(\int_{E_N} (1 + |\xi|^2)^{1-\theta} |\hat{f}(\xi)|^2 d\xi\right)^{1/2} \left(\int_{E_N} (1 + |\xi|^2)^{\theta} |\hat{h}(\xi)|^2 d\xi\right)^{1/2}.$$

En utilisant le lemme 2.9, on obtient de (4.5)

(4.6) $$\left|\int_\Omega \bar{a} \frac{\partial f}{\partial x_i} \bar{g}\, dx\right| \leqslant c_2 |f|_{W_2^{(1-\theta)}(E_N)} |h|_{W_2^{(\theta)}(\Omega)}.$$

Soit maintenant $\tilde{f}$ un autre prolongement de la restriction de f sur Ω sur E_N tel que $\tilde{f} \in W_2^{(1-\theta)}(E_N)$. On peut trouver $\tilde{f}_n \to \tilde{f}$ dans $W_2^{(1-\theta)}(E_N)$, $\tilde{f}_n \in \mathfrak{D}(E_N)$.

Par (4.6) on obtient

(4.7) $$\left|\int_\Omega a \frac{\partial \tilde{f}_n}{\partial x_i} \bar{g}\, dx\right| \leqslant c_2 |\tilde{f}_n|_{W_2^{(1-\theta)}(E_N)} |h|_{W_2^{(\theta)}(\Omega)}.$$

On a

(4.8) $$\lim_{n\to\infty} \int_\Omega \bar{a} \frac{\partial \tilde{f}_n}{\partial x_i} \bar{g}\, dx = \int_\Omega \bar{a} \frac{\partial f}{\partial x_i} \bar{g}\, dx.$$

En effet : d'après le lemme 2.10 : $W_2^{(\theta)}(\Omega) = \overset{\circ}{W}{}_2^{(\theta)}(\Omega)$, alors à chaque $\varepsilon > 0$ on peut trouver $h_\varepsilon \in \mathfrak{D}(\Omega)$ de sorte que $|h_\varepsilon - h|_{W_2^{(\theta)}(\Omega)} < \varepsilon$. On a alors

$$\lim_{n\to\infty} \int_\Omega \frac{\partial \tilde{f}_n}{\partial x_i} \bar{h}_\varepsilon dx = -\lim_{n\to\infty} \int_\Omega \tilde{f}_n \frac{\partial \bar{h}_\varepsilon}{\partial x_i} dx = -\int_\Omega f \frac{\partial \bar{h}_\varepsilon}{\partial x_i} dx = \int_\Omega \frac{\partial f}{\partial x_i} \bar{h}_\varepsilon dx,$$ d'où

$$\lim_{n\to\infty} \left|\int_\Omega \frac{\partial f_n}{\partial x_i} (\bar{h}_\varepsilon - \bar{h})\, dx\right| \leqslant c_2 |\tilde{f}|_{W_2^{(1-\theta)}(E_N)} \varepsilon,$$ d'où (4.8). On obtient alors

de (4.7)

(4.9) $$\left|\int_\Omega \bar{a}\frac{\partial f}{\partial x_i}\bar{g}dx\right| \leqslant c_2 |\tilde{f}|_{W_2^{(1-\theta)}(E_N)} |h|_{W_2^{(\theta)}(\Omega)};$$

$\tilde{f}$ étant un prolongement quelconque de f, on a $\left|\int_\Omega \bar{a}\frac{\partial f}{\partial x_i}\bar{g}dx\right| \leqslant$ $\leqslant c_2 |f|_{W_2^{(1-\theta)}} |h|_{W_2^{(\theta)}(\Omega)}$, d'où le lemme.

On désigne par $\mathcal{E}(\overline{\Omega})$ l'ensemble des fonctions continues avec toutes leurs dérivées sur la fermeture de Ω.

Soit maintenant $|\theta| < 1/2$ et supposons les coefficients de (4.1) : (4.10) $\varkappa$-höldériens avec $\varkappa > |\theta|$ pour $|i| = k$, $|j| \leqslant k$ si $0 < \theta < 1/2$ et pour $|i| \leqslant k$, $|j| = k$ si $-1/2 < \theta < 0$.

On définit la forme (4.1) sur $W_2^{(k-\theta)}(\Omega) \times W_2^{(k+\theta)}(\Omega)$ à l'aide du lemme 4.1 en la définissant d'abord sur $\mathfrak{D}_\Omega(E_N) \times \mathfrak{D}_\Omega(E_N)$.

Nous démontrerons maintenant le

THÉORÈME 2. *Soit Ω un domaine à frontière lipschitzienne. Soit $|\theta| < 1/2$. Alors la forme* (4.1), *satisfaisant* (4.2), (4.10) *est sesquilinéaire sur* $W_2^{(k-\theta)}(\Omega) \times W_2^{(k+\theta)}(\Omega)$. *Si* $|\theta|$ *est assez petit et* λ *assez grand, la forme est coercive sur* $\mathring{W}_2^{(k-\theta)}(\Omega) \times \mathring{W}_2^{(k+\theta)}(\Omega)$; *si les coefficients* a_{ij} *pour* $|i| = |j| = k$ *sont réels, la forme* 4.1 *est coercive pour tous* $|\theta| < 1/2$ *et* λ *assez grand.*

Démonstration. Pour simplifier, on omettra l'index λ chez $B_\lambda(v, u)$. Soit $z \in \overline{\Omega}$; on désigne par $B'_z(v,u) = \int_\Omega \sum_{|i|=|j|=k} \bar{a}_{ij}(z) D^i v D^j \bar{u}\, dx + {}$ $+ \lambda \int_\Omega v\bar{u}dx$. Par $B'(v, u)$ on désigne $\int_\Omega \sum_{|i|,|j|=k} \bar{a}_{ij} D^i v D^j \bar{u} dx + \lambda \int_\Omega v\bar{u}dx$.

L'assertion de notre théorème est bien connue pour $\theta = 0$ (cf. L. Garding [7]) ; nous ne répéterons pas sa démonstration dans ce cas. Considérons alors le cas du $|\theta| \neq 0$. Soit $u \in \mathfrak{D}(\Omega_N)$. On définit $Q(u) = w$, $w \in L_2(E_N)$, en posant $\hat{w}(\xi) = \hat{u}(\xi)(1 + |\xi|^2)^\theta$. On a évidemment $w \in W_2^{(m)}(E_N)$ pour $m \geqslant 0$. Pour $z \in \overline{\Omega}$, $\lambda \geqslant 2$: $\mathcal{R}e \int_{E_N} \sum_{|i|,|j|=k} \bar{a}_{ij}(z) D^i w D^j \bar{u} dx + \mathcal{R}e\,\lambda \int_{E_N} w\bar{u}dx =$ $= \mathcal{R}e\,(1/2\pi)^N \left(\int_{E_N} \sum_{|i|,|j|=k} \bar{a}_{ij}(z)\, \xi^{i+j} (1 + |\xi|^2)^\theta |\hat{u}(\xi)|^2 d\xi + \lambda \int_{E_N} (1 + {}\right.$ $\left.+ |\xi|^2)^\theta |\hat{u}(\xi)|^2 d\xi\right)$, d'où, en vertu de (3.21),

(4.11) $$\mathcal{R}e \int_{E_N} \sum_{|i|,|j|=k} \bar{a}_{ij}(z) D^i w D^j \bar{u} dx + \mathcal{R}e\, \lambda \int_{E_N} w\bar{u}dx \geqslant$$ $$\geqslant c_1 |u|^2_{*W_2^{(k+\theta)}(\Omega)} + c_1 (\lambda - 1)^{1/2} |u|^2_{L_2(\Omega)},$$

où c_1 ne dépend pas de z. On a

(4.12) $$|w|_{*W_2^{(k-\theta)}(E_N)} \leqslant c_2 |u|_{*W_2^{(k+\theta)}(\Omega)}.$$

Revenons pour le moment à la définition de U_r dans (2.14). En prenant $\alpha' < \alpha$ et $\beta' < \beta$ de sorte que $|\alpha - \alpha'|$, $|\beta - \beta'|$ soit assez petit, le système des U'_r recouvre aussi $\dot{\Omega}$.

Soit maintenant $u \in \mathfrak{D}(U'_r)$. Sans restreindre la généralité, on peut supposer chaque fonction a_r définie et lipschitzienne sur E_{N-1}, nulle au voisinage de l'infini. Désignons le demi-espace en question Ω_r. D'après le théorème 1, il existe des transformations linéaires et continues de $\overset{\circ}{W}{}_2^{(k+\theta)}(\Omega_r)$ dans $\overset{\circ}{W}{}_2^{(k-\theta)}(\Omega_r)$, soit Q_r, $r = 1, 2, \ldots M$, de sorte que pour $w = Q_r(u)$:

$$|Q_r(u)|_{*W_2^{(k-\theta)}(\Omega_r)} \leqslant c_3 |u|_{*W_2^{(k+\theta)}(\Omega)}, \tag{4.13}$$

$$\mathcal{R}e \int_{\Omega_r} \sum_{|i|,\, |j|=k} \bar{a}_{ij}(z) D^i w D^j \bar{u} dx + \mathcal{R}e\, \lambda \int_{\Omega_r} w\bar{u} dx \geqslant \tag{4.14}$$
$$\geqslant c_4 |u|^2_{*W_2^{(k+\theta)}(\Omega)} + c_4 (\lambda - 1)^{1/2} |u|^2_{L_2(\Omega)},$$

où c_4 est indépendant de z. Soit $c_5 = \min(c_1, c_4)$, $c_6 = \max(c_2, c_3)$. On désigne encore par $C(z, \rho)$ la boule centrée au point z et du rayon ρ. En vertu de la condition (4.10), on peut trouver $\rho_0 > 0$ si petit que pour $v, u \in \mathfrak{D}(\Omega) \cap C(z, (2/3)\rho)$ on a avec $\rho \leqslant \rho_0$

$$\left| \mathcal{R}e \int_{\Omega \cap C(z,(2/3)\rho)} \sum_{|i|,\, |j|=k} (\bar{a}_{ij}(x) - \bar{a}_{ij}(z)) D^i v D^j \bar{u} dx \right| \leqslant \tag{4.15}$$
$$\leqslant c_7 |v|_{*W_2^{(k-\theta)}(\Omega)} |u|_{*W_2^{(k+\theta)}(\Omega)} + c_8(\rho) |v|_{W_2^{([k-\theta])}(\Omega)} |u|_{*W_2^{(k+\theta)}(\Omega)} +$$
$$+ c_9(\rho) |v|_{*W_2^{(k-\theta)}(\Omega)} |u|_{W_2^{([k+\theta])}(\Omega)},$$

avec

$$c_6 c_7 < c_5/2. \tag{4.16}$$

Pour fixer les idées, soit $0 < \theta$ (le cas $\theta < 0$ se démontre de la même manière). Désignons pour i, j fixes $a_{ij}(x) - a_{ij}(z) = b(x)$. On a d'après (2.17) :

$$|b D^j u|^2_{*W_2^{(\theta)}(\Omega)} \leqslant 2 \int_\Omega \int_\Omega \frac{|b(x)|^2 |D^j u(x) - D^j u(y)|^2}{|x - y|^{N+2\theta}} dx dy +$$
$$+ 2 \int_\Omega \int_\Omega \frac{|D^j u(y)|^2 |b(x) - b(y)|^2}{|x - y|^{N+2\theta}} dx dy \leqq$$
$$\leqq 2 \cdot \omega^2(\rho) \int_\Omega \int_\Omega \frac{|D^j u(x) - D^j u(y)|^2}{|x - y|^{N+2\theta}} dx dy +$$
$$+ 4M^2 \int_{\Omega - C(z,\rho)} \int_{\Omega \cap C(z,\rho^2/3)} \frac{|D^j u(y)|^2}{|x - y|^{N+2\theta}} dx dy +$$
$$+ 2 \int_\Omega \int_\Omega \frac{|D^j u(y)|^2 |b(x) - b(y)|^2}{|x - y|^{N+2\theta}} dx dy \leqslant$$
$$\leqslant 2\omega^2(\rho) |D^j u|^2_{W_2^{(\theta)}(\Omega)} + c_{10}(\rho) |D^j u|^2_{L_2(\Omega)},$$

$\omega(\rho)$ étant le module de la continuité de la fonction b. ρ_0 fixé, on peut trouver $\rho \leqslant \rho_0$ et des points $z_1, z_2, \ldots, z_l$ de sorte que les boules $C(z_s, \rho/3)$ recouvrent $\bar{\Omega}$; sans restreindre la généralité, on peut supposer soit $C(z_i, \rho) \subset \Omega$, soit $C(z_i, \rho) \subset U_r'$ pour r choisi convenablement. Soient $h_s \in \mathfrak{D}(C(z_s, \rho/3))$, des fonctions réelles, $0 \leqslant h_s(x) \leqslant 1$, telles que $x \in \bar{\Omega} \Rightarrow \sum_{s=1}^{l} h_s(x) = 1$. Soit maintenant $u \in \mathfrak{D}(\Omega)$ et désignons par $u_s = uh_s$.

Le premier cas: $C(z_s, \rho) \subset \Omega$. Soit $\psi_s \in \mathfrak{D}(C(z_s, \rho\, 2/3))$ telle que $0 \leqslant \psi_s(x) \leqslant 1$, $\psi_s(x) = 1$ pour $x \in C(z_s, \rho/3)$. Posons $P_s(u_s) = \psi_s Q(u_s) = v_s$. On a

$$\mathcal{R}e\, B'_{z_s}(v_s, u_s) \geqslant c_5 |u_s|^2_{*W_2^{(k+\theta)}(\Omega)} + c_5(\lambda - 1)^{1/2} |u_s|^2_{L_2(\Omega)}. \tag{4.17}$$

De plus, on peut trouver constantes c_{11} et c_{12} de sorte que

$$|v_s|_{*W_2^{(k+\theta)}(\Omega)} \leqslant c_{11} |u_s|_{*W_2^{(k+\theta)}(\Omega)} + c_{12} |u_s|_{W_2^{([k+\theta])}(\Omega)}, \tag{4.18}$$

où $c_{11} > c_2$ mais telle que $c_{11} c_7 < c_5/2$. En effet, on a pour $\varepsilon > 0$, arbitrairement petit,

$$|v_s|^2_{*W_2^{(k-\theta)}(\Omega)} \leqslant (1+\varepsilon) |Q_s(u_s)|^2_{*W_2^{(k-\theta)}(\Omega)} + \mu(\varepsilon) |Q(u_s)|^2_{W_2^{([k-\theta])}(\Omega)}. \tag{4.19}$$

Pour le voir, il suffit de démontrer pour $0 < m < 1$ l'inégalité

$$|\omega \psi_s|^2_{*W_2^{(m)}(\Omega)} \leqslant (1+\varepsilon) |\omega|^2_{*W_2^{(m)}(\Omega)} + \nu(\varepsilon) |\omega|^2_{L_2(\Omega)}, \tag{4.20}$$

ω étant de $\mathcal{E}(\bar{\Omega})$. On a $\displaystyle\int_\Omega \int_\Omega \frac{|\omega(x)\psi_s(x) - \omega(y)\psi_s(y)|^2}{|x-y|^{N+m}}\, dx dy \leqslant$

$$\leqslant (1+\varepsilon) \int_\Omega \int_\Omega \frac{|\omega(x) - \omega(y)|^2}{|x-y|^{N+m}}\, dx dy +$$

$$+ (1 + 1/\varepsilon) \int_\Omega \int_\Omega |\omega(y)|^2 \frac{|\psi_s(x) - \psi_s(y)|^2}{|x-y|^{N+m}}\, dx dy,$$

d'où (4.20). D'après la définition de Q, on a $|Q(u_s)|^2_{W_2^{([k-\theta])}(\Omega)} \leqslant$

$$\leqslant |Q(u_s)|^2_{W_2^{([k-\theta])}(E_N)} \leqslant c_{13} \int_{E_N} (1 + |\xi|^2)^{[k-\theta]+2\theta} |\hat{u}_s(\xi)|^2\, d\xi =$$

$$= c_{13} |u_s|^2_{W_2^{([k-\theta]+2\theta)}(E_N)}.$$

Il s'ensuit de (4.12), (4.19) avec $\varepsilon > 0$ assez petit

$$|v_s|^2_{*W_2^{(k-\theta)}(\Omega)} \leqslant (1+\varepsilon) c_2^2 |u_s|^2_{*W_2^{(k+\theta)}(\Omega)} + \mu(\varepsilon) c_{13} |u_s|^2_{W_2^{([k-\theta]+2\theta)}(E_N)}. \tag{4.21}$$

En utilisant le lemme 2.13, tenant compte du fait que pour $\theta \neq 0 \Rightarrow$ $\Rightarrow [k - \theta] + 2\theta < k + \theta$, on obtient

$$(4.21') \quad \mu(\varepsilon)\, c_{13} |u_s|^2_{W_2^{([k-\theta]+2\theta)}(E_N)} \leqslant \varepsilon c_2^2 |u_s|^2_{*W_2^{(k+\theta)}(\Omega)} + c_{14} |u_s|^2_{W_2^{([k+\theta])}(\Omega)},$$

d'où, avec (4.21), on tire finalement (4.18). En utilisant (4.15), (4.17), (4.18), (4.21′) on vient à l'inégalité

$$(4.22) \quad \mathcal{Re}\, B(v_s, u_s) \geqslant (c_5/3)\, |u_s|^2_{*W_2^{(k+\theta)}(\Omega)} - c_{15}\, |u_s|^2_{W_2^{([k+\theta])}(\Omega)} + $$
$$+ c_5 (\lambda - 1)^{1/2} |u_s|^2_{L_2(\Omega)}.$$

Le deuxième cas : $C(z_s, \rho) \not\subset \Omega$. Soit $C(z_s, \rho) \subset U'_r$. On désigne de nouveau par $v_s = \psi_s Q(u_s)$, ψ_s étant une fonction réelle, $0 \leqslant \psi_s(x) \leqslant$ $\leqslant 1$, $\psi_s \in \mathfrak{D}(C(z_s, \rho 2/3))$, $\psi_s(x) = 1$ pour $x \in C(z_s, \rho/3)$. Notons que $v_s \in \mathfrak{D}(U'_r)$. En raisonnant comme au cas précédent, on trouve c_{16} et c_{17}, de sorte que $c_{16} > c_3$, $c_{16} c_{17} < c_5/2$ et telles que

$$(4.23) \quad |v_s|_{*W_2^{(k-\theta)}(\Omega)} \leqslant c_{16} |u_s|_{*W_2^{(k+\theta)}(\Omega)} + c_{17} |u_s|_{W_2^{([k+\theta])}(\Omega)}$$

et on arrive à l'inégalité

$$(4.24) \quad \mathcal{Re}\, B(v_s, u_s) \geqslant (c_5/3) |u_s|^2_{*W_2^{(k+\theta)}(\Omega)} - c_{18} |u_s|^2_{W_2^{([k+\theta])}(\Omega)} + $$
$$+ c_5 (\lambda - 1)^{1/2} |u_s|^2_{L_2(\Omega)}.$$

On a maintenant pour $s = 1, 2, \ldots l$

$$(4.25) \quad \mathcal{Re}\, B(v_s h_s, u) \geqslant (c_5/3) |u_s|^2_{*W_2^{(k+\theta)}(\Omega)} - c_{17} |u_s|^2_{W_2^{([k+\theta])}(\Omega)} + $$
$$+ c_5 (\lambda - 1)^{1/2} |u_s|^2_{L_2(\Omega)} - c_{20} |u|_{*W_2^{(k+\theta)}(\Omega)} |u|_{W_2^{([k+\theta])}(\Omega)}.$$

En effet, on obtient sans difficulté que $|B(v_s, h_s u) - B(v_s h_s, u)| \leqslant$ $\leqslant c_{20} \cdot |u|_{W_2^{([k+\theta])}(\Omega)} |u|_{*W_2^{(k+\theta)}(\Omega)}$, d'où (4.25). On a évidemment

$$(4.26) \quad \sum_{s=1}^{l} |u_s|^2_{W_2^{(k+\theta)}(\Omega)} \geqslant c_{21} |u|^2_{W_2^{(k+\theta)}},$$

d'où $\mathcal{Re}\, B(\sum_{s=1}^{l} v_s h_s, u) \geqslant (c_5/3) \cdot c_{21} |u|^2_{W_2^{(k+\theta)}(\Omega)} + c_5 c_{21} (\lambda - 1)^{1/2} |u|^2_{L_2(\Omega)} -$ $- c_{22} |u|^2_{W_2^{([k+\theta])}(\Omega)} - c_{23} |u|_{W_2^{(k+\theta)}(\Omega)} |u|_{W_2^{([k+\theta])}(\Omega)}$; en utilisant le lemme 2.13, on obtient pour λ assez grand l'assertion du théorème.

5. PROBLÈME DE DIRICHLET

On considère dans Ω (un domaine à frontière lipschitzienne) la forme 4.1, jouissant de (4.2) et (4.10). Soient, pour $|\theta| < 1/2$, $u_0 \in W_2^{(k+\theta)}(\Omega)$, $f \in W_2^{(\theta-k)}(\Omega)$. On dit d'une fonction de $W_2^{(k+\theta)}(\Omega)$, soit u, qu'elle résout le problème de Dirichlet : $Au = f$ dans Ω (où A est l'opérateur (1.1)), $u = u_0$, $\dfrac{\partial u}{\partial \nu} = \dfrac{\partial u_0}{\partial \nu}, \ldots, \dfrac{\partial^{k-1} u}{\partial \nu^{k-1}} = \dfrac{\partial^{k-1} u_0}{\partial \nu^{k-1}}$ sur Ω au sens généralisé $\left(\dfrac{\partial}{\partial \nu}\right.$ désignant la dérivée dans la direction de la normale extérieure$\left.\right)$, si pour chaque $\varphi \in \mathfrak{D}(\Omega)$

(5.1) $$A(\varphi, u) = \bar{f}(\varphi), \quad (\bar{f}(\varphi) = \overline{f(\bar{\varphi})})$$

et si

(5.2) $$u - u_0 \in \overset{\circ}{W}{}_2^{(k+\theta)}(\Omega).$$

Pour simplifier, on supposera dans cette partie que la forme $A(v, u)$ est $\overset{\circ}{W}{}_2^{(k)}(\Omega)$ elliptique, à savoir :

(5.3) $$v \in \overset{\circ}{W}{}_2^{(k)}(\Omega) \Rightarrow |A(v, v)| \geqslant c\, |v|^2_{W_2^{(k)}(\Omega)} \;{}^{*)}$$

On démontre sans difficulté :

LEMME 5.1. *La forme sesquilinéaire $B(v,u)$ sur $H_1 \times H_2$, deux espaces de Hilbert, soit coercive. Alors chaque fonctionnelle f sur H_1 peut être représentée d'une manière unique sous la forme $B(v, u) = f(v)$. On a*

(5.4) $$|u|_{H_2} \leqslant c\, |f|.$$

En effet, en désignant par $(\ ,\ )_{H_1}$ le produit scalaire dans H_1, on a, par le théorème de Riesz : $B(v, u) = (v, Z(u))_{H_1}$, Z étant une transformation linéaire et continue de H_2 dans H_1. On a par (1.3) $\sup\limits_{|v|_{H_1} \leqslant 1} |B(v, u)| = |Z(u)|_{H_1} \geqslant c_1 |u|_{H_2}$, d'où Z est une transformation simple, et $Z(H_2)$ est fermé dans H_1. D'autre part, en vertu de l'inégalité (1.3.b.), on a $Z(H_2) = H_1$, d'où l'assertion. On a maintenant

THÉORÈME 3. *Soit Ω un domaine borné à frontière lipschitzienne. L'opérateur A avec sa forme sesquilinéaire $A(v, u)$ satisfasse les hypothèses (5.3), (4.1), (4.2), (4.10). Soit $|\theta| < 1/2$, un nombre assez petit au cas général et $|\theta| < 1/2$ seulement si les coefficients a_{ij} pour $|i| = |j| = k$ sont réels. Alors pour chaque $f \in W_2^{(\theta-k)}(\Omega)$, $u_0 \in W_2^{(k+\theta)}(\Omega)$, il existe précisément une solution du problème de Dirichlet, soit u, et on a*

(5.5) $$|u|_{W_2^{(k+\theta)}(\Omega)} \leqslant c\, [|f|^2_{W_2^{(\theta-k)}(\Omega)} + |u_0|^2_{W_2^{(k+\theta)}(\Omega)}]^{1/2}.$$

Démonstration. Naturellement, on considère θ, pour lesquels théorème 2 a lieu. Soit $\theta \geqslant 0$. On a dans ce cas $f \in W_2^{(\theta-k)}(\Omega) \subset W_2^{(-k)}(\Omega)$, $u_0 \in W_2^{(k+\theta)}(\Omega) \subset W_2^{(k)}(\Omega)$, les inclusions étant algébriques et topologiques.

---

*) Sans supposer (5.3) il vaut l'alternative de Fredholm.

En posant $\overline{F}(v) = \overline{f}(v) - A(v, u_0)$, on obtient, tenant compte du lemme 5.1 et de (5.3), l'assertion du théorème pour $\theta = 0$ et l'inégalité

$$(5.6) \qquad |u|_{W_2^{(k)}(\Omega)} \leqslant c_1[|f|^2_{W_2^{(-k)}(\Omega)} + |u_0|^2_{W_2^{(k)}(\Omega)}]^{1/2}.$$

On désigne de nouveau $B_\lambda(v, u) = \int_\Omega \sum_{|i|, |j| \leqslant k} \overline{a}_{ij} D^i v D^j \overline{u} dx + \lambda \int_\Omega v\overline{u} dx$. Soit λ si grand pour que le théorème 2 ait lieu. On prend λ si grand pour que le théorème 2 ait lieu aussi pour $\theta = 0$. On a pour $v \in \overset{\circ}{W}{}_2^{(k)}(\Omega)$:

$$(5.7) \qquad B_\lambda(v, u) = \overline{f}(v) - \lambda(v, u),$$

$$(5.8) \qquad u - u_0 \in \overset{\circ}{W}{}_2^{(k)}(\Omega).$$

Soit maintenant $u^* \in W_2^{(k+\theta)}(\Omega)$ telle que

$$(5.9) \qquad \varphi \in \mathfrak{D}(\Omega) \Rightarrow B_\lambda(\varphi, u^*) = \overline{f}(\varphi) - \lambda(\varphi, u),$$

$$(5.10) \qquad u^* - u_0 \in \overset{\circ}{W}{}_2^{(k+\theta)}(\Omega).$$

Il existe une seule solution de ce problème et on a

$$(5.11) \qquad |u^*|_{W_2^{(k+\theta)}(\Omega)} \leqslant c_2[|f|^2_{W_2^{(\theta-k)}(\Omega)} + |u_0|_{W_2^{(k+\theta)}(\Omega)}]^{1/2}.$$

En effet, d'après le lemme 5.1, il existe précisément un élément de $\overset{\circ}{W}{}_2^{(k+\theta)}(\Omega)$, soit w, de sorte que pour $v \in \overset{\circ}{W}{}_2^{(k-\theta)}(\Omega)$, on a $B_\lambda(v, w) = \overline{f}(v) - \lambda(v, u) - B_\lambda(v, u_0)$.

Il vaut

$$(5.12) \qquad |w|_{W_2^{(k+\theta)}} \leqslant c_3[|f|^2_{W_2^{(\theta-k)}(\Omega)} + |u|^2_{L_2(\Omega)} + |u_0|^2_{W_2^{(k+\theta)}(\Omega)}]^{1/2}.$$

Évidemment, l'élément $w + u_0$ est une solution du problème. D'après (1.3 a), elle est unique. Il s'ensuit de (5.9) et (5.10) que u^* satisfait à (5.7), (5.8) ; u étant unique, on a $u = u^*$ et l'inégalité (5.5) (c'est (5.11)). La solution du problème est unique ; cela découle de l'unicité de la solution pour $\theta = 0$. Soit maintenant $\theta < 0$. On prend de nouveau λ si grand que le théorème 2 vaut pour θ et 0. On trouve comme au premier cas une solution unique du problème $(A + \lambda) u = f$ dans Ω, $u - u_0 \in \overset{\circ}{W}{}_2^{(k+\theta)}(\Omega)$ et on a

$$(5.13) \qquad |u|_{W_2^{(k+\theta)}(\Omega)} \leqslant c_4[|f|^2_{W_2^{(\theta-k)}(\Omega)} + |u_0|^2_{W_2^{(k+\theta)}(\Omega)}]^{1/2}.$$

Soit maintenant $h \in W_2^{(k)}(\Omega)$ la solution du problème $Ah = \lambda u$ dans Ω, $h \in \overset{\circ}{W}{}_2^{(k)}(\Omega)$; il existe une seule solution de ce problème et on a

$$(5.14) \qquad |h|_{W_2^{(k)}}(\Omega) \leqslant c_5|u|_{L_2(\Omega)}.$$

Évidemment, la fonction $u + h$ résout notre problème et on a, en vertu de (5.13) et (5.14), l'inégalité (5.5). La solution u est unique. En effet, soit $Au = 0$ dans Ω, $u \in \overset{\circ}{W}{}_2^{(k+\theta)}(\Omega)$. On a $(A + \lambda) u = \lambda u$, dans Ω ; la solution du problème $(A + \lambda) u = f$ dans $\Omega, f \in L_2(\Omega)$, $u \in \overset{\circ}{W}{}_2^{(k+\theta)}(\Omega)$, étant unique, on a $u \in \overset{\circ}{W}{}_2^{(k)}(\Omega)$, parce qu'il existe une seule solution de ce problème pour $\theta = 0$. Ayant $Au = 0$ dans Ω et $u \in \overset{\circ}{W}{}_2^{(k)}(\Omega)$, on obtient $u = 0$, la solution du dernier problème étant unique, d'où le théorème.

L'interprétation du théorème 3 dépend du cas si $\theta < 0$ ou > 0. Au premier cas, le théorème 3 nous donne l'existence et l'unicité de la solution pour les données moins régulières ; on obtient la solution sans l'intégrale de Dirichlet finie. Au deuxième cas, la méthode variationnelle est applicable ; le théorème 3 peut être compris comme un théorème sur la régularité de la solution.

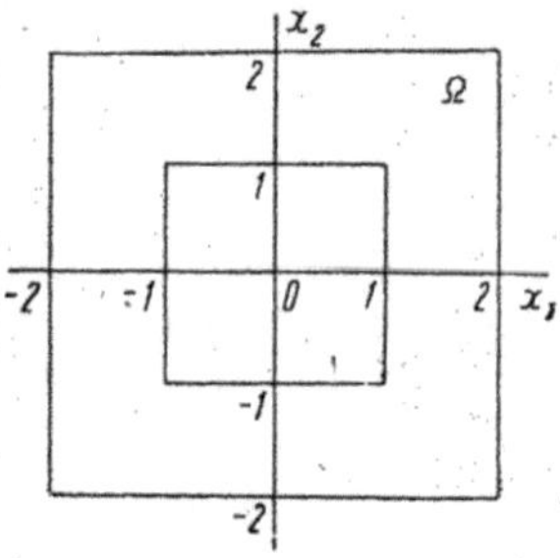

Fig. 1

Exemple 1. On considère un domaine plan (cf. figure 1). On cherche la solution du problème biharmonique, pour les conditions aux limites engendrées par la fonction $u_0(x_1, x_2) = -(x_1^2/2)\, \varphi(x)$, où $\varphi(x) \in \mathfrak{D}(E_2)$, égale à 1 au voisinage du contour extérieur de Ω et à 0 au voisinage du contour intérieur de Ω. (Ce problème correspond au problème de l'élasticité plan, étant donné un cadre, chargé par des forces normales constantes sur les côtés $x_2 = \pm 1$). On a évidemment $u_0 \in W_2^{(2+\theta)}(\Omega)$ pour $0 \leqslant \theta < 1/2$, alors on peut appliquer le théorème 3 et l'on obtient la solution du problème : $\Delta^2 u = 0$ dans Ω, $u = u_0$, $\dfrac{\partial u}{\partial \nu} = \dfrac{\partial u_0}{\partial \nu}$ sur $\dot{\Omega}$, appartenant à $W_2^{(2+\theta)}(\Omega)$. On sait bien que la fonction u est indéfiniment continûment différentiable dans Ω et que l'équation est satisfaite au sens classique (cf. p. ex. L. Nirenberg [13]). Il s'ensuit du théorème 3 que les premières dérivées de u sont μ-höldériennes dans $\overline{\Omega}$ avec $\mu < 1/2$. En effet, considérons p. ex. $\dfrac{\partial u}{\partial x_1} = g$. On peut prolonger g sur E_2 de sorte que $g \in W_2^{(1+\theta)}(E_2)$. On a alors $|g(x) - g(y)| \leqslant (1/2\pi)^2 \left| \int_{E_2} (e^{I(x,\xi)} - e^{I(y,\xi)})\, \hat{g}(\xi)\, d\xi \right|$. Soit $\mu < \theta < 1/2$. On obtient

$$|g(x) - g(y)| \leqslant c_1 \int_{E_2} |e^{I(x,\xi)} - e^{I(y,\xi)}|^{\mu} |\hat{g}(\xi)|\, d\xi \leqslant c_2 |x - y|^{\mu} \int_{E_2} |\xi|^{\mu} (1 +$$

$$+ |\xi|^2)^{(1+\theta)/2} \cdot (1 + |\xi|^2)^{-(1+\theta)/2} |\hat{g}(\xi)|\, d\xi \leqslant c_2 |x - y|^{\mu} \left(\int_{E_2} |\hat{g}(\xi)|^2 (1 + |\xi|^2)^{1+\theta} d\xi \right)^{1/2} \cdot$$

$$\cdot \left(\int_{E_2} (1 + |\xi|^2)^{-(1+\theta)} |\xi|^{2\mu}\, d\xi \right)^{1/2}, \text{ d'où l'assertion.}$$

Voilà une conséquence du théorème 3 :

CONSÉQUENCE. *Supposons que le théorème 3 a lieu pour tous* $|\theta| < 1/2$. *Soit* $[2N/(1+N)] < p < 2$, $u_0 \in W_p^{(k)}(\Omega)$, *et soit* f *une fonctionnelle sur* $\overset{\circ}{W}{}_p^{(k)}(\Omega)$.

Alors il existe précisément une solution du problème de Dirichlet et appartient à $W_2^{(k-\theta)}(\Omega)$ *pour* $\theta > N(1/p - 1/2)$ *en vertu du fait que* $W_p^{(k)}(\Omega) \subset W_2^{(k-\theta)}(\Omega)$ *algébriquement et topologiquement.*

Démonstration. Considérons d'abord les dérivées de l'ordre $\leqslant k-1$. D'après le théorème de l'immersion de S. L. Sobolev (cf. E. Gagliardo [6]), on a $W_p^{(k)}(\Omega) \subset W_2^{(k-1)}(\Omega)$ algébriquement et topologiquement. En effet, l'on sait que $W_p^{(1)}(\Omega) \subset L_q(\Omega)$ algébriquement et topologiquement pour $1/q = 1/p - 1/N$ si $p < N$. Il suffit alors de démontrer que $W_2^{(1-\theta)}(\Omega) \subset W_p^{(1)}(\Omega)$ algébriquement et topologiquement. Utilisant la partition de l'unité, engendrée par (2.15), la transformation locale des cartes (cf. (2.9)) et la construction du prolongement donné par (2.8), on obtient sans difficulté un opérateur-prolongement des fonctions de $W_p^{(1)}(\Omega)$ sur E_N, linéaire et borné de $W_p^{(1)}(\Omega)$ dans $W_p^{(1)}(E_N)$. Cela étant, utilisant le théorème de Young-Hausdorf, on obtient pour $f \in W_p^{(1)}(E_N)$: $\hat{f}(\xi)(1 + |\xi|^2)^{1/2} \in L_{p'}(E_N)$, $1/p + 1/p' = 1$, la transformation $f(x) \to \hat{f}(\xi)(1 + |\xi|^2)^{1/2}$ étant continue de $W_p^{(1)}(E_N)$ dans $L_{p'}(E_N)$. On en tire $\int_{E_N} |\hat{f}(\xi)|^2 (1 + |\xi|^2)^{1-\theta} d\xi \leqslant \left(\int_{E_N} |\hat{f}(\xi)|^{p'} (1 + |\xi|^2)^{p'/2} d\xi\right)^{2/p'}$, $\left(\int_{E_N} (1 + |\xi|^2)^{-\theta p/(2-p)} d\xi\right)^{(2-p)/p}$, d'où l'assertion.

Exemple 2. Finissons le présent article par une remarque montrant que notre résultat, au sens suivant, est le meilleur possible. Soit $1/2 < \theta \leqslant 1$. Considérons le domaine plan Ω, exprimé aux cartes polaires par $0 \leqslant r < 1$, $|\vartheta| < \pi - \varepsilon$, $\varepsilon > 0$. Soit $1 \leqslant p < 2$. Utilisant la représentation conforme du cercle unité sur ce domaine, on voit qu'on peut trouver $\varepsilon > 0$ assez petit et une suite des polynômes harmoniques, soit $\{u_n\}$, telle que $|u_n|_{L_p(\dot{\Omega})} <$ const. et $\lim_{n\to\infty} |u_n|_{L_2(\Omega)} = \infty$. (On utilise les théorèmes de l'immersion (cf. p. ex. E. Gagliardo [6]), les théorèmes sur l'approximation d'une fonction holomorphe par des polynômes). Cela étant, supposons que l'inégalité (5.5) a lieu pour θ considéré. Soit u la solution du problème $\Delta u = f$ dans Ω, $f \in L_2(\Omega)$, $u \in \mathring{W}_2^{(1)}(\Omega)$. On doit avoir alors l'inégalité $|u|_{W_2^{(1+\theta)}\Omega} \leqslant c_1 |f|_{L_2(\Omega)}$, d'où, en vertu des théorèmes de l'immersion pour k fractionnaire (on y parvient par la transformation de Fourier), on obtiendrait $\sup_{x \in \dot{\Omega}} \left|\frac{\partial u}{\partial \nu}\right| \leqslant c_2 |f|_{L_2(\Omega)}$. Utilisant la formule de Green, on parvient à l'inégalité $|u_n|_{L_2(\Omega)} \leqslant c_3 |u_n|_{L_p(\dot{\Omega})}$ ce qui est une contradiction.

BIBLIOGRAPHIE

1. S. AGMON, *The coerciveness problem for integro-differential forms.* J. Anal. Math., 1958, **6**, 183–223.
2. S. AGMON, A. DOUGLIS, L. NIRENBERG, *Estimates near the boundary for solutions of elliptic partial differential equations satisfying general boundary conditions.* I. Comm. Pure Appl. Math., 1959, **12**, 623–727.

3. N. ARONSZAJN, *On coercive integro-differential quadratic forms*. University of Kansas, 1954, 94–106.
4. N. ARONSZAJN, K. T. SMITH, *Theory of Bessel Potentials*. Ann. Inst. Fourier, 1961, **11**, 385–476.
5. E. GAGLIARDO, *Caratterizzazione delle tracce sulla frontiera relative ad alcune classi di funzioni in n variabili*. Rend. Sem. Mat. Padova, 1957, **27**, 284–305.
6. E. GAGLIARDO, *Proprietà di alcune classi di funzioni in più variabili*. Ricerche di Matematica, 1958, **7**, 102–137.
7. L. GARDING, *Dirichlet's problem for linear elliptic partial differential equations*. Math. Scand., 1953, **1**, 55–72.
8. J. L. LIONS, *Problèmes aux limites en théorie des distributions*. Acta Mathematica, 1955, **94**, 13–153.
9. J. L. LIONS, E. MAGENES, *Problèmes aux limites non homogènes. II*. Annal. Inst. Fourier, 1961, **11**, 137–178; *V*, Ann. Scuola Norm. Sup. Pisa, ser. III, 1962, **16**, *1*, 1–44.
10. J. L. LIONS, E. MAGENES, *Problèmes aux limites non homogènes. II*, cf. [9]; *III*, Ann. Scuola Norm. Sup. Pisa, ser. III, 1961, **15**, *1–2*, 39–101.
11. J. L. LIONS, E. MAGENES, *Problèmes aux limites non homogènes. IV*. Ann. Scuola Norm. Sup. Pisa, ser. III, 1961, **15**, *4*, 311–326.
12. J. NEČAS, *Sur une méthode pour résoudre les équations aux dérivées partielles du type elliptique, voisine de la variationnelle*. Ann. Scuola Norm. Sup. Pisa, ser. III, 1962, **16**, *4*, 305–326.
13. L. NIRENBERG, *Remarks on strongly elliptic partial differential equations*. Comm. Pure Appl. Math., 1955, **8**, 649–675.
14. L. N. SLOBODĚTSKIJ, *Les espaces de S. L. Sobolev de l'ordre fractionnaire et leurs applications*. Dokl. Akad. Naouk, 1957, **118**, 243–246.
15. S. L. SOBOLEV, *Quelques applications de l'analyse fonctionnelle dans la physique mathématique*. Leningrad, 1950.
16. M. I. VIŠIK, *Sur les systèmes fortement elliptiques*. Mat. Sbornik, 1951, **29**, 615–676.

L'APPLICATION DE L'ÉGALITÉ DE RELLICH
SUR LES
SYSTÈMES ELLIPTIQUES DU DEUXIÈME ORDRE ;

PAR JINDŘICH NEČAS
(Prague).

Dans ce travail, on étudie les systèmes d'équations aux dérivées partielles fortement elliptiques du deuxième ordre. On démontre quelques théorèmes sur la régularité de la solution du problème de Dirichlet, spécialement ceux, fondés sur l'égalité de Rellich. Par le procédé dual, on parvient à résoudre le problème de Dirichlet sous des hypothèses faibles pour les données sur la frontière. On suppose seulement qu'elles sont de carrés sommables. La solution ainsi obtenue n'a pas, en général, l'intégrale de Dirichlet bornée. Le point essentiel du travail est que les domaines considérés ne sont que lipschitziens.

On développe des idées des travaux ([2] à [5]) de l'auteur utilisant la technique de L. E. Payne et H. F. Weinberger [7].

1. LEMMES. — On rappelle dans ce chapitre brièvement les lemmes nécessaires pour la suite. Leurs démonstrations sont contenues dans les travaux [2] et [5].

On désigne par E_n l'espace euclidien avec les cartes $[x_1, x_2, ..., x_n]$.

Un domaine borné Ω sera dit appartenir à la classe $\mathfrak{N}^{(k),\mu}$ si :

1° Il existe m systèmes de cartes dans E_n, soient $[x_{r1}, x_{r2}, ..., x_{rn-1}, x_{rn}]$ (en bref $[X_r, x_{rn}]$), et m fonctions a_r de sorte qu'on puisse représenter chaque point de la frontière sous la forme $[X_r, a_r(X_r)]$. Les fonctions a_r appartiennent à l'espace $C^{(k),\mu}(\overline{\Delta}_r)$

© Gauthier-Villars 1965. We tried to trace the rights holder. If someone has legitimate claims, please contact the publisher. Reprinted from: Nečas, J., L'application de l'égalité de Rellich sur les systèmes elliptiques du deuxième ordre, Journal de Mathématiques Pures et Appliqueés 44, 133-147.

(l'espace des fonctions dont les dérivées jusqu'à l'ordre k sont μ höldériennes), où Δ_r signifie la boule $\Delta_r = |X_r| < \alpha$;

2° Il existe un nombre $\beta \leqq 1$ tel que les points

$$[X_r, x_{rn}], \qquad |X_r| < \alpha, \qquad a_r(X_r) - \beta < x_{rn} < a_r(X_r)$$

soient à l'intérieur de Ω, tandis que les points

$$[X_r, x_{rn}], \ |X_r| < \alpha, \qquad a_r(X_r) < x_{rn} < a_r(X_r) + \beta$$

sont à l'extérieur de Ω.

Par $\dot{\Omega}$, on désigne la frontière de Ω.

LEMME 1.1. — *Soit* Ω *du type* $\mathfrak{N}^{(0),1}$. *Soit* $p \geqq 1$. *Alors il existe une suite de sous-domaines* Ω_s,

$$\overline{\Omega}_s \subset \Omega_{s+1} \subset \Omega, \qquad \lim_{s\to\infty} \Omega_s = \Omega$$

(*au sens que pour chaque compact* $K \subset \Omega$ *il existe* s_0 *de sorte que* $s \geqq s_0 \Rightarrow K \subset \Omega_s$), Ω_s *sont du type* $\mathfrak{N}^\infty$; *en désignant par* a_{rs} *les fonctions représentant les frontières de* Ω_s *on a* : a_{rs}, *sont définies dans les mêmes boules* Δ_r, $r = 1, \ldots, m$, *sont uniformément lipschitziennes et*

$$\lim_{s\to\infty} \int_{\Delta_r} \left| \frac{\partial a_r}{\partial x_{ri}} - \frac{\partial a_{rs}}{\partial x_{ri}} \right|^p dX_r = 0 \qquad (i = 1, 2, \ldots, n-1, \ r = 1, 2, \ldots, m).$$

On définit

$$U_r = \mathop{\mathcal{E}}_{X} (X = [X_r, x_{rn}], \ |X_r| < \alpha, \ a_r(X_r) - \beta < x_{rn} < a_r(X_r) + \beta),$$
$$V_r = \mathop{\mathcal{E}}_{X} (X = [X_r, x_{rn}], \ |X_r| < \alpha, \ a_r(X_r) - \beta < x_{rn} < a_r(X_r)).$$

Il existe une partition de l'unité telle que

$$X \in \Omega \ \Rightarrow \ 1 = \sum_{r=1}^{m+1} \varphi_r(X),$$

les φ_r sont indéfiniment continûment différentiables, $0 \leqq \varphi_r(X) \leqq 1$, pour $r = 1, \ldots, m$, les supports des φ_r sont contenus dans U_r, pour $m+1$ dans Ω.

On utilise désormais souvent U_r, V_r, φ_r sans répéter leurs définitions.

On désigne par $\mathcal{E}(\overline{\Omega})$ l'espace des fonctions réelles, indéfiniment continûment différentiables dans Ω, continues avec toutes leurs dérivées dans $\overline{\Omega}$.

On désigne par $\mathcal{D}(\Omega)$ le sous-espace de $\mathcal{E}(\overline{\Omega})$ des fonctions à support compact dans Ω.

Soit $k \geqq 0$ un entier. On désigne par $W_2^{(k)}(\Omega)$ l'espace des fonctions réelles qui sont avec toutes leurs dérivées (au sens des distributions) jusqu'à l'ordre k de carré sommable. L'espace $W_2^{(k)}(\Omega)$ est l'espace de Hilbert pour le produit scalaire

$$(1.1) \qquad (u, v) = \sum_{|i| \leq k} \int_\Omega \frac{\partial^{|i|} u}{\partial x_1^{i_1} \ldots \partial x_n^{i_n}} \frac{\partial^{|i|} v}{\partial x_1^{i_1} \ldots \partial x_n^{i_n}} d\Omega.$$

On désigne par $\mathring{W}_2^{(k)}(\Omega) = \overline{\mathcal{D}(\Omega)}$ où la fermeture est définie moyennant la norme de $W_2^{(k)}(\Omega)$.

Par $W_2^{(-k)}(\Omega)$ on désigne le dual de $\mathring{W}_2^{(k)}(\Omega)$.

Lemme 1.2. — *Soit* $\Omega \in \mathfrak{N}^{(0)}$.

Alors $\overline{\mathcal{E}(\overline{\Omega})} = W_2^{(k)}(\Omega)$. [*La fermeture est définie moyennant la norme de* $W_2^{(k)}(\Omega)$.]

Lemme 1.3. — *Soit* $\Omega \in \mathfrak{N}^{(0)}$. *Il existe précisément une transformation* Z, *linéaire et continue de* $W_2^{(1)}(\Omega)$ *dans* $W_2^{(0)}(\dot{\Omega})$ *de sorte que*

$$u \in \mathcal{E}(\overline{\Omega}) \quad \Rightarrow \quad Z(u) = u.$$

On désigne par $W_2^{(\frac{1}{2})}(\dot{\Omega})$ l'espace des traces des fonctions de $W_2^{(1)}(\Omega)$. On munit $W_2^{(\frac{1}{2})}(\dot{\Omega})$ de la topologie de $W_2^{(1)}(\Omega)/\mathring{W}_2^{(1)}(\Omega)$.

Lemme 1.4. — *Soit* $\Omega \in \mathfrak{N}^{(0),1}$.

Alors $W_2^{(\frac{1}{2})}(\dot{\Omega})$ *est dense dans* $W_2^{(0)}(\dot{\Omega})$.

Lemme 1.5. — *Soit* $\Omega \in \mathfrak{N}^{(0),1}$.

Alors

$$u \in W_2^{(1)}(\Omega),\ Z(u) = 0 \iff u \in \mathring{W}_2^{(1)}(\Omega).$$

2. Les systèmes elliptiques du deuxième ordre. Problème de Dirichlet, l'intégrale de Dirichlet bornée. — On considère le système suivant pour les fonctions inconnues $u_1, u_2, \ldots, u_N$:

$$(2.1)\qquad -\frac{\partial}{\partial x_i}\left(a_{ij}^{\alpha\beta}\frac{\partial}{\partial x_j}u_\beta\right) + b^{\alpha\beta}u_\beta = f_\alpha \qquad (\alpha, \beta = 1, \ldots, N).$$

La convention usuelle de sommation est utilisée. On désigne par $u = (u_1, u_2, \ldots, u_N)$.

On suppose les coefficients $a_{ij}^{\alpha\beta} \in C^{(0),1}(\overline{\Omega})$ [1], $b^{\alpha\beta}$ mesurables, bornés. On suppose dans la suite que toutes ou quelques-unes des hypothèses suivantes sont vérifiées (cela dépend du cas) :

$$(2.2)\qquad \left\{\begin{array}{l} [\xi_1, \xi_2, \ldots, \xi_n], [\eta_1, \eta_2, \ldots, \eta_n] \in E_n \\ \Rightarrow (a_{ij}^{\alpha\beta}\xi_i\xi_j)\eta_\alpha\eta_\beta \geq c\left(\sum_{i=1}^{n}\xi_i^2\right)\left(\sum_{j=1}^{N}\eta_j^2\right) \quad [2] \\ \text{(ellipticité forte)}, \end{array}\right.$$

$$(2.3)\qquad a_{ij}^{\alpha\beta} = a_{ji}^{\alpha\beta},$$

$$(2.4)\qquad a_{ij}^{\alpha\beta} = a_{ij}^{\beta\alpha},$$

$$(2.5)\qquad \left\{\begin{array}{l} u \in \left[\mathring{W}_2^{(1)}(\Omega)\right]^N \quad [3] \\ \Rightarrow \displaystyle\int_\Omega \left(a_{ij}^{\alpha\beta}\frac{\partial u_\alpha}{\partial x_i}\frac{\partial u_\beta}{\partial x_j} + b^{\alpha\beta}u_\alpha u_\beta\right)d\Omega \geq c\,|u|^2_{[W_2^{(1)}(\Omega)]^N} \\ \text{(ellipticité pour le problème de Dirichlet)}. \end{array}\right.$$

On utilisera dans la suite la notation suivante :

$$\frac{\partial}{\partial x_i}\left(a_{ij}^{\alpha\beta}\frac{\partial}{\partial x_j}u_\beta\right) + b^{\alpha\beta}u_\beta = A_\alpha u.$$

---

(1) Pour la section 2, on peut évidemment affaiblir les hypothèses sur la régularité des $a_{ij}^{\alpha\beta}$.

(2) On va désigner dans la suite la plupart des constantes par le même symbole c.

(3) On utilisera dans la suite la notation suivante : soit B l'espace de Banach. On désigne par $[B]_N$ le produit $\underbrace{B \times B \times \ldots \times B}_{N \text{ fois}}$ avec la norme $|u_1| + \ldots + |u_N|$.

On écrira

$$f=(f_1, f_2, \ldots, f_N), \qquad \ldots .$$

Les lemmes suivants sont bien connus (pour la démonstration, *cf.* L. Nirenberg [6]) :

LEMME 2.1. — *Si* $a_{ij}^{\alpha\beta}$ *sont constantes, alors* (2.2) *et l'hypothèse*

$$[\eta_1, \eta_2, \ldots, \eta_N] \in E_N \Rightarrow b^{\alpha\beta}\eta_\alpha \eta_\beta \geqq 0 \Rightarrow (2.5).$$

LEMME 2.2. — *Si* $a_{ij}^{\alpha\beta}$ *sont continues dans* $\overline{\Omega}$, *alors* (2.2) *entraîne*

$$\int_\Omega \left(a_{ij}^{\alpha\beta} \frac{\partial u_\alpha}{\partial x_i} \frac{\partial u_\beta}{\partial x_j} + b^{\alpha\beta} u_\alpha u_\beta \right) d\Omega + \lambda |u|^2_{[W_2^{(0)}(\Omega)]^N} \geqq c |u|^2_{[W_2^{(1)}(\Omega)]^N}$$

pour $\lambda > 0$ *assez grand.*

On utilisera dans la suite la notation suivante :

$$\int_\Omega \left(a_{ij}^{\alpha\beta} \frac{dv_\alpha}{dx_i} \frac{\partial u_\beta}{\partial x_j} + b^{\alpha\beta} v_\alpha u_\beta \right) d\Omega = B(v, u).$$

Remarque. — Le système classique de l'élasticité est fortement elliptique.

Position du problème de Dirichlet : on se donne

$$f \in [W_2^{(-1)}(\Omega)]^N, \qquad w \in [W_2^{(1)}(\Omega)]^N.$$

On cherche $u \in [W_2^{(1)}(\Omega)]^N$ de sorte que

$$(2.6) \qquad v \in \left[\mathring{W}_2^{(1)}(\Omega)\right]^N \Rightarrow B(v, u) = f_\alpha(v_\alpha) = f(v),$$

$$(2.7) \qquad u - w \in \left[\mathring{W}_2^{(1)}(\Omega)\right]^N.$$

THÉORÈME 1. — *Soit* Ω *un domaine borné, le système* (2.1) *satisfaisant seulement à* (2.5). *Soit*

$$f \in [W_2^{(-1)}(\Omega)]^N, \qquad w \in [W_2^{(1)}(\Omega)]^N.$$

Alors il existe une solution du problème de Dirichlet et une seule (on dira désormais du P. D.) et l'on a

$$(2.8) \qquad |u|^2_{[W_2^{(1)}(\Omega)]^N} \leqq c\left[|w|^2_{[W_2^{(1)}(\Omega)]^N} + |f|^2_{[W_2^{(-1)}(\Omega)]^N} \right].$$

Démonstration. — La forme bilinéaire $B(v, u)$ satisfait les hypothèses du théorème de Lax et Milgram pour

$$v = (v_1, \ldots, v_N), \qquad u = (u_1, \ldots, u_N), \qquad v, u \in \left[\mathring{W}_2^{(1)}(\Omega)\right]^N \quad \text{(cf. [6])}.$$

Il existe alors une solution ω et une seule de l'équation $B(v, \omega) = f(v) - B(v, w)$. En posant $u = \omega + w$, on obtient l'assertion du théorème.

Théorème 2. — *Soit* $\Omega \in \mathfrak{N}^{(0),1}$, *le système satisfaisant seulement* (2.5). *Soit*

$$f \in [W_2^{(-1)}(\Omega)]^N, \qquad g \in \left[W_2^{(\frac{1}{2})}(\dot{\Omega})\right]^N,$$

Alors il existe précisément une solution faible au sens de (2.6) *du système* (2.1), *soit* u *de* $[W_2^{(1)}(\Omega)]^N$, *telle que* $Z(u_\alpha) = g_\alpha$ *sur* Ω. *On a*

$$(2.9) \qquad |u|^2_{[W_2^{(1)}(\Omega)]^N} \leq c\left[|g|^2_{\left[W_2^{(\frac{1}{2})}(\dot{\Omega})\right]^N} + |f|^2_{[W_2^{(-1)}(\Omega)]^N}\right].$$

La démonstration est une conséquence immédiate du théorème 1 et du lemme 1.5.

Théorème 3. — *Soit* Ω *borné, le système* (2.1) *satisfaisant seulement à* (2.5). *Soit*

$$f \in [W_2^{(-1)}(\Omega)], \qquad w \in [W_2^{(1)}(\Omega)]^N.$$

Soit $\Omega_s \subset \Omega$, $\lim_{s \to \infty} \Omega_s = \Omega$ *au sens mentionné ci-dessus. Ayant*

$$f \in [W_2^{(-1)}(\Omega_s)]^N, \qquad w \in [W_2^{(1)}(\Omega_s)]^N,$$

soit u^s (*resp.* u) *la solution du P. D. correspondant sur* Ω_s (*resp.* Ω). *On prolonge* u^s *sur* Ω *en posant* $u^s = w$ *hors de* Ω_s.

Alors on a $u^s \to u$ *dans* $[W_2^{(1)}(\Omega)]^N$.

Démonstration. — On introduit dans $\left[\mathring{W}_2^{(1)}(\Omega)\right]^N$ le produit scalaire par

$$(2.10) \qquad (v, u) = (v_\alpha, u_\alpha)_{W_2^{(1)}(\Omega)}.$$

Évidemment, sans restreindre la généralité on peut se borner au cas où $w = 0$. Pour chaque $v \in \left[\mathring{W}_2^{(1)}(\Omega)\right]^N$ on définit une transformation T linéaire et bornée par

$$B(v, u) = (T(v), u) \qquad \text{pour} \quad u \in \left[\mathring{W}_2^{(1)}(\Omega)\right]^N.$$

Il suit de (2.5) que T est simple et sur $\left[\mathring{W}_2^{(1)}(\Omega)\right]^N$. Soit $\varphi \in [\mathcal{D}(\Omega)]^N$. On a évidemment

$$B(\varphi, u^s) = (T(\varphi), u^s) \quad \to \quad B(\varphi, u) = (T(\varphi), u),$$

alors u^s tend vers u faiblement. On a

$$B(u - u^s, u - u^s) = B(u, u) - B(u, u^s) = B(u, u) - (Z(u), u^s) \to 0,$$

alors (2.5) entraîne que $u^s \to u$ fortement. La démonstration est par là achevée.

Nous démontrerons maintenant quelques théorèmes sur la régularité de la solution faible qui sont importants pour la suite. La technique des démonstrations est celle du travail [6].

Théorème 4. — *Soit Ω un domaine borné. On considère le système* (2.1) *qui satisfait seulement à la condition* (2.5) *pour chaque sous-domaine de Ω. On suppose $a_{ij}^{\alpha\beta} \in C^{(0),1}(\Omega')$, où Ω' est le sous-domaine arbitraire de Ω pour lequel $\overline{\Omega}' \subset \Omega$.* (*On utilisera souvent dans la suite cette notation.*) *Soit $f \in [W_2^{(0)}(\Omega)']^N$. Soit u la solution faible du système et soit*

$$u \in [W_2^{(1)}(\Omega')]^N.$$

Alors $u \in [W_2^{(2)}(\Omega')]^N$ et l'on a

$$|u|^2_{[W_2^{(2)}(\Omega'')]^N} \leq c(\Omega', \Omega'') \left[|f|^2_{[W_2^{(0)}(\Omega')]^N} + |u|^2_{[W_2^{(1)}(\Omega')]^N} \right].$$

Les équations du système (2.1) *sont satisfaites presque partout.*

Démonstration. — Soit $\psi \in \mathcal{D}(\Omega')$ choisie de sorte que $\psi(X) = 1$ pour $X \in \Omega''$. On désigne par $v = u\psi = (u_1\psi, \ldots, u_N\psi)$. Par un calcul immédiat on voit que v résout P. D. avec $w = 0$ et avec le terme absolu donné par

$$g_\alpha = f_\alpha \psi - \frac{\partial}{\partial x_i}\left(a_{ij}^{\alpha\beta} \frac{\partial \psi}{\partial x_j} u_\beta\right) - a_{ij}^{\alpha\beta} \frac{\partial \psi}{\partial x_i} \frac{\partial u_\beta}{\partial x_j}. \tag{2.11}$$

Soit $H = [\underbrace{0, \ldots, 0}_{k-1 \text{ fois}}, h, 0, \ldots, 0]$ le vecteur de E_n choisi de manière que

$$X + tH \in \Omega \quad \text{si} \quad X \in \Omega', \quad -2 \leq t \leq 2.$$

On considère le quotient

$$\frac{v(X+H) - v(X)}{h} = D_h^k v(X).$$

Soit $w \in [\overset{\circ}{W}{}_2^{(1)}(\Omega')]^N$. On a

$$\begin{aligned}
(2.12) \quad B(w, D_h^k v) = & \int_\Omega a_{ij}^{\alpha\beta}(X) \frac{\partial w_\alpha(X)}{\partial x_i} \frac{\partial D_h^k v_\beta(X)}{\partial x_j} dX \\
& + \int_\Omega b^{\alpha\beta}(X) w_\alpha(X) D_h^k v_\beta(X) dX \\
= & \int_\Omega a_{ij}(X-H) \frac{\partial w_\alpha(X-H)}{\partial x_i} \frac{1}{h} \frac{\partial v_\beta(X)}{\partial x_j} dX \\
& - \int_\Omega a_{ij}^{\alpha\beta}(X) \frac{\partial w_\alpha(X)}{\partial x_i} \frac{1}{h} \frac{\partial v_\beta(X)}{\partial x_j} dX \\
& + \int_\Omega b^{\alpha\beta}(X) w_\alpha(X) \frac{v_\alpha(X+H)}{h} dX \\
& - \int_\Omega b^{\alpha\beta}(X) w_\alpha(X) \frac{v_\alpha(X)}{h} dX \\
= & B(D_{-h}^k w, v) + \int_\Omega \frac{a_{ij}^{\alpha\beta}(X-H) - a_{ij}^{\alpha\beta}(X)}{h} \\
& \times \frac{\partial w_\alpha(X-H)}{\partial x_i} \frac{\partial v_\beta(X)}{\partial x_j} dX \\
& + \int_\Omega b^{\alpha\beta}(X) w_\alpha(X) \frac{1}{h} [v_\alpha(X+H) - v_\alpha(X)] dX \\
& - \int_\Omega b^{\alpha\beta}(X) \frac{1}{h} [w_\alpha(X-H) - w_\alpha(X)] v_\beta(X) dX.
\end{aligned}$$

Utilisant le fait que

$$\varphi \in \mathcal{D}(\Omega) \Rightarrow |D_h^k \varphi|_{W_2^{(0)}(\Omega)} \leq c |\varphi|_{W_2^{(1)}(\Omega)},$$

on obtient

$$(2.13) \quad |B(w, D_h^k v)| \leq c[|w|_{[W_2^{(1)}(\Omega)]^N} |v|_{[W_2^{(1)}(\Omega)]^N} + |g|_{[W_2^{(0)}(\Omega)]^N} |w|_{[W_2^{(1)}(\Omega)]^N}].$$

Posant $w = D_h^k v$, utilisant (2.13) et (2.5) on obtient

$$(2.14)\qquad |D_h^k v|^2_{[W_2^{(1)}(\Omega)]^N} \leq c\left[|g|^2_{[W_2^{(0)}(\Omega)]^N} + |v|^2_{[W_2^{(1)}(\Omega)]^N}\right].$$

On a

$$D_h^k v \to \frac{\partial v}{\partial x_k} \quad \text{dans} \quad [W_2^{(0)}(\Omega)]^N,$$

alors on obtient que $v \in [W_2^{(2)}(\Omega)]^N$ et l'inégalité

$$(2.15)\qquad |v|^2_{[W_2^{(2)}(\Omega)]^N} \leq c\left[|g|^2_{[W_2^{(0)}(\Omega)]^N} + |v|^2_{[W_2^{(1)}(\Omega)]^N}\right],$$

d'où l'assertion.

Théorème 5. — *Soit $\Omega \in \mathfrak{N}^{(1),1}$. On considère le système* (2.1) *en supposant la validité de* (2.5) *et* (2.2). *On suppose*

$$f \in [W_2^{(0)}(\Omega)]^N, \qquad w \in [W_2^{(2)}(\Omega)]^N.$$

Alors la solution faible u du P. D. appartient à $[W_2^{(2)}(\Omega)]^N$ et l'on a

$$(2.16)\qquad |u|^2_{[W_2^{(2)}(\Omega)]^N} \leq c\left[|f|^2_{[W_2^{(0)}(\Omega)]^N} + |w|^2_{[W_2^{(2)}(\Omega)]^N}\right].$$

Démonstration. — Par le procédé de la démonstration du théorème 1, on voit qu'on peut se borner au cas $w = 0$. On utilise la partition de l'unité signalée à la section 1. On pose $u^r = u\varphi_r$ $(r = 1, \ldots, m+1)$. En vertu du théorème précédent, on a (2.16) pour u^{m+1}. Soit alors $r \leq m$. L'élément u^r est la solution du P. D. dans le domaine U_r avec $w = 0$ et avec le terme absolu donné par (2.11) où figure φ_r au lieu de ψ.

On transforme le système X aux cartes $[X_r, x_{rn}]$, ce qui ne change pas des hypothèses du théorème. Soit K_r le cylindre

$$\left[\sum_{i=1}^{n-1} y_{ri}^2\right]^{\frac{1}{2}} < \alpha, \qquad -\beta < y_{rn} < 0$$

et soit $T(Y) = X$ la transformation de ce cylindre sur V_r définie par

$$(2.17)\qquad X_r = Y_r, \qquad x_{rn} = y_{rn} + a_r(X_r)$$

Tenant compte de $\Omega \in \mathfrak{N}^{(1),1}$, on obtient deux constantes $0 < c_1 \leq c_2$ de sorte que pour $u \in [\mathcal{E}(\overline{\Omega})]^N \Rightarrow$

$$c_1 \,|\, u \,|_{[W_2^{(2)}(U_r)]^N} \leq |\, u \,|_{[W_2^{(2)}(K_r)]^N} \leq c_2 \,|\, u \,|_{[W_2^{(2)}(U_r)]^N}. \tag{2.18}$$

On transforme de nouveau le système (2.1) aux cartes Y, ce qui ne change pas les hypothèses de notre théorème, spécialement (2.5) et (2.2). En utilisant maintenant le même procédé comme dans la démonstration du théorème 4, prenant H dans l'hyperplan $y_n = 0$, on obtient que toutes les dérivées du deuxième ordre des fonctions u_α^r sont de carrés sommables sur K_r, sauf les dérivées du type $\dfrac{\partial^2 u_\alpha^r}{\partial y_{rn}^2}$. En vertu de (2.2), on peut calculer ces dérivées du système (2.1) et l'on obtient par là qu'elles sont de carrés sommables dans K_r. Tenant compte de (2.18) on achève la démonstration.

3. L'égalité de Rellich. — Soit $\Omega \in \mathfrak{N}^{(1),1}$, $u \in [\mathcal{E}(\overline{\Omega})]^N$. Soit

$$a_{ij}^{\alpha\beta} \in C^{(0),1}(\overline{\Omega}), \qquad h \in [C^{(1)}(\overline{\Omega})]^N.$$

On suppose (2.2)-(2.5) ([1]). Alors on a l'identité

$$\begin{aligned} \frac{\partial}{\partial x_k}\left[\left(h_k a_{ij}^{\alpha\beta} - 2 h_i a_{kj}^{\alpha\beta}\right)\frac{\partial u_\alpha}{\partial x_i}\frac{\partial u_\beta}{\partial x_j}\right] &= \left(\frac{\partial h_k}{\partial x_k} a_{ij}^{\alpha\beta} + h_k \frac{\partial a_{ij}^{\alpha\beta}}{\partial x_k} - 2\frac{\partial h_i}{\partial x_k} a_{kj}^{\alpha\beta}\right)\frac{\partial u_\alpha}{\partial x_i}\frac{\partial u_\beta}{\partial x_j} \\ &\quad + 2 h_i \frac{\partial u_\alpha}{\partial x_i} A_\alpha u - 2 h_i \frac{\partial u_\alpha}{\partial x_i} b^{\alpha\beta} u_\beta. \end{aligned} \tag{3.1}$$

La formule de Green nous donne

$$\begin{aligned} \int_{\dot\Omega} \left(h_k a_{ij}^{\alpha\beta} - 2 h_i a_{kj}^{\alpha\beta}\right)\frac{\partial u_\alpha}{\partial x_i}\frac{\partial u_\beta}{\partial x_j} \nu_k \, dS &= \int_\Omega b_{ij}^{\alpha\beta} \frac{\partial u_\alpha}{\partial x_i}\frac{\partial u_\beta}{\partial x_j} d\Omega + 2\int_\Omega h_i \frac{\partial u_\alpha}{\partial x_i} A_\alpha u \, d\Omega \\ &\quad - 2\int_\Omega h_i \frac{\partial u_\alpha}{\partial x_i} b^{\alpha\beta} u_\beta \, d\Omega. \end{aligned} \tag{3.2}$$

([1]) La condition (2.4) est satisfaite pour le système de l'élasticité dans le cas général. Pour les autres systèmes, on peut s'y ramener par des transformations linéaires des inconnues u_α et par des transformations régulières des cartes.

Ici

$$b_{ij}^{\alpha\beta} = \left(\frac{\partial h_k}{\partial x_k} a_{ij}^{\alpha\beta} + h_k \frac{\partial a_{ij}^{\alpha\beta}}{\partial x_k} - 2 \frac{\partial h_i}{\partial x_k} a_{kj}^{\alpha\beta}\right).$$

Soit $\Omega \in \mathfrak{N}^{(0),1}$, Ω_s les sous-domaines du lemme 1.2. Soit u^s la solution du P. D. $A_\alpha u^s = f_\alpha$ dans Ω_s,

$$f \in [W_2^{(0)}(\Omega)]^N, \qquad u^s \in \left[\overset{0}{W}{}_2^{(1)}(\Omega_s)\right]^N.$$

En vertu du théorème 5 et du lemme 1.2, (3.2) est valable pour u^s. Évidemment, le vecteur $h_k \nu_k a_{ij}^{\alpha\beta} - h_i \nu_k a_{kj}^{\alpha\beta}$ pour α, β, j fixes est orthogonal à la normale extérieure. Ayant alors

$$\int_{\dot{\Omega}_s} (h_k \nu_k a_{ij}^{\alpha\beta} - h_i \nu_k a_{kj}^{\alpha\beta}) \frac{\partial u_\alpha^s}{\partial x_i} \frac{\partial u_\beta^s}{\partial x_j} dS = 0,$$

on en tire, en tenant compte de (3.2),

$$-\int_{\dot{\Omega}_s} h_k \nu_k a_{ij}^{\alpha\beta} \frac{\partial u_\alpha^s}{\partial x_i} \frac{\partial u_\beta^s}{\partial x_j} dS = \int_{\Omega_s} b_{ij}^{\alpha\beta} \frac{\partial u_\alpha^s}{\partial x_i} \frac{\partial u_\beta^s}{\partial x_j} d\Omega_s$$
$$+ 2 \int_{\Omega_s} h_i \frac{\partial u_\alpha^s}{\partial x_i} A_\alpha u \, d\Omega_s - 2 \int_{\Omega_s} h_i \frac{\partial u_\alpha^s}{\partial x_i} b^{\alpha\beta} u_\beta \, d\Omega_s.$$

Pour les domaines en question, on peut trouver h de sorte que pour s assez grand on ait $h_i \nu_i \geq c > 0$ *cf.* [1]). Tenant compte du théorème 1 on parvient à l'inégalité

$$\int_{\dot{\Omega}_s} a_{ij}^{\alpha\beta} \frac{\partial u_\alpha^s}{\partial x_i} \frac{\partial u_\beta^s}{\partial x_j} dS \leq c \, |f|^2_{[W_2^{(0)}(\Omega_s)]^N}. \tag{3.3}$$

On pose

$$\left(\frac{\partial u}{\partial \nu}\right)_\alpha = a_{ij}^{\alpha\beta} \frac{\partial u_\beta}{\partial x_j} \nu_i.$$

Soit maintenant $c^{\alpha\beta}$ la matrice inverse de la matrice $a_{ij}^{\alpha\beta} \nu_i \nu_j$ (pour $X \in \dot{\Omega}_s$). On a

$$\int_{\dot{\Omega}_s} \left(c^{\alpha\beta} \left(\frac{\partial u^s}{\partial \nu}\right)_\alpha \left(\frac{\partial u^s}{\partial \nu}\right)_\beta - a_{ij}^{\alpha\beta} \frac{\partial u_\alpha^s}{\partial x_i} \frac{\partial u_\beta^s}{\partial x_j}\right) dS = 0. \tag{3.4}$$

En effet,

$$c^{\alpha\beta}\left(\frac{\partial u^s}{\partial \nu}\right)_\alpha\left(\frac{\partial u^s}{\partial \nu}\right)_\beta - a_{ij}^{\alpha\beta}\frac{\partial u_\alpha^s}{\partial x_i}\frac{\partial u_\beta^s}{\partial x_j}$$

$$= c^{\alpha\beta}a_{kl}^{\alpha\gamma}\frac{\partial u_\gamma^s}{\partial x_l}\nu_k a_{mn}^{\beta\delta}\frac{\partial u_\delta^s}{\partial x_n}\nu_m - a_{ln}^{\delta\gamma}\frac{\partial u_\delta^s}{\partial x_n}\frac{\partial u_\gamma^s}{\partial x_l}$$

$$= \left(c^{\alpha\beta}a_{kl}^{\alpha\gamma}\nu_k a_{mn}^{\beta\delta}\nu_m - a_{ln}^{\gamma\delta}\right)\frac{\partial u_\gamma^s}{\partial x_l}\frac{\partial u_\delta^s}{\partial x_n}.$$

Pour γ, l, δ fixes, le vecteur $c^{\alpha\beta}a_{kl}^{\alpha\gamma}\nu_k a_{mn}^{\beta\delta}\nu_n - a_{ln}^{\delta\gamma}$ est orthogonal à la normale extérieure, d'où (3.4). On a

$$c^{\alpha\beta}\left(\frac{\partial u^s}{\partial \nu}\right)_\alpha\left(\frac{\partial u^s}{\partial \nu}\right)_\beta \geq c\sum_{\alpha=1}^{N}\left(\frac{\partial u^s}{\partial \nu}\right)_\alpha^2,$$

alors on tire de (3.3) en posant $\frac{\partial u}{\partial \nu} = \left(\left(\frac{\partial u}{\partial \nu}\right)_1, \ldots, \left(\frac{\partial u}{\partial \nu}\right)_N\right)$:

$$(3.5) \qquad \left|\frac{\partial u^s}{\partial \nu}\right|^2_{[W_2^{(0)}(\dot{\Omega}_s)]^N} \leq c\,|f|^2_{[W_2^{(0)}(\Omega_s)]^N} \quad (^5).$$

Soit maintenant $\psi \in [\mathcal{D}(\Delta_r)]^N$ (⁶). Soit $\omega \in \mathcal{D}(U_r)$ choisie de sorte que pour s assez grand on ait $\psi_\alpha(X_r) = \psi_\alpha(X_r)\,\omega(_r, a_{rs}(X_r X))$. On définit $v = \psi\omega$. La formule de Green nous donne

$$\int_{\Omega_s} f_\alpha v_\alpha\, d\Omega = -\int_{\dot{\Omega}_s}\left(\frac{\partial u^s}{\partial \nu}\right)_\alpha v_\alpha\, dS + \int_{\Omega_s} a_{ij}^{\alpha\beta}\frac{\partial v_\alpha}{\partial x_i}\frac{\partial u_\beta^s}{\partial x_j}\,d\Omega + \int_{\Omega_s} b^{\alpha\beta}v_\alpha u_\beta^s\, d\Omega.$$

Tenant compte du théorème 3, on obtient que

$$\lim_{s\to\infty}\int_{\dot{\Omega}_s}\left(\frac{\partial u^s}{\partial \nu}\right)_\alpha \psi_\alpha\left(1+\sum_{i=1}^{n-1}\left(\frac{\partial a_{rs}}{\partial x_{ri}}\right)^2\right)^{\frac{1}{2}} dX_r$$

est finie. En vertu du lemme 1.1, on aura aussi

$$\lim_{s\to\infty}\int_{\dot{\Omega}_s}\left(\frac{\partial u^s}{\partial \nu}\right)_\alpha \psi_\alpha\left(1+\sum_{i=1}^{n-1}\left(\frac{\partial a_r}{\partial x_{ri}}\right)^2\right)^{\frac{1}{2}} dX_r.$$

(⁵) Ici c ne dépend pas de s.

(⁶) *Cf.* la définition des domaines du type $\mathfrak{N}^{(h),\mu}$.

Les éléments $\psi\left(1+\sum_{i=1}^{n-1}\left(\frac{\partial a_r}{\partial x_{ri}}\right)^2\right)^{\frac{1}{2}}$ forment un ensemble dense dans $[W_2^{(0)}(\Delta_r)]^N$. On a alors :

LEMME 3.1. — *Soit* $\Omega\in\mathfrak{N}^{(0),1}$, A_α *le système* (2.1). *On suppose* (2.2)-(2.5). *Soit* Ω_s *du lemme* 1.1, $f[\in W_2^{(0)}(\Omega)]^N$. *Soit* u^s *la solution du P. D.* $A_\alpha u=f_\alpha$ *dans* Ω_s, $u_\alpha^s=0$ *sur* $\dot\Omega_s$.

Alors

$$u^s\in[W_2^{(2)}(\Omega_s)]^N,\qquad \frac{\partial u^s}{\partial\nu}\in[W_2^{(0)}(\dot\Omega_s)]^N$$

et, par projection sur l'hyperplan $x_{rn}=0$, *à* $[W_2^{(0)}(\Delta_r)]^N$, $r=1, \ldots, m$. *On a*

$$\left(\frac{\partial u^s}{\partial\nu}\right)_\alpha\rightharpoonup w_\alpha^r$$

où $\rightharpoonup$ *signifie la convergence faible dans* $W_2^{(0)}(\Delta_r)$.

Définissons

$$(3.6)\qquad \left(\frac{\partial u}{\partial\nu}\right)_\alpha=\sum_{r=1}^{m} w_\alpha^r\varphi_r;$$

on a :

THÉORÈME 6. — *Soit* $\Omega\in\mathfrak{N}^{(0),1}$, A_α *le système* (2.1). *On suppose* (2.2)-(2.5). *Soit* $f\in[W_2^{(0)}(\Omega)]^N$. *Soit u la solution du P. D.* : $A_\alpha u=f_\alpha$ *dans* Ω, $u_\alpha=0$ *sur* $\dot\Omega$. *Soit*

$$v\in[W_2^{(1)}(\Omega)]^N.$$

Alors à chaque solution u on fait correspondre linéairement $\left(\frac{\partial u}{\partial\nu}\right)_\alpha$ *par* (3.6), *on a*

$$(3.7)\qquad \left|\frac{\partial u}{\partial\nu}\right|^2_{[W_2^{(0)}(\dot\Omega)]^N}\leq c\,|f|^2_{[W_2^{(0)}(\Omega)]^N},$$

$$(3.8)\qquad \int_\Omega v_\alpha f_\alpha\,d\Omega=-\int_{\dot\Omega}v_\alpha\left(\frac{\partial u}{\partial\nu}\right)_\alpha dS+B(v, u).$$

Démonstration. — (3.7) est une conséquence immédiate de (2.5). Pour chaque r on a

$$(3.9)\qquad \int_{\Omega_s}v_\alpha\varphi_r f_\alpha\,d\Omega_s=-\int_{\dot\Omega_s}v_\alpha\varphi_r\left(\frac{\partial u^s}{\partial\nu}\right)_\alpha dS+B(v, u^s).$$

En vertu du lemme 2.1 on obtient

$$\int_{\Omega} v_\alpha \varphi_r f_\alpha \, d\Omega = -\int_{\dot{\Omega}} v_\alpha \varphi_r w_\alpha^r \, dS + B(v, u). \tag{3.10}$$

L'addition des équations (3.10) pour $r = 1, \ldots, m$ achève la démonstration.

Remarque. — Il suit immédiatement de (3.8) que la définition de $\frac{\partial u}{\partial \nu}$ est unique.

4. Le procédé dual. — Soit v la solution du P. D. : $A_\alpha v = 0$ dans Ω, $v_\alpha = h_\alpha$ sur $\dot{\Omega}$ où $h \in \left[W_2^{(\frac{1}{2})}(\dot{\Omega})\right]^N$. Alors on a, tenant compte de (3.8),

$$\int_{\Omega} v_\alpha f_\alpha \, d\Omega = -\int_{\dot{\Omega}} h_\alpha \left(\frac{\partial u}{\partial \nu}\right)_\alpha dS. \tag{4.1}$$

De (3.7) et (4.1) on tire l'inégalité duale

$$|v|^2_{[W_2^{(0)}(\Omega)]^N} \leq c\,|h|^2_{[W_2^{(0)}(\dot{\Omega})]^N}. \tag{4.2}$$

En vertu du lemme 1.4 on parvient au

Théorème 7. — *Soit* $\Omega \in \mathfrak{N}^{(0),1}$, A_α *le système* (2.1). *On suppose* (2.2)-(2.5).

Alors il existe précisément une transformation R, *linéaire et continue de* $[W_2^{(0)}(\dot{\Omega})]^N$ *dans* $[W_2^{(0)}(\Omega)]^N$ *de sorte que pour* $h \in \left[W_2^{(\frac{1}{2})}(\dot{\Omega})\right]^N$, $R(h)$ *est la solution du P. D.* $A_\alpha v = 0$ *dans* Ω, $v_\alpha = h_\alpha$ *sur* $\dot{\Omega}$ *au sens du théorème* 2.

Remarque. — On peut démontrer que la solution du P. D. obtenue par le théorème 7 est régulière dans Ω : on a $v_\alpha \in W_2^{(2)}(\Omega')$ et $A_\alpha v = 0$ presque partout. Les coefficients $a_{ij}^{\alpha\beta}$, $b^{\alpha\beta}$ étant plus réguliers, on obtient une solution satisfaisant au système au sens classique.

Comme conséquences immédiates du lemme 3.1 on obtient les divers théorèmes d'unicité :

Théorème 8. — *Soit* $\Omega \in \mathfrak{N}^{(0),1}$, *soit* A_α *le système* (2.1) *satisfaisant* (2.2)-(2.5). *Soit* $u \in \left[W^{(1)}_{\frac{2n}{n-1}}(\Omega')\right]^N$ *et soit* u *la solution faible des équations* (2.1) *dans* Ω. *Soit* $u_\alpha = 0$ *sur* $\dot{\Omega}$ *au sens que*

$$\lim_{s\to\infty} \int_{\Delta_r} (u_\alpha(X_r, a_{rs}(X_r)))^2\, dX_r = 0 \qquad \text{pour } r = 1, \ldots, m.$$

(*Cf.* définition des domaines du type $\mathfrak{N}^{(k),\mu}$.)

Alors $u = 0$.

Démonstration. — Il suit des théorèmes d'immersion de Sobolev (*cf.* E. Gagliardo [1]) que $u_\alpha \in W_2^{(0)}(\dot{\Omega}_s)$, d'où l'assertion.

Théorème 9. — *Soit* $\Omega \in \mathfrak{N}^{(0),1}$, A_α *le système* (2.1) *satisfaisant* (2.2)-(2.5). *Soit* $u \in \left[\mathring{W}^{(1)}_{\frac{2n}{n-1}}(\Omega)\right]^N$. $\left(\mathring{W}^{(1)}_{\frac{2n}{n-1}}(\Omega) = \overline{\mathcal{D}(\Omega)}\right)$. *Soit* $A_\alpha u = 0$ *dans* Ω *faiblement.*

Alors $u = 0$.

Démonstration. — $u \in \left[\mathring{W}^{(1)}_{\frac{2n}{n-1}}(\Omega)\right]^N$ entraîne que $u_\alpha = 0$ au sens du théorème précédent.

BIBLIOGRAPHIE.

[1] E. Gagliardo, *Proprietà di alcune classi di funzioni in più variabili* (*Ricerche di Matematica*, t. 7, 1958, p. 102-137).

[2] J. Nečas, *O rešenijach elliptičeskich differencialnych uravnĕnij v častnych proizvodnych vtorovo porjadka s nĕograničennym intĕgralom Dirichle* (*Czechoslovak Math. J.*, t. 10, 1960, p. 283-298).

[3] J. Nečas, *Sur le problème de Dirichlet pour l'équation aux dérivées partielles du quatrième ordre du type elliptique* (*Rend. Sem. Mat. Univ. Padova*, t. 31, 1961, p. 198-231).

[4] J. Nečas, *Sur les équations différentielles aux dérivées partielles du type elliptique du deuxième ordre* (*Czechoslovak Math. J.*, (sous presse).

[5] J. Nečas, *Ob oblastach tipa* $\mathfrak{N}$ (*Czechoslovak Math. J.*, t. 12, 1962, p. 274-287).

[6] L. Nirenberg, *Remarks on Strongly Elliptic Partial Differential Equations* (*Comm. Pure Appl. Math.*, t. 8, 1955, p. 649-675).

[7] L. E. Payne et H. F. Weinberger, *New bounds for solutions of second order elliptic partial differential equations* (*Pacific. J. Math.*, t. 8, n° 3, 1958, p. 551-573).

(Manuscrit reçu le 4 mars 1963.)

On Inequalities of Korn's Type

I. Boundary-Value Problems for Elliptic Systems of Partial Differential Equations

IVAN HLAVÁČEK & JINDŘICH NEČAS

Communicated by R. FINN

Preface

If one derives the fundamental properties of solutions of the boundary-value problems for elliptic systems of partial differential equations using a variational approach, as was done, for example, in the book [1] of J. NEČAS, one realizes the important role of the *coerciveness* inequality (compare also [2]). Part I of this work presents some consequences of the coerciveness inequality, and yields the existence, uniqueness and continuous dependence of the solution upon the given data.

In Part II, the general theory of Part I is applied to the linear theory of three-dimensional elasticity, where inequalities resulting from coerciveness have been called KORN's inequalities and have been studied by KORN [3], FRIEDRICHS [4], EYDUS [5], PAYNE & WEINBERGER [6], GOBERT [7] and others. We present here new proofs for the previous results and extend them to some mixed boundary-value problems with more general boundary conditions, in which displacements and tractions are prescribed in the normal and tangential directions to the boundary. The cases of elastic supports are also considered.

1. Notation. Preliminary Definitions

Let a Lipschitz* region Ω in N-dimensional Euclidian space E_N and positive integers m and κ_s, $s=1, 2, \dots m$, be given. $L_2(\Omega)$ will denote the space of real functions which are square-integrable on Ω (in the Lebesgue-sense). $W_2^{(k)}(\Omega)$ denotes the subspace of $L_2(\Omega)$ consisting of functions whose derivatives up to order k, in the sense of distributions, are in $L_2(\Omega)$.

Let us introduce the scalar product on $W_2^{(k)}(\Omega)$ by means of

$$(v, u)_k = \sum_{|\alpha| \leq k} \int_\Omega D^\alpha v \, D^\alpha u \, dX, \tag{1.1}$$

* We call a region $\Omega \subset E_N$ Lipschitz if it is bounded and its boundary Γ has the following properties: a) to each point $X \in \Gamma$ an open hypersphere S_X about X exists, such that the intersection $S_X \cap \Gamma$ may be described by means of a Lipschitz function, and b) $S_X \cap \Gamma$ divides S_X into exterior and interior parts with respect to Ω.

© Springer-Verlag 1970. Reprinted with kind permission from: Hlaváček, I. & Nečas, J., On inequalities of Korn's type. I. Boundary-value problems for elliptic system of partial differential equations, Archive of Rational Mechanics and Analysis 36, 305-311.

where

$$D^\alpha \equiv \frac{\partial^{|\alpha|}}{\partial x_1^{\alpha_1} \dots \partial x_N^{\alpha_N}}.$$

Let W denote the Cartesian product $\prod_{s=1}^{m} W_2^{(\kappa_s)}(\Omega)$, and let two bilinear forms $A(v, u)$, $a(v, u)$ be given on $W \times W$:

$$A(v, u) = \int_\Omega \sum_{r,s=1}^{m} \sum_{\substack{|i| \leq \kappa_r \\ |j| \leq \kappa_s}} a_{ij}^{rs} D^i v_r D^j u_s \, dX \tag{1.2}$$

where a_{ij}^{rs} are bounded and measurable on Ω, and

$$a(v, u) = \int_\Gamma \sum_{r,s=1}^{m} \sum_{i=0}^{\kappa_r - 1} \sum_{|\alpha| \leq \kappa_s - 1} b_{i\alpha}^{rs} \frac{\partial^i v_r}{\partial n^i} D^\alpha u_s \, d\Gamma \tag{1.3}$$

where $b_{i\alpha}^{rs}$ are bounded and measurable on the boundary Γ. Furthermore, let $\mathscr{D}(\Omega)$ be the space of real functions with compact support in Ω which are infinitely differentiable, let $\mathring{W}_2^{(k)}(\Omega)$ be the closure of $\mathscr{D}(\Omega)$ in $W_2^{(k)}(\Omega)$, and let $\mathring{W} = \prod_{s=1}^{m} \mathring{W}_2^{(\kappa_s)}(\Omega)$.

Let V be a closed subspace of W, such that $\mathring{W} \subset V \subset W$. Define

$$f(v) = \int_\Omega \sum_{s=1}^{m} f_s v_s \, dX$$

where $f_s \in L_2(\Omega)$. Let g be a linear continuous functional on V such that $g(v) = 0$ for $v \in \mathring{W}$. Moreover, let $u_0 \in W$.

We say that $u \in W$ is a *weak solution of the boundary-value problem* if

$$u - u_0 \in V \tag{1.4}$$

and if for each $v \in V$ the relation

$$A(v, u) + a(v, u) = f(v) + g(v) \tag{1.5}$$

holds.

Let the operators $N_l v$, mapping W into $L_2(\Omega)$, $l = 1, \dots h$, be given in the form

$$N_l v = \sum_{r=1}^{m} \sum_{|\alpha| \leq \kappa_r} n_{lr\alpha} D^\alpha v_r$$

where $n_{lr\alpha}$ are bounded and measurable on Ω. Suppose that these operators form a *coercive system* on* W, i.e.,

$$v \in W \Rightarrow \sum_{l=1}^{h} \|N_l v\|_{L_2}^2 + \sum_{r=1}^{m} \|v_r\|_{L_2}^2 \geq c \|v\|_W^2, \quad c > 0. \tag{1.6}$$

We assume furthermore that

$$v \in W \Rightarrow A(v, v) \geq c \sum_{l=1}^{h} \|N_l v\|_{L_2}^2, \tag{1.7}$$

$$v \in W \Rightarrow a(v, v) \geq 0. \tag{1.8}$$

* See the theorems of Section 3 for necessary and sufficient conditions for coerciveness.

2. Fundamental Properties of the Weak Solution of the Boundary-Value Problem

Let us put

$$\mathscr{P}=\left\{v\in V,\ \sum_{l=1}^{h}\|N_l v\|_{L_2}^2+a(v,v)=0\right\}$$

and

$$((v,u))=A(v,u)+a(v,u),$$

and denote by $V/\mathscr{P}$ the factor-space of classes $\tilde{v}=\{v+p,\ v\in V,\ p\in\mathscr{P}\}$ with the norm

$$\|\tilde{v}\|_{V/\mathscr{P}}=\inf_{p\in\mathscr{P}}\|v+p\|_V.$$

Theorem 2.1. *Let the relations*

$$\begin{aligned}A(\tilde{v},\tilde{u})&=A(v,u),\\ a(\tilde{v},\tilde{u})&=a(v,u),\end{aligned}\qquad v\in\tilde{v},\quad u\in\tilde{u},\quad \tilde{v},\tilde{u}\in V/\mathscr{P},$$

define two bilinear forms on $V/\mathscr{P}\times V/\mathscr{P}$. *Let* (1.6), (1.7) *and* (1.8) *hold. Then*

$$((\tilde{v},\tilde{v}))\geqq c_1\left(\sum_{l=1}^{h}\|N_l\tilde{v}\|_{L_2}^2+a(\tilde{v},\tilde{v})\right)\geqq c_2\|\tilde{v}\|_{V/\mathscr{P}}^2 \tag{2.1}$$

holds for each $\tilde{v}\in V/\mathscr{P}$.

Proof. Suppose the contrary holds. Then there exists a sequence of elements $\tilde{v}\in V/\mathscr{P}$ such that

$$\|\tilde{v}_n\|_{V/\mathscr{P}}=1 \tag{2.2}$$

and

$$\sum_{l=1}^{h}\|N_l\tilde{v}_n\|_{L_2}^2+a(\tilde{v}_n,\tilde{v}_n)<\frac{1}{n}. \tag{2.3}$$

Let $V=K\oplus\mathscr{P}$ be the orthogonal decomposition of V by means of the scalar product (see (1.1))

$$\sum_{s=1}^{m}(v_s,u_s)_{\kappa_s}.$$

Denote by v_K the orthogonal projections of v on K. We have

$$\|\tilde{v}_n\|_{V/\mathscr{P}}=\|v_{nK}\|_V.$$

As the immersion of $W_2^{(1)}(\Omega)$ into $L_2(\Omega)$ and $L_2(\Gamma)$, respectively, is compact and V is weakly compact, we can choose a subsequence of v_{nK} (denoted again by v_{nK}) such that $v_{nK}\to v_K$ (weak convergence), $v_{nK}\to v_K$ in $\prod_{s=1}^{m}W_2^{(\kappa_s-1)}(\Omega)$ and $a(v_{nK},v_{nK})\to a(v_K,v_K)$. Then (2.3) implies $v_K\in\mathscr{P}$ and consequently $v_K=0$. By virtue of (2.3) and (1.6), $v_{nK}\to 0$ in V, which contradicts (2.2).

Theorem 2.2. *Let the form* $((v,u))$ *define the bilinear form* $((\tilde{v},\tilde{u}))$ *for* $\tilde{v}\in W/\mathscr{P}$, $\tilde{u}\in W/\mathscr{P}$, $v\in\tilde{v}$, $u\in\tilde{u}$. *Let* (1.6)–(1.8) *hold. Then a necessary and sufficient condition for the existence of a weak solution of the boundary-value problem is*

$$p\in\mathscr{P}\ \Rightarrow\ f(p)+g(p)=0. \tag{2.4}$$

In this case the solution is determined except for an element $\boldsymbol{p}\in\mathscr{P}$. *Moreover*

$$(2.5) \qquad \|\tilde{\boldsymbol{u}}\|_{W/\mathscr{P}} \leqq c(\|\boldsymbol{u}_0\|_W + \|f\|_{[L_2]^m} + \|g\|_{V'}).$$

Let $(\boldsymbol{v}, \boldsymbol{u})_W$ *and* $(\boldsymbol{v}, \boldsymbol{u})_V$ *be scalar products in* W *and* V, *supplying norms equivalent to*

$$\left[\sum_{s=1}^{m} (v_s, v_s)_{K_s}\right]^{\frac{1}{2}}$$

in W *and* V, *respectively. Let* $\boldsymbol{Q}$ *and* $\boldsymbol{R}$ *denote the orthogonal complements of* $\mathscr{P}$ *in* W *and* V, *respectively, by means of these scalar products, and let* $\boldsymbol{u}_Q$, $\boldsymbol{u}_R$ *be the corresponding orthogonal projections of* $\boldsymbol{u}$.

Then we may choose $\boldsymbol{u}_Q$ *and* $\boldsymbol{u}' = \boldsymbol{u}_0 + (\boldsymbol{u} - \boldsymbol{u}_0)_R$, *respectively, as the representatives of* $\tilde{\boldsymbol{u}}$, *and*

$$(2.6) \qquad \|\boldsymbol{u}_Q\|_W + \|\boldsymbol{u}'\|_W \leqq c(\|\boldsymbol{u}_0\|_W + \|f\|_{[L_2]^m} + \|g\|_{V'}).$$

Proof. The necessity of (2.4) follows from (1.5). If (2.4) is satisfied, let us seek a $\tilde{\boldsymbol{w}}\in V/\mathscr{P}$ such that

$$(2.7) \qquad ((\tilde{\boldsymbol{v}}, \tilde{\boldsymbol{w}})) = F(\tilde{\boldsymbol{v}})$$

holds for $\tilde{\boldsymbol{v}}\in V/\mathscr{P}$, where

$$F(\tilde{\boldsymbol{v}}) = f(\boldsymbol{v}) + g(\boldsymbol{v}) - ((\boldsymbol{v}, \boldsymbol{u}_0)).$$

The condition (2.1) enables us to use the Lax-Milgram theorem. Hence there exists a unique element $\tilde{\boldsymbol{w}}$ which satisfies (2.7). Evidently, any element $\boldsymbol{u}$ of the class $\tilde{\boldsymbol{u}} = \tilde{\boldsymbol{u}}_0 + \tilde{\boldsymbol{w}}$ is a solution of the problem. The remaining part of the theorem is obvious.

Remark 1. Let us take $(\boldsymbol{v}, \boldsymbol{u})_V = (\boldsymbol{v}, \boldsymbol{u})_W$ for $\boldsymbol{v}$, $\boldsymbol{u}\in V$. Then applying this scalar product to the orthogonal decomposition of W, we obtain

$$W = \mathscr{P} \oplus Q = \mathscr{P} \oplus R \oplus D$$

and consequently,

$$\boldsymbol{u}' = \boldsymbol{u}_0 + \boldsymbol{u}_R - \boldsymbol{u}_{0R} = \boldsymbol{u}_{0D} + \boldsymbol{u}_{0\mathscr{P}} + \boldsymbol{u} - \boldsymbol{u}_D - \boldsymbol{u}_{\mathscr{P}} = \boldsymbol{u}_Q + \boldsymbol{u}_{0\mathscr{P}} + (\boldsymbol{u}_0 - \boldsymbol{u})_D = \boldsymbol{u}_Q + \boldsymbol{u}_{0\mathscr{P}}.$$

Remark 2. Put

$$\mathscr{P}_A = \left\{\boldsymbol{v}\in V,\ \sum_{l=1}^{h} \|N_l \boldsymbol{v}\|_{L_2}^2 = 0\right\}.$$

Let

$$A(\boldsymbol{v}, \boldsymbol{u}) = A(\boldsymbol{u}, \boldsymbol{v}),$$

$$0 \leqq A(\boldsymbol{v}, \boldsymbol{v}) \leqq c \sum_{l=1}^{h} \|N_l \boldsymbol{v}\|_{L_2}^2$$

hold for $\boldsymbol{v}\in W$. Then a necessary condition for the existence of a solution is

$$(2.8) \qquad \boldsymbol{p}\in\mathscr{P}_A \Rightarrow a(\boldsymbol{p}, \boldsymbol{u}) = f(\boldsymbol{p}) + g(\boldsymbol{p}).$$

In fact, the assertion follows from (1.5) when we note that the Schwarz inequality for $A(v, u)$ implies $A(p, u)=0$.

Theorem 2.3. *Let* (1.6)–(1.8) *hold and let* $p_i \in V'$, $i=1, 2, \dots r$ *be a system of linear continuous functionals on* V *such that*

$$v\in V,\ \sum_{l=1}^{h} \|N_l v\|_{L_2}^2 + a(v, v) + \sum_{i=1}^{r} (p_i(v))^2 = 0 \Rightarrow v = 0. \tag{2.9}$$

Then

$$\sum_{l=1}^{h} \|N_l v\|_{L_2}^2 + a(v, v) + \sum_{i=1}^{r} (p_i(v))^2 \geqq c\,\|v\|_W^2 \tag{2.10}$$

holds for each $v\in V$.

Proof. The method is analogous to that of Theorem 2.1. Suppose the contrary holds. Then elements $v_n \in V$ exist such that $\|v_n\|_V = 1$, $v_n \rightharpoonup v$ in V, $v_n \to v$ in $\prod_{s=1}^{m} W_2^{(\kappa_s - 1)}(\Omega)$, $a(v_n, v_n) \to a(v, v)$, and

$$\sum_{l=1}^{h} \|N_l v_n\|_{L_2}^2 + a(v_n, v_n) + \sum_{i=1}^{r} (p_i(v_n))^2 \leqq \frac{1}{n}. \tag{2.11}$$

Hence by transition to the limit $n\to\infty$, $v=0$ follows. (2.11) and (1.6) yield $v_n \to 0$ in V, and we have a contradiction.

Remark 3. Let the hypotheses of Theorem 2.3 hold. Put

$$V_p = \left\{ v\in V, \sum_{i=1}^{r} (p_i(v))^2 = 0 \right\}.$$

Then

$$v\in V_p \Rightarrow ((v, v)) \geqq c_1 \left(\sum_{l=1}^{h} \|N_l v\|_{L_2}^2 + a(v, v) \right) \geqq c_2 \|v\|_W^2 .$$

Remark 4. Let the hypotheses of Theorem 2.3 hold. Furthermore, let the system p_i be linearly independent on $\mathscr{P}$, i.e., let $\sum_{i=1}^{r} \alpha_i p_i(p) = 0$ for each $p\in\mathscr{P}$ imply $\sum_{i=1}^{r} \alpha_i^2 = 0$. Let us choose in Theorem 2.2

$$(v, u)_V = \sum_{l=1}^{h} \int_\Omega N_l v\, N_l u\, dX + \tfrac{1}{2}\big(a(v, u) + a(u, v)\big) + \sum_{i=1}^{r} p_i(v)\, p_i(u).$$

Then

$$R = V_p .$$

Indeed, let $v\in R$. Then $(v, p)_V = 0$ for $p\in\mathscr{P}$. Schwarz's inequality implies that

$$\sum_{l=1}^{h} \int_\Omega N_l v\, N_l p\, dX + \tfrac{1}{2}\big(a(v, p) + a(p, v)\big) = 0; \tag{2.12}$$

consequently,

$$\sum_{i=1}^{r} p_i(v)\, p_i(p) = 0$$

for every $p \in \mathscr{P}$. Hence

$$\sum_{i=1}^{r} (p_i(v))^2 = 0$$

and therefore $v \in V_p$.

Conversely, let $v \in V_p$. Then (2.12) yields $(v, p)_V = 0$ for every $p \in \mathscr{P}$, and therefore $v \in R$.

3. Coerciveness of the System $N_l v$

Let us consider operators $N_l v$ having the form

$$N_l v = \sum_{s=1}^{m} \sum_{|\alpha| \leq \kappa_s} n_{l_s \alpha} D^\alpha v_s, \tag{3.1}$$

and let us define

$$N_{l_s} \xi = \sum_{|\alpha| = \kappa_s} n_{l_s \alpha}(X) \xi_1^{\alpha_1} \xi_2^{\alpha_2} \ldots \xi_N^{\alpha_N}. \tag{3.2}$$

NECAS ([1], [2]) proved and BESOV generalized ([8]) the following results.

Theorem 3.1. *Let $n_{l s \alpha}$ be constants for $|\alpha| = \kappa_s$. Then the system $N_l v$ is coercive on W if and only if*

$$\text{the rank of the matrix } (N_{l_s} \xi) = m \tag{3.3}$$

for each $\xi \in C_N$, $\xi \neq 0$, where C_N denotes complex N-dimensional space.

Theorem 3.2. *Let $n_{l s \alpha} \in C(\overline{\Omega})$* for $|\alpha| = \kappa_s$. Then the system $N_l v$ is coercive on W if and only if* (3.3) *holds for $\xi \in E_N$ if $X \in \Omega$, and for $\xi \in C_N$ if $X \in \Gamma$.*

Example. Let $m = N$, $\kappa_1 = \kappa_2 = \cdots = \kappa_m = 1$. Let the system $N_l v$ be defined by the operators $\left(\frac{\partial v_i}{\partial x_j} + \frac{\partial v_j}{\partial x_i}\right)$, $i, j = 1, \ldots N$, $i \neq j$ and $\frac{\partial v_i}{\partial x_i}$, $i = 1, \ldots N$. If we order these operators as follows:

$$\frac{\partial v_1}{\partial x_1}, \frac{\partial v_1}{\partial x_2} + \frac{\partial v_2}{\partial x_1}, \frac{\partial v_1}{\partial x_3} + \frac{\partial v_3}{\partial x_1}, \ldots \frac{\partial v_1}{\partial x_N} + \frac{\partial v_N}{\partial x_1},$$

$$\frac{\partial v_2}{\partial x_2}, \frac{\partial v_2}{\partial x_3} + \frac{\partial v_3}{\partial x_2}, \ldots \frac{\partial v_2}{\partial x_N} + \frac{\partial v_N}{\partial x_2}, \frac{\partial v_3}{\partial x_3}, \ldots \frac{\partial v_N}{\partial x_N},$$

then the transposed matrix $N_{l s} \xi$ will have the form

$$\begin{pmatrix}
\xi_1, & \xi_2, & \xi_3, & & \ldots \xi_N, & 0, & & & \ldots 0, & & & \ldots & & \ldots 0 \\
0, & \xi_1, & 0, & & \ldots 0, & \xi_2, & \xi_3, & & \ldots \xi_N, & 0, & 0, & \ldots & & 0 \\
0, & 0, & \xi_1, & 0, & 0, & 0, & \xi_2, & 0 & \ldots 0, & \xi_3, & \xi_4, & \ldots \xi_N, & 0, & \ldots 0 \\
0, & 0, & 0, & \xi_1 & 0, & 0, & 0, & \xi_2 & \ldots 0, & 0, & \xi_3, & 0, \ldots 0, & \xi_4, \ldots & \\
\vdots & \vdots & \vdots & \ddots & \vdots & & & \ddots & \vdots & \vdots & \vdots & \ddots & & \\
0, & 0, & & & \ldots \xi_1, & 0, & 0, & & \xi_2, & 0, & 0, & \ldots \xi_3, & & \ldots
\end{pmatrix}.$$

Obviously, the rank of $(N_{l s} \xi)$ is equal to N for each $\xi \in C_N$, $\xi \neq 0$. Hence the system $N_l v$ is coercive on W.

* $C(\overline{\Omega})$ denotes the space of functions, which are continuous on Ω and have continuous extensions to Γ.

References

1. NEČAS, J., Les méthodes directes en théorie des équations elliptiques. Prague: Academia 1967.
2. NEČAS, J., Sur les normes équivalentes dans $W_p^{(k)}(\Omega)$ et sur la coercitivité des formes formellement positives. Les presses de l'Université de Montréal, Janvier 1966, 102—128.
3. KORN, A., Über einige Ungleichungen, welche in der Theorie der elastischen und elektrischen Schwingungen eine Rolle spielen. Bull. Int. Cracovie Akad. Umiejet, Classe des sc. math. et nat. (1909).
4. FRIEDRICHS, K. O., On the boundary-value problems of the theory of elasticity and Korn's inequality. Annals of Math. **48**, 2 (1947).
5. EYDUS, D. M., On the mixed problem of the theory of elasticity. Dokl. A. N. SSSR **76**, 2 (1951) [Russian].
6. PAYNE, L. E., & H. F. WEINBERGER, On Korn's inequality. Arch. Rational Mech. Anal. **8**, 89—98 (1961).
7. GOBERT, J., Une inégalité fondamentale de la théorie de l'élasticité. Bull. Soc. Royale des Sciences de Liège 3—4 (1962).
8. BESOV, O. V., On the coerciveness in the anisotropic Sobolev space. Matem. Sbornik **73**, 115, 4, 585—599 (1967) [Russian].

Matematický ústav ČSAV
Žitná 25
Prague 1, Czechoslovakia

(Received July 18, 1968)

On Inequalities of Korn's Type

II. Applications to Linear Elasticity

IVAN HLAVÁČEK & JINDŘICH NEČAS

Communicated by R. FINN

Preface

In Part I [1] we studied a general mixed boundary-value problem using the recent theory of linear elliptic partial differential equations formulated by NEČAS [2]. We determined there certain general conditions for the V-ellipticity of the corresponding bilinear form of internal energy, i.e., an inequality of KORN's type, yielding the existence, uniqueness, and continuous dependence of the solution on the given data. In the present article we illustrate the general theory by applying it to the classical linear theory of elasticity. The main effort is devoted to cases in which the contact conditions and conditions of elastic support on the boundary do not suffice to eliminate the possibility of rigid body motions. Some sufficient conditions for the opposite case to occur are derived for a general shape of the (piecewise smooth) boundary.

1. The General Mixed Boundary-Value Problem

Let an elastic body occupy a Lipschitz region Ω (see [1]) in three-dimensional Euclidean space. Let the body forces $\boldsymbol{K}\in[L_2(\Omega)]^3$ and surface tractions $\boldsymbol{P}\in[L_2(\Gamma)]^3$ be prescribed, and assume that surface displacements are given by means of a vector-function $\boldsymbol{u}_0\in[W_2^{(1)}(\Omega)]^3$.

Suppose that the strain-displacement relations

$$\varepsilon_{ik}=\tfrac{1}{2}(u_{i,k}+u_{k,i}), \tag{1}$$

the stress-strain relations (generalized Hooke's law)

$$\tau_{ik}=c_{iklm}\varepsilon_{lm} \tag{2}$$

and the stress equations of equilibrium

$$\tau_{ik,k}+K_i=0 \tag{3}$$

hold on Ω. Here u_i, ε_{ik} and τ_{ik} designate respectively the rectangular Cartesian components of the displacement vector $\boldsymbol{u}$, the strain tensor ε and the stress tensor τ. We use the convention that a repeated suffix implies summation over the range 1, 2, 3, and we write $u_{i,k}=\partial u_i/\partial x_k$. The elastic coefficients $c_{iklm}(X)$ are assumed to be measurable and bounded on Ω, and to satisfy the symmetry relations

$$c_{iklm}=c_{kilm}=c_{lmik}. \tag{4}$$

© Springer-Verlag 1970. Reprinted with kind permission from: Hlaváček, I. & Nečas, J., On inequalities of Korn's type. II. Applications to linear elasticity, Archive of Rational Mechanics and Analysis 36, 312-334.

Moreover, we suppose that

$$c_{iklm}(X)\,\varepsilon_{ik}\,\varepsilon_{lm} \geqq \mu_0 \sum_{i,k=1}^{3} \varepsilon_{ik}^2 \tag{5}$$

for every symmetric tensor ε_{ik} at each point $X \in \Omega$ where a positive constant μ_0 is independent of X.

On the boundary Γ of the region Ω the boundary conditions are prescribed in the form of linear combinations of the displacements and surface traction components, thus

$$\begin{aligned} A_n u_n + B_n T_n &= C_n, \\ A_t \boldsymbol{u}_t + B_t \boldsymbol{T}_t &= \boldsymbol{C}_t, \end{aligned} \tag{6}$$

where the suffixes n or t denote the components of vectors $\boldsymbol{u}$ and $\boldsymbol{T}$ with components $T_i = \tau_{ik} n_k$ in the direction of the unit outward normal $\boldsymbol{n}$ or of the tangential plane to Γ, respectively, that is

$$\begin{aligned} u_n &= u_k n_k, & u_{tj} &= u_j - u_i n_i n_j, \\ T_n &= \tau_{ik} n_i n_k, & T_{tj} &= \tau_{jk} n_k - \tau_{ik} n_i n_k n_j. \end{aligned} \tag{7}$$

The coefficients A_n, A_t are piecewise constant functions on Γ whose values are either 0 or 1, while B_n and B_t are bounded measurable functions on Γ, such that

$$\begin{aligned} B_n \geqq \beta_n > 0 \quad &\text{or} \quad B_n = 0, \\ B_t \geqq \beta_t > 0 \quad &\text{or} \quad B_t = 0 \end{aligned} \tag{8}$$

with β_n, β_t constant. We suppose that

$$A_n + B_n > 0, \qquad A_t + B_t > 0 \tag{9}$$

holds for almost every point* $X \in \Gamma$.

Let us introduce the following point sets:

$$\begin{aligned} \mathscr{A}_n &= \{X \in \Gamma, B_n = 0\}, \qquad \mathscr{B}_n = \{X \in \Gamma, B_n > 0\}, \\ \mathscr{A}_t &= \{X \in \Gamma, B_t = 0\}, \qquad \mathscr{B}_t = \{X \in \Gamma, B_t > 0\}, \\ \Gamma_n &= \{X \in \Gamma, A_n B_n > 0\}, \\ \Gamma_t &= \{X \in \Gamma, A_t B_t > 0\}. \end{aligned}$$

A set $G \subset \Gamma$ will be called *open in* Γ if for any point $X_0 \in G$ there exists an $\eta > 0$ such that each $X \in \Gamma$, which satisfies $|X - X_0| < \eta$, belongs to G.**

Suppose that the sets $\mathscr{A}_n$, $\mathscr{A}_t$, $\mathscr{B}_n$, $\mathscr{B}_t$, Γ_n, Γ_t are either empty, or open in Γ. Furthermore, let a vector-function $\boldsymbol{u}_0 \in [W_2^{(1)}(\Omega)]^3$ be given, defining functions $C_n \in L_2(\mathscr{A}_n)$ and $\boldsymbol{C}_t \in [L_2(\mathscr{A}_t)]^3$ on $\mathscr{A}_n$ and $\mathscr{A}_t$ by means of (6), that is, by means of the relations

$$u_{0n} = C_n \quad \text{on} \quad \mathscr{A}_n, \qquad \boldsymbol{u}_{0t} = \boldsymbol{C}_t \quad \text{on} \quad \mathscr{A}_t.$$

* I.e., for $X \in \Gamma \dot{-} \mathscr{N}$, where $\mathscr{N}$ is a set of surface measure zero.

** $|X| = \left(\sum_{i=1}^{3} x_i^2\right)^{\frac{1}{2}}$.

C_n is defined on $\mathscr{B}_n$ by means of the given surface traction $\boldsymbol{P}$, namely

$$C_n = B_n P_n \in L_2(\mathscr{B}_n),$$

and $\boldsymbol{C}_t$ is defined on $\mathscr{B}_t$ by means of

$$\boldsymbol{C}_t = B_t \boldsymbol{P}_t \in [L_2(\mathscr{B}_t)]^3.$$

Inserting (1) and (2) into (3) and making use of the symmetry condition (4), we obtain Lamé's system of equilibrium equations

$$(c_{iklm} u_{l,m})_{,k} + K_i = 0.$$

If moreover we express the surface stress components T_n, $\boldsymbol{T}_t$ in terms of $\boldsymbol{u}$, using (7), (1) and (2), we obtain from (6) the general mixed boundary-value problem of the linear theory of elasticity, formulated in terms of the displacements.

Let us consider this problem in the notation of Part I. We have $m = N = 3$, $\kappa_s = 1$ for $s = 1, 2, 3$; also V is the space of functions $\boldsymbol{u} \in [W_2^{(1)}(\Omega)]^3$ which satisfy the boundary conditions $u_n = 0$ on $\mathscr{A}_n$, $\boldsymbol{u}_t = \boldsymbol{0}$ on $\mathscr{A}_t$.

By a *solution* we mean a function $\boldsymbol{u} \in [W_2^{(1)}(\Omega)]^3$ such that, first of all, $\boldsymbol{u} - \boldsymbol{u}_0 \in V$, (or $\boldsymbol{u} - \boldsymbol{u}_0 \in V_p$, according to Remark 4 of Part I) and, secondly,

$$\text{(10)} \qquad \boldsymbol{v} \in V \Rightarrow A(\boldsymbol{v}, \boldsymbol{u}) + a(\boldsymbol{v}, \boldsymbol{u}) = \int_\Omega K_i v_i \, dX + \int_{\mathscr{B}_n} P_n v_n \, dS + \int_{\mathscr{B}_t} \boldsymbol{P}_t \cdot \boldsymbol{v}_t \, dS.$$

Here

$$A(\boldsymbol{v}, \boldsymbol{u}) = \int_\Omega c_{iklm} v_{i,k} u_{l,m} \, dX = A(\boldsymbol{u}, \boldsymbol{v}),$$

with

$$\text{(11)} \qquad A(\boldsymbol{v}, \boldsymbol{v}) \geqq \mu_0 \sum_{i,k=1}^{3} \int_\Omega \varepsilon_{ik}^2(\boldsymbol{v}) \, dX$$

according to (4) and (5). Furthermore

$$\text{(11')} \qquad a(\boldsymbol{v}, \boldsymbol{u}) = \int_{\Gamma_n} \frac{1}{B_n} v_n u_n \, dS + \int_{\Gamma_t} \frac{1}{B_t} \boldsymbol{v}_t \cdot \boldsymbol{u}_t \, dS$$

where

$$\text{(12)} \qquad a(\boldsymbol{v}, \boldsymbol{v}) \geqq 0$$

according to (8); consequently, inequality (1.8) holds.* From (11), the inequality (1.7) follows with the choice

$$N_1 \boldsymbol{v} = v_{1,1}, \quad N_2 \boldsymbol{v} = \frac{1}{\sqrt{2}}(v_{1,2} + v_{2,1}), \quad N_3 \boldsymbol{v} = \frac{1}{\sqrt{2}}(v_{1,3} + v_{3,1}),$$

$$N_4 \boldsymbol{v} = v_{2,2}, \quad N_5 \boldsymbol{v} = \frac{1}{\sqrt{2}}(v_{2,3} + v_{3,2}), \quad N_6 \boldsymbol{v} = v_{3,3}.$$

In fact, in this case $h = 6$ and

$$\sum_{l=1}^{h} \|N_l \boldsymbol{v}\|_{L_2(\Omega)}^2 = \sum_{i,k=1}^{3} \int_\Omega \varepsilon_{ik}^2(\boldsymbol{v}) \, dX,$$

* The formulas of Part I will be denoted by (1.1), (1.2), …, (2.1), (2.2), …, and so forth.

and (11) implies

$$A(\boldsymbol{v},\boldsymbol{v}) \geqq \mu_0 \sum_{l=1}^{h} \|N_l \boldsymbol{v}\|^2_{L_2(\Omega)}$$

as required.

The validity of (1.6) follows from Theorem 3.1 of Part I (see also the Example there). Finally, the functionals on the right-hand side of (10) satisfy the assumptions of Part I because

$$K_i \sim f_s \in L_2(\Omega); \quad P_n n_s, P_{tj}, P_{tj} n_s n_i \in L_2(\Gamma).$$

We now establish the important

Lemma II.1. *Let*

$$\boldsymbol{v} \in [W_2^{(1)}(\Omega)]^3, \quad \sum_{i,k=1}^{3} \int_\Omega \varepsilon_{ik}^2(\boldsymbol{v})\, dX = 0.$$

Then $\boldsymbol{v}$ *has the form*

$$\boldsymbol{v} = \boldsymbol{a} + \boldsymbol{b} \times \boldsymbol{r}$$

where $\boldsymbol{a}, \boldsymbol{b}$ *are constant vectors,* $\boldsymbol{r}$ *denotes the radius vector, and* $\times$ *denotes the vector product.*

Proof. Let us regularize all components of $\boldsymbol{v}$ by means of the function

$$\omega(X,h) = \begin{cases} \exp\left(|X|^2/(|X|^2 - h^2)\right) & \text{for } |X| < h, \\ 0 & \text{for } |X| \geqq h, \end{cases}$$

that is, we introduce a new vector field $\boldsymbol{v}^h$ with components

$$v_i^h(X) = \frac{1}{\kappa h^3} \int_\Omega \omega(X - Y, h)\, v_i(Y)\, dY,$$

where

$$\kappa = \int_{|X|<1} \omega(X,1)\, dX.$$

The derivatives of $\boldsymbol{v}^h$ satisfy the relations

$$(13) \qquad 2\varepsilon_{lm}(\boldsymbol{v}^h) = v_{l,m}^h + v_{m,l}^h = (v_{l,m} + v_{m,l})^h = 0, \qquad l, m = 1, 2, 3,$$

in every region Ω^*, $\bar{\Omega}^* \subset \Omega$, if h is sufficiently small ($h < \operatorname{dist}(\bar{\Omega}^*, \Gamma)$). The functions $v_i^h(X)$ have all derivatives continuous in $\bar{\Omega}$. Differentiating (13), we can show that all derivatives of the second order vanish, except the derivatives of the type $v_{i,jk}^h$, where $i \neq j \neq k$. Furthermore, all derivatives of the third order vanish, and therefore the v_i^h have the form

$$v_i^h = a_i^h + b_{ij}^h x_j + c_{ijk}^h x_j x_k, \qquad i,j,k = 1,2,3; \; i \neq j < k.$$

The coefficients b_{ij}^h, c_{ijk}^h must satisfy the conditions which follow by substitution of this expression into (13). Hence we obtain

$$b_{12}^h = -b_{21}^h = -b_3^h, \qquad b_{13}^h = -b_{31}^h = b_2^h, \qquad b_{23}^h = -b_{32}^h = -b_1^h$$

and the system

$$c_{123}^h + c_{213}^h = 0,$$
$$c_{123}^h + c_{312}^h = 0,$$
$$c_{213}^h + c_{312}^h = 0,$$

which has only the trivial solution. Consequently,

$$\begin{aligned} v_1^h &= a_1^h - b_3^h x_2 + b_2^h x_3, \\ v_2^h &= a_2^h - b_1^h x_3 + b_3^h x_1, \qquad (14) \\ v_3^h &= a_3^h - b_2^h x_1 + b_1^h x_2, \end{aligned}$$

which is equivalent to

$$\boldsymbol{v}^h = \boldsymbol{a}^h + \boldsymbol{b}^h \times \boldsymbol{r}.$$

As v_i^h converge to v_i in $L_2(\Omega)$ for $h \to 0$, and since finite dimensional subsets are closed in $L_2(\Omega)$, (14) holds even for $\boldsymbol{v}(X)$ in every interior subset Ω^*, $\overline{\Omega^*} \subset \Omega$. Hence

$$\boldsymbol{v} = \boldsymbol{a} + \boldsymbol{b} \times \boldsymbol{r} \qquad \text{for } X \in \Omega.$$

Remark II.1. If the function $\boldsymbol{v}$ has the form $\boldsymbol{v} = \boldsymbol{a} + \boldsymbol{b} \times \boldsymbol{r}$ in a Cartesian frame X with radius vector $\boldsymbol{r}(x_1, x_2, x_3)$, then it has similar form in every other Cartesian frame X' with radius vector $\boldsymbol{R}(x_1', x_2', x_3')$. In fact, we have

$$\boldsymbol{r}(X) = \boldsymbol{r}_0 + \boldsymbol{R}(X')$$

where $\boldsymbol{r}_0$ is the radius vector of the origin of X' in the frame X, and therefore

$$\boldsymbol{v}(X) = \boldsymbol{a} + \boldsymbol{b} \times (\boldsymbol{r}_0 + \boldsymbol{R}(X')) = \boldsymbol{a} + \boldsymbol{b} \times \boldsymbol{r}_0 + \boldsymbol{b} \times \boldsymbol{R}(X') = \boldsymbol{v}(X').$$

2. Displacement and Traction Boundary-Value Problems

In this section we shall treat the fundamental boundary-value problems in which either displacements or surface tractions are prescribed on the whole boundary. Let us start with a brief discussion of the

First Boundary-Value Problem (Dirichlet's Type)

Here $\Gamma = \mathscr{A}_u = \mathscr{A}_t$,* $a(\boldsymbol{v}, \boldsymbol{u}) \equiv 0$, $V_p = V$ and the system of functionals $p_i(\boldsymbol{v})$ is vacuous $(r = 0)$; $\mathscr{P} = \{\boldsymbol{0}\}$. Indeed, if $\boldsymbol{v} \in V$ (i.e., $\boldsymbol{v} = \boldsymbol{0}$ on Γ) and $\sum_{i,k=1}^{3} \|\varepsilon_{ik}(\boldsymbol{v})\|_{L_2(\Omega)}^2 = 0$, then according to Lemma II.1 and Remark II.1 $\boldsymbol{v}$ has the form $\boldsymbol{v} = \boldsymbol{a} + \boldsymbol{b} \times \boldsymbol{r}$ in every Cartesian frame. Choosing the coordinate axes along principal axes of inertia of the surface Γ, we obtain from the conditions

$$\int_\Gamma v_i \, dS = 0 \quad \text{and} \quad \int_\Gamma x_j v_i \, dS = 0,$$

that $a_i = 0$ and $b_k = 0$, respectively; consequently $\boldsymbol{v} = \boldsymbol{0}$. The condition (2.9) is therefore satisfied even without introducing any functional $p_i(\boldsymbol{v})$.

* Except for a set of surface measure zero.

Theorem 2.3 implies that KORN's inequality

(15) $$\int_\Omega c_{iklm} v_{i,k} v_{l,m}\, dX \geqq \mu_0 \sum_{i,k=1}^{3} \int_\Omega \varepsilon_{ik}^2(\boldsymbol{v})\, dX \geqq c \|\boldsymbol{v}\|^2_{[W_2^{(1)}(\Omega)]^3}$$

holds for each $\boldsymbol{v} \in V$. According to Theorem 2.2 *the boundary-value problem has precisely one solution* $\boldsymbol{u} \in \boldsymbol{u}_0 \oplus V$, whereupon

$$\|\boldsymbol{u}\|_{[W_2^{(1)}(\Omega)]^3} \leqq c(\|\boldsymbol{u}_0\|_{[W_2^{(1)}(\Omega)]^3} + \|\boldsymbol{K}\|_{[L_2(\Omega)]^3}).$$

Second Boundary-Value Problem (Neumann's Type)

In this case

$$\Gamma = \mathscr{B}_n = \mathscr{B}_t,^* \qquad \Gamma_n = \Gamma_t = \phi.$$

Again $a(\boldsymbol{v}, \boldsymbol{u}) \equiv 0$; furthermore, $\boldsymbol{u}_0 = \boldsymbol{0}$, $V = [W_2^{(1)}(\Omega)]^3$. On the basis of Remark 3 in Part I, we derive

Theorem II.1. *Let*

$$p_i \in ([W_2^{(1)}(\Omega)]^3)', \qquad i = 1, 2, \dots 6,$$

be a system of linear continuous functionals on $[W_2^{(1)}(\Omega)]^3$ *such that if* $\boldsymbol{v} = \boldsymbol{a} + \boldsymbol{b} \times \boldsymbol{r}$ *and* $\sum_{i=1}^{6} (p_i(\boldsymbol{v}))^2 = 0$, *then* $\boldsymbol{v} = \boldsymbol{0}$. *Let* V_p *be the subspace of functions* $\boldsymbol{v} \in [W_2^{(1)}(\Omega)]^3$, *for which* $\sum_{i=1}^{6} (p_i(\boldsymbol{v}))^2 = 0$. *Then* KORN's *inequality* (15) *holds for each* $\boldsymbol{v} \in V_p$.

Proof. If

$$\sum_{i,k=1}^{3} \|\varepsilon_{ik}(\boldsymbol{v})\|^2_{L_2(\Omega)} + \sum_{i=1}^{6} (p_i(\boldsymbol{v}))^2 = 0,$$

then $\|\varepsilon_{ik}(\boldsymbol{v})\|_{L_2(\Omega)} = 0$ for all $i, k = 1, 2, 3$, and by virtue of Lemma II.1,

$$\boldsymbol{v} = \boldsymbol{a} + \boldsymbol{b} \times \boldsymbol{r}.$$

We also have $\sum_{i=1}^{6} (p_i(\boldsymbol{v}))^2 = 0$, so that the supposition implies $\boldsymbol{v} = \boldsymbol{0}$; the condition (2.9) is satisfied and from Remark 3, (15) follows.

The functionals of Theorem II.1 may be chosen in many different forms, for instance

(16) $$\begin{aligned} p_i(\boldsymbol{v}) &= \int_{\Omega^*} v_i\, dX && (i = 1, 2, 3) \\ p_j(\boldsymbol{v}) &= \int_{\Omega^*} (\boldsymbol{r} \times \boldsymbol{v})_{j-3}\, dX && (j = 4, 5, 6); \end{aligned}$$

(17) $$\begin{aligned} p_i(\boldsymbol{v}) &= \int_{\Omega^*} v_i\, dX && (i = 1, 2, 3) \\ p_j(\boldsymbol{v}) &= \int_{\Omega^*} (\operatorname{rot} \boldsymbol{v})_{j-3}\, dX && (j = 4, 5, 6); \end{aligned}$$

---

* Except for a set of surface measure zero.

$$p_i(\boldsymbol{v}) = \int_{\Gamma^*} v_i\, dS \qquad (i=1,2,3)$$

(18)

$$p_j(\boldsymbol{v}) = \int_{\Gamma^*} (\boldsymbol{r}\times\boldsymbol{v})_{j-3}\, dS \qquad (j=4,5,6).$$

Here $\Omega^*\subset\Omega$ denotes an arbitrary measurable set with positive volume measure; Γ^* denotes an arbitrary open part of the boundary Γ' of a Lipschitz region $\Omega'\subset\Omega$. In particular, we may choose $\Omega^*=\Omega$, $\Gamma^*\subset\Gamma$.

Lemma II.2. *Systems* (16), (17) *and* (18) *satisfy the assumptions of Theorem* II.1.

Proof. First we can verify that for all the systems (16), (17) and (18) the conditions $p_i(\boldsymbol{v})=0$, $i=1, 2, \dots 6$, are invariant with respect to Cartesian rectangular frame changes. This property follows for $i=1,2,3$ by insertion of the transformation formula

$$v'_j = c_i^j\, v_i,$$

where c_i^j are coefficients of the (orthonormal) matrix of the transformation, v'_j are coordinates in the new, and v_i coordinates in the old Cartesian system. In fact, we have

$$\int_{M'} v'_j\, dM' = c_i^j \int_M v_i\, dM = 0,$$

where M denotes Ω^* and Γ^*, respectively, and M' the corresponding image of the set M. Similarly,

$$\int_{\Omega^{*\prime}} (\operatorname{rot}\boldsymbol{v})'_k\, dX' = c_i^k \int_{\Omega^*} (\operatorname{rot}\boldsymbol{v})_i\, dX = 0.$$

Inserting $\boldsymbol{r}(X)=\boldsymbol{r}_0+\boldsymbol{R}(X')$ (see Remark II.1) into $p_j(\boldsymbol{v})=0$, $j=4,5,6$, in (16) and (18), respectively, we obtain

$$\int_M (\boldsymbol{r}_0+\boldsymbol{R})\times\boldsymbol{v}\, dM = \int_M \boldsymbol{r}_0\times\boldsymbol{v}\, dM + \int_{M'} (\boldsymbol{R}\times\boldsymbol{v})\, dM' = \boldsymbol{0}.$$

Moreover,

$$\int_M \boldsymbol{r}_0\times\boldsymbol{v}\, dM = \boldsymbol{r}_0\times\int_M \boldsymbol{v}\, dM = \boldsymbol{0};$$

therefore

$$\int_{M'} (\boldsymbol{R}\times\boldsymbol{v})\, dM' = \boldsymbol{0}.$$

Now let us choose the coordinate axes of the coordinate system X' along the principal axes of inertia of the set M. By virtue of the above-mentioned invariance, from the conditions $p_i(\boldsymbol{v})=0$, $i=1,2,\dots 6$, in the original coordinate system, the same conditions follow in the new coordinate system. Inserting $\boldsymbol{v}=\boldsymbol{a}'+\boldsymbol{b}\times\boldsymbol{R}(X')$ into these conditions, we obtain $\boldsymbol{a}'=\boldsymbol{0}$ from the first three and $\boldsymbol{b}=\boldsymbol{0}$ from the remaining three.

The functionals $p_i(\boldsymbol{v})$ are linear and bounded on $[W_2^{(1)}(\Omega)]^3$. This fact can be observed immediately for (16) and (17), because $\Omega^*\subset\Omega$ and Ω is bounded. For (18) we use the continuity of the embedding of $[W_2^{(1)}(\Omega)]^3$ into $[L_2(\Gamma^*)]^3$.

Remark II.2. KORN's inequality (15) for functions which are continuously differentiable on $\bar{\Omega}$ and which annul the functionals (17), was proved by EYDUS (see reference [5] in Part I) on the basis of some results of FRIEDRICHS ([4] in Part I).

The subspace $\mathscr{P}$ of Theorem 2.2 is, according to Lemma II.1, the linear manifold of functions $\boldsymbol{v}=\boldsymbol{a}+\boldsymbol{b}\times\boldsymbol{r}$. The bilinear form

$$((\tilde{\boldsymbol{v}},\tilde{\boldsymbol{u}}))=A(\tilde{\boldsymbol{v}},\tilde{\boldsymbol{u}})=\int_\Omega c_{iklm}\,\varepsilon_{ik}(\tilde{\boldsymbol{v}})\,\varepsilon_{lm}(\tilde{\boldsymbol{u}})\,dX$$

is bounded on the Cartesian product $[W_2^{(1)}(\Omega)]^3/\mathscr{P}\times[W_2^{(1)}(\Omega)]^3/\mathscr{P}$.

Theorem II.2. *Necessary and sufficient conditions for the existence of a solution of the second boundary-value problem are the total equilibrium conditions*

$$\begin{gathered}\int_\Omega \boldsymbol{K}\,dX+\int_\Gamma \boldsymbol{P}\,dS=\boldsymbol{0},\\ \int_\Omega(\boldsymbol{K}\times\boldsymbol{r})\,dX+\int_\Gamma(\boldsymbol{P}\times\boldsymbol{r})\,dS=\boldsymbol{0}.\end{gathered}\tag{19}$$

If they are satisfied, the solution is determined except for a vector-polynomial of the form

$$\boldsymbol{p}=\boldsymbol{a}+\boldsymbol{b}\times\boldsymbol{r},$$

i.e., the solutions form precisely one class $\tilde{\boldsymbol{u}}\in[W_2^{(1)}(\Omega)]^3/\mathscr{P}$.

If we choose the orthogonal projection $\boldsymbol{u}_Q$ *of an arbitrary representative* $\boldsymbol{u}\in\tilde{\boldsymbol{u}}$ *of the class* $\tilde{\boldsymbol{u}}$ *of solutions into the orthogonal complement* $\boldsymbol{Q}$ *of the subspace* $\mathscr{P}$ *in* $[W_2^{(1)}(\Omega)]^3$,

$$[W_2^{(1)}(\Omega)]^3=\mathscr{P}\oplus\boldsymbol{Q},$$

by means of a scalar product, supplying an equivalent norm in $[W_2^{(1)}(\Omega)]^3$, *then the inequality*

$$\|\boldsymbol{u}_Q\|_{[W_2^{(1)}(\Omega)]^3}\leqq\text{const}\left(\|\boldsymbol{K}\|_{[L_2(\Omega)]^3}+\|\boldsymbol{P}\|_{[L_2(\Gamma)]^3}\right)\tag{20}$$

holds.

Proof. The condition (2.4) is equivalent to six conditions

$$\int_\Omega \boldsymbol{K}\cdot\boldsymbol{p}^{(s)}\,dX+\int_\Gamma \boldsymbol{P}\cdot\boldsymbol{p}^{(s)}\,dS=0,\qquad s=1,2,\dots 6,\tag{21}$$

where

$$\begin{aligned}&\boldsymbol{p}^{(1)}=\boldsymbol{i}, &&\boldsymbol{p}^{(2)}=\boldsymbol{j}, &&\boldsymbol{p}^{(3)}=\boldsymbol{k},\\ &\boldsymbol{p}^{(4)}=\boldsymbol{i}\times\boldsymbol{r}, &&\boldsymbol{p}^{(5)}=\boldsymbol{j}\times\boldsymbol{r}, &&\boldsymbol{p}^{(6)}=\boldsymbol{k}\times\boldsymbol{r},\end{aligned}$$

and $\boldsymbol{i},\boldsymbol{j},\boldsymbol{k}$ are the unit vectors of the rectangular Cartesian base of coordinates. In fact, the vectors $\boldsymbol{p}^{(s)}$ form a basis for the subspace $\mathscr{P}$. The relations (21) are equivalent to (19). The rest of the assertion to be proved is an immediate consequence of Theorem 2.2.

Theorem II.3. *Let the system of functionals* p_i *be linearly independent on* $\mathscr{P}$ *and let it satisfy the suppositions of Theorem* II.1. *Then the solution of the second boundary-value problem exists in* V_p, *if and only if the conditions of the total equilibrium* (19) *are met. The solution is unique in* V_p *and the inequality* (20) *holds for it.*

Proof. The system of functionals satisfying the suppositions of Theorem II.1 meet the condition (2.9), and (19) are equivalent to the condition (2.4), as was shown in the proofs of Theorems II.1 and II.2. Hence the assertion follows with use of Theorem 2.2 and Remark 4.

Corollary II.1. *In the space V_p, defined by means of the functionals* (16), (17) *or* (18), *precisely one solution to the second boundary-value problem exists and inequality* (20) *holds, if and only if the conditions of the total equilibrium* (19) *are met.*

Proof. With regard to Lemma II.2 it remains only to prove that the functionals are linearly independent on $\mathscr{P}$. First we shall consider the types (16) and (18). Let

$$\sum_{i=1}^{6} \alpha_i p_i(\boldsymbol{p}) = 0 \tag{22}$$

hold for each $\boldsymbol{p} \in \mathscr{P}$. Let $\boldsymbol{\alpha} = (\alpha_1, \alpha_2, \alpha_3)$, $\boldsymbol{\beta} = (\alpha_4, \alpha_5, \alpha_6)$. Then (22) may be re-written in the form

$$\int_M [\boldsymbol{\alpha} \cdot \boldsymbol{p} + \boldsymbol{\beta} \cdot (\boldsymbol{r} \times \boldsymbol{p})]\, dM = 0$$

or

$$\int_M \boldsymbol{p} \cdot [\boldsymbol{\alpha} + \boldsymbol{\beta} \times \boldsymbol{r}]\, dM = 0 .$$

Choosing $\boldsymbol{p} = \boldsymbol{\alpha} + \boldsymbol{\beta} \times \boldsymbol{r}$, we obtain conditions

$$\boldsymbol{\alpha} + \boldsymbol{\beta} \times \boldsymbol{r} = \boldsymbol{0}$$

for each $X \in M$, which represent a system of three linear equations for x_1, x_2, x_3. If $|\boldsymbol{\alpha}|^2 + |\boldsymbol{\beta}|^2 > 0$, only the following cases would be possible: (a) the solution of this system does not exist, (b) the solutions correspond with one straight-line. By virtue of the suppositions on the set M we obtain a contradiction; therefore $\boldsymbol{\alpha} = \boldsymbol{\beta} = \boldsymbol{0}$.

For the type (17) we choose $\boldsymbol{p} = \boldsymbol{p}^{(s)}$, $s = 1, \ldots, 6$ (see the proof of Theorem II.2) and insert this into (22). Thus $\alpha_i = 0$ follows immediately for $i = 1, 2, \ldots 6$.

3. Mixed Boundary-Value Problem

Next let us consider some particular cases of the conditions (6), namely

$$\Gamma = \Gamma_u \cup \Gamma_P \cup \Gamma_K \cup \Gamma_V \cup \Gamma_n \cup \Gamma_t \cup \mathscr{N}, \qquad \operatorname{mes} \mathscr{N} = 0,^*$$

where Γ_u, Γ_P, Γ_K, Γ_V, $\Gamma_n \cup \Gamma_t$ are mutually disjoint sets, which are either empty or open in Γ. On Γ_u a displacement vector $\boldsymbol{u}_0$ is prescribed; on Γ_P surface tractions $\boldsymbol{P}$; on Γ_K the contact conditions

$$u_n = 0, \qquad \boldsymbol{T}_t = 0$$

are given; on Γ_V the conditions, corresponding for example with the torsion problem of a cylinder on its end, i.e.,

$$\boldsymbol{u}_t = \boldsymbol{u}_{0t}, \qquad T_n = 0,$$

* i.e., $\mathscr{N}$ is a set of surface measure zero.

are prescribed, and on Γ_n, Γ_t we have the conditions of elastic supports:

$$(23)\qquad \begin{aligned} u_n &= -B_n T_n, & \boldsymbol{T}_t &= \boldsymbol{0} \quad \text{on } \Gamma_n, \\ \boldsymbol{u}_t &= -B_t \boldsymbol{T}_t, & T_n &= 0 \quad \text{on } \Gamma_t, \quad \text{where } \Gamma_n \cap \Gamma_t = \phi. \end{aligned}$$

Consequently

$$\mathscr{A}_n = \Gamma_u \cup \Gamma_K, \qquad \mathscr{A}_t = \Gamma_u \cup \Gamma_V,$$

$$\mathscr{B}_n = \Gamma_P \cup \Gamma_V \cup \Gamma_n \cup \Gamma_t, \qquad \mathscr{B}_t = \Gamma_P \cup \Gamma_K \cup \Gamma_n \cup \Gamma_t.$$

Clearly, the conditions (23) do not express all the possibilities of elastic supports of the type (6). Thus, for example, further kinds of elastic supports are given by

$$(23')\qquad \begin{aligned} u_n &= -B_n T_n, & \boldsymbol{u}_t &= \boldsymbol{0} \quad \text{on } \Gamma_n, \\ \boldsymbol{u}_t &= -B_t \boldsymbol{T}_t, & u_n &= 0 \quad \text{on } \Gamma_t \end{aligned}$$

or

$$(23'')\qquad \begin{aligned} u_n &= -B_n T_n, \\ \boldsymbol{u}_t &= -B_t \boldsymbol{T}_t \quad \text{on } \Gamma_n = \Gamma_t \end{aligned}$$

or by means of nonhomogeneous conditions of the above-mentioned kinds. For (23′) we have

$$\mathscr{A}_n = \Gamma_u \cup \Gamma_K \cup \Gamma_t, \qquad \mathscr{A}_t = \Gamma_u \cup \Gamma_V \cup \Gamma_n,$$

$$\mathscr{B}_n = \Gamma_P \cup \Gamma_V \cup \Gamma_n, \qquad \mathscr{B}_t = \Gamma_P \cup \Gamma_K \cup \Gamma_t$$

and for (23″)

$$\mathscr{A}_n = \Gamma_u \cup \Gamma_K, \qquad \mathscr{A}_t = \Gamma_u \cup \Gamma_V,$$

$$\mathscr{B}_n = \Gamma_P \cup \Gamma_V \cup \Gamma_n, \qquad \mathscr{B}_t = \Gamma_P \cup \Gamma_K \cup \Gamma_n.$$

We shall distinguish the boundary-value problems according to the character of $\mathscr{A}_n \cup \mathscr{A}_t \cup \Gamma_n \cup \Gamma_t$. An important case occurs, when homogeneous conditions on $\mathscr{A}_n \cup \mathscr{A}_t$ and $a(\boldsymbol{v}, \boldsymbol{v}) = 0$ together exclude the possibility of rigid body motions, i.e., when

$$(24)\qquad \big(\boldsymbol{v} \in V,\ \boldsymbol{v} = \boldsymbol{a} + \boldsymbol{b} \times \boldsymbol{r},\ a(\boldsymbol{v}, \boldsymbol{v}) = 0\big) \Rightarrow \boldsymbol{v} = \boldsymbol{0}.$$

In the following we shall obtain some sufficient conditions for (24). The simplest of them is

Lemma II.3. *If Γ_u is a set open in Γ, then* (24) *holds.*

Proof. As $\boldsymbol{a} + \boldsymbol{b} \times \boldsymbol{r} = \boldsymbol{0}$ on an open set in Γ, the rank of the matrix (compare (14))

$$(25)\qquad \begin{pmatrix} 0 & b_3 & -b_2 & a_1 \\ -b_3 & 0 & b_1 & a_2 \\ b_2 & -b_1 & 0 & a_3 \end{pmatrix}$$

is at most one. Then all the minor determinants of the second degree vanish and consequently $b_1 = b_2 = b_3 = 0$. By insertion of this result into the condition $\boldsymbol{v} = \boldsymbol{0}$ on Γ_u, the result $a_1 = a_2 = a_3 = 0$ follows.

Lemma II.4. *Let $\boldsymbol{v} = \boldsymbol{a} + \boldsymbol{b} \times \boldsymbol{r}$ and $v_n = 0$ on an open part F of the plane $x_3 = \text{const}$. Then $\boldsymbol{v} = b_1 \boldsymbol{i} + a_2 \boldsymbol{j} + b_3 \boldsymbol{k} \times \boldsymbol{r}$.*

The proof follows immediately from the condition

$$v_n = \pm \boldsymbol{k} \cdot (\boldsymbol{a} + \boldsymbol{b} \times \boldsymbol{r}) = \pm (a_3 + b_2 x_1 - b_1 x_2) = 0 \quad \text{on } F.$$

Lemma II.5. *Let $\boldsymbol{v} = \boldsymbol{a} + \boldsymbol{b} \times \boldsymbol{r}$ and $v_n = 0$ on an open part F of a circular cylinder with axis x_3. Then*

$$\boldsymbol{v} = a_3 \boldsymbol{k} + b_3 \boldsymbol{k} \times \boldsymbol{r}.$$

Proof. Introducing cylindrical coordinates

$$x_1 = \rho \cos \varphi, \qquad x_2 = \rho \sin \varphi, \qquad x_3 = z,$$

the condition $v_n = 0$ on F reads

$$a_1 \cos \varphi + a_2 \sin \varphi + (b_2 \cos \varphi - b_1 \sin \varphi) z = 0,$$

which implies

$$a_1 = a_2 = b_1 = b_2 = 0$$

(compare also the proof of Lemma II.7).

Lemma II.6. *Let $\boldsymbol{v} = \boldsymbol{a} + \boldsymbol{b} \times \boldsymbol{r}$ and $v_n = 0$ on an open part F of a spherical surface with the centre in the origin. Then*

$$\boldsymbol{v} = \boldsymbol{b} \times \boldsymbol{r}.$$

Proof. On F there holds

$$v_n = C \boldsymbol{r} \cdot (\boldsymbol{a} + \boldsymbol{b} \times \boldsymbol{r}) = 0,$$

i.e., $\boldsymbol{a} \cdot \boldsymbol{r} = 0$ and as F is open, $\boldsymbol{a} = \boldsymbol{0}$.

Lemma II.7. *Let $\boldsymbol{v} = \boldsymbol{a} + \boldsymbol{b} \times \boldsymbol{r}$ and $v_n = 0$ on an open part F of a smooth general surface of rotation, excluding planes, circular cylinders and spheres. Then, if the axis of rotation is identified with the axis x_3,*

$$\boldsymbol{v} = b_3 \boldsymbol{k} \times \boldsymbol{r}.$$

Proof. Introducing cylindrical coordinates, we may write

$$v_n = a_\rho n_\rho + a_z n_z + b_\varphi (z n_\rho - \rho n_z), \tag{26}$$

where

$$a_\rho = \boldsymbol{a} \cdot \boldsymbol{e}_\rho = a_1 \cos \varphi + a_2 \sin \varphi,$$

$$a_z = \boldsymbol{a} \cdot \boldsymbol{k} = a_3,$$

$$b_\varphi = \boldsymbol{b} \cdot \boldsymbol{e}_\varphi = -b_1 \sin \varphi + b_2 \cos \varphi,$$

$\boldsymbol{e}_\rho$ and $\boldsymbol{e}_\varphi$ are unit vectors in the radial and tangential directions, respectively,

$$n_\rho = \cos(\boldsymbol{n}, \boldsymbol{e}_\rho), \qquad n_z = \cos(\boldsymbol{n}, \boldsymbol{k}).$$

Let the part F of the surface be given by

$$\rho = \rho(z) \quad \text{or} \quad z = z(\rho),$$

where $d\rho/dz$ and $dz/d\rho$ are continuous on bounded intervals $z \in \mathcal{O}_z \subset (-\infty, \infty)$ and $\rho \in \mathcal{O}_\rho \subset \langle 0, \infty)$, respectively. As F is open in the surface of rotation, we may suppose that all points $(\varphi, z, \rho(z))$ and $(\varphi, z(\rho), \rho)$, respectively, where $\varphi \in I_\varphi$, $z \in \mathcal{O}_z$ and $\rho \in \mathcal{O}_\rho$ (I_φ means an open interval), belong to F. We have

$$n_\rho = \pm(1+\rho'^2)^{-\frac{1}{2}}, \quad n_z = \mp\rho'(1+\rho'^2)^{-\frac{1}{2}}$$

and

$$n_\rho = \pm z'(1+z'^2)^{-\frac{1}{2}}, \quad n_z = \mp(1+z'^2)^{-\frac{1}{2}},$$

respectively, and the condition $v_n = 0$ may be expressed according to (26) in the forms

$$(27') \quad a_1 \cos\varphi + a_2 \sin\varphi - a_3 \rho'(z) + (b_2 \cos\varphi - b_1 \sin\varphi)\big(z + \rho(z)\rho'(z)\big) = 0$$

and

$$(27'') \quad (a_1 \cos\varphi + a_2 \sin\varphi) z'(\rho) - a_3 + (b_2 \cos\varphi - b_1 \sin\varphi)\big(\rho + z(\rho) z'(\rho)\big) = 0.$$

From the linear independence of the functions 1, $\cos\varphi$, $\sin\varphi$ on the interval I_φ and from (27′) follows

$$a_3 \rho'(z) = 0,$$
$$a_1 + b_2\big(z + \rho(z)\rho'(z)\big) = 0,$$
$$a_2 - b_1\big(z + \rho(z)\rho'(z)\big) = 0$$

for all $z \in \mathcal{O}_z$. As F is not cylindrical, $\rho'(z) \neq 0$, and therefore $a_3 = 0$. Let the functions 1, $z + \rho(z)\rho'(z)$ be linearly dependent on $\mathcal{O}_z$. Then the differential equation

$$\rho \frac{d\rho}{dz} + z + C = 0$$

(C=const.) holds on $\mathcal{O}_z$, whose solutions are circles

$$(28) \qquad \rho^2 + (z+C)^2 = D$$

generating spherical surfaces. Hence the above-mentioned functions are linearly independent, and consequently

$$a_1 = a_2 = b_1 = b_2 = 0.$$

From (27″) follows similarly

$$a_3 = 0,$$
$$a_1 z'(\rho) + b_2\big(\rho + z(\rho) z'(\rho)\big) = 0,$$
$$a_2 z'(\rho) - b_1\big(\rho + z(\rho) z'(\rho)\big) = 0$$

for all $\rho \in \mathcal{O}_\rho$. Let the functions $z'(\rho)$, $\rho + z'(\rho) z(\rho)$ be linearly dependent on $\mathcal{O}_\rho$. As the planes orthogonal to the axis of rotation are excluded, $z'(\rho) \neq 0$ and we may write

$$\rho + z \frac{dz}{d\rho} + C_1 \frac{dz}{d\rho} = 0$$

(C_1 =const.) for $\rho \in \mathcal{O}_\rho$. The solutions of this equation are again circles (28), generating spheres. Hence the linear independence of the above-mentioned functions follows and consequently, $a_1 = a_2 = b_1 = b_2 = 0$.

Lemma II.8. *Let $\boldsymbol{v}=\boldsymbol{a}+\boldsymbol{b}\times\boldsymbol{r}$ and $v_n=0$ on an open part F of a smooth general cylindrical surface parallel with the axis x_3, excluding planes and circular cylinders. Then*

$$\boldsymbol{v}=a_3\,\boldsymbol{k}.$$

Proof. Let F be given by

$$x_2=f(x_1) \quad \text{or} \quad x_1=g(x_2),$$

where f and g are continuously differentiable on bounded open intervals $x_1\in\mathcal{O}_1$ and $x_2\in\mathcal{O}_2$, respectively. As F is open in the cylindrical surface, we may suppose that all points $(x_1, f(x_1), x_3)$ and $(g(x_2), x_2, x_3)$, respectively, where $x_1\in\mathcal{O}_1$, $x_2\in\mathcal{O}_2$ and $x_3\in I_3$ (I_3 denotes an open interval), belong to F.

The condition $v_n=0$ on F may be written in the forms

$$a_1+b_2x_3+g'(x_2)(-a_2+b_1x_3)-b_3\big(x_2+g(x_2)\,g'(x_2)\big)=0$$

and

$$-a_2+b_1x_3+f'(x_1)(a_1+b_2x_3)-b_3\big(x_1+f(x_1)\,f'(x_1)\big)=0.$$

Similarly to the proof of Lemma II.7 we may prove that the functions 1, g', $x_2+g\,g'$ and $1, f'$, x_1+ff' are linearly independent on $\mathcal{O}_2$ and $\mathcal{O}_1$, respectively, and consequently $a_1=a_2=b_1=b_2=b_3=0$, $\boldsymbol{v}=a_3\boldsymbol{k}$.

In fact, if these functions are linearly dependent, there holds

$$\alpha_1+\alpha_2\,g'+\alpha_3(x_2+g\,g')=0 \qquad \text{for } x_2\in\mathcal{O}_2.$$

If the functions $1, g'(x_2)$ are linearly dependent, we would have

$$g'+C=0,$$

yielding the straight lines $g(x_2)=x_1=C\,x_2+\bar{C}$, generating planes. Thus the functions 1, $g'(x_2)$ are linearly independent and $\alpha_3\neq 0$, so that we obtain the differential equation

$$g\,g'+x_2+C_1\,g'+C_2=0,$$

whose solutions are circles

$$(x_1+C_1)^2+(x_2+C_2)^2=C_3,$$

generating the circular cylinders. The same procedure may be used for $x_2=f(x_1)$.

Lemma II.9. *Let $\boldsymbol{v}=\boldsymbol{a}+\boldsymbol{b}\times\boldsymbol{r}$ and $v_n=0$ on an open part F of a smooth helicoidal surface with the axis x_3, which is given by*

$$x_1=\rho\cos\varphi, \qquad x_2=\rho\sin\varphi, \qquad x_3=f(\rho)+h\,\varphi, \tag{29}$$

where ρ, φ are cylindrical coordinates, $h\neq 0$ and $f(\rho)$ is continuously differentiable in an interval $\mathcal{O}_\rho\subset\langle 0,\infty)$. Then

$$\boldsymbol{v}=b_3(h\,\boldsymbol{k}+\boldsymbol{k}\times\boldsymbol{r}).$$

Proof. The components of the unit normal to the surface (29) have the form

$$n_1 = \frac{1}{\Delta}(h \sin\varphi - \rho f'(\rho)\cos\varphi),$$

$$n_2 = -\frac{1}{\Delta}(h\cos\varphi + \rho f'(\rho)\sin\varphi), \qquad n_3 = \frac{\rho}{\Delta},$$

with $\Delta = (h^2 + \rho^2(1+f'^2))^{\frac{1}{2}}$, $\rho \in \mathcal{O}_\rho$, $\varphi \in I_\varphi$, (I_φ is an open interval).

The condition $v_n = 0$ on F may be rewritten in the form

$$\begin{aligned}\rho(a_3 - h b_3) + [-h a_2 - b_2 \rho(\rho + f f') + b_1 h f - a_1 \rho f']\cos\varphi \\ + [h a_1 + b_1 \rho(\rho + f f') + b_2 h f - a_2 \rho f']\sin\varphi \\ + (h b_1 - h b_2 \rho f')\varphi\cos\varphi + (h b_2 + h b_1 \rho f')\varphi \sin\varphi = 0.\end{aligned}$$

As the functions 1, $\cos\varphi$, $\sin\varphi$, $\varphi\cos\varphi$, $\varphi\sin\varphi$ are linearly independent on I_φ, their coefficients must vanish for all $\rho \in \mathcal{O}_\rho$. Thus we derive, for example, the system

$$h(b_1 - b_2 \rho f'(\rho)) = 0,$$
$$h(b_2 + b_1 \rho f'(\rho)) = 0,$$

whose determinant is

$$h(1 + \rho^2 f'^2(\rho)) \neq 0;$$

consequently $b_1 = b_2 = 0$. Quite analogously, from the system

$$h a_2 + a_1 \rho f'(\rho) = 0,$$
$$h a_1 - a_2 \rho f'(\rho) = 0,$$

$a_1 = a_2 = 0$ follows, and finally we obtain $a_3 = h b_3$. Hence

$$\boldsymbol{v} = a_3 \boldsymbol{k} + b_3 \boldsymbol{k} \times \boldsymbol{r} = h b_3 \boldsymbol{k} + b_3 \boldsymbol{k} \times \boldsymbol{r}.$$

Henceforth, let Γ_u be empty. In order to determine the subspace $\mathscr{P}$ of Theorems 2.1, 2.2, let us consider besides $\boldsymbol{v} \in \boldsymbol{V}$, $\boldsymbol{v} = \boldsymbol{a} + \boldsymbol{b} \times \boldsymbol{r}$ the condition $a(\boldsymbol{v}, \boldsymbol{v}) = 0$. Having regard to (11′), there follows $v_n = 0$ on Γ_n, $\boldsymbol{v}_t = \boldsymbol{0}$ on Γ_t. Thus the condition $v_n = 0$ is to be satisfied by $\boldsymbol{v} \in \mathscr{P}$ on $\mathscr{A}_n \cup \Gamma_n$, the condition $\boldsymbol{v}_t = \boldsymbol{0}$ on $\mathscr{A}_t \cup \Gamma_t$. As the general analysis, based on both these boundary conditions, would be too complicated, we shall restrict our considerations to the first condition $v_n = 0$.

Let us always suppose that $\mathscr{A}_n \cup \Gamma_n$ is a sum of a finite number of portions F_m of smooth surfaces,

$$\mathscr{A}_n \cup \Gamma_n = \bigcup_{m=1}^{M} F_m,$$

such that each F_m is open in Γ and can be described by

$$f_m(x_1, x_2, x_3) = 0,$$

where f_m is continuously differentiable and

$$\sum_{i=1}^{3} \left(\frac{\partial f_m}{\partial x_i}\right)^2 > 0 \tag{30}$$

for $X \in F_m$.

In the following we shall investigate the number and kind of the "degrees of freedom" of the given rigid body, under prescribed homogeneous boundary conditions for displacement on $\mathscr{A}_n$, ($\Gamma_u = \phi$), and replacing the originally elastic supports on Γ_n by "stiffened" ones, for which $v_n = 0$ on Γ_n. In other words, we seek the most general form of the function $\boldsymbol{v} = \boldsymbol{a} + \boldsymbol{b} \times \boldsymbol{r}$, restricted by means of the condition $v_n = 0$ on $\mathscr{A}_n \cup \Gamma_n$. Thus we are omitting the influence of the condition $\boldsymbol{v}_t = \boldsymbol{0}$ on $\mathscr{A}_t \cup \Gamma_t$, which must be studied separately and superposed upon the results of the analysis mentioned above.

Theorem II.4. *Let at least one of the F_m contain a smooth portion F_0, open in Γ, such that*

(a) *it does not belong to any helicoidal surface, surface of rotation or cylinder,**

(b) *each open part of F_0 has the property* (a).

Then

$$\{A \equiv (\boldsymbol{v} = \boldsymbol{a} + \boldsymbol{b} \times \boldsymbol{r},\ v_n = 0 \ \ on\ \ \mathscr{A}_n \cup \Gamma_n)\} \ \Rightarrow\ \boldsymbol{v} = \boldsymbol{0}. \tag{31}$$

Proof. Let $|\boldsymbol{b}| > 0$. Choose the rectangular system (x, y, z) so that $\boldsymbol{b} = b_3 \boldsymbol{k}$. The condition $v_n = 0$ on F_0 yields

$$(a_1 - b_3 y)\frac{\partial \tilde{f}}{\partial x} + (a_2 + b_3 x)\frac{\partial \tilde{f}}{\partial y} + a_3 \frac{\partial \tilde{f}}{\partial z} = 0. \tag{F}$$

Let us consider the corresponding system

$$\frac{dx}{a_1 - b_3 y} = \frac{dy}{a_2 + b_3 x} = \frac{dz}{a_3}.$$

By transition to the cylindrical coordinates

$$x = -\frac{a_2}{b_3} + \rho \cos\varphi, \qquad y = \frac{a_1}{b_3} + \rho \sin\varphi, \qquad z = z$$

we find two first integrals

$$\rho = C_1, \qquad z - \frac{a_3}{b_3}\varphi = C_2.$$

Therefore, the general solution of (F) has the form

$$\Phi\left(\rho, z - \frac{a_3}{b_3}\varphi\right) = 0$$

where $\Phi(\xi, \eta)$ is an arbitrary differentiable function. From the relation

$$f_m(x_1, x_2, x_3) = \tilde{f}\,(x(\rho, \varphi), y(\rho, \varphi), z) = \Phi\left(\rho, z - \frac{a_3}{b_3}\varphi\right)$$

---

* Here the term helicoidal surface is meant in the sense of (29), surface of rotation in the general sense, including planes (orthogonal to the axis of rotation), circular cylinders and spheres, and the term cylinder is meant also in the general sense, including planes and circular cylinders.

we infer that the first derivatives of Φ are continuous and

$$\left(\frac{\partial\phi}{\partial\xi}\right)^2+\left(\frac{\partial\phi}{\partial\eta}\right)^2>0$$

for $\xi\neq 0$. Therefore, two and only two cases are possible:

(i) An open part of F_0 may be expressed by

$$\rho=\nu\left(z-\frac{a_3}{b_3}\varphi\right)$$

where $\nu(\eta)$ is continuously differentiable. If $d\nu/d\eta=0$ in an interval, then $\rho=$const., yielding a part of a circular cylinder. If $d\nu/d\eta\neq 0$ in an interval, then we may write inversely

$$z=\frac{a_3}{b_3}\varphi+\mu(\rho)$$

where $\mu(\rho)$ is continuously differentiable, yielding a part of either a helicoidal surface (for $a_3\neq 0$) or a surface of rotation (for $a_3=0$).

(ii) An open part of F_0 may be expressed by

$$z-\frac{a_3}{b_3}\varphi=\mu(\rho)\,.$$

This case leads again to a part of a helicoidal surface or of a surface of rotation.

Second, let $\boldsymbol{b}=\boldsymbol{0}$, $|\boldsymbol{a}|>0$. By an analogous method we find that the general solution of (F) represents general cylinders, parallel to the vector $\boldsymbol{a}$.

Thus in all cases, we obtain a contradiction with the suppositions (a) and (b); consequently $\boldsymbol{a}=\boldsymbol{b}=\boldsymbol{0}$, $\boldsymbol{v}=\boldsymbol{0}$.

Theorem II.5. *Let F_1 be a part of a surface of rotation with the axis x_3 (excluding planes, circular cylinders and spheres as in Lemma* II.7). *Then either* (31) *holds or*

$$A\;\Rightarrow\;\boldsymbol{v}=b_3\,\boldsymbol{k}\times\boldsymbol{r}\,.$$

(31) *fails to hold if and only if either all the remaining F_m belong to surfaces of rotation with the axis x_3 (including cylindrical or spherical surfaces and planes $x_3=$const.) or $\mathscr{A}_n\cup\Gamma_n=F_1$.*

Theorem II.6. *Let F_1 be a part of a general cylinder, parallel with the axis x_3 (excluding planes and circular cylinders as in Lemma* II.8). *Then either* (31) *holds or*

$$A\;\Rightarrow\;\boldsymbol{v}=a_3\,\boldsymbol{k}\,.$$

(31) *fails to hold if and only if either all the remaining F_m belong to cylinders parallel with the x_3 axis (including planes and circular cylinders), or $\mathscr{A}_n\cup\Gamma_n=F_1$.*

Theorem II.7. *Let F_1 be a part of a helicoidal surface with the axis x_3 (in the sense of* (29)). *Then either* (31) *holds or*

$$A\;\Rightarrow\;\boldsymbol{v}=h\,b_3\,\boldsymbol{k}+b_3\,\boldsymbol{k}\times\boldsymbol{r}\,.$$

(31) *fails to hold if and only if either all the remaining* F_m *consist of portions* of helicoidal surfaces, described by means of the relation* (29) *with the same* h *and with the axis* x_3, *and circular cylinders with the axis* x_3, *or* $\mathscr{A}_n \cup \Gamma_n = F_1$.

Theorem II.8. *Let all* F_m *belong to planes, circular cylinders and spheres. Let at least one* F_m *be a part of a circular cylinder with the axis* x_3. *If*

α) *all* F_m *are parts of cylinders with the common axis* x_3, *then*

$$A \Rightarrow \boldsymbol{v} = a_3 \boldsymbol{k} + b_3 \boldsymbol{k} \times \boldsymbol{r};$$

β) *if all* F_m *consist of portions of circular cylinders with the axis* x_3, *spherical surfaces with the centre on the* x_3*-axis and planes orthogonal to the* x_3*-axis, and at least one* F_m *is not cylindrical, then*

$$A \Rightarrow \boldsymbol{v} = b_3 \boldsymbol{k} \times \boldsymbol{r};$$

γ) *if all* F_m *consist of portions of planes and circular cylinders parallel with the* x_3*-axis, and at least one* F_m *is not a cylinder with axis* x_3, *then*

$$A \Rightarrow \boldsymbol{v} = a_3 \boldsymbol{k}.$$

In all other cases (different from α, β, γ) (31) *holds.*

Theorem II.9. *Let all* F_m *belong to planes and spherical surfaces. Let at least one* F_m *be a part of a sphere with the centre at the origin. If*

α) *all* F_m *are parts of spheres with the centre at the origin, then*

$$A \Rightarrow \boldsymbol{v} = \boldsymbol{b} \times \boldsymbol{r};$$

β) *if all* F_m *belong to spheres with the centres on the* x_3*-axis and to planes* $x_3 = \text{const.}$, *and at least one* F_m *is not spherical with the centre at the origin, then*

$$A \Rightarrow \boldsymbol{v} = b_3 \boldsymbol{k} \times \boldsymbol{r}.$$

In all other cases (different from α, β) (31) *holds.*

Theorem II.10. *Let all* F_m *be parts of planes. If*

α) *all* F_m *are orthogonal to the* x_3*-axis, then*

$$A \Rightarrow \boldsymbol{v} = a_1 \boldsymbol{i} + a_2 \boldsymbol{j} + b_3 \boldsymbol{k} \times \boldsymbol{r};$$

β) *if all* F_m *are parallel with the* x_3*-axis and at least two of them intersect each other, then*

$$A \Rightarrow \boldsymbol{v} = a_3 \boldsymbol{k}.$$

In all other cases (different from α, β) (31) *holds.*

Remark II.3. It is evident that the supposition on the position of F_m with respect to the x_3-axis (or to the origin) cannot mean a loss of generality. In fact,

---

* We say that F_m consists of portions S_k, if

$$F_m = \bigcup_k S_k \cup \mathscr{N},$$

where $\mathscr{N}$ is a set of surface measure zero, S_k are mutually disjoint, open in Γ.

by means of a suitable change of coordinate system, it is possible to reach a position in which the x_3-axis either represents the axis of rotation or is parallel or orthogonal to F_m (see also Remark II.1).

Proofs of Theorems II.5–II.9 follow easily from Theorem II.4 and Lemmas II.4–II.9 with use of a counterpart of Lemma II.3. As an example, let us prove Theorem II.5: According to Lemma II.7 we have $\boldsymbol{v}=b_3\boldsymbol{k}\times\boldsymbol{r}$ and the question arises, whether b_3 vanishes or need not vanish. Comparing the function $b_3\boldsymbol{k}\times\boldsymbol{r}$ with the following functions

$$b_3'\boldsymbol{k}'\times(\boldsymbol{r}-\boldsymbol{r}_0),\quad a_3'\boldsymbol{k}',\quad h\,b_3'\boldsymbol{k}'+b_3'\boldsymbol{k}'\times(\boldsymbol{r}-\boldsymbol{r}_0),$$

$$a_3'\boldsymbol{k}'+b_3'\boldsymbol{k}'\times(\boldsymbol{r}-\boldsymbol{r}_0),\quad \boldsymbol{b}'\times(\boldsymbol{r}-\boldsymbol{r}_0),\quad a_1'\boldsymbol{i}'+a_2'\boldsymbol{j}'+b_3'\boldsymbol{k}'\times(\boldsymbol{r}-\boldsymbol{r}_0),$$

using a counterpart of Lemma II.3 and Theorem II.4, we obtain the assertion to be proved.

Theorem II.11. *Let* (24) *hold. Then the generalized Korn inequality*

$$\begin{aligned}(32)\quad &\int_\Omega c_{iklm}v_{i,k}v_{l,m}\,dX+\int_{\Gamma_n}\frac{1}{B_n}v_n^2\,dS+\int_{\Gamma_t}\frac{1}{B_t}\boldsymbol{v}_t^2\,dS\\ &\geqq c_1\left[\sum_{i,k=1}^{3}\int_\Omega\varepsilon_{ik}^2(\boldsymbol{v})\,dX+\int_{\Gamma_n}\frac{1}{B_n}v_n^2\,dS+\int_{\Gamma_t}\frac{1}{B_t}\boldsymbol{v}_t^2\,dS\right]\geqq c_2\|\boldsymbol{v}\|^2_{[W_2^{(1)}(\Omega)]^3}\end{aligned}$$

holds for every $\boldsymbol{v}\in V$, *one and only one solution of the mixed boundary-value problem exists, and the inequality*

$$(33)\quad \|\boldsymbol{u}\|_{[W_2^{(1)}(\Omega)]^3}\leqq c(\|\boldsymbol{u}_0\|_{[W_2^{(1)}(\Omega)]^3}+\|\boldsymbol{K}\|_{[L_2(\Omega)]^3}+\|P_n\|_{L_2(\mathscr{B}_n)}+\|\boldsymbol{P}_t\|_{[L_2(\mathscr{B}_t)]^3})$$

holds for this solution.

Proof. (24) implies that $\mathscr{P}=\{\boldsymbol{0}\}$ according to the definition in Section 2 of Part I. Then (32) follows from Theorem 2.1; existence, uniqueness and inequality (33) for the solution is a consequence of Theorem 2.2.

Theorem II.12. *Let* Γ_u *be non-empty or let* $\mathscr{A}_n\cup\Gamma_n$ *be such that it satisfies any of the conditions of Theorems* II.4–II.10, *sufficient for* (31). *Then the generalized Korn's inequality* (32), *existence and uniqueness of a solution* $\boldsymbol{u}\in\boldsymbol{u}_0\oplus V$ *of the mixed boundary-value problem and inequality* (33) *hold.*

Proof. If Γ_u is non-empty, it represents an open set in Γ (see the assumptions at the beginning of this section) and by virtue of Lemma II.3 (24) holds. If Γ_u is empty, then one of the Theorems II.4–II.10 implies (31). Using, moreover, the definition of the space V and (11′), we obtain (24). Finally, we apply Theorem II.11.

Remark II.4. It is necessary to mention that even if Γ_u is empty and (31) fails to hold, (24) may be valid, if the "degrees of freedom" of Theorems II.5–II.10 are eliminated by the condition $\boldsymbol{v}_t=\boldsymbol{0}$ on $\Gamma_t\cup\mathscr{A}_t$. We shall not discuss these possibilities here.

Theorem II.13. *Let* $\Gamma_u=\phi$ *and* $\Gamma_n\cup\Gamma_K\cup\Gamma_t\cup\Gamma_V$ *be such that*

$$\boldsymbol{v}\in V,\quad \boldsymbol{v}=\boldsymbol{a}+\boldsymbol{b}\times\boldsymbol{r},\quad v_n=0\ \text{on}\ \Gamma_n,\quad \boldsymbol{v}_t=\boldsymbol{0}\ \text{on}\ \Gamma_t$$

holds precisely if $\boldsymbol{v}=\boldsymbol{p}^{(\sigma)}$, *where six different cases* ($\sigma=$a, b, c, d, e, f) *occur:*

a) $$\boldsymbol{p}^{(a)}=b_3\,\boldsymbol{k}\times\boldsymbol{r},$$

b) $$\boldsymbol{p}^{(b)}=a_3\,\boldsymbol{k},$$

c) $$\boldsymbol{p}^{(c)}=h\,b_3\,\boldsymbol{k}+b_3\,\boldsymbol{k}\times\boldsymbol{r},$$

d) $$\boldsymbol{p}^{(d)}=a_3\,\boldsymbol{k}+b_3\,\boldsymbol{k}\times\boldsymbol{r},$$

e) $$\boldsymbol{p}^{(e)}=\boldsymbol{b}\times\boldsymbol{r},$$

f) $$\boldsymbol{p}^{(f)}=a_1\,\boldsymbol{i}+a_2\,\boldsymbol{j}+b_3\,\boldsymbol{k}\times\boldsymbol{r}$$

(with arbitrary coefficients a_i, b_i*).*

Then the necessary and sufficient conditions for the existence of a solution to the mixed boundary-value problem are the total equilibrium conditions, namely, in case

a) (34) $$\int_\Omega (x_1 K_2 - x_2 K_1)\,dX + \int_{\mathscr{B}_n} P_n(n_2 x_1 - n_1 x_2)\,dS + \int_{\mathscr{B}_t} (x_1 P_{t2} - x_2 P_{t1})\,dS = 0;$$

b) (35) $$\int_\Omega K_3\,dX + \int_{\mathscr{B}_n} P_n n_3\,dS + \int_{\mathscr{B}_t} P_{t3}\,dS = 0;$$

c) $$\int_\Omega (x_1 K_2 - x_2 K_1)\,dX + h\int_\Omega K_3\,dX + \int_{\mathscr{B}_n} P_n(n_2 x_1 - n_1 x_2)\,dS + h\int_{\mathscr{B}_n} P_n n_3\,dS + \int_{\mathscr{B}_t} (x_1 P_{t2} - x_2 P_{t1})\,dS + h\int_{\mathscr{B}_t} P_{t3}\,dS = 0;$$

d) (34) *and* (35);

e) $$\int_\Omega (\boldsymbol{K}\times\boldsymbol{r})\,dX + \int_{\mathscr{B}_n} P_n(\boldsymbol{n}\times\boldsymbol{r})\,dS + \int_{\mathscr{B}_t} (\boldsymbol{P}_t\times\boldsymbol{r})\,dS = 0;$$

f) $$\int_\Omega K_j\,dX + \int_{\mathscr{B}_n} P_n n_j\,dS + \int_{\mathscr{B}_t} P_{tj}\,dS = 0, \quad (j=1,2), \quad \text{and} \quad (34).$$

If these conditions are satisfied, the solution is determined except for a polynomial $\boldsymbol{p}^{(\sigma)}$, *i.e., the solutions form precisely one class* $\tilde{\boldsymbol{u}}^{(\sigma)} \in [W_2^{(1)}(\Omega)]^3/\mathscr{P}^{(\sigma)}$, *where* $\mathscr{P}^{(\sigma)}$ *denotes the linear manifold of all polynomials of the form* $\boldsymbol{p}^{(\sigma)}$.

If we choose the orthogonal projection $\boldsymbol{u}_Q^{(\sigma)}$ *of an arbitrary representative* $\boldsymbol{u}\in\tilde{\boldsymbol{u}}^{(\sigma)}$ *of the class* $\tilde{\boldsymbol{u}}^{(\sigma)}$ *of solutions into the orthogonal complement* $\boldsymbol{Q}^{(\sigma)}$ *of the subspace* $\mathscr{P}^{(\sigma)}$ *in* $[W_2^{(1)}(\Omega)]^3$,

$$[W_2^{(1)}(\Omega)]^3 = \mathscr{P}^{(\sigma)} \oplus \boldsymbol{Q}^{(\sigma)},$$

by means of a scalar product, providing an equivalent norm in $[W_2^{(1)}(\Omega)]^3$, *or if we set*

$$\hat{\boldsymbol{u}}^{(\sigma)} = \boldsymbol{u}_0 + \boldsymbol{w}_{\boldsymbol{K}}^{(\sigma)},$$

where $\boldsymbol{w}_{\boldsymbol{K}}^{(\sigma)}$ *is the orthogonal projection of the difference* $\boldsymbol{u}-\boldsymbol{u}_0$ *into the orthogonal complement* $\boldsymbol{R}^{(\sigma)}$ *of the subspace* $\mathscr{P}^{(\sigma)}$ *in* $\boldsymbol{V}$,

$$\boldsymbol{V} = \mathscr{P}^{(\sigma)} \oplus \boldsymbol{R}^{(\sigma)},$$

by means of a scalar product, providing an equivalent norm in V, *then* $u_Q^{(\sigma)}$ *and* $\hat{u}^{(\sigma)}$ *respectively, being determined uniquely, satisfy the inequality* (33).

The proof follows from Theorem 2.2 in a manner similar to the proof of Theorem II.2, taking for the basis of the subspace $\mathscr{P}^{(\sigma)}$ in case a) $p^{(6)}$, b) $p^{(3)}$, c) $h p^{(3)}+p^{(6)}$, d) $p^{(3)}$, $p^{(6)}$, e) $p^{(4)}$, $p^{(5)}$, $p^{(6)}$, f) $p^{(1)}$, $p^{(2)}$, $p^{(6)}$, where the notation of the above-mentioned proof is used. The bilinear form

$$((\tilde{v}, \tilde{u}))=A(\tilde{v}, \tilde{u})+a(\tilde{v}, \tilde{u})$$

is bounded on $[W_2^{(1)}(\Omega)]^3/\mathscr{P}^{(\sigma)} \times [W_2^{(1)}(\Omega)]^3/\mathscr{P}^{(\sigma)}$. This fact can be proved easily.

Remark II.5. Let Γ_u be empty. If the solution u exists, then it must satisfy the integral equilibrium condition

$$\int_{\Gamma_n} \frac{1}{B_n} u_n v_n \, dS + \int_{\Gamma_t} \frac{1}{B_t} u_t \cdot v_t \, dS = \int_{\Omega} K \cdot v \, dX + \int_{\mathscr{B}_n} P_n v_n \, dS + \int_{\mathscr{B}_t} P_t \cdot v_t \, dS \tag{36}$$

for all $v=a+b\times r$ consistent with the conditions

$$v_n=0 \text{ on } \mathscr{A}_n, \quad v_t=0 \text{ on } \mathscr{A}_t, \quad \text{i.e.,} \quad v\in V.$$

Let, for example, $\mathscr{A}_n=\mathscr{A}_t=\phi$, $\Gamma=\Gamma_P\cup\Gamma_n\cup\Gamma_t$; then (36) is equivalent to the six conditions, which follow by inserting $v=p^{(l)}$, $l=1, 2, \ldots 6$: $(s=1, 2, 3)$

$$\int_{\Gamma_n} \frac{1}{B_n} n_s u_n \, dS + \int_{\Gamma_t} \frac{1}{B_t} (u_s - n_s u_n) \, dS = \int_{\Omega} K_s \, dX + \int_{\mathscr{B}_n} P_n n_s \, dS + \int_{\mathscr{B}_t} P_{ts} \, dS,$$

$$\begin{aligned} &\int_{\Gamma_n} \frac{1}{B_n} (n\times i_s)\cdot r u_n \, dS + \int_{\Gamma_t} \frac{1}{B_t} (u_t \times i_s)\cdot r \, dS \\ &\quad = \int_{\Omega} (K\times i_s)\cdot r \, dX + \int_{\mathscr{B}_n} (n\times i_s)\cdot r P_n \, dS + \int_{\mathscr{B}_t} (P_t\times i_s)\cdot r \, dS. \end{aligned}$$

As the form $A(v, u)$ satisfies the assumptions of Remark 2 in Part I, the assertion follows immediately.

Theorem II.14. *Let the assumptions of Theorem* II.13 *be satisfied. Let the linear functionals* $p_1^{(a)}, p_1^{(b)}, p_1^{(c)}, p_1^{(d)}, p_2^{(d)}, p_1^{(e)}, p_2^{(e)}, p_3^{(e)}, p_1^{(f)}, p_2^{(f)}, p_3^{(f)}$ *be continuous on* V *and such that*

a) $$p_1^{(a)}(p^{(a)})=0 \Rightarrow b_3=0,$$

b) $$p_1^{(b)}(p^{(b)})=0 \Rightarrow a_3=0,$$

c) $$p_1^{(c)}(p^{(c)})=0 \Rightarrow b_3=0,$$

d) $$\sum_{i=1}^{2} [p_i^{(d)}(p^{(d)})]^2=0 \Rightarrow a_3=b_3=0,$$

e) $$\sum_{i=1}^{3} [p_i^{(e)}(p^{(e)})]^2=0 \Rightarrow b=0,$$

f) $$\sum_{i=1}^{3} [p_i^{(f)}(p^{(f)})]^2=0 \Rightarrow a_1=a_2=b_3=0.$$

Moreover, let each of the systems $\{p_i^{(\sigma)}\}$ *be linearly independent on* $\mathscr{P}^{(\sigma)}$, $\sigma = a, b, c, d, e, f$ *(according to the definition of Remark* 4 *in Part* I). *Put*

$$V_p^{(\sigma)} = \{v \in V, \sum_i [p_i^{(\sigma)}(v)]^2 = 0\},$$

where $i=1$ *for* $\sigma = a, b, c$; $i=1, 2$ *for* $\sigma = d$ *and* $i=1, 2, 3$ *for* $\sigma = e, f$.

Then the generalized Korn inequality (32) *holds for* $v \in V_p^{(\sigma)}$, *and the solution of the mixed problem exists if and only if the conditions of total equilibrium from Theorem* II.13 *are satisfied. This solution is unique in* $u_0 \oplus V_p^{(\sigma)}$ *and the inequality* (33) *holds for it.*

The proof is an immediate consequence of Theorem 2.2, Remark 3 and 4 of Section 2 in Part I.

The choice of functionals $p_i^{(\sigma)}$ is, of course, not unique. In the following lemma we show some possibilities, analogous to the formulas (16), (17), (18), for the second boundary-value problem.

Lemma II.10. *Let us set in the cases*

a) (37a) $$p_1^{(a)}(v) = \int_{\Omega^*} (x_1 v_2 - x_2 v_1)\, dX,$$

(38a) $$p_1^{(a)}(v) = \int_{\Omega^*} (v_{2,1} - v_{1,2})\, dX,$$

(39a) $$p_1^{(a)}(v) = \int_{\Gamma^*} (x_1 v_2 - x_2 v_1)\, d\Gamma;$$

b) (37b) $$p_1^{(b)}(v) = \int_{\Omega^*} v_3\, dX,$$

(38b) $$p_1^{(b)}(v) = \int_{\Gamma^*} v_3\, d\Gamma;$$

c) (37c) $$p_1^{(c)}(v) = \int_{\Omega^*} (x_1 v_2 - x_2 v_1)\, dX,$$

(38c) $$p_1^{(c)}(v) = \int_{\Omega^*} (v_{2,1} - v_{1,2})\, dX,$$

(39c) $$p_1^{(c)}(v) = \int_{\Gamma^*} (x_1 v_2 - x_2 v_1)\, dS,$$

(40c) $$p_1^{(c)}(v) = \int_{\Omega^*} v_3\, dX,$$

(41c) $$p_1^{(c)}(v) = \int_{\Gamma^*} v_3\, dS;$$

d) (37d) $$p_1^{(d)}(v) = \int_{\Omega^*} v_3\, dX, \quad p_2^{(d)}(v) = \int_{\Omega^*} (x_1 v_2 - x_2 v_1)\, dX,$$

(38d) $$p_1^{(d)}(v) = \int_{\Omega^*} v_3\, dX, \quad p_2^{(d)}(v) = \int_{\Omega^*} (v_{2,1} - v_{1,2})\, dX,$$

(39d) $$p_1^{(d)}(v) = \int_{\Gamma^*} v_3\, dS, \quad p_2^{(d)}(v) = \int_{\Gamma^*} (x_1 v_2 - x_2 v_1)\, dS;$$

e) (37e) $$p_i^{(e)}(\boldsymbol{v}) = \int_{\Omega^*} (\boldsymbol{r}' \times \boldsymbol{v})_i \, dX',$$

(38e) $$p_i^{(e)}(\boldsymbol{v}) = \int_{\Omega^*} (\operatorname{rot} \boldsymbol{v})_i \, dX,$$

(39e) $$p_i^{(e)}(\boldsymbol{v}) = \int_{\Gamma^*} (\boldsymbol{r}' \times \boldsymbol{v})_i \, dS, \quad (i=1,2,3);$$

f) (37f) $$p_i^{(f)}(\boldsymbol{v}) = \int_{\Omega^*} v_i \, dX, \quad (i=1,2), \quad p_3^{(f)}(\boldsymbol{v}) = \int_{\Omega^*} (x_1 v_2 - x_2 v_1) \, dX,$$

(38f) $$p_i^{(f)}(\boldsymbol{v}) = \int_{\Omega^*} v_i \, dX, \quad (i=1,2), \quad p_3^{(f)}(\boldsymbol{v}) = \int_{\Omega^*} (v_{2,1} - v_{1,2}) \, dX,$$

(39f) $$p_i^{(f)}(\boldsymbol{v}) = \int_{\Gamma^*} v_i \, dS, \quad (i=1,2), \quad p_3^{(f)}(\boldsymbol{v}) = \int_{\Gamma^*} (x_1 v_2 - x_2 v_1) \, dS,$$

where $\Omega^* \subset \Omega$ is an arbitrary measurable set with positive volume measure, Γ^* denotes an arbitrary open part of the boundary Γ' of a Lipschitz region $\Omega' \subset \Omega$. In particular, we may choose $\Omega^* = \Omega$, $\Gamma^* = \Gamma$. In (37e) *and* (39e) *the coordinate frame is formed by the principal axes of inertia of the sets Ω^* and Γ^*, respectively.*

Then the functionals (37a)–(39f) *satisfy the hypotheses of Theorem* II.14.

Proof. The cases a)–d) can be verified easily by inserting into $p_i^{(\sigma)}(\boldsymbol{p}^{(\sigma)}) = 0$. In cases (37e) and (39e) we use the fact that

$$\int_M x_i' \, dM = 0 \quad \text{and} \quad \int_M x_i' x_j' \, dM = 0$$

for $i \neq j$ ($M = \Omega^*$ and Γ^*, respectively), to obtain $b_i = 0$. By substitution $\operatorname{rot}(\boldsymbol{b} \times \boldsymbol{r}) = 2\boldsymbol{b}$ into (38e) we derive $\boldsymbol{b} = \boldsymbol{0}$. In case f) let us translate the coordinate frame into the centroid of the set M, so that

$$\boldsymbol{p}^{(f)} = a_1' \, \boldsymbol{i} + a_2' \, \boldsymbol{j} + b_3 \, \boldsymbol{k} \times \boldsymbol{r}',$$

where $\boldsymbol{r}'$ denotes the radius vector relative to the centroid. Then from $p_i^{(f)}(\boldsymbol{p}^{(f)}) = 0$ ($i = 1, 2$), $a_1' = a_2' = 0$ follows immediately and $p_3^{(f)}(\boldsymbol{p}^{(f)}) = 0$ implies $b_3 = 0$.

It remains to prove that the systems of functionals $\{p_i^{(d)}\}$, $\{p_i^{(e)}\}$ and $\{p_i^{(f)}\}$ are linearly independent on the corresponding $\mathscr{P}^{(\sigma)}$. In cases (37d), (38d), (39d), let us suppose that

$$\alpha_1 \, p_1^{(d)}(\boldsymbol{p}) + \alpha_2 \, p_2^{(d)}(\boldsymbol{p}) = 0$$

for all $\boldsymbol{p} = a_3 \boldsymbol{k} + b_3 \boldsymbol{k} \times \boldsymbol{r}$. Inserting $a_3 = 1$, $b_3 = 0$ and $a_3 = 0$, $b_3 = 1$, we obtain $\alpha_1 = 0$ and $\alpha_2 = 0$.

In cases (37e), (39e) we may write

$$\sum_{i=1}^{3} \alpha_i \, p_i^{(e)}(\boldsymbol{p}) = \int_M \boldsymbol{\alpha} \cdot (\boldsymbol{r}' \times \boldsymbol{p}) \, dM = 0,$$

where $\boldsymbol{\alpha} = (\alpha_1, \alpha_2, \alpha_3)$ is to be understood as a vector. Hence

$$\int_M \boldsymbol{p} \cdot (\boldsymbol{\alpha} \times \boldsymbol{r}') \, dM = 0.$$

If we choose $\boldsymbol{p}=\boldsymbol{\alpha}\times\boldsymbol{r}=\boldsymbol{\alpha}\times\boldsymbol{r}'+\boldsymbol{\alpha}\times\boldsymbol{r}_0$, where $\boldsymbol{r}(X)=\boldsymbol{r}'(X')+\boldsymbol{r}_0$, and $\boldsymbol{r}'$ is the radius vector relative to the principal axes of inertia of M, we obtain

$$\int_M |\boldsymbol{\alpha}\times\boldsymbol{r}'|^2\,dM=0.$$

Consequently $\boldsymbol{\alpha}\times\boldsymbol{r}'=\boldsymbol{0}$ on M, which represents a system of three linear equations for x'_1, x'_2, x'_3. If $|\boldsymbol{\alpha}|>0$, the solutions of this system define a straight-line. By virtue of the hypotheses on the set M we obtain a contradiction and therefore $\boldsymbol{\alpha}=\boldsymbol{0}$.

In the functionals (38e) it suffices to insert consecutively $\boldsymbol{p}=\boldsymbol{i}\times\boldsymbol{r}$, $\boldsymbol{p}=\boldsymbol{j}\times\boldsymbol{r}$, $\boldsymbol{p}=\boldsymbol{k}\times\boldsymbol{r}$ into the relation

$$\sum_{i=1}^{3}\int_{\Omega^*}\alpha_i(\operatorname{rot}\boldsymbol{p})_i\,dX=0,$$

to obtain $\alpha_i=0$, $i=1, 2, 3$.

For (37f), (39f) we may write

$$\sum_{i=1}^{3}\alpha_i\,p_i^{(f)}(\boldsymbol{p})=\int_M[\alpha_1\,p_1+\alpha_2\,p_2+\alpha_3\,\boldsymbol{k}\cdot(\boldsymbol{r}\times\boldsymbol{p})]\,dM=0$$

or

$$\int_M \boldsymbol{p}\cdot[\alpha_1\,\boldsymbol{i}+\alpha_2\boldsymbol{j}+\alpha_3\,\boldsymbol{k}\times\boldsymbol{r}]\,dM=0.$$

Choosing $\boldsymbol{p}=\alpha_1\,\boldsymbol{i}+\alpha_2\boldsymbol{j}+\alpha_3\boldsymbol{k}\times\boldsymbol{r}$, we have $\boldsymbol{p}=\boldsymbol{0}$ on M, which leads to a system of two linear equations

$$\alpha_1-\alpha_3\,x_2=0,$$

$$\alpha_2+\alpha_3\,x_1=0,$$

the solutions of which yield a single straight-line for $\alpha_3\neq 0$. By virtue of the hypotheses on the set M, we conclude $\alpha_3=0$ and then also $\alpha_1=\alpha_2=0$.

In the functionals (38f) it suffices to insert consecutively $\boldsymbol{p}=\boldsymbol{i}$, $\boldsymbol{p}=\boldsymbol{j}$, $\boldsymbol{p}=\boldsymbol{k}\times\boldsymbol{r}$, to obtain $\alpha_1=\alpha_2=0$ and $\alpha_3=0$.

References

1. HLAVÁČEK, I., & J. NEČAS, On inequalities of Korn's type. I. Boundary-value problems for elliptic systems of partial differential equations. Arch. Rational Mech. Anal. **36**, 305–311 (1970).
2. NEČAS, J., Les méthodes directes en théorie des équations elliptiques. Prague: Academia 1967.

Matematický ústav ČSAV
Žitná 25
Prague 1, Czechoslovakia

(Received July 18, 1968)

Chapter 6
Commentary

6.1 Linear Elliptic Operators

Alois Kufner

Abstract The achievements of Jindřich Nečas in the field of linear (elliptic) partial differential equations are described. Some most important results are pointed out.

The interest of J. Nečas in linear problems started in connection with his collaboration with Ivo Babuška and others in solving problems from practical applications, such as dam building, see I. Babuška's contribution. Some of his first results were relatively standard, but already the paper about the biharmonic problem in an infinite cone shows that he was able to identify problems which became quite popular later (BVPs on domains with corners and edges—see the work of Olejnik, Mazya and others).

The next period was devoted to the investigation of the so-called weak (variational) solutions. This topic was quite new and needed a deeper investigation of a class of new function spaces (the so-called Sobolev spaces) and their properties. A survey of achievements of J. Nečas in this new area has been presented in Chap. 2 of his book *Les Méthodes Directes en Théorie des Equations Elliptiques*. In connection with the problems of existence of such weak solutions, which is closely connected with the concepts of ellipticity and coercivity, he wrote papers [JN-C28, JN-C23], but his main contribution to the area is in my opinion exactly in the following two papers [JN-C17, C18]. For the investigation of Sobolev spaces it was necessary to solve some technical problems, in particular the description of the domain considered. A very useful class of domains was the class of domains with the cone condition, and J. Nečas introduced in his paper [JN-C17] a very practical technical apparatus for describing the boundary of such domains. His description was a starting point of a more detailed investigation of the properties of Sobolev spaces, and this description was used not only in the book mentioned, but also

© Springer Basel 2015

Š. Nečasová et al. (eds.), *Selected Works of Jindřich Nečas*, Advances in Mathematical Fluid Mechanics, DOI 10.1007/978-3-0348-0231-4_6

by numerous other authors (let us mention for example the book *Function spaces*, see [KJF]). Nečas's description (called by him domains of type $\mathcal{N}$) continues to be useful. The second paper, [C18], *Sur une méthode pour résoudre les équations aux derivées partielles du type elliptique, voisine de la variationelle*, appeared in the prestigious Annali di Pisa, and this paper again shows that J. Nečas was able to look far into the future. His contribution in this paper is in the extension of the theory of weak solutions to weighted Sobolev spaces. Weighted spaces appeared earlier, but mainly in connection with unbounded domains. The description of the behaviour of the solution at infinity is realized via an appropriate weight function—here a pioneering work was done by L.D. Kudrjavcev. It also appeared in connection with the degeneration or singularity of differential operators (see a number of Russian authors), or sometimes in connection with the geometry of domain boundaries. In the paper [C18] J. Nečas investigated standard domains and operators, for which it is possible to use standard Sobolev spaces, and he also started to look for a weak solution in certain weighted spaces, which allow one to obtain a better characterization of properties of the solution. (It should be mentioned that behind his efforts was the problem of regularity of the solution, although it is not so explicitly expressed in [C18].)

The application of weighted spaces led to the necessity to investigate bilinear forms (which are important for definition of the weak solution) not on the product $H \times H$ (with H representing some Hilbert space), but on the product $H_1 \times H_2$ of two (different) Hilbert spaces. Therefore it was necessary to extend and generalize the celebrated Lax-Milgram Lemma to this non-symmetric case. This is by my opinion one of the most important contributions of J. Nečas to the linear theory. Many authors continued in his approach and weighted spaces became an important and useful tool.

All results of J. Nečas, connected with the linear theory, have been included into the book *Les Méthodes Directes en Théorie des Equations Elliptiques* and therefore it is not necessary to describe in detail every single paper. In this book a strong interest in the problem of the regularity of the weak solution can already be detected. Later on, REGULARITY and NONLINEARITY have been the main focuses of the work of J. Nečas.

Nečas' Works Quoted Above

The references [Cxx] refer to the respective papers listed in Chap. 2 "Complete Bibliography of Jindřich Nečas". The references [JN-Cyy] refer to papers [Cyy] that are reproduced in Chap. 5 "Selected papers on linear elliptic operators".

[C18] J. Nečas, *Sur une méthode pour résoudre les équations aux dérivées partielles du type elliptique, voisine de la variationnelle*, Ann. Scuola Norm. Sup. Pisa (3), 1962, vol. 16, pp. 305–326.

[JN-C17] J. Nečas, *On domains of type $\mathcal{N}$*, Czechoslovak Math. J., 1962, vol. 12 (87), pp. 274–287. In Russian.

[JN-C23] J. Nečas, *Sur la coercivité des formes sesquilinéaires, elliptiques*, Rev. Roumaine Math. Pures Appl., 1964, vol. 9, pp. 47–69.

[JN-C28] J. Nečas, *Sur les normes equivalentes dans* W_p^k *et sur la coercivité des formes formellement positives*, Sem. math. sup., Université Montréal, 1966, pp. 102–128.

References

[KJF] A. Kufner, O. John, S. Fučík, Function spaces, Monographs and Textbooks on Mechanics of Solids and Fluids; Mechanics: Analysis. Noordhoff International Publishing, Leyden; Academia, Prague, 1977.

Alois Kufner
Institute of Mathematics, Academy of Sciences, Žitná 25, 11567 Prague 1, Czech Republic
email: kufner@math.cas.cz

Part III
Regularity Theory

Chapter 7
Selected Papers on Regularity Theory

Commentationes Mathematicae Universitatis Carolinae
8,2 (1967)

SUR L'APPARTENANCE DANS LA CLASSE $C^{(k),\mu}$ DES SOLUTIONS VARIATIONNELLES DES ÉQUATIONS ELLIPTIQUES NON-LINÉAIRES DE L'ORDRE 2 k EN DEUX DIMENSIONS

Jindřich NEČAS, Praha

(Communication préalable)

Il est considéré la régularité à l'intérieur du domaine plan Ω de la solution du problème variationnel

$$\text{(1)} \quad \min_{u-u_0\in \overset{\circ}{W}{}_m^{(k)}} \left[\int_\Omega \mathcal{F}(x, D^\alpha u)\,dx - \int_\Omega \sum_{|i|\le k} D^i u\, f_i\, dx \right] .$$

Ici $W_m^{(k)}(\Omega)$ est l'espace de Sobolev des fonctions réelles, dont les dérivées au sens des distributions jusqu'à l'ordre k sont de m-ème puissance sommable sur Ω, $1 < m < \infty$, et $\overset{\circ}{W}{}_m^{(k)}(\Omega)$ est la fermeture en $W_m^{(k)}(\Omega)$ de l'espace $\mathcal{D}(\Omega)$ des fonctions indéfiniment différentiables à support compact. La notation usuelle $D^\alpha = \dfrac{\partial^{|\alpha|}}{\partial x_1^{\alpha_1}\dots\partial x_N^{\alpha_N}}$ est utilisée.

Si $k = 1$, l'appartenance de la solution u à la classe des fonctions, dont les premières dérivées sont μ-höldériennes dans Ω ou même dans la fermeture de Ω : $\overline{\Omega}$, est démontrée sous différentes hypothèses (dont les substentielles sont les mêmes et correspondent aux notres) dans plusieurs travaux, dont nous citons: C.B. Morrey [4], E.R. Buley [1], O.A. Ladyženskaja, N.N. Uralceva [3]; ici la dimension de Ω

© Faculty of Mathematics and Physics of Charles University 1967. Reprinted with kind permission from: Nečas, J., Sur l'appartenance dans la classe $C^{(k),\mu}$ des solutions variationnelles des équations elliptiques non-linéaires de l'ordre $2k$ en deux dimensions, Commentationes Mathematicae Universitatis Carolinae 8, 209-217.

est $N \geq 2$. Tous ces travaux utilisent les résultats de De Giorgi - Nash, cf. De Giorgi [2]. La situation pour N = 2 (et k = 1) étant moins compliquée, C.B. Morrey a démontré ce résultat plus avant au travail [5] pour le cas m = 2 . Pour $k \geq 1$, $N = 2$ et $m = 2$, ce résultat a été annoncé par J. Nečas au travail [6].

Dans ce travail est démontré un théorème sur la régularité des solutions faibles des équations elliptiques linéaires pour la dimension $N \geq 2$ (un analogue du théorème de De Giorgi pour N = 2) qui permet en liaison avec une technique assez fine de démontrer pour le cas N = 2 que la solution faible u du problème (1) appartient dans la classe $C^{(k),\mu}(\overline{\Omega}')$ (la classe des fonctions μ -höldériennes sur $\overline{\Omega}'$ avec leurs dérivées jusqu'à l'ordre k). Ici $\Omega' \subset \overline{\Omega}' \subset \Omega$, $\mu = \mu(\Omega')$.

Le problème (1) correspond au problème de Dirichlet. Un raisonement simple montre que la question sur la régularité dans l'intérieur de Ω des solutions d'autres problèmes aux limites se ramène facilement à celle pour le problème (1). Remarquons que nous n'avons pas taché de chercher les conditions les plus générales pour $F(x, \xi_\alpha)$, f_i pour ne pas compliquer la situation.

<u>Hypothèses et notation</u>: Le domaine Ω soit à frontière $\partial\Omega$ lipschitzienne. Par c , on note les différentes constantes positives.

La fonction réelle $\mathcal{F}(x, \xi_\alpha)$ soit définie pour $x \in \overline{\Omega}$, $|\xi_\alpha| < \infty$, $|\alpha| \leq k$, continue ici. On suppose l'existence d'un sousensemble M des indices

$|\alpha| \leq k$, contenant tous les indices $|\alpha| = k$, tel que

(2) $|F(x, \mathcal{J}_\alpha)| \leq c(1 + \sum_{\alpha \in M} |\mathcal{J}_\alpha|)^m$, $1 < m < \infty$,

(3) $F(x, \mathcal{J}_\alpha) \geq c_1 (1 + \sum_{\alpha \in M} |\mathcal{J}_\alpha|)^m - c_2$.

Si $m = 2$, on suppose encore l'existence des dérivées continues $\frac{\partial F}{\partial \mathcal{J}_i}, \frac{\partial^2 F}{\partial \mathcal{J}_i \partial \mathcal{J}_j}, \frac{\partial F}{\partial x_\ell}, \frac{\partial^2 F}{\partial \mathcal{J}_i \partial x_\ell}$ dans le domaine de la définition de $F(x, \mathcal{J}_\alpha)$ et

(4) $|\frac{\partial F}{\partial \mathcal{J}_i}(x, \mathcal{J}_\alpha)| \leq c(1 + \sum_{\alpha \in M} |\mathcal{J}_\alpha|)$,

(5) $|\frac{\partial^2 F}{\partial \mathcal{J}_i \partial x_\ell}(x, \mathcal{J}_\alpha)| \leq c(1 + \sum_{\alpha \in M} |\mathcal{J}_\alpha|)$,

(6) $|\frac{\partial^2 F}{\partial \mathcal{J}_i \partial \mathcal{J}_j}(x, \mathcal{J}_\alpha)| \leq c$,

(7) $c_3 \sum_{|i| = k} \xi_i^2 \leq \sum_{|i| = |j| = k} \frac{\partial^2 F}{\partial \mathcal{J}_i \partial \mathcal{J}_j} \xi_i \xi_j \leq c_4 \sum_{|i| = k} \xi_i^2$,

(8) $\sum_{|i|, |j| \leq k} \frac{\partial^2 F}{\partial \mathcal{J}_i \mathcal{J}_j} \xi_i \xi_j > 0$.

Si $m > 2$, posons $\sigma = [\frac{m}{2}]$ (partie entière), $h = \frac{m-2}{\sigma}$ et supposons l'existence d'une fonction $F(x, \mathcal{J}_\alpha, \lambda_1, \lambda_2, \ldots \lambda_\sigma)$ définie pour $x \in \bar{\Omega}$, $|\mathcal{J}_\alpha| < \infty$, $|\alpha| \leq k$, $0 \leq \lambda_\ell \leq 1$, $\ell = 1, 2, \ldots \sigma$, continue avec les dérivées $\frac{\partial F}{\partial \mathcal{J}_i}, \frac{\partial F}{\partial x_\ell}, \frac{\partial^2 F}{\partial \mathcal{J}_i \partial x_\ell}, \frac{\partial^2 F}{\partial \mathcal{J}_i \partial \mathcal{J}_j}$ au domaine de définition et telle que $F(x, \mathcal{J}_\alpha, 0, \ldots 0) =$

$= F(x, \zeta_\alpha)$ (une homotopie).

Soit $0 \leq \tau \leq \sigma$ et supposons:

(9) $$| F(x, \zeta_\alpha, \lambda_1, \lambda_2, \dots \lambda_\tau, 0, \dots 0)| \leq c(1 + \sum_{\alpha \in M} |\zeta_\alpha|)^m \cdot \prod_{\ell=1}^{\tau} \cdot (1 + \lambda_\ell \sum_{\alpha \in M} |\zeta_\alpha|)^{-h} ,$$

(10) $$F(x, \zeta_\alpha, \lambda_1, \lambda_2, \dots \lambda_\tau, 0, \dots 0) \geq c_1 + \sum_{\alpha \in M} |\zeta_\alpha|)^m \cdot \prod_{\ell=1}^{\tau} (1 + \lambda_\ell \sum_{\alpha \in M} |\zeta_\alpha|^{-h} - c_2 ,$$

(11) $$| \frac{\partial F}{\partial \zeta_i} (x, \zeta_\alpha, \lambda_1, \lambda_2, \dots \lambda_\tau, 0, \dots 0)| \leq c (1 + \sum_{\alpha \in M} |\zeta_\alpha|)^{m-1} \cdot \prod_{\ell=1}^{\tau} (1 + \lambda_\ell \sum_{\alpha \in M} |\zeta_\alpha|)^{-h} ,$$

(12) $$| \frac{\partial^2 F}{\partial \zeta_i \partial x_\ell} (x, \zeta_\alpha, \lambda_1, \dots \lambda_\tau, 0, \dots 0)| \leq c (1 + \sum_{\alpha \in M} |\zeta_\alpha|)^{m-1} \cdot \prod_{\ell=1}^{\tau} (1 + \lambda_\ell \sum_{\alpha \in M} |\zeta_\alpha|)^{-h} ,$$

(13) $$c_3 (1 + \sum_{\alpha \in M} |\zeta_\alpha|)^{m-2} \prod_{\ell=1}^{\tau} (1 + \lambda_\ell \sum_{\alpha \in M} |\zeta_\alpha|)^{-h} \sum_{|i|=k} \xi_i^2 \leq$$
$$\leq \sum_{|i|=|j|=k} \frac{\partial^2 F}{\partial \zeta_i \partial \zeta_j} (x, \zeta_\alpha, \lambda_1, \dots \lambda_\tau, 0, \dots 0) \xi_i \xi_j \leq$$
$$\leq c_4 (1 + \sum_{\alpha \in M} |\zeta_\alpha|)^{m-2} \prod_{\ell=1}^{\tau} (1 + \lambda_\ell \sum_{\alpha \in M} |\zeta_\alpha|)^{-h} \sum_{|i|=k} \xi_i^2 ,$$

(14) $$| \frac{\partial^2 F}{\partial \zeta_i \partial \zeta_j} (x, \zeta_\alpha, \lambda_1, \dots \lambda_\tau, 0, \dots 0)| \leq$$
$$\leq c (1 + \sum_{\alpha \in M} |\zeta_\alpha|)^{m-2} \prod_{\ell=1}^{\tau} (1 + \lambda_\ell \sum_{\alpha \in M} |\zeta_\alpha|)^{-h} ,$$

(15) $$\sum_{|i|, |j| \leq k} \frac{\partial^2 F}{\partial \zeta_i \partial \zeta_j} (x, \zeta_\alpha, \lambda_1, \dots \lambda_\tau, 0, \dots 0) \xi_i \xi_j \geq 0 .$$

Si $1 < m < 2$, on suppose l'existence d'une fonction $F(x, \zeta_\alpha, \lambda)$, définie pour $x \in \bar{\Omega}$, $|\zeta_\alpha| < \infty$, $|\alpha| \leq k$, $0 \leq \lambda \leq 1$, continue avec les dérivées $\frac{\partial F}{\partial \zeta_i}, \frac{\partial F}{\partial x_\ell}, \frac{\partial^2 F}{\partial \zeta_i \partial \zeta_j}, \frac{\partial^2 F}{\partial \zeta_i \partial x_\ell}$ au domaine de définition, telle que $F(x, \zeta_\alpha, 0) = F(x, \zeta_\alpha)$.

On suppose encore:

(16) $$|F(x, \zeta_\alpha, \lambda)| \leq c(1+\sum_{\alpha \in M} |\zeta_\alpha|)^m (1+\lambda \sum_{\alpha \in M} |\zeta_\alpha|)^{2-m},$$

(17) $$F(x, \zeta_\alpha, \lambda) \geq c_1 (1+\sum_{\alpha \in M} |\zeta_\alpha|)^m (1+\lambda \sum_{\alpha \in M} |\zeta_\alpha|)^{2-m} - c_2,$$

(18) $$\left|\frac{\partial F}{\partial \zeta_i}(x, \zeta_\alpha, \lambda)\right| \leq c(1+\sum_{\alpha \in M} |\zeta_\alpha|)^{m-1} (1+\lambda \sum_{\alpha \in M} |\zeta_\alpha|)^{2-m},$$

(19) $$\left|\frac{\partial^2 F}{\partial \zeta_i \partial x_\ell}(x, \zeta_\alpha, \lambda)\right| \leq c(1+\sum_{\alpha \in M} |\zeta_\alpha|)^{m-1} (1+\lambda \sum_{\alpha \in M} |\zeta_\alpha|)^{2-m},$$

(20) $$\left|\frac{\partial^2 F}{\partial \zeta_i \partial \zeta_j}(x, \zeta_\alpha, \lambda)\right| \leq c(1+\sum_{\alpha \in M} |\zeta_\alpha|)^{m-2} (1+\lambda \sum_{\alpha \in M} |\zeta_\alpha|)^{2-m},$$

(21) $$\begin{aligned} &c_3 (1+\sum_{\alpha \in M} |\zeta_\alpha|)^{m-2} (1+\lambda \sum_{\alpha \in M} |\zeta_\alpha|)^{2-m} \sum_{|i|=k} \xi_i^2 \leq \\ &\leq \sum_{|i|=|j|=k} \frac{\partial^2 F}{\partial \zeta_i \partial \zeta_j}(x, \zeta_\alpha, \lambda)\, \xi_i \xi_j \leq \\ &\leq c_4 (1+\sum_{\alpha \in M} |\zeta_\alpha|)^{m-2} (1+\lambda \sum_{\alpha \in M} |\zeta_\alpha|)^{2-m}, \end{aligned}$$

(22) $$\sum_{|i|,|j| \leq k} \frac{\partial^2 F}{\partial \zeta_i \partial \zeta_j}(x, \zeta_\alpha, \lambda)\, \xi_i \xi_j \geq 0 .$$

Pour la fonction u_0 de (1) on suppose:

(23) $$u_0 \in W_m^{(k)}(\Omega) \text{ pour } m \geq 2,$$

(24) $$u_0 \in W_2^{(k)}(\Omega) \text{ pour } m < 2,$$

en ce qui concerne f_i de (1), on suppose:

(25) $m \geq 2 : f_i \in L_2(\Omega), \ |i| \leq k ,$

(26) $m \geq 2 : \int_\Omega \rho^{kp_0} \sum_{|i|=k} \left| \frac{\partial f_i}{\partial x_l} \right|^{p_0} dx \leq c , \ 2 < p_0 ,$

(27) $m < 2 : f_i \in L_{\frac{m}{m-1}}(\Omega), \ |i| \leq k ,$

(28) $m < 2 : \int_\Omega \rho^{k\frac{m}{m-1}} \sum_{|i|=k} \left| \frac{\partial f_i}{\partial x_l} \right|^{\frac{m}{m-1}} dx \leq c ;$

ici $\rho(x) = distance\,(x, \partial\Omega)$.

<u>Exemple</u>. Soit $F(x, \xi_\alpha) = (1 + \sum_{\alpha \in M} \xi_\alpha^2)^{\frac{m}{2}} , \ 1 < m < \infty .$
Alors les hypothèses sur $F(x, \xi_\alpha)$ sont satisfaites, si nous posons pour m ≥ 2 :

$$F(x, \xi_\alpha, \lambda_1, \ldots \lambda_\tau) = (1 + \sum_{\alpha \in M} \xi_\alpha^2)^{\frac{m}{2}} \prod_{l=1}^{\tau} (1 + \lambda_l \sum_{\alpha \in M} \xi_\alpha^2)^{-\frac{h}{2}} ,$$

pour $1 < m \leq 2$:

$$F(x, \xi_\alpha, \lambda) = (1 + \sum_{\alpha \in M} \xi_\alpha^2)^{\frac{m}{2}} (1 + \lambda \sum_{\alpha \in M} \xi_\alpha^2)^{1-\frac{m}{2}} .$$

<u>Résultats</u>:

<u>Théorème 1</u>. (Facile et connu, cf. F.E. Browder [8], J.Nečas [7].) Il existe une solution unique du problème (1) et satisfait à l'équation:

(29) pour $\varphi \in \mathcal{D}(\Omega)$

$$\int_\Omega \sum_{|i| \leq k} \frac{\partial F}{\partial \xi_i}(x, D^\alpha u) D^i \varphi \, dx = \int_\Omega \sum_{|i| \leq k} D^i \varphi \, f_i \, dx , \ (N \geq 2) .$$

<u>Théorème 2</u>. Soit $K_d = \{x \in E_N , |x| < d\}$, $N \geq 2$ et $u \in \mathring{W}_2^{(k)}(K_d)$ une solution faible de l'équation

$$\sum_{|i|=|j|=k} D^i(A_{ij} D^j u) = \sum_{|i|=k} D^i f_i , \qquad \text{où } A_{ij} \in L_\infty(K_d) ,$$

$A_{ij} = A_{ji}$,

(30) $c_1 \sum_{|i|=k} \xi_i^2 \le \sum_{|i|=|j|=k} A_{ij} \xi_i \xi_j \le c_2 \sum_{|i|=k} \xi_i^2$,

$f_i \in L_p(K_d)$, $|i| = k$, $p \geq 2$. Alors il existe des constantes c_3, $c_4 > 1$ telles que

$$\Big(\int_{K_d} \sum_{|i|=k} |D^i u|^p dx\Big)^{\frac{1}{p}} \le \frac{2}{c_1} c_3^{\frac{n-2}{2}} \Big(\int_{K_d} \sum_{|i|=k} |f_i|^p dx\Big)^{\frac{1}{p}}$$

si $2 \le p \le n$, n arbitraire mais fixe, $c_3 = c_3(n)$,

$c_4 = c_4(n)$, et si $p\Big(1 - \dfrac{\log \dfrac{1-\frac{1}{2}\frac{c_1}{c_2}}{1-\frac{c_1}{c_2}}}{\log c_4}\Big) \le 2$.

Théorème 3. (Cela généralise les résultats du travail de J. Nečas [7].) Soit u la solution de (1). Alors pour $2 < m < \infty$:

$$\int_\Omega \varphi^{2k} \Big(1 + \sum_{\alpha \le M} |D^\alpha u|\Big)^{m-2} \sum_{|i| \le k+1} (D^i u(x))^2 dx \le c, \quad N \geq 2 .$$

Théorème 4. (Cf. le résultat de J. Nečas [6] .) Soit m = 2 , u la solution de (1), N = 2 . Alors il existe $\infty > p_1 > 2$ tel que pour $|i| = k + 1$

$$\| D^i u \|_{L_{p_1}(L(x_0))} \le c\, d^{-(1+k)} ,$$

où $L(x_0) = \{x \in \Omega, |x - x_0| < \frac{1}{2} d, 2d = \text{distance}(x_0, \partial\Omega)\}$. La fonction $D^i u \in C^{(0),\mu}(\overline{\Omega}')$, $\Omega' \subset \overline{\Omega}' \subset \Omega$, $\mu = 1 - \frac{2}{p_1}$, $|i| = k$.

Théorème 5. Soit m > 2, N = 2 , u la solution du problème (1). Alors si $\Omega_d = \{x \in \Omega, \text{dist}(x, \partial\Omega) > d\}$:

$$\sup_{x \in \Omega_d} \sum_{|\alpha| \le k} |D^\alpha u(x)| \le c\, d^{-\frac{2(1+k)}{1-\frac{4}{2}}} .$$

Posons $A_d = 1 + \sup_{x \in \Omega_d} \sum_{|\alpha| \le k} |D^\alpha u(x)|$. Il existe

$0 < c_1 \le 1$ de sorte que si $p = 2 + c_1 A_d^{2-m}$,

alors pour $|\alpha| = k + 1$:

$$\| D^{\alpha} u \|_{L_p(x_0)} \le c d^{-(1+k)[1+\frac{2(1+k)(m-2)}{1-\frac{m}{2}}]} .$$

Enfin, nous avons pour $|\alpha| = 1 + k$:

$$\| D^{\alpha} u \|_{C^{(0),\mu}(\Omega_{2d})} \le c(\Omega_{2d}) \quad \text{avec} \quad \mu = 1 - \frac{2}{p} .$$

<u>Théorème 6</u>. Soit $1 < m < 2$, $N = 2$, u la solution du problème (1). Alors

$$\sup_{x \in \Omega_d} \sum_{|\alpha| \le k} | D^{\alpha} u(x) | \le c d^{-\frac{2(1+k)}{m-1}} .$$

Il existe $0 < c_1 \le 1$ de sorte que si $p = 2 + c_1 A_d^{m-2}$, alors pour $|\alpha| = 1 + k$:

$$\| D^{\alpha} u \|_{L_p(x_0)} \le c d^{-(1+k)[1+\frac{4(2-m)}{m-1}]} .$$

Enfin, pour $|\alpha| = 1 + k$

$$\| D^{\alpha} u \|_{C^{(0),\mu}(\Omega_{2d})} \le c(\Omega_{2d}) \quad \text{avec} \quad \mu = 1 - \frac{2}{p} .$$

B i b l i o g r a p h i e

[1] E.R. BULEY: The differentiability of solutions of certain variational problems for multiple integrals, Technical Report 16(1960),Univ.Berkeley.

[2] E.De GIORGI: Sulla differenziabilità e l'analyticità delle estremali degli integrali multipli regolari, Mem.Accad.Sci.Torino,3(1957),25-43.

[3] O.A. LADYŽENSKAJA, N.N. URALCEVA: Linějnyje i kvazilinějnyje uravněnija elliptičeskogo tipa, Izd.Nauka,Moskva 1964.

[4] C.B. MORREY: Quelques résultats récents du calcul des variations, Colloque CNRS,Paris 1962.

[5] C.B. MORREY: Multiple integral problems in the calculus of variations and related topics, Univ.of California Publ.1(1943),1-130.

[6] J. NEČAS: On the existence and regularity of solutions of nonlinear elliptic equations,Proceedings of Equadiff II,Bratislava 1966.

[7] J. NEČAS: Sur la méthode variationnelle pour les équations aux dérivées partielles non linéaires du type elliptique;l'éxistence et la régularité des solutions,Comment.Math.Univ. Carolinae 7,3(1966),301-317.

[8] F.E. BROWDER: Variational methods for nonlinear elliptic eigenvalue problems,Bull.Amer.Math. Soc.71(1965),176-183.

(Received January 13,1967)

COMMENTATIONES MATHEMATICAE UNIVERSITATIS CAROLINAE
20, 1 (1979)

ON THE REGULARITY OF WEAK SOLUTIONS TO NONLINEAR ELLIPTIC SYSTEMS VIA LIOUVILLE'S TYPE PROPERTY

M. GIAQUINTA, J. NEČAS

Abstract: Let u be a weak solution with bounded gradient of a nonlinear elliptic system. In the present paper it is proved that the first derivatives of u are Hölder-continuous if the system satisfies a Liouville's type condition. This condition, roughly speaking, means that every solution defined on the whole R^n to the system and with bounded gradient is a polynomial of at most first degree.

Key words: Regularity, weak solution, nonlinear elliptic system, Liouville's property, Sobolev space.

AMS: 35J60

---

§ 1. Introduction. Let $\Omega \subset R^n$, $n \geqq 2$, be a bounded domain and let us consider a nonlinear elliptic system

$$(1.1) \quad -\frac{\partial}{\partial x_i}[a_i^r(x,u,\nabla u)] + a^r(x,u,\nabla u) = -\frac{\partial f_i^r}{\partial x_i} + f^r,$$

$r = 1,2,\dots,m$, where $u \in [W^{1,\infty}(\Omega)]^m$, ∇u is the set of the derivatives $\frac{\partial u_s}{\partial x_j}$, $a_i^r(x,\xi,\eta)$, $\frac{\partial a_i^r}{\partial x_\ell}$, $\frac{\partial a_i^r}{\partial \xi_s}$, $\frac{\partial a_i^r}{\partial \eta_j^s}$ are continuous functions on $\Omega \times R^m \times R^{nm}$, $f_i^r \in W^{1,p}(\Omega)$, $f^r \in W^{1,\frac{p}{2}}(\Omega)$, $p > n$,

$$(1.2) \quad \frac{\partial a_i^r}{\partial \eta_j^s}(x,\xi,\eta)\eta_i^r\eta_j^s > 0 \text{ for } \eta \neq 0,$$

and the summation convention is used.

Here $W^{k,p}(\Omega)$ denotes, as usual, the Sobolev space of

© Faculty of Mathematics and Physics of Charles University 1979. Reprinted with kind permission from: Giaquinta, M. & Nečas, J., On the regularity of weak solutions to nonlinear elliptic systems via Liouville's type property. Commentationes Mathematicae Universitatis Carolinae 20, 111-121.

$L^p(\Omega)$ functions whose derivatives up to order k are also $L^p(\Omega)$ functions.

We say that (1.1).(1.2) is a <u>regular system</u> (R) if a weak solution u belongs to the space $[C^{1,\alpha}(\Omega)]^m$, where, of course, $C^{1,\alpha}(\Omega)$ is the space of continuously differentiable functions in Ω whose derivatives are locally α-Hölder continuous,

The history of the regularity problem is described in the book by O.A. Ladyženskaja, N.N. Ural'ceva [1], in the paper by Ch.B. Morrey [2] abd elsewhere. It is well known from the result of E. De Giorgi [3] that, for m = 1, the single equation (1.1),(1.2) is regular. By virtue of a counter example of J. Nečas [4], there exist systems (1.1),(1.2) which are not regular for $n \geqq 5$; this question is still open for n = = 3,4. Sufficient conditions for the regularity are also of interest, see M. Giaquinta [5], J. Nečas [6]. Since the examples of the regularity are $[W^{1,\infty}(R^n)]^m$ solutions to a system

$$(1.3) \qquad -\frac{\partial}{\partial x_i}[a_i^r(\nabla u)] = 0$$

of the type

$$(1.4) \qquad |x - x_0|\, g\left(\frac{x-x_0}{|x-x_0|}\right),$$

and in virtue of a trivial fact that a $C^1(R^n)$ vector function of the type (1.4) is a polynomial of at most first degree, we see that the regularity implies weak Liouville's property: we say that the system (1.1),(1.2) has <u>weak Liouville's property</u> (WL), if for every $x^0 \in \Omega$, $\xi \in R^m$, every function v with a bounded gradient of the type (1.4), solving in R^n the system

$$(1.5) \qquad - \frac{\partial}{\partial x_i}[a_i^r(x^\circ, \xi, \nabla v)] = 0,$$

is a polynomial (more exactly, a vector of polynomials) of at most first degree. We speak about <u>Liouville's property</u> (L) if the same is true without supposing (1.4).

We prove in this paper by the "partial regularity" method, see Ch. B. Morrey [7], E. Giusti, M. Miranda [8], E. Giusti [9], M. Giaquinta [5], that (L) $\Longrightarrow$ (R). In this connection 31 relations can be thought of between (R), (WL), (L) (some are trivial), especially (WL) $\overset{?}{\Longrightarrow}$ (R), (R) $\overset{?}{\Longrightarrow}$ (L).

Considering the solutions to (1.3) in the form (1.4), we can get, see J. Nečas [6], that, for m = 1, $n \geqq 2$, we have (WL). Because there is still some hope that for n = 3,4 we get (R) for the systems (1.1),(1.2) it is not unthinkable that we have the property (L) for n = 3,4, which would be a way how to prove this conjecture.

Clearly there are many other interesting questions, as, for example, how to avoid the condition $u \in [W^{1,\infty}(\Omega)]^m$; this seems to be possible via some growth conditions.

We also prove (in § 3 of this paper) an easy result that for the systems (1.1).(1.2) and for n = 2, the property (L) is satisfied. So we get once more the known result that for n = 2 we have (R).

§ 2. <u>Lemmas</u>. Let us first introduce some notation: put

$$u_{x_0,R} = \frac{1}{\operatorname{mes} B_R(x_0)} \int_{B_R(x_0)} u(x)dx,$$

where $B_R(x_0)$ is the ball with the center x_0 and the radius R and

$$U(x_0,R) = R^{-n} \int_{B_R(x_0)} |u(x) - u_{x_0,R}|^2\, dx.$$

Let us mention the result of S. Campanato [10]: if

$$u \in [W^{1,2}_{loc}(B(0,1)) \cap L^2(B(0,1))]^m$$

is a weak solution to the equation with constant coefficients

$$\int_{B(0,1)} b^{hk}_{ij} D_i u_h D_j \psi_k\, dx = 0, \qquad \forall \psi \in [\mathcal{D}(B(0,1))]^m,$$

then for every $0 < \rho < 1$ we have

(*) $U(0,\rho) \leq c\,\rho^2\, U(0,1)$, where c depends on $\max|b^{hk}_{ij}|$

and on the constant α of ellipticity:

$$b^{hk}_{ij}\,\eta^h_i\,\eta^k_j \geq \alpha|\eta|^2 .$$

First we get a modification of the main lemma from [8], [9],[5].

<u>Lemma 2.1.</u> Let $v \in [L^\infty(\Omega)]^N \cap [W^{1,2}(\Omega)]^N$ be a weak solution to the system

$$(2.1) \quad \int_\Omega [A^{hk}_{ij}(x,v) D_i v_h D_j \varphi_k + A^{hk}_i(x,v) D_i v_h \varphi_k]dx =$$

$$= \int_\Omega [g^k_j D_j \varphi_k + g^k \varphi_k]dx,$$

where $\Omega \subset R^n$ is a bounded domain, $A^{hk}_{ij}(x,\xi)$, $A^{hk}_i(x,\xi)$ are continuous functions in $\overline{\Omega} \times R^N$, $g^h_j \in L_p(\Omega)$, $g^k \in L_{\frac{p}{2}}(\Omega)$, $p > n$,

$$(2.2) \qquad A^{hk}_{ij}(x,\xi)\,\eta^h_i\,\eta^k_j > 0 \text{ for } \eta \neq 0.$$

If $x_0 \in \Omega$ and $R \leq \operatorname{dist}(x_0,\partial\Omega)$ we put $v = v^* + w$, where $w \in H^1_0(B(x_0,R))$ is a solution to

$$(2.2') \quad \int_{B(x_0,R)} [A^{hk}_{ij}(x,v) D_i w_h D_j \varphi_k + A^{hk}_i(x,v) D_i w_h \varphi_k]dx =$$

$$= \int_{B(x_0,R)} [g_j^k D_h \varphi_k + g^k \varphi_k] dx.$$

Then for every τ, $0 < \tau < 1$, there exist $\varepsilon_0 = \varepsilon_0(\tau, |v|_\infty)$, $R_0 = R_0(\tau, |v|_\infty)$ such that if $R \leq \min(R_0, \operatorname{dist}(x_0, \partial\Omega))$ and if

$$V^*(x_0, R) < \varepsilon_0^2 \tag{2.3}$$

then

$$V^*(x_0, \tau R) \leq 2c\, \tau^2\, V^*(x_0, R), \tag{2.4}$$

where the constant c is from ($*$).

<u>Proof</u>. Let us suppose the contrary. Then $\exists \tau$, $x_\nu \in \Omega$, $\varepsilon_\nu \longrightarrow 0$, $R_\nu \longrightarrow 0$, $v^\nu \in [H^1(\Omega)]^N$, $|v^\nu|_{L_\infty} \leq |v|_{L_\infty}$, such that $V^{*(\nu)}(x_\nu, R_\nu) = \varepsilon_\nu^2$, $V^{*(\nu)}(x_\nu, \tau R_\nu) > 2c\, \tau^2\, \varepsilon_\nu^2$. Put $x = x_\nu + R_\nu y$, $s^\nu(y) = \varepsilon_\nu^{-1} [v^{*\nu}(x_\nu + R_\nu y) - v^{*\nu}_{x_\nu, R_\nu}]$. We have $\int_{B(0,1)} |s^\nu(y)|^2 dy = S^\nu(0,1) = 1$,

$$S^\nu(0, \tau) > 2c\, \tau^2. \tag{2.5}$$

Put further $t^\nu(y) = \omega^\nu(x_\nu + R_\nu y)$ ($v^\nu = v^{*\nu} + \omega^\nu$). Then we can suppose $x_\nu \longrightarrow x_0 \in \bar{\Omega}$, $s^\nu \rightharpoonup s$ in $L_2(B(0,1))$, $\varepsilon_\nu s^\nu(y) \longrightarrow 0$ almost everywhere in $B(0,1)$. We have

$$v^\nu(x_\nu + R_\nu y) = s^\nu(y)\varepsilon_\nu + v^{*\nu}_{x_\nu, R_\nu} + t^\nu(y). \tag{2.6}$$

Since

$$\int_{B(x_\nu, R_\nu)} |\omega^\nu|^2 dx \leq c_1 R_\nu^2 \int_{B(x_\nu, R_\nu)} D_i \omega_h^\nu D_i \omega_h^\nu dx, \tag{2.7}$$

we first get that (2.2) is uniquely solvable for R_ν small enough. We further get from (2.7) and (2.2) that

$$\int_{B(0,1)} |t^\nu(y)|^2 dy \leq c_2 R^{2(1-\frac{n}{p})}, \tag{2.8}$$

so we can also suppose that $t^{\nu}(y) \longrightarrow 0$ almost everywhere. Hence from (2.6) it follows that we can suppose $v^{*\nu}_{x_\nu, R_\nu} \longrightarrow$ $\longrightarrow \xi \in R^N$ and therefore

$$A^{hk}_{ij}(x_\nu + R_\nu y, s^\nu(y)\varepsilon_\nu + v^{*\nu}_{x_\nu, R_\nu} + t^\nu(y)) \longrightarrow A^{hk}_{ij}(x^\circ, \xi)$$

almost everywhere in B(0,1). Hence we get that $s^\nu \longrightarrow s$ in $[W^{1,2}_{loc}(B(0,1))]^N$ and that

$$(2.9) \qquad \int_{B(0,1)} A^{hk}_{ij}(x^\circ, \xi) D_i s_h D_j \psi_k \, dy = 0$$

$$\forall \psi \in [\mathcal{D}(B(0,1))]^N.$$

Thus we have

$$(2.10) \qquad S(0,\tau) \leq c\,\tau^2 \, S(0,1) \leq c\,\tau^2,$$

which is a contradiction with

$$(2.11) \qquad S(0,\tau) > 2c\,\tau^2$$

obtained from (2.5).

<u>Lemma 2.2.</u> Under the conditions of Lemma 2.1, for every point $x_o \in \Omega$ such that $V^*(x_o, R) < \varepsilon_o^2$, there exists a $B(x_o, R_1) \subset \Omega$ such that $v \in C^{\alpha}(\overline{B(x_o, R_1)})$ with $\alpha = \min(\frac{1}{2}, 1 - \frac{n}{p})$.

<u>Proof</u>. We get by a standard argument that if $\sigma > 0$ is small enough, $|\bar{x} - x_o| < \sigma$, and $R_{\bar{x}} = R - |\bar{x} - x_o|$, then $V^*(\bar{x}, R_{\bar{x}}) < \varepsilon_o^2$. If $v = v^* + \omega$ in $B(\bar{x}, R_{\bar{x}})$, we first have

$$(2.12) \qquad \int_{B(\bar{x}, R_{\bar{x}})} |\omega|^2 dx \leq c_1 R^2_{\bar{x}} \int_{B(\bar{x}, R_{\bar{x}})} |D\omega|^2 dx \leq$$

$$\leq c_2 R^2_{\bar{x}} R_{\bar{x}}^{n(1-\frac{2}{p})} \Big[\sum \Big[\int_B |f^r_i|^p dx \Big]^{\frac{2}{p}} +$$

$$+ \sum \Big(\int_B |f^r|^{\frac{p}{2}} dx \Big)^{\frac{4}{p}} \Big] \leq c_3 R_{\bar{x}}^{2+n-\frac{2n}{p}} .$$

Thus

$$(2.13)\qquad V(\bar{x}, \tau R_{\bar{x}}) \leq 2V^*(\bar{x}, \tau R_{\bar{x}}) + 2\Omega(\bar{x}, \tau R_{\bar{x}}) \leq$$

$$\leq 4c\,\tau^2 V^*(\bar{x}, R_{\bar{x}}) + 2c_3 R_{\bar{x}}^{2(1-\frac{n}{p})}\ \tau^{-n} \leq$$

$$\leq 8c\,\tau^2 V(\bar{x}, R_{\bar{x}}) + 8c\,\tau^2 c_3 R_{\bar{x}}^{2(1-\frac{n}{p})} + 2c_3 \tau^{-n} R_{\bar{x}}^{2(1-\frac{n}{p})}.$$

Choose $\tau \in (0,1)$ such that $8c\tau = \varkappa \leq 1$ and small enough. We get from (2.13) that

$$(2.14)\qquad V(\bar{x}, \tau R_{\bar{x}}) \leq 8c\,\tau^2\, V(\bar{x}, R_{\bar{x}}) + c_4\, R_{\bar{x}}^{2(1-\frac{n}{p})}.$$

For k being a positive integer, we get from (2.14) that

$$(2.15)\qquad V(\bar{x}, \tau^k R_{\bar{x}}) \leq \tau^k\, V(\bar{x}, R_{\bar{x}}) +$$

$$+ R_{\bar{x}}^{2(1-\frac{n}{p})}\, c_4\, \frac{(\varkappa\tau)^k + \tau^{k2(1-\frac{n}{p})}}{|\varkappa\tau - \tau^{2(1-\frac{n}{p})}|}.$$

If $0 < \rho < R - \delta$ and if we choose k such that $\tau^{k+1} R_{\bar{x}} < \rho \leq \tau^k R_{\bar{x}}$, we get $\tau^n V(\bar{x}, \rho) \leq \left(\frac{\rho}{\tau^k R_{\bar{x}}}\right)^n V(\bar{x}, \rho) \leq V(\bar{x}, \tau^k R_{\bar{x}}) \leq$ $\leq \frac{\rho}{R_{\bar{x}}\tau} V(\bar{x}, R_{\bar{x}}) + c_4 R_{\bar{x}}^{2(1-\frac{n}{p})} \dfrac{\frac{\rho}{R_{\bar{x}}\tau} + \left(\frac{\rho}{R_{\bar{x}}\tau}\right)^{2(1-\frac{n}{p})}}{|\varkappa\tau - \tau^{2(1-\frac{n}{p})}|}$, and using [10], we get the result, q.e.d.

§ 3. Main results

Theorem 1. Let $u \in [W^{1,\infty}(\Omega)]^m$ be a weak solution to (1.1) and let the conditions on a_i^r, a^r, f_i^r, f^r, Ω, mentioned in § 1, be fulfilled. Let the system (1.1) satisfy the Liouville's property, i.e., for $\forall x^\circ \in \Omega$ and $\forall \xi \in R^m$ the only solution to (1.5) defined in the whole R^n and possessing a bounded gradient is a polynomial of at most first degree.

Then $u \in [C^{1,\alpha}(\Omega)]^m$, $\alpha = \min(\frac{1}{2}, 1 - \frac{n}{p})$.

<u>Proof.</u> Let $x^0 \in \Omega$. Put $u_R(y) = \frac{1}{R}[u(x^0 + Ry) - u(x^0)]$, $x^0 + Ry = x$. If O is the image of Ω we have

$$(3.1) \quad \int_O [a_i^r(x^0+Ry,\ Ru_R(y)+u(x^0),\ \nabla_y u_R(y)) \frac{\partial \psi_r(y)}{\partial y_i} +$$

$$+ a^r(x^0+Ry,\ Ru_R(y) + u(x^0),\ \nabla_y u_R(y)) R \psi_r(y)]\, dy =$$

$$= \int_O [f_i^r(x^0+Ry) \frac{\partial \psi_r}{\partial y_i}(y) + f^r(x^0+Ry) R \psi_r(y)]\, dy.$$

Let $B(0,a) \subset O$. We get in a standard way that

$$(3.2) \qquad \int_{B(0,a)} | D^2 u_R(y)|^2 \, dy \leqq c(a).$$

Hence we can choose $R_k \longrightarrow 0$ in such a way, that $u_{R_k} \longrightarrow p$ in $[W^{1,2}(B(0,a))]^m$ $\forall a > 0$. Thus $p \in [W^{1,\infty}(R^n)]^m$ and it is a weak solution to

$$(3.3) \quad \int_{R^n} a_i^r(x^0, u(x^0),\ \nabla_y p) \frac{\partial \psi_r}{\partial y_i}\, dy = 0$$

$$\forall \psi \in [\mathcal{D}(R^n)]^m.$$

Therefore, by assumption, p is a polynomial of at most first degree. So we have

$$(3.3') \quad 0 \longleftarrow \int_{B(0,1)} | Du_{R_k}(y) - Dp|^2 \, dy = R_k^{-n} \int_{B(x^0, R_k)} | Du(x) - Dp|^2 dx.$$

If $'$ is the $\frac{\partial}{\partial x_t}$ derivative we get from (1.1) the equation in variations

$$(3.4) \int_\Omega [\frac{\partial a_i^r}{\partial \frac{\partial u_s}{\partial x_j}} \frac{\partial u_s'}{\partial x_j} \frac{\partial \varphi_r}{\partial x_i} + \frac{\partial a_i^r}{\partial u_s} u_s' \frac{\partial \varphi_r}{\partial x_i} + \frac{\partial a_i^r}{\partial x_t} \frac{\partial \varphi_r}{\partial x_i} +$$

$$+ \frac{\partial a^r}{\partial \frac{\partial u_s}{\partial x_j}} \frac{\partial u_s}{\partial x_j} \varphi_r + \frac{\partial a^r}{\partial u_s} u'_s \varphi_r + \frac{\partial a^r}{\partial x_t} \varphi_r] dx =$$

$$= \int_\Omega [f_i^{r'} \frac{\partial \varphi_r}{\partial x_i} + f^{r'} \varphi_r] dx .$$

Writing (3.4) for every $\frac{\partial}{\partial x_t}$, $t = 1,2,\dots,n$, removing the terms $\frac{\partial a_i^r}{\partial u_s} u'_s \frac{\partial \varphi_r}{\partial x_i}$, $\frac{\partial a_i^r}{\partial x_t} \frac{\partial \varphi_r}{\partial x_i}$, $\frac{\partial a^r}{\partial u_s} u'_s \varphi_r$, $\frac{\partial a^r}{\partial x_t} \varphi_r$ to the right-hand side of (3.4), and denoting by v_κ the derivatives $\frac{\partial u_s}{\partial x_t}$, we get, with $\frac{\partial a_i^r}{\partial \frac{\partial u_s}{\partial x_j}}(x,u(x),v) \equiv b_{ij}^{rs}(x,v)$ (and the same with a_i^r), a system of the type (2.1). The result follows from Lemmas 2.1, 2.2 and from (3.3′), because, in decomposing $v = v^* + \omega$ on $B(x^o,R)$ as in Lemma 2.1, we have $\Omega(x^o,R) \longrightarrow 0$ for $R \longrightarrow 0$, as above, so $V^*(x^o,R) \longrightarrow 0$, q.e.d.

<u>Theorem 2</u>. Let us consider the system (1.1),(1.2). Let n be the dimension of the space, n = 2. Then (L) is satisfied.

<u>Proof</u>. Let $v \in [W^{1,\infty}(R^2)]^m$ be a weak solution to the equation

$$(3.5) \qquad \int_{R^2} a_i^r(x^o, \xi, \nabla v) \frac{\partial \psi_r}{\partial y_i} dy = 0.$$

Let $T > 0$ and let $\eta \in \mathcal{D}(B(0,2T))$, $0 \leq \eta \leq 1$, $\eta = 1$ in $B(0,T)$, $|D_i \eta| \leq \frac{c_1}{T}$. We get the equation in variations

$$(3.6) \qquad \int_{R^2} \frac{\partial a_i^r}{\partial \frac{\partial v_s}{\partial x_j}}(x^o, \xi, \nabla v) \frac{\partial v'_s}{\partial y_j} \cdot \frac{\partial \psi_r}{\partial y_i} dy = 0.$$

Putting $\psi_r = v'_r \eta^2$, we get from (3.6), using the boundedness of the gradient, that

(3.7) $\int_{B(0,2T)} |Dv'|^2 \eta^2 \, dy \leq c_2.$

Hence $\int_{R^2} |Dv'|^2 \, dy < \infty$. But there exists $\psi^n \in [\mathcal{D}(R^2)]^m$ such that $D\psi^n \to Dv'$ in $[L^2(R^2)]^{2m}$ (and there exists $\Lambda^n \in R^m$ such that $\Lambda^n + \psi^n \to v'$ in $[L^2_{loc}(R^n)]^m$. Hence

$$\int_{R^2} \frac{\partial a_i^{\kappa}}{\partial \frac{\partial v_s}{\partial x_j}}(x^0, \xi, \nabla v) \frac{\partial v'_s}{\partial y_j} \frac{\partial v'_{\kappa}}{\partial y_i} \, dy = 0$$

and thus v is a polynomial of at most first degree.

References

[1] O.A. LADYŽENSKAJA, N.N. URAL'CEVA: Linejnye i kvazilinejnye uravnenija elliptičeskogo tipa, Moscow (1973), 2-nd edition,

[2] Ch.B. MORREY: Differentiability theorems for weak solutions of nonlinear elliptic differential equations, BAMS, Vol. 75(1969), 684-705.

[3] E. De GIORGI: Sulla differenziabilità e analiticità delle estremali degli integrali multipli regolari, Mem. Acad. Sci. Torino Cl. Sci. Fis. Mat. Nat. (3),3(1957), 25-43.

[4] J. NEČAS: Example of an irregular solution to a nonlinear elliptic system with analytic coefficients and conditions for regularity, Theory of Nonlinear Operators, Abhandlungen der Akademie der Wissenschaften der DDR, Jahrg. 1977, Nr. 1N, 197-206.

[5] M. GIAQUINTA: Sistemi ellittici non lineari, Convegno su: Sistemi ellittici non lineari ed applicazioni, Università di Ferrara, Editrice Universitaria, 1978,

[6] J. NEČAS: On the regularity of weak solutions to variational equations and inequalities for nonline-

ar second order elliptic systems, Proceedings of Equadiff IV, Prague 1977, to appear in Springer 1979,

[7] Ch.B. MORREY: Partial regularity results for nonlinear elliptic systems, Journ. Math. and Mech. 17(1968),

[8] E. GIUSTI; M. MIRANDA: Sulla regolarità delle soluzioni deboli di una classe di sistemi ellittici quasi lineari, Arch. Rat. Mech. and Anal. 31(1968),

[9] E. GIUSTI: Regolarità parziale delle soluzioni di sistemi ellittici quasi lineari di ordine arbitrario, Ann. Scuola Norm. Sup. Pisa 23(1969),

[10] S. CAMPANATO: Equazioni ellitiche del II ordine e spazi $L^{2,\lambda}$, Ann. Mat. Pura e Appl. 69(1965),

Università di Ferrara,
Ferrara
Italia

Mat.-fyz. fakulta Karlovy Univ.
Malostranské nám. 25,
Praha - Malá Strana
Československo

(Oblatum 6.11. 1978)

On the regularity of weak solutions to non linear elliptic systems of partial differential equations

By *M. Giaquinta* at Firenze and *J. Nečas* at Praha

The aim of this paper is to discuss the regularity (partial and everywhere) of the weak solutions of second order non linear elliptic systems with particular emphasis on the everywhere regularity.

We shall restrict ourselves to second order elliptic systems in divergence form

$$(0.1)\qquad -\frac{\partial}{\partial x_i} a_i^r(x, u, \nabla u) + a^r(x, u, \nabla u) = -\frac{\partial f_i^r}{\partial x_i} + f^r, \qquad r = 1, \dots, m$$

where $u(x) = (u_1(x), \dots, u_m(x))$ and x belongs to a domain Ω of R^n; we shall use the summation convention over repeated indices and the standard notations D_r for $\frac{\partial}{\partial x_r}$, $|\nabla u|^2$ for $D_r u\, D_r u$.

We shall suppose that coefficients in (0.1) are continuously differentiable in $\bar{\Omega} \times R^m \times R^{nm}$ and satisfy the *strong ellipticity* condition:

$$(0.2)\qquad \frac{\partial a_i^r(x, \xi, \eta)}{\partial \eta_j^s} \zeta_i^r \zeta_j^s > 0 \quad \text{for} \quad \zeta \neq 0.$$

A weak solution to system (0.1) is a vector function from $[W^{1,p}(\Omega)]^m$ satisfying (0.1) in the weak sense, where $W^{1,p}$ is the usual Sobolev space. For the sake of simplicity we shall write in the sequel $W^{1,p}(\Omega)$ also for vector valued functions.

Such a solution is regular if it belongs to $C^{1,\alpha}(\Omega)$; we shall refer to the property that 'every weak solution to (0.1) is regular' as 'property (R)'.

For the history of the regularity problem, that is, the history of the 19-th Hilbert problem, we refer to Ch. B. Morrey [18], [19], O. A. Ladyzenskaya and N. N. Ural'tseva [16] and, as for what concerns mainly systems, to Morrey [20], [21], Giusti and Miranda [14], [15], Giaquinta [7], Nečas [25], Giaquinta and Modica [10]. We only mention that if we get $C^{1,\alpha}$ solutions then $C^{l,\alpha}$ regularity for $1 < l \leqq +\infty$ can be obtained from the theory of linear equations. Moreover we mention that for $n = 2$, $m \geqq 1$ the regularity was proved by Ch. B. Morrey [18]; while for $m = 1$ and $n \geqq 2$ we have the famous result due to de Giorgi [4] and Nash [22].

0075-4102/80/0316-0011$02.00
Copyright by Walter de Gruyter & Co.

© Walter de Gruyter 1980. Reprinted with kind permission from: Giaquinta, M. & Necaš, J., On the regularity of weak solutions to nonlinear elliptic systems of partial differential equations, Journal für die reine und angewandte Mathematik, 316:140-159.

For $n \geqq 3$, $m > 1$ counterexamples to the regularity were given by Giusti and Miranda [14], Mazja [17], Arakcejev [1], and Giaquinta [6]. But all these counterexamples are essentially quasi-linear elliptic systems of the type

$$-\frac{\partial}{\partial x_i}\left[A_{ij}^{rs}(u)\frac{\partial u_s}{\partial x_j}\right]=0$$

with solutions in $L^\infty(R^n) \cap W^{1,2}_{\mathrm{loc}}(R^n)$ of the form $u=\frac{x}{|x|}$. On the other hand for completely non linear systems of the type

$$-\frac{\partial}{\partial x_i}a_i^r(\nabla u)=0$$

which are very important, for example, in the continuum mechanics, counterexamples have been given by J. Nečas [24], [25], who presented solutions in R^n, but with $n \geqq 5$, which are uniformly Lipschitz-continuous, i.e. $|\nabla u| \leqq c < +\infty$ in R^n, but not regular solutions.

In section 4 we shall present a new counterexample in dimension $n \geqq 3$, due to the second author, and precisely a strongly elliptic system of equations, with infinitely differentiable coefficients of the form

$$-\frac{\partial}{\partial x_i}a_i^{kl}(\nabla u)=0 \qquad k, l=1,\dots,n \tag{0. 3}$$

with the solution $u=(u_{hk})$

$$u_{hk}=\frac{x_h x_k}{|x|}-\frac{1}{n}|x|\delta_{hk}$$

which has bounded gradient in R^n.

Equations (0. 3) are not yet Euler's equations and $m=\frac{n^2+n}{2}-1$ due to the symmetry of the matrix u_{hk} and the condition $u_{kk}=0$. However from the former counterexamples we see that in general the 19-th Hilbert problem has a negative answer.

The above mentioned example shows that there exist systems of the type (0. 3) with solution v in the whole R^n satisfying $|\nabla v| \leqq c < +\infty$ which are not polynomial (vector polynomial) of at most first degree. So one is led to the following definition.

Definition. The system (0. 1) satisfies the Liouville's property (L) if for every $x_0 \in \Omega$ and every $\xi \in R^n$ the only solutions in R^n to

$$-\frac{\partial}{\partial x_i}a_i^r(x^0, \xi, \nabla v)=0 \tag{0. 4}$$

with bounded gradient, $|\nabla v| \leqq c < +\infty$, are polynomials of at most first degree.

We shall see in section 2, compare also Giaquinta-Nečas [11], that $(L) \Rightarrow (R)$. In section 3 and also 2 we give a few conditions which grant that property (L) and then (R) holds, while in section 1 we discuss a few results concerning the partial regularity of weak solutions.

1. Partial regularity

In this section we want to present and discuss a few results concerning the partial regularity of weak solutions of non linear elliptic systems.

We shall first consider quasilinear systems of the type

$$(1.1)\qquad -\frac{\partial}{\partial x_i}\left[A_{ij}^{rs}(x,v)\frac{\partial v_s}{\partial x_j}\right]+A_j^{rs}(x,v)\frac{\partial v_s}{\partial x_j}=-\frac{\partial g_i^r}{\partial x_i}+g^r,\qquad r=1,\dots,m$$

where $x\in\Omega$ bounded domain of $\mathbb{R}^n$, the coefficients $A_{ij}^{rs}(x,v)$, $A_j^{rs}(x,v)$ are continuous functions in $\bar\Omega\times\mathbb{R}^m$, $g_i^r\in L^p(\Omega)$, $g^r\in L^{p/2}(\Omega)$ $p>n$ and we shall suppose that the *strong ellipticity* condition holds:

$$(1.2)\qquad A_{ij}^{rs}(x,\xi)\,\zeta_i^r\zeta_j^s>0\quad\text{for}\quad \zeta\neq 0.$$

Note that for $|\xi|\leqq M$ and $x_0\in\Omega$ we have

$$A_{ij}^{rs}(x^0,\xi)\,A_{ij}^{rs}(x^0,\xi)\leqq N,\quad A_{ij}^{rs}(x^0,\xi)\,\zeta_i^r\zeta_j^s\geqq \nu|\zeta|^2$$

where N, ν are positive constants depending on M.

We shall restrict ourselves to consider bounded weak solutions to system (1.1), i.e. vector functions $v\in W^{1,2}(\Omega)\cap L^\infty(\Omega)$ satisfying

$$\int_\Omega [A_{ij}^{rs}(x,v)\,D_jv_sD_i\psi_r+A_j^{rs}D_jv_s\cdot\psi_r]\,dx=\int_\Omega[g^r\psi_r+g_i^rD_i\psi_r]\,dx\qquad\forall\psi\in\mathscr{D}(\Omega).$$

If v is such a solution, $x^0\in\Omega$ and $R<\operatorname{dist}(x^0,\partial\Omega)$ be small enough, the linear elliptic system

$$\int_{B_R(x^0)}[A_{ij}^{rs}(x,v)\,D_jw_s^RD_i\varphi_r+A_j^{rs}(x,v)\,D_jw_s^R\varphi_r]\,dx=\int_{B_R(x^0)}[g_i^rD_i\varphi_r+g^r\varphi_r]\,dx\qquad\forall\varphi\in\mathscr{D}(B_R(x_0))$$

is uniquely solvable in $W_0^{1,2}(B_R(x^0))$ ($B_R(x_0)$ being the ball of center x_0 and radius R). We shall set then

$$(1.3)\qquad v=v^R+w^R.$$

Moreover for a vector valued function u we shall define

$$u_{x^0,R}=\fint_{B_R(x^0)}u\,dx=\frac{1}{\operatorname{meas}B_R(x^0)}\int_{B_R(x^0)}u(x)\,dx$$

and

$$U(x_0,R)=R^{-n}\int_{B_R(x^0)}|u(x)-u_{x^0,R}|^2\,dx.$$

With the former assumptions and notations we now are ready to state the main results of this section.

Proposition 1. 1. *Let $v \in W^{1,2}(\Omega) \cap L^\infty(\Omega)$ be a weak solution to system* (1. 1). *Then for every $x^0 \in \Omega$ such that*

$$\liminf_{R \to 0^+} V^R(x^0, R) = 0$$

there exists a ball $B_{R_1}(x^0) \subset \Omega$ such that $v \in C^\alpha(B_{R_1}(x^0))$ with

$$0 < \alpha < \varkappa = \frac{1}{2} \wedge \left(1 - \frac{n}{p}\right)$$

and hence with $\alpha = 1 - \dfrac{n}{p}$.

Theorem 1. 2. *Let $v \in W^{1,2}(\Omega) \cap L^\infty(\Omega)$ be a weak solution to system* (1. 1). *Then v is Hölder continuous on an open set $\Omega_0 \subset \Omega$ with Hölder exponent $1 - \dfrac{n}{p}$. Moreover for the singular set $\Omega \setminus \Omega_0$ we have*

$$\Omega \setminus \Omega_0 \subset \{x \in \Omega : \liminf_{R \to 0^+} \int_{B_R(x)} |v(y) - v_{x,R}|^2 dy > 0\}$$

and

$$\mathcal{H}^{n-2}(\Omega \setminus \Omega_0) = 0$$

where $\mathcal{H}^{n-2}$ denotes the $(n-2)$-dimensional Hausdorff measure.

The main argument we shall use in proving the former results is one of reduction to absurd contained in lemma 1. 4 and connected with a device introduced by F. J. Almgren and E. de Giorgi in order to study the regularity of minimal surfaces. So our proof appears as a slight modification of the analogous ones in C. B. Morrey [20], E. Giusti, M. Miranda [15], E. Giusti [13], [12]. And in fact theorem 1. 2 is essentially contained in Giusti-Miranda [15], Giusti [13], [12]. See also [8] for a direct argument.

Let b_{ij}^{rs}; $r, s = 1, \ldots, m$; $i, j = 1, \ldots, n$ be constant and let $u \in W^{1,2}_{\text{loc}}(B_1(0)) \cap L^2(B_1(0))$ be a weak solution to

$$\int_{B_1(0)} b_{ij}^{rs} D_j u_s D_i \psi_r dx = 0 \qquad \forall \psi \in \mathcal{D}(B_1(0)). \tag{1. 4}$$

Set

$$N = b_{ij}^{rs} b_{ij}^{rs}$$

and suppose system (1. 4) strongly elliptic, i.e.

$$b_{ij}^{rs} \eta_i^r \eta_j^s \geqq \nu |\eta|^2 \qquad \forall \eta;\ \nu > 0.$$

We have the following result, see [3].

Lemma 1. 3. *Let u be a weak solution of system* (1. 4). *Then*

$$U(0, \varrho) \leqq c\varrho^2\, U(0, 1) \tag{1. 5}$$

for $0 \leqq \varrho \leqq 1$, where c is a constant depending only on N and ν.

Proof. We get from (1. 4)

$$\int_{B_{1/2}(0)} |\nabla u|^2 dx \leqq c_1 \int_{B_1(0)} |u-u_{0,1}|^2 dx = c_2\, U(0,1)$$

and from the regularity of solution u in the interior of $B_1(0)$

$$\max_{B_{1/4}(0)} |\nabla u|^2 \leqq c_3 \int_{B_{1/2}(0)} |\nabla u|^2;$$

hence

$$t^{-n} \int_{B_t(0)} |\nabla u|^2 dx \leqq c_4 \int_{B_{1/2}(0)} |\nabla u|^2 dx$$

for $0<t<\frac{1}{4}$. Therefore, taking into account Poincaré's inequality

$$t^{-2}\, U(0,t) \leqq c_5 t^{-n} \int_{B_t} |\nabla u|^2 dx$$

we obtain the result for $0<\varrho<\frac{1}{4}$. As (1. 5) is obvious for $\frac{1}{4}\leqq \varrho \leqq 1$, the proof is complete. q.e.d.

Lemma 1. 4. *Let v be a weak solution of system (1. 1), $\|v\|_\infty \leqq M$ and $0<\tau<1$. There exist $\varepsilon_0=\varepsilon_0(\tau, M)$, $R_0=R_0(\tau, M)$ such that if $R\leqq R_0 \wedge \operatorname{dist}(x^0, \partial\Omega)$ $x^0\in\Omega$ and if*

(1. 6) $$V^R(x^0, R) < \varepsilon_0^2$$

then

(1. 7) $$V^R(x^0, \tau R) \leqq 2c\tau^2 V^R(x^0, R)$$

where c is the constant in (1. 5) for the family of coefficients

$$A_{ij}^{rs}(x^0, \xi),\ x^0\in\Omega,\ |\xi|_{R^m} \leqq \|v\|_{L^\infty(\Omega)}.$$

Proof. Let us suppose the contrary. Then there exist $\tau\in\,]0,1[$, $x^\nu\in\Omega$, $\varepsilon^\nu\to 0$, $R_\nu\to 0$ and solutions $v^\nu\in W^{1,2}(\Omega)\cap L^\infty(\Omega)$ $\|v^\nu\|_\infty\leqq M$ to system (1. 1) such that

$$V^{\nu R_\nu}(x^\nu, R_\nu) = \varepsilon_\nu^2$$

and

$$V^{\nu R_\nu}(x^\nu, \tau R_\nu) > 2c\tau^2\varepsilon_\nu^2.$$

Set

$$x = x^\nu + R_\nu y,\quad s^\nu(y) = \varepsilon_\nu^{-1}[v^{\nu R_\nu}(x^\nu + R_\nu y) - v^{\nu R_\nu}_{x^\nu, R_\nu}].$$

We have

(1. 8) $$S^\nu(0,1) = \int_{B_1(0)} |s^\nu(y)|^2 dy = \varepsilon_\nu^{-2}\, V^{\nu R_\nu}(x_\nu, R_\nu) = 1,$$

(1. 9) $$\begin{aligned} S^\nu(0,\tau) &= \tau^{-n}\int_{B_\tau(0)} |s^\nu(y) - s^\nu_{0,\tau}|^2 dy = \tau^{-n} R_\nu^{-n}\varepsilon_\nu^{-2} \int_{B_{\tau R_\nu}(x^\nu)} |v^{\nu R_\nu}(x) - v^{\nu R_\nu}_{x^\nu, R_\nu} - \varepsilon_\nu s^\nu_{0,\tau}|^2 dx \\ &\geqq \varepsilon_\nu^{-2}\, V^{\nu, R_\nu}(x^\nu, \tau R_\nu) > 2c\tau^2. \end{aligned}$$

So that we can suppose that $x^\nu \to x^0$, $s^\nu(y) \to s(y)$ in $L^2(B_1(0))$ and $\varepsilon_\nu s^\nu(y) \to 0$ almost everywhere. Let $t^\nu(y) = w^{\nu R_\nu}(x^\nu + R_\nu y)$. We have by definition

$$v^\nu(x^\nu + R_\nu y) = s^\nu(y) \cdot \varepsilon_\nu + v^{\nu R_\nu}_{x^\nu, R_\nu} + t^\nu(y)$$

and

$$\int_{B_1(0)} |t^\nu(y)|^2 \, dy \leqq \operatorname{const} R_\nu^{2\varkappa},$$

$\varkappa = 1 - \dfrac{n}{p}$ for $n \geqq 3$, $0 < \varkappa < 1 - \dfrac{2}{p}$ arbitrary for $n = 2$.

Thus we can suppose that $t^\nu \to 0$ almost everywhere in $B_1(0)$ and $v^{\nu R_\nu}_{x^\nu, R_\nu} \to \xi$ with clearly $|\xi| \leqq M$; hence

$$A^{rs}_{ij}\left(x^\nu + R_\nu y, \varepsilon^\nu s^\nu(y) + v^{\nu R_\nu}_{x^\nu, R_\nu} + t^\nu(y)\right) \to A^{rs}_{ij}(x^0, \xi)$$

almost everywhere.

Set now

$$B^{rt\nu}_{ij}(y) = A^{rt}_{ij}\left(x^\nu + R_\nu y, v^\nu(x^\nu + R_\nu y)\right), \quad B^{rt\nu}_{j}(y) = A^{rt}_{j}\left(x^\nu + R_\nu y, v^\nu(x^\nu + R_\nu y)\right).$$

We have

$$\int_{B_1(0)} [B^{rt\nu}_{ij}(y) \, D_j s^\nu_t D_i \psi_r + R_\nu B^{rt}_j(y) \, D_j s^\nu_t \cdot \psi_r] \, dy = 0 \qquad \forall \psi \in \mathscr{D}(B_1(0)).$$

Therefore we get, taking into account (1.8), (1.9), that $s^\nu(y)$ converges in $L^2(B_1(0))$ and weakly in $W^{1,2}_{\mathrm{loc}}(B_1(0))$ to $s(y)$ which is solution of the system

$$\int_{B_1(0)} A^{rt}_{ij}(x^0, \xi) \, D_j s_t D_i \psi_r \, dx = 0 \qquad \forall \psi \in \mathscr{D}(\Omega).$$

Moreover

$$S(0, 1) = 1, \quad S(0, \tau) \geqq 2c\tau^2$$

which contradict lemma 1.3. q.e.d.

Proof of proposition 1.1. Choose $0 < \tau < 1$ in such a way that $\sqrt{2c}\sqrt{\tau} < \frac{1}{2}$, where c is the constant in (1.7). Let $R_1 \leqq R_0 \wedge \operatorname{dist}(x^0, \partial\Omega) \wedge 1$ be such that

$$V^{R_1}(x^0, R_1) < \varepsilon_0^2$$

where ε_0 and R_0 are the constants in lemma 1.4, we have

$$V^{R_1}(x^0, \tau R_1) \leqq 2c\tau^2 \, V^{R_1}(x^0, R_1);$$

hence

$$\sqrt{V^{\tau R_1}(x^0, \tau R_1)} \leqq \sqrt{2c}\,\tau \sqrt{V^{R_1}(x^0, R_1)} + \gamma_2^{1/2} R_1^\varkappa (\tau^\varkappa + \tau^{1-\frac{n}{2}}) \leqq \frac{\varepsilon_0}{2} + \gamma_2^{1/2} R_1^\varkappa (\tau^\varkappa + \tau^{1-\frac{n}{2}})$$

because for any $R < \operatorname{dist}(x^0, \partial\Omega)$

$$\text{(1.10)} \qquad \int_{B_R(x^0)} |w^R|^2 \, dx \leqq \gamma_1 R^2 \int_{B_R(x^0)} |\nabla w^R|^2 \, dx \leqq \gamma_2 R^{2+n-\frac{2n}{p}}.$$

Therefore we get

$$\sqrt{V^{\tau R_1}(x^0, \tau R_1)} \leqq \tau^{\varkappa} \sqrt{V^{R_1}(x^0, R_1)} + \gamma_3 R_1^{\varkappa} \tag{1.11}$$

and, if $\gamma_2^{1/2} R_1^{\varkappa}(\tau^{\varkappa} + \tau^{1-\frac{n}{2}}) < \frac{\varepsilon_0}{2}$,

$$V^{\tau R_1}(x^0, \tau R_1) < \varepsilon_0^2.$$

So that we can reiterate (1.11) getting for every integer k

$$\sqrt{V^{\tau^k R_1}(x^0, \tau^k R_1)} \leqq \tau^{k\varkappa} \sqrt{V^{R_1}(x^0, R_1)} + k\gamma_3 \tau^{\varkappa(k-1)} R_1^{\varkappa}$$

and hence

$$\sqrt{V^{\tau^k R_1}(x^0, \tau^k R_1)} \leqq \gamma_4 \tau^{\alpha k}. \tag{1.12}$$

From (1.12) and (1.10) now easily follows the estimate

$$\sqrt{V(x^0, \tau^k R_1)} \leqq \gamma_5 \tau^{\alpha k}. \tag{1.13}$$

But once having $V^{R_1}(x^0, R_1) < \varepsilon_0$, if δ is small enough we have for $|x - x^0| < \delta$, $R_x = R_1 - |x - x^0|$

$$V^{R_x}(x_1 R_x) < \varepsilon_0^2$$

because of the continuous dependence of w^{R_x} and $V^{R_x}(x_1 R_x)$ on x. Let now $0 < \varrho < R - \delta$ and let us choose k in such a way that $\tau^{k+1} R_x < \varrho < \tau^k R_x$; then we get

$$\tau^n V(x, \varrho) \leqq \left(\frac{\varrho}{\tau^k R_x}\right)^n V(x, \varrho) \leqq V(x, \tau^k R_x) \leqq \gamma_5^2 \left(\frac{\varrho}{R_x \tau}\right)^{2\alpha}$$

which implies that $v \in C^{\alpha}(B_{R_1}(x^0))$, see [2]. q.e.d.

Proof of theorem 1. 2. The proof of the first part of the statement is contained in the proof of proposition 1. 1; anyway it follows from proposition 1. 1 and (1. 10).

The second part is a consequence of the following result due to E. Giusti [12]: if $f \in L^1_{\text{loc}}$ then

$$\mathscr{H}^{n-q}\{x \in \Omega : \liminf_{R \to 0^+} R^{q-n} \int_{B_R(x)} |f| dy > 0\} = 0.$$

In fact

$$R^{-n} \int_{B_R(x)} |v(y) - v_{x_1 R}|^2 \leqq R^{2-n} \int_{B_R(x)} |\nabla v|^2. \qquad \text{q.e.d.}$$

Remark 1. 5. As from the Hölder inequality we have

$$(R^{2-n} \int_{B_R} |\nabla v|^2 dx)^{1/2} \leqq (R^{q-n} \int_{B_R} |\nabla v|^q dx)^{1/q}$$

we see that one can improve the estimate of the Hausdorff dimension of the singular set once it is known that $v \in W^{1,p}(\Omega)$, $p \geqq 2$: in this case in fact we would have

$$\mathscr{H}^{n-p}(\Omega \setminus \Omega_0) = 0.$$

We end this section with a few remarks on the partial regularity of weak solutions (not necessarily bounded) of non linear systems.

As our main interest is in studying solutions of systems (0. 1) with bounded gradient, we have restricted ourselves to $W^{1,2} \cap L^\infty$ solutions of system (1. 1). In fact the former statements and proofs extend easily to $W^{1,2}$ solutions: that is, the restriction of boundedness can be dropped, compare [20], [15], [13]. Moreover by the same technique one can study the partial regularity of solutions of systems of the type

$$(1.14) \qquad -D_i[A_{ij}^{rs}(x,u)\,D_j u_s] = B_r(x,u,\nabla u) \qquad r = 1 - m$$

but with a linear growth with respect to the gradient of u on $B(x,u,\nabla u)$, compare [13]. In these cases instead of the inclusion in theorem 1. 1 the following occurs:

$$\Omega \setminus \Omega_0 \subset \{x \in \Omega \colon \liminf_{R \to 0^+} R^{-n} \int_{B_R(x)} |v(y) - v_{x,R}|^2 dy > 0\} \cup \{x \in \Omega \colon \sup_R |v_{x,R}| = +\infty\}.$$

It is to be noted that the above argument turns out to be unsuccessful when carried over to treat systems of the type (1. 14) with a quadratic growth on B with respect to the gradient or to treat general non linear systems

$$(1.15) \qquad -D_i A_i^r(x,u,\nabla u) = B^r(x,u,\nabla u)$$

without the assumption of the boundedness of the gradient, under the *natural* growth conditions which follow:

$$\text{(i)} \qquad \begin{aligned} &|A_i^r(x,u,\nabla u)| \leqq L(1+|\nabla u|),\\ &|A_{iu_s}^r(x,u,\nabla u)| \leqq L(1+|\nabla u|),\\ &|A_{ip_j^s}^r(x,u,p)| \leqq L, \end{aligned}$$

the strong ellipticity condition

$$\text{(ii)} \qquad A_{ip_j^s}^r(x,u,p)\,\xi_i^r \xi_j^s \geqq \nu|\xi|^2 \qquad \forall \xi \in \mathbb{R}^{nm};\ \nu > 0$$

and

$$\text{(iii)} \qquad |B(x,u,p)| \leqq a|p|^2 + b.$$

Recently M. Giaquinta-E. Giusti [8] introduced a new technique of a direct type, which can be extended so as to show regularity also for quasilinear systems like (1. 14) with right-hand side of quadratic growth; see also [9] for sharper and more general results.

By strenghthening the technique of [8], M. Giaquinta-G. Modica [10] were then able to show partial regularity of solutions of nonlinear systems of the type (1. 15) under the natural conditions (i), (ii), (iii).

For example we have, see [10], the results:

Let $u \in W^{1,2}(\Omega)$ be a weak solution to system

$$D_i A_i^r(x,u,\nabla u) = 0 \qquad 1 \leqq r \leqq m.$$

Suppose assumptions (i), (ii) *are satisfied. Then the first derivatives of u are Hölder-continuous on an open set* Ω_0. *Moreover*

$$\Omega\setminus\Omega_0\subset\Sigma_1\cup\Sigma_2$$

where

$$\Sigma_1=\{x\in\Omega:\ \liminf_{R\to 0^+} R^{-n}\int_{B_R(x)}|\nabla u-(\nabla u)_{x,R}|^2>0\},$$

$$\Sigma_2=\{x\in\Omega:\ \sup_R\left(|u_{x,R}|+|(\nabla u)_{x,R}|\right)=+\infty\}.$$

In particular

$$\operatorname{meas}(\Omega\setminus\Omega_0)=0.$$

While in the general case (1. 15) under the assumptions (i), (ii), (iii) the same results hold for weak solutions $u\in W^{1,2}\cap L^\infty$ which moreover satisfy the condition $2aM<\nu$ where $\|u\|_\infty\leqq M$.

We refer to [10] for a discussion of the assumptions and results.

2. Regularity via Liouville's condition

Let us consider the non linear system

$$\text{(2. 1)}\qquad -\frac{\partial}{\partial x_i}a_i^r(x,u,\nabla u)+a^r(x,u,\nabla u)=-\frac{\partial f_i^r}{\partial x_i}+f^r,\qquad r=1,\dots,m$$

where $x\in\Omega$ open set of R^n, a_i^r, a^r are continuously differentiable functions, $f_i^r\in W^{1,p}(\Omega)$, $f^r\in W^{1,p/2}(\Omega)$ $p>n$ and the ellipticity condition holds:

$$\text{(2. 2)}\qquad \frac{\partial a_i^r(x,\xi,\eta)}{\partial\eta_j^s}\zeta_i^r\zeta_j^s>0\quad\text{for}\quad \zeta\neq 0.$$

We shall only consider $W^{1,\infty}(\Omega)$ solutions. A weak $W^{1,2}(\Omega)$ solution is in $W^{1,\infty}_{\text{loc}}$ under some asymptotical conditions for the behaviour of $\frac{\partial a_i^r}{\partial\eta_j^s}(x,\xi,\eta)$ in η, see [27].

Then we have, compare [11], the following:

Theorem 2. 1. *Let* $u\in W^{1,2}(\Omega)$ *be a weak solution to system* (2. 1) *and* $|\nabla u|\leqq c<+\infty$. *Suppose that system* (2. 1) *satisfies Liouville's condition* (*L*). *Then* $u\in C^{1,1-\frac{n}{p}}(\Omega)$.

Proof. Let $x^0\in\Omega$. Set

$$u_R(y)=\frac{1}{R}\left[u(x^0+Ry)-u(x^0)\right],\quad x=x^0+Ry$$

and denote by O the image of Ω through this transformation. We have

$$\int_O\left[a_i^r(x^0+Ry,Ru_R(y)+u(x^0),\nabla_y u_R(y))\frac{\partial\psi_r}{\partial y_i}+a^r(x^0+Ry,Ru_R(y)+u(x^0),\nabla_y u_R(y))\,R\cdot\psi_r\right]dy$$

$$\text{(2. 3)}\qquad =\int_O\left[f_i^r(x^0+Ry)\frac{\partial\psi_r}{\partial y_i}+Rf^r(x_0+Ry)\,\psi_r(y)\right]dy.$$

From (2. 3) by a standard method we get that $u_R \in W^{2,2}_{\text{loc}}(\mathrm{O})$ and that $v_R = \dfrac{\partial u_R}{\partial y_l}$ satisfies the equation in variations

$$\int_{\mathrm{O}} \left[\frac{\partial a_i^r}{\partial \eta_j^s} \frac{\partial v_{Rs}}{\partial y_j} \frac{\partial \psi_r}{\partial y_i} + \frac{\partial a_i^r}{\partial x_l} \cdot R \cdot \frac{\partial \psi_r}{\partial y_i} + \frac{\partial a_i^r}{\partial \xi_s} R \cdot \frac{\partial \psi_r}{\partial y_i} v_{Rs} + \frac{\partial a^r}{\partial \eta_j^s} R \cdot \psi_r \frac{\partial v_{Rs}}{\partial y_j} \right.$$

(2. 4)
$$\left. + \frac{\partial a^r}{\partial x_l} R \psi_r + \frac{\partial a^r}{\partial \xi_s} R \psi_r v_{Rs} \right] dy = \int_{\mathrm{O}} \left[\frac{\partial f_i^r}{\partial x_l} R \cdot \frac{\partial \psi_r}{\partial y_i} + \frac{\partial f^r}{\partial x_l} R^2 \cdot \psi_r \right] dy.$$

In a standard way we get now from (2. 4)

$$\int_{B_a(0)} \sum_{|i|=2} |D^i u_R|^2 \, dy \leq c(a)$$

provided $B_{2a}(0) \subset \mathrm{O}$. Therefore we can choose $R_k \to 0$ such that $u_{R_k} \to p$ in

$$W^{1,2}(B_a(0)) \qquad \forall a > 0.$$

p is then defined in the whole R^n, $|\nabla p| \leq c < +\infty$ and it is a weak solution of the system

$$\int_{R^n} a_i^r(x^0, u(x^0), \nabla p) \frac{\partial \psi_r}{\partial y_i} \, dy = 0 \qquad \forall \psi \in \mathcal{D}(R^n)$$

so that, by the assumption (L), p is a polynomial of at most first degree, and thus

(2. 5)
$$0 \leftarrow \int_{B_1(0)} |\nabla_y u_{R_k} - \nabla_y p|^2 \, dy = R_k^{-n} \int_{B_{R_k}(x^0)} |\nabla_x u - \nabla_x p|^2 \, dx.$$

Let us consider now the equation in variation for system (2. 1). We get as before, if we denote by u' the derivative $\dfrac{\partial u}{\partial x_l}$, that

$$\int_{\Omega} \left[\frac{\partial a_i^r}{\partial \eta_j^s} \frac{\partial u_s'}{\partial x_j} \frac{\partial \varphi_r}{\partial x_i} + \frac{\partial a_i^r}{\partial \xi_s} \frac{\partial \varphi_r}{\partial x_i} u_s' + \frac{\partial a_i^r}{\partial x_l} \frac{\partial \varphi_r}{\partial x_i} + \frac{\partial a^r}{\partial \eta_j^s} \frac{\partial u_s'}{\partial x_j} \varphi_r \right.$$

(2. 6)
$$\left. + \frac{\partial a^r}{\partial u_s} u_s' \cdot \varphi_r + \frac{\partial a^r}{\partial x_l} \varphi_r \right] dx = \int_{\Omega} \left[f_i^{r'} \frac{\partial \varphi_r}{\partial x_i} + f^{r'} \varphi_r \right] dx.$$

If we move the terms $\dfrac{\partial a_i^r}{\partial \xi_s} \dfrac{\partial \varphi_r}{\partial x_i} u_s'$, $\dfrac{\partial a_i^r}{\partial x_l} \dfrac{\partial \varphi_r}{\partial x_i}$, $\dfrac{\partial a^r}{\partial \xi_s} \varphi_r u_s'$, $\dfrac{\partial a^r}{\partial x_l} \varphi_r$ to the right-hand side of (2. 6), and denote by U the vector function whose components are $\dfrac{\partial U_s}{\partial x_l} = U_s^l$, system (2. 6) can be rewritten as

$$\int_{\Omega} \left[A_{ij}^{rs} \delta_{kl} \frac{\partial U_s^l}{\partial x_j} \frac{\partial \varphi_r^k}{\partial x_i} + A_j^{rs} \delta_{kl} \frac{\partial u_s^l}{\partial x_j} \varphi_r^k \right] dx = \int_{\Omega} \left[g_i^{rk} \frac{\partial \varphi_r^k}{\partial x_i} + g^{rk} \varphi_r^k \right] dx$$

where

$$A_{ij}^{rs}(x, U) = \frac{\partial a_i^r}{\partial \eta_j^s}(x, u, U), \quad A_j^{rs}(x, U) = \frac{\partial a^r}{\partial \eta_j^s}(x, u, U).$$

Therefore system (2. 6) appears as a system of type (1. 1) and because of (2. 5) we can use theorem 1. 2 and obtain the result. q.e.d.

In the sequel of this section we want to make a few consideration about the validity of Liouville's condition. The first proposition which follows provides a different proof of a well known result of regularity in two dimension, see also [18], [23]. The second one gives a sufficient condition in dimension $n=3, 4$ in order that (L) holds, compare [5], while in the third proposition we show that (L) holds for *equations* in every dimension.

Proposition 2. 2. *Let us consider system* (2. 1) *and let us suppose* $n=2$. *Then condition* (L) *is satisfied.*

Proof. Let $v \in W^{1,2}_{\mathrm{loc}}(\mathbb{R}^2)$, $|\nabla v| \leqq c < +\infty$ be a solution in $\mathbb{R}^2$ of

$$\int_{\mathbb{R}^2} a_i^r(x^0, \xi, \nabla v) \frac{\partial \varphi_r}{\partial x_i} dx = 0 \qquad \forall \varphi \in \mathscr{D}(\mathbb{R}^2). \tag{2. 7}$$

The equation in variations is

$$\int_{\mathbb{R}^2} \frac{\partial a_i^r}{\partial \eta_j^s}(x^0, \xi, \nabla v) \frac{\partial v_s'}{\partial x_j} \frac{\partial \varphi_r}{\partial x_i} dx = 0 \qquad \forall \varphi \in \mathscr{D}(\mathbb{R}^2). \tag{2. 8}$$

Putting $\varphi_r = v_r' \eta^2$, $\eta \in \mathscr{D}(B_{2T}(0))$, $0 \leqq \eta \leqq 1$, $\eta \equiv 1$ in $B_T(0)$, $|\nabla \eta| \leqq \frac{c_1}{T}$, we easily get

$$\int_{\mathbb{R}^2} \eta^2 |\nabla v'|^2 dx \leqq c_2 \sup_{x \in \mathbb{R}^2} |\nabla v|.$$

Hence there exist $\varphi_k \in \mathscr{D}(\mathbb{R}^2)$ such that $\nabla \varphi_k \to \nabla v'$ in $L^2(\mathbb{R}^2)$ (and there exist $\Lambda_k \in \mathbb{R}^m$ such that $\Lambda_k + \varphi_k \to v'$ in $L^2_{\mathrm{loc}}(\mathbb{R}^2)$). Therefore from (2. 8) we get

$$\int_{\mathbb{R}^2} \frac{\partial a_i^r}{\partial \eta_j^s}(x^0, \xi, \nabla v) \frac{\partial v_s'}{\partial x_j} \frac{\partial v_r'}{\partial x_i} dx = 0$$

which gives the result. q.e.d.

Proposition 2. 3. *Let* $v \in W^{1,2}_{\mathrm{loc}}(\mathbb{R}^n)$, $|\nabla v| \leqq c < +\infty$ *be a solution of system* (2. 7). *Suppose* $3 \leqq n \leqq 4$ *and that there exists a polynomial of degree* $\leqq 1$ *such that*

$$|v - p| \leqq c < +\infty. \tag{2. 9}$$

Then v *is a polynomial of at most first degree.*

Proof. Putting $\varphi = (v-p)'\eta^4$, η as before, in (2. 8) we get

$$0 = \int_{n} \frac{\partial a_i^r}{\partial \eta_j^s} \frac{\partial}{\partial x_i}(v_r - p_r)' \frac{\partial}{\partial x_j}(v_s - p_s)' \eta^4 dx + \int_{n} \frac{\partial a_i^r}{\partial \eta_j^s}(v_r - p_r)' 4\eta^3 \frac{\partial \eta}{\partial x_i} \frac{\partial}{\partial x_j}(v_s - p_s)' dx.$$

On the other hand we have

$$\frac{1}{T^2} \int_{B_{2T}(0)} \frac{\partial}{\partial x_l}(v_r - p_r) \frac{\partial}{\partial x_l}(v_r - p_r)\eta^2 dx \overset{\text{def.}}{=} \frac{I}{T^2} = -\frac{1}{T^2} \int_{B_{2T}(0)} (v_r - p_r)\, \Delta(v_r - p_r)\eta^2$$

$$- \frac{1}{T^2} \int_{B_{2T}(0)} (v_r - p_r) \frac{\partial}{\partial x_l}(v_r - p_r)\, 2\eta \frac{\partial \eta}{\partial x_l} dx.$$

Hence in virtue of (2. 9) we have

$$\frac{I}{T^2} \leqq \frac{c_1}{T^2} T^{n/2} J^{1/2} + \frac{c_1}{T^3} T^{n/2} I^{1/2}$$

where

$$J = \int_{B_{2T}(0)} \frac{\partial}{\partial x_i} (v_r - p_r)' \frac{\partial}{\partial x_i} (v_r - p_r)' \eta^u \, dx.$$

Thus it follows

$$\frac{I}{T^2} \leqq c_2(\varepsilon) T^{n-4} + \varepsilon J, \quad J \leqq \frac{1}{2} J + c_3 \frac{I}{T^2}$$

which yield

$$\int_{R^n} \frac{\partial (v_r - p_r)'}{\partial x_i} \frac{\partial (v_r - p_r)'}{\partial x_i} dx < +\infty.$$

Since now for $n \geqq 3$ if $\nabla h \in L^2(R^n)$ there exist $\varphi^k \in \mathscr{D}(R^n)$ such that $\nabla \varphi^k \to \nabla h$ in $L^2(R^n)$ and $\varphi^k \to h$ in $L^2_{\mathrm{loc}}(R^n)$ we obtain the result. q.e.d.

Proposition 2. 4. *Assume* $m = 1$. *Then system* (2. 1) *satisfies Liouville's condition.*

Proof. Let $v \in W^{1,2}_{\mathrm{loc}}(R^n)$, $|\nabla v| \leqq c < +\infty$ be a solution of

$$-\frac{\partial}{\partial x_i} a_i(x^0, \xi, \nabla v) = 0,$$

we see that $v \in W^{2,2}_{\mathrm{loc}}(R^n)$ and

$$\int_{R^n} \frac{\partial a_i}{\partial \eta_j} (x^0, \xi, \nabla v) \frac{\partial v'}{\partial x_j} \frac{\partial \varphi}{\partial x_i} dx = 0$$

holds. Now remember that if $u \in W^{1,2}(\Omega)$, $u \geqq 0$ is a solution of

$$-\frac{\partial}{\partial x_i} \left(a_{ij} \frac{\partial u}{\partial x_j} \right) = 0 \quad \text{in} \quad \Omega$$

and $B_{4R} \subset \Omega$ the Harnack inequality

$$\sup_{B_R(y)} u(x) \leqq c \inf_{B_R(y)} u(x) \tag{2. 10}$$

holds (c depends only on the ellipticity constant and $\sup |a_{ij}(x)|$). Set

$$M(R) = \sup_{B_R(0)} v', \quad m(R) = \inf_{B_R(0)} v', \quad \omega(R) = M(R) - m(R).$$

Applying (2. 10) for $u = M(R) - v'$ we then get

$$M(R) - m\left(\frac{R}{4}\right) \leqq c \left[M(R) - M\left(\frac{R}{4}\right) \right]$$

while applying (2. 10) for $u=v'-m(R)$ we have

$$M\left(\frac{R}{4}\right)-m(R)\leqq c\left[m\left(\frac{R}{4}\right)-m(R)\right]$$

which yield

$$\omega\left(\frac{R}{4}\right)\leqq\frac{c-1}{c+1}\,\omega(R)\leqq\frac{c-1}{c+1}\,\omega(\infty),$$

that is

$$\omega(\infty)\leqq\frac{c+1}{c-1}\,\omega(\infty),\quad \text{i.e.}\quad \omega(\infty)=0. \qquad \text{q.e.d.}$$

3. Condition number

In this section we shall consider systems of the type (2. 1) for which $\frac{\partial a_i^r}{\partial\eta_j^s}$ are bounded in $\bar{\Omega}\times\mathbb{R}^m\times\mathbb{R}^{nm}$ and symmetric i.e.

$$\frac{\partial a_i^r}{\partial\eta_j^s}=\frac{\partial a_j^s}{\partial\eta_i^r}\overset{\text{def}}{=}a_{ij}^{rs}.$$

We shall also suppose the strong ellipticity condition

$$\lambda_1|\zeta|^2\leqq a_{ij}^{rs}(x^0,\xi,\eta)\,\zeta_i^r\zeta_j^s\leqq\lambda_2|\zeta|^2. \tag{3. 1}$$

Let us remark that condition (3. 1) and Liouville's condition (L) are invariant under a change of coordinates of the type

$$x^*=Ax,\quad v^*=Bv$$

where A and B are respectively $n\times n$ and $m\times m$ regular matrices (x^0, ξ, fixed). So we can define

$$k=\inf_{x^0,\xi}\sup_{A,B}\frac{\lambda_1}{\lambda_2}. \tag{3. 2}$$

The main result of this section is the following theorem.

Theorem 3. 1. *Let* $|\nabla v|\leqq c<+\infty$ *be a weak solution in* $\mathbb{R}^n$, $n\geqq 3$, *to system*

$$\int_{\mathbb{R}^n}a_i^r(x^0,\xi,\nabla v)\,\frac{\partial\varphi_r}{\partial x_i}\,dx=0. \tag{3. 3}$$

Suppose

$$k>k_0=\frac{\sqrt{1+\dfrac{(n-2)^2}{n-1}}-1}{\sqrt{1+\dfrac{(n-2)^2}{n-1}}+1}. \tag{3. 4}$$

Then v is a polynomial of at most the first degree. That is, system (2. 1) satisfies Liouville's condition.

The proof of theorem 3. 1 which grants the regularity of weak solution of system (2. 1) under the quite precise condition (3. 4) uses the estimate stated in the following lemma which refers to Laplace operator, compare [25].

Lemma 3. 2. *Suppose $v \in L^\infty(R^n) \cap W^{1,2}_{\mathrm{loc}}(R^n) \cap C^1(B_\delta(0))$ for some $\delta > 0$ be a solution of*

$$-\Delta v = -\frac{\partial g_i}{\partial x_i} \tag{3. 5}$$

where $g_i \in L^2_{\mathrm{loc}}(R^n) \cap L^\infty(B_\delta(0))$. Moreover suppose

$$\int_{R^n} D_i v\, D_i v\, r^{-\lambda}\, dx < +\infty, \quad \int_{R^n} g_i g_i\, r^{-\lambda}\, dx < +\infty \tag{3. 6}$$

where $n-2 < \lambda < n$, $n \geqq 3$ and finally suppose $v(0) = 0$.

Then we have

$$\int_{R^n} \frac{\partial v}{\partial x_i}\frac{\partial v}{\partial x_i} r^{-\lambda}\, dx \leqq \alpha(\lambda)\left(1 + \frac{(n-2)^2}{n-1}\right) \int_{R^n} g_i g_i\, r^{-\lambda}\, dx \tag{3. 7}$$

where $\alpha(\lambda) \to 1$ as $\lambda \to (n-2)$.

Proof. Let us introduce polar coordinates $r, \theta_1, \ldots, \theta_{n-1}$, $0 < r < \infty$, $0 < \theta_i < \pi$, $i = 1, \ldots, n-2$, $0 < \theta_{n-1} < 2\pi$.

$x_1 = r\cos\theta_1$, $x_2 = r\sin\theta_1\cos\theta_2, \ldots, x_{n-1} = r\sin\theta_1\sin\theta_2\cdots\sin\theta_{n-2}\cos\theta_{n-1}$,

$x_n = r\sin\theta_1\sin\theta_2\cdots\sin\theta_{n-1}$.

Let us define the elementary differential operators

$$\partial_1 v = \frac{\partial v}{\partial r}, \quad \partial_2 v = \frac{1}{r}\frac{\partial v}{\partial\theta_1}, \quad \partial_3 v = \frac{\partial v}{\partial\theta_2}\frac{1}{r\sin\theta_1}, \ldots, \partial_n v = \frac{\partial v}{\partial\theta_{n-1}}\frac{1}{r\sin\theta_1\cdots\sin\theta_{n-2}},$$

$$\bar\partial_2 h = \frac{\partial h}{\partial\theta_1}, \quad \bar\partial_3 h = \frac{\partial h}{\partial\theta_2}\frac{1}{\sin\theta_1}, \ldots, \bar\partial_n h = \frac{\partial h}{\partial\theta_{n-1}}\frac{1}{\sin\theta_1\cdots\sin\theta_{n-2}}, \quad \bar\partial h = (\bar\partial_2 h, \ldots, \bar\partial_n h).$$

Then we have

$$\partial_i v = a_{ij}\frac{\partial v}{\partial x_j}$$

where a_{ij} is an orthonormal matrix, and equation (3. 5) can be written

$$\int_0^\infty \int_S \partial_i v\, \partial_i \varphi\, r^{n-1}\, dr\, ds = \int_0^\infty \int_S m_i\, \partial_i\varphi\, r^{n-1}\, dr\, ds \tag{3. 8}$$

with $m_i = a_{ij} g_j$, S being the boundary of the unitary ball. Let $w \in \varepsilon(S)$ define

$$\varphi(r, \theta) = r^{p'-1} w(\theta)$$

where

$$p' = p - (n-3), \quad p = x' + iy, \quad y \in R, \quad -1 < x' < x, \quad x = -\frac{1}{2}[\lambda - (n-2)].$$

Now in (3. 8) instead of φ put $\psi(x)=\eta_R(r)\,\varphi(r,\theta)$ where $\eta_R\in\mathscr{D}((0,2R))$, $0\leqq\eta_R\leqq 1$, $\eta(r)\equiv 1$ for $\frac{1}{R}\leqq r\leqq R$, $|\eta_R'|\leqq R$ for $0\leqq r<\frac{1}{R}$, $|\eta_R'|\leqq\frac{1}{R}$ for $R\leqq r\leqq 2R$. Since

$$\left|\frac{\partial v}{\partial x_i}\right|\leqq c<+\infty$$

in $B_\delta(0)$ and because of (3. 6) we can let $R\to\infty$ and we get (3. 8) with $\varphi=r^{p'-1}w(\theta)$ and absolutely converging integrals. Denoting now by

$$V(p,\theta)=\int_0^\infty r^{p-1}v(r,\theta)\,dr,\quad M_i(p,\theta)=\int_0^\infty r^p m_i(r,\theta)\,dr$$

Mellin's transform of v and m_i, we obtain from (3. 8)

$$\text{(3. 9)}\qquad \int_S[-p(p'-1)Vw+\bar{\partial}_iV\,\bar{\partial}_iw]\,ds=\int_S[M_1(p'-1)w+M_{i'}\bar{\partial}_{i'}w]\,ds$$

where i' takes values from 2 to n.

Clearly we can now take $w\in W^{1,2}(s)$ $(V\in W^{1,2}(s))$ where $W^{1,2}(s)$ is the Sobolev space over complex numbers with the scalar product

$$\int_S(u\bar{v}+\bar{\partial}_iu\,\bar{\partial}_i\bar{v})\,ds.$$

Let us now decompose $W^{1,2}(s)=W_1\oplus W_2$, $L^2(s)=L_1\oplus L_2$, $W_1=L_1=\{\text{constants}\}$ and

$$V=V_1+V_2,\quad M_1=M_1^1+M_1^2$$

be the subordinate decompositions. If we put

$$w=-\frac{\bar{p}}{p'-1}\bar{V}_1$$

in (3. 9), we obtain

$$\text{(3. 10)}\qquad \int_S|p|^2\,|V_1|^2\,ds\leqq\int_S|M_1'|^2\,dS.$$

On the other hand for every function w from W_2 we have

$$(n-1)\int_S|w|^2\,ds\leqq\int_S\bar{\partial}_{i'}w\,\bar{\partial}_{i'}w\,ds$$

so that putting in (3. 9) $w=\bar{V}_2$, taking the real part in (3. 9) and setting

$$J=\int_S[|p|^2\,|V_2|^2+\bar{\partial}_{i'}V_2\,\bar{\partial}_{i'}\bar{V}_2]\,ds,$$

we get

$$\left(1-\frac{x'(x'-(n-2)+x'^2)}{n-1}\right)J\leqq\Big(\int_S(|M_1^2|^2+M_{i'}\bar{M}_{i'})\,ds\Big)^{1/2}\cdot\Big(\int_S(|p'-1|\,|V_2|^2+\bar{\partial}_{i'}V_2\,\bar{\partial}_{i'}\bar{V}_2)\,ds\Big)^{1/2}$$

$$\text{(3. 11)}\qquad \leqq\left(1+\frac{(n-2)^2}{n-1}-\frac{2x'(n-2)}{n-1}\right)^{1/2}J^{1/2}M^{1/2}.$$

And putting together (3. 10), (3. 11) we obtain

$$\int_S (|p|^2 |V|^2 + \bar{\partial}_{i'} V \bar{\partial}_{i'} \bar{V}) ds \leqq \frac{1 + \frac{(n-2)^2}{n-1} - \frac{2x'(n-2)}{n-1}}{\left[1 - \frac{x'(x'-(n-2)) + x'^2}{n-1}\right]^2} \int_S M_i \bar{M}_i. \tag{3. 12}$$

If we now set $\lambda' = -2x' + (n-2)$, we find, using Parseval's identities

$$\frac{1}{2\pi} \int_{-\infty}^{+\infty} (|p|^2 |V|^2 + \bar{\partial}_{i'} V \partial_{i'} \bar{V}) dy = \int_0^\infty (\partial_i v \, \partial_i v) \, r^{-\lambda' + n - 1} dr,$$

$$\frac{1}{2\pi} \int_{-\infty}^{+\infty} M_i \bar{M}_i dy = \int_0^\infty m_i m_i r^{-\lambda' + n - 1} dr$$

which together with (3. 12) imply the result. q.e.d.

Proof of theorem 3. 1. Because of the partial regularity results in section 1, we can suppose that in a neighbourhood of 0 the function v is twice differentiable, and, if v' as usual denotes a derivative $\frac{\partial v}{\partial x_l}$, that $v'(0) = 0$.

As $|v'(x)| \leqq c < +\infty$ we have for $\lambda > n-2$

$$\int_{R^n} |v'|^2 r^{-\lambda-2} dx < +\infty,$$

so that from the equation in variations

$$\int_{{}_n} \frac{\partial a_i^r}{\partial \eta_j^s} (x^0, \xi, \nabla v) \frac{\partial v_s'}{\partial x_j} \frac{\partial \varphi_r}{\partial x_i} dx = 0$$

putting $\varphi_r = v' r^{-\lambda} \eta_R^2(r)$ ($\eta_R(r)$ being the function in the proof of lemma 3. 2) we get

$$\int_{R^n} \eta_R^2 \frac{\partial v_r'}{\partial x_i} \frac{\partial v_r'}{\partial x_i} r^{-\lambda} \leqq c < +\infty$$

where c does not depend on R. So

$$\int_{{}_n} \frac{\partial v_r'}{\partial x_i} \frac{\partial v_r'}{\partial x_i} r^{-\lambda} dx \leqq c < +\infty$$

and we can use for v' lemma 3. 2 as follows. Set

$$\gamma = \frac{2}{\lambda_1 + \lambda_2}, \quad g_i^r = \frac{\partial v_r'}{\partial x_i} - \gamma a_{ij}^{rs} \frac{\partial v_s'}{\partial x_j}.$$

Then we obviously have

$$\int D_i v_r' D_i \varphi_r = \int g_i^r D_i \varphi_r$$

and

$$\int_{R^n} g_i^r g_i^r r^{-\lambda} dx \leqq \left(\frac{\lambda_2 - \lambda_1}{\lambda_2 + \lambda_1}\right)^2 \int_{R^n} \frac{\partial v_r'}{\partial x_i} \frac{\partial v_r'}{\partial x_i} r^{-\lambda} dx.$$

Therefore from lemma 3. 2 we get

$$\int_{R^n} \frac{\partial v_r'}{\partial x_i}\frac{\partial v_r'}{\partial x_i} r^{-\lambda} = I \leqq \alpha(\lambda)\left(1+\frac{(n-2)^2}{n-1}\right)\left(\frac{\lambda_2-\lambda_1}{\lambda_2+\lambda_1}\right)^2 I.$$

But

$$\left(1+\frac{(n-2)^2}{n-1}\right)\frac{1-\dfrac{\lambda_1}{\lambda_2}}{1+\dfrac{\lambda_1}{\lambda_2}} < \left(1+\frac{(n-2)^2}{n-1}\right)\frac{1-k_0}{1+k_0} = 1$$

so that $I=0$. q.e.d.

4. A counterexample

In this section we present an elliptic system of the type

$$-\frac{\partial}{\partial x_i} a_i^r(\nabla v) = 0$$

with smooth coefficients a_i^r satisfying

$$\left|\frac{\partial a_i^r(\eta)}{\partial \eta_j^s}\right| \leqq L < +\infty, \quad \frac{\partial a_i^r(\eta)}{\partial \eta_j^s}\zeta_i^r\zeta_j^s \geqq \nu|\xi|^2 \qquad \nu>0 \tag{4. 1}$$

and having as solution in the whole R^n, $n \geqq 3$, the vector function u of components

$$u_{ij} = \frac{x_i x_j}{|x|} - \frac{1}{n}\delta_{ij}|x|. \tag{4. 2}$$

Clearly $|\nabla u| \leqq c < +\infty$, u is not a polynomial and is not regular.

Systems of this kind were constructed only for $n \geqq 5$ in [24], [25] by J. Nečas.

First, let us remark that it is sufficient to construct coefficients a_i^r satisfying (4. 1) in the domain $\{|\eta| \leqq c\}$; in fact we can then change them smoothly outside of $|\eta| \leqq c$ in such a way that (4. 1) still holds.

Set

$$\nabla_i u = \frac{\partial u_{ij}}{\partial x_j}, \quad \nabla_{ijk} u = \frac{\partial u_{ij}}{\partial x_k} + \gamma[\nabla_i u\delta_{jk} + \nabla_j u\delta_{ik} + \nabla_k u\delta_{ij}], \quad \gamma \leqq 0, \quad \delta_i u = \frac{\partial u_{jj}}{\partial x_i}.$$

Then it is not difficult to show that the vector valued function in (4. 2) satisfies the non linear system

$$\begin{aligned}&\int_{R^n} \nabla_{ijk}u\,\nabla_{ijk}\varphi\,dx + \delta\int_{R^n}\nabla_i u\,\delta_i\varphi\,dx + \beta\int_{R^n}\delta_i u\,\delta_i\varphi\,dx \\ &\qquad + \lambda\int_{R^n}\nabla_i u\,\nabla_j u\,\nabla_k u\,\nabla_{ijk}\varphi\,dx + \bar{\nu}\int_{R^n}(\nabla_i u\,\nabla_i u)^2\,\nabla_j u\,\nabla_j\varphi\,dx = 0\end{aligned} \tag{4. 3}$$

provided λ and δ are chosen as follows

$$\lambda=\frac{n+1+4\gamma(n-\mu)+3\gamma^2(n+2)(n-\mu)+\bar{v}(n-\mu)^5}{(n-1)(n-\mu)^3-3\gamma(n-\mu)^3},$$

$$\delta=-\frac{1+\mu+4\gamma(n-\mu)+(n-1)(n-\mu)\gamma+3\lambda\gamma(n-\mu)^3+3\gamma^2 n(n-\mu)+6\gamma^2(n-\mu)+\bar{v}(n-\mu)^5}{(n-1)(n-\mu)},$$

$$\mu=\frac{1}{n}$$

and β is free.

Set now

$$\tilde{\eta}_k^{ij}=\tilde{\eta}_k^{ji},\quad \tilde{v}_i=\tilde{\eta}_j^{ij},\quad \tilde{\mu}_i=\tilde{\eta}_i^{jj},\quad \zeta_{ijk}=\tilde{\eta}_k^{ij}+\gamma[\tilde{v}_i\delta_{jk}+\tilde{v}_j\delta_{ik}+\tilde{v}_k\delta_{ij}]$$

then

$$|\zeta|^2=|\eta|^2+[4\gamma+3(n-2)\gamma^2]\,|\tilde{v}|^2+2\gamma(\tilde{v},\tilde{\mu}).$$

If $3(n-2)\gamma^2+4\gamma<0$, then using

$$|\tilde{v}|\leqq\sqrt{n}\,|\tilde{\eta}|$$

we find that the condition

$$1+3n(n+2)\gamma^2+4n\gamma<0$$

with $\gamma\leqq 0$ for $2\leqq n\leqq 5$ implies $|\tilde{\zeta}|^2\geqq c|\eta|^2+2\gamma|\tilde{v}|\,|\tilde{\mu}|$, $c>0$. For $n\geqq 6$ we shall choose $\gamma=0$.

And we see that the ellipticity condition in (4. 1) will be satisfied if

$$|\tilde{\zeta}|^2-|\delta|\,|\tilde{v}|\,|\tilde{\mu}|+\beta|\tilde{\mu}|^2-|\lambda|\,|\tilde{\zeta}|\sqrt{3}\,[|\tilde{v}|^2\,|v|^4+2(v,\tilde{v})^2\,|v|^2]^{1/2}$$
$$+\bar{v}[|\tilde{v}|^2\,|v|^4+4(v,\tilde{v})^2\,|v|^2]\geqq c_1|\tilde{\zeta}|^2\qquad c_1>0,$$

and this will be true, setting

$$\varkappa^2=|\tilde{v}|^2\,|v|^4+2(v,\tilde{v})\,|v|^2,$$

if for some $\bar{v}>0$

$$|\tilde{\zeta}|^2-\sqrt{3}\,|\lambda|\,|\tilde{\zeta}|\varkappa+\bar{v}\varkappa^2>c_2|\tilde{\zeta}|^2\qquad c_2>0 \tag{4. 4}$$

because $\beta>0$ can be chosen arbitrarily.

Now (4. 4) is satisfied if

$$\bar{v}-\frac{3}{4}\lambda^2>0. \tag{4. 5}$$

Set

$$V=\bar{v}(n-\mu)^5,\quad A(n)=n+1+4\gamma(n-\mu)+3\gamma^2(n+2)(n-\mu),\quad B(n)=(n-\mu)(n-1-3\gamma)^2.$$

We easily see that (4. 5) is equivalent to

$$(4.6) \qquad -\frac{3}{4B(n)}V^2+V\left[1-\frac{3}{2}\frac{A(n)}{B(n)}\right]-\frac{3}{4}\frac{A(n)^2}{B(n)}>0;$$

thus it is sufficient that

$$3\frac{A(n)}{B(n)}<1$$

which is equivalent to

$$(4.7) \qquad 9\gamma^2(n+1)(n-1)+6\gamma(n-1)(n+1)+3n-(n-1)^3<0.$$

For $n=2$, (4. 7) is impossible (it must be so!). In order to satisfy (4. 7) for $n\geqq 4$ it is sufficient to choose $\gamma=0$ and for $n=3$ to choose γ in the interval

$$\frac{-4-\sqrt{14}}{12}<\gamma<\frac{-4+\sqrt{14}}{2}<0.$$

Finally from (4. 7) we get the possible choices of V from the interval $]V_1, V_2[$, where V_i are the roots of (4. 6); we have $0<V_1<V_2$.

References

[1] *S. A. Arakčejev*, O hladkosti obobščennych rešenij někotorovo klassa kvazilinějnych elliptičeskich uravněnij, Věstnik Moskovsk. Univ. **1** (1975), 49—56.
[2] *S. Campanato*, Proprietà di hölderianità di alcune classi di funzioni, Ann. S.N.S. Pisa **17** (1963), 175—188.
[3] *S. Campanato*, Equazioni ellittiche del II ordine e spazi $\mathcal{L}^{2,\lambda}$, Ann. Mat. Pura e Appl. **69** (1965), 321—382.
[4] *E. de Giorgi*, Sulla differenziabilità e l'analiticità delle estremali degli integrali multipli regolari, Mem. Accad. Sci. Torino **3** (1957), 25—43.
[5] *J. Frehse*, Essential selfadjointness of singular elliptic operators, preprint Universität Bonn 1977.
[6] *M. Giaquinta*, A counter-example to the boundary regularity of solutions to elliptic quasilinear systems, Manuscripta Math. **24** (1978), 217—220.
[7] *M. Giaquinta*, Sistemi ellittici non lineari, teoria della regolarità, to appear in Boll. U.M.I. (1979).
[8] *M. Giaquinta, E. Giusti*, Nonlinear elliptic systems with quadratic growth, Manuscripta Math. **24** (1978), 323—349.
[9] *M. Giaquinta, G. Modica*, Regularity results for some classes of higher order non linear elliptic systems, J. reine angew. Math. **311/312** (1979), 145—169.
[10] *M. Giaquinta, G. Modica*, Almost-everywhere regularity results for solutions of non linear elliptic systems, Manuscripta Math. **28** (1979), 109—158.
[11] *M. Giaquinta, J. Nečas*, On the regularity of weak solutions to nonlinear elliptic systems via Liouville's type property, Comm. Math. Univ. Carolinae **20** (1979), 111—121.
[12] *E. Giusti*, Precisazione delle funzioni di $H^{1,p}$ e singolarità delle soluzioni deboli di sistemi ellittici non lineari, Boll. U.M.I. **2** (1969), 71—76.
[13] *E. Giusti*, Regolarità parziale delle soluzioni di sistemi ellittici quasi lineari di ordine arbitrario, Ann. S.N.S. Pisa **23** (1969), 115—141.
[14] *E. Giusti, M. Miranda*, Un esempio di soluzioni discontinue per un problema di minimo relativo ad un integrale regolare del calcolo delle variazioni, Boll. U.M.I. **2** (1968), 1—8.
[15] *E. Giusti, M. Miranda*, Sulla regolarità delle soluzioni deboli di sistemi ellittici quasilineari, Arch. Rat. Mech. and Anal. **31** (1968), 173—184.

[16] *O. A. Ladyzenskaya, N. N. Ural'tseva*, Linear and quasilinear elliptic equations, New York 1968.

[17] *V. G. Mazja*, Priměry něregularnych rešenij kvazilinějnych elliptičeskich uravněnij e analitičeskimi koefficientami, Funk. Analiz. **2** vyp. 3 (1968).

[18] *C. B. Morrey*, Existence and differentiabilty theorems for the solutions of variational problems for multiple integrals, Bull. A.M.S. **46** (1940), 439—458.

[19] *C. B. Morrey*, Multiple integrals in the Calculus of Variations, Berlin-Heidelberg-New York 1966.

[20] *C. B. Morrey*, Partial regularity results for non linear elliptic systems, J. Math. and Mech. **17** (1968), 649—670.

[21] *C. B. Morrey*, Differentiability theorems for non linear elliptic equations, Actes Congrès intern. Math. Nice (1970).

[22] *J. Nash*, Continuity of the solutions of parabolic and elliptic equations, Amer. J. Math. **80** (1958), 931—954.

[23] *J. Nečas*, Sur la régularité des solutions variationnelles des équations elliptiques non-lineaires d'ordre $2k$ en deux dimensions, Ann. S.N.S. Pisa **21** (1967), 427—457.

[24] *J. Nečas*, Example of an irregular solution to a non linear elliptic system with analitic coefficients and conditions for regularity, in: Theory of Nonlinear Operators, Abhandlungen Akad. der Wissenschaften der DDR (1977).

[25] *J. Nečas*, On the regularity of weak solutions to variational equations and inequalities for non linear second order elliptic systems, Equadiff IV, Praha, Lecture Notes n. 703.

[26] *J. Nečas, J. Stara*, Principio di massimo per i sistemi ellittici quasilineari non diagonali, Boll. U.M.I. **6** (1972).

Università degli Studi di Firenze, Istituto di Matematica Applicata, Via di S. Marta 3, I-50139 Firenze

Matematicky ústav ČSAV, Žitná 25, 11565 Praha 1, CSSR

Eingegangen 28. Juni 1979

COMMENTATIONES MATHEMATICAE UNIVERSITATIS CAROLINAE

21, 1 (1980)

COUNTEREXAMPLE TO THE REGULARITY OF WEAK SOLUTION OF ELLIPTIC SYSTEMS

J. NEČAS, O. JOHN, J. STARÁ

Abstract: In the paper there will be given an example of nonlinear elliptic system

$$(1)\quad \sum_{i=1}^{n} D^i(a_r^i(\text{grad } u)) = 0,\ r = 1,\dots,m \text{ on } \Omega = \{x\in R_n,\ |x|<1\},\ u = u_0 \text{ on } \partial\Omega,$$

having analytic coefficients and unique solution with discontinuous but bounded first derivatives even in dimensions n = 3,4. (For n = 5 an example of considered type was constructed by J. Nečas (see [10]).

In the introduction we give a brief survey of the problem of regularity and counterexamples. In Chapter 1 there will be studied the counterexample mentioned above. In Chapter 2 we add same calculations omitted in Chapter 1 in details.

Key words: Regularity, elliptic systems

Classification: 35J60, 35D10

Introduction. The problem of regularity (or analyticity) of weak solutions of nonlinear elliptic systems can be traced to the beginning of this century - to the 19. D. Hilbert's problem and can be expressed by the question: Supposing a_r^i and u_0 in (1) to be analytic, is the weak solution u also analytic function? The history of this problem is described in several books and papers (see [5],[6],

© Faculty of Mathematics and Physics of Charles University 1980. Reprinted with kind permission from: Nečas, J., John, O. & J. Stará, J., Counterexample to the regularity of weak solution of elliptic systems, Commentationes Mathematicae Universitatis Carolinae 21, 145-154.

[4]), hence we will mention here only some crucial points. Very soon - in 1939 - the problem was solved positively for systems of equations of second order in plane by Ch.B. Morrey. Very important further step was made by E.De Giorgi and J. Nash in 1957. They proved regularity of solution of one equation of second order in the space R_n of arbitrarily high dimension n. Another positive result was proved by one of the authors (J. Nečas - in 1967) for equations of arbitrarily high order in plane. Almost immediately there appeared counterexamples (E.De Giorgi - 1968, E. Giusti, M. Miranda - 1968), showing that the situation of one equation of second order in R_n or of systems of arbitrary order in plane is in some sense exceptional and that there exist systems with analytic coefficients whose solutions are not even continuous (x). Unfortunately, these counterexamples have some disadvantages:

(i) They have analytic coefficients, they are naturally defined on Sobolev spaces W_2^1, but the corresponding operators are not differentiable on this space.

(ii) For low dimensions (which play the most important role in physics) it is unclear, whether the irregular solution is unique or if, perhaps, there could exist another regular solution of system in question (xx).

----

x) are bounded and have unbounded gradients.

xx) i.e. the typical quasilinear system

$$\sum_{i,j,s=1}^{m} D^i(A_{rs}^{ij}(u)\ D^j u_s) = 0 \text{ for } r = 1,\dots,n.$$

In 1977 J. Nečas constructed a counterexample without these disadvantages and working in all dimensioms $n \geqq 5$, but the problem in $n = 3,4$ still remained unsolved. The aim of this paper is to give a counterexample with unique irregular solution in R_n with dimensions $n \geq 3$.

Chapter 1

1.1. Notation. Let $\Omega = \{x \in R_n; |x| < 1\}$, $u : \Omega \to R_{n^2}$; $u = \{u_{ij}\}_{i,j=1}^n$.

Let us denote

$$D^k u_{ij} = \frac{\partial u_{ij}}{\partial x_k}, \quad \delta_{ij} \text{ the Cronecker symbol,}$$

$$\nabla_j u = \sum_{i=1}^n D^i u_{ji}, \quad \|\nabla u\|^2 = \sum_{j=1}^n (\nabla_j u)^2, \quad (\nabla u, \nabla v) = \sum_{j=1}^n \nabla_j u \, \nabla_j v,$$

for a fixed real number γ let

$$\nabla_{ijk} u = D^k u_{ij} + \gamma(\delta_{ij} \nabla_k u + \delta_{ik} \nabla_j u + \delta_{jk} \nabla_i u),$$

$$\|\delta u\|^2 = \sum_{i,j,k=1}^n (\nabla_{ijk} u)^2, \quad (\delta u, \delta v) = \sum_{i,j,k=1}^n \nabla_{ijk} u \, \nabla_{ijk} v.$$

1.2. System and its solution. Let γ, λ, ν be real numbers. We shall consider the system

$$(2) \quad \sum_{k=1}^n D^k \{D^k u_{ij} + \gamma(\delta_{ij} \nabla_k u + \delta_{ik} D^j u_{kk}) + \lambda \nabla_i u \, \nabla_j u \, \nabla_k u \, [1 + \|\nabla u\|^2]^{-1} + \delta_{ik} \nabla_j u \{\gamma(4 + 3\gamma(n+2)) + 3\gamma\lambda \|\nabla u\|^2 [1 + \|\nabla u\|^2]^{-1} + \nu \|\nabla u\|^4 [1 + \|\nabla u\|^2]^{-2}\}\} = 0.$$

We shall prove that the function $u = \{u_{ij}\}_{i,j=1}^n$

(3) $$u_{ij}(x) = x_i x_j |x|^{-1} - \frac{1}{n}|x|\delta_{ij}$$

is a weak solution of the Dirichlet boundary problem for the system (2), i.e. for every infinitely differentiable function φ with compact support in Ω the equality

(4) $$\langle Au,\varphi\rangle = \int_\Omega \{(\nabla_{ijk}u + \lambda\nabla_i u\,\nabla_j u\,\nabla_k u[1+\|\nabla u\|^2]^{-1})\nabla_{ijk}\varphi + \nu\|\nabla u\|^4[1+\|\nabla u\|^2]^{-2}(\nabla u,\nabla\varphi)\} = 0$$

holds, if the numbers λ, ν, γ satisfy the following conditions

(5) $$\lambda = [1 + (n - \tfrac{1}{n})^2](n - \tfrac{1}{n})^{-2}(\tfrac{1}{n-1} - \gamma),$$

(6) $$\nu = -(n - \tfrac{1}{n})^{-5}\{3\gamma^2(n+1)(n - \tfrac{1}{n}) + \gamma(n^2 + 3n + 2) + 1 + \tfrac{1}{n}\}\times[1 + (n - \tfrac{1}{n})^2]^2 .$$

1.3. <u>Unicity of the solution</u>. Unicity of the solution is an immediate consequence of the following inequality

(7) $$\langle DA(u)\varphi,\varphi\rangle \geq C\|\varphi\|^2_{[W_2^1]^{n^2}},$$

holding for a positive coonstant C and a class of test functions φ, and which implies that the operator A is strongly monotone. In fact, as it is proved in 2.3, we establish an algebraic condition of monotonicity, i.e. the integrand of $\langle DA(u)\varphi,\varphi\rangle$ is greater than

$$(1 - \tfrac{3}{4}\tfrac{\lambda^2}{\nu})\|\delta\varphi\|^2.$$

We have

$$\langle DA(u)\varphi,\varphi\rangle = \int_\Omega\{\|\delta\varphi\|^2 + \lambda\{\{\nabla_i u\,\nabla_j u\nabla_k\varphi + \nabla_i u\,\nabla_j\varphi\,\nabla_k u + \nabla_i\varphi\,\nabla_j u\,\nabla_k u\}[1+\|\nabla u\|^2]^{-1} + \nabla_i u\,\nabla_j u\,\nabla_k u\,(-2(\nabla u,\nabla\varphi)$$

$$[1 + \|\nabla u\|^2]^{-2}\} \nabla_{ijk}\varphi + \nu\{\{4 \|\nabla u\|^2 (\nabla u, \nabla\varphi)^2 +$$

$$+ \|\nabla u\|^4 \|\nabla\varphi\|^2\}[1 + \|\nabla u\|^2]^{-2} -$$

$$- 4 \|\nabla u\|^4 (\nabla u, \nabla\varphi)^2 [1 + \|\nabla u\|^2]^{-3}\}\}.$$

Estimating the second term by Hölder inequality we get

$$(8) \qquad \langle DA(u)\varphi, \varphi\rangle \geq \int_\Omega (1 - \frac{3\lambda^2}{4\nu}) \|\delta\varphi\|^2.$$

Taking in consideration the symmetry of the solution and of the system in the indices i, j, we can choose the test function φ symmetric in i, j, too. Thus it suffices to consider the test functions φ with only $m = \frac{1}{2} n (n + 1)$ different components. For such functions φ we get (supposing that $\gamma <$ < 0)

$$(9) \qquad \| \|\delta\varphi\| \|^2_{L_2} \geq (1 - \frac{n\gamma}{4 + 3\gamma(n+2)}) \|\varphi\|^2_{[\frac{e1}{2}]n^2} \cdot$$

Summarizing the inequalities (8) and (9) we obtain (7) with a constant C which is positive if

$$1 - \frac{n\gamma}{4 + 3\gamma(n+2)} > 0, \text{ which implies the inequality}$$

$$\gamma < \frac{-2}{n + 3} ;$$

and if

$$1 - \frac{3\lambda^2}{4\nu} > 0.$$

The second condition implies that $\gamma \in (\gamma_1, \gamma_2)$ where (for $n = 3$)

$$\gamma_i = \frac{-27 \pm 2\sqrt{42}}{102}$$

is approximately $\gamma_1 = - 0{,}39$, $\gamma_2 = - 0{,}13$. Analogous numerical results show that the counterexample works in dimensions $n = 4,5$. For higher dimensions the function

$$\gamma_1(n) + \frac{2}{n+3}$$

is a decreasing function of variable n. It proves that for all $n \geq 3$ we can choose γ so that the function u given by (3) is the unique solution of the system (2) with analytic coefficients and linear growth. Moreover, u is the solution of the Dirichlet boundary problem with analytic boundary condition u_o.

Chapter 2

2.1. Deduction of $\lambda(\gamma)$ and $\nu(\gamma)$. The system (4)' can be written in the form

$$(10) \qquad \int \{\Phi_{ijk} \nabla_{ijk}\varphi + \Psi_i \nabla_i\varphi\} dx = 0, \quad \varphi_{ij} = \varphi_{ji} \in \mathcal{D}.$$

By means of the Gauss formula we deduce from (10) the system in a strong form

$$(11) \qquad D^k \Phi_{ijk} + \gamma D^j(\Phi_{kki} + \Phi_{kik} + \Phi_{ikk}) + D^j \Psi_i = 0,$$

$$(i,j = 1,\dots,n),$$

remembering that

$$(12) \qquad \begin{aligned} \Phi_{ijk} &= \nabla_{ijk}u + \lambda[1 + \|\nabla u\|^2]^{-1} \nabla_i u \, \nabla_j u \, \nabla_k u, \\ \Psi_i &= \nu[1 + \|\nabla u\|^2]^{-2} \|\nabla u\|^4 \, \nabla_i u. \end{aligned}$$

We want to choose the parameters ν, λ, γ in such a way that the function

$$(13) \qquad u_{ij}(x) = \frac{x_i x_j}{|x|} - \frac{1}{n} \delta_{ij}|x|, \quad (i,j = 1,\dots,n)$$

would be the solution of (11).

After tedious but not difficult calculations we get the following expressions for Φ and Ψ (see (12)) in case of function u given by (13):

$$(14)\qquad \begin{aligned} \Phi_{ijk} &= |x|^{-1}[a(\delta_{ik}x_j + \delta_{jk}x_i) + b\,\delta_{ij}x_k] + c\,\frac{x_i x_j x_k}{|x|^3}, \\ \Psi_i &= (n - \tfrac{1}{n})^5[1 + (n - \tfrac{1}{n})^2]^{-2}\,\frac{x_i}{|x|}, \end{aligned}$$

where

$$(15)\qquad a = 1 + \gamma(n - \tfrac{1}{n}),\quad b = -\tfrac{1}{n} + \gamma(n - \tfrac{1}{n}),$$

$$c = \lambda[1 + (n - \tfrac{1}{n})^2]^{-1}(n - \tfrac{1}{n})^3 - 1.$$

Substituting (14) and (15) into (11) and differentiating we put the coefficients of $\delta_{ij}|x|^{-1}$ and $x_i x_j |x|^{-3}$ equal zero. Thus we obtain

$$(16)\qquad \begin{aligned} &2a + (n-1)b + \nu[1 + (n - \tfrac{1}{n})^2]^{-2}(n - \tfrac{1}{n})^5 + \\ &+ \gamma[2(2 + n)a + (2 + n)b + 3c] = 0, \\ &-2a + (n-1)c - \nu[1 + (n - \tfrac{1}{n})^2]^{-2}(n - \tfrac{1}{n})^5 - \\ &- \gamma[2(2 + n)a + (2 + n)b + 3c] = 0. \end{aligned}$$

From here it follows immediately that

$$(17)\qquad \lambda = [1 + (n - \tfrac{1}{n})^2]\,(n - \tfrac{1}{n})^{-2}\left(\frac{1}{n-1} - \gamma\right),$$

$$(18)\qquad \nu = -(n - \tfrac{1}{n})^{-5}\{3\gamma^2(n+1)(n - \tfrac{1}{n}) + \gamma(n^2 + 3n + 2) + (1 + \tfrac{1}{n})\}\times[1 + (n - \tfrac{1}{n})^2]^2.$$

2.2. <u>Equivalent norms</u>. We are to formulate sufficient conditions on parameter γ under which there exists a constant $c_\gamma > 0$ such that

$$(19)\qquad \|\delta u\|^2 \geqq c_\gamma \|Du\|^2$$

where $\|Du\|^2 = \sum_{i,j,k} (D^k u_{ij})^2$, $\|\delta u\|^2 = \sum_{i,j,k} (\nabla_{ijk} u)^2$.

It is

$$\|\delta u\|^2 = \sum_{i,j,k} [D^k u_{ij} + \gamma(\delta_{ij} \nabla_k u + \delta_{ik} \nabla_j u + \delta_{jk} \nabla_i u)]^2 \geqq$$

$$\geqq (\text{supposing that } \gamma < 0) \geqq \|Du\|^2 + 2\gamma \sum_{i,k} |D^k u_{ii}|\, |\nabla_k u| +$$

$$+ [4\gamma + 3\gamma^2(n + 2)] \|\nabla u\|^2 \geqq \|Du\|^2 + 2\gamma \sqrt{n} \|Du\| \, \|\nabla u\| +$$

$$+ [4\gamma + 3\gamma^2(n + 2)] \|\nabla u\|^2 = [4\gamma + 3\gamma^2(n + 2)] \{\|\nabla u\| +$$

$$+ (\gamma \sqrt{n} \|Du\|) [4\gamma + 3\gamma^2(n + 2)]^{-1}\}^2 +$$

$$+ (1 - (\gamma n)[4 + 3\gamma(n+2)]^{-1}) \|Du\|^2.$$

It is easy to see that for

$$\gamma < -\frac{2}{n+3} \tag{20}$$

it takes place (19) with $c_\gamma = 1 - (\gamma n)[4 + 3\gamma(n+2)]^{-1} > 0$.

2.3. <u>Monotonicity condition</u>. Let us suppose that $\lambda > 0$, $\nu > 0$. (From (17) it follows that $\lambda > 0$ is implied by the condition (20).) Putting

$$\overline{\nabla_i u} = \nabla_i u (1 + \|\nabla u\|^2)^{\frac{1}{2}} \tag{21}$$

and denoting by $I(\varphi)$ the integrand of $\langle DA(u)\varphi, \varphi\rangle$ we have

$$I(\varphi) = \|\delta\varphi\|^2 + \tag{22}$$

$$+ \lambda \{[\overline{\nabla_i u}\, \overline{\nabla_j u}\, \nabla_k \varphi + \overline{\nabla_i u}\, \nabla_j \varphi\, \overline{\nabla_k u} + \nabla_i \varphi\, \overline{\nabla_j u}\, \overline{\nabla_k u} -$$

$$- 2(\overline{\nabla u}, \nabla\varphi)\, \overline{\nabla_i u}\, \overline{\nabla_j u}\, \overline{\nabla_k u}]\, \nabla_{ijk}\varphi\} + \nu\{4 \|\overline{\nabla u}\|^2 (\overline{\nabla u}, \nabla\varphi)^2 +$$

$$+ \|\overline{\nabla u}\|^4 \|\nabla\varphi\|^2 - 4 \cdot \|\overline{\nabla u}\|^4 (\overline{\nabla u}, \nabla\varphi)^2\}.$$

The expression standing by λ in figure brackets can be estimated by means of Hölder inequality as follows:

$|\{[\overline{\nabla_i u}\,\overline{\nabla_j u}\,\nabla_k\varphi + \ldots]\nabla_{ijk}\varphi\}| \leq \{\sum_{i,j,k}[\overline{\nabla_i u}\,\overline{\nabla_j u}\,\nabla_k\varphi +$
$+\ldots]^2\}^{\frac{1}{2}} = \|\delta\varphi\|\{3\|\overline{\nabla u}\|^4\|\nabla\varphi\|^2 + 2\|\overline{\nabla u}\|^2$
$(\overline{\nabla u}, \nabla\varphi)^2[3 - 6\|\overline{\nabla u}\|^2 + 2\|\overline{\nabla u}\|^4]\}^{\frac{1}{2}} \leq$ (using the fact that $0 \leq \|\overline{\nabla u}\| < 1$) $\leq \|\delta\varphi\|\{3[\|\overline{\nabla u}\|^4\|\nabla\varphi\|^2 +$
$+ 2\|\overline{\nabla u}\|^2(\overline{\nabla u}, \nabla\varphi)^2(1 - \|\overline{\nabla u}\|^2)]\}^{\frac{1}{2}}$.

Using the estimate in (22) and putting

$$Q^2 = \|\overline{\nabla u}\|^4\|\nabla\varphi\|^2 + 2\|\overline{\nabla u}\|^2(\overline{\nabla u}, \nabla\varphi)^2(1 - \|\overline{\nabla u}\|^2)$$

we obtain

$$(23) \qquad I(\varphi) \geq \|\delta\varphi\|^2 - \sqrt{3}\lambda Q\|\delta\varphi\| + \nu Q^2 \geq \|\delta\varphi\|^2\left(1 - \frac{3\lambda^2}{4\nu}\right).$$

Let now

$$(24) \qquad 4\nu > 3\lambda^2$$

and let (20) hold. Then $I(\varphi) \geq c_\gamma^* \|D\varphi\|^2$ with $c_\gamma^* > 0$ and so the monotonicity of the operator A defined by (4) takes place.

References

[1] De GIORGI E.: Sulla differenziabilità e l'analicità delle estremali degli integrali multipli regolari, Mem. Acad. Sci. Torino 3(1957), 25-43, Matematika 4(1960), 25-38.

[2] De GIORGI E.: Un esempio di estremali discontinue per un problema variazionale di tipo ellittico, Boll. Unione Mat. Italiana 1(1968), 135-138.

[3] GIUSTI E., MIRANDA M.: Un esempio di soluzioni discontinue per un problema di minimo relativo ad un integrale regolare del calcolo delle variazioni, Boll. Unione Mat. Italiana (4)1(1968),219-227.

[4] KOŠELEV A.I.: Regularnosť rešenij kvasilinějnych eliptičeskich sistem, Uspechi mat. nauk 23(1978), 3-49.

[5] LADYŽENSKAJA O.A., URAL'CEVA N.N.: Linějnyje i kvasilinějnyje uravněnija eliptičeskogo tipa, Nauka, Moskva 1964.

[6] MORREY Ch.B.: Differentiability theorems for weak solutions of nonlinear elliptic differential equations, Bull. Am. Math. Soc. 75(1969), 684-705.

[7] MORREY Ch.B.: Existence and differentiability theorems for the solutions of variational problems for multiple integrals, Bull. Am. Math. Soc. 46 (1940), 439-458.

[8] NASH J.: Continuity of solutions of parabolic and elliptic equations, Amer. J. Math. 80(1950),931-954.

[9] NEČAS J.: Sur la régularité des solutions variationelles des équations elliptiques non-linéaires d'ordre 2k en deux dimensions, Annali Sc. Norm. Sup. Pisa 21(1967), 427-457.

[10] NEČAS J.: Example of an irregular solution to a nonlinear elliptic system with analytic coefficients and conditions for regularity, Abh. Deutsch. Akad. Wiss. Berlin Kl. Math. Phys. Tech.,1977, 1N,197-206.

Matemat. fyz. fakulta
Universita Karlova
Malostran. nám. 1
11000 Praha 1
Československo

Matemat. fyz. fakulta
Universita Karlova
Sokolovská 83
18600 Praha 8
Československo

Mat. fyz. fakulta
Universita Karlova
Sokolovská 83
18600 Praha 8
Československo

(Oblatum 20.6. 1979)

COMMENTATIONES MATHEMATICAE UNIVERSITATIS CAROLINAE
23,4 (1982)

A LIOUVILLE THEOREM FOR NONLINEAR ELLIPTIC SYSTEMS WITH ISOTROPIC NONLINEARITIES

P. L. LIONS, J. NEČAS and I. NETUKA

__Abstract__: We show that if $u=(u_1,\dots,u_m)$ is a solution with bounded gradient in $\mathbb{R}^n$ of an elliptic system of the form:

$$-\frac{\partial}{\partial x_i}\left(a_{ij}(|\nabla u|^2)\frac{\partial u_\alpha}{\partial x_j}\right)=0,\ 1\le\alpha\le m,$$

then each u_α is an affine function on $\mathbb{R}^n$.

__Key words__: elliptic systems, Liouville theorem, regularity, Harnack inequality.

AMS: Primary 35J60, 35D10,
Secondary 35G20

I Introduction:

We consider here a nonlinear second-order elliptic system of the following form:

$$(1)\qquad -\frac{\partial}{\partial x_i}\left(a_{ij}(|\nabla u|^2)\frac{\partial u_\alpha}{\partial x_j}\right)=0 \text{ in } \mathbb{R}^n,\ u=(u_1,\dots,u_m),\quad 1\le\alpha\le m.$$

Throughout all the paper we will assume that $a_{ij}\in C^1(\mathbb{R})$ (for $1\le i,j\le n$) and that (1) is <u>very strongly elliptic</u> in the sense that for every η and $\xi\ne 0$

$$(2)\qquad a_{ij}(|\eta|^2)\,\xi_i^\alpha\xi_j^\alpha+2\,a'_{ik}(|\eta|^2)\,\eta_i^\alpha\eta_j^\beta\xi_i^\alpha\xi_j^\beta>0$$

We prove below that if <u>u has a bounded gradient on $\mathbb{R}^n$, then</u>

© Faculty of Mathematics and Physics of Charles University 1982. Reprinted with kind permission from: Lions, P.-L., Nečas, J. & Netuka, I., A Liouville theorem for nonlinear elliptic systems with isotropic nonlinearities, Commentationes Mathematicae Universitatis Carolinae 23, 645-655.

each component u_α of u is affine on $\mathbb{R}^n$.

This result is clearly a Liouville type theorem. Let us explain now how this result is related to various facts from nonlinear second-order elliptic systems theory. To this end, let us consider a general second order elliptic system:

$$(3) \qquad -\frac{\partial}{\partial x_i}(a_i^\alpha(x,u,\nabla u)) + a^\alpha(x,u,\nabla u) = f^\alpha(x) \text{ in } \Omega$$

where $1 \leq \alpha \leq m$, $u = (u_1,\dots,u_m)$ and Ω is a bounded domain in $\mathbb{R}^n$. The very strong ellipticity of the system (3) is expressed by the following condition:

$$(4) \qquad \frac{\partial a_i^\alpha}{\partial \gamma_j^\beta}(x,\xi,\gamma)\ \xi_i^\alpha \xi_j^\beta > 0, \qquad \xi \neq 0.$$

Of course, when (3) reduces to (1), (4) is nothing else than (2). Assuming that u is a Lipschitz solution of (3), one may ask the following natural (and fundamental) question: is u of class C^1 or even $C^{1,\mu}$ (for some $\mu \in (0,1)$) ?

As shown by M.Giaquinta and J.Nečas [2] , this regularity question turns out to be, in some sense, equivalent to the following Liouville type condition: (3) is said to satisfy the Liouville condition (in short $L(\mathbb{R}^n)$) provided the following implication holds: for all $x^0 \in \Omega$, $\xi \in \mathbb{R}^m$, if $v = (v_1,\dots,v_m)$ is a solution with bounded gradient of

$$(3') \qquad -\frac{\partial}{\partial x_i}(a_i^\alpha(x^0,\xi,\nabla v)) = 0 \quad \text{in } \mathbb{R}^n,$$

then each v is affine on $\mathbb{R}^n$. More precisely, in [2] it is proved that if the system (3) (where we assume (4) with a_i^α, $a^\alpha \in C^1(\bar\Omega \times \mathbb{R}^m \times \mathbb{R}^{mn})$) satisfies $L(\mathbb{R}^n)$ and $p > n$, then for every $\gamma > 0$ and every compact set $K \subset \Omega$ there is $c(\gamma,K) < \infty$ such that

$$(5) \qquad \| u_\alpha \|_{C^{1,\mu}(K)} \leq c(\nu, K), \qquad 1 \leq \alpha \leq m,$$

with $\mu = 1 - (n/p)$, whenever $f^\alpha \in L^p(\Omega)$ and u is a Lipschitz solution of (3) such that

$$\| u \|_{[W^{1,\infty}(\Omega)]^m} + \| f \|_{[L^p(\Omega)]^m} \leq \nu .$$

Conversely, in some sense, $L(\mathbb{R}^n)$ is a consequence of regularity results of the form (5) - see J.Nečas [6],[7] or M.Giaquinta [1].

Therefore the Liouville result we prove in this paper immediately yields the $C^{1,\mu}$ regularity for special systems of form:

$$(6) \qquad - \frac{\partial}{\partial x_i} (a_{ij}(x,u,|\nabla u|^2) \frac{\partial u_\alpha}{\partial x_j}) + a^\alpha (x,u,\nabla u) = f^\alpha(x) \text{ in } \Omega$$

(for $1 \leq \alpha \leq m$). At this point, we want to point out that this regularity result (a consequence of our result and an equivalent when a_{ij} depend on $|\nabla u|^2$ only) was established by P.A.Ivert [4] in a generalization of deep results due to K.Uhlenbeck [8]. Thus, in some sense, the result we present here is not new and could be derived from Uhlenbeck - Ivert results. On the other hand, our method of proof is quite different from those of [4], [8] and, we believe, much simpler. Let us also mention that it is straightforward to adapt our method of proof to show directly the $C^{1,\mu}$ regularity result (looking, roughly speaking, at little balls instead of large balls).

Let us conclude this introduction by a few words on our method of proof. In section II below, we present a general result on nonlinear elliptic systems which implies in particular that, if we denote by $\omega = \nabla u$, we have: there is $\epsilon_0 > 0$ such that if $\Phi_\omega(R) < \epsilon_0$, then for every $\rho \in (0,R)$

$$(7) \qquad \Phi_\omega(\rho) \leq c_0 \, \Phi_\omega(R)$$

where C_o depends only on $\|\omega\|_{L^\infty(B_R)}$ and where for a vector valued function g we denote:

$$\Phi_g(\rho) = \frac{1}{\rho^n} \int_{B_\rho} |g(x) - (g)^\rho|^2 \, dx,$$

$$(g)^\rho = (1/|B_\rho|) \int_{B_\rho} g(x)dx.$$

By an easy use of Poincaré inequality, we see that in order to conclude (using(7)) we just need to show that $\omega = \nabla u$ has the so-called Saint-Venant property:

$$\lim_{R\to\infty} R^{-n+2} \int_{B_R} |\nabla \omega(x)|^2 \, dx = 0 \quad . \tag{8}$$

The main idea used to prove (7) goes back to a fundamental lemma of E.Giusti - see e.g. [2] .

Next, in section III, we state and prove a Liouville type theorem. This is done by remarking - following [4],[8] - that $|\nabla u|^2 = w$ satisfies:

$$-\frac{\partial}{\partial x_i} (A_{ij} \frac{\partial w}{\partial x_j}) + \alpha |D^2 u|^2 \leq 0 \quad \text{in } \mathbb{R}^n \tag{9}$$

for some $\alpha > 0$, and for some uniformly elliptic coefficients A_{ij}. Using this inequality and a Harnack type inequality proved in D.Gilbarg and N.S.Trudinger [3] (for example), we show that (8) holds and thus ω is constant.

The authors wish to thank P.A.Ivert for useful discussions and for a careful reading of our manuscript.

II A general result on quasilinear elliptic systems:

In this section we consider a solution $\omega = (\omega_1, \ldots, \omega_N)$ of

$$(10) \qquad -\frac{\partial}{\partial x_i}\left[A_{ij}^{\alpha\beta}(\omega)\frac{\partial \omega_\beta}{\partial x_j}\right] = 0 \quad \text{in } \mathbb{R}^n, \quad \alpha = 1,\dots,N,$$

where $A_{ij}^{\alpha\beta}$ are continuous on $\mathbb{R}^m$ and where the ellipticity condition

$$(11) \qquad A_{ij}^{\alpha\beta}(\xi)\,\xi_i^\alpha \xi_j^\beta > 0 \quad \text{for } \xi \neq 0$$

holds.

Theorem II. 1: Let $R > 0$, let ω be a bounded solution of (10) in $(H^1(B_R))^N$ and let us assume that (11) holds. We denote $\mu = \|\omega\|_{L^\infty(B_R)}$. Then there exist $\varepsilon_0 > 0$, $C_0 > 0$ such that the following statement holds:

if $\qquad \Phi_\omega(R) \leq \varepsilon_0^2$,

then $\qquad \Phi_\omega(\rho) \leq C_0\, \Phi_\omega(R)$

whenever $\rho \in (0,R)$. In addition ε_0, C_0 depend only on μ and on the ellipticity constants in (11).

Before giving the proof of Theorem II.1, let us mention the Corollary II.1: Let ω be a bounded solution of (10) in $(H^1_{loc}(\mathbb{R}^n))^N$ satisfying the Saint-Venant property

$$\lim_{R\to\infty} R^{-n+2} \int_{B_R} |\nabla \omega(x)|^2\, dx = 0,$$

and let us assume that (11) holds. Then ω is a constant vector.

Proof: Observe that we have by Poincaré inequality:

$$(12) \qquad R^{-n} \int_{B_R} |\omega(x) - (\omega)^R|^2\, dx \leq c_1 R^{-n+2} \int_{B_R} |\nabla \omega(x)|^2\, dx.$$

(Here and below $c_1, c_2, \dots$ denote various positive contents independent of R, ω, u.) Thus we see that (8) implies: $\lim_{R\to\infty} \Phi_\omega(R) = 0$. Therefore by Theorem II.1, $\Phi_\omega(\rho) = 0$ for all $\rho > 0$ and the proof is complete.

Proof of Theorem II.1: First of all, in view of (11), there exists $\nu > 0$ such that for every ξ and $|\xi| \leq \mu$ we have

$$[A_{ij}^{\alpha\beta}(\xi)\, A_{ij}^{\alpha\beta}(\xi)]^{1/2} \leq \frac{1}{\nu}, \quad A_{ij}^{\alpha\beta}(\xi)\, \xi_i^{\alpha}\, \xi_j^{\beta} \geq \nu\, |\xi|^2.$$

Let us also recall that it is known (see e.g.[2]) that there exists c_2 ($= c_2(\mu, \nu)$) such that we have:

$$\text{(13)} \qquad \Phi_\omega(\tau) \leq c_2 \tau^2 .\, \Phi_\omega(1), \qquad 0 < \tau \leq 1,$$

if ω is a solution of the system:

$$-\frac{\partial}{\partial y_i}\left(A_{ij}^{\alpha\beta}(\xi)\frac{\partial \omega_\beta}{\partial y_j}\right) = 0 \ \text{ in } B_1$$

where $|\xi| \leq \mu$.

Next, let $\tau \in (0,1)$. We are first going to prove that there exist $\varepsilon_0 = \varepsilon_0(\mu, \tau, \nu) > 0$ such that

$$\text{(14)} \qquad \Phi_\omega(\tau) \leq 2\, c_2\, \tau^2\, \Phi_\omega(1)$$

where ω solves (10) and satisfies: $\|\omega\|_{L^\infty(B_1)} \leq \mu$, $\Phi_\omega(1) \leq \varepsilon_0^2$.

Let us argue by contradiction and let us thus assume that there exists a sequence $(\omega^n)_{n\geq 1}$ of solutions of (10) satisfying:

$$\text{(15)} \quad \|\omega^n\|_{L^\infty(B_1)} \leq \mu, \ \{\Phi_{\omega^n}(1)\}^{1/2} = \varepsilon_n \longrightarrow 0,$$
$$\Phi_{\omega^n}(\tau) > 2c_2\, \tau^2\, \varepsilon_n^2.$$

To simplify notations, we will use indifferently the notations $\Phi_{\omega^n}(\tau)$ or $\Phi(\omega^n, \tau)$. We then set: $\sigma^n = \frac{1}{\varepsilon_n}[\omega^n - (\omega^n)^1]$.

Obviously we have:

$$\text{(16)} \qquad \int_{B_1} |\sigma^n(x)|^2 dx = 1; \ \Phi(\sigma^n, \tau) > 2c_2 \tau^2;$$

$$(17) \qquad - \frac{\partial}{\partial x_i} \left(A_{ij}^{\alpha\beta}(\omega^n) \frac{\partial \sigma_\beta^n}{\partial x_j}\right) = 0 .$$

Without loss of generality we may assume that:

$\sigma^n \longrightarrow \sigma$ weakly in $(L^2(B_1))^N$, $\varepsilon_n \sigma^n \longrightarrow 0$ in $(L^2(B_1))^N$ and a.e.,

for some $\sigma \in (L^2(B_1))^N$. In addition, in wiew of (16): $\Phi_\sigma(1) \leq 1$. Furthermore, recalling that we have:

$$\omega^n = \varepsilon_n \sigma^n + (\omega^n)^1, \quad \|\omega^n\|_{L^\infty(B_1)} \leq \mu ,$$

we see that $|(\omega^n)^1| \leq \mu$ and $\omega^n - (\omega^n)^1 \longrightarrow 0$ a.e. Since we may assume without loss of generality that $(\omega^n)^1 \longrightarrow \xi$ ($|\xi| \leq \mu$), we finally deduce: $\omega^n \longrightarrow \xi$. a.e. .

Next, we obtain from (16) and (17):

$$(18) \qquad \int_{B_k} |\nabla \sigma^n(y)|^2 \, dy \leq C(k) \quad \text{for} \quad k \in (0,1),$$

thus we may suppose that $\sigma^n \longrightarrow \sigma$ weakly in $(H^1(B_k))^N$ (for all $k < 1$). Thus, passing to the limit in (17), we get:

$$- \frac{\partial}{\partial x_i} \left(A_{ij}^{\alpha\beta}(\xi) \frac{\partial \sigma_\beta}{\partial x_j}\right) = 0 \quad \text{in} \quad B_1 .$$

In addition, since $\sigma_n \longrightarrow \sigma$ in $(L^2(B_k))^N$ (for all $k < 1$), we deduce from (16): $\Phi(\sigma, \tau) \geq 2\, c_2 \tau^2 \geq 2 c_2 \tau^2 \Phi(\sigma, 1)$. This contradicts (13) and the contradiction shows our claim.

Let us choose now $\tau \in (0,1)$ satisfying: $2 c_2 \tau^2 \leq 1$. Given $\rho \in (0,1)$, let $k \geq 0$ be the integer such that: $\tau^{k+1} \leq \rho < \tau^k$. Now, if ω solves (10) and satisfies: $\|\omega\|_{L^\infty(B_1)} \leq \mu$, $\Phi_\omega(1) \leq \varepsilon_0^2$, we have in view of (14):

$$\tau^n \rho^{-n} \int_{B_\rho} |\omega - (\omega)^\rho|^2 \, dx \leq (\rho / \tau^k)^n \rho^{-n} \int_{B_\rho} |\omega - (\omega)^\rho|^2 \, dx \leq$$

$$\leq (\tau^k)^{-n} \int_{B_\varrho} |\omega - (\omega)^{\tau^k}|^2 \, dx \leq (\tau^k)^{-n} \int_{B_{\tau^k}} |\omega - (\omega)^{\tau^k}|^2 \, dx \leq$$

$$\leq \int_{B_1} |\omega - (\omega)^1|^2 \, dx;$$

that is, we proved: $\Phi_\omega(\varrho) \leq \tau^{-n} \Phi_\omega(1)$.

The proof of Theorem II.1 is easily completed by considering the function $\tilde{\omega}(x) = \omega(x/R)$.

Remark II.1: We now show how the preceding results are related to the system (1): indeed, if $u \in (H^2_{loc}(\mathbb{R}^n))^m$ is a solution of (1) then, for $1 \leq k \leq n$, $\frac{\partial u_\alpha}{\partial x_k}$ satisfies:

$$-\frac{\partial}{\partial x_i}\left[A^{\alpha\beta}_{ij}(\nabla u)\frac{\partial}{\partial x_j}\left(\frac{\partial u_\beta}{\partial x_k}\right)\right] = 0 \quad \text{in } \mathbb{R}^n, \quad 1 \leq \alpha \leq m,$$

where $A^{\alpha\beta}_{ij}(\nabla u) = a_{ij}(|\nabla u|^2)\,\delta_{\alpha\beta} + 2\,a'(|\nabla u|^2)\frac{\partial u_\alpha}{\partial x_\ell}\frac{\partial u_\beta}{\partial x_j}$.

Thus $\omega = \nabla u$ satisfies a system of the form (10) and (11) is a consequence of (2).

III The main result:

Let $u = (u_1, \dots, u_m)$ be a solution of (1):

$$-\frac{\partial}{\partial x_i}\left(a_{ij}(|\nabla u|^2)\frac{\partial u_\alpha}{\partial x_j}\right) = 0 \quad \text{in } \mathbb{R}^n, \quad 1 \leq \alpha \leq m.$$

Theorem III.1: We assume the ellipticity condition (2) and $\nabla u \in (L^\infty(\mathbb{R}^n))^{nm}$. Then each component u_α of u is affine on $\mathbb{R}^n$.

Proof: Standard arguments yield $u \in W^{2,2}_{loc}(\mathbb{R}^n)$; cf. [7] or [1]. In view of the results of the preceding section and of Remark II.1, it is enough to show:

$$\lim_{R \to \infty} R^{-n+2} \int_{B_R} |D^2 u|^2 \, dx = 0. \tag{19}$$

In order to prove (19), we first observe that an easy computation yields:

$$- \frac{\partial}{\partial x_i} \left[A_{ij}(\nabla u) \frac{\partial}{\partial x_j}(|\nabla u|^2) \right] + a_{ij} \frac{\partial^2 u_\alpha}{\partial x_i \partial x_k} \frac{\partial^2 u_\alpha}{\partial x_j \partial x_k} +$$

(20)

$$+ 2 a'_{ik} \frac{\partial u_\alpha}{\partial x_k} \frac{\partial u_\beta}{\partial x_j} \frac{\partial^2 u_\alpha}{\partial x_i \partial x_s} \frac{\partial^2 u_\beta}{\partial x_j \partial x_s} = 0 \quad ,$$

where $A_{ij}(\nabla u) = \frac{1}{2} a_{ij}(|\nabla u|^2) + a'_{ik}(|\nabla u|^2) \frac{\partial u_\alpha}{\partial x_k} \frac{\partial u_\alpha}{\partial x_j}$

In view of (2), we see that (for more details, see [4])

(21) $\exists \nu > 0, \ \forall \xi \in \mathbb{R}^n, \ A_{ij}(\nabla u(x)) \xi_i \xi_j \geq \nu |\xi|^2$,

$$\{A_{ij}(\nabla u(x)) A_{ij}(\nabla u(x))\}^{1/2} \leq \frac{1}{\nu} \text{ a.e. in } \mathbb{R}^n$$

and (20) implies:

(22) $$- \frac{\partial}{\partial x_i} (A_{ij}(\nabla u) \frac{\partial}{\partial x_j} (|\nabla u|^2)) + \alpha |D^2 u|^2 \leq 0 \text{ in } \mathbb{R}^n,$$

for some $\alpha > 0$. We denote $M = \| |\nabla u|^2 \|_{L^\infty(\mathbb{R}^n)}$.

We are now going to prove:

(23) $$R^{-n+2} \int_{B_{R/2}} |D^2 u|^2 \, dx \leq c_3 R^{-n} \int_{B_{2R}} (M - |\nabla u|^2) dx.$$

To this end we introduce $\eta \in H_0^1(B_{2R})$, the solution of:

(24) $$- \frac{\partial}{\partial x_i} (A_{ji} \frac{\partial \eta}{\partial x_j}) = \frac{1}{R^2} \text{ in } B_{2R} \quad .$$

Standard results yield: $\eta \geq 0$ in B_{2R} and

(25) $\|\eta\|_{L^\infty(B_{2R})} \leq c \qquad \operatorname{infess}_{B_{R/2}} \eta \geq c_5 > 0.$

Then multiplying (22) by η^2 and using (24), (25), we deduce:

$$c_6 \int_{B_{R/2}} |D^2u(x)|^2\,dx \le \int_{B_{2R}} A_{ij} \frac{\partial \eta^2}{\partial x_i} \frac{\partial}{\partial x_j} (M-|\nabla u|^2)\,dx \le$$

$$\le 2 \int_{B_{2R}} \frac{1}{R^2}\, \eta\, (M-|\nabla u|^2)\,dx - 2 \int_{B_{2R}} A_{ij} \frac{\partial \eta}{\partial x_i} \frac{\partial \eta}{\partial x_j} (M-|\nabla u|^2)dx$$

and this yields:

$$\int_{B_{R/2}} |D^2u(x)|^2\,dx \le \frac{c_7}{R^2} \int_{B_{2R}} (M-|\nabla u|^2)dx$$

and (23) is proved.

To conclude, we see that (19) follows from (23), applying the following lemma to $w = |\nabla u|^2$, $\alpha_{ij}(x) = A_{ij}(\nabla u(x))$.

<u>Lemma III.1</u>: Let $w \in H^1_{loc}(\mathbb{R}^n) \cap L^\infty(\mathbb{R}^n)$ satisfy:

$$-\frac{\partial}{\partial x_i}\left(\alpha_{ij}(x) \frac{\partial w}{\partial x_j}\right) \le 0 \quad \text{in } \mathbb{R}^n$$

where $\alpha_{ij} \in L^\infty(\mathbb{R}^n)$ satisfy:

$$\{\alpha_{ij}(x)\,\alpha_{ij}(x)\}^{1/2} \le \frac{1}{\nu}\,, \quad \alpha_{ij}(x)\,\xi_i\,\xi_j \ge \nu|\xi|^2 \quad \forall \xi \in \mathbb{R}^n,$$

a.e.in $\mathbb{R}^n$

for some $\nu > 0$. If $M = \sup\mathrm{ess}_{\mathbb{R}^n}\, w$, then we have:

$$(26) \qquad \lim_{R\to\infty} (1/|B_R|) \int_{B_R} w(x)dx = M.$$

<u>Proof</u>: This lemma is proved by the use of a weak Harnack inequality (cf.[3], for example) which implies:

$$(27) \qquad R^{-n} \int_{B_{2R}} z(x)\,dx \le c_8 \inf_{B_R}\mathrm{ess}\; z$$

with $z = M - w$. Now if we let $R \to \infty$, we obtain (26) since $\inf\limits_{B_R} \mathrm{ess}\, z \longrightarrow \inf\limits_{\mathbb{R}^n} \mathrm{ess}\, z = 0$; and $z \geq 0$ a.e.in R^n.

References

[1] M.Giaquinta: Multiple integrals in the calculus of variations and nonlinear elliptic systems, Universität Bonn, Sonderforschungsbereich 72, Preprint No. 443.

[2] M.Giaquinta and J.Nečas: On the regularity of weak solutions to nonlinear elliptic systems of partial differential equations, J.Reine Angew.Math.316(1980), 140 - 159.

[3] D.Gilbarg and N.S.Trudinger: Elliptic partial differential equations of second order, Springer-Verlag, New York, 1977.

[4] P.A.Ivert: Regularitätsuntersuchungen von Lösungen elliptischer Systeme vom quasilinearen Differentialgleichungen zweiter Ordnung, Manuscripta Math.30(1970),53-88.

[5] J.Nečas: Les méthodes directes en théorie des équations elliptiques, Academia, Prague, 1967.

[6] J.Nečas: A necessary and sufficient condition for the regularity of weak solutions to nonlinear elliptic systems of partial differential equations, Abh.Akad.Wiss.DDR, 1981.

[7] J.Nečas: Régularité des solutions faibles d'équations elliptiques nonlinéaires; applications à l'élasticité, Université Paris VI, Analyse numérique et fonctionnelle,1981.

[8] K.Uhlenbeck: Regularity for a class of non-linear elliptic systems, Acta Math.138(1977), 219-240.

CEREMADE
Université Paris IX,Dauphine
Place De-Lattre-de-Tassigny
75 775 Paris, Cedex 16
France

Matematicko-fyzikální fakulta
Univerzita Karlova
Malostranské nám.25
110 00 Praha 1
Československo

Matematicko-fyzikální fakulta
Univerzita Karlova
Sokolovská 83
186 00 Praha 8
Československo

(Oblatum 7.7. 1982)

PACIFIC JOURNAL OF MATHEMATICS
Vol. 99, No. 1, 1982

ON THE REGULARITY UP TO THE BOUNDARY FOR SECOND ORDER NONLINEAR ELLIPTIC SYSTEMS

M. GIAQUINTA, J. NEČAS, O. JOHN, AND J. STARÁ

It is proved the regularity up to the boundary of the uniformly Lipschitz-continuous weak solutions of a boundary value problem for the elliptic system

$$-D_i a_i^r(x, u, Du) + \bar{a}^r(x, u, Du) = f^r\,; \qquad r = 1, \cdots, m \tag{1.1}$$

from the Liouville properties of the system.

In (1.1) $u = \{u^r\}_{r=1,\cdots,m}$ is a vector function and $Du = \{D_i u^r\}_{\substack{i=1,\cdots,n\\ r=1,\cdots,m}}$ is its gradient. We write $D_i u^r = \partial u^r/\partial x_i$ and the summation convention is used throughout the paper.). We follow the ideas of our previous work (see [1-4]) where interior regularity was shown to be equivalent (in some sense) to the Liouville property (L) (see Definition 2.2). In the present paper, regularity up to the boundary is shown to be, essentially, equivalent to the previous (L) together with a certain "boundary" Liouville property (L^+) (see Definition 2.3).

2. Notation and assumptions. Let $\boldsymbol{R}^n$ be an n-dimensional Euclidean space; for $x = (x_1, \cdots, x_{n-1}, x_n) = (x', x_n) \in \boldsymbol{R}^n$ let $|x| = \max\{|x_i|; i = 1, \cdots, n\}$; further let $\boldsymbol{R}^n_+ = \{x \in \boldsymbol{R}^n; x_n > 0\}$; $\Omega = \{x \in \boldsymbol{R}^n_+; |x| < 1\}$; $\Gamma = \{x \in \boldsymbol{R}^n; |x'| < 1; x_n = 0\}$; $B(x_0, R) = \{x \in \Omega; |x - x_0| < R\}$; $\Gamma(x_0, R) = \overline{B(x_0, R)} \cap \Gamma$.

Let us denote

$$a(x, u, Du) = \{a_i^r(x, u, Du)\}_{\substack{i=1,\cdots,n\\ r=1,\cdots,m}}$$

$$\bar{a}(x, u, Du) = \{\bar{a}^r(x, u, Du)\}_{r=1,\cdots,m}$$

$$f(x) = \{f^r(x)\}_{r=1,\cdots,m}\,,$$

where a, $\bar{a}$ are once continuously differentiable functions on $\bar{\Omega} \times \boldsymbol{R}^m \times \boldsymbol{R}^{nm}$, and $f \in [W^{1,p/2}(\Omega)]^m$ for some p, $p > n$.

REMARK. In what follows we omit the notation of the Cartesian product. So we write $f \in W^{1,p}(\Omega)$ instead of $f \in [W^{1,p}(\Omega)]^{mn}$ etc.

In this notation the system (1.1) can be rewritten as

$$-\operatorname{div}(a(x, u, Du)) + \bar{a}(x, u, Du) = f(x) \tag{2.1}$$

on Ω. We suppose that the strong ellipticity condition holds:

$$\frac{\partial a_i^r}{\partial \eta_j^s}(x, \xi, \eta)\zeta_i^r \zeta_j^s > 0 \tag{2.2}$$

© Pacific Journal of Mathematics 1982. Reprinted with kind permission from Mathematical Sciences Publishers from: Giaquinta, M., Nečas, J., John, O. & Stará, J., On the regularity up to the boundary for second order nonlinear elliptic systems, Pacific Journal of Mathematics 99:1-17.

for every $\zeta \neq 0$ and each $(x, \xi, \eta) \in \bar{\Omega} \times \boldsymbol{R}^m \times \boldsymbol{R}^{nm}$.

To describe the boundary conditions we introduce two disjoint sets M, N of positive integers such that $M \cup N = \{1, \cdots, m\}$ (both the cases $M = \emptyset$ and $N = \emptyset$ being admissible). Let $\{b_{rs}\}_{r \in M, s \in N}$ be the set of real constants. The stable boundary operators B_r $(r \in M)$ are given by the formulas

$$B_r u = u^r - \sum_{s \in N} b_{rs} u_s .$$

Put

$$C = \{c_{rs}\}_{r,s=1,\cdots,m} , \quad \text{where} \quad c_{rs} = \begin{cases} -b_{rs} , & r \in M,\ s \in N , \\ \delta_{rs} , & r, s \in M , \\ 0 , & r \in N , \end{cases}$$

$$C^* = \{c^*_{rs}\}_{r,s=1,\cdots,m} , \quad \text{where} \quad c^*_{rs} = \begin{cases} b_{sr} , & r \in N,\ s \in M . \\ -\delta_{rs} , & r, s \in N , \\ 0 , & r \in M , \end{cases} \tag{2.3}$$

$$\mathscr{F}(x, u, Du) = \{a^r_n(x, u, Du) + h^r(x, u) - g^r(x) - f^r_n(x)\}_{r=1,\cdots,m} ,$$

where h and g are given functions; $h \in C^1(\bar{\Gamma} \times \boldsymbol{R}^m)$, $g \in W^{1,\infty}(\Gamma)$.

Let, finally, $u_0 = \{u^r_0\}_{r=1,\cdots,m}$ be a given function from $W^{2,p}(\Omega)$. We consider the following boundary value problem for the system (2.1) (in its classical formulation):

$$\begin{aligned} &C(u - u_0) = 0 \quad \text{on} \quad \Gamma , \\ &C^* \mathscr{F}(x, u, Du) = 0 \quad \text{on} \quad \Gamma , \\ &u - u_0 = 0 \quad \text{on} \quad \partial\Omega \backslash \Gamma . \end{aligned} \tag{2.4}$$

Denote the scalar product in $\boldsymbol{R}^n$ as well as in $\boldsymbol{R}^{nm}$ by (,) and put

$$V = \{v \in W^{1,2}(\Omega) ; \quad Cv = 0 \quad \text{on} \quad \Gamma ; \quad v = 0 \quad \text{on} \quad \partial\Omega \backslash \Gamma\} . \tag{2.5}$$

A function $u \in W^{1,2}(\Omega)$ is said to be a weak solution of the problem (2.1), (2.4) if

(2.6) (i) $u - u_0 \in V$,

(ii) for each $\varphi \in V$, it holds

$$\int_\Omega \{(a, D\varphi) + (\bar{a}, \varphi) - (f, \varphi)\} dx = \int_\Gamma (h - g, \varphi) dx' .$$

(Let us rewrite for once the equation (2.6) (ii) in a more detailed form:

(2.6) (ii)′

$$\int_\Omega \{a^r_i(x, u(x), Du(x)) D_i \varphi^r(x) + \bar{a}^r(x, u(x), Du(x)) \varphi^r(x) - f^r(x) \varphi^r(x)\} dx$$

$$= \int_\Gamma \{h^r(x, u(x)) - g^r(x)\} \varphi^r(x) dx' .$$

Let us now formulate the regularity of the problem and the Liouville conditions.

DEFINITION 2.1 (R). We say that the problem (2.1), (2.4) is regular (and denote this property by (R)) if for each weak solution u of this problem for which $Du \in L_\infty(\Omega)$, the gradient Du is locally α-Hölder continuous on $\Omega \cup \Gamma$, and for each Ω' for which $\Omega' \subset \Omega \cup \Gamma$ it holds

$$\|\nabla u\|_{C^\alpha(\bar{\Omega}')} \leqq C ,$$

where the constant C depends on $\|\nabla u\|_{L_\infty(\Omega)}$, Ω' and the data of the problem.

DEFINITION 2.2 (L). We say that the system (2.1) satisfies the Liouville condition (L) if for each $x_0 \in \Omega$ and each $\xi \in \boldsymbol{R}^m$ the solution $u \in W^{1,2}_{\text{loc}}(\boldsymbol{R}^n)$ of the equation

$$\int_{R^n} (a(x_0, \xi, Du), D\varphi)dx = 0 \qquad \forall \varphi \in C_0^\infty(R^n) \tag{2.7}$$

for which $Du \in L_\infty(\boldsymbol{R}^n)$ is a polynomial of at most the first degree.

DEFINITION 2.3 (L^+). Write $Z = \{\varphi \in C_0^\infty(\boldsymbol{R}^n);\ C\varphi = 0 \text{ on } \{x \in \boldsymbol{R}^n;\ x_n = 0\}\}$. We say that the problem (2.1), (2.4) satisfies the Liouville condition (L^+) if for each $x_0 \in \Gamma$; $\xi \in \boldsymbol{R}^m$; $d \in \boldsymbol{R}^m$ the solution $u \in W^{1,2}_{\text{loc}}(\overline{\boldsymbol{R}^n_+})$ of the equation

$$\int_{R^n_+} (a(x_0, \xi, Du), D\varphi)dx = \int_{\{x \in R^n;\ x_n = 0\}} (d, \varphi)dx' \qquad \forall \varphi \in Z \tag{2.8}$$

is the polynomial of at most the first degree, provided that Cu is a polynomial of at most the first degree on $\{x \in \boldsymbol{R}^n;\ x_n = 0\}$ and $Du \in L_\infty(\boldsymbol{R}^n_+)$.

Our paper contains the proof that (roughly speaking): (2.1), (2.4) is regular iff (L) and (L^+) hold simultaneously. The necessity of the Liouville conditions is proved in §3 with the definition of the regularity being slightly changed. In §4 the proof of the implication $(L) \wedge (L^+) \Rightarrow (R)$ is given.

3. The necessity of Liouville conditions. Considering the definitions 2.1–2.3 we conclude that the property (R) concerns one fixed problem (2.1), (2.4) whilst the Liouville conditions (L) and (L^+) refer to a system of problems (2.8). We do not know whether the implication $(R) \Rightarrow (L) \wedge (L^+)$ holds. To obtain the implication of this type we modify at first the definition of regularity.

DEFINITION 3.1 (R'). Let for each $x_0 \in \overline{\boldsymbol{R}^n_+}$, $\xi \in \boldsymbol{R}^m$, $d \in \boldsymbol{R}^m$ and for each solution u of (2.8) for which $Du \in L_\infty(\boldsymbol{R}^n_+)$ and Cu is a polynomial of at most the first degree on $\{x \in \boldsymbol{R}^n; x_n = 0\}$ there exists $T > 0$ such that u belongs to the space $C^{1,\alpha}(\overline{B(0, T)})$ with $\alpha = \min\{1/2, 1 - n/p\}$ and

$$\|u\|_{C^{1,\alpha}} \leqq C , \tag{3.1}$$

where C and T depend only on $\|Du\|_{L_\infty}$, $|u(0)|$ and the data of the problem.

THEOREM 3.1. $(R') \Rightarrow (L^+)$.

Proof. Suppose $x_0 = 0$. The function $u_R(y) = (1/R)u(Ry)$ solves (2.8). Further $\|Du_R\|_{L_\infty} = \|Du\|_{L_\infty}$ and for $R > 1$ the values $|u_R(0)|$ and Cu_R are bounded by the same constants as the corresponding values of u. Thus u_R $(R > 1)$ satisfies (3.1) with the constant independent of R. Let $x \in \boldsymbol{R}^n_+$; $TR \geqq |x|$; $Ry = x$. According to (3.1) we get

$$|D_y u_R(y) - D_y u_R(0)| \leqq C|y|^\alpha \leqq C\left|\frac{x}{R}\right|^\alpha ,$$

hence

$$|D_x u(x) - D_x u(0)| \leqq C\frac{|x|^\alpha}{R^\alpha}$$

and it implies that $D_x u(x) = D_x u(0)$ for $R \to \infty$.

Let us mention that the necessity of the condition (L) was proved in [4].

4. Sufficiency of the Liouville conditions. Put for an arbitrary vector function $f = \{f^r\}_{r=1,\cdots,s}$,

$$\begin{cases} F(x_0, R) = R^{2-n} \displaystyle\int_{B(x_0,R)} \sum_{r=1}^{s} \sum_{i=1}^{n} |D_i f^r(x)|^2 dx \quad \text{and} \\ VB(x_0, R) = \{u \in W^{1,2}(B(x_0, R)); Cu = 0 \quad \text{on} \quad \Gamma(x_0, R) \quad \text{and} \\ \qquad u = 0 \quad \text{on} \quad \partial B(x_0, R)\backslash\Gamma(x_0, R)\} . \end{cases} \tag{4.1}$$

The following notation will be used in Lemma 4.1 only:

$$B(x_0, t) = \{x \in \boldsymbol{R}^n_+;\ |x - x_0| < t\} ; \qquad t \in [0, 1] .$$

Let

$$\Gamma = \{x \in \boldsymbol{R}^n;\ |x'| < 1; x_n = 0\} .$$

With the so defined $B(x_0, t)$ the symbols $F(x_0, t)$ and $VB(x_0, t)$ have the same meaning as in (4.1).

LEMMA 4.1. *Let* $\boldsymbol{B} = \{B^{ij}_{rs}\}_{\substack{i,j=1,\cdots,n\\ r,s=1,\cdots,m}}$ *be a real matrix such that*

$$\exists \mathscr{H} > 0 \qquad \forall \eta \in R^{nm} \qquad B^{ij}_{rs}\eta^i_r\eta^j_s \geqq \mathscr{H}|\eta|^2 . \tag{4.2}$$

Then there exists a positive number K *such that for every* $x_0 = (0, \cdots, 0, q)$, $(q \in [0, 1])$ *for each* $v \in W^{1,2}(B(x_0, 1))$ *for which*

$$\boldsymbol{C}v = 0 \quad \text{on} \quad \Gamma \tag{4.3}$$

and which solves the system

$$\int_{B(x_0,1)} (\boldsymbol{B}Dv, D\varphi)dx = 0 \qquad \forall \varphi \in VB(x_0, 1) , \tag{4.4}$$

and for every $t \in]0, 1/2[$, *the inequalities*

$$V(x_0, t) \leqq Kt^2 V(x_0, 1) , \tag{4.5}$$

$$t^{-n}\int_{B(x_0,t)} |v(x) - P_t|^2 dx \leqq Kt^2 \int_{B(x_0,1)} |v(x) - Q|^2 dx \tag{4.6}$$

hold, where Q *is an arbitrary vector such that* $\boldsymbol{C}Q = 0$ *and* P_t *is either a value* $v(\tilde{x})$ *at an arbitrary point* $\tilde{x} \in \overline{B(x_0, t)}$ *or an integral mean value of* v *over any connected subset of* $\overline{B(x_0, t)}$.

Proof. Let $k \in \boldsymbol{N}$ be such that $W^{2,k}(G) \subset C^1(\bar{G})$ for a bounded domain $G \subset \boldsymbol{R}^n$. Let $1 = t_0 > t_1 > \quad > t_k = 1/2$ be an equidistant subdivision of the segment [1/2, 1].

Let $\Phi \in C^\infty(\overline{B(x_0, 1)})$; $\operatorname{supp} \Phi \subset B(x_0, (t_0 + t_1)/2)$; $0 \leqq \Phi \leqq 1$, $\Phi \equiv 1$ on $B(x_0, t_1)$; $|D\Phi| \leqq C/(t_0 - t_1)$.

Let $\boldsymbol{C}Q = 0$ and put

$$\varphi = \Phi^2(v - Q)$$

in (4.4). By usual calculations we obtain (denoting in what follows all the constants by C)

$$\int_{B(x_0,t_1)} |Dv|^2 dx \leqq C\int_{B(x_0,1)} |v - Q|^2 dx \tag{4.7}$$

If $\overline{B(x_0, t_1)} \subset B(x_0, 1)$ we use the fact that all the derivatives up to the order k solve the system (4.4) and we get finally the estimate

$$\int_{B(x_0,1/2)} |D^k v|^2 dx \leqq C\int_{B(x_0,1)} |v - Q|^2 dx . \tag{4.8}$$

If $B(x_0, t_1)$ reaches up to the boundary, only the tangent derivatives $D_j v$ $(j = 1, \cdots, n-1)$ of the solution v solve again the boundary value problem. For them we get

$$\int_{B(x_0,t_2)} |D(D_j v)|^2 dx \leqq C \int_{B(x_0,t_1)} |Dv|^2 dx . \tag{4.8'}$$

The second normal derivative can be expressed from the equation

$$B_{rs}^{ij} D_{ij} v^s = 0 ; \qquad r = 1, \cdots, m ,$$

which holds a.e. in $B(x_0, 1)$. Advancing this process up to the estimate of the derivatives of the kth order we obtain (using the Sobolev imbedding theorem)

$$\max{}^2\left\{|Dv(x)|; x \in B\left(x_0, \frac{1}{2}\right)\right\} \leqq C \int_{B(x_0, 1)} |v - Q|^2 dx . \tag{4.9}$$

Let now $t \in]0, 1/2]$, $x, \tilde{x} \in \overline{B(x_0, tx)}$. Then

$$|v(x) - v(\tilde{x})|^2 \leqq Ct^2 \max{}^2\left\{|Dv(x)|; x \in \overline{B\left(x_0, \frac{1}{2}\right)}\right\} = C \int_{B(x_0,1)} |v - Q|^2 dx .$$

Let us recall that the constant C does not depend on the position of the point x_0 satisfying the assumptions of Lemma 4.1. Its value will be needed in the next text; because of an easier quotation we denote it by K. Integrating the last inequality over $B(x_0, t)$, we get (4.6) for the case $P_t = v(\tilde{x})$ with $\tilde{x} \in B(x_0, t)$. The case of P_t being an integral mean value can be reduced to the previous one by means of the integral mean value theorem.

To prove the inequality (4.5) we start with the estimate (4.8′) and applying the same method as before, we obtain

$$\left(\max\left\{|Dv(x)|; x \in \overline{B\left(x_0, \frac{1}{2}\right)}\right\}\right)^2 = C \int_{B(x_0,1)} |Dv|^2 dx . \tag{4.10}$$

The inequality (4.5) is an immediate consequence of (4.10).

The main result of this section is the following

THEOREM 4.2. *Let* (L) *and* (L^+) *be satisfied. Let* $u \in W^{1,2}(\Omega)$ *with the gradient* $Du \in L_\infty(\Omega)$ *be a weak solution of the problem* (2.1), (2.4). *Then* Du *is* α*-Hölder continuous on* $\Omega \cup \Gamma$ *with* $\alpha = \min(1/2, 1 - n/p)$ *and for every domain* Ω' *such that* $\bar{\Omega}' \subset \Omega \cup \Gamma$ *the inequality holds*:

$$\begin{aligned} \|u\|_{C^{1,\alpha}(\bar{\Omega}')} \leqq C(&\|Du\|_{L_\infty}, \|u_0\|_{W^{2,p}}, \|f\|_{W^{1,p/2}} , \\ &\|g\|_{L_\infty}, \operatorname{dist}(\bar{\Omega}', \boldsymbol{R}_+^n \backslash \Omega)) . \end{aligned} \tag{4.11}$$

Schema of the proof of the Theorem 4.2. In Lemma 4.8 we shall prove that Du belongs to certain Morrey-Campanato space and use then embedding of this space into $C^{1,\alpha}$.

For the case $\bar{\Omega}' \subset \Omega$ it follows from the condition (L). We can prove it by the method described in [4] modifying it slightly.

For the case $\bar{\Omega}' \cap \Gamma \neq \emptyset$ more substantial modifications of the method are needed. Denoting tangent derivatives as $\omega_l^r = D_l u^r$; $r = 1, \cdots, m$; $l = 1, \cdots, n-1$, we decompose them on $B(x, R)$ as

$$\omega = v + w$$

in the following way:

(i) The function w solves the linearized equation in variations (see (4.14)) and satisfies the nonhomogeneous boundary conditions

$$Cw_l = CD_l u_0 \quad \text{on} \quad \Gamma(x, R) ,$$

$$w_l - D_l u_0 = 0 \quad \text{on} \quad \partial B(x, R) \backslash \Gamma(x, R) ,$$

$l = 1, \cdots, n-1$. The L_2-norm of Dw can be easily estimated (see Lemma 4.3).

(ii) The second component $v = \omega - w$ solves the homogeneous linearized equation (4.15) and satisfies the homogeneous boundary conditions $Cv_l = 0$ on $\Gamma(x, R)$ and nonhomogeneous boundary conditions $v_l = \omega_l - D_l u_0$ on $\partial B(x, R) \backslash \Gamma(x, R)$, $l = 1, \cdots, n-1$.

In Lemma 4.4 we shall prove that, starting with sufficiently small oscillations of v on $B(x, R)$ we can describe how they decrease on $B(x, \tau R)$, $(\tau \in (0, 1))$.

The Liouville condition (L^+) yields, for each $x_0 \in \Gamma$, the fact that

$$\liminf_{R \to 0+} V(x_0, R) = 0 . \qquad \text{(See Theorem 4.5.)}$$

Combining this result together with the estimates of v and w, we obtain the assertion of Theorem 4.2.

First we shall describe more precisely the decomposition of ω.

Let u be a solution of (2.6) with $Du \in L_\infty(\Omega)$. Using the finite difference technique, we prove that $u \in W_{\text{loc}}^{2,2}(\Omega)$ and that each component ω_l of the tangent gradient ω solves the equation

$$\begin{aligned} \int_\Omega \Big\{ \Big(\frac{\partial a}{\partial \eta} D\omega_l + \frac{\partial a}{\partial \xi} \omega_l + \frac{\partial a}{\partial x_l}, D\varphi \Big) + \Big(\frac{\partial \bar{a}}{\partial \eta} D\omega_l \\ + \frac{\partial \bar{a}}{\partial \xi} \omega_l + \frac{\partial \bar{a}}{\partial x_l}, \varphi \Big) \Big\} dx = \int_\Omega \Big(\frac{\partial f}{\partial x_l}, \varphi \Big) dx \\ + \int_\Gamma \Big\{ \Big(\frac{\partial h}{\partial \xi} \omega_l + \frac{\partial h}{\partial x_l} - \frac{\partial g}{\partial x_l}, \varphi \Big) \Big\} dx' , \qquad \forall \varphi \in V . \end{aligned} \tag{4.12}$$

Moreover, $C(\omega_l - D_l u_0) = 0$ on Γ.

Let $x_0 \in \Omega \cup \Gamma$; $R > 0$ and $x_{0n} \leqq R$ (i.e., $\Gamma(x_0, R) \neq \emptyset$). Define $w = \{w^r\}_{r=1,\cdots,m} \in W^{1,2}(B(x_0, R))$ as a unique weak solution of the problem

$$w_l - D_l u_0 \in VB(x_0, R)\,, \tag{4.13}$$

$$\forall \varphi \in VB(x_0, R)$$

$$\begin{aligned}
&\int_{B(x_0,R)} \left\{\left(\frac{\partial a}{\partial \eta}(x, u, Du)Dw_l, D\varphi\right) + \left(\frac{\partial \bar{a}}{\partial \eta}(x, u, Du)Dw_l, \varphi\right)\right\}dx \\
&\quad = -\int_{B(x_0,R)} \left\{\left(\frac{\partial a}{\partial \xi}\omega_l + \frac{\partial a}{\partial x_l}, D\varphi\right) + \left(\frac{\partial \bar{a}}{\partial \xi}\omega_l + \frac{\partial \bar{a}}{\partial x_l}, \varphi\right)dx\right. \\
&\qquad + \int_{B(x_0,R)} \left(\frac{\partial f}{\partial x_l}, \varphi\right)dx + \int_{\Gamma(x_0,R)} \left\{\frac{\partial h}{\partial \xi}\omega_l + \frac{\partial h}{\partial x_l} + \frac{\partial g}{\partial x_l}, \varphi\right\}dx'\,.
\end{aligned} \tag{4.14}$$

The relations (4.12) and (4.13), (4.14) imply that (defining $v_l = \omega_l - w_l$) the component v_l solves the equation

$$\forall \varphi \in VB(x_0, R)$$

$$\int_{B(x_0,R)} \left\{\left(\frac{\partial a}{\partial \eta}(x, u, Du)Dv_l, D\varphi\right) + \left(\frac{\partial \bar{a}}{\partial \eta}(x, u, Du)Dv_l, \varphi\right)dx = 0\,,\right. \tag{4.15}$$

and satisfies the boundary conditions

$$\begin{aligned}
&Cv_l = 0 \quad \text{on} \quad \Gamma(x_0, R) \\
&v_l = \omega_l - D_l u_0 \quad \text{on} \quad \partial B(x_0, R)\backslash\Gamma(x_0, R)\,.
\end{aligned} \tag{4.16}$$

The components v_l and w_l depend on the choice of x_0 and R. We shall denote them by $v = \{v_l\}_{l=1,\cdots,n-1}$, omitting to express the dependence on x_0 and R if not necessary.

Taking into account the assumptions on the coefficients, the right-hand side, the boundary conditions, and the solution u ($Du \in L_\infty(\Omega)$), we get easily that the problem (4.13), (4.14) can be rewritten as follows:

$$w - w_0 \in VB(x_0, R)\,; \tag{4.17}$$

$$\begin{aligned}
&\int_{B(x_0,R)} \{(\boldsymbol{A}Dw, D\varphi) + (\bar{\boldsymbol{A}}Dw, \varphi)\}dx \\
&\quad = \int_{B(x_0,R)} (\boldsymbol{F}, \varphi)dx + \int_{\Gamma(x_0,R)} (\boldsymbol{H}, \varphi)dx' \qquad \forall \varphi \in VB(x_0, R)\,,
\end{aligned} \tag{4.18}$$

where

$$\begin{aligned}
&(1)\quad \boldsymbol{A} = \left\{\frac{\partial a_i^r}{\partial \eta_j^s}\delta_{kl}\right\} \in L_\infty(B(x_0, R))\,, \\
&\qquad (\boldsymbol{A}\eta, \eta) \geq \mathscr{H}|\eta|^2 \qquad \forall \eta \in R^{nm(n-1)}\,, \\
&(2)\quad \bar{\boldsymbol{A}} = \left\{\frac{\partial \bar{a}^r}{\partial \eta_j^s}\right\} \in L_\infty(B(x_0, R))\,, \\
&(3)\quad w_0 = D_l u_0 \in W^{1,p}(B(x_0, R))\,, \\
&(4)\quad \boldsymbol{F} \in L_{p/2}(B(x_0, R))\,, \\
&(5)\quad \boldsymbol{H} \in L_\infty(\Gamma(x_0, R))\,.
\end{aligned} \tag{4.19}$$

(Let us remind here that ω_l is bounded on Ω and (as it solves the system 4.12) it belongs to the space $W^{1,2}_{\mathrm{loc}}(\Omega \cup \Gamma)$; thus it has a well defined trace on Γ and, since $\|Du\|_{L_\infty} \leqq C$, we have $\|\omega_l\|_{L_\infty(\Gamma)} \leqq C$, too.)

Putting $\varphi = w - w_0$ and using the assumptions (4.19), we get

LEMMA 4.3. *There exist a positive constant C and a positive radius R_0 such that, for every $R \in (0, R_0)$ and for every solution $w \in W^{1,2}(B(x_0, R))$ of the problem* (4.17), (4.18) *satisfying* (4.19), *the inequality*

$$\|Dw\|_{L_2(B(x_0,R))} \leqq CR^{n(1/2-1/p)} \tag{4.20}$$

holds.

The local behaviour of the oscillations of the second component v is shown in the next lemma.

LEMMA 4.4. *For every Ω'; $\bar{\Omega}' \subset \Omega \cup \Gamma$, for every positive C and each $\tau \in (0, 1)$, there exist a positive ε and R_0 such that, for every solution u of the problem* (2.6) *with $\|Du\|_{L_\infty} \leqq C$, for every $x_0 \in \Omega'$ and $R \in]0, \min(R_0, 1 - |x_0|)[$, the implication*

$$V(x_0, R) < \varepsilon^2 \Longrightarrow V(x_0, \tau R) < 2K\tau^2 V(x_0, R) \tag{4.21}$$

holds.

(*Here*

$$V(x_0, R) = R^{2-n} \int_{B(x_0,R)} |Dv|^2 dx\,, \qquad V(x_0, \tau R) = (\tau R)^{2-n} \int_{B(x_0,\tau R)} |Dv|^2 dx$$

and, in both the expressions, v is the component of the decomposition of ω on $B(x_0, R)$). K is the maximal of the constants from Lemma 4.1, corresponding to

$$B^{rs}_{ij} = \frac{\partial a^r_i}{\partial \xi^s_j}(x_0, \zeta, \xi)\,,$$

$x_0 \in \bar{\Omega}$; $|\xi| \leqq C$ and the upper bound for ζ derived as the upper bound for weak solution of the problem (2.6) *for which $\|Du\|_{L_\infty(\Omega)} \leqq C$.*

Proof. Suppose that the assertion of the lemma does not hold. Then there exist $C \in (0, \infty)$, $\tau \in]0, 1[$, sequences $\varepsilon_\nu \searrow 0$, $R_\nu \searrow 0$, $x_\nu \to x_0 \in \Omega \cup \Gamma$ and u_ν; $\|Du_\nu\|_{L_\infty} \leqq C$, such that

$$V(x_\nu, R_\nu) = \varepsilon^2_\nu \tag{4.22}$$

and simultaneously

$$V(x_\nu, \tau R_\nu) > 2K\tau^2 V(x_\nu, R_\nu) . \tag{4.23}$$

If $x_0 \in \Omega$, all $\overline{B(x_\nu, R_\nu)} \subset \Omega$ for ν sufficiently large and the proof is substantially the same as in [4]. A similar situation occurs if $x_0 \in \Gamma$ but $R_\nu < x_{\nu n}$ for infinitely many indices ν (i.e. the closed sets $\overline{B(x_\nu, R_\nu)} \subset \Omega$). In what follows, $x_{\nu j}$ will denote the jth component of the vector x_ν (i.e., $x_\nu = \{x_{\nu j}\}_{j=0}^m$). The same notation will be used for sequences u_ν, v_ν etc.

Suppose that $x_\nu \to x_0 \in \Gamma$ and $x_{\nu n} \leqq R_\nu$. Using the decomposition $\omega_{\nu l} = w_{\nu l} + v_{\nu l}$ on $B(x_\nu, R_\nu)$ and estimating Dv_ν by (4.22) and Dw_ν by Lemma 4.3 we get

$$\| D\omega_{\nu l} \|^2_{L_2(B(x_\nu, R_\nu))} \leqq c(R_\nu^{n(1-2/p)} + \varepsilon_\nu^2 R_\nu^{n-2}) , \qquad l = 1, \cdots, n-1 . \tag{4.24}$$

The second normal derivatives $\partial^2 u_\nu / \partial x_n^2$ can be expressed from the equation

$$\frac{\partial a_i^r}{\partial \eta_j^s} D_{i+j} u^s + \frac{\partial a_i^r}{\partial \xi_s} D_i u^s + \frac{\partial a_i^r}{\partial x_i} + \bar{a}^r - f^r = 0 ; \qquad r = 1, \cdots, m ,$$

which is satisfied a.e. on Ω. Thus we get

$$\| D^2 u_\nu \|^2_{L_2(B(x_\nu, R_\nu))} \leqq C(R_\nu^{n(1-2/p)} + \varepsilon_\nu^2 R_\nu^{n-2}) . \tag{4.25}$$

Put

$$\mathscr{H}_{\nu l} = \frac{1}{\text{meas}_{n-1} \Gamma(x_\nu, R_\nu)} \int_{\Gamma(x_\nu, R_\nu)} v_{\nu l}(y) dy' , \tag{4.26}$$

$$\mathscr{H}_\nu = \{\mathscr{H}_{\nu l}\}_{l=1,\cdots,n-1} ,$$

$$\psi_\nu \colon y \longrightarrow x , \tag{4.27}$$

where

$$x_i = x_{\nu i} + R_\nu y^i \qquad i = 1, \cdots, n-1 ,$$

$$x_n = R_\nu y_n ,$$

$$a_\nu = 1 + \frac{x_{\nu n}}{R_\nu} \in \langle 1, 2 \rangle . \tag{4.28}$$

Then the substitution ψ_ν^{-1} transforms the sets $B(x_\nu, R_\nu)$ into

$$B_\nu = \{y \in \boldsymbol{R}^n ; |y_i| < 1 \text{ for } i = 1, \cdots, n-1,\ 0 < y_n < a_\nu\} \tag{4.29}$$

and the sets $\Gamma(x_\nu, R_\nu)$ into

$$\Gamma_0 = \{y \in \boldsymbol{R}^n ; |y_i| < 1 \text{ for } i = 1, \cdots, n-1,\ y_n = 0\} . \tag{4.30}$$

Moreover, put

$$B_{\nu,\tau} = \psi_\nu^{-1}(B(x_\nu, \tau R_\nu)) .$$

Defining

$$s_\nu(y) = \frac{1}{\varepsilon_\nu}\{v_\nu(\psi_\nu(y)) - \mathscr{H}_\nu\}\,, \tag{4.31}$$

we get from (4.22), (4.23) that

$$S_\nu = \int_{B_\nu} |Ds_\nu|^2 dy = 1\,, \tag{4.32}$$

$$S_{\nu,\tau} = \int_{B_{\nu,\tau}} |Ds_\nu|^2 dy > 2K\tau^2\,. \tag{4.33}$$

Applying the following type of Poincaré's inequality to s_ν and using (4.32), we obtain

$$\|s_\nu\|_{W^{1,2}(B_\nu)} \leqq C\,. \tag{4.34}$$

Poincaré's inequality. There exists $C > 0$ such that for each $\nu \in N$ and for each $f \in W^{1,2}(B_\nu)$

$$\int_{B_\nu}\left[f(y) - \frac{1}{\mathrm{meas}_{n-1}\,\Gamma_0}\int_{\Gamma_0} f(z)dz'\right]^2 dy \leqq C\int_{B_\nu} |Df|^2 dy \tag{4.35}$$

holds.

In what follows, we dare to pass to a suitable subsequence without notice and without changing the notation.

We distinguish two cases

$$\begin{aligned} &\text{(a)} \quad a_\nu \searrow a_0;\ \bigcap_{\nu\in N} B_\nu \supset B_0 \\ &\qquad\qquad = \{y \in \boldsymbol{R}^n;\ |y_i| < 1;\ i = 1, \cdots, n-1,\ 0 < y_n < a_0\}\,; \\ &\text{(b)} \quad a_\nu \nearrow a_0;\ \bigcup_{\nu\in N} B_\nu = B_0\,. \end{aligned} \tag{4.36}$$

We shall prove that $\{s_\nu\}$ converges on B_0 to a function s solving the system with constant coefficients and such boundary conditions that Lemma 4.1 can be applied to s. Then the passage to the limit in the relations (4.32), (4.33) gives the contradiction.

From (4.34) we can conclude that there is a function $s \in W^{1,2}(B_0)$ such that

$$s_\nu \longrightarrow s \quad \text{and} \quad \varepsilon_\nu s_\nu \longrightarrow 0 \quad \text{a.e. on } B_0 \tag{4.37}$$

and

$$\begin{aligned} &\text{(a)} \quad s_\nu \longrightarrow s \text{ in } W^{1,2}(B_0)\,, \quad \text{weakly} \\ &\text{(b)} \quad s_\nu \longrightarrow s \text{ in } W^{1,2}(G_0) \text{ weakly for each} \\ &\qquad\qquad G;\ \bar{G} \subset \bar{B}_0\backslash\{y \in \boldsymbol{R}^n;\ y_n = a_0\}\,. \end{aligned} \tag{4.37'}$$

Taking into account the definition of s_ν (see (4.31)), we get

(4.38) $$\omega_\nu(\psi_\nu(y)) = \varepsilon_\nu s_\nu(y) + \mathscr{H}_\nu + t_\nu(y) ,$$

where

(4.39) $$t_\nu(y) = w_\nu(\psi_\nu(y)) .$$

The boundedness of ω together with Lemma 4.3 and (4.37) yield the existence of a constant vector $\sigma = \{\sigma_l^r\}_{\substack{l=1,\cdots,n-1\\ r=1,\cdots,m}}$ such that $\mathscr{H}_{\nu l} \to \sigma_l$ and

(4.40) $$\omega_{\nu l}(\psi_\nu(y)) \longrightarrow D_l u_0(x_0) + \sigma_l \qquad \text{a.e. on } B_0 .$$

A similar technique may be used for the normal derivative. Put

$$\mathscr{H}_{\nu n} = \frac{1}{\operatorname{meas}_{n-1} \Gamma(x_\nu, R_\nu)} \int_{\Gamma(x_\nu, R_\nu)} D_n u_\nu(x', x_n) dx' .$$

By Poincaré's inequality and (4.25) we get

(4.41) $$\| D_n u_\nu(\psi_\nu(y)) - \mathscr{H}_{\nu n} \|^2_{L_2(B_\nu)} \leqq C(R_\nu^{2(1-n/p)} + \varepsilon_\nu^2) \longrightarrow 0 .$$

This and the boundedness of Du imply the boundedness of the sequence $\mathscr{H}_{\nu n}$ and thus the existence of such a constant vector $\xi_n = \{\xi_n^r\}_{r=1,\cdots,m}$ that

(4.42) $$D_n u_\nu(\psi_\nu(y)) \longrightarrow \xi_n \qquad \text{a.e. on } B_0 .$$

Put $\xi = \{\xi_l^r\}_{\substack{r=1,\cdots,m\\ l=1,\cdots,n}}$; $\xi_l^r = D_l u_0^r(x_0) + \sigma_l^r$ for $r = 1, \cdots, m$ and $l = 1, \cdots, n-1$. Then (4.40) and (4.42) give

(4.43) $$D_l u_\nu^r(\psi_\nu(y)) \longrightarrow \xi_l^r \text{ a.e. on } B_0 \text{ for } r = 1, \cdots, m, \ l = 1, \cdots, n ,$$

and the norm of ξ is bounded by the same constant as the L_∞-norm of Du_ν.

Deduce now the equation for s:
Substituting $x = \psi_\nu(y)$ into (4.15) and using (4.31), we obtain

(4.44) $$\int_{B_\nu} \{(MDs_\nu, D\varphi) + R_\nu(\bar{M}Ds_\nu, \varphi)\} dy = 0 \qquad \forall \nu \in N, \ \forall \varphi \in VB_\nu ;$$

where

$$M = \{m_{ij}^{rs}(\nu, y)\}_{\substack{r,s=1,\cdots,m\\ i,j=1,\cdots,n}} , \qquad m_{ij}^{rs}(\nu, y) = \frac{\partial a_i^r}{\partial \eta_j^s}(\psi_\nu(y), u_\nu(\psi_\nu(y)), Du_\nu(\psi_\nu(y))) ,$$

$$\bar{M} = \{\bar{m}_{ij}^r(\nu, y)\}_{\substack{r=1,\cdots,m\\ i,j=1,\cdots,n}} , \qquad \bar{m}_{ij}^r(\nu, y) = \frac{\partial \bar{a}^r}{\partial \eta_j^s}(\psi_\nu(y), u_\nu(\psi_\nu(y)), Du_\nu(\psi_\nu(y))) .$$

Taking into account that $\psi_\nu(y) \to x_0$ on B_0, $u_\nu(\psi_\nu(y)) \to \zeta$ on B_0 and $Du_\nu(\psi_\nu(y)) \to \xi$ a.e. on B_0, we can conclude that

(4.45) $$m_{ij}^{rs}(\nu, y) \longrightarrow B_{ij}^{rs} = \frac{\partial a_i^r}{\partial \eta_j^s}(x_0, \zeta, \xi) \qquad \text{a.e. on } B_0$$

and passing to the limit in (4.44), we get finally

(4.46) $$\int_{B_0} (\boldsymbol{B}Ds, D\varphi)dy = 0 \qquad \forall \varphi \in VB_0 .$$

According to (4.16), $Cv_\nu = 0$ on $\Gamma(x_\nu, R_\nu)$ for each $\nu \in N$, hence $Cs_\nu = 0$ on Γ_0 for each $\nu \in N$ and

(4.47) $$Cs = 0 \quad \text{on} \quad \Gamma_0 .$$

Thus the function s solves the boundary value problem of the type required in Lemma 4.1 and

(4.48) $$S_\tau = \tau^{2-n} \int_{B_{0,\tau}} |Ds|^2 dy \leqq K\tau^2 \int_{B_0} |Ds|^2 dy = K\tau^2 S ,$$

where $B_{0,\tau} = \{y \in \boldsymbol{R}^n; |y_i| < \tau \text{ for } i = 1, \cdots, n-1, |y_n - a_0 + 1| < \tau\}$, and K is the constant described in Lemma 4.4.

The weak lower semicontinuity of the functional $\vartheta: s \to \int_{B_0} |Ds|^2 dy$ together with (4.37), (4.32) gives

(4.49) $$S = \int_{B_0} |Ds|^2 dy \leqq 1 .$$

To get the contradiction it is sufficient to prove that

$$S_{\nu,\tau} \longrightarrow S_\tau = \tau^{2-n} \int_{B_{0,\tau}} |Ds|^2 dy .$$

We shall prove (by the choice of a test function) that $Ds_\nu \to Ds$ in $L_{2,\text{loc}}(B_0)$. Let us sketch the choice for the case (a): Take $\nu_0 \in N$ so large that

$$B_{\nu_0,\tau} \subset \left\{ y \in \boldsymbol{R}^n; y_n < a_0 - 1 + \frac{\tau + 1}{2} \right\} ;$$

let $\Phi \in C_\infty(\bar{R}_+^n)$; supp $\Phi \subset B_0 \cup \Gamma_0$; $\Phi \equiv 1$ on $\bigcup_{\nu \geqq \nu_0} B_{\nu,\tau}$; Then $\varphi = (s_\nu - s)\Phi^2$ (prolonged by zero if necessary) is an admissible test function for both (4.44) and (4.46). Therefore

(4.50) $$\int_{B_0} \{(\boldsymbol{M}Ds_\nu, D(s_\nu - s))\Phi^2 + 2\Phi(\boldsymbol{M}Ds_\nu, (s_\nu - s)D\Phi) + (\bar{\boldsymbol{M}}Ds_\nu, s_\nu - s)\Phi^2\}dy = 0 ,$$

(4.51) $$\int_{B_0} \{(\boldsymbol{B}Ds, D(s_\nu - s))\Phi^2 + 2\Phi(\boldsymbol{B}Ds, (s_\nu - s))D\Phi\}dy = 0 .$$

Finally, using the ellipticity condition

$$
\begin{aligned}
&\int_{B_0} \Phi^2 |D(s_\nu - s)|^2 dy \leqq \frac{1}{\mathscr{H}} \int_{B_0} \Phi^2(MD(s_\nu - s), D(s_\nu - s)) dy \\
(4.52) \qquad & = \frac{1}{\mathscr{H}} \left\{ \int_{B_0} \Phi^2(MDs_\nu, D(s_\nu - s)) dy - \int_{B_0} \Phi^2(MDs, D(s_\nu - s)) dy \right\} .
\end{aligned}
$$

Now we can estimate the first integral on the right hand side of (4.52) from (4.50) and the second one from (4.51) and we get

$$
(4.53) \qquad \int_{B_0} \Phi^2 |D(s_\nu - s)|^2 dy \longrightarrow 0 .
$$

To bound the difference $S_{\nu,\tau} - S_\tau$ we write

$$
\begin{aligned}
|S_{\nu,\tau} - S_\tau| \leqq \tau^{2-n} \Big\{ & \int_{B_0} \Phi^2(Ds_\nu^2 - Ds^2) dy + \int_{B_{0,\tau} \setminus B_{\nu,\tau}} |Ds|^2 dy \\
& + \int_{B_{\nu,\tau} \setminus B_{0,\tau}} |Ds|^2 dy \Big\} .
\end{aligned}
$$

Here the first integral on the right hand side tends to zero by (4.53) and the second and third ones because of the uniform absolute continuity. Thus

$$
(4.54) \qquad S_\tau = \lim_{\nu \to \infty} S_{\nu,\tau} \geqq 2K\tau^2 \geqq 2K\tau^2 S ,
$$

which contradicts (4.48).

THEOREM 4.5. *Let the system* (2.1), (2.4) *satisfies the condition* (L^+), *let* u *be a weak solution of* (2.1), (2.4) *for which* $Du \in L_\infty(\Omega)$. *Then for each* $x_0 \in \Gamma$, *there exists a sequence* $R_\nu \searrow 0$ *such that*

$$
(4.55) \qquad \lim_{\nu \to \infty} Z(x_0, R_\nu) = 0 ,
$$

where $z = Du - Du_0$. (*For* $Z(x_0, R_\nu)$ *see* (4.1)).

Proof. Be $x_0 \in \Gamma$; $0 < R < \operatorname{dist}(x_0, \partial\Omega \setminus \Gamma)$. Put

$$
(4.56) \qquad y = y(x) = \frac{x - x_0}{R} ,
$$

$$
(4.57) \qquad u_R(y) = \frac{u(x_0 + Ry) - u(x_0)}{R} .
$$

Then $y(B(x_0, R)) = B(0, 1)$. Put $0_R = y(\Omega)$. For each T, let $R(T)$ be such a positive radius that it is $B(0, T) \subset 0_R$ for $R < R(T)$.

In the following part of the proof we use the fact that for every $T > 0$ the set of second gradients $\{D^2 u_R; R < R(T)\}$ is bounded in $L_2(B(0, T))$. More precisely, it holds

LEMMA 4.6. *Let u be a solution of the problem* (2.1), (2.4) *for which $Du \in L_\infty(\Omega)$. Then for each $x_0 \in \Gamma$ and for every T, there exist $R(T)$ and C such that*

$$\| D^2 u_R \|_{L_2(B(0,T))} \leqq C \qquad \forall R < R(T) \ . \tag{4.58}$$

The value of C depends on $\| Du \|_{L_\infty}$, $\| f \|_{W^{1,p/2}}$, $\| u_0 \|_{W^{2,p}}$, $\| g \|_{W^{1,\infty}}$, T, and dist $(x_0, \partial\Omega \backslash \Gamma)$.

Proof of the Lemma 4.6 is standard: using the finite difference technique and Nirenberg's lemma, we get the estimates for $D_{ij}u_R$, $ij \neq nn$. The bound for $D_{nn}u_R$ can be obtained by means of the equation in variations, which is valid a.e. on $B(0, T)$, and which enables us to express $D_{nn}u_R$ through the other second derivatives which we had estimated before.

Returning to the proof of Theorem 4.5, we see that the set $\{Du_R; R < R(T)\}$ is bounded in $L_\infty(\Omega)$-it follows from the assumption $Du \in L_\infty(\Omega)$ and the simple equality

$$\frac{\partial u_R(y)}{\partial y_i} = \frac{\partial u}{\partial x_i}(x_0 + Ry) \ .$$

Taking into account that $u_R(0) = 0$, we get finally the boundedness of the set $\{u_R; R < R(T)\}$ in $W^{2,2}(B(0, T))$. The compactness of the imbedding of $W^{2,2}(B(0, T))$ into $W^{1,2}(B(0, T))$ allows us to choose a sequence R_ν, $R_\nu \searrow 0$, such that $u_{R_\nu} \to p$ in $W^{1,2}(B(0, T))$, and, using the diagonal process, also

$$\begin{aligned} &\lim_{\nu\to\infty} u_{R_\nu} = p \quad \text{in} \quad W^{1,2}_{\text{loc}}(\boldsymbol{R}^n_+) \ , \\ &\lim_{\nu\to\infty} Du_{R_\nu} = Dp \quad \text{a.e. on } \boldsymbol{R}^n_+ \ . \end{aligned} \tag{4.59}$$

Deduce now the equation for the limit function p: To this end we substitute (4.56) and (4.57) into (2.6); after the passage to the limit we obtain

$$\int_{\boldsymbol{R}^n_+} (a(x_0, \xi, Dp(y)), D\varphi(y))dy = \int_{\{y \in \boldsymbol{R}^n; y_n = 0\}} (d, \varphi(y))dy' \ . \tag{4.60}$$

Using the theorem on traces and (4.59), we get

$$\lim_{\nu\to\infty} \boldsymbol{C}u_{R_\nu} = \boldsymbol{C}p \quad \text{a.e. on} \quad \{y \in \boldsymbol{R}^n; y_n = 0\} \ .$$

The transformed boundary conditions give

$$\boldsymbol{C}u_{R_\nu} = \boldsymbol{C}\Big(\frac{u_0(x_0 + R_\nu y) - u_0(x_0)}{R_\nu}\Big) \ ,$$

but $u_0 \in C_1(\bar{\Omega})$, hence

$$u_{0,R_\nu} = R_\nu^{-1}(u_0(x_0 + R_\nu y) - u_0(x_0)) \longrightarrow y_1 D_1 u_0(x_0) + \cdots + y_n D_n u_0(x_0) ,$$

so that Cp is a polynomial of at most the first degree on $\{y \in \boldsymbol{R}^n; y_n = 0\}$. The condition (L^+) implies that p is a polynomial of at most the first degree on $\boldsymbol{R}^n_+$.

Because of (4.59) and the fact that Dp is a constant vector, we have that

$$D(x_0, R_\nu) = \int_{B(0,1)} |Du_{R_\nu}(y) - (Du_{R_\nu})_{0,1}|^2 dy \longrightarrow 0$$

here $(Du)_{0,1}$ is the integral mean value of Du, i.e.

$$(Du)_{0,1} = \frac{1}{\text{meas}_n B(0, 1)} \int_{B(0,1)} Du dy .$$

After an easy calculation ($u_0 \in W^{2,p}$ with $p > n$) we obtain that also

$$\tilde{Z}(x_0, R) = \int_{B(0,1)} |D[u_{R_\nu} - u_{0,R_\nu}](y) - (D[u_{R_\nu} - u_{0,R_\nu}])_{0,1}|^2 dy \longrightarrow 0 . \tag{4.16}$$

The following lemma shows the relations between Z and $\tilde{Z}$.

LEMMA 4.7. *Let the notation of the preceding lemma be preserved. Then there exists constants $\gamma > 0$, $\gamma_1 > 0$ such that for each point $x_0 \in \Gamma$, $R < \text{dist}(x_0, \partial\Omega \backslash \Gamma)$, the estimate*

$$Z\left(x_0, \frac{R}{2}\right) \leqq \gamma \tilde{Z}(x_0, R) + \gamma_1 \| D^2 u_0 \|^2_{L_p} R^{2(1-n/p)} \tag{4.62}$$

holds.

The proof of this lemma is similar to that of Lemma 4.1 — we insert a suitable test function of the type $\Phi^2(\omega - \omega_0 - c)$ (here $\omega_0 = \{D_i u_0\}_{i=1,\cdots,n-1}$; c is a constant vector satisfying the condition $Cc = 0$) into the equation in variations.

From (4.61) and (4.62) the assertion (4.55) of Theorem 4.5 follows.

To finish the proof of Theorem (4.2) it remains to observe that the difference between $Z(x_0, R)$ and $V(x_0, R)$ is small for small R thanks to the assumption $u_0 \in W^{2,p}(\Omega)$ and to use the same procedure as in [2], proof of Proposition 1.1 for the estimates of tangential derivatives. As for the second normal derivative, we repeat the estimates of (4.25). In such a way we get that the whole gradient belongs to the Morrey-Campanato space and thus $u \in C^{1,\alpha}(\overline{B(x_0, R_1)})$

with some R_1 sufficienty small.

REMARK. With some modification the same method can be used to prove the analogous theorems for any bounded domain with sufficiently smooth boundary.

REFERENCES

1. M. Giaquinta and J. Nečas, *On the regularity of weak solutions to nonlinear elliptic systems via Liouville's type property*, CMUC 20, **1** (1979), 111-121.
2. ———, *On the regularity of weak solutions to nonlinear elliptic systems of partial differential equations*, J. reine angew. Math., **316** (1980), 140-159.
3. J. Nečas, *On the regularity of weak solutions to nonlinear elliptic systems of partial differential equations*, to appear in Editrice Tecnica Scientifiea, Pisa.
4. ———, *A Necessary and sufficient condition for the regularity of weak solutions to nonlinear elliptic systems of partial differential equations*, to appear in Abhandlungen Akad. der Wissensehaften der DDR.

Received November 17, 1980.

CHARLES UNIVERSITY
1, MALOSTRANSKE n. 25
11800 PRAHA 1, CZECHOSLOVAKIA

On Regularity of Solutions of Nonlinear Parabolic Systems

JINDŘICH NEČAS - VLADIMÍR ŠVERÁK

1. - Introduction

It is well-known that the full regularity of the elliptic systems

$$D_\alpha A_i^\alpha(Du) = 0$$

in two dimensions can (under standard assumptions) be proved by using $W^{1,2+\delta}$-estimates for linear elliptic systems with L^∞ coefficients. (See, for example, M. Giaquinta [2]). The purpose of this paper is to show that a similar method can be used when dealing with nonlinear parabolic systems

$$\frac{\partial u_i}{\partial t} = D_\alpha A_i^\alpha(Du).$$

The idea is to show that $\frac{\partial u}{\partial t}$ is bounded in $L^\infty(-T,0;L^{2+\delta}(\Omega))$ and then apply the theory of elliptic systems. The required estimate is obtained by using estimates for solutions of linear parabolic systems with L^∞-coefficients. (See Lemma 1). In the two-dimensional case we get full regularity.

2. - Preliminaries

Let $n \geq 2$, $N \geq 1$. We shall be dealing with open sets $Q = \Omega \times (-T,0) \subset \mathbb{R}^{n+1}$, where Ω is a bounded domain in $\mathbb{R}^n$ and $T > 0$. A typical point of $\mathbb{R}^{n+1}$ is denoted by $z = (x,t), x \in \mathbb{R}^n, t \in \mathbb{R}$.

For $\delta > 0$ we let

$$\Omega_\delta = \{x \in \Omega, \ \text{dist}\,(x, \partial\Omega) > \delta\}$$

Pervenuto alla Redazione il 16 Febbraio 1990.

© Annali della Scuola Normale Superiore 1991. Reprinted with kind permission from: Nečas, J. & Šverák, V., On regularity of solutions of nonlinear parabolic systems, Annali della Scuola Normale Superiore di Pisa - Classe di Scienze 18:1-11.

and

$$Q_\delta = \Omega_\delta \times (-T+\delta, 0).$$

For $x \in \mathbb{R}^n$ and $\rho > 0$ we define

$$B_{x,\rho} = \{y \in \mathbb{R}^n, |x-y| < \rho\}.$$

If $a, b \in \mathbb{R}$, we denote by $a \wedge b$ the minimum of the two numbers.

The Sobolev spaces W_p^k, $\overset{\circ}{W}{}_p^k$ are defined in the standard way.

The space $L^2(-T,0;W_2^1(\Omega))$ is denoted by $W_2^{1,0}(Q)$. The norm $[\cdot]_{2,Q}$ on $W_2^{1,0}(Q)$ is defined by

$$[u]_{2,Q} = \{\int_Q (|u|^2 + \sum_{i=1}^n |D_i u|^2)\}^{\frac{1}{2}}.$$

The spaces $L^\infty(-T,0;L^p(\Omega))$, $p \geq 1$ will be denoted by $L^{p,\infty}(Q)$ and the corresponding norm is denoted by $\|\cdot\|_{p,\infty,Q}$.

The usual L^p-norm is denoted by $\|\cdot\|_{p,Q}$.

Let us consider the nonlinear parabolic system

$$\frac{\partial u_i}{\partial t} - D_\alpha A_i^\alpha(Du) = 0. \qquad (i = 1,2,\ldots,N) \tag{1}$$

where $u = (u_1,\ldots,u_N), Du = (D_\alpha u_i)_{1\leq i\leq N, 1\leq\alpha\leq n} = (\frac{\partial u_i}{\partial x_\alpha})_{1\leq i\leq N,1\leq\alpha\leq n}$ is the gradient matrix of u and the summation over repeated indexes is understood.

We shall suppose that the functions A_i^α have continuous derivatives satisfying

$$\left\{\sum_{i,j}\sum_{\alpha,\beta} |\frac{\partial A_i^\alpha}{\partial \xi_\beta^j}(\xi)|^2\right\}^{\frac{1}{2}} \leq M \tag{2}$$

and

$$\frac{\partial A_i^\alpha}{\partial \xi_\beta^j}(\xi)\pi_\alpha^i \pi_\beta^j \geq \nu|\pi|^2, \qquad \nu > 0,$$

for every $\xi, \pi \in \mathbb{R}^{nN}$.

(Of course, for higher regularity results we have to assume higher smoothness of A_i^α).

By a weak solution of (1) we mean a function $u \in W_2^{1,0}(Q)$ satisfying

$$\int_Q \left(u_i \frac{\partial \varphi_i}{\partial t} - A_i^\alpha(Du) D_\alpha \varphi_i\right) dz = 0$$

for every $\varphi \in \overset{\circ}{W}{}_2^1(Q)$.

We shall also be dealing with linear strongly parabolic systems

$$\frac{\partial u_i}{\partial t} - D_\alpha a_{ij}^{\alpha\beta} D_\beta u_j = 0 \qquad (i = 1, \ldots, N) \tag{4}$$

where $a_{ij}^{\alpha\beta} = a_{ij}^{\alpha\beta}(z)$ are L^∞-functions in Q satisfying for almost every $z \in \Omega$ the conditions

$$\left\{ \sum_{i,j} \sum_{\alpha,\beta} |a_{ij}^{\alpha\beta}|^2 \right\}^{\frac{1}{2}} \leq M \tag{2'}$$

and

$$a_{ij}^{\alpha\beta} \xi_\beta^j \xi_\alpha^i \geq \nu |\xi|^2 \tag{3'}$$

for every $\xi \in \mathbb{R}^{nN}$. By a weak solution of (4) we mean a function $u \in W_2^{1,0}(Q)$ satisfying

$$\int_Q \left(u_i \frac{\partial \varphi_i}{\partial t} - a_{ij}^{\alpha\beta} D_\beta u_j D_\alpha \varphi_i \right) dz = 0 \tag{$*$}$$

for every $\varphi \in \overset{\circ}{W}{}_2^1(Q)$.

We shall use the following well-known results.

(i) If u is a weak solution of (1) or (4), then u is continuous in time with respect to the L^2-norm. More precisely, if $\Omega' \subset\subset \Omega$, then the map $t \to u(\cdot, t)$ from $(-T, 0)$ into $L^2(\Omega')$ is continuous. (See, for example, O.A. Ladyzenskaja, V.A. Solonnikov, N.N. Ural'ceva [4], Chap. 3, Lemma 4.3).

(ii) We have the imbedding

$$L^{2,\infty}(Q) \cap W_2^{1,0}(Q) \hookrightarrow L^{q_0}(Q)$$

where

$$q_0 = \begin{cases} \dfrac{2(n+2)}{n}, & \text{if } n > 2 \\ \text{is any number } \in [1, 4) & \text{if } n = 2. \end{cases}$$

(See, for example, O.A. Ladyzenskaja, V.A. Solonnikov, N.N. Ural'ceva [4], Chap. 2).

We denote by c_i various constants. The value of these constants can depend on ν, M, Ω, T, n and N. The dependence on additional parameters will be indicated.

3. - L^p-estimates

The first statement of the following Lemma is well known (see, for example, O.A. Ladyzenskaja, V.A. Solonnikov, N.N. Ural'ceva [4], Chap. 3).

The second statement will be used for the $L^\infty(-T,0;L^{2+\delta}(\Omega))$-estimate mentioned in the introduction.

LEMMA 1. *Let u be a weak solution of the linear system (4). Then, for any $\delta > 0$,*

(i)

$$u \in L^{2,\infty}(Q_\delta) \cap W_2^{1,0}(Q_\delta)$$

and

$$||u||_{2,\infty,Q_\delta} + [u]_{2,Q_\delta} \leq c_1(\delta)||u||_{2,Q}.$$

(ii) *For every $p \in [2,(2+\frac{\nu}{NM}) \wedge q_0)$ the function u belongs to $L^{p,\infty}(Q_\delta)$ and*

$$||u||_{p,\infty,Q_\delta} \leq c_2(\delta,p)||u||_{2,Q}.$$

PROOF. Let $\gamma \geq 1$ and let $k > 0$ be such that $meas\{z \in Q, |u(z)| = k\} = 0$. Define $g_k : [0,\infty) \to \mathbb{R}$ by

$$g_k(t) = \begin{cases} t^\gamma, & \text{if } 0 \leq t \leq k, \\ k^\gamma + \gamma k^{\gamma-1}(t-k), & \text{if } t \geq k. \end{cases}$$

Clearly $g_k'(t) = \gamma(t \wedge k)^{\gamma-1}$ and

$$g_k''(t) = \begin{cases} \gamma(\gamma-1)t^{\gamma-2}, & \text{if } 0 < t < k, \\ 0, & \text{if } t > k. \end{cases}$$

Define also the function $\eta^k : \mathbb{R}^N \to \mathbb{R}$ by

$$\eta^k(u) = g_k(|u|^2)$$

We have

$$\eta^k_{u_i}(u) = \frac{\partial \eta^k}{\partial u_i}(u) = 2\gamma(|u|^2 \wedge k)^{\gamma-1}u_i$$

$$\eta^k_{u_i u_j}(u) = \frac{\partial^2 \eta^k}{\partial u_i \partial u_j}(u) = 2\gamma(|u|^2 \wedge k)^{\gamma-1}(\delta_{ij} + 2(\gamma-1)d_{ij}(u)).$$

In the second formula we assume $|u|^2 \neq k$ and

$$d_{ij}(u) = \begin{cases} 0, & \text{if } |u|^2 > k, \\ \frac{u_i u_j}{|u|^2}, & \text{if } |u|^2 < k. \end{cases}$$

Let ω_ϵ be a family of symmetric mollifying functions satisfying

$\omega_\epsilon \in \mathcal{D}(\mathbb{R}),\ \omega_\epsilon(t) \geq 0,$

$\omega_\epsilon(t) = \omega_\epsilon(-t),$

support $\omega_\epsilon \subset (\epsilon, -\epsilon),$

$\int_{\mathbf{R}} \omega_\epsilon = 1.$

For $f \in L^1(Q)$ let us denote by $(f)_\epsilon$ the function defined a.e. in Q by

$$(f)_\epsilon(x,t) = \int_{\mathbf{R}} f(x, t-s)ds.$$

(We extend f by zero outside Q). Let $\psi \in \overset{\circ}{W}{}_2^1(\Omega \times (-T+\epsilon, -\epsilon))$. Following E. Giusti, M. Giaquinta [3] we set $\varphi = (\psi)_\epsilon$ in $(*)$ and we see that

$$\int_Q \frac{\partial (u_i)_\epsilon}{\partial t}\psi_i dz = -\int_Q (a_{ij}^{\alpha\beta} D_\beta u_j)_\epsilon D_\alpha \psi_i dz. \tag{5}$$

Let $\theta \in \mathcal{D}(\Omega)$ with $0 \leq \theta \leq 1$ and $\theta = 1$ on Ω_δ and let $\rho \in \mathcal{D}(-T, 0)$, $\rho \geq 0$. For ϵ sufficiently small we can use (5) with

$$\psi_i = \eta_{u_i}^k(u_\epsilon)\theta^2\rho^2$$

to get

$$\begin{aligned}\int_Q \left(\frac{\partial}{\partial t}\eta^k(u_\epsilon)\right)\theta^2\rho^2 dz = &-\int_Q \left(a_{ij}^{\alpha\beta} D_\beta u_j\right)_\epsilon D_\alpha(\eta_{u_i}^k(u_\epsilon))\theta^2\rho^2 dz\\ &-\int_Q (a_{ij}^{\alpha\beta} D_\beta u_j)_\epsilon \eta_{u_i}^k(u_\epsilon) 2\theta D_\alpha\theta\rho^2 dz.\end{aligned}$$

Integrating by parts on the left-hand side, letting $\epsilon \to 0$ and then using the chain rule for the derivative $D_\alpha(\eta_{u_i}^k(u))$ (which is legal) we see that

$$\begin{aligned}\int_Q \eta^k(u)\theta^2(\rho^2)' dz = &-\int_Q a_{ij}^{\alpha\beta} D_\beta u_j \eta_{u_i u_l}^k(u) D_\alpha u_l \theta^2\rho^2 dz\\ &-\int_Q a_{ij}^{\alpha\beta} D_\beta u_j \eta_{u_i}^k(u) 2\theta D_\alpha\theta\rho^2 dz.\end{aligned} \tag{6}$$

Since $|d_{ij}| \leq 1$ we see that if $0 \leq 2(\gamma-1) < \frac{\nu}{NM}$ then the matrix $\tilde{a}_{lj}^{\alpha\beta} = a_{ij}^{\alpha\beta}(\delta_{il} + 2(\gamma-1)d_{il}(u))$ satisfies the condition (2') with ν replaced by $\nu_1 = \nu - 2(\gamma-1)NM$.

We can estimate the right-hand side of (6) by

$$\nu_1 \int_Q 2\gamma(|u|^2 \wedge k)^{\gamma-1}|Du|^2\theta^2\rho^2 dz$$

$$+2M\left\{\int_Q |Du|^2\theta^2\rho^2 2\gamma(|u|^2 \wedge k)^{\gamma-1}dz\right\}^{\frac{1}{2}}$$

$$\times\left\{\int_Q |u|^2 2\gamma(|u|^2 \wedge k)^{\gamma-1}|D\theta|^2\rho^2 dz\right\}^{\frac{1}{2}}$$

$$\leq -\frac{\nu_1}{2}\int_Q 2\gamma(|u|^2 \wedge k)^{\gamma-1}|Du|^2\theta^2\rho^2 dz$$

$$+\frac{4M^2}{\nu_1}\int_Q |u|^2 2\gamma(|u|^2 \wedge k)^{\gamma-1}|D\theta|^2\rho^2 dz.$$

Let $t_1 \in (-T+\delta, 0)$. As we have remarked in Section 2, under our assumptions the function $t \to u(\cdot, t)$ is continuous mapping of $(-T, 0)$ into $L^2(\Omega_\delta)$. Hence we can use (6) with ρ defined by

$$\rho^2(t) = \begin{cases} 0 & \text{if } t \in (-T, -T+\frac{\delta}{2}) \\ \frac{2}{\delta}(t+T-\frac{\delta}{2}) & \text{if } t \in (-T+\frac{\delta}{2}, -T+\delta) \\ 1 & \text{if } t \in (-T+\delta, t_1) \\ 0 & \text{if } t \in (t_1, 0). \end{cases}$$

We get

$$\int_Q \eta^k(u(x,t_1))\theta(x)dx + \frac{\nu_1}{2}\int_Q |Du|^2\theta^2\rho^2 2\gamma(|u|^2 \wedge k)^{\gamma-1}dz$$

$$\leq c_3\int_Q |u|^2 2\gamma(|u|^2 \wedge k)^{\gamma-1}|D\theta|^2\rho^2 dz + \int_{\Omega\times(-T,t_1)} (\rho^2)'\eta^k(u)\theta^2 dz.$$

Letting $\gamma = 1$ we get (i).

We can use (i) and the imbedding

$$L^{2,\infty}(Q) \cap W_2^{1,0}(Q) \hookrightarrow L^{q_0}(Q)$$

to infer that $\|u\|_{q_0,Q_\delta} \leq c_4(\delta)\|u\|_{2,Q}$. Using this and letting $k \to \infty$ in (7) we get (ii) with $p = 2\gamma$.

LEMMA 2. *Let u be a weak solution of the nonlinear system* (1). *Then* $u \in W_2^1(Q_\delta)$, *the derivatives* $D_i u$, $i = 1, \dots, n$ *and* $D_{n+1}u = \frac{\partial u}{\partial t}$ *belong to the space* $L^{p,\infty}(Q_\delta) \cap W_2^{1,0}(Q_\delta)$ *and for each* $i = 1, \dots, n, n+1$

$$\|D_i u\|_{p,\infty,Q_\delta} + [D_i u]_{2,Q_\delta} \le [u]_{2,Q}.$$

PROOF. As above, we denote by Du the vector $(D_1 u, \dots, D_n u) \in \mathbb{R}^{nN}$. let us fix an index r, $1 \le r \le n+1$ and let $e_r \in \mathbb{R}^n \times \mathbb{R}$ be the r-th vector of the canonical basis. Let $\delta' > 0$. For $0 < h < \delta'$ let

$$u_h(z) = h^{-1}[u(z) - u(z - he_r)].$$

Define the functions $a_{hij}^{\alpha\beta} \in L^\infty(Q_{\delta'})$ for a.e. $z \in Q_{\delta'}$ by

$$a_{hij}^{\alpha\beta}(z) = \int_0^1 A_{i,\xi_\beta^j}^\alpha (Du(z) - hDu_h(z) + \tau h Du_h(z)) d\tau.$$

It is not difficult to see that u_h is the weak solution of the linear system

$$\frac{\partial u_{hi}}{\partial t} - D_\alpha a_{hij}^{\alpha\beta} D_\beta u_{hj} = 0$$

in $Q_{\delta'}$. The functions $a_{hij}^{\alpha\beta}$ clearly satisfy the conditions (2') and (3'). Hence, by Lemma 1

$$\begin{aligned} &\|u_h\|_{p,\infty,Q_{2\delta'}} \le c_6(\delta', p)\|u_h\|_{2,Q_{\delta'}} \\ &\|Du_h\|_{2,Q_{2\delta'}} \le c_7(\delta')\|u_h\|_{2,Q_{\delta'}}. \end{aligned} \tag{8}$$

Suppose first $1 \le r \le n$. In this case the difference is taken in the direction of the space variables. Since $u \in W_2^{1,0}(Q)$, we have

$$\|u_h\|_{2,Q_{\delta'}} \le \|D_r u\|_{2,Q}. \tag{9}$$

Using Nirenberg's Lemma we see from (8) that $Du \in W_2^{1,0}(Q_{\delta'})$ and

$$\begin{aligned} &\|D_r u\|_{p,\infty,Q_{2\delta'}} \le c_6(\delta', p)\|D_r u\|_{2,Q_{\delta'}} \\ &\|DD_r u\|_{2,Q_{2\delta'}} \le c_7(\delta')\|D_r u\|_{2,Q_{\delta'}} \end{aligned} \tag{10}$$

for every $1 \le r \le n$. Now let $r = n+1$. Following S. Campanato [1] we notice that we can use equation (1) and the L^2-estimate of $D_\alpha D_\beta u$ obtained above to infer that $\frac{\partial u}{\partial t} \in L^2(Q_{2\delta'})$ and

$$\left\|\frac{\partial u}{\partial t}\right\|_{2,Q_{2\delta'}} \le c_8(\delta')\|Du\|_{2,Q}. \tag{11}$$

Now we can use (8) with Q replaced by $Q_{2\delta'}$ and using (11) we get by the same argument as above

$$\|\frac{\partial u}{\partial t}\|_{p,\infty,Q_{4\delta'}} \leq c_9(\delta', p)\|Du\|_{2,Q}$$

$$\|D\frac{\partial u}{\partial t}\|_{2,Q_{4\delta'}} \leq c_{9'}(\delta')\|Du\|_{2,Q}.$$

The proof is finished.

THEOREM 1. *Let u be a weak solution of the system (1) and let p be the exponent from Lemma 1. The for each $\delta > 0$*

$$\frac{\partial u}{\partial t} \in L^{p,\infty}(Q_\delta)$$

and

$$u \in L^\infty(-T+\delta, 0; W_q^2(\Omega_\delta))$$

for some $q = q(\nu, M, p, \delta)$ with $2 < q < p$. Moreover

$$\|u\|_{L^\infty(-T+\delta,0;W_q^2(\Omega_\delta))} + \|\frac{\partial u}{\partial t}\|_{p,\infty,Q_\delta} \leq c_{10}(\delta, p, q)\|u\|_{2,Q}.$$

PROOF. Let $\delta' > 0$. We notice that u can be considered as a weak solution of the linear system (4) with

$$a_{ij}^{\alpha\beta}(z) = \int_0^1 A_{i,\xi_\beta^j}^{\alpha}(\tau Du(z))d\tau.$$

(See, for example S. Campanato [1]). Using this and Lemma 1 we get estimates for the norms $\|u\|_{p,\infty,Q_{\delta'}}$ and $[u]_{2,Q_{\delta'}}$. Now we can use Lemma 2 to get estimates of the norms $\|Du\|_{p,\infty,Q_{2\delta'}}$, $\|\frac{\partial u}{\partial t}\|_{p,\infty,Q_{2\delta'}}$. Lemma 2 also implies $D_\alpha D_\beta u \in L^2(Q_{2\delta'})$, $(0 \leq \alpha,\ \beta \leq n)$. We see that equation (1) is satisfied pointwise almost everywhere in $Q_{2\delta'}$ and that for almost every $t \in (-T+2\delta', 0)$ the function $u(\cdot, t)$ belongs to $W_2^2(\Omega_{2\delta'})$ and is the weak solution of the elliptic system

$$D_\alpha A_i^\alpha(Dv) = \frac{\partial u_i}{\partial t}$$

in $\Omega_{2\delta'}$. We can now use well-known L^p-estimates for elliptic systems (see Lemma 3 below). The proof is finished.

LEMMA 3. *Let $p > 2$ and let $g \in L^p(\Omega)$. Let $u \in W_2^1(\Omega)$ be a weak solution of the elliptic system*

$$D_\alpha A_i^\alpha(Du) = g_i \qquad i = 1, \ldots, n \tag{12}$$

Then there exists $q = q(\nu, M, p) > 2$ *such that* $u \in W^2_{q,loc}(\Omega)$. *Moreover, for every* $\delta > 0$

$$\|u\|_{W^2_q(\Omega_\delta)} \le c_{11}(\nu, M, p, q, \delta)(\|u\|_{W^1_2(\Omega)} + \|g\|_{p,\Omega}).$$

PROOF. Using the standard difference quotient technique, it is not difficult to verify that the following computations are legal.

Let $1 \le s \le n$. We let $v = D_s u$ and take the s-th derivative of (12). We get

$$D_\alpha a^{\alpha\beta}_{ij} D_\beta v_j = D_s g_i \tag{13}$$

where $a^{\alpha\beta}_{ij}(z) = A^\alpha_{i,\xi^j_\beta}(Du(x))$.

This implies

$$\frac{\nu}{2} \int_\Omega \varsigma^2 |Dv|^2 dx \le c_{12}(\nu, M) \int_\Omega (|v|^2 |D\varsigma|^2 + |g|^2) dx \tag{14}$$

for every $\varsigma \in \mathcal{D}(\Omega)$ (Cacciopoli's inequality). The required estimate can now be obtained by using the technique of reverse Hölder inequalities. (See, for example, M. Giaquinta [2], Chap. 5, Theorem 2.2). The proof is finished.

COROLLARY. *Let the assumptions of Theorem 1 be satisfied.*

(i) *If* $n \le 4$, *then* u *is Hölder continuous in* Q.

(ii) *If* $n \le 2$, *then* Du *is Hölder continuous in* Q.

(iii) *If* $n \le 2$ *and the functions* A^α_i *are smooth, then the solution* u *is smooth.*

REMARK. If $n \ge 3$, then Du may not be continuous. Examples are provided by nonregular solutions of elliptic systems. These can be found in J. Nečas [5].

PROOF OF THE COROLLARY. Let $\delta > 0$.

(i) Since $W^2_q(\Omega_{\delta/2}) \hookrightarrow C^{0,\alpha}(\Omega_\delta)$ with $\alpha = (2 - \frac{n}{q}) \wedge 1$, we have $u \in L^\infty(-T + \delta, 0; C^{0,\alpha}(\Omega_\delta))$.

Since we have also $\frac{\partial u}{\partial t} \in L^2(Q_\delta)$, u is Hölder continuous by Lemma 4 below.

(ii) In this case we have $W^2_q(\Omega_{\delta/2}) \hookrightarrow C^{1,\beta}(\Omega_\delta)$ $\beta = 1 - \frac{n}{q}$.

Hence $Du \in L^\infty(-T + \delta, 0; C^{0,\beta}(\Omega_\delta))$. Using the Hölder continuity of u it is easy to see that in fact $Du(\cdot, t) \in C^{0,\beta}(\Omega_\delta)$ for every $t \in (-T + \delta, 0)$, the $C^{0,\beta}$-norm being bounded independently of t.

Now we can use Lemma 3.1, Chap. 2 from O.A. Ladyzenskaja, V.A. Solonnikov, N.N. Ural'ceva [4] to infer that Du is Hölder continuous in Q_δ.

(iii) The higher regularity follows in the standard way from the theory of linear equations.

LEMMA 4. *Let $\alpha > 0$, $q > 1$, $\delta > 0$ and suppose*

$$u \in L^\infty(-T,0;C^{0,\alpha}(\Omega)) \text{ and } \frac{\partial u}{\partial t} \in L^q(Q).$$

Denote $K_1 = \|u\|_{L^\infty(-T,0;C^{0,\alpha}(\Omega))}$, $K_2 = \|\frac{\partial u}{\partial t}\|_{q,Q}$. *Then there exists* $K = K(K_1,K_2,\delta)$ *such that*

$$|\tilde{u}(x,t_1) - \tilde{u}(x,t_2)| \leq K|t_1 - t_2|^\beta$$

for every $x \in \Omega_\delta$ *and every* t_1, $t_2 \in (-T,0)$, *where* $\beta = \frac{\alpha/q'}{\alpha + n/q}$, $q' = \frac{q}{q-1}$ *and* $\tilde{u}$ *is a suitable representative of* u.

PROOF. Suppose first that u is continuous. Let $x \in \Omega_\delta$ and let $0 < \rho < \delta$. Define

$$w_\rho(t) = \frac{1}{|B_{x,\rho}|}\int_{B_{x,\rho}} u(y,t)dy.$$

It is easy to see that w'_ρ is bounded in $L^q(-T,0)$ by $c_{13}\rho^{-\frac{n}{q}}K_2$.

Let $t_1, t_2 \in (-T,0)$. We can write

$$\begin{aligned}
&|u(x,t_1) - u(x,t_2)| \\
&\quad \leq |u(x,t_1) - w_\rho(t_1) + w_\rho(t_1) - w_\rho(t_2) + w_\rho(t_2) - u(x,t_2)| \\
&\quad \leq 2K_1\rho^\alpha + c_{12}K_2\rho^{-\frac{n}{q}}|t_1 - t_2|^{\frac{1}{q'}}.
\end{aligned}$$

The proof is easily finished by using this inequality with $\rho = |t_1 - t_2|^{\frac{\beta}{\alpha}}$.

REMARK. It is not difficult to see that if the boundary of Ω is sufficiently regular (say, lipshitzian), then K can be chosen independent of δ.

Acknowledgement

We thank Oldřich John, Jan Malý and Jana Stará for helpful discussions.

REFERENCES

[1] S. CAMPANATO, *On the Nonlinear Parabolic Systems in Divergence Form. Hölder Continuity and Partial Hölder Continuity of the Solutions*. Ann. Mat. Pura Appl., **137** (1984).

[2] M. GIAQUINTA, *Multiple Integrals in the Calculus of Variations and Nonlinear Elliptic Systems*, Princeton University Press, 1983.

[3] M. GIAQUINTA - E. GIUSTI, *Partial Regularity for the Solutions to Nonlinear Parabolic Systems*, Ann. Mat. Pura Appl., **97** (1973).

[4] O.A. LADYZENSKAJA - V.A. SOLONNIKOV - N.N. URAL'CEVA, *Linear and Quasilinear Equations of Parabolic Type*, Moscow, 1966.

[5] J. NEČAS, *Introduction to the Theory of Nonlinear Elliptic Equations*, Teubner-Texte zur Math. **52**, Leipzig, 1983.

MFF UK
Sokolovská 83
18600 PRAHA 8
Czechoslovakia

Example of an irregular solution to a nonlinear elliptic system with analytic coefficients and conditions for regularity

Jindrich Nečas, Praha

§ 1. Introduction

The problem on the regularity of a weak solution to a nonlinear elliptic system of the second-order with analytic coefficients was negatively solved by the example of E. Giusti, M. Miranda [1] (see also other examples of the same type in V. G. Mazja [2], S. A. Arakčejev [3]) who have given for the dimension $n \geqq 3$, $0 \in \Omega \subset R_n$ and for the vector function $u = (u_1, \ldots, u_n)$ a functional

$$\Phi(u) = \int_\Omega F(u, \partial u)\, dx \tag{1.1}$$

with analytic function $F(u, \partial u)$, whose critical point is a discontinuous function with components $\frac{x_i}{r}$. The function $F(u, \partial u)$ depends on u in a substantial way and the functional (1.1) has not the second differential.

In this paper, in § 2, we give a functional for $n \geqq 3$, $0 \in \Omega \subset R_n$,

$$\Phi(u) = \int_\Omega F(\partial u)\, dx, \tag{1.2}$$

with an analytic function F, satisfying the growth conditions:

$$|D^l F| \leqq c_l \frac{|\partial u|^2}{(1 + |\partial u|)^l}, \qquad l = 0, 1, \ldots, \tag{1.3}$$

where u is an n^2 dimensional vector function. It is $\frac{\partial F}{\partial \xi_k{}^{ij}}(0) = 0$ and for n enough large

$$\frac{\partial^2 F}{\partial \xi_k{}^{ij} \partial \xi_\gamma{}^{\alpha\beta}}(\xi)\, \eta_k{}^{ij} \eta_\gamma{}^{\alpha\beta} \geqq c|\eta|^2, \qquad c > 0. \tag{1.4}$$

The functional (1.2) is on the Sobolev space $[W^{1,2}]^{n^2}$ twice continuously differentiable and for n enough large the second differential is positive definite. The critical point of (1.2) is the function u^0 with components $u_{ij}^0 = \frac{x_i x_j}{r}$. We give also an example of a second-order elliptic system

$$\int_\Omega a_k{}^{ij}(\partial u) \frac{\partial \varphi_{ij}}{\partial x_k}\, dx = 0, \tag{1.5}$$

© Akademie-Verlag 1977. Reprinted with kind permission from Wiley-VCH Verlag: Nečas, J., Example of irregular solution to a nonlinear elliptic system with analytic coefficients and conditions for regularity, Theory of Nonlinear Operators, Abhandlungen der Akademie der Wissenschaften der DDR 1N, 197-206.

with analytic $a_k^{ij}(\partial u)$, $a_k^{ij}(0) = 0$, satisfying the ellipticity condition for $n \geqq 3$:

$$a_k^{ij}(\xi)\, \xi_k^{ij} \geqq c\, |\xi|^2, \qquad c > 0 \tag{1.6}$$

and the strong monotony condition for $n \geqq 5$:

$$\frac{\partial a_k^{ij}(\xi)}{\partial \xi_\gamma^{\alpha\beta}}\, \eta_k^{ij} \eta_\gamma^{\alpha\beta} \geqq c|\eta|^2, \qquad c > 0. \tag{1.7}$$

A weak solution of (1.5) is also the above defined function u^0.

In § 3, we give an "a priori" estimate for the regularity of a critical point to the functional

$$\Phi(u) = \int_\Omega [F(\partial u) - (f, \partial u)]\, dx, \tag{1.8}$$

where $\Omega \subset R_3$, $\partial\Omega$ is enough smooth, $u_1 = (u, u_2, \ldots, u_m)$ and ∂u is the jet of the first order derivatives. The function $F(\xi)$ is defined in a bounded ball K_a, $F \in C^{(4)}(\overline{K}_a)$. Let $f \in [W^{2,2}]^{3m}$, $\bar{u} \in [W^{3,2}]^m$ and let $u \in [W^{3,2}]^m$ be a critical point of (1.8), such that $\forall x \in \overline{\Omega}: |\partial u(x)| < r \leqq a$ and let $u = \bar{u}$ on $\partial\Omega$. Let us suppose for $|\xi| < r$:

$$\frac{\partial^2 F}{\partial \xi_i^r\, \partial \xi_j^s}(\xi)\, \eta_i^r \eta_j^s \geqq c(r)\, |\eta|^2, \qquad c(r) > 0, \tag{1.9}$$

$$\frac{\partial^4 F}{\partial \xi_i^r\, \partial \xi_j^s\, \partial \xi_k^t\, \partial \xi_l^v}\, \eta_i^r \eta_j^s \eta_k^t \eta_l^v \leqq T(r) \sum_{r,i} (\eta_i^r)^4, \tag{1.10}$$

$$c(r) - 3r^2 T(r) > 0 \quad \text{(the regularity condition)}, \tag{1.11}$$

then

$$\|u\|_{3,2} \leqq A(r)\left[1 + \|\bar{u}\|_{3,2}^2 + \|f\|_{2,2}^2\right].$$

This estimate enables us to find classical solutions of problems with weak solutions, or of problems where the existence of weak solutions is not proved, in the form of solution's path when the data's paths $f(t)$, $\bar{u}(t)$ are prescribed. For details see the paper by J. Nečas, J. Stará, R. Švarc [4].

In § 4, constructive aspects of the method are briefly discussed and the second problem to the finite elasticity is defined as an example, where the theory of monotone operators fail.

§ 2. Examples for irregular solutions

Let $0 \in \Omega \subset R_n$, where Ω is a bounded domain. Put $m = n^2$ and let us look for $u \in [W^{1,2}]^m$, such that

$$u - u^0 \in [W_0^{1,2}]^m \quad \text{(functions with zero traces on } \partial\Omega\text{)}, \tag{2.1}$$

where $u_{ij}^0 = \dfrac{x_i x_j}{r}$, $i, j = 1, 2, \ldots, n$, and where u is the weak solution to the elliptic system: $\forall \varphi \in [\mathcal{D}(\Omega)]^m$

$$\int_\Omega a_k^{ij}(\partial u)\, \frac{\partial \varphi_{ij}}{\partial x_k} = 0. \tag{2.2}$$

Let

$$\Phi(u) = \int_\Omega \left[\frac{1}{2}\frac{\partial u_{ij}}{\partial x_k}\frac{\partial u_{ij}}{\partial x_k} + \frac{\mu}{2}\frac{\partial u_{ij}}{\partial x_i}\frac{\partial u_{kk}}{\partial x_j} + \lambda \frac{\partial u_{ij}}{\partial x_i}\frac{\partial u_{k\alpha}}{\partial x_k}\frac{\partial u_{l\beta}}{\partial x_l}\frac{\partial u_{j\alpha}}{\partial x_\beta}\left(1 + \frac{\partial u_{ab}}{\partial x_c}\frac{\partial u_{ab}}{\partial x_c}\right)^{-1}\right] dx \equiv \int_\Omega F(\partial u)\, dx, \quad (2.3)$$

where $\lambda = 2\,\dfrac{n^3 - 1}{n(n-1)(n^3 - n + 1)}$, $\mu = -\dfrac{4 + \lambda n}{n^2 - n + 1}$.

We have (1.7), u^0 is a critical point of (2.3), $\dfrac{\partial F}{\partial \xi_k{}^{ij}}(0) = 0$, and for n enough large (1.4) holds. In order to see the last statement, put $\nu^i = \xi_j{}^{ji}$, $\nu_\eta{}^i = \eta_j{}^{ji}$, $M^i = \xi_i{}^{jj}$, $N_i = \eta_i{}^{jj}$ and note $V = \dfrac{\partial u_{ij}}{\partial x_i}\dfrac{\partial u_{k\alpha}}{\partial x_k}\dfrac{\partial u_{l\beta}}{\partial x_l}\dfrac{\partial u_{j\alpha}}{\partial x_\beta}$, $\dfrac{\partial u_{ab}}{\partial x_l}\dfrac{\partial u_{ab}}{\partial x_l} = U$, $\xi_k{}^{ij}\eta_k{}^{ij} = [\xi, \eta]$, $\nu^i\nu_\eta{}^j = (\nu, \nu_\eta)$. We get

$$\begin{aligned} D^2F(\xi, \eta, \eta) &= |\eta|^2 + \mu(\nu_\eta, M_\eta) \\ &\quad + \lambda D^2V(\xi, \eta, \eta)(1 + U)^{-1} - 4\lambda DV(\xi, \eta)(1 + U)^{-2}[\xi, \eta] \\ &\quad + 8\lambda V(1 + U)^{-3}[\xi, \eta]^2 - 2\lambda V(1 + U)^{-2}|\eta|^2. \end{aligned} \quad (2.4)$$

Because $|\nu| \leqq \sqrt{n}\,|\xi|$, $|\nu_\eta| \leqq \sqrt{n}\,|\eta|$, …, we get from (2.4)

$$D^2F(\xi, \eta, \eta) \geqq \left(1 - \frac{c}{\sqrt{n}}\right)|\eta|^2. \quad (2.5)$$

The construction of the function F and the estimate (2.5) are crude. In order to get better estimations for the dimension where "the theorem on the regularity" is not valid, let us consider the elliptic system

$$\int_\Omega \left[\frac{\partial u_{ij}}{\partial x_k}\frac{\partial \varphi_{ij}}{\partial x_k} + a_1 \frac{\partial u_{ik}}{\partial x_i}\frac{\partial \varphi_{jk}}{\partial x_j} + a_2 \frac{\partial u_{ii}}{\partial x_k}\frac{\partial \varphi_{jj}}{\partial x_k} + \mu \frac{\partial u_{ij}}{\partial x_i}\frac{\partial \varphi_{kk}}{\partial x_j} + \lambda \frac{\partial u_{ij}}{\partial x_i}\frac{\partial u_{\alpha\beta}}{\partial x_\alpha}\frac{\partial u_{k\gamma}}{\partial x_k}\frac{\partial \varphi_{j\beta}}{\partial x_j}\left(1 + \frac{\partial u_{ab}}{\partial x_a}\frac{\partial u_{cb}}{\partial x_c}\right)^{-1}\right] dx \equiv \int_\Omega f(\partial u, \partial\varphi)\, dx. \quad (2.6)$$

Put $a_1 = \dfrac{n+1}{n}$, $a_2 = \dfrac{n+3}{n-1}$, $\lambda = 2\,\dfrac{(n+1)(n^2+1)}{n^3(n-1)}$, $\mu = -2\,\dfrac{n+3}{n(n-1)}$, $u_{ij}^0 = \dfrac{x_i x_j}{r}$. We get (2.6) for $u = u^0$. For $n \geqq 3$, we get

$$\begin{aligned} f(\xi, \xi) &\geqq |\xi|^2 + \frac{n+1}{n}|\nu|^2 + \frac{n+3}{n-1}|M|^2 - |\mu|\,|\nu|\,|M| - \lambda\frac{|\nu|^3\,|\xi|}{1 + |\nu|^2} \\ &\geqq |\xi|^2 + \frac{n^3 - 2n - 3}{n^2(n-1)}|\nu|^2 - \frac{2(n+1)(n^2+1)}{n^3(n-1)}|\nu|\,|\xi|. \end{aligned} \quad (2.7)$$

It follows that for $n \geqq 3$ the last form is positive definite.

To obtain (1.7), we can see that the integral $I = \int_\Omega \frac{\partial u_{ij}}{\partial x_i} \frac{\partial u_{\alpha\beta}}{\partial x_\alpha} \frac{\partial u_{k\gamma}}{\partial x_k} \frac{\partial \varphi_{j\beta}}{\partial x_\gamma} dx$ is independent with respect to a permutation of the indices j, β, γ. Put $\eta^{i,j,k} = \frac{1}{3} (\eta_k^{ji} + \eta_k^{ji} + \eta_j^{ik})$, $V(\xi, \eta) = \xi_i^{ij} \xi_\alpha^{\alpha\beta} \xi_k^{k\gamma} \eta_\gamma^{j\beta} (1 + \xi_a^{ab} \xi_c^{ab})^{-1}$. We have

$$DV(\xi, \eta, \eta) = 3\nu_\eta^{j} \nu^\beta \nu^\gamma \eta^{j\beta\gamma} (1 + |\nu|^2)^{-1} - 2\nu^j \nu^\beta \nu^\gamma \eta^{j\beta\gamma} (1 + |\nu|^2)^{-2} (\nu, \nu_\eta). \tag{2.8}$$

Put $\nu' = \frac{\nu}{(1 + |\nu|^2)^{\frac{1}{2}}}, \quad \mu^j = \nu'^\beta \nu'^\gamma \eta^{j\beta\gamma}$.

We have to estimate the term

$$W \equiv |3(\nu_\eta, \eta) - 2(\nu', \mu)(\nu', \nu_\eta)|.$$

By the projection of the vector ν' on the plane generated by ν_η, μ (we let the details to the reader), we get finally

$$W \leqq 3 |\eta| |\nu_\eta|. \tag{2.9}$$

Hence

$$Df(\xi, \eta, \eta) \geqq |\eta|^2 + \frac{n+1}{n} |\nu_\eta|^2 + \frac{n+3}{n-1} |M_\eta|^2 - |\mu| |\nu_\eta| |M_\eta| - 3\lambda |\nu_\eta| |\eta|$$

$$\geqq |\eta|^2 + \frac{n^3 - 2n - 3}{n^2(n-1)} |\nu_\eta|^2 - \frac{6(n+1)(n^2+1)}{n^3(n-1)} |\nu_\eta| |\eta|, \tag{2.10}$$

which implies (1.7) for $n \geqq 5$.

We have also

$$|f(\xi, \eta)| \leqq c|\xi| |\eta|, \tag{2.11}$$

for

$$k \geqq 1 : |D_\xi^k f(\xi, \eta)| \leqq c_k |\eta| (1 + |\nu|)^{1-k}. \tag{2.12}$$

§ 3. Conditions for regularity

Let us consider $\Omega \subset R_3$ with $\partial\Omega$ enough smooth and let $F(\xi)$, where $\xi = \{\xi_i^r, r = 1, 2, \ldots, m, i = 1, 2, 3\}$, be a real function defined on the ball $K_a = \{\xi, |\xi| < a\}$. Let $F \in C^{(4)}(\overline{K}_a)$, $\frac{\partial F}{\partial \xi_i^r}(0) = 0$. Let $f \in [W^{2,2}]^{3m}$, $\tilde{u} \in [W^{3,2}]^m$. Let us have $u \in [W^{3,2}]^m$ such that $u = \tilde{u}$ on $\partial\Omega$, $\|u\|_{1,\infty} \equiv \max_{x \in \overline{\Omega}} \left(\frac{\partial u_r}{\partial x_i}(x), \frac{\partial u_r}{\partial x_i}(x) \right)^{\frac{1}{2}} < a$, and such that it is a solution to the elliptic system:

$$\forall \varphi \in [W_0^{1,2}]^m$$

$$\int_\Omega \frac{\partial F}{\partial \xi_i^r}(\partial u) \frac{\partial \varphi_r}{\partial x_i} dx = \int_\Omega f_i^r \frac{\partial \varphi_r}{\partial x_i} dx. \tag{3.1}$$

Theorem 3.1. *Let u be a solution of the system* (3.1), *satisfying the boundary condition $\tilde{u}$, let $\|u\|_{1,\infty} < r < \varrho \leqq a$ and let for $|\xi| < r$ the conditions* (1.9), (1.10), (1.11) *be valid. Then there exists a continuous function $d(r)$ for $0 \leqq r < \varrho$ such that*

$$\|u\|_{3,2} \leqq d(r) \left(1 + \|f\|_{2,2}^2 + \|\tilde{u}\|_{3,2}^2\right). \tag{3.2}$$

If $\lim_{r\to\varrho} c(r) = \alpha > 0$ and $\lim_{r\to\varrho} c(r) - 3T(r)\, r^2 = \beta > 0$, then $d(r)$ is bounded on the interval $[0, \varrho[$.

We refer the reader for details to the paper [4]. We sketch the proof in this paper for $\tilde{u} = 0$ and $\Omega = R_3^+$, the halfspace.

Theorem 3.2. *Let $\Omega = R_3^+$, let u be a solution of* (3.1) *with zero boundary data and let the conditions of Theorem* 3.1 *be satisfied. Then there exists a continuous function $c_1(r)$ in the interval $[0, \varrho]$, such that*

$$\|u\|_{3,2} \leqq c_1(r) \left(c(r)\right)^{-4} \left[c(r) - 3T(r)\, r^2\right]^{-1} \left(\|f\|_{2,2} + \|f\|_{2,2}^2\right). \tag{3.3}$$

Proof. For the halfspace R_3^+ we use the following norms for $W^{k,2} \cap W_0^{1,2}$ spaces $(k \geqq 1)$:

$$\|u\|_{k,2} = \left(\int_{R_3^+} \sum_{1\leqq|\alpha|\leqq k} (D^\alpha u)^2\, dx\right)^{\frac{1}{2}}.$$

Let $\varphi \in [\mathscr{D}]^m$ and let φ' be the derivative $\dfrac{\partial\varphi}{\partial x_1}$ or $\dfrac{\partial\varphi}{\partial x_2}$. We have

$$\int_{R_3^+} \frac{\partial F}{\partial \xi_i^r} \frac{\partial \varphi_r^{(4)}}{\partial x_i}\, dx = \int_{R_3^+} (f_i^r)'' \frac{\partial \varphi_r''}{\partial x_i}\, dx = \int_{R_3^+} \frac{\partial^2 F}{\partial \xi_i^r\, \partial \xi_j^s} \frac{\partial u_s''}{\partial x_j} \frac{\partial \varphi_r''}{\partial x_i}\, dx$$
$$+ \int_{R_3^+} \frac{\partial^3 F}{\partial \xi_i^r\, \partial \xi_j^s\, \partial \xi_k^t} \frac{\partial u_s'}{\partial x_j} \frac{\partial u_t'}{\partial x_k} \frac{\partial \varphi_r}{\partial x_i}\, dx. \tag{3.4}$$

Passing to the corresponding limit, we can put $\varphi = u$. We first get

$$\int_{R_3^+} \frac{\partial^3 F}{\partial \xi_i^r\, \partial \xi_j^s\, \partial \xi_k^t} \frac{\partial u_r''}{\partial x_i} \frac{\partial u_s'}{\partial x_j} \frac{\partial u_t'}{\partial x_k}\, dx$$
$$= -\frac{1}{3} \int_{R_3^+} \frac{\partial^4 F}{\partial \xi_i^r\, \partial \xi_j^s\, \partial \xi_k^t\, \partial \xi_l^v} \frac{\partial u_r'}{\partial x_i} \frac{\partial u_s'}{\partial x_j} \frac{\partial u_t'}{\partial x_k} \frac{\partial u_v'}{\partial x_l}\, dx. \tag{3.5}$$

In the following we use a special case of the L. Nirenberg's interpolation inequality, see L. Nirenberg [5], which we can get by a simple integration by parts:

$$\text{for } v \in \mathscr{D}(-\infty, \infty), \quad \int_{-\infty}^{\infty} (v')^4\, dx \leqq 9 \max_{x\in R_1} |v(x)|^2 \int_{-\infty}^{\infty} (v'')^2\, dx. \tag{3.6}$$

From (3.6) and (1.10) we obtain

$$-\frac{1}{3}\int_{R_3^+}\frac{\partial^4 F}{\partial\xi_i^r\,\partial\xi_j^s\,\partial\xi_k^t\,\partial\xi_l^v}\,\frac{\partial u_r'}{\partial x_i}\,\frac{\partial u_s'}{\partial x_j}\,\frac{\partial u_t'}{\partial x_k}\,\frac{\partial u_v'}{\partial x_l}\,dx$$

$$\geqq -3T(r)\,r^2\int_{R_3^+}\frac{\partial u_r''}{\partial x_i}\,\frac{\partial u_r''}{\partial x_i}\,dx. \tag{3.7}$$

It follows from (3.4), (3.5), (3.7) and (1.11) that

$$\left(c(r) - 3T(r)\,r^2\right)\|u''\|_{1,2} \leqq \|f\|_{2,2}. \tag{3.8}$$

In virtue of (3.6), we get

$$\int_{R_3^+}\sum_{r,i}\left(\frac{\partial u_r'}{\partial x_i}\right)^4 dx \leqq \frac{9r^2}{\left(c(r) - 3T(r)\,r^2\right)^2}\,\|f\|_{2,2}^2, \tag{3.9}$$

hence

$$\left\|\left(\frac{\partial F}{\partial\xi_i^r}\right)'\right\|_{0,4} \leqq \frac{M(r)}{\left(c(r) - 3T(r)\,r^2\right)^{\frac{1}{2}}}\,\|f\|_{2,2}^{\frac{1}{2}}. \tag{3.10}$$

But

$$\sup_{\varphi\in[\mathcal{D}]^3,\,\|\varphi\|_{0,\frac{4}{3}}\leq 1}\left|\int_{R_3^+}\frac{\partial}{\partial x_3}\left(\frac{\partial F}{\partial\xi_3^r}\right)\varphi_r\,dx\right|$$

$$\leqq \sup\left[\left|\int_{R_3^+}\frac{\partial}{\partial x_1}\left(\frac{\partial F}{\partial\xi_1^r}\right)\varphi_r\,dx\right| + \left|\int_{R_3^+}\frac{\partial}{\partial x_2}\left(\frac{\partial F}{\partial\xi_2^r}\right)\varphi_r\,dx\right| + \left|\int_{\Omega} f_i^r\,\frac{\partial\varphi_r}{\partial x_i}\,dx\right|\right]$$

$$\leqq \frac{N(r)}{\left(c(r) - 3T(r)\,r^2\right)^{\frac{1}{2}}}\,\|f\|_{2,2}^{\frac{1}{2}} + c\,\|f\|_{2,2}. \tag{3.11}$$

We can calculate the derivatives $\dfrac{\partial^2 u_s}{\partial x_3^2}$, $s = 1, 2, \ldots, m$, from the elliptic system

$$-\frac{\partial^2 F}{\partial\xi_i^r\,\partial\xi_j^s}\,\frac{\partial^2 u_s}{\partial x_i\,\partial x_j} = -\frac{\partial f_i^r}{\partial x_i}, \qquad r = 1, 2, \ldots, m \tag{3.12}$$

and we get

$$\left\|\frac{\partial^2 u_s}{\partial x_3^2}\right\|_{0,4} \leqq \frac{O(r)}{c(r)\left[c(r) - 3T(r)\,r^2\right]^{\frac{1}{2}}}\left[\|f\|_{2,2}^{\frac{1}{2}} + \|f\|_{2,2}\right]. \tag{3.13}$$

If we differentiate (3.12) with respect to x_1 or x_2, we obtain

$$-\frac{\partial^2 F}{\partial\xi_i^r\,\partial\xi_j^s}\,\frac{\partial^2 u_s'}{\partial x_i\,\partial x_j} = -\frac{\partial f_i^{r\prime}}{\partial x_i} + \frac{\partial^3 F}{\partial\xi_i^r\,\partial\xi_j^s\,\partial\xi_k^t}\,\frac{\partial u_t'}{\partial x_k}\,\frac{\partial^2 u_s}{\partial x_i\,\partial x_j}, \tag{3.14}$$

hence

$$\|u'\|_{2,2} \leqq \frac{Q(r)}{c(r)^3\,[c(r) - 3T(r)\,r^2]}\left(\|f\|_{2,2} + \|f\|_{2,2}^2\right). \tag{3.15}$$

Once more differentiating (3.12) with respect to x_3, we get the inequality (3.2), q.e.d.

Let data paths be given: $\tilde{u} \in C([0, T], [W^{3,2}]^m)$, $f \in C([0, T], [W^{2,2}]^{3m})$, $\tilde{u}(0) = 0$, $f(0) = 0$, and let us look for $u(t) \in [W^{3,2}]^m$ such that, $\forall t$ from the below precised interval, $u(t)$ solves the equation (3.1) with the boundary condition $\tilde{u}(t) = u(t)$ and such that $\|u(t)\|_{1,\infty} < \varrho \leqq a$. We have

Theorem 3.3. *Let the conditions* (1.9) *and* (1.11) *be satisfied for all* $r < \varrho$. *Then there exists a unique solution path* $u(t)$ *and we have:*

$$u \in C([0, T], [W^{3,2}]^m) \tag{3.16}$$

or

$$u \in C([0, t_{\mathrm{cr}}[, [W^{3,2}]^m), \tag{3.17}$$

where $t_{\mathrm{cr}} \leqq T$ *and, in the case* (3.17), *we have*

$$\lim_{t \to t_{\mathrm{cr}}} \min_{x \in \bar{\Omega}} |\partial u(x, t) - \partial K_\varrho| = 0. \tag{3.18}$$

If $\lim\limits_{r \to \varrho} c(r) = \alpha > 0$ *and* $\lim\limits_{r \to \varrho} \left(c(r) - 3r^2 T(r)\right) = \beta > 0$, *then*

$$\sup_{t \in [0, t_{\mathrm{cr}}[} \|u(t)\|_{3,2} < \infty. \tag{3.19}$$

The reader can easily see that Theorem 3.3 follows readily from Theorem 3.1 and from the implicit function theorem applied to our problem. We sketch the situation for $\Omega = R_3{}^+$:

Theorem 3.4. *Let* $u_0(t) \in [W^{3,2}(R_3{}^+)]^m \cap [W_0{}^{1,2}(R_3{}^+)]^m$ *be a solution of the system* (3.1) *for some* t_0 *and let* $U(u(t_0))$ *be some small neighbourhood of* $u(t_0)$. *On* $U(u(t_0)) \times (t_0 - \varepsilon, t_0 + \varepsilon)$, *let us define the operator* A *to* $[W^{2,2}(R_3{}^+)]^{3m}/S$

$$\int\limits_{R_3^+} \frac{\partial F}{\partial \xi_i{}^r}\left(\partial u(t)\right) \frac{\partial \varphi_r}{\partial x_i}\, dx = \int\limits_{R_3^+} g_i{}^r \frac{\partial \varphi_r}{\partial x_i}; \tag{3.20}$$

here

$$S = \left\{ g \,\middle|\, \int\limits_{R_3^+} g_i{}^r \frac{\partial \varphi_r}{\partial x_i}\, dx = 0 \right\}.$$

Then the operator $A(u, t)$ *is continuous, has a continuous derivative* $A_u{}'(u, t)$, *invertible at the point* $[u(t_0), t_0]$.

Proof. We shall sketch the proof of the invertibility of $A_u{}'$. But this is equivalent to the proof that the system

$$\int\limits_{R_3^+} \frac{\partial^2 F}{\partial \xi_i{}^r\, \partial \xi_j{}^s}\left(\partial u(t_0)\right) \frac{\partial v_s}{\partial x_j} \frac{\partial \varphi_r}{\partial x_i}\, dx = \int\limits_{R_3^+} h_i{}^r \frac{\partial \varphi_r}{\partial x_i}\, dx \tag{3.21}$$

has a unique solution in $[W^{3,2}]^m \cap [W_0^{1,2}]^m$. But the condition (1.9) implies the unique solvability of (3.21) in $[W_0^{1,2}]^m$. We have to prove that such a solution $v \in [W^{3,2}]^m$. This we can do by investigating the integral

$$\int_{R_3^+} \frac{\partial^2 F}{\partial \xi_i^r \, \partial \xi_j^s} \left(\partial u(t_0)\right) \frac{\partial v_s}{\partial x_j} \frac{\partial \varphi_r^{(4)}}{\partial x_i} \, dx \tag{3.22}$$

and by following the analogous steps as in the proof of Theorem 3.2. For the usual $W^{l,2}$ norms $\|f\|_{l,2} = \left(\int_{R_3^+} \sum_{|\alpha| \leq l} (D^\alpha f)^2 \, dx \right)^{\frac{1}{2}}$ we use the inequalities

$$\|f\|_{k,2} \leqq \varepsilon \|f\|_{l,2} + c_\varepsilon \|f\|_{0,2} \tag{3.23}$$

for $k < l$, which we get easily in R_3 by the Fourier transform and in R_3^+ by the extension of f from R_3^+ to R_3. For details see, for example, J. Nečas [6]. Using the inequality (3.6), we get first

$$\|u\|_{2,4} \leqq c \, \|u\|_{3,2} \tag{3.24}$$

and then

$$\|v''\|_{1,2} \leqq \varepsilon \, \|v\|_{3,2} + c_\varepsilon \|h\|_{2,2} \,. \tag{3.25}$$

The next step is the inequality

$$\|v\|_{2,4} \leqq \varepsilon \|v\|_{3,2} + d_\varepsilon \|h\|_{2,2} \tag{3.26}$$

and finally, we obtain the inequality

$$\|v\|_{3,2} \leqq c \|h\|_{2,2} \,. \tag{3.27}$$

(3.27) is an "a priori" inequality, whose validity for the solution v can be obtained by the standard homotopy argument.

§ 4. Constructive aspects and finite elasticity

Let t_0 be an attainable parameter from Theorem 3.2. The path $u(t)$ can be understood as an embedding process, connecting $t = 0$, where $u(0) = 0$, with $u(t_0)$. For details see H. Schwetlick [7]. Instead of looking for the whole path, we can calculate $u(0)$, $u(t_1)$, $u(t_2), \ldots, u(t_0)$ with $t_i - t_{i-1}$ enough small, using Newton's method. The conditions for the applicability of this method are, roughly speaking, the same as for the implicit function theorem, see Theorem 3.4, provided that the paths $\bar{u}(t)$ and $f(t)$ are in $C^{(1)}$. Let us underline the fact that the conditions for the applicability of Newton's method are not satisfied for the weak solutions of nonlinear elliptic systems of the type (3.1).

Let us define the place boundary value problem for the finite elasticity. For details see K. Washizu [8].

Let $\Omega \subset R_3$ be a bounded domain with $\partial\Omega$ enough smooth, the considered body before the deformation Ω is supposed to play the role of a reference domain. Let u be the displacement vector and

$$e_{ij} = \frac{1}{2}\left(\frac{\partial u_i}{\partial x_j} + \frac{\partial u_j}{\partial x_i} + \frac{\partial u_k}{\partial x_i}\frac{\partial u_k}{\partial x_j}\right) \tag{4.1}$$

the finite strain-tensor. Let σ^i be the stress vectors acting on the surface, which before the deformation had the normal parallel to the x_i axis, and let σ^i be defined with respect to the unit area of the surface before the deformation. Let P be the vector of body forces with respect to the unit volume before the deformation. In every point of the body, we have the equations of equilibrium

$$\frac{\partial \sigma^i}{\partial x_i} + P = 0. \tag{4.2}$$

Let $y = x + u(x)$, and let us define a tensor $\sigma^{\lambda\mu}$ by

$$\sigma^\lambda = \sigma^{\lambda\mu}\frac{\partial y}{\partial x_\mu}. \tag{4.3}$$

The moment equation of equilibrium gives $\sigma^{\mu\lambda} = \sigma^{\lambda\mu}$.

In some domain of the strain tensor, let the Hooke law be defined by

$$\sigma^{\mu\lambda} = \frac{\partial A}{\partial e_{\mu\lambda}}. \tag{4.4}$$

On $\partial\Omega$ let the displacement $\tilde{u}$ be described. Let us suppose that the body forces depend only on x (and not on $u(x)$). From (4.2), (4.3) and (4.4) we then get the system of finite elasticity:

$$\frac{\partial}{\partial x_i}\left[\frac{\partial A}{\partial e_{ij}}\left(\delta_{\alpha j} + \frac{\partial u_\alpha}{\partial x_j}\right)\right] + P_\alpha = 0. \tag{4.5}$$

(4.5) are the Euler equations for the functional of the total potential energy

$$\Phi(u) = \int_\Omega A\big(e_{ij}(u)\big)\,dx - \int_\Omega u_i P_i\,dx. \tag{4.6}$$

If we put $F(\partial u) = A\big(e_{ij}(u)\big)$, we have an example of the system (3.1). The functional (4.6) is not positive definite and not lower weakly semicontinuous in the energy space (which is, for a linear Hooke law, $[W^{1,4}]^3$!), see the examples from the paper by J. Nečas [9]. The conditions (1.9) can be defined only locally in the space $\|u\|_{1,\infty}$ because $c(r) > 0$ gives the condition of the local stability of the system (and the instability can occur). The paths $\tilde{u}(t)$, $f(t)$ express the loading paths.

If the conditions (1.9), (1.11) are satisfied in the domain of definition for the function $F(\partial u)$, the solution path $u(t)$ exists up to a point where it loses its physical meaning.

Bibliography

[1] Giusti, E., M. Miranda, Un esempio di soluzioni discontinue per un problema di minimo relativo ad un integrale regolare del calcolo delle variazioni. Boll. Un. Mat. Ital. (4) 2, 1–8 (1969).

[2] Мазья, В. Г., Примеры нерегулярных решений квазилинейных эллиптических уравнений с аналитическими коэффициентами. Функц. Анализ **2**, 3, 53–57 (1968).
[3] Аракчеев, С. А., О гладкости обобщенных решений некоторого класса квазилинейных эллиптических уравнений. Вестник Московск. Унив. **30**, 1, 49–56 (1975).
[4] Nečas, J., J. Stará, R. Švarc, Classical solution to second-order nonlinear elliptic system in R_3 (to appear).
[5] Nirenberg, L., An extended interpolation inequality. Ann. Scuola Norm. Sup. Pisa **20**, 733 to 737 (1966).
[6] Nečas, J., Les méthodes directes en théorie des équations elliptiques. Academia, Prague 1967.
[7] Schwetlick, H., Continuation methods for solving nonlinear equations (to appear in the volume: Mathematical Models and Numerical Methods Semester, edited by S. Banach international mathematical center).
[8] Washizu, K., Variational methods in elasticity and plasticity. Pergamon Press, 1968.
[9] Nečas, J., Theory of locally monotone operators modeled on the finite displacement theory for hyperelasticity (to appear).

Address of the author:

Mathematical Institute of the Czechoslovak Academy of Sciences
Žitná 25
11567 Praha 1
Czechoslovakia

Chapter 8
Commentaries

8.1 Regularity for Nonlinear Equations and Systems

Jana Stará and Oldřich John

Abstract In the following article we restrict ourselves to a short review of J. Nečas' achievements in studying the smoothness properties of solutions to elliptic and parabolic systems. Besides the positive results it includes also a view on construction of counterexamples which demonstrate that standard assumptions on a functional of the calculus of variations cannot guarantee smoothness of minimizers.

A significant part of J. Nečas' work in regularity for nonlinear problems is closely related to classical questions going back to Hilbert's 19th and 20th problems. These concern existence and regularity of functions minimizing (under suitable conditions) variational integrals of the form

$$\int_\Omega F(x, u, Du)\, dx \tag{I}$$

and their higher-order analogues. J. Nečas started to actively work on these problems around 1965. One of the factors which turned his attention in this direction was a series of lectures by F. Browder at a summer school in Montreal held in 1965. J. Nečas' first important result in this field, published in [JN-C30, C33] can be viewed as a highly non-trivial generalization of a famous result of C.B. Morrey from 1938. Morrey proved full regularity for minimizers of (I) in dimension 2 (under appropriate assumptions). In [JN-C30, C33] J. Nečas generalized this result to a large class of functionals of super linear growth depending on higher-order derivatives. His basic tool is a higher-order version of a well-known linear $L^{2+\epsilon}$-estimate due to N.G. Meyers (1963) (see [MEY]), which is also proved in [C33] using Agmon–Douglis–Nirenberg L^p estimates for solutions of a Laplace equation

© Springer Basel 2015

Š. Nečasová et al. (eds.), *Selected Works of Jindřich Nečas*, Advances in Mathematical Fluid Mechanics, DOI 10.1007/978-3-0348-0231-4_8

(see [ADN]). Nečas' method of proof has found applications in other situations, for example in regularity problems concerning stationary solutions of certain fluid-dynamics models. An important milestone in the study of regularity for integrals (I) in dimensions higher than two is the famous 1957 results of E. De Giorgi and J. Nash that, under appropriate assumptions, minimizers of (I) are smooth in the case of a scalar unknown function u. The scalar case is proved by an estimate for solutions of linear equations with measurable coefficients. Some additional subtle issues, such as optimal assumptions about dependence of F on u were clarified later by Ladyzhenskaya and Uralceva [LU]. In general, these works left open the case of vector-valued unknown functions u. In 1968, E. De Giorgi and V. Mazja independently constructed examples showing that the critical estimate may fail for the corresponding linear equations with vector-valued unknowns. These examples were later refined by E. Guisti and M. Miranda (see [GM]). However, the original problem of regularity of minimizers of (I) remained open until it was settled by J. Nečas in 1977. Nečas showed that in dimensions five and higher one can construct analytic functions F which do not depend on x and u, are uniformly convex in Du, have bounded second derivatives, and the integral (I) admits singular minimizers. We remark that the convexity of F guarantees uniqueness of solutions to natural boundary value problems in the class of weak solutions. It also implies that weak solutions can be identified with minimizers.[1]

Nečas' counter-example remains as astonishing today as it was when it first appeared. Further results related to this problem can be found in [C177, MA, SOU, SJ]. Many basic questions concerning minimizers of (I) for vector-valued u remain open. For example, there are no known examples of non-smooth minimizers of functionals (I) for functions u mapping a three-dimensional domain into R^3 (under appropriate assumptions on F, of course). For some recent developments see for example [SY] and [MS].

In 1979, J. Nečas together with M. Giaquinta pointed out (in [JN-C96, JN-C100]) an interesting connection between regularity of nonlinear elliptic systems and a Liouville-type condition. For example, in the context of variational integral (I) with $F(x, u, Du) = F(Du)$, the connection is that regularity of minimizers of (I) is equivalent, modulo technicalities, to the absence of non-trivial global Lipschitz solutions of the corresponding Euler–Lagrange equation. (Theorems which are conceptually similar appeared first in the theory of minimal surfaces, see for example [GI] for an overview.) A similar connection holds for boundary regularity, too (see [JN-C114]).

The Liouville condition can be hard to check. Nevertheless, the theorem is important from a conceptual point of view. Also, it was used by J. Nečas, I. Netuka and P. L. Lions to give new proofs of regularity results of K. Uhlenbeck for $F(x, u, DU) = f(|Du|)$ (see [JN-C113]), and J. Nečas used it to give a different proof of regularity results of A.I. Koshelev addressing the case when $Du \rightarrow$

[1]Later was given an example of an elliptic system of equations with unique singular solution of Dirichlet boundary value problem working in dimensions 3,4 in [JN-C104].

$F(x, u, Du)$ is not too far from a quadratic form. Many basic problems related to these ideas remain open.

For a short period of time, J. Nečas also paid attention to regularity for nonlinear parabolic systems. Even in this area his contribution cannot be overlooked. In 1991, he studied (with V. Šverák) a problem of regularity of weak solutions of the system

$$\frac{\partial u}{\partial t} = div(A(\nabla u)), \quad u = (u_1(z), u_2(z), \ldots, u_N(z)), \quad z = (x_1, x_2 \ldots x_n, t),$$

where A is a uniformly monotone mapping.[2]

In [JN-C156], which soon became well-known among specialists, Hölder continuity of the weak solutions is proved for $n \leq 4$ and Hölder continuity of the gradient is proved for $n \leq 2$ (implying full regularity in that case).

The method of proof is based on a non-trivial generalization of a well-known linear $L^{2+\epsilon}$ estimate of N. G. Meyers (1963) (see [MEY]) to the parabolic case. The new linear parabolic estimate enables the authors to move the time derivative to the right-hand side and treat the system by known elliptic techniques.

We remark that it still seems to be an open problem whether, for $n = 2$, one can improve the $L^{2+\epsilon}$ estimate for linear systems with bounded measurable coefficients to a Hölder estimate. (This would immediately give full regularity for the nonlinear problem.) It is known that this cannot work for $n = 3$, due to elliptic counter-examples.

Nečas' Works Quoted Above

The references [Cxx] refer to the respective papers listed in Chapter 2 "Complete Bibliography of Jindřich Nečas". The references [JN-Cyy] refer to papers [Cyy] that are reproduced in Chapter 7 "Selected Papers on Regularity Theory".

[C33] J. Nečas. Sur la régularité des solutions variationelles des équations elliptiques non-linéaires d'ordre $2k$ en deux dimensions. *Ann. Scuola Norm. Sup. Pisa (3)*, 21:427–457, 1967.

[C177] Wenge Hao, S. Leonardi, and J. Nečas. An example of irregular solution to a nonlinear Euler-Lagrange elliptic system with real analytic coefficients. *Ann. Scuola Norm. Sup. Pisa Cl. Sci. (4)*, 23(1):57–67, 1996.

[JN-C30] J. Nečas. Sur l'appartenance dans la classe $C^{(k),\mu}$ des solutions variationnelles des équations elliptiques non-linéaires de l'ordre $2k$ en deux dimensions. *Comment. Math. Univ. Carolinae*, 8:209–217, 1967.

[JN-C96] M. Giaquinta and J. Nečas. On the regularity of weak solutions to nonlinear elliptic systems via Liouville's type property. *Comment. Math. Univ. Carolin.*, 20(1):111–121, 1979.

[JN-C100] M. Giaquinta and J. Nečas. On the regularity of weak solutions to nonlinear elliptic systems of partial differential equations. *J. Reine Angew. Math.*, 316:140–159, 1980.

[2] When A merely satisfies the Legendre–Hadamard condition, regularity can completely fail even for solutions independent of t, see [MS].

[JN-C104] J. Nečas, O. John, and J. Stará. Counterexample to the regularity of weak solution of elliptic systems. *Comment. Math. Univ. Carolin.*, 21(1):145–154, 1980.

[JN-C113] P.-L. Lions, J. Nečas, and I. Netuka. A Liouville theorem for nonlinear elliptic systems with isotropic nonlinearities. *Comment. Math. Univ. Carolin.*, 23(4):645–655, 1982.

[JN-C114] M. Giaquinta, J. Nečas, O. John, and J. Stará. On the regularity up to the boundary for second order nonlinear elliptic systems. *Pacific J. Math.*, 99(1):1–17, 1982.

[JN-C156] J. Nečas and V. Šverák. On regularity of solutions of nonlinear parabolic systems. *Ann. Scuola Norm. Sup. Pisa Cl. Sci. (4)*, 18(1):1–11, 1991.

References

[ADN] S. Agmon, A. Douglis, L. Nirenberg. Estimates near the boundary for solutions to elliptic PDE's satisfying general boundary conditions I,II. *Comm. Pure Appl. Math.*12:623–727, 1959; 17:35–92, 1964.

[GI] E. Giusti. *Minimal surface and functions of bounded variation*. Birkhauser, Basel, 1984 Monographs in Mathematics

[GM] E. Giusti, M. Miranda. Un esempio di soluzioni discontinue per un problema di minimo relativo ad un integrale regolare del calcolo delle variazioni *Boll.U.M.I.* 2:1–8, 1968

[LU] O.A. Ladyzhenskaya, N.N.Uralceva. *Linear and quasilinear elliptic equations* Academic Press, New York, 1968

[MA] J. Malý. Nonisolated singularities of solutions to a quasilinear elliptic system *Comment. Math. Univ. Carolinae* 29(3):421–426,1988

[MEY] N.G. Meyers. On L_p estimates for the gradient of solutions of second order elliptic divergence equations *Ann. Scuola Norm. Sup. Pisa* ,17: 189–206,1963

[MS] S. Müller, V. Šverák. Convex integration for Lipschitz functions and counterexamples to regularity *Ann. Math.* (2)158: 1–23,2003

[SOU] J. Souček. Nonisolated singularities of solutions to a quasilinear elliptic systems *Comment. Math. Univ. Carolinae* 25:273–281, 1984

[SJ] J. Stará, O. John On some regularity and nonregularity results for solutions to parabolic systems *Le Matematiche* LV 2000-Supplemento n.2: 145–163, 2000

[SY] V. Šverák, X. Yan. Non-Lipschitz minimizers of smooth strongly convex functionals *Proc. Natl.Acad.Sci.USA* 99:15269–15276,2002

Jana Stará · Oldřich John
Faculty of Mathematics and Physics, Charles University in Prague, Sokolovská 83, 18675 Prague 8, Czech Republic
email: stara@karlin.mff.cuni.cz; john@karlin.mff.cuni.cz

8.2 Sketches of Regularity Theory from the Twentieth Century and the Work of Jindřich Nečas

Jan Kristensen and Giuseppe Mingione

Abstract The work of Jindřich Nečas on Partial Differential Equations is well-known and profound. In this survey article we discuss aspects of his work in the context of Regularity Theory.

8.2.1 Jindřich Nečas' Work in Regularity Theory

When reviewing the main aspects of general regularity theory for elliptic and parabolic problems, as it has evolved during the second part of the past century, it is remarkable to note how it has developed along the dichotomy between regularity and singularities. In turn, this can be understood by looking at a closely related dichotomy, namely the one between scalar and vectorial problems (meaning problems with one and many unknown real-valued functions, respectively). While, after the pioneering contributions of Almgren, De Giorgi, Morrey, Nash related to Hilbert's 19th problem, several mathematicians devoted themselves to the study of one of the two aspects of the above dichotomies, the work of Nečas interestingly moved along the borderline between the two aspects in, what turned out to be, a particularly fruitful way. Indeed, while investigating vectorial problems—where, as we shall see, singularities naturally occur—he was particularly interested in those conditions that ensure everywhere regularity, thereby revealing interesting connections between certain vectorial structures and properties enjoyed by solutions to scalar problems. On the other hand, moving from such a viewpoint, he developed a deep insight into the nature of singularities of solutions to vectorial problems that allowed him to construct fundamental counterexamples and to open the way to later important developments. In this paper we shall try to describe those aspects of regularity theory for elliptic and parabolic problems that are related to the work of Jindřich Nečas, and, on the other hand, those parts of Nečas' multifaceted work that mostly relate to general regularity theory. In doing this we shall also take the opportunity to describe some of the results of a group of well-known and appreciated analysts who scientifically grew up in Prague under the general guidance of Nečas. Indeed, an important aspect of Nečas' professional activity was devoted to build what is nowadays recognized as the Prague school of PDEs. Nečas' deep insight, kindness, enthusiasm and exceptional abilities, allowed him to attract many of the most talented young Czech mathematicians and to train a group that subsequently has produced an impressive number of important and fundamental results over the last 40 years.

Of course, and unavoidably, we shall by no means aim at completeness in describing general aspects of regularity theory nor in the description of the work of Nečas. Rather, we shall, more modestly, provide some personal viewpoints on some of the main issues of regularity theory relevant in this context, and in particular on the work of Jindřich Nečas, a master to the memory of whom this manuscript is dedicated.

8.2.2 Regularity and Singularities

After the first results in low dimensions due to a distinguished group of mathematicians including Agmon, Bernstein, Caccioppoli, Douglis, Morrey, Nirenberg and Stampacchia, the breakthrough allowing for the proof of everywhere regularity came with the work of De Giorgi [18], followed by the one of Nash [80]. De Giorgi's brilliant insights—with new proofs of Moser [77, 78]—eventually opened the way to the general nonlinear theory, as described in [21, 38, 58, 63]. As for the vectorial case, the theory has been shaped again by the initial insight of De Giorgi [19] in the context of minimal surfaces theory. Indeed, methods inspired by those of De Giorgi proved to be fundamental in establishing the so-called partial regularity theory both for elliptic systems and variational integrals (see [34, 39, 71]).

In this section we give a rapid overview of those general aspects of regularity theory which help in appreciating and understanding the relevance and the historical context of Nečas' work in regularity theory. For more detailed and exhaustive information on regularity theory we refer to the classical books [10, 21, 34, 38, 39, 58, 59, 75], and to the more recent survey paper [71]. To clarify the role of the regularity theory in the more general context of general PDEs methods we recommend that the reader takes a look at [11].

8.2.2.1 Overview of General Regularity Theory

We shall consider variational integrals of the type

$$\mathcal{F}(v, A) := \int_A F(x, v, Dv)\, dx \tag{8.1}$$

defined for Sobolev maps $v \in W^{1,p}_{\mathrm{loc}}(\Omega, \mathbb{R}^N)$, and open sets A whose closure is compact and contained in Ω. Here $n \geq 2$, $N \geq 1$, Ω is a bounded open set in $\mathbb{R}^n$, $p \geq 1$, and $F\colon \Omega \times \mathbb{R}^N \times \mathbb{R}^{Nn} \to \mathbb{R}$ is an integrand, for simplicity assumed to be measurable with respect to the first variable, and continuous with respect to the remaining two. We shall also write $\mathcal{F} \equiv \mathcal{F}(v) \equiv \mathcal{F}(v, \Omega)$. A local minimizer of the functional $\mathcal{F}$ is a map $u \in W^{1,p}_{\mathrm{loc}}(\Omega, \mathbb{R}^N)$ such that $\mathcal{F}(u, A) \leq \mathcal{F}(v, A)$, whenever $A \subset\subset \Omega$ and $u - v \in W^{1,p}_0(A, \mathbb{R}^N)$. As for equations and systems, we shall consider those of quasilinear type, i.e.

$$\operatorname{div} a(x, u, Du) = b(x, u, Du)\ , \tag{8.2}$$

where $a\colon \Omega \times \mathbb{R}^N \times \mathbb{R}^{Nn} \to \mathbb{R}^{Nn}$ and $b\colon \Omega \times \mathbb{R}^N \times \mathbb{R}^{Nn} \to \mathbb{R}^N$ are vector fields, again assumed to be measurable with respect to the first variable, and continuous with respect to the remaining two. Indeed, when the integrand $F(x,v,z)$ of $\mathcal{F}$ in (8.1) is regular enough, minimizers of $\mathcal{F}$ solve the so-called Euler–Lagrange system associated to $\mathcal{F}$

$$\operatorname{div} F_z(x,u,Du) = F_v(x,u,Du)\,, \tag{8.3}$$

which turns out to be elliptic provided $F(x,v,z)$ satisfies suitable convexity assumptions with respect to z, see [34, Chaps. 1 and 2]. The symbol F_z denotes of course the partial derivative of F with respect to the gradient variable z. In the following, when referring to nonlinear elliptic systems, we shall for simplicity treat the simpler cases such as

$$\operatorname{div} a(x,u,Du) = 0\,. \tag{8.4}$$

This is not too restrictive, as, following a more recent approach, in several regularity respects, minimizers of functionals (8.1) can often be treated without appealing to their Euler–Lagrange systems (8.3).

As mentioned above, the key regularity results of De Giorgi, Nash and Moser opened the way to the regularity theory of solutions to nonlinear problems, which initially belonged only to Sobolev spaces. Let us consider a linear elliptic equation in divergence form with bounded and measurable coefficients:

$$D_i(a_{i,j}(x)D_j u) = 0\,,$$

that is, in its weak formulation

$$\int_\Omega a_{i,j}(x) D_j u D_i \varphi \, dx = 0\,, \qquad \forall\, \varphi \in C_c^\infty(\Omega)\,, \tag{8.5}$$

where as usual the summation convention of summing over repeated indices is in force. The coefficient matrix $\{a_{i,j}(x)\}$ is assumed to be real and symmetric, with bounded and measurable entries satisfying the boundedness and ellipticity conditions:

$$|a_{i,j}(x)| \le L\,, \qquad a_{i,j}(x)\lambda_i\lambda_j \ge \nu|\lambda|^2\,, \tag{8.6}$$

for almost every $x \in \Omega$, and every $\lambda \in \mathbb{R}^n$, while, here as in the following we shall consider $0 < \nu \le L < \infty$ as constants. We then have the following

Theorem 8.1 (De Giorgi). *Let $u \in W^{1,2}(\Omega)$ be a weak solution to (8.5), under the assumptions (8.6). Then there exists a positive number $\alpha \equiv \alpha(n, L/\nu) > 0$ such that $u \in C^{0,\alpha}_{\mathrm{loc}}(\Omega)$.*

Under the additional assumption that the solution is bounded the previous result was independently obtained by Nash [80], who also treated the parabolic case (a case that can be treated by De Giorgi's methods as well, as shown in [59]). Later on a different proof was given by Moser [77,78]. It is interesting to note that De Giorgi's proof is so robust that it equally well applies to nonlinear equations. In particular, the result extends to more general quasilinear equations of the type in (8.4) under the following growth and monotonicity assumptions:

$$|a(x,v,z)| \le L(1+|z|)^{p-1}\,, \qquad \nu|z|^p - L \le \langle a(x,v,z), z\rangle\,, \tag{8.7}$$

for every $x \in \Omega$, $v \in \mathbb{R}$, and $z \in \mathbb{R}^n$, with $p > 1$ and $0 < \nu \le L < \infty$, a notation we shall keep in the sequel.

Theorem 8.2. *Let $u \in W^{1,p}(\Omega)$ be a weak solution to (8.4), under the assumptions (8.7). Then there exists a positive number $\alpha \equiv \alpha(n,p,L/\nu) > 0$ such that $u \in C^{0,\alpha}_{\mathrm{loc}}(\Omega)$.*

We refer to [21, 59] for the analogous results in the parabolic context, including equations of the type $u_t - \operatorname{div} a(x,u,Du) = 0$.

Concerning minimizers of functionals, as already mentioned above, a key point is that a direct appeal to minimality allows us to by-pass the use of the Euler–Lagrange system (8.3). This is crucial: in fact, under the assumptions (8.12) below, that are natural and standard in the Calculus of Variations, the Euler–Lagrange system might not even exist as F_u does not make sense. Let us give a brief outlook to some of the results that can be obtained without directly using (8.3), but rather using minimality and the coercivity and growth conditions

$$\nu|z|^p \le F(x,y,z) \le L(1+|z|)^p\,. \tag{8.8}$$

A first result is the following, that was first observed in [31] in a special case, and then carried out in [35] in full generality.

Theorem 8.3 ([31,35]). *Let $u \in W^{1,p}_{\mathrm{loc}}(\Omega)$ be a local minimizer of the functional $\mathcal{F}$, under the assumptions (8.8). Then there exists a positive number $\alpha \equiv \alpha(n,p,L/\nu)$ such that $u \in C^{0,\alpha}_{\mathrm{loc}}(\Omega)$.*

The remarkable fact in the previous result is that no pointwise smoothness or convexity properties are required on the integrand $F(\cdot)$ to ensure the local Hölder continuity of minimizers.

Proving higher regularity for minimizers and weak solutions requires correspondingly higher smoothness of the involved integrands and systems, respectively. The focal point of regularity is the continuity of the gradient; starting from this it is indeed possible to prove higher regularity—up to analyticity—by standard bootstrap methods. Therefore we shall here confine ourselves to describe those results dealing with $C^{1,\alpha}$-regularity. As far as equations and systems are concerned we shall consider the following set of assumptions:

$$\begin{cases} z \mapsto a(x,v,z) \quad \text{is} \quad C^1 \quad \text{and} \quad (x,v,z) \mapsto a_z(x,v,z) \quad \text{is continuous} \\ |a(x,v,z)| \le L(1+|z|)^{p-1} \\ \nu(\mu^2+|z|^2)^{\frac{p-2}{2}}|\lambda|^2 \le \langle a_z(x,v,z)\lambda,\lambda\rangle \le L(\mu^2+|z|^2)^{\frac{p-2}{2}}|\lambda|^2 \\ |a(x,u,z)-a(y,v,z)| \le L\omega(|x-y|+|u-v|)(1+|z|)^{p-1}, \end{cases} \tag{8.9}$$

for all $x, y \in \Omega$, $u, v \in \mathbb{R}$ and $z, \lambda \in \mathbb{R}^n$, where $\mu \in [0,1]$ is a fixed constant and $\omega\colon \mathbb{R}^+ \to [0,1]$ is a nondecreasing function such that $\omega(0) = 0$ used to measure the degree of continuity with respect to the coefficients (x, v) (often called a "modulus of continuity"). The exponent p satisfies $1 < p < \infty$. Concerning the modulus of continuity, we impose one of the following two conditions,

$$\omega(\varrho) \le \varrho^\alpha \tag{8.10}$$

for some $\alpha \in (0,1)$, or merely,

$$\lim_{\varrho\to 0} \omega(\varrho) = 0\,. \tag{8.11}$$

The former corresponds to Hölder continuous dependence of $a(\cdot)$ with respect to the coefficients (x, v), while assuming only the latter means that such a dependence is uniformly continuous.

The corresponding assumptions for functionals:

$$\begin{cases} z \mapsto F(x,v,z) \quad \text{is} \quad C^2 \quad \text{and} \quad (x,v,z) \mapsto F_{zz}(x,v,z) \quad \text{is continuous} \\ \nu|z|^p \le F(x,v,z) \le L(1+|z|)^p \\ \nu(\mu^2+|z|^2)^{\frac{p-2}{2}}|\lambda|^2 \le \langle F_{zz}(x,v,z)\lambda,\lambda\rangle \le L(\mu^2+|z|^2)^{\frac{p-2}{2}}|\lambda|^2 \\ |F(x,u,z)-F(y,v,z)| \le L\omega(|x-y|+|u-v|)(1+|z|)^p, \end{cases} \tag{8.12}$$

in a similar notation to that employed for (8.9). The parameter μ serves to distinguish the non-degenerate case ($\mu > 0$) from the degenerate one ($\mu = 0$). In the latter case we for instance encounter functionals such as

$$v \mapsto \int_\Omega c(x,v)|Dv|^p\,dx$$

whose Euler–Lagrange system for $c(\cdot) = 1$ is the celebrated p-Laplacean system

$$\operatorname{div}(|Du|^{p-2}Du) = 0. \tag{8.13}$$

For this system, regularity was first proved in [96] and [95], in the scalar and in the vectorial cases, respectively. Observe again that the assumptions considered in (8.12) do not allow us to reduce the treatment of minima to that of solutions to quasilinear systems via (8.3), in particular because the integrands may be non-differentiable with respect to v. We refer to [50] for more details. The available regularity results in the case of equations can be summarized as follows:

Theorem 8.4. *Let $u \in W^{1,p}(\Omega)$ be a weak solution to (8.4), under the assumptions (8.9).*

- *[22, 36, 64, 96] If (8.10) holds for some $\alpha \in (0,1)$ then Du is locally Hölder continuous in Ω.*
- *[64] If (8.11) holds then we have that $u \in C^{0,\alpha}_{\mathrm{loc}}(\Omega)$ for every $\alpha \in (0,1)$.*

The corresponding theorem for minima is

Theorem 8.5. *Let $u \in W^{1,p}(\Omega)$ be a local minimizer to the functional in (8.1), under the assumptions (8.12).*

- *[36, 44, 64] If (8.10) holds for some $\alpha \in (0,1)$, then Du is locally Hölder continuous in Ω.*
- *[64] If (8.11) holds, then we have that $u \in C^{0,\alpha}_{\mathrm{loc}}(\Omega)$ for every $\alpha \in (0,1)$.*

Notice that in the previous theorem one cannot expect that the Hölder continuity exponent of the gradient is precisely the exponent α in (8.10), because the previous results also cover the degenerate case $\mu = 0$. Indeed, already in the case of the functional (8.13) minimizers do not in general belong to $C^{1,\alpha}$ for every $\alpha < 1$, as initially shown again by Uraltseva; see [61] for further counterexamples.

Theorems 8.4 and 8.5 virtually summarize the efforts of several authors, starting from the fundamental techniques developed in [18, 58, 77, 78, 91] over a period of almost 30 years, and the references given here mainly concern those papers where the statement can be found in a form which is essentially the optimal one stated here.

When passing to the vectorial case $N > 1$ the situation drastically changes, and solutions are in general no longer everywhere regular. In a sense this could be expected; indeed, for instance several multidimensional vectorial variational problems considered stem from physical situations where singularities naturally occur (for instance cavitation phenomena in nonlinear elasticity). The first examples of vectorial problems featuring singular solutions were given by De Giorgi [20] and Maz'ya [65]. The next section will be devoted to this aspect. What is possible to prove in the vectorial case is instead partial regularity, that is *the regularity of minima and weak solutions outside a negligible closed subset* referred to as the *singular set*. Besides proving such partial regularity one would like to find size (that is, dimensional) estimates and structural properties of the singular set; for this we refer to Sect. 8.2.2.2 below. Partial regularity results and methods are essentially based on linearization techniques around points—the so-called regular ones—where the oscillations of solutions or of their gradients, are a priori known to be suitably small. In fact, they are nothing but the Lebesgue points of solutions or their gradients, and the almost everywhere regularity results stem from the fact that the Lebesgue points indeed form a set of full measure (see Remark 8.1 below).

The linearization techniques employed to prove such partial regularity theorems were pioneered by De Giorgi in his fundamental work on minimal surfaces [19] and eventually by Almgren [3] in the setting of Geometric Measure Theory. Eventually, these methods were transferred to the case of elliptic systems, and the first results can be retrieved in the papers of Giusti and Miranda [41] and Morrey [76]. In the following we report on a couple of theorems which summarize the recent status of partial regularity for general functionals and systems.

Theorem 8.6. *Let $u \in W^{1,p}(\Omega, \mathbb{R}^N)$ be a weak solution to the system (8.4), under the assumptions (8.9) with $\mu > 0$.*

- *[25, 37, 43] If (8.10) holds for some $\alpha \in (0, 1)$, then there exists an open subset $\Omega_u \subset \Omega$ such that*

$$|\Omega \setminus \Omega_u| = 0 \qquad \textit{and} \qquad Du \in C^{0,\alpha}_{\mathrm{loc}}(\Omega_u, \mathbb{R}^{Nn}).$$

- *[30] If (8.11) holds, then there exists an open subset $\Omega_u \subset \Omega$ such that*

$$|\Omega \setminus \Omega_u| = 0 \quad \textit{and} \quad u \in C^{0,\alpha}_{\mathrm{loc}}(\Omega_u, \mathbb{R}^N) \textit{ holds for every } \alpha \in (0, 1).$$

Again, the corresponding results for minimizers are given in the following:

Theorem 8.7. *Let $u \in W^{1,p}(\Omega, \mathbb{R}^N)$ be a local minimizer of the functional (8.1), under the assumptions (8.12) with $\mu > 0$.*

- *[36, 44] If (8.10) holds for some $\alpha \in (0, 1)$, then there exists an open subset $\Omega_u \subset \Omega$ such that*

$$|\Omega \setminus \Omega_u| = 0 \qquad \textit{and} \qquad Du \in C^{0,\alpha/2}_{\mathrm{loc}}(\Omega_u, \mathbb{R}^{Nn}).$$

- *[30] If (8.11) holds then there exists an open subset $\Omega_u \subset \Omega$ such that*

$$|\Omega \setminus \Omega_u| = 0 \quad \textit{and} \quad u \in C^{0,\alpha}_{\mathrm{loc}}(\Omega_u, \mathbb{R}^N) \textit{ holds for every } \alpha \in (0, 1).$$

Notice that the results in [30] are optimal in the sense that when the dependence on coefficients is merely continuous partial Hölder continuity of u holds, as shown in the last two theorems, but that of the gradient fails in the sense that examples can be constructed, and already in the one-dimensional case, of minima u that fail to be differentiable on a dense set, as eventually shown in [42].

Partial regularity extends to the boundary, under suitable restrictions on the structure of the system/functional, and on the Hölder continuity exponent α, which in general is assumed to be larger than $1/2$. Initially treated only in the case of systems of the type

$$\operatorname{div}(A(x, u)Du) = 0\,, \tag{8.14}$$

a general theory of boundary regularity for quasilinear systems as in (8.4) and functionals as in (8.1) has been developed only recently in the papers [26, 52], to which we refer for results and details.

Remark 8.1. For the following it is important to remark that in Theorems 8.6 and 8.7 the regular points $x_0 \in \Omega$—i.e., those at which *the solution is close enough* to an affine map, meaning that for some $M > 0$ it holds that

$$\fint_{B_R(x_0)} |Du|^2 \, dx \le M \quad \text{and} \quad \fint_{B_R(x_0)} |Du - (Du)_{B_R(x_0)}|^2 \, dx \le \varepsilon(M)\,. \qquad (8.15)$$

When considering simpler systems of the type (8.14), conditions in (8.15) can be relaxed in

$$\fint_{B_R(x_0)} |u|^2 \, dx \le M \quad \text{and} \quad \fint_{B_R(x_0)} |u - (u)_{B_R(x_0)}|^2 \, dx \le \varepsilon(M)\,. \qquad (8.16)$$

8.2.2.2 Higher Integrability and Singular Sets

One of the most important and versatile regularity properties of weak solutions to elliptic vectorial problems, is the so-called higher integrability. In turn, this reveals itself to be a crucial starting point for establishing further regularity properties of solutions, like for instance the everywhere Hölder continuity in low dimensions, and more generally the inference of estimates on the singular sets of solutions.

The main theorem can be summarized as follows:

Theorem 8.8 ([28, 37]). *Let $u \in W^{1,p}(\Omega, \mathbb{R}^N)$ be a weak solution to the system (8.4), under the assumptions (8.7). Then there exists an exponent $s(n, p, L/\nu) > p$ such that $Du \in L^s_{\text{loc}}(\Omega, \mathbb{R}^{Nn})$.*

As for local minimizers, after a preliminary result in [5] obtained under additional assumptions, the following general theorem was obtained in [35].

Theorem 8.9 ([5,35]). *Let $u \in W^{1,p}(\Omega, \mathbb{R}^N)$ be a local minimizer of the functional $\mathcal{F}$, under the assumption (8.8). Then there exists an exponent $s(n, p, L/\nu) > p$ such that $Du \in L^s_{\text{loc}}(\Omega, \mathbb{R}^{Nn})$.*

Both proofs of Theorems 8.8 and 8.9 rely on the fundamental self-improving integrability properties established in [33] (the so-called "Gehring's lemma" [45]). We also recall that the non-trivial parabolic extension of Theorem 8.8 has been obtained in [47].

Theorems 8.6–8.7 immediately raise the question of the size of the singular set $\Omega \setminus \Omega_u$, for instance in terms of its Hausdorff dimension $\dim_{\mathcal{H}}(\Omega \setminus \Omega_u)$.

Theorem 8.10 ([69, 70]). *Let $u \in W^{1,p}(\Omega, \mathbb{R}^N)$ be a weak solution to the system (8.4), under the assumptions (8.9) with $\mu > 0$. Then*

$$\dim_{\mathcal{H}}(\Omega \setminus \Omega_u) \le n - \min\{2\alpha, s - p\}\,, \qquad (8.17)$$

where $s > p$ is the higher integrability exponent appearing in Theorem 8.8. Moreover if $n \leq p+2$, or if the vector field $a(\cdot)$ is independent of u, then

$$\dim_{\mathcal{H}}(\Omega \setminus \Omega_u) \leq n - 2\alpha\,. \tag{8.18}$$

The corresponding theorem for minimizers is given by

Theorem 8.11 ([50]). *Let $u \in W^{1,p}(\Omega, \mathbb{R}^N)$ be a local minimizer of the functional (8.1), under the assumptions (8.9) with $\mu > 0$. Then*

$$\dim_{\mathcal{H}}(\Omega \setminus \Omega_u) \leq n - \min\{\alpha, s-p\}\,, \tag{8.19}$$

where $s > p$ is the higher integrability exponent appearing in Theorem 8.9. Moreover if $n \leq p+2$, then

$$\dim_{\mathcal{H}}(\Omega \setminus \Omega_u) \leq n - \alpha\,. \tag{8.20}$$

Further results can be obtained under additional structure conditions; for instance we report the following theorem, that concerns integrands with measurable coefficients:

Theorem 8.12 ([50]). *Let $u \in W^{1,p}(\Omega, \mathbb{R}^N)$ be a local minimizer of the functional (8.1)*

$$v \mapsto \int_\Omega f(x, Dv) + g(x, v)\, dx$$

where $f(\cdot)$ satisfies assumptions (8.12) and $g(\cdot)$ is a bounded, Carathéodory regular, being α-holder continuous in second argument. Then

$$\dim_{\mathcal{H}}(\Omega \setminus \Omega_u) \leq n - \alpha\,.$$

Remark 8.2. We notice that better dimensional bounds appear in the case $n \leq p+2$. This is in turn linked to the fact that in low dimensions elliptic systems have additional regularity properties. See Sect. 8.2.4 below.

Remark 8.3. In the vectorial case, a natural condition that replaces convexity in order to obtain existence theorems is quasiconvexity, originally introduced by Morrey in [74]. This condition is essentially equivalent to sequential semicontinuity in the natural weak topologies of Sobolev spaces (see [1, 74]) and is also used in mathematical materials science and nonlinear elasticity (see [6]). Also in the case of quasiconvexity it is possible to prove partial regularity theorems and singular sets estimates (see [29, 51]), while an essential difference occurs between minimizers (regular outside a negligible closed set) and solutions to the related Euler–Lagrange systems, which in general do not enjoy partial regularity properties (see [79]).

Remark 8.4. Several of the results in this and in the previous section have, up to a certain extent, a parabolic analog; we refer to [27] for a panoramic view and for a list of references.

8.2.3 Singularities: Nečas' Insight and Heritage

Nečas gave striking and important contributions towards the understanding of the problem of singularities of minima and of solutions to nonlinear elliptic systems. In this section we are going to review his work in the historical context, eventually describing some developments of his work, mainly due to his collaborators and former students.

In the following we often make use of the summation convention (summing over repeated indices).

8.2.3.1 De Giorgi's Example

In [20] De Giorgi considered the following quadratic-type functional with discontinuous coefficients:

$$\mathcal{DG}(u) := \int_{B_1} |Du|^2 + \left[(n-2) \sum_{i=1}^{n} D_i u^i + n \sum_{i,j=1}^{n} \frac{x_i x_j}{|x|^2} D_i u^j \right]^2 dx \,, \qquad n = N \,. \tag{8.21}$$

For $n \geq 3$ the map

$$u(x) := \frac{x}{|x|^\alpha} \,, \qquad \alpha := \frac{n}{2}\left[1 - \frac{1}{\sqrt{(2n-2)^2+1}} \right] \,, \tag{8.22}$$

belongs to $W^{1,2}(B_1, \mathbb{R}^n)$, and minimizes $\mathcal{DG}$. We observe that it is necessary to take discontinuous dependence on x in the integrand, otherwise one can use perturbation methods to show that the solution is locally Hölder continuous with any exponent $\alpha < 1$. No such example is possible when $n = 2$, since in this case Hölder continuity of solutions, for some exponent $\alpha < 1$, follows by Sobolev's embedding theorem and the higher integrability afforded by Theorem 8.9 below. We remark that independently, Maz'ya [65] found examples of higher-order elliptic equations, with real analytic coefficients, but with discontinuous solutions.

8.2.3.2 Giusti and Miranda's Example

The main point in De Giorgi's example is that the integrand considered exhibits coefficients with a singularity at the origin; on the other hand they are allowed to depend only on the space variable x. Instead, when the coefficients depend on the solution itself Giusti and Miranda showed in [40] that the dependence can be even analytic. They considered the quadratic-type functional

$$\mathcal{GM}(v) := \int_{B_1} A^{ij}_{\alpha\beta}(v) D_j v^\alpha D_i v^\beta \, dx \,, \qquad n = N \,, \tag{8.23}$$

with

$$A^{ij}_{\alpha\beta}(v) := \delta_{i,j}\delta_{\alpha,\beta} + \left[\delta_{\alpha,i} + \frac{4}{n-2}\cdot\frac{v_\alpha v_i}{1+|v|^2}\right]\cdot\left[\delta_{\beta,j} + \frac{4}{n-2}\cdot\frac{v_\beta v_j}{1+|v|^2}\right] .$$

Here $\delta_{i,j}$ denotes the usual Kronecker's symbol. For $n > 2$ sufficiently large, the discontinuous map

$$u(x) := \frac{x}{|x|} , \tag{8.24}$$

minimizes $\mathcal{GM}$. Similar examples also work for quasilinear systems of the type

$$\int_\Omega A^{ij}_{\alpha\beta}(u) D_j u^\alpha D_i \varphi^\beta \, dx = 0 , \qquad \forall\, \varphi \in C^\infty_c(\Omega, \mathbb{R}^n) . \tag{8.25}$$

The key to understand why in this example real analytic coefficients are allowed, when compared to the example of De Giorgi in Sect. 8.2.3.1, is that if we think of the identification $B^{ij}_{\alpha\beta}(x) \equiv A^{ij}_{\alpha\beta}(u(x))$, then the map $x \mapsto B^{ij}_{\alpha\beta}(x)$ can be discontinuous. Moreover, the example of Giusti and Miranda is built in a way that when one composes the solution in (8.24) with the matrix $a(v)$ one obtains a structure similar to the one of De Giorgi's counterexample, and this is the key to obtain the counterexample.

8.2.3.3 Nečas' Example

In the previous examples the singularity of minimizers occur by the peculiar way the coefficients x or v mix-up with the components of the gradient variable z. It was then a very challenging open problem to understand whether it was possible to construct examples of singular minimizer to regular variational integrals

$$v \mapsto \int_\Omega F(Dv)\, dx \tag{8.26}$$

with convex and analytic dependence on the gradient variable, and satisfying for instance assumptions (8.12) (together with similar growth and ellipticity bounds on the higher order derivatives of $F(\cdot)$) with $p = 2$. The quest for a such a high regularity of the integrand is clearly linked to the traditional line of thought that connects regularity theory to Hilbert's 19th problem, that is indeed concerned with the regularity of solutions to variational problems involving analytic integrands.

A major breakthrough was achieved by Nečas initially in [JN-C86], providing an example of such a minimizer with a discontinuous (but still bounded) gradient. In [JN-C86] Nečas provided actually two examples. The first is about a real analytic variational integral as described above with a singular minimizer $u\colon \Omega\,(\subset \mathbb{R}^n) \to \mathbb{R}^{n^2}$ for large n. The second concerns an elliptic system of the type

$$\operatorname{div} a(Du) = 0 \tag{8.27}$$

again with a real analytic vector field $a\colon \mathbb{R}^{n^3} \to \mathbb{R}^{n^3}$ satisfying assumptions (8.9) for the case $p = 2$, and without (x, v)-dependence, having a non-C^1 weak solution for $n \geq 5$. The basic idea of Nečas is to start with the mapping that should be the minimizer and then construct the integrand F, or the structure of the vector field $a(\cdot)$, that makes it happen. His candidate minimizer/solution is the mapping

$$u = (u^{ij}), \quad u^{ij}(x) := \frac{x_i x_j}{|x|}, \tag{8.28}$$

which exhibits a high gradient anisotropy ("large sets of gradients") and then he constructs an integrand such that (8.31) becomes a (unique) minimizer/solution. It is rather clear that the approach of Nečas has an intrinsic *combinatorial nature* that will influence subsequent developments. As a matter of fact, in his constructions, Nečas always used the map (8.28), and a closely related one, see (8.31) below, as a starting candidate to be a solution/minimizer, and then *constructed suitable integrals and systems* to conclude with the counterexample. The necessity of using a map with a large set of gradients, and consequently to use systems exhibiting a wide anisotropy (indeed in Nečas words) goes back to De Giorgi's ideas. Indeed, on one hand systems which are too close to be diagonal, as for instance those considered in Sect. 8.2.4 below, typically force solutions to be regular. On the other hand, solutions whose image is close to the image of a single real-valued function are always regular. Indeed, when considering a simpler system of the type

$$\operatorname{div}(A(x,u)Du) = 0$$

it happens that if in a given ball $B_R(x_0)$ the range of values of the solution is close to that of a real-valued function in the sense that

$$⨍_{B_R(x_0)} |u - \xi|^2 \, dx - ⨍_{B_R(x_0)} |\langle u - \xi, \xi_0 \rangle|^2 \, dx \quad \text{is small}$$

for suitable vectors ξ, ξ_0 with $|\xi_0| = 1$, then no discontinuity appears in a neighborhood of x_0 (see [34, Chap. 4, Theorem 3.1]).

We remark that the occurrence of maps as for instance the ones in (8.24) and (8.31) is also typical when studying for instance the formation of singularities in nonlinear elasticity (for instance, cavitation). As a matter of fact, the examples above admit an interpretation in the context of elasticity (see for instance [85]).

The astonishing results in [JN-C86] still left some natural questions open, and we emphasize in particular the following two (recall we refer to assumptions (8.9) with $p = 2$):

- The allowed dimensions to consider in order to find singular minimizers/solutions are described in the following scheme:

$$\begin{cases} u \notin L^\infty_{\text{loc}} & \text{if } n \geq 5 \\ Du \notin L^\infty_{\text{loc}} & \text{if } n \geq 3\,. \end{cases} \tag{8.29}$$

- The counterexamples in [JN-C86] fail to be C^1 but are still Lipschitz regular, so the work in [JN-C86] still leaves open the possibility that minimizers, and perhaps even weak solutions to quasilinear systems, could be locally Lipschitz.

Notice that the scheme above is motivated by the fact that solutions to (8.27) under assumptions (8.9) (for $\mu > 0$ and $p = 2$) have the property

$$u \in W^{2,q}_{\text{loc}} \qquad \text{for some } q > 2\,,$$

and therefore the Sobolev–Morrey embedding theorem implies that u and Du are locally Hölder continuous when $n \leq 4$ and $n = 2$, respectively. See Sect. 8.2.4 below for more on these types of results.

For these reasons Nečas still kept on thinking about the problem of singularities and over the years he produced a number of further examples. In doing this he stimulated a rather large and comprehensive study of singularities at Prague, that in turn involved several well-known mathematicians including John, Malý, Souček, Stará. Later on, fundamental contributions would be given by Šverák; see Sect. 8.2.3.5 below.

To reveal the *combinatorial complexity* of Nečas' approach we shall now report in detail on a couple of constructions. In the paper [JN-C104], with John and Stará, the authors gave an example of a system of the type (8.27) exhibiting a non-C^1 solution for $n \geq 3$. Specifically, the system considered in [JN-C104] is

$$\begin{aligned} \sum_{k=1}^{n} D_k \Big\{ & D_k u^{i,j} + \gamma \left(\delta_{ij} \nabla_k u + \delta_{ik} D_j u^{k,k}\right) + \lambda \nabla_i u \nabla_j u \nabla_k u \left[1 + |\nabla u^2|\right]^{-1} \\ & + \delta_{ik} \nabla_j u \left\{\gamma(4 + 3\gamma(n+2)) + 3\gamma\lambda |\nabla u|^2 [1 + |\nabla u|^2]^{-1} \right. \\ & \left. + \nu |\nabla u|^4 [1 + |\nabla u|^2]^{-2}\right\}\Big\} = 0\,. \end{aligned} \tag{8.30}$$

The solution is this time provided by the map

$$u = (u^{ij}), \quad u^{i,j}(x) := \frac{x_i x_j}{|x|} - \frac{\delta_{ij}|x|}{n}\,, \tag{8.31}$$

where $\delta_{i,j}$ denotes the usual Kronecker's symbol. As for the operators and constants appearing in the previous systems, we have

$$\nabla_j u := D_i u^{j,i}\,, \qquad \nabla_{ijk} u = D_k u^{i,j} + \gamma \left(\delta_{ij} \nabla_k u + \delta_{ik} \nabla_j u + \delta_{jk} \nabla_i u\right)$$

and the numbers λ, ν satisfy

$$\lambda := \left[1+\left(n-\frac{1}{n}\right)^2\right]\left(n-\frac{1}{n}\right)^{-2}\left(\frac{1}{n-1}-\gamma\right)$$

and

$$\nu = -\left(n-\frac{1}{n}\right)^{-5}\left\{3\gamma^2(n+1)\left(n-\frac{1}{n}\right)+\gamma(n^2+3n+3)+1+\frac{1}{n}\right\}\cdot \\ \cdot\left[1+\left(n-\frac{1}{n}\right)^2\right]^2$$

for a suitable choice of γ. We notice that further examples developed by Nečas are in the joint paper [JN-C100] with Giaquinta, and in the lecture notes [B8].

While the paper [JN-C104] gave an answer to the dimensional problem for the case of systems, it still remained the problem of lowering the dimension on the case of functionals. An interesting contribution is the one Nečas provided in [C177] together with Hao and Leonardi, where, in the case $n \geq 5$, once again the map (8.31) is shown to be a minimizer of a regular variational integral. Specifically, in [C177] the authors considered a functional satisfying (8.12) with $p = 4$ given as

$$\begin{aligned} v \mapsto \frac{\alpha}{4}\int_{B_1}|Dv|^4\,dx + \frac{\alpha_0}{4}\int_{B_1}|Dv|^2\,dx + \frac{\alpha_1}{4}\int_{B_1}\nabla_m v\nabla_m v\nabla_{ijk}v\nabla_{ijk}v\,dx \\ +\frac{\alpha_2}{4}\int_{B_1}(\nabla_{ijk}v\nabla_{ijk}v)^2\,dx + \frac{\alpha_3}{4}\int_{B_1}(\nabla_i v\nabla_i v)^2\,dx \\ +\alpha_4\int_{B_1}\nabla_i v\nabla_j v\nabla_k v\nabla_{ijk}v\,dx + \alpha_5\int_{B_1}\nabla_i v\tau_i v\,dx + \frac{\alpha_6}{2}\int_{B_1}\tau_i v\tau_i v\,dx \end{aligned} \tag{8.32}$$

where the operators ∇ and τ are this time defined as follows:

$$\tau_i v := D_i v^{j,j}\,, \qquad \nabla_i v := \frac{1}{2}\left(D_j v^{i,j} + D_j v^{j,i} + \tau_i v\right),$$

$$\begin{aligned} \nabla_{ijk}v := \frac{1}{6}\left(D_k v^{i,j} + D_j v^{k,i} + D_i v^{j,k} + D_k v^{k,j} + D_j v^{i,k}\right) \\ +\frac{2n-1}{3(n^2-1)}\left(\delta_{ij}\nabla_k v + \delta_{ki}\nabla_j v + \delta_{jk}\nabla_i v\right) \end{aligned}$$

where as usual δ_{ij} is Kronecker's symbol. As for the values of the constants involved we have

$$\alpha := \frac{n}{(2n^2-n-1)^3}, \qquad \alpha_0 := \left(\frac{1}{n^2-1}\right)^2,$$

$$\alpha_1 := \begin{cases} \dfrac{n^4}{(n^2+2)(n^2-1)} & \text{if } n \le 50 \\ 0 & \text{if } n > 50, \end{cases}$$

$$\alpha_2 := n^2 - 1 - \left(\frac{n^2-1}{n}\right)^2 \alpha_1,$$

$$\alpha_3 := \left(\frac{n}{n^2-1}\right)^3 \left[-\alpha_1 \frac{n^2-1}{n} + 3n + n(n^2-n+4)(\beta_1+\beta_2) - \beta_3(n+1)\right],$$

$$\alpha_4 := \left(\frac{n}{n^2-1}\right)(1+\beta_1+\beta_2)n,$$

$$\alpha_5 := \begin{cases} \dfrac{n-4+\beta\beta_3(n+2)}{2\beta(n-1)} & \text{if } n \le 50 \\ \dfrac{1+\beta\beta_3(n+2)}{2\beta(n-1)} & \text{if } n > 50, \end{cases}$$

$$\alpha_6 := \frac{3\alpha_5^2}{\alpha_0} n^3,$$

$$\beta := \frac{1-n}{n(n^3-n^2-n+1)}, \qquad \beta_1 := -\beta(n+2)\beta_3,$$

$$\beta_2 := 2\beta(n-1)\alpha_5, \qquad \beta_3 := \alpha_0 + \left(\frac{1}{2n^2-n-1}\right)^2.$$

The map in (8.31) minimizes the functional in (8.32) in its Dirichlet class. Notice that although assumptions in (8.12) are not formally satisfied for $p = 2$ and $\mu > 0$, this fact is not really relevant. Indeed, as the gradient of the map in (8.31) is bounded the growth of the functional (which is formally of polynomial type of order 4) becomes irrelevant, while the related Euler–Lagrange system becomes non-degenerate when considered on the map (8.31) (see [C177, p. 62]). Notice also that again in the case considered in [C177] the dimensional bound considered when $n \ge 5$ is not optimal, instead $n \ge 3$ being the one to hope for. Indeed, when considering minimizers of functionals under assumptions (8.12) we have in general

$$|Du|^{(p-2)/2} Du \in W^{1,2}_{\text{loc}}, \tag{8.33}$$

see for instance [12, 13, 70], that again implies, via imbedding, that Du is locally Hölder continuous when $n = 2$.

8.2.3.4 Wild Singularities and the Prague School

The reader will have noticed that all of the above mentioned examples of singular weak solutions and minimizers are in fact *partially regular*, as they are smooth away from a point singularity at the origin. One is left to wonder if De Giorgi's theorem in the vectorial case could be replaced instead by some kind of partial regularity result for weak solutions of linear elliptic systems. To set the scene we let the coefficients $A^{ij}_{\alpha\beta}\colon \Omega \to \mathbb{R}$ be bounded measurable functions defined on a nonempty open subset Ω of $\mathbb{R}^n$ satisfying the symmetry condition $A^{ij}_{\alpha\beta} = A^{ji}_{\beta\alpha}$ almost everywhere. We further suppose they satisfy

$$\nu|\xi|^2 \leq A^{ij}_{\alpha\beta}(x)\xi^{\alpha}_i\xi^{\beta}_j \leq L|\xi|^2 \tag{8.34}$$

for all $\xi \in \mathbb{R}^{nN}$ and almost all $x \in \Omega$, where the summation convention is used (summing from 1 to N over α, β, and from 1 to n over i, j), and where $0 < \nu \leq L$ are constants. We consider maps $u \in W^{1,2}_{\mathrm{loc}}(\Omega, \mathbb{R}^N)$ satisfying

$$D_i\left(A^{ij}_{\alpha\beta}(x)D_j u^{\beta}\right) = 0, \quad \alpha = 1, \ldots, N, \tag{8.35}$$

in the distributional sense on Ω. Now if the coefficients $A^{ij}_{\alpha\beta}$ are continuous, then classical results imply that u is locally Hölder continuous [23,76]. In fact, as a closer inspection of the proofs reveal, if the coefficients $A^{ij}_{\alpha\beta}$ are continuous at the point x_0, then u will be Hölder continuous near x_0, see for instance [34, Theorem 3.1]. The issue here is the general vectorial case n, $N \geq 2$, where, as remarked above, weak solutions can be unbounded, and hence in particular fail to be Hölder continuous, when $n = N \geq 3$. In the cases $n = 2$, $N \geq 2$, weak solutions are again Hölder continuous [68, 72, 73], see also Sect. 8.2.6 for related works of Nečas. In the case of *complex* elliptic equations (8.34)–(8.35), so for $n \geq 3$, $N = 2$, examples of unbounded weak solutions have been given recently by Frehse [32]. See also [66] for earlier examples in dimensions $n \geq 5$, $N = 2$.

The issue regarding partial regularity of weak solutions was settled in the negative by Souček in [92] for $n = N \geq 3$: there exist systems (8.34)–(8.35) with bounded weak solutions that are discontinuous on a dense set. Hence partial regularity, as understood here, is excluded for such systems. We emphasize that discontinuity of the coefficients $A^{ij}_{\alpha\beta}(x)$ is not equivalent to the existence of singular weak solutions, and that the relation is more subtle, see [2,62] for interesting results in this regard. Souček's construction of an elliptic system with a singular weak solution is based on the following remarkable fact:

Lemma 8.1 ([92]). *Let $u \in W^{1,2}_{\mathrm{loc}}(\Omega, \mathbb{R}^N)$ and $\sigma = (\sigma^i_\alpha) \in L^2_{\mathrm{loc}}(\Omega, \mathbb{R}^{nN})$ be row-wise solenoidal, $D_i\sigma^i_\alpha = 0$ in the sense of distributions on Ω for each $\alpha = 1, \ldots, n$. Assume that*

$$\langle\sigma, Du\rangle \geq \lambda|Du|^2 \quad \textit{and} \quad |\sigma|^2 \leq \mu^2|Du|^2$$

hold almost everywhere on Ω, where $0 < \lambda \leq \mu$ are positive constants, and the angular brackets denote the usual inner product on $\mathbb{R}^{Nn}$ with corresponding norm $|\cdot|$. Then u is a weak solution to a linear elliptic system as in (8.34)–(8.35), with ellipticity ratio

$$\frac{\nu}{L} = \frac{\frac{\mu}{\lambda} - \sqrt{\frac{\mu^2}{\lambda^2} - 1}}{\frac{\mu}{\lambda} + \sqrt{\frac{\mu^2}{\lambda^2} - 1}}. \tag{8.36}$$

The proof of this result is so elegant and short that we report it here.

Proof. For $\theta \in (0, \lambda)$ define

$$A_{\alpha\beta}^{ij} = \theta \delta_{\alpha\beta} \delta_{ij} + \frac{(\sigma_\alpha^i - \theta D_i u^\alpha)(\sigma_\beta^j - \theta D_j u^\beta)}{\langle \sigma - \theta Du, Du \rangle}.$$

For $\xi \in \mathbb{R}^{nN}$ we have for almost all x,

$$A_{\alpha\beta}^{ij} \xi_i^\alpha \xi_j^\beta = \theta |\xi|^2 + \frac{\langle \sigma - \theta Du, \xi \rangle}{\langle \sigma - \theta Du, Du \rangle} \geq \theta |\xi|^2$$

and

$$A_{\alpha\beta}^{ij} \xi_i^\alpha \xi_j^\beta \leq |\xi|^2 \Big(\theta + \frac{|\sigma - \theta Du|^2}{\langle \sigma - \theta Du, Du \rangle} \Big) \leq \frac{\mu^2 - \theta\lambda}{\lambda - \theta} |\xi|^2.$$

Choosing $\theta = (\mu^2 - \mu\sqrt{\mu^2 - \lambda^2})/\lambda$ the result follows. □

Using Lemma 8.1 in connection with a delicate and precise construction, John, Malý and Stará established in [46] the following result:

Theorem 8.13 ([46]). *Let $n = N \geq 3$, and assume $0 < \nu \leq L$ satisfy*

$$\frac{\nu}{L} < K(n) := \frac{\sqrt{1 + \frac{(n-2)^2}{n-1}} - 1}{\sqrt{1 + \frac{(n-2)^2}{n-1}} + 1}. \tag{8.37}$$

Then for any given F_σ subset F of Ω there exists a linear elliptic system (8.34)–(8.35) that has a bounded solution $u \in W_{\mathrm{loc}}^{1,2}(\Omega, \mathbb{R}^N)$ which is essentially discontinuous on F and essentially continuous on its complement $\Omega \setminus F$.

It is remarkable that when the ellipticity ratio strictly violates (8.37), meaning that $\frac{\nu}{L} > K(n)$, then Koshelev has shown in [48] that any weak solution of (8.34)–(8.35) must be locally Hölder continuous. Koshelev also observed that the everywhere Hölder continuity of weak solutions is lost when (8.37) holds and $n = N \geq 3$. Theorem 8.13 shows in particular that a weak solution to a system (8.34)–(8.35) with ellipticity ratio satisfying (8.37) can be *everywhere* essentially discontinuous when $n = N \geq 3$, just like a generic $W^{1,2}_{\mathrm{loc}}$-Sobolev map.

Finally, let us remark that in the vectorial case the natural ellipticity condition for linear elliptic systems is the Legendre–Hadamard condition. It amounts to requiring the lower bound in (8.34) to hold only when $\xi \in \mathbb{R}^{Nn} \simeq \mathbb{R}^{N\times n}$ is a matrix whose rank equals one—we can state this equivalently as

$$\nu|a|^2|b|^2 \leq A^{ij}_{\alpha\beta}(x)a^\alpha b_i a^\beta b_j \tag{8.38}$$

for all $a \in \mathbb{R}^N$, $b \in \mathbb{R}^n$ and almost all $x \in \Omega$. Whereas the lower bound in (8.34) at the level of frozen integrands (fixed x) corresponds to convexity of $\xi \mapsto A^{ij}_{\alpha\beta}(x)\xi^\alpha_i\xi^\beta_j - \nu|\xi|^2$, the Legendre–Hadamard condition (8.38) corresponds to *rank-one convexity*. Since these frozen integrands are quadratic forms, a classical result, proved by use of the Fourier transformation, gives that they are in fact *quasiconvex* in the sense of Morrey [74].

For weak solutions to Legendre–Hadamard elliptic linear systems with bounded measurable coefficients even Meyers' higher integrability result [68] fails in general, and this is already observed in the low-dimensional case $n = N = 2$. We refer to the seminal paper [79] for further discussion and details, including the failure of partial regularity for extremals of strongly quasiconvex integrals.

8.2.3.5 The Constructions of Šverák and Yan

We return to the closely related regularity problem for minimizers of regular variational integrals

$$v \mapsto \int_\Omega F(Dv(x))\,dx, \tag{8.39}$$

where the integrand $F\colon \mathbb{R}^{Nn} \to \mathbb{R}$ is assumed C^∞ smooth, and

$$\nu|\xi|^2 \leq F_{z^\alpha_i z^\beta_j}(z)\xi^\alpha_i\xi^\beta_j \leq L|\xi|^2 \tag{8.40}$$

holds for all $z, \xi \in \mathbb{R}^{Nn}$, where $0 < \nu \leq L$ are given constants. Let us recapitulate: in dimensions $n = 2$, and any N, a minimizer is smooth, and in dimensions $n \leq 4$, and any N, a minimizer is locally Hölder continuous, while in dimensions $n \geq 5$, $N \geq \frac{1}{2}n(n+1) - 1$, Nečas showed us that minimizers can be Lipschitz but not C^1. There is a gap between theory and examples, that in fact puzzled the experts for several years. In [93, 94] Šverák and Yan clarified the situation, and, in particular, they showed that in dimensions $n = 3$, $N = 5$ minimizers of (8.39)–(8.40) can be non-Lipschitz, while in dimensions $n = 5$, $N = 14$ they can be even unbounded. This constituted a major and definitive progress on the regularity problem. As it relates to the constructions of Nečas let us discuss their results and methods a bit further. The minimizers of Šverák and Yan are related to those of Nečas as they have the form $u\colon B(0,1) \subset \mathbb{R}^n \to \mathbb{R}^N$ with

$$u = (u^{ij}), \quad u^{ij}(x) := \frac{1}{|x|^{\varepsilon}}\Big(\frac{x_i x_j}{|x|} - \frac{\delta_{ij}|x|}{n}\Big) \tag{8.41}$$

for a suitable choice of the parameter $\varepsilon = \varepsilon(n)$. Considering u as a mapping into the symmetric trace-free $n \times n$ matrices we have $N = \frac{n(n+1)}{2} - 1$. As in the constructions of De Giorgi and Nečas they start with the mapping that they want to be a minimizer, in this case (8.41), and then they construct/show existence of the integrand F. The novelty is in their construction of F that loosely stated relies on the notion of a *Null Lagrangian*. Recall that a smooth integrand $L\colon \mathbb{R}^{N\times n} \simeq \mathbb{R}^{Nn} \to \mathbb{R}$ is a Null Lagrangian (see [6, 15, 74, 75]) if for any Lipschitz mapping $v\colon \Omega \subset \mathbb{R}^n \to \mathbb{R}^N$ we have that

$$D_i\Big(L_{z_i^\alpha}(Dv)\Big) = 0$$

in the sense of distributions on Ω for each $\alpha = 1, \dots, N$ (so the Euler–Lagrange system for L is satisfied by all suitably regular mappings). Define the set $K = \{Du(x) : x \in B(0,1)\} \subset \mathbb{R}^{Nn}$ with the map u defined at (8.41). The approach of Šverák and Yan is now based on the observation that the construction will be complete if we can find a Null Lagrangian L and an integrand F satisfying (8.40) and so $F_{z_i^\alpha} = L_{z_i^\alpha}$ on K. To achieve this the authors invoke results and methods from representation theory, and we refer the interested reader to the original papers for the details. Instead let us quote their main result here.

Theorem 8.14 ([94]). *The mapping $u\colon B(0,1) \subset \mathbb{R}^n \to \mathbb{R}^N$ defined at (8.41) and with $N = \frac{1}{2}n(n+1) - 1$ is a minimizer of a regular variational integral (8.39)–(8.40) provided that*

$$\varepsilon < \frac{n + 1 - \sqrt{\frac{3(n+1)}{n-1}}}{1 + \sqrt{\frac{3(n+1)}{n-1}}}.$$

In particular we record that for $n \geq 3$ and $N = \frac{1}{2}n(n+1) - 1$ there exists a regular variational integral (8.39)–(8.40) with a minimizer $u\colon B(0,1) \subset \mathbb{R}^n \to \mathbb{R}^N$ satisfying

$$u \notin W^{2,p}_{\mathrm{loc}} \quad \text{for} \quad p \geq p_2 := \frac{n}{n+2}\Big(1 + \sqrt{\frac{3(n+1)}{n-1}}\Big),$$

$$u \notin W^{1,p}_{\mathrm{loc}} \quad \text{for} \quad p \geq p_1 := \frac{n\Big(1 + \sqrt{\frac{3(n+1)}{n-1}}\Big)}{n + 1 - \sqrt{\frac{3(n+1)}{n-1}}},$$

and, for $n \geq 5$,

$$u \notin L^p_{\rm loc} \quad \textit{for} \quad p \geq p_0 := \frac{n}{n+1}\left(1 + \sqrt{\tfrac{3(n+1)}{n-1}}\right).$$

Finally we mention that Šverák and Yan [94] also construct a non-Lipschitz minimizer in dimensions $n = 4$, $N = 3$ by considering a suitable complex version of the mapping (8.41) and, which, as it turns out, is related to the so-called Hopf fibration $\mathbb{S}^3 \to \mathbb{S}^2$.

It remains to find the precise higher integrability exponents for minimizers of regular variational integrals (8.39)–(8.40). Dimension-free higher integrability improvement has been established in [49], where it is shown that minimizers of (8.39)–(8.40) belong to $W^{2,p}_{\rm loc}$ for all $p \leq 2 + \frac{\nu}{50L}$ regardless of the dimensions n, N.

8.2.4 Conditions for Everywhere Regularity

Nečas always insisted on the importance of deriving conditions under which systems have everywhere regular weak solutions (say locally Hölder continuous, or of class $C^{1,\alpha}_{\rm loc}$), and indeed, this is still today a major issue in the theory of elliptic systems. There are several research directions in this respect, but all of them have a common root: the attempt to reduce certain vectorial issues to scalar-type situations or to linear situations, where solutions are known to be everywhere regular. Let us try to identify a few of the main ones:

- Perturbation methods. The original system is "close" to another one, for instance a linear one, whose solutions are regular. Since ellipticity means also stability under perturbations, solutions to the original system must be regular by comparison with the ones of the close regular system
- Structure conditions. Additional structure properties make the system not anisotropic enough to develop singularities: the eigenvalues of its linearization are not too sparse, roughly. Anisotropic structure as for instance the one in (8.30) are ruled out
- Certain scalar type properties of solutions are assumed to hold. Solutions are a priori known, either as a direct assumption, or as an assumption on the system, to share some relevant features of solutions to single equations. This is used to prove their regularity. This is an indirect way to look at regularity since a priori conditional assumptions are shown to force regularity
- Low-dimensional arguments. The level of anisotropicity of the system in low dimensions is obviously limited; then regularity simply follows via higher differentiability of solutions and Sobolev–Morrey embedding. This is a classical path commonly taken in the times preceding DeGiorgi's–Nash–Moser multidimensional approach. In Sect. 8.2.6 below we are going to say more on these aspects

Let us briefly comment on the above approaches, premising that all have important points in common. As for perturbation methods, these are classical and originate in the proof of Schauder estimates and the use of freezing methods of Korn type. The main idea is that the system in question is not far from a linear elliptic system, whose solutions are in turn regular. Related methods are those employed by Cordes, who assumed that the eigenvalues of the system are not too sparse. We shall not dwell further on these aspects, but instead refer the interested reader to [48] for conditions regarding regularity and perturbation methods. We only remark that such an approach was often pursued by former students and collaborators of Nečas [16, 17].

As for structure conditions that prevent the formation of singularities, the most important contribution in this respect is arguably the fundamental work of Uhlenbeck [95]. The main assumption in [95] prescribes, when referred to functionals, that the integrand depends on the gradient only via its Euclidean length:

$$F(x,v,z) \equiv F(z) = g(|z|)\,, \tag{8.42}$$

for a suitable function $g\colon [0,\infty) \to [0,\infty)$, such that (8.12) are still satisfied. So, the dependence of the gradient must occur directly via the Euclidean norm $|Du|$, making the functional "less anisotropic" and, ultimately ruling out singularities of minima. The most classical instance of such a structure is given in (8.13). In fact, when passing to the so-called second variation of the functional in (8.13) (that in a sense corresponds to looking at the linearized system) we find

$$D^2F(Du) = |Du|^{p-2}\left[I + (p-2)\frac{Du \otimes Du}{|Du|^2}\right]. \tag{8.43}$$

In other words, D^2F is almost diagonal—it differs from a diagonal system by a positive definite bilinear form—the eigenvalues are not sparse, and they are proportional to $|Du|^{p-2}$. The connections with the perturbation approaches are now evident.

When considering systems, the analog of the structure in (8.42) is given by

$$a_i^\alpha(x,v,z) \equiv a_i^\alpha(z) = h(|z|)z_i^\alpha \qquad \alpha \in \{1,\dots,N\}\,, \quad i \in \{1,\dots,n\}\,, \tag{8.44}$$

that is sometimes called "Uhlenbeck structure". We refer to the book [9] for more results in this direction. Notice how the structure in (8.43) and (8.44) are very far from the ones considered for building counterexamples, as the one in (8.30).

Additional, less explicit structures, allowing for everywhere regularity, can be explicitly given in terms of the second variation of the functional; see for instance [67], where various regularity results for solutions to systems, such as (8.4), are considered and proved in connection to the behavior of the so-called indicator function

$$\mathcal{I}(u) := \sum_{i,j} \frac{u^i u^j}{|u|^2} Du^j a^i(x,u,Du)\,.$$

This particularly relates to the work of Nečas as suitable assumptions of the sign of the previous indicator function essentially rules out the occurrence of solutions such as those in (8.31).

As for additional properties that resemble the scalar behavior, one has already been encountered in Sect. 8.2.3, in the discussion before (8.29). Another one, considered in the papers [JN-C96, JN-C100], regards the impact that Liouville type properties have on solutions, and was the object of intensive investigation of Nečas; see also [B8]. We shall dedicate the next section to this theme.

Finally, the low-dimensional cases. Everywhere regularity of solutions is here essentially linked to the possibility of using higher integrability and/or higher differentiability properties of solutions, and then, via embedding, the desired regularity follows. This is a classical, and historically speaking, one of the oldest paths taken in regularity theory (low-dimensional cases were essentially the only ones treatable before De Giorgi–Nash–Moser theory). For instance, under the assumptions, (8.9) and (8.12), respectively, solutions to systems of the type (8.27), and consequently minima of integrals such as in (8.26) satisfy

$$V(Du) := (1 + |Du|^2)^{(p-2)/2} Du \in W^{1,q}_{\mathrm{loc}} \tag{8.45}$$

for some $q > 2$, which in turn depends essentially only on the ellipticity parameters ν and L. The proof of this fact goes in two steps; one first uses the difference quotients method to prove that $V(Du) \in W^{1,2}_{\mathrm{loc}}$ together with the local estimate

$$\fint_{B_{r/2}} |DV(Du)|^2\, dx \le \frac{c}{r^2} \fint_{B_r} |V(Du) - (V(Du))_{B_r}|^2\, dx$$

whenever $B_r \subset \Omega$ is a ball. Then the Poincaré–Sobolev inequality yields

$$\left(\fint_{B_{r/2}} |DV(Du)|^2\, dx \right)^{1/2} \le c \left(\fint_{B_r} |DV(Du)|^t\, dx \right)^{1/t}$$

where $t < 2$ is such that taking its Sobolev conjugate we obtain 2. The last estimate is a so-called reverse-Hölder inequality with increasing support that allows us to apply Gehring's lemma [39, 45] and to deduce the existence of some higher integrability exponent $q > 2$ such that

$$\left(\fint_{B_{r/2}} |DV(Du)|^q\, dx \right)^{1/q} \le c \left(\fint_{B_r} |DV(Du)|^t\, dx \right)^{1/t} \tag{8.46}$$

holds whenever $B_r \subset \Omega$. Therefore (8.45) follows; in turn Sobolev's embedding theorem allows us to deduce that u is locally Hölder continuous in Ω provided $n \le p + 2$. The procedure we have just described—that can be extended to higher order systems as for instance shown in [12]—does not carry over to systems and functionals, such as those at (8.1), (8.2) and (8.4), partly because the coefficients are supposed to be just Hölder continuous. However, even when the coefficients are smooth and globally Lipschitz continuous minimizers of functionals (8.4) might not be twice differentiable. Nevertheless, a key point, observed by Campanato in a series of papers from the 1980s—see again [12] and for instance [13]—is that the regularity estimates as (8.46) are sufficient for perturbation methods to prove general partial regularity theorems for complete functionals and systems. Indeed, for the variational case this approach has been followed in [36, 43, 44]. In turn, as shown for instance in [51, 70], these results have strong impact on the singular sets of solutions: the origin of the condition $n < p + 2$ considered in Theorems 8.10 and 8.11 stems exactly from this.

Let us point out that regularity proofs based on higher integrability and Sobolev's embedding theorem in the two-dimensional case originate in the works of Bojarski [10] and Meyers [68], where linear elliptic equations were treated. Early regularity papers of Nečas moved along this line, and we shall dedicate Sect. 8.2.6 to Nečas's work on low dimensional problems.

8.2.5 *Everywhere Regularity via Liouville Properties*

Here we shall deal with those works of Nečas [JN-C96, JN-C100, JN-C113, JN-C114] considering the relation between Liouville type properties of systems, and the everywhere regularity of solutions. For this it is interesting to recall the related historical context; after the examples of singular solutions given since the end of the 1960s (some of them described in Sect. 8.2.3) the main issue was—and in several respects still is—to find additional structure conditions on systems that rule out singularities. One major contribution was the previously mentioned work of Uhlenbeck [95], while various research directions have been already described in Sect. 8.2.4. Nečas pursued a path aimed at exploiting up to which extent *significant properties of solutions to scalar equations force everywhere regularity when a priori assumed to hold in the vectorial case*. One of the key properties of harmonic functions is given by the Liouville theorem: a harmonic function which is defined in the whole of $\mathbb{R}^n$ and having a bounded gradient is necessarily affine. The work of Giaquinta and Nečas revealed that assuming a priori the validity of such a property for weak solutions to elliptic systems forces everywhere regularity and implies several other known results. We shall present a model result valid for simpler systems of the type (8.4), referring to [JN-C100, B8] for more results in this direction.

Definition 8.1. Consider a vector field $a\colon \Omega \times \mathbb{R}^n \times \mathbb{R}^{Nn} \to \mathbb{R}^{Nn}$ satisfying assumptions (8.9) for $p = 2$. The vector field $a(\cdot)$ is said to have the Liouville property (L) iff whenever $v \in W^{1,\infty}(\mathbb{R}^n, \mathbb{R}^N) \cap W^{1,2}(\mathbb{R}^n, \mathbb{R}^N)$ is a solution of the system

$$\operatorname{div} a(x_0, \xi, Dv) = 0$$

for each fixed $x_0 \in \mathbb{R}^n$ and $\xi \in \mathbb{R}^N$, then v is an affine map.

The main result is now:

Theorem 8.15 ([JN-C96, JN-C100]). *Let $u \in W^{1,\infty}(\Omega, \mathbb{R}^N)$ be a solution to (8.4) under assumptions (8.9) for $p = 2$ and with $\omega(\varrho) \le \varrho$ (differentiability with respect to coeffcients). Assume that $a(\cdot)$ satisfies the Liouville property* (L). *Then Du is locally Hölder continuous in Ω.*

This theorem looks both at the past and at the future. On one hand, by looking at its proof, it goes along the traditional line of applying ideas from the theory of minimal surfaces to the theory of nonlinear elliptic systems. On the other hand it predates several papers where the link between global and local properties of solutions have been established, and in turn, local regularity results have been proved by using global properties of solutions.

It is interesting to give a sketch of the proof of Theorem 8.15, that proceeds via a blow-up argument. Specifically, one considers the rescaled functions

$$u_r(y) := \frac{u(x_0 + ry) - u(x_0)}{r} \tag{8.47}$$

which solve rescaled systems

$$\operatorname{div} a_r(y, u_r, Du_r) = 0 \tag{8.48}$$

where

$$a_r(y, v, z) := a(x_0 + ry, u(x_0) + rv, z)\,.$$

Here x_0 is an interior point of Ω and $r < \operatorname{dist}(x_0, \partial\Omega) =: d$. Notice that the functions u_r solve (8.48) in the balls $B_{d/r}$, that get larger and larger when $r \to 0$. Coefficients are now differentiable, and therefore, differentiating (8.48) yields

$$\operatorname{div}\left(\partial_z a_r(y, u_r, Du_r) D^2 u_r\right) = \text{“additional controllable terms”}\,. \tag{8.49}$$

The standard Caccioppoli estimate for a system as (8.49) gives a uniform—with respect to r—bound of the type

$$\int_{B_{d/(2r)}(0)} |D^2 u_r|^2 \, dy \le c\,.$$

In turn this implies—up to a non-relabeled subsequence—strong convergence of the gradients $Du_r \to DP$ in L^2, where P solves a system of the type

$$\operatorname{div} a(x_0, \xi, DP) = 0$$

in $\mathbb{R}^n$. Moreover as Du is bounded, also DP is bounded, and the assumed Liouville property gives that P is affine, i.e. $P(y) = \langle \eta, y \rangle + \eta_0$, for some $\eta \in \mathbb{R}^{Nn}$ and $\eta_0 \in \mathbb{R}^N$. By scaling back to balls $B_r(x_0)$ the strong convergence implies

$$\fint_{B_r(x_0)} |Du - \eta|^2 \, dx \ \text{ is small}$$

and in turn, this implies that the excess functional

$$\fint_{B_r(x_0)} |Du - (Du)_{B_r(x_0)}|^2 \, dx \ \text{ is small}\,, \tag{8.50}$$

provided we take in turn r suitably small. Now, observe that if we let $v^{\alpha,s} := D_s u^\alpha$, then v solves a system of the type

$$\operatorname{div}(\tilde{A}(x, v)Dv) = \text{“controllable terms”}\,.$$

In turn—see the content of Remark 8.1 and in particular (8.16)—the conditions for $v \equiv Du$ to be Hölder continuous in a neighborhood of x_0 prescribes that the quantity

$$\fint_{B_r(x_0)} |v - (v)_{B_r(x_0)}|^2 \, dx$$

be suitably small for some $r > 0$, a condition which is now satisfied by (8.50). This completes the proof of Theorem 8.15 as the choice of the point x_0 was arbitrary. We refer to [JN-C100, B8] for the details.

The interest in the previous theorem also relies on the fact that it allows us to draw a unified picture between typical blow-up methods from vectorial problems and theory of minimal surfaces, and certain scalar type properties of solutions. In turn, in [JN-C100] the authors also showed that using this theorem one can give new proofs of several typical regularity results that hold in special situations. For instance the classical Cordes' type condition involving nearness between eigenvalues, and the two-dimensional case (where solutions are regular) still allow for the Liouville property (L) to be satisfied. Therefore another proof of regularity follows applying Theorem 8.15.

Other interesting results can be proved using these methods. For instance, in the paper [JN-C114] the authors are able to prove everywhere boundary regularity of Lipschitz solutions to Dirichlet problems, once again provided Liouville property holds. This involves a typical “boundarization” of the argument employed to prove Theorem 8.15.

8.2.6 Two Low-Dimensional Theorems

As described in Sect. 8.2.4, low-dimensional contexts allow for everywhere regularity, both for elliptic systems and higher-order nonlinear equations, via basic higher differentiability and integrability properties and Sobolev–Morrey embedding. Nečas often pursued this path and here we want to recall a couple of results that stand out from his works in this respect. They reveal a certain continuity in his methodological approach starting from the 1967 paper [C33] and ending with the 1992 one [JN-C156]; incidentally (or maybe not) both of the papers were published in the *Annali della Scuola Normale Superiore di Pisa, Classe di Scienze.*

8.2.6.1 Higher-Order Equations ([C33])

To better frame Nečas paper [C33] we have to go back to the late 1960s and recall the historical situation. DeGiorgi–Nash–Moser theory was still in its early days, still raising hopes for an extension to the general vectorial case (and therefore to higher-order equations). As seen in Sect. 8.2.3, these hopes disappeared after the examples given in [20,40,65], but before, at the time of Nečas' paper [C33], several people were actively devoted to obtain an analog of DeGiorgi–Nash–Moser theory in the vectorial case. On the other side, Gehring's lemma and higher integrability, a nowadays basic tool to prove regularity in low dimensions, was not available then. Let us briefly recall the "modern approach", indeed via higher integrability. We want to look for the gradient spatial continuity of solutions u to a general higher-order elliptic system, or of a minimizer of a general integral functional depending on higher-order derivatives as

$$v \mapsto \int_\Omega F(D^k v)\,dx$$

defined on $W^{k,p}(\Omega,\mathbb{R}^N)$, and satisfying the growth conditions as in (8.8) (now z is the variable indicating the highest order derivatives). Then in two dimensions one can again appeal to Gehring's lemma and higher integrability as described in Sect. 8.2.2.2 and Theorem 8.9. These give, in the present situation, that

$$D^k u \in L^q\,, \qquad \text{for some } q > p$$

holds for minima u (and also for solutions to elliptic systems of order $2k$), and actually this result holds in every dimension n. Therefore, if $p \geq 2$ (but also if $|p-2|$ is sufficiently small) then $D^{k-1}u$ is locally Hölder continuous when $n = 2$, simply by the Sobolev–Morrey embedding theorem. Moreover, if further differentiability is assumed as in (8.12) for functionals of the type

$$v \mapsto \int_\Omega F(D^k v)\,dx$$

defined on $W^{k,p}(\Omega,\mathbb{R}^N)$ where $p > 1$, then, similarly to Sect. 8.2.4 (see (8.45)) one obtains that

$$V(D^k u) = (1 + |D^k u|^2)^{(p-2)/2} D^k u \in W^{1,q}_{\mathrm{loc}}\,, \qquad \text{for some } q > 2 \tag{8.51}$$

and then the local Hölder continuity of $V(D^k u)$, and hence of $D^k u$, follows in the case $n = 2$. So far, this is the current approach; as said, all this was not available before the higher integrability results originating from the work of Gehring [33] were proved. In the paper [C33] Nečas gave the first everywhere regularity results for minimizers of higher-order functionals of the type

$$v \mapsto \int_\Omega \left(F(x, D^{[k]}v) + \sum_{|i|\le k} D^i u f_i \right) dx \tag{8.52}$$

$\Omega \subset \mathbb{R}^2$ (here i is a multi-index whose length $|i|$ is at most k and $D^{[k]}v$ denotes the set of all partial derivatives of order less than or equal to k). The integrand $F(\cdot)$ and the assigned functions f_i satisfy suitable smoothness assumptions. In particular, the integrand F satisfies (8.12) with $\mu = 1$, when obviously recast for the higher-order case under consideration. The proof of Nečas goes in three steps, and we shall summarize it in the simpler case of quadratic growth, $p = 2$.

Step 1: Differentiation. This is essentially the derivation of (8.51).

Step 2: Higher-order Meyers estimates. Nečas considers a general linear elliptic higher-order system with measurable coefficients of the type

$$\int_\Omega \langle A(x) D^k w, D^k \varphi \rangle \, dx = \int_\Omega \sum_{|i|\le k} D^i w g_i \, dx \qquad \forall \ \varphi \in C^\infty(\Omega, \mathbb{R}^N) \tag{8.53}$$

with ellipticity bounds

$$\nu |\xi|^2 \le \langle A(x)\xi, \xi \rangle \le L|\xi|^2 .$$

Under such assumptions Nečas proves that there exists a universal number $q > 2$, depending only on $n, L/\nu$ such that

$$w \in W^{k,2} \qquad \text{and} \qquad g \in L^q \Longrightarrow w \in W^{k,q} .$$

This result extends the classical one of Meyers [68] valid for second-order elliptic equations. As in [68], the proof is based on the version of Calderón–Zygmund estimates for general systems due to Agmon et al. [4] and a careful perturbation argument.

Step 3: Conclusion. The last step uses the result of Step 1 to differentiate the Euler–Lagrange systems relative to the functional (8.51) (assumptions have to be made to allow for the derivation of such a system). This gives that every partial derivative of the original minimizer u is a solution of a system of the type (8.53) and therefore we have that $u \in W^{k+1,q}$ for some $q > 2$. Notice that the functions g_i considered here now involve non-maximal order derivatives of u, and therefore some kind of bootstrap argument is needed here to apply the result of Step 2. At this stage the Hölder continuity of $D^k u$ follows via the Sobolev–Morrey embedding theorem since we are in the two-dimensional case, and the proof is complete.

As specified at the beginning, the result in [C33] was obtained when there was still hope for developing a theory of everywhere regularity for systems and higher order equations, in the style of De Giorgi–Nash–Moser's theory. Shortly after, the examples in [20, 40, 65] would show that Nečas' result was essentially the best possible.

We finally observe that the paper [C33] appears as a synthesis of many of the remarkable technical abilities that Nečas developed till then, and that are well documented in his famous book [A1]. His broad approach to elliptic equations eventually inspired several people from the Prague school to develop related theoretical issues. We would like, in this respect, to mention the book [53], that became a standard reference over the years, and that highlights many of the ideas Nečas had developed at the time.

8.2.6.2 General Parabolic Systems ([JN-C156])

A by now classical result of Nečas and Šverák asserts the everywhere (spatial) gradient Hölder continuity of solutions to parabolic systems of the type

$$u_t - \operatorname{div} a(Du) = 0 \tag{8.54}$$

when the space dimension is 2. The system is considered in a cylindrical domain $\Omega_T = \Omega \times (-T, 0)$, where $\Omega \subset \mathbb{R}^n$ is a bounded domain and $n = 2$. Here the vector field $a(\cdot)$ is assumed to satisfy (8.9) for $p = 2$ and no dependence on (x, t, u) occurs. The above regularity result is actually a consequence of a more general higher integrability result for $(u_t, D^2 u)$ valid also in higher dimensions. The elliptic analog of the result of Nečas and Šverák is classical; as explained in Sect. 8.2.4, and also in the preceding one, for solutions to an elliptic systems as in (8.27) are such that $V(Du) \equiv Du \in W^{1,q}$ for some $q > 2$; in two dimensions this implies everywhere Hölder continuity Du via Sobolev–Morrey embedding. In the case of (8.54) we still have, and in every dimension, that second spatial derivatives are higher integrable $D^2 u \in L^q$, but this is no longer sufficient to establish the everywhere Hölder continuity via Sobolev embedding, since time derivatives are missing. The idea of Nečas and Šverák is then to use elliptic theory slicewise to deduce the uniform-in-time higher differentiability/integrability of the spatial gradient in the following sense:

$$\sup_{t_1 \le t \le t_2} \int_{\Omega'} |D^2 u(x,t)|^q \, dx < \infty \,, \qquad \text{for some } q > 2 \,, \tag{8.55}$$

whenever $\Omega' \Subset \Omega$ and $-T < t_1 < t_2 < 0$. The previous result then implies the required regularity via slicewise (elliptic) Sobolev embedding and standard parabolic arguments. The proof in [JN-C156] is very elegant in that it combines some known regularity results, with a clever boot-strap method. We think it deserves to be sketched in the following lines.

Step 1. The system in (8.54) can be differentiated in space, thereby getting that each component of the spatial gradient in turn again solves a linear parabolic systems with measurable coefficients

$$v_t - \operatorname{div}(A(x,t)Dv) = 0 \tag{8.56}$$

for

$$v = u_{x_s}, \qquad A(x,t) = a_z(Du(x,t)), \qquad s = 1, \ldots, n,$$

and moreover

$$D^2u, u_t \in L^2. \tag{8.57}$$

These facts belong to the classical theory of parabolic systems in divergence form of the type (8.54), see for instance [14].

Step 2. Another classical fact states that, when considering a general energy solution to (8.54), the map

$$t \in (-T, 0) \mapsto u(\cdot, t) \in L^2(\Omega) \tag{8.58}$$

is continuous and therefore locally bounded in L^2. The *crucial observation* in [JN-C156] is that it is possible to improve this result by proving that actually the map in (8.58) is locally bounded in L^{q_0} for some $q_0 > 2$, depending only on $n, N, L/\nu$. Summarizing:

$$\sup_{t_1 \le t \le t_2} \int_{\Omega'} |v(x,t)|^{q_0}\, dx < c \left(\int_{\Omega_T} |v(x,t)|^2\, dx\, dt \right)^{q_0/2},$$

whenever $\Omega' \Subset \Omega$ and $-T < t_1 < t_2 < 0$; the constant c of course depends on these objects. The proof given in [JN-C156] uses the stability of vectorial truncation for small power-type perturbations.

Step 3. As shown in [JN-C156], we can use (8.57) and the result of Step 2 (applied to a difference quotient version of the original system) to deduce

$$u_{tt} - \operatorname{div}(A(x,t)Du_t) = 0. \tag{8.59}$$

Again the result of Step 2 to the systems in (8.56)–(8.59) ($v = Du, u_t$) it follows that

$$\sup_{t_1 \le t \le t_2} \int_{\Omega'} (|u_t(x,t)|^{q_0} + |Du(x,t)|^{q_0})\, dx < \infty .$$

This is the crucial estimate, as now we can read the original parabolic system as a slicewise elliptic system

$$\operatorname{div} a(Dw) = g\,, \qquad g = u_t\,, \qquad w(x) = u(x,t)$$

for $t \in (-T, 0)$, and with a higher integrable right-hand side

$$\sup_{t_1 \le t \le t_2} \int_{\Omega'} |g|^{q_0}\, dx < \infty .$$

It is now possible to prove a "non-homogeneous" variant of the higher integrability theorem explained in Sect. 8.2.4 which asserts the following extension to (8.46):

$$\left(\fint_{B_{r/2}} |D^2 u(x,t)|^q\, dx \right)^{1/q} \le c \left(\fint_{B_r} |D^2 u(x,t)|^2\, dx \right)^{1/2} + c \left(\fint_{B_r} |g(x,t)|^{q_0}\, dx \right)^{1/q_0}$$

for some $q \in (2, q_0)$, whenever $B_r \subset \Omega$ (see for instance [34]). Since the previous inequality is uniform in time (8.55) follows via a standard covering argument.

8.2.7 The Late Style

Like other artists, Nečas had a "late style" too [88]. The late Nečas was very interested in fluid-dynamics, and in the basic problem of regularity of solutions to the Navier–Stokes system. Besides this, by starting from some old ideas of Olga Ladyzhenskaya, he studied regularity and existence problems for solutions to the systems of so-called non-Newtonian fluids. These incompressible fluids are indeed modelled by PDEs where the constitutive law shows a nonlinear dependence on the symmetric part of the velocity gradient. In dealing with such issue Nečas employed all the strength of the analytical techniques he had developed in so many years dedicated to the study of regularity problems.

More generally, during his life Nečas was always interested in applications of the PDE theory he was dealing with, and this transversal approach came probably partly through the influence that his Ph.D. advisor, Ivo Babuška, had on him. Nečas's interests covered a very broad range of topics and results: elasto-plasticity [A4]; the theory of multipolar viscous fluids, characterized by having a stress tensor depending also on higher-order derivatives of the gradient velocity (see [JN-C155]); the existence theorems for transonic flows via the method of entropic compactification and the method of viscosity [C121, A11, C135].

In this section we shall briefly recall some of his results which are mostly linked to regularity and existence issues related to linear and nonlinear fluid-dynamics; these occupied the last part of his working life.

8.2.7.1 Leray's Conjecture

A striking result obtained in [JN-C176] responds in the negative to a longstanding open problem posed by Leray [60]. Specifically, and considering the Navier–Stokes system

$$\begin{cases} u_t - \nu\Delta u + (u\cdot D)u + D\pi = 0 \\ \operatorname{div} u = 0 \end{cases} \quad \text{in} \quad \mathbb{R}^3 \times (t_1, t_2)\ , \tag{8.60}$$

Leray proposed to prove the existence of irregular solutions by looking for a self-similar map (satisfying certain natural energy inequalities, see [60]), that is, a solution u of the special type

$$u(x,t) = \frac{1}{\sqrt{2a(T-t)}} U\left(\frac{x}{\sqrt{2a(T-t)}}\right),$$

where $a,\ T > 0$ are parameters. Assuming that u satisfies (8.60) and the usual energy inequalities, one sees that

$$\int_{\mathbb{R}^3} |U|^3\, dx < \infty$$

and that the stationary system

$$\begin{cases} -\nu\Delta U + aU + a(y\cdot D)U + a(U\cdot D)U + D\pi = 0 \\ \operatorname{div} U = 0 \end{cases} \quad \text{in} \quad \mathbb{R}^3 \tag{8.61}$$

is satisfied. In [JN-C176] Nečas, Růžička and Šverák prove that the only possible solution U to (8.61) is the zero one, therefore ruling out the possibility to construct singular solutions to the Navier–Stokes system via self-similar solutions. The paper is a cornerstone in the modern theory of Mathematical Fluid Dynamics.

8.2.7.2 Non-Newtonian Fluid Dynamics

The story starts with Olga Ladyzhenskaya, who, observing that for the Navier–Stokes system the basic unique solvability and regularity questions were still open, proposed to use a different model, eventually allowing for the existence and uniqueness of solutions. As it is well known, the difficulty in treating the Navier–Stokes system (8.60) lies in the fact that it becomes a systems with a critical nonlinearity in three dimensions, and this means that the convective term $(u \cdot D)u$ cannot be controlled by the Laplacean (viscosity). Expressed equivalently, no gain is obtained by scaling the system. To overcome this lack of coercivity Ladyzhenskaya proposed to replace the Lapacean by another monotone operator, featuring more coercivity (ellipticity) so that the convective term becomes now subcritical in the sense that it can be controlled by the new elliptic (viscosity) term on the left-hand side. Specifically, the new system to be considered is now of the type

$$\begin{cases} u_t - \operatorname{div} \sigma + (u \cdot D)u + D\pi = 0 \\ \operatorname{div} u = 0 \end{cases} \quad \text{in} \quad \mathbb{R}^3 \times (t_1, t_2) \tag{8.62}$$

where, in general, σ is a nonlinear, monotone vector field of the symmetric part of the gradient $D(u)$. A basic example is given by

$$\sigma \approx (1 + |D(u)|^2)^{(p-2)/2} D(u)\,, \qquad p > 1\,.$$

For $p > 2$ the system in (8.62) is no longer critical and there is hope for proving regularity and uniqueness, and, needless to say, existence. The first results concerning existence of solutions for certain values of $p > 2$ were obtained by Ladyzhenskaya [54–57], who proved existence and uniqueness of solutions (under suitable boundary conditions) for $p > 5/2$; a first improvement of Ladyzhenskaya 's results was eventually obtained in [25]. These questions were later tackled by Nečas, his students and collaborators in the 1990s, who proved a number of existence and regularity results. We recall the foundational papers [JN-C167,JN-C165,JN-C184], where the authors gave a series of results concerning the solvability of the problem (8.62), with different assumptions on σ and on the type of the data considered (for instance periodic boundary conditions and non-slippery side conditions are considered). One of the main features in the work of the authors is that they consider the subquadratic case $p < 2$, for which the treatment becomes even more difficult as the system exhibits even less coercivity than (8.60). The case $p \in (1, 2)$ is interesting from the applicative viewpoint as the ("hardening") exponent p actually serves to describe some relevant physical properties of the fluid in question as for instance the viscosity (see [A12] and [87]). Accordingly, various types of solutions are considered: when p is large enough, the coercivity of σ is in turn strong enough and existence and uniqueness of usual weak energy solutions can be proved. For intermediate values of p (crossing the value $p = 2$) existence, but not uniqueness of weak energy solutions, can be still proved; when p becomes too close to one then the authors prove existence in a very weak sense, using the notion of (Young) measure-valued solutions, already adopted in the Calculus of Variations and in Mathematical Materials Science (see for instance [7]). The basics of this approach is also described in the book [A12], where several of the techniques introduced by Nečas and his collaborators are discussed. In turn, several other papers by different researchers followed. In particular, a generalization of the model above prescribing a dependence of the viscosity rate on the space-time variables, for instance of the type

$$\sigma = (s^2 + |D(u)|^2)^{\frac{p(x,t)-2}{2}} D(u)\,, \qquad 1 < \gamma_1 \le p(x,t) \le \gamma_2 < \infty\,,$$

was introduced and studied by Rajagopal and Růžička [86,87] to model the so-called electrorheological fluids. It eventually became the source of several theoretical and applied studies.

All in all, the interest of Nečas in fluid-dynamics stimulated several people to the study of several aspects of the theory, and nowadays in Prague there is an active and well-recognized school of Applied Mathematics, see [8, 24, 81–84, 90, C180].

In his working activity on fluid dynamics, we finally want to recall that Nečas was one of the founders, and actually one of the leading figures, of a very successful series of summer schools on "Mathematical Theory in Fluid Mechanics", typically held every 2 years in Paseky. The school, nowadays managed by E. Feireisl, J. Málek, A. Novotný, M. Pokorný, M. Rokyta and M. Růžička, has become over the years a classical appointment for researchers in Fluid Dynamics.

Acknowledgements The authors are supported by the ERC grant 207573 "Vectorial Problems" and by EPSRC Science and Innovation Award EP/E035027/1. The second-named author also acknowledge the hospitality and support from the OxPDE Centre at the Mathematical Institute - Oxford, in March 2011.

Nečas' Works Quoted Above

The references [Axx], [Bxx], [Cxx] refer to the respective papers listed in Chapter 2 "Complete Bibliography of Jindřich Nečas". The references [JN-Cyy] refer to papers [Cyy] that are reproduced either in Chapter 7 "Selected Papers on Regularity Theory" or in Chapter 13 "Selected Papers on Fluid Mechanics".

[A1] Nečas, J.: *Les méthodes directes en théorie des équations elliptiques.* Masson, Paris; Academia, Prague 1967.

[A4] Hlaváček, I. & Nečas, J.: *Mathematical theory of elastic and elasto-plastic bodies: an introduction.* Studies in Applied Mechanics, 3. Elsevier, Amsterdam-New York, 1980.

[A11] Nečas, J.: *Écoulements de fluide: compacité par entropie.* RMA: Research Notes in Applied Mathematics, 10. Masson, Paris, 1989.

[A12] Málek, J. & Nečas, J. & Rokyta, M. & Růžička, M.: *Weak and measure-valued solutions to evolutionary PDEs.* Applied Mathematics and Mathematical Computation, 13. Chapman & Hall, London, 1996.

[B8] Nečas, J.: *On the regularity of weak solutions to nonlinear elliptic systems of partial differential equations.* Scuola Normale Superiore, Pisa. 1979.

[C33] Nečas, J.: Sur la régularité des solutions variationelles des équations elliptiques non-linéaires d'ordre $2k$ en deux dimensions. (French) *Ann. Scu. Norm. Sup. Pisa (III)* 21 (1967), 427–457.

[C121] Feistauer, M. & Mandel, J. & Nečas, J.: Entropy regularization of the transonic potential flow problem. *Comment. Math. Univ. Carolin.* 25 (1984), 431–443.

[C135] Feistauer, M. & Nečas, J.: Viscosity method in a transonic flow. *Comm. Partial Differential Equations* 13 (1988), 775–812.

[C175] Nečas, J. & Růžička, M. & Šverák, V.: Sur une remarque de J. Leray concernant la construction de solutions singulres des équations de Navier-Stokes. *C. R. Acad. Sci. Paris Sér. I Math.* 323 (1996), 245–249.

[C177] Hao, W. & Leonardi, S. & Nečas, J.: An example of irregular solution to a nonlinear Euler-Lagrange elliptic system with real analytic coefficients. *Ann. Scuola Norm. Sup. Pisa Cl. Sci. (IV)* 23 (1996), 57–67.

[C180] Leonardi, S. & Málek, J. & Pokorný,M. & Nečas, J.: On axially symmetric flows in $\mathbb{R}^3$. *Z. Anal. Anwendungen* 18 (1999), 639–649.

[JN-C86] Nečas, J.: Example of an irregular solution to a nonlinear elliptic system with analytic coefficients and conditions for regularity. *Theory of nonlinear operators* (Proc. Fourth Internat. Summer School, Acad. Sci., Berlin, 1975), pp. 197–206.

[JN-C96] Giaquinta, M. & Nečas, J.: On the regularity of weak solutions to nonlinear elliptic systems via Liouville's type property. *Comment. Math. Univ. Carolin.* 20 (1979), 111–121.

[JN-C100] Giaquinta, M. & Nečas, J.: On the regularity of weak solutions to nonlinear elliptic systems of partial differential equations. *J. Reine Angew. Math. (Crelles J.)* 316 (1980), 140–159.

[JN-C104] Nečas, J. & John, O. & Stará, J.: Counterexample to the regularity of weak solution of elliptic systems. *Comment. Math. Univ. Carolin.* 21 (1980), 145–154.

[JN-C113] Lions, P. L. & Nečas, J. & Netuka, I.: A Liouville theorem for nonlinear elliptic systems with isotropic nonlinearities. *Comment. Math. Univ. Carolin.* 23 (1982), 645–655.

[JN-C114] Giaquinta, M. & Nečas, J. & John, O. & Stará, J.: On the regularity up to the boundary for second order nonlinear elliptic systems. *Pacific J. Math.* 99 (1982), 1–17.

[JN-C155] Nečas, J. & Šilhavý, M.: Multipolar viscous fluids. *Quart. Appl. Math.* 49 (1991), 247–265.

[JN-C156] Nečas, J. & Šverák, V.: On regularity of solutions of nonlinear parabolic systems. *Ann. Scuola Norm. Sup. Pisa Cl. Sci. (IV)* 18 (1991), 1–11.

[JN-C165] Malek, J. & Nečas, J. & Růžička, M.: On the non-Newtonian incompressible fluids. *Math. Models Methods Appl. Sci.* 3 (1993), 35–63.

[JN-C167] Bellout, H. & Bloom, F. & Nečas, J.: Young measure-valued solutions for non-Newtonian incompressible fluids. *Comm. Partial Differential Equations* 19 (1994), no. 11–12, 1763–1803

[JN-C176] Nečas, J. & Růžička, M. & Šverák, V.: On Leray's self-similar solutions of the Navier-Stokes equations. *Acta Math.* 176 (1996), 283–294.

[JN-C184] Málek, J. & Nečas, J. & Růžička, M.: On weak solutions to a class of non-Newtonian incompressible fluids in bounded three-dimensional domains: the case $p \geq 2$. *Adv. Differential Equations* 6 (2001), 257–302.

References

[1] Acerbi, E. & Fusco, N.: Semicontinuity problems in the calculus of variations. *Arch. Ration. Mech. Anal.* 86 (1984), 125–145.

[2] Acquistapace, P.: On BMO regularity for linear elliptic systems. *Ann. Mat. Pure Appl. (4)* **161** (1992), 231–269.

[3] Almgren, F. J., Jr.: Some interior regularity theorems for minimal surfaces and an extension of Bernstein's theorem. *Ann. of Math. (2)* 84 (1966), 277–292.

[4] Agmon, S. & Douglis, A. & Nirenberg, L.: Estimates near the boundary for solutions of elliptic partial differential equations satisfying general boundary conditions. I. *Comm. Pure Appl. Math.* 12 (1959), 623–727.

[5] Attouch, H. & Sbordone, C.: Asymptotic limits for perturbed functionals of calculus of variations. *Ricerche Mat.* 29 (1980), 85–124.
[6] Ball, J.: Convexity conditions and existence theorems in nonlinear elasticity. *Arch. Ration. Mech. Anal.* 63 (1976/77), 337–403.
[7] Ball, J.: A version of the fundamental theorem for Young measures. *PDEs and continuum models of phase transitions (Nice, 1988), 207–215.* Lecture Notes in Phys., 344, Springer, Berlin, 1989.
[8] Beirao da Veiga, H.: On the Ladyzhenskaya-Smagorinsky turbulence model of the Navier-Stokes equations in smooth domains. The regularity problem. *J. Eur. Math. Soc. (JEMS)* 11 (2009), 127–167.
[9] Bensoussan, A. & Frehse, J.: *Regularity results for nonlinear elliptic systems and applications.* Applied Mathematical Sciences, 151. Springer-Verlag, Berlin, 2002.
[10] Bojarski, B. V.: Generalized solutions of a system of differential equations of first order and of elliptic type with discontinuous coefficients. (Russian) *Mat. Sb. N.S.* 43(85) (1957), 451–503.
[11] Brezis, H. & Browder, F.: Partial differential equations in the 20th century. *Adv. Math.* 135 (1998), 76–144.
[12] Campanato, S.: Hölder continuity of the solutions of some non-linear elliptic systems. *Adv. Math.* 48 (1983), 16–43.
[13] Campanato, S.: Elliptic systems with nonlinearity q greater or equal to two. Regularity of the solution of the Dirichlet problem. *Ann. Mat. Pura Appl. (IV)* 147 (1987), 117–150.
[14] Campanato, S.: L^p regularity and partial Hölder continuity for solutions of second order parabolic systems with strictly controlled growth. *Ann. Mat. Pura Appl. (IV)* 128 (1981), 287–316.
[15] Dacorogna, B.: *Direct Methods in the Calculus of Variations*, Second Edition, Applied Math. Sciences, Springer, 2008.
[16] Daněček, J. & Viszus, E.: $C^{0,\gamma}$-regularity for vector-valued minimizers of quasilinear functionals. *NoDEA Nonlinear Differential Equations Appl.* 16 (2009), 189–211.
[17] Daněček, J. & John, O. & Stará, J.: Interior $C^{1,\gamma}$-regularity for weak solutions of nonlinear second order parabolic systems. *Comm. Partial Differential Equations* 31 (2006), 1085–1098.
[18] De Giorgi, E.: Sulla differenziabilità e l'analiticità delle estremali degli integrali multipli regolari. *Mem. Accad. Sci. Torino Cl. Sci. Fis. Mat. Nat. (3)* 125 3 (1957), 25–43.
[19] De Giorgi, E.: Frontiere orientate di misura minima. *Seminario di Matematica della Scuola Normale Superiore di Pisa, 1960–61.*
[20] De Giorgi, E.: Un esempio di estremali discontinue per un problema variazionale di tipo ellittico. *Boll. Un. Mat. Ital.* (IV) 1 (1968), 135–137.
[21] DiBenedetto, E.: *Degenerate parabolic equations.* Universitext. Springer-Verlag, New York, 1993.
[22] DiBenedetto, E.: $C^{1+\alpha}$ local regularity of weak solutions of degenerate elliptic equations. *Nonlinear Anal.* 7 (1983), 827–850.
[23] Douglis, A. & Nirenberg, L.: Interior estimates for elliptic systems of partial differential equations. *Comm. Pure Appl. Math.* 8 (1955), 503–538.
[24] Du, Q. & Gunzburger, M. D.: Analysis of a Ladyzhenskaya model for incompressible viscous flow. *J. Math. Anal. Appl.* 155 (1991), 21–45.
[25] Duzaar, F. & Grotowski, J.F.: Optimal interior partial regularity for nonlinear elliptic systems: the method of A-harmonic approximation. *manuscripta math.* 103 (2000), 267–298.
[26] Duzaar, F. & Kristensen, J. & Mingione, G.: The existence of regular boundary points for non-linear elliptic systems. *J. reine ang. Math. (Crelles J.)* 602 (2007), 17–58.
[27] Duzaar, F. & Mingione, G. & Steffen, K.: Parabolic systems with polynomial growth and regularity. *Mem. Amer. Math. Soc.* 212 (2011).
[28] Elcrat, A. & Meyers, N.: Some results on regularity for solutions of non-linear elliptic systems and quasi-regular functions. *Duke Math. J.* 42 (1975), 121–136.

[29] Evans, L.C.: Quasiconvexity and partial regularity in the calculus of variations. *Arch. Ration. Mech. Anal.* 95 (1986), 227–252.
[30] Foss, M. & Mingione, G.: Partial continuity for elliptic problems. *Ann. Inst. H. Poincaré, Anal. Non Lineaire* 25 (2008), 471–503.
[31] Frehse, J.: A note on the Hölder continuity of solutions of variational problems. *Abh. Math. Sem. Univ. Hamburg* 43 (1975), 59–63.
[32] Frehse, J.: An irregular complex valued solution to a scalar uniformly elliptic equation. *Calc. Var. Partial Differential Equations* 33 (2008), 263–266.
[33] Gehring, F. W.: The L^p-integrability of the partial derivatives of a quasiconformal mapping. *Acta Math.* 130 (1973), 265–277.
[34] Giaquinta, M.: *Multiple integrals in the calculus of variations and nonlinear elliptic systems.* Annals of Mathematics Studies, 105. Princeton University Press, Princeton, NJ, 1983.
[35] Giaquinta, M. & Giusti, E.: On the regularity of the minima of variational integrals. *Acta Math.* 148 (1982), 31–46.
[36] Giaquinta, M. & Giusti, E.: Differentiability of minima of nondifferentiable functionals. *Invent. Math.* 72 (1983), 285–298.
[37] Giaquinta, M. & Modica, G.: Almost-everywhere regularity for solutions of nonlinear elliptic systems. *manuscripta math.* 28 (1979), 109–158.
[38] Gilbarg, D. & Trudinger, N.S.: *Elliptic partial differential equations of second order.* Grundlehren der Mathematischen Wissenschaften, Vol. 224. Springer-Verlag, Berlin-New York, 1977.
[39] Giusti, E.: *Direct methods in the calculus of variations.* World Scientific Publishing Co., Inc., River Edge, NJ, 2003.
[40] Giusti, E. & Miranda, M.: Un esempio di soluzioni discontinue per un problema di minimo relativo ad un integrale regolare del calcolo delle variazioni. *Boll. Un. Mat. Ital. (IV)* 1 (1968), 219–226.
[41] Giusti, E. & Miranda, M.: Sulla regolarità delle soluzioni deboli di una classe di sistemi ellittici quasi-lineari. *Arch. Ration. Mech. Anal.* 31 (1968/1969), 173–184.
[42] Gratwick, R. & Preiss, D.: A One-Dimensional Variational problem with continuous lagrangian and singular minimizer. *Arch. Ration. Mech. Anal.* (2011).
[43] Ivert, P.A.: Regularittsuntersuchungen von Lösungen elliptischer Systeme von quasilinearen Differentialgleichungen zweiter Ordnung. *manuscripta math.* 30 (1979/80), 53–88.
[44] Ivert, P.A.: Partial regularity of vector valued functions minimizing variational integrals. *Preprint Univ. Bonn, 1982.*
[45] Iwaniec, T.: The Gehring lemma. *Quasiconformal mappings and analysis (Ann Arbor, MI, 1995)*, 181–204, Springer, New York, 1998.
[46] John, O. & Malý, J. & Starà, J.: Nowhere continuous solutions to elliptic systems. *Comment. Math. Univ. Carolin.* 30 (1989), 33–43.
[47] Kinnunen, J. & Lewis, J. L.: Higher integrability for parabolic systems of p-Laplacian type. *Duke Math. J.* 102 (2000), 253–271.
[48] Koshelev, A.: *Regularity problem for quasilinear elliptic and parabolic systems.* Lecture Notes in Mathematics, 1614. Springer-Verlag, Berlin, 1995.
[49] Kristensen, J. & Melcher, C.: Regularity in oscillatory nonlinear elliptic systems. *Math. Z.* **260** (2008), 813–847.
[50] Kristensen, J. & Mingione, G.: The singular set of minima of integral functionals. *Arch. Ration. Mech. Anal.* 180 (2006), 331–398.
[51] Kristensen, J. & Mingione, G.: The singular set of Lipschitzian minima of multiple integrals. *Arch. Ration. Mech. Anal.* 184 (2007), 341–369.
[52] Kristensen, J. & Mingione, G.: Boundary regularity in variational problems. *Arch. Ration. Mech. Anal.* 198 (2010), 369–455.
[53] Kufner, A. & John, O. & Fučík, S.: *Function spaces.* Noordhoff International Publishing, Leyden; Academia, Prague, 1977.

[54] Ladyzhenskaya, O.A.: On nonlinear problems of continuum mechanics. *Proc. Internat. Congr. Math. (Moscow, 1966)*. Nauka, Moscow, 1968, pp. 560– 573; English transl. in *Amer. Math. Soc. Transl. (2)*, 70 (1968).

[55] Ladyzhenskaya, O.A.: New equations for the description of motion of viscous incompressible fluids and global solvability of boundary value problemes for them. *Trudy Mat. Inst. Steklov* 102 (1967), 85–104; English transl. in *Proc. Steklov Inst. Math.* 102 (1967).

[56] Ladyzhenskaya, O.A.: On some modifications of the Navier-Stokes equations for large gradients of velocity. *Zap. Nauchn. Sem. Leningrad. Otdel. Mat. Inst. Steklov (LOMI)* 7 (1968), 126–154; English transl. in *Sem. Math. V. A. Steklov Math. Inst. Leningrad* 7 (1968).

[57] Ladyzhenskaya, O.A.: *Mathematical problems in the dynamics of a viscous incompressible fluid*, 2-nd rev. aug. ed., Nauka, Moscow, 1970; English transl. of 1st ed., The mathematical theory of viscous incompressible flow, Cordon and Breach, New York, 1963.

[58] Ladyzhenskaya, O.A. & Ural'tseva, N.N.: *Linear and quasilinear elliptic equations.* Academic Press, New York-London 1968

[59] Ladyzhenskaya, O.A. & Solonnikov, V.A. & Ural'tseva, N.N.: *Linear and Quasilinear Equations of Parabolic Type.* Translations of Mathematical Monographs, vol. 23, American Mathematical Society, 1967.

[60] Leray, J.: Sur le mouvement d'un liquide visqueux emplissant l'espace. *Acta Math.* 63 (1934), 193–248.

[61] Lewis, J.: Smoothness of certain degenerate elliptic equations. *Proc. Amer. Math. Soc.* 80 (1980), 259–265.

[62] Li, Y.Y. & Nirenberg, L.: Estimates for elliptic systems from composite material. *Comm. Pure Appl. Math.* **56**(7) (2003), 892–925.

[63] Malý, J. & Ziemer, W.P.: *Fine regularity of solutions of elliptic partial differential equations.* Mathematical Surveys and Monographs, 51. American Mathematical Society, Providence, RI, 1997.

[64] Manfredi, J.J.: Regularity of the gradient for a class of nonlinear possibly degenerate elliptic equations. *Ph.D. Thesis*. University of Washington, St. Louis.

[65] Maz'ya, V. G.: Examples of nonregular solutions of quasilinear elliptic equations with analytic coefficients. (Russian) *Funkcional. Anal. i Pril.* 2 (1968), 53–57.

[66] Maz'ja, V. G. & Nasarov, S.A. & Plamenewski, B.A.: *Asymptotische Theorie Elliptischer Randwertaufgaben in Singulär Gestörten Gebieten*, Akademie–Verlag, Berlin 1991.

[67] Meier, M.: Boundedness and integrability properties of weak solutions of quasilinear elliptic systems. *J. Reine Angew. Math. (Crelles J.)* 333 (1982), 191–220.

[68] Meyers, M.: An L^p-estimate for the gradient of solutions to second order elliptic divergence equations. *Ann. Scuola Norm. Sup. Pisa (III)* 17 (1963), 189–206.

[69] Mingione, G.: The singular set of solutions to non-differentiable elliptic systems. *Arch. Ration. Mech. Anal.* 166 (2003), 287–301.

[70] Mingione, G.: Bounds for the singular set of solutions to non linear elliptic systems. *Calc. Var. & PDE* 18 (2003), 373–400.

[71] Mingione, G.: Regularity of minima: an invitation to the dark side of the Calculus of Variations. *Applications of Mathematics* 51 (2006), 355–425.

[72] Morrey, C.B.: On the solutions of quasi-linear elliptic partial differential equations. *Trans. Amer. Math. Soc.* **43** (1938), no. 1, 126–166.

[73] Morrey, C.B.: Second elliptic systems of differential equations. *Ann. of Math. Studies No.* **33** (1954), 101–159.

[74] Morrey, C.B.: Quasi-convexity and the lower semicontinuity of multiple integrals. *Pacific J. Math.* 2 (1952), 25–53.

[75] Morrey, C.B.: *Multiple integrals in the calculus of variations.* Springer-Verlag, New York, 1966.

[76] Morrey, C.B.: Partial regularity results for non-linear elliptic systems. *J. Math. Mech.* 17 1967/1968 649–670.

[77] Moser, J.: A new proof of De Giorgi's theorem concerning the regularity problem for elliptic differential equations. *Comm. Pure Appl. Math.* 13 (1960), 457–468.

[78] Moser, J.: On Harnack's theorem for elliptic differential equations. *Comm. Pure Appl. Math.* 14 (1961), 577–591.
[79] Müller, S. & Šverák, V.: Convex integration for Lipschitz mappings and counterexamples to regularity. *Ann. of Math. (II)* 157 (2003), 715–742.
[80] Nash, J.: Continuity of solutions of parabolic and elliptic equations. *Amer. J. Math.* 80 (1958), 931–954.
[81] Nečasová, Š. & Medvidová, M.: Bipolar barotropic non-Newtonian fluid. *Comment. Math. Univ. Carolin.* 35 (1994), 467–483.
[82] Nečasová, Š. & Penel, P.: Incompressible non-Newtonian fluids: time asymptotic behaviour of weak solutions. *Math. Methods Appl. Sci.* 29 (2006), 1615–1630.
[83] Nečasová, Š. & Penel, P.: L^2 decay for weak solution to equations of non-Newtonian incompressible fluids in the whole space. *Nonlinear Anal. TMA* 47 (2001), 4181–4192.
[84] Neustupa, J.: Partial regularity of weak solutions to the Navier-Stokes equations in the class $L^\infty(0,T;L^3(\Omega)^3)$. *J. Math. Fluid Mech.* 1 (1999), 309–325.
[85] Podio Guidugli, P.: De Giorgi's counterexample in elasticity. *Quart. Appl. Math.* 34 (1976/77), 411–419.
[86] Rajagopal, K.R. & Růžička, M.: Mathematical modeling of electrorheological materials. *Continuum Mechanics and Thermodynamics* 13 (2001), 59–78.
[87] Růžička, M.: *Electrorheological fluids: modeling and mathematical theory.* Lecture Notes in Mathematics, 1748. Springer-Verlag, Berlin, 2000.
[88] Said, E.: *On the late style: literature and music against the grain.* Pantheon Books, 2006.
[89] Seregin, G. A.: Interior regularity for solutions to the modified Navier-Stokes equations. *J. Math. Fluid Mech.* 1 (1999), 235–281.
[90] Seregin, G. A. & Šverák, V.: Navier-Stokes equations with lower bounds on the pressure. *Arch. Ration. Mech. Anal.* 163 (2002), 65–86.
[91] Serrin, J.: Local behavior of solutions of quasi-linear equations. *Acta Math.* 111 (1964), 247–302.
[92] Souček, J.: Singular solution to linear elliptic systems. *Comment. Math. Univ. Carolinae* 25 (1984), 273–281.
[93] Šverák, V. & Yan, X.: A singular minimizer of a smooth strongly convex functional in three dimensions. *Calc. Var. Partial Differential Equations* 10 (2000), 213–221.
[94] Šverák, V. & Yan, X.: Non-Lipschitz minimizers of smooth uniformly convex variational integrals. *Proc. Natl. Acad. Sci. USA* 99/24 (2002), 15269–15276.
[95] Uhlenbeck, K.: Regularity for a class of non-linear elliptic systems. *Acta Math.* 138 (1977), 219–240.
[96] Ural'tseva, N.N.: Degenerate quasilinear elliptic systems. *Zap. Na. Sem. Leningrad. Otdel. Mat. Inst. Steklov. (LOMI)* 7 (1968), 184–222.

Jan Kristensen
Mathematical Institute, University of Oxford, 24-29 St. Giles',
Oxford OX1 3LB, UK
email: jan.kristensen@maths.ox.ac.uk; kristens@maths.ox.ac.uk

Giuseppe Mingione
Dipartimento di Matematica, Università di Parma, Parco Area delle
Scienze 53/a, Campus, 43100 Parma, Italy
email: giuseppe.mingione@unipr.it

Part IV
Nonlinear Functional Analysis, Calculus of Variations

Chapter 9
Selected Papers on Nonlinear Functional Analysis and Calculus of Variations

SUR L'ALTERNATIVE DE FREDHOLM POUR LES OPÉRATEURS NON-LINÉAIRES AVEC APPLICATIONS AUX PROBLÈMES AUX LIMITES

JINDŘICH NEČAS

§ 1. Introduction.

La théorie des opérateurs monotons s'est montrée une base pour la solution des problèmes aux limites pour les équations non-linéaires aux dérivées partielles du type elliptique.

On considère un opérateur T d'un espace de Banach B, complex, réfléxif, dans son antidual B^*, en désignant par (Tu, v) la dualité entre B^* et B. Le problème fondamental : sous quelles hypothèses $T(B) = B^*$. Deux hypothèses suivantes sont importantes :

(1.1) monotonie : $\quad \operatorname{Re}(Tv - Tu, v - u) \geq 0,$

(1.2) coercitivité : $\quad \lim\limits_{\|u\| \to \infty} |(Tu, u)| \cdot \|u\|^{-1} = \infty .$

Si encore T est continu de chaque sous-espace de dimension finie dans B^* faible, on a $T(B) = B^*$, cf. le théorème 1 de ce travail ; si B est réel et séparable et $\lim\limits_{\|u\| \to \infty} (Tu, u) . \|u\|^{-1} = \infty$, cf. J. Leray, J. L. Lions [1]; sans séparabilité cf. F. E. Browder [2] et J. Nečas [6].

Si B est un espace de Hilbert et T un opérateur linéaire, continu et (v, u) le produit scalaire, on peut se donner encore un opérateur S de H dans H, complètement continu, et de considérer la question sur la surjectivité de l'opérateur $A_\lambda = T - \lambda S$, $\lambda \in C_1$. L'alternative de Fredholm a

Pervenuto alla Redazione il 31 Agosto 1968.

© Annali della Scuola Normale Superiore 1969. Reprinted with kind permission from: Nečas , J ., Sur l'alternative de Fredholm pour les opérateurs non-linéaires avec applications aux problèmes aux limites, Annali della Scuola Normale Superiore di Pisa - Classe di Scienze 23, 331-345.

lieu : ou bien λ est une valeur propre, ou bien il existe un opérateur inverse. La structure des valeurs propres est bien connue et on connaît aussi l'image de $T - \lambda S$ pour λ, une valeur propre. La condition (1.2) est remplacée dans ler applications linéaires du type elliptique par l'inégalité de Gårding. C'est l'idée principale qu'on a suivie en considérant les opérateurs non-linéaires.

Sous certaines conditions sur la monotonie et la coercivité de T, sous la condition que S est continu de la topologie faible dans la topologie forte et encore sous certaines conditions, dont un cas particulier très important est $\varkappa$-homogénéité de T et S : pour $\mu>0$, $u \in B$, $T(\mu u)=\mu^{\varkappa} T(u)$, $S(\mu u)=\mu^{\varkappa} S(u)$, nous avons obtenu un analogue de l'alternative de Fredholm : s'il n'existe pas u de B, $u \neq \theta$, telle que $Tu - \lambda Su = \theta$, alors $T - \lambda S$ est surjectif. On démontre aussi qu'il existe un sous-espace F de B^*, de dimension finie, tel que chaque f de B^* s'écrit sous la forme $f = f_1 + f_2$, où $f_1 \in (T - \lambda S)(B)$ et $f_2 \in F$.

EXEMPLES : Dans les exemples qui suivront, on désigne par

$$\overset{\circ}{W}{}_m^{(1)}(\Omega) = \left\{ u \,\middle|\, \int_\Omega \left(\sum_{i=1}^N \left| \frac{\partial u}{\partial x_i} \right|^m + |u|^m \right) dx < \infty,\ u = 0 \quad \text{sur } \partial\Omega \right\},$$

$\partial\Omega$ étant la frontière d'un domaine Ω de R_N.

EXEMPLE 1. On cherche u de $\overset{\circ}{W}{}_2^{(1)}(\Omega)$, telle que pour $v \in \overset{\circ}{W}{}_2^{(1)}(\Omega)$:

$$\int_\Omega \sum_{i=1}^N \frac{\partial u}{\partial x_i} \frac{\partial \bar{v}}{\partial x_i} dx - \lambda \int_\Omega \frac{|u|^s}{1+|u|^s} u \bar{v}\, dx = \int_\Omega f \bar{v}\, dx,$$

formellement : $u = 0$ sur $\partial\Omega$, $-\Delta u - \lambda u \dfrac{|u|^s}{1+|u|^s} = f$ dans Ω. $f \in L_2(\Omega)$, $s > 0$. Alors pour chaque f de L_2, il existe une solution si $\lambda \neq \lambda_k$, où $\{\lambda_k\}_{k=1}^\infty$ est le spectre pour le problème de Dirichlet et l'opérateur de Laplace.

EXEMPLE 2. On cherche u de $\overset{\circ}{W}{}_4^{(1)}(\Omega)$, telle que pour chaque v de $\overset{\circ}{W}{}_4^{(1)}(\Omega)$:

$$\int_\Omega \sum_{i=1}^N \left| \frac{\partial u}{\partial x_i} \right|^2 \frac{\partial u}{\partial x_i} \frac{\partial \bar{v}}{\partial x_i} dx - \lambda \int_\Omega |u|^2 u \bar{v}\, dx = \int_\Omega f \bar{v}\, dx,$$

formellement $u = 0$ sur $\partial\Omega$, $-\sum_{i=1}^N \dfrac{\partial}{\partial x_i} \left[\left| \dfrac{\partial u}{\partial x_i} \right|^2 \dfrac{\partial u}{\partial x_i} \right] - \lambda |u|^2 u = f$ dans Ω. Alors pour chaque f de $L_{\frac{4}{3}}(\Omega)$, il existe une solution, s'il n'existe pas u de $\overset{\circ}{W}{}^{(1)}_4(\Omega)$, $u \neq \theta$, telle que pour chaque $v \in \overset{\circ}{W}{}^{(1)}_4(\Omega)$:

$$\int_\Omega \sum_{i=1}^{N} \left|\frac{\partial u}{\partial x_i}\right|^2 \frac{\partial u}{\partial x_i}\,\frac{\partial \bar v}{\partial x_i}\,dx - \lambda \int_\Omega |u|^2\, u\, \bar v\, dx = 0.$$

Exemple 3. Soit $1 < m < \infty$. On cherche u de $\overset{\circ}{W}{}^{(1)}_m(\Omega)$, telle que pour chaque v de $\overset{\circ}{W}{}^{(1)}_m(\Omega)$ on a

$$\int_\Omega \sum_{i=1}^{N} \left(1 + \sum_{j=1}^{N} \left|\frac{\partial u}{\partial x_j}\right|^2\right)^{\frac{m}{2}-1} \cdot \frac{\partial u}{\partial x_i}\,\frac{\partial \bar v}{\partial x_i}\cdot dx - \lambda \int_\Omega (1 + |u|^2)^{\frac{m}{2}-1} \cdot u \bar v\, dx = \int_\Omega f \bar v\, dx.$$

Formellement $u = 0$ sur $\partial\Omega$,

$$-\sum_{j=1}^{N} \frac{\partial}{\partial x_i}\left[\left(1 + \sum_{j=1}^{N} \left|\frac{\partial u}{\partial x_j}\right|^2\right)^{\frac{m}{2}-1} \frac{\partial u}{\partial x_i}\right] - \lambda (1 + |u|^2)^{\frac{m}{2}-1} u = f \text{ dans } \Omega.$$

Alors pour chaque f de $L_{\frac{m}{m-1}}$, il existe une solution, s'il n'existe pas u de $\overset{\circ}{W}{}^{(1)}_m(\Omega)$, $u \neq \theta$, telle que pour v de $\overset{\circ}{W}{}^{(1)}_m(\Omega)$:

$$\int_\Omega \sum_{i=1}^{N} \left(\sum_{j=1}^{N} \left|\frac{\partial u}{\partial x_j}\right|^2\right)^{\frac{m}{2}-1} \frac{\partial u}{\partial x_i}\,\frac{\partial \bar v}{\partial x_i}\,dx - \lambda \int_\Omega |u|^{m-2} \cdot u \cdot \bar v \cdot dx = 0.$$

§ 2. Une généralisation du théorème de Leray-Lions.

Nous considérons dans la suite les espaces de Banach B, complex. Par B^*, on désigne l'antidual de B, à savoir l'espace des fonctionelles antilinéaires. On désigne par (f, u) la dualité entre B^* et B.

Hypothèse 1. *Soit A, un opérateur borné, continu de tout sous-espace de B de dimension finie dans B^* faible. A soit coercitif:*

(2.1) $$\lim_{\|u\| \to \infty} |(Au, u)| \cdot \|u\|^{-1} = \infty.$$

Hypothèse 2. *Il existe une application bornée de $B \times B$ dans B^*, soit $A(v, u)$, telle que $A(u, u) = A(u)$ et telle que*

(2.2) $\begin{cases} \forall\, u \text{ de } B, \text{ l'application } v \longrightarrow A(v, u) \text{ est hemicontinue (continue de toute droite de } B \text{ dans } B^* \text{ faible) et telle que pour } u, v \text{ de } B, \\ \operatorname{Re}(A(u, u) - A(v, u), u - v) \geq 0, \end{cases}$

(2.3) $\begin{cases} \text{si } u_n \longrightarrow u \text{ (convergence faible) et si } (A(u_n, u_n) - A(u, u_n), \\ u_n - u) \longrightarrow 0, \text{ alors pour } v \text{ de } B : A(v, u_n) \longrightarrow A(v, u) \text{ dans } B^*, \end{cases}$

(2.4) *si $u_n \longrightarrow u$ et $A(v, u_n) \longrightarrow v^*$ dans B^*, alors $(A(v, u_n), u_n) \longrightarrow (v^*, u)$.*

Nous dirons que $A(B) = B^*$ régulièrement, si de $f \in B^*$, $\|f\| \leq R < \infty$, suit $\|u\| \leq r < \infty$, u étant la solution de $A(u) = f$.

Théorème 1. ***Soit B un espace de Banach réflexif, complex, de dimension ≥ 2 et A une application de B dans B^*, satisfaisfaisant aux hypothèses 1 et 2. Alors $A(B) = B^*$ règuliérement.***

Démonstration. Soit $F \subset B$ un sous-ensemble de dimension finie, F^* son antidual et désignons par $(f, v)_F$ la dualité entre F^* et F. Définissons A_F de F dans F^* par $u, v \in F \Longrightarrow (A_F v, u)_F = (Av, u)$. Si l'on introduit dan F et F^* une base biorthonormale, soit e_i, e_j^*, $i, j = 1, 2, \ldots \varkappa$, alors u de F s'écrit comme $u = \sum_{j=1}^{\varkappa} x_i e_i$, f de F^* s'écrit comme $f = \sum_{j=1}^{\varkappa} y_j e_j^*$ et on peut identifier F et F^* avec l'espace euclidien $C_\varkappa$, au produit scalaire $(x, y) = \sum_{i=1}^{\varkappa} x_i \bar{y}_i$ et l'opérateur A_F avec un opérateur R, continu de $C_\varkappa$ dans lui même, defini par $(A_F v, u) = (Rx, y)_{C_\varkappa}$, où $v = \sum_{i=1}^{\varkappa} x_i e_i$, $u = \sum_{i=1}^{\varkappa} y_i e_i$.

On a

$$\lim_{\|x\| \to \infty} |(Rx, x)_{C_\varkappa}| \, \|x\|^{-1} = \infty . \tag{2.5}$$

Soit $f \in B^*$, $(f_F, v)_F = (f, v)$ et soit $\eta \in C_\varkappa$ avec $f_F = \sum_{i=1}^{\varkappa} \eta_i e_j^*$. Considérons encore $Ux = Rx - \eta$ qui satisfait aussi à (2.5). Il s'agit de trouver $x \in C_\varkappa$ de sorte que $Ux = 0$. Soit $\varrho > 0$ si grand pour que $|(Ux, x)| \neq 0$ pour $|x| = \varrho$. Sans restreindre la généralité, on peut supposer $\varkappa \geq 2$. Dans ce cas, on peut trouver $\arg(Ux, x)$ de sorte qu'il résulte continue sur la sphère $x| = \varrho$. Posons $V(x) = \exp\left(-i \cdot \frac{\|x\|}{\varrho} \cdot \arg\left(U \frac{x}{\|x\|} \varrho, \frac{x}{\|x\|} \varrho\right)\right) \cdot U(x)$; c'est

une application continue de C_n dans lui même. Sur $|x| = \varrho$, on a $(V(x), x) > 0$, alors on peut trouver $\varepsilon > 0$ si petit pour que la transformation $x - \varepsilon V(x)$ transforme la boule $|x| \leq \varrho$ dans elle même. D'après le théorème de Brouwer, il existe un point fixe x_F, dans cette boule, alors $V(x_F) = 0$, d'où $U(x_F) = 0$. Désignons par u_F le point correspondant dans F. Nous avons alors pour v de F:

$$(A_F u_F, v)_F = (A u_F, v) = (f, v). \tag{2.6}$$

Il suit de (2.1) qu'il existe une constante $c > 0$ telle que pour chaque F, $\|u_F\| \leq c$. Désignons par Λ l'ensemble de tous les sous-espaces de B de dimension finie et ≥ 2 et par $B_{F_0} = \bigcup_{\Lambda \ni F \supset F_0} u_F$. Soit $\overline{B}_{F_0}$ la fermeture de B_{F_0} dans la topologie faible. La famille de $\overline{B}_{F_0}$ possède la propriété des intersections finies. Il en suit que $\bigcap_{F_0 \in \Lambda} \overline{B}_{F_0} \neq \theta$, donc il existe $u_0 \in \bigcap_{F_0 \in \Lambda} \overline{B}_{F_0}$. Soit v de B et $F_0 \in \Lambda$ contenant v et u_0. Il existe maintenant une suite u_n de B_{F_0}, telle que $u_n \rightharpoonup u$. Comme u_n est une suite bornée, $A(u_0, u_n)$ l'est aussi, alors, on peut supposer que $A(u_0, u_n) \rightharpoonup u^*$. Vérifions que

$$(A(u_n, u_n) - A(u_0, u_n), u_n - u_0) \to 0. \tag{2.7}$$

En effet, $(A(u_n, u_n), u_n) = (f, u_n) \to (f, u_0)$, $(A(u_n, u_n), u_0) = (u_0, f)$. D'après (2.4): $(A(u_0, u_n), u_n) \to (u^*, u_0)$ et enfin $(A(u_0, u_n), u_0) \to (u^*, u_0)$; tout cela entraîne (2.7). Alors pour v en question, compte tenu de (2.3), $A(v, u_n) \rightharpoonup A(v, u_0)$. Il s'en suit de (2.4) que $(A(v, u_n), u_n) \to (A(v, u_0), u_0)$. Soit $C_n = (A(u_n, u_n) - A(v, u_n), u_n - v)$. On a $C_n \to (f - A(v, u_0), u_0 - v)$. Mais d'après (2.2), Re $C_n \geq 0$, donc on a

$$\operatorname{Re}(f - A(v, u_0), u_0 - v) \geq 0 \tag{2.8}$$

pour chaque v de B. Si nous posons $v = u_0 - \lambda w$, w de B, $\lambda > 0$, il suit de (2.8) que $\operatorname{Re}(f - A(u_0 - \lambda w, u_0), w) \geq 0$ et en faisant tendre $\lambda \to 0$, il en suit que $\operatorname{Re}(f - A(u_0), w) \geq 0$, d'où $A(u_0) = f$; pour chaque boule $K_\varrho = \{\|f\| \leq \varrho\}$, il suit de (2.1) que $A^{-1}(K_\varrho)$ est bornée, d'où l'assertion.

§ 3. Alternative.

Soit $T(u)$ un opérateur du théorème 1, tel que

$$(T(u), u) \geq 0. \tag{3.1}$$

Hypothèse 3. *Soit $S(u)$ un opérateur de B dans B^*, continu « de la topologie faible dans la topologie forte » : si $u_n \rightharpoonup u$, alors $S(u_n) \to S(u)$ dans B^*. Soit encore*

$$(3.2) \qquad (S(u), u) \geq 0.$$

Désignons par $A_\lambda(u) = T(u) - \lambda S(u)$. Évidemment, il suit du théorème 1 :

Conséquence 1. *Pour $\lambda \leq 0$, $A_\lambda(B) = B^*$ régulièrement.*

Théorème 2. *Soit $\lambda = \lambda_1 + i\lambda_2$ et $A_\lambda = T - \lambda S$ avec T satisfaisant aux hypothèses 1 et 2 et S à l'hypothèse 3. Supposons encore (3.1). Alors si $\lambda_2 \neq 0$, l'opérateur A_λ est régulièrement surjectif.*

Démonstration. Il suffit de démontrer (2.1). Mais (3.2)' : $|(A_\lambda u, u)|^2 = (Tu, u)^2 + \lambda_1^2 (Su, u)^2 - 2\lambda_1 (Tu, u)(Su, u) + \lambda_2^2 (Su, u)^2 \geq (Tu, u)^2 (1 - \varepsilon\lambda_1) + \left(\lambda_1^2 + \lambda_2^2 - \frac{\lambda_1}{\varepsilon}\right)(Su, u)^2$ pour chaque $\varepsilon > 0$.

Si nous prenons $0 < \varepsilon < \frac{1}{\lambda_1}$ mais assez proche de $1/\lambda_1$, nous obtenons le résultat.

Nous avons un lemme important :

Lemme 1. *Soit T un opérateur borné de B dans B^*, satisfaisant à l'hypothèse 2. Alors si $u_n \rightharpoonup u$ et $T(u_n) \to g$, il en suit que $g = T(u)$.*

Démonstration. On peut supposer que $T(u, u_n) \rightharpoonup u^*$; d'après (2.4), $(T(u, u_n), u_n) \to (u^*, u)$, d'où $(T(u_n, u_n) - T(u, u_n), u_n - u) \to 0$, d'où d'apres (2.3) pour chaque v de B : $T(v, u_n) \rightharpoonup T(v, u)$. D'après (2.2) $\operatorname{Re}(T(u_n, u_n) - T(v, u_n), u_n - v) \geq 0$; si $n \to \infty$, on obtient de (2.4) que $(T(v, u_n), u_n) \to (T(v, u), u)$, d'où $\lim_{n\to\infty} \operatorname{Re}(T(u_n, u_n) - T(v, u_n), u_n - v) = \operatorname{Re}(g - T(v, u), u - v) \geq 0$.

Si nous posons $v = u - \lambda w$, $\lambda > 0$, w de B, nous obtenons $\operatorname{Re}(g - T(u - \lambda w, u), w) \geq 0$ et en faisant tendre $\lambda \to 0$, on obtient $\operatorname{Re}(g - T(u), w) \geq 0$, d'où $g = T(u)$.

Nous dirons qu'un opérateur N de B dans B^* est $\varkappa > 0$ homogène, si pour chaque $0 < \mu$, u de B : $N(\mu u) = \mu^\varkappa N(u)$.

Nous dirons qu'un opérateur N de B dans B^* est $\varkappa > 0$ quasi-homogène, s'il existe un opérateur N_0 de B dans B^*, $\varkappa$-homogène, tel que : si $w_n \rightharpoonup w$ et si les nombres $\mu_n > 0$, $\mu_n \to 0$ et si $\mu_n^\varkappa N\left(\frac{w_n}{\mu_n}\right) \to g$, alors

$$(3.2)'' \qquad \lim_{n\to\infty} \mu_n^\varkappa N\left(\frac{w_n}{\mu_n}\right) = N_0(w).$$

Nous dirons que N est fortement $\varkappa > 0$ quasi-homogène si (3.2)″ est valable pour chaque $\mu_n \to 0$, $\mu_n > 0$, $w_n \rightharpoonup w$. (Dans ce cas N_0 est $\varkappa$-homogène automatiquement).

PROPOSITION 1. *Soit S un opérateur de B dans B^* continu « de la topologie faible dans la topologie forte ». Si S est $\varkappa$-homogène, il est $\varkappa$ quasi-homogène fortement avec $S_0 = S$.*

En effet $\mu_n^\varkappa S\left(\dfrac{w_n}{\mu_n}\right) = S(w_n) \to S(w)$.

PROPOSITION 2. *Soit T un opérateur de B dans B^*, satisfaisant à l'hypothèse 2. Alors s'il est $\varkappa$ homogène, il est $\varkappa$ quasi-homogène avec $T_0 = T$.*

En effet $\mu_n^\varkappa T\left(\dfrac{w_n}{\mu_n}\right) = T(w_n)$. Maintenant, on utilise le lemme 1.

PPROPOSITION 3. *Si T est un opárateur de B dans B^* satisfaisant à* (2.1) *et s'il est $\varkappa$-homogène, on a*

$$|(T(u), u)| \geq c \| u \|^{1+\varkappa}, \; c > 0.$$

En effet, pour $\| u \| = \varrho$, avec ϱ assez grand, $|(T(u), u)| \geq c_1 > 0$. On a pour u de B

$$|(T(u), u)| = \left|\left(T\left(\frac{\| u \|}{\varrho}\frac{u}{\| u \|}\varrho\right), \frac{\| u \|}{\varrho}\frac{u}{\| u \|}\varrho\right)\right| \geq \left(\frac{\| u \|}{\varrho}\right)^{1+\varkappa} c_1,$$

d'où l'assertion.

THÉORÈME 3. (*alternative*) *Soient T et S les opérateurs satisfaisant aux hypothèses 1 et 2 et à l'hypothèse 3 respectivement. Soit T $\varkappa$-quasi-homogène et S fortement $\varkappa$-quasi-homogène. Supposons encore*

$$(T(u), u) \geq c \| u \|^{1+\varkappa}, \; c > 0. \tag{3.3}$$

Soit $\lambda > 0$. Alors

(i) *ou bien il existe $u \neq \theta$, « un élément propre » tel que $T_0(u) - \lambda S_0(u) = \theta$*

(ii) *ou bien il n'existe pas ; dans ce cas $A_\lambda = T - \lambda S$ est régulièrement surjectif.*

DÉMONSTRATION. D'après le théorème 2, si nous posons $\lambda^* = \lambda + i\lambda_2$, $\lambda_2 > 0$, nous obtenous que A_{λ^*} est régulièrement surjectif. Considérons $\lambda > 0$

qui n'est pas une valeur propre et soit $\lambda^* = \lambda + i\lambda_2$, $0 < \lambda_2 \leq 1$. J'affirme que $A_{\lambda^*}^{-1}(K_R)$ est uniformément borné par rapport à $0 < \lambda_2 \leq 1$. Supposons que ceci n'est pas vrai. Dans ce cas, il existe f_n de B^*, $\| f_n \| \leq R$, $\lambda_{2n} \to 0$ et u_n de B, telles que $A_{\lambda_n^*}(u_n) = f_n$, $\| u_n \| \to \infty$. ($\lambda_{2n} \to 0$ en vertu du théorème 2, où on a démontré de plus l'inégalité (3.2)'). Soit $v_n = \frac{u_n}{\| u_n \|}$. On peut supposer que $v_n \rightharpoonup v$. On a

$$\| u_n \|^{-\varkappa} A_{\lambda_n^*}(u_n) = \| u_n \|^{-\varkappa} A_{\lambda_n^*}(\| u_n \| v_n) = \| u_n \|^{-\varkappa} f_n \tag{3.4}$$

et d'autre part, ètant

$$\| u_n \|^{-\varkappa} S(\| u_n \| v_n) \to S_0(v), \tag{3.5}$$

on a $\| u_n \|^{-\varkappa} T(\| u_n \| v_n) \to \lambda S_0(v)$,

donc $\| u_n \|^{-\varkappa} T(\| u_n \| v_n) \to T_0(v)$. On a alors $T_0(v) - \lambda S_0(v) = \theta$.

D'autre part, $\| u_n \|^{-\varkappa}(T(\| u_n \| v_n), v_n) = \| u_n \|^{-1-\varkappa}(T(v_n \| u_n \|), v_n \| u_n \|) \geq$ $\geq c > 0$, et étant $\lim_{n\to\infty} \| u_n \|^{-\varkappa}(T(\| u_n \| v_n), v_n) = (T_0(v), v)$, on a $(T_0(v), v) > 0$, donc $v \neq \theta$. Mais c'est une contradiction.

Soit alors f de B^* et u_{λ^*} de K_ϱ, les solutions de $A_{\lambda^*}(u) = f$, pour $\lambda^* = \lambda + i\lambda_2$, $\lambda_2 > 0$. Soit $\lambda_{2n} \to 0$ et désignons $u_{\lambda_n^*} = u_n$. On peut supposer que $u_n \rightharpoonup u$. Mais $A_{\lambda_n^*}(u_n) = T(u_n) - \lambda_n^* S(u_n) = f$, d'où $T(u_n) \to f + \lambda$. Su, alors d'après le lemme 1, $Tu = f + \lambda Su$, d'où l'assertion.

PROPOSITION 4. *Sous les hypothèses du théorème 3, les valeurs propres sont réels, positives.*

En effet, si λ est une valeur propre, $(T_0(u), u) = \lambda(S_0(u), u)$, $(S_0(u), u) \neq 0$, d'où $\lambda = \frac{(T_0(u), u)}{(S_0(u), u)}$.

§ 4. Image des applications $T - \lambda S$.

Dans ce paragraphe, nous donnerons quelques résultats simples sur l'image de $T - \lambda S$ au cas général, alors aussi au cas, ou λ est une valeur propre de $T_0 - \lambda S_0$. Ces résultats généralisent dans un sens les résultats de F. F. Browder [4].

Nous supposerons pour l'opérateur T l'hypothèse 2 et la coercitivité dans une forme un peu plus forte :

$$(4.1)\qquad |(T(u)\,u)| \geq c_1\,\|u\|^{1+\varkappa} - c_2\,,\ c_1\,,\ c_2 > 0, \quad \varkappa > 0.$$

Pour l'opérateur S, nous supposerons :

Hypothèse 4. *Il existe G, un espace de Banach, tel que $B \subset G$ algébriquement et topologiquement et tel que l'application identique de B dans G est compacte ; S est continue de G dans B^*.*

Désignons par $\|u\|_0$ la norme dans G.

Hypothèse 5. *Il existe une application complètement continu, soit S_1 de G dans B^*, $\varkappa$-homogène, et telle que*

$$(4.2)\qquad \|S(u) - S_1(u)\| \leq c\,(1 + |u|)^{\varkappa'}, \qquad 0 \leq \varkappa' < \varkappa.$$

Nous avons le lemme suivant :

Lemme 2. *Soit S_1 une application complètement continue de G dans B, $\varkappa$-homogène. Alors à chaque $\varepsilon > 0$, il existe une application complètement continue, $\varkappa$-homogène, soit R, de G dans un sous-espace F de B de dimension finie, telle que pour u de G :*

$$(4.3)\qquad \|S_1(u) - R(u)\| \leq \varepsilon\,\|u\|_0^{\varkappa}.$$

Démonstration. L'application S_1 étant complètement continue de G, l'image de la sphère $K = \{\|u\|_0 = 1\}$: $S_1(K)$ est compacte. Donc, il existe un $\varepsilon > 0$ réseau dans $S_1(K)$, soit $f_1, f_2, \dots, f_p$ tel que si $f \in S_1(K)$, alors il existe f_i tel que $\|f - f_i\| < \varepsilon$. Soit F le sous-espace engendré par f_j, $j = 1, 2, \dots, p$. Posons pour u de K :

$$\mu_i(u) = \varepsilon - \|S_1(u) - f_i\|\,, \quad \text{si} \quad \|S_1(u) - f_i\| \leq \varepsilon \quad \text{et} \quad \mu_i(u) = 0$$

si $\|S_1(u) - f_i\| > \varepsilon$. Posons $P(u) = \dfrac{\sum_{i=1}^{p} \mu_i(u) f_i}{\sum_{=1}^{p} \mu_i(u)}$. Enfin soit $R(u) = \|u_0\|^{\varkappa} \cdot$ $\cdot P\left(\dfrac{u}{\|u\|_0}\right)$. On obtient aisément (4.3), d'où l'assertion.

Théorème 4. *Soit T une application de B dans B^* satisfaisant à l'hypothèse 2, à l'inégalité (4.1) et S l'opérateur complètement continu de G dans*

B^* satisfaisant aux hypothèses 4 et 5. Alors il existe un opérateur R, complètement continu de G dans B^*, $\varkappa$-homogène, avec image dans $F \subset B^*$, un espace de dimension finie, tel que $T - \lambda S = T_0 - \lambda R$, où T_0 satisfait à l'hypothèse 2, et à l'inégalité (4.1). Il en suit que chaque élément f de B^* peut s'écrire comme $f = f_1 + f_2$, où $f_1 \in (T - \lambda S)(B)$ et $f_2 \in F$.

DÉMONSTRATION. $T - \lambda S = T - \lambda(S - S_1) - \lambda S_1$. L'opérateur $T_1 = T - \lambda(S - S_1)$ satisfait à l'hypothèse 2 et à (4.1): l'hypothèse 2 est évidente en vertu de la continuité complète de S de G dans B^*. Mais

$$|(T_1(u), u)| = |(T - \lambda(S - S_1)(u), u)| \geq c_1 \|u\|^{1+\varkappa} - c_2 -$$
$$- |\lambda| (1 + \|u\|)^{\varkappa'} \|u\| \geq c_3 \|u\|^{1+\varkappa} - c_4 ,$$

d'où l'énoncé. Pour S_1, on applique le lemme 2 et on construit R. On obtient $T - \lambda S = T_1 - \lambda(S_1 - R) - \lambda R$. Si nous posons $T_0 = T_1 - \lambda(S_1 - R)$, on a de nouveau l'hypothèse 2 en vertu de la continuité complète de R de G dans B^*. Pour obtenir (4.1), on a $|(T_0(u), u)| \geq c_3 \|u\|^{1+\varkappa} - c_4 - |\lambda| \varepsilon c_5 \|u\|^{1+\varkappa}$, où c_5 est la norme de l'application identique de G dans B. Il suffit maintenant de prendre $\varepsilon < \frac{c_3}{c_5 |\lambda|}$. Si $f \in B^*$, il existe u de B tel que $T_0(u) = f$. Posons $f_2 = \lambda R(u)$ et $f_1 = (T - \lambda S)(u)$, d'où la démonstration.

§ 5. Applications aux problèmes aux limites.

D'abord, nous démontrerons encore un lemme facile, utile dans la suite.

LEMME 3. *Soient T_n, $n = 1, 2, \ldots$, une suite des opérateurs de B dans B^*, monotone:*

$$\mathrm{Re}\,(T_n(u) - T_n(v), u - v) \geq 0 \tag{5.1}$$

et tels que

$$\lim_{n \to \infty} T_n(u) = T(u) \text{ dans } B^* \text{ pour chaque } u \text{ de } B. \tag{5.2}$$

Soit l'opérateur T hemicontinu. Alors si $u_n \rightharpoonup u$ et si $T_n(u_n) = g$, alors $T(u) = g$.

En effet, pour chaque v de B: $0 \leq \mathrm{Re}\,(T_n(u_n) - T_n(v), u_n - v) \to \mathrm{Re}\,(g - T(v), u - v)$. Si nous posons $v = u - \lambda w$, $\lambda > 0$, nous obtenons $\mathrm{Re}\,(g - T(u - \lambda w), w) \geq 0$ et par le passage à la limite $\lambda \to 0$: $\mathrm{Re}\,(g - T(u), w) \geq 0$, d'où le résultat.

Dans la suite, on va se bormer aux applications simples de la théorie ci-dessus. Le lecteur pourrait faire les généralisations lui même. Pour la littérature, cf. J. Leray, J. L. Lions [1], J. Nečas [3].

Soit Ω un domaine borné de l'espace euclidien R_N, à frontière lipschitzienne. On désigne par $\mathcal{D}(\Omega)$ l'espace des fonctions complexes, à support compact dans Ω. La notation usuelle $D^i = \frac{\partial^{|i|}}{\partial x_1^{i_1} \dots \partial x_N^{i_N}}$ est utilisée. On introduit comme d'habitude l'espace $W_m^{(k)}(\Omega)$ avec la norme $\| u \|_{W_m^{(k)}} = \left(\int_\Omega \sum_{|i| \leq k} | D^i u |^m dx \right)^{\frac{1}{m}}$. On désigne encore par $\mathring{W}_m^{(k)}(\Omega)$ le fermeture de $\mathcal{D}(\Omega)$ dans $W_m^{(k)}(\Omega)$.

On définit les fonctions $a_i(x, \zeta_j)$ pour x de Ω, $|\zeta_j| < \infty$, $|i| \leq k$, $|j| \leq k$. On suppose que $a_i(x, \zeta_j)$ sont mesurables en x pour tous ζ_j, ζ_j fixés et continues en ζ_j pour presque tout x (condition de Carathéodory). On suppose encore

$$|a_i(x, \zeta_j)| \leq c\,(1 + \sum_{|j| \leq k} |\zeta_j|)^{m-1}, \quad 1 < m < \infty. \tag{5.3}$$

Il est connu, cf. par exemple M. M. Vajnberg [5]: l'opérateur $a_i(x, D^j u)$ résulte continu de $W_m^{(k)}(\Omega)$ dans $L_{m'}(\Omega)$, $\frac{1}{m} + \frac{1}{m'} = 1$.

Soit f de $L_{m'}(\Omega)$, u_0 de $W_m^{(k)}(\Omega)$. On s'occupera du problème de Dirichlet (pour les autres problèmes cf. J. Nečas [3]). Une fonction u de $W_m^{(k)}(\Omega)$ est dite une solution faible du problème de Dirichlet:

$$\sum_{|i| \leq k} (-1)^{|i|} D^i (a_i(x, D^j u)) = f \quad \text{dans} \quad \Omega, \tag{5.4}$$

$$\frac{\partial^l u}{\partial n^l} = \frac{\partial^l u_0}{\partial n^l} \quad l = 0, 1, \dots, k-1 \quad \text{sur} \quad \partial\Omega \tag{5.5}$$

$\left(\frac{\partial}{\partial n} \right.$ étant la dérivée selon la nomale $\left. \text{extérieure} \right)$ si pour chaque v de $\mathring{W}_m^{(k)}(\Omega)$:

$$\int_\Omega \sum_{|i| \leq k} a_i(x, D^j u) \cdot D^i \bar{v} \cdot dx = \int_\Omega f \bar{v}\, dx, \tag{5.6}$$

$$u - u_0 \in \mathring{W}_m^{(k)}(\Omega). \tag{5.7}$$

Désignons encore par $a_i(x, \zeta_\alpha, \zeta_\beta) = a_i(x, \zeta_j)$, où $|\alpha| = k$, $|\beta| < k$. Supposons

(5.8) $$\sum_{|i| \leq k} a_i(x, \zeta_j)\bar{\zeta}_i \geq c_1 \sum_{|i|=k} |\zeta_i|^m,$$

(5.9) $$\operatorname{Re} \sum_{|i|=k} (a_i(x, \xi_\alpha, \zeta_\beta) - a_i(x, \eta_\alpha, \zeta_\beta))(\bar{\xi}_i - \bar{\eta}_i) > 0$$

pour $\xi \neq \eta$,

(5.10) $$\sum_{|i|=k} a_i(x, \xi_\alpha, \eta_\beta)\bar{\xi}_i \geq c_3 \sum_{|i|=k} |\xi_i|^m - c_4.$$

Posons

$$(T(u, v), w) \overset{\text{df}}{=} \int_\Omega \sum_{|i|=k} a_i(x, D^\alpha u_0 + D^\alpha u, D^\beta u_0 + D^\beta v) D^i \bar{w}\, dx +$$

$$+ \int_\Omega \sum_{|i|<k} a_i(x, D^j u_0 + D^j v) \cdot D^i \bar{w}\, dx.$$

Il est démontré dans J. Leray, J. L. Lions [1] ou dans J. Nečas [3] que l'opérateur $T(u, v)$ ainsi défini satisfait aux hypothèses du théorème 3. Soient $b_i(x, \zeta_j)$ définies comme $a_i(x, \zeta_j)$ mais pour $|i| < k$, $|j| < k$. On suppose

(5.11) $$|b_i(x, \zeta_j)| \leq c\,(1 + \sum_{|j|<k} |\zeta_j|)^{m-1},$$

(5.12) $$\sum_{|i| \leq k} b_i(x, \zeta_j)\bar{\zeta}_i \geq 0$$

et posons

(5.12)′ $$(S(u), v) \overset{\text{df}}{=} \int_\Omega \sum_{|i|<k} b_i(x, D^j u) D^i \bar{v}\, dx.$$

Compte tenu de ce que l'application identique de $W_m^{(k)}(\Omega)$ dans $W_m^{(k-1)}(\Omega)$ est complètement continue, on a l'hypothèse 3. Si nous supposons que pour chaque $\mu > 0$, $\varkappa > 0$ fixé, $\varkappa = m - 1$:

(5.13) $$\begin{cases} a_i(x, \mu\zeta_j) = \mu^\varkappa \cdot a_i(x, \zeta_j), \\ b_i(x, \mu\zeta_j) = \mu^\varkappa\, b_i(x, \zeta_j), \end{cases}$$

nous pouvons utiliser les propositions 2 et 3. Dans ces conditions, nous avons :

THÉORÈME 5. *On suppose les conditions* (5.3), (5.8), (5.13), $u_0 \equiv \theta$ *de* (5.7). *Alors si* $\operatorname{Im} \lambda \neq 0, \lambda \leq 0$ *ou si* $\lambda > 0$ *et dans le dernier cas s'il n'existe pas* $u \neq \theta$ *dans* $\overset{\circ}{W}{}_m^{(k)}(\Omega)$ *telle que pour* v *de* $\overset{\circ}{W}{}_m^{(k)}(\Omega)$: $\int_\Omega \sum_{|i| \leq k} a_i(x, D^j u) D^i \bar{v}\, dx -$ $- \lambda \int_\Omega \sum_{|i| < k} b_i(x, D^j u) D^i \bar{v}\, dx = 0$, *alors pour chaque* f *de* $L_{m'}(\Omega)$ (*à plus forte raison pour* f *de* $(\overset{\circ}{W}{}_m^{(k)}(\Omega))^*$) *il existe une solution faible* u *de* $W_m^{(k)}(\Omega)$ *de l'équation*

$$(5.13)' \qquad \sum_{|i| \leq k} (-1)^{|i|} D^i(a_i(x, D^j u)) - \lambda \sum_{|i| < k} (-1)^{|i|} D^i(b_i(x, D^j u)) = f$$

dans Ω. (*Cf. exemple* 2).

Supposons maintenant pour $\varkappa = m - 1$ et $0 < \mu \leq 1$ que

$$(5.14) \qquad \begin{cases} \mu^{m-1} a_i\left(x, \dfrac{\zeta_i}{\mu}\right) = a_i(x, \zeta_j, \mu), \\[2ex] \mu^{m-1} b_i\left(x, \dfrac{\zeta_i}{\mu}\right) = b_i(x, \zeta_j, \mu), \end{cases}$$

où $a_i(x, \zeta_j, \mu)$, $b_i(x, \zeta_j, \mu)$ à côté de la condition de Carathéodory sont pour presque tout x de Ω et pour tous ζ_j continues en μ dans l'intervalle $0 \leq \mu \leq 1$. Supposons

$$(5.15) \qquad \begin{aligned} |a_i(x, \zeta_j, \mu)| &\leq c\,(1 + \sum_{|j| \leq k} |\zeta_j|)^{m-1}, \\ |b_i(x, \zeta_j, \mu)| &\leq c\,(1 + \sum_{|j| < k} |\zeta_j|)^{m-1}. \end{aligned}$$

Pour fixer les idées, sopposons encore que pour $\mu > 0$:

$$(5.16) \qquad \operatorname{Re} \sum_{|i| \leq k} (a_i(x, \xi_\alpha, \mu) - a_i(x, \eta_\alpha, \mu)) (\bar{\xi}_i - \bar{\eta}_i) \geq 0.$$

Définissons :

$$(5.17) \qquad (T(u), v) \overset{\mathrm{df}}{=} \int_\Omega \sum_{|i| \leq k} a_i(x, D^j u) D^i \bar{v}\, dx$$

et $S(u)$ comme audessus. Nous avons :

LEMME 4. *L'opérateur T, défini par* (5.17), *sous les conditions* (5.3), (5.14), (5.15) *et* (5.16) *est $m-1$ quasi-homogène avec $T_0(u)$ défini par* $(T_0(u), v) \overset{\mathrm{df}}{=}$ $\overset{\mathrm{df}}{=} \int_\Omega \sum_{|i| \leq k} a_i(x, D^j u, 0) D^i \bar{v}\, dx$ *et* $B = \overset{\circ}{W}{}_m^{(k)}(\Omega)$.

DÉMONSTRATION. Définissons $T_\mu(u)$ par

$$(T_\mu(u), v) = \int_\Omega \sum_{|i| \leq k} a_i(x, D^j u, \mu) D^i \bar{v}\, dx$$

et soit $\mu_n \to 0$, $u_n \to u$ et $T_{\mu_n}(u_n) \to g$. Pour appliquer le lemme précédent, on voit de (5.16) que (5.1) est satisfaite. L'opérateur T_0 est continu. Il faut encore voir (5.2). Mais $a_i(x, D^j u, \mu_n) \to a_i(x, D^j u, 0)$ presque partout et en vertu de (5.15) :

$$\lim_{n \to \infty} \int_\Omega | a_i(x, D^j u, \mu_n) - a_i(x, D^j u, 0) |^{m'} dx = 0,$$

d'où (5.2), donc le lemme 3 s'applique, d'où l'assertion. On a aussi :

LEMME 5. *L'operateur S, défini par* (5.12)', *sous les conditions* (5.14), (5.15) *est fortement $m-1$ quasi-homogène.*

En effet, il suffit de voir que pour $\mu_n \to 0$, $u_n \rightharpoonup u$:

$$\lim_{n \to \infty} \int_\Omega | b_i(x, D^j u_n, \mu_n) - b_i(x, D^j u, 0) |^{m'} = 0 . \tag{5.17}$$

Mais $b_i(x, D^j u_n(x), \mu_n) \to b_i(x, D^j u(x), 0)$ presque partout et les intégrales (5.17) sont en vertu du théorème de l'immersion : $W_m^{(k)}(\Omega) \subset W_q^{(k-1)}(\Omega)$, $q > m$, algébriquement et topologiquement, absolument équicontinues. Il suffit d'appliquer le théorème de Jegorov.

Nous avons alors, en vertu des Lemmes 4 et 5 :

THÉORÈME 6. *On suppose les hypothèses* (5.3), (5.8), (5.11), (5.12), (5.14), (5.15), (5.16), $u_0 = \theta$ *de* (5.7). *Alors si* $\mathrm{Im}\, \lambda \neq 0$ *ou si* $\lambda \leq 0$ *ou si* $\lambda > 0$ *et si dans le dernier cas il n'existe pas* $u \neq \theta$ *dans* $\overset{\circ}{W}{}_m^{(k)}(\Omega)$ *telle que pour* $\overset{\circ}{W}{}_m^{(k)}(\Omega)$

$$\int_\Omega \sum_{|i| \leq k} a_i(x, D^j u, 0) D \bar{v}\, dx - \lambda \int_\Omega \sum_{|i| < k} b_j(x, D^j u, 0) \cdot D_i \bar{v}\, dx = 0,$$

alors pour chaque f de $L_{m'}(\Omega)((\overset{\circ}{W}{}_m^{(k)}(\Omega))^*)$, il existe une solution de l'équation (5.13)'.

En effet, il suffit de définir $T(u, v) = T(u)$, où $T(u)$ est défini par (5.17). $T(u, v)$ satisfait aux hypothèses du théorème 3. Cf. les exemples 1 et 3.

REMARQUE. On voit aisément que le cas du problème de Dirichlet non-homogène se résout par le théorème 6, où l'on peut remplacer la condition 3.3 par la condition $(T(u), u) \geq c_1 \| u \|^{1+\varkappa} - c_3 \| u \|^{1+\varkappa''}$, $-1 \leq \varkappa'' < \varkappa$.

Institut Mathématique
Zitná 25, Praha 1
Tchécoslovaquie

BIBLIOGRAPHIE

[1] J. LERAY, J. L. LIONS : *Quelques résultats de Višik sur les problèmes elliptiques non linéaires par les méthodes de Minty-Browder*, Bull. Soc. Math. France 93 (1965), 97-107.

[2] F. E. BROWDER : *Problèmes non-linéaires, Séminaire de mathématiques supérieures*, Montréal 1966.

[3] J. NEČAS : *Les équations elliptiques non linéaires*, Cours d'été 1967, Tchécoslovaquie, à paraître dans Czechoslovac mathematical journ. 1969.

[4] F. E. BROWDER : *Variational boundary value problems for quasi-linear elliptic equations*, II, Proc. Nat. Acad. Sciences, Vol. 50, n. 4, 592-598, (1963).

[5] M. M. VAJNBERG : *Variacionnyje metody issledovanija nělinějnych operatorov*, Moskva 1956.

[6] J. NEČAS : *Sur la régularité des solutions faibles des équations elliptiques non linéaires*, Sémin. équat. part., Collège de France, 1968.

Časopis pro pěstování matematiky, roč. 97 (1972), Praha

FREDHOLM ALTERNATIVE FOR NONLINEAR OPERATORS AND APPLICATIONS TO PARTIAL DIFFERENTIAL EQUATIONS AND INTEGRAL EQUATIONS

JINDŘICH NEČAS, Praha*)

(Received April 15, 1970)

1. Introduction. The problem of solving a nonlinear boundary value problem or an integral equation can be reduced often to the following abstract one: find a solution u of $Tu = f$, where T is a mapping from a real, reflexive Banach space B to its dual B^*.

Example 1. Let Ω be a bounded domain with Lipschitz boundary $\partial\Omega$ and let $a_i(x, \xi_0, \xi_1, \ldots, \xi_n)$, $i = 0, 1, \ldots, n$, be continuous functions in $\bar{\Omega} \times R_{n+1}$, satisfying growth conditions

$$|a_i(x, \xi)| \leqq c(1 + |\xi|)^{m-1}, \tag{1.1}$$

where $1 < m < \infty$. Let $f_i \in L_{m'}(\Omega)$, $1/m' + 1/m = 1$, $i = 0, \ldots, n$. By $W_m^{(1)}(\Omega)$ we denote the well-known Sobolev space of real L_m functions whose first derivatives are also L_m functions. $W_m^{(1)}(\Omega)$ is a Banach space with the norm $\|u\|_{W_m(1)} = (\int_\Omega (|u|^m + \sum_{i=1}^n |\partial u/\partial x_i|^m) \,dx)^{1/m}$ and is separable. $W_m^{(1)}(\Omega)$ is also reflexive as the closed subspace of $[L_m]^{n+1}$. Let $\mathring{W}_m^{(1)}(\Omega)$ be the closure of $D(\Omega)$, the space of infinitely differentiable functions with compact support, in the space $W_m^{(1)}(\Omega)$. We have to find $u \in \mathring{W}_m^{(1)}(\Omega)$ such that for any $v \in \mathring{W}_m^{(1)}(\Omega)$

$$\int_\Omega \left(\sum_{i=1}^n \frac{\partial v}{\partial x_i} a_i \left(x, u, \frac{\partial u}{\partial x_1}, \ldots, \frac{\partial u}{\partial x_n}\right) + v a_0 \left(x, u, \frac{\partial u}{\partial x_1}, \ldots, \frac{\partial u}{\partial x_n}\right)\right) dx = \tag{1.2}$$
$$= \int_\Omega v f_0 \,dx - \int_\Omega \sum_{i=1}^n \frac{\partial v}{\partial x_i} f_i \,dx .$$

*) Lecture held on the Chicago area applied mathematics seminar.

© Institute of Mathematics, Academy of Sciences of the Czech Republic 1972. Reprinted with kind permission from: Nečas, J., Fredholm alternative for nonlinear operators and applications to partial differential equations and integral equations, asopis pro pstovn matematiky, 97, 65-71, http://dml.cz.

The function u is called weak solution of the differential equation

$$(1.3)\quad -\sum_{i=1}^{n} \frac{\partial}{\partial x_i}\left(a_i\left(x, u, \frac{\partial u}{\partial x_1}, \ldots, \frac{\partial u}{\partial x_n}\right)\right) + a\left(x, u, \frac{\partial u}{\partial x_1}, \ldots, \frac{\partial u}{\partial x_n}\right) = f + \sum_{i=1}^{n} \frac{\partial f_i}{\partial x_i}$$

in Ω, satisfying on the boundary the condition $u = 0$.

Denoting by (w^*, u) the pairing between B^* and B, we can define an operator T: $: B \to B^*$, putting

$$(Tu, v) \overset{\mathrm{df}}{=} \int_{\Omega}\left(\sum_{i=1}^{n} a_i\left(x, u, \frac{\partial u}{\partial x_1}, \ldots, \frac{\partial u}{\partial x_n}\right)\frac{\partial v}{\partial x_i} + a_0\left(x, u, \frac{\partial u}{\partial x_1}, \ldots, \frac{\partial u}{\partial x_n}\right)v\right) \mathrm{d}x\,.$$

Because $\int_\Omega f_0 v \,\mathrm{d}x - \int_\Omega \sum_{i=1}^{n} f_i(\partial v/\partial x_i)\,\mathrm{d}x = (f, v)$, the equation (1.2) is reduced to the problem of solving the equation $Tu = f$.

Example 2. Let us consider the Hammerstein's integral equation

$$(1.4)\qquad u(x) - \lambda \int_M K(x, y) f(y, u(y))\,\mathrm{d}y = w(x)\,,$$

where the solution is supposed in $L_2(M)$, M being a compact subset of R_n, $w \in L_2(M)$, $f(y, u)$ is a continuous function on $M \times R_1$, satisfying the growth condition $|f(y, u)| \leqq c(1 + |u|)$. We suppose $\int_M \int_M K^2(x, y)\,\mathrm{d}x\,\mathrm{d}y < \infty$. If $(Tu)(x) \overset{\mathrm{df}}{=} u(x) - \lambda \int_M K(x, y) f(y, u(y))\,\mathrm{d}y$, then $T: L_2(M) \to L_2(M)$ and the problem is reduced to the solution of $Tu = w$.

2. Borsuk type theorem. A mapping T is said to be bounded if the image of bounded set is bounded and it is said to be demicontinuous, if from $u_n \to u$ (strong convergence) follows $Tu_n \rightharpoonup Tu$ (weak convergence).

Theorem 1. *Let $T: B \to B^*$, where B is a reflexive space, be a bounded, demicontinuous mapping. Let $T_t(u) = T(u) - t\,T(-u)$ for $0 \leqq t \leqq 1$. Let for $0 \leqq t \leqq 1$, the condition (S) be satisfied:*

$$(2.1)\qquad \text{if}\quad u_n \rightharpoonup u \quad\text{and}\quad (T_t(u_n) - T_t(u), u_n - u) \to 0\,,$$

then $u_n \to u$, and for $f \in B^*$ the condition

$$(2.2)\qquad T_t u - (1 - t) f \neq 0 \quad \text{for}\quad \|u\| = R > 0\,,\quad 0 \leqq t \leqq 1\,.$$

Then there exists a solution of $Tu = f$.

Let us remark first that the above solution is unique if, for example, the operator T is strictly monotone: $u \neq v \Rightarrow (Tu - Tv, u - v) > 0$.

Theorems as above are based on the concept of monotone operators, and there is a large amount of literature on this subject, compare, for example, M. I. Višik [11],

F. E. Browder [1], J. Leray, J. L. Lions [6], G. J. Minty [7]. The concept using Borsuk's theorem was recently used in the paper of D. G. de Figueiredo, Ch. P. Gupta [3] and elsewhere.

The main ideas of the proof of Theorem 1: First, if $B = R_n$, then the degree $(T_1(u), B(0, R), 0)$ is an odd integer by Borsuk's theorem, hence by homotopy, this is true for $T(u) - f$, hence, there exists $\|u\| < R$ such that $Tu = f$. If $F \subset B$ is a finite dimensional subspace of B and ψ_F is the injection of $T \to B$, ψ_F^* being its dual mapping, then for $T_F \overset{df}{=} \psi_F^* T \psi_F$, it can be proved by contradiction existence of a F such that if $F' \supset F$, then $T_{F'}(u) - tT_{F'}(-u) - (1 - t)\psi_{F'}^* f \neq 0$ for $\|u\| = R$, $u \in F'$, $0 \leq t \leq 1$. Hence for every $F' \supset F$, there exists $u_{F'} \in F'$ such that $T_{F'}u_{F'} = \psi_{F'}^* f$. Let us put $M_{F'} = \{u_{F''} \mid F'' \supset F'\}$. The set of $M_{F'}$ has finite intersection property. If $\overline{M}_{F'}$ is the closure in the weak topology, then $\bigcap_{F'} \overline{M}_{F'} \ni u$. If $w, u \in F'$ for F' such chosen, then there exists $u_n \in M_{F'}$, $u_n \rightharpoonup u$ and because of $\lim_{n\to\infty} (Tu_n - Tu, u_n - u) = \lim_{n\to\infty} (Tu_n, u_n - u) = \lim_{n\to\infty} (f, u_n - u) = 0$, ($(Tu_n, u_n - u) = (f, u_n - u)$ follows from the definition of $T_{F'}$) the condition (2.1) implies $u_n \to u$, what, in virtue of the demicontinuity of T, gives the result. We have clearly:

Consequence 1. *If the operator T is coercive:*

$$\lim_{\|u\|\to\infty} \frac{(Tu, u)}{\|u\|} = \infty\,, \quad then \quad T(B) = B^*\,.$$

This is because $(T_t u, u) \geq c(\|u\|)\,\|u\|$, with $c(s) \to \infty$ for $s \to \infty$.

Consequence 2. *If the conditions of theorem 1 are satisfied and T is odd: $T(-u) = -T(u)$ and if T is weakly coercive: $\lim_{\|u\|\to\infty} \|Tu\| = \infty$, then $T(B) = B^*$.*

Let us consider the following class of operators: first if for $\varkappa > 0$ and every $t > 0$: $A(tu) = t^\varkappa A(u)$, then A is called $\varkappa$-homogeneous.

An operator S is asymptotically zero if for $\varkappa > 0$ $\lim_{\|u\|\to\infty} \|Su\|/\|u\|^\varkappa = 0$.

We have the following Fredholm alternative:

Theorem 2. *Let $T = A + S$, where A is demicontinuous, $\varkappa$-homogeneous, satisfies the condition (S) (i.e. if $u_n \rightharpoonup$ and $(A(u_n) - A(u), u_n - u) \to 0$, then $u_n \to u$), S is demicontinuous, asymptotically zero (with the same $\varkappa$ as for A) and T is bounded odd and satisfies the condition (S). Then the range of T is all of B^* if $Au = 0 \Rightarrow u = 0$. In this case, for every solution,*

$$\|u\| \leq c(1 + \|f\|^{1/\varkappa})\,. \tag{2.3}$$

If (2.3) is true for every solution, then $Au = 0 \Rightarrow u = 0$.

Theorems of this type are recent. It seems the first paper is due to S. I. POCHOŽAJEV [10] and to the author [8]. For further results, compare F. E. BROWDER [2] and the forthcoming paper of J. NEČAS [9]; compare also M. KUČERA [5].

Proof of Theorem 2:

(i) If (2.3) is true and there exists $u_0 \neq 0$ such that $Au_0 = 0$, then for $u = tu_0$:

$$\|u_0\| \leq c\left(\frac{1}{t} + \frac{\|S(u_0 t)\|^{1/\varkappa}}{t\|u_0\|}\right)\|u_0\| \to 0$$

which is a contradiction.

(ii) Let $Au = 0 \Rightarrow u = 0$. Then (2.3) is true: if not, there exists a sequence $\|u_n\| \to \to \infty$ such that

$$(2.4) \qquad \|u_n\|^\varkappa > n(1 + \|Tu_n\|) \quad \text{and putting} \quad v_n = \frac{u_n}{\|u_n\|},$$

we can suppose $v_n \rightharpoonup v$ and we obtain from (2.4) $Av_n \to 0$ and using (S) condition: $v_n \to v$, hence $\|v\| = 1$ and $Av = 0$ which is a contradiction.

(iii) (2.3) implies (2.2), and (2.1) is satisfied because $T_t(u) = (1 + t)\, T(u)$.

3. Back to the applications. Let us remark first that it is only a question of introducing enough of indices to treat general systems instead of one partial differential equation as we will do; there is no essential difference.

I) We consider first the problem:

$$(3.1) \qquad -\sum_{i,j=1}^{n} \frac{\partial}{\partial x_i}\left(a_{ij}\frac{\partial u}{\partial x_j}\right) - \lambda a_0\left(x, u, \frac{\partial u}{\partial x_1}, \ldots, \frac{\partial u}{\partial x_n}\right) = f_0(x) + \sum_{i=1}^{n} \frac{\partial f_i}{\partial x_i}$$

with $a_{ij} \in L_\infty(\Omega)$, $\sum_{i,j=1}^{n} a_{ij}\xi_i\xi_j \geq c|\xi|^2$. Let us suppose

$$(3.2) \qquad \left|\frac{1}{t}\, a_0(x, t\xi) - \sum_{i=0}^{n} b_i(x)\,\xi_i\right| \leq c(t)\,[(\sum_{i=0}^{n} \xi_i^2)^{1/2} + 1]$$

with $c(t) \to 0$ for $t \to \infty$, $b_i \in L_\infty(\Omega)$. The condition (3.2) implies immediately that $Ru \stackrel{\mathrm{df}}{=} a_0(x, u, \partial u/\partial x_1, \ldots, \partial u/\partial x_n) - \sum_{i=1}^{n} b_i(x)\,\partial u/\partial x_i - b_0(x)\,u$ satisfies the condition $\lim_{\|u\|\to\infty} \|Ru\|_{L_2}/\|u\|_{W_2^{(1)}} = 0$. Supposing $a_0(x, -\xi) = -a_0(x, \xi)$ and defining

$$(Au, v) \stackrel{\mathrm{df}}{=} \int_\Omega \sum_{i,j=1}^{n} a_{ij}\frac{\partial u}{\partial x_j}\frac{\partial v}{\partial x_i}\,\mathrm{d}x\,, \quad (Su, v) \stackrel{\mathrm{df}}{=} -\lambda\int_\Omega a_0\left(x, u, \frac{\partial u}{\partial x_1}, \ldots, \frac{\partial u}{\partial x_n}\right) v\,\mathrm{d}x\,,$$

we obtain in virtue of the fact the imbedding $W_2^{(1)}(\Omega) \to L_2(\Omega)$ is completely continuous, that Su is a completely continuous operator from $\mathring{W}_2^{(1)} \to (\mathring{W}_2^{(1)})^*$. Because A

and S above defined satisfy with $\varkappa = 1$ the conditions of the theorem 2, this altogether gives by theorem 2 this result:

For every $f_i \in L_2(\Omega)$, $i = 0, \ldots, n$, there exists a solution of (3.1) with $u = 0$ on $\partial\Omega$ and for every solution, we have $\|u\|_{W_2^{(1)}} \leqq c(1 + \|f_0\|_{L_2} + \sum_{i=1}^{n} \|f_i\|_{L_2})$ if and only if λ is not an eigenvalue for the linear problem

$$-\sum_{i,j=1}^{n} \frac{\partial}{\partial x_i}\left(a_{ij}\frac{\partial u}{\partial x_j}\right) - \lambda \sum_{i=1}^{n}\left(b_i(x)\frac{\partial u}{\partial x_i} + b_0(x)\,u\right) = 0\,, \quad u \in \mathring{W}_2^{(1)}(\Omega)\,.$$

II) If we consider a nonlinear problem with $m \neq 2$, then we suppose:

$$\left|\frac{a_i(x, t\xi)}{t^{m-1}} - A_i(x, \xi)\right| \leqq c_i(t)\,(1 + |\xi|^{m-1})\,, \quad i = 0, \ldots, n\,, \tag{3.3}$$

where $A_i(x, \xi)$ satisfy the conditions (1.1) and $c_i(t) \to 0$ for $t \to \infty$. Let $A_i(x, \xi)$ and $a_i(x, \xi)$ be odd in ξ and $A_i(x, t\xi) = t^{m-1}A_i(x, \xi)$, $t > 0$. We shall suppose for $a_i(x, \xi)$ and $A_i(x, \xi)$ the conditions (we write them only for A_i): if $[\xi_1, \ldots, \xi_n] \neq [\xi'_1, \ldots, \xi'_n]$ then

$$\sum_{i=1}^{n} (A_i(x, \xi_0, \xi_1, \ldots, \xi_n) - A_i(x, \xi_0, \xi'_1, \ldots, \xi'_n))\,(\xi_i - \xi'_i) > 0 \tag{3.4}$$

and

$$\sum_{i=1}^{n} A_i(x, \xi_0, \xi_1, \ldots, \xi_n)\,\xi_i \geqq c_1 \sum_{i=1}^{n} |\xi_i|^m - c_2|\xi_0|^m\,. \tag{3.5}$$

For to apply theorem 2, we can easily verify (for details compare J. Leray, J. L. Lions [6]) the hypothesis eventually with the exception of the condition (S): for to see this, let $u_k \rightharpoonup u$ in $\mathring{W}_m^{(1)}(\Omega)$. We have first by the complete continuity of the imbedding $W_m^{(1)}(\Omega) \to L_m(\Omega)$: $u_k \to u$ in $L_m(\Omega)$. Choosing a subsequence, if necessary, still noted u_k, we have $u_k(x) \to u(x)$ almost everywhere. By hypothesis,

$$\lim_{k\to\infty} \int_\Omega \sum_{i=1}^{n} \left(a_i\left(x, u_k, \frac{\partial u_k}{\partial x_1}, \ldots, \frac{\partial u_k}{\partial x_n}\right) - a_i\left(x, u, \frac{\partial u}{\partial x_1}, \ldots, \frac{\partial u}{\partial x_n}\right)\right)$$
$$\left(\frac{\partial u_k}{\partial x_i} - \frac{\partial u}{\partial x_i}\right) \mathrm{d}x + \int_\Omega \left(a_0\left(x, u_k, \frac{\partial u_k}{\partial x_1}, \ldots, \frac{\partial u_k}{\partial x_n}\right) - a_0\left(x, u, \frac{\partial u}{\partial x_1}, \ldots, \frac{\partial u}{\partial x_n}\right)\right)$$
$$(u_k - u)\,\mathrm{d}x = 0\,.$$

The second member tends to zero, hence also the first, but in virtue of (3.4) putting

$$f_k(x) = \sum_{i=1}^{n} \left(a_i\left(x, u_k, \frac{\partial u_k}{\partial x_1}, \ldots, \frac{\partial u_k}{\partial x_n}\right) - a_i\left(x, u_k, \frac{\partial u}{\partial x_1}, \ldots, \frac{\partial u}{\partial x_n}\right)\left(\frac{\partial u_k}{\partial x_i} - \frac{\partial u}{\partial x_i}\right)\right),$$

we have $f_k(x) \geqq 0$ and $\int_\Omega f_k(x)\,dx \to 0$. If necessary, for a subsequence, still noted f_k, $f_k(x) \to 0$ almost everywhere. This implies $(\partial u_k/\partial x_i)(x) \to (\partial u/\partial x_i)(x)$ almost everywhere. But (3.5) gives uniform continuity of the integrals $\int_M \sum_{i=1}^n |\partial u_k/\partial x_i|^m\,dx$ with respect to k. This implies $\partial u_k/\partial x_i \to \partial u/\partial x_i$ in $L_m(\Omega)$ for the original sequence.

Hence we obtain: the conditions (1.1), (3.3)–(3.5) being satisfied, there exists a solution $u \in W_m(\Omega)$ of (1.2), and every solution is such that

$$\|u\|_{W_m^{(1)}} \leqq c\Big(1 + \sum_{i=0}^{n} \|f_i\|_{L_{m'}}^{1/(m-1)}\Big) \tag{3.6}$$

if and only if $u = 0$

is the only solution of (1.2) for $f_i = 0$ and the coefficients A_i.

III) As far as the integral equation (1.4), although there is a lot of possible generalizations, we shall consider the condition:

$$\left|\frac{1}{t} f(y, tu) - a(y)\,u\right| \leqq c(t)\,(1 + |u|) \quad \text{with} \quad c(t) \to 0 \tag{3.7}$$

for $t \to \infty$ and $a \in L_\infty(M)$. The operators from $L_2(M) \to L_2(M)$ defined by

$$\int_M K(x, y)\,f(y, u(y))\,dy\,, \quad \int_M K(x, y)\,a(y)\,u(y)\,dy \tag{3.8}$$

are completely continuous. If we have

$$f(y, -u) = -f(y, u)\,, \tag{3.9}$$

we can immediately apply Theorem 2. But it is easy to see that we can immediately apply Theorem 1 in virtue of the complete continuity of (3.8) without (3.9). We obtain:

The equation (1.4) provided (3.7) has a solution for every $w \in L_2(M)$ and for every solution holds $\|u\|_{L_2} \leqq c(1 + \|w\|_{L_2})$ if and only if λ is not an eigenvalue for the linear equation

$$u(x) - \lambda \int_M K(x, y)\,a(y)\,u(y)\,dy = 0\,.$$

This result is very close to the corresponding result of M. A. Krasnoselskij [4].

Bibliography

[1] *F. E. Browder:* "Existence and uniqueness theorems for solutions of non-linear boundary value problems", Proc. Symposia on Appl. Math. Amer. Math. Soc. *17* (1965), 24–49.

[2] *F. E. Browder:* "Existence theorems for non-linear partial differential equations", Proc. Amer. Math. Soc. 1968 Summer Institute in global Analysis (to appear).

[3] *D. G. de Figueiredo, Ch. P. Gupta:* "Borsuk type theorems for non-linear non-compact mappings in Banach space", to appear.

[4] *M. A. Krasnoselskij:* "Topological methods in the theory of non-linear integral equations", Pergamon Press, N. Y., 1964.

[5] *M. Kučera:* "Fredholm alternative for non-linear operators", thesis 1969, Charles University, Prague.

[6] *J. Leray, J. L. Lions:* "Quelques résultats de Višik sur les problèmes elliptiques non linéaires par les méthodes de Minty-Browder", Bull. Soc. Math. France *93* (1965), 97—107.

[7] *G. J. Minty:* "Monotone (non-linear) operators in Hilbert space", Duke Math. J. *29* (1962), 341—346.

[8] *J. Nečas:* "Sur l'alternative de Fredholm pour les opérateurs non linéaires avec applications aux problèmes aux limites", Annali Scuola Norm. Sup. Pisa, *XXIII* (1969), 331—345.

[9] *J. Nečas:* "Remark on the Fredholm alternative for non-linear operators with application to non-linear integral equation of generalized Hammerstein type", to appear.

[10] *S. I. Pochožajev:* "On the solvability of non-linear equations involving odd operators", Functional Analysis and Appl. (Russian), *1* (1967), 66—73.

[11] *M. I. Višik:* "Quasilinear strongly elliptic system of differential equations having divergence form", (Russian), Trudy Mosk. Mat. Obšč. *12* (1963), 125—184.

Author's address: Praha 8 - Karlín, Sokolovská 83 (Matematicko-fyzikální fakulta UK).

Czechoslovak Mathematical Journal, 24 (99) 1974, Praha

APPLICATION OF ROTHE'S METHOD TO ABSTRACT PARABOLIC EQUATIONS

JINDŘICH NEČAS, Praha

(Received August 25, 1973)

Introduction. The method of Rothe, introduced by Rothe in his paper [1], for the investigation of parabolic equations, was used by many authors, see for example O. A. LADYŽENSKAJA [2], A. M. ILJIN, A. S. KALAŠNIKOV, O. A. OLEJNIK [3]. Recently, a paper of K. REKTORYS [4] and papers of J. KAČUR [5] led the author to apply this method on the initial value problem for abstract parabolic equations $\mathrm{d}u/\mathrm{d}t + A(u(t)) = f(t)$ with A a nonlinear operator. The author followed the paper of F. E. BROWDER [6], where the theory of semigroups is used and generalised his result for nonvanishing $f(t)$; see also H. BRÉZIS [7]. The method of Rothe consists in replacing the equation in fixed points $t_j = jh$, by $(u(t_j) - u(t_{j-1}))/h + A(u(t_j)) = f(t_j)$. The nonlinear operator A is supposed monotone in its domain $D(A) \subset H$, where H is a real Hilbert space, and such that Range $(A + I) = H$. If $u_0 \in D(A)$ and if f is a continuous function from the interval $\langle 0, T\rangle$ to H and of bounded variation, then a unique solution of the problem exists in the sense precised below and the piecewise linear functions or piecewise constant functions inset in points t_j tend uniformly in the interval $\langle 0, T\rangle$ for $h \to 0$ to the solution.

Assertions and main theorem. Let the operator A be monotone in $D(A)$:

$$(1) \qquad (Au - Av, u - v) \geqq 0 \; \forall \, u, v \in D(A)$$

and let us suppose that

$$(2) \qquad (A + I)(D(A)) = H .$$

Let u_0 be in $D(A)$ and f in $C(\langle 0, T\rangle, H)$ with bounded variation in the interval $\langle 0, T\rangle$. By a solution of the problem

$$(3) \qquad \frac{\mathrm{d}u}{\mathrm{d}t} + A(u(t)) = f(t)\,, \quad 0 \leqq t \leqq T\,, \quad u(0) = u_0\,,$$

we mean a strongly continuous function $u : \langle 0, T\rangle \to H$, weakly differentiable to H for $t \in \langle 0, T\rangle$, such that $u(t) \in D(A)$ for $t \in \langle 0, T\rangle$, $A(u(t))$ is weakly continuous,

© Institute of Mathematics, Academy of Sciences of the Czech Republic 1974. Reprinted with kind permission from: Nečas, J., Application of Rothe's method to abstract parabolic equations, Czechoslovak Mathematical Journal 24, 496-500, http://dml.cz.

$u(t)$ satisfies the equation (3) for $t \in \langle 0, T \rangle$ and the initial condition in the obvious sens. (See [6].)

Put for $\lambda > 0 : A_\lambda = A + \lambda I$. We obtain as in [6]:

Assertion 1. *For $\lambda > 0$, Range $(A_\lambda) = H$ and $\|A_\lambda^{-1}x - A_\lambda^{-1}y\| \leqq \lambda^{-1}\|x - y\|$.*

Proof. Let $u, v \in D(A)$. Then $(A_\lambda u - A_\lambda v, u - v) \geqq \lambda\|u - v\|^2$, hence $\|u - v\| \leqq \leqq \lambda^{-1}\|A_\lambda u - A_\lambda v\|$. If for some $\lambda_0 > 0$ the Range $(A_{\lambda_0}) = H$, then $A_{\lambda_0}^{-1}$ is defined and is Lipschitzian with Lipschitz constant λ_0^{-1}. To solve the equation $A_\lambda u = y$, put $u = A_{\lambda_0}^{-1}x$. We obtain the equation

$$x + (\lambda - \lambda_0)\, A_{\lambda_0}^{-1}x = y\,, \tag{4}$$

whose solution gives a solution of the original equation. If $|\lambda - \lambda_0|\,\lambda_0^{-1} < 1$, then by the theorem on contractive mappings, the equation (4) has a unique solution, q.e.d.

Let $h = T/n$, $t_j = jh$, $j = 0, 1, \ldots, n$ and put $z_0 = u_0$, $Az_1 + (z_1 - z_0)/h = f(t_1)$, $A(z_j) + (z_j - z_{j-1})/h = f(t_j)$, $j = 1, 2, \ldots, n$. It follows from the Assertion 1 that z_j are uniquely determined. We have

$$(Az_1 - Az_0,\, z_1 - z_0) + \frac{\|z_1 - z_0\|^2}{h} = (f(t_1),\, z_1 - z_0) - (Az_0, z_1 - z_0)\,,$$

hence

$$h^{-1}\|z_1 - z_0\| \leqq \|f(t_1)\| + \|Az_0\| \tag{5}$$

and similarly for $j \geqq 2$: $(Az_j - Az_{j-1}, z_j - z_{j-1}) + \|z_j - z_{j-1}\|^2/h = (f(t_j), z_j - z_{j-1}) - (f(t_{j-1}), z_j - z_{j-1}) + h^{-1}(z_{j-1} - z_{j-2}, z_j - z_{j-1})$, so

$$h^{-1}\|z_j - z_{j-1}\| \leqq \|f(t_j) - f(t_{j-1})\| + h^{-1}\|z_{j-1} - z_{j-2}\|\,. \tag{6}$$

We obtain from (5) and (6)

$$h^{-1}\|z_j - z_{j-1}\| \leqq \operatorname*{Var}_{\langle 0,T\rangle} f + \operatorname*{Max}_{\langle 0,T\rangle} \|f(t)\| + \|Az_0\|\,. \tag{7}$$

Put in $[0, T]$, $z^n(t) = z_{j-1} + h^{-1}(t - t_{j-1})(z_j - z_{j-1}) = h^{-1}(h - (t - t_{j-1}))\,.\, z_{j-1} + h^{-1}(t - t_{j-1})\, z_j$ for $t_{j-1} \leqq t \leqq t_j$. It follows from (5) and (7)

Assertion 2.

$$\|z^n(\tau) - z^n(\mu)\| \leqq |\tau - \mu|\,(\operatorname*{Var}_{\langle 0,T\rangle} f + \operatorname*{Max}_{\langle 0,T\rangle} \|f(t)\| + \|Az_0\|)\,. \tag{8}$$

Let us define $x^n(t)$ by $x^n(0) = u_0$, $x^n(t) = z_j$ for $(j - 1)\,h < t \leqq jh$. In the same manner we define $f^n(t)$. It follows from the construction above that

$$\int_0^t A(x^n)(\tau))\,d\tau + z^n(t) = \int_0^t f^n(\tau)\,d\tau + u_0 \tag{9}$$

and for the derivatives from the left:

$$A(x^n(t)) + \frac{\mathrm{d}^- z^n}{\mathrm{d}t} = f(t)\,. \tag{10}$$

It follows from the definition of the functions $x^n(t)$ and $f^n(t)$ that for all $t \in \langle 0, T\rangle$, $f^n(t) \to f(t)$ (strong convergence), uniformly in $\langle 0, T\rangle$, and

$$x^n(t) - z^n(t) \to 0 \quad \text{uniformly in} \quad \langle 0, T\rangle\,. \tag{11}$$

Assertion 3. $x^n(t) \to u(t)$ *uniformly in* $\langle 0, T\rangle$.

Proof. We have

$$\frac{\mathrm{d}^-}{\mathrm{d}t}\|z^m(t) - z^n(t)\|^2 = 2\left(\frac{\mathrm{d}^- z^m}{\mathrm{d}t}(t) - \frac{\mathrm{d}^- z^n}{\mathrm{d}t}(t), \right. \tag{12}$$

$$\left. z^m(t) - z^n(t)\right) = 2(f^m(t) - A(x^m(t)) - f^n(t) + A(x^n(t)), z^m(t) - z^n(t)\Big)\,.$$

It follows from (8) and (10) that

$$\|A(x^n(t))\| \leqq 2 \operatorname*{Max}_{\langle 0,T\rangle} \|f(t)\| + \operatorname*{Var}_{\langle 0,T\rangle} f + \|Az_0\|\,, \tag{13}$$

hence (11), (12) and (13) give

$$\frac{\mathrm{d}^-}{\mathrm{d}t}\|z^m(t) - z^n(t)\|^2 \leqq \varepsilon_{m,n} + 2(f^m(t) - A(x^m(t)) - f^n(t) +$$

$$+ A(x^n(t)), x^m(t) - x^n(t)) \leqq \varepsilon_{m,n} + 2\|f^m(t) - f^n(t)\|\,\|x^m(t) - x^n(t)\|\,,$$

where $\varepsilon_{m,n} \to 0$. Because $\|z^m(t) - z^n(t)\| \geqq 0$ and $\|z^m(0) - z^n(0)\| = 0$, it follows that with some $\bar{\varepsilon}_{m,n} \to 0 : \|z^m(t) - z^n(t)\| \leqq \bar{\varepsilon}_{m,n}t$, q.e.d.

We prove as in the paper [6] the following Assertions 4 and 5:

Assertion 4. A *is maximal monotone, i.e. if* $(Av - w, v - u) \geqq 0$ *for all* v *in* $D(A)$, *then* $u \in D(A)$ *and* $Au = w$.

Proof. We have $(Av - w + v - u, v - u) \geqq 0$. For fixed $z \in H$, $t > 0$, put $v_t = A_1^{-1}(u + w + tz)$. It follows that $Av_t + v_t - u - w = tz$, hence $t(z, v_t - u) \geqq$ $\geqq 0$ so $(z, v_t - u) \geqq 0$. Letting $t \to 0+ : v_t \to A_1^{-1}(u + w)$, hence $(z, A_1^{-1}(u + w) -$ $- u) \geqq 0$ so $u = A_1^{-1}(u + w)$, $u \in D(A)$, and $Au = w$, q.e.d.

Assertion 5. *If* $u_j \to u$ *and* $Au_j \rightharpoonup w$ (weak convergence) *then* $u \in D(A)$ *and* $Au = w$.

Proof. For $v \in D(A) : (Av - Au_j, v - u_j) \geqq 0$. From this follows taking the limit as $j \to \infty$ that $(Av - w, v - u) \geqq 0$, hence by the Assertion 4, $u \in D(A)$ and $Au = w$, q.e.d.

Assertion 6.

$$(14) \qquad \|u(t) - u(\tau)\| \leqq |t - \tau| \left(\operatorname*{Var}_{\langle 0,T\rangle} f + \operatorname*{Max}_{\langle 0,T\rangle} \|f(t)\| + \|Au_0\|\right).$$

Proof. (14) follows from (8), (11) and the Assertion 3.

Assertion 7. *$A(x^n(t)) \rightharpoonup A(u(t))$ and $A(u(t))$ is weakly continuous.*

Proof. We use estimations (13), (14) and Assertion 3 and 5.

Theorem. *Let A be a nonlinear operator from the domain $D(A) \subset H$ to H. Let (1) and (2) be satisfied. Let $u_0 \in D(A)$, $f(t)$ be continuous on $\langle 0, T\rangle$ with values in H and of bounded variation on $\langle 0, T\rangle$. Then there exists a unique solution $u(t)$ of the initial problem (3) and $u(t)$ is a Lipschitz continuous function satisfying (14). For the functions x^n, z^n constructed by the Rothe's method (8), (9), (10), (13) are valid. Also*

$$(15) \qquad z^n(t) \to u(t) \quad \text{uniformly in} \quad \langle 0, T\rangle\,,$$

$$(16) \qquad A\, x^n(t) \rightharpoonup A(u(t))\,.$$

If u, v are two solutions corresponding to the f, u_0, g, v_0 respectively, then

$$(17) \qquad \operatorname*{Max}_{\langle 0,T\rangle} \|u(t) - v(t)\| \leqq 2\int_0^T \|f(t) - g(t)\|\, dt + \|u(0) - v(0)\|\,.$$

If $f(t) \equiv 0$, then $\|A(u(t))\|$ is non-increasing.

Proof. First we prove (17) from which follows also uniqueness. We have

$$\frac{d}{dt}\|u(t) - v(t)\|^2 = 2(f(t) - g(t) - A(u(t)) + A(v(t))\,,$$

$$u(t) - v(t)) \leqq 2\|f(t) - g(t)\|\;\|u(t) - v(t)\|\,,$$

so

$$\frac{d}{dt}\|u(t) - v(t)\|^2 \leqq 2(\operatorname*{Max}_{\langle 0,T\rangle} \|u(\tau) - v(\tau)\|)\;\|f(t) - g(t)\|\,,$$

hence

$$\|u(t) - v(t)\|^2 \leqq \|u(0) - v(0)\|^2 + 2\operatorname*{Max}_{\langle 0,T\rangle} \|u(\tau) - v(\tau)\| \cdot \int_0^T \|f(\tau) - g(\tau)\|\, d\tau\,,$$

from which (17) follows. Let us consider first sequences $x^n(t)$, $z^n(t)$ introduced above.

It follows from (9) that for every $v \in H$

(18) $$\int_0^t (A(x^n(\tau)), v)\,d\tau + (z^n(t), v) = \int_0^t (f^n(\tau), v)\,d\tau + (u_0, v).$$

Hence from the Assertions 3,7 and from (11), it follows for $n \to \infty$:

(19) $$\int_0^t (A(u(\tau)), v)\,d\tau + (u(t), v) = \int_0^t (f(\tau), v)\,d\tau + (u_0, v),$$

hence it follows from (19) that $u(t)$ is weakly differentiable, because of the Assertions 7, and that the equation (3) is satisfied. (15) and (16) are evident.

From (13) and the Assertion 7 we obtain for $f(t) \equiv 0$

(20) $$\|A(u(t))\| \leqq \|Au_0\|,$$

hence from the uniqueness

(21) $$\|A(u(t))\| \leqq \|A(u(s))\|, \quad t \geqq s,$$

q.e.d.

Remark. If $f(t) \equiv 0$, then clearly for two solutions corresponding to u_0, v_0:

$$\frac{d}{dt} \|u(t) - v(t)\|^2 = 2(u'(t) - v'(t), u(t) - v(t)) =$$
$$= 2(A(u(t)) - A(v(t)), u(t) - v(t)) \leqq 0,$$

hence the transition operator $U(t)$ defined by $U(t)\,u(0) = u(t)$ is nonexpansive.

References

[1] *E. Rothe:* Zweidimensionale parabolische Randwertaufgaben als Grenzfall eindimensionaler Randwertaufgaben. Math. Ann. *102* (1930), 650—670.

[2] *O. A. Ладыженская:* Решение в целом первой краевой задачи для квазилинейных параболических уравнений ДАН СССР *107* (1956), 636—639.

[3] *A. M. Ильин, A. C. Калашников, O. A. Олейник:* Линейные уравнения второво порядка параболического типа, УНМ *17*, vyp. 3 (1962), 3—146.

[4] *K. Rektorys:* On application of direct variational methods to the solution of parabolic boundary value problems of arbitrary order in the space variables, Czech. Math. J. *21* (*96*) 1971, 318—339.

[5] *J. Kačur:* Method of Rothe and nonlinear parabolic boundary value problems of arbitrary order I, II, to appear in Czech. Math. J.,

[6] *F. E. Browder:* Existence theorems for nonlinear partial differential equations, Global Analysis, Proc. Symp. Pure Math. Vol. *16*, Amer. Math. Soc. (1970) 1—60.

[7] *H. Brézis:* Opérateurs Maximaux Monotones, Mathematics Studies, North-Holland, 1973.

Author's address: 115 67 Praha 1, Žitná 25, ČSSR (Matematický ústav ČSAV).

Math. Nachr. 93, 21–31 (1979)

On a Class of Nonlinear Initial Value Problems in Hilbert Spaces

By K. Gröger of Berlin and J. Nečas of Prague

(Eingegangen am 13. 7. 1978)

Introduction

In this paper we shall deal with initial value problems of the form

$$\text{(A)} \qquad (Eu)' + \partial\varphi(u) \ni f, \quad u(0) = a ,$$

where E is a strongly monotone and Lipschitzian mapping from a Hilbert space H into itself and φ is a lower semicontinuous convex functional on H such that $\varphi \circ E^{-1}$ is also convex. The paper was initiated by results of Trávníček [5]. Starting from certain nonlinear constitutive relations in plasticity Trávníček formulated and solved initial value problems which are closely related to problems of type (A).

The paper consists of 3 sections. In Section 1 we introduce some notations and we formulate precisely the class of problems we are interested in. In Section 2 we prove an existence-uniqueness result. Section 3 is devoted to regularity properties of the solution to (A) which can be proved under additional assumptions on the data of the problem.

Our method of proof differs from that used by Trávníček [5]. We use the fact that problem (A) can be approximated by problems of the type

$$E'(u(t_{k-1}))\, u'(t) + \partial\varphi(u(t)) \ni f(t) \quad \text{for a.e.} \quad t \in]t_{k-1}, t_k[, \quad k = 1, \dots, n ,$$
$$u(0) = a, \ u \text{ continuous on } [0, T],$$

where $0 = t_0 < t_1 < \dots < t_n = T$ is a subdivision of the time interval.

Let us mention that results on initial value problems of the type

$$(Au)' + Bu \ni f, \quad u(0) = a ,$$

were obtained also by Barbu [1] and Gajewski-Zacharias [3], but the assumptions of each of these papers are quite different from those used here.

1. Preliminaries

Let H be a real Hilbert space. By $(\cdot\,,\cdot)$ and $|\cdot|$ we denote the scalar product and the norm of H, respectively. Let E be an operator such that

© Wiley-VCH Verlag 1979. Reprinted with kind permission from: Gröger, K. & Nečas, J., On a class of nonlinear initial value problems in Hilbert spaces, Mathematische Nachrichten, 93:21-31.

$$(1)\quad \begin{cases} E\colon H \to H \text{ is a Fréchet-differentiable potential operator;} \\ \forall x_1,\ \forall x_2 \in H\colon (Ex_1 - Ex_2, x_1 - x_2) \geqq m\ |x_1 - x_2|^2,\ m > 0, \\ \qquad |Ex_1 - Ex_2| \leqq M\ |x_1 - x_2|; \\ E'\colon H \to \mathfrak{L}(H; H) \text{ is Lipschitzian on bounded sets.} \end{cases}$$

We shall denote the inverse operator E^{-1} by F. From (1) it follows:

$F\colon H \to H$ is a Fréchet-differentiable potential operator;

$$\forall y_1,\ \forall y_2 \in H\colon (Fy_1 - Fy_2, y_1 - y_2) \geqq \frac{1}{M} |y_1 - y_2|^2,$$

$$|Fy_1 - Fy_2| \leqq \frac{1}{m} |y_1 - y_2|;$$

$F'\colon H \to \mathfrak{L}(H; H)$ is Lipschitzian on bounded sets.

Moreover,

$$\forall x,\ \forall a \in H\colon \|E'(x)\|_{\mathfrak{L}(H;H)} \leqq M,\ (E'(x)\, a, a) \geqq m\ |a|^2,\ E'(x) \text{ is symmetric;}$$

$$\forall y,\ \forall b \in H\colon \|F'(y)\|_{\mathfrak{L}(H;H)} \leqq \frac{1}{m},\ (F'(y)\, b, b) \geqq \frac{1}{M} |b|^2,\ F'(y) \text{ is symmetric.}$$

We assume that we are given a functional φ such that

$$(2)\quad \begin{cases} \varphi\colon H \to]-\infty, +\infty] \text{ is proper, convex and lower semicontinuous;} \\ \psi := \varphi \circ F \text{ is convex}. \end{cases}$$

Remark 1. If (1) and (2) hold then ψ is proper and lower semicontinuous.

Remark 2. If C is a nonempty convex closed subset of H and $\varphi = I_C$ (where I_C denotes the indicator functional of C, see Brezis [2]), then $\psi = I_C \circ F = I_{EC}$. Therefore, ψ is convex if and only if EC is convex.

If χ is any convex lower semicontinuous functional we denote by $D(\chi)$ its domain, by $\partial\chi$ its subdifferential and by $D(\partial\chi)$ the domain of $\partial\chi$ (see Brezis [2]).

Lemma 1. *If* (1) *and* (2) *are satisfied then*

$$\forall x \in H\colon \partial\varphi(x) = E'(x)\ \partial\psi(Ex), \quad \forall y \in H\colon \partial\psi(y) = F'(y)\ \partial\varphi(Fy).$$

Proof. Let $f \in \partial\varphi(x)$. Then for every $y \in H$ and $t \in]0, 1]$

$$\begin{aligned} \psi(y) - \psi(Ex) &\geqq \frac{1}{t} (\psi\ (ty + (1-t)\, Ex) - \psi(Ex)) \\ &= \frac{1}{t} (\varphi(F(ty + (1-t)\, Ex)) - \varphi(x)) \\ &\geqq \frac{1}{t} (f, F\ (ty + (1-t)\, Ex) - x) \\ &= \left(f, \frac{1}{t} (F\ (Ex + t\ (y - Ex)) - FEx)\right). \end{aligned}$$

Letting $t \downarrow 0$ we obtain

$$\psi(y) - \psi(Ex) \geqq (f, F'(Ex)\ (y - Ex)) = (F'(Ex)\, f, y - Ex).$$

Hence $(E'(x))^{-1}f = F'(Ex)\,f \in \partial\psi(Ex)$ and $\partial\varphi(x) \subset E'(x)\,\partial\psi(Ex)$. In the same way we find $\partial\psi(y) \subset F'(y)\,\partial\varphi(Fy)$. Combining these results we obtain the assertion of the lemma.

Let $S =]0, T]$ be a finite interval of time and let $\bar{S} = [0, T]$. We use the notations $L^p(S; H)$ and $C(\bar{S}; H)$ with their usual meaning. We set

$$AC(\bar{S}; H) := \{u : \bar{S} \to H \mid u \text{ is absolutely continuous from } \bar{S} \text{ into } H\},$$
$$AC(S; H) := \{u : S \to H \mid \forall \delta \in S : u \text{ is absolutely cont. from } [\delta, T] \text{ into } H\}.$$

If $u \in L^1(S; H)$ we denote by u' the derivative of u in the sense of distributions on $]0, T[$ with values in H. By $H^1(S; H)$ we denote the Hilbert space of all $u \in L^2(S; H)$ such that $u' \in L^2(S; H)$ (equipped with its usual scalar product). For $u : S \to H$ we introduce $Eu : S \to H$ by

$$\forall t \in S : (Eu)(t) := Eu(t)\,,$$

i.e., we consider E also as a mapping defined on (H-valued) functions of time. Similarly we shall deal with other operators.

In the following sections we shall consider problems of the type

$$\text{(A)} \qquad (Eu)'(t) + \partial\varphi(u(t)) \ni f(t) \quad \text{for a.e. } t \in S,\ u(0) = a,\ u \in C(\bar{S}; H) \cap AC(S; H),$$

assuming that $a \in H$ and $f : S \to H$ are given.

Remark 3. Using Lemma 1 and the relation

$$\forall u \in AC(S; H) : (Eu)'(t) = E'(u(t))\,u'(t) \quad \text{for a.e. } t \in S$$

it is easy to check that under the assumptions (1), (2) a function u is a solution to (A) if and only if $v := Eu$ is a solution to

$$\text{(B)} \qquad \begin{cases} (Fv)'(t) + \partial\psi(v(t)) \ni F'(v(t))\,f(t) & \text{for a.e. } t \in S\,, \\ v(0) = Ea, \quad v \in C(\bar{S}; H) \cap AC(S; H)\,. \end{cases}$$

2. Main results

Theorem 1. *Let* (1), (2) *be satisfied and let* $a_i \in H$, $f_i \in L^1(S; H)$, $i = 1, 2$. *If*

$$(Eu_i)'(t) + \partial\varphi(u_i(t)) \ni f_i(t) \quad \text{for a.e. } t \in S\,,$$
$$u_i(0) = a_i, \quad u_i \in C(\bar{S}; H) \cap AC(S; H), \quad i = 1, 2\,,$$

then

$$(3) \qquad \|u_1 - u_2\|_{C(\bar{S}; H)} \leqq r\,(\max\{|a_i| + \|f_i\|_{L^1(S;H)}\})\,(|a_1 - a_2| + \|f_1 - f_2\|_{L^1(S;H)})\,,$$

where r denotes an increasing real valued function defined on $\mathbf{R}^+$.

Proof. 1) Let $v_i := Eu_i$, $i = 1, 2$. Then (cf. Remark 3)

$$\left.\begin{aligned} v_i'(t) + \partial\varphi(u_i(t)) &\ni f_i(t), \\ u_i'(t) + \partial\psi(v_i(t)) &\ni F'(v_i(t))\,f_i(t) \end{aligned}\right\} \quad \text{for a.e. } t \in S, \quad i = 1, 2.$$

We choose $a_0 \in H$, $f_0 \in H$ such that $f_0 \in \partial\varphi(a_0)$. Then $F'(Ea_0)\, f_0 \in \partial\psi(Ea_0)$. Using the monotonicity of $\partial\varphi$ and $\partial\psi$ we obtain for $0 < t_0 \leqq t \leqq T$:

$$\int_{t_0}^{t} (v_i'(s) - f_i(s) + f_0,\, u_i(s) - a_0)\, ds \leqq 0\,,$$

$$\int_{t_0}^{t} (u_i'(s) - F'(v_i(s))\, f_i(s) + F'(Ea_0)\, f_0,\, v_i(s) - Ea_0)\, ds \leqq 0\,.$$

Adding these inequalities and letting $t_0 \downarrow 0$ we find

$$(v_i(t) - Ea_0,\, u_i(t) - a_0) - (Ea_i - Ea_0,\, a_i - a_0)$$
$$\leqq \left(1 + \frac{M}{m}\right) (\|f_i\|_{L^1(S;H)} + T\, |f_0|)\, (\|u_i\|_{C(S;H)} + |a_0|)\,.$$

In view of the properties of E this implies that

$$\|v_i\|_{C(S;H)} \leqq r_0\, (|a_i| + \|f_i\|_{L^1(S;H)} + 1)\,,$$

where r_0 is a suitable constant independent of a_i and f_i.

2) Using once more the monotonicity of $\partial\varphi$ and $\partial\psi$ we obtain for $0 < t_0 \leqq t \leqq T$:

$$\int_{t_0}^{t} ((v_1' - v_2')\,(s),\, (u_1 - u_2)\,(s))\, ds \leqq \int_{t_0}^{t} ((f_1 - f_2)\,(s),\, (u_1 - u_2)\,(s))\, ds\,,$$

$$\int_{t_0}^{t} ((u_1' - u_2')\,(s),\, (v_1 - v_2)\,(s))\, ds$$
$$\leqq \int_{t_0}^{t} (F'(v_1(s))\, f_1(s) - F'(v_2(s))\, f_2(s),\, (v_1 - v_2)\,(s))\, ds\,.$$

Adding these inequalities and letting $t_0 \downarrow 0$ we find

$$((v_1 - v_2)\,(t),\, (u_1 - u_2)\,(t)) - (Ea_1 - Ea_2,\, a_1 - a_2)$$
$$\leqq \int_0^t \{((f_1 - f_2)\,(s),\, (u_1 - u_2)\,(s) + F'(v_1(s))\,(v_1 - v_2)\,(s))$$
$$+ ((F'(v_1(s)) - F'(v_2(s)))\, f_2(s),\, (v_1 - v_2)\,(s))\}\, ds\,.$$

If we denote by K the Lipschitz constant of F' on the ball

$$\{y \in H \mid |y| \leqq r_0 \max_{i=1,2} (|a_i| + \|f_i\|_{L^1(S;H)} + 1)\}$$

(cf. the first step of the proof) we have

$$((v_1 - v_2)\,(t),\, (u_1 - u_2)\,(t)) - (Ea_1 - Ea_2,\, a_1 - a_2)$$
$$\leqq \left(1 + \frac{M}{m}\right) \|f_1 - f_2\|_{L^1(S;H)}\, \|u_1 - u_2\|_{C(S;H)} + K \int_0^t |f_2(s)| \cdot |(v_1 - v_2)\,(s)|^2\, ds\,.$$

Thus

$$m\, |(u_1 - u_2)\,(t)|^2 \leqq M\, |a_1 - a_2|^2 + \left(1 + \frac{M}{m}\right) \|f_1 - f_2\|_{L^1(S;H)}\, \|u_1 - u_2\|_{C(S;H)}$$
$$+ KM^2 \int_0^t |f_2(s)| \cdot |(u_1 - u_2)\,(s)|^2\, ds\,;$$

In view of GRONWALL's Lemma this implies that

$$m\,|(u_1-u_2)(t)|^2 \leqq \exp\left(\frac{KM^2}{m}\|f_2\|_{L^1(S;H)}\right)$$
$$\times\left(M\,|a_1-a_2|^2+\left(1+\frac{M}{m}\right)\|f_1-f_2\|_{L^1(S;H)}\|u_1-u_2\|_{C(\bar S;H)}\right).$$

From this inequality the assertion follows by elementary estimations.

Theorem 2. *Let* (1), (2) *be satisfied and let* $a\in D(\varphi)$, $f\in L^2(S;H)$. *Then the problem* (A) *has a unique solution* u. *This solution has the properties*

$$u\in H^1(S;H),\quad \varphi(u(\cdot))\in AC(\bar S;H)\,.$$

Proof. 1) For every natural number n we define $t_n^k := \frac{kT}{n}$, $k=0,\dots,n$, and we consider the following problem approximating (A):

$$(\mathrm{A}_n)\qquad \begin{cases} E'(u_n(t_n^{k-1}))\,u_n'(t)+\partial\varphi(u_n(t))\ni f(t) & \text{for a.e. } t\in\,]t_n^{k-1},t_n^k[,\\ k=1,\dots,n,\quad u_n(0)=a,\quad u_n\in H^1(S;H)\,. \end{cases}$$

It is easy to prove that (A_n) is uniquely solvable. One has to consider the problem successively on the time intervals $[t_n^{k-1},t_n^k]$, $k=1,\dots,n$, and to use the scalar product $(E'(u_n(t_n^{k-1}))\,x,y)$ in $L^2([t_n^{k-1},t_n^k];H)$ (cf. BREZIS [2]).

2) Multiplying the approximating evolution equation by u_n' we obtain (cf. BREZIS [2], Lemma 3.3):

$$m\,\|u_n'\|^2_{L^2(S;H)}+\varphi(u_n(T))-\varphi(a)\leqq\|f\|_{L^2(S;H)}\|u_n'\|_{L^2(S;H)}\,.$$

Because φ has an affine minorante this implies that

$$m\,\|u_n'\|^2_{L^2(S;H)}\leqq c_1(|u_n(T)|+1)+\|f\|_{L^2(S;H)}\|u_n'\|_{L^2(S;H)}$$
$$\leqq c_1\Big(\Big|\int_S u_n'(s)\,ds+a\Big|+1\Big)+\|f\|_{L^2(S;H)}\|u_n'\|_{L^2(S;H)}$$
$$\leqq c_2\,(\|u_n'\|_{L^2(S;H)}+1),\quad c_1,c_2 \text{ independent of } n\,.$$

Since $u_n(0)=a$ we obtain

$$(4)\qquad \sup_n\|u_n\|_{H^1(S;H)}<\infty\,.$$

3) Let

$$f_n(t):=f(t)+(E'(u_n(t))-E'(u_n(t_n^{k-1})))\,u_n'(t)\quad\text{if}\quad t\in\,]t_n^{k-1},t_n^k]\,.$$

Since

$$|u_n(t)-u_n(t_n^{k-1})|\leqq\int_{t_n^{k-1}}^{t_n^k}|u_n'(s)|\,ds\leqq\sqrt{\frac{T}{n}}\,\|u_n'\|_{L^2(S;H)}\quad\text{if}\quad t\in[t_n^{k-1},t_n^k]\,,$$

we have (cf. (1) and (4))

$$(5)\qquad \lim_{n\to\infty}\|f_n-f\|_{L^2(S;H)}=0\,.$$

Moreover,

(6) $$(Eu_n)'(t)+\partial\varphi(u_n(t))\ni f_n(t) \quad \text{for a.e.} \quad t\in S,\ u_n(0)=a,\ u_n\in H^1(S;H)\,.$$

In view of Theorem 1 from (5) and (6) it follows the existence of a function $u\in C(\bar{S};H)$ such that $u_n\to u$ in $C(\bar{S};H)$. Because of (1) and (4) this implies $Eu_n\to Eu$ in $C(\bar{S};H)$ and $u_n\rightharpoonup u$, $Eu_n\rightharpoonup Eu$ in $H^1(S;H)$. The operator $\partial\varphi$ considered as a mapping from $L^2(S;H)$ into itself is maximal monotone. Therefore from $u_n\to u$ in $L^2(S;H)$, $f_n-(Eu_n)'\rightharpoonup f-(Eu)'$ in $L^2(S;H)$, $f_n-(Eu_n)'\in\partial\varphi(u_n)$ it follows $f-(Eu)'\in\partial\varphi(u)$. Hence u is a solution to (A) belonging to $H^1(S;H)$. The absolute continuity of $\varphi(u(\cdot))$ follows from Lemma 3.3 in [2].
4) The uniqueness of the solution is a consequence of Theorem 1.

Theorem 3. *Let* (1), (2) *be satisfied and let* $a\in\overline{D(\varphi)}$, $f\in L^2(S;H)$. *Then the problem* (A) *has a unique solution* u. *This solution has the properties*

$$\int_S t\,|u'(t)|^2\,dt<\infty,\quad \varphi(u(\cdot))\in AC(S;H)\,.$$

Proof. Let $a_n\in D(\varphi)$ be such that $a_n\to a$ in H and let u_n denote the solution to (A) corresponding to the data a_n and f (cf. Theorem 2). Moreover, let $v_n:=Eu_n$. Then $u_n\to u$, $v_n\to Eu$ in $C(\bar{S};H)$ (cf. Theorem 1) and

(7) $$\begin{cases}\int_S t(v_n'(t),u_n'(t))\,dt+T\varphi(u_n(T))-\int_S\varphi(u_n(t))\,dt=\int_S t(f(t),u_n'(t))\,dt,\\ \int_S t(u_n'(t),v_n'(t))\,dt+T\psi(v_n(T))-\int_S\psi(v_n(t))\,dt\\ =\int_S t(F'(v_n(t))\,f(t),v_n'(t))\,dt\,.\end{cases}$$

In view of the convexity of φ and ψ we have

$$\begin{aligned}&\int_S \{\varphi(u_n(T))-\varphi(u_n(t))+\psi(v_n(T))-\psi(v_n(t))\}\,dt\\ &\geqq\int_S \{(f(t)-v_n'(t),u_n(T)-u_n(t))+(F'(v_n(t))\,f(t)-u_n'(t),v_n(T)-v_n(t))\}\,dt\\ &=(v_n(T)-Ea_n,u_n(T)-a_n)\\ &\quad+\int_S\{(f(t),u_n(T)-u_n(t))+(F'(v_n(t))\,f(t),v_n(T)-v_n(t))\}\,dt\geqq -c\,,\end{aligned}$$

where c is independent of n. Therefore (cf. (7))

$$2m\int_S t\,|u_n'(t)|^2\,dt-c\leqq\int_S t\,|f(t)|\left(1+\frac{M}{m}\right)|u_n'(t)|\,dt\,.$$

This is possible only if

(8) $$\sup_n\int_S t\,|u_n'(t)|^2\,dt<\infty\,.$$

Hence for every $\delta\in S$

$$u_n\to u \text{ in } L^2([\delta,T];H),\quad f-(Eu_n)'\rightharpoonup f-(Eu)' \text{ in } L^2([\delta,T];H)\,.$$

Because $\partial\varphi$ considered as a mapping from $L^2([\delta,T];H)$ into itself is maximal monotone this implies that

$$f(t)-(Eu)'(t)\in\partial\varphi(u(t)) \quad \text{for a.e.} \quad t\in S\,.$$

In view of (8)

$$\int_S t\,|u'(t)|^2\,dt<\infty\,.$$

Therefore u is a solution to (A). It is unique because of Theorem 1. The relation $\varphi(u(\cdot))\in AC(S;H)$ follows from Lemma 3.3 in [2]. This completes the proof.

Remark 4. If f satisfies the condition

$$\forall y_1,\ \forall y_2\in H: F'(y_1)\,f(t)=F'(y_2)\,f(t) \quad \text{for a.e.} \quad t\in S\,, \tag{9}$$

then in the preceding theorems we may replace the Lipschitz continuity of E' on bounded sets by the weaker condition that E' is uniformly continuous on bounded sets. In that case instead of (3) we could prove

$$\|u_1-u_2\|_{C(S;H)}\leqq\sqrt{\frac{M}{m}}\,|a_1-a_2|+\frac{1}{2m}\left(1+\frac{M}{m}\right)\|f_1-f_2\|_{L^1(S;H)}\,.$$

The condition (9) is satisfied for instance if

$$H=H_1\times H_2,\quad F[y_1,y_2]=[F_1y_1,F_2y_2]\,,$$
$$F_1\in\mathfrak{L}(H_1;H_1),\quad \forall t\in S: f(t)=[f_1(t),0]\,.$$

Problems, where these conditions are met, arise in the theory of plasticity (see [4]).

Remark 5. For every $a\in\overline{D(\varphi)}$ let u_a denote the solution to

$$(Eu)'(t)+\partial\varphi(u(t))\ni 0 \quad \text{for a.e.} \quad t\in S,\ u(0)=a,\ u\in C(\bar S;H)\cap AC(S;H)\,.$$

Because we can choose the length T of the time interval S arbitrarily large, we may assume that u_a is defined on $\boldsymbol{R}^+$. Evidently, by

$$Q(t)\,a:=u_a(t) \quad \text{for} \quad t\in\boldsymbol{R}^+,\ a\in\overline{D(\varphi)}\,,$$

there is defined a semigroup Q on $\overline{D(\varphi)}$. From the proof of Theorem 1 it follows that for every $t\in\boldsymbol{R}^+$ we have

$$\forall a_1,\ \forall a_2\in H: \big(EQ(t)\,a_1-EQ(t)\,a_2,\ Q(t)\,a_1-Q(t)\,a_2\big)$$
$$\leqq(Ea_1-Ea_2,\ a_1-a_2)\,.$$

This can be interpreted as a certain contractivity property of the semigroup operators $Q(t)$. The last inequality proves that $Q(t)$ is Lipschitzian with the Lipschitz constant $\frac{M}{m}$.

Remark 6. From Theorem 1 – Theorem 3 it follows that the mapping $[a,f]\mapsto u$, where u denotes the solution to (A), can be extended from $\overline{D(\varphi)}\times L^2(S;H)$ to a continuous mapping R from $\overline{D(\varphi)} \times L^1(S; H)$ into $C(\bar{S}; H)$. $R[a, f]$ is called the weak solution to (A) corresponding to the data $a \in \overline{D(\varphi)}$, $f \in L^1(S; H)$.

Remark 7. Most of the previous results can be generalized to the case that the functional φ depends (in a sufficiently nice way) on time.

3. Regularity results

Theorem 4. *Suppose that* (1), (2) *are valid and that* $a \in D(\partial\varphi)$, $f \in AC(\bar{S}; H)$. *Then the solution* u *to* (A) *is* Lipschitz*ian and for every* $t \in [0, T[$ *we have:*

$$(10)\qquad \begin{cases} u(t) \in D(\partial\varphi)\,; \text{ the right derivative } \dfrac{d^+u}{dt}(t) \text{ exists;} \\ E'(u(t))\dfrac{d^+u}{dt}(t) \in f(t) - \partial\varphi(u(t)), \\ \left(E'(u(t))\dfrac{d^+u}{dt}(t), \dfrac{d^+u}{dt}(t)\right) = \inf\limits_{y \in \partial\varphi(u(t))} \left(f(t) - y, (E'(u(t)))^{-1}(f(t) - y)\right). \end{cases}$$

Moreover, $\dfrac{d^+u}{dt}$ *is continuous from the right.*

Proof. 1) Let $y \in \partial\varphi(a)$ and $\bar{u}(t) = a$, $\bar{f}(t) = y$, $t \in \bar{S}$. Then $(E\bar{u})' + \partial\varphi(\bar{u}) \ni \bar{f}$. Setting $v := Eu$, $\bar{v} := E\bar{u}$ we obtain as in the proof of Theorem 1

$$\begin{aligned} &((v - \bar{v})(t), (u - \bar{u})(t)) \\ &\leqq \int_0^t \{((f - \bar{f})(s), (u - \bar{u})(s)) + (F'(v(s))f(s) - F'(\bar{v}(s))\bar{f}(s), (v - \bar{v})(s))\}\, ds \\ &= \int_0^t \{(f(s) - y, u(s) - a) + (F'(Eu(s))f(s) - F'(Ea)y, Eu(s) - Ea)\}\, ds\,. \end{aligned}$$

Consequently,

$$|u(t) - a|^2 \leqq c_1 \int_0^t |u(s) - a|\, ds, \quad c_1 \text{ independent of } t\,.$$

This implies (cf. Brezis [2], Lemma A.5)

$$(11)\qquad \forall t \in S\colon |u(t) - a| \leqq c_1 t\,.$$

2) Let $t \in [0, T[$ and $h > 0$ such that $t + h \leqq T$. Due to the monotonicity of $\partial\varphi$ and $\partial\psi$

$$\int_0^t (v'(s+h) - v'(s) - f(s+h) + f(s), u(s+h) - u(s))\, ds \leqq 0\,,$$

$$\begin{aligned} &\int_0^t (u'(s+h) - u'(s) - F'(v(s+h))f(s+h) \\ &+ F'(v(s))f(s), v(s+h) - v(s))\, ds \leqq 0\,. \end{aligned}$$

Adding these inequalities we obtain

$$(v\,(t+h)-v(t),\,u\,(t+h)-u(t))-(v(h)-v(0),\,u(h)-u(0))$$
$$\leqq \int_0^t \{(f\,(s+h)-f(s),\,u\,(s+h)-u(s))+((F'(v\,(s+h))-F'(v(s)))\,f(s)$$
$$+F'(v\,(s+h))\,(f\,(s+h)-f(s)),\,v\,(s+h)-v(s))\}\,ds\,.$$

Hence

$$m\,|u\,(t+h)-u(t)|^2 \leqq M\,|u(h)-a|^2$$
$$+\int_0^t \left\{|f\,(s+h)-f(s)|\cdot|u\,(s+h)-u(s)|\right.$$
$$\left.\times\left(1+\frac{M}{m}\right)+KM^2\,|f(s)|\cdot|u\,(s+h)-u(s)|^2\right\}ds\,,$$

where K denotes the LIPSCHITZ constant of F' on the ball $\{y\in H|\ |y|\leqq\|v\|_{C(S;H)}\}$. Using GRONWALL's Lemma and Lemma A.5 of [2] we find

$$|u\,(t+h)-u(t)|\leqq c_2\left(|u(h)-a|+\int_0^t |f\,(s+h)-f(s)|\,ds\right),\quad c_2=\text{const}.$$

In view of (11) and the absolute continuity of f this implies (cf. [2], Prop. A.5):

(12) $\qquad |u\,(t+h)-u(t)|\leqq c_3 h,\quad c_3$ independent of t and h.

Thus u is LIPSCHITZian and $u'\in L^\infty(S;H)$.

3) Let $t_0\in[0,T[$ be fixed and let $S_0:=[t_0,T]$, $L:=(E'(u(t_0)))^{\frac{1}{2}}$, $w:=Lu$ and

$$\forall t\in S_0\colon g(t):=L^{-1}(f(t)-(E'(u(t))-E'(u(t_0)))\,u'(t))\,.$$

Evidently, $g\in L^\infty(S_0;H)$ and g is continuous from the right at t_0. It is easy to see that

$$w'(t)+L^{-1}\partial\varphi L^{-1}w(t)\ni g(t)\quad\text{for a.e.}\quad t\in S_0,\ w(t_0)=Lu(t_0),\ w\in H^1(S_0;H).$$

The operator $L^{-1}\partial\varphi L^{-1}$ is maximal monotone. Therefore (cf. (12) and [2], Th. 3.5)

$$w(t_0)\in D(L^{-1}\partial\varphi L^{-1}),\ \frac{d^+w}{dt}(t_0)\ \text{exists},$$

$$\frac{d^+w}{dt}(t_0)=(g(t_0)-L^{-1}\partial\varphi L^{-1}w(t_0))^0\,,$$

where C^0 denotes the unique element of C of minimal norm if $C\subset H$ is closed and convex. From these results it follows (10).

4) Using the same notation as before we obtain similarly as in the second step of the proof for every $t\in[t_0,T[$ and $h>0$ such that $t+h\leqq T$:

$$(v\,(t+h)-v(t),\,u\,(t+h)-u(t))-(v\,(t_0+h)-v(t_0),\,u\,(t_0+h)-u(t_0))$$

$$\leqq c_4 \int_{t_0}^{t} \{|f(s+h)-f(s)| \cdot |u(s+h)-u(s)| + |u(s+h)-u(s)|^2\}\, ds$$

$$\leqq c_5 h \int_{t_0}^{t} \{|f(s+h)-f(s)| + h\}\, ds\,.$$

Dividing by h^2 and letting $h \downarrow 0$ we find (cf. BREZIS [2], Prop. A.5):

$$\left(E'(u(t))\frac{d^+u}{dt}(t), \frac{d^+u}{dt}(t)\right) - \left(E'(u(t_0))\frac{d^+u}{dt}(t_0), \frac{d^+u}{dt}(t_0)\right)$$

$$\leqq c_5 \int_{t_0}^{t} \{|f'(s)|+1\}\, ds\,.$$

Hence

$$\overline{\lim_{t \downarrow t_0}} \left(E'(u(t))\frac{d^+u}{dt}(t), \frac{d^+u}{dt}(t)\right) \leqq \left(E'(u(t_0))\frac{d^+u}{dt}(t_0), \frac{d^+u}{dt}(t_0)\right). \tag{13}$$

Let (t_n) be a sequence such that $t_n \downarrow t_0$, $t_n < T$ and $\frac{d^+u}{dt}(t_n) \rightharpoonup x$ in H. Because $u \in C(\bar{S}; H)$ we have $E'(u(t_n))\frac{d^+u}{dt}(t_n) \rightharpoonup E'(u(t_0))\,x$ and $u(t_n) \to u(t_0)$ in H. Since f is continuous and $\partial\varphi$ is maximal monotone this implies that

$$E'(u(t_0))\,x \in f(t_0) - \partial\varphi(u(t_0))\,. \tag{14}$$

In view of

$$\begin{aligned}(E'(u(t_0))\,x, x) &\leqq \varliminf_{n \to \infty} \left(E'(u(t_0))\frac{d^+u}{dt}(t_n), \frac{d^+u}{dt}(t_n)\right)\\ &= \varliminf_{n \to \infty} \left(E'(u(t_n))\frac{d^+u}{dt}(t_n), \frac{d^+u}{dt}(t_n)\right)\\ &\leqq \left(E'(u(t_0))\frac{d^+u}{dt}(t_0), \frac{d^+u}{dt}(t_0)\right)\end{aligned}$$

(cf. (13)) it follows from (10) and (14) that $x = \frac{d^+u}{dt}(t_0)$. Hence we have $\frac{d^+u}{dt}(t) \to \frac{d^+u}{dt}(t_0)$ as $t \downarrow t_0$ (weak convergence and convergence of the norm $\|z\| = (E'(u(t_0))\,z, z)^{\frac{1}{2}}$). This completes the proof of Theorem 4.

Corollary. *Let* (1), (2) *be satisfied and let* $a \in \overline{D(\varphi)}$, $f \in AC(S; H) \cap L^2(S; H)$. *Then the solution* u *to* (A) *satisfies* (10) *for every* $t \in]0, T[$ *and* $\frac{d^+u}{dt}$ *is continuous from the right on* $]0, T[$.

Proof. Since $f(t)-(Eu)'(t)\in\partial\varphi(u(t))$ for a.e. $t\in S$ we have $u(t)\in D(\partial\varphi)$ for a.e. $t\in S$. Let $t_n\downarrow 0$ such that $a_n:=u(t_n)\in D(\partial\varphi)$. Then we may apply Theorem 4 to the problem

$$(Eu)'(t)+\partial\varphi(u(t))\ni f(t) \quad \text{for a.e.} \quad t\in[t_n, T],$$
$$u(t_n)=a_n, \quad u\in H^1([t_n, T]; H).$$

This yields the assertions of the corollary.

Remark 8. It is easy to show that the solution u to (A) satisfies the condition

$$\sup_{t\in]0,T[} t\left|\frac{d^+u}{dt}(t)\right|^2<\infty$$

if $a\in D(\varphi)$ and $f\in AC(\bar{S}; H)$.

References

[1] V. Barbu, Existence theorems for a class of two points boundary problems. J. Diff. Eq. **17**, 236–257 (1975).

[2] H. Brezis, Opérateurs maximaux monotones et semi-groupes de contractions dans les espaces de Hilbert. Math. Studies **5**, North Holland 1973.

[3] H. Gajewski und K. Zacharias, Über eine Klasse nichtlinearer Gleichungen im Hilbertraum. J. Math. Anal. Appl. **44**, 71–87 (1973).

[4] K. Gröger, J. Nečas and L. Trávníček, Dynamic deformation processes of elastic-plastic systems. ZAMM **59**, 567–572 (1979).

[5] L. Trávníček, Existence a jednoznačnost řešení hranicního problému pro elasticko-plastické materialy. Thesis, Prague 1977.

Akademie der Wissenschaften der DDR
Zentralinstitut für Mathematik und Mechanik
DDR – 108 Berlin
Mohrenstraße 39

Československa Akademie Věd
Matematický Ústav
Praha 1
Žitná ulice 25

ON THE CONSTRUCTION OF LUSTERNIK-SCHNIRELMANN CRITICAL VALUES WITH APPLICATION TO BIFURCATION PROBLEMS

JINDRICH NEČAS, ARI LEHTONEN AND PEKKA NEITTAANMÄKI

ABSTRACT. An iterative method to construct Lusternik-Schnirelmann critical values is presented. Examples of its use to obtain numerical solutions to nonlinear eigenvalue problems and their bifurcation branches are given.

1. INTRODUCTION

Existence theorems for nonlinear eigenvalue problems of the form

$$g'(x) = \mu x, \tag{1.1}$$

where g is a functional on a Hilbert space H and $g'(x)$ the corresponding gradient, are considered in many papers. The existence theorems are based on the existence of a critical vector with respect to the manifold $S = \{x \in H : \|x\| = 1\}$. Very general results on nonlinear eigenvalue problems in Banach spaces can be found in H. Amann [**1**] and in E. Zeidler [**11**], which also contains an extensive bibliography on critical point theories.

Iterative methods for the construction of all Lusternik-Schnirelmann critical values and critical vectors has been presented by J. Nečas [**7**] and by A. Kratochvíl and J. Nečas [**4**],[**5**].

In this paper we shall give an extension of the method used in [**5**] to study the eigenvalue problem (1.1). Also, we give examples of problems that have been tested numerically.

2. ITERATIVE CONSTRUCTION OF THE FIRST LUSTERNIK-SCHNIRELMANN CRITICAL VALUE

Let H be real Hilbert space with inner product $(\cdot|\cdot)$ and norm $\|\cdot\|$. Furthermore, we set $S = \{x \in H : \|x\| = 1\}$. Let g be a continuously differentiable functional on H such that the derivative g' is strongly continuous, i.e. for each sequence $(x_n)_{n=1}^{\infty} \subset H$ converging weakly to $x_0 \in H$, the sequence $(g'(x_n))_{n=1}^{\infty}$ converges to $g'(x_0)$.

Assume that the following conditions are fulfilled:

$$g(0) = 0, \quad g'(0) = 0; \tag{2.1}$$

$$\text{if } g(x) \neq 0 \text{ and } \|x\| \leq 1 \text{ then } g'(x) \neq 0; \tag{2.2}$$

there exists a constant $c > 0$ such that

$$(g'(x) - g'(y)|x - y) \geq -c\|x - y\|^2 \tag{2.3}$$

© Taylor & Francis 1987. Reprinted by permission from: Nečas, J., Lehtonen, A. & Neittaanmäki, P., On the construction of Lusternik-Schnirelmann critical values with application to bifurcation problems, Applicable Analysis, 25, 253-268.

holds for all $x, y \in H$, $\|x\| \le 1$, $\|y\| \le 1$.

For a given $x \in S$ we divide $g'(x)$ into the tangential and normal components, i.e. $g'(x) = Tg'(x) + Ng'(x)$, where

$$Tg'(x) = g'(x) - \frac{(g'(x)|x)}{\|x\|^2}x. \tag{2.4}$$

THEOREM 2.1. *Let the above hypothesis be satisfied. Assume that $x_1 \in S$ and $g(x_1) > 0$. Let $0 < \theta < 1/c$. Define a sequence $(x_n)_{n=1}^{\infty}$ by*

$$x_{n+1} = \frac{x_n + \theta g'(x_n)}{\|x_n + \theta g'(x_n)\|}. \tag{2.5}$$

Then there exists a subsequence of $(x_n)_{n=1}^{\infty}$ converging to a solution of (1.1). Hence, there exists a point $x_0 \in S$ and a number $\mu \in \mathbf{R}$ such that

$$g'(x_0) = \mu x_0. \tag{2.6}$$

PROOF: First, note that $x \perp Tg'(x)$ for all $x \in S$. Put $\mu_n = (g'(x_n)|x_n)$ and $r_n = \|x_n + \theta g'(x_n)\|$. Then $r_n^2 = \|(1+\theta\mu_n)x_n + \theta Tg'(x_n)\|^2 = (1+\theta\mu_n)^2 + \theta^2\|Tg'(x_n)\|^2$.

From (2.3) we get $\mu_n \ge -c$. Since $0 < \theta < 1/c$ we have $r_n \ge 1 - \theta c > 0$. Futhermore, from (2.5) we obtain the following identity

$$\begin{aligned}(g'(x_n)|x_{n+1} - x_n) &= \frac{1}{\theta}(r_n x_{n+1} - x_n | x_{n+1} - x_n)\\ &= \frac{1}{\theta}(r_n - (x_n|x_{n+1}) - r_n(x_{n+1}|x_n) + 1)\\ &= \frac{1}{2\theta}(1 + r_n)\|x_{n+1} - x_n\|^2.\end{aligned} \tag{2.7}$$

Next, we shall show that the sequence $(g(x_n))_{n=1}^{\infty}$ is increasing. For this we note that using (2.3) we obtain

$$\begin{aligned}g(y) - g(x) &= \int_0^1 (g'(x + t(y-x))|y - x)\, dt\\ &= \int_0^1 (g'(x + t(y-x)) - g'(x)|y - x)\, dt + (g'(x)|y - x)\\ &\ge -\frac{c}{2}\|y - x\|^2 + (g'(x)|y - x).\end{aligned}$$

Using this and (2.7) we get

$$g(x_{n+1}) - g(x_n) \ge (1/\theta - c)\|x_{n+1} - x_n\|^2. \tag{2.8}$$

Hence the sequence $(g(x_n))_{n=1}^{\infty}$ is increasing. Because of the strong continuity of g' the functional g is bounded on S. Therefore, the limit $\lim_{n\to\infty} g(x_n)$ exists and

$$\lim_{n\to\infty} \|x_{n+1} - x_n\| = 0.$$

Since the sequence $(x_n)_{n=1}^{\infty}$ is bounded it has a subsequence (we use the same notation for each subsequence) converging weakly to some $x_0 \in H$. The strong continuity of g' yields $r_n^2 = 1 + 2\theta(g'(x_n)|x_n) + \theta^2\|g'(x_n)\|^2 \to 1 + 2\theta(g'(x_0)|x_0) + \theta^2\|g'(x_0)\|^2 =: r_0$.

Since $x_{n+1} = 1/r_n(x_n + \theta g'(x_n))$ we get $(r_0 - 1)x_0 = \theta g'(x_0)$. Furthermore, as $g(x_0) \geq g(x_n) > 0$, we obtain from (2.1) and (2.2), $g'(x_0) \neq 0$. Therefore $r_0 \neq 1$. To show that the sequence $(x_n)_{n=1}^{\infty}$ converges to x_0 we write (2.5) as

$$(1 - r_n)x_n = r_n(x_{n+1} - x_n) - \theta g'(x_n). \tag{2.9}$$

Here the right hand side converges in norm to $-\theta g'(x_0)$ and the left hand side to $1 - r_0$. Since $1 - r_0 \neq 0$, we obtain the desired convergence from (2.9).

COROLLARY 2.2. *Assume that insted of (2.3) the inequality*

$$(g'(x) - g'(y)|x - y) \leq c\|x - y\|^2 \tag{2.10}$$

holds for some $c > 0$ and for all $x, y \in H$, $\|x\| \leq 1$, $\|y\| \leq 1$, and that $x_1 \in S$ is such that $g(x_1) < 0$. Let $-1/c < \theta < 0$. Define the sequence $(x_n)_{n=1}^{\infty}$ by (2.5). Then there exists a subsequence of $(x_n)_{n=1}^{\infty}$ converging to a solution of (2.6).

PROOF: We can apply Theorem 2.1 to $-g$ and $-\theta$. The sequence $(g(x_n))_{n=1}^{\infty}$ is now decreasing.

3. HIGHER ORDER CRITICAL POINTS

To study higher order critical poins we recall some definitions concerning the Lusternik-Schnirelmann theory in an infinite dimensional Hilbert space.

We use the notion of the order of a set rather than the category or genus, cf. [**2**]. Let K be a symmetric closed set in H. We say that ord $K = 0$ if K is empty; that ord $K = 1$ if $K = K_1 \cup K_2$, where the K_i are closed subsets of K and neither K_1 nor K_2 contains antipodal poins. In general, ord $K = n$ if $K = \cup_{i=1}^{n+1} K_i$, where the K_i are closed subsets of K not containing antipodal points and n is the least possible number. Finally, ord $K = \infty$ if no such n exists.

Let $V_k^{\pm}$ denote the set of all symmetric compact subsets K of S such that ord $K \geq k$ and $\pm g(x) > 0$ on K. Denote

$$\pm\gamma_k^{\pm} = \begin{cases} \sup_{K \in V_k^{\pm}} \min_{x \in K} \pm g(x), \text{ if } V_k^{\pm} \neq \emptyset \\ 0, \text{ if } V_k^{\pm} = \emptyset \end{cases} \tag{3.1}$$

The fundamental theorem of the Lusternik-Schnirelmann theory states that there exists a sequence of critical points $x_k^{\pm}$ of g such that $g(x_k^{\pm}) = \gamma_k^{\pm}$, $\pm\gamma_k^{\pm} \searrow 0$ and $x_k^{\pm} \to 0$ weakly. For a proof see [**1**] or [**11**, Ch. 44]. In case $g(x) > 0$ on H a proof has been given in [**7**]. There the critical points are found as limits $\lim_{t\to\infty} x(t)$, where $x(t)$ is the solution of the differential equation on S

$$\dot{x} = Tg'(x), \qquad x(0) = x_0 \in S.$$

Our method using a discrete iteration seems to be more stable with respect to variations in x_0 than this one.

Let the assumptions of Theorem 2.1 hold for a functional g. Furthermore, assume that g is even on S, i.e. $g(-x) = g(x)$ when $\|x\| = 1$. We assume that g has positive critical values. Let γ_1^+ and γ_2^+ be the first and second Lusternik-Schnirelmann critical values of $g|_S$, $\gamma_1^+ > \gamma_2^+$. Let there exist a positive constant ε such that there are no critical values in the interval $]\gamma_2^+ - \varepsilon, \gamma_2^+[$. Let K_1 be a compact symmetric subset of S such that ord $K_1 \geq 2$, $g(x) > 0$ on K_1 and

$$\gamma_2^+ - \varepsilon < \min_{x \in K_1} g(x) < \gamma_2^+. \tag{3.2}$$

We denote by φ the function used to define the iteration in (2.5), i.e.

$$\varphi(x) = \frac{x + \theta g'(x)}{\|x + \theta g'(x)\|}. \tag{3.3}$$

Then φ is a well-defined, odd continuous map $S \to S$. Choose $x_1 \in K_1$, and put $x_{n+1} = \varphi(x_n) = \varphi^n(x_1)$, where φ^n denotes the n-fold composition $\varphi \circ \cdots \circ \varphi$.

Let $x_n^{(0)}$ be a vector from K_1 such that

$$\min_{x \in K_1} g(\varphi^n(x)) = g(\varphi^n(x_n^{(0)})) \tag{3.4}$$

for any integer n.

THEOREM 3.1. *Let the above assumptions be fulfilled. Then the following assertions hold:*

(i) $\lim_{n\to\infty} g(\varphi^n(x_n^{(0)})) = \gamma_2^+$;

(ii) *there exists* $x^{(0)} \in K_1$ *such that*

$$\lim_{n\to\infty} g(\varphi^n(x^{(0)})) = \gamma_2^+;$$

(iii) *there exists a subsequence of* $(x_n^{(0)})_{n=1}^\infty$ *converging to* $x^{(0)}$

(iv) *for each* $x^{(0)}$ *satisfying (ii) there exists a subsequence of* $(\varphi^n(x^{(0)}))_{n=1}^\infty$ *converging to some* x_0 *such that*

$$g'(x_0) = \mu x_0$$

PROOF: From the proof Theorem 2.1 (see (2.8)) we have

$$g(\varphi^n(x)) \leq g(\varphi^{n+1}(x)), \tag{3.5}$$

for each $x \in K_1$. Since φ is an odd continuous operator from S to S we have

$$\text{ord } \varphi^n(K_1) \geq \text{ord } K_1 \geq 2.$$

From this, the definition of γ_2^+ and (3.5) we get

$$\lim_{n\to\infty} \min_{x\in K_1} g(\varphi^n(x)) \leq \gamma_2^+, \tag{3.6}$$

Let us denote

$$\kappa = \lim_{n\to\infty} \min_{x\in K_1} g(\varphi^n(x)) = \lim_{n\to\infty} g(\varphi^n(x_n^{(0)})). \tag{3.7}$$

By compactness there exists a subsequence of $(x_n^{(0)})_{n=1}^{\infty}$ converging to $x^{(0)} \in K_1$. According to Theorem 2.1,

$$\lim_{n\to\infty} g(\varphi^n(x^{(0)}))$$

is a critical value of g. Since there are no critical values in the interval $]\gamma_2^+ - \varepsilon, \gamma_2^+[$, we obtain

$$\lim_{n\to\infty} g(\varphi^n(x^{(0)})) \geq \gamma_2^+.$$

Take $\eta > 0$. Then by the definition of $x_n^{(0)}$ and the convergence $x_n^{(0)} \to x^{(0)}$ there exist numbers n_0 and n_1 such that

$$g(\varphi^{n_0}(x_n^{(0)})) \geq \gamma_2^+ - \eta, \tag{3.8}$$

for any integer $n \geq n_1$.

According to (3.5) and (3.8)

$$g(\varphi^n(x_n^{(0)})) \geq g(\varphi^{n_0}(x_n^{(0)})) \geq \gamma_2^+ - \eta,$$

for any $n \geq \max\{n_0, n_1\}$ large enough. Therefore

$$\kappa = \lim_{n\to\infty} g(\varphi^n(x_n^{(0)})) \geq \gamma_2^+ - \eta,$$

and thus (i) follows.

The rest of the proof follows as in Theorem 2.1.

In a similar way we obtain the following result

COROLLARY 3.2. *Let the assumptions of Theorem 3.1 be fulfilled. Let*

$$\gamma_1^+ \geq \cdots \geq \gamma_k^+ > \gamma_{k+1}^+ = \cdots = \gamma_{k+l}^+ > \gamma_{k+l}^+$$

be the positive Lusternik-Schnirelmann values of $g|_S$.

Let there exist a constant $\varepsilon > 0$ such that there are no critical values in the interval $]\gamma_{k+l}^+ - \varepsilon, \gamma_{k+l}^+[$.

Let K_1 be a compact symmetric subset of S such that

$$\text{ord } K_1 \geq k+1$$

$$\gamma_{k+l}^{+} - \varepsilon < \min_{x \in K_1} g(x) < \gamma_{k+l}^{+}.$$

For $x \in K_1$ let the sequence $(x_n^{(0)})_{n=1}^{\infty}$ be defined by (3.4). Then

$$\lim_{n \to \infty} g(\varphi^n(x_n^{(0)})) = \gamma_{k+l}^{+},$$

and there exists a point $x^{(0)} \in K_1$ such that

$$\lim_{n \to \infty} g(\varphi^n(x^{(0)})) = \gamma_{k+l}^{+}.$$

Moreover, the assertions (iii) and (iv) of Theorem 3.1 hold.

REMARK 3.3. Assume that instead of (2.3) the condition (2.10) holds and that θ is chosen as in Corollary 2.2. Then, with obvious modifications in Theorem 3.1 and Corollary 3.2 we can also construct negative critical values and corresponding critical points.

4. Bifurcation of solutions

We will use the previous results to find bifurcation branches for the equation (1.1). Let $g : H \to \mathbf{R}$ be a given function and for $r > 0$ define $g_r : H \to \mathbf{R}, g_r(x) = g(rx)$. Then $g_r'(x) = \mu x$ if and only if $g'(x_r) = \mu_r x_r$, where $x_r = rx$ and $\mu_r = \mu r^{-2}$. It is immediately verified that if the second derivative $g''(0)$ exists and $(\mu_0, 0)$ is a bifurcation point of (1.1), then μ_0 is an eigenvalue of $g''(0)$.

To calculate solutions for equation (1.1) in Section 6 we will use the following method: Let $\mu_0 \neq 0$ be an eigenvalue and x_1 an eigenvector of $g''(0)$. Using Theorems 2.1 and 3.1 and Corollary 2.2 solve the equation $g_r'(x) = \mu x, x \in S$. Then (μ_r, x_r) gives a bifurcation branch of (1.1).

In order to verify that this method can be applied we need the following result due to Krasnoselskii:

THEOREM 4.1. *Let $B : H \to H$ be a linear, continuous and symmetric operator with $\mathcal{R}(B)$ closed and $0 < \dim \mathcal{N}(B) < \infty$. Let N be a continuously differentiable potential operator on a neighbourhood of the origin such that $\|N(x)\|/\|x\| \to 0$ as $x \to 0$. Then the point $(0,0)$ is a bifurcation point of the equation*

$$Bx + N(x) = \varepsilon x. \tag{4.1}$$

A proof of this result can be found in [**11**, Ch. 45]; see also [**2**, App. 1].

Now we can prove

THEOREM 4.2. *Let $g : H \to \mathbf{R}$ be a twice continuously differentiable function on a neighbourhood of the origin. Assume that g' is strongly continuously and that $g''(0)$ has an eigenvalue $\mu_0 \neq 0$. Then the point $(\mu_0, 0)$ is a bifurcation point of (1.1).*

PROOF: We can write equation (1.1) in the form

$$\mu_0 x - g''(0)x - (g'(x) - g''(0)x) = (\mu_0 - \mu)x.$$

Denote $\varepsilon = \mu_0 - \mu$, $B = \mu_0 \mathbf{1} - g''(0)$ and $N(x) = -g'(x) + g''(0)x$. From the strong continuity of g' it follows that $g''(0)$ is compact. Therefore $\mathcal{R}(B)$ is closed and $0 < \dim \mathcal{N}(B) < \infty$, cf. [**10**, Ch. X.5]. Since g'' is continuous, we have $\|N(x)\|/\|x\| \to 0$ as $x \to 0$. Hence the conditions of Theorem 4.1 are fulfilled and the result follows.

5. Application to differential equations

We will apply the previous results to the following two point boundary value problem

$$\begin{cases} -u'' = \lambda G'(u) \\ u(a) = u(b) = 0 \end{cases} \tag{5.1}$$

on a bounded interval $]a, b[\subset \mathbf{R}$, where $G : \mathbf{R} \to \mathbf{R}$ is a given continuously differentiable function and $G'(u)$ denotes the function $t \mapsto G'(u(t))$. For partial differential equations the situation is analogous and will not handled in this connection.

We use the inner product $(u|v) = (u'|v')_{L^2(a,b)}$ in the Sobolev space $H = H_0^1(a, b)$. Problem (5.1) is then equivalent to finding $u \in H$ such that

$$(G'(u)|v)_{L^2(a,b)} = \mu\,(u|v) \qquad \text{for all } v \in H, \tag{5.2}$$

where $\mu = 1/\lambda$.

We define $g : H \to \mathbf{R}$ by

$$g(u) = \int_a^b G(u(t))\,dt.$$

Then the derivative of g is given by

$$(g'(u)|v) = \int_a^b G'(u(t))v(t)\,dt \qquad \text{for all } v \in H.$$

Since H embeds compactly into the space of continuous functions on the inteval $[a, b]$, the derivative g' is strongly continuous.

We assume that $G(0) = G'(0) = 0$. In order to fulfil condition (2.2) we assume that $G'(r)$ does not vanish on any interval. If we assume that G admits a continuous second derivative, then g'' exists and is given by

$$g''(u)(h,k) = \int_a^b G''(u(t))h(t)k(t)\,dt.$$

It is easily verified that condition (2.3) holds for $c = ((b-a)/2)^2 \sup\{G''(r) : |r| \leq \sqrt{(b-a)/2}\}$.

Therefore, the linear eigenvalue problem $g''(0)u = \mu_0 u$ is equivalent to the boundary value problem

$$\begin{cases} -\mu_0 u'' = G''(0)\,u \\ u(a) = u(b) = 0. \end{cases} \tag{5.3}$$

If $G''(0) \neq 0$ then problem (5.3) has nonzero eigenvalues and we can apply Theorem 4.2 to prove that they give bifurcation branches for problem (5.1).

6. Numerical examples

We shall consider four examples where the method presented above is applied in connection of the finite element method. The problem (5.1) is solved by FEM with piecewise linear elements. We divide the interval $[a, b]$ into $n(h)$ subintervals $[a_i, a_{i+1}]$, $h = a_{i+1} - a_i$. Let $(Q_1, \ldots, Q_{n(h)-1})^T$ be the vector of the nodal values of the finite element solution. Then the approximation of the problem (5.1) reads

$$AQ = \lambda G'(Q),$$

where A is the stiffness matrix

$$A = \frac{1}{h} \begin{pmatrix} 2 & -1 & 0 & \ldots & 0 & 0 \\ -1 & 2 & -1 & \ldots & 0 & 0 \\ \vdots & \vdots & \vdots & \ddots & \vdots & \vdots \\ 0 & 0 & 0 & \ldots & 2 & 1 \\ 0 & 0 & 0 & \ldots & 1 & 2 \end{pmatrix}$$

and $G'(Q) = h\,(G'(Q_1), \ldots, G'(Q_{n(h)-1}))^T$.

The norms appeared above are computed here in the discrete form

$$\|Q\| = \sqrt{(Q|AQ)}.$$

The authors are indebted to Mr. T. Männikkö for his assistance in numerical tests.

Example 6.1. Consider the nonlinear eigenvalue problem

$$\begin{cases} -u''(t) = \lambda \sin u(t), & t \in]0,1[\\ u(0) = u(1) = 0. \end{cases} \tag{6.1}$$

The simple eigenvalues of the linearized problem for (6.1) are $\lambda_i = i^2\pi^2$ and the corresponding eigensolutions are $\phi_i(t) = \sin \pi i t$. Consequently, the symmetric bifurcation points are $(\lambda_i, 0)$.

In Figures 6.1 – 6.3 we see the numerical reults obtained with discretization parameter $h = 1/32$ for the first, second and third bifurcation branches.

Fig. 6.1. First bifurcation branch.

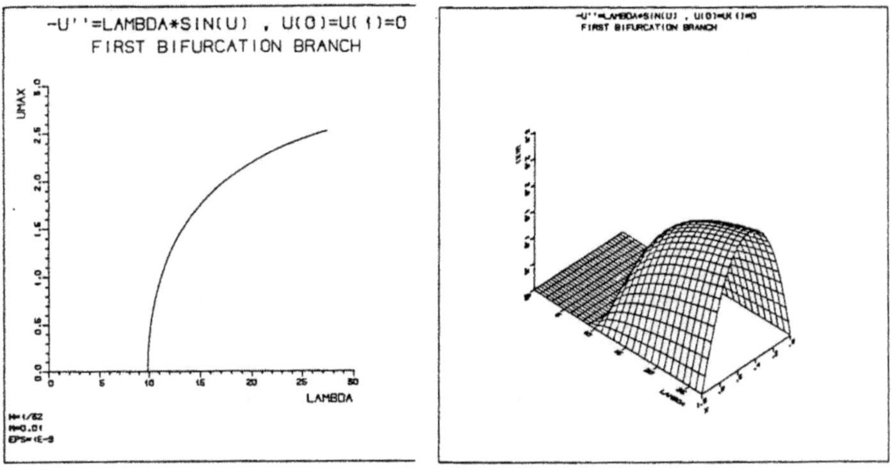

Fig. 6.2. Second bifurcation branch.

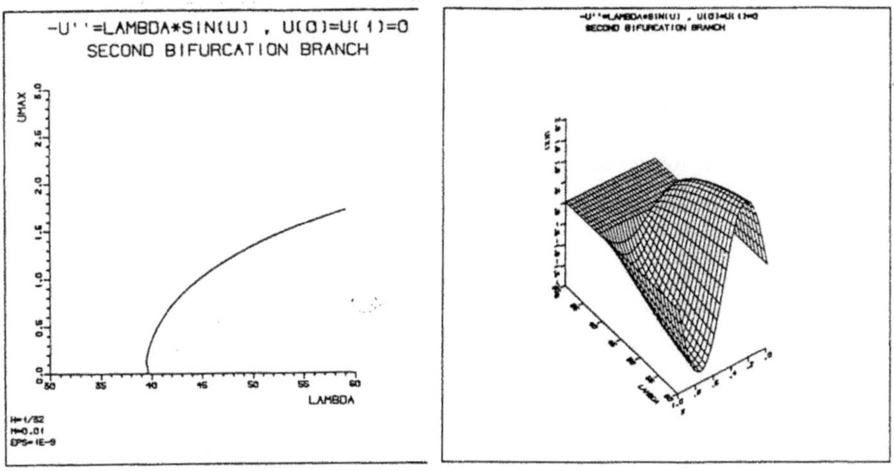

Fig. 6.3. Third bifurcation branch.

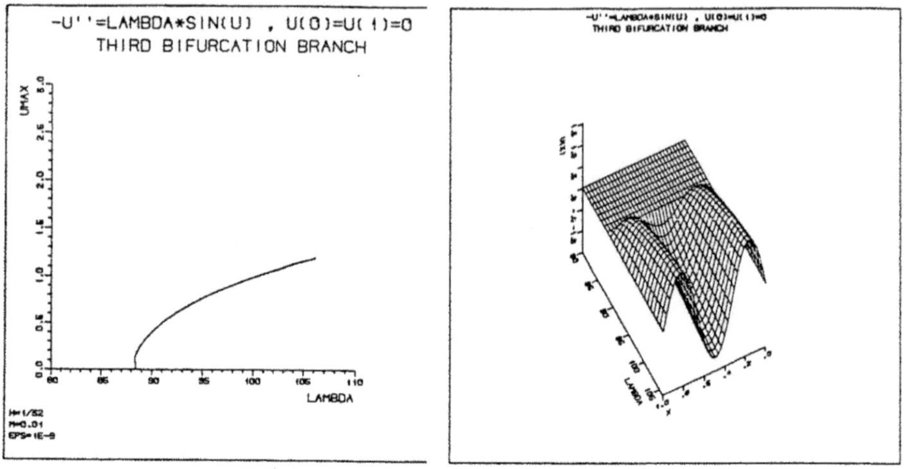

Example 6.2. Consider the nonlinear eigenvalue problem

$$\text{(6.2)} \qquad \begin{cases} -u''(t) = \lambda(u(t))^5, & t \in]0,1[\\ u(0) = u(1) = 0. \end{cases}$$

In Figures 6.4 – 6.5 we see the numerical results with $h = 1/32$ for first and second bifurcation branches.

Fig. 6.4. First bifurcation branch.

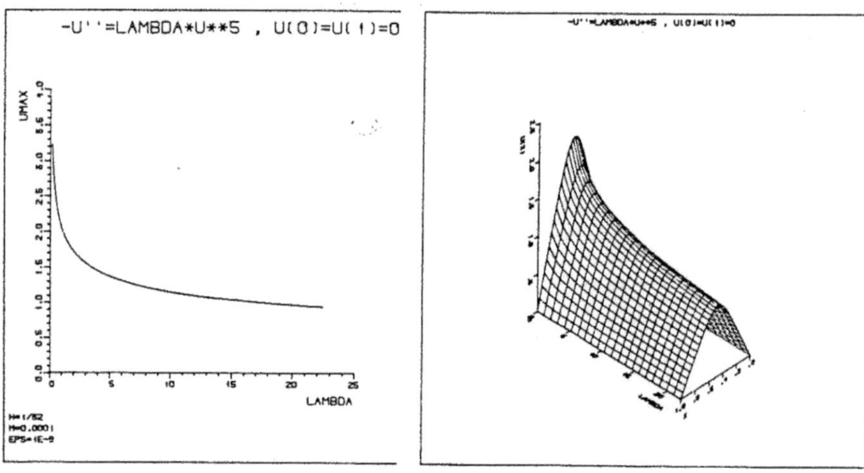

Fig. 6.5. Second bifurcation branch.

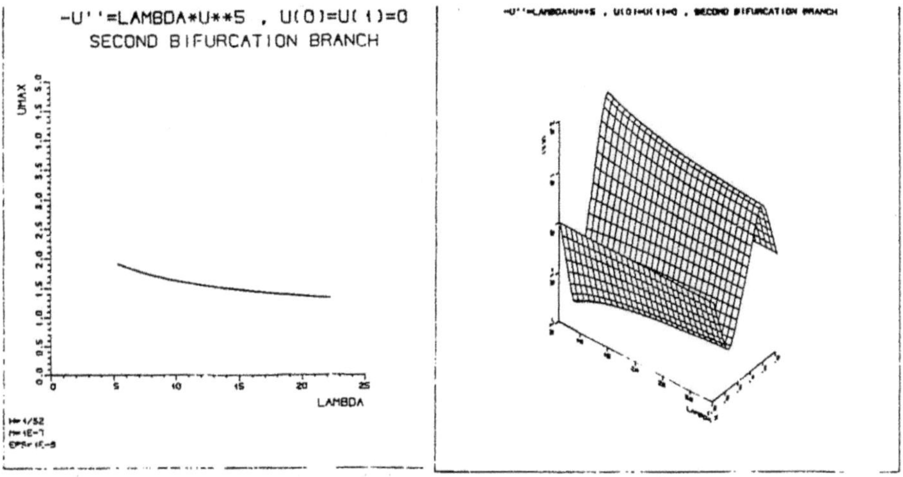

Example 6.3. Consider the nonlinear eigenvalue problem

$$(6.3)\qquad \begin{cases} -u''(t) = \lambda\big(u(t)(1-\sin u(t)) + (u(t))^3\big) & t \in]0,1[\\ u(0) = u(1) = 0. \end{cases}$$

In Figure 6.6 we see the behaviour of $\max_{t\in]0,1[} |u(t,\lambda)|$ for $\lambda \in]0,15[$, $h = 1/32$. At point $\lambda \approx 12.7$ a turning point occurs.

Fig. 6.6. First bifurcation branch.

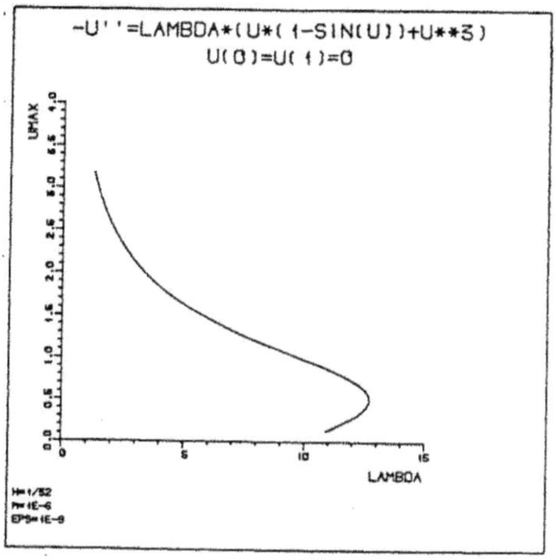

Example 6.4. Consider the nonlinear eigenvalue problem

$$\text{(6.4)} \qquad \begin{cases} -u''(t) = \exp\big(u(t)/(1+Du(t))\big), & t \in]-1,1[\\ u(-1) = u(1) = 0, \end{cases}$$

where $D \geq 0$.

Figures 6.7 –6.9 show the numerical results with $h = 1/32$ for various parameters D. In Figure 6.7 the illustration of $\max_{t\in]-1,1[} |u(t,\lambda)|$ for $\lambda \in]0,2[$ with $D = 0$ and $D = 0.1$. Figure 6.8 shows the result with $D = 0.2$ and $D = 0.3$.

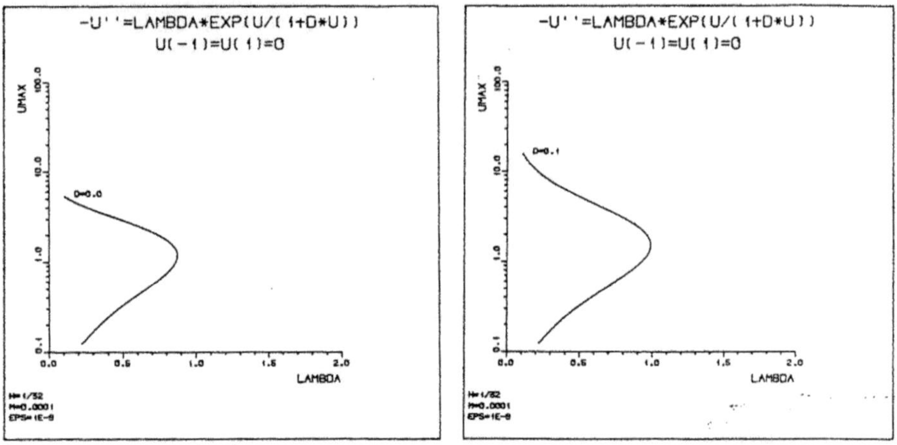

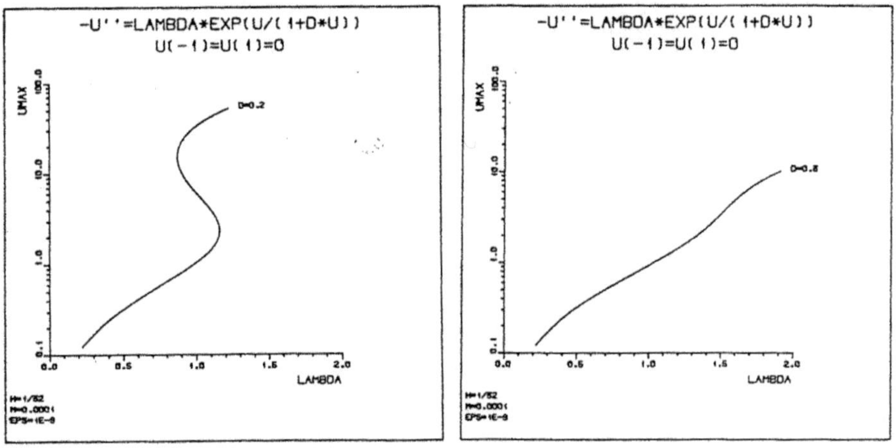

As s summary of above results we can say that by the method presented here we were able to find the bifurcation point or turning point with a relatively good accuracy already with a rough discretization parameter. Those examples are often used in literature to test numerical methods for nonlinear problems. Our results agree well with those given in literature, cf. [3], [6], [8].

The method works of course for nonlinear eigenvalue problems for PDE's. About these application will be reported in a forthcoming paper.

References

1. H. Amann, *Lusternik-Schnirelmann Theory and Non-Linear Eigenvalue Problems*, Math. Ann. **199** (1972), 55–72.
2. S. Fučik, J. Nečas, J. Souček and V. Souček, "Spectral Analysis of Nonlinear Operators", Lecture Notes in Mathematics 346, Springer-Verlag, Berlin-Heidelberg-New York, 1973.
3. R.Glowinski, H.Keller and L.Reinhart, *Continuation-conjugate gradient methods for least square solution of nonlinear boundary value problems*, INRIA, Rapports de Recherche **141** (1982).
4. A.Kratochvíl and J. Nečas, *Secant Modulus Method for the Construction of a Solution of Nonlinear Eigenvalue Problems*, Boll. Un. Mat. Ital. B(5) **16** (1979), 694–710.
5. ______, *Gradient Methods for the Construction of Lusternik-Schnirelmann Critical Values*, RAIRO Anal. Numér. **14** (1980), 43–54.
6. M.D.Mittelman and H.Weber (ed.), "Bifurcation Problems and Their Numerical Solution", ISNM 54, Birkhäuser, Basel, 1980.
7. J. Nečas, *An Approximate Method of Finding Critical Poins of Even Functionals*, Proc. Steklov Inst. Math. **134** (1975), 267–272.
8. P.Neittaanmäki and K.Ruotsalainen, *On the numerical solution of the bifurcation problem for the sine-Gordon equation*, Arab. J. Math. (1986) (to appear).
9. L.Nirenberg, "Topics in Nonlinear Functional Analysis", Lecture Note, Courant Institute of Mathematical Sciences, New York University, 1974.
10. K. Yosida, "Functional Analysis", Fourth Edition, Grundlehren der mathematischen Wissenschaften 123, Springer-Verlag, New York-Berlin-Heidelberg-Tokyo, 1974.
11. E. Zeidler, "Nonlinear Functional Analysis and its Applications III. Variational Methods and Optimization", Springer-Verlag, New York-Berlin-Heidelberg-Tokyo, 1985.

Keywords and phrases. Lusternik-Schnirelmann -theory, bifurcation theory, differential equations
1980 Mathematics subject classifications (Amer. Math. Soc.): Primary:58E05,65L15; Secondary:34B15,47H15
Current address (J. Nečas) : Charles University, Faculty of Mathematics and Physics, KAM MFF UK Malostranske 2/25, CS-11800 Prague, Czechoslovakia
University of Jyväskylä, Department of Mathematics, SF-40100 Jyväskylä, Finland

Chapter 10
Commentary

10.1 Nonlinear Functional Analysis, Calculus of Variations

Pavel Drábek

Abstract This contribution is focused on the research of Professor Jindřich Nečas dedicated to nonlinear eigenvalue problems, theory of monotone operators and calculus of variations. Several Necas's works are mentioned here as milestones of the development of these fields.

Professor Jindřich Nečas was one of the most distinguished Czech mathematicians of the second half of the twentieth century. His first publication registered by MathSciNet goes back to 1956, the last one is dated 2003. His "author profile" on MathSciNet registers 202 of his publications with 67 co-authors and more than 1,500 citations of his works. He contributed to many branches of mathematical analysis and its applications. His profound papers in several directions of mathematics were essential impulses for the further development of these subjects. His charismatic personality influenced many young mathematicians not only in the Czech Republic but also in many countries all over the world. I had the privilege to be a regular visitor of his seminar in partial differential equations (PDEs) at the Mathematical Institute of the Academy of Sciences in Prague during the period of the 1970s and 1980s. Professor Nečas was one of the top world experts in PDEs, in particular, in "modern methods" in PDEs. One of the most favourite tools of J. Nečas to study qualitative properties of PDEs were nonlinear functional analysis and calculus of variations. The most significant publications of J. Nečas in this direction are listed in [JN-C40, JN-C53, JN-C73, JN-C94, JN-C130].

The Fredholm alternative is one of the main tools in the study of solvability properties of linear operator equations. It was initially invented in the context of linear integral equations by Ivar Fredholm at the beginning of the twentieth century. The proof of its general operator version is a nice demonstration of the power of

© Springer Basel 2015

Š. Nečasová et al. (eds.), *Selected Works of Jindřich Nečas*, Advances in Mathematical Fluid Mechanics, DOI 10.1007/978-3-0348-0231-4_10

abstract linear functional analysis. However, if the linearity of an operator equation is violated, this proof does not work at all. Several mathematicians in the 1960s observed that even if the operator lacks the additivity but preserves homogeneity (even in some generalized form), the corresponding operator equation inherits some of the properties indicated by the linear Fredholm alternative. We would like to mention S.I. Pohozaev and J. Nečas at this stage. Nečas' paper [JN-C40] published in Annali della Scuola Normale Superiore di Pisa in 1969 is a pioneering work in this direction. In this paper J. Nečas generalizes the famous Leray–Lions Theorem. Moreover, for the first time in the literature he presents the Fredholm alternative for a nonlinear operator equation. Roughly speaking, under suitable assumptions on T and S, either $T - \lambda S$ is an invertible operator or else the spectral parameter is an eigenvalue in the sense that there exists $u \neq 0$ satisfying $T(u) - \lambda S(u) = 0$. The final section of [JN-C40] is devoted to the applications to boundary value problems for PDEs. Nečas formulates here the assumptions on the data of the problem so that the abstract version of a nonlinear Fredholm alternative yields the existence of weak solutions for a wide class of quasilinear elliptic PDEs on bounded domains. Further results and applications in this direction are presented in Nečas' paper [JN-C40] published in Mathematica Bohemica (Časopis pro pěstování matematiky) in 1972.

It is important to mention here that J. Nečas introduced this subject also to his students and colleagues. Let us mention here S. Fučík and M. Kučera among others. The results of their Theses had been published in series of papers and were influenced by the ideas of their advisor. J. Nečas' scientific life in the 1960s and 1970s was closely connected with the study of nonlinear operator equations. This effort culminated in Springer Lecture Notes published by S. Fučík, J. Nečas, J. Souček and V. Souček in 1973. This book is a classical reference book in the field of nonlinear analysis and continues to be quite frequently cited.

In contrast with works [JN-C40, JN-C53] which are associated with stationary problems, the papers [JN-C73] and [JN-C94] deal with evolution parabolic equations. J. Nečas' work [JN-C73] published in the Czechoslovak Mathematical Journal in 1974 generalizes Rothe's original idea of the time discretization to abstract parabolic equations. The basic idea consists in solving a series of nonlinear stationary problems using the theory of monotone operators separately on each fixed time level. Then the convergence has to be proved for fining down the step of the time discretization. A different approach to evolution problems is used in a joint paper of J. Nečas and German mathematician K. Gröger [JN-C94] published in Mathematische Nachrichten in 1979. Existence and uniqueness of the weak solution is proved here using the methods of calculus of variations. The last section of this paper deals with regularity of the weak solution. This was another favourite topic of J. Nečas, in particular, in the late 1970s and 1980s.

The last paper in this series, [JN-C130], was published by J. Nečas jointly with two Finish mathematicians, A. Lehtonen and P. Neittaanmäki, in Applicable Analysis in 1987. This paper is devoted to the study of nonlinear eigenvalue problems and related bifurcation problems. The authors use the calculus of variations on manifolds as their tool to approach the problem. In order to get the critical points of higher energy they apply the Lusternik–Schnirelmann theory. Abstract results are than applied to the second-order nonlinear boundary value problems for ordinary differential equations. The results are illustrated by numerical simulations since the paper was written already when the first efficient computers were available in order to support the theoretical mathematical research.

Everybody who was close to professor Jindřich Nečas, saw him working or just talking about the problems of everyday life, had to be affected by the strong charisma of his personality. For this reason he has many successors not only in the field of nonlinear analysis and calculus of variations but also in other branches of mathematics where he left his deep trace.

Nečas' Works Quoted Above

The references [Cxx] refer to the respectice papers listed in Chapter 2 "Complete Bibliography of Jindřich Nečas". The references [JN-Cyy] refer to papers [Cyy] that are reproduced in Chapter 9 "Selected Papers on Nonlinear Functional Analysis and Calculus of Variations".

[JN-C40] J. Nečas, *Sur l'alternative de Fredholm pour les opérateurs non-linéaires avec applications aux problèmes aux limites*, Ann. Scuola Norm. Sup. Pisa (3), 1969, vol. 23, pp. 331–345.

[JN-C53] J. Nečas, *Fredholm alternative for nonlinear operators and applications to partial differential equations and integral equations*, Časopis Pěst. Mat., 1972, vol. 97, pp. 65–71.

[JN-C73] J. Nečas, *Application of Rothe's method to abstract parabolic equations*, Czechoslovak Math. J., 1974, vol. 24 (99), pp. 496–500.

[JN-C94] K. Gröger and J. Nečas, *On a class of nonlinear initial value problems in Hilbert spaces*, Math. Nachr., 1979, vol. 93, pp. 21–31.

[JN-C130] J. Nečas, A. Lehtonen, and P. Neittaanmäki, *On the construction of Lusternik-Schnirelmann critical values with application to bifurcation problems*, Appl. Anal., 1987, vol. 25, pp. 253–268.

Pavel Drábek
Department of Mathematics, University of West Bohemia, Univerzitní 22,
306 14 Plzeň, Czech Republic
email: pdrabek@kma.zcu.cz

Part V
Solid Mechanics, Elasticity, Thermoelasticity, and Plasticity

Chapter 11
Selected Papers on Solid Mechanics, Elasticity, Thermoelasticity, and Plasticity

ZAMM 59, 567 – 572 (1979)

K. Gröger / J. Nečas / L. Trávníček

Dynamic Deformation Processes of Elastic-Plastic Systems

Die Arbeit befaßt sich mit dynamischen Prozessen von elastisch-plastischen Systemen, deren Verhalten durch Modelle beschrieben wird, die von internen Zustandsvariablen und internen Kräften Gebrauch machen. Es wird ein Existenz- und Einzigkeitsergebnis bezüglich der Verschiebungen, der Verzerrungen, der Spannungen, der internen Zustandsvariablen und der internen Kräfte bewiesen, und es wird gezeigt, daß diese Größen von den Anfangswerten und den äußeren Kräften stetig abhängen.

This paper deals with dynamic processes of elastic-plastic systems the behaviour of which is described by models that make use of internal state variables and internal forces. It is proved an existence-uniqueness result with respect to displacements, strains, stresses, internal state variables and internal forces, and it is shown that these values depend continuously on initial values and external forces.

В этой работе исследованы динамические процессы упруго-пластических систем, поведение которых описывается моделями, которые используют внутренние переменные состояния и внутренные силы. Доказано существование и единственность для перемещений, деформаций, напряжений, внутренних переменных состояния и внутренних сил и показана непрерывная зависимость этих величин от начальных данных и венешних сил.

Introduction

In this paper we shall deal with some evolution equations arising in the theory of plasticity. More precisely, we shall be concerned with *dynamic deformation processes of elastic-plastic materials* the behaviour of which can be described by models proposed by Nguyen [10] and Halphen and Nguyen [5]. These models include internal state variables and internal forces which are in a certain sense dual to internal state variables. Quasi-static deformation processes governed by Halphen-Nguyen's constitutive relations were investigated from the mathematical point of view by Nečas and Trávníček [9] and Trávníček [11]. Gröger [3] discussed dynamic and quasi-static deformation processes assuming that the relation between internal state variables and internal forces is linear whereas in Trávnjček's paper [11] this relation is allowed to be nonlinear. In the present paper we shall deal with dynamic processes under assumptions on the constitutive relations which are slightly weaker than those used by Trávníček [11] in the quasi-static case.

The paper consists of two sections. In Section 1 we introduce some notations and notions and we formulate precisely the class of problems we are interested in. Following Moreau [8] we do this in a rather general way. In Section 2 we prove that the problems formulated in Section 1 are well posed. More precisely, we prove an existence-uniqueness result with respect to displacements, elastic and plastic strains, stresses, internal state variables and internal forces and we show that these values depend continuously on the given data, i. e. on external forces and initial values.

1. Preliminaries

Most of the notions and notations to be introduced here were used already in [3]. Let U, Y, Z be Hilbert spaces the elements of which can be considered as configurations (displacements), strains and internal parameters, respectively. The elements of the dual spaces U', Y', Z' can be considered as forces, stresses and internal forces, respectively. We shall identify Y and Y' as well as Z and Z'. Moreover, we assume U to be densely and continuously imbedded into U' and we set $H := [U, U']_{1/2}$. (The definition of the interpolation space $[U, U']_{1/2}$ can be found in Lions and Magenes [7].) By $\langle \cdot , \cdot \rangle$ we shall denote the dual pairing of U' and U.

Let $S =]0, T]$ be a finite interval of time and let $\bar{S} = [0, T]$. If X is a Hiblert space we shall use the notations $L^p(S; X)$ and $C(\bar{S}; X)$ with their usual meaning. We set

$$AC(\bar{S}; X) := \{u : \bar{S} \to X \mid u \text{ is absolutely continuous from } \bar{S} \text{ into } X\}.$$

If $u \in L^1(S; X)$ we denote by u' the derivative of u in the sense of distributions on $]0, T[$ with values in X. By $H^1(S; X)$ we denote the Hilbert space of all $u \in L^2(S; X)$ such that $u' \in L^2(S; X)$ (equipped with its usual scalar product). A mapping from $\bar{S}$ into X will be called a *process*; for example we shall call σ a *stress process*, if σ maps $\bar{S}$ into Y. If X_1, X_2 are two Hilbert spaces we denote by $\mathcal{L}(X_1; X_2)$ the space of all bounded linear mappings from X_1 into X_2.

We assume that the strain process $\varepsilon : \bar{S} \to Y$ corresponding to a configuration process $u : \bar{S} \to U$ is given by

$$\forall t \in \bar{S}:\ \varepsilon(t) = Ku(t) + g(t)\,, \tag{1}$$

where $g : \bar{S} \to Y$ is a given function and

$$\text{(H1)} \quad K \in \mathcal{L}(U; Y), \quad \forall u \in U:\ \|Ku\|_Y \geqq m\,\|u\|_U\,,\ m > 0\,.$$

In many concrete situations the last inequality becomes an inequality of Korn's type (cf. Hlaváček and Nečas [6]). The function g can represent internal deformations or motions of the reference configuration. The force process $f : \bar{S} \to U'$ corresponding to a stress process $\sigma : \bar{S} \to Y$ is given by

$$\forall t \in \bar{S}:\ f(t) = K^*\sigma(t)\,,$$

© Wiley-VCH Verlag 1979. Reprinted with kind permission from: Gröger, K., Nečas, J., & Trávníček, L., Dynamic deformation processes of elasticplastic systems, Zeitschrift für Angewandte Mathematische Mechanik, 59:567-572.

where $K^* \in \mathcal{L}(Y; U')$ denotes the adjoint operator of K. This is a consequence of the principle of virtual work (cf. MOREAU [8]). Therefore, if $f: \bar{S} \to U'$ is a given process of external forces, the condition of dynamic equilibrium is

$$\forall t \in \bar{S}: u''(t) + K^*\sigma(t) = f(t) . \tag{2}$$

For the sake of simplicity we have assumed that the system under consideration is of constant "density" 1.

Let $\varepsilon, \sigma, a, \alpha$ be simultaneous processes of strain, stress, internal state variable and internal force of the system, respectively. Following HALPHEN and NGUYEN [5] we assume that the following constitutive relations hold:

$$\forall t \in \bar{S}: \varepsilon(t) = e(t) + p(t),\ \sigma(t) = Ae(t),\ \alpha(t) = Ba(t),\ \big(p'(t), -a'(t)\big) \in \partial\varphi\big(\sigma(t), \alpha(t)\big) , \tag{3}$$

where

(H2) $A \in \mathcal{L}(Y; Y)$ is symmetric and positive definite;

(H3) $B: Z \to Z$ is a FRÉCHET differentiable strongly montone and LIPSCHITZIAN potential operator such that $B': Z \to \mathcal{L}(Z; Z)$ is uniformly continuous on bounded sets;

(H4) $\varphi: Y \times Z \to]-\infty, +\infty]$ is a proper convex and lower semicontinuous functional such that $(e, a) \mapsto \varphi(Ae, Ba)$ is convex on $Y \times Z$ and $\|y\|_Y \leqq \text{const}\,(\|b\|_Z + 1)$ if $(y, b) \in \partial\varphi(\sigma, \alpha)$ ($\partial\varphi$ denotes the subdifferential of φ, see BRÉZIS [1]).

The functions e and p are called *elastic* and *plastic strain*, respectively. The operator A represents a linear HOOKE law whereas B characterizes the (possibly nonlinear) relationship between internal state variables and internal forces. The functional φ represents the "plasticity properties" of the system. The last condition imposed on φ is satisfied if φ describes materials with hardening (cf. the examples given in [3], Remark 2 and Remark 3). We shall denote by ψ the functional $(e, a) \mapsto \varphi(Ae, Ba)$ on $Y \times Z$ which is convex in view of (H4). We define $E: Y \times Z \to Y \times Z$ by

$$\forall (\sigma, \alpha) \in Y \times Z: E(\sigma, \alpha) := (A^{-1}\sigma, B^{-1}\alpha) . \tag{4}$$

Then (cf. [4], Lemma 1)

$$\forall (e, a) \in Y \times Z: \partial\psi(e, a) = (E^{-1})'(e, a)\, \partial\varphi\big(E^{-1}(e, a)\big) . \tag{5}$$

If $u: \bar{S} \to U$ we introduce $Ku: \bar{S} \to Y$ by

$$\forall t \in \bar{S}: (Ku)(t) = Ku(t) ,$$

i.e., we consider K also as a mapping defined on configuration processes. Similarly we shall deal with other operators. Moreover, we shall not distinguish processes which are equal for almost every $t \in S$.

For the time being we assume that we are given $f \in L^1(S; H)$, $g \in AC(\bar{S}; H)$, $u_0 \in U$, $v_0 \in U$, $\sigma_0 \in Y$, $\alpha_0 \in Z$. We are interested in the dynamic process determined by the processes f and g of force and deformation and by the initial values $u_0, v_0, \sigma_0, \alpha_0$ of displacement, velocity, stress and internal force, respectively. More precisely, we are looking for functions u, σ, α such that

$$(u, u', \sigma, \alpha) \in AC(\bar{S}; U \times H \times Y \times Z),\ (u, u', \sigma, \alpha)(0) = (u_0, v_0, \sigma_0, \alpha_0),\ u'' + K^*\sigma = f,$$

$$(A^{-1}\sigma', (B^{-1}\alpha)') + \partial\varphi(\sigma, \alpha) \ni (Ku' + g', 0) .$$

The last relation is obtained easily from (3) and (1) by elimination of ε and p. Introducing $v := u'$ as a new unknown function and using the operator E (cf. (4)) we can write this problem as follows

$$\text{(P)}\quad \begin{cases} (v, \sigma, \alpha) \in AC(\bar{S}; H \times Y \times Z),\ v \in L^1(S; U),\ (v, \sigma, \alpha)(0) = (v_0, \sigma_0, \alpha_0), \\ v' + K^*\sigma = f,\ (E(\sigma, \alpha))' + \partial\varphi(\sigma, \alpha) \in (Kv + g', 0) . \end{cases}$$

It is this problem we are concerned with in the next section. Most of our results will be proved under the following assumptions with respect to the data of problem (P):

$$\text{(H5)}\quad \begin{cases} f \in AC(\bar{S}; H),\quad g' \in AC(\bar{S}; Y) , \\ v_0 \in U,\ \sigma_0 \in Y,\ \alpha_0 \in Z,\ K^*\sigma_0 \in H,\ \partial\varphi(\sigma_0, \alpha_0) \neq \emptyset . \end{cases}$$

In order to simplify notations we shall denote by X the cartesian product $H \times Y \times Z$.

2. Results

Theorem 1: *Let the hypotheses* (H1) — (H4) *be satisfied. Moreover, let* $f_i \in L^1(S; H)$, $g_i' \in L^1(S; Y)$, $i = 1, 2$, *and let*

$$\begin{aligned} &(v_i, \sigma_i, \alpha_i) \in AC(\bar{S}; X),\ v_i \in L^1(S; U) , \\ &v_i' + K^*\sigma_i = f_i,\ \big(E(\sigma_i, \alpha_i)\big)' + \partial\varphi(\sigma_i, \alpha_i) \ni (Kv_i + g_i', 0), \quad i = 1, 2. \end{aligned} \tag{6}$$

Then

$$\|(v_1 - v_2, \sigma_1 - \sigma_2, \alpha_1 - \alpha_2)\|_{C(\bar{S}; X)} \leqq c\, \|(v_1 - v_2, \sigma_1 - \sigma_2, \alpha_1 - \alpha_2)(0)\|_X + \|f_1 - f_2\|_{L^1(S; H)} + \|g_1' - g_2'\|_{L^1(S; Y)}) ,$$

where c is a constant depending only on E.

Proof: Let $(e_i, a_i) := E(\sigma_i, \alpha_i) = (A^{-1}\sigma_i, B^{-1}\alpha_i)$, $i = 1, 2$, and let

$$(f, g, v, \sigma, \alpha, e, a) := (f_1 - f_2, g_1 - g_2, v_1 - v_2, \sigma_1 - \sigma_2, \alpha_1 - \alpha_2, e_1 - e_2, a_1 - a_2).$$

Then $(e_i', a_i') = (E(\sigma_i, \alpha_i))' = E'(\sigma_i, \alpha_i)(\sigma_i', \alpha_i')$ and (cf. (5), (6))

$$(e_i', a_i') + \partial\varphi(\sigma_i, \alpha_i) \in (Kv_i + g_i', 0), (\sigma_i', \alpha_i') + \partial\psi(e_i, a_i) \ni (AKv_i + Ag_i', 0).$$

Due to the monotonicity of $\partial\varphi$ and $\partial\psi$ this implies for every $t \in S$

$$0 \geqq \int_0^t \{((e' - Kv - g')(s), \sigma(s))_Y + (a'(s), \alpha(s))_Z + ((\sigma' - AKv - Ag')(s), e(s))_Y + (\alpha'(s), a(s))_Z\}\, ds =$$

$$= (e(t), \sigma(t))_Y - (e(0), \sigma(0))_Y + (a(t), \alpha(t))_Z - (a(0), \alpha(0))_Z - 2\int_0^t ((Kv + g')(s), \sigma(s))_Y\, ds.$$

On the other hand from (6) it follows

$$0 = 2\int_0^t \langle (v' + K^*\sigma - f)(s), v(s)\rangle\, ds = \|v(t)\|_H^2 - \|v(0)\|_H^2 + 2\int_0^t \{(Kv(s), \sigma(s))_Y - (f(s), v(s))_H\}\, ds.$$

Adding the last relations we obtain

$$\|v(t)\|_H^2 + (e(t), \sigma(t))_Y + (a(t), \alpha(t))_Z \leqq \|v(0)\|_H^2 + (e(0), \sigma(0))_Y + (a(0), \alpha(0))_Z + 2\int_0^t \{(g'(s), \sigma(s))_Y +$$
$$+ f((s), v(s))_H\}\, ds.$$

Using the definition of e, a and the properties of E we find

$$\|(v, \sigma, \alpha)(t)\|_X^2 \leqq c_0(\|(v, \sigma, \alpha)(0)\|_X^2 + 2\int_0^t (\|g'(s)\|_Y + \|f(s)\|_H)\, \|(v, \sigma, \alpha)(s)\|_X\, ds).$$

Hence (cf. Brézis [1], Lemma A.5)

$$\|(v, \sigma, \alpha)(t)\|_X \leqq \sqrt{c_0}\, \|(v, \sigma, \alpha)(0)\|_X + c_0\int_0^t \{\|g'(s)\|_Y + \|f(s)\|_H\}\, ds.$$

This completes the proof.

Theorem 2: *If the hypotheses* (H1) – (H5) *are satisfied then the problem* (P) *has a unique solution* (v, σ, α). *This solution has the properties*

$$(v', \sigma', \alpha') \in L^\infty(S; X), \qquad v \in L^\infty(S; U), \tag{8}$$

and for every $t \in [0, T[$

the right derivative $\frac{d^+(v, \sigma, \alpha)}{dt}(t)$ *exists in* X,

$$\left(A^{-1}\frac{d^+\sigma}{dt}(t), (B^{-1})'(\alpha(t))\frac{d^+\alpha}{dt}(t)\right) + \partial\varphi(\sigma(t), \alpha(t)) \ni (Kv(t) + g'(t), 0), \frac{d^+v}{dt}(t) + K^*\sigma(t) = f(t),$$

$$\left(A^{-1}\frac{d^+\sigma}{dt}(t), \frac{d^+\sigma}{dt}(t)\right)_Y + \left((B^{-1})'(\alpha(t))\frac{d^+\alpha}{dt}(t), \frac{d^+\alpha}{dt}(t)\right)_Z = \tag{9}$$

$$\inf_{(y,b)\in\partial\varphi(\sigma(t),\alpha(t))} \{(Kv(t) + g'(t) - y, A(Kv(t) + g'(t) - y))_Y + (b, B'(B^{-1}\alpha(t))\, b)_Z\}.$$

Moreover, $\frac{d^+(v, \sigma, \alpha)}{dt} : [0, T[\to X$ *is continuous from the right.*

The proof of this theorem is based on the following lemmas.

Lemma 1: *For every natural number n the problem*

$$(v_n, \sigma_n, \alpha_n) \in H^1(S; U \times Y \times Z), (v_n, \sigma_n, \alpha_n)(0) = (v_0, \sigma_0, \alpha_0),$$

$$\frac{1}{n^2}K^*Kv_n' + v_n' + K^*\sigma_n = f, \quad (E(\sigma_n, \alpha_n))' + \partial\varphi(\sigma_n, \alpha_n) \in (Kv_n + g', 0) \tag{10}$$

is uniquely solvable.

Proof: For any given $v \in L^2(S; U)$ there exists a unique (σ, α) such that

$$(E(\sigma, \alpha))' + \partial\varphi(\sigma, \alpha) \in (Kv + g', 0), (\sigma, \alpha)(0) = (\sigma_0, \alpha_0), (\sigma, \alpha) \in H^1(S; Y \times Z) \tag{11}$$

(see [4], Th. 2 and Rem. 4). Introducing $Q: L^2(S; U) \to L^2(S; U')$ by

$$Qv := K^*\sigma \quad \text{if} \quad (\sigma, \alpha) \text{ satisfies } (11)$$

we can write (10) as follows:

$$\frac{1}{n^2}K^*Kv_n' + v_n' + Qv_n = f, \quad v_n(0) = v_0, v_n \in H^1(S; U). \tag{12}$$

The operator Q is LIPSCHITZian (cf. [4], Th. 1 and Rem. 4) and it is a so called VOLTERRA operator (cf. [2], Ch. V). In view of (H1) the operator $K^*K \in \mathscr{L}(U; U')$ is positive definite. The existence and uniqueness of a solution to (12) follows from results on pseudo-parabolic equations (see [2], Ch. V).

Lemma 2: *The solution* $(v_n, \sigma_n, \alpha_n)$ *to* (10) *has the properties*

$$v_n' \in C(\overline{S}; U), \quad (\sigma_n', \alpha_n') \in L^\infty(S; Y \times Z); \tag{13}$$

$$\forall t \in [0, T[: \frac{\mathrm{d}^+(\sigma_n, \alpha_n)}{\mathrm{d}t}(t) \quad \text{exists in } Y \times Z;$$

$$\left(E'(\sigma_0, \alpha_0)\frac{\mathrm{d}^+(\sigma_n, \alpha_n)}{\mathrm{d}t}(0), \frac{\mathrm{d}^+(\sigma_n, \alpha_n)}{\mathrm{d}t}(0)\right)_{Y\times Z} = \inf_{w\in(Kv_0+g'(0),\,0)-\partial\varphi(\sigma_0,\alpha_0)} (w, (E'(\sigma_0, \alpha_0))^{-1}w)_{Y\times Z}. \tag{14}$$

Proof: From (12) it follows immediately that $v_n' \in C(\overline{S}; U)$. Therefore we may apply Theorem 4 of [4] to problem (11) with v_n instead of v. The remaining statements of Lemma 2 follow from this theorem.

Lemma 3: *Let* $(v_n, \sigma_n, \alpha_n)$ *be the solution to* (10) $(n = 1, 2, \ldots)$. *Then*

$$\sup_n \left\{\|(v_n', \sigma_n', \alpha_n')\|_{L^\infty(S;X)} + \frac{1}{n}\|Kv_n'\|_{L^2(S;Y)} + \|Kv_n\|_{L^\infty(S;Y)}\right\} < \infty.$$

Proof: Let $e_n := A^{-1}\sigma_n$, $a_n := B^{-1}\alpha_n$. Then

$$(e_n', a_n') + \partial\varphi(\sigma_n, \alpha_n) \ni (Kv_n + g', 0), \quad (\sigma_n', \alpha_n') + \partial\psi(e_n, a_n) \ni (AKv_n + Ag', 0).$$

Let $t \in [0, T[$ and $h > 0$ such that $t + h \leqq T$. If w is any function defined on $\overline{S}$ we denote by w_h the function defined on $[0, T - h]$ by

$$w_h(s) := w(s + h) - w(s).$$

Due to the monotonicity of $\partial\varphi$ and $\partial\psi$ we find

$$\int_0^t \{((e_{nh}' - Kv_{nh} - g_h')(s), \sigma_{nh}(s))_Y + (a_{nh}'(s), \alpha_{nh}(s))_Z +$$
$$+ ((\sigma_{nh}' - AKv_{nh} - Ag_h')(s), e_{nh}(s))_Y + (\alpha_{nh}'(s), a_{nh}(s))_Z\} \,\mathrm{d}s \geqq 0.$$

Furthermore,

$$2\int_0^t \left\langle\left(\frac{1}{n^2}K^*Kv_{nh}' + v_{nh}' + K^*\sigma_{nh} - f_h\right)(s), v_{nh}(s)\right\rangle \mathrm{d}s = 0.$$

Adding the last relations we obtain

$$(e_{nh}(t), \sigma_{nh}(t))_Y + (a_{nh}(t), \alpha_{nh}(t))_Z + \frac{1}{n^2}\|Kv_{nh}(t)\|_Y^2 + \|v_{nh}(t)\|_H^2 \leqq$$

$$\leqq (e_{nh}(0), \sigma_{nh}(0))_Y + (a_{nh}(0), \alpha_{nh}(0))_Z + \frac{1}{n^2}\|Kv_{nh}(0)\|_Y^2 + \|v_{nh}(0)\|_H^2$$

$$+ 2\int_0^t \{(g_h'(s), \sigma_{nh}(s))_Y + (f_h(s), v_{nh}(s))_H\} \,\mathrm{d}s.$$

Hence

$$\|(v_n, \sigma_n, \alpha_n)_h(t)\|_X^2 + \frac{1}{n^2}\|Kv_{nh}(t)\|_Y^2 \leqq$$

$$\leqq c_1\left(\|(v_n, \sigma_n, \alpha_n)_h(0)\|_X^2 + \frac{1}{n^2}\|Kv_{nh}(0)\|_Y^2 + \int_0^t \{\|g_h'(s)\|_Y \|\sigma_{nh}(s)\|_Y + \|f_h(s)\|_H \|v_{nh}(s)\|_H\} \,\mathrm{d}s\right).$$

This implies (cf. BRÉZIS [1], Lemma A.5)

$$\|(v_n, \sigma_n, \alpha_n)_h(t)\|_X + \frac{1}{n}\|Kv_{nh}(t)\|_Y \leqq$$

$$\leqq c_2\left(\|(v_n, \sigma_n, \alpha_n)_h(0)\|_X + \frac{1}{n}\|Kv_{nh}(0)\|_Y + \int_0^t \{\|g_h'(s)\|_Y + \|f_h(s)\|_H\} \,\mathrm{d}s\right).$$

Dividing by h and letting $h \downarrow 0$ we obtain (cf. [1], Prop. A.5)

$$\left\|\frac{\mathrm{d}^+(v_n, \sigma_n, \alpha_n)}{\mathrm{d}t}(t)\right\|_X + \frac{1}{n}\left\|K\frac{\mathrm{d}^+v_n}{\mathrm{d}t}(t)\right\|_Y \leqq c_3\left(\left\|\frac{\mathrm{d}^+(v_n, \sigma_n, \alpha_n)}{\mathrm{d}t}(0)\right\|_X + \frac{1}{n}\left\|K\frac{\mathrm{d}^+v_n}{\mathrm{d}t}(0)\right\|_Y + 1\right). \tag{15}$$

Since

$$\frac{1}{n^2}K^*K\frac{\mathrm{d}^+v_n}{\mathrm{d}t}(0) + \frac{\mathrm{d}^+v_n}{\mathrm{d}t}(0) + K^*\sigma_0 = f(0)$$

(cf. (10), (13)) we have

$$\frac{1}{n^2}\left\|K\frac{\mathrm{d}^+v_n}{\mathrm{d}t}(0)\right\|_Y^2 + \left\|\frac{\mathrm{d}^+v_n}{\mathrm{d}t}(0)\right\|_H^2 = \left\langle\frac{1}{n^2}K^*K\frac{\mathrm{d}^+v_n}{\mathrm{d}t}(0) + \frac{\mathrm{d}^+v_n}{\mathrm{d}t}(0), \frac{\mathrm{d}^+v_n}{\mathrm{d}t}(0)\right\rangle =$$

$$= \left(f(0) - K^*\sigma_0, \frac{\mathrm{d}^+v_n}{\mathrm{d}t}(0)\right)_H \leqq c_4\left\|\frac{\mathrm{d}^+v_n}{\mathrm{d}t}(0)\right\|_H.$$

Thus

$$\frac{1}{n}\left\|K\frac{\mathrm{d}^+v_n}{\mathrm{d}t}(0)\right\|_Y + \left\|\frac{\mathrm{d}^+v_n}{\mathrm{d}t}(0)\right\|_H \leqq c_5. \tag{16}$$

From (14) — (16) it follows

$$\left\|\frac{\mathrm{d}^+(v_n, \sigma_n, \alpha_n)}{\mathrm{d}t}(t)\right\|_X + \frac{1}{n}\left\|K\frac{\mathrm{d}^+v_n}{\mathrm{d}t}(t)\right\|_Y \leqq c_6. \tag{17}$$

This implies (cf. (H3))

$$\sup_n \|a_n'\|_{L^\infty(S;Z)} < \infty.$$

Using (H4) we obtain

$$\sup_n \|e_n' - Kv_n - g'\|_{L^\infty(S;Y)} < \infty.$$

Since

$$\sup_n \|e_n'\|_{L^\infty(S;Y)} = \sup_n \|A^{-1}\sigma_n'\|_{L^\infty(S;Y)}$$

(cf. (17)) this is possible only if

$$\sup_n \|Kv_n\|_{L^\infty(S;Y)} < \infty. \tag{18}$$

The results (17) and (18) complete the proof of Lemma 3.

Lemma 4: *The sequence* $(v_n, \sigma_n, \alpha_n)$, $n = 1, 2, \ldots$, *of solutions to* (10) *converges in* $C(\bar{S}; X)$.

Proof: For every $t \in \bar{S}$ we have (cf. (12))

$$0 = 2\int_0^t \left\langle\left(\frac{1}{n^2}K^*Kv_n' + v_n' + Qv_n - \frac{1}{m^2}K^*Kv_m' - v_m' - Qv_m\right)(s), (v_n - v_m)(s)\right\rangle \mathrm{d}s \geqq$$

$$\geqq \|(v_n - v_m)(t)\|_H^2 + 2\int_0^t \langle K^*(\sigma_n - \sigma_m)(s), (v_n - v_m)(s)\rangle \,\mathrm{d}s$$

$$- 2\left(\frac{1}{n^2}\|Kv_n'\|_{L^2(S;Y)} + \frac{1}{m^2}\|Kv_m'\|_{L^2(S;Y)}\right)\|K(v_n - v_m)\|_{L^2(S;Y)}.$$

The last term in this inequality tends to 0 as $n, m \to \infty$ (cf. Lemma 3). Furthermore (cf. (7))

$$2\int_0^t \langle K^*(\sigma_n - \sigma_m)(s), (v_n - v_m)(s)\rangle \,\mathrm{d}s = 2\int_0^t ((\sigma_n - \sigma_m)(s), K(v_n - v_m)(s))_Y \,\mathrm{d}s \geqq$$

$$\geqq ((e_n - e_m)(t), (\sigma_n - \sigma_m)(t))_Y + ((a_n - a_m)(t), (\alpha_n - \alpha_m)(t))_Z$$

$$\geqq c_0(\|(\sigma_n - \sigma_m)(t)\|_Y^2 + \|(\alpha_n - \alpha_m)(t)\|_Z^2), \quad c_0 > 0.$$

Combining this result and the preceding inequality we obtain the assertion of the Lemma.

Proof of Theorem 2:

1) Let us denote by (v, σ, α) the limit of the sequence $(v_n, \sigma_n, \alpha_n)$ the existence of which is guaranteed by Lemma 4. From Lemma 3 and (H1) it follows immediately that (v, σ, α) satisfies (8).

2) From (10), Lemma 3 and Lemma 4 it follows

$$v' + K^*\sigma = f, \quad (v, \sigma, \alpha)(0) = (v_0, \sigma_0, \alpha_0). \tag{19}$$

3) Lemma 4 implies that

$$(e_n, a_n) = E(\sigma_n, \alpha_n) \to E(\sigma, \alpha) \quad \text{in} \quad C(\bar{S}; Y \times Z). \tag{20}$$

In view of Lemma 3

$$\sup_n \|(E(\sigma_n, \alpha_n))'\|_{L^\infty(S; Y \times Z)} < \infty . \tag{21}$$

From (20), (21) it follows

$$(E(\sigma_n, \alpha_n))' \rightharpoonup (E(\sigma, \alpha))' \quad \text{in} \quad L^2(S; Y \times Z).$$

Furthermore (cf. Lemma 3, Lemma 4)

$$Kv_n \rightharpoonup Kv \quad \text{in} \quad L^2(S; Y).$$

Thus

$$(Kv_n + g', 0) - (E(\sigma_n, \alpha_n))' \rightharpoonup (Kv + g\cdot, 0) - (E(\sigma, \alpha))' \quad \text{in} \quad L^2(S; Y \times Z),$$

$$(\sigma_n, \alpha_n) \to (\sigma, \alpha) \quad \text{in} \quad L^2(S; Y \times Z), \quad (Kv_n + g', 0) - (E(\sigma_n, \alpha_n))' \in \partial\varphi(\sigma_n, \alpha_n).$$

Because $\partial\varphi$ (considered as a multivalued mapping in $L^2(S; Y \times Z)$) is maximal monotone these relations imply that

$$(Kv + g', 0) - (E(\sigma, \alpha))' \in \partial\varphi(\sigma, \alpha). \tag{22}$$

From (8), (19) and (22) it follows that (v, σ, α) is a solution to problem (P).

4) The uniqueness of the solution is a consequence of Theorem 1.

5) The assertions with respect to the right derivative $\frac{d^+(v. \sigma, \alpha)}{dt}$ can be proved in the same way as the corresponding assertions in Theorem 4 of [4]. One has to use the fact that by

$$M(v. \sigma, \alpha) := \{(K^*\sigma, -Kv + y, b) \mid v \in U, \; K^*\sigma \in H, \; (y, b) \in \partial\varphi(\sigma, \alpha)\}$$

there is defined a maximal monotone mapping in X (cf. [3], Lemma 2).

This completes the proof of Theorem 2.

Remark: Let $D := \{(\sigma_0, \alpha_0) \in Y \times Z \mid K^*\sigma_0 \in H, \; \partial\varphi(\sigma_0, \alpha_0) \neq \emptyset\}$ and let $\bar{D}$ denote the closure of D in $Y \times Z$. Theorem 1 and Theorem 2 show that the mapping

$$(f, g', v_0, \sigma_0, \alpha_0) \to (v, \sigma, \alpha), \text{ where } (v, \sigma, \alpha) \text{ is the solution to (P)},$$

can be extended from the set of data satisfying (H5) to a Lipschitzian mapping R from $L^1(S; H) \times L^1(S; Y) \times H \times \bar{D}$ into $C(\bar{S}; X)$. $R(f, g', v_0, \sigma_0, \alpha_0)$ is called the *weak solution to* (P) corresponding to the data $(f, g', v_0, \sigma_0, \alpha_0)$.

References

1 Brézis, H., Opérateurs maximaux monotones et semi-groupes de contractions dans les espaces de Hilbert, North Holland Math. Studies 5, 1973.
2 Gajewski, H., Gröger, K., Zacharias, K., Nichtlineare Operatorgleichungen und Operatordifferentialgleichungen, Berlin 1974.
3 Gröger, K., Evolution equations in the theory of plasticity, Proc. Fifth Summer School on Nonlinear Operators, Berlin 1977, pp. 97–107.
4 Gröger, K., Nečas, J., On a class of nonlinear initial value problems in Hilbert spaces, Math. Nachr. (to appear).
5 Halphen, B., Nguyen, Q. S., Sur les matériaux standards généralisés, J. Mécanique 14, 39–63 (1975).
6 Hlaváček, I., Nečas, J., On inequalities of Korn's Type. I, II, Arch. Rat. Mech. Analysis 36, 305–311, 312–334 (1970).
7 Lions, J.-L., Magenes, E., Problèmès aux limites non homogenes et applications, Paris 1968.
8 Moreau, J. J., On unilateral constraints, friction and plasticity. In: New Variational Techniques in Mathematical Physics, CIME, II Ciclo 1973, Ed. G. Capriz, G. Stampacchia, 1974, pp. 175–322.
9 Nečas, J., Trávníček, L., Evolutionary variational inequalities and applications in the plasticity. Apl. Mat. (to appear).
10 Nguyen, Q. S., Matériaux élastoplastiques écrouissables. Distribution de la contrainte dans une evolution quasi-statique, Arch. Mech. Stosow. 25, 695–702 (1973).
11 Trávníček, L., Existence a jednoznačnost řešení hraničního problému pro elasticko-plastické materialy. Thesis, Prague 1977

Eingereicht: 9. 8. 1978

Anschriften: Dr. K. Gröger, Zentralinstitut für Mathematik und Mechanik der Akademie der Wissenschaften der DDR, DDR 108 Berlin, Mohrenstr. 39; Dr. J. Nečas, Dr. L. Trávníček, Department of Theoretical Physics, Faculty of Mathematics and Physics of the Charles University, Ke Karlovu 3, ČS 121 16 Praha 2, ČSSR

Bollettino U. M. I.
(5) **17**-B (1980), 796-811

On the Solution of the Variational Inequality to the Signorini Problem with Small Friction.

Jindřich Nečas - Jiří Jarušek - Jaroslav Haslinger (Prague)

Sunto. – *Si dimostra l'esistenza della soluzione del problema di Signorini con attrito in una striscia in R^2 per i coefficienti d'attrito abbastanza piccoli e si dà una valutazione precisa di essi.*

1. – Introduction.

Unilateral contact problems of Signorini's type stimulated the development of the theory of variational inequalities, see e.g. G. Fichera [1], G. Duvaut-J. L. Lions [2], J. Nečas [3] etc. In this paper we shall restrict ourselves to a simple model-situation of a plane domain in the form of a strip. The authors had no informations about any theoretical result concerning the Signorini problem with friction. This problem is formulated also in the book [2] as an open problem.

Let $\Omega = \{x \in R^2; -\infty < x_1 < +\infty, 0 < x_2 < 1\}$ and let us look for the displacement vector $u = (u_1, u_2)$. We put

$$e_{ij}(u) = \frac{1}{2}\left(\frac{\partial u_i}{\partial x_j} + \frac{\partial u_j}{\partial x_i}\right), \quad i, j = 1, 2$$

and we consider the simplest Hook's law, the relation between strain tensor e_{ij} and stress tensor:

$$\tau_{ij} = \lambda e_{kk}\delta_{ij} + 2\mu e_{ij}, \tag{1.1}$$

where $\lambda \geq 0$, $\mu > 0$ are constants.

For the stress tensor the equilibrium conditions must be satisfied:

$$-\frac{\partial \tau_{ij}}{\partial x_j} = f_i, \quad i = 1, 2. \tag{1.2}$$

© Unione Matematica Italiana 1980. Reprinted with kind permission from: Nečas, J., Jarušek, J. & Haslinger, J., On the solution of the variational inequality to the Signorini problem with small friction, Bollettino dell'Unione Matematica Italiana 17, 1980, 796-811.

For simplicity, we prescribe on

$$\Gamma_u = \{x \in R^2; -\infty < x_1 < +\infty, x_2 = 1\}$$

the displacement, i.e.

$$u = u^0 . \tag{1.3}$$

Put $\Gamma_c = \{x \in R^2; -\infty < x_1 < +\infty, x_2 = 0\}$. Let us denote $u_n = u_i n_i$, where n_i are components of the outer normal to $\partial\Omega$, $u_t = u_i t_i$ where t_i are components of the tangential vector orienting $\partial\Omega$ in the positive direction ($n_1 = 0, n_2 = -1, t_1 = 1, t_2 = 0$ on Γ_c). Put $T_n = \tau_{ij} n_i n_j$, $T_t = \tau_{ij} t_i n_j$. The unilateral contact on Γ_c with the coefficient of the friction $\mathcal{F}$ is as follows:

$$u_n \leqq 0 \qquad T_n \leqq 0 \qquad u_n T_n = 0 \tag{1.4}$$

$$T_t \leqq \mathcal{F}|T_n| , \qquad (\mathcal{F}|T_n| - |T_t|) u_t = 0 , \qquad u_t T_t \leqq 0$$

on Γ_c.

A standard idea for the search of sequence of successive approximations is the following: Let T_t^0 be known on Γ_c. We look for the solution of the classical Signorini problem: $u_n^1 \leqq 0$, $T_n^1 \leqq 0$, $u_n^1 T_n^1 = 0$, $T_t^1 = T_t^0$ on Γ_c. Then we look for the problem with friction: $T_n^2 = T_n^1$, $|T_t^2| \leqq \mathcal{F}|T_n^1|$ $(\mathcal{F}|T_n^1| - |T_t^2|) u_t^2 = 0$, $u_t^2 T_t^2 \leqq 0$ on Γ_c etc. We can abbreviate this scheme, in the second step looking for the solution satisfying $u_n^2 \leqq 0$, $T_n^2 \leqq 0$, $u_n^2 T_n^2 = 0$, $|T_t^2| \leqq \mathcal{F}|T|_n^1$, $(\mathcal{F}|T_n^1| - |T_t^2|) u_t^2 = 0$, $u_t^2 T_t^2 \leqq 0$ on Γ_c for in advance known T_n^1. So we get a mapping Φ: $T_n^1 \to T_n^2$ and we can look for a fixed point of Φ. We shall solve this problem proving that the map Φ maps the set of all non-positive distributions of the space $H^{-\frac{1}{2}+\alpha}(\Gamma_c)$ into itself for $0 < \alpha < \frac{1}{4}$ and is weakly continuous for these α. For to get a bounded convex set mapped into itself we need some smallness of $\mathcal{F}$; therefore we shall do some careful estimations for to get estimations for $\mathcal{F}$ to discover when our theorem is valid.

Let us introduce

$$a(u, v) \equiv \int_\Omega \lambda \operatorname{div} u \cdot \operatorname{div} v + 2\mu e_{ij}(u) e_{ij}(v)\, dx \tag{1.5}$$

and a convex set

$$K \equiv \{v \in (H^1(\Omega))^2, v_n \leqq 0 \text{ on } \Gamma_c, v = u_0 \text{ on } \Gamma_u\}$$

and the space $\mathcal{H} = (H^1(\Omega))^2$. Here $H^1(\Omega)$ is the Sobolev space of square integrable functions with square integrable first derivatives on Ω. u^0 is a given function, here for instance of $H^{\frac{1}{2}+\alpha}(\Gamma_u)$. By usual formal reasoning, we note $T_n^1 = g_n$, u^2 as u. We look for $u \in K$, such that for each $v \in K$ the following inequality

$$a(u, v-u) + \int_{\Gamma_c} \mathcal{F}|g_n|(|v_t| - |u_t|)\,ds \geq (f, v-u) \tag{1.6}$$

holds. We suppose $f \in (L_2(\Omega))^2$. Let us remark that in general for $g_n \in H^{-\frac{1}{2}+\alpha}(\Gamma_c)$, $0 \leqq \alpha < \frac{1}{2}$ the operation $|g_n|$ is meaningless. But it is sufficient to consider only $g_n \leqq 0$. Let us define

$$\int_{\Gamma_c} T_n v_n\,ds = a(u, v) - (f, v) \tag{1.7}$$

for $v = 0$ on Γ_u, $v_t = 0$ on Γ_c. Thus the fixed point is iff $T_n = g_n$.

2. – The problem with prescribed g_n.

We shall look in detail at the solution of the inequality (1.6), where instead of $|g_n|$ we put $-g_n$. As usually, $H^{\frac{1}{2}}(\Gamma_c)$ is a space of traces of functions from $H^1(\Omega)$ and $H^{-\frac{1}{2}}(\Gamma_c)$ its dual, see for example the book by J. Nečas [4].

We shall look for $u \in K$, such that for each $v \in K$

$$a(u, v-u) - \langle \mathcal{F}g_n, |v_t| - |u_t| \rangle \geq (f, v-u) \tag{2.1}$$

where $\langle \cdot,\cdot \rangle$ is the $L_2(\Gamma_c)$ pairing and $\mathcal{F} \in \mathcal{D}(R^1)$. We shall suppose $u^0 \equiv 0$ on Γ_c. Because $g_n \in H^{-\frac{1}{2}}(\Gamma_c)$ iff $\left(\int_{-\infty}^{+\infty} |\hat{g}_n|^2 (1+|\xi|)^{-1}\,d\xi\right)^{\frac{1}{2}} < +\infty$ which is an equivalent norm (see (3.1) and what follows), where $\hat{g}_n = F(g_n)$ and F is the Fourier transform, we can define $\hat{g}_n^m(\xi) = \hat{g}_n(\xi)$ for $|\xi| < m$ and $\hat{g}_n^m(\xi) = 0$ for $|\xi| \geq m$ and we get $g_n^m \in L_2(\Gamma_c)$ and $g_n^m \to g_n$ in $H^{-\frac{1}{2}}(\Gamma_c)$. Clearly a solution u^m of (2.1), where instead of «g_n» we put g_n^m, can be found as

$$\min_{v \in K} \tfrac{1}{2} a(v, v) - \langle \mathcal{F}g_n^m, |v_t| \rangle - (f, v)\,. \tag{2.2}$$

We put $J(v) = \frac{1}{2}a(v, v) - \langle \mathcal{F}g_n^m, |v_t| \rangle - (f, v)$. $J(v)$ is weakly lower semicontinuous and coercive (if $v_m \rightharpoonup v$ (weak convergence) then

$\lim_{m\to+\infty} J(v_m) \geqq J(v)$; $\lim_{\substack{\|v\|\to+\infty\\ v\in K}} J(v) = +\infty$). The solution of (2.1) is uniquely determined: if w_1, w_2 are two solutions, then

$$a(w_2, w_1 - w_2) - \langle \mathcal{F} g_n^m, |w_{t,1}| - |w_{t,2}|\rangle \geqq (f, w_1 - w_2),$$

$$a(w_1, w_2 - w_1) - \langle \mathcal{F} g_n^m, |w_{t,2}| - |w_{t,1}|\rangle \geqq (f, w_2 - w_1)$$

hence $a(w_2 - w_1, w_2 - w_1) \leqq 0$. It is very well known that for every $v \in \mathcal{H}$, $v = 0$ on Γ_u the Korn inequality

$$a(v, v) \geqq c\|v\|^2_{\mathcal{H}} \tag{2.3}$$

is valid, therefore $w_1 = w_2$. We shall come back to (2.3) later on. We used also the well known fact that for $0 < a < +\infty$ the imbedding $H^{\frac{1}{2}}((-a, a)) \to L_2((-a, a))$ is compact, see S. Fučík, O. John, A. Kufner [6].

As above we get

$$\begin{aligned} a(u^{m_2} - u^{m_1}, u^{m_2} - u^{m_1}) &\leqq \langle \mathcal{F} g_n^{m_2} - \mathcal{F} g_n^{m_1}, |u_t^{m_2}| - |u_t^{m_1}|\rangle \leqq \\ &\leqq c\|g_n^{m_2} - g_n^{m_1}\|_{-\frac{1}{2},\Gamma_c}[\|u^{m_1}\|_{\mathcal{H}} + \|u^{m_2}\|_{\mathcal{H}}], \end{aligned} \tag{2.4}$$

where $\|\cdot\|_{-\frac{1}{2},\Gamma_c}$ is the norm of the space $H^{-\frac{1}{2}}(\Gamma_c)$. If $g_n^m \to g_n$ in $H^{\frac{1}{2}}(\Gamma_c)$, then $u^m \to u$ in $\mathcal{H}$ because of the estimation $\|u^m\|_{\mathcal{H}} \leqq \leqq c[\|f\|_{(L_2(\Omega))^2} + \|u^0\|_{\mathcal{H}} + \|g_n^m\|_{-\frac{1}{2},\Gamma_c}]$ which can be obtained by putting $v = u^0$ into (2.1). Clearly $\langle \mathcal{F} g_n^m, |u_t^m|\rangle \to \langle \mathcal{F} g, |u_t|\rangle$. So we get

LEMMA 2.1. – *There is a unique solution $u \in K$ of (2.1) and*

$$\|u\|_{\mathcal{H}} \leqq c[\|u^0\|_{\mathcal{H}} + \|g_n\|_{-\frac{1}{2},\Gamma_c} + \|f\|_{(L_2(\Omega))^2}]. \tag{2.5}$$

For $g_n \leqq 0$ and fixed f we have

$$\|u\|_{\mathcal{H}} \leqq c\|u^0\|_{\mathcal{H}} \tag{2.6}$$

The mapping $g_n \mapsto u$ is continuous, we have also $T_n(u) \leqq 0$.

PROOF. – The estimation (2.5) follows immediately from (2.1), if we put $v = u^0$. The continuity of the mapping $g_n \mapsto u$ follows from (2.4). If $g_n \leqq 0$ and putting in (2.1) $v = u^0$ again, we get $a(u, u^0 - u) + \langle \mathcal{F} g_n, |u_t|\rangle \geqq (f, u^0 - u)$, hence $a(u, u) \leqq a(u, u^0) + + \langle \mathcal{F} g_n, |u_t|\rangle + (f, u) - (f, u^0)$, what implies (2.5) and because of

$\|f\|_{(L_2(\Omega))^2} \leqq c_1 \|u^0\|_{\mathcal{H}}$ for fixed f and $\langle \mathcal{F} g_n, |u_t| \rangle \leqq 0$ because $\mathcal{F} \geqq 0$ —see (1.4)—we have (2.6). If we put $v = u + w$ with $w_t = 0$, $w_n \leqq 0$ on Γ_c and $w = 0$ on Γ_u, we get from (2.1) and (1.7) $\langle T_n(u), w_n \rangle \geqq 0$, therefore $T_n(u) \leqq 0$.

3. – Some statements on the norms in $H^\alpha(R^1)$ and the introduction of auxiliary spaces.

It is very well known, that for $0 < k < 1$ the norm in the space $H^k(R^1)$ can be introduced as

$$\|w\|^2_{k,R^1} = \int_{-\infty}^{+\infty} w^2\,dx + \int_{-\infty}^{+\infty}\int_{-\infty}^{+\infty} \frac{\big(w(x+h) - w(x)\big)^2}{|h|^{1+2k}}\,dx\,dh\,, \tag{3.1}$$

see for example [4]. Let us denote

$$\|w\|'^2_{k,R^1} = \int_{-\infty}^{+\infty}\int_{-\infty}^{+\infty} \frac{\big(w(x+h) - w(x)\big)^2}{|h|^{1+2k}}\,dx\,dh\,.$$

Put $w_h = w(x - h)$. We have

LEMMA 3.1. – *Let* $0 < \alpha < k$, $0 < \beta < k$, $\alpha + \beta = k < 1$. *Then*

$$\int_{-\infty}^{+\infty} \frac{\|w_{-h} - w\|'^2_{\alpha,R^1}}{|h|^{1+2\beta}}\,dh = \|w\|'^2_{k,R^1} \cdot d(\alpha, \beta)\,, \tag{3.2}$$

where

$$d(\alpha, \beta) = \frac{2\pi c(\beta)\, c(\alpha)}{c(k)}\,, \quad c(\alpha) = \frac{2^{1-2\alpha}}{\pi}\int_{-\infty}^{+\infty} \frac{\sin^2 t}{|t|^{1+2\alpha}}\,dt\,.$$

PROOF. – For $\beta > 0$ we have

$$\int_{-\infty}^{+\infty} \frac{\|w_{-h} - w\|^2_{0,R^1}}{|h|^{1+2\beta}}\,dh = \frac{1}{2\pi}\int_{-\infty}^{+\infty} |h|^{-1-2\beta}\,dh \int_{-\infty}^{+\infty} |\hat{w}(\xi)(1 - \exp[ih\xi])|^2\,d\xi = \tag{3.3}$$

$$= \frac{2^{1-2\beta}}{\pi}\int_{-\infty}^{+\infty} \frac{\sin^2 t}{|t|^{1+2\beta}}\,dt \cdot \int_{-\infty}^{+\infty} |\hat{w}(\xi)|^2 |\xi|^{2\beta}\,d\xi = c(\beta)\cdot\int_{-\infty}^{+\infty} |\hat{w}(\xi)|^2 |\xi|^{2\beta}\,d\xi\,,$$

hence

$$(3.4)\qquad \int_{-\infty}^{+\infty} \frac{\|w_{-h}-w\|'^2_{\alpha,R^1}}{|h|^{1+2\beta}}\,dh = 2\pi c(\alpha)\cdot\frac{1}{2\pi}\int_{-\infty}^{+\infty}|h|^{-1-2\beta}\,dh\cdot$$

$$\cdot\int_{-\infty}^{+\infty}|\hat{w}(\xi)|^2|1-\exp[ih\xi]|^2\cdot|\xi|^{2\alpha}\,d\xi = 2\pi c(\alpha)\,c(\beta)\int_{-\infty}^{+\infty}|\hat{w}(\xi)|^2\cdot$$

$$\cdot|\xi|^{2k}\,d\xi = \frac{2\pi c(\alpha)\,c(\beta)}{c(k)}\,\|w\|'^2_{k,R^1}\,.$$

Using the same technique, we get

LEMMA 3.2 - *Let* $-1<\alpha<0,\ 0<\beta<1,\ \alpha+\beta=k<0$. *Let for* $k<0$

$$(3.5)\qquad \|w\|^2_{k,R^1} = \frac{1}{2\pi}\int_{-\infty}^{+\infty}|\hat{w}(\xi)|^2(1+2\pi c(-k)|\xi|^{-2k})^{-1}\,d\xi\,,$$

where for $r>0$ $c(r)$ *is the same as in the preceding lemma.*

Then

$$\int_{-\infty}^{+\infty}|\tilde{w}(\xi)|^2(1+2\pi c(-\alpha)|\xi|^{-2\alpha})^{-1}|\xi|^{2\beta}d\xi = (c(\beta))^{-1}\int_{-\infty}^{+\infty}|h|^{-1-2\beta}\|w_{-h}-w\|^2_{\alpha,R^1}dh.$$

Let us introduce on Ω the spaces $H^1_\alpha\subset H^1$, $0<\alpha<\frac{1}{2}$ such that

$$(3.6)\quad \|w\|^2_{1,\alpha}\equiv\|w\|_{H^1_\alpha(\Omega)}=\int_{-\infty}^{+\infty}\int_0^1\int_{-\infty}^{+\infty}\sum_{i=1}^{2}\left[\frac{\partial w}{\partial x_i}(x_1+h,x_2)-\frac{\partial w}{\partial x_i}(x_1,x_2)\right]^2\cdot$$

$$\cdot|h|^{-1-2\alpha}\,dx_1\,dx_2\,dh + \|w\|^2_{H^1(\Omega)}\,.$$

As above we get

LEMMA 3.3. -

$$\|w\|^2_{1,\alpha} = c(\alpha)\int_{-\infty}^{+\infty}\int_0^1\left(\left|\frac{\partial\tilde{w}}{\partial x_2}\right|^2+|\tilde{w}\xi_1|^2\right)|\xi_1|^{2\alpha}\cdot d\xi_1\,dx_2 + \|w\|^2_{H^1(\Omega)}\,,$$

where $\tilde{w}$ *denote the Fourier transform of* w *in the transformation with respect to the first variable and* $c(\alpha)$ *is from lemma 3.1.*

For $0<\alpha<1$ we introduce the space $H^1_{-\alpha}(\Omega)$ as the subspace $L_2(\Omega)$ consisting of all functions, such that for their one-dimensional Fourier transforms $\tilde{w}(\xi_1, x_2)$ we have

$$(3.7)\qquad \|w\|^2_{1,-\alpha} = \frac{1}{2\pi}\int_{-\infty}^{+\infty}\int_0^1\left(\left|\frac{\partial\tilde{w}}{\partial x_2}\right|^2 + |\tilde{w}\xi_1|^2\right)(1+|\xi_1|^{2\alpha}\cdot 2\pi c(\alpha))^{-1} + + |\tilde{w}|^2\, d\xi_1\, dx_2 < +\infty\,.$$

Because of imbeding $H^1(\Omega)$ into $H^{\frac{1}{2}}(\Gamma_c)$, we have immediately:

LEMMA 3.4. – *For* $0\leq\alpha\leq\frac{1}{2}$ $\|w(\cdot,0)\|_{\frac{1}{2}+\alpha,R^1}\leq c\|w\|_{1,\alpha}$.

LEMMA 3.5. – *Let* $0\leq\alpha<\frac{1}{2}$. *Then*

$$(3.8)\qquad \|w(\cdot,0)\|_{\frac{1}{2}-\alpha,R^1}\leq c\|w\|_{1,-\alpha}\,.$$

PROOF. – Let $\eta(x_2)=1$ for $0\leq x_2<\frac{1}{2}$, $\eta(x_2)=0$ for $x_2\geq\frac{3}{4}$,

$$\eta\in C^\infty(R^1)\,.$$

Put $\tilde{v}(\xi_1,x_2)=\eta(x_2)\tilde{w}(\xi_1,x_2)$ for $x_2\geq 0$ and $\tilde{v}(\xi_1,x_2)=\tilde{v}(\xi_1,-x_2)$ for $x_2\leq 0$.

We get $v\in H^1_{-\alpha}(R^2)$,

$$(3.9)\qquad \|v\|_{1,-\alpha,R^2}\leq c\|w\|_{1,-\alpha,\Omega}$$

and the Fourier transform in both variables

$$(3.10)\qquad \int_{R^2}|\hat{v}|^2(1+|\xi|^2)(1+|\xi_1|^{2\alpha})^{-1}\,d\xi_1\,d\xi_2\leq c\|v\|^2_{1,-\alpha,R^2}\,.$$

But

$$\tilde{w}(\xi_1,0)=\frac{1}{2\pi}\int_{-\infty}^{+\infty}\hat{v}(\xi_1,\xi_2)\,d\xi_2$$

(the inverse transform with respect to the second variable), hence

$$(3.11)\qquad \int_{-\infty}^{+\infty}|\tilde{w}(\xi_1,0)|^2(1+|\xi_1|^2)^{\frac{1}{2}-\alpha}\,d\xi_1\leq\frac{1}{4\pi}\int_{-\infty}^{+\infty}\int_{-\infty}^{+\infty}|\hat{v}|^2(1+|\xi|^2)\cdot \cdot(1+\xi_1^2)^{-\alpha}\,d\xi_1\,d\xi_2$$

because

$$\int_{-\infty}^{+\infty}(1+|\xi|^2)^{-1}\,d\xi_2=(1+\xi_1^2)^{-\frac{1}{2}}\pi$$

and the left term in (3.11) is an equivalent norm in $H^{\frac{1}{2}+\alpha}(R^1)$.

LEMMA 3.6. – *Let $0\leq\alpha<\frac{1}{2}$. Then there exists a continuous linear map E from $H^{\frac{1}{2}-\alpha}(R^1)$ into $H^1_{-\alpha}(\Omega)$ such that if $g\in H^{\frac{1}{2}-\alpha}(R^1)$ and $Eg=f$, then $\hat{g}(\xi_1)=\tilde{f}(\xi_1,0)$, $\tilde{f}(\xi_1,1)=0$ and $\|Eg\|_{1,-\alpha}\leq$ $\leq c\|g\|_{\frac{1}{2}-\alpha,R^1}$.*

PROOF. – Put $\tilde{f}(\xi_1,x_2)=\exp\left[-(1+|\xi_1|)x_2\right]g(\xi_1)\eta(x_2)$ with η of the preceding proof.

4. – Regularity of the solution of the inequality (2.1) and the existence of the solution of the variational inequality of the problem with friction.

THEOREM 4.1. – *Let $0<\alpha<\frac{1}{4}$ and let $g_n\in H^{-\frac{1}{2}+\alpha}(R^1)$,*

$$u^0\in\left(H^1_\alpha(\Omega)\right)^2,\qquad f\in\left(L_2(\Omega)\right)^2.$$

Then for the solution of (2.1) we have

$$\|u\|_{(H^1_\alpha(\Omega))^2}\leq c(\alpha)\left(\|g_n\|_{-\frac{1}{2}+\alpha,R^1}+\|u^0\|_{(H^1_\alpha(\Omega))^2}+\|f\|_{(L_2(\Omega))^2}\right).\tag{4.1}$$

PROOF. – Put in (2.1) for $v=u_{-h}-u^0_{-h}+u^0$. We get

$$\begin{aligned}a(u,u_{-h}-u^0_{-h}+u^0-u)&-\\&-\langle\mathcal{F}g_n,|u_t|_{-h}-|u_t|\rangle\geq(f,u_{-h}-u^0_{-h}+u^0-u).\end{aligned}\tag{4.2}$$

After the shift of (2.1) we get

$$a(u_{-h},v_{-h}-u_{-h})-\langle\mathcal{F}_{-h}(g_n)_{-h},|v_t|_{-h}-|u_t|_{-h}\rangle\geq(f_{-h},v_{-h}-u_{-h})\tag{4.3}$$

and by putting $v=u_h-u^0_h+u^0$ we get

$$\begin{aligned}a(u_{-h},u-u^0+u^0_{-h}-u_{-h})-\langle\mathcal{F}_{-h}(g_n)_{-h},|u_t|-|u_t|_{-h}\rangle&\geq\\\geq-(f_{-h},u_{-h}-u^0_{-h}+u^0-u)&,\end{aligned}\tag{4.4}$$

hence it follows from (4.2) and (4.4)

$$(4.5)\qquad a(u_{-h}-u,\,u_{-h}-u)\leq a(u_{-h}-u,\,u^0_{-h}-u^0)+ \\ +\langle \mathcal{F}_{-h}(g_n)_{-h}-\mathcal{F}g_n,\,|u_t|_{-h}-|u_t|\rangle-\big(f-f_{-h},\,u_{-h}-u^0_{-h}-(u-u^0)\big)\,.$$

So we get, using lemmas from § 3, (2.6) and the Korn inequality

$$(4.6)\qquad \|u\|^2_{(H^1_\alpha(\Omega))^2}\leq c_1[\|u\|_{(H^1_\alpha(\Omega))^2}\cdot\|u^0\|_{(H^1_\alpha(\Omega))^2}+\|g_n\|_{-\frac12,R^1}\cdot \\ \cdot\|u\|_{\mathcal{H}}]+c_2(\alpha)\bigg(\int\limits_{-\infty}^{+\infty}\|(g_n)_{-h}-g_n\|^2_{-\frac12,R^1}\cdot|h|^{-1-2\alpha}\,dh\bigg)^{\frac12}\cdot \\ \cdot\bigg(\int\limits_{-\infty}^{+\infty}\|\,|u_t|_{-h}-|u_t|\,\|_{\frac12,\Gamma_c}|h|^{-1-2\alpha}\,dh\bigg)^{\frac12}+\int\limits_{-\infty}^{+\infty}|h|^{-1-2\alpha}(f_{-h}- \\ -f,\,u_{-h}-u^0_{-h}-(u-u^0))\,dh\leq c_1(\|u\|_{(H^1_\alpha(\Omega))^2}\|u^0\|_{(H^1_\alpha(\Omega))^2}+ \\ +\|g_n\|_{-\frac12,R^1}\|u^0\|_{\mathcal{H}})+c_3(\alpha)[\|g_n\|_{-\frac12+\alpha,R^1}\|\,|u_t|\,\|_{\frac12+\alpha,\Gamma_c}+\|f\|_{(L_2(\Omega))^2}\cdot\|u^0\|_{\mathcal{H}}]\,.$$

Bearing in mind $\|\,|u_t|\,\|_{\frac12+\alpha,\Gamma_c}\leq\|u_t\|_{\frac12+\alpha,\Gamma_c}$ we get by the usual method the inequality (4.1).

THEOREM 4.2. – *For the map $\Phi\colon g_n\mapsto T_n$ defined by the solution of* (2.1) *we have*

$$(4.7)\qquad \|\Phi(g_n)\|_{-\frac12+\alpha,R^1}=c(\alpha,\mathcal{F})\|g_n\|_{-\frac12+\alpha,R^1}+c_1\|u^0\|_{(H^1_\alpha(\Omega))^2}+c_2\|f\|_{(L_2(\Omega))^2}\,.$$

The map Φ is weakly continuous.

PROOF. – (4.7) follows from theorem 4.1, (1.7) and lemma 3.6. Let $g^m_n\rightharpoonup g_n$ in $H^{-\frac12+\alpha}(R^1)$. Because of the condition $\mathcal{F}\in\mathfrak{D}(R^1)$ we have $\mathcal{F}g^m_n\to\mathcal{F}g_n$ in $H^{-\frac12}(R^1)$, hence from lemma 2.1 and the Korn inequality $T_n(u^m)\to T_n(u)$ in $H^{-\frac12}(R^1)$. From (4.7) and the theorem of Eberlein-Šmuljan-Kaplanski $T_n(u^m)\rightharpoonup T_n(u)$ in $H^{-\frac12+\alpha}(R^1)$.

THEOREM 4.3. – *Let $0<\alpha<\frac14$, $u^0\in(H^1_\alpha(\Omega))^2$, $f\in(L_2(\Omega))^2$ and $c(\alpha,\mathcal{F})<1$ (see* (4.7)*). Then there exists a solution of the problem with friction, i.e. there exists $u\in K$ such that for each $v\in K$*

$$(4.8)\qquad a(u,v-u)-\langle\mathcal{F}T_n(u),\,|v_t|-|u_t|\rangle\geq(f,v-u)\,,$$

where $T_n(u)$ is defined in (1.7).

PROOF. – Because $c(\alpha,\mathcal{F})<1$, there exists $r>0$ such that Φ maps the set $M_r=B_{(r)}\cap C^-$ in $H^{-\frac12+\alpha}(R^1)$ into itself (where $B_{(r)}$

is the ball with radius r and C^- is the dual cone of non-positive distributions). Because the map is weakly continuous and M_r is weakly compact, Φ has a fixed point (thanks to Tichonov's fixed point theorem—see Day [5]).

5. – Estimation of the coefficient of friction.

In this part, we shall find a more exact estimate of the constant $c(\alpha, \mathcal{F})$ of the theorem 4.2. Firstly, if we introduce the norm in $H^{\frac{1}{2}+\alpha}(R^1)$, $-\frac{1}{2} < \alpha < \frac{1}{2}$, by (3.1), by the lemma 3.1 we get

$$\|f\|^2_{\frac{1}{2}+\alpha,R^1} = \frac{1}{2\pi} \int_{-\infty}^{+\infty} |\hat{f}(\xi_1)|^2 (1 + 2^{-2\alpha} \lambda_{\alpha+\frac{1}{2}} |\xi_1|^{2\alpha+1})\, d\xi_1 , \tag{5.1}$$

where

$$\lambda_k = 2 \int_{-\infty}^{+\infty} \frac{\sin^2 t}{|t|^{1+2k}}\, dt , \quad 0 < k < 1 .$$

So we have

$$\|f\|^2_{-\frac{1}{2}+\alpha,R^1} = \frac{1}{2\pi} \int_{-\infty}^{+\infty} |\hat{f}(\xi_1)|^2 (1 + 2^{2\alpha} \lambda_{\frac{1}{2}-\alpha} |\xi_1|^{1-2\alpha})^{-1}\, d\xi_1 , \tag{5.2}$$

LEMMA 5.1. – *For $u \in \mathcal{H} = (H^1(\Omega))^2$, $u = 0$ on Γ_u we have the Korn inequality*

$$\|u\|^2_{\mathcal{H}} \leq c a(u, u) . \tag{5.3}$$

PROOF. – For simplicity, let us consider $\Gamma_c = \{x \in R^2, x_1 \in R^1, x_2 = \pi\}$, Γ_u for $x_2 = 0$. Put $u_1(x_1, x_2) = -u_1(x_1, -x_2)$, $u_2(x_1, x_2) = u_2(x_1, -x_2)$ for $-\pi < x_2 \leq 0$. Put

$$a_k = \frac{1}{\sqrt{\pi}} \int_{-\pi}^{\pi} \tilde{u}_1(\xi_1, x_2) \sin kx_2\, dx_2 , \qquad k = 1, 2, \ldots$$

$$b_k = \frac{1}{\sqrt{\pi}} \int_{-\pi}^{\pi} \tilde{u}_2(\xi_1, x_2) \cos kx_2\, dx_2 , \qquad k = 1, 2, \ldots$$

$$b_0 = \frac{1}{\sqrt{2\pi}} \int_{-\pi}^{\pi} \tilde{u}_2(\xi_1, x_2)\, dx_2 , \qquad a_0 = 0 .$$

We have

$$(5.4)\qquad \int_{-\infty}^{+\infty}\int_{-\pi}^{\pi}\left(\frac{\partial u_1}{\partial x_1}\right)^2+\frac{1}{2}\left(\frac{\partial u_1}{\partial x_2}+\frac{\partial u_2}{\partial x_1}\right)^2+\left(\frac{\partial u_2}{\partial x_2}\right)^2 dx_1\,dx_2 =$$

$$= 2\int_{-\infty}^{+\infty}\int_{0}^{\pi}\left(\frac{\partial u_1}{\partial x_1}\right)^2+\frac{1}{2}\left(\frac{\partial u_1}{\partial x_2}+\frac{\partial u_2}{\partial x_2}\right)^2+\left(\frac{\partial u_2}{\partial x^2}\right)^2 dx_1\,dx_2\,.$$

Hence

$$(5.5)\qquad \int_{-\infty}^{+\infty}\int_{-\pi}^{\pi}\left(\frac{\partial u_1}{\partial x_1}\right)^2+\frac{1}{2}\left(\frac{\partial u_1}{\partial x_2}+\frac{\partial u_2}{\partial x_1}\right)^2+\left(\frac{\partial u_2}{\partial x_2}\right)^2 dx_1\,dx_2 =$$

$$=\frac{1}{2\pi}\int_{-\infty}^{+\infty}\sum_{k=0}^{+\infty}\left[\xi_1^2|a_k|^2+k^2|b_k|^2+\frac{1}{2}\left(k^2|a_k|^2+\xi_1^2|b_k|^2+2k\xi_1\operatorname{Re} ia_k\bar{b}_k\right)\right]d\xi_1\geqq$$

$$\geqq\frac{1}{2\pi}\int_{-\infty}^{+\infty}\frac{1}{2}\sum_{k=0}^{+\infty}\left[\xi_1^2|a_k|^2+k^2|a_k|^2+\xi_1^2|b_k|^2+k^2|b_k|^2\right]dx_1=$$

$$=\int_{\Omega}\left(\frac{\partial u_1}{\partial x_1}\right)^2+\left(\frac{\partial u_1}{\partial x_2}\right)^2+\left(\frac{\partial u_2}{\partial x_2}\right)^2+\left(\frac{\partial u_2}{\partial x_1}\right)^2 dx\,.$$

But we have

$$(5.6)\qquad \int_{\Omega}u_1^2+u_2^2\,dx\leqq c_1\int_{\Omega}\left(\frac{\partial u_1}{\partial x_1}\right)^2+\left(\frac{\partial u_1}{\partial x_2}\right)^2+\left(\frac{\partial u_2}{\partial x_1}\right)^2+\left(\frac{\partial u_2}{\partial x_2}\right)^2 dx$$

because $u=0$ for $x_2=0$.

Let us start to investigate more closely some terms of the inequality (4.5). We have (using $\mathcal{F}_{-h}(g_n)_{-h}-\mathcal{F}g_n=(\mathcal{F}_{-h}-\mathcal{F})(g_n)_{-h}+$ $+\mathcal{F}\cdot((g_n)_{-h}-g_n)$):

$$(5.7)\qquad |\langle(\mathcal{F}_{-h}-\mathcal{F})(g_n)_{-h},\,|u_t|_{-h}-|u_t|\rangle\leqq$$

$$\leqq\|g_n\|_{-\frac{1}{2},R^1}\|(\mathcal{F}_{-h}-\mathcal{F})|(u_t|_{-h}-|u_t|)\|_{\frac{1}{2},R^1}\,.$$

By denoting $v=|u_t|_{-h}-|u_t|$, we get

$$(5.8)\qquad \|(\mathcal{F}_{-h}-\mathcal{F})v\|_{\frac{1}{2},R^1}\leqq|h|\|v\|_{\frac{1}{2},R^1}\varkappa(h)\,,$$

where $\varkappa(h)\in L_\infty(R^1)$, $\varkappa(h)\leqq c|h|^{-1}$ for $|h|\geqq h_0>0$.

Furthermore, from lemma 3.1 and 3.2 we have

$$(5.9)\quad \int_{-\infty}^{+\infty} |\langle \mathcal{F}((g_n)_{-h} - g_n),\ |u_t|_{-h} - |u_t|\rangle| \cdot |h|^{-1-2\alpha}\, dh \leqq$$

$$\leqq \|\mathcal{F}\|_\infty \left(\int_{-\infty}^{+\infty} |h|^{-1-2\alpha} \cdot \|(g_n)_{-h} - g_n\|_{-\frac{1}{2},R^1}\, dh \right)^{\frac{1}{2}} \cdot$$

$$\cdot \left(\int_{-\infty}^{+\infty} |h|^{-1-2\alpha} \|(u_t)_{-h} - u_t\|_{\frac{1}{2},R^1}\, dh \right)^{\frac{1}{2}} \leqq$$

$$\leqq \left(\sqrt{2} \cdot \|u_t\|_{\frac{1}{2}+\alpha} \cdot (\lambda_{\frac{1}{2}} \lambda_\alpha)^{\frac{1}{2}} \lambda_{\alpha+\frac{1}{2}}^{-\frac{1}{2}} + k_1 \|u\|_{\mathcal{H}}\right) \cdot \|\mathcal{F}\|_\infty \cdot$$

$$\cdot \left(\frac{2^{1-2\alpha}}{\pi} \lambda_\alpha \int_{-\infty}^{+\infty} |\hat{g}_n(\xi)|^2 (1 + \lambda_{\frac{1}{2}} |\xi|)^{-1} \cdot |\xi|^{-2\alpha}\, d\xi \right)^{\frac{1}{2}}.$$

We have

$$(1 + \lambda_{\frac{1}{2}}|\xi|)^{-1} \cdot |\xi|^{2\alpha} = (1 + 2^{2\alpha} \lambda_{\frac{1}{2}-\alpha} |\xi|^{1-2\alpha})^{-1} \cdot$$

$$\cdot \left[|\xi|^{2\alpha} + 2^{2\alpha} \lambda_{\frac{1}{2}-\alpha} |\xi|\right] \cdot \left[1 + \lambda_{\frac{1}{2}}|\xi|\right]^{-1} = (1 + 2^{2\alpha} \lambda_{\frac{1}{2}-\alpha} |\xi|^{1-2\alpha})^{-1} \cdot$$

$$\cdot \left[\left[2^{2\alpha} \lambda_{\frac{1}{2}-\alpha} \cdot \lambda_{\frac{1}{2}}^{-1}\right] + \left(|\xi|^{2\alpha} \lambda_{\frac{1}{2}} - 2^{2\alpha} \lambda_{\frac{1}{2}-\alpha}\right) \cdot \left[\lambda_{\frac{1}{2}}(1 + |\xi| \lambda_{\frac{1}{2}})\right]^{-1}\right].$$

So it is easy to see (bearing in mind also (2.6)) that for $0 < \alpha < \frac{1}{4}$:

$$(5.10)\quad \int_{-\infty}^{+\infty} |\langle \mathcal{F}((g_n)_{-h} - g_n),\ |u_t|_{-h} - |u_t|\rangle |h|^{-1-2\alpha}\, dh \leqq$$

$$\leqq \|\mathcal{F}\|_\infty 2 \lambda_\alpha \lambda_{\frac{1}{2}+\alpha}^{-\frac{1}{2}} \cdot \lambda_{\frac{1}{2}-\alpha}^{\frac{1}{2}} \left[\|u_t\|_{\frac{1}{2}+\alpha, \Gamma_c} + k_1 \|u^0\|_{\mathcal{H}}\right] \cdot$$

$$\cdot \left[\|g_n\|_{\frac{1}{2}+\alpha, \Gamma_c} + k_2 \|u^0\|_{\mathcal{H}}\right].$$

In the following, we consider as in lemma 5.1 Γ_u with $x_2 = 0$, Γ_c with $x_2 = \pi$. Let us denote $y = u - u^0$. Similarly as in lemma 5.1 we extend y onto the strip $R^1 \times (-\pi, \pi)$. Let us denote

$$G(y) = \lambda \left| i\xi_1 \tilde{y}_1 + \frac{\partial \tilde{y}_2}{\partial x_2} \right|^2 + 2\mu \left(\xi_1^2 |\tilde{y}_1|^2 + \frac{1}{2} \left| \frac{\partial \tilde{y}_1}{\partial x_2} + \tilde{y}_2 i\xi_1 \right|^2 + \left| \frac{\partial \tilde{y}_2}{\partial x_2} \right|^2 \right).$$

Analogously we get

$$(5.11)\quad \frac{2^{1-2\alpha} \lambda_\alpha}{2\pi} \int_{-\infty}^{+\infty} \int_0^\pi G(y) |\xi_1|^{2\alpha}\, d\xi_1\, dx_2 \geqq \frac{2^{1-2\alpha} \lambda_\alpha}{2\pi} \mu \int_{-\infty}^{+\infty} \int_0^\pi \Big[|\xi_1 \tilde{y}_1|^2 +$$

$$+\left|\frac{\partial \tilde{y}_1}{\partial x_2}\right|^2 + |\xi_1 \tilde{y}_2|^2 + \left|\frac{\partial \tilde{y}_2}{\partial x_2}\right|^2\Bigg] |\xi_1|^{2\alpha} d\xi_1 dx_2 =$$

$$= \mu \int_{-\infty}^{+\infty} \int_0^{\pi} \int_{-\infty}^{+\infty} \sum_{i=1}^{2} \sum_{j=1}^{2} \left(\frac{\partial y_i}{\partial x_j}(x_1 + h, x_2) - \frac{\partial y_i}{\partial x_j}(x_1, x_2)\right)^2 \cdot |h|^{-1-2\alpha} dx_1 dx_2 dh .$$

It is easy to see that the function

$$\tilde{w} = \frac{\operatorname{sh} \xi_1 x_2}{\operatorname{sh} \xi_1 \pi} \tilde{y}_1(\xi_1, \pi)$$

is the one-dimensional Fourier transform of the unique solution of the following problem:

$$\Delta w = 0 \text{ in } \Omega , \quad w \in H^1(\Omega) \tag{5.12}$$

$$w(x_1, 0) = 0 , \quad w(x_1, \pi) = y_1(x_1, \pi)$$

Hence clearly

$$\int_{-\infty}^{+\infty} \int_0^{\pi} \sum_{i=1}^{2} \left(\frac{\partial y_1}{\partial x_i}(x_1 + h, x_2) - \frac{\partial y_1}{\partial x_i}(x_1, x_2)\right)^2 dx_1 dx_2 \geqq$$

$$\geqq \int_{-\infty}^{+\infty} \int_0^{\pi} \sum_{j=1}^{2} \left(\frac{\partial w}{\partial x_i}(x_1 + h, x_2) - \frac{\partial w}{\partial x_i}(x_1, x_2)\right)^2 dx_1 dx_2 .$$

Hence

$$\frac{2^{1-2\alpha}}{2\pi} \lambda_\alpha \int_{-\infty}^{+\infty} \int_0^{\pi} G(y) |\xi_1|^{2\alpha} d\xi_1 dx_2 \geqq \frac{2^{1-2\alpha} \lambda_\alpha}{2\pi} \mu \cdot \tag{5.13}$$

$$\cdot \int_{-\infty}^{+\infty} \int_0^{\pi} \left[|\xi_1 \cdot \tilde{w}|^2 + \left|\frac{\partial \tilde{w}}{\partial x_2}\right|^2 \right] |\xi_1|^{2\alpha} d\xi_1 dx_2 = \frac{2^{1-2\alpha} \lambda_\alpha}{2\pi} \mu \cdot$$

$$\cdot \int_{-\infty}^{+\infty} |\xi_1|^{2+2\alpha} \operatorname{sh}^{-2} \xi_1 \pi \; |\tilde{w}(\xi_1, \pi)|^2 d\xi_1 \cdot \int_0^{\pi} \operatorname{sh}^2 x_2 \xi_1 +$$

$$+ \operatorname{ch}^2 x_2 \xi_1 dx_2 \geqq \frac{2^{1-2\alpha} \lambda_\alpha}{2\pi} \mu \int_{-\infty}^{+\infty} |\xi_1|^{1+2\alpha} |\tilde{y}(\xi_1, \pi)|^2 d\xi_1 \geqq$$

$$\geqq \frac{2\mu\lambda_\alpha}{\lambda_{\frac{1}{2}+\alpha}} \|y_1\|^2_{\frac{1}{2}+\alpha, \Gamma_c} - c \|y_1\|^2_{0, \Gamma} .$$

Using considerations analogous to those in the proof of lemma 5.1, we can see that for every $u \in (H^1_\alpha(\Omega))^2$ and $v \in (H^1_{-\alpha}(\Omega))^2$ the following inequality holds:

$$a(u, v) = \frac{1}{2\pi} \int_{-\infty}^{+\infty} \int_0^{\pi} \lambda \left(i\xi_1 \tilde{u}_1 + \frac{\partial \tilde{u}_2}{\partial x_2}\right) \overline{\left(i\xi_1 \tilde{v}_1 + \frac{\partial \tilde{v}_2}{\partial x_2}\right)} + \tag{5.14}$$

$$+ 2\mu \left[\xi_1^2 \tilde{u}_1 \bar{\tilde{v}}_1 + \frac{1}{2}\left(\frac{\partial \tilde{u}_1}{\partial x_2} + i\xi_1 \tilde{u}_2\right) \cdot \overline{\left(\frac{\partial \tilde{v}_1}{\partial x_2} + i\xi_1 \tilde{v}_2\right)} + \frac{\partial \tilde{u}_2}{\partial x_2} \cdot \overline{\frac{\partial \tilde{v}_2}{\partial x_2}}\right] d\xi_1 \, dx_2 \leqq$$

$$\leqq \frac{1}{(2\pi)^{\frac{1}{2}}} \left(\int_{-\infty}^{+\infty} \int_0^{\pi} G(u)(1 + |\xi_1|^{2\alpha}) \, d\xi_1 \, dx_2 \right)^{\frac{1}{2}} \cdot$$

$$\cdot \frac{1}{(2\pi)^{\frac{1}{2}}} \left(\int_{-\infty}^{+\infty} \int_0^{\pi} G(v)(1 + |\xi_1|^{2\alpha})^{-1} \, d\xi_1 \, dx_2 \right)^{\frac{1}{2}} = B(u) \cdot C(v) .$$

For every $v_n \in H^{\frac{1}{2}-\alpha}(\Gamma_c)$ we can make a « harmonic » extension onto the strip Ω analogously as in (5.12). We can denote this extension ω and put $v = (0, \omega)$. From (1.7) we have

$$\|T_n\|_{-\frac{1}{2}+\alpha, \Gamma_c} \leqq \sup_{\|v_n\|_{-\frac{1}{2}+\alpha,\Gamma_c} \leqslant 1} B(u) \, C(v) + (f, v) . \tag{5.15}$$

For $v = (0, \omega)$ we have

$$C^2(v) = \frac{1}{2\pi} \int_{-\infty}^{+\infty} \int_0^{\pi} \left[(\lambda + 2\mu) \left|\frac{\partial \tilde{\omega}}{\partial x_2}\right|^2 + \mu \xi_1^2 |\tilde{\omega}|^2\right] (1 + |\xi_1|^{2\alpha})^{-1} \, d\xi_1 \, dx_2 \leqq$$

$$\leqq \left(\frac{\lambda}{2} + \frac{3}{2}\mu\right) \left[\|v_n\|^2_{\frac{1}{2}-\alpha,\Gamma_c} \lambda^{-1}_{\frac{1}{2}-\alpha} 2^{-2\alpha} + c_1 \|v_n\|^2_{0,\Gamma_c}\right] .$$

If we put for $\tilde{\omega}(\xi)$ the extension of

$$\hat{T}_n(\xi)(1 + \lambda_{\frac{1}{2}-\alpha} 2^{2\alpha} |\xi_1|^{1-2\alpha})^{-1} \cdot \|\hat{T}_n(\xi)(1 + \lambda_{\frac{1}{2}-\alpha} 2^{2\alpha} |\xi_1|^{1-2\alpha})^{-1}\|^{-1}_{\frac{1}{2}-\alpha,\Gamma_c} ,$$

we get for $0 < \alpha \leqq \frac{1}{4}$ (using also (2.6) and denoting $\|\cdot\|_0$ the norm in $(L_2(\Omega))^2$ and $\|\cdot\|_{1,\alpha}$ the norm in $(H^1_\alpha(\Omega))^2$)

$$\|T_n\|_{-\frac{1}{2}+\alpha,\Gamma_c} \leqq B(u) \left(\frac{\lambda}{2} + \frac{3}{2}\mu\right)^{\frac{1}{2}} \lambda^{-\frac{1}{2}}_{\frac{1}{2}-\alpha} 2^{-\alpha} + c_2 \frac{\|u^0\|_{\mathcal{K}}}{\|T_n\|_{-\frac{1}{2}+\alpha,\Gamma_c}} + \tag{5.16}$$

$$+ c_3 \|f\|_0 \frac{\|u^0\|_{\mathcal{K}}}{\|T_n\|_{-\frac{1}{2}+\alpha,\Gamma_c}} .$$

Using the inequality $z \leqq p + \sum_{i=1}^{k} \sqrt{q_i}$ which is valid for

$$0 \leqq z \leqq p + \sum_{i=1}^{k} \frac{q_i}{z}, \quad p \geqq 0, \; q_i \geqq 0 \text{ for } i = 1, \dots, k$$

and the Cauchy inequality, we get for every $\varepsilon > 0$:

$$\|T_n\|_{-\frac{1}{2}+\alpha,\Gamma_c} \leqq \left(\frac{3}{2}\mu + \frac{\lambda}{2}\right)^{\frac{1}{2}} \lambda_{\frac{1}{2}-\alpha}^{-\frac{1}{2}} 2^{-\alpha}(1+\varepsilon) B(u) + c_4(\varepsilon) \cdot \cdot \left[\|u^0\|_{1,\alpha} + \|f\|_0\right]. \tag{5.17}$$

From (5.14) and (2.6) we can see that

$$\left| B^2(u)\, 2^{1-2\alpha} \lambda_\alpha - \int_{-\infty}^{+\infty} |h|^{-1-2\alpha} a(u_{-h} - u, u_{-h} - u)\, dh \right| \leqq c_5(\alpha) \|u^0\|_{1,-\alpha}^2 .$$

Hence from (5.9), (5.10) and analogously to (4.6) we have for every $\varepsilon > 0$ (thanks to the validity of the inequality $\|g_n\|_{-\frac{1}{2},\Gamma_c} \|u_t\|_{\frac{1}{2},\Gamma_c} \leqq \leqq \|g_n\|_{-\frac{1}{2},\Gamma_c} \|u^0\|_{\mathcal{H}}$)

$$2^{1-2\alpha} \lambda_\alpha B^2(u) \leqq (1+\varepsilon) 2 \lambda_\alpha \lambda_{\frac{1}{2}+\alpha}^{-\frac{1}{2}} \lambda_{\frac{1}{2}-\alpha}^{\frac{1}{2}} \|\mathcal{F}\|_\infty \|g_n\|_{-\frac{1}{2}+\alpha,\Gamma_c} \cdot \cdot \|u_t\|_{\frac{1}{2}+\alpha,\Gamma_c} + c_5\left[\|u^0\|_{1,\alpha}^2 + \|f\|_0^2\right]. \tag{5.18}$$

But from (5.13) and (5.14) we obtain

$$\|u_t\|_{\frac{1}{2}+\alpha,\Gamma_c}^2 \leqq \frac{\lambda_{\frac{1}{2}+\alpha}}{2\lambda_\alpha \mu} \int_{-\infty}^{+\infty} |h|^{-1-2\alpha} a(u_{-h} - u, u_{-h} - u)\, dh + + c_7 \|u^0\|_{1,\alpha}^2 \leqq 2^{-2\alpha} \mu^{-1} \lambda_{\frac{1}{2}+\alpha} B^2(u) + c_8(\alpha) \|u^0\|_{1,\alpha}^2 . \tag{5.19}$$

Hence, from (5.18) and (5.19) it follows for every $\varepsilon > 0$.

$$B(u) \leqq (1+\varepsilon) 2^\alpha \lambda_{\frac{1}{2}-\alpha}^{\frac{1}{2}} \mu^{-\frac{1}{2}} \|\mathcal{F}\|_\infty \|g_n\|_{-\frac{1}{2}+\alpha,\Gamma_c} + + c_9(\alpha, \varepsilon)\left[\|u^0\|_{1,\alpha} + \|f\|_0\right]. \tag{5.20}$$

To sum up (5.17) and (5.20), we finally obtain ($\varepsilon > 0$ is an arbitrary number)

$$\|T_n\|_{-\frac{1}{2}+\alpha,\Gamma_c} \leqq (1+\varepsilon)\left(\frac{3\mu}{2} + \frac{\lambda}{2}\right)^{\frac{1}{2}} \mu^{-\frac{1}{2}} \|\mathcal{F}\|_\infty \|g_n\|_{-\frac{1}{2}+\alpha,\Gamma_c} + + c_{10}(\alpha, \varepsilon)\left[\|u^0\|_{1,\alpha} + \|f\|_0\right]. \tag{5.21}$$

THEOREM 5.1. – Let

$$\mathcal{F} \in \mathfrak{D}(R^1), \qquad \max_{x\in R^1} |\mathcal{F}(x)| < \sqrt{\frac{2\mu}{\lambda + 3\mu}}.$$

Then there exists a solution of the problem with friction.

REMARK. – Put

$$\mu = \frac{E}{2(1+\sigma)}, \quad \lambda = \frac{E\sigma}{(1+\sigma)(1-2\sigma)}.$$

Then

$$\sqrt{\frac{2\mu}{3\mu+\lambda}} = \sqrt{\frac{2-4\sigma}{3-4\sigma}}.$$

BIBLIOGRAPHY

[1] G. FICHERA, *Existence theorem in elasticity. Boundary value problems of elasticity with unilateral constraints*, Springer-V., 1972.

[2] G. DUVAUT - J. L. LIONS, *Les inéquations en mécanique et en physique*, Dunod, Paris, 1972.

[3] J. NEČAS, *On regularity of solutions to nonlinear variational inequalities for second-order elliptic systems*, Rend. Mat., (6) **8**, 2 (1975), 481-498.

[4] J. NEČAS, *Les méthodes directes en théorie des équations elliptiques*, Academia, Prague, 1967.

[5] M. M. DAY, *Normed linear spaces*, Springer-V., Berlin-Göttingen-Heidelberg, 1958.

[6] A. KUFNER - O. JOHN - S. FUČÍK, *Function spaces*, Academia, Prague, 1977.

Pervenuta alla Segreteria dell' U. M. I.
il 18 gennaio 1979

"Monograf" - Bologna - Via Collamarini 5

Unilateral Problems in Nonlinear, Three-Dimensional Elasticity

PHILIPPE G. CIARLET & JINDŘICH NEČAS

Communicated by S. ANTMAN

Résumé. En 1977, JOHN BALL a obtenu une série de résultats d'existence très importants en hyperélasticité. L'objet de cet article est l'extension de ses résultats dans deux directions, premièrement en considérant des conditions aux limites unilatérales correspondant au contact sans frottement, deuxièmement en admettant une contrainte de "blocage" dans l'ensemble des déformations admissibles.

Abstract. In 1977 JOHN BALL obtained a series of very important existence results in hyperelasticity. The objective of the present paper is to extend his results in two directions, first by considering unilateral boundary conditions corresponding to contact without friction, secondly by allowing "locking" constraints in the set of admissible deformations.

1. Introduction

To begin with, we give a very brief review of non-linear three-dimensional elasticity with "classical" boundary conditions. For details, see *e.g.* CIARLET [1983, 1985], GERMAIN [1972], GURTIN [1981], MARSDEN & HUGHES [1983], TRUESDELL & NOLL [1965], WANG & TRUESDELL [1973]. Let us first specify the various notations we shall use, concerning notably vectors, matrices, and functions spaces (in what follows, Latin indices take their values in the set $\{1, 2, 3\}$ and the repeated index convention is used):

$\boldsymbol{u} = (u_i)$:	vector with components u_i,
$\boldsymbol{A} = (A_{ij})$:	matrix with elements A_{ij} (i: row index, j: column index),
$\boldsymbol{u} \cdot \boldsymbol{v} = u_i v_i$:	Euclidean inner product,
$\lvert \boldsymbol{u} \rvert = \sqrt{\boldsymbol{u} \cdot \boldsymbol{u}}$:	Euclidean vector norm,
$\boldsymbol{A} : \boldsymbol{B} = A_{ij} B_{ij}$:	matrix inner product,

© Springer-Verlag 1985. Reprinted with kind permission from: Ciarlet, P.G. & Nečas, J., Unilateral Problems in Nonlinear, Three-Dimensional Elasticity, Archive of Rational Mechanics and Analysis 87, 319-338.

$\|A\| = \sqrt{A:A}$: matrix norm associated with the matrix inner product,

$\boldsymbol{adj}\, A$: adjugate of the matrix A = transpose of the cofactor matrix ($\boldsymbol{adj}\, A = (\det A)\, A^{-1}$ if A is regular),

$\iota(A) = (\operatorname{tr} A, \operatorname{tr} \boldsymbol{adj}\, A, \det A)$: set of the principal invariants of a matrix of order 3,

$\mathbb{M}^3$: set of all matrices of order 3,

$\mathbb{M}^3_+ = \{F \in \mathbb{M}^3; \det F > 0\}$,

$\mathbb{S}^3 = \{F \in \mathbb{M}^3; F = F^T\}$,

$\mathbb{O}^3_+ = \{F \in \mathbb{M}^3; FF^T = F^TF = I, \det F = +1\}$,

$\partial_i = \dfrac{\partial}{\partial x_i}$: usual partial derivatives,

$\nabla v = (\partial_i v) \in \mathbb{R}^3$: gradient of a mapping $v: \Omega \subset \mathbb{R}^3 \to \mathbb{R}$,

$\nabla \boldsymbol{\phi} = (\partial_j \phi_i) \in \mathbb{M}^3$: gradient of a mapping $\boldsymbol{\phi}: \Omega \subset \mathbb{R}^3 \to \mathbb{R}^3$,

$\operatorname{div} T = (\partial_j T_{ij}) \in \mathbb{R}^3$: divergence of a tensor field $T: \Omega \subset \mathbb{R}^3 \to \mathbb{M}^3$,

$\dfrac{\partial W}{\partial F}(F) = \dfrac{\partial W}{\partial F_{ij}}(F) \in \mathbb{M}^3$: gradient of a mapping $W: \Omega \subset \mathbb{M}^3 \to \mathbb{R}$,

dx: volume element in $\mathbb{R}^3$,

da: area element in $\mathbb{R}^3$,

$\boldsymbol{L}^p(\Omega) = (L^p(\Omega))^n$ for some integer $n \geqq 2$,

$|\cdot|_{0,p,\Omega}$: norm of the space $L^p(\Omega)$ or $\boldsymbol{L}^p(\Omega)$,

$\boldsymbol{W}^{1,p}(\Omega) = (W^{1,p}(\Omega))^n$ for some integer $n \geqq 2$,

$\|\cdot\|_{1,p,\Omega}$: norm of the space $W^{1,p}(\Omega)$ or $\boldsymbol{W}^{1,p}(\Omega)$,

$|v|_{1,p,\Omega} = \left\{\int_\Omega |\nabla v|^p \, dx\right\}^{1/p}$: semi-norm in the space $W^{1,p}(\Omega)$,

$|\boldsymbol{\phi}|_{1,p,\Omega} = \left\{\int_\Omega \|\nabla \boldsymbol{\phi}\|^p \, dx\right\}^{1/p}$: semi-norm in the space $\boldsymbol{W}^{1,p}(\Omega)$.

We denote by c_1, c_2, *etc.*, constants that are independent of the various functions found in a given inequality.

Let Ω be a bounded open connected subset of $\mathbb{R}^3$ with a boundary Γ that is Lipschitz continuous (in the sense of NEČAS [1967]) and let Γ_0 and Γ_1 be disjoint (relatively) open subsets of Γ, such that area $\{\Gamma - (\Gamma_0 \cup \Gamma_1)\} = 0$. The set $\bar{\Omega}$ is the *reference configuration* occupied by a *homogeneous hyperelastic body* in the absence of any applied force. With the assumption that all applied forces are *dead loads*, the corresponding *mixed displacement-traction problem* classically consists in finding the stationary points of the *total energy*

(1.1) $$I(\psi) = \int_\Omega W(\nabla \psi)\,dx - \left\{ \int_\Omega f \cdot \psi\,dx + \int_{\Gamma_1} g \cdot \psi\,da \right\}$$

when the *admissible deformations* $\psi: \bar{\Omega} \to \mathbb{R}^3$ (whose smoothness will be specified later) satisfy the *constraints*

(1.2) $$\det \nabla \psi > 0 \quad \text{in } \Omega,$$

(1.3) $$\psi = \phi_0 \quad \text{on } \Gamma_0.$$

In (1.1), the real-valued function W, called the *stored energy function*, should be defined for all admissible deformations. Taking into account the *principle of material frame-indifference*, one can show that the associated hyperelastic material is isotropic if and only if there is a function $\tilde{W}$ such that

(1.4) $$W(F) = \tilde{W}(\iota(F^T F)) \text{ for all } F \in \mathbb{M}^3_+.$$

Remark. The extension to *non-homogeneous* materials, whose stored energy function also depends on the material points $x \in \bar{\Omega}$, offers no difficulty, but for simplicity we consider only homogeneous materials here. □

We assume that the reference configuration is a *natural state*, *i.e.*, that there is no stress in the absence of deformation. In this case the terms of lowest order in the expansion of the stored energy function of an isotropic hyperelastic homogeneous body for "small" deformations are of a specific form, *viz.*

(1.5) $$W(F) = \frac{\lambda}{2}(\operatorname{tr} E)^2 + \mu \operatorname{tr}(E^2) + o(\|E\|^2), \quad \text{with } E = \tfrac{1}{2}(F^T F - I),$$

for all $F \in \mathbb{M}^3_+$. In (1.5), the numbers λ and μ are the *Lamé constants* of the elastic material under consideration. Experimental evidence shows that they satisfy the inequalities

(1.6) $$\lambda > 0, \quad \mu > 0,$$

for real materials. When $F = \nabla \phi$ and ϕ is an admissible deformation, the matrix $F^T F$ (often denoted by C) is called the *right Cauchy-Green strain tensor*, while the matrix E is called the *Green-St. Venant strain tensor*. We shall see (*cf.* Theorems 1.1 and 1.2) that additional properties of the stored energy function (polyconvexity, coerciveness, appropriate behavior as $\det F \to 0^+$) also play a crucial role.

The given vector fields $f: \Omega \to \mathbb{R}^3$ and $g: \Gamma_1 \to \mathbb{R}^3$, respectively, measure the densities per unit volume and per unit surface of the reference configuration respectively, of the *applied forces*. Condition (1.2) ensures that the admissible deformations are *orientation-preserving* and *locally invertible* (at least if they are smooth enough). The *boundary condition of place* (1.3) means that the *displacement*

$$u \stackrel{\text{def}}{=} \phi - id$$

is imposed on the portion Γ_0 of the boundary.

If an admissible deformation ϕ is a stationary point of the total energy, it satisfies (at least formally) the associated *mixed displacement-traction boundary value problem*

$$-\operatorname{div} \hat{T}(\nabla\phi) = f \quad \text{in } \Omega, \tag{1.7}$$

$$\phi = \phi_0 \quad \text{on } \Gamma_0, \tag{1.8}$$

$$\hat{T}(\nabla\phi)\, n = g \quad \text{on } \Gamma_1, \tag{1.9}$$

where the mapping

$$\hat{T}: F \in \mathbb{M}^3_+ \to \hat{T}(F) \stackrel{\text{def}}{=} \frac{\partial W}{\partial F}(F) \in \mathbb{M}^3 \tag{1.10}$$

denotes the *response function* associated with the *first Piola-Kirchhoff stress tensor*

$$T(x) = \hat{T}(\nabla\phi(x)), \quad x \in \Omega, \tag{1.11}$$

and n denotes the unit ($|n| = 1$) outer normal vector along the boundary Γ.

In the special case where $\Gamma_1 = \phi$, we have a *pure displacement problem*, and in the special case where $\Gamma_0 = \phi$, we have a *pure traction problem*. In this case, it is immediate from eqs. (1.7) and (1.9) that the given applied forces should satisfy the *condition of equilibrium*:

$$\int_\Omega f\, dx + \int_\Gamma g\, da = 0. \tag{1.12}$$

JOHN BALL [1977] has established the existence of *absolute minima* of the total energy (which are thus particular stationary points of the energy) in situations which include in particular the boundary value problems considered above. The problem of finding reasonable conditions under which such minima are weak solutions of the equations is still open. To put our extensions in their proper perspective, let us first restate the results of JOHN BALL in the form appropriate to our subsequent purposes. We begin with the *mixed displacement-traction* and the *pure displacement problems*.

Theorem 1.1 (BALL [1977, Theorem 7.6]). *Let there be given a stored energy function* $W: \mathbb{M}^3_+ \to \mathbb{R}$ *that satisfies the following assumptions*:

(i) *Polyconvexity*: *There exists a convex function* $\mathbb{W}: \mathbb{M}^3 \times \mathbb{M}^3 \times]0, +\infty[\to \mathbb{R}$ *such that*

$$W(F) = \mathbb{W}(F, \operatorname{adj} F, \det F) \text{ for all } F \in \mathbb{M}^3_+. \tag{1.13}$$

(ii) *Behavior as* $\det F \to 0^+$:

$$\left.\begin{aligned} &F_n \to F \text{ in } \mathbb{M}^3, \\ &H_n \to H \text{ in } \mathbb{M}^3, \\ &\delta_n \to 0 \text{ in }]0, +\infty[\end{aligned}\right\} \text{ imply } \mathbb{W}(F_n, H_n, \delta_n) \to +\infty. \tag{1.14}$$

(iii) *Coerciveness*: *There exist* $a \in \mathbb{R}$, $b > 0$, $p \geqq 2$, $q \geqq \frac{p}{p-1}$, $r > 1$, *such that*

$$\mathbb{W}(\boldsymbol{F}, \boldsymbol{H}, \delta) \geqq a + b(\|\boldsymbol{F}\|^p + \|\boldsymbol{H}\|^q + \delta^r) \text{ for all } (\boldsymbol{F}, \boldsymbol{H}, \delta) \in \mathbb{M}^3 \times \mathbb{M}^3 \times]0, +\infty[. \tag{1.15}$$

Let $\boldsymbol{f} \in \boldsymbol{L}^{\varrho}(\Omega)$ *and* $\boldsymbol{g} \in \boldsymbol{L}^{\sigma}(\Gamma_1)$ *be such that the linear forms*

$$\boldsymbol{\psi} \in \boldsymbol{W}^{1,p}(\Omega) \to \int_\Omega \boldsymbol{f} \cdot \boldsymbol{\psi}\, dx \quad \text{and} \quad \boldsymbol{\psi} \in \boldsymbol{W}^{1,p}(\Omega) \to \int_{\Gamma_1} \boldsymbol{g} \cdot \boldsymbol{\psi}\, da \tag{1.16}$$

are continuous. On the assumtion that the area $\Gamma_0 > 0$, *let* $\boldsymbol{\phi}_0 \in \boldsymbol{W}^{1,p}(\Omega)$ *be given in such away that the set*

$$U = \{\boldsymbol{\psi} \in \boldsymbol{W}^{1,p}(\Omega); \quad \mathbf{adj}\, \nabla\boldsymbol{\psi} \in \boldsymbol{L}^q(\Omega), \ \det \nabla\boldsymbol{\psi} \in L^r(\Omega), \ \det \nabla\boldsymbol{\psi} > 0 \text{ a.e. in } \Omega, \ \boldsymbol{\psi} = \boldsymbol{\phi}_0 \text{ on } \Gamma_0\} \tag{1.17}$$

is not empty. Consider finally the total energy functional defined in (1.1), *and assume that*

$$\inf_{\boldsymbol{\psi} \in U} I(\boldsymbol{\psi}) < +\infty. \tag{1.18}$$

Then the problem: *Find* $\boldsymbol{\phi}$ *such that*

$$\boldsymbol{\phi} \in U \quad \text{and} \quad I(\boldsymbol{\phi}) = \inf_{\boldsymbol{\psi} \in U} I(\boldsymbol{\psi}) \tag{1.19}$$

has at least one solution. □

In the case of the *pure traction problem*, an additional condition has to be included in the definition of the set U, notably in order to preserve the coerciveness of the total energy. From a mechanical viewpoint, this condition removes the indeterminacy due to rigid body translations, which do not alter the equilibrium.

Theorem 1.2 (Ball [1977, Theorem 7.10]). *Let there be given a stored energy function which satisfies assumptions* (1.13), (1.14), (1.15), *let* $\Gamma = \Gamma_1$, *and let there be given applied forces which satisfy the condition of equilibrium* (1.12) *and whose associated linear forms* (1.16) *are continuous. Given a constant vector* $\boldsymbol{e} \in \mathbb{R}^3$, *assume that the set*

$$U(\boldsymbol{e}) = \left\{\boldsymbol{\psi} \in \boldsymbol{W}^{1,p}(\Omega); \mathbf{adj}\, \nabla\boldsymbol{\psi} \in \boldsymbol{L}^q(\Omega), \ \det \nabla\boldsymbol{\psi} \in L^r(\Omega), \ \det \nabla\boldsymbol{\psi} > 0 \text{ a.e. in } \Omega, \quad \int_\Omega \boldsymbol{\psi}\, dx = \boldsymbol{e}\right\} \tag{1.20}$$

is not empty, and let the total energy be defined as in (1.1) *with* $\Gamma = \Gamma_1$. *Finally, assume that* $\inf_{\boldsymbol{\psi} \in U} I(\boldsymbol{\psi}) < +\infty$.

Then the problem: Find ϕ such that

$$\phi \in U(e) \quad \text{and} \quad I(\phi) = \inf_{\psi \in U(e)} I(\psi) \tag{1.21}$$

has at least one solution. □

The following result shows that it is possible to find a class of *very simple* stored energy functions of the form (1.4) which satisfy assumptions (1.13), (1.14), (1.15) used in both Theorems 1.1 and 1.2, and whose expansion in terms of the Green-St. Venant strain tensor $\boldsymbol{E}$ agrees with the expansion (1.5) for any given Lame constants λ, μ satisfying (1.6). For a closely related result, see NEČAS [1983].

Theorem 1.3 (CIARLET & GEYMONAT [1982]). *Given two constants $\lambda > 0$ and $\mu > 0$, it is possible to find a stored energy function of the form*

$$W(\boldsymbol{F}) = a\,\|\boldsymbol{F}\|^2 + b\,\|\boldsymbol{adj}\,\boldsymbol{F}\|^2 + c\,(\det \boldsymbol{F})^2 - d\,\mathrm{Log}\,(\det \boldsymbol{F}), \quad \boldsymbol{F} \in \mathbb{M}^3_+, \tag{1.22}$$

with $a > 0$, $b > 0$, $c > 0$, $d > 0$, that satisfies

$$W(\boldsymbol{F}) = \tfrac{1}{2}\lambda\,(\mathrm{tr}\,\boldsymbol{E})^2 + \mu\,\mathrm{tr}\,(\boldsymbol{E}^2) + o(\|\boldsymbol{E}\|^2), \quad \boldsymbol{E} = \tfrac{1}{2}(\boldsymbol{F}^T\boldsymbol{F} - \boldsymbol{I}), \quad \boldsymbol{F} \in \mathbb{M}^3_+. \tag{1.23}$$

Sketch of the proof. Expanding the expressions

$$\|\boldsymbol{F}\|^2 = \mathrm{tr}\,(\boldsymbol{I} + 2\boldsymbol{E}), \quad \|\boldsymbol{adj}\,\boldsymbol{F}\|^2 = \mathrm{tr}\,\{\boldsymbol{adj}\,(\boldsymbol{I} + 2\boldsymbol{E})\}, \quad \det \boldsymbol{F} = \{\det(\boldsymbol{I} + 2\boldsymbol{E})\}^{\frac{1}{2}}$$

around $\boldsymbol{E} = \boldsymbol{0}$, and writing *a priori* the stored energy function in the form

$$W(\boldsymbol{F}) = a\,\|\boldsymbol{F}\|^2 + b\,\|\boldsymbol{adj}\,\boldsymbol{F}\|^2 + \Gamma(\det \boldsymbol{F}),$$

with $a \in \mathbb{R}$, $b \in \mathbb{R}$, and $\Gamma :]0, +\infty[\to \mathbb{R}$ a convex function twice differentiable at the point 1, we find

$$W(\boldsymbol{F}) = (3a + 3b + \Gamma(1)) + (2a + 4b + \Gamma'(1))\,\mathrm{tr}\,\boldsymbol{E}$$
$$- (2b + \Gamma'(1))\,\mathrm{tr}\,(\boldsymbol{E}^2) + (2b + \tfrac{1}{2}(\Gamma'(1) + \Gamma''(1)))\,(\mathrm{tr}\,\boldsymbol{E})^2 + o(\|\boldsymbol{E}\|)^2.$$

Hence in order that this expansion agree with (1.23), we must have

$$2a + 4b + \Gamma'(1) = 0, \quad 2b + \Gamma'(1) = -\mu, \quad 2b + \tfrac{1}{2}(\Gamma'(1) + \Gamma''(1)) = \tfrac{1}{2}\lambda,$$

and these equations must be solved in such a way that $a > 0$, $b > 0$, $\Gamma''(1) \geqq 0$ (Γ is a convex function). It is then easily verified that for any $\theta \in]0, 1[$,

$$\Gamma'(1) = -\mu(1 + \theta), \quad \Gamma''(1) = \lambda\theta + (1 - \theta)(\lambda + \mu),$$

is an admissible solution of this system. Then for $0 < \theta < \dfrac{\lambda}{2\mu}$, one can find $c > 0$, $d > 0$ such that the particular convex function $\Gamma(\delta) = c\,\delta^2 - d\,\mathrm{Log}\,\delta$ has matching values $\Gamma'(1)$ and $\Gamma''(1)$. □

Let us verify that such a stored energy function satisfies the assumptions of both. Theorems 1.1 and 1.2:

(i) The polyconvexity condition (1.13) is satisfied, since both functions $\boldsymbol{F} \in \mathbb{M}^3_+ \to \|\boldsymbol{F}\|^2$ and $\delta \in]0, +\infty[\to \Gamma(\delta)$ are convex (the corresponding function $\mathbb{W}$ in (1.13) is $\mathbb{W}(\boldsymbol{F}, \boldsymbol{H}, \delta) = a\,\|\boldsymbol{F}\|^2 + b\,\|\boldsymbol{H}\|^2 + \Gamma(\delta)$; this is why we wished $a > 0$, $b > 0$, and $c > 0$, $d > 0$ in $\Gamma(\delta) = c\,\delta^2 - d\,\mathrm{Log}\,\delta$).

(ii) The behavior (1.14) as $\det F \to 0^+$ is taken care of by the inequality $d > 0$.
(iii) The coerciveness condition (1.15) holds with $p = q = r = 2$.

Remark. The stored energy function of (1.22) extends to compressible materials the *Mooney-Rivlin* stored energy function of the incompressible case. It is a special case of the general stored energy functions proposed by OGDEN [1972] in the form

$$W(F) = \sum_{i=1}^{M} a_i \operatorname{tr} \{(F^T F)^{\alpha_i/2}\} + \sum_{j=1}^{n} b_j \operatorname{tr}\{adj\,(F^T F)^{\beta_j/2}\} + \Gamma(\det F)$$

(recall that $\operatorname{tr} F^T F = \|F\|^2$, $\operatorname{tr} adj\, F^T F = \|adj\, F\|^2$). Such a function W is polyconvex if $a_i > 0$, $b_j > 0$, $\alpha_i \geqq 1$, $\beta_j \geqq 1$ and $\Gamma:]0, +\infty[\to \mathbb{R}$ is a convex function. The simplest stored energy function whose expansion agrees with (1.23) is of course given by

$$W(F) = \tfrac{1}{2}\lambda (\operatorname{tr} E)^2 + \mu \operatorname{tr} (E^2).$$

It corresponds to what are called *St. Venant-Kirchhoff materials*, which are often used in computations. However, such a stored energy function is *not* polyconvex (*cf.* NEČAS [1983]). □

We extend the results of Theorem 1.2 in two ways: First, to the boundary conditions $\phi = \phi_0$ on Γ_0 and $\hat{T}(\nabla\phi)\, n = g$ on Γ_1 will be added a *unilateral boundary condition* of the form

$$\phi \notin C \text{ on } \Gamma_2,$$

where Γ_2 is another portion of the boundary Γ and C is some given set in $\mathbb{R}^3$. As a formal interpretation shows (*cf.* Sect. 2), such a boundary condition corresponds to *contact without friction*. Secondly, a *locking constraint* in the form

$$L(\nabla\phi) \leqq 0 \text{ a.e. in } \Omega,$$

will be also allowed in the set of admissible deformations. Examples and a formal interpretation are given in Section 3. Finally in Section 4, we prove two existence theorems corresponding to different combinations of boundary conditions.

The results of this paper were announced in CIARLET & NEČAS [1984].

2. Unilateral Boundary Conditions

Throughout this section and the next one, we assume that all data and unknowns are smooth enough to justify all the computations involved.

Let C be an open subset (which may be unbounded) of $\mathbb{R}^3$, let Ω be a bounded open connected subset of $\mathbb{R}^3$ with boundary Γ, and let Γ_0, Γ_1, Γ_2 be disjoint (relatively) open subsets of Γ such that area $\{\Gamma - (\Gamma_0 \cup \Gamma_1 \cup \Gamma_2)\} = 0$. Let the set of *admissible deformations* be of the form

(2.1)

$$U = \{\psi : \bar{\Omega} \to \mathbb{R}^3;\ \det \nabla\psi > 0 \text{ in } \Omega,\ \psi = \phi_0 \text{ on } \Gamma_0,\ \psi(x) \notin C \text{ for } x \in \Gamma_2\},$$

the smoothness of the admissible deformations ψ being purposely left unspecified, and let the total energy be of the form

$$I(\psi) = \int_\Omega W(\nabla\phi)\,dx - \left\{\int_\Omega f\cdot\psi\,dx + \int_{\Gamma_1} g\cdot\psi\,da\right\}, \tag{2.2}$$

where the stored energy function W and the fields (f) and (g) are as in Sect. 1.

Theorem 2.1. *A solution* ϕ *of the minimization problem*

$$\phi\in U,\quad I(\phi)=\inf_{\psi\in U} I(\psi), \tag{2.3}$$

where the set U *and the functional* I *are as in* (2.1)–(2.2), *is, at least formally, a solution of the following boundary value problem:*

$$-\text{div}\,\hat{T}(\nabla\phi) = f \quad\text{in}\quad \Omega,\quad\text{with}\quad \hat{T}(F) = \frac{\partial W}{\partial F}(F)\ \text{for}\ F\in \mathbb{M}^3_+, \tag{2.4}$$

$$\phi = \phi_0 \quad\text{on}\ \Gamma_0, \tag{2.5}$$

$$\hat{T}(\nabla\phi)\,n = g \quad\text{on}\ \Gamma_1, \tag{2.6}$$

$$\left.\begin{aligned} &\hat{T}(\nabla\phi)\,n = 0 \quad\text{if}\ \phi\notin\bar{C}, &(2.7)\\ &\hat{T}(\nabla\phi)\,n = \alpha n',\quad \alpha\leqq 0,\ \text{if}\ \phi\in\partial C, &(2.8)\end{aligned}\right\}\ \text{on}\ \Gamma_2,$$

where n *denotes the unit outer normal vector along the surface* Γ *at the point* x *and where* n' *denotes the unit outer normal vector along the surface* $\phi(\Gamma)$ *at the point* $\phi(x)$.

Proof. We shall repeatedly use the following Green's formula:

$$\int_\Omega \hat{T}(\nabla\phi):\nabla\theta\,dx = -\int_\Omega \text{div}\,\hat{T}(\nabla\phi)\cdot\theta\,dx + \int_\Gamma \hat{T}(\nabla\phi)\,n\cdot\theta\,da, \tag{2.9}$$

valid for smooth enough functions $\theta:\Omega\to\mathbb{R}^3$. Assume first that θ vanishes in a neighborhood of $\Gamma_0\cup\Gamma_2$. There exists $\varepsilon_0=\varepsilon_0(\theta)>0$ such that $\phi^\varepsilon=\phi+\varepsilon\theta\in U$ for $|\varepsilon|\leqq\varepsilon_0$. Stating that $I(\phi^\varepsilon)\geqq I(\phi)$, and using Green's formula (2.9), we find that

$$\varepsilon\left\{\int_\Omega\{-\text{div}\,\hat{T}(\nabla\phi)-f\}\cdot\theta\,dx + \int_{\Gamma_1}\{\hat{T}(\nabla\phi)\,n-g\}\cdot\theta\,da + O(\varepsilon)\right\}\geqq 0$$

for all $|\varepsilon|\leqq\varepsilon_0$. From such inequalities, we classically derive equations (2.4) and (2.6). The boundary condition (2.5) follows from the definition of the set U.

Next, let $x\in\Gamma_2$ be such that $\phi(x)\notin C$. It is easily seen that there is an $r>0$ such that, given any function $\theta:\bar{\Omega}\to\mathbb{R}^3$ with support in the ball $B(x;r)$, there is an $\varepsilon_0=\varepsilon_0(\theta)>0$ such that $\phi^\varepsilon=\phi+\varepsilon\phi\in U$ for $|\varepsilon|\leqq\varepsilon_0$. Stating again that $I(\phi^\varepsilon)\geqq I(\phi)$ for such functions ϕ^ε, using Green's formula

(2.9) and taking into account equations (2.4), we find that

$$\varepsilon\left\{\int_{\Gamma_2} \hat{T}(\nabla\boldsymbol{\phi})\,\boldsymbol{n}\cdot\boldsymbol{\theta}\,da + O(\varepsilon)\right\} \geqq 0$$

for all $|\varepsilon| \leqq \varepsilon_0$, from which we deduce the boundary condition (2.7).

Finally, let $\tilde{x} \in \Gamma_2$ be such that $\boldsymbol{\phi}(\tilde{x}) \in \partial C$ and assume that the surfaces ∂C and $\boldsymbol{\phi}(\Gamma)$ have the same tangent space at the point $\boldsymbol{\phi}(\tilde{x})$. Let $(\boldsymbol{t}_1', \boldsymbol{t}_2')$ be a smooth field of linearly independent tangent vectors to the surface $\boldsymbol{\phi}(\Gamma)$, with $|\boldsymbol{t}_1'| = |\boldsymbol{t}_2'| = 1$. Let $B(\tilde{x};\varrho)$ be such that $B(\tilde{x};\varrho) \cap \Gamma \subset \Gamma_2$. Given any functions $\theta_1, \theta_2 : \bar{\Omega} \to \mathbb{R}$ with supports in $B(\tilde{x};\varrho)$, there exists $\varepsilon_0 = \varepsilon_0(\theta_1, \theta_2) > 0$ and there are functions $\lambda_1^\varepsilon, \lambda_2^\varepsilon : \bar{\Omega} \to \mathbb{R}$ with supports in $B(\tilde{x};\varrho)$ such that, for $\alpha = 1, 2$,

$$\begin{cases} \boldsymbol{\phi}_\alpha^\varepsilon = \boldsymbol{\phi} + \varepsilon(\theta_\alpha \boldsymbol{t}_\alpha' + \lambda_\alpha^\varepsilon \boldsymbol{n}') \in U \quad \text{for} \quad |\varepsilon| \leqq \varepsilon_0, \\ |\lambda^\varepsilon(x)| = O(\varepsilon), \end{cases}$$

where $\boldsymbol{n}'$ denotes the unit outer normal vector along the surface $\boldsymbol{\phi}(\Gamma)$. Proceeding as before, we find that

$$\varepsilon\left\{\int_{\Gamma_2} \theta_\alpha(\hat{T}(\nabla\boldsymbol{\phi})\,\boldsymbol{n}\cdot\boldsymbol{t}_\alpha')\,da + O(\varepsilon)\right\} \geqq 0,$$

and thus we must have $\hat{T}(\nabla\boldsymbol{\phi})\,\boldsymbol{n}\cdot\boldsymbol{t}_\alpha' = 0$ for $\alpha = 1, 2$, *i.e.*, the vector $\hat{T}(\nabla\boldsymbol{\phi})\,\boldsymbol{n}$ must be parallel to $\boldsymbol{n}'$ at those points $\tilde{x} \in \Gamma_2$ where $\boldsymbol{\phi}(\tilde{x}) \in \partial C$ (we already know that $\hat{T}(\nabla\boldsymbol{\phi})\,\boldsymbol{n} = \boldsymbol{0}$ at those points $x \in \Gamma_2$ where $\boldsymbol{\phi}(x) \notin C$).

Given any function $\theta : \bar{\Omega} \to \mathbb{R}$ with support in $B(\tilde{x};\varrho)$, and which is $\geqq 0$, there is an $\varepsilon_0 = \varepsilon_0(\theta) > 0$ such that

$$\boldsymbol{\phi}^\varepsilon = \boldsymbol{\phi} - \varepsilon\theta\boldsymbol{n}' \in U \quad \text{for} \quad 0 \leqq \varepsilon \leqq \varepsilon_0.$$

Thus

$$\varepsilon\left\{\int_{\Gamma_2} \theta\{\hat{T}(\nabla\boldsymbol{\phi})\,\boldsymbol{n}\cdot\boldsymbol{n}'\}\,da + O(\varepsilon)\right\} \leqq 0 \quad \text{for} \quad 0 \leqq \varepsilon \leqq \varepsilon_0,$$

which forces the normal component $\{\hat{T}(\nabla\boldsymbol{\phi})\,\boldsymbol{n}\cdot\boldsymbol{n}'\}$ of the vector $\hat{T}(\nabla\boldsymbol{\phi})\,\boldsymbol{n}$ to be $\leqq 0$. Since we have just shown that this vector is in fact parallel to $\boldsymbol{n}'$, the boundary condition (2.8) is satisfied. □

To *interpret* the boundary conditions (2.7)–(2.8), we recall that the first Piola-Kirchhoff stress tensor $\hat{\boldsymbol{T}} = \hat{T}(\nabla\boldsymbol{\phi})$ of (1.11), the normal vector $\boldsymbol{n}$ and the area element da at a point $x \in \Gamma$, and the corresponding *Cauchy stress tensor* $\boldsymbol{T}'$, normal vector $\boldsymbol{n}'$, and area element da' at the point $\boldsymbol{\phi}(x) \in \boldsymbol{\phi}(\Gamma)$, are related by

$$\boldsymbol{T}\boldsymbol{n}\,da = \boldsymbol{T}'\boldsymbol{n}'\,da'. \tag{2.10}$$

In particular, the vectors $\boldsymbol{T}\boldsymbol{n}$ and $\boldsymbol{T}'\boldsymbol{n}'$ are parallel, and so the boundary conditions (2.7) and (2.8) can also be written as (*cf.* Fig. 2.1)

$$\left.\begin{aligned} &\boldsymbol{T}'\boldsymbol{n}' = \boldsymbol{0} \quad \text{if } \boldsymbol{\phi} \notin C, && (2.11)\\ &\boldsymbol{T}'\boldsymbol{n}' = \alpha'\boldsymbol{n}', \quad \alpha' \leqq 0, \quad \text{if} \quad \boldsymbol{\phi} \in \partial C, && (2.12) \end{aligned}\right\} \quad \text{on} \quad \boldsymbol{\phi}(\Gamma_2).$$

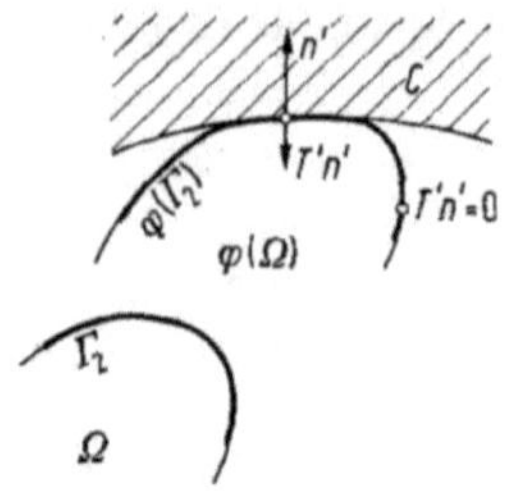

Fig. 2.1. At those points where the deformed boundary $\varphi(\Gamma_2)$ and the obstacle C have a common tangent plane, the Cauchy stress vector $\boldsymbol{T}'\boldsymbol{n}'$ is directed along the outer normal to the obstacle (contact without friction)

Since the *Cauchy stress vector* $\boldsymbol{t}' = \boldsymbol{T}'\boldsymbol{n}'$ measures the density of the surface force per unit area of the deformed surface $\boldsymbol{\phi}(\Gamma)$ and since we have shown that it is normal to the deformed surface $\boldsymbol{\phi}(\Gamma)$, the physical interpretation of the boundary conditions (2.7)–(2.8) as a model for *contact without friction* is now clear.

For *linearly elastic* materials, various *unilateral* problems, including contact with or without friction, have been extensively studied, notably by DUVAUT & LIONS [1972], FICHERA [1972], NEČAS & HLAVÁČEK [1981], HLAVÁČEK, HASLINGER, NEČAS & LOVÍŠEK [1982].

3. Locking Constraints

PRAGER [1957] has considered materials with locking constraints: Roughly speaking, such a material is linearly elastic for "small" strains, while it becomes "locked" when some measure of the strains reaches a certain level, *i.e.*, no more deformations can occur during the loading process at those points where the material is "locked". A *locking constraint* thus takes the form of an inequality

$$L(\nabla\boldsymbol{\phi}) \leqq 0 \quad \text{in } \Omega, \tag{3.1}$$

where the *locking function* $L: \mathbb{M}^3_+ \to \mathbb{R}$ is given. If such a constraint is taken into consideration, our problem is to minimize the total energy (2.2) over a *set of admissible deformations* of the form:

$$U = \{\boldsymbol{\psi}: \bar{\Omega} \to \mathbb{R}^3;\ \det \nabla\boldsymbol{\psi} > 0 \text{ in } \Omega,\ L(\nabla\boldsymbol{\psi}) \leqq 0 \text{ in } \Omega,\ \boldsymbol{\psi} = \boldsymbol{\phi}_0 \text{ on } \Gamma_0,\ \boldsymbol{\psi} \notin C \text{ on } \Gamma_2\}. \tag{3.2}$$

Since the additional constraint takes the form of an inequality which may become an equality (by contrast with the inequality $\det \nabla\boldsymbol{\psi} > 0$ in Ω), its formal interpretation is classical in constrained optimization: It consists in replacing the stress tensor $\hat{\boldsymbol{T}}(\nabla\boldsymbol{\phi}) = \frac{\partial W}{\partial \boldsymbol{F}}(\nabla\boldsymbol{\phi})$ in the equilibrium equations (2.4) and (2.6) by

$$\hat{\boldsymbol{T}}(\nabla\boldsymbol{\phi}) = \frac{\partial W}{\partial \boldsymbol{F}}(\nabla\boldsymbol{\phi}) + \lambda \frac{\partial L}{\partial \boldsymbol{F}}(\nabla\boldsymbol{\phi}), \tag{3.3}$$

where the *generalized Kuhn-Tucker multiplier* $\lambda : \bar{\Omega} \to \mathbb{R}$ satisfies

$$(3.4)\qquad \begin{cases} \lambda = 0 \text{ at those points where } L(\nabla\boldsymbol{\phi}) < 0, \\ \lambda \geqq 0 \text{ elsewhere.} \end{cases}$$

Remark. By contrast, *plasticity* is taken into account by inequalities involving the *stress tensor* instead of the deformation gradient $\nabla\boldsymbol{\phi}$. In this case, a major difficulty lies in the definition of a suitable dual functional (see GUO ZHONG-HENG [1980]). □

The locking constraints introduced by W. PRAGER are of the form

$$(3.5)\qquad L(\nabla\boldsymbol{\phi}) = l(\varepsilon(\nabla\boldsymbol{u})) \leqq 0 \text{ in } \Omega, \quad \nabla\boldsymbol{\phi} = \boldsymbol{I} + \nabla\boldsymbol{u},$$

where $l : \mathbb{S}^3 \to \mathbb{R}$ is a convex real-valued function satisfying $l(\boldsymbol{0}) < 0$, and

$$(3.6)\qquad \varepsilon(\nabla\boldsymbol{u}) = \tfrac{1}{2}(\nabla\boldsymbol{u}^T + \nabla\boldsymbol{u}) = \boldsymbol{E}(\nabla\boldsymbol{\phi}) - \tfrac{1}{2}\nabla\boldsymbol{u}^T\nabla\boldsymbol{u}, \quad \nabla\boldsymbol{\phi} = \boldsymbol{I} + \nabla\boldsymbol{u},$$

denotes the *linearized strain tensor*. In this respect, DUVAUT & LIONS [1972, p. 269] have obtained existence results for materials whose stored energy function is quadratic and positive definite with respect to the linearized strain tensor, *i.e.*, their results apply to *linearly* elastic models and to locking functions such as

$$(3.7)\qquad l(\varepsilon) = \varepsilon' : \varepsilon' - \nu, \quad \nu > 0,$$

where

$$(3.8)\qquad \varepsilon' = \varepsilon - \tfrac{1}{3}(\operatorname{tr} \varepsilon)\, \boldsymbol{I}$$

denotes the *deviator* of the linearized strain tensor. Within the limited framework of such first-order approximations (which are not frame-invariant), such a locking function is natural in the sense that the number $\varepsilon'(\nabla\boldsymbol{u}) : \varepsilon'(\nabla\boldsymbol{u})$ represents *the intensity of linearized shear strains*: Arguing as did NEČAS & HLAVÁČEK [1981, p. 27–28], one can show that it is an *average of the maximum shear strains*.

Since we are concerned here with nonlinear elasticity, it is one step further to replace in (3.7) the linearized strain tensor $\varepsilon(\nabla\boldsymbol{u})$ by the "full" strain tensor $\boldsymbol{E}(\nabla\boldsymbol{\phi})$. In this fashion, we obtain a locking function of the form

$$(3.9)\qquad \Lambda(\nabla\boldsymbol{\phi}) = l(\boldsymbol{E}(\nabla\boldsymbol{\phi})) = \boldsymbol{E}'(\nabla\boldsymbol{\phi}) : \boldsymbol{E}'(\nabla\boldsymbol{\phi}) - \nu, \quad \nu > 0,$$

where

$$(3.10)\qquad \boldsymbol{E}'(\nabla\boldsymbol{\phi}) = \boldsymbol{E}(\nabla\boldsymbol{\phi}) - \tfrac{1}{3}(\operatorname{tr} \boldsymbol{E}(\nabla\boldsymbol{\phi}))\, \boldsymbol{I}$$

denotes the *deviator of the Green-St. Venant strain tensor* $\boldsymbol{E}(\nabla\boldsymbol{\phi})$.

For reasons that will become apparent in the next section, we need locking functions $L : \mathbb{M}^3_+ \to \mathbb{R}$ which are *polyconvex* in the sense of BALL [1977]: there exists a *convex* function $\mathbb{L} : \mathbb{M}^3 \times \mathbb{M}^3 \times]0, +\infty[\to \mathbb{R}$ such that

$$(3.11)\qquad L(\boldsymbol{F}) = \mathbb{L}(\boldsymbol{F}, \boldsymbol{adj}\, \boldsymbol{F}, \det \boldsymbol{F}) \quad \text{for all} \quad \boldsymbol{F} \in \mathbb{M}^3_+$$

(compare with (1.13)). In this direction, the following result is reminiscent of Theorem 1.3:

Theorem 3.1. *There exists a convex function* $\mathbb{L}: \mathbb{M}^3 \times \mathbb{M}^3 \times]0, +\infty[\to \mathbb{R}$ *such that*

$$(3.12)\qquad \begin{aligned} L(F) &= \mathbb{L}(F, \operatorname{adj} F, \det F) = -\nu + E' : E' + o(\|E\|^2), \\ E &= \tfrac{1}{2}(F^T F - I), \quad E' = E - \tfrac{1}{3}(\operatorname{tr} E)\, I. \end{aligned}$$

Proof. A simple computation shows that

$$E' : E' = -\tfrac{1}{3}(\operatorname{tr} E)^2 + \operatorname{tr}(E^2).$$

Thus we have an expression of the form (1.23) with $\lambda = -\frac{2}{3}$, $\mu = 1$. Let us look *a priori* for a function L in the form (compare with (1.22)):

$$L(F) = -\nu + a\,\|F\|^2 + b\,\|\operatorname{adj} F\|^2 + \Gamma(\det F).$$

If we can find $a > 0$, $b > 0$ and a convex function $\Gamma:]0, +\infty[\to \mathbb{R}$ such that condition (3.12) is fulfilled, the theorem is proved. The argument that was used to prove Theorem 1.3 shows that the inequalities $a > 0$, $b > 0$, $\Gamma''(1) \geqq 0$ are satisfied if

$$\Gamma'(1) = -(1 + \theta), \quad \Gamma''(1) = \tfrac{1}{3} - \theta, \quad \text{for any} \quad \theta \in]0, \tfrac{1}{3}].$$

It then suffices to choose *any* convex function Γ with matching values $\Gamma'(1)$ and $\Gamma''(1)$. For example, let $\theta = \frac{1}{3}$ and $\Gamma(\delta) = -\frac{4}{3}\delta$. □

Remarks. (1) By contrast with Theorem 1.3, where the condition $\lim_{\delta \to 0^+} \Gamma(\delta) = +\infty$ on the stored energy function plays a crucial role in the existence theory, it is no longer needed here in regard to the locking function (this follows from the proof of Theorem 4.1).

(2) It is natural to require that the stored energy function and the locking function both be *frame-indifferent*, or *objective*. A necessary and sufficient condition that this property hold is that such a function, say G, satisfy

$$G(QF) = G(Q) \quad \text{for all } F \in \mathbb{M}^3_+, \quad Q \in \mathbb{Q}^3_+.$$

Clearly such a condition is satisfied by the examples given in Theorems 1.3 and 3.1.

(3) If the function L agrees with the function $(-\nu + E' : E')$ to within terms of second order in the variable E, it agrees *a fortiori* with the function $(-\nu + \varepsilon' : \varepsilon')$ to within the second order in the variable ε in an expansion with respect to the deformation gradient.

(4) The subsequent analysis applies even if *several* locking constraints are simultaneously taken into account. □

4. Results on Existence

We begin by considering problems with $\Gamma_0 \neq \phi$ which include unilateral boundary conditions and locking constraints of the form considered in Sect. 2 and 3. *The following theorem extends Theorem 1.1*:

Theorem 4.1. *Let there be given a stored energy function* $W: \mathbb{M}^3_+ \to \mathbb{R}$ *which satisfies assumptions* (1.13), (1.14), (1.15), *and let there be given a locking function*

$L: \mathbb{M}^3_+ \to \mathbb{R}$ which is polyconvex in the sense of (3.11). *Let Ω be a bounded open connected subset of $\mathbb{R}^3$ with a Lipschitz continuous boundary Γ, and let Γ_0, Γ_1, Γ_2 be disjoint (relatively) open subsets of Γ, with*

$$\text{(4.1)} \qquad \text{area } \Gamma_0 > 0, \text{ and area } \{\Gamma - (\Gamma_0 \cup \Gamma_1 \cup \Gamma_2\} = 0.$$

Let $f \in \boldsymbol{L}^\varrho(\Omega)$ and $g \in \boldsymbol{L}^\sigma(\Gamma_1)$ be such that the linear forms of (1.16) *are continuous. Let C be an open subset of $\mathbb{R}^3$, and let $\boldsymbol{\phi}_0 \in \boldsymbol{W}^{1,p}(\Omega)$ be such that the set*

$$\text{(4.2)} \qquad U = \{\boldsymbol{\psi} \in \boldsymbol{W}^{1,p}(\Omega);\ \boldsymbol{adj}\,\nabla\boldsymbol{\psi} \in \boldsymbol{L}^q(\Omega),\ \det\nabla\boldsymbol{\psi} \in L^r(\Omega),$$

$$\det\nabla\boldsymbol{\psi} > 0 \text{ a.e. in } \Omega,\ L(\nabla\boldsymbol{\psi}) \leqq 0 \text{ a.e. in } \Omega,\ \boldsymbol{\psi} = \boldsymbol{\phi}_0 \text{ on } \Gamma_0,\ \boldsymbol{\psi} \notin C \text{ on } \Gamma_2\}$$

is not empty. Consider finally the functional I defined by (1.1) *and assume that $\inf_{\psi \in U} I(\boldsymbol{\psi}) < +\infty$.*

Then the problem: Find $\boldsymbol{\phi}$ such that

$$\text{(4.3)} \qquad \boldsymbol{\phi} \in U \quad \text{and} \quad I(\boldsymbol{\phi}) = \inf_{\psi \in U} I(\boldsymbol{\psi})$$

has at least one solution.

Proof. We first remark that the total energy $I(\boldsymbol{\psi})$ is well defined for functions $\boldsymbol{\psi} \in U$. Since the stored energy function is polyconvex, the function $x \in \Omega \to \mathbb{W}(\nabla\boldsymbol{\phi}(x), \boldsymbol{adj}\,\nabla\boldsymbol{\phi}(x), \det\nabla\boldsymbol{\phi}(x))$ is measurable; since $\mathbb{W} \geqq a$ by (1.15) and since Ω is bounded, the integral $\int_\Omega W(\nabla\boldsymbol{\phi}(x))\,dx$ is well defined in the interval $[a \text{ meas } \Omega, +\infty]$. Next we extend the domain of definition of the function $\mathbb{W}$ by letting

$$\text{(4.4)} \qquad \mathbb{W}(\boldsymbol{F}, \boldsymbol{H}, \delta) = +\infty \quad \text{for } \boldsymbol{F} \in \mathbb{M}^3, \quad \boldsymbol{H} \in \mathbb{M}^3, \quad \delta \leqq 0,$$

so that by (1.14) it is equivalent to effect the minimization with the extended function W which is also convex and continuous.

Using assumption (4.1) and arguing as in Nečas [1967, Theorem 7.9, p. 116], we can show that

$$\text{(4.5)} \qquad \int_\Omega |\boldsymbol{\psi}|^p\,dx \leqq c_1 \left\{ |\boldsymbol{\psi}|^p_{1,p,\Omega} + \left| \int_{\Gamma_0} \boldsymbol{\psi}\,da \right|^p \right\} \quad \text{for all } \boldsymbol{\psi} \in \boldsymbol{W}^{1,p}(\Omega).$$

From this inequality, the boundary condition $\boldsymbol{\psi} = \boldsymbol{\phi}_0$ on Γ_0, the coerciveness inequality (1.15), and the continuity of the linear forms (1.16), it follows that

$$\text{(4.6)} \quad I(\boldsymbol{\psi}) \geqq c_2 + c_3(\|\boldsymbol{\psi}\|^p_{1,p,\Omega} + |\boldsymbol{adj}\,\nabla\boldsymbol{\psi}|^q_{0,q,\Omega} + |\det\nabla\boldsymbol{\psi}|^r_{0,r,\Omega}),\ c_3 > 0,$$

for all $\boldsymbol{\psi} \in \boldsymbol{W}^{1,p}(\Omega)$ with $\boldsymbol{adj}\,\nabla\boldsymbol{\psi} \in \boldsymbol{L}^q(\Omega)$, $\det\nabla\boldsymbol{\psi} \in L^r(\Omega)$.

Then let $(\boldsymbol{\phi}^k)$ with $\boldsymbol{\phi}^k \in U$ be an infimizing sequence for the functional I. By inequality (4.6), the sequence $(\boldsymbol{\phi}^k, \boldsymbol{adj}\,\nabla\boldsymbol{\phi}^k, \det\nabla\boldsymbol{\phi}^k)$ is bounded in the space $\boldsymbol{W}^{1,p}(\Omega) \times \boldsymbol{L}^q(\Omega) \times L^r(\Omega)$. Since this space is reflexive (recall that the numbers p, q, r are > 1 by assumption (1.15), there exists a subsequence, again denoted by $(\boldsymbol{\phi}^k)$, such that

(4.7)

$$\boldsymbol{\phi}^k \rightharpoonup \boldsymbol{\phi} \text{ in } \boldsymbol{W}^{1,p}(\Omega), \quad \boldsymbol{adj}\,\nabla\boldsymbol{\phi}^k \rightharpoonup \boldsymbol{H} \text{ in } \boldsymbol{L}^q(\Omega), \quad \det\nabla\boldsymbol{\phi}^k \rightharpoonup \delta \text{ in } L^r(\Omega),$$

where $\rightharpoonup$ denotes weak convergence. Besides, one has

$$H = \boldsymbol{adj}\,\nabla\boldsymbol{\phi}, \qquad \delta = \det \nabla\boldsymbol{\phi}, \tag{4.8}$$

by the results of BALL [1977, Theorem 6.2]. Since the extended function $\mathbb{W}$ is convex and continuous and since the linear forms of (1.16) are continuous, the functional

$$(\boldsymbol{\psi}, \boldsymbol{adj}\,\nabla\boldsymbol{\psi}, \det \nabla\boldsymbol{\psi}) \in \boldsymbol{W}^{1,p}(\Omega) \times \boldsymbol{L}^q(\Omega) \times L^r(\Omega) \to I(\boldsymbol{\psi})$$

is weakly lower semi-continuous and therefore

$$I(\boldsymbol{\phi}) \leqq \liminf I(\boldsymbol{\phi}^k) < +\infty. \tag{4.9}$$

In order to finish, we must verify that the function $\boldsymbol{\phi}$ belongs to the set $\boldsymbol{U}$. Since $I(\boldsymbol{\phi}) < +\infty$ we first deduce that $\det \nabla\boldsymbol{\psi} > 0$ (so far the arguments and those of BALL [1977] are identical). Since the trace operator is completely continuous from the space $W^{1,p}(\Omega)$ into the space $L^p(\Gamma)$, it next follows that $\boldsymbol{\phi} = \boldsymbol{\phi}_0$ on Γ_0 and $\boldsymbol{\phi} \notin C$ on Γ_2 (C is open). Finally, by MAZUR's theorem, convex combinations with the following strong convergence properties for $k \to \infty$ exist:

$$\left.\begin{aligned}
&\sum_{l=0}^{k} \alpha_{lk}\boldsymbol{\psi}^l \to \boldsymbol{\psi} \quad \text{in} \quad \boldsymbol{W}^{1,p}(\Omega),\\
&\sum_{l=0}^{k} \alpha_{lk}\,\boldsymbol{adj}\,\nabla\boldsymbol{\psi}^l \to \boldsymbol{adj}\,\nabla\boldsymbol{\psi} \quad \text{in} \quad \boldsymbol{L}^q(\Omega),\\
&\sum_{l=0}^{k} \alpha_{lk} \det \nabla\boldsymbol{\psi}^l \to \det \nabla\boldsymbol{\psi} \quad \text{in} \quad L^r(\Omega),
\end{aligned}\right\} \text{with } \sum_{l=0}^{k} \alpha_{lk} = 1,\ \alpha_{lk} \geqq 0,\ 0 \leqq l \leqq k.$$

These sequences have subsequences, which may be assumed to be indexed by the same integers k', which converge pointwise almost everywhere to their respective limits. Since the function $\mathbb{L}: \mathbb{M}^3 \times \mathbb{M}^3 \times]0, +\infty[\to \mathbb{R}$ is convex (by the assumed polyconvexity of the locking function L), it is continuous. Thus passing to the limit, we obtain

$$\begin{aligned}
L(\nabla\boldsymbol{\psi}(x)) &= \mathbb{L}(\nabla\boldsymbol{\psi}(x), \boldsymbol{adj}\,\nabla\boldsymbol{\psi}(x), \det \nabla\boldsymbol{\psi}(x))\\
&= \lim_{k'\to\infty} \left(\sum_{l=0}^{k'} \alpha_{lk'}\boldsymbol{\psi}^l(x), \sum_{l=0}^{k'} \alpha_{lk'}\,\boldsymbol{adj}\,\nabla\boldsymbol{\psi}^l(x), \det \nabla\boldsymbol{\psi}^l(x)\right)\\
&\leqq \limsup_{k'\to\infty} \sum_{l=0}^{k'} \alpha_{lk'}, \mathbb{L}(\nabla\boldsymbol{\psi}^l(x)\ \boldsymbol{adj}\,\nabla\boldsymbol{\psi}^l(x), \det \nabla\boldsymbol{\psi}^l(x)) \leqq 0
\end{aligned} \tag{4.10}$$

for almost all points $x \in \Omega$, and the proof is complete. □

The case where $\Gamma_0 = \Gamma_2 = \phi$ has been treated by J. BALL (*cf.* Theorem 1.2). Adding a locking constraint in the set $\boldsymbol{U}(\boldsymbol{e})$ of (1.20) presents no difficulty, since it can be dealt with as in Theorem 4.1.

The case $\Gamma_0 = \phi$ and $\Gamma_2 \neq \phi$ is more interesting: In particular, we shall be able to substantiate mathematically in this case the following natural but vague idea: *If the applied forces are "opposite" or "at worst" perpendicular to directions where the body can "escape", then we expect solutions to exist, without imposing an additional condition such as* $\int_\Omega \boldsymbol{\psi}\, dx = \boldsymbol{e}$ *in* (1.20). To make this idea more precise, assume the body is subject only to its weight and that the set C is an "infinite" goblet as suggested in Fig. 4.1.

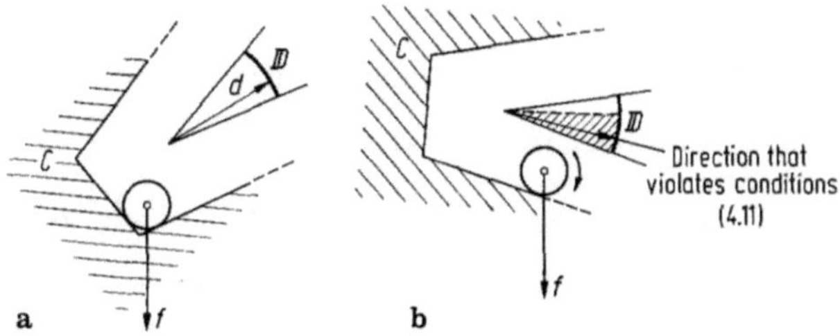

Fig. 4.1. (a) All possible "directions of escape"" $\boldsymbol{d}$ make an obtuse angle with the direction of the applied force $\boldsymbol{f}$. (b) Some "directions of escape" $\boldsymbol{d}$ make an acute angle with the direction of the applied force $\boldsymbol{f}$: There is no equilibrium position (unless some additional condition is imposed)

Then we see that if $\boldsymbol{d}$ is a non-zero vector such that, if $\boldsymbol{\phi}$ is *any* solution of the minimization problem, then $\boldsymbol{\phi}^\varepsilon = \boldsymbol{\phi} + \varepsilon\, \boldsymbol{d}$ is also *admissible* for $\varepsilon \in [0, \varepsilon_0]$ with $\varepsilon_0 = \varepsilon_0(\boldsymbol{d}) > 0$, the inequalities $I(\boldsymbol{\phi}^\varepsilon) \geqq I(\boldsymbol{\phi})$, $0 \leqq \varepsilon \leqq \varepsilon_0$, imply

(4.11) $$\int_\Omega \boldsymbol{f} \cdot \boldsymbol{d}\, dx + \int_\Gamma \boldsymbol{g} \cdot \boldsymbol{d}\, da \leqq 0.$$

In other words, equilibrium positions can occur only if inequalities (4.11) hold for all possible "directions of escape" $\boldsymbol{d}$ (*cf.* Fig. 4.1), which make up an appropriate set $\mathbb{D}$ of "dangerous directions" (*cf.* (4.13)).

Remark. In our proof of existence we shall need the stronger assumption that the inequalities (4.11) are *strict* for all vectors $\boldsymbol{d} \in \mathbb{D}$. □

Motivated essentially by the proof of the next existence theorem, but also by the previous example, we set the following *definition*: Let the set of admissible deformations be of the form (all data being as before):

(4.12) $$\boldsymbol{U} = \{\boldsymbol{\psi} \in \boldsymbol{W}^{1,p}(\Omega);\ \boldsymbol{adj}\, \nabla \boldsymbol{\psi} \in \boldsymbol{L}^q(\Omega),\ \det \nabla \boldsymbol{\psi} \in L^r(\Omega),$$
$$\det \nabla \boldsymbol{\psi} > 0 \ \text{ a.e. in } \Omega,\quad L(\nabla \boldsymbol{\psi}) \leqq 0 \ \text{ a.e. in } \Omega,\ \boldsymbol{\psi} \notin C \ \text{ on } \Gamma_2\}.$$

Then the associated *set of dangerous directions* is defined by

(4.13)
$$\mathbb{D} = \mathrm{cl}\left\{\boldsymbol{d} \in \mathbb{R}^3;\ |\boldsymbol{d}| = 1,\ \text{there exist } \boldsymbol{\phi}^k \in \boldsymbol{U} \text{ such that } \boldsymbol{d} = \lim \frac{\overline{\boldsymbol{\Phi}}^k}{|\overline{\boldsymbol{\Phi}}^k|},\right.$$
$$\left.\lim |\overline{\boldsymbol{\Phi}}^k| = +\infty\right\},$$

where cl denotes the closure, and $\overline{\boldsymbol{\psi}}$ denotes the average

$$(4.14)\qquad \overline{\boldsymbol{\psi}} = \frac{1}{\int_{\Gamma_2} da} \int_{\Gamma_2} \boldsymbol{\psi}\, da \quad \text{for any} \quad \boldsymbol{\psi} \in \boldsymbol{L}^1(\Gamma_2).$$

Since $\overline{\boldsymbol{\psi}}$ is a *vector* of $\mathbb{R}^3$, it can be identified with a constant *function* in the space $\boldsymbol{W}^{1,p}(\Omega)$, which will be denoted by the same symbol $\overline{\boldsymbol{\psi}}$. Thus the relations $\boldsymbol{d} = \lim \frac{\overline{\boldsymbol{\phi}}^k}{|\overline{\boldsymbol{\phi}}^k|}$ or $\lim |\overline{\boldsymbol{\phi}}^k| = +\infty$ in (4.13) may be also understood as convergences in $\boldsymbol{W}^{1,p}(\Omega)$. In the same vein, notice that there is a constant c_4 such that

$$(4.15)\qquad c_4 > 0 \quad\text{and}\quad c_4^{-1}\,|\boldsymbol{d}| \leqq \|\boldsymbol{d}\|_{1,p,\Omega} \leqq c_4\,|\boldsymbol{d}| \quad \text{for all} \quad \boldsymbol{d} \in \mathbb{R}^3.$$

In the light of the above definition, let us review some *examples*: Suppose first that *the set* $\mathbb{R}^3 - C$ *is bounded.* Then *the set of dangerous directions is empty,* since we cannot have $\lim \|\overline{\boldsymbol{\psi}}^k\|_{1,p,\Omega} = +\infty$ with $\boldsymbol{\psi}^k \in \boldsymbol{U}$: For if $\boldsymbol{\psi} \in \boldsymbol{U}$, then $\boldsymbol{\psi} \notin C$ on Γ_2 by definition and thus

$$|\overline{\boldsymbol{\psi}}| \leqq \sqrt{3}\,|\boldsymbol{\psi}|_{0,\infty,\Gamma_2} \leqq c_5 \quad \text{for all} \quad \boldsymbol{\psi} \in \boldsymbol{U}.$$

Suppose that $\mathbb{R}^3 - C$ is unbounded and that there is a closed convex cone $\mathbb{C}_0$ with vertex $\boldsymbol{a}_0$ such that

$$\mathbb{R}^3 - C \subset \mathbb{C}_0$$

and that $\mathbb{C}_0$ is not a half space. If $\boldsymbol{\psi} \in \boldsymbol{U}$, then $\boldsymbol{\psi} \notin C$ on Γ_2 by definition and so $\boldsymbol{\psi} \in \mathbb{C}_0$ on Γ_2 whence,

$$\boldsymbol{\psi} \in \boldsymbol{U} \Rightarrow \overline{\boldsymbol{\psi}} \in \mathbb{C}_0.$$

The relations $\boldsymbol{d} = \lim \frac{\overline{\boldsymbol{\phi}}^k}{|\overline{\boldsymbol{\phi}}^k|}$ with $|\boldsymbol{d}| = 1$ and $\lim |\overline{\boldsymbol{\phi}}^k| = +\infty$ with $\boldsymbol{\phi}^k \in \boldsymbol{U}$ force the vector $\boldsymbol{d}$ to belong to the cone $\{\mathbb{C}_0 - \boldsymbol{a}_0\}$, *i.e.*, the cone $\mathbb{C}_0$ translated by the vector $-\boldsymbol{a}_0$. Therefore we have in this case

$$\mathbb{D} \subset \partial B(0; 1) \cap \{\mathbb{C}_0 - \boldsymbol{a}_0\}.$$

Suppose that $\mathbb{R}^3 - C$ is unbounded and that there is a closed convex cone $\mathbb{C}_1$ with a non-empty interior such that

$$\mathbb{C}_1 \subset \mathbb{R}^3 - C.$$

Denoting by $\boldsymbol{a}_1$ the vertex of the cone $\mathbb{C}_1$, we have in this case

$$\partial B(0; 1) \cap \{\mathbb{C}_1 - \boldsymbol{a}_1\} \subset \mathbb{D}.$$

To see this, let $\boldsymbol{d} \in \operatorname{int} \mathbb{C}_1$ with $|\boldsymbol{d}| = 1$. Then the particular functions $\boldsymbol{\phi}^k(x) = x + k\,\boldsymbol{d}$ for $k \geqq 0$ large enough, can be used in definition (4.13).

Observe that *if there is a closed convex cone* $\mathbb{C}$ *with a non-empty interior, and which is not a half space, such that* (cf. Fig. 4.2)

$$\mathbb{C} = \mathbb{C}_0 - \boldsymbol{a}_0 = \mathbb{C}_1 - \boldsymbol{a}_1,$$

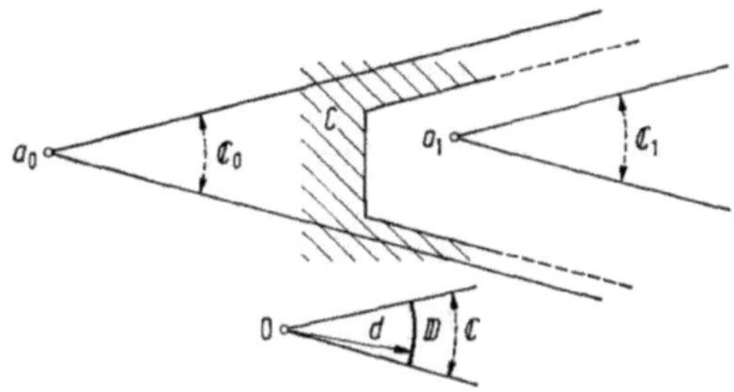

Fig. 4.2. A situation in which the set $\mathbb{D}$ of dangerous directions can be completely identified

this instance is one in which *the set* $\mathbb{D}$ *of dangerous directions can be completely identified*, with

$$\mathbb{D} = \partial B(0; 1) \cap \mathbb{C}.$$

This was the case for instance in the problem considered in Fig. 4.1.

Let us now turn to the theorem of existence.

Theorem 4.2. *Let there be given a stored energy function* $W = \mathbb{M}^3_+ \to \mathbb{R}$ *which satisfies assumptions* (1.13), (1.14), (1.15), *and let there be given a locking function* $L: \mathbb{M}^3_+ \to \mathbb{R}$ *which is polyconvex in the sense of* (3.11). *Let* Ω *be a bounded open connected subset of* $\mathbb{R}^3$ *with a Lipschitz continuous boundary* Γ, *and let* Γ_1 *and* Γ_2 *be disjoint (relatively) open subsets of* Γ, *with*

$$\text{(4.16)} \qquad \text{area } \Gamma_2 > 0 \text{ and area } \{\Gamma - (\Gamma_1 \cup \Gamma_2)\} = 0.$$

Let $\boldsymbol{f} \in L^\varrho(\Omega)$ *and* $\boldsymbol{g} \in L^\sigma(\Gamma_1)$ *be such that the linear forms of* (1.16) *are continuous. Let* C *be an open subset of* $\mathbb{R}^3$, *and assume that the set*

$$\text{(4.17)} \qquad U = \{\boldsymbol{\psi} \in \boldsymbol{W}^{1,p}(\Omega);\ \boldsymbol{adj}\, \nabla\boldsymbol{\psi} \in \boldsymbol{L}^q(\Omega),\ \det \nabla\boldsymbol{\psi} \in L^r(\Omega),$$
$$\det \nabla\boldsymbol{\psi} > 0 \text{ a.e. in } \Omega,\ L(\nabla\boldsymbol{\psi}) \leqq 0 \text{ a.e. in } \Omega,\ \boldsymbol{\psi} \notin C \text{ on } \Gamma_2\}$$

is not empty. Assume also that the applied force densities satisfy

$$\text{(4.18)} \qquad \int_\Omega \boldsymbol{f} \cdot \boldsymbol{d}\, dx + \int_{\Gamma_1} \boldsymbol{g} \cdot \boldsymbol{d}\, da < 0 \quad \text{for all} \quad \boldsymbol{d} \in \mathbb{D},$$

where the set of dangerous directions $\mathbb{D}$ *is defined as in* (4.13). *Consider finally the functional* I *of* (1.1) *and assume that* $\inf_{\boldsymbol{\psi}\in U} I(\boldsymbol{\psi}) < +\infty$.

Then the problem: Find $\boldsymbol{\phi}$ *such that*

$$\text{(4.19)} \qquad \boldsymbol{\phi} \in U \quad \text{and} \quad I(\boldsymbol{\phi}) = \inf_{\boldsymbol{\psi}\in U} I(\boldsymbol{\psi}),$$

has at least one solution.

Proof. The parts of the proof which are common to the proof of Theorem 4.1 are omitted. From the coerciveness inequality (1.15), we deduce that

$$\text{(4.20)} \qquad I(\boldsymbol{\psi}) \geqq c_6(|\boldsymbol{\psi}|^p_{1,p,\Omega} + |\boldsymbol{adj}\, \nabla\boldsymbol{\psi}|^q_{0,q,\Omega} + |\det \nabla\boldsymbol{\psi}|^r_{0,r,\Omega}) + c_7$$
$$- \left\{ \int_\Omega \boldsymbol{f} \cdot \boldsymbol{\psi}\, dx + \int_\Gamma \boldsymbol{g} \cdot \boldsymbol{\psi}\, da \right\},$$

with $c_6 > 0$, for all $\boldsymbol{\psi} \in \boldsymbol{W}^{1,p}(\Omega)$ with $\boldsymbol{adj}\, \nabla\boldsymbol{\psi} \in \boldsymbol{L}^q(\Omega)$, $\det \nabla\boldsymbol{\psi} \in L^r(\Omega)$. The difficulty here, due to the lack of the boundary condition $\boldsymbol{\psi} = \boldsymbol{\varphi}_0$ on Γ_0, is that *we cannot derive from* (4.20) *an inequality of the form* (4.6). We shall thus resort to another argument, by showing that there is a constant $\alpha > 0$ with the following property:

Let $(\boldsymbol{\phi}^k)$ *with* $\boldsymbol{\phi}^k \in \boldsymbol{U}$ *be such that*

$$\lim \|\boldsymbol{\phi}^k\|_{1,p,\Omega} = +\infty. \tag{4.21}$$

Then

$$\liminf \frac{I(\boldsymbol{\phi}^k)}{\|\boldsymbol{\phi}^k\|_{1,p,\Omega}} = \alpha > 0 \tag{4.22}$$

(the constant α does not depend on the particular sequence chosen). To prove this, assume it is not true. Then there would exist a sequence $(\boldsymbol{\phi}^k)$ with

(4.23)

$$\boldsymbol{\phi}^k \in \boldsymbol{U}, \quad \lim \|\boldsymbol{\phi}^k\|_{1,p,\Omega} = +\infty, \quad I(\boldsymbol{\phi}^k) = \alpha^k \|\boldsymbol{\phi}^k\|_{1,p,\Omega} \text{ with } \lim \alpha^k = \alpha \leqq 0.$$

From inequality (4.20) we deduce

$$\alpha^k \|\boldsymbol{\phi}^k\|_{1,p,\Omega} \geqq c_6 |\boldsymbol{\phi}^k|^p_{1,p,\Omega} + c_7 - \left\{\int_\Omega \boldsymbol{f}\cdot\boldsymbol{\phi}^k\,dx + \int_{\Gamma_1} \boldsymbol{g}\cdot\boldsymbol{\phi}^k\,da\right\},$$

and this inequality may be rewritten as

$$\alpha^k \geqq c_6 \|\boldsymbol{\phi}^k\|^{p-1}_{1,p,\Omega} |\tilde{\boldsymbol{\phi}}^k|^p_{1,p,\Omega} - \left\{\int_\Omega \boldsymbol{f}\cdot\tilde{\boldsymbol{\phi}}^k\,dx + \int_{\Gamma_1} \boldsymbol{g}\cdot\tilde{\boldsymbol{\phi}}^k\,da\right\}, \tag{4.24}$$

with

$$\tilde{\boldsymbol{\phi}}^k = \frac{\boldsymbol{\phi}^k}{\|\boldsymbol{\phi}^k\|_{1,p,\Omega}}. \tag{4.25}$$

From the relations $c_6 > 0$, $\lim \alpha^k = 0$, $\lim \|\boldsymbol{\phi}^k\|_{1,p,\Omega} = +\infty$, $p > 1$, and $\|\tilde{\boldsymbol{\phi}}^k\|_{1,p,\Omega} \leqq 1$, we deduce that

$$\lim |\tilde{\boldsymbol{\phi}}^k|_{1,p,\Omega} = 0. \tag{4.26}$$

Arguing as in (4.5), we also have

$$\|\boldsymbol{\psi}\|^p_{1,p,\Omega} \leqq c_7 \left\{ |\boldsymbol{\psi}|^p_{1,p,\Omega} + \left|\int_{\Gamma_2} \boldsymbol{\psi}\,da\right|^p \right\} \text{ for all } \boldsymbol{\psi} \in \boldsymbol{W}^{1,p}(\Omega). \tag{4.27}$$

Applying this inequality to the particular functions $\|\boldsymbol{\phi}^k\|^{-1}_{1,p,\Omega}(\boldsymbol{\phi}^k - \bar{\boldsymbol{\phi}}^k)$ with $\bar{\boldsymbol{\phi}}^k = \frac{1}{\int_{\Gamma_2} da} \int_{\Gamma_2} \boldsymbol{\phi}^k\,da$ (*cf.* (4.14)), we obtain

$$\frac{\|\boldsymbol{\phi}^k - \bar{\boldsymbol{\phi}}^k\|_{1,p,\Omega}}{\|\boldsymbol{\phi}^k\|_{1,p,\Omega}} \leqq c_8^{1/p} |\tilde{\boldsymbol{\phi}}^k|_{1,p,\Omega}, \tag{4.28}$$

and thus by (4.26),

$$\lim \frac{\|\boldsymbol{\phi}^k - \bar{\boldsymbol{\phi}}^k\|_{1,p,\Omega}}{\|\boldsymbol{\phi}^k\|_{1,p,\Omega}} = 0. \tag{4.29}$$

Since

$$(4.30)\qquad \left|\frac{\|\overline{\boldsymbol{\phi}}^k\|_{1,p,\Omega}}{\|\boldsymbol{\phi}^k\|_{1,p,\Omega}}-1\right|\leq\frac{\|\boldsymbol{\phi}^k-\overline{\boldsymbol{\phi}}^k\|_{1,p,\Omega}}{\|\boldsymbol{\phi}^k\|_{1,p,\Omega}},$$

we conclude from (4.23), (4.28), and (4.29), that

$$(4.31)\qquad \lim\|\overline{\boldsymbol{\phi}}^k\|_{1,p,\Omega}=+\infty.$$

With inequalities (4.15), we conclude that the sequence $(\boldsymbol{\phi}^k)$ is a "candidate" for defining an element of the set $\mathbb{D}$ of dangerous directions, as definined in (4.13). Since the constant functions

$$\boldsymbol{\delta}^k=\frac{\overline{\boldsymbol{\phi}}^k}{\|\boldsymbol{\phi}^k\|_{1,p,\Omega}}$$

are bounded in $W^{1,p}(\Omega)$ independently of k (by (4.29) and (4.30)), there is a subsequence (indexed by the same integers k, for convenience) such that

$$\lim\boldsymbol{\delta}^k=\lambda\,\boldsymbol{d}\ \text{in both}\ \mathbb{R}^3\ \text{and}\ W^{1,p}(\Omega)\ \text{with}\ \boldsymbol{d}\in\mathbb{D}\ \text{and}\ \lambda>0,$$

and thus we also have

$$\lim\tilde{\boldsymbol{\phi}}^k=\lambda\,\boldsymbol{d}\ \text{in}\ W^{1,p}(\Omega),$$

by (4.26) and (4.28). From inequality (4.24) we deduce

$$\lambda\left\{\int_\Omega\boldsymbol{f}\cdot\boldsymbol{d}\,dx+\int_{\Gamma_1}\boldsymbol{g}\cdot\boldsymbol{d}\,da\right\}=\lim\left\{\int_\Omega\boldsymbol{f}\cdot\boldsymbol{\phi}^k\,dx+\int_{\Gamma_1}\boldsymbol{g}\cdot\boldsymbol{\phi}^k\,da\right\}\geqq 0.$$

But this inequality contradicts assumption (4.18), and thus relation (4.21) is proved.

It is now easy to conclude: Let $(\boldsymbol{\phi}^k)$ with $\boldsymbol{\phi}^k\in\boldsymbol{U}$ be an infimizing sequence for the functional I. The above result shows that the sequence $(\boldsymbol{\phi}^k)$ is bounded in the space $W^{1,p}(\Omega)$. From inequality (4.20), it then follows that the sequence $((\boldsymbol{adj}\,\nabla\boldsymbol{\phi}^k,\det\nabla\boldsymbol{\phi}^k))$ is bounded in the space $L^q(\Omega)\times L^r(\Omega)$. The proof then proceeds along the same lines as the proof of Theorem 4.1. □

Remark. We could have equally well defined the set of dangerous directions as in (4.13) but with the additional condition that the sequence $\boldsymbol{\phi}^k\in\boldsymbol{U}$ satisfy

$$\lim\frac{\|\boldsymbol{\phi}^k-\overline{\boldsymbol{\phi}}^k\|_{1,p,\Omega}}{\|\overline{\boldsymbol{\phi}}^k\|_{1,p,\Omega}}=0,$$

i.e., a relation that we *proved* in Theorem 4.2 to hold automatically for any sequence satisfying (4.21). This new set of dangerous directions, which is smaller in general, nevertheless coincides with that of (4.13) in the situations considered in Fig. 4.1 or 4.2. □

Inequalities (4.11) and (4.18) thus appear as the natural extension to nonlinear elasticity of analogous inequalities, whose relation to existence theory for what is called the *Signorini problem* in *linearized elasticity* has been established by DUVAUT & LIONS [1972, p. 149] and FICHERA [1972, p. 413].

Acknowledgement. The authors are indebted to STUART S. ANTMAN for several useful suggestions.

References

BALL, J. M. [1977]: Convexity conditions and existence theorems in nonlinear elasticity, Arch. Rational Mech. Anal. **63**, 337–403.

CIARLET, P. G. [1983]: *Lectures on Three-Dimensional Elasticity*, Tata Institute Lectures on Mathematics, Springer-Verlag, Berlin.

CIARLET, P. G. [1985]: *Topics in Mathematical Elasticity*, Vol. 1, North-Holland, Amsterdam.

CIARLET, P. G. & GEYMONAT, G. [1982]: Sur les lois de comportement en élasticité non-linéaire compressible, C. R. Acad. Sci. Paris Sér. A **295**, 423–426.

CIARLET, P. G. & NEČAS, J. [1984]: Problèmes unilatéraux en élasticité non linéaire tridimensionnelle, C. R. Acad. Sci. Paris, Sér. A 298, 189–192.

DUVAUT, G. & LIONS, J. L. [1972]: *Les Inéquations en Mécanique et en Physique*, Dunod, Paris.

FICHERA, G. [1972]: Boundary value problems of elasticity with unilateral constraints, *Handbuch der Physik*, Vol. VIa/2, 391–424, Springer-Verlag, Berlin, Heidelberg, New York.

GERMAIN, P. [1972]: *Mécanique des Milieux Continus*, Tome 1, Masson, Paris.

GUO ZHONG-HENG [1980]: The unified theory of variational principles in nonlinear elasticity, Arch. Mech. **32**, 577–596.

GURTIN, M. E. [1981]: *Introduction to Continuum Mechanics*, Academic Press, New York.

HLAVÁČEK, I.; HASLINGER, J.; NEČAS, J. & LOVÍŠEK, J. [1982]: *Solution of Variational Inequalities in Mechanics*, Alfa, Bratislava.

MARSDEN, J. E. & HUGHES, T. J. R. [1983]: *Mathematical Foundations of Elasticity*, Prentice-Hall, Englewood Cliffs.

NEČAS, J. [1967]: *Les Méthodes Directes en Théorie des Equations Elliptiques*, Masson, Paris.

NEČAS, J. [1983]: *Introduction to the Theory of Nonlinear Equations*, Teubner Texte zur Mathematik, Band 52, Leipzig.

NEČAS, J. & HLAVÁČEK, I. [1981]: *Mathematical Theory of Elastic and Elasto-Plastic Bodies: An Introduction*, Elsevier, Amsterdam.

OGDEN, R. W. [1972]: Large deformation isotropic elasticity: On the correlation of theory and experiment for compressible rubberlike solids, Proc. Roy. Soc. London A **328**, 567–583.

PRAGER, W. [1957]: On ideal locking materials, Transaction of the Society of Rheology **1**, 169–175.

TRUESDELL, C. & NOLL, W. [1965]: The non-linear field theories of mechanics, *Handbuch der Physik*, Vol. III/3, 1–602.

WANG, C.-C. & TRUESDELL, C. [1973]: *Introduction to Rational Elasticity*, Noordhoff, Groningen.

Université Pierre et Marie Curie
Paris

Charles University
Prague

(Received May 16, 1984)

Injectivity and Self-Contact in Nonlinear Elasticity

PHILIPPE G. CIARLET & JINDŘICH NEČAS

Communicated by S. ANTMAN

Abstract

Let Ω be a bounded open connected subset of $\mathbb{R}^3$ with a sufficiently smooth boundary. The additional condition $\int_\Omega \det \nabla\boldsymbol{\psi}\, dx \leqq \operatorname{vol} \boldsymbol{\psi}(\Omega)$ is imposed on the admissible deformations $\boldsymbol{\psi} : \bar{\Omega} \to \mathbb{R}^3$ of a hyperelastic body whose reference configuration is $\bar{\Omega}$. We show that the associated minimization problem provides a mathematical model for matter to come into frictionless contact with itself but not interpenetrate. We also extend J. BALL's theorems on existence to this case by establishing the existence of a minimizer of the energy in the space $W^{1,p}(\Omega;\mathbb{R}^3)$, $p > 3$, that is injective almost everywhere.

1. Introduction

Theorems on injectivity in nonlinear three-dimensional elasticity seem to have been confined so far to the *pure displacement problem*, for which the deformation is specified *a priori* on the entire boundary of the reference configuration. For this problem, global injectivity has been established under the assumption that the applied body force density is small enough in the space $L^p(\Omega, \mathbb{R}^3)$, $p > 3$, by VALENT [1979, 1982] and by CIARLET & DESTUYNDER [1979, Th. 2.2] within the existence theory based on the implicit function theorem. Global injectivity has also been established by BALL [1981 a] for hyperelastic bodies whose associated energy "grows sufficiently fast for extreme strains", within the existence theory based on minimization of the energy as developed by BALL [1977]. In these works, injectivity is obtained as a consequence of the following fact: If a smooth enough mapping $\boldsymbol{\varphi} : \bar{\Omega} \to \mathbb{R}^3$, where Ω is a bounded, open, connected subset of $\mathbb{R}^3$, is orientation-preserving ($\det \nabla\boldsymbol{\varphi} > 0$ in $\bar{\Omega}$) and is equal to an injective mapping $\boldsymbol{\varphi}_0 : \bar{\Omega} \to \mathbb{R}^3$ on the boundary of the set Ω, then it is also injective on $\bar{\Omega}$ (a more precise statement is given in Theorem 3 of this paper; BALL [1981 a, Theorem 1]

© Springer-Verlag 1987. Reprinted with kind permission from: Ciarlet, P.G. & Nečas, J., Injectivity and Self-Contact in Nonlinear Elasticity, Archive of Rational Mechanics and Analysis 97, 171-188.

also gives a subtle extension of this result to mapping φ in the Sobolev space $W^{1,p}(\Omega; \mathbb{R}^3)$, $p > 3$).

We propose here a different approach, which rests upon the additional constraint

$$(1.1) \qquad \int_\Omega \det \nabla\varphi \, dx \leq \operatorname{vol} \varphi(\Omega)$$

on the minimizers $\varphi : \bar{\Omega} \to \mathbb{R}^3$ of the energy, and more generally on all admissible deformations (*cf.* (2.2)). Our idea (*cf.* Sect. 2) is to combine (1.1) with the following classical formula in integration theory:

$$(1.2) \qquad \int_{\varphi(\Omega)} \operatorname{card} \varphi^{-1}(x') \, dx' = \int_\Omega \det \nabla\varphi \, dx$$

(see, *e.g.*, RADO & REICHELDERFER [1955], SCHWARTZ [1967, p. 676], SMITH [1983, Chap. 16]). In Section 3 we combine (1.1) with its extension to mappings $\varphi : \bar{\Omega} \to \mathbb{R}^3$ with values in the Sobolev spaces $W^{1,p}(\Omega; \mathbb{R}^3)$, $p > 3$ (*cf.* MARCUS & MIZEL [1973] and VODOP'YANOV, GOL'DSHTEIN & RESHETNYAK [1979]). In this fashion, we are able to show also that minimizers of the energy are injective almost everywhere when the displacement is prescribed only on a portion of the boundary of the set Ω while the traction is prescribed on the rest. Moreover, for these mixed problems we even allow *unilateral boundary conditions* on the remaining portion of the boundary, in a form that generalizes those introduced in CIARLET & NEČAS [1985a].

More specifically, we first show that when the constraint (1.1) is imposed on the admissible deformations, the associated minimization problem is a *mathematical model of frictionless self-contact without interpenetration of matter* (Theorem 4). Secondly we show that the constraint (1.1) "holds in the weak limit" when infimizing sequences of the energy are considered, so that the minimizers are injective almost everywhere (Theorem 5). This existence theorem generalizes those of BALL [1977], BALL, CURRIE & OLVER [1981], and CIARLET & NEČAS [1985a].

That formula (1.2) can be used to prove the injectivity almost everywhere of the minimizers of the energy in nonlinear three-dimensional elasticity had already been observed by BALL [1981a], who used it in conjunction with a delicate *a priori* estimate on the minimizer (*cf.* BALL [*ibid*, (1.3)]), which is a consequence of suitable assumptions on growth of the stored-energy function. As noted earlier, however, this approach was restricted to pure displacement problems.

The conclusions of this paper were announced in CIARLET & NEČAS [1985b].

Finally, let us specify the various notations that we shall use, concerning notably sets, vectors, matrices, and function space. (In what follows, Latin indices have values in the set $\{1, 2, 3\}$, and the repeated index convention is used):

(α^k): an infinite sequence $(\alpha^0, \alpha^1, \ldots, \alpha^k, \ldots)$,
card E: number of elements of a set E,
$\varphi(A) = \{x' \in F; x' = \varphi(x)$ for at least one $x \in A\}$: direct image of $A \subset E$ by a mapping $\varphi : E \to F$,
$\varphi^{-1}(B) = \{x \in E; \varphi(x) \in B\}$: inverse image of $B \subset F$ by a mapping $\varphi : E \to F$,
$A - B = \{x \in A; x \notin B\}$: complement of $B \subset E$ relative to $A \subset E$,

$\text{int}\, A$: interior of the set A,
$\bar{A}$: closure of the set A,
∂A: boundary of the set A,
$\text{ext}\, A$: exterior of the set A ($\text{ext}\, A = \text{int}\,(E - A)$ if $A \subset E$),
$B_r(x) = \{y \in X;\, |y - x| < r\}$: open ball in a vector space X normed by $|\cdot|$,
$\boldsymbol{u} \cdot \boldsymbol{v} = u_i v_i$: Euclidean inner product of the vectors $\boldsymbol{u} = (u_i)$ and $\boldsymbol{v} = (v_i)$,
$|\boldsymbol{u}| = \sqrt{\boldsymbol{u} \cdot \boldsymbol{u}}$: Euclidean norm of the vector $\boldsymbol{u}$,
$\boldsymbol{A} : \boldsymbol{B} = A_{ij} B_{ij}$: matrix inner product of the matrices $\boldsymbol{A} = (A_{ij})$ and $\boldsymbol{B} = (B_{ij})$,
$\|\boldsymbol{A}\| = \sqrt{\boldsymbol{A} : \boldsymbol{A}}$: associated norm of the matrix $\boldsymbol{A}$,
$\det \boldsymbol{A}$: determinant of the matrix $\boldsymbol{A}$,
$\mathbf{Cof}\, \boldsymbol{A}$: cofactor matrix of the matrix A ($\text{Cof}\, A = (\det A)\, A^{-T}$ if A is regular),
$\mathbb{M}^3$: set of all matrices of order 3,
$\mathbb{M}^3_+ = \{\boldsymbol{F} \in \mathbb{M}^3;\, \det \boldsymbol{F} > 0\}$,
$\nabla \boldsymbol{\varphi} = \left(\frac{\partial \varphi_i}{\partial x_j}\right) \in \mathbb{M}^3$: gradient of a mapping $\boldsymbol{\varphi}: A \subset \mathbb{R}^3 \to \mathbb{R}^3$,
$\mathbf{div}\, \boldsymbol{T} = (\partial_j T_{ij}) \in \mathbb{R}^3$: divergence of a tensor field $\boldsymbol{T}: A \subset \mathbb{R}^3 \to \mathbb{R}^3$,
$\frac{\partial W}{\partial \boldsymbol{F}}(\boldsymbol{F}) = \left(\frac{\partial W}{\partial F_{ij}}(\boldsymbol{F})\right) \in \mathbb{M}^3$: gradient of a mapping $W: A \subset \mathbb{M}^3 \to \mathbb{R}$,
dx: volume element in $\mathbb{R}^3$,
$\text{vol}\, A = \int_A dx$: volume of a dx-measurable set $A \subset \mathbb{R}^3$,
da: area element in $\mathbb{R}^3$,
$\text{area}\, \Delta = \int_\Delta da$: area of a da-measurable set $\Delta \subset \mathbb{R}^3$,
$\mathscr{C}^m(A)$: space of functions $\varphi: A \to \mathbb{R}$ which are m times continuously differentiable on $A \subset \mathbb{R}^3$,
$\mathscr{C}^m(A; \mathbb{R}^n) = \{\boldsymbol{\varphi} = (\varphi_i): A \to \mathbb{R}^n;\, \varphi_i \in \mathscr{C}^m(A)$ for all $i = 1, \ldots, n\}$,
$\mathscr{D}(\Omega)$: space of functions $\theta \in \mathscr{C}^\infty(\Omega)$ with compact support in Ω (Ω open in $\mathbb{R}^n$),
$\mathscr{D}(\Omega; \mathbb{R}^n) = \{\boldsymbol{\theta} = (\theta_i): \Omega \to \mathbb{R}^n;\, \theta_i \in \mathscr{D}(\Omega)$ for all $i = 1, \ldots, n\}$,
$L^p(\Omega; \mathbb{R}^n) = \{v = (v_i): \Omega \to \mathbb{R}^n;\, v_i \in L^p(\Omega)$ for all $i = 1, \ldots, n\}$,
$|\cdot|_{0,p,\Omega}$: norm of the spaces $L^p(\Omega)$ and $L^p(\Omega; \mathbb{R}^n)$,
$W^{1,p}(\Omega; \mathbb{R}^n) = \{\boldsymbol{v} = (v_i): \Omega \to \mathbb{R}^n;\, v_i \in W^{1,p}(\Omega)$ for all $i = 1, \ldots, n\}$,
$\|\cdot\|_{1,p,\Omega}$: norm of the spaces $W^{1,p}(\Omega)$ and $W^{1,p}(\Omega; \mathbb{R}^n)$.

2. The Mathematical Model

Let Ω be a bounded, open, connected subset of $\mathbb{R}^3$, whose boundary $\partial\Omega = \Gamma$ is Lipschitz continuous in the sense of Nečas [1967], and let Γ_0 and Γ_1 be disjoint relatively open subsets of Γ such that $\text{area}\,\{\Gamma - (\Gamma_0 \cup \Gamma_1)\} = 0$. Either of the sets Γ_0, Γ_1 may be empty.

The set $\bar{\Omega}$ is the *reference configuration* of a *homogeneous hyperelastic body* that is subjected to an applied body force with density $\boldsymbol{f}: \Omega \to \mathbb{R}^3$ per unit mass, (We assume for simplicity that the field $\boldsymbol{f}$ is independent of the deformation, *i.e.*, that the body force is a dead load). If in a given situation, a particle occupying the point $x \in \bar{\Omega}$ in the reference configuration moves to a point represented by a vector $\varphi(x)$, the corresponding mapping $\varphi : \bar{\Omega} \to \mathbb{R}^3$, which is called a *deformation*, is the unknown of our problem. Let $W: \mathbb{M}_+^3 \to \mathbb{R}$ denote the stored-energy function of the material constituting the body. Then classically the *mixed displacement-traction problem* seeks to find particular solution φ as minimizers of the *total energy*

$$I(\psi) = \int_\Omega W(\nabla\psi)\, dx - \int_\Omega f \cdot \psi\, dx, \tag{2.1}$$

when the trial functions $\psi : \bar{\Omega} \to \mathbb{R}^3$ vary in the set of all smooth enough mappings $\psi : \bar{\Omega} \to \mathbb{R}^3$ that *preserve orientation, namely* $\det \nabla\psi > 0$ in $\bar{\Omega}$, and satisfy the *boundary condition* of place $\psi = \varphi_0$ on Γ_0, where $\varphi_0 : \Gamma_0 \to \mathbb{R}^3$ is a prescribed function. For details, see *e.g.* CIARLET [1985, 1987], GERMAIN [1972], GURTIN [1981], HANYGA [1985], MARSDEN & HUGHES [1983], TRUESDELL & NOLL [1965], WANG & TRUESDELL [1973].

Remark. We could handle also the case where a nonzero surface force is applied on a subset of Γ_1, but since such an extension merely produces additional technicalities, we ignore it. □

Our purpose is to add *two new conditions* to be satisfied by the trial functions ψ; one is a *confinement condition* in the form of the inclusion $\psi(\bar{\Omega}) \subset B$ (*cf.* NOLL [1978]), which extends the generalized unilateral constraints "$\psi \in B$ on a prescribed portion of Γ" introduced by CIARLET & NEČAS [1985a]. The other one takes the form of the inequality

$$\int_\Omega \det \nabla\psi\, dx \leqq \operatorname{vol} \psi(\Omega), \tag{2.2}$$

whose effect is to ensure the *injectivity almost everywhere of the minimizer over the set* $\bar{\Omega}$ (*cf.* Theorems 1, 4, 5). More specifically, let B be an arbitrary *closed subset of* $\mathbb{R}^3$ (possibly unbounded). To avoid unnecessary technicalities, we assume that $\varphi_0(\Gamma_0)$ is contained in ∂B. We then define the *set of admissible deformations*:

$$\Phi = \Big\{\psi : \bar{\Omega} \to \mathbb{R}^3;\ \det \nabla\psi > 0 \text{ in } \bar{\Omega},\ \psi = \varphi_0 \text{ on } \Gamma_0,$$
$$\psi(\bar{\Omega}) \subset B,\ \int_\Omega \det \nabla\psi\, dx \leqq \operatorname{vol} \psi(\Omega)\Big\}, \tag{2.3}$$

the smoothness of the admissible deformations ψ being purposely left unspecified at this stage. Notice that the closure of the set $\mathbb{R}^3 - B$ acts as an *obstacle*.

We begin by establishing useful properties, notably the injectivity on the open set Ω, of sufficiently smooth admissible deformations satisfying inequality (2.2). The definition of a boundary of class $\mathscr{C}^1$ used here is that of NEČAS [1967] if the boundary is bounded. If the boundary is unbounded (as may be the case for the set ∂B), the definition is simply extended to allow an infinite number of local coordinate systems.

Theorem 1. *Let* $\varphi \in \mathscr{C}^1(\overline{\Omega}; \mathbb{R}^3)$ *be such that*

$$\det \nabla\varphi > 0 \text{ in } \overline{\Omega}, \tag{2.4}$$

$$\int_\Omega \det \nabla\varphi \, dx \leq \operatorname{vol} \varphi(\Omega). \tag{2.5}$$

Then the mapping φ *is injective on the set* Ω.

Proof. By the classical result (1.2) from the theory of integration,

$$\int_{\varphi(\Omega)} \operatorname{card} \varphi^{-1}(x') \, dx' = \int_\Omega \det \nabla\varphi(x) \, dx \tag{2.6}$$

whenever one of the two integrals exists (as is the case here since $\det \nabla\varphi \in \mathscr{C}^0(\overline{\Omega})$ by assumption). Let x and y be two distinct points in the set Ω such that $\varphi(x) = \varphi(y)$. Since the matrices $\nabla\varphi(x)$ and $\nabla\varphi(y)$ are invertible, the implicit function theorem provides open sets U, V, W' such that

$$x \in U \subset \Omega, y \in V \subset \Omega, \quad U \cap V = \emptyset, \quad \varphi(x) = \varphi(y) \in W' \subset \varphi(\Omega),$$

$$\varphi : U \to W' \text{ and } \varphi : V \to W' \text{ are } \mathscr{C}^1\text{-diffeomorphisms}.$$

Hence

$$\operatorname{card} \varphi^{-1}(x') \geq 2 \text{ for all } x' \in W'.$$

Since $\operatorname{vol} W' > 0$, it follows from (2.6) that

$$\operatorname{vol} \varphi(\Omega) = \int_{\varphi(\Omega)} dx' < \int_{\varphi(\Omega)} \operatorname{card} \varphi^{-1}(x') \, dx' = \int_\Omega \det \nabla\varphi(x) \, dx,$$

but this contradicts (2.5). Hence the mapping φ is injective in Ω. □

Theorem 2. *Let* B *be a closed subset of* $\mathbb{R}^3$. *We assume that the boundaries,* ∂B *and* $\partial\Omega$ *are of class* $\mathscr{C}^1$. *Let* $\varphi \in \mathscr{C}^1(\overline{\Omega}; \mathbb{R}^3)$ *be a mapping that satisfies* (2.4), (2.5), *and*

$$\varphi(\overline{\Omega}) \subset B. \tag{2.7}$$

Then

$$\varphi(\Omega) \subset \operatorname{int} B. \tag{2.8}$$

If

$$x' \in \partial B \cap \varphi(\partial\Omega), \quad \text{then} \quad \operatorname{card} \varphi^{-1}(x') = 1, \tag{2.9}$$

and the surfaces ∂B *and* $\varphi(\partial\Omega)$ *have the same tangent plane at the point* x'. *If*

$$x' \in \varphi(\partial\Omega) \cap \operatorname{int} B, \text{ then } \operatorname{card} \varphi^{-1}(x') = 1 \text{ or } 2. \tag{2.10}$$

If $\varphi^{-1}(x') = \{x, y\}$, *then* $x \in \partial\Omega$, $y \in \partial\Omega$, *and the outer normal vectors to the surface* $\varphi(\partial\Omega)$ *at the points* $\varphi(x) = \varphi(y)$ *have opposite directions. (Consequently the tangent planes coincide.)*

Proof. Given any point $x \in \Omega$, the implicit function theorem provides an open set $U \subset \Omega$ containing x such that $\varphi: U \to \varphi(U)$ is a $\mathscr{C}^1$-diffeomorphism. The set $\varphi(U)$, which is open (by the theorem on invariance of domains), contains $\varphi(x)$, and is contained in the set B by assumption (2.7). Hence $\varphi(x) \in \text{int } B$, which proves (2.8).

Next let $x' \in \partial B \cap \varphi(\partial\Omega)$, and let $x \in \partial\Omega$ be such that $\varphi(x) = x'$. We must show that $\varphi^{-1}(x') = \{x\}$. Since the boundary $\partial\Omega$ is of class $\mathscr{C}^1$ and since the mapping φ is in the space $\mathscr{C}^1(\bar{\Omega}; \mathbb{R}^3)$, we can find an open set $A \subset \mathbb{R}^3$ containing x and an extension (still denoted by) $\varphi \in \mathscr{C}^1(\bar{\Omega} \cup A; \mathbb{R}^3)$ such that $\det \nabla\varphi > 0$ in $\bar{\Omega} \cup A$. This extension can be constructed using the extension techniques for smooth boundaries described by NEČAS [1967]. Then by the implicit function theorem, there are open sets $U \subset A$ and U' containing x and x' respectively, such that $\varphi: U \to U'$ is a $\mathscr{C}^1$-diffeomorphism.

If the surfaces ∂B and $\varphi(\partial\Omega)$ had different tangent planes at the point x', we should have $\varphi(\Omega) \cap \partial B \neq \emptyset$, in contradiction with (2.8). (To see this, let $\boldsymbol{n}'$ denote the unit outer normal vector to the surface ∂B at the point x', and consider a plane containing the vector $\boldsymbol{n}'$ and intersecting the two tangent planes along two distict lines). Hence the tangent planes coincide and the unit outer normal vectors to the surfaces ∂B and $\varphi(\partial\Omega)$ necessarily have the same direction by (2.7).

Assume that the set $\varphi^{-1}(x')$ contains another point $y \neq x$, which also belongs to $\partial\Omega$ by (2.8). Using the implicit function theorem, we can find open sets U_x, U_y, U' respectively containing x, y, x' such that $U_x \cap U_y = \emptyset$ and such that $\varphi: U_x \cap \Omega \to U' \cap \varphi(\Omega)$ and $\varphi: U_y \cap \Omega \to U' \cap \varphi(\Omega)$ are both $\mathscr{C}^1$-diffeomorphisms. This would imply that the mapping φ is not one-to-one on Ω, in contradiction with Theorem 1. Hence the set $\varphi^{-1}(x')$ contains exactly one point.

Let $x' \in \varphi(\partial\Omega) \cap \text{int } B$ be such $\text{card } \varphi^{-1}(x') \geqq 2$. Then there are an $x \in \partial\Omega$ and a $y \in \bar{\Omega}$ such that $x \neq y$ and $x' = \varphi(x) = \varphi(y)$. If $y \in \Omega$, the above argument shows that the mapping φ is not one-to-one on Ω, in contradiction with Theorem 1; we thus conclude that $y \in \partial\Omega$.

Next, there are open sets V and U' respectively containing y and x' such that $\varphi: V \cap \bar{\Omega} \to U' \cap \varphi(\bar{\Omega})$ is a $\mathscr{C}^1$-diffeomorphism. It then suffices to reproduce the above arguments, the set $U' \cap \varphi(\Omega)$ playing the rôle of the set ext $\boldsymbol{B}$: This proves that x is the only point other than y in the inverse image $\varphi^{-1}(x')$ and that the tangent planes to the surface $\varphi(\partial\Omega)$ at the points $\varphi(x)$ and $\varphi(y)$ coincide and have opposite outer normal vectors. □

Remarks. (1) Theorem 1 still holds under the weaker assumption that $\det \nabla\varphi > 0$ in Ω. The stronger assumption that $\det \nabla\varphi > 0$ in $\bar{\Omega}$ is needed in Theorem 2, however, in order to ensure that the extension of φ over the open set denoted A in the proof still satisfies $\det \nabla\varphi > 0$ on A.

(2) The assumptions of regularity in Theorem 2 are essential. In particular, it is easy to construct counterexamples to relations (2.9) and (2.10) corresponding to less smooth data $\partial\Omega$, ∂B and φ. □

Finally, we also record the following theorem, which will be needed to show that the "variations" considered in the proof of Theorem 4 are injective; its proof relies on properties of the topological degree (*cf.* CIARLET [1987, Chapter 5] or BALL [1981a, Theorem 1], where it is extended to mappings $\psi \in W^{1,p}(U;\mathbb{R}^n)$, $p > n$; see also the closely related result of MEISTERS & OLECH [1963]).

Theorem 3. *Let U be a bounded, open, connected subset of $\mathbb{R}^n$ such that $\operatorname{int}\bar{U} = U$, let $\varphi \in \mathscr{C}^0(\bar{U}, \mathbb{R}^n)$ be an injective mapping, and let $\psi \in \mathscr{C}^0(\bar{U};\mathbb{R}^n) \cup \mathscr{C}^1(U;\mathbb{R}^n)$ be a mapping that satisfies*

$$\det \nabla\psi(x) > 0 \quad \textit{for all } x \in U,$$

$$\psi(x) = \varphi(x) \quad \textit{for all } x \in \partial U.$$

Then $\psi : \bar{U} \to \psi(\bar{U})$ is a homeomorphism (in particular, the mapping ψ is injective on $\bar{U}$), the mapping $\psi : U \to \psi(U)$ is a $\mathscr{C}^1$-diffeomorphism, and finally,

$$\psi(U) = \varphi(U), \psi(\bar{U}) = \varphi(\bar{U}). \quad \square$$

We are now in a position to give a formal interpretation of a minimizer of the total energy (2.1) over the set of admissible deformations (2.3).

Theorem 4. *Let the boundaries ∂B and $\partial\Omega = \Gamma$ be smooth enough and let φ be a smooth enough solution of the minimization problem:*

$$\varphi \in \boldsymbol{\Phi}, \quad I(\varphi) = \inf_{\psi \in \boldsymbol{\Phi}} I(\psi), \tag{2.11}$$

where the functional I and the set $\boldsymbol{\Phi}$ are defined as in (2.1) and (2.3). Then the minimizer φ is injective on Ω and is a solution of the following boundary-value problem:

$$-\mathbf{div}\, \boldsymbol{T}(\nabla\varphi) = \boldsymbol{f} \quad \textit{in } \Omega, \quad \textit{with} \quad \boldsymbol{T}(\boldsymbol{F}) = \frac{\partial W}{\partial \boldsymbol{F}}(\boldsymbol{F}) \quad \textit{for all } \boldsymbol{F} \in \mathbb{M}^3_+, \tag{2.12}$$

$$\varphi = \varphi_0 \quad \textit{on } \Gamma_0, \tag{2.13}$$

$$\boldsymbol{T}(\nabla\varphi(x))\, \boldsymbol{n}(x) = \lambda(x)\, \boldsymbol{n}'(x) \quad \textit{with } \lambda(x) \leq 0, \textit{ for } x \in \Gamma_1; \tag{2.14}$$

the last equation corresponds to one of the following situations (cf. Figure 1):

$$\varphi(x) \in \operatorname{int} B \quad \textit{and} \quad \varphi^{-1}(\varphi(x)) = \{x\} \textit{ whence } \lambda(x) = 0 \ (\textit{case } (\mathrm{a})), \tag{2.15}$$

$$\varphi(x) \in \partial B \textit{ whence } \varphi^{-1}(\varphi(x)) = \{x\} \ (\textit{case } (\mathrm{b})), \tag{2.16}$$

$$\varphi(x) \in \operatorname{int} B \quad \textit{and} \quad \varphi^{-1}(\varphi(x)) = \{x, y\} \quad \textit{with } y \in \Gamma_1 \textit{ whence} \\ \boldsymbol{n}'(x) + \boldsymbol{n}'(y) = 0 \quad \textit{and} \quad \lambda(x)\, da(x) = \lambda(y)\, da(y) \ (\textit{case } (\mathrm{c})), \tag{2.17}$$

where $\boldsymbol{n}(z)$ and $\boldsymbol{n}'(z)$ denote the unit outer normal vectors along the surfaces Γ and $\varphi(\Gamma)$ at the points z and $\varphi(z)$, respectively, and $da(z)$ denotes the differential area along Γ at the point z.

Proof. The proof consists in extending the argument given by CIARLET & NEČAS [1985a] by showing that *adequate "variations" $\varphi^\varepsilon = \varphi + \varepsilon\boldsymbol{\theta}$ around a minimizer φ of the energy can be constructed in such a way as to satisfy not only* *the conditions* $\det \nabla \varphi^\varepsilon > 0$ *in* $\bar{\Omega}$, $\varphi^\varepsilon = \varphi_0$ *on* Γ_0, $\varphi^\varepsilon(\bar{\Omega}) \subset B$, *but also the additional constraint* (2.2), *or equivalently, that they be injective over the set* Ω (by Theorem 1; the converse is clear).

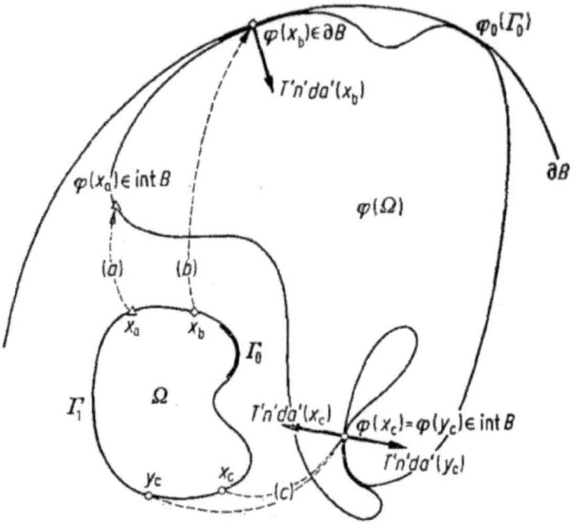

Fig. 1. The different types of boundary conditions along Γ_1 described in Theorem 4. The differential surface force $\boldsymbol{T}'\boldsymbol{n}'\,da'$ vanishes in case (a), and is normal to and directed away from the obstacle in case (b) (contact without friction with the obstacle); in case (c), the differential surface forces $\boldsymbol{T}'(x_c)\,\boldsymbol{n}'(x_c)\,da'(x_c)$ and $\boldsymbol{T}'(y_c)\,\boldsymbol{n}'(y_c)\,da'(y_c)$ are normal to the surface $\boldsymbol{\varphi}(\Gamma_1)$ and directed inward, and take opposite values (self-contact without friction)

(i) Let $\boldsymbol{\varphi}$ be a smooth enough solution of the minimization problem (2.11), let x be a point in Ω, and let $B_r(x)$ be an open ball centered at x and such that $\overline{B_r(x)} \subset \Omega$. Given any function $\boldsymbol{\theta} \in \mathscr{D}(B_r(x); \mathbb{R}^3)$ and any $\varepsilon \in \mathbb{R}$, let $\boldsymbol{\varphi}^\varepsilon = \boldsymbol{\varphi} + \varepsilon\boldsymbol{\theta}$. Clearly, $\boldsymbol{\varphi}^\varepsilon = \boldsymbol{\varphi}^0$ on Γ_0, and there is an $\varepsilon_0 = \varepsilon_0(\boldsymbol{\theta}) > 0$ such that $\det \nabla\boldsymbol{\varphi}^\varepsilon > 0$ in $\bar{\Omega}$ and $\boldsymbol{\varphi}^\varepsilon(\Omega) \subset B$ if $|\varepsilon| \leqq \varepsilon_0$.

Applying Theorem 3 with $U = B_r(x)$ and $\boldsymbol{\psi} = \boldsymbol{\varphi}^\varepsilon$ for $|\varepsilon| \leqq \varepsilon_0$, we find that $\boldsymbol{\varphi}^\varepsilon$ is injective on $\overline{B_r(x)}$ and that $\boldsymbol{\varphi}^\varepsilon(B_r(x)) = \boldsymbol{\varphi}(B_r(x))$. (By Theorem 1, the mapping $\boldsymbol{\varphi}$ is injective on the set Ω, which contains the set $\overline{B_r(x)}$ by assumption). The injectivity on Ω follows by observing that $\boldsymbol{\varphi}^\varepsilon$ is injective on $\Omega - B_r(x)$ (since $\boldsymbol{\varphi} = \boldsymbol{\varphi}^\varepsilon$ on $\Omega - B_r(x)$), and that

$$\boldsymbol{\varphi}^\varepsilon(\Omega - B_r(x)) \cap \boldsymbol{\varphi}^\varepsilon(B_r(x)) = \boldsymbol{\varphi}(\Omega - B_r(x)) \cap \boldsymbol{\varphi}(B_r(x)) = \emptyset.$$

Applying Green's formula

$$(2.18) \qquad \int_\Omega \boldsymbol{T}(\nabla\boldsymbol{\varphi}) : \nabla\boldsymbol{\theta}\, dx = -\int_\Omega \operatorname{\mathbf{div}} \boldsymbol{T}(\nabla\boldsymbol{\varphi}) \cdot \boldsymbol{\theta}\, dx + \int_\Gamma \boldsymbol{T}(\nabla\boldsymbol{\varphi})\,\boldsymbol{n} \cdot \boldsymbol{\theta}\, da$$

to the inequality $I(\varphi^\varepsilon) \geq I(\varphi)$, we find that

$$\varepsilon \int_\Omega \{-\mathbf{div}\, T(\nabla\varphi) - f\} \cdot \theta \, dx \geqq 0 \quad \text{for all } |\varepsilon| \leqq \varepsilon_0.$$

From these inequalities, we derive equations (2.12). The boundary condition (2.13) is satisfied by definition of the set Φ.

(ii) Let $x \in \Gamma_1$ be such that $\varphi(x) \in \operatorname{int} B$ and $\operatorname{card} \varphi^{-1}(\varphi(x)) = 1$ (case (a)). We can find an open set $V_x \subset \mathbb{R}^3$ containing x and a smooth extension of the mapping φ, which is denoted by the same letter φ, such that $\det \nabla\varphi > 0$ in $\bar{\Omega} \cup V_x$. We first prove that *there is an open ball* $B_r(x)$ *such that*

$$(2.19) \qquad B_r(x) \subset V_x \quad \textit{and} \quad \varphi : \Omega \cup B_r(x) \to \mathbb{R}^3 \quad \textit{is injective.}$$

By the implicit function theorem, there is an $r_0 > 0$ such that the mapping $\varphi : B_{r_0}(x) \to \mathbb{R}^3$ is injective. We also recall that we have proved in Theorem 1 that the mapping φ is injective on Ω. If relation (2.19) were false, there would be sequences (x^k) and (y^k) such that

$$x^k \in B_{1/k}(x), \quad y^k \in \Omega - B_{1/k}(x), \quad \varphi(x^k) = \varphi(y^k) \quad \text{for all } k \geqq 1.$$

Since $\bar{\Omega}$ is bounded, there is a subsequence (y^l) that converges to a point $y \in \bar{\Omega}$. Since $\lim_{l\to\infty} x^l = x$, we have

$$\varphi(x) = \lim_{l\to\infty} \varphi(x^l) = \lim_{l\to\infty} \varphi(y^l) = \varphi(y),$$

and thus the assumption that $\operatorname{card} \varphi^{-1}(\varphi(x)) = 1$ implies that $x = y$. Hence the points x^l and y^l belong to the ball $B_{r_0}(x)$ for l large enough, which contradicts the injectivity of φ on $B_{r_0}(x)$, and relation (2.19) is proved.

After further reducing the radius r is necessary, we can assume that $B_r(x) \cap \Gamma_0 = \emptyset$, that $\varphi(\bar{\Omega} \cup \overline{B_r(x)}) \subset B$ (we assume here that $\varphi(x) \in \operatorname{int} B$), and finally that φ is injective on $\Omega \cup \overline{B_r(x)}$ (this last property is needed in the ensuing argument).

Given any function $\theta \in \mathcal{D}(B_r(x); \mathbb{R}^3)$ and any $\varepsilon \in \mathbb{R}$, let $\varphi^\varepsilon = \varphi + \varepsilon\theta$. Clearly $\varphi^\varepsilon = \varphi^0$ on Γ_0 and there is an $\varepsilon_0 = \varepsilon_0(\theta) > 0$ such that $\det \nabla\varphi^\varepsilon > 0$ in $\bar{\Omega}$ and $\varphi^\varepsilon(\bar{\Omega}) \subset B$ if $|\varepsilon| \leqq \varepsilon_0$. Then the injectivity of the mapping φ^ε on the set $\Omega \cup B_r(x)$, hence *a fortiori* on the set Ω, follows from the injectivity of the mapping φ on $\Omega \cup B_r(x)$ by the same argument as in part (i).

Beginning again with the inequality $I(\varphi^\varepsilon) \geqq I(\varphi)$ for such functions φ^ε and using Green's formula (2.18), we find that

$$\varepsilon \left\{ \int_{\Gamma_1} T(\nabla\varphi)\, \boldsymbol{n} \cdot \theta \, da + O(\varepsilon) \right\} \geqq 0$$

for all $|\varepsilon| \leqq \varepsilon_0$, from which we deduce that the boundary condition (2.14) holds with $\lambda(x) = 0$. Hence the implication (2.15) is proved.

(iii) Let $x \in \Gamma_1$ be such that $\varphi(x) \in \partial B$, so that $\operatorname{card} \varphi^{-1}(\varphi(x)) = 1$ by Theorem 2 (case (b)). Arguing as in part (ii), we can find a ball $B_r(x)$ and an extension φ such that $\varphi : \Omega \cup B_r(x) \to \mathbb{R}^3$ is injective.

Let (t_1', t_2', t_3') be a smooth field of linearly independent vectors defined in a neighborhood V' of the point $\varphi(x)$ containing the open set $\varphi(B_r(x))$ (after reducing r if necessary) with the following properties:

(2.20) $$|t_1'(z)| = |t_2'(z)| = |t_3'(z)| = 1 \quad \text{for all } z \in V',$$

$t_1'(\varphi(y))$ and $t_2'(\varphi(y))$ span the tangent plane at the surface $\partial\varphi(\Omega)$ at all points $\varphi(y) \in \partial\varphi(\Omega) \cap V'$,

$t_3'(\varphi(y))$ is the unit outer normal vector $n'(\varphi(y))$ to the surface $\partial\varphi(\Omega)$ at all points $\varphi(y) \in \partial\varphi(\Omega) \cap V'$.

Given any functions $\theta_1, \theta_2 \in \mathscr{D}(B_r(x))$, there is an $\varepsilon_0 = \varepsilon_0(\theta_1, \theta_2) > 0$ and there are functions $\lambda_1^\varepsilon, \lambda_2^\varepsilon : \bar{\Omega} \to \mathbb{R}$ such that

(2.21) $$\sup \lambda_\alpha^\varepsilon \subset B_r(x), \quad \sup_{y \in B_r(x)} |\lambda_\alpha^\varepsilon(y)| = O(\varepsilon) \text{ for all } |\varepsilon| \leqq \varepsilon_0, \quad \alpha = 1, 2,$$

and such that the functions

(2.22) $$\varphi_\alpha^\varepsilon = \varphi + \varepsilon(\theta_\alpha t_\alpha' \circ \varphi + \lambda_\alpha^\varepsilon t_3' \circ \varphi), \quad \alpha = 1, 2,$$

satisfy

(2.23)

$$\varphi_\alpha^\varepsilon(\bar{\Omega}) \subset B \quad \text{and} \quad \det \nabla\varphi_\alpha^\varepsilon > 0 \quad \text{in } \bar{\Omega} \text{ for all } |\varepsilon| \leqq \varepsilon_0, \quad \alpha = 1, 2.$$

Hence the same argument as in part (ii) shows that the functions $\varphi_\alpha^\varepsilon$ are injective on the set Ω. Another application of Green's formula (2.18) then shows that

$$\varepsilon\left\{\int_{\Gamma_1} \theta_\alpha\{T(\nabla\varphi)\, n \cdot (t_\alpha' \circ \varphi)\}\, da + O(\varepsilon)\right\} \geqq 0 \quad \text{for all } |\varepsilon| \leqq \varepsilon_0,$$

and thus we must have $T(\nabla\varphi)\, n \cdot t_\alpha' = 0$ for $\alpha = 1, 2$, *i.e., the vector* $T(\nabla\varphi)\, n$ *must be parallel to the unit outer normal vector* n' *along* Γ_1 *at the point* x.

We can similarly prove that, given any function $\theta \in \mathscr{D}(B_r(x))$ that is $\geqq 0$, there is an $\varepsilon_0 = \varepsilon_0(\theta) > 0$ such that

(2.24) $$\varphi^\varepsilon = \varphi + \varepsilon\theta t_3' \circ \varphi \in \Phi \text{ for } -\varepsilon_0 \leqq \varepsilon \leqq 0.$$

Observing that $t_3' = n'$ along Γ_1 and using Green's formula, we obtain the inequality

$$\varepsilon\left\{\int_{\Gamma_1} \theta\{\hat{T}(\nabla\varphi)\, n \cdot (n' \circ \varphi)\}\, da + O(\varepsilon)\right\} \geqq 0 \quad \text{for } -\varepsilon_0 \leqq \varepsilon \leqq 0,$$

which shows that *the vector* $\hat{T}(\nabla\varphi)\, n$ *must be of the form* $\lambda n'$ *with* $\lambda \leqq 0$ (we already know that $\hat{T}(\nabla\varphi)\, n$ is parallel to n'). Hence relation (2.14) is proved in this case.

(iv) Let $x \in \Gamma_1$ be such that $\varphi^{-1}(\varphi(x)) = \{x, y\}$ with $y \in \Gamma_1$ and $y \neq x$ (case (c)). By Theorem 2, the tangent planes to the surface $\varphi(\partial\Omega)$ at the points $\varphi(x)$ and $\varphi(y)$ coincide, and their unit outer normal vectors have opposite directions. By the implicit function theorem, there are disjoint open balls $B_{2r}(x)$ and $B_r(y)$ such that the restrictions of φ to both sets $B_{2r}(x) \cap \bar{\Omega}$ and $B_r(y) \cap \bar{\Omega}$ are $\mathscr{C}^1$-diffeomorphisms. Moreover,

$$\varphi(B_{2r}(x) \cap \Omega) \cap \varphi(B_r(y) \cap \Omega) = \emptyset,$$

since φ is injective on Ω by Theorem 1. Then we can argue as in part (iii): We first consider a smooth field of linearly independent vectors in a neighborhood V' of x' that satisfies (2.20), with $\varphi(\Gamma_1 \cap B_{2r}(x))$ playing the rôle of the set $\partial\varphi(\Omega) \cap V'$ in (2.20); then we consider variations of the form (2.22), with functions $\theta_1, \theta_2 \in \mathscr{D}(B_r(x))$. This means that the set $\varphi(B_r(y) \cap \bar{\Omega})$ is kept fixed, while variations of the form (2.22) with supports in $B_r(x) \cap \bar{\Omega}$ are allowed. Such variations are injective on the set $\overline{B_{2r}(x)} \cap \Omega$ if $|\varepsilon|$ is small enough. Moreover, we can find functions $\lambda_1^\varepsilon, \lambda_2^\varepsilon : \bar{\Omega} \to \mathbb{R}$ such that $\sup_{z \in B_r(x)} |\lambda_\alpha^\varepsilon(z)| = O(\varepsilon)$ and

$$\varphi_\alpha^\varepsilon(B_r(y) \cap \Omega) \cap \varphi_\alpha(B_r(y) \cap \Omega) = \emptyset.$$

(Doing this is analogous to proving that $\varphi_\alpha^\varepsilon(\bar{\Omega}) \subset B$ in (2.23)). Hence these variations $\varphi_\alpha^\varepsilon$ are injective on Ω. Considering next variations of the form (2.24), and arguing as in step (iii), we conclude that

$$T(\nabla\varphi(x))\, \boldsymbol{n}(x) = \lambda(x)\, \boldsymbol{n}'(x) \quad \text{with } \lambda'(x) \leqq 0,$$

and relation (2.14) is proved in this case too.

Finally, let $r' > 0$ be so chosen that

$$\varphi^{-1}(B_{r'}(x') \cap \varphi(\bar{\Omega})) \subset (B_r(x) \cap \bar{\Omega}) \cup (B_r(y) \cap \bar{\Omega}),$$

let a vector field (t_3') be defined as in (2.20) in a neighborhood of x' that contains the ball $B_{r'}(x')$, with $\varphi(\Gamma_1 \cap B_r(x))$ playing the rôle of the set $\partial\varphi(\Omega) \cap V'$ in (2.20), and let $\eta \in \mathscr{D}(B_{r'}(x'); \mathbb{R}^3)$ be given. Then for $|\varepsilon|$ small enough, the variations φ^ε defined by

$$\varphi^\varepsilon(z) = \varphi(z) + \varepsilon\eta(\varphi(z))\, t_3'(\varphi(z)) \quad \text{for } z \in \bar{\Omega},$$

are injective. (This is seen by invoking the same kinds of arguments as before.) Using the inequality $I(\varphi^\varepsilon) \geqq I(\varphi)$, we obtain

$$\varepsilon \Bigg(\int_{\Gamma_1 \cap B_r(x)} (\eta \circ \varphi) \{T(\nabla\varphi)\, \boldsymbol{n} \cdot (t_3' \circ \varphi)\}\, da$$
$$+ \int_{\Gamma_1 \cap B_r(x)} (\eta \circ \varphi) \{T(\nabla\varphi)\, \boldsymbol{n} \cdot (t_3' \circ \varphi)\}\, da + O(\varepsilon) \Bigg) \geqq 0$$

for $|\varepsilon|$ small enough. Using the relations

$$t_3'(\varphi(z)) = n'(z) \quad \text{for } z \in \Gamma_1 \cap B_r(x),$$

$$n'(u) + n'(v) = 0 \quad \text{if } u \in \Gamma_1 \cap B_r(x), \quad v \in \Gamma_1 \cap B_r(y), \quad \text{and } \varphi(u) = \varphi(v),$$

$$T(\nabla\varphi(u)) = \lambda(u)\, n'(u) \quad \text{for } u \in \Gamma_1,$$

$$\lambda(u) = 0 \quad \text{if } u \in \Gamma_1 \cap B_r(x), \quad \text{and } \varphi^{-1}(\varphi(u)) = \{u\},$$

we obtain

$$\int_{\Gamma_1 \cap B_r(x)} (\eta \circ \varphi)\, \lambda\, da + \int_{\Gamma_1 \cap B_r(y)} (\eta \circ \varphi)\, \lambda\, da = 0,$$

and since $(\theta \circ \varphi)(x) = (\theta \circ \varphi)(y)$, we conclude that $\lambda(x)\, da(x) = \lambda(y)\, da(y)$, and thus relations (2.17) are proved. □

Remark. Had we not assumed that $\varphi_0(\Gamma_0)$ was part of the boundary ∂B, we should still have to consider the case where $x \in \Gamma_1$ is such that $\varphi^{-1}(\varphi(x)) = \{x, y\}$ with $y \in \Gamma_0$. This situation could then be treated like case (c). □

To interpret the boundary conditions (2.14)–(2.17) found along Γ_1, we recall that the *first Piola-Kirchhoff stress tensor* $T(\nabla\varphi(x)) = \frac{\partial W}{\partial F}(\nabla\varphi(x))$ (*cf.* (2.12)) at a point x of the boundary Γ of the reference configuration $\bar{\Omega}$ and the *Cauchy stress tensor* $T'(x)$ at the corresponding point $\varphi(x)$ of the boundary of the deformed configuration $\varphi(\bar{\Omega})$ are related by

$$T(\nabla\varphi(x))\, n(x)\, da(x) = T'(x)\, n'(x)\, da'(x), \tag{2.25}$$

where $n(x)$, $n'(x)$, $da(x)$ are defined as in Theorem 4, and $da'(x)$ denotes the differential area along $\partial\varphi(\bar{\Omega})$ at the point $\varphi(x)$. Hence the boundary condition (2.14) can be also written as

$$T'(x)\, n'(x) = \lambda'(x)\, n'(x) \quad \text{with } \lambda'(x) \leqq 0 \quad \text{for } x \in \Gamma_1. \tag{2.26}$$

Since the *Cauchy stress vector* $T'(x)\, n'(x)$ measures the intensity of the surface force per unit area of the boundary of the deformed configuration, the various situations considered in (2.15)–(2.17) can be interpreted as follows:

Case (*a*). $\varphi(x) \in \operatorname{int} B$ *and* $\varphi^{-1}(\varphi(x)) = \{x\}$ (*cf.* (2.15)). There is neither contact with the boundary of the obstacle nor self-contact, and the differential surface force $T(x)\, n(x)\, da(x)$ vanishes, as when null traction acts upon the boundary.

Case (*b*). $\varphi(x) \in \partial B$ (*cf.* (2.16)). There is contact with the obstacle, and the differential surface force $T(x)\, n(x)\, da(x)$ is normal to and directed away from the obstacle: This is a model of frictionless *contact with the obstacle*.

Case (c). *There are two distinct points* $x, y \in \Gamma_1$ *such that* $\varphi(x) = \varphi(y)$ (*cf.* (2.17)). Then

$$n'(x) + n'(y) = 0, \quad \lambda(x)\, da(x) = \lambda(y)\, da(y),$$

and thus

$$T'(x)\,n'(x)\,da'(x) + T'(y)\,n'(y)\,da'(y) = 0 \tag{2.27}$$

as a consequence of (2.25). Hence the differential surface forces $T'(x)\,n'(x)\,da'(x)$ and $T'(y)\,n'(y])\,da'(y)$ are normal to and directed inside the surface $\partial\varphi(\Gamma_1)$, and they take opposite values: This is a *model of frictionless self-contact.*

Remark. The function $\lambda: \Gamma_1 \to \mathbb{R}$ defined by equations (2.14)–(2.17), or the associated function $\lambda': \Gamma_1 \to \mathbb{R}$ defined in (2.26), can be thought of as a *Kuhn & Tucker multiplier* associated with the constraint $\varphi(\bar{\Omega}) \subset B$ in cases (a) and (b) and with the constraint $\int_\Omega \det \nabla\varphi\, dx \leqq \operatorname{vol} \varphi(\Omega)$ in case (c). Observe also that the function λ' can be interpreted in mechanical terms as the *reaction of contact.* □

3. Results on Existence

Condition (2.2) is of interest also in that it requires the energy functional to be weakly lower semi-continuous with respect to infimizing sequences of functions. This leads to the following extension of the existence results of Ball [1977].

Theorem 5. *Let the boundary* $\Gamma = \partial\Omega$ *be Lipschitz-continuous, and let* da-meas $\Gamma_0 > 0$. *Let a stored-energy function* $W: \mathbb{M}^3_+ \to \mathbb{R}$ *be given that satisfies the following assumptions:*

(i) *Polyconvexity: There is a convex function* $\mathbb{W}: \mathbb{M}^3 \times \mathbb{M}^3 \times]0, +\infty[\to \mathbb{R}$ *such that*

$$W(F) = \mathbb{W}(F, \operatorname{Cof} F, \det F) \quad \text{for all } F \in \mathbb{M}^3_+. \tag{3.1}$$

(ii) *Behavior as* $\det F \to 0^+$:

$$W(F) \to +\infty \quad \text{as } \det F \to 0^+. \tag{3.2}$$

(iii) *Coerciveness: There are constants* $a > 0$, $b \in \mathbb{R}$, $p > 0$, $q \geqq \frac{p}{p-1}$, $r > 1$, *such that*

$$W(F) \geqq a(\|F\|^p + \|\operatorname{Cof} F\|^q + (\det F)^r) + b \quad \text{for all } F \in \mathbb{M}^3_+. \tag{3.3}$$

Let the density f *of the applied body force be such that the linear form* $\psi \in W^{1,p}(\Omega) \to \int f \cdot \psi\, dx$ *is continuous, and let* $\varphi_0 \in W^{1,p}(\Omega)$ *be given in such a way that the set*

$$\begin{aligned}\Phi = \{\psi \in W^{1,p}(\Omega; \mathbb{R}^3);\ &\operatorname{Cof} \nabla\psi \in L^q(\Omega; \mathbb{M}^3),\ \det \nabla\psi \in L^r(\Omega),\\ &\det \nabla\psi > 0 \quad \text{a.e. in } \Omega,\\ &\psi = \varphi_0 \text{ on } \Gamma_0,\quad \int_\Omega \det \nabla\psi\, dx \leqq \operatorname{vol} \psi(\Omega)\}\end{aligned} \tag{3.4}$$

is not empty. Finally, assume that

$$\inf_{\psi \in \Phi} I(\psi) < +\infty, \tag{3.5}$$

where the total energy I is defined as in (2.1).

Then the problem of finding a φ such that

$$\varphi \in \Phi \quad \textit{and} \quad I(\varphi) = \inf_{\psi \in \Phi} I(\psi) \tag{3.6}$$

has at least one solution. Moreover, the mapping $\varphi : \bar{\Omega} \to \mathbb{R}^3$ is injective almost everywhere in the sense that

$$\operatorname{card} \varphi^{-1}(x') = 1 \quad \textit{for almost all } x' \in \varphi(\bar{\Omega}). \tag{3.7}$$

Proof. We only sketch those parts of the proof which are the same as in BALL [1977]. First, the total energy $I(\psi)$ is well defined for functions $\psi \in \Phi$. Next, (3.3) requires that there be numbers $\alpha > 0$ and $\beta \in \mathbb{R}$ such that

$$I(\psi) \geqq \alpha(\|\psi\|^p_{1,p,\Omega} + |\mathbf{Cof}\, \nabla\psi|^q_{0,q,\Omega} + |\det \nabla\psi|^r_{0,r,\Omega}) + \beta \textit{ for all } \psi \in \Phi. \tag{3.8}$$

(The assumption that da-meas $\Gamma_0 \geqq 0$ is used here). Now let (φ^l) with $\varphi^l \in \Phi$ be an infimizing sequence for the functional I. By (3.8), the sequence $(\varphi^l, \mathbf{Cof}\, \nabla\varphi^l, \det \nabla\varphi^l)$ is bounded in the reflexive space $W^{1,p}(\Omega; \mathbb{R}^3) \times L^q(\Omega; \mathbb{M}^3) \times L^r(\Omega)$. Thus there is a subsequence $(\varphi^k, \mathbf{Cof}\, \nabla\varphi^k, \det \nabla\varphi^k)$ that converges weakly in this space to an element of the form $(\varphi, \mathbf{Cof}\, \nabla\varphi, \det \nabla\varphi)$ satisfying $\varphi = \varphi_0$ on Γ_0 (*cf.* BALL [1977, Theorem 6.2]). The behavior of the stored-energy function as $\det F \to 0^+$ (cf. (3.2)) then implies that $\det \nabla\varphi > 0$ almost everywhere in Ω (BALL, CURRIE & OLVER [1981]). In order to conclude that $\varphi \in \Phi$, we must therefore show that $\int_\Omega \det \nabla\varphi \, dx \leqq \operatorname{vol} \varphi(\Omega)$.

We first note that, by a theorem of MARCUS & MIZEL [1973], the combined asumptions that Ω is a bounded open subset of $\mathbb{R}^3$ and that $p > 3$ imply that any function $\varphi \in W^{1,p}(\Omega; \mathbb{R}^3)$ maps sets of zero volume (*i.e.*, of zero dx-measure) into sets of zero volume and maps dx-measurable subsets of $\mathbb{R}^3$ into dx-measurable subsets of $\mathbb{R}^3$. Since the volume of a Lipschitz continuous boundary such as $\partial\Omega$ is zero, it follows that $\operatorname{vol} \varphi(\partial\Omega) = 0$, and consequently that $\operatorname{vol} \varphi(\Omega) = \operatorname{vol} \varphi(\bar{\Omega})$. We shall take advantage of this observation in (3.14) below.

Again beacuse $p > 3$ and Ω is a bounded open subset of $\mathbb{R}^3$ with a Lischitz-continuous boundary, the Sobolev space $W^{1,p}(\Omega)$ is compactly imbedded in the space $\mathscr{C}^0(\bar{\Omega})$ by the Kondrašov theorem (*cf.* NEČAS [1967, p. 107]). Hence the sequence (φ^k) converges uniformly to φ, *i.e.*, in the space $\mathscr{C}^0(\bar{\Omega}; \mathbb{R}^3)$. Let $\varepsilon > 0$ be arbitrary. Since the set $\varphi(\bar{\Omega})$ is compact, whence measurable, a classical property of Lebesgue measure provides an open set O_ε such that

$$\varphi(\bar{\Omega}) \subset O_\varepsilon, \quad \text{and} \quad \operatorname{vol}(O_\varepsilon - \varphi(\bar{\Omega})) < \varepsilon. \tag{3.9}$$

Let us next show that there is a number $\delta = \delta(\varepsilon) > 0$ such that

$$\bigcup_{x \in \varphi(\bar{\Omega})} B_\delta(x) \subset O_\varepsilon. \tag{3.10}$$

Otherwise for a certain $\varepsilon > 0$, we could find sequences (x^m), (y^m), (δ^m) satisfying

$$x^m \in \varphi(\bar{\Omega}), \quad y^m \notin O_\varepsilon, \quad |x^m - y^m| < \delta^m \quad \text{and} \lim_{m\to\infty} \delta^m = 0.$$

Since the set $\varphi(\bar{\Omega})$ is compact, there is a subsequence (x^n) that converges to an element $x \in \varphi(\bar{\Omega})$, and therefore the subsequence (y^n) also converges to the same element x since $\lim_{n\to\infty} |y^n - x^n| = 0$. But the limit of the sequence (y^n) should also belong to the closed set $\mathbb{R}^3 - O_\varepsilon$, which is impossible by virtue of the inclusion of (3.9). Hence (3.10) is established. As a consequence, there is an integer $k_0 = k_0(\delta) = k_0(\varepsilon)$ such that

$$\varphi^k(\bar{\Omega}) \subset O_\varepsilon \quad \text{for all } k \geqq k_0, \tag{3.11}$$

since the sequence (φ^k) converges uniformly to φ. Hence the fact that $\varphi^k \in \Phi$ and the inclusion (3.11) together imply that

$$\int_\Omega \det \nabla\varphi^k \, dx \leqq \operatorname{vol} \varphi^k(\bar{\Omega}) \leqq \operatorname{vol} O_\varepsilon \quad \text{for all } k \geqq k_0. \tag{3.12}$$

Since the sequence $(\det \nabla\varphi^k)$ converges weakly to $\det \nabla\varphi$ in the space $L^r(\Omega)$, we conclude that

$$\int_\Omega \det \nabla\varphi \, dx = \lim_{k\to\infty} \int_\Omega \det \nabla\varphi^k \, dx \leqq \operatorname{vol} O_\varepsilon, \tag{3.13}$$

and since $\operatorname{vol} O_\varepsilon = \operatorname{vol} \varphi(\bar{\Omega}) + \operatorname{vol} (O_\varepsilon - \varphi(\bar{\Omega}))$, we infer from (3.9), (3.13), and the arbitrariness of ε that

$$\int_\Omega \det \nabla\varphi \, dx \leqq \operatorname{vol} \varphi(\bar{\Omega}) = \operatorname{vol} \varphi(\Omega), \tag{3.14}$$

and thus we have proved that $\varphi \in \Phi$.

It is now easy to conclude the proof of existence: The assumed polyconvexity (3.1) of the stored-energy function implies that

$$I(\varphi) \leqq \liminf_{k\to\infty} I(\varphi^k),$$

and the existence of at least one minimizer φ of the total energy is thus established.

Let us now prove that any mapping $\varphi \in \Phi$, hence in particular a minimizer of the total energy, is injective almost everywhere, in the sense of (3.7). A theorem of Marcus & Mizel [1973] shows that if Ω is a bounded open subset of $\mathbb{R}^3$ and if $\varphi \in W^{1,p}(\Omega; \mathbb{R}^3)$ with $p > 3$, then

$$\int_\Omega |\det \nabla\varphi| \, dx = \int_{\varphi(\Omega)} \operatorname{card} \{\varphi^{-1}(x')\} \, dx', \tag{3.15}$$

whenever one of the two sides is meaningful. (Relation (3.15) is nothing but the extension to mappings with values in Sobolev spaces of the analogous relation (2.6) for smooth maps). Using (3.15) and the definition of the set Φ, we obtain

$$\operatorname{vol} \varphi(\Omega) = \int_{\varphi(\Omega)} dx' \leqq \int_{\varphi(\Omega)} \operatorname{card} \{\varphi^{-1}(x')\} dx' = \int_\Omega \det \nabla\varphi \, dx \leqq \operatorname{vol} \varphi(\Omega),$$

which shows that

$$\operatorname{card}\{\varphi^{-1}(x')\} = 1 \text{ for almost all } x' \in \varphi(\Omega),$$

and the proof is complete. □

4. Concluding Remarks

(i) For the pure displacement problem there is no need to impose the additional constraint $\int_\Omega \det \nabla\boldsymbol{\psi}\, dx \leqq \operatorname{vol} \boldsymbol{\psi}(\Omega)$ on the admissible deformation $\boldsymbol{\psi}$, provided the boundary function $\boldsymbol{\varphi}_0 : \Gamma \to \mathbb{R}^3$ and the stored-energy function satisfy suitable hypotheses, since the injectivity of the minimizers of the energy is automatic in this case (*cf.* BALL [1981 a, Theorem 3]).

(ii) The link between the invertibility of equilibrium solutions in non-linear elasticity (Thence in particular of the minimizers of the energy) and the behavior of the stored-energy function as $\det \boldsymbol{F} \to 0^+$ (*cf.* (3.2)) has been studied for one-dimensional problems by ANTMAN [1976], ANTMAN & BREZIS [1978], and BALL [1981 b].

(iii) As shown in CIARLET & GEYMONAT [1982], it is always possible to construct stored-energy functions that satisfy the assumption (3.1)–(3.3) of Theorem 5, and whose expansions in terms of the Green-St. Venant strain tensor $\boldsymbol{E} = \frac{1}{2}(\boldsymbol{F}^T\boldsymbol{F} - \boldsymbol{I})$ are of the form

$$W(\boldsymbol{F}) = \frac{\lambda}{2}(\operatorname{tr} \boldsymbol{E})^2 + \mu \operatorname{tr}(\boldsymbol{E}^2) + o(\|\boldsymbol{E}\|^2),$$

for arbitrary Lamé constants $\lambda > 0$ and $\mu > 0$. Such stored-energy functions are special cases of the general stored-energy functions proposed by OGDEN [1972].

(iv) Using the notion of "*set of dangerous directions*" introduced in CIARLET & NEČAS [1985 a], we could treat also the case where $\Gamma_0 = \emptyset$, provided the averaged body forces make obtuse angles with the "dangerous directions". It is worth noticing that this last condition is always satisfied if the set $\boldsymbol{B}$ is bounded, in which case the set of dangerous directions is empty.

(v) Finally, we could also consider an analogous problem, but where *several non-interpenetrating bodies* are treated simultaneously. In this case the set Ω has a finite number of connected components, and the constitutive stored-energy functions may differ from one connected component to another.

References

ANTMAN, S. S. [1976]: Ordinary differential equations of nonlinear elasticity II: Existence and regularity theory for conservative boundary value problems, Arch. Rational Mech. Anal. **61**, 353–393.

Antman, S. S., & H. Brezis [1978]: The existence of orientation-preserving deformations in nonlinear elasticity, in *Nonlinear Analysis and Mechanics: Heriot-Watt Symposium, Vol. II* (R. J. Knops, Editor), pp. 1–29, Pitman, London.

Ball, J. M. [1977]: Convexity conditions and existence theorems in nonlinear elasticity, Arch. Rational Mech. Anal. **63**, 337–403.

Ball, J. M. [1981a]: Global invertibility of Sobolev functions and the interpenetration of matter, Proc. Royal Soc. Edinburgh **88 A**, 315–328.

Ball, J. M. [1981b]: Remarques sur l'existence et la régularité des solutions d'élastostatique nonlinéaire, in *Recent Contributions to Nonlinear Partial Differential Equations*, pp. 50–62, Pitman, Boston.

Ball, J. M., J. C. Currie & P. J. Olver [1981]: Null Lagrangians, weak continuity and variational problems of arbitrary order, J. Functional Analysis **41**, 135–174.

Ciarlet, P. G. [1985]: *Elasticité Tridimensionnelle*, Masson, Paris.

Ciarlet, P. G. [1987]: *Mathematical Elasticity*, Vol. 1, North-Holland, Amsterdam.

Ciarlet, P. G., & P. Destuynder [1979]: A justification of a nonlinear model in plate theory, Comput. Meth. Applied Mech. Engrg. **17/18**, 227–258.

Ciarlet, P. G., & G. Geymonat [1982]: Sur les lois de comportement en élasticité non-linéaire compressible, C. R. Acad. Sci. Paris Sér. II **295**, 423–426.

Ciarlet, P. G., & J. Nečas [1985a]: Unilateral problems in nonlinear, three-dimensional elasticity, Arch. Rational Mech. Anal. **87**, 319–338.

Ciarlet, P. G., & J. Nečas [1985b]: Injectivité presque partout, auto-contact, et non-interpénétrabilité en élasticité non linéaire tridimensionnelle, C. R. Acad. Sci. Paris Sér. I **301**, 621–624.

Germain, P. [1972]: *Mécanique des Milieux Continus*, Tome 1, Masson, Paris.

Gurtin, M. E. [1981]: *Introduction to Continuum Mechanics*, Academic Press, New York.

Hanyga, A. [1985]: *Mathematical Theory of Non-Linear Elasticity*, Polish Scientific Publishers, Warszawa, and Ellis Horwood, Chichester.

Marcus, M., & V. J. Mizel [1973]: Transformations by functions in Sobolev spaces and lower semicontinuity for parametric variational problems, Bull. Amer. Math. Soc. **79**, 790–795.

Marsden, J. E., & T. J. R. Hughes [1983]: *Mathematical Foundations of Elasticity*, Prentice-Hall, Englewood Cliffs.

Meisters, G. H., & C. Olech [1973]: Locally one-to-one mappings and a classical theorem on Schlicht functions, Duke Math. J. **30**, 63–80.

Nečas, J. [1967]: *Les Méthodes Directes en Théorie des Equations Elliptiques*, Masson, Paris.

Noll, W. [1978]: A general framework for problems in the statics of finite elasticity, in *Contemporary Developments in Continuum Mechanics and Partial Differential Equations*, pp. 363–387, North-Holland Mathematics Studies, Amsterdam.

Ogden, R. W. [1972]: Large deformation isotropic elasticity: On the correlation of theory and experiment for compressible rubberlike solids, Proc. Roy. Soc. London A **328**, 567–583.

Rado, T., & P. V. Reichelderfer [1955]: *Continuous Transformations in Analysis*, Springer-Verlag, Berlin.

Schwartz, L. [1967]: *Cours d'Analyse*, Hermann, Paris.

Smith, K. T. [1983]: *Primer of Modern Analysis*, Springer-Verlag, New York, Second Edition.

Truesdell, C., & W. Noll [1965]: The non-linear field theories of mechanics, *Handbuch der Physik*, Vol. III/3, 1–602.

Valent, T. [1979]: Teoremi di esistenza e unicità in elastostatica finita, Rend. Sem. Mat. Univ. Padova **60**, 165–181.

Valent, T. [1982]: Local theorems of existence and uniqueness in finite elastostatics, in *Finite Elasticity* (D. E. Carlson & R. T. Shield, Editors), pp. 401–421.

Vodop'yanov, S. K., V. M. Gol'dshtein, & Yu. G. Reshetnyak [1979]: On geometric properties of functions with generalized first derivatives, Russian Math. Surveys **34**, 19–74.

Wang, C. -C., & C. Truesdell [1973]: *Introduction to Rational Elasticity*. Noordhoff, Groningen.

Université Pierre, et Marie Curie, Paris
and
Charles University
Prague

(Received February 28, 1986)

SIAM J. MATH. ANAL.
Vol. 24, No. 1, pp. 36-45, January 1993

© 1993 Society for Industrial and Applied Mathematics
003

EXISTENCE OF GLOBAL WEAK SOLUTIONS TO THE DYNAMICAL PROBLEM FOR A THREE-DIMENSIONAL ELASTIC BODY WITH SINGULAR MEMORY*

HAMID BELLOUT†, FREDERICK BLOOM†‡, AND JINDRICH NECAS§

Abstract. A three-dimensional elastic body with memory is considered, for which the memory term is generated by a singular but integrable kernel; the existence of a global weak solution to an associated initial-boundary value problem is established by constructing Galerkin approximations and deriving suitable energy estimates.

Key words. singular viscoelasticity, energy estimates

AMS(MOS) subject classifications. 35Q99, 45K05, 73F15

1. Introduction. In spite of intense efforts by many outstanding mathematicians, most questions concerning the global existence of both weak and classical solutions to the mixed problem, with large data, for the nonlinear wave equation governing the evolution of the displacement vector **u** in an elastic body, i.e.,

$$\rho\frac{\partial^2 u_i}{\partial t^2}-\frac{\partial}{\partial x_j}\left(\frac{\partial W}{\partial e_{ij}}\right)=\rho f_i, \qquad i=1,2,3, \tag{1.1}$$

remain open. Equations (1.1) are set in an open domain $\Omega\subseteq R^n$, $n\geqq 2$, with $e_{ij}=\frac{1}{2}((\partial u_i/\partial x_j)+(\partial u_j/\partial x_i))$ the components of the infinitesimal strain tensor. The evolution equation (1.1) reflects only the presence of an instantaneous elastic response governed by the stress tensor that applies in the case of small strains, namely,

$$\sigma^I_{ij}=\frac{\partial W}{\partial e_{ij}}, \tag{1.2}$$

where $W=W(e_{ij})$, the stored energy function, is supposed (as it will be in this paper) to be twice continuously differentiable, with bounded second derivatives, and positive definite, i.e., for some $\alpha>0$ and all symmetric $n\times n$ matrices,

$$\frac{\partial^2 W}{\partial e_{ij}\partial e_{kl}}\,\xi_{ij}\xi_{kl}\geqq\hat{\alpha}\,\|\boldsymbol{\xi}\|^2. \tag{1.3}$$

We also suppose that $W(0)=0$ and $(\partial W/\partial e_{ij})(0)=0$.

For dynamic nonlinear elasticity problems of the type (1.1) in one-space dimension it is well known, e.g., Lax [16], Klainerman and Majda [17], that solutions develop singularities in finite time even for data that are small and compactly supported; existence of weak solutions for such problems follows from the recent work of DiPerna [18] and is based on compensated compactness arguments. For spatial dimension $n\geqq 2$ there are many partial results for the dynamic problem, e.g., local (in time) existence of classical solutions [19] and formation of singularities in finite time; excellent surveys of progress to date on problems involving but small and finite strain may be found, for example, in the expository paper [20] and the text [21].

* Received by the editors May 7, 1990; accepted for publication (in revised form) March 30, 1992.

† Department of Mathematical Sciences, Northern Illinois University, DeKalb, Illinois 60115.

‡ The research of this author was partially supported by National Science Foundation grant DMS-8823037.

§ Mathematical Institute, Charles University, Prague, Czechoslovakia. This author was on leave during Fall, 1989, at the Department of Mathematical Sciences, Northern Illinois University, DeKalb, Illinois 60115.

© Society for Industrial and Applied Mathematics 1993. Reprinted with kind permission from: Bellout, H., Bloom, F. & Nečas, J., Existence of global weak solutions to the dynamical problem for a three-dimensional elastic body with singular memory, SIAM Journal on Mathematical Analysis 24:36-45.

In this paper we consider an elastic material with memory for which the Cauchy stress has the form

$$\sigma_{ij} = \sigma_{ij}^I + \sigma_{ij}^M. \tag{1.4}$$

In (1.4), σ_{ij}^I is still given by (1.2), while

$$\sigma_{ij}^M = -\int_0^t h(t-\tau)\frac{\partial V}{\partial e_{ij}}(\mathbf{e}(\tau)-\mathbf{e}(t))\,d\tau, \tag{1.5}$$

with $V = V(\mathbf{e})$ an energy function possessing bounded second derivatives and positive definite, i.e., for some $\beta > 0$

$$\frac{\partial^2 V}{\partial e_{ij}\partial e_{kl}}\xi_{ij}\xi_{kl} \geqq \beta\|\boldsymbol{\xi}\|^2 \tag{1.6}$$

for all symmetric $n \times n$ matrices $\boldsymbol{\xi}$; as was the case with the stored energy function W, we also assume that $V(0) = 0$ and $(\partial V/\partial e_{ij})(0) = 0$. The kernel $h(t)$ in (1.5) is singular but integrable and assumed to have (a nonlinear version of) a form due to Boltzman, namely, $h(t) = \lambda h_0(t)$ with $h_0(t) = t^{-\nu}a(t) + r(t), 0 < \nu < \frac{1}{2}$, and $\lambda > 0$. Both $a(t)$ and $r(t)$ are sufficiently smooth, decreasing, nonnegative functions on $[0, \infty)$ with $a(t) > 0$ in some neighborhood of the origin. We suppose, additionally, that $\int_0^\infty h(\tau)\,d\tau < \infty$. Moreover we take the density $\rho \equiv 1$, which opens the door to the possibility of using a Galerkin approximation scheme, and we employ throughout the Langrangian description of the motion relative to the Cartesian coordinates x_i.

Elastic materials with singular memories of the type described by (1.5) have been considered in several recent pieces of work, e.g., Londen [1], Renardy [2], Hrusa and Renardy [3], Hannsgen and Wheeler [4], Hrusa and Renardy [5], Hrusa and Renardy [6], and Renardy [7]. In [1] the author establishes the existence of weak solutions to an abstract integrodifferential equation; the result applies to the one-dimensional viscoelastic model equation

$$u_{tt} = \phi(u_x)_x + \int_{-\infty}^t a'(t-\tau)\psi(u_x(x,\tau))_x\,d\tau + f(x,t), \tag{1.7}$$

provided $\phi = \psi$ and the memory function has a singularity that is stronger than logarithmic. In [2] it is shown that certain singular kernels do not permit the propagation of singularities and have a smoothing effect; these ideas were further expounded upon in [3], all within a one-dimensional framework. The authors in [4] consider the constant coefficient linear problem on a bounded domain and show that the evolution operator is compact for positive time if and only if the kernel of the integral operator is singular; results such as these, of course, point towards the fact that the models with singular kernels should have nicer existence properties than those with regular kernels. In [5] local and global existence in time (for small data) is established for the history-value problem associated with (1.7) when $a'(t) \sim -t^{-\alpha}, 0 < \alpha < 1$, as $t \to 0^+$; the equation (1.7) is approximated by equations with regular kernels, and energy estimates are used to prove convergence of the corresponding approximate solutions. The work in [6] differs from that in [5] in that $\alpha(t)$ is allowed to have an integrable singularity at $t = 0$, and local (but not global) existence of regular solutions for small data is established. In [7] the author provides an alternative existence proof based on coercive properties of the linearized problem; global existence for small data and local existence for large data (of smooth solutions) is proved for $a(t)$ having a singularity at $t = 0$ at least as strong as a negative power of t. We note, in passing, that viscoelastic response, of the Boltzmann variety, governed by kernels that are singular at zero, has been noted by several investigators in the area of polymer physics, e.g., Doi and Edwards [8], Rouse [9], Zimm [10], and Laun [11].

The method that we use in the work reported here is very close in spirit to that employed recently by Milota, Nečas, and Šverák [12], where the scalar equation on $\Omega \subset R^n$,

$$u_{tt} - \frac{\partial}{\partial x_i}(a_i(\nabla u)) + \int_0^t h(t-\tau)\Delta u(\mathbf{x}, \tau)\, d\tau = f, \tag{1.8}$$

is studied, with $h(t) = e^{-\alpha t} t^{-\nu}$, $\alpha > 0$, $0 < \nu < 1$. After the work was completed we received a copy of a preprint of [13], where a model viscoelastic problem similar to the one considered here is treated for a scalar displacement u (as opposed to our vector $\mathbf{u}$). In [13] analogous results to those obtained here are produced by use of an entirely different approximation scheme, one that involves adding the term $\varepsilon \Delta u$ to the left-hand side of (2.2) and studying the limit of u^ε as $\varepsilon \to 0$ (i.e., by use of the viscosity method).

We note here that the work in [13] is somewhat flawed: from Lemma A.2 in [13], and the estimate $\|\mathbf{w}^n - \mathbf{v}^n\|_* \leqq \sqrt{\varepsilon_n}\, C$, therein, it follows only that $\|J_s \mathbf{v}^n - \mathbf{v}^n\|_* \leqq \omega(s) + \sqrt{\varepsilon_n}\, \tilde{C}$ for some $\tilde{C} > 0$ and not that $\|J_s \mathbf{v}^n - \mathbf{v}^n\|_* \leqq \omega(s)$, uniformly in n, for some ω: $[0, 1] \to [0, \infty)$, with $\omega(s) \downarrow 0$ as $\sigma \downarrow 0$, as stated by the author in equation (3.26) of [13]. The lemma following (3.27) is then correct, but vacuous, as one of the hypotheses, i.e., (3.26) of [13], is not satisfied. If (3.26) in [13] is replaced by the somewhat weaker statement alluded to above, a lemma corresponding to that which follows (3.27) in [13] may be stated in a manner sufficiently strong, so as to salvage the rest of the argument in § 3 of [13].

Remarks.

1. It is well known (e.g., Coleman [14]) that, for a viscoelastic body of the type under consideration here, the following condition must be satisfied: in an isothermal process starting from equilibrium, the integral of the stress power around a closed path in strain space must be nonnegative. In the present situation this condition is equivalent to the statement that

$$\int_{t_0}^{t_1} \sigma_{ij} \dot{e}_{ij}\, dt \geqq 0 \quad \forall e_{ij} \in C^1([t_0, t_1]), \quad e_{ij}(x, t_0) = e_{ij}(x, t_1), \tag{1.9}$$

where $x \in \Omega$. For such a closed strain path, of course, $\int_{t_0}^{t_1} (\partial W/\partial e_{ij})(\mathbf{e}(t))\dot{e}_{ij}(t)\, dt = W(\mathbf{e}(t))|_{t_0}^{t_1} = 0$; if we take a closed strain path $e_{ij} \in C^1([t_0, t_1])$ and extend it to the interval $[0, t_0)$ by setting $e_{ij}(x, t) = e_{ij}(x, t_0)$, for $0 \leqq t < t_0$, $x \in \Omega$, then we may compute that

$$\begin{aligned}
&\int_{t_0}^{t_1} \dot{e}_{ij}(t)\left(\int_0^t h(t-\tau)\frac{\partial V}{\partial e_{ij}}(\mathbf{e}(\tau) - \mathbf{e}(t))\, d\tau\right) dt \\
&= \int_0^{t_1} \dot{e}_{ij}(t)\left(\int_0^t h(t-\tau)\frac{\partial V}{\partial e_{ij}}(\mathbf{e}(\tau) - \mathbf{e}(t))\, d\tau\right) dt \\
&= \int_0^{t_1} \left(\int_0^t h(t-\tau)\frac{\partial}{\partial t} V(\mathbf{e}(\tau) - \mathbf{e}(t))\, d\tau\right) dt \\
&= \int_0^{t_1} d\tau \int_\tau^{t_1} h(t-\tau)\frac{\partial}{\partial t} V(\mathbf{e}(\tau) - \mathbf{e}(t))\, dt \\
&= \int_0^{t_1} h(t_1-\tau) V(\mathbf{e}(\tau) - \mathbf{e}(t))\, d\tau \\
&\quad - \int_0^{t_1} d\tau \int_\tau^{t_1} \frac{\partial h}{\partial t}(t-\tau) V(\mathbf{e}(\tau) - \mathbf{e}(t))\, dt \geqq 0,
\end{aligned} \tag{1.10}$$

and (1.9) is, therefore, satisfied.

2. We emphasize here that the operator $(\partial/\partial x_j)\sigma_{ij}^M$ is continuous in the same topology, i.e., in $L^2((0, T); W^{1,2}(\Omega; R^n))$ as $(\partial/\partial x_j)\sigma_{ij}^I$ and is even compact in time; this result will be established in the analysis that follows. A key point to be noted is that the memory part of the stress tensor σ_{ij}^M allows us to establish an estimate of the form

$$\|\mathbf{u}\|_{W^{\nu/2,2}((0,T);\, W^{1,2}(\Omega;R^n))} \leqq C(T), \tag{1.11}$$

for any $T>0$, which, together with various energy estimates, and an interpolation lemma, yields an estimate of the type

$$\|\mathbf{u}\|_{W^{(1+\nu/2)\alpha,2}((0,T);\, W^{\alpha,2}(\Omega;R^n))} \leqq C'(T) \tag{1.12}$$

for $1/(1+\nu/2)<\alpha<1$. The estimates (1.11), (1.12) are key ingredients in our global existence result.

2. Galerkin approximations and basic estimates. We assume in all that follows that Ω is a bounded domain, with Lipschitz continuous boundary $\partial\Omega$, and begin by choosing a basis $\{\mathbf{w}^i\}_{i=1}^{\infty}$ in the space $W_0^{1,2}(\Omega; R^n)$; a natural choice for such a basis would consist of the set of eigenfunctions for the problem

$$\frac{\partial}{\partial x_j}\left(\frac{\partial^2 W}{\partial e_{ij}\partial e_{kl}}(0)\frac{\partial u_k}{\partial x_l}\right)+\lambda u_i=0 \quad \text{in } \Omega, \tag{2.1}$$
$$\mathbf{u}\in W_0^{1,2}(\Omega;\, R^n).$$

In any case we suppose that the basis chosen is orthonormal in the space $L^2(\Omega; R^n)$.

DEFINITION 2.1. A weak solution to the mixed problem

$$\ddot{u}_i-\frac{\partial}{\partial x_j}\sigma_{ij}^I-\frac{\partial}{\partial x_j}\sigma_{ij}^M=f_i \quad \text{in } Q=\Omega\times(0,\infty), \tag{2.2}$$

$$\mathbf{u}=0 \quad \text{on } \partial\Omega\times(0,\infty), \tag{2.3}$$

$$\mathbf{u}(0,\cdot)=\mathbf{u}^0, \qquad \ddot{\mathbf{u}}(0,\cdot)=\mathbf{u}^1 \tag{2.4}$$

with $u^0\in W_0^{1,2}(\Omega;\mathbb{R}^n)$, $\mathbf{u}^1\in L^2(\Omega;\mathbb{R}^n)$, is a function $\mathbf{u}\in L^\infty((0,\infty); W_0^{1,2}(\Omega;\mathbb{R}^n))$, for which $\dot{\mathbf{u}}\in L^\infty((0,\infty); L^2(\Omega; R^n))$, $\ddot{u}\in L^\infty((0,\infty); W^{-1,2}(\Omega; R^n))$ and such that for almost all $T>0$, and all $\mathbf{v}\in W_0^{1,2}(\Omega; R^n)$,

$$\int_{\Omega_r}\ddot{u}_i v_i\,d\mathbf{x}+\int_{\Omega T}\sigma_{ij}^I e_{ij}(\mathbf{v})\,d\mathbf{x}+\int_{\Omega_T}\sigma_{ij}^M e_{ij}(\mathbf{v})\,d\mathbf{x}=\int_{\Omega_T} f_i v_i\,d\mathbf{x}. \tag{2.5}$$

In (2.2), (2.5)

$$\mathbf{f}\in L^\infty((0,\infty); L^2(\Omega;\mathbb{R}^n))\cap L^2((0,\infty); L^2(\Omega; R^n)), \tag{2.6}$$

while $\Omega_T=(0, T)\times\Omega$. For integer k, the $W^{k,2}$ are, of course, the Sobolev spaces of L^2 functions with square integrable derivatives up to order k, while the zero subscript denotes those functions with vanishing traces on $\partial\Omega$ up to order $k-1$; via duality $W^{-k,2}=(W_0^{k,2})^*$. Spaces of fractional derivatives will be introduced below.

If we construct Galerkin approximations of the solution of (2.2)–(2.4) of the form $\mathbf{u}^n=\sum_{k=1}^n e_i(t)\mathbf{w}^i(\mathbf{x})$, we then find that for all $\mathbf{w}^k$, $k=1,2,\ldots,n$ and all $t>0$,

$$\begin{aligned}&\int_\Omega \ddot{u}_i^n w_i^k\,d\mathbf{x}+\int_n\frac{\partial W}{\partial e_{ij}}(\mathbf{e}(\mathbf{u}^n))e_{ij}(\mathbf{w}^k)\,d\mathbf{x}\\&-\int_\Omega\left(\int_0^t h(t-\tau)\frac{\partial V}{\partial e_{ij}}(\mathbf{e}(u^n(\tau))-\mathbf{e}(u^n(t)))\,d\tau\right)e_{ij}(\mathbf{w}^k)\,d\mathbf{x}=\int_\Omega f_i w_i^k\,d\mathbf{x},\end{aligned} \tag{2.7}$$

$$\int_{\Omega t} u_i^n(0,\cdot)w_i^k\,d\mathbf{x}=\int_{\Omega} u_i^0 w_i^k\,d\mathbf{x}, \tag{2.8}$$

$$\int_{\Omega} \dot u_i(0,\cdot)w_i^k\,d\mathbf{x}=\int_{\Omega} u_i^1 w_i^k\,d\mathbf{x}. \tag{2.9}$$

The problem (2.7)-(2.9) clearly possesses a unique solution on some interval $[0, t_0]$, and for this solution we may state the following.

LEMMA 2.1. *For any $T, 0\leqq T<T_\infty$, T_∞ being the maximal time of existence of the solution of (2.7)-(2.9), the solution $\mathbf{u}^n$ satisfies, for some $c_1>0$,*

$$\begin{aligned}\frac{1}{2}\int_{\Omega_T}\|\dot{\mathbf{u}}^n\|^2\,d\mathbf{x}+\int_{\Omega_T}W(\mathbf{e}(\mathbf{u}^n))\,d\mathbf{x}+c_1\int_{\Omega}\int_{M_{T,\delta}}\frac{\|\mathbf{e}(\mathbf{u}^n(\tau))-\mathbf{e}(\mathbf{u}^n(t))\|^2}{|t-\tau|^{1+\nu}}\,d\mathbf{x}\,dt\,d\tau\\ \leqq\int_0^T\int_{\Omega} f_i u_i^n\,d\mathbf{x}\,dt+\frac{1}{2}\int_{\Omega}\|\dot{\mathbf{u}}^n\|^2 d\mathbf{x}+\int_{\Omega}W(\mathbf{e}(\mathbf{u}^n))\,d\mathbf{x},\end{aligned} \tag{2.10}$$

where

$$M_{T,\delta}=\{(t,\tau)|0<t<T, 0<\tau<T, t-\delta<\tau<t+\delta, \delta>0\}.$$

Proof. We replace $\mathbf{w}^k$ in (2.7) by $\dot{\mathbf{u}}^n$ and then integrate the equation over $[0, T]$; there results

$$\begin{aligned}&\frac{1}{2}\int_{\Omega_T}\|\dot{\mathbf{u}}\|^2\,d\mathbf{x}-\frac{1}{2}\int_{\Omega}\|\dot{\mathbf{u}}^n(0)\|\,d\mathbf{x}+\int_{\Omega_T}W(\mathbf{e}(\mathbf{u}^n))\,d\mathbf{x}\\ &\quad-\int_{\Omega}W(\mathbf{e}(\mathbf{u}^n(0)))\,d\mathbf{x}\\ &\quad-\int_0^T dt\int_{\Omega}\left(\int_0^t h(t-\tau)\frac{\partial V}{\partial e_{ij}}(\mathbf{e}(\mathbf{u}^n(\tau))-\mathbf{e}(\mathbf{u}^n(t)))\,d\tau\right)\cdot\dot e_{ij}(\mathbf{u}^n(t))\,d\mathbf{x}\\ &=\int_0^t dt\int_{\Omega} f_i\dot u_i^n\,d\mathbf{x}\,dt.\end{aligned} \tag{2.11}$$

Denoting by $\nu=(\nu_t, \nu_\tau)$ the outer normal to ∂M_T, where

$$M_T=\{(t,\tau)|0<t<\tau, 0<t<\tau\},$$

we compute that for some $c_1>0$,

$$\begin{aligned}&-\int_{M_T}H(t-\tau)\frac{\partial}{\partial t}(e_{ij}(\mathbf{u}^n(t))-e_{ij}(\mathbf{u}^n(\tau)))\frac{\partial V}{\partial e_{ij}}(\mathbf{e}(u^n(\tau)-\mathbf{e}(u^n(t)))\,dt\,d\tau\\ &=\int_{M_T}h(t-\tau)V(\mathbf{e}(\mathbf{u}^n(\tau))-\mathbf{e}(\mathbf{u}^n(t)))\,dt\,d\tau\\ &=\int_{\partial M_T}h(t-\tau)\frac{\partial}{\partial t}V(\mathbf{e}(\mathbf{u}^n(\tau))-\mathbf{e}(\mathbf{u}^n(t)))\nu_t\,ds\\ &\quad-\int_{M_T}h(t-\tau)V(\mathbf{e}(\mathbf{u}^n(\tau))-\mathbf{e}(\mathbf{u}^n(t)))\,dt\,d\tau\\ &\geqq c_1\lambda\int_{M_{T,\delta}}\frac{\|\mathbf{e}(\mathbf{u}^n(\tau))-\mathbf{e}(\mathbf{u}^n(t))\|^2}{|t-\tau|^{1+\nu}}\,dt\,d\tau.\end{aligned} \tag{2.12}$$

Combining (2.11) with (2.12) we find the estimate (2.10). □

In what follows we will assume that besides (2.6), $\mathbf{f}$ satisfies the condition

$$\int_\Omega \int_0^\infty \int_0^\infty \frac{\|\mathbf{f}(\tau)-\mathbf{f}(t)\|^2}{|\tau-t|^{1+\nu}}\,dt\,d\tau\,d\mathbf{x}<\infty. \tag{2.13}$$

It follows from Lemma 2.1 that the Galerkin approximants $\mathbf{u}^n$ are defined on the entire interval $[0,\infty]$. We now make the following definition.

DEFINITION 2.2. Let B be a Banach space, $\mathbf{f}(t)\in B$, $0<t<T$, and $0<\mu<1$. Then we define

$$W^{\mu,2}((0,T);B)=\left\{\mathbf{f}\in L^2((0T);B)\,\middle|\,\int_0^T \|\mathbf{f}(t)\|_B^2\,dt+\int_0^T\int_0^T \frac{\|\mathbf{f}(t)-\mathbf{f}(\tau)\|_B^2}{|t-\tau|^{1+2\mu}} \equiv \|\mathbf{f}\|^2_{W^{\mu},2}<\infty\right\}, \tag{2.14}$$

and we have the following.

LEMMA 2.2. *For each $T>0$, $\exists C^*(T)>0$ such that the Galerkin approximants $\mathbf{u}^n$ satisfy*

$$\|\dot{\mathbf{u}}^n\|_{W^{\nu/2,2}((0,T);W^{-1,2}(\Omega,\mathbb{R}^n))}\le C^*(T). \tag{2.15}$$

Proof. Let $\boldsymbol{\phi}\in W_0^{1,2}(\Omega;\mathbb{R}^n)$, and let R_n be the projection operator mapping $W_0^{1,2}(\Omega;\mathbb{R}^n)$ to the subspace spanned by $(\mathbf{w}^1,\dots,\mathbf{w}^n)$. Then

$$\begin{aligned}\|\ddot{\mathbf{u}}(\tau)-\ddot{\mathbf{u}}^n(t)\|_{W^{-1,2}}&=\sup_{\|\boldsymbol{\phi}\|_{W^{1,2}}\le 1}\int_\Omega (\ddot{u}_i^n(\tau)-\ddot{u}_i(t))\phi_i\,d\mathbf{x}\\&=\sup_{\|\boldsymbol{\phi}\|_{W^{1,2}}\le 1}\int_\Omega (\ddot{\mathbf{u}}_i^n(\tau)-\ddot{\mathbf{u}}_i(t))(R_n\boldsymbol{\phi})_i\,d\mathbf{x}.\end{aligned} \tag{2.16}$$

We now employ (2.7) so as to estimate the right-hand side of (2.16); it is sufficient to look at the term in (2.7) generated by the memory portion of the stress σ_{ij}^M, as the estimates for the other terms follow in a straightforward fashion. We have

$$\begin{aligned}&\sup_{\|\boldsymbol{\phi}\|_{W^{1,2},\le 1}}\bigg|\int_\Omega e_{ij}(R_n\boldsymbol{\phi})\,d\mathbf{x}\int_0^{t_1} h(t_1-\tau_1)\frac{\partial V}{\partial e_{ij}}(\mathbf{e}(\mathbf{u}^n(\tau_1))-\mathbf{e}(\mathbf{u}^n(t_1)))\,d\tau_1\\&\qquad-\int_\Omega e_{ij}(R_n\boldsymbol{\phi})\,d\mathbf{x}\int_0^{t_2} h\left((t_2-\tau_2)\frac{\partial V}{\partial e_{ij}}\right)\mathbf{e}(\mathbf{u}^n(\tau_2)-\mathbf{e}(\mathbf{u}^n(t_2)))\,d\tau_2\bigg|\\&\le c_1\bigg\{\int_\Omega\sum_{i,j=1}^n\bigg[\int_0^{t_1} h(t_1-\tau_1)\frac{\partial V}{\partial e_{ij}}(\mathbf{e}(\mathbf{u}^n(\tau_1))-\mathbf{e}(\mathbf{u}^n(t_1)))\,d\tau_1\\&\qquad-\int_0^{t_2} h(t_2-\tau_2)\frac{\partial V}{\partial e_{ij}}(\mathbf{e}(\mathbf{u}^n(\tau_2))-\mathbf{e}(\mathbf{u}^n(t_2)))\,d\tau_2\bigg]^2 d\mathbf{x}\bigg\}^{1/2}\end{aligned} \tag{2.17}$$

for some $c_1>0$. For $0<t_2<t_1<T$, we may write the terms within the square bracket, on the right-hand side of the above estimate, in the form

$$\begin{aligned}&\int_0^{t_2} h(s)\left[\frac{\partial V}{\partial e_{ij}}(\mathbf{e}(\mathbf{u}^n(t_1-s))-\mathbf{e}(\mathbf{u}^n(t_1)))-\frac{\partial V}{\partial e_{ij}}(\mathbf{e}(\mathbf{u}^n(t_2-s))-\mathbf{e}((\mathbf{u}^n(t_2)))\right]ds\\&\quad+\int_{t_2}^{t} h(s)\frac{\partial V}{\partial e_{ij}}(\mathbf{e}(\mathbf{u}^n(t_1-s))-\mathbf{e}(\mathbf{u}^n(t_1)))\,ds.\end{aligned} \tag{2.18}$$

Using Schwarz's inequality on the first integral in this last expression we find that

$$\int_\Omega d\mathbf{x} \int_0^T dt_1 \int_0^{t_1} \frac{dt_2}{|t_1-t_2|^{1+\nu}} \int_0^{t_2} \left[\frac{\partial V}{\partial e_{ij}}(\mathbf{e}(\mathbf{u}^n(t_1-s))-\mathbf{e}(\mathbf{u}^n(t_1))) - \frac{\partial V}{\partial e_{ij}}(\mathbf{e}(\mathbf{u}^n(t_2-s)-\mathbf{e}(\mathbf{u}^n(t_2-s)-\mathbf{e}(\mathbf{u}^n(t_2)))\right]^2 ds$$

$$\text{(2.19)} \quad \leqq c_2(T)\int_\Omega d\mathbf{x}\int_0^T dt_1 \int_0^{t_1} \frac{dt_2}{|t_1-t_2|^{1+\nu}} \int_0^{t_2} [\|\mathbf{e}(\mathbf{u}^n(t_1-s))-\mathbf{e}(\mathbf{u}^n(t_2-s))\|^2 + \|\mathbf{e}(\mathbf{u}^n(t_1))-\mathbf{e}(\mathbf{u}^n(t_2))\|^2]\, ds$$

$$\leqq c_3(T)\int_0^T\int_0^T\int_\Omega \frac{\|\mathbf{e}(\mathbf{u}^n(t_1))-\mathbf{e}(\mathbf{u}^n(t_2))\|^2}{|t_1-t_2|^{1+\nu}}\, dt_1\, dt_2\, d\mathbf{x},$$

while for the second integral in (2.18) we have the estimates

$$\left[\int_{t_2}^{t_1} h(s)\frac{\partial V}{\partial e_{ij}}(\mathbf{e}(\mathbf{u}^n(t_1-s))-\mathbf{e}(\mathbf{u}^n(t_1)))\, ds\right]^2$$

$$\text{(2.20)} \quad \leqq c_4(T)\int_{t_2}^{t_1} h^2(s)\, ds \int_{t_2}^{t_1} \|\mathbf{e}(\mathbf{u}^n(t_1-s))-\mathbf{e}(\mathbf{u}^n(t_1))\|^2\, ds$$

$$\leqq c_5(T)[(t_1-t_2)+(t_1-t_2)^{1-2\nu}]\int_{t_2}^{t_1} \|\mathbf{e}(\mathbf{u}^n(t_1-s)-\mathbf{e}(\mathbf{u}^n(t_1))\|^2\, ds.$$

However,

$$\text{(2.21)} \quad \int_\Omega d\mathbf{x}\int_0^T dt_1\int_0^{t_1}\frac{dt_2}{|t_1-t_2|^{1+\nu}}[(t_1-t_2)+(t_1-t_2)^{1-2\nu}] \times \int_{t_2}^{t_1} \|\mathbf{e}(\mathbf{u}^n(t_1-s))-\mathbf{e}(\mathbf{u}^n(t_1))\|^2\, ds \leqq c_6(T)\sup_{0\leqq t\leqq T}\|\mathbf{u}^n(t)\|^2_{W^{1,2}}.$$

So for some $c_\eta(T)>0$,

$$\text{(2.22)} \quad \int_\Omega d\mathbf{x}\int_0^T dt_1\int_0^{t_1}\frac{dt_2}{|t_1-t_2|^{1+\nu}}\left[\int_{t_2}^{t_1} h(s)\frac{\partial V}{\partial e_{ij}}(\mathbf{e}(\mathbf{u}^n(t_1-s))-\mathbf{e}(\mathbf{u}^n(t_1)))\, ds\right]^2 \leqq c_\tau(T)\sup_{0\leqq t\leqq T}\|\mathbf{u}^n(t)\|^2_{W^{1,2}}.$$

Combining (2.16), (2.17), (2.18), (2.19), and (2.22), with the obvious estimates generated by the terms in (2.7), other than the one engendered by σ_{ij}^M, and using the definition (2.14), with $\mu=\nu/2$, we are led to (2.15).

3. An interpolation lemma. Let v^i be an orthonormal basis in $L^2(\Omega)$ formed by the eigenfunctions for the problem

$$\text{(3.1)} \quad \Delta v^i + \lambda v^i = 0, \quad \text{in } \Omega; \qquad v^i \in W_0^{1,2}(\Omega).$$

Let $0\leqq \eta \leqq \frac{3}{2}$, $-1\leqq\theta\leqq 1$, and consider an element $v\in L^2((0,T); W^{-1,2}(\Omega))$, $T>0$. Then v may be expanded in a double Fourier series

$$\text{(3.2)} \quad v = \sum_{i=0}^{\infty}\sum_{j=1}^{\infty} C_{ij}h_i(t)v^i(x),$$

where $h_0(t)=1/\sqrt{T}$, $h_i(t)=\sqrt{2/T}\cos(i\pi/T)t$. We define

$$W^{\eta,2}((0,T);\,W_0^{\theta,2}(\Omega)) \tag{3.3}$$
$$=\left\{u\in L^2((0,T);\,W^{-1,2}(\Omega))\,\Big|\,\sum_{i=0}^{\infty}\sum_{j=1}^{\infty} c_{ij}^2(1+i^2)^\eta\lambda_j^\theta\equiv\|\mathbf{u}\|_{\eta,\theta}^2<\infty\right\}.$$

For $1<\eta<\frac{3}{2}$, a norm equivalent to $\|\mathbf{u}\|_{\eta,\theta}$ is given by

$$\left[\|\mathbf{u}\|^2_{L^2((0,T);\,W^{-1,2}(\Omega))}+\int_0^T\int_0^T\frac{\|\dot{\mathbf{u}}(t)-\dot{\mathbf{u}}(\tau)\|^2_{W_0^{\theta,2}}}{|t-\tau|^{1+2(\eta-1)}}\,d\tau\,dt\right]^{1/2}. \tag{3.4}$$

For the spaces defined by (3.3) we may state the following.

LEMMA 3.1. *Let* $0\leqq\alpha\leqq 1$, $0\leqq\eta<\frac{3}{2}$. *Then there exists* $C>0$ *such that*

$$\|\mathbf{v}\|_{W^{\alpha\eta,2}}((0,T);\,W^{-\alpha,2}(\Omega))\leqq C\|\mathbf{v}\|^{1-\alpha}_{L^2((0,T);L^2(\Omega))}\|\mathbf{v}\|^{\alpha}_{W^{\eta,2}((0,T);W^{-1,2}(\Omega))}. \tag{3.5}$$

Proof. We compute directly that

$$\begin{aligned}\|\mathbf{v}\|_{W^{\alpha\eta,2}}&((0,T);\,W^{-\alpha,2}(\Omega))\\ &=\sum_{i,j}c_{ij}^2(1+i^2)^{\alpha,\eta}\lambda_j^{-\alpha}\\ &\leqq\sum_{i,j}c_{ij}^{2\alpha}(1+i^2)^{\alpha\eta}\lambda_j^{-\alpha}c_{ij}^{2(1-\alpha)}\\ &\leqq\left(\sum_{i,j}c_{ij}^2(1+i^2)^\eta\lambda_j^{-1}\right)^{\alpha}\left(\sum_{i,j}c_{ij}^2\right)^{1-\alpha}.\end{aligned}$$

□

4. Existence of a weak solution. We begin by first proving that the operator $(\partial/\partial x_j)\sigma_{ij}^M$ is Lipschitz continuous, as indicated in the remarks of § 1; this result is the essential content of the following theorem.

THEOREM 4.1. *There exists* $p>0$, *independent of* T, *such that*

$$\begin{aligned}&\|\sigma_{ij}^M(\mathbf{u}^1)-\sigma_{ij}^M(\mathbf{u}^2)\|_{L^2((0,T);L^2(\Omega;R^{n^2}))}\\ &\qquad\leqq\lambda p\|\mathbf{e}(\mathbf{u}^1)-\mathbf{e}(\mathbf{u}^2)\|_{L^2((0,T);L^2(\Omega;R^{n^2}))}.\end{aligned} \tag{4.1}$$

Proof. A direct computation yields the following series of estimates:

$$\begin{aligned}&\int_0^T dt\int_\Omega(\sigma_{ij}^M(\mathbf{u}^1))-(\sigma_{ij}^M(\mathbf{u}^2))^2\,d\mathbf{x}\\ &\leqq d_1\int_0^T dt\int_\Omega\left[\int_0^t(h(t-\tau))[\|\mathbf{e}(\mathbf{u}^1(\tau))-\mathbf{e}(\mathbf{u}^2(\tau))\|+\|\mathbf{e}(\mathbf{u}^1(t))-\mathbf{e}(\mathbf{u}^2(t))\|]\,d\tau\right]^2d\mathbf{x}\\ &\leqq d_2\int_0^T dt\int_\Omega\|\mathbf{e}(\mathbf{u}^1(t))-\mathbf{e}(\mathbf{u}^2(t))\|^2\,d\mathbf{x}\\ &\qquad+d_3\int_0^T dt\int_\Omega\left(\int_0^t h(t-\tau)\,d\tau\right)\left(\int_0^t h(t+\tau)\|\mathbf{e}(\mathbf{u}^1(\tau))-\mathbf{e}(\mathbf{u}^2(\tau))\|^2\,d\tau\right)d\mathbf{x}\\ &\leqq d_4\int_0^T dt\int_\Omega\|\mathbf{e}(\mathbf{u}^1(t))-\mathbf{e}(\mathbf{u}^2(t))\|^2\,d\mathbf{x}\\ &\qquad+d_5\int_0^T d\tau\int_\Omega\|\mathbf{e}(\mathbf{u}^1(\tau))-\mathbf{e}(\mathbf{u}^2(\tau))\|^2\,d\mathbf{x}\int_\tau^T h(t-\tau)\,dt\\ &\leqq d_6\int_0^T dt\int_\Omega\|\mathbf{e}(\mathbf{u}^1(t))-\mathbf{e}(\mathbf{u}^2(t))\|^2\,d\mathbf{x}.\end{aligned} \tag{4.2}$$

□

THEOREM 4.2 (Global existence of a weak solution). *The initial boundary value problem* (2.2)–(2.4) *possesses, under the conditions* (1.2)–(1.6), *with* $W(0) = V(0) = (\partial W/\partial e_{ij})(0) = (\partial V/\partial e_{ij})(0) = 0$, $\mathbf{u}^0 \in W_0^{1,2}(\Omega; R^n)$, $\mathbf{u}^1 \in L^2(\Omega; \mathbb{R}^n)$, *and* (2.6), (2.13), *a weak solution on* $(0, \infty)$ *for* $\lambda p < \hat{\alpha}$ *and* $\nu < \frac{1}{2}$. *The solution* $\mathbf{u}$ *satisfies*

(4.3a) (i) $\mathbf{u} \in L^\infty((0,\infty); W_0^{1,2}(\Omega; R^n))$,

(4.3b) (ii) $\dot{\mathbf{u}} \in L^\infty((0,\infty); L^2(\Omega; \mathbb{R}^n))$,

(4.3c) (iii) $\mathbf{u} \in W^{\nu/2,2}((0, T); W_0^{1,2}(\Omega; R^n)) \quad \forall T > 0$,

(4.3d) (iv) $\mathbf{u} \in W^{\nu/2,2}((0, T); W^{-1,2}(\Omega; \mathbb{R}^n)) \cap L^2((0, T); W^{-1/(1+(\nu/2))}(\Omega; \mathbb{R}^n)) \quad \forall T > 0$.

Remark. The condition $\lambda p < \hat{\alpha}$ yields the result that the operator $(\partial/\partial x_j)\sigma_{ij} = (\partial/\partial x_j)\sigma_{ij}^I + (\partial/\partial x_j)\sigma_{ij}^M$ is strongly monotone.

Proof. Let $1/(1+(\nu/2)) < \alpha < 1$. Then it follows from our interpolation Lemma 3.1, with $v = \dot{u}_i^n$, that $\dot{u}_i^n$ is a bounded sequence in $W^{(1+(\nu/2))\alpha,2}((0, T); W^{-\alpha,2}(\Omega))$, which, in turn, is compactly embedded in $W^{1,2}((0, T); W^{-1,2}(\Omega))$. Therefore, we may choose a subsequence $\mathbf{u}^{(n_k)}$, such that for each $T > 0$, and some $\mathbf{u}$ satisfying (4.3a–d),

(4.4a) $\mathbf{u}^{n_k} \to \mathbf{u}$, in $L^2((0, T); W_0^{1,2}(\Omega; \mathbb{R}^n))$,

(4.4b) $\dot{\mathbf{u}}^{n_k} \to \dot{\mathbf{u}}$, in $L^2((0, T); L^2(\Omega; \mathbb{R}^n))$,

(4.4c) $\mathbf{u}^{n_k} \to \mathbf{u}$, in $W^{\nu/2,2}((0, T); W^{1,2}(\Omega; \mathbb{R}^n))$,

(4.4d) $\ddot{\mathbf{u}}^{n_k} \to \ddot{\mathbf{u}}$, in $W^{\nu/2,2}((0, T); W^{-1,2}(\Omega; \mathbb{R}^n))$,

(4.4e) $\ddot{\mathbf{u}}^{n_k} \to \ddot{\mathbf{u}}$, in $L^2((0, T); W^{-1,2}(\Omega; R^n))$.

Now, let P_n be the projection operator from $L^2((0, T); W_0^{1,2}(\Omega; \mathbb{R}^n))$ to the space spanned by the vectors $c_j(t)\mathbf{w}^j(t)$, where, for $j = 1, \ldots, n$, the $c_j \in L^2((0, T))$. From (2.7) we obtain

$$\begin{aligned}
&\int_0^T \int_\Omega \ddot{u}_i^{n_k}((\ddot{u}_i^{n_k} - P_{n_k}\mathbf{u})_i)\, d\mathbf{x}\, dt \\
&+ \int_0^T \int_\Omega \frac{\partial W}{\partial e_{ij}}(\mathbf{e}(\mathbf{u}^{n_k}))e_{ij}(\mathbf{u}^{n_k} - P_{n_k}\mathbf{u})\, d\mathbf{x}\, dt \\
&- \int_0^T \int_\Omega \left[\int_0^t h(t-\tau)\frac{\partial V}{\partial e_{ij}}(\mathbf{e}(\mathbf{u}^{n_k}(\tau)) - \mathbf{e}(\mathbf{u}^{n_k}(t)))\, d\tau\right] \cdot e_{ij}(\mathbf{u}^{n_k} - P_{n_k}\mathbf{u})\, d\mathbf{x}\, dt \\
&= \int_0^T \int_\Omega f_i(u_i^{n_k} - (P_{n_k}\mathbf{u})_i)\, d\mathbf{x}\, dt.
\end{aligned} \tag{4.5}$$

For $\mathbf{u} \in W_0^{1,2}(\Omega; \mathbb{R}^n)$, Korn's inequality, i.e.,

$$\int_\Omega e_{ij}(\mathbf{u})e_{ij}(\mathbf{u})\, d\mathbf{x} \geqq \kappa \|\mathbf{u}\|^2_{W^{1,2}(\Omega;\mathbb{R}^n)}, \tag{4.6}$$

holds for some $\kappa > 0$ (see, e.g., [15]). Now we know that as $n_k \to \infty$

$$P_{n_k}\mathbf{u} - \mathbf{u}, \quad \text{in } L^2((0, T); W_0^{1,2}(\Omega; R^n)). \tag{4.7}$$

Therefore, both

$$\int_0^T \int_\Omega \frac{\partial W}{\partial e_{ij}}(\mathbf{e}(P_{n_k}\mathbf{u}))e_{ij}(\mathbf{u}^{n_k} - P_{n_k}\mathbf{u})\, d\mathbf{x}\, dt \to 0 \tag{4.8}$$

and

$$\int_0^T \int_\Omega \left[\int_0^t h(t-\tau)\frac{\partial V}{\partial e_{ij}}(\mathbf{e}(P_{n_k}\mathbf{u})(\tau)) - \mathbf{e}(P_{n_k}\mathbf{u}(t))\,d\tau\right] e_{ij}(\mathbf{u}^{n_k} - P_{n_k}\mathbf{u})\,d\mathbf{x}\,dt \to 0 \tag{4.9}$$

as $n_k \to \infty$. By virtue, therefore, of (4.8), (4.9), (4.4e), and (4.1), with $\lambda p < \alpha$, we find that as $n_k \to \infty$

$$\mathbf{u}^{n^k} \to \mathbf{u}, \text{ in } L^2((0, T);\ W_0^{1,2}(\Omega; \mathbb{R}^n)), \tag{4.10}$$

and the existence of the required global weak solution follows as a direct consequence of (2.7). □

REFERENCES

[1] S. O. LONDEN, *An existence result for a Volterra equation in a Banach space*, Trans. Amer. Math. Soc., 235 (1978), pp. 285–304.

[2] M. RENARDY, *Some remarks on the propagation and non-propagation of discontinuities in linearly viscoelastic liquids*, Rheol. Acta, 21 (1982), pp. 251–254.

[3] W. J. HRUSA AND M. RENARDY, *On wave propagation in linear viscoelasticity*, Quart. Appl. Math., 43 (1985), pp. 237–254.

[4] K. B. HANNSGEN AND R. L. WHEELER, *Behavior of the solutions of a Volterra equation as a parameter tends to infinity*, J. Integral Equations, 7 (1984), pp. 229–237.

[5] W. J. HRUSA AND M. RENARDY, *On a class of quasilinear partial integrodifferential equations with singular kernels*, J. Differential Equations, 64 (1986), pp. 195–220.

[6] ———, *A model equation for viscoelasticity with a strongly singular kernel*, SIAM J. Math. Anal, 19 (1988), pp. 257–269.

[7] M. RENARDY, *Coercive estimates and existence of solutions for a model of one-dimensional viscoelasticity with a nonintegrable memory function*, J. Integral Equations Appl., to appear.

[8] M. DOI AND S. F. EDWARDS, *Dynamics of concentrated polymer systems*, J. Chem. Soc. Faraday, 74 (1978), pp. 1789–1832; 75 (1979), pp. 38–54.

[9] P. E. ROUSE, *A theory of the linear viscoelastic properties of dilute solutions of coiling polymers*, J. Chem. Phys. 21 (1953), pp. 1271–1280.

[10] B. H. ZIMM, *Dynamics of polymer molecules in dilute solutions: viscoelasticity, flow birefringince, and dielectric loss*, J. Chem. Phys, 24 (1956), pp. 269–278.

[11] H. M. LAUN, *Description of the nonlinear shear behavior of a low density polyethylene melt by means of an experimentally determined strain dependent memory function*, Rheol. Acta, 17 (1978), pp. 1–15.

[12] J. MILOTA, J. NEČAS, AND V. ŠVEŔAK, *On weak solutions to a viscoelasticity model*, Comment. Math. Univ. Carolin., 31 (1990), pp. 557–565.

[13] H. ENGLER, *Weak solutions of a class of quasilinear hyperbolic integro-differential equations describing viscoelastic materials*, Arch. Rational Mech. Anal., 113 (1991), pp. 1–38.

[14] B. D. COLEMAN, *On thermodynamics, strain impulses, and viscoelasticity*, Arch. Rational Mech. Anal., 17 (1964), pp. 1–46.

[15] J. NEČAS AND I. HLAVÁČEK, *Mathematical Theory of Elastic and Elastico-Plastic Bodies*, Elsevier, New York, 1981.

[16] P. D. LAX, *Development of singularities of solutions of nonlinear hyperbolic partial differential equations*, J. Math. Phys., 5 (1964), pp. 611–613.

[17] S. KLAINERMAN AND A. MAJDA, *Formation of singularities for wave equations including the nonlinear vibrating string*, Comm. Pure Appl. Math, 33 (1980), pp. 241–263.

[18] R. DIPERNA, *Convergence of approximate solutions to conservation laws*, Arch. Rational Mech. Anal., 82 (1983), pp. 27–70.

[19] T. J. R. HUGHES, T. KATO, AND T. E. MARSDEN, *Well-posed quasi-linear hyperbolic systems with applications to nonlinear elastodynamics and general relativity*, Arch. Rational Mech. Anal., 63 (1977), pp. 273–294.

[20] S. S. ANTMAN, *The influence of elasticity on analysis: modern developments*, Bull. Amer. Math. Soc., 9 (1983), pp. 267–291.

[21] J. E. MARSDEN AND T. J. R. HUGHES, *Mathematical Foundations of Elasticity*, Prentice-Hall, Englewood Cliffs, NJ, 1983.

Math. Ann. 299, 275–291 (1994)

Mathematische Annalen

© Springer-Verlag 1994

Existence of global weak solutions for a class of quasilinear hyperbolic integro-differential equations describing visco-elastic materials

Hamid Bellout[1] **and Jindrich Nečas**[2]

[1] Department of Mathematical Sciences, Northern Illinois University, DeKalb, IL 60115, USA
[2] Mathematical Institute, Charles University, Prague, Czech Republic

Received January 16, 1993; in revised form September 22, 1993

Mathematics Subject Classification (1991): 35Q99, 45K05, 73F15

1 Introduction

We are interested in the existence of a global solution to the Cauchy-Dirichlet problem, for large data, for the nonlinear wave equation

$$\frac{\partial^2 u_i}{\partial t^2} - \frac{\partial \sigma_{ij}}{\partial x_j} = f_i \quad \text{in} \quad (0,\infty) \times \Omega . \tag{1.1}$$

This equation governs the evolution of the displacement vector $\mathbf{u}$ in the elastic body $\Omega \subset R^N$, $N \geqq 2$. $\sigma_{ij} = \sigma_{ij}(\mathbf{e})$, where $\mathbf{e}$ is the infinitesimal stress tensor whose components are $e_{ij} = \frac{1}{2}\left(\frac{\partial u_i}{\partial x_j} + \frac{\partial u_j}{\partial x_i}\right)$.

Here the Cauchy stress tensor will be assumed to take the form

$$\sigma_{ij} = \sigma^I_{ij} + \sigma^M_{ij} ,$$

where σ^I_{ij} is the part corresponding to an instantaneous elastic response and σ^M_{ij} is the part of the stress tensor expressing the memory of the elastic body in the form due to Boltzmann [26, p. 4].

We will assume that

$$\sigma^I_{ij} = \frac{\partial W}{\partial e_{ij}}; \qquad e_{ij} = \frac{1}{2}\left(\frac{\partial u_i}{\partial x_j} + \frac{\partial u_j}{\partial x_i}\right). \tag{1.2}$$

The stored energy function W is supposed continuously differentiable, and such that, $W(\mathbf{o}) = 0$, $\dfrac{\partial W(\mathbf{o})}{\partial e_{ij}} = 0$. Also, $\exists p \in [2,\infty)$ and $\exists \alpha_1 > 0$, $\exists \alpha_2 > 0$ $\exists \alpha_3 > 0$ such that

$$|\sigma^I_{ij}(\mathbf{e})| \leqq \alpha_1(1 + |\mathbf{e}|^{p-1}), \tag{1.3}$$

$$(\sigma^I_{ij}(\mathbf{e}(\mathbf{u})) - \sigma^I_{ij}(\mathbf{e}(\mathbf{v}))) \cdot (\varepsilon_{ij}(\mathbf{u}) - e_{ij}(\mathbf{v})) \geqq \alpha_2(1 + |\mathbf{e}(\mathbf{u})|^{p-2} + |\mathbf{e}(\mathbf{v})|^{p-2})\,|\mathbf{e}(\mathbf{u}) - \mathbf{e}(\mathbf{v})|^2 + \alpha_3|\mathbf{e}(\mathbf{u}) - \mathbf{e}(\mathbf{v})|^p . \tag{1.4}$$

© Springer-Verlag 1994. Reprinted with kind permission from: Bellout, H. & Nečas, J., Existence of global weak solutions for a class of quasilinear hyperbolic integro-differential equations describing visco-elastic materials, Mathematische Annalen 299, 275-291.

A crucial role is plaed by the memory part σ_{ij}^M which is of the type

$$\sigma_{ij}^M = -\int_0^t h(t-\tau)\frac{\partial V}{\partial e_{ij}}(\mathbf{e}(\tau)-\mathbf{e}(t))d\tau\,, \tag{1.5}$$

where $h(t) = \gamma h_0(t)$ and $h_0(t) = t^{-\nu}a(t) + r(t)$, $0 < \nu < 1$, $\gamma > 0$. $a(t)$ and $r(t)$ are decreasing, nonnegative, smooth functions in $[0,\infty)$. Our proof requires only that $a(t) > 0$ in the neighborhood of the origin. However, in order to simplify the exposition we will assume that $a(0) > 0$ and $a(t) \geqq 0$. We suppose also that $\int_0^\infty h(\tau)d\tau < \infty$.

The stored energy function V is assumed to be twice continuously differentiable and such that $V(\mathbf{o}) = 0$, $\frac{\partial V}{\partial e_{ij}}(\mathbf{o}) = 0$. Furthermore $\exists q \in [2,\infty)$, $\alpha > 0$, $\beta > 0$ such that for all symmetric $N \times N$ matrices ξ_{ij}.

$$V(\mathbf{e}) \geqq \alpha|\mathbf{e}|^q \tag{1.6}$$

$$0 \leqq \frac{\partial^2 V(\mathbf{e})}{\partial e_{ij}\partial e_{kl}}\xi_{ij}\xi_{kl} \tag{1.7}$$

$$\left|\frac{\partial^2 V(\mathbf{e})}{\partial e_{ij}\partial e_{kl}}\right| \leqq \beta|\mathbf{e}|^{q-2}\,. \tag{1.8}$$

We prove the existence of a global weak solution, for large data, to the Cauchy-Dirichlet problem associated to the partial differential Eq. (1.1) provided that p and q satisfy the relations

$$p \geqq 2\,, \qquad q \leqq 2\,, \tag{1.9}$$

$$p\left[1 - \frac{\nu}{2} - \frac{\nu}{N}\right] + \nu < q \leqq \frac{p}{2} + 1\,. \tag{1.10}$$

The values of p and q which satisfy the conditions above, for $N = 3$, are shown in the graphs.

Elastic materials with singular memories have been considered by several authors, e.g., Renardy, [24], Hrusa and Renardy [13, 15], Hannsgen and Wheeler [14], Engler [9], Londen [19], Milota et al. [21].

For the special case $p = q = 2$ in (1.3), (1.4), (1.6), (1.8), the existence of a global weak solution, for the higher space dimension case ($N \geqq 2$), was established in [4] by Bellout et al. The proof of [4] uses fully and profoundly the special nature of the case $p = q = 2$. The proof differs in a substantial way from that of [4] because of the nonquadratic growth and some basic new ideas are used: first there is the problem with the Schauder basis described below and also with the control of the memory term which is no longer dominated by the instantaneous elastic response.

To the best of our knowledge all of the previously known global existence results of weak solutions to hyperbolic integro-differential equations were proved either for the easier 1 space dimension case or for the special case of quadratic growth of the storage functions V and W (i.e., $p = q = 2$ in (1.3), (1.4), (1.6), (1.8)). A detailed discussion of the previously known results is provided in the introduction of the recent paper [4].

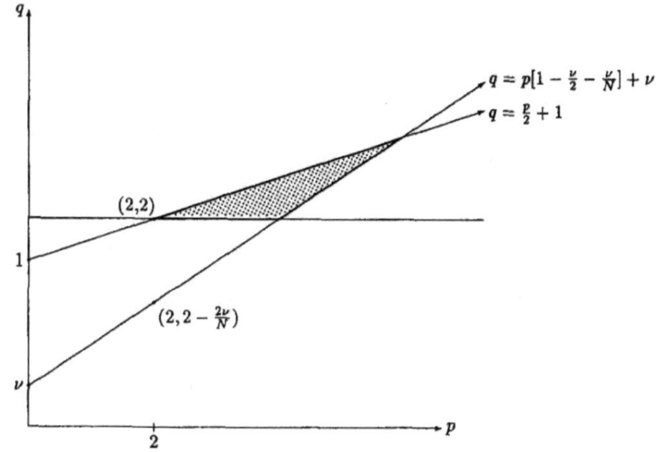

Fig. 1 For $\nu < \frac{3}{5}$

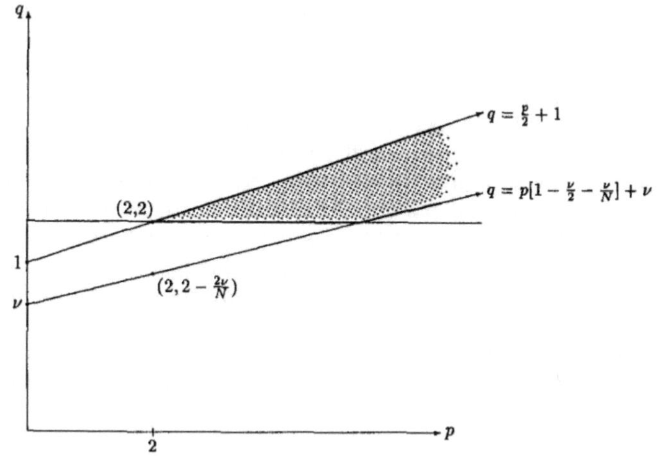

Fig. 2 For $\nu > \frac{3}{5}$

For a more complete review of related work as well as a discussion of the physical origin of this problem see the survey paper of Antman [2]; see also [4, 9], the monograph of Renardy et al. [26], and the references therein.

The existense of a solution is established via the Galerkin method. The blueprint for applying this method (first used in [12] for non-linear evolution equations) is fairly simple and well known. The typical essential (often the only) difficulty in applying it to nonlinear equaitons resides in showing that the nonlinear terms (involving the approximate solution) of the equation (in our particular case, $\sigma(\mathbf{u}_m)$) converge to the corresponding term ($\sigma(\mathbf{u})$). An extremely effective "trick" (see [18, Sect. 6.4 and Sect. 1.7]), which has been in use for several decades, consists in showing that $\mathbf{u}_m''$ ($\mathbf{u}_m'$ for the parabolic case) is regular in time ($\mathbf{u}_m'' \in L^2(0, T; W^{-1,2}(\Omega))$ for example) and using compactness results of the type of Aubin's lemma. (See Theorem 5.1 of [18] for example.)

The usual method used for deriving the necessary regularity in time (of $\mathbf{u}''_m$) requires that the Schauder basis used in the Galerkin approximation consist of functions which are *mutually orthogonal in* $L^2(\Omega)$. (See Sect. 1.7 and Sect. 6.4 of [18].) The existence of a Schauder basis with this property in the context of the Hilbertien L^2 structure of $W^{s,2}$ is well known and standard (for example the eigenfunctions of the Laplace equation with homogeneous Dirichlet boundary condition make up such a basis).

The existence of a Schauder basis for $W^{1,p}$, $p \neq 2$, with the above property seems to be an open question. In particular it does not seem to be known whether the eigenfunctions of the Laplacian make a basis of this space or not. The existence of a Schauder basis for $W^{s,p}(\Omega)p \neq 2$, with the above property has been established in the particular case where Ω is a cube (see [6]).

The existence of a Schauder basis for $W^{s,p}(\Omega)$, $p \neq 2$, for a general domain Ω was proved in [10], but it is not clear that the elements of this basis are mutually orthogonal in L^2.

It should be mentioned that the only cases of use of the "trick" described above that we are acquainted with were done within the framework of L^2 structures, i.e., $W^{s,2}$.

We introduce here a variation on the "trick" mentioned above, namely that for an appropriate basis (the existence of which is proved in [3]) there exists an operator Q_m such that $Q_m(\mathbf{u}_m)''$ has the in time regularity sought. This method may be the most important result in the paper.

The other notable difficulty here has to do with the presence of the memory part σ^M. The presence of σ^M provides us with some regularity in time of the approximate solution, which is critical in proving the existence of a solution. On the other hand, it is not clear, a priori, how its presence affects the monotonicity of the operator σ. Contrary to the case $p = q = 2$ of [4], here σ^I no longer dominates the memory part σ^M. However, it turns out that the non-monotone part of σ^M is dominated by σ^I so that for γ small enough, the operator σ is monotone. This is the object of Lemma 1.2.

The conditions (1.2)–(1.4) reflect the assumptions of small deformations and nonlinear constitutive behavior. This point of view is common and a classical exmaple is the theory of elast-plasticity with Hencky-Mise's relation

$$\sigma_{ij} = \left(- \tfrac{2}{3}\, \mu(\Gamma)\right) \delta_{ij} e_{ll} + 2\mu(\Gamma) e_{ij}\,,$$

where $\Gamma = a_{ij}a_{ij}$ and $a_{ij} = e_{ij} - \frac{1}{3}\,\delta_{ij}e_{ll}$. In the case of large deformations and non-linear Hooke's law, the mathematical difficulties seem to be insurmountable; in fact, even in the steady state case, the equilibrium equation has not yet been proven to have a rigorous mathematical sense.

The special form of the memory term (1.5), instead of the general formula

$$\sigma^M_{ij} = - \int_0^t K_{ij}(t-\tau, (\mathbf{e}(\tau) - \mathbf{e}(t))\,d\tau\,,$$

was introduced for the sake of simplicity. This special form of the memory term has been considered, in the unidimensional as well as the multidimensional case, by several authors (see, for example [9, 4, 26] and the references therein).

2 Some auxiliary results

We will use the standard notation $W^{s,r}(\Omega)$, $W_0^{s,r}(\Omega)$ to denote the usual Sobolev spaces $W^{s,r}(\Omega; R^N)$, $W_0^{s,r}(\Omega; R^N)$, (see [1, 18] for example).

Definition 2.1. *Let B be a Banach space, $u(t) \in B$, $0 < t < T$ we define:*
- *for $r \geqq 1$, $L^r(0,T;B) = \{u(t) \in B \mid \|u(t)\|_B \in L^r(0,T)\}$.*
- *for $r \in (1,\infty)$, $0 < \mu < 1$,*

$$W^{\mu,r}(0,T;B) = \left\{ u \in L^r(0,T;B) \,\middle|\, \int_0^T \|u(t)\|_B^r \,dt + \int_0^T\int_0^T \frac{\|u(t)-u(\tau)\|_B^r}{\|t-\tau\|^{1+r\mu}} \right.$$
$$\left. \equiv \|u\|^r_{W^{\mu,r}(0,T;B)} < \infty \right\}. \tag{2.1}$$

- for $r \in (1,\infty)$, $0 < \mu < \dfrac{1}{r}$ we denote the dual of $W^{\mu,r}(0,T;B)$ by $W^{-\mu,r'}(0,T;B^*)$.

Various definitions of the spaces $W^{s,r}(0,T;B)$ exist in the literature (see [5, 11, 22, 28]) and various results about the properties of functions in these spaces have been proved using these different definitions. The compactness result of the next proposition may already be known but we were unable to find a reference (based on or consistent with the above definition) for it. For this reason and for the convenience of the reader we will prove it here.

We start with a compactness result.

Proposition 2.1. *Let the sequence $\mathbf{f}_n$ be such that:*
1. *$\mathbf{f}_n$ is uniformly bounded in $L^2(0,T;L^2(\Omega))$;*
2. *$\dfrac{\partial \mathbf{f}_n}{\partial t}$ is uniformly bounded in $L^\infty(0,T;W^{-1,p'}(\Omega))$.*

Then $\forall\lambda \in (0,1)$; $\forall\delta > 0$ such $\lambda - \delta > 0$

a. *$\mathbf{f}_n$ is a compact sequence in $W^{\lambda-\delta,2}(0,T;W^{-\lambda-\delta,s}(\Omega))$.*

b. *$\dfrac{\partial \mathbf{f}_n}{\partial t}$ is a compact sequence in $W^{-1+\lambda-\delta,2}(0,T;W^{-\lambda-\delta,s}(\Omega))$,*

where s satisfies $\dfrac{1}{s} = \dfrac{\lambda}{p'} + \dfrac{1-\lambda}{2}$.

Proof. Part a. To prove part a, it is enough to show that the sequence $\mathbf{f}_n$ has a subsequence, denoted again by $\mathbf{f}_n$, which converges in the strong topology of $W^{\lambda-\delta,2}(0,T;W^{-\lambda-\delta,s}(\Omega))$. Without loss of generality we can assume that the sequence $\mathbf{f}_n$ converges weakly to 0 in $L^2(0,T;L^2(\Omega))$. From assumptions 1 and 2 it follows by Aubin's theorem (see [18]) that $\forall\varepsilon > 0$, $\mathbf{f}_n$ converges stronlgy to 0 in $L^2(0,T;W^{-\varepsilon,2}(\Omega))$.

From standard results of interpolation theory (see [29, p. 186] for example) we have that for $\lambda \in (0,1)$, $\delta > 0$, and $\varepsilon = \dfrac{\delta}{1-\lambda}$, $\exists c > 0$ such that $\forall n, \forall t, \forall\tau$;

$$\|\mathbf{f}_n(t) - \mathbf{f}_n(\tau)\|_{W^{-\lambda-\delta,s}(\Omega)} \leqq c\|\mathbf{f}_n(t) - \mathbf{f}_n(\tau)\|^{\lambda}_{W^{-1,p'}(\Omega)}$$
$$\times \|\mathbf{f}_n(t) - \mathbf{f}_n(\tau)\|^{1-\lambda}_{W^{-\varepsilon,2}(\Omega)},$$

where $\frac{1}{s} = \frac{\lambda}{p'} + \frac{1-\lambda}{2}$. Hence,

$$\int_0^T \int_0^T \frac{\|\mathbf{f}_n(t) - \mathbf{f}_n(\tau)\|^2_{W^{-\lambda-\delta,s}(\Omega)}}{|t-\tau|^{1+2(\lambda-\delta)}}\, dt\, d\tau$$

$$\leqq c \sup_{0\leqq t\leqq T,\, 0\leqq \tau\leqq T} \left\| \frac{\mathbf{f}_n(t) - \mathbf{f}_n(\tau)}{t-\tau} \right\|^{2\lambda}_{W^{-1,p'}(\Omega)}$$

$$\times \int_0^T \int_0^T \frac{\|\mathbf{f}_n(t) - \mathbf{f}_n(\tau)\|^{2(1-\lambda)}_{W^{-\varepsilon,2}(\Omega)}}{|t-\tau|^{1-2\delta}}\, dt\, d\tau$$

$$\leqq c \left(\left\| \frac{\partial \mathbf{f}_n}{\partial t} \right\|^2_{L^\infty(0,T;W^{-1,p'}(\Omega))} \right)^\lambda \left(\int_0^T \|\mathbf{f}_n(t)\|^2_{W^{-\varepsilon,2}(\Omega)} dt \right)^{1-\lambda}. \tag{2.2}$$

The Holder inequality, and the fact that $\int_0^T \int_0^T |t-\tau|^{(1-\theta)}\, dt\, d\tau$ is finite for $\theta \in (0,1)$, were used above.

Hence,

$$\|\mathbf{f}_n\|^2_{W^{\lambda-\delta,2}(0,T;W^{-\lambda-\delta,s}(\Omega))} \leqq c \left(\left\| \frac{\partial \mathbf{f}_n}{\partial t} \right\|^1_{L^\infty(0,T;W^{-1,p'}(\Omega))} \right)^\lambda$$

$$\times \left(\|\mathbf{f}_n\|^2_{L^2(0,T;W^{-\varepsilon,2}(\Omega))} \right)^{(1-\lambda)2}, \tag{2.3}$$

from which part a of the proposition follows immediately.

Part b. Using that for $r \in (1,\infty)$, $0 < \mu < \frac{1}{r}$, $W_0^{\mu,r}(0,T;B) \equiv W^{\mu,r}(0,T;B)$ and that $\{W_0^{1,r}(0,T;B)\}^* \equiv \left\{ \frac{\partial u}{\partial t} \,\middle|\, u \in L^{r'}(0,T;B) \right\}$ part b of the proposition becomes a simple consequence of part a, and (3.a) p. 69 of [29].

Next, we prove a general integral inequality which will be used to show that σ is a monotone operator.

Lemma 2.1. *Under the assumptions* (1.3)–(1.8) *and assuming that* $p \geqq 2q-2$ *we have that*

$$\int_0^T \int_{\Omega_t} (\sigma^M_{ij}(\mathbf{u}) - \sigma^M_{ij}(\mathbf{v}))(e_{ij}(\mathbf{u}) - e_{ij}(\mathbf{v}))\, dx\, dt$$

$$\geqq -c\gamma \int_0^T \int_{\Omega_t} (\sigma^I_{ij}(\mathbf{u}) - \sigma^I_{ij}(\mathbf{v}))(e_{ij}(\mathbf{u}) - e_{ij}(\mathbf{v}))\, dx\, dt\,, \tag{2.4}$$

where $c = \frac{\beta}{\alpha_2} \int_0^\infty h_0(s)\, ds \cdot \min(1; 2^{q-3})$.

Proof.

$$
\begin{aligned}
&(\sigma_{ij}^M(\mathbf{u}) - \sigma_{ij}^M(\mathbf{v}))(e_{ij}(\mathbf{u}) - e_{ij}(\mathbf{v}))\\
&= -\int_0^t h(t-\tau)\left(\frac{\partial V(\mathbf{e}(\mathbf{u}(\tau)) - \mathbf{e}(\mathbf{u}(t)))}{\partial e_{ij}} - \frac{\partial V(\mathbf{e}(\mathbf{v}(\tau)) - \mathbf{e}(\mathbf{v}(t)))}{\partial e_{ij}}\right) d\tau\\
&\times (e_{ij}(\mathbf{u}(t)) - e_{ij}(\mathbf{v}(t))).
\end{aligned}
\tag{2.5}
$$

Using that

$$
\begin{aligned}
&\left(\frac{\partial V(\mathbf{e}(\mathbf{u}(\tau)) - \mathbf{e}(\mathbf{u}(t)))}{\partial e_{ij}} - \frac{\partial V(\mathbf{e}(\mathbf{v}(\tau)) - \mathbf{e}(\mathbf{v}(t)))}{\partial e_{ij}}\right)\\
&= \int_0^1 \Bigg(\frac{\partial^2 V(\mathbf{e}(\mathbf{v}(\tau)) - \mathbf{e}(\mathbf{v}(t)) + \lambda(\mathbf{e}(\mathbf{u}(\tau)) - \mathbf{e}(\mathbf{u}(t)) - \mathbf{e}(\mathbf{v}(\tau)) + \mathbf{e}(\mathbf{v}(t))))}{\partial e_{ij}\partial e_{kl}}\\
&\times (e_{kl}(\mathbf{u}(\tau)) - e_{kl}(\mathbf{u}(t)) - e_{kl}(\mathbf{v}(\tau)) + e_{kl}(\mathbf{v}(t)))\Bigg) d\lambda,
\end{aligned}
\tag{2.6}
$$

we find that

$$
\begin{aligned}
&(\sigma_{ij}^M(\mathbf{u}) - \sigma_{ij}^M(\mathbf{v}))(e_{ij}(\mathbf{u}) - e_{ij}(\mathbf{v}))\\
&= -\int_0^t h(t-\tau)\int_0^1\left(\frac{\partial^2 V}{\partial e_{ij}\partial e_{kl}}\right) d\lambda\\
&\times (e_{kl}(\mathbf{u}(\tau)) - e_{kl}(\mathbf{v}(\tau)))(e_{ij}(\mathbf{u}(t)) - e_{ij}(\mathbf{v}(t)))\,d\tau\\
&+ \int_0^t h(t-\tau)\int_0^1\left(\frac{\partial^2 V}{\partial e_{ij}\partial e_{kl}}\right) d\lambda\\
&\times (e_{kl}(\mathbf{u}(t)) - e_{kl}(\mathbf{v}(t)))(e_{ij}(\mathbf{u}(t)) - e_{ij}(\mathbf{v}(t)))\,d\tau.
\end{aligned}
\tag{2.7}
$$

By (1.7) the second term in the right side of equality above is nonnegative.

Using (1.8) to estimate from below the first term in the right side of equality (2.7) we find that

$$
\begin{aligned}
&\int_0^T\int_{\Omega_t}(\sigma_{ij}^M(\mathbf{u}) - \sigma_{ij}^M(\mathbf{v}))(e_{ij}(\mathbf{u}) - e_{ij}(\mathbf{v}))\,dx\,dt\\
&\geqq -\beta c_1\int_0^T\int_{\Omega_t}\int_0^t h(t-\tau)\\
&\times [|\mathbf{e}(\mathbf{u}(\tau))|^{q-2} + |\mathbf{e}(\mathbf{u}(t))|^{q-2} + |\mathbf{e}(\mathbf{v}(\tau))|^{q-2} + |\mathbf{e}(\mathbf{v}(t))|^{q-2}]\\
&\times |\mathbf{e}(\mathbf{u}(\tau)) - \mathbf{e}(\mathbf{v}(\tau))|\,|\mathbf{e}(\mathbf{u}(t)) - \mathbf{e}(\mathbf{v}(t))|\,d\tau\,dx\,dt,
\end{aligned}
\tag{2.8}
$$

where $c_1 = \frac{1}{2}\min(1; 2^{q-3})$.

Multiplying inside the square brackets we find 4 terms. We will treat the first such term; the others are similar.

$$I_1 = \int_0^T \int_{\Omega_t} \int_0^t h(t-\tau)\, |\mathbf{e}(\mathbf{u}(\tau))|^{q-2}\, |\mathbf{e}(\mathbf{u}(\tau)) - \mathbf{e}(\mathbf{v}(\tau))|$$

$$\times\, |\mathbf{e}(\mathbf{u}(t)) - \mathbf{e}(\mathbf{v}(t))|\, d\tau\, dx\, dt$$

$$\leq \int_0^T \int_{\Omega_t} \left(\int_0^t h(t-\tau)\, |\mathbf{e}(\mathbf{u}(\tau))|^{2q-4}\, |\mathbf{e}(\mathbf{u}(\tau)) - \mathbf{e}(\mathbf{v}(\tau))|^2\, d\tau \right)^{1/2}$$

$$\times \left(\int_0^t h(t-\tau)\, d\tau \right)^{1/2} |\mathbf{e}(\mathbf{u}(t)) - \mathbf{e}(\mathbf{v}(t))|\, dt$$

$$\leq \left(\int_0^T \int_{\Omega_t} \int_0^t h(t-\tau)\, |\mathbf{e}(\mathbf{u}(\tau))|^{2q-4}\, |\mathbf{e}(\mathbf{u}(\tau)) - \mathbf{e}(\mathbf{v}(\tau))|^2\, d\tau\, dx\, dt \right)^{1/2}$$

$$\times \left(\int_0^T \int_{\Omega_t} \left(\int_0^t h(t-\tau)\, d\tau \right)^2 |\mathbf{e}(\mathbf{u}(t)) - \mathbf{e}(\mathbf{v}(t))|^2\, dx\, dt \right)^{1/2}$$

$$\leq \frac{1}{2} \int_0^T \int_{\Omega_t} \int_0^t h(t-\tau)\, |\mathbf{e}(\mathbf{u}(\tau))|^{2q-4}\, |\mathbf{e}(\mathbf{u}(\tau)) - \mathbf{e}(\mathbf{v}(\tau))|^2\, d\tau\, dx\, dt$$

$$+ \frac{1}{2} \int_0^T \int_{\Omega_t} \left(\int_0^t h(t-\tau)\, d\tau \right)^2 |\mathbf{e}(\mathbf{u}(t) - \mathbf{e}(\mathbf{v}(t))|^2\, dx\, dt\,.$$

Changing the order of integration in the term before last and using that h is integrable we find

$$I_1 \leq \left(\int_0^\infty h(s)\, ds \right) \int_0^T \int_{\Omega_\tau} |\mathbf{e}(\mathbf{u}(\tau))|^{2q-4}\, |\mathbf{e}(\mathbf{u}(\tau)) - \mathbf{e}(\mathbf{v}(\tau))|^2\, dx\, d\tau$$

$$+ \left(\int_0^\infty h(s)\, ds \right) \int_0^T \int_{\Omega_t} |\mathbf{e}(\mathbf{u}(t)) - \mathbf{e}(\mathbf{v}(t))|^2\, dx\, dt\,. \tag{2.9}$$

Proceeding similarly we find that I_2 satisfies the same estimate. Also, I_3 and I_4 satisfy the same estimate with $|\mathbf{e}(\mathbf{v}(\tau))|^{2q-4}$ instead of $|\mathbf{e}(\mathbf{u}(\tau))|^{2q-4}$, which, after grouping, yields

$$\int_0^T \int_{\Omega_t} (\sigma_{ij}^M(\mathbf{u}) - \sigma_{ij}^M(\mathbf{v}))(e_{ij}(\mathbf{u}) - e_{ij}(\mathbf{v}))\,dx\,dt$$

$$\geqq -c\gamma \int_0^T \int_{\Omega_t} (1 + |\mathbf{e}(\mathbf{u})|^{2q-4} + |\mathbf{e}(\mathbf{v})|^{2q-4})(|\mathbf{e}(\mathbf{u}) - \mathbf{e}(\mathbf{v})|^2)\,dx\,dt\,.$$

The lemma is then a direct consequence of (1.4).

We recall also the well known Korn's inequality, which we will need.

Lemma 2.2. *$\forall p \in (1, \infty)$, there exists $c > 0$ such that for any $\mathbf{u} \in W_0^{1,p}(\Omega)$,*

$$\|\mathbf{e}(\mathbf{u})\|_{L^p(\Omega)} \geqq \|\mathbf{u}\|_{W_0^{1,p}(\Omega)}\,.$$

Proof. See [23], for example.

3 The main result

We begin with

Definition 3.1. *A* weak solution *to the initial value problem*

$$\frac{\partial^2 u_i}{\partial t^2} - \frac{\partial}{\partial x_j}\sigma_{ij}^I - \frac{\partial}{\partial x_j}\sigma_{ij}^M = f_i\,, \quad \text{in} \quad (0,\infty)\times\Omega \tag{3.1}$$

$$\mathbf{u} = 0\,, \quad \text{on} \quad (0,\infty)\times\partial\Omega \tag{3.2}$$

$$\mathbf{u}(0,\cdot) = \mathbf{u}^0\,, \qquad \mathbf{u}_t(0,\cdot) = \mathbf{u}^1 \tag{3.3}$$

with $\mathbf{u}^0 \in W_0^{1,p}(\Omega; R^N)$, $\mathbf{u}^1 \in L^2(\Omega; R^N)$, is a function $\mathbf{u} \in L^\infty_{\text{loc}}([0;\infty); W_0^{1,p}(\Omega; R^N))$, for which $\mathbf{u}_t \in L^\infty_{\text{loc}}([0,\infty); L^2(\Omega; R^N))$, $\mathbf{u}_{tt} \in L^\infty_{\text{loc}}([0,\infty); W^{-1,p'}(\Omega; R^N))$ and such that for all $T > 0$, and all $\mathbf{v} \in W_0^{1,p}(\Omega; R^N)$

$$\int_{\Omega_T} \frac{\partial^2 u_i}{\partial t^2} v_i dx + \int_{\Omega_T} \sigma_{ij}^I e_{ij}(\mathbf{v})\,dx + \int_{\Omega_T} \sigma_{ij}^M e_{ij}(\mathbf{v})\,dx = \int_{\Omega_T} f_i v_i\,dx \tag{3.4}$$

where $\Omega_T = \{T\}\times\Omega$.

In (3.1) *and* (3.4)

$$\mathbf{f} \in L^\infty((0,\infty); L^2(\Omega; R^N)) \cap L^2((0,\infty); L^2(\Omega; R^N))\,. \tag{3.5}$$

Theorem 3.1. *Under the assumptions* (1.2)–(1.8) *and*

- *$\gamma c < 1$ where $c = \dfrac{\beta}{\alpha_2}\int_0^\infty h_0(s)\,ds \cdot \min(1; 2^{q-3})$,*
- *$p \geqq 2$, $q \geqq 2$, and*

$$p\left[1 - \frac{\nu}{2} - \frac{\nu}{N}\right] + \nu < 1 \leqq \frac{p}{2} + 1\,, \tag{3.6}$$

- *$\mathbf{u}^0 \in W_0^{1,p}(\Omega; R^N)$, $\mathbf{u}^1 \in L^2(\Omega; R^N)$,* *the problem* (3.1)–(3.3) *has a weak global solution* $\mathbf{u}$. *Moreover* $\forall T > 0$ *there exists* C *such that*

1. $|\mathbf{u}|_{L^\infty(0,T;W_0^{1,p}(\Omega))} \leqq C$
2. $\left\|\frac{\partial \mathbf{u}}{\partial t}\right\|_{L^\infty(0,T;L^2(\Omega))} \leqq C$
3. $\left\|\frac{\partial^2 \mathbf{u}}{\partial t^2}\right\|_{L^\infty(0,T;W^{-1,p'}(\Omega))} \leqq C$
4. $\|\mathbf{u}\|_{W^{\frac{\nu}{q},q}(0,T;W_0^{1,q}(\Omega))} \leqq C.$

The solution $\mathbf{u}$ will be constructed using the Galerkin approximation method.

First we introduce some finite dimensional spaces where the Galerkin approximate is to be defined. These spaces will be constructed using a certain Schauder basis of $W_0^{1,p}$.

For this purpose we recall that $(-\Delta+1)^{1/2}$ defines an isomorphism from $W_0^{1,p}(\Omega)$ into L^p and set $L = (-\Delta + 1)^{1/2}$. We can then state:

Proposition 3.1. *Let* p_0, p_1 *be real numbers such that* $1 < p_0 < p_1 < \infty$.

There exists a sequence of functions $\mathbf{w}^k$, $k = 1, \ldots, \infty$, *which forms a Schauder basis for* $W_0^{1,p}(\Omega; R^N)$ $\forall p \in (p_0, p_1)$ *and satisfies*

1. $\mathbf{w}^k \in C^\infty \cap W_0^{1,2}(\Omega)$
2. $\int_\Omega L\mathbf{w}^j \cdot L\mathbf{w}^i \, dx = \delta_{ij}$
3. *if* $f \in W_0^{1,p}(\Omega)$; $\alpha_i = \int_\Omega Lf \cdot L\mathbf{w}^i \, dx$ *and* $P_n(f) = \sum_{k=1}^{n} \alpha_k \mathbf{w}^k$ *then* $P_n(f)$

converges to f *and the norm of* P_n *in* $\mathscr{L}(W_0^{1,p}(\Omega); W_0^{1,p}(\Omega))$ *is bounded uniformly in* n *and* $p \in (p_0, p_1)$.

Proof. See [3].

In order to state our next result we need the following notations. Denote by E_n the closure in L^2 of the space spanned by the subsequence $\{\mathbf{w}^i\}_{i>n}$ and by Q_n, the L^2 projection onto the space $E_n^\perp$ (in L^2).

Lemma 3.2. *If* $a_{ij} = \int_\Omega Q_n(\mathbf{w}^i) \cdot Q_n(\mathbf{w}^j) dx$, *for* $i \leqq n$ *and* $j \leqq n$, *then the matrix* $(a_{ij})_{ij}$ *is positive definite and invertible.*

Proof. See [3].

Since $\mathbf{u}^0 \in W^{1,p}$ and $\mathbf{u}^1 \in W^{1,2}$ there exist sequences $\lambda_i^k, \bar{\lambda}_i^k$ such that

$$\lim_{n\to\infty} \left\|\mathbf{u}^0 - \sum_{k=1}^{n} \lambda_i^k w_i^k\right\|_{W^{1,p}} = 0 \tag{3.7}$$

$$\lim_{n\to\infty} \left\|\mathbf{u}^1 - \sum_{k=1}^{n} \bar{\lambda}_i^k w_i^k\right\|_{W^{1,2}} = 0. \tag{3.8}$$

For n fixed consider the problem of existence and uniqueness of $\mathbf{u}^n = \sum_{k=1}^{n} c^{n,k}(t)\mathbf{w}^k$, such that $c^{n,k} \in C^1$ and $\forall \mathbf{w} \in \operatorname{span}\{\mathbf{w}^1, \ldots, \mathbf{w}^n\}$

$$\int_{\Omega_t} \frac{\partial^2 Q_n(u_i^n)}{\partial t^2} w_i + \int_{\Omega_t} \sigma_{ij}^I(\mathbf{e}(\mathbf{u}^n)) e_{ij}(\mathbf{w})\, dx$$

$$- \int_{\Omega_t} \sigma_{ij}^M(\mathbf{e}(\mathbf{u}^n)) e_{ij}(\mathbf{w})\, dx = \int_{\Omega_t} Q_n(f_i) w_i\, dx\,, \tag{3.9}$$

$$c_i^{n,k}(0) = \lambda_i^k \tag{3.10}$$

$$\frac{dc_i^{n,k}}{dt}(0) = \bar{\lambda}_i^k\,. \tag{3.11}$$

Since Q_n is linear and $\int Q_n(\mathbf{w}^k)\mathbf{w}^j\, dx = \int Q_n(\mathbf{w}^k) Q_n(\mathbf{w}^j)\, dx$, it follows from Lemma 3.2 that this system of ordinary differential equations is not degenerate. Therefore, by the Picard theorem, $\exists t_0 > 0$ such that problem (3.9)–3.11) has a unique solution on the interval $(0, t_0)$. We intend to show that in fact the solution exists on $[0, +\infty)$. This is a direct consequence of the following lemma.

Lemma 3.3. *$\exists M > 0$ such that for any solution $\mathbf{u}^n$ to problem (3.9)–(3.11), and any $0 < t < t_0$*

$$\int_{\Omega_t} \left| \frac{\partial Q_n(u_i^n)}{\partial t} \right|^2 dx + \int_{\Omega_t} W(\mathbf{e}(\mathbf{u}^n))\, dx$$

$$+ \int_0^{t_0} \int_0^{t_0} \int_{\Omega}^{0} \frac{|\mathbf{e}(\mathbf{u}^n(\tau)) - \mathbf{e}(\mathbf{u}^n(t))|^q}{|t - \tau|^{1+\nu}}\, dx\, dt\, d\tau$$

$$\leqq M\,. \tag{3.12}$$

Proof. Set $\mathbf{w} = \mathbf{w}_t^n$ in (3.9). Integrating this equality over $(0, T)$ and summing over i we find:

$$\frac{1}{2} \int_{\Omega_T} \left| \frac{\partial Q_n(\mathbf{u}^n)}{\partial t} \right|^2 dx - \frac{1}{2} \int_{\Omega_0} \left| \frac{\partial Q_n(\mathbf{u}^n)}{\partial t} \right|^2 dx$$

$$+ \int_{\Omega_T} W(\mathbf{e}(\mathbf{u}^n))\, dx - \int_{\Omega_0} W(\mathbf{e}(\mathbf{u}^n))\, dx$$

$$\times \int_0^T \int_{\Omega} \int_0^t h(t - \tau) \frac{\partial V}{\partial e_{ij}} (\mathbf{e}(\mathbf{u}^n(\tau)) - \mathbf{e}(\mathbf{u}^n(t))) \frac{\partial}{\partial t} (e_{ij}(\mathbf{u}^n(t)))\, d\tau\, dx\, dt$$

$$= \int_0^T \int_{\Omega} Q_n(f_i) \frac{\partial u_i^n}{\partial t}\, dx\, dt\ . \tag{3.13}$$

We will first estimate the last term in the left-hand side of (3.13) from below.

Since

$$\frac{\partial V}{\partial e_{ij}}(\mathbf{e}(\mathbf{u}^n(\tau)) - \mathbf{e}(\mathbf{u}^n(t)))\frac{\partial}{\partial t}(e_{ij}(\mathbf{u}^n(t))) = -\frac{\partial}{\partial t}V(\mathbf{e}\mathbf{u}^n(\tau) - \mathbf{e}(\mathbf{u}^n(t))),$$

we can rewrite the last term on the left-hand side of (3.13) as:

$$\begin{aligned}
&-\int_0^T\int_\Omega\int_0^t h(t-\tau)\frac{\partial V}{\partial e_{ij}}(\mathbf{e}(\mathbf{u}^n(\tau)) - \mathbf{e}(\mathbf{u}^n(t)))\frac{\partial}{\partial t}e_{ij}(\mathbf{u}^n(t))\,d\tau\,dx\,dt \\
&= \int_\Omega\int_0^T\int_0^t h(t-\tau)\frac{\partial V}{\partial t}(\mathbf{e}(\mathbf{u}^n(e)) - \mathbf{e}(\mathbf{u}^n(t)))\,d\tau\,dt\,dx \\
&= \lim_{\delta\to 0+}\int_\Omega\int_0^T\int_0^T Y(t-\tau-\delta)h(t-\tau) \\
&\quad\times\frac{\partial V}{\partial t}(\mathbf{e}(\mathbf{u}^n(\tau)) - \mathbf{e}(\mathbf{u}^n(t)))\,d\tau\,dt\,dx \\
&= \lim_{\delta\to 0+} I_\delta
\end{aligned}\tag{3.14}$$

where Y is the usual Heavyside function.

Integrating by parts with respect to t we find

$$\begin{aligned}
I_\delta = &-\int_\Omega\int_0^T\int_0^T Y(t-\tau-\delta)h'(t-\tau)V(\mathbf{e}(\mathbf{u}^n(\tau)) - \mathbf{e}(\mathbf{u}^n(t)))\,d\tau\,dt\,dx \\
&-\int_\Omega\int_0^T h(\delta)V(\mathbf{e}(\mathbf{u}^n(\tau)) - \mathbf{e}(\mathbf{u}^n(\tau+\delta)))\,dt\,dx \\
&+\int_\Omega\int_0^{T-\delta} h(T-\tau-\delta)V(\mathbf{e}(\mathbf{u}^n(\tau)) - \mathbf{e}(\mathbf{u}^n(T)))\,d\tau \\
\equiv\ & I_\delta^1 + I_\delta^2 + I_\delta^3 .
\end{aligned}\tag{3.15}$$

From the assumptions on h and V it can be easily seen that $I_\delta^3 \geqq 0$, and that for some positive constant c

$$I_\delta^1 \geqq c\int_\Omega\int_0^T\int_0^T Y(t-\tau-\delta)\frac{|\mathbf{e}(\mathbf{u}^n(\tau)) - \mathbf{e}(\mathbf{u}^n(t))|^q}{|t-\tau|^{1+\nu}}\,d\tau\,dt\,dx .$$

Since $\nu < 1$ it follows from the smoothness of $\mathbf{u}^n(\tau)$ and V that

$$\lim_{\delta\to 0} I_\delta^2 = 0 .$$

Consequently,

$$\lim_{\delta\to 0} I_\delta \geqq \lim_{\delta\to 0} I_\delta^1 \geqq c \int_\Omega \int_0^T \int_0^T Y(t-\tau)\, \frac{|\mathbf{e}(\mathbf{u}^n(\tau)) - \mathbf{e}(\mathbf{u}^n(t))|^q}{|t-\tau|^{1+\nu}}\, d\tau\, dt\, dx$$

$$\geqq \frac{1}{2} c \int_\Omega \int_0^T \int_0^T Y(t-\tau)\, \frac{|\mathbf{e}(\mathbf{u}^n(\tau)) - \mathbf{e}(\mathbf{u}^n(t))|^q}{|t-\tau|^{1+\nu}}\, d\tau\, dt\, dx\,.$$

We deduce from (3.13)

$$\begin{aligned}
&\int_{\Omega_T} \left|\frac{\partial Q_n(\mathbf{u}^n)}{\partial t}\right|^2 dx + \int_{\Omega_T} W(\mathbf{e}(\mathbf{u}^n))\, dx \\
&\quad + \frac{c}{2} \int_\Omega \int_0^T \int_0^T \frac{|\mathbf{e}(\mathbf{u}^n(\tau)) - \mathbf{e}(\mathbf{u}^n(t))|^q}{|t-\tau|^{1+\nu}}\, d\tau\, dt\, dx \\
&\leqq \int_\Omega \int_0^T Q_n(f_i)\, \frac{\partial u_i^n}{\partial t}\, dx\, dt + \int_{\Omega_0} \left|\frac{\partial \mathbf{u}^n}{\partial t}\right|^2 dx \\
&\quad + \int_{\Omega_0} W(\mathbf{e}(\mathbf{u}^n))\, dx\,.
\end{aligned} \tag{3.16}$$

The terms $\int\limits_{\Omega_0} \left|\frac{\partial \mathbf{u}^n}{\partial t}\right|^2 dx$ and $\int\limits_{\Omega_0} W(\mathbf{e}(\mathbf{u}^n))\, dx$ are bounded by constants which depend only on the initial conditions $\mathbf{u}^0$ and $\mathbf{u}^1$.

Also,

$$\left|\int_0^T Q_n(f_i)\, \frac{\partial u_i^n}{\partial t}\, dx\, dt\right| \leqq \|Q_n(f_i)\|_{L^2(0,T;L^2(\Omega))} \cdot \left\|\frac{Q_n(\partial u_i^n)}{\partial t}\right\|_{L^2(0,T;L^2(\Omega))}. \tag{3.17}$$

The lemma is then deduced from (3.16) and (3.17) in a standard manner.

Lemma 3.4. *There exist a constant $M > 0$ such that $\forall n$*

$$\|Q_n(\mathbf{u}_{tt}^n)\|_{L^\infty(0,T;W^{-1,p'}(\Omega;R^n))} \leqq M\,. \tag{3.18}$$

Proof. By design, $Q_n(\mathbf{u}^n)$ is orthogonal to E_n. Therefore, $\forall \mathbf{w} \in W_0^{1,p}(\Omega)$, $\mathbf{w} = \sum\limits_{i=1}^{\infty} \lambda_i \mathbf{w}^i$ there holds

$$\int_\Omega Q_n(\mathbf{u}_{tt}^n) \cdot \mathbf{w}\, dx = \int_\Omega Q_n(\mathbf{u}_{tt}^n) \cdot P_n(\mathbf{w})\, dx = \int_\Omega Q_n(\mathbf{u}_{tt}^n) \cdot \sum_{i=1}^n \lambda_i \mathbf{w}^i\, dx\,. \tag{3.19}$$

Also, from the definition of $\mathbf{u}^n$ we have that

$$\int\limits_{\Omega_0} \frac{\partial^2 Q_n(u_i^n)}{\partial t^2} P_n(\mathbf{w})\,dx + \int\limits_{\Omega_t} \sigma_{ij}^I(\mathbf{e}(\mathbf{u}^n))e_{ij}(P_n(\mathbf{w}))\,dx$$
$$- \int\limits_{\Omega_t} \sigma_{ij}^M(\mathbf{e}(\mathbf{u}^n))e_{ij}(P_n(\mathbf{w}))\,dx = \int\limits_{\Omega_t} Q_n(f_i)P_n(\mathbf{w})_i\,dx\,. \tag{3.20}$$

Hence,

$$\int\limits_{\Omega_t} \frac{\partial^2 Q_n(u_i^n)}{\partial t^2} \mathbf{w})\,dx = -\int\limits_{\Omega_t} \sigma_{ij}^I(\mathbf{e}(\mathbf{u}^n))e_{ij}(P_n(\mathbf{w}))\,dx$$
$$+ \int\limits_{\Omega_t} \sigma_{ij}^M(\mathbf{e}(\mathbf{u}^n))e_{ij}(P_n(\mathbf{w}))\,dx + \int\limits_{\Omega_t} Q_n(f_i)P_n(\mathbf{w})_i\,dx\,, \tag{3.21}$$

and

$$\left|\int\limits_{\Omega_t} \frac{\partial^2 Q_n(u_i^n)}{\partial t^2} \mathbf{w}\,dx\right| \leqq M\|P_n(\mathbf{w})\|_{W_0^{1,p}(\Omega)} \leqq C\|\mathbf{w}\|_{W_0^{1,p}(\Omega)}\,,$$

which proves the lemma.

From the estimates (3.12) (3.18) it follows that there exist a subsequence, denoted again by $\mathbf{u}^n$, and a function $\mathbf{u}$ such that

$$\mathbf{u}^n \to \mathbf{u} \text{ weakly in } L^2(0,T;W_0^{1,p}(\Omega)) \tag{3.22}$$

$$\frac{\partial Q_n \mathbf{u}^{(n)}}{\partial t} \to \frac{\partial \mathbf{u}}{\partial t} \text{ weakly in } L^2(0,T;L^2(\Omega)) \tag{3.23}$$

$$\frac{\partial^2 Q_n \mathbf{u}^{(n)}}{\partial t^2} \to \frac{\partial^2 \mathbf{u}}{\partial t^2} \text{ weakly in } L^2(0,T;W^{-1,p'}(\Omega)) \tag{3.24}$$

$$\mathbf{u}^n \to \mathbf{u} \text{ weakly in } W^{\frac{\nu}{q},q}(0,T;W_0^{1,q}(\Omega))\,. \tag{3.25}$$

Lemma 3.5. *Let* $\mathbf{u}$ *be the weak limit of the sequence* $\mathbf{u}^n$. *Then* $\forall T > 0$

$$\lim_{n\to\infty} \int\limits_0^T \int\limits_{\Omega_0} \frac{\partial^2 Q_n(u_i^n)}{\partial t^2} \cdot (u_i^n - (P_n(\mathbf{u}))_i)\,dx\,dt = 0\,.$$

Proof. For $\lambda \in [0,1]$, $\delta > 0$ and $\frac{1}{s} = \frac{\lambda}{p'} + \frac{1-\lambda}{2}$, by Proposition 2.1, (3.12) and (3.18), $\frac{\partial Q_n(u_i^n)}{\partial t}$ converges strongly in $W^{\lambda-\delta,2}(0,T;W_{(\Omega)}^{-\lambda-\delta,s})$ to $\frac{\partial(u_i)}{\partial t}$ and

$$\frac{\partial^2 Q_n(\mathbf{u}^n)}{\partial t^2} \to \frac{\partial^2 \mathbf{u}}{\partial t^2} \text{ strongly in } W^{\lambda-\delta-1,2}(0,T;W^{-\lambda-\delta,s}(\Omega))\,. \tag{3.26}$$

Also, from (3.25) and Propositon 3.1 it easily follows that $\forall \delta_1 > 0$,

$$u_i^n - P_n(\mathbf{u})_i \to 0 \text{ weakly in } W^{\frac{\nu}{q}-\delta_1,2}(0,T;W_0^{1,q}(\Omega))\,. \tag{3.27}$$

Using the duality product, it follows that the lemma is proved, provided $W^{\lambda-\delta-1,2}(0, T; W_{(\Omega)}^{-\lambda-\delta,s})$ is continuously embedded in the dual of $W^{\frac{\nu}{q}-\delta_1,2}(0,T;W_0^{1,q}(\Omega))$. Since $\frac{\nu}{q}-\delta_1<\frac{1}{2}$, this is the case whenever the following conditions are satisfied:

$$\frac{\nu}{q}-\delta_1+(\lambda-1-\delta)\geqq 0 \tag{3.28}$$

$$W_0^{1,q}(\Omega,T)\hookrightarrow W_0^{\lambda+\delta,s'}(\Omega)\,. \tag{3.29}$$

Since δ and δ_1 are only required to be positive the above conditons will be met if $\exists\lambda\in(0,1)$ such that

$$\frac{\nu}{q}+(\lambda-1)>0\,, \tag{3.30}$$

$$1-\frac{N}{q}\geqq\lambda-\frac{N}{s'}\,. \tag{3.31}$$

Using the definitions of s' the last condition means that

$$1-N\left[\frac{1}{q}-1+\lambda-\frac{\lambda}{p}+\frac{1-\lambda}{2}\right]>\lambda\,. \tag{3.32}$$

Elementary calculations show that whenever

$$p\left[1-\frac{\nu}{2}-\frac{\nu}{N}\right]+\nu<q \tag{3.33}$$

it is possible to find a value $\lambda\in(0,1)$ which satisfies both conditons (3.32) and (3.30).

In order to prove the theorem by simply letting n go to ∞ in (3.9) we need know that the nonlinear terms $\sigma(\mathbf{u}^n)$ converge to the corresponding terms $\sigma(\mathbf{u})$. Due to the nonlinear nature of σ the weak convergence results above do not guarantee this. The object of the next lemma is to remedy this situation.

Lemma 3.6.

$$\mathbf{u}^n\to\mathbf{u}\ \textit{strongly in}\ L^2(0,T;W_0^{1,p}(\Omega))\,. \tag{3.34}$$

Proof. In order to prove this lemma we will combine (3.22)–(3.25), (3.26) and the fact that $\mathbf{u}^n$ is a *solution* of (3.9), and exploit the elliptic regularity of the partial differential equation.

For this purpose, using $(\mathbf{u}^n-P_n(\mathbf{u}))$ as the test function in (3.9) we find:

$$\begin{aligned}&\int_0^T\int_{\Omega_0}\frac{\partial^2 Q_n(u_i^n)}{\partial t^2}\cdot(u_i^n-(P_n(\mathbf{u}))_i)\,dx\,dt+\int_0^t\int_{\Omega_0}\sigma_{ij}^I(\mathbf{e}(\mathbf{u}^n))\cdot(e_{ij}(\mathbf{u}^n)-e_{ij}(P_n(\mathbf{u})))\\ &\quad-\int_0^T\int_{\Omega_t}\sigma_{ij}^M(\mathbf{e}(\mathbf{u}^n))\cdot(e_{ij}(\mathbf{u}^n)-e_{ij}(P_n(\mathbf{u})))\,dx\,dt\\ &\quad-\int_0^T\int_{\Omega_t}Q_n(f_i)(u_i^n(t)-(P_n(\mathbf{u}(t)))_i)\,dx\,d\tau\\ &\equiv I_1^{(n)}+I_2^{(n)}+I_3^{(n)}+I_4^{(n)}=0\,.\end{aligned} \tag{3.35}$$

Hence

$$\int_0^T\int_{\Omega_0}(\sigma^I_{ij}(\mathbf{e}(\mathbf{u}^n)) - \sigma^I_{ij}(\mathbf{e}(P_n(\mathbf{u})))) \cdot (e_{ij}(\mathbf{u}^n) - e_{ij}(P_n(\mathbf{u})))$$

$$- \int_0^T\int_{\Omega_0}(\sigma^M_{ij}(\mathbf{e}(\mathbf{u}^n)) - \sigma^M_{ij}(\mathbf{e}(P_n(\mathbf{u})))) \cdot (e_{ij}(\mathbf{u}^n) - e_{ij}(P_n(\mathbf{u})))\,dx\,dt$$

$$= -I_1^{(n)} - I_4^{(n)} - \int_0^t\int_{\Omega_0}\sigma^I_{ij}(\mathbf{e}(P_n(\mathbf{u}))) \cdot (e_{ij}(\mathbf{u}^n) - e_{ij}(P_n(\mathbf{u})))$$

$$+ \int_0^T\int_{\Omega_0}\sigma^M_{ij}(\mathbf{e}(P_n(\mathbf{u}))) \cdot (e_{ij}(\mathbf{u}^n) - e_{ij}(P_n(\mathbf{u})))\,dx\,dt$$

$$\equiv -I_1^{(n)} - I_4^{(n)} - I_5^{(n)} - I_6^{(n)}, \tag{3.36}$$

where the terms $I_5^{(n)}, I_6^{(n)}$ have been subtracted from both sides of (3.36).

From Lemma 2.1 it then follows that for γ small enough there exists a positive number δ that

$$\sigma\int_0^T\int_{\Omega_0}(\sigma^I_{ij}(\mathbf{e}(\mathbf{u}^n)) - \sigma^I_{ij}(\mathbf{e}(P_n(\mathbf{u}))) \cdot (e_{ij}(\mathbf{u}^n) - e_{ij}(P_n(\mathbf{u}))) \leqq -I_1^{(n)} - I_4^{(n)} - I_5^{(n)} - I_6^{(n)}. \tag{3.37}$$

Since $\mathbf{u}^n - P_n(\mathbf{u})$ converges weakly to 0 in $L^2(0,T;W_0^{1,p}(\Omega))$ and $P_n(\mathbf{u})$ converges strongly to $\mathbf{u}$ in $L^2(0,T;W_0^{1,p}(\Omega))$ it immediately follows that $I_4^{(n)}$, $I_5^{(n)}$, and $I_6^{(n)}$ go to 0 as n goes to ∞.

Also, by the previous lemma we have that the $\lim_{n\to\infty} I_1^{(n)} = 0$.

Therefore,

$$\lim_{n\to\infty}\sigma\int_0^T\int_{\Omega_0}(\sigma^I_{ij}(\mathbf{e}(\mathbf{u}^n)) - \sigma^I_{ij}(\mathbf{e}(P_n(\mathbf{u})))) \cdot (e_{ij}(\mathbf{u}^n) - e_{ij}(P_n(\mathbf{u}))) = 0,$$

which proves the Lemma.

Proof of the Theorem 3.1. Letting n go to infinity in (3.9), we then obtain in the limit that $\mathbf{u}$ is a weak solution of (3.1)–(3.3) in the sense of definition (3.1).

Remark. It can easily be verified that for the storage functions $W(\mathbf{e}) = c(1 + e_{ij}e_{ij})^{p/2} - c$, $V(\mathbf{e}) = c(1 + e_{ij}e_{ij})^{q/2} - c$ all of the assumptions of Theorem 3.1 are satisfied.

References

1. Adams, R.A.: Sobolev spaces. New York: Academic Press 1975
2. Antman, S.S.: The influence of elasticity on analysis: modern developments. Bull Am. Math. Soc. **9**, 267–291 (1983)
3. Bellout, H.: On a special Schauder basis for the Sobolev spaces $W_0^{1,p}(\Omega)$. Ill. J. Math. (to appear)

4. Bellout, H., Bloom, F., Nečas, J.: Existence of global solutions to the dynamical problem for a three dimensional elastic body with singular memory. SIAM J. Math. Anal. **24**, 46–58 (1993)
5. Besov, O.V., Il'in, V.P., Nikol'skij, S.M.: Integral representations of functions and embedding theorems. Washington DC: Winston 1978/79
6. Ciesielski, Z., Domsta, J.: Construction of an orthonormal basis in $C^m(I^d)$ and $W_p^m(I^d)$. Stud. Math. **41**, 211–224 (1972)
7. Coleman, B.D.: On thermodynamics, strain impulses, and viscoelasticity. Arch. Ration. Mech. Anal. **17**, 1–46 (1964)
8. Doi, M., Edwards, S.F.: Dynamics of concentrated polymer systems. J. Chem. Soc. Faraday **74**, 1789–1832 (1978); **75**, 38–54 (1979)
9. Engler, H.: Weak solutions of a class of quasilinear hyperbolic integro-differential equations describing viscoelastic materials. Arch. Ration. Mech. Anal. **113**, 1–38 (1991)
10. Fucik, S., John, O., Nečas, J.: On the existence of Schauder basis in Sobolev spaces. Commentat. Math. Univ. Carol. **13**, 163–175 (1972)
11. Grisvard, P.: Commutativite de deux foncteurs d'interpolation et applications. J. Math. Pures Appl. **45**, 143–290 (1966)
12. Hopf, E.: Über die Anfangswertaufgabe für die hydrodynamischen Grundgleichungen. Math. Nachr. **4**, 213–231 (1951)
13. Hrusa, W.J., Renardy, M.: On wave propagation in linear viscoelasticity. Q. Appl. Math. **43**, 237–254 (1985)
14. Hannsgen, K.B., Wheeler, R.L.: Behavior of the solutions of a Volterra equation as a parameter tends to infinity. J. Integral Equations **7**, 229–237 (1984)
15. Hrusa, W.J., Renardy, M.: On a class of quasilinear partial integrodifferential equations with singular kernels. J. Differ. Equations **64**, 195–220 (1986)
16. Hrusa, W.J., Renardy, M.: A model equation for viscoelasticity with a strongly singular kernel. SIAM J. Math. Anal. **19**, 257–269 (1988)
17. Laun, H.M.: Description of the nonlinear shear behavior of a low density polyethylene melt by means of an experimentally determined strain dependent memory function. Rheol. Acta **17**, 1–15 (1978)
18. Lions, J.L.: Quelques méthodes de resolution des problémes aux limites nonlinéaires. Paris: Dunrod Gauthier-Villars 1969
19. Londen, S.O.: An existence result for a Volterra equation in a Banach space. Trans. Am. Math. Soc. **235**, 285–304 (1978)
20. Marsden, J.E., Hughes, T.J.R.: Mathematical foundations of elasticity. Englewood Cliffs, NJ: Prentice-Hall 1983
21. Milota, J. Nečas, J., Śveŕak, V.: On weak solutions to a viscoelasticity model. Commentat. Math. Univ. Carol. **31**, 557–565 (1990)
22. Muramatu, T.: On Besov spaces and Sobolev spaces of generalised functions defined on a general regions. Publ. Res. Inst. Math. Sci. **9**, 325–396 (1974)
23. Nečas, J., Hlavăček, I.: Mathematical theory of elastic and elastico-plastic bodies. New York: Elsevier 1981
24. Renardy, M.: Some remarks on the propagation and non-propagation of discontinuities in linearly viscoelastic liquids. Rheol. Acta **21**, 251–254 (1982)
25. Renardy, M.: Coercive estimates and existence of solutions for a model of one-dimensional viscoelasticity with a nonintegrable memory functions. J. Integral. Equations Appl. (to appear)
26. Renardy, M., Hrusa, W., Nohel, J.: Mathematical problems in viscoelasticity. (Pitman Monogr. Surv. Pure Appl. Math., vol. 35) Harlow: Longman and New York: Wiley 1987
27. Rouse, P.E.: A theory of the lienar viscoelastic properties of dilute solutions of coiling polymers. J. Chem. Phys. **21**, 1271–1280 (1953)
28. Schmeisser, H., Triebel, H.: Topics in Fourier analysis and function spaces. New York: Wiley 1987
29. Triebel, H.: Interpolation theory, function spaces, differential operator. Amsterdam: North-Holland 1978
30. Zimm, B.H.: Dynamics of polymer molecues in dilute solutions: Viscoelasticity, flow birefringence, and dielectric loss. J. Chem. Phys. **24**, 269–278 (1956)

Chapter 12
Commentaries

12.1 Mechanics of Solids

Ivan Hlaváček

From In Memoriam Jindřich Nečas, Matematica Bohemica, 124, 2004, 4, 421–446.[1]

Abstract We devote this article to the part of scientific life of J. Nečas, when he was working on the elasticity problem.

In 1967, J. Nečas founded the Continuum Mechanics Seminar which attracted a wide audience from Charles University, Czech Technical University, and various institutes of the Czech Academy of Sciences. The audience would typically include students, mathematicians, and engineers. He also lectured on continuum mechanics at the Faculty of Mathematics and Physics of Charles University. These lectures, together with the seminar, were the starting point of Nečas' next successful book [A4], written with I. Hlaváček. The book contains many original results of Nečas and his collaborators. Among J. Nečas' favorite topics were unilateral problems in elasticity. These can be modelled by variational inequalities. This is another area where Nečas made important contributions. Together with I. Hlaváček, J. Haslinger and J. Lovíšek he wrote the monograph [A5]. Besides providing an excellent introduction to the subject, it also gives an exposition of original results of Nečas and his collaborators. Of these we should mention at least [JN-C101], where subtle existence results for a model of quasistatic unilateral contact with Coulomb friction are proved, and [C76], where semicoercive unilateral contact problems are discussed and solved. Monograph [A5] later appeared in Russian [A8] and English [A10].

[1] ©Institute of Mathematics, Academy of Sciences of the Czech Republic 2004. Reprinted with kind permission.

© Springer Basel 2015
Š. Nečasová et al. (eds.), *Selected Works of Jindřich Nečas*, Advances in Mathematical Fluid Mechanics, DOI 10.1007/978-3-0348-0231-4_12

Jindřich Nečas also made important contributions to the mathematical analysis of models describing the yielding of plastic materials. He proved existence and uniqueness of solutions for models of perfectly plastic bodies and models with isotropic or kinematic hardening. (The penalty method proved to be a very effective tool here.) He enthusiastically embraced the 1977 existence theory of J. Ball, and together with P.G. Ciarlet studied its extensions to unilateral contact problems (see [C122, C124, JN-C125]). He studied the dynamics of elastoplastic bodies and proved existence and uniqueness for the corresponding PDEs in [JN-C95]. Later, he turned his attention to nonlinear thermoelasticity [C147, C152, C144]. In [C147], he studied the problem of hardening in the framework of problems with moving boundary. In [C146, C161] he analyzed viscoelastic materials and incompressible multipolar fluids. We must also mention the contributions of J. Nečas to numerical methods suitable for problems in solid mechanics. His proofs of important theoretical results were often based on methods that suggested a good numerical approximation. Kačanov's (secant modules) method [C64] with its applications in the nonlinear theory of plastic deformations [C119] and Galerkin's method [A12] are good examples. Although Nečas' work in continuum mechanics concentrated mostly on nonlinear problems, he also made important contributions to linear theory. For example, papers [JN-C43, JN-C44] give a simple algebraic criterion which is equivalent to a coercivity condition for quite general systems of linear PDEs. This result generalizes Korn's inequality known from linear elasticity and can be used to obtain ellipticity for other boundary value problems arising in applications. We would also like to mention important contribution [A13].

Nečas' Works Quoted Above

The references [Axx] and [Cxx] refer to the respective papers listed in Chapter 2 "Complete Bibliography of J. Nečas". The references [JN-Cyy] refer to papers [Cyy] that are reproduced either in Chapter 11 "Selected Papers on Solid Mechanics, Elasticity, Thermoelasticity, and Plasticity" or in Chapter 5 "Selected Papers on Linear Elliptic Operators".

[A4] J. Nečas, I. Hlaváček: Mathematical Theory of Elastic and Elasto-Plastic Bodies: An Introduction. Studies in Applied Mechanics vol. 3, Elsevier, Amsterdam, 1981; Czech edition in SNTL, Praha, 1983.

[A5] I. Hlaváček, J. Haslinger, J. Nečas, J. Lovíšek: Riešenie variačných nerovností v mechanike. Alfa, Bratislava, 1982 (In Slovak.); for Russian edition see [A8] and for English edition [A10].

[A8] I. Hlaváček, J. Haslinger, J. Nečas, J. Lovíšek: Reshenie variatsionnykh neravenstv v mekhanike. Mir, Moskva, 1986; Russian translation of [A5].

[A10] I. Hlaváček, J. Haslinger, J. Nečas, J. Lovíšek: Solution of Variational Inequalities in Mechanics. Applied Mathematical Sciences vol. 66, Springer, Verlag, New York, 1988; English translation of [A5].

[A12] J. Málek, J. Nečas, M. Rokyta, M. Růžička: Weak and Measure-Valued Solutions to Evolutionary PDEs. Applied Mathematics and Mathematical Computation vol. 13, Chapman & Hall, London, 1996.

[A13] J. Haslinger, I. Hlaváček, and J. Nečas. Numerical methods for unilateral problems in solid mechanics. *Handbook of numerical analysis*, 4, 313–485, 1996.

[C64] S. Fučík, A. Kratochvíl, J. Nečas: Kačanov-Galerkin method. Comment. Math. Univ. Carolin. 14 (1973), 651–659.
[C76] J. Nečas: On regularity of solutions to nonlinear variational inequalities for second-order elliptic systems. Rend. Mat. 8 (1975), 481–498.
[C119] J. Nečas, I. Hlaváček: Solution of Signorini's contact problem in the deformation theory of plasticity by secant modules method. Apl. Mat. 28 (1983), 199–214.
[C122] Ph. G. Ciarlet, J. Nečas: Problèmes unilatéraux en élasticité non linéaire tridimensionnelle. C. R. Acad. Sci. Paris Sér. I Math. 298 (1984), 189–192.
[C124] Ph. G. Ciarlet, J. Nečas: Injectivité presque partout, auto-contact, et non-interpénétrabilité en élasticité non-linéaire tridimensionnelle. C. R. Acad. Sci. Paris Sér. I Math. 301 (1985), 621–624.
[C144] J. Nečas, A. Novotný, V. Šverák: Uniqueness of solutions to the systems for thermoelastic bodies with strong viscosity. Math. Nachr. 149 (1990), 319–324.
[C146] J. Milota, J. Nečas, V.Šverák: On weak solutions to a viscoelasticity model. Comment. Math. Univ. Carolin. 31 (1990), 557–565.
[C147] J. Nečas, T. Roubíček: Approximation of a nonlinear thermoelastic problem witha moving boundary via a fixed-domain method. Apl. Mat. 35 (1990), 361–372.
[C152] J. Nečas, M. Růžička: A dynamic problem of thermoelasticity. Z. Anal. Anwend. 10 (1991), 357–368.
[C161] J. Nečas, M.Růžička: Global solution to the incompressible viscous-multipolar material problem. J. Elasticity 29 (1992), 175–202.
[JN-C43] I. Hlaváček, J. Nečas: On inequalities of Korn's type. I. Boundary-value problems for elliptic system of partial differential equations. Arch. Ration. Mech. Anal. 36 (1970), 305–311.
[JN-C44] I. Hlaváček, J. Nečas: On inequalities of Korn's type. II. Applications to linear elasticity. Arch. Ration. Mech. Anal. 36 (1970), 312–334.
[JN-C95] K. Gröger, J. Nečas, L.Trávníček: Dynamic deformation processes of elastic-plastic systems. Z. Angew. Math. Mech. 59 (1979), 567–572.
[JN-C101] J. Nečas, J. Jarušek, J. Haslinger: On the solution of the variational inequality to the Signorini problem with small friction. Boll. Un. Mat. Ital. B 17 (1980), 796–811.
[JN-C125] Ph. G. Ciarlet, J. Nečas: Unilateral problems in nonlinear, three-dimensional elasticity. Arch. Ration. Mech. Anal. 87 (1985), 319–338.

Ivan Hlaváček
Academy of Sciences, Institute of Mathematics, Žitná 25, 11567 Prague 1,
Czech Republic
email: hlavacek@math.cas.cz

12.2 The Contribution of Jindřich Nečas to the Development of the Theory of Solvability of Contact Problems

Jiří Jarušek

Abstract The article describes the main contribution of J. Nečas on the solvability of contact problems.

His main contribution started with the pioneering paper [JN-C95], where the solvability of Signorini's problem with Coulomb friction was solved via regularity of solutions to a suitable class of approximate solutions. The impact of this paper is clearly seen from its citation in more than 150 works of different authors from many different countries. As his contemporary Ph.D. student I developed his idea in two papers [9, 10]. This original idea was related with the use of the Tichonov fixed point theorem. Later it was combined with the penalty approximation which enable one to include also the dependence of the coefficient of friction on the solution (to distinguish the "slip" and "stick" coefficient; it does not make much sense to expect a jump between them, because this idea increases the possible multiplicity of solutions only, but a continuous transition between them is physically reasonable). This was used in the papers [6, 7] where improved and probably optimal bounds for the admissible magnitude of the coefficient of friction were calculated. Also in the results [1,5] for the solvability of the quasistatic contact problem with Coulomb friction it plays its crucial role.

The dynamic version of the problem has been solved for the contact condition formulated in velocities only, cf. e.g. [14] (there are versions including thermal effects or some kind of plasticity). The reason is that the Nečas idea works here only if a sufficient regularity of velocities can be proved.

His contribution continued in papers [C162] and [JN-C131] with the main content published almost simultaneously. His idea to employ an L_2 estimate of the penalty term in the domain contact together with some viscosity of the material in the first paper written for a membrane led to new results in solvability of dynamic contact problems devoted to different models of plates with the physically optimal contact condition formulated in displacements. Let me mention at least [2, 11] and [3], although the last one is written for purely elastic von Kármán plates. In boundary contacts the dual estimate of the acceleration can be obtained more straightforwardly (cf. e.g. [13]), but its importance was seen in [C162] at first.

The idea of singular memory introduced in [JN-C131] led to a parallel series of papers of solvability of dynamic contact problems for membranes and plates made of such material, cf. e.g. [12] and Sect. 4 of [4] which are still unsolvable for purely elastic material.

With the exception of the special result by Lebeau and Schatzman [16] and the unique paper of Kim [15], all important existing results about solvability of two- or three-dimensional unilateral dynamic contact problems have been essentially influenced by the two mentioned papers of J. Nečas.

A part of these and further results are contained in the monograph [8]. To honour the Nečas contribution to the development of solvability of contact problems, it has been dedicated to him.

The frequent citation of a great part of the works mentioned here shows the importance of the Nečas contribution as well.

Nečas' Works Quoted Above

The references [Cxx] refer to the respectice papers listed in Chapter 2 "Complete Bibliography of Jindřich Nečas". The references [JN-Cyy] refer to papers [Cyy] that are reproduced in Chapter 11 "Selected Papers on Solid Mechanics, Elasticity, Thermoelasticity, and Plasticity".

[C162] J. Jarušek, J. Málek, J. Nečas, and V. Šverák, *Variational inequality for a viscous drum vibrating in the presence of an obstacle*, Rend. Mat. Appl. (7), 1993, vol. 12, pp. 943–958.

[JN-C95] K. Gröger, J. Nečas, and L. Trávníček, *Dynamic deformation processes of elastic-plastic systems*, Z. Angew. Math. Mech., 1979, vol. 59, pp. 567–572.

[JN-C131] Ph. G. Ciarlet and J. Nečas, *Injectivity and self-contact in nonlinear elasticity*, Arch. Rational Mech. Anal., 1987, vol. 97, pp. 171–188.

Bibliography

[1] L.-E. ANDERSSON: Existence result for quasistatic contact problems with Coulomb friction. *Appl. Math. Optimiz.* **42** (2000), 169–202.

[2] I. BOCK AND J. JARUŠEK: Solvability of dynamic contact problems for elastic von Kármán plates. *SIAM J. Math. Anal.* **41** (1) (2009), 37–45.

[3] I. BOCK AND J. JARUŠEK: Dynamic contact problem for a bridge modeled by a viscoelastic full von Kármán system. *Z. Angew. Math. Phys.* **61** (5) (2010), 865–876.

[4] I. BOCK AND J. JARUŠEK: Unilateral dynamic contact problem for viscoelastic Reissner-Mindlin plates. *Nonlin. Anal., Theory Meth. Appl* **74** (12), (2011), 4192–4202.

[5] R. ROCCA & M. COCOU: Existence and approximation of a solution to quasistatic Signorini problem with local friction. *Int. J.Engrg. Sci.* **39** (11), (2001), 1233–1255.

[6] C. ECK AND J. JARUŠEK: Existence results for the static contact problem with Coulomb–friction. *Math. Models Meth. Appl. Sci.* **8** (1998), 445–468.

[7] C. ECK AND J. JARUŠEK: Existence results for the semicoercive static contact problem with Coulomb–friction. *Nonlin. Anal., Theory Meth. Appl.* **42** (2000) 961–976.

[8] C. ECK, J. JARUŠEK AND M. KRBEC: *Unilateral Contact Problems in Mechanics. Variational Methods and Existence Theorems.* Monographs & Textbooks in Pure & Appl. Math. No. 270 (ISBN 1-57444-629-0). Chapman & Hall/CRC (Taylor & Francis Group), Boca Raton – London – New York – Singapore 2005.

[9] J. JARUŠEK: Contact problems with bounded friction. Coercive case. *Czech. Math. J.* **33** (108) (1983), 237–261.

[10] J. JARUŠEK: Contact problems with bounded friction. Semicoercive case. *Czech. Math. J.* **34** (109) (1984), 619–629.

[11] J. JARUŠEK: Solvability of unilateral hyperbolic problems involving viscoelasticity via penalization. **In:** Proc. Int. Conf. Diff. Eq. & Math. Modelling held at Varenna 1992 (R. Salvi, ed.). *S.A.A.C.M.* **3**, 2 (1993), 129–140.
[12] J. JARUŠEK: Solvability of the variational inequality for a drum with a memory vibrating in the presence of an obstacle. *Boll. Unione Mat. Ital.* **7** (8–A) (1994), 113–122.
[13] J. JARUŠEK: Dynamical contact problems with given friction for viscoelastic bodies. *Czech. Math. J.* **46** (1996), 475–487.
[14] J. JARUŠEK AND C. ECK: Dynamic contact problems with small Coulomb friction for viscoelastic bodies. Existence of solutions. *Math. Models Meth. Appl. Sci.* **9** (1999), 11–34.
[15] J. U. KIM: A boundary thin obstacle problem for a wave equation. *Comm. Part. Diff. Equations* **14**(849), (1989), 1001–1025.
[16] G. LEBEAU AND M. SCHATZMAN: A wave problem in a half-space with a unilateral constraint at the boundary. *J. Diff. Eq.* **53** (1984), 309–361.

Jiří Jarušek
Academy of Sciences, Institute of Mathematics, Žitná 25, 11567 Prague 1,
Czech Republic
email: jarusek@math.cas.cz

Part VI
Fluid Mechanics

Chapter 13
Selected Papers on Fluid Mechanics

manuscripta math. 61, 417 - 428 (1988)

manuscripta mathematica
© Springer-Verlag 1988

REMARKS ON THE SOLVABILITY OF TRANSONIC FLOW PROBLEMS

Miloslav Feistauer and Jindřich Nečas

The paper is devoted to the study of the viscosity method and upwinding in the transonic flow problems. We weaken the conditions from [6,7] under which we can obtain a nonviscous compressible transonic flow as a limit of viscous solutions, if the viscosity and heat conductivity tend to zero. Further, we give the justification of the density upwinding applied often in the calculation of transonic potential flows.

INTRODUCTION

In [4,5,6,7,9] we studied the questions of the solvability of transonic flows and showed also the connection with numerical approaches from [2, 8]. There are two goals of this paper:

1) To generalize the results from [6,7] by weakening the conditions and assumptions in the viscosity method. The starting point is the solution of a compressible, perfect, viscous, conductive gas flow. We let to vanish the viscosity and thermal conductivity constants and look for the limit of the density, temperature and velocity under such weak assumptions which yield a limit solution satisfying the continuity and Euler equations and in some sense the energy equation and permit strong shocks in the velocity and density and even cavitation effects on sets of positive measure.

2) To justify the density upwinding methods (very popular in the numerical modeling of transonic potential flows

© Springer-Verlag 1988. Reprinted with kind permission from: Feistauer, M. & Nečas, J., Remarks on the solvability of transonic flow problems, manuscripta mathematica 61, 417-428.

- see e. g. [1, 2, 8]) and to show that they lead to entropy compactification inequalities.

1. VISCOSITY METHOD

Let $\Omega \subset R^N$ (N = 2 or 3) be a bounded domain with a Lipschitz-continuous boundary $\partial\Omega$. We denote ρ-density, p-pressure, T-temperature, S-entropy, $v = (v_1,\ldots,v_N)$-velocity, c_p, c_v- specific heats, μ-viscosity, k-thermal conductivity, $R = c_p - c_v$ - gas constant, $\kappa = c_p/c_v$ - adiabatic constant, $n = (n_1,\ldots,n_N)$ - unit outer normal to $\partial\Omega$. ($\mu, k, c_p, c_v, R>0, \kappa>1$.)

Stationary flow of a compressible, perfect, viscous, conductive gas in the domain Ω is governed by the following system:

(1.1) $\quad p = R\rho T$ (state equation)

(1.2) a) $\frac{\partial}{\partial x_i}(\rho v_i) = 0$ in Ω (continuity equation)

b) $\rho v_i n_i = g$ on $\partial\Omega$,

c) $\int_{\partial\Omega} g\,ds = 0,$

(1.3) a) $\rho v_j \frac{\partial v_i}{\partial x_j} + \frac{\partial p}{\partial x_i} = 2\mu \frac{\partial}{\partial x_j} e_{ij}(v) - \frac{2}{3}\mu \frac{\partial}{\partial x_i} \operatorname{div} v,$

$e_{ij}(v) = \frac{1}{2}\left(\frac{\partial v_i}{\partial x_j} + \frac{\partial v_j}{\partial x_i}\right)$ in Ω

(Navier-Stokes equations),

b) $v = v^o$ on $\partial\Omega$,

(1.4) a) $T\frac{\partial}{\partial x_i}(\rho S v_i) = k\Delta T + 2\mu e_{ij}(v)e_{ij}(v) - \frac{2}{3}\mu(\operatorname{div} v)^2$ in Ω (energy equation),

b) $\frac{\partial T}{\partial n} = h$ on $\partial\Omega$,

c) $S = c_v \ln \frac{T}{\rho^{\kappa-1}}.$

We use the summation convention over repeated indices. By $L^p(\Omega)$, $L^p(\partial\Omega)$ and $W^{k,p}(\Omega)$, $W_0^{k,p}(\Omega)$ we denote the Lebesgue and Sobolev spaces, respectively, and put $L^p(\Omega;R^N) = L^p(\Omega) \times \ldots \times L^p(\Omega)$, $W^{k,p}(\Omega;R^N) = W^{k,p}(\Omega) \times \ldots \times W^{k,p}(\Omega)$ (N-times) etc. The letter c will denote a positive generic constant independent of μ, k, with generally different values at different places.

We assume that the above problem (1.1) - (1.4) has a weak solution satisfying the conditions

(1.5) $\rho \in W^{1,2}(\Omega), \quad 0 < \tilde{c}(\mu,k) \le \rho(x) \le \tilde{\rho}_1 < +\infty, \quad x \in \Omega,$

(1.6) $v \in W^{1,2}(\Omega;R^N), \quad |v| \le K,$

(1.7) $g \in L^\infty(\partial\Omega), \quad h \in L^1(\partial\Omega), \quad \|g\|_{L^\infty(\partial\Omega)}, \quad \|h\|_{L^1(\partial\Omega)} \le K,$

(1.8) $T \in W^{1,2}(\Omega), \quad 0 < T_0 \le T,$

(1.9) $\int_{\partial\Omega} \ln \frac{T}{T_o}\, ds \le K,$

and equations

(1.10) $\int_\Omega \rho v_i \frac{\partial\phi}{\partial x_i} dx = \int_{\partial\Omega} g\phi ds \qquad \forall\ \phi \in W^{1,2}(\Omega),$

(1.11) $\int_\Omega \rho v_j \frac{\partial v_i}{\partial x_j}\phi_i dx = \int_\Omega p\frac{\partial\phi_i}{\partial x_i} dx + \frac{2}{3}\mu\int_\Omega \frac{\partial v_j}{\partial x_j}\,\frac{\partial\phi_i}{\partial x_i} dx -$

$- 2\mu\int_\Omega e_{ij}(v)e_{ij}(\phi)dx \qquad \forall\ \phi \in W_0^{1,2}(\Omega;R^N),$

$v = v^o \quad \text{on } \partial\Omega,$

(1.12) $-\int_\Omega \rho v_i S\frac{\partial\psi}{\partial x_i} dx + \int_{\partial\Omega} Sg\psi ds =$

$= -k\int_\Omega \frac{\nabla T}{T}\nabla\psi dx + k\int_\Omega \frac{|\nabla T|^2}{T^2}\psi dx + k\int_{\partial\Omega}\frac{h}{T}\psi ds + \int_\Omega \frac{E(v)}{T}\psi dx$

$\forall\ \psi \in W^{1,2}(\Omega) \cap L^\infty(\Omega),$

where $E(v) = 2\mu e_{ij}(v)e_{ij}(v) - \frac{2}{3}\mu(\operatorname{div} v)^2 \ge 0$ is the so-called dissipation. We assume that $\tilde{\rho}_1$, T_O, $K > 0$ are constants independent of μ and k. The constant $\tilde{c}(\mu,k) > 0$ can tend to zero, if μ, $k \to 0$.

In the following we shall suppose that $k = \beta\mu$, with a constant $\beta > 0$ independent of μ and k, and

(1.13) $\left|\frac{1}{\mu}\int_{\partial\Omega} Sgds\right| \le K.$

By the same way as in [6, Theorem 3.3] we get

(1.14) $\int_\Omega \frac{|\nabla T|^2}{T^2} dx + \frac{1}{\mu}\int_\Omega \frac{E(v)}{T} dx \le c,$

and provided

(1.15) $\|v^o\|_{W^{1,2}(\Omega;R^N)} + \|v^o\|_{L^\infty(\Omega;R^N)} \le K,$

we also have ([6, Theorem 3.9] or [7, Theorem 2.23])

(1.16) $\int_\Omega \frac{|\nabla v|^2}{T} dx \le c.$

Let us consider $\mu_n \to 0$, $\mu_n > 0$ and a sequence of solutions $\{\rho_n, T_n, p_n, S_n, v^n\}$ to (1.10) - (1.12) with properties (1.5) - (1.9), where $\mu := \mu_n$, $k := \beta\mu_n$. With respect to

(1.8) and the Bepo-Levi definition of $W^{1,2}(\Omega)$ we have $\ln T_n \in W^{1,2}(\Omega)$ and $\nabla \ln T_n = \nabla T_n / T_n$. Moreover, taking into account (1.14) and (1.9) we conclude that

$$\|\ln T_n\|_{W^{1,2}(\Omega)} \le c \qquad \forall\, n = 1,2,\dots$$

In view of (1.5) and (1.6), $\{\rho_n\}$ and $\{v^n\}$ are bounded in $L^2(\Omega)$ and $L^2(\Omega;R^N)$, respectively. Now, on the basis of the reflexivity of the spaces $L^2(\Omega)$ and $W^{1,2}(\Omega)$ and the compact imbedding $W^{1,2}(\Omega) \hookrightarrow\hookrightarrow L^2(\Omega)$ we can choose a sequence of solutions satisfying

(1.17)
a) $\rho_n \rightharpoonup \rho$ weakly in $L^2(\Omega)$
b) $\ln T_n \rightharpoonup \ln T$ weakly in $W^{1,2}(\Omega)$,
c) $\ln T_n \to \ln T$ strongly in $L^2(\Omega)$,
d) $\ln T_n \to \ln T$ almost everywhere in Ω (or briefly, a. e. in Ω),
e) $v^n \rightharpoonup v$ weakly in $L^2(\Omega;R^N)$.

1.18. LEMMA. $v^n \to v$ *strongly in* $L^2(\Omega;R^N)$.

Proof. If we put $w^n = \dfrac{v^n}{\sqrt{T_n}}$, then

$$\frac{\partial w_i^n}{\partial x_j} = \frac{\partial v_i^n}{\partial x_j}\frac{1}{\sqrt{T_n}} - \frac{1}{2}\frac{1}{T_n^{3/2}} v_i^n \frac{\partial T_n}{\partial x_j}.$$

This, (1.6), (1.8), (1.4) and (1.16) give the estimate

$$\|w^n\|_{W^{1,2}(\Omega;R^N)} \le c.$$

Hence (possibly for a subsequence) $w^n \to w$ a. e. in Ω. In view of (1.17,d), $\sqrt{T_n} \to \sqrt{T}$ a. e. and therefore also $v^n \to v^* = w\sqrt{T}$ a. e. in Ω. Thus, $v^n \to v^*$ strongly in $L^2(\Omega;R^N)$, since $|v^n| \le K$. Now by (1.17,e), $v = v^*$. ⌑

1.19. LEMMA. $\rho_n \to \rho$ *strongly in* $L^2(\Omega)$.

Proof of this lemma is a slight modification of the proof of Theorem 3.2 from [7]. Let us put $\phi = \frac{\psi}{T_n}$ in (1.11) for arbitrary $\psi \in W_0^{1,2}(\Omega;R^N) \cap L^\infty(\Omega;R^N)$. Then

$$(1.20)\quad -\overbrace{\int_\Omega \rho_n v_i^n v_j^n \frac{1}{T_n}\frac{\partial \psi_i}{\partial x_j}dx}^{1.} + \overbrace{\int_\Omega \rho_n v_i^n v_j^n \frac{1}{T_n^2}\frac{\partial T_n}{\partial x_j}\psi_i dx}^{2.} =$$

$$= \overbrace{R\int_\Omega \rho_n \frac{\partial \psi_i}{\partial x_i}dx}^{3.} - \overbrace{R\int_\Omega \rho_n \frac{\partial T_n}{\partial x_i}\frac{1}{T_n}\psi_i dx}^{4.} + \overbrace{\frac{2}{3}\mu_n \int_\Omega \frac{\partial v_j^n}{\partial x_j}\frac{1}{T_n}\frac{\partial \psi_i}{\partial x_i}dx}^{5.} -$$

$$- \underbrace{\frac{2}{3}\mu_n \int_\Omega \frac{\partial v_j^n}{\partial x_j} \frac{1}{T_n^2} \frac{\partial T_n}{\partial x_i}\psi_i dx}_{6.} - \underbrace{2\mu_n \int_\Omega e_{ij}(v^n) e_{ij}(\psi) \frac{1}{T_n} dx}_{7.} +$$

$$+ \underbrace{2\mu_n \int_\Omega e_{ij}(v^n)\psi_i \frac{\partial T_n}{\partial x_j} \frac{1}{T_n^2} dx}_{8.}.$$

It is necessary to estimate the terms 5. - 8. in (1.20), the rest of the proof is the same as in [7].

Using (1.8) and (1.16) we have

$$(1.21) \qquad \mu_n \left| \int_\Omega \frac{\partial v_j^n}{\partial x_j} \frac{1}{T_n} \frac{\partial \psi_i}{\partial x_i} dx \right| \le c\mu_n \int_\Omega \frac{|\nabla v^n|}{\sqrt{T_n}} \left|\frac{\partial \psi_i}{\partial x_i}\right| dx \le$$

$$\le c\mu_n \left(\int_\Omega \frac{|\nabla v^n|^2}{T_n} dx\right)^{\frac{1}{2}} \|\psi\|_{W^{1,2}(\Omega;R^N)} \le c\mu_n \|\psi\|_{W^{1,2}(\Omega;R^N)}.$$

Similarly for the 7th term. Further, by (1.8), (1.14) and (1.16),

$$(1.22) \qquad \mu_n \left| \int_\Omega \frac{\partial v_j^n}{\partial x_j} \frac{1}{T_n^2} \frac{\partial T_n}{\partial x_i}\psi_i dx \right| \le$$

$$\le c\mu_n \left(\int_\Omega \frac{|\nabla v^n|^2}{T_n} dx\right)^{\frac{1}{2}} \left(\int_\Omega \frac{|\nabla T_n|^2}{T_n^2} dx\right)^{\frac{1}{2}} \|\psi\|_{L^\infty(\Omega;R^N)} \le$$

$$\le c\mu_n \|\psi\|_{L^\infty(\Omega;R^N)}.$$

The same estimate holds for the 8th term in (1.20). ∎

We choose eventually a subsequence such that

$$(1.23) \qquad \rho_n \to \rho, \quad v^n \to v, \quad T_n \to T \quad \text{a. e. in } \Omega.$$

1.24. THEOREM. *Let*

$$S(x) = c_v \ln \frac{T(x)}{\rho(x)^{\kappa-1}}, \quad H(x) = \rho(x)S(x), \quad \textit{if } \rho(x) > 0,\ x \in \Omega,$$

$$S(x) = \infty, \qquad H(x) = 0, \quad \textit{if } \rho(x) = 0,\ x \in \Omega.$$

Then ρ, T, v, S *satisfy* $0 \le \rho \le \tilde{\rho}_1$, $T \ge T_0$, $|v| \le K$ *and form a weak solution of the nonviscous conservation law equations in the following sense:*

$$(1.25) \qquad \int_\Omega \rho v_i \frac{\partial \phi}{\partial x_i} dx = \int_{\partial\Omega} g_0 \phi ds \qquad \forall\ \phi \in W^{1,2}(\Omega),$$

$$(1.26) \qquad -\int_\Omega \rho v_i v_j \left(\frac{1}{T} \frac{\partial \psi_i}{\partial x_j} - \frac{\partial T}{\partial x_j} \frac{1}{T^2}\psi_i\right) dx =$$

$$= R\int_\Omega \rho \frac{\partial \psi_i}{\partial x_i} dx - R\int_\Omega \rho \frac{1}{T} \frac{\partial T}{\partial x_i}\psi_i dx \qquad \forall\ \psi \in W_0^{1,2}(\Omega;R^N),$$

(1.27) $\int_\Omega v_i H \frac{\partial \psi}{\partial x_i} dx = \int_{\partial\Omega} S_0 g \psi ds \quad \forall\ \psi \in W^{1,2}(\Omega) \cap L^\infty(\Omega),$

provided

(1.28) $S_n \to S_0$ *in* $L^1(\partial\Omega)$ *and* $g_n \to g_0$ *a. e. in* $\partial\Omega$.

Proof. It is necessary to carry out the limit process $n \to +\infty$, $\mu_n \to 0+$ in the system (1.10) - (1.12).

In the continuity equation (1.10), i. e.

$$\int_\Omega \rho_n v_i^n \frac{\partial \phi}{\partial x_i} dx = \int_{\partial\Omega} g_n \phi ds \qquad \forall\ \phi \in W^{1,2}(\Omega)$$

we apply Lebesgue's theorem using (1.5), (1.6) and (1.23).

Now we shall deal with the particular terms 1. - 8. of the Navier-Stokes equations (1.20). In view of the density of the set $\mathcal{D}(\Omega;R^N)$ (of infinitely differentiable functions with compact supports in Ω) in the space $W_0^{1,2}(\Omega;R^N)$ it is sufficient to consider test functions $\psi \in \mathcal{D}(\Omega;R^N)$ only.

In the proof of Lemma 1.19 we have already shown that the terms 5. - 8. tend to zero - cf. (1.21) and (1.22). Let us treat the terms 1. - 4. of (1.20):

1. As (1.5), (1.6), (1.8), (1.23) hold, we can use Lebesgue's theorem in the first term.

2. Let us denote

$$\alpha_n = \frac{1}{T_n}\frac{\partial T_n}{\partial x_j}, \qquad \alpha = \frac{1}{T}\frac{\partial T}{\partial x_j},$$

$$\beta_n = \rho_n v_i^n v_j^n \frac{1}{T_n}\psi_i, \qquad \beta = \rho v_i v_j \frac{1}{T}\psi_i.$$

In view (1.17,b), $\alpha_n \to \alpha$ weakly in $L^2(\Omega)$. Easily we find out that $\beta_n \to \beta$ strongly in $L^2(\Omega)$. Hence,

$$\int_\Omega \alpha_n \beta_n dx \to \int_\Omega \alpha\beta dx,$$

which proves the limit process in the second term.

3. Concerning the third term in (1.20), we get the result from 1.19.

4. The limit process in the fourth term is a consequence of (1.17,b) and 1.19.

Hence, (1.26) is proved.

Now let us devote ourselves to the energy equation (1.12), i. e.

(1.29) $$-\overbrace{\int_\Omega \rho_n v_i^n S_n \frac{\partial \psi}{\partial x_i}}^{1.} + \overbrace{\int_{\partial\Omega} S_n g_n \psi ds}^{2.} =$$

$$= \overbrace{-k_n \int_\Omega \frac{\nabla T_n}{T_n} \nabla\psi dx}^{3.} + \overbrace{k_n \int_\Omega \frac{|\nabla T_n|^2}{T_n^2} \psi dx}^{4.} +$$

$$+ \overbrace{k_n \int_{\partial\Omega} \frac{h_n}{T_n} \psi ds}^{5.} + \overbrace{\int_\Omega \frac{E(v^n)}{T_n} \psi dx}^{6.}$$

As a test function we can consider an arbitrary $\psi \in C^\infty(\bar{\Omega})$.

1. In order to prove the limit in the first term, i. e.

$$(1.30) \qquad \int_\Omega \rho_n v_i^n S_n \frac{\partial \psi}{\partial x_i} dx \to \int_\Omega v_i H \frac{\partial \psi}{\partial x_i} dx,$$

we shall use the following elementary assertion:

<u>1.31. LEMMA.</u> *Let* $a_n \to a$ *a. e. in* Ω, a_n *be measurable,* $|a_n| \le c$, $b_n \to b$ *in* $L^1(\Omega)$. *Then* $a_n b_n \to ab$ *in* $L^1(\Omega)$. ⌑

Now let us put

$$(1.32) \qquad H_n = \rho_n S_n = \rho_n c_v \ln \frac{T_n}{\rho_n^{\kappa-1}} = c_v \rho_n \ln T_n - c_v(\kappa-1)\rho_n \ln \rho_n.$$

By 1.31,

$$(1.33) \qquad \rho_n \ln T_n \to \rho \ln T \qquad \text{in } L_1(\Omega)$$

($\rho_n \to \rho$ a. e. in Ω, ρ_n is bounded and $\ln T_n \to \ln T$ in $L^1(\Omega)$).

Further, if we take into account that the function $\tau \ln \tau$, $\tau \in (0, \tilde{\rho}_1]$ is bounded and $\tau \ln \tau \to 0$, if $\tau \to 0+$, we conclude that there exist a constant $c > 0$ and a function Q such that

$$|\rho_n \ln \rho_n| \le c \qquad \text{a. e. in } \Omega \quad \forall\ n = 1,2,\dots,$$

$$\rho_n \ln \rho_n \to Q \qquad \text{a. e. in } \Omega.$$

($Q(x) = 0$, if $\rho(x) = 0$.) Hence,

$$(1.34) \qquad \rho_n \ln \rho_n \to Q \qquad \text{in } L^1(\Omega).$$

By (1.32) - (1.34) we have $H_n \to \Phi = c_v\ \rho \ln T + c_v(\kappa-1)Q$ in $L^1(\Omega)$. On the other hand, $\Phi(x) = 0$, if $\rho(x) = 0$ and $\Phi(x) = c_v\ \rho(x) \ln T(x) + c_v(\kappa-1)\rho(x) \ln \rho(x)$, if $\rho(x) \ne 0$. Hence $\Phi = H$ and

$$(1.35) \qquad H_n \to H \qquad \text{in } L^1(\Omega).$$

Now, (1.30) is an easy consequence of (1.6), (1.23), (1.35), Lemma 1.31 and the boundedness of $\partial\psi/\partial x_i$.

2. In virtue of (1.7), (1.28), (1.35) and Lemma 1.31, where we substitute $\Omega := \partial\Omega$, we get

$$\int_{\partial\Omega} S_n g_n \psi ds \to \int_{\partial\Omega} S_o g_o \psi ds.$$

3. By (1.14) and the Cauchy inequality we have

$$k_n \int_\Omega \frac{\nabla T_n}{T_n} \nabla\psi dx \le ck_n \Big(\int_\Omega \frac{|\nabla T_n|^2}{T_n^2} dx\Big)^{\frac{1}{2}} \le ck_n \to 0.$$

4. In the fourth and sixth terms we use again (1.14).

5. Finally, by (1.7) - (1.8),

$$k_n \int_{\partial\Omega} \frac{h_n}{T_n} \psi ds \le ck_n \to 0.$$

Summarizing the above results we get (1.27). ⌑

2. UPWINDING

Let us consider a *transonic potential flow* described by the velocity potential equation

(2.1) $$-\frac{\partial}{\partial x_i}\Big(\rho(|\nabla u|^2)\frac{\partial u}{\partial x_i}\Big) = 0 \quad \text{in } \Omega.$$

Here Ω is a simply connected bounded domain in R^N with $\partial\Omega$ Lipschitz-continuous, u is the velocity potential and the density ρ depends on $s = |\nabla u|^2$ in the following way ([5]):

(2.2)
a) ρ, ρ' are continuous in $[0,+\infty)$,

b) $\rho(s) = \rho_o\big(1-\frac{\kappa-1}{2a_o^2}s\big)^{\frac{1}{\kappa-1}}$ for $s \in [0,s^*]$ with

c) $s^* \in \big(\frac{2a_o^2}{\kappa+1}, \frac{2a_o^2}{\kappa-1}\big)$ (s^* is close to $\frac{2a_o^2}{\kappa-1}$),

d) $0 < \rho_\infty \le \rho(s) \le \rho_o < +\infty$,

e) $-\infty < -\delta \le \rho'(s) \le 0$, $\delta > 0$,

f) $|\rho(s)-\alpha|\sqrt{s} \le c$ with some c, $\alpha > 0$.

To (2.1) we add the boundary condition

(2.3) $$\rho(|\nabla u|^2)\frac{\partial u}{\partial n} = g \qquad \text{on } \partial\Omega,$$

where $g \in L^\infty(\partial\Omega)$ and

(2.4) $$\int_{\partial\Omega} g ds = 0.$$

If we put

(2.5) $$V = \{u \in W^{1,2}(\Omega);\ \int_\Omega u dx = 0\},$$

then the problem (2.1), (2.3) has the following weak formulation: Find $u \in V$ such that

(2.6) $$\langle A(u),v\rangle = \langle g,v\rangle \qquad \forall\ v \in V,$$

where $A: W^{1,2}(\Omega) \to (W^{1,2}(\Omega))^*$, $g \in (W^{1,2}(\Omega))^*$, $\langle .,. \rangle$ denotes the duality between the space $W^{1,2}(\Omega)$ and its dual $(W^{1,2}(\Omega))^*$ and

$$(2.7)\qquad \langle A(u),v\rangle = \int_\Omega \rho(|\nabla u|^2)\nabla u\ \nabla v\ dx,$$
$$\langle g,v\rangle = \int_{\partial\Omega} gv ds.$$

In [4,5] we showed that in the study of problem (2.6) a suitable *entropy condition* plays a crucial role. We can use e. g. the entropy condition proposed by Bristeau et al. in [2] (see also [8])

$$\Delta u \le K$$

with some $K > 0$, i. e.

$$-\int_\Omega \nabla u\ \nabla\phi dx \le K\int_\Omega \phi dx$$
$$\forall\ \phi\in \mathcal{D}^+(\Omega) = \{v\in \mathcal{D}(\Omega);\ v \ge 0 \text{ in } \Omega\},$$

which was used with success in numerical calculations. Many authors use the artificial viscosity method introduced into numerical schemes by the upwinding of the density (see e. g. [1,3,8,12]). It means that the value of ρ at a given point P is replaced by its approximation equal to the value of ρ at a suitable point P' in the upstream direction to P.

We shall prove the following theorem:

2.8. Theorem. *Let* $\{u_n\}_{n=1}^\infty$ *be a regular generic sequence to problem* (2.6). *I. e.,*

$$(2.9)\qquad u_n\in W^{2,2}(\Omega)\cap V,$$

$$(2.10)\qquad \langle A(u_n),v\rangle = \langle g,v\rangle + \langle F_n,v\rangle \qquad \forall\ v\in V$$

and

$$(2.11)\qquad F_n \to 0 \qquad\qquad \text{in } (W^{1,2}(\Omega))^*.$$

Further, let

$$(2.12)\qquad f_n \overset{\text{def.}}{=} \left[\Delta u_n + 2\frac{\rho'(|\nabla u_n|^2)}{\rho(|\nabla u_n|^2)}\frac{\partial u_n}{\partial x_k}\frac{\partial u_n}{\partial x_i}\frac{\partial^2 u_n}{\partial x_k\partial x_i}\right]^+ \rightharpoonup f$$
weakly in $L^2(\Omega)$,

$$(2.13)\qquad q_n \overset{\text{def.}}{=} \left[2\frac{\rho'(|\nabla u_n|^2)}{\rho(|\nabla u_n|^2)}\frac{\partial u_n}{\partial x_k}\frac{\partial u_n}{\partial x_i}\frac{\partial^2 u_n}{\partial x_k\partial x_i}\right]^- \rightharpoonup q$$
weakly in $L^2(\Omega)$.

Then

$$(2.14)\qquad \|u_n\|_{W^{1,2}(\Omega)} \le c$$

and provided $u_n \rightharpoonup u$ *weakly in* $W^{1,2}(\Omega)$, *we have*

(2.15) $\quad u_n \to u \quad$ *strongly in* $W^{1,2}(\Omega)$

and u *is a solution of the transonic flow problem* (2.6), *entropic in the sense*

(2.16) $\quad \Delta u \le f + q.$

<u>Proof.</u> From (2.2) we get the coercivity of the operator A:

(2.17) $\quad \langle A(u),u\rangle \ge c_1\|u\|^2_{W^{1,2}(\Omega)} \qquad \forall\, u \in V$

with $c_1 > 0$. This and (2.10) - (2.11) yield the estimate

$$0 \ge c_1\|u_n\|^2_{W^{1,2}(\Omega)} - \|g+F_n\|_{V^*}\|u_n\|_{W^{1,2}(\Omega)} \ge$$
$$\ge c_1\|u_n\|^2_{W^{1,2}(\Omega)} - c\|u_n\|_{W^{1,2}(\Omega)},$$

which implies (2.14).

Similarly as in the proof of Theorem 4.30 from [5] we write

(2.18) $\quad u_n = u_n^1 + u_n^2, \qquad u_n^1,\ u_n^2 \in V,$

$$\alpha\int_\Omega \nabla u_n^1 \nabla v\,dx = \int_\Omega (\alpha-\rho(|\nabla u_n|^2))\nabla u_n \nabla v\,dx + \langle g,v\rangle \quad \forall\, v \in V,$$
$$\alpha\int_\Omega \nabla u_n^2 \nabla v\,dx = \langle F_n,v\rangle \qquad \forall\, v \in V.$$

By (2.11),

(2.19) $\quad u_n^2 \to 0 \quad$ in $W^{1,2}(\Omega)$

and thus,

(2.20) $\quad u_n^1 \rightharpoonup u \quad$ in $W^{1,2}(\Omega)$.

Now, by [10], adopted in [12] to the Neumann problem,

(2.21) $\quad \|u_n^1\|_{W^{1,p_2}(\Omega)} \le c \qquad \forall\, n = 1,2,\ldots$

with some $p_2 > 2$ and $c > 0$. From this, (2.20) and the well--known fact that a bounded, closed, convex set is weakly closed we get

(2.22) $\quad \|u\|_{W^{1,p_2}(\Omega)} \le c.$

Let us define functionals $G,\ G_n \in (W^{1,2}(\Omega))^*$:

(2.23) $\quad \langle G_n,v\rangle = \int_\Omega (f_n+q_n)v\,dx + \int_\Omega \nabla u_n\, \nabla v\,dx,$

$$\langle G,v\rangle = \int_\Omega (f+q)v\,dx + \int_\Omega \nabla u\, \nabla v\,dx,$$
$$v \in W^{1,2}(\Omega).$$

By the definitions (2.12), (2.13) of f_n and q_n we find out that

(2.24) $\Delta u_n \le f_n + q_n$

and thus,

(2.25) $-\int_\Omega \nabla u_n \nabla v dx \le \int_\Omega (f_n+q_n) v dx \qquad \forall\, v \in \mathcal{D}^+(\Omega)$.

Actually, denoting

$$\theta_n = 2\frac{\rho'(|\nabla u_n|^2)}{\rho(|\nabla u_n|^2)} \frac{\partial u_n}{\partial x_k} \frac{\partial u_n}{\partial x_i} \frac{\partial^2 u_n}{\partial x_k \partial x_i},$$

we have

$$\Delta u_n = (\Delta u_n+\theta_n) - \theta_n = (\Delta u_n+\theta_n)^+ - (\Delta u_n+\theta_n)^- - \theta_n \le$$
$$\le (\Delta u_n+\theta_n)^+ + \theta_n^- = f_n + q_n.$$

Relation (2.25) is equivalent to

(2.26) $\langle G_n, v\rangle \ge 0 \qquad \forall\, v \in \mathcal{D}^+(\Omega)$.

Moreover, $G_n \rightharpoonup G$ in $(W^{1,2}(\Omega))^*$. Using the stronger variant of the Murat's famous theorem ([11]) proved in [4] we see that

(2.27) $G_n \to G \qquad$ strongly in $(W^{1,p_2}(\Omega))^*$.

Now let us put $h_n = u_n^1 - u$. Then, by (2.10), (2.18), (2.23), (2.11) - (2.13), (2.27) and the implication "$h_n \rightharpoonup 0$ in $W^{1,2}(\Omega) \Rightarrow h_n \to 0$ in $L^2(\Omega)$",

$$\alpha\|h_n\|^2_{W^{1,2}(\Omega)} = \alpha\langle G_n, h_n\rangle - \int_\Omega \alpha(f_n+q_n) h_n dx -$$
$$- \alpha\langle G, h_n\rangle + \int_\Omega \alpha(f+q) h_n dx - \langle F_n, h_n\rangle \to 0.$$

Hence, $h_n \to 0$ in $W^{1,2}(\Omega)$ and, in virtue of (2.19), $u_n \to u$ in $W^{1,2}(\Omega)$.

Taking into account the demicontinuity of A: "$u_n \to u$ in $W^{1,2}(\Omega) \Rightarrow A(u^n) \rightharpoonup A(u)$ in $(W^{1,2}(\Omega))^*$", we conclude from (2.10) and (2.11) that u is a solution of problem (2.6). Finally, in view of (2.26) and (2.27),

$$\langle G, v\rangle \ge 0 \qquad \forall\, v \in \mathcal{D}^+(\Omega),$$

which is (2.16). ⌑

<u>2.28. REMARK.</u> Condition (2.12) represents the correctness of the approximate solvability of the transonic equation in $W^{2,2}(\Omega)$ norm with respect to the positive part of the residual. (2.13) means that the density does not have the jumps prohibited by the physical entropy condition.

REFERENCES

[1] AMARA, M., JOLY, P., THOMAS, J. M.: A mixed finite element method for solving transonic flow equations. Comp. Meth. Appl. Mech. Eng. 39, 1-18(1983)

[2] BRISTEAU, M. O., GLOWINSKI, R., PERIAUX, J., PERRIER, P., PIRONNEAU, O.: On the numerical solution of nonlinear problems in fluid dynamics by least squares and finite element methods. (I) Least square formulations and conjugate gradient solution of the continuous problems. Comp. Meth. Appl. Mech. Eng. 17/18, 619-657 (1979)

[3] DECONINCK, H.: The Numerical Computation of Transonic Potential Flows (Thesis). Vrije Universiteit Brussel (1983)

[4] FEISTAUER, M., MANDEL, J., NEČAS, J.: Entropy regularization of the transonic potential flow problem. CMUC 25 (3), 431-443 (1984)

[5] FEISTAUER, M., NEČAS, J.: On the solvability of transonic potential flow problems. Zeitschrift für Anal. und ihre Anwendungen, Bd. 4(4), 305-329 (1985)

[6] FEISTAUER, M., NEČAS, J.: On the solution of transonic flows with weak shocks. CMUC 27, 4, 791-804 (1986)

[7] FEISTAUER, M., NEČAS, J.: Viscosity method in a transonic flow. Comm. PDE (to appear)

[8] GLOWINSKI, R.: Numerical Methods for Nonlinear Variational Problems. Springer-Verlag, New York - Berlin - Heidelberg - Tokyo (1984)

[9] MANDEL, J., NEČAS, J.: Convergence of finite elements for transonic potential flows. SIAM J. Numer. Anal. (to appear)

[10] MEYERS, N. G.: An L^p-estimate for the gradient of solutions of second order elliptic divergence equations. Ann. Scuola Norm. Sup. Pisa (3) 17, 189-206 (1963)

[11] MURAT, F.: L'injection du cône positif de H^{-1} dans $W^{-1/q}$ est compacte pour tout $q < 2$. J. Math. Pures Appl. (9) 60, 309-322 (1981)

[12] NEČAS, J.: Compacité par entropie et ecoulement de fluides. L'Université Pierre et Marie Curie, Paris VI (1988)

Miloslav Feistauer and Jindřich Nečas
Faculty of Mathematics and Physics
Charles University Prague
Sokolovská 83
186 00 Praha 8
Czechoslovakia

(Received December 4, 1987)

GLOBAL SOLUTION TO THE COMPRESIBLE ISOTHERMAL MULTIPOLAR FLUID

J. Nečas, A. Novotný, M. Šilhavý

Authors addresses : J. Nečas, Math. Phys. Faculty, Charles University,Sokolovská 83, 186 00 Praha 8,Czechoslovakia
A. Novotný, Institute of Physics of the Atmosphere Cz. Acad. Sci.,Boční II,1401, 141 31 Praha 4, Czechoslovakia
M. Šilhavý, Math. Institute Cz. Acad. Sci.,Žitná 25,115 67 Praha 1,Czechoslovakia

Abstract: The global existence of weak solutions of the initial boundary value problem in bounded domains to the system of partial differential equations for viscous compressible isothermal bipolar and multipolar fluids is proved. Some other properties as cavitation, regularity up to the strong solution and uniqueness are discussed.

May 1989

Classification 35Q, 76N
Keywords: isothermal motion,compressible fluid, viscous fluid, multipolar fluid, weak solution, existence, uniqueness, initial boundary value problem.

1

© Elsevier 1991. Reprinted with kind permission from: J. Nečas, A. Novotný & M. Šilhavý), Global solution to the compressible isothermal multipolar fluid, Journal of Mathematical Analysis and Applications 162, 223-241.

I. Introduction

The paper of M. Padula [10] concerns the same topic for the classical ideal gas and two-dimensional space domains. Padula's paper is pioneer, especially the use of Orlicz spaces chacterizing the finite entropy and theorems of the compensated compactness type. Nevertheless, there are some places not clear for authors.

In this paper we follow the ideas of the work by J. Nečas, M. Feistauer, V. Šverák [1] inspired by [10]. The main step is the study of multipolar fluids. The physical backround is studied in the paper by J. Nečas, M. Šilhavý [8], where higher stress tensors to constitution laws are introduced. There is proved that it is possible (and we shall work like) to satisfy all thermodynamical laws. Higher stress tensors imply the use of higher derivatives of the velocity field. This point of view exprimes some space nonlocality and seems also better to describe the turbulence phenomena. It is interesting that the proof of the existence of the central manifold to the incompressible fluid requires in fact higher stress tensors, see C. Foias, G. R. Sell, R. Temam [2].

We prove a global existence of Hopf solutions to the bipolar fluid under general initial data and volume forces in the time cylinder $(0,T)\times\Omega$ with $T>0$ and $\Omega \subset R^N$, $N=2$ or 3 provided the temperature $\Theta=\Theta_0=\text{const.}>0$. In the same spirit, the general multipolar gas can be treated. We concern also the problem of cavitation, regularity up to the strong solution and uniqueness. In all these studies we look for the lowest multipolarity.

2

In the present case only one new stress tensor is needed such that the momentum equations are of the fourth order. So we handle a bipolar fluid. The corresponding stress-strain relations are supposed to be linear.

II. Formulation of the problem

We suppose the classical state equation

(2.1) $$p = R\,\rho\,\theta \quad ,$$

where p, ρ, θ are pressure, density and temperature respectively and R is the universal gas constant. The isothermal process implies

(2.2) $$p = \beta\,\rho \quad , \quad \beta = \text{const.}>0.$$

We denote as usually by v the velocity vector, hence the continuity equation has its standard form

(2.3) $$\frac{\partial \rho}{\partial t} + \frac{\partial}{\partial x_j}(\rho v_j) = 0.$$

In (2.3) as well as throughout the paper we use summation convention.

A standard symmetric stress tensor τ_{ij} is considered such that

(2.4) $$\tau_{ij} = -p\,\delta_{ij} + \tau^{d}_{ij} \quad ;$$

its power on an elementary surface dS with outer normal ν is

(2.5) $$\tau_{ij}\, v_i\, \nu_j\, dS \, .$$

We consider a further 3-order stress tensor τ^{d}_{ijk}, whose power on an elementary surface dS with outer normal ν is

(2.6) $$\tau^{d}_{ijk}\, \frac{\partial v_i}{\partial x_j}\, \nu_k\, dS \, .$$

The general linear form for τ^{d}_{ij} (with coefficients depending on the temperature θ only and therefore constant in our case) provided τ^{d}_{ij} is symmetric reads

$$(2.7)\qquad \tau^d_{ij}(v) = \gamma \frac{\partial v_l}{\partial x_l}\delta_{ij} + 2\mu e_{ij} - \gamma_1\Delta \frac{\partial v_l}{\partial x_l}\delta_{ij} -$$

$$- 2\mu_1 \Delta e_{ij} + \gamma_2 \frac{\partial^2}{\partial x_i \partial x_j}\left(\frac{\partial v_l}{\partial x_l}\right),$$

see [8] (Δ denotes as usually Laplace operator). We shall suppose $\gamma \geq (-2/3)\mu$, $\mu > 0$, $\gamma_1 > (-2/3)\mu_1$, $\mu_1 > 0$, $\gamma_2 = 0$, $2e_{ij} = \frac{\partial v_i}{\partial x_j} + \frac{\partial v_j}{\partial x_i}$.

For the stress tensor τ^d_{ijk} we require the symmetry in i, j ; then the general form, according to [8] is

$$(2.8)\qquad \tau^d_{ijk}(v) = 2\mu_1 \frac{\partial}{\partial x_k} e_{ij} + \gamma_1 \delta_{ij} \frac{\partial}{\partial x_k} e_{ll} + \gamma_3 \delta_{ij} \Delta v_k +$$

$$+ \gamma_4 \delta_{ik} \Delta v_j + \gamma_4 \delta_{jk} \Delta v_i + \gamma_5 \delta_{ik} \frac{\partial}{\partial x_j} e_{ll} +$$

$$+ \gamma_5 \delta_{jk} \frac{\partial}{\partial x_i} e_{ll} + \gamma_6 \frac{\partial^2}{\partial x_i \partial x_j} v_k + \gamma_7 \frac{\partial^2}{\partial x_j \partial x_k} v_i +$$

$$+ \gamma_7 \frac{\partial^2}{\partial x_i \partial x_k} v_j .$$

We shall restrict ourselves to the case $\gamma_3 = \gamma_4 = \gamma_5 = \gamma_6 = \gamma_7 = 0$. The Clasius-Duhem inequality implies (see [3])

$$(2.9)\qquad \tau^d_{ij}(v) e_{ij} + \tau^d_{ijk}(v) \frac{\partial^2 v_i}{\partial x_j \partial x_k} + \frac{\partial}{\partial x_k}(\tau^d_{ijk}(v)) \frac{\partial v_i}{\partial x_j} \geq 0 ;$$

this is satisfied in our example, we shall work with and also the corresponding Korn's inequality will be satisfied, see later.

Let $\Omega \subset R^N$, N = 2 or 3 be a bounded domain with a boundary smooth enough and let $I=(0,T)$, $Q_T = I \times \Omega$ be the time cylinder. Let F be the density of volume forces. The momentum equations combined with (2.3) give

$$(2.10)\qquad \frac{\partial}{\partial t}(\rho v_i) + \frac{\partial}{\partial x_j}\left(\rho v_i v_j + \beta\rho\delta_{ij} - \tau^d_{ij}(v)\right) = \rho F_i .$$

Besides initial conditions for v and ρ we suppose v=0 on $(0,T) \times \partial\Omega$. The further boundary condition is in the general case

(2.11) $\quad \tau^{d}_{ijk}(v)\, \nu_j\, \nu_k = 0 \quad$ on $(0,T) \times \Omega$.

First let us suppose that we consider a solution smooth enough and $\rho>0$ in $\overline{Q}_T$. Then we get

2.12. Theorem

$$\int_{\Omega_t} \rho \, dx - \int_{\Omega_0} \rho \, dx = 0,$$

$$\frac{1}{2} \int_{\Omega_t} \rho |v|^2 \, dx - \frac{1}{2} \int_{\Omega_0} \rho |v|^2 \, dx + \int_{Q_t} (\gamma \, (e_{ll})^2 + 2\mu \, e_{ij} e_{ij} +$$

$$+ \gamma_1 \frac{\partial}{\partial x_k}(e_{ll}) \frac{\partial}{\partial x_k}(e_{ll}) + 2\mu_1 \frac{\partial}{\partial x_k}(e_{ij}) \frac{\partial}{\partial x_k}(e_{ij})) \, d\tau \, dx +$$

$$+ \beta \int_{\Omega_t} \rho \ln \rho \, dx - \beta \int_{\Omega_0} \rho \ln\rho \, dx = \int_{Q_t} \rho \, F_i \, v_i \, dx \, d\tau ,$$

$$t \in (0,T\rangle \, , \quad Q_t = (0,t) \times \Omega \, .$$

For usual notation for Sobolev, Orlicz and Bochner spaces see for example A. Kufner, O. John, S. Fučík [3]. The equalities 2.12 show the function spaces, we shall work in:

(2.13) $\quad v \in L^2 \, (I, \, W^{2,2}(\Omega, R^N)),$

(2.14) $\quad \rho \in L^\infty \, (I, \, L_\psi(\Omega)) \quad , \; \rho \geq 0$.

We denote by $\psi(t)=(1+t)\ln(1+t)-t$, $\Phi(t)=e^t-t-1$ resp. $\psi_{1/2}(t)$, $\Phi_2(t)=e^{t^2}-1$ the couple of Young complementary functions and by $L_\psi(\Omega), L_\Phi(\Omega)$ resp. $L_{\psi_{1/2}}(\Omega)$, $L_{\Phi_2}(\Omega)$ the Orlicz spaces of functions for which the Luxemburg norm

(2.15) $\quad \| u \|_{L_f(\Omega)} = \inf_h \{ h>0; \int_\Omega f \left(\frac{|u(x)|}{h}\right) dx \leq 1 \} < +\infty$,

where f stands for $\psi, \Phi, \psi_{1/2}, \Phi_2$.
Let $\mathcal{B}(\Omega)$ be the set of all bounded measurable functions defined on Ω. Let us denote further by $C_f(\Omega)$ the closure of $\mathcal{B}(\Omega)$ in $L_f(\Omega)$. We have (see [3])

$$(2.16) \qquad (C_\Phi(\Omega))^* = L_\psi(\Omega)\ ,(C_{\Phi_2}(\Omega))^* = L_{\psi_{1/2}}(\Omega),$$

$$(C_\psi(\Omega))^* = L_\Phi(\Omega),\ (C_{\psi_{1/2}}(\Omega))^* = L_{\Phi_2}(\Omega)\ .$$

Of course, $C_f(\Omega)$ are separable Banach spaces and $\mathcal{E}_0^\infty(\Omega)$ is dense in every $C_f(\Omega)$. We realize that $\psi, \psi_{1/2}$ satisfiy Δ_2 condition; hence

$$(2.17) \qquad C_\psi(\Omega) = L_\psi(\Omega)\ ,\ C_{\psi_{1/2}}(\Omega) = L_{\psi_{1/2}}(\Omega)$$

The weak formulation to the equation (2.10) will also be useful. It reads

$$(2.18) \qquad - \int_{Q_T} v_i \frac{\partial w_i}{\partial t}\, dx\, dt - \int_\Omega \rho_0\, v_{0i} w_i(0)\, dx +$$

$$+ \int_0^T ((v,w))\, dt - \int_{Q_T} (\rho\, v_i v_j + \beta\rho\delta_{ij}) \frac{\partial w_i}{\partial x_j}\, dx\, dt =$$

$$= \int_{Q_T} \rho\, F_i\, w_i\, dx\, dt$$

for every $w \in \mathcal{E}^\infty(\overline{Q}_T, R^N)$, $w(t) \in W_0^{1,2}(\Omega, R^N)$, $t \in I$, $w(T) = 0$; v_0 is the initial condition for v; in our situation, for $v, w \in V = (W^{2,2}(\Omega, R^N) \cap W_0^{1,2}(\Omega, R^N)$, $((v,w))$ is defined by (3.1) in the following section.

III. A modified Galerkin method

Let us denote $V = W^{2,2}(\Omega, R^N) \cap W_0^{1,2}(\Omega, R^N)$ and for $v, w \in V$ let

$$(3.1) \qquad ((v,w)) = \int_\Omega (\gamma\, e_{ii}(v)\, e_{kk}(w) + 2\mu\, e_{ij}(v)\, e_{ij}(w) +$$

$$+ \gamma_1 \frac{\partial}{\partial x_k} e_{ii}(v) \frac{\partial}{\partial x_k} e_{ss}(w) + 2\mu_1 \frac{\partial}{\partial x_k} e_{ij}(v) \frac{\partial}{\partial x_k} e_{ij}(w))\, dx\ .$$

From the coercivness of deformations (see e.g. J.Nečas, I.Hlaváček [7]) and from the very strong ellipticity of the bilinear form $((v,w))$ it follows for $v \in V$

$$(3.2) \qquad ((v,v)) \geq k_2 \int_\Omega (|v|^2 + |\nabla v|^2 + |\nabla_2 v|^2)\, dx\ , \quad k_2>0$$

(as usually ∇ resp. ∇_2 denotes first resp. second gradient).

Let $\{w^k\}_{k=1}^{+\infty}$ be a complete orthonormal system of eigenfunctions in V, solutions to the following eigenproblem in V

$$(3.3) \qquad ((v,w)) = \lambda\ (v,w) \text{ for every } v \in V\ ,\ w \in V\ ,$$

where $(v,w) = \int_\Omega v_i\ w_i\ dx$. We have

$$(3.4) \qquad ((v,w^k)) = \lambda_k\ (v,w^k)\ \ ;\ k = 1,2,\dots;\ \ 0<\lambda_1\leq\lambda_2\leq\dots,$$
$$((w^k,w^l)) = \delta_{kl}\,.$$

From the regularity of solutions to the linear elliptic problem (6.1) (see appendix) we get

$$(3.5) \qquad w^k \in C^\infty(\bar{\Omega},R^N).$$

Let us put for $w \in L^2(\Omega,R^N)$

$$(3.6) \qquad P_m w = \sum_{k=1}^{m} \lambda_k (w^k,w)\ w^k\ .$$

If $L^2_m = \text{span}\ \{w^1,\dots,w^m\}$ in $L^2(\Omega,R^N)$, $V_m = \text{span}\ \{w^1,\dots,w^m\}$ in V, then P_m is an orthogonal projector of L^2 onto L^2_m and of V onto V_m.

By (6.1) one defines the operator $\mathcal{A}$: $((v,w)) = (\mathcal{A}v,w)$ for every $w \in V$. Its definition domain is denoted by $D(\mathcal{A})$, of course $W_0^{4,2}(\Omega,R^N) \subset D(\mathcal{A})$. It is the consequence of Theorem 6.1 (see appendix) that

$$(3.7) \qquad \| v \|_{W^{4,2}(\Omega,R^N)} \leq k_3 \| \mathcal{A}v \|_{L^2(\Omega,R^N)} \qquad (k_3>0)$$

for every $v \in D(\mathcal{A})$, hence

$$\| P_m v \|_{W^{4,2}(\Omega,R^N)} \leq k_3 \| \mathcal{A}\, P_m v \|_{L^2(\Omega,R^N)} \leq k_3 \| P_m A\ v \|_{L^2(\Omega,R^N)} \leq$$
$$\leq k_4 \| v \|_{W^{4,2}(\Omega,R^N)} \qquad (k_4>0) \text{ for every } v \in W_0^{4,2}(\Omega,R^N).$$

Thus, due to the interpolation theorem it holds for every $v \in W_0^{3,2}(\Omega, R^N)$

(3.8) $$\| P_m v \|_{W^{3,2}(\Omega,R^N)} \le k_5 \| v \|_{W^{3,2}(\Omega,R^N)} \quad (k_5>0).$$

Let $c_i \in \mathcal{C}^1(\bar{I})$ and let us put $v^m(t,x) = \sum_{i=1}^{m} c_i(t) w^i(x)$.

Let us look for $\rho_m \in \mathcal{C}^1(\bar{Q}_T)$ such that

(3.9) $$\frac{\partial \rho_m}{\partial t} + \frac{\partial}{\partial x_i}(\rho_m v_i^m) = 0 .$$

We shall suppose throughout the paper

(3.10) $$\rho_m(0,x) = \rho_0(x) \in \mathcal{C}^1(\bar{\Omega}) , \quad \rho_0(x) > 0 \text{ in } \bar{\Omega} .$$

Let

(3.11) $$\dot{x}^m(t) = v^m(t, x^m(t)) , \quad x^m(0) = y , \quad y \in \bar{\Omega} .$$

For every $t \in \bar{I}$, $y \longrightarrow x^m(t)$ is a diffeomorphism of $\bar{\Omega}$ onto $\bar{\Omega}$.

For $\sigma_m = \ln \rho_m$ we have

(3.12) $$\frac{d}{dt}(\sigma_m(t, x^m(t))) = - \frac{\partial}{\partial x_i} v_i^m(t, x^m).$$

Hence

(3.13) $$\rho_m(t,x) = \rho_0(y) \exp\left(-\int_0^t \frac{\partial}{\partial x_i} v_i^m(\tau, x^m(\tau)) \, d\tau \right),$$

where $x = x^m(t)$, $y = x^m(0)$. Let us look now for $\bar{v}^m$ such that for every $t \in I$:

(3.14) $$\int_\Omega \left(\rho_m \frac{\partial \bar{v}_i^m}{\partial t} + \rho_m v_j^m \frac{\partial \bar{v}_i^m}{\partial x_j} + \beta \frac{\partial \rho_m}{\partial x_i} - \rho_m F_i\right) w_i^k \, dx =$$

$$= - ((\bar{v}^m, w^k)) , \quad k = 1,2,\ldots,m.$$

The initial conditions let be given by

(3.15) $$\int_\Omega \bar{c}_j(0) \, w_k^j \, w_k^i \, dx = \int_\Omega \bar{v}_k(0,x) \, w_k^i(x) \, dx .$$

We shall suppose in the sequel

(3.16) $$v_0(x) = \bar{v}(0,x) \in L^2(\Omega, R^N), \quad F_i \in L^2(I, L^\infty(\Omega)).$$

Because $\det(\int_\Omega \rho_m w_i^k w_i^r \, dx) \neq 0$, we can solve (3.14), (3.15) uniquely in I. We get the standard estimate

$$(3.17)\quad \frac{1}{2}\int_{\Omega_t}\rho_m|\bar{v}^m|^2\,dx - \frac{1}{2}\int_{\Omega}\rho_0|\bar{v}_0^m|^2\,dx - \beta\int_{Q_t}\rho_m\frac{\partial}{\partial x_i}\bar{v}_i^m\,dx\,d\tau -$$

$$-\int_{Q_t}\rho\,F_i\bar{v}_i^{\,m}\,d\tau\,dx + \int_0^t((\bar{v}^m,\bar{v}^m))\,d\tau = 0\ ,\quad \bar{v}_0^m=\bar{v}^m(0).$$

If we start with $c_i(t)$ in the ball

$$(3.18)\quad \max_{[0,\alpha]}|c_i(t)-c_i(0)|\le 1\ ;\ i=1,2,\ldots,m\ ,$$

we get

$$(3.19)\quad \max_{[0,\alpha]}|\bar{c}_i(t)-c_i(0)|\le 1\ ,\ i=1,\ldots,m\ ;$$

$$(3.20)\quad \max_{[0,\alpha]}|\dot{\bar{c}}_i|\le k_6(\alpha)\quad ,\quad k_6>0$$

provided α is small enough. So applying the Schauder fixed point theorem, we get on $[0,\alpha]$ $\bar{c}_i=c_i$. But for such solution we get

$$(3.21)\quad \int_{\Omega_t}\rho_m\,dx - \int_{\Omega}\rho_0\,dx = 0\ ,$$

$$\frac{1}{2}\int_{\Omega_t}\rho_m\,|\bar{v}^m|^2\,dx - \frac{1}{2}\int_{\Omega}\rho_0\,|\bar{v}_0^m|^2\,dx + \int_0^t((\bar{v}^m,\bar{v}^m))d\tau +$$

$$+\beta\int_{\Omega_t}(1+\rho_m)\ln(1+\rho_m)\,dx - \beta\int_{\Omega}(1+\rho_0)\ln(1+\rho_0)\,dx =$$

$$= \int_{Q_t}\rho_m\,F_i\,\bar{v}_i^m\,dxd\tau + \beta\int_{Q_t}\ln(1+\rho_m)\frac{\partial\bar{v}_j^m}{\partial x_j}\,dxd\tau\ .$$

We have

$$\int_{Q_t}\rho_m F_i\bar{v}_i^m\,dxd\tau\le k_7\|F_i\|_{L^2(I_t,L^\infty(\Omega))}\|\bar{v}^m\|_{L^2(I_t,V)}\|\rho^m\|_{L^\infty(I_t,L^1(\Omega))},$$

$$\int_{Q_t}\ln(1+\rho_m)\frac{\partial\bar{v}_i^m}{\partial x_j}\,dxd\tau\le k_7\Big(\int_{Q_t}(1+\rho_m)\ln(1+\rho_m)\Big)^{1/2}\|\bar{v}^m\|_{L^2(I_t,V)},\ k_7>0,$$

$\|\quad\|_V=((\quad,\quad))^{1/2},\ I_t=(0,t),\ t\in(0,\alpha]$.

Using Young inequality at the r.h.s. of (3.21) and applying Gronwall lemma, we get

3.22 Lemma

$$\int_{\Omega_t} (\tfrac{1}{2}\rho_m\ |\overline{v}^m|^2 + (1+\rho_m)\ \ln(1+\rho_m))\ dx -$$

$$- \int_{\Omega} (\tfrac{1}{2}\rho_0|v_0|^2+(1+\rho_0)\ln(1+\rho_0))dx + \int_0^t ((\overline{v}^m,\overline{v}^m))d\tau \le k_8\ ,k_8>0.$$

This implies we can continue with α to T . For more detailed proof see for example [9].

IV. The limit proces

4.1 Lemma

Let B be Banach space, B_i (i=0,1) separable reflexive Banach spaces. Let $B_0 \subset\subset B \subset B_1$ ($\subset\subset$ denotes compact imbedding) , $1<p_i<\infty$. Let $W = \{v,\ v \in L^{p_0}(I,B_0),\ \frac{\partial v}{\partial t} \in L^{p_1}(I,B_1)\}$.
Then $W \subset\subset L^{p_0}(I,B)$.

For proof we refer to Lions [5].

Main theorem

Let (2.2), (2.7), (3.10), (3.16) be satisfied. Then there exists

(4.2) $\rho \in L^{\infty}(I, L\psi(\Omega))$, $\rho \geq 0$ a.e in Q_T ,

(4.3) $v \in L^2(I, W^{2,2}(\Omega, R^N) \cap W_0^{1,2}(\Omega, R^N))$,

(4.4) $\frac{\partial \rho}{\partial t} \in L^2(I, W^{-3,2}(\Omega))$,

(4.5) $\frac{\partial}{\partial t}(\rho\, v) \in L^2(I, W^{-3,2}(\Omega, R^N))$

satisfying (2.3),(2.10) in the sense of distribution and such that (2.18) holds. Besides

(4.6) $\| \rho \|_{L^{\infty}(I, L_1(\Omega))} \leq \int_{\Omega_0} \rho_0 \, dx$

(4.7) $\frac{1}{2} \| \rho |v|^2 \|_{L^{\infty}(I, L^1(\Omega))} + \| v \|^2_{L^2(I, W^{2,2}(\Omega, R^N))} +$

$+ \beta \operatorname{supess}_I \int_{\Omega_t} (1+\rho)\ln(1+\rho)\, dx \leq k_9 (1 + \sum_{i=1}^{N} \|F_i\|_{L^2(I, L_{\infty}(\Omega))})^2$,

$k_9 > 0$.

Proof.

Let $0 \leq k \leq 2$ and let $W^{k,2}(\Omega)$, $W_0^{k,2}(\Omega)$ be the usual Sobolev spaces with fractional derivatives, see e.g. [3]; let $V^k = \overline{V}$, where the closure is taken in $W^{k,2}(\Omega, R^N)$; of course the traces are zero only for $k > 1/2$. Because

(4.8) $W_0^{2,2}(\Omega) \subset\subset W_0^{k_1,2}(\Omega) \subset\subset W_0^{k_2,2}(\Omega) \subset C_{\Phi}(\Omega) \subset C_{\Phi_2}(\Omega)$,

for $N/2 < k_2 < k_1 < 2$,

we have

(4.9) $L_{\psi_{1/2}}(\Omega) \subset L_{\psi}(\Omega) \subset W^{-k_2,2}(\Omega) \subset\subset W^{-k_1,2}(\Omega) \subset\subset W^{-2,2}(\Omega)$;

of course $W^{k,2}(\Omega) \subset C_{\psi}(\Omega)$, hence $L_{\psi}(\Omega) \subset (W^{k,2}(\Omega))^*$ ($(N/2) < k$)
It folows from the interpolation theorem that

(4.10) $\sup_{\substack{v \in V^k \\ \| v \|_{W^{k,2}(\Omega, R^N)} \leq 1}} \| P_m v \|_{W^{k,2}(\Omega, R^N)} \leq k_{10}$, $k_{10} > 0$, ($0 \leq k \leq 2$)

and

(4.11) $$\sup_{\substack{v \in (V^k)^* \\ \| v \|_{(V^k)^*} \le 1}} \| P_m^* v \|_{(V^k)^*} \le k_{11} \quad , k_{11}>0 \; (\; 0 \le k \le 2 \;)$$

(P_m^* is the dual operator to P_m) .

4.12 Lemma

Let $0 \le k < 2$. Then for every $\varepsilon>0$, l_0 exists such that for $l \ge l_0$ $\| u - P_l u \|_{W^{k,2}(\Omega,R^N)} < \varepsilon$ provided $\| u \|_{W^{2,2}(\Omega,R^N)} \le 1$.

Proof.

Let us suppose the contrary. Then there exist P_{l_j}, $l_j \longrightarrow +\infty$ and $\| u_{l_j} \|_{W^{2,2}(\Omega,R^N)} \le 1$ such that $\| u_{l_j} - P_{l_j} u_{l_j} \|_{W^{k,2}(\Omega,R^N)} \ge$ $\ge \varepsilon_0 \ge \varepsilon$. Because $W^{2,2}(\Omega,R^N) \subset\subset W^{k,2}(\Omega,R^N)$, we can suppose $u_{l_j} \longrightarrow u$ strongly in $W^{k,2}(\Omega,R^N)$; hence $P_{l_j} u_{l_j} \longrightarrow u$ strongly in $W^{k,2}(\Omega,R^N)$, what is contradictory.

Let ρ_m , v^m be an approximative solution from the section III. There exist subsequences (denoted $\{\rho_m\}_{m=1}^{+\infty}$, $\{v^m\}_{m=1}^{+\infty}$ again) such that $\rho_m \longrightarrow \rho$ $*$-weakly in $L^2(I,L_\psi(\Omega)) = L^2(I,(C_\Phi\,(\Omega)^*)$;

$$v^m \longrightarrow v \quad , \quad \frac{\partial}{\partial x_i} v^m \longrightarrow \frac{\partial}{\partial x_i} v \quad , \quad \frac{\partial^2 v^m}{\partial x_i \partial x_j} \longrightarrow \frac{\partial^2 v}{\partial x_j \partial x_i} \quad (i,j = 1,\dots,N)$$

weakly in $L^2(Q_T)$, $v^m \longrightarrow v$ $*$-weakly in $L^2(I,L_\Phi(\Omega)),L^2(I,L_{\Phi_2}(\Omega))$.

Due to 3.22 and (4.9)

(4.12) $$\| \rho_m \|_{L^\infty(I,W^{-k,2}(\Omega))} \le k_{12} \quad , \; k_{12}>0 \; ,$$
$$(N/2)<k\le 2.$$

For N=2,3

(4.13) $$\| v^m \|_{C^0(\bar\Omega,R^N)} \le k_{13} \| v^m \|_{W^{2,2}(\Omega,R^N)} \quad , \; k_{13}>0 \; ;$$

hence

(4.14) $$\| \rho_m v^m \|_{L^2(I,L_\psi(\Omega,R^N))} \le k_{14} \quad , \; k_{14}>0 \; .$$

It follows from (3.9)

(4.15) $\| \frac{\partial \rho_m}{\partial t} \|_{L^2(I,W^{-3,2}(\Omega,R^N)} \le k_{15}$, $k_{15}>0$.

Due to 4.1 $\rho_m \longrightarrow \rho$ strongly in $L^2(I,W^{-2,2}(\Omega))$; hence due to

(4.14) $\rho_m v^m \longrightarrow \rho v$ $*$-weakly in $L^2(I,L_\psi(\Omega))$.

By (4.9), (4.14) we get

(4.16) $\| \rho_m v^m \|_{L^2(I,W^{-k,2}(\Omega,R^N)} \le k_{16}$, $k_{16}>0$ $((N/2)<k\le 2)$;

hence by (4.11)

(4.17) $\| P^*_m(\rho_m v^m) \|_{L^2(I,W^{-k,2}(\Omega,R^N))} \le k_{17}$, $k_{17}>0$.

According to 3.22 $\rho_m |v^m|^2$ is bounded in $L^\infty(I,L^1(\Omega))$; therefore

(4.18) $\| \rho_m |v^m|^2 \|_{L^2(I,W^{-2,2}(\Omega,R^N))} \le k_{18}$, $k_{18}>0$.

By (3.14), (3.8), 3.22, (4.12), (4.18)

(4.19) $\| \frac{\partial}{\partial t}(P^*_m \rho_m v^m) \|_{L^2(I,W^{-3,2}(\Omega,R^N))} \le k_{19}$, $k_{19}>0$

holds. So by 4.1 $P^*_m(\rho_m v^m) \longrightarrow a$ strongly in $L^2(I,W^{-2,2}(\Omega,R^N))$ (eventually for a subsequence).

Let $w \in L^2(I,W_0^{2,2}(\Omega,R^N))$. Because of 4.12 for m large enough

(4.21) $\int_0^t \| P_m w - w \|^2_{W^{k,2}(\Omega,R^N)} d\tau \le \varepsilon^2 \int_0^t \| w \|^2_{W^{2,2}(\Omega,R^N)} d\tau$

hence for $\int_0^t \| w \|^2_{W^{2,2}(\Omega,R^N)} d\tau \le 1$, it follows uniformly with respect to w

(4.22) $\lim_{m\to\infty} \int_0^T \int_\Omega (P^*_m(\rho_m v^m_i) - \rho_m v^m_i) w_i \, dx \, d\tau = 0$.

Therefore $\rho_m v^m$ is a Cauchy sequence in $L^2(I,W^{-2,2}(\Omega,R^N))$ and $\rho_m v^m \longrightarrow a$ strongly in $L^2(I,W^{-2,2}(\Omega,R^N))$. But $\rho_m v^m \longrightarrow \rho v$ in $\mathcal{D}'(Q_T)$ in the sense of distributions , hence $a = \rho v$. Therefore, due to (4.18)

(4.23) $\rho_m v_i^m v_j^m \longrightarrow \rho v_i v_j$ weakly in $L^2(I, W^{-2,2}(\Omega, R^N))$.

Due to the estimate

$\int_\Omega |\rho_m v_j^m| u \, dx \le (\int_\Omega \rho_m |v^m|^2 dx)^{\frac{1}{2}} (\int_\Omega \rho_m u^2 \, dx)^{\frac{1}{2}}$ which holds for every $u \in L_{\Phi_2}(\Omega)$ and due to (4.13), we get

(4.24) $\| \rho_m v_i^m v_j^m \|_{L^2(I, L_{\psi_{1/2}}(\Omega))} \le k_{20}$, $k_{20} > 0$.

Therefore $\rho_m v_i^m v_j^m \longrightarrow \rho v_i v_j$ *-weakly in $L^2(I, L_{\psi_{1/2}}(\Omega))$.

It follows from (3.14) that for every $\varphi \in \mathcal{E}^\infty(\overline{Q}_T, R^N)$ satisfying $\varphi(t) \in V_m$ for every $t \in [0,T]$ and $\varphi(T) = 0$, it holds

$$\int_{Q_T} \rho v_i \frac{\partial \varphi_i}{\partial t} \, dxdt + \int_{Q_T} \rho v_i v_j \frac{\partial \varphi_j}{\partial x_j} \, dxdt + \beta \int_{Q_T} \rho \frac{\partial \varphi_j}{\partial x_j} =$$

$$= \int_0^T ((v, \varphi)) \, dt - \int_{Q_T} \rho F_i \varphi_i \, dx \, dt - \int_\Omega \rho_0 v_{0i} \varphi_i \, dx .$$

Due to the density arguments (2.18) holds and (2.10) is satisfied in the sense of distributions. The continuity equation is obviously satisfied in the sense of distributions.

V. Higher polarity and qualitative properties of the solutions.

In the case of a bipolar gas, we got for the limit density $\rho \ge 0$ only, i.e. in general, there can be a set of positive measure in Q_T such that $\rho = 0$ here, this means that there is a strong cavitation. For this reason and also because of uniqueness of the solution, it is worth to consider k-polar gas (k=3,4,...). We refer the reader to the work [8].

In our situation we shall consider on $V = W^{k,2}(\Omega,R^N) \cap$ $\cap W_0^{1,2}(\Omega,R^N)$ a symmetric V-coercive bilinear form

$$(5.1) \qquad ((v,w)) = \int_\Omega \sum_{l=1}^{k} A^l_{iji_1\ldots i_l j_1\ldots j_l} \frac{\partial^l v_i}{\partial x_{i_1}\ldots\partial x_{i_l}} \frac{\partial^l w_j}{\partial x_{j_1}\ldots\partial x_{j_l}} dx,$$

where $A^l_{iji_1\ldots i_l j_1\ldots j_l}$ are constants, for l=1 there are only combinations of $e_{ij}(v)$, $e_{ij}(w)$; we suppose that $A^l_{iji_1\ldots i_l j_1\ldots j_l}$ are symmetric under the permutation of indexes $(i_1\ldots i_l)$.$(j_1\ldots j_l)$. Of course, we suppose for $v \in V$:

$$(5.2) \qquad ((v,v)) \geq \alpha_1 \| v \|^2_{W^{k,2}(\Omega,R^N)}, \quad \alpha_1>0.$$

This follows for example from the following conditions

$$(5.3) \qquad A^1_{iji_1j_1}\frac{\partial v_i}{\partial x_{i_1}} \frac{\partial v_j}{\partial x_{j_1}} \geq \alpha_2 e_{ij}(v)\, e_{ij}(v), \quad \alpha_2>0 ,$$

$$(5.4) \qquad \sum_{l=2}^{k} A^l_{iji_1\ldots i_l j_1\ldots j_l} J^i_{i_1\ldots i_l} J^j_{j_1\ldots j_l} \geq$$

$$\geq \alpha_2 \sum_{l=2}^{k} J^i_{i_1\ldots i_l} J^i_{i_1\ldots i_l} \quad \text{for every real vector } (J^i_{i_1\ldots i_l})$$

$$i,i_1,\ldots,i_l = 1,\ldots,N .$$

In our situation

$$(T^d_{ij}),_j = \sum_{l=1}^{k} (-1)^{l+1} A^l_{iji_1\ldots i_l j_1\ldots j_l} \frac{\partial^{2l} v_j}{\partial x_{i_1}\ldots\partial x_{i_l}\partial x_{j_1}\ldots\partial x_{j_l}} .$$

We consider v=0 on $(0,T)\times\partial\Omega$ and unstable boundary conditions given by

$$\sum_{l=1}^{k}\sum_{s=0}^{l-1}\int_{\partial\Omega}(-1)^{s+1}A^l_{iji_1\ldots i_l j_1\ldots j_l} \frac{\partial^{l+s} v_j}{\partial x_{i_1}\ldots\partial x_{i_l}\partial x_{j_1}\ldots\partial x_{j_s}} \nu_{j_{s+1}}$$

$$\frac{\partial^{l-s-1} w_i}{\partial x_{j_{s+2}}\ldots\partial x_{j_l}} dS = 0 \quad \text{for every } w \in C^\infty(\bar\Omega,R^N) \cap W_0^{1,2}(\Omega,R^N).$$

5.5 Theorem

Let $k=3$ and (2.2), (3.10), (3.16), (5.1)-(5.2) be satisfied. Then there exist ρ, v

(5.6) $\rho \in L^{\infty}(Q_T)$, $\rho \geq \gamma > 0$ in Q_T ,

(5.7) $v \in L^2(I, W^{3,2}(\Omega, R^N) \cap W_0^{1,2}(\Omega, R^N))$,

(5.8) $\dfrac{\partial \rho}{\partial t} \in L^2(I, W^{-1,2}(\Omega))$,

(5.9) $\dfrac{\partial}{\partial t}(\rho\, v) \in L^2(I, W^{-3,2}(\Omega, R^N))$

such that (2.3) is satisfied in the sense of distributions and also in the sens of duality in $L^2(I, W^{-1,2}(\Omega))$ and (2.10) in the sense of distributions and also in the weak sense, i.e. (2.18) is fullfiled . Besides (4.6), (4.7) hold and

(5.10) $\min_{\bar{\Omega}} \rho_0(x) \exp(-k_{20} t^{1/2}) \leq \rho(t,x) \leq \max_{\bar{\Omega}} \rho_0(x) \exp(+k_{20} t^{1/2})$,

$k_{20} > 0$, $t \in I$.

5.11 Theorem

Let $k \geq 4$ and (2.2),(3.10),(3.16),(5.1)-(5.2) be satisfied. Let $\rho_0 \in \mathcal{C}^{k-3}(\bar{\Omega})$. Then there exist ρ, v

(5.12) $\dfrac{\partial^s \rho}{\partial x_1^{s_1} \ldots \partial x_N^{s_N}} \in L^{\infty}(Q_T)$ for $0 \leq s \leq k-3$, $s = s_1 + \ldots s_N$,

(5.13) $\dfrac{\partial^s}{\partial x_1^{s_1} \ldots \partial x_N^{s_N}} \dfrac{\partial \rho}{\partial t} \in L^2(I, L^{\infty}(\Omega))$ for $0 \leq s \leq k-4$,

(5.14) $v \in L^2(I, W^{k,2}(\Omega, R^N) \cap W_0^{1,2}(\Omega, R^N))$,

(5.15) $\dfrac{\partial}{\partial t}(\rho v) \in L^2(I, W^{-k,2}(\Omega, R^N))$

such that (2.3),(2.10) are satisfied in the sense of distrubutions, (2.3) also almost everywhere and (2.10) in the weak sense, i.e. (2.18) holds.

Proof of 5.5 and 5.II.

We take w^k ($k=1,2,\dots$) from (3.4). We define P_m by (3.6). We get ρ_m, v^m as before. Of course (3.21) and Lemma 3.22 hold. Due to

$$(5.16)\qquad \| v^m(t) \|_{C^{k-2}(\overline{\Omega},R^N)} \le k_{21} \| v^m(t) \|_{W^{k,2}(\Omega,R^N)},$$

it follows from (3.13) that

$$(5.17)\qquad (\min_{\overline{\Omega}}\rho_0(x))\exp(-\int_0^t k_{21}\| v^m(\tau) \|_{W^{k,2}(\Omega,R^N)}d\tau) \le \rho^m(t,x)\le$$

$$(\max_{\overline{\Omega}} \rho_0(x))\exp(+\int_0^t k_{21}\| v^m(\tau) \|_{W^{k,2}(\Omega,R^N)}d\tau .$$

Using (3.11), 3.22, (5.16) and Gronwall lemma, we obtain

$$(5.18)\qquad \max_{\overline{Q}_T} \left| \frac{\partial^s x_i}{\partial y_1^{s1}\dots\partial y_N^{sN}}(t,y) \right| \le k_{22} ,$$

where $0\le s\le k-3$, $k_{22}>0$ and $(s1,\dots,sN)$ is any multiindex such that $s=s1+\dots sN$. Due to

$$(5.19)\qquad \rho_m(t,x(t,y)) \det \left(\frac{\partial x_i}{\partial y_j}\right) = \rho_0(y)$$

also

$$(5.20)\qquad \max_{\overline{Q}_T} \left| \frac{\partial^s y_i}{\partial x_1^{s1}\dots\partial x_N^{sN}}(t,x) \right| \le k_{23} ,\ k_{23}>0 ,\ s \le k-3 .$$

It follows from (5.20) and (3.13) that

$$(5.21)\qquad \| \rho_m \|_{C^0(\overline{I},C^s(\overline{\Omega}))} \le k_{24},\ k_{24} > 0$$

provided $s \le k-3$. Now from (3.3) one gets

$$(5.22)\qquad \left\| \frac{\partial\rho_m}{\partial t} \right\|_{L^2(I,C^{s-1}(\overline{\Omega}))} \le k_{25} ,\ k_{25} > 0\ (\ s\le k-3, k\ge 4)$$

or

$$\left\| \frac{\partial\rho_m}{\partial t} \right\|_{L^2(I,W^{-1,2}(\Omega,R^N))} \le k_{25} \text{ for } k=3 .$$

In any case, we can suppose $\rho_{m_l} \longrightarrow \rho$ weakly in $L^2(Q_T)$ and $\rho_{m_l} \longrightarrow \rho$ strongly in $L^2(I,W^{-1,2}(\Omega))$. We have

(5.23) $\quad \| \rho_m v^m \|_{L^2(Q_T,R^N)} \le k_{26}$, $k_{26}>0$,

therefore $\rho_m v^m \longrightarrow \rho v$ weakly in $L^2(Q_T,R^N)$ at least for a chosen subsequence . It follows from (3.14) with $\bar{v}^m = v^m$

(5.24) $\quad \| \frac{d}{dt} P^*_m(\rho_m v^m) \|_{L^2(I,V^*)} \le k_{27}$, $k_{27}>0$,

hence for a subsequence if necessary $P^*_m(\rho_m v^m) \longrightarrow a$ strongly in $L^2(I,W^{-1,2}(\Omega,R^N))$. But we have for every $\varepsilon>0$
$\| P_m w - w \|_{L^2(\Omega,R^N)} \le \varepsilon$ provided $w \in W^{1,2}_0(\Omega,R^N)$,
$\| w \|_{W^{1,2}(\Omega,R^N)} \le 1$ and m is large enough (see 4.12) ; so
$\rho_m v^m \longrightarrow a = \rho v$ strongly in $L^2(I,W^{-1,2}(\Omega,R^N))$.
We have the estimate

(5.25) $\quad \| \rho_m |v^m|^2 \|_{L^2(Q_T)} \le k_{28}$, $k_{28}>0$;

therefore $\rho_m v^m_i v^m_j \longrightarrow \rho v_i v_j$ weakly in $L^2(Q_T)$ at least for a chosen subsequence . The other is obvious. The proof of 5.5,5.11 is finished.

5.26 Theorem

Let $k\ge 4$. Let the conditions of the theorem 5.5 be satisfied and let further $v_0 \in V$. Then for the solution it holds

(5.27) $\quad \int_0^T \int_\Omega |\frac{\partial v}{\partial t}|^2 \, dxdt + \sup_{\bar{I}} \| v(t) \|^2_{W^{k,2}(\Omega,R^N)} \le k_{29}, k_{29}>0,$

(5.28) $\quad v \in L^2(I,\dot{W}^{2k,2}(\Omega,R^N))$.

The equation (2.10) is fulfilled a.e. in Q_T.

Proof.

For $\bar{v}^m = v^m$,we use in (3.14) $\frac{\partial v^m}{\partial t}$ for the test function.

Because of 3.22 and (5.17), $\int_{\Omega_t} |v^m|^2 \, dx \le k_{30}$ ($k_{30}>0$), hence

using (5.12), 3.22, (5.16), Holder and Young inequality, we get

(5.29) $\quad \int_0^T \int_\Omega |\frac{\partial v^m}{\partial t}|^2 \, dxdt + \sup_{\bar{I}} \| v^m(t) \|^2_{W^{k,2}(\Omega,R^N)} \le k_{31}$,

($k_{31}>0$).

Now, (5.28) follows from the regularity to the elliptic systems $((v,w)) = \int_\Omega g_i w_i \, dx$, where $g \in L^2(\Omega, R^N)$ (see 6.1)

5.30 Theorem

Let the conditions of the theorems 5.11, 5.26 be satisfied . Then in the set of solutions satisfying 5.11, 5.23 there exists at most one solution to the problem (2.3),(2.10).

Proof.

Let $(\rho, v), (\bar\rho, \bar v)$ be two solutions to the problem. Then for $(\xi, w) = (\rho - \bar\rho , v - \bar v)$ it holds

$$(5.31) \qquad \frac{\partial \xi}{\partial t} = - \xi \frac{\partial v_i}{\partial x_i} - \bar\rho \frac{\partial w_i}{\partial x_i} - \frac{\partial \xi}{\partial x_i} v_i - \frac{\partial \bar\rho}{\partial x_i} w_i$$

$$(5.32) \qquad \int_{\Omega_t} \bar\rho \frac{\partial w_i}{\partial t} \varphi_i \, dx + ((w, \varphi)) =$$

$$= - \int_{\Omega_t} (\xi \frac{\partial v_i}{\partial t} + \xi v_j \frac{\partial v_i}{\partial x_j} + \bar\rho w_j \frac{\partial v_i}{\partial x_j} + \bar\rho \bar v_j \frac{\partial w_i}{\partial x_j}) \varphi_i \, dx +$$

$$+ \int_{\Omega_t} \xi \frac{\partial \varphi_i}{\partial x_i} \, dx + \int_{\Omega_t} \xi F_i \varphi_i \, dx$$

for every $\varphi \in V$ a.e. in I .

We multiply (5.31) by ξ and integrate over Ω ; after some computation we get the estimate

$$(5.33) \qquad \frac{d}{dt} (\| \xi(t) \|_{L^2(\Omega)})^2 \le a_1(t) \| \xi(t) \|^2_{L^2(\Omega)} +$$

$$+ \frac{1}{2} ((w, w)) ,$$

where $a_1 \le k_{32} (((v,v)) + (\|\bar\rho\|_{L^\infty(\Omega)} + \|\bar\rho\|_{W^{1,2}(\Omega)})^2)$, $k_{32} > 0$; hence $a_1 \in L^1(I)$.

We put in (5.32) $\varphi = w$. We get

$$(5.34) \quad \frac{d}{dt}\int_{\Omega_t} \tfrac{1}{2}\bar\rho |w|^2\, dx \quad + \quad ((w(t),w(t))) \quad =$$

$$= \; - \int_{\Omega_t} \Big(\frac{\partial v_i}{\partial t} w_i \xi + \xi v_j w_i \frac{\partial v_i}{\partial x_j} + \bar\rho w_i w_j \frac{\partial v_i}{\partial x_j} + \bar\rho \bar v_j \frac{\partial w_i}{\partial x_j} w_i \Big)\, dx \; +$$

$$+ \int_{\Omega_t} \Big(\xi \frac{\partial w_j}{\partial x_j} + \frac{1}{2}\frac{\partial \bar\rho}{\partial t}|w|^2 \Big)\, dx \quad \text{a.e. in } I \; .$$

Using Holder and Young inequalities, r.h.s. of (5.34) can be estimated by

$$a_2(t) \; (\; \| \; w(t) \; \|^2_{L^2(\Omega,R^N)} \; + \; \| \; \xi(t) \; \|^2_{L^2(\Omega)} \;) \; + \; \tfrac{1}{2} \; ((w(t),w(t))),$$

where

$$a_2 \; \le \; k_{33}\{ \; (1+\| \frac{\partial v}{\partial t} \|_{L^2(\Omega,R^N)}+\| \; v \; \|_{W^{k,2}(\Omega,R^N)}+\| \; \bar v \; \|_{W^{k,2}(\Omega,R^N)} \;)^2$$

$$(\; \| \; \bar\rho \; \|_{L^\infty(\Omega)} \; + \; \| \; \frac{\partial \bar\rho}{\partial t} \; \|_{L^\infty(\Omega)} \; + \; \| \; \bar\rho \; \|_{W^{1,2}(\Omega)} \; +1)^2 \; \} \; , k_{33} \; > \; 0 \quad .$$

From (5.27),(5.12), (2.3) we have $\frac{\partial \bar\rho}{\partial t} \in L^\infty(Q_T)$, hence $a_2 \in L^1(I)$.

Due to (5.6)

$$k_{34}\| \; w \; \|^2_{L^2(\Omega,R^N)} \; \le \; \int_\Omega \; \bar\rho |w|^2 \; dx \quad \le \; k_{35}\| \; w \; \|^2_{L^2(\Omega,R^N)} \qquad (0<k_{34}<k_{35}).$$

Hence from (5.34)

$$(5.35) \qquad \frac{d}{dt}\| \; w(t)\|^2_{L^2(\Omega,R^N)} \quad \le$$

$$\le \; a_2(t) \; (\; \| \; w(t) \; \|^2_{L^2(\Omega,R^N)} \quad + \quad \| \; \xi(t) \; \|^2_{L^2(\Omega)} \;) \; .$$

We add (5.33),(5.35) and apply Gronwall lemma . Because of $\xi(0) = 0$, $w(0) = 0$, $\xi = 0$ and $w = 0$ a.e. in Q_T . The proof is finished.

VI. Appendix

For the construction of the basis for Galerkin method we have used the following regularity property to the weak solution of the elliptic problem

(6.1) $\quad u \in V \ , \ f \in L^2(\Omega, R^N),$

$$((\, v, u \,)) = \int_\Omega v_i f_i \ dx \quad \text{for every } v \in V \ .$$

6.2 Theorem

Let $u \in V$ be a solution to (6.1) . Then $u \in W^{2k,2}(\Omega, R^N)$ and

(6.3) $\quad \| u \|_{W^{2k,2}(\Omega, R^N)} \leq c \, \| f \|_{L^2(\Omega, R^N)} \ , \quad c>0 \ .$

The detailed proof can be constructed according to the procedure from J. Nečas [6]. We shall restrict ourselves to the case $\Omega = R^N_+ = \{ \, x \in R^N \ , \ x_N > 0 \, \}$ provided the solution $u \in W^{2k,2}(R^N_+, R^N)$. For exact proof one has to use differences instead of derivatives. For general domain $\Omega \subset R^N$ one uses in the usual way the partition of unity to replace the original problem by the problem in R^N_+ .

Idea of the proof.

We shall proceed by induction. Let us suppose, that for $2k-1 \geq l \geq k$ we have

(6.4) $\quad \| u \|_{W^{l,2}(R^N_+, R^N)} \leq c \, \| f \|_{L^2(R^N_+, R^N)} \ .$

First, let $v \in V \cap W^{k+1,2}(R^N_+, R^N)$. Then $\frac{\partial v}{\partial x_j} \in V$, $j=1,2,\dots,N-1$.

So from (6.1) it follows

(6.5) $\quad - ((v, \frac{\partial u}{\partial x_j})) = \int_{R^N} \frac{\partial v_i}{\partial x_j} f_i \ dx \ .$

Especially for $v = \frac{\partial u}{\partial x_j}$

(6.6) $$\left(\left(\frac{\partial u}{\partial x_j} , \frac{\partial u}{\partial x_j} \right)\right) = \int_{R^N_+} \frac{\partial}{\partial x_j} \frac{\partial u_i}{\partial x_j} f_i \, dx \quad ,$$

hence

(6.7) $$\left\| \frac{\partial u}{\partial x_j} \right\|_{W^{k,2}(R^N_+, R^N)} \le c \, \| f \|_{L^2(R^N_+, R^N)} .$$

Let $\varphi \in \mathcal{D}(R^N_+, R^N)$. We have

(6.8) $$- \left(\left(\varphi , \frac{\partial u}{\partial x_N} \right)\right) = \int_{R^N_+} \frac{\partial \varphi_i}{\partial x_N} f_i \, dx .$$

Put $(0,\dots,0,\psi,0,\dots,0)$, where ψ is on the j-th position in the vector ; of course $\psi \in \mathcal{D}(R^N_+)$. So we get by the theorem about negatives norms, see J. Nečas [6]

(6.9) $$\Big\| A^k_{ij \underbrace{N\dots N}_{k\text{-times}} \underbrace{N\dots N}_{k\text{-times}}} \frac{\partial^{k+1} u_i}{\partial x_N^{k+1}} \Big\|_{L^2(R^N_+)} \le c$$

$(j=1,2,\dots,N)$.

Therefore in virtue of (5.4) , $\det (A^k_{ij \underbrace{N\dots N}_{k\text{-times}} \underbrace{N\dots N}_{k\text{-times}}}) \neq 0$

and we have

(6.10) $$\| u \|_{W^{k+1,2}(R^N_+, R^N)} \le c \, \| f \|_{L^2(R^N_+, R^N)} .$$

Let $\frac{\partial^{l+1-k}}{\partial x_{i_1} \dots \partial x_{i_{l+1-k}}}$ be some derivative for

$i_1, \dots, i_{l+1-k} \le N-1$. Then we have as before

(6.11) $$\left(\left(\frac{\partial^{l+1-k} u}{\partial x_{i_1} \dots \partial x_{i_{l+1-k}}} , \frac{\partial^{l+1-k} u}{\partial x_{i_1} \dots \partial x_{i_{l+1-k}}} \right)\right) =$$

$$= (-1)^{l+1} \int_{R^N_+} \frac{\partial^{2(l+1-k)} u_i}{\partial x^2_{i_1} \dots \partial x^2_{i_{l+1-k}}} f_i \, dx \quad .$$

Hence $$\left\| \frac{\partial^{l+1-k} u}{\partial x_{i_1} \dots \partial x_{i_{l+1-k}}} \right\|_{W^{k,2}(R^N_+, R^N)} \le c \, \| f \|_{L^2(R^N_+, R^N)} .$$

Let $\dfrac{\partial^{l+1-k}}{\partial x_N \partial x_{i_1} \ldots \partial x_{i_{l-k}}}$ with $i_s \le N-1$ $(s=1,\ldots,l-k)$.

Then we get from (6.1) for $\varphi \in \mathcal{D}(R^N_+, R^N)$

$$\left(\left(\frac{\partial^{l+1-k} u}{\partial x_N \partial x_{i_1} \ldots \partial x_{i_{l-k}}}, \varphi \right)\right) = (-1)^{l+1} \int_{R^N_+} \frac{\partial^{l+1-k} \varphi_i}{\partial x_N \partial x_{i_1} \ldots \partial x_{i_{l-k}}} f_i \, dx.$$

So we get as before

$$\| A^k_{ij} \underbrace{N\ldots N}_{k\text{-times}} \underbrace{N\ldots N}_{k\text{-times}} \frac{\partial^{l+1} u_i}{\partial x_N^{k+1} \partial x_{i_1} \ldots \partial x_{i_{l-k}}} \|_{L^2(R^N_+, R^N)} \le$$

$$\le c \| f \|_{L^2(R^N_+, R^N)} \quad ; \text{ hence}$$

$$\| \frac{\partial^{l+1} u}{\partial x_{i_1} \ldots \partial x_{i_{l-k}} \partial x_N^{k+1}} \|_{L^2(R^N_+, R^N)} \le c \| f \|_{L^2(R^N_+, R^N)} \quad . \tag{6.12}$$

Further step is to consider $\dfrac{\partial^{l+1}}{\partial x_{i_1} \ldots \partial x_{i_{l-k-1}} \partial x_N^{k+2}}$. By the same reasoning we get

$$\| \frac{\partial^{l+1} u}{\partial x_{i_1} \ldots \partial x_{i_{l-k-1}} \partial x_N^{k+2}} \|_{L^2(R^N_+, R^N)} \le c \| f \|_{L^2(R^N_+, R^N)} \quad .$$

References

[1] M.Feistauer, J.Nećas, V.Šverák: On the weak compactness of the equations of compressible flow, preprint MFF UK Prague(1988), to appear in Applicable Analysis

[2] C.Foias, G.R.Sell, R.Temam: Inertial manifolds for nonlinear evolution equations, Preprint series #234, Inst. Math. Appl. Univ. Minnesota (1986)

[3] A.Kufner, O.John, S.Fučík: Functional spaces, Academia Prague (1977)

[4] J.L.Lions, E.Magenes: Problèmes aux limites non homogenes et application, Dunod Paris (1968)

[5] J.L.Lions: Quelques méthodes de résolution des problèmes aux limites non linéaires, Dunod Paris (1969)

[6] J.Nečas: Les méthodes directes en la théorie des équations elliptiques, Academia Prague (1967)

[7] J.Nečas, I.Hlaváček: Introduction into the theory of elastic and elastoplastic bodies, SNTL Prague (1983) (in Czech)

[8] J.Nečas, M.Šilhavý: Viscous multipolar fluids, to appear

[9] J.Neustupa, A.Novotný: Global weak solvability to the regularized viscous heat conductive flow, to appear

[10] M.Padula: Existence of global solutions for 2-dimensional viscous compressible flow, J. Funct. Anal. 69(1986)

COMMUN. IN PARTIAL DIFFERENTIAL EQUATIONS, 16(2&3), 197-220 (1991)

SOME QUALITATIVE PROPERTIES OF THE VISCOUS COMPRESSIBLE HEAT CONDUCTIVE MULTIPOLAR FLUID

J. Nečas, A. Novotný

J. Nečas, Mathematical Faculty, Charles University, Sokolovská 83, 186 00 Praha 8, Czechoslovakia

A. Novotný, Inst. Phys. Atm. Cz. Acad. Sci., Boční II 1401, 141 31 Praha 4, Czechoslovakia

I. Introduction

Our paper [4] has been devoted to the solvability of the initial boundary value problem in bounded domains in two or three dimensions to the evolution equations of viscous multipolar compressible heat conductive fluid. Physical background to this problem is discussed in [6]. We were able to prove solvability for the polarity $k \geq 4$.

We will generelize these results to the case $k \geq 3$. We will use the same method. The crucial point is a more detailed study of the characteristics to the continuity equation, which makes possible to improve estimates of the density and consequently to prove existence of strong solutions. Further we deal with the uniqueness of these solutions. In the last section we derive some important estimates, energy and entropy

Copyright © 1991 by Marcel Dekker, Inc.

© Taylor & Francis 1991. Reprinted with kind permission from: Nečas, J. & A. Novotný, A., Some qualitative properties of the viscous compressible heat conductive multipolar fluid, Communications in Partial Differential Equations, 16, 197-220.

estimates included, which has to be fulfilled by any solution.

First, let us give the formulation of the problem. For more detailed discussion see [4], [6].

The state equation for ideal fluid reads

$$(1.1) \qquad p = R\rho\Theta ,$$

where p is the pressure, R the universal gas constant, ρ the density and Θ the temperature. We consider the standard stress tensor

$$(1.2) \qquad \tau_{ij} = - p\delta_{ij} + \overset{V}{\tau}_{ij}$$

and higher order stress tensors

$$(1.3) \qquad \overset{V}{\tau}_{ii_1 \dots i_m j} \; ; \; 1 \le m \le k \; , \; k \ge 3$$

(the tensor $\overset{V}{\tau}_{ii_1 \dots i_k j}$ is equal to zero by definition), whose power on an elementary surface dS with an outer normal ν is

$$\overset{V}{\tau}_{ii_1 \dots i_m j} \frac{\partial^m v_i}{\partial x_{i_1} \dots \partial x_{i_m}} \nu_j \; dS \; .$$

The tensors $\overset{V}{\tau}_{ij}, \overset{V}{\tau}_{ii_1 \dots i_m j}$; $1 \le m \le k-1$ depend linearly on the gradients of the velocity field up to the order 2k-1 (for the first gradient there is only the dependence on $e_{ij}(v) = \frac{1}{2}(\frac{\partial v_i}{\partial x_j} + \frac{\partial v_j}{\partial x_i})$); the coefficients of linear dependence are real constants such that $\overset{V}{\tau}_{ij}$, $\frac{\partial}{\partial x_j}\overset{V}{\tau}_{ii_1 j}$ are symmetric and $\overset{V}{\tau}_{ii_1 \dots i_m j}$ are symmetric in $i_1, \dots, i_m$. Moreover, due to the Clausius-Duhem inequality we suppose that

$$(1.4) \qquad \overset{V}{\tau}_{ij}(v)e_{ij}(v) + \\ + \sum_{m=1}^{k-1} \frac{\partial}{\partial x_j}(\overset{V}{\tau}_{ii_1 \dots i_m j}(v) \frac{\partial^m v_i}{\partial x_{i_1} \dots \partial x_{i_m}}) \ge 0 \; .$$

Let $V = W_0^{1,2}(\Omega,R^N) \cap W^{k,2}(\Omega,R^N)$. We assume that there exists a symmetric bilinear V-coercive form $((v,w))$

$$(1.5) \qquad ((v,w)) = $$

$$= \int_\Omega (\sum_{m=1}^{k} A^m_{iji_1\dots i_m j_1\dots j_m} \frac{\partial^m v_i}{\partial x_{i_1}\dots\partial x_{i_m}} \frac{\partial^m w_j}{\partial x_{j_1}\dots\partial x_{j_m}})dx ,$$

$$(1.6) \qquad ((v,v)) \geq \alpha \| v \|^2_{W^{k,2}(\Omega,R^N)} , \quad \alpha > 0$$

such that for $v,w \in V \cap W^{2k,2}(\Omega,R^N)$ it holds

$$(1.7) \qquad ((v,w)) = - \int_\Omega \frac{\partial}{\partial x_j}(\tau^V_{ij}(v))w_i\, dx + [[v,w]] ,$$

where $[[v,w]] = \sum_{m=1}^{k-1} \int_{\partial\Omega} \tau^V_{ii_1\dots i_m j}(v) \frac{\partial^m w_i}{\partial x_{i_1}\dots\partial x_{i_m}} \nu_j\, dS$.

Assumption (1.7) is very restrictive one and it implies further conditions on the coefficients of the linear dependence of $\tau^V_{ii_1\dots i_m j}$ $(m = 0,1,\dots,k-1)$ on the gradients of velocity field. An example of multipolar fluid, which is consistent with thermodynamics and satisfies (1.7) is given at the end of this section. We define

$$<<v,w>> = \tau^V_{ij}(v)e_{ij}(w) +$$

$$+ \sum_{m=1}^{k-1} \frac{\partial}{\partial x_j}(\tau^V_{ii_1\dots i_m j}(v)\frac{\partial^m w_i}{\partial x_{i_1}\dots\partial x_{i_m}}) .$$

It follows from (1.4) that the form $<<v,w>>$ contents the derivatives of v resp. w at most of order k.

We consider the following system of partial differential equations for density ρ, velocity v and temperature θ given by the continuity equation

$$(1.8) \qquad \frac{\partial\rho}{\partial t} + \frac{\partial}{\partial x_i}(\rho v_i) = 0 ,$$

the momentum equations

$$(1.9)\quad \frac{\partial}{\partial t}(\rho v_i) + \frac{\partial}{\partial x_j}(\rho v_i v_j + p(\rho,\theta)\delta_{ij} - \tau^V_{ij}(v)) = \rho F_i \ ,$$

and the energy equation

$$(1.10)\quad \frac{\partial}{\partial t}(c_v \rho\theta) + \frac{\partial}{\partial x_j}(c_v \rho\theta v_j) + p(\rho,\theta)\frac{\partial v_j}{\partial x_j} - \lambda\frac{\partial^2\theta}{\partial x_j \partial x_j} =$$

$$= \langle\langle v, v\rangle\rangle \ ,$$

where c_V (the specific heat at constant volume) and λ (the heat conductivity) are positive constants and F denotes the density of external forces. This system is to be solved in a bounded domain $\Omega \subset R^N$ $(N = 2,3)$ with the smooth boundary $\partial\Omega$ in the time cylinder $Q_T = I\times\Omega$ $(I = (0,T)\ ,\ T > 0)$. Besides initial conditions

$$(1.11)\qquad \rho(0) = \rho_0 \ ,\ v(0) = v_0 \ ,\ \theta(0) = \theta_0 \ ,$$

we consider boundary conditions

$$(1.12)\qquad v = 0 \ ,\ \frac{\partial\theta}{\partial\nu} = 0 \text{ on } I \times \partial\Omega$$

(ν is the outer normal to $\partial\Omega$) and

$$(1.13)\qquad [[v,w]] = 0 \text{ on } I \times \partial\Omega$$

for every $w \in V \cap W^{2k,2}(\Omega,R^N)$. The weak formulation of (1.9), (1.10) reads

$$(1.14)\qquad \int_0^T \int_\Omega \frac{\partial}{\partial t}(\rho v_i)w_i\,dxdt -$$

$$- \int_0^T\int_\Omega(\rho v_i v_j\frac{\partial w_i}{\partial x_j} + p(\rho,\theta)\frac{\partial w_j}{\partial x_j})dxdt +$$

$$+ \int_0^T((v,w))dt = \int_0^T \int_\Omega \rho F_i w_i\,dxdt$$

for every $w \in L^2(I,V)$;

$$(1.15)\quad - c_v\int_0^T \int_\Omega \rho\theta\frac{\partial\eta}{\partial t}dxdt - c_v\int_\Omega \rho_0\theta_0\eta(0)dx + \lambda\int_0^T\int_\Omega \frac{\partial\theta}{\partial x_j}\frac{\partial\eta}{\partial x_j}dxdt - c_v\int_0^T\int_\Omega \rho\theta v_j\frac{\partial\eta}{\partial x_j}dxdt + \int_0^T\int_\Omega p(\rho,\theta)\frac{\partial v_j}{\partial x_j}\eta dxdt = \int_0^T\int_\Omega \langle\langle v,v\rangle\rangle\eta dxdt$$

for every $\eta \in \mathcal{E}^\infty(\bar{Q}_T)$, $\eta(T) = 0$.

1.16 Example

We give one example of viscous multipolar fluid compatible with thermodynamics , which satisfies (1.4),(1.7). We consider ideal gas, e.g. the state equation (1.1) holds. We introduce specific internal energy and entropy

$$(1.17)\qquad e(\rho,\theta) = c_v\theta ,$$

$$(1.18)\qquad \varkappa(\rho,\theta) = c_v\ln\theta - R\ln\rho$$

and stress tensors

$$(1.19)\qquad \tau_{ii_1\ldots i_m j} = \tau^E_{ii_1\ldots i_m j} + \tau^V_{ii_1\ldots i_m j} ;$$

$m = 0,\ldots,k$, where

$$(1.20)\qquad \tau^E_{ij}(\rho,\theta) = -R\rho\theta\delta_{ij} , \quad \tau^E_{ii_1\ldots i_m j} = 0$$

for $1 \le m \le k$ and

$$(1.21)\qquad \tau^V_{ij}(v) = \sum_{r=0}^{k-1} (-1)^r\Delta^r q^r_{ij} ,$$

$$\tau^V_{ii_1\ldots i_m j}(v) = \mathrm{Sym}\Big(\sum_{r=m}^{k-1} (-1)^{r+m}\Delta^{r-m}\frac{\partial^m q^r_{im}}{\partial x_{i_1}\ldots\partial x_{i_{m-1}}\partial x_j}\Big) ,$$

$$1 \le m \le k-1 \ , \ \tau^{V}_{i i_1 \dots i_k j} = 0 \ ;$$

Sym denotes the symmetrization with respect to $i_1, \dots, i_m$; $q^r_{ij} = \varkappa_r \frac{\partial v_l}{\partial x_l}\delta_{ij} + 2\mu_r e_{ij}(v)$; $\varkappa_s, \mu_s$ $(s = 0,\dots,k-1)$ are constants such that $\varkappa_r \ge -\frac{2}{3}\mu_r$, $\mu_r > 0$ $(r = 0,\dots,k-2)$ and $\varkappa_{k-1} > -\frac{2}{3}\mu_{k-1}$, $\mu_{k-1} > 0$.

It is only a subject of standard computation to verify that the fluid described by the constitutive equations (1.1),(1.17)-(1.21) satisfies (1.4) , (1.7) with

$$((v,w)) = \int_\Omega \sum_{m=0}^{k-1} \frac{\partial^m q^m_{ij}(v)}{\partial x_{i_1} \dots \partial x_{i_m}} \frac{\partial^m e_{ij}(w)}{\partial x_{i_1} \dots \partial x_{i_m}} dx \ .$$

Consequently Clausius-Duhem inequality is satisfied (see [6]).

II. Modified Galerkin method

The orthonormal basis $\{w^r\}_{r=1}^{+\infty}$ in V is given by solving the following eigenvalue problem

$$(2.1) \qquad ((v,w^r)) = \lambda_r \int_\Omega v_i w^r_i dx \ ,$$

for every $v \in V$ $(0 < \lambda_1 \le \lambda_2 \le \dots)$. From the regularity to the elliptic systems (see [3], [5])

$$(2.2) \qquad w^r \in \mathcal{C}^\infty(\bar\Omega, R^N)$$

Let $\gamma = (\gamma_1,\dots,\gamma_n) \in \mathcal{C}^1(\bar I, R^n)$ $(n = 1,2,\dots)$ and put $v^n(t) = \sum_{r=1}^{n} \gamma_r(t) w^r$. Let

$$(2.3) \qquad \rho_0 \in \mathcal{C}^d(\bar\Omega) \ , \ \rho_0 > 0 \text{ in } \bar\Omega \ , \ d = 1,2,\dots \ ,$$

$$(2.4) \qquad v_0 \in L^2(\Omega, R^N) \ ,$$

(2.5) $\qquad \theta_0 \in L^2(\Omega)\ ,\ \theta_0 > 0$ a.e. in Ω ,

(2.6) $\qquad F \in L^\infty(Q_T, R^N)$,

We refer to [4] that there exists a solution $\rho^n \in \mathcal{C}^1(\overline{Q}_T) \cap \mathcal{C}^1(\overline{I}, \mathcal{C}^d(\overline{\Omega}))$, $\rho^n > 0$ in $\overline{Q}_T$, $v^n(t) = \sum_{r=1}^{n} \gamma_r(t) w^r$, $\gamma = (\gamma_1, \dots, \gamma_n) \in \mathcal{C}^1(\overline{I}, R^n)$, $\theta^n \in \mathcal{C}^1(\overline{Q}_T) \cap \mathcal{C}^0(\overline{I}, \mathcal{C}^2(\overline{\Omega}))$, $\theta^n \geq 0$ in $\overline{Q}_T$ satisfying the equations

(2.7) $$\frac{\partial \rho^n}{\partial t} + \frac{\partial}{\partial x_i}(\rho^n v_i^n) = 0 ,$$

(2.8) $$\frac{\partial}{\partial t}(c_v \rho^n \theta^n) + \frac{\partial}{\partial x_j}(c_v \rho^n \theta^n v_j^n) - \lambda \frac{\partial^2 \theta^n}{\partial x_j \partial x_j} = - R\rho^n \theta^n \frac{\partial v_j^n}{\partial x_j} + \langle\langle v^n, v^n \rangle\rangle ,$$

(2.9) $$\int_\Omega (\rho^n \frac{\partial v_i^n}{\partial t} + \rho^n v_j^n \frac{\partial v_i^n}{\partial x_j} + R\frac{\partial}{\partial x_j}(\rho^n \theta^n) - \rho^n F_i) w_i^r \, dx = - ((v^n, w^r)) \ ; \ r = 1, \dots, n ,$$

(2.10) $\qquad \rho^n(0) = \rho_0$,

(2.11) $\qquad \theta^n(0) = \theta_0^n \ ,\ \dfrac{\partial \theta^n}{\partial \nu} = 0$ in $I \times \partial\Omega$,

where $\theta_0^n \in \mathcal{C}^2(\overline{\Omega})$, $\theta_0^n > 0$ in $\overline{\Omega}$, $\dfrac{\partial \theta_0^n}{\partial \nu} = 0$ on $\partial\Omega$, $\theta_0^n \longrightarrow \theta_0$ strongly in $L^2(\Omega)$,

(2.12) $$\int_\Omega v_i^n(0) w_i^r \, dx = \int_\Omega v_{0i} w_i^r \, dx$$

such that

(2.13) $$\rho^n(t,x) = \rho_0(y) \exp(-\int_0^t \frac{\partial}{\partial x_j} v_j^n(\tau, x^n(\tau)) d\tau) ,$$

where $x = x^n(t)$, $y = x^n(0)$ (see (3.2) in the next chapter),

(2.14) $\| \rho^n |v^n|^2 \|_{L^\infty(I,L^1(\Omega))} \le K_1$, $K_1 > 0$,

(2.15) $\| \rho^n\theta^n \|_{L^\infty(I;L^1(\Omega))} \le K_1$,

$\theta^n \ge 0$ in Q_T ,

(2.16) $\| v^n \|_{L^2(I,W^{k,2}(\Omega,R^N))} \le K_1$,

(2.17) $(\min_{x\in\bar\Omega} \rho_0(x))\exp(-c_1 t^{\frac{1}{2}}) \le \rho^n(t,x) \le$

$\le (\max_{x\in\bar\Omega} \rho_0(x))\exp(+c_1 t^{\frac{1}{2}})$, $c_1 > 0$, $(t,x) \in \bar Q_T$,

$0 < \varepsilon < \rho^n(t,x) < K_1$,

(2.18) $\| v^n \|_{L^\infty(I,L^2(\Omega,R^N))} \le K_1$.

III. Estimates of approximations

First, we shall deal with the continuity equation (2.7). For $\sigma^n = \ln\rho^n$ it reads

(3.1) $$\frac{d\sigma^n}{dt}(t,x^n(t,y)) = -\frac{\partial v_j}{\partial x_j}(t,x^n(t,y)) ,$$

where $x^n(t,.)$ is for every $t \in \bar I$ a diffeomorphism of $\bar\Omega$ onto $\bar\Omega$ given by solving the problem of characteristics

(3.2) $$\frac{dx^n}{dt}(t) = v^n(t,x^n) , \quad x^n(0) = y \in \bar\Omega .$$

Using Gronwall lemma, (3.2) and the imbedding

(3.3) $\| v^n(t) \|_{\mathcal{C}^1(\bar\Omega,R^N)} \le c\| v^n(t) \|_{W^{3,2}(\Omega,R^N)}$ $(c > 0)$

for every $t \in \bar I$, we get

(3.4) $\| \frac{\partial x_r^n}{\partial y_s} \|_{L^\infty(Q_T)} \le K_2 \ (K_2 > 0)$, $r,s = 1,\dots,N$.

Put $\hat{\sigma}^n(t,y) = \sigma^n(t,x^n(t,y))$, $\hat{\rho}^n(t,y) = \rho^n(t,x^n(t,y))$,
$z(t,y) = \frac{\partial v_j}{\partial x_j}(t,x^n(t,y))$.

The continuity equation in Lagrange coordinates reads

(3.5) $\hat{\rho}^n(t,y)\, G(t,y) = \rho_0(y)$,

where $G(t,y) = \det(\frac{\partial x_r^n}{\partial y_s})$. Due to (3.5), (2.17)

(3.6) $0 < \varepsilon_1 < G \le K_3 \ (K_3 > 0)$

and hence

(3.7) $\| \frac{\partial y_s}{\partial x_r^n} \|_{L^\infty(Q_T)} \le K_4 \ (K_4 > 0)$.

The equation (3.1) can be rewritten

(3.8) $\frac{\partial \hat{\sigma}^n}{\partial t}(t,y) = - z(t,y)$.

Using substitution theorem, we get

(3.9) $\int_\Omega | \frac{\partial z}{\partial y_s} |^q dy = \int_\Omega | \frac{\partial^2 v_j}{\partial x_j \partial x_r} \frac{\partial x_r}{\partial y_s} |^q G^{-1} dx$.

We know that $W^{3,2}(\Omega) \subset W^{2,q}(\Omega)$ for $1 \le q \le 6$ $(N = 3)$ or $1 \le q < +\infty$ $(N = 2)$, hence

(3.10) $\| \frac{\partial \hat{\sigma}^n}{\partial t} \|_{L^2(I,W^{1,q}(\Omega))} \le K_5$, $K_5 > 0$,

(3.11) $\| \hat{\sigma}^n \|_{C^{0,\alpha}(\bar{I},W^{1,q}(\Omega))} \le K_5$, $0 < \alpha < \frac{1}{2}$.

Of course $\sigma^n(t,x) = \hat{\sigma}^n(t,y(t,x))$. Due to the substitution theorem

$\int_\Omega | \frac{\partial \sigma^n}{\partial x_j} |^q dx = \int_\Omega | \frac{\partial \hat{\sigma}^n}{\partial y_r} \frac{\partial y_r}{\partial x_j} |^q G dy$;

hence $\| \ln \rho^n \|_{L^\infty(I,W^{1,q}(\Omega))} \le K_6 \ (K_6 > 0)$, $1 \le q \le 6$ $(N = 3)$ or $1 \le q < +\infty$ $(N = 2)$ and due to (2.17) also

(3.12) $\| \rho^n \|_{L^\infty(I,W^{1,q}(\Omega))} \le K_7$, $K_7 > 0$.

The continuity equation implies

$$(3.13) \qquad \left\| \frac{\partial \rho^n}{\partial t} \right\|_{L^2(I,L^q(\Omega))} \le K_8 \ , \ K_8 > 0 \ .$$

Hence $\| \rho^n \|_{C^{0,\alpha}(\overline{I},L^q(\Omega))} \le K_9 \ (K_9 > 0\ ) \ , \ 0 < \alpha < \frac{1}{2}$.

For $k \ge 3$ we get from (3.2)

$$(3.14) \qquad \left\| \frac{\partial^s x_i}{\partial y_1^{s1} \ldots \partial y_N^{sN}} \right\|_{L^\infty(Q_T)} \le K_{10} \ (K_{10} > 0)$$

for any multiindex $(s1,\ldots,sN)$, $s = s1 + \ldots + sN$, $0 \le s \le k-2$ and due to (3.6) for the inverse function $y(t,.)$

$$(3.15) \qquad \left\| \frac{\partial^s y_i}{\partial x_1^{s1} \ldots \partial x_N^{sN}} \right\|_{L^\infty(Q_T)} \le K_{11} \ , \ K_{11} > 0 \ ,$$

$$s = s1 + \ldots + sN, \ 0 \le s \le k-2 \ .$$

Using (3.14), (3.15), the imbedding $W^{k,2}(\Omega,R^N) \subset \subset W^{k-1,q}(\overline{\Omega},R^N)$ for $1 \le q \le 6$ $(N = 3)$, $1 \le q < +\infty$ $(N = 2)$, (3.6), one obtains from (3.8) by substitution theorem

$$(3.16) \qquad \left\| \frac{\partial \hat{\sigma}^n}{\partial t} \right\|_{L^2(I,W^{k-2,q}(\Omega))} \le K_{12} \ , \ K_{12} > 0 \ .$$

Due to (2.3)

$$(3.17) \qquad \| \hat{\sigma}^n \|_{C^{0,\alpha}(\overline{I},W^{p,q}(\Omega)} \le K_{12} \ ,$$

where $p = \min(d,k-2)$ and $1 \le q \le 6$ $(N=3)$ or $1 \le q < +\infty$ $(N=2)$; hence the following lemma holds

3.18 Lemma

Let $k \ge 3$, $p = \min(d,k-2)$, $1 \le q < +\infty$ *if* $N = 2$,$1 \le q \le 6$ *if* $N = 3$, $0 < \alpha < \frac{1}{2}$. *Then*

$$(3.19) \qquad \| \rho^n \|_{L^\infty(I,W^{p,q}(\Omega)} \le K_{13} \ , \ K_{13} > 0 \ ,$$

$$(3.20) \qquad \left\| \frac{\partial \rho^n}{\partial t} \right\|_{L^2(I,W^{p-1,q}(\Omega))} \le K_{13} \ ,$$

$$(3.21) \qquad \| \rho^n \|_{C^{0,\alpha}(\overline{I},W^{p-1,q}(\Omega))} \le K_{13} \ , \ 0 < \alpha < \frac{1}{2} \ .$$

Further we will proceed as in [4].

3.22 Lemma

Let $v_0 \in W^{k,2}(\Omega,R^N) \cap W_0^{1,2}(\Omega,R^N)$. *Then*

$$(3.23)\qquad \left\| \frac{\partial v^n}{\partial t} \right\|^2_{L^2(Q_T,R^N)} + \sup_{[0,T]} ((v^n,v^n)) \le$$

$$\le K_{14}(1 + \| \theta^n \|^2_{L^2(I,W^{1,2}(\Omega))}),\ K_{14} > 0 .$$

$$(3.24)\qquad \| v^n \|^2_{L^2(I,W^{2k,2}(\Omega,R^N))} \le$$

$$\le K_{14}(1 + \| \theta^n \|^2_{L^2(I,W^{1,2}(\Omega))}) .$$

Proof. We use in (2.9) the test function $\frac{\partial v^n}{\partial t}$; after some computation applying 3.18, (2.16), (2.18), we get (3.23). Due to the regularity to the elliptic systems (see [3],[5]) one gets (3.24)■

3.25 Lemma

$$\| \theta^n \|^2_{L^2(I,W^{1,2}(\Omega))} + \| \theta^n \|^2_{L^\infty(I,L^2(\Omega))} \le K_{15} ,$$
$K_{15} > 0$.

Proof. We multiply (2.8) by θ^n and integrate over Ω

$$(3.26)\qquad \frac{\partial}{\partial t}(\tfrac{1}{2}c_v\int_\Omega \rho^n(\theta^n)^2 dx) + \lambda\int_\Omega \frac{\partial\theta^n}{\partial x_j}\frac{\partial\theta^n}{\partial x_j}dx =$$

$$= R\int_\Omega \rho^n\frac{\partial v_j^n}{\partial x_j}(\theta^n)^2 dx + \int_\Omega \langle\langle v^n,v^n\rangle\rangle\theta^n dx .$$

Due to

$$(3.27)\qquad \| v^n \|_{\mathcal{C}^k(\bar\Omega,R^N)} \le \varepsilon_1 \| v^n \|_{W^{2k,2}(\Omega,R^N)} +$$

$$+ \bar K_1(\varepsilon_1)\| v^n \|_{L^2(\Omega,R^N)}$$

for every $\varepsilon_1 > 0$ ($\bar K_1(\varepsilon_1) > 0$) (see e.g. Lions [2]),

$$(3.28)\quad \Big(\int_{\Omega_t} ((v^n,v^n))^2 dx\Big)^{\frac{1}{2}} \| \theta^n \|_{L^2(\Omega)} \le$$

$$\le c_2\varepsilon_1 \| v^n(t) \|_{W^{k,2}(\Omega,R^N)} \| v^n(t) \|_{W^{2k,2}(\Omega,R^N)} \| \theta^n(t) \|_{L^2(\Omega)} +$$

$$+ \overline{K}_2(\varepsilon_1) \| \theta^n(t) \|_{L^2(\Omega)} , \ c_2, \ \overline{K}_2(\varepsilon_1) > 0 .$$

Hence according to 3.22 and Young inequality the expression on the r.h.s. is bounded by

$$(3.29)\quad \overline{K}_3(\varepsilon_1) \| v^n(t) \|^2_{W^{k,2}(\Omega,R^N)} \| \theta^n \|^2_{L^2(\Omega)} +$$

$$+ \varepsilon_1 \| \theta^n \|^2_{W^{1,2}(\Omega)} + \overline{K}_3(\varepsilon_1) , \ \overline{K}_3(\varepsilon_1) > 0 .$$

Thus 3.25 holds by Gronwall's lemma∎

Due to 3.22,3.25 we have

3.30. Lemma

Let $v_0 \in W^{k,2}(\Omega,R^N) \cap W_0^{1,2}(\Omega,R^N)$. *Then*

$$(3.31)\quad \Big\| \frac{\partial v^n}{\partial t} \Big\|^2_{L^2(Q_T,R^N)} + \sup_{[0,T]} ((v^n,v^n)) \le K_{16} ,$$

$$(3.32)\quad \| v^n \|_{L^2(I,W^{2k,2}(\Omega,R^N))} \le K_{16} , \ K_{16} > 0 .$$

3.33 Lemma

$$(3.34)\quad \Big\| \frac{\partial}{\partial t}(\rho^n\theta^n) \Big\|_{L^2(I,W^{-1,2}(\Omega,R^N))} \le K_{17} ,$$

$$(3.35)\quad \| \rho^n\theta^n \|_{L^2(I,W^{1,2}(\Omega,R^N))} \le K_{17} , \ K_{17} > 0 .$$

Proof. Estimate (3.35) is consequence of (3.19) and (3.25); (3.34) follows from (2.8)∎

Now, we improve the estimate of $\frac{\partial \rho^n}{\partial t}$. Due to (2.7), 3.30 it holds

3.36 Lemma

Let $k \ge 3$, $1 \le q < +\infty$ *if* $N = 2$, $1 \le q \le 6$ *if* $N = 3$, $0 < \alpha < \frac{1}{2}$. *Then*

(3.37) $$\| \frac{\partial \rho^n}{\partial t} \|_{L^\infty(I,W^{p-1,q}(\Omega))} \le K_{18} , K_{18} > 0 ,$$

$p = \min(d,k-2)$. *If* $k-2 < d$, *then also*

(3.38) $$\| \rho^n \|_{L^\infty(I,W^{p,q}(\Omega))} \le K_{18} ,$$

(3.39) $$\| \frac{\partial \rho^n}{\partial t} \|_{L^2(I,W^{p-1,q}(\Omega))} \le K_{18} ,$$

(3.40) $$\| \rho^n \|_{\mathcal{C}^{0,\alpha}(\bar{I},W^{p-1,q}(\Omega))} \le K_{18} ,$$

where $p = \min(d,2k-2)$.

3.41 Lemma

Let $\theta_o \in W^{1,2}(\Omega)$. *Then*

$$\| \frac{\partial \theta^n}{\partial t} \|^2_{L^2(Q_T)} + \| \theta^n \|^2_{L^\infty(I,W^{1,2}(\Omega))} \le K_{19} , K_{19} > 0 ,$$

$$\| \theta^n \|_{L^2(I,W^{2,2}(\Omega))} \le K_{19} .$$

Proof. Let us choose a sequence $\theta_o^n \in \mathcal{C}^2(\bar{\Omega})$, $\theta_o^n > 0$, $\theta_o^n \longrightarrow \theta_o$ strongly in $W^{1,2}(\Omega)$, $\frac{\partial \theta_o^n}{\partial \nu} = 0$ on $\partial\Omega$. Multiplying (2.8) by $\frac{\partial \theta^n}{\partial t}$ and integrating over $Q_t = (0,t) \times \Omega$, $t \in I$, we get

$$(3.42)\qquad \int_{Q_t} \rho^n \left(\frac{\partial\theta^n}{\partial t}\right)^2 dx d\tau + \lambda \int_{\Omega_t} \frac{\partial\theta^n}{\partial x_j} \frac{\partial\theta^n}{\partial x_j} dx =$$

$$= - \int_{Q_t} \rho^n v_j^n \frac{\partial\theta^n}{\partial x_j} \frac{\partial\theta^n}{\partial t} dx d\tau - \int_{Q_t} R\rho^n \theta^n \frac{\partial v_j^n}{\partial x_j} \frac{\partial\theta^n}{\partial t} dx d\tau +$$

$$+ \int_{Q_t} \langle\langle v^n, v^n \rangle\rangle \frac{\partial\theta^n}{\partial t} dx d\tau + \lambda \int_{\Omega} \frac{\partial\theta_0^n}{\partial x_j} \frac{\partial\theta_0^n}{\partial x} dx .$$

But r.h.s. of (3.42) is bounded by $K_{20}(1 + \| \frac{\partial\theta^n}{\partial t} \|_{L^2(Q_t)})$ $(K_{20} > 0)$. Applying Young's inequality, we verify the first part of Lemma 3.41. Second part is consequence of the regularity properties to the elliptic equations (see [5]).

IV. Limiting process

4.1 Lemma

Let the assumptions (2.3)-(2.6), 3.30 be satisfied. Then one can choose a subsequence of $\{(\rho^n, v^n, \theta^n)\}_{n=1}^{+\infty}$ *(denoted* $\{(\rho^n, v^n, \theta^n)\}_{n=1}^{+\infty}$ *again) such that*

i) $\rho^n \longrightarrow \rho$ *strongly in* $L^p(Q_T)$ $, 1 < p < +\infty$,
$\rho > \varepsilon > 0$ *a.e. in* Q_T ;

ii) $v^n \longrightarrow v$ *strongly in* $L^2(I, W^{2k-1,2}(\Omega, R^N))$,
$v^n \longrightarrow v$ *strongly in* $L^p(I, W^{k-1,2}(\Omega, R^N))$,
$D^i v^n \longrightarrow D^i v$ *weakly in* $L^2(Q_T, R^N)$ $(i = 1, \dots, 2k)$;

iii) $D^i \theta^n \longrightarrow D^i \theta$ *weakly in* $L^2(Q_T)$ $(i = 0,1)$;

iv) $\rho^n\theta^n \longrightarrow \rho\theta$ *strongly in* $L^2(Q_T)$,
$\theta \geq 0$ *a.e. in* Q ;

v) $\frac{\partial\rho^n}{\partial t} \longrightarrow \frac{\partial\rho}{\partial t}$ *weakly in* $L^2(Q_T)$;

vi) $\frac{\partial v^n}{\partial t} \longrightarrow \frac{\partial v}{\partial t}$ *weakly in* $L^2(Q_T,R^N)$;

vii) $\rho^n v^n \longrightarrow \rho v$ *strongly in* $L^2(Q_T,R^N)$;

viii) $\rho^n v_i^n v_j^n \longrightarrow \rho v_i v_j$ *weakly in* $L^2(Q_T)$;

ix) $\rho^n \theta^n \frac{\partial v_j^n}{\partial x_j} \longrightarrow \rho\theta\frac{\partial v_j}{\partial x_j}$ *weakly in* $L^2(Q_T)$;

x) $\int_0^T\int_\Omega \langle\langle v^n, v^n\rangle\rangle\phi dxdt \longrightarrow \int_0^T\int_\Omega \langle\langle v,v\rangle\rangle\phi dxdt$

for every $\phi \in \mathcal{C}^1(\overline{Q}_T)$.

Proof. Let us recall one assertion called Lions lemma, see e.g. [2],[8]. Let B_0, B, B_1 be Banach spaces, B_0, B_1 reflexive such that $B_0 \subset\subset B \subset B_1$ ($\subset\subset$ is compact imbedding). Let $1 < p_0, p_1 < +\infty$. Then
$L^{p_0}(I,B) \subset\subset \{ g;\ g\in L^{p_0}(I,B_0)\ ,\ \frac{\partial g}{\partial t} \in L^{p_1}(I,B_1)\ \}$.

i) First assertion follows directly from Lions lemma; we put $B_0 = W^{1,6}(\Omega)$, $B = L^p(\Omega)$, $B_1 = L^q(\Omega)$, $1 < p < +\infty$, $p_0 = p$, $p_1 = 2$, $1 \le q \le 6$ $(N = 3)$, $1 \le q < +\infty$ $(N = 2)$. Second assertion follows from (2.17).

ii) First two assertions hold due to Lions lemma with $B_0 = W^{2k,2}(\Omega,R^N)$, $B = W^{2k-1,2}(\Omega,R^N)$, $B_1 = L^2(\Omega,R^N)$, $p_0 = p_1 = 2$ resp. $B_0 = W^{k,2}(\Omega,R^N)$, $B = W^{k-1,2}(\Omega,R^N)$, $B_1 = L^2(\Omega,R^N)$, $p_0 = p$, $p_1 = 2$. Last assertion ii) follows from the boundedness of $\{v^n\}_{n=1}^{+\infty}$ in $L^2(I,W^{2k,2}(\Omega,R^N))$.

iii) holds due to 3.25.

iv) Strong convergence $\rho^n\theta^n \longrightarrow a$ in $L^2(Q_T)$ is consequence of Lions lemma with $B_0 = W^{1,2}(\Omega)$, $B = L^2(\Omega)$, $B_1 = W^{-1,2}(\Omega)$, $p_0 = p_1 = 2$; $a = \rho\theta$ due to i), iii); $\theta \ge 0$ a.e. in Q_T due to 2.15.

v) resp. vi) follow directly from (3.20) resp. (3.31). vii), viii) are consequence of i), ii); ix) is consequence of iv), ii) and x) follows from ii)■

Due to Lemma 4.1, we can pass to the limit in (2.7), (2.8), (2.9). We get the following theorems:

4.2 Theorem *(weak solutions)*

a) Let $k \ge 3$, $\rho_0 \in \mathcal{C}^d(\bar\Omega)$, $d = 1,2,\dots$, $\rho_0 > 0$ *in* $\bar\Omega$, $\theta_0 \in L^2(\Omega)$, $\theta_0 > 0$ *a.e. in* Ω, $v_0 \in W^{k,2}(\Omega,R^N) \cap \cap W_0^{1,2}(\Omega,R)^N$. *Let* $1 \le q < +\infty$ $(N = 2)$ *or* $1 \le q \le 6$ $(N = 3)$, $0 < \alpha < \frac{1}{2}$. *Then there exists* (ρ, v, θ)

$$\rho \in L^\infty(I, W^{p,q}(\Omega)) \cap \mathcal{C}^{0,\alpha}(\bar I, W^{p-1,q}(\Omega)), \tag{4.3}$$

$$\rho > \varepsilon \text{ a.e. in } Q_T \text{ for some } \varepsilon > 0,$$

$$\frac{\partial \rho}{\partial t} \in L^\infty(I, W^{p-1,q}(\Omega)), \tag{4.4}$$

where $p = \min(d, k-2)$,

$$v \in L^\infty(I, W^{k,2}(\Omega,R^N) \cap W_0^{1,2}(\Omega,R^N)) \cap \cap L^2(I, W^{2k,2}(\Omega,R^N)), \tag{4.5}$$

$$\frac{\partial v}{\partial t} \in L^2(Q_T, R^N), \tag{4.6}$$

$$\theta \in L^\infty(I, L^2(\Omega)) \cap L^2(I, W^{1,2}(\Omega)), \tag{4.7}$$

$$\theta \ge 0 \text{ a.e. in } Q_T,$$

such that (1.8), (1.9) hold a.e. in Q_T *and (1.15) is fulfilled.*

b) If $d > k-2$, *then*

$$\rho \in L^\infty(I, W^{p,q}(\Omega)) \cap \mathcal{C}^{0,\alpha}(\bar I, W^{p-1,q}(\Omega)), \tag{4.8}$$

(4.9) $$\frac{\partial\rho}{\partial t} \in L^2(I, W^{p-1,q}(\Omega)) ,$$

where p = min (d,2k−2).

4.10 Theorem *(strong solutions)*

Let the assumptions of Theorem 4.2 a) *be satisfied and let* $\theta_0 \in W^{1,2}(\Omega)$. *Then there exists* (ρ, v, θ) *satisfying* (4.3)−(4.7) *and*

(4.11) $$\theta \in L^2(I, W^{2,2}(\Omega)) \cap L^\infty(I, W^{1,2}(\Omega)) ,$$

(4.12) $$\frac{\partial\theta}{\partial t} \in L^2(Q_T) ,$$

such that the equations (1.8)−(1.10) *are satisfied a.e. in* Q_T.

Proof. The existence of strong solution was proved in 4.2. Clearly, Lemma 4.1 is fulfilled. Due to 3.41 $\theta \in L^2(I, W^{1,2}(\Omega))$, $\frac{\partial\theta}{\partial t} \in L^2(Q_T)$. The other is obvious∎

V. Uniqueness

Our aim in this section is to prove the following theorem.

5.1 Theorem

Let the assumptions of Theorem 4.10 *with* d ≥ 2 *be satisfied. Then in the class* (4.3)−(4.9), (4.11), (4.12) *there exists at most one solution* (ρ, v, θ) *satisfying* (1.8)−(1.10) *a.e. in* Q_T.

Proof. Let (ρ, v, θ), $(\bar\rho, \bar v, \bar\theta)$ be two solutions with the same initial conditions. Then $(\xi, w, \eta) = (\rho-\bar\rho, v-\bar v, \theta-\bar\theta)$ satisfies

$$(5.2)\qquad \frac{\partial}{\partial t}\xi + \frac{\partial}{\partial x_j}(\xi v_j) + \frac{\partial}{\partial x_j}(\bar\rho w_j) = 0 \text{ a.e. in } Q_T ,$$

$$(5.3)\quad \bar\rho\frac{\partial}{\partial t}w_i + \xi v_j\frac{\partial v_i}{\partial x_j} + \bar\rho w_j\frac{\partial v_i}{\partial x_j} + \bar\rho\bar v_j\frac{\partial w_i}{\partial x_j} - \frac{\partial}{\partial x_j}(\tau^V_{ij}(w)) +$$

$$+ R\frac{\partial}{\partial x_i}(\xi\theta) + R\frac{\partial}{\partial x_i}(\bar\rho\eta) + \xi\frac{\partial v_i}{\partial t} = \xi F_i \text{ a.e. in } Q_T ,$$

$$(5.4)\qquad c_V\bar\rho\frac{\partial\eta}{\partial t} + c_V\xi v_j\frac{\partial\theta}{\partial x_j} + c_V\bar\rho v_j\frac{\partial\eta}{\partial x_j} + c_V\bar\rho\frac{\partial\bar\theta}{\partial x_j}w_j -$$

$$- \lambda\frac{\partial^2\eta}{\partial x_j\partial x_j} + R\theta\frac{\partial v_j}{\partial x_j}\xi + R\bar\rho\frac{\partial v_j}{\partial x_j}\eta + R\bar\rho\bar\theta\frac{\partial w_j}{\partial x_j} + c_V\xi\frac{\partial\theta}{\partial t} =$$

$$= \langle\langle v,v\rangle\rangle - \langle\langle\bar v,\bar v\rangle\rangle \text{ a.e. in } Q_T .$$

From (5.2) one obtains the estimates

$$(5.5)\quad \frac{\partial}{\partial t}\| \xi(t) \|^2_{W^{1,2}(\Omega)} \le \bar K_4(\varepsilon_1)a_1(t)\| \xi(t) \|^2_{W^{1,2}(\Omega)} +$$

$$+ \varepsilon_1\| w(t) \|^2_{W^{k,2}(\Omega,R^N)} , \quad \bar K_4(\varepsilon_1) > 0 ,$$

where $a_1 = (1 + \| w \|_{W^{k,2}(\Omega,R^N)} + \| \bar\rho \|_{W^{2,6}(\Omega)})^2$ (hence $a_1 \in L^\infty(I)$), which holds for every $\varepsilon_1 > 0$. From (5.3) we get

$$(5.6)\qquad \frac{\partial}{\partial t}(\int_{\Omega_t} \bar\rho|w|^2dx) + \| w(t) \|^2_{W^{k,2}(\Omega,R^N)} \le$$

$$\le \bar K_5(\varepsilon_1)a(t)(\| \xi(t) \|^2_{W^{1,2}(\Omega)}) + \| w(t)\|^2_{L^2(\Omega,R^N)} +$$

$$+ \| \eta(t) \|^2_{L^2(\Omega)}) + \varepsilon_1\| w(t)\|^2_{W^{k,2}(\Omega,R^N)}$$

($\bar K_5(\varepsilon_1) > 0$) for a.e. $t \in I$ and every $\varepsilon_1 > 0$, where

$$a = (1 + \| \frac{\partial\bar\rho}{\partial t} \|_{L^6(\Omega)} + \| \bar\rho \|_{W^{1,6}(\Omega)} + \| \rho \|_{W^{1,6}(\Omega)})^2$$

$$(1+\| \bar v \|_{W^{2k,2}(\Omega,R^N)}+\| v \|_{W^{2k,2}(\Omega,R^N)}+\| \frac{\partial v}{\partial t} \|_{L^2(\Omega,R^N)}+$$

$$+ \| \frac{\partial\theta}{\partial t} \|_{L^2(\Omega,R^N)})^2 (1+ \| \bar\theta \|_{W^{1,2}(\Omega)}+\| \theta \|_{W^{1,2}(\Omega)})^2,$$

hence $a \in L^1(I)$.

We multiply (5.4) by η and integrate over Ω . Due to (2.4), $\langle\langle v,v\rangle\rangle = \sum \mathcal{L}_{(s1,\dots,sN)}(v)\mathcal{L}_{(s1,\dots,sN)}(v)$, where $\mathcal{L}_{(s1,\dots,sN)}(v)$ is a linear combination of e_{ij} , $\frac{\partial^2 v_i}{\partial x_{i_1}\partial x_{i_2}}$, ... , $\frac{\partial^k v_i}{\partial x_{i_1}\dots\partial x_{i_k}}$ and the sum is taken over all multiindexes $(s_1,\dots,s_N)$, $s_1 + \dots + s_N \le k$. Thus r. h. s. of (5.4) is equal to

$$\sum \mathcal{L}_{(s1,\dots,sN)}(w)\mathcal{L}_{(s1,\dots,sN)}(v + \bar{v}) .$$

After some computation we get the estimate

$$(5.7)\quad \frac{\partial}{\partial t}\left(\int_{\Omega_t} \rho\eta^2 dx\right) + \| \nabla\eta(t) \|^2_{L^2(\Omega)} \le$$

$$\le \bar{K}_6(\varepsilon_1)a(t)(\| \xi(t) \|^2_{W^{1,2}(\Omega)} + \| w(t) \|^2_{L^2(\Omega,R^N)} +$$

$$+ \| \eta(t) \|^2_{L^2(\Omega)}) + \varepsilon_1(\| w(t) \|^2_{W^{k,2}(\Omega,R^N)} +$$

$$+ \| \nabla\eta(t) \|^2_{L^2(\Omega)})$$

$(K_6(\varepsilon_1) > 0)$ for a.e. $t \in I$ and every $\varepsilon_1 > 0$ (∇ denotes the gradient with respect to space variables). We take into consideration that $\int_\Omega \bar{\rho}|w|^2 dx$ resp. $\int_\Omega \bar{\rho}\eta^2 dx$ are equivalent norms in $L^2(\Omega,R^N)$ resp. $L^2(\Omega)$; hence due to (5.5), (5.6), (5.7) and Gronwall's lemma $\xi=0$, $w=0$, $\eta=0$ a.e. in I . The proof is complete∎

VI. Energy and entropy estimates

In this section we derive some important estimates. For simplicity, put F=0. Generalization is easy. We shall start with the following lemma.

6.1 Lemma

Let $\{(\rho^n,u^n,\theta^n)\}_{n=1}^{+\infty}$ *be a sequence of solutions to* (2.7)-(2.12). *Let the assumptions* 4.10 *be satisfied and let* $\Theta_0 \in C^2(\bar{\Omega})$, $\Theta_0 > 0$ *in* $\bar{\Omega}$, $\frac{\partial\Theta_0}{\partial\nu} = 0$ *on* $\partial\Omega$. *We put* $\Theta_0^n = \Theta_0$. *Then*

i) $\Theta^n > 0$ *in* $\bar{Q}_T$;

ii) *there exists a subsequence (denoted* $\{(\rho^n, u^n, \Theta^n)\}_{n=1}^{+\infty}$ *again) such that*

ii)$_1$ $1/\Theta^n \longrightarrow 1/\Theta$ *strongly in* $L^p(Q_T)$, $1 < p < +\infty$;

ii)$_2$ $\int_\Omega \rho^n |\ln\Theta^n|\, dx \longrightarrow \int_\Omega \rho|\ln\Theta|dx$ *for a.e.* $t \in I$;

ii)$_3$ $\int_0^t \int_\Omega (\langle\langle v^n, v^n\rangle\rangle/\Theta^n) dx d\tau \longrightarrow \int_0^t \int_\Omega (\langle\langle v, v\rangle\rangle/\Theta)\, dx d\tau$, $t \in (0,T]$;

ii)$_4$ $(\ln\Theta^n)_{,i} \longrightarrow (\ln\Theta)_{,i}$ *weakly in* $L^4(Q_T)$.

Proof. Due to the smoothness of Θ^n we can find a minimal number σ, $0 < \sigma \le T$ such that $\min_{\bar{\Omega}} \Theta^n(\sigma,x) = 0$. Hence $\Theta(t,x) > 0$ in Q_σ. Put $\vartheta = \ln \Theta^n$. We derive from (2.8)

$$c_v\rho^n\frac{\partial\vartheta}{\partial t} + c_v\rho^n v_j^n\frac{\partial\vartheta}{\partial x_j} - \lambda\frac{\partial^2\vartheta}{\partial x_j\partial x_j} = \lambda(1/\Theta^n)^2\,\frac{\partial\Theta^n}{\partial x_j}\frac{\partial\Theta^n}{\partial x_j} +$$

$$+ \langle\langle v^n, v^n\rangle\rangle/\Theta^n - R\rho^n\frac{\partial v_j^n}{\partial x_j}$$

For $t < \sigma$, put $\vartheta = \vartheta_1 + \vartheta_2$, where $\vartheta_1(0) = \ln \Theta_0^n$, $\frac{\partial\vartheta_1}{\partial\nu} = 0$ on $(0,\sigma) \times \partial\Omega$, $c_v\rho^n\frac{\partial\vartheta_1}{\partial t} + c_v\rho^n v_j^n\frac{\partial\vartheta_1}{\partial x_j} - \lambda\frac{\partial^2\vartheta_1}{\partial x_j\partial x_j} =$ $= \lambda(1/\Theta^n)^2\frac{\partial\Theta^n}{\partial x_j}\frac{\partial\Theta^n}{\partial x_j} + \langle\langle v^n, v^n\rangle\rangle/\Theta^n$ and $\vartheta_2(0) = 0$, $\frac{\partial\vartheta_2}{\partial\nu} = 0$ on $(o,\sigma) \times \partial\Omega$,

$$c_v\rho^n\frac{\partial\vartheta_2}{\partial t} + c_v\rho^n v_j^n\frac{\partial\vartheta_2}{\partial x_j} - \lambda\frac{\partial^2\vartheta_2}{\partial x_j\partial x_j} = - R\rho^n\frac{\partial v_j^n}{\partial x_j} .$$

According to the maximum principle $\vartheta_1 \geq \ln \Theta_0^n$ and ϑ_2 can be continuously extended on $\overline{Q}_\sigma$. This is contradiction.

ii)$_1$ We multiply (2.8) by $1/(r(\Theta^n)^{r+1}$, $r > 0$ and integrate over Q_t, $t \in I$. We get

$$(6.2)\quad c_v\int_{\Omega_t}(\rho^n/(\Theta^n)^r)dx + \frac{1}{r}\int_0^t\int_\Omega(\langle\langle v^n, v^n\rangle\rangle/(\Theta^n)^{r+1})dxd\tau +$$
$$+ \frac{r+1}{r}\int_0^t\int_\Omega\left(\frac{\partial\Theta^n}{\partial x_j}\frac{\partial\Theta^n}{\partial x_j}/(\Theta^n)^{r+2}\right)dxd\tau =$$
$$= \frac{1}{r}\int_0^t\int_\Omega\left(R\rho^n\frac{\partial v_j^n}{\partial x_j}/(\Theta^n)^r\right)dxd\tau + c_v\int_\Omega(\rho_0^n/(\Theta_0^n)^r)dx .$$

From (6.2), one obtains

$$\| (\Theta^n)^{-\sqrt{r}} \|_{L^\infty(I,L^2(\Omega))} \leq K_{21} , \quad K_{21} > 0 .$$

Due to Lemma 4.1, i), iv), $\Theta^n \longrightarrow \Theta$ a e. in Q_T ; hence $1/\Theta^n \longrightarrow 1/\Theta$ strongly in $L^p(Q_T)$, $p > 1$.

ii)$_2$ $\ln\Theta^n(t) \longrightarrow \ln\Theta(t)$ and $\rho^n(t) \longrightarrow \rho(t)$ a.e. in Ω for a.e. $t \in I$. The sequence $\rho^n(t)\ln\Theta^n(t)$ is bounded in $L^2(\Omega)$ for a.e. $t \in I$. Thus

$\rho^n(t)|\ln\Theta^n(t)| \longrightarrow \rho(t)|\ln\Theta(t)|$ weakly in $L^2(\Omega)$.

ii)$_3$ holds due to ii)$_1$ and Lemma 4.1 ii).

ii)$_4$ is consequence of ii)$_1$ and iii) in Lemma 4.1∎

We will prove the following theorem

6.3 Theorem

Let (ρ, u, Θ) be a strong solution to the system (1.8)-(1.13) and let the assumptions of Lemma 6.1 *be satisfied. Then it holds*

$$(6.4)\qquad \int_\Omega \rho dx = \int_\Omega \rho_0 dx ,$$

$$(6.5)\qquad \frac{1}{2}\int_{\Omega_t}\rho|v|^2dx + c_v\int_{\Omega_t}\rho\Theta dx \leq \frac{1}{2}\int_\Omega\rho_0|v_0|^2dx +$$
$$+ c_v\int_\Omega\rho_0\Theta_0dx ,$$

$$(6.6)\qquad R\int_{\Omega_t}\rho|\ln\rho|dx+\int_0^t\int_\Omega(\langle\langle v,v\rangle\rangle/\theta)dxd\tau+$$

$$+\lambda\int_0^t\int_\Omega\frac{\partial\ln\theta}{\partial x_i}\frac{\partial\ln\theta}{\partial x_i}dxd\tau\le\tfrac{1}{2}\int_\Omega\rho_0|v_0|^2dx+$$

$$+2\tfrac{R}{e}m(\Omega)-R\int_\Omega\rho_0\ln\rho_0dx-c_v\int_\Omega\rho_0\ln\theta_0dx+c_v\int_\Omega\rho_0\theta_0dx\ ,$$

$$(6.7)\qquad c_v\int_{\Omega_t}\rho|\ln\theta|dx\le\tfrac{R}{e}m(\Omega)+R\int_\Omega\rho_0\ln\rho_0dx+$$

$$+2c_v\int_\Omega\rho_0\theta_0dx+\int_\Omega\rho_0|v_0|^2dx-c_v\int_\Omega\rho_0\ln\theta_0dxdt$$

for a.e. $t\in I$ *(m denotes the Lebesgue measure).*

Proof. Equalities (6.4),(6.5) represent the conservation of mass and energy respectively. They are derived in the standard way (see e.g. [7],[9]).

Multiplying (2.8) by $1/\theta^n$, integrating over Q_t $(t\in I)$ and using the identity

$$(6.8)\quad -\int_0^t\int_\Omega R\rho^n\frac{\partial v_j^n}{\partial x_j}dxd\tau=\int_{\Omega_t}R\rho^n\ln\rho^n dx-\int_\Omega R\rho_0\ln\rho_0dx$$

(which is a consequence of the equation (2.7)), we arrive at

$$(6.9)\qquad c_v\int_{\Omega_t}\rho^n\ln\theta^n dx=\int_{\Omega_t}R\rho^n\ln\rho^n dx-\int_\Omega R\rho_0\ln\rho_0dx+$$

$$+\int_0^t\int_\Omega(\langle\langle v^n,v^n\rangle\rangle/\theta^n)dxd\tau+$$

$$+\lambda\int_0^t\int_\Omega\frac{\partial\ln\theta^n}{\partial x_i}\frac{\partial\ln\theta^n}{\partial x_i}dxd\tau+c_v\int_\Omega\rho_0\ln\theta_0dx\ .$$

We know that $\ln\theta<\theta,\rho\ln\rho=\rho(\ln\rho)^{+}-\rho(\ln\rho)^{-}$, $\ln\theta=(\ln\theta)^{+}-(\ln\theta)^{-}$ (a^{+}, a^{-} are positive resp. negative parts of a) and $\rho(\ln\rho)^{-}\le\frac{1}{e}$. Using these facts we arrive at

$$(6.10)\quad R\int_\Omega \rho^n|\ln\rho^n|dx + \int_0^t\int_\Omega \langle\langle v^n, v^n\rangle\rangle/\theta^n)dxd\tau +$$

$$+\lambda\int_0^t\int_\Omega \frac{\partial\ln\theta^n}{\partial x_i}\frac{\partial\ln\theta^n}{\partial x_i}dxd\tau \le \tfrac{1}{2}\int_\Omega \rho_0|v_0|^2dx + 2\tfrac{R}{e}m(\Omega) +$$

$$+ R\int_\Omega \rho_0\ln\rho_0 dx - c_v\int_\Omega \rho_0\ln\theta_0 dx + c_v\int_\Omega \rho_0\theta_0 dx ,$$

$$(6.11)\quad c_v\int_{\Omega_t} \rho^n|\ln\theta^n|dx \le \tfrac{R}{e}m(\Omega) + R\int_\Omega \rho_0\ln\rho_0 dx +$$

$$+ 2c_v\int_\Omega \rho_0\theta_0 dx + \int_\Omega \rho_0|v_0|^2dx - c_v\int_\Omega \rho_0\ln\theta_0 dxdt .$$

According to Fatou's lemma

$$(6.12)\quad \liminf_{n\to+\infty} \|(\ln\theta^n)_{,i}\|_{L^2(Q_T)} \ge \|(\ln\theta)_{,i}\|_{L^2(Q_T)} .$$

We use 4.1, 6.1 and (6.12). We can pass to the limit in (6.10), (6.11) and we get (6.6), (6.7). Theorem 6.3 is proved■

6.13 Consequence

Let the assumptions of 6.3 be satisfied and let $\theta_0 > 1$ *in* $\overline{\Omega}$. *Then*

$$(6.14)\qquad 1/\theta \in L^\infty(Q_T) .$$

Proof. For $\theta_0 > 1$ we get from (6.2) that $1/\theta^n$ is uniformly bounded in $L^\infty(I, L^r(\Omega))$ $(1 < r < +\infty)$ with respect to r. Hence (6.14) holds■

REFERENCES

[1] A.Friedman: Partial differential equations of parabolic type, Prentice-Hall, Englewood Cliffs, N.J., 1964.

[2] J.L.Lions: Quelques méthodes de résolution des problèmes aux limites nonlinéaires, Dunod, Paris 1969.

[3] J. Nečas: Les méthodes directes en la théorie des equations elliptiques, Academia, Prague, 1967.

[4] J. Nečas, A. Novotný, M. Šilhavý: Global solution to the ideal compressible multipolar heat conductive fluid, Comment. Mat. Univ. Carolinae **30**,3 (1989), 551-564.

[5] J. Nečas, A. Novotný, M. Šilhavý: Global solution to the viscous compressible isothermal multipolar fluid, to appear in J. Math. Anal. Appl.

[6] J. Nečas, M. Šilhavý: Viscous multipolar fluids, to appear Quart. Appl. Math.

[7] J. Neustupa, A. Novotný: Global solvability to the regularized viscous compressible flow, to appear in Aplikace Matematiky.

[8] R. Temam: Navier-Stokes equations, Amsterdam-New York-Oxford, 1979.

[9] A. Valli: An existence theorem for compressible viscous fluids, Ann. Mat. Pura Appl. **130** (1982), 197-213.

Received August 1989

Revised December 1990

QUARTERLY OF APPLIED MATHEMATICS
VOLUME XLIX, NUMBER 2
JUNE 1991, PAGES 247-265

MULTIPOLAR VISCOUS FLUIDS

By

J. NEČAS (*Charles University, Prague, Czechoslovakia*)

AND

M. ŠILHAVÝ (*Czechoslovak Academy of Sciences, Prague, Czechoslovakia*)

Abstract. The purpose of the paper is to develop a thermodynamic theory of constitutive equations of multipolar viscous fluids. The restrictions which the principle of material frame-indifference and the Clausius-Duhem inequality place on the constitutive equations are derived. Explicit forms of the viscous stresses are obtained for linear viscous fluids.

1. Introduction. In the classical Navier-Stokes theory, the viscosity of fluids is modelled by the dependence of the stress on the first spatial gradients of velocity. It is well known that the corresponding mathematical theory contains a number of unsolved problems; in particular, there is no adequate existence theory for compressible flows of such fluids. These problems have led the present authors to believe that a stronger mechanism of dissipation and viscosity, namely the dependence of the stress on the higher gradients of velocity, must occur in the flows of viscous fluids. In a work in progress, the authors together with A. Novotný [1] show that a satisfactory existence theory can be developed for materials of this type (see also Nečas and Růžička [2] for an analogous treatment of viscous incompressible solids).

In materials in which the higher gradients of velocity and the higher gradients of deformation influence the response, the rate of work of internal forces cannot be expected to be only the product of the usual second order stress tensor with the first gradient of velocity; instead of this, a more general expression must be assumed containing additionally the sum of products of higher order multipolar stress tensors with the higher gradients of velocity. Otherwise such materials cannot be compatible with the Clausius-Duhem inequality. The general theory of multipolar materials is due to Green and Rivlin [3, 4]. However, they consider only the constitutive equations of elastic, nonviscous materials. (See also Bucháček [5] for the theory of multipolar materials with fading memory.)

The purpose of this paper is to develop a thermodynamic theory of constitutive equations of multipolar viscous fluids within the framework of the theory of Green and Rivlin. The postulated constitutive equations express the Helmholtz free

Received April 19, 1990.

©1991 Brown University

© Brown University 1991. Reprinted with kind permission from Walter Freiberger (Managing Ed.) from: Nečas, J. & Silhavy, M., Multipolar viscous fluids, Quarterly of Applied Mathematics, 49:247-265.

energy, entropy, heat flux vector, and the multipolar stress tensors as functions of the following variables: the density and its gradients up to a fixed order, gradients of velocity up to a fixed order, the temperature, and the gradient of temperature. We derive the general restrictions which the principle of material frame-indifference and the Clausius-Duhem inequality place on the constitutive functions of the fluid. Then we restrict our attention to linear fluids for which the constitutive quantities depend on the gradients of velocity and temperature linearly, with the coefficients independent of temperature and gradients of density. Using the representation theorems for isotropic linear functions, we obtain explicit forms of the viscous stresses. The corresponding scalar coefficients in front of the gradients of velocity in these expressions generalize (and include as special cases) the classical viscosities of the fluid. As in the classical case, the Clausius-Duhem inequality gives the nonnegativeness of the viscous work which, in the strengthened form, plays a crucial role in the existence theory [1, 2].

2. Balance equations and the Clausius-Duhem inequality. We refer to Green and Rivlin [3, 4] for a detailed exposition of the thermomechanics of multipolar bodies.

We use the spatial description of the processes of the fluid. The fields associated with the processes are functions of the actual position $x = (x_i)$ $(i = 1, 2, 3)$ of the points of the fluid and of time t. Let $N \geq 1$ be an integer. A thermodynamic process of a multipolar fluid R of grade N is a collection of $8 + N$ functions of position and time: $v, \theta, \rho, e, \eta, b, r, q, T^{(k)}$ $(k = 0, 1, \ldots, N-1)$, whose interpretation and tensorial nature is as follows:

- $v = (v_i)$ is the velocity field,
- θ is the field of positive absolute temperature,
- ρ is the actual density,
- e is the specific internal energy,
- η is the specific entropy,
- $b = (b_i)$ is the specific external body force,
- r is the rate of the external communication of heat to the body,
- $q = (q_i)$ is the heat flux vector,
- $T^{(k)} = (T^{(k)}_{jj_1\cdots j_k i})$ is the spatial multipolar stress tensor of order $k+2$, $k = 0, 1, \ldots, N-1$; it is assumed that $T^{(k)}$ is symmetric in the indices $j_1 \cdots j_k$. The symmetry is motivated by the fact that the tensor $T^{(k)}$ enters the basic equations only through its product with the spatial gradient of velocity which has the same type of symmetry (cf. (2.2) below).

It is also assumed that any function occurring in this paper is as many times continuously differentiable as needed to make the expressions meaningful. Each process satisfies the equations of balance of mass, energy, linear and angular momentum, and the Clausius-Duhem inequality. Their local forms read as follows:

$$\dot{\rho} + \rho v_{i,i} = 0 \quad \text{or} \quad \partial \rho / \partial t + (\rho v_i)_{,i} = 0, \tag{2.1}$$

$$\rho\left(e+\frac{1}{2}v_iv_i\right)^{\cdot} = \left(-q_i+\sum_{k=0}^{N-1} T^{(k)}_{jj_1\cdots j_k i}v_{j,j_1\cdots j_k}\right)_{,i} + \rho b_i v_i + \rho r, \tag{2.2}$$

$$\rho\dot{v}_j = T^{(0)}_{ji,i} + \rho b_j, \tag{2.3}$$

$$\rho(\varepsilon_{jkp}x_kv_p)^{\cdot} = (\varepsilon_{jkp}x_kT^{(0)}_{pi} + \varepsilon_{jkp}T^{(1)}_{pki})_{,i} + \rho\varepsilon_{jkp}x_kb_p, \tag{2.4}$$

$$\rho\dot{\eta} \geq -\left(\frac{q_i}{\theta}\right)_{,i} + \rho\frac{r}{\theta}. \tag{2.5}$$

Here a superimposed dot denotes the material time derivative, and a comma followed by an index (or several indices) denotes the partial derivative with respect to the coordinate corresponding to the index (or indices). Green and Rivlin [3, 4] show that the equations of balance of linear and angular momenta follow from the equation of balance of energy and the principle of material frame-indifference. We also note that the higher order stresses $T^{(1)}, T^{(2)}, \ldots$ do not enter the equation of balance of linear momentum while only the first two stresses $T^{(0)}, T^{(1)}$ enter the equation of balance of angular momentum.

Routine manipulations with the balance equations give the reduced forms of the balance equations:

$$\rho\dot{e} = \sum_{k=0}^{N-1}(T^{(k)}_{jj_1\cdots j_k i} + T^{(k+1)}_{jj_1\cdots j_k ip,p})v_{j,j_1\cdots j_k i} - q_{i,i} + \rho r, \tag{2.6}$$

$$\varepsilon_{ijk}(T^{(0)}_{kj} + T^{(1)}_{kjp,p}) = 0, \tag{2.7}$$

$$\rho\theta\dot{\eta} \geq \rho\dot{e} - \sum_{k=0}^{N-1}(T^{(k)}_{jj_1\cdots j_k i} + T^{(k+1)}_{jj_1\cdots j_k ip,p})v_{j,j_1\cdots j_k i} + q_i\theta_{,i}/\theta. \tag{2.8}$$

In (2.6) and in what follows we use the convention that

$$T^{(k)} = 0 \quad \text{for } k \geq N. \tag{2.9}$$

Another equivalent form of the Clausius-Duhem inequality is the dissipation inequality which reads

$$\rho\dot{\psi} \leq \sum_{k=0}^{N-1}(T^{(k)}_{jj_1\cdots j_k i} + T^{(k+1)}_{jj_1\cdots j_k ip,p})v_{j,j_1\cdots j_k i} - \rho\eta\dot{\theta} - q_i\theta_{,i}/\theta, \tag{2.10}$$

where

$$\psi = e - \theta\eta \tag{2.11}$$

is the Helmholtz free energy.

3. Constitutive equations of multipolar viscous fluids. We postulate the constitutive equations in the following form:

$$f = f(\rho, \nabla\rho, \ldots, \nabla^{M-1}\rho, \nabla v, \ldots, \nabla^K v, \theta, \nabla\theta), \tag{3.1}$$

where $M, K \geq 1$ are prescribed positive integers, and f stands for any of the functions $e, \eta, \psi, q, T^{(k)}, k = 0, 1, \ldots, N-1$. The functions occurring on the right-hand side of (3.1), i.e., the functions $e, \eta, \psi, q, T^{(k)}$, are given smooth functions of their arguments defined on the natural domain given by the physical interpretation of the arguments; the functions e, η, ψ naturally satisfy the relation (2.11). The values of f on the left-hand side of (3.1) as well as the arguments on the right-hand side of (3.1) are evaluated at (x, t). The postulated form of the constitutive equations can be motivated by the general definition of fluids in terms of the symmetry group of the material. (See Noll [6], Cross [7], Samohýl [8], and Gurtin, Vianello, and Williams [9] for various particular cases.) We assume that the integers N, M, K take their least possible values. If N, M, K are so chosen, then the material is called a multipolar viscous fluid of type (N, M, K). If $M > 1$ or $K > 1$, then the material is nonsimple in the sense of Truesdell and Noll [10].

In addition to the symmetry which has already been taken into account, we shall use two general principles to restrict the form of the constitutive functions: the principle of material frame-indifference and the requirement that the Clausius-Duhem inequality be satisfied in every process. We shall take these two principles in turn and combine them with the idea of linearization with respect to the nonequilibrium parameters $\nabla v, \ldots, \nabla^K v, \nabla\theta$.

Consider a change of frame of the form

$$\overline{x}_i = Q_{ij}(t)x_j + c_i(t), \tag{3.2}$$

where $Q(t) = (Q_{ij}(t))$ is a time-dependent orthogonal matrix,

$$Q_{ij}(t)Q_{ik}(t) = \delta_{jk}, \tag{3.3}$$

and $c(t) = (c_i(t))$ is a time-dependent vector. The principle of material frame-indifference postulates that under the changes of frame (3.2) the quantities θ, ρ, e, η, r, q, and $T^{(k)}$ transform in the following way:

$$\overline{\theta} = \theta, \quad \overline{\eta} = \eta, \quad \overline{e} = e, \quad \overline{\rho} = \rho, \quad \overline{r} = r, \tag{3.4}$$

$$\overline{q}_i = Q_{ij}q_j, \tag{3.5}$$

$$\overline{T}^{(k)}_{j_1\cdots j_{k+2}} = Q_{j_1m_1}\cdots Q_{j_{k+2}m_{k+2}}T^{(k)}_{m_1\cdots m_{k+2}}. \tag{3.6}$$

The quantities on the two sides of the above equations are evaluated at the same material point; since the spatial description is used, the arguments on the left-hand side are $(\overline{x}, t)$, while the arguments on the right-hand side are (x, t). Consequently, we have the following transformation laws for the gradients of density, velocity, and temperature:

$$\overline{\rho}_{,i_1\cdots i_k} = Q_{i_1j_1}\cdots Q_{i_kj_k}\rho_{,j_1\cdots j_k}, \tag{3.7}$$

$$\overline{v}_{i,j} = Q_{il}Q_{jm}v_{l,m} + W_{ij}, \tag{3.8}$$

$$W_{ij} = \dot{Q}_{im}Q_{jm}, \qquad W_{ij} = -W_{ji}, \tag{3.9}$$

$$\overline{v}_{i,j_1\cdots j_k} = Q_{il}Q_{j_1m_1}\cdots Q_{j_km_k}v_{l,m_1\cdots m_k} \qquad (k \geq 2), \tag{3.10}$$

$$\overline{\theta}_{,i} = Q_{ij}\theta_{,j}. \tag{3.11}$$

Let us also note that in view of (3.8) and (3.9), the symmetric part $D = (D_{ij})$ of the gradient of velocity,

$$D_{ij} = \tfrac{1}{2}(v_{i,j} + v_{j,i}), \tag{3.12}$$

transforms as

$$\overline{D}_{ij} = Q_{il}Q_{jm}D_{lm}. \tag{3.13}$$

PROPOSITION 3.1. A multipolar viscous fluid of type (N, M, K) satisfies the principle of material frame-indifference if and only if the following two conditions are satisfied:

(1) The functions $e, \eta, \psi, q, T^{(k)}$ depend on the first spatial gradient of velocity only through its symmetric part D, i.e.,

$$\begin{aligned} &f(\rho, \nabla\rho, \dots, \nabla^{M-1}\rho, \nabla v, \dots, \nabla^K v, \theta, \nabla\theta) \\ &\quad = f(\rho, \nabla\rho, \dots, \nabla^{M-1}\rho, D, \nabla^2 v, \dots, \nabla^K v, \theta, \nabla\theta), \end{aligned} \tag{3.14}$$

where f stands for any of the functions $e, \eta, \psi, q, T^{(k)}$.

(2) The constitutive functions $e, \eta, \psi, q, T^{(k)}$ are isotropic scalar-, vector-, or tensor-valued functions of the scalar, vector, or tensor arguments ρ, $\nabla\rho, \dots, \nabla^{M-1}\rho$, $D, \nabla^2 v, \dots, \nabla^K v, \theta, \nabla\theta$.

The proof is omitted. We only note that the fact that the response can depend on the first gradient of velocity only through its symmetric part is related with the occurrence of the skew tensor W in the transformation law (3.8) and this in turn depends on the admittance of the time-dependent rotations in the statement of the principle of material frame-indifference (cf., e.g., Truesdell [11]). Assertion (2) means that, for instance, the constitutive function $T^{(k)}$ satisfies the functional equation of the form

$$\begin{aligned} &T^{(k)}_{j_1\cdots j_{k+2}}(\overline{\rho}, \nabla\overline{\rho}, \dots, \nabla^{M-1}\overline{\rho}, \overline{D}, \nabla^2\overline{v}, \dots, \nabla^K\overline{v}, \overline{\theta}, \nabla\overline{\theta}) \\ &\quad = Q_{j_1m_1}\cdots Q_{j_{k+2}m_{k+2}} T^{(k)}_{m_1\cdots m_{k+2}}(\rho, \nabla\rho, \dots, \nabla^{M-1}\rho, D, \nabla^2 v, \dots, \nabla^K v, \theta, \nabla\theta). \end{aligned} \tag{3.15}$$

Using this equation and a similar one for q, we shall derive explicit expressions for the functions q and $T^{(k)}$ (see Sec. 5).

4. The consequences of the Clausius-Duhem inequality. We now demand, following Coleman and Noll [12], that the Clausius-Duhem inequality be satisfied in every process compatible with the constitutive equations and the equations of balance of energy and linear momentum. These contain the external sources r and b respectively and it is essential to admit that r and b can be arbitrary functions of x, t. In view of the form of the constitutive equations this means that all possible motions and evolutions of the absolute temperature are admitted.

To facilitate the statement of the restrictions which the Clausius-Duhem inequality places on the constitutive functions, we introduce the following quantities and functions:

- the equilibrium part of the multipolar stress $T^{(k,E)}$,

$$\begin{aligned} &T^{(k,E)}(\rho, \nabla\rho, \dots, \nabla^{M-1}\rho, \theta) \\ &\quad = T^{(k)}(\rho, \nabla\rho, \dots, \nabla^{M-1}\rho, 0, \dots, 0, \theta, 0) \end{aligned} \tag{4.1}$$

$(k = 0, 1, \dots)$, as the multipolar stress corresponding to the zero values of the gradients of velocity and the zero value of the gradient of temperature;

- the viscous part of the multipolar stress $T^{(k,V)}$,

$$\begin{aligned} T^{(k,V)}&(\rho, \nabla\rho, \dots, \nabla^{M-1}\rho, \nabla v, \dots, \nabla^K v, \theta, \nabla\theta) \\ &= T^{(k)}(\rho, \nabla\rho, \dots, \nabla^{M-1}\rho, \nabla v, \dots, \nabla^K v, \theta, \nabla\theta) \\ &- T^{(k,E)}(\rho, \nabla\rho, \dots, \nabla^{M-1}\rho, \theta), \end{aligned} \tag{4.2}$$

$k = 0, 1, \dots;$
- the variable σ is defined by

$$\sigma = \ln\rho, \tag{4.3}$$

so that the dependence of the constitutive quantities on ρ and its gradients can be replaced by the dependence on σ and its gradients.

We say that a multipolar viscous fluid R satisfies the second law of thermodynamics if the Clausius-Duhem inequality (2.5) holds in every process compatible with the constitutive equations (3.1) and the equations of balance of mass (2.1), energy (2.2), and linear momentum (2.3).

THEOREM 4.1. A multipolar viscous fluid of type (N, M, K) satisfies the second law of thermodynamics if and only if the following two conditions are satisfied in every process:

(1) The generalized Gibbs equation,

$$\rho\dot\psi = -\rho\eta\dot\theta + \sum_{k=0}^{N-1}(T^{(k,E)}_{jj_1\cdots j_k i} + T^{(k+1,E)}_{jj_1\cdots j_k ip,p})v_{j,j_1\cdots j_k i} - \sum_{k=1}^{N-1}\frac{\partial}{\partial\theta}T^{(k,E)}_{jj_1\cdots j_k p}\theta_{,p}v_{j,j_1\cdots j_k}. \tag{4.4}$$

(2) The residual dissipation inequality,

$$\sum_{k=0}^{N-1}(T^{(k,V)}_{jj_1\cdots j_k i} + T^{(k+1,V)}_{jj_1\cdots j_k ip,p})v_{j,j_1\cdots j_k i} + \sum_{k=1}^{N-1}\frac{\partial}{\partial\theta}T^{(k,E)}_{jj_1\cdots j_k p}\theta_{,p}v_{j,j_1\cdots j_k} - q_i\theta_{,i}/\theta \ge 0. \tag{4.5}$$

The Clausius-Duhem inequality or equivalently the dissipation inequality thus splits into the generalized Gibbs equation and the residual dissipation inequality. Further consequences of the generalized Gibbs equation will be given in Theorem 4.2, below, while the consequences of the residual dissipation inequality will be analyzed in the next section.

THEOREM 4.2. If a multipolar viscous fluid of type (N, M, K) satisfies the second law of thermodynamics then the following two assertions hold:

(1) The constitutive functions ψ, η, e are independent of the gradients of velocity and of the gradient of temperature, i.e.,

$$f(\rho, \nabla\rho, \dots, \nabla^{M-1}\rho, \nabla v, \dots, \nabla^K v, \theta, \nabla\theta) = f(\rho, \nabla\rho, \dots, \nabla^{M-1}\rho, \theta) \tag{4.6}$$

throughout the domain of the constitutive functions, where f stands for any of the functions ψ, η, e.

(2) The entropy relation

$$\eta = -\frac{\partial\psi}{\partial\theta} \tag{4.7}$$

and the generalized stress relations

$$\mathrm{Sym}(T^{(k,E)}_{jj_1\cdots j_k i} + T^{(k+1,E)}_{jj_1\cdots j_k ip,p} - \frac{\partial}{\partial\theta}T^{(k+1,E)}_{jj_1\cdots j_k ip}\theta_{,p}) = -\rho\,\mathrm{Sym}\left(\frac{\partial\psi}{\partial\sigma_{j_1\cdots j_k}}\delta_{ij}\right), \tag{4.8}$$

$k = 0, 1, \dots,$ hold throughout the domain of the constitutive functions; here the constitutive functions are taken as functions of σ and its gradients, and Sym denotes the symmetrization with respect to the indices $j_1 \cdots j_k i$.

In (4.8) we have used our earlier convention that for $k \geq N$, we have set $T^{(k)} = 0$ and hence also, as a consequence of this, $T^{(k,E)} = T^{(k,V)} = 0$ (cf. (2.9), (4.1), (4.2)).

Proof of Theorems 4.1 *and* 4.2. We shall use the dissipation inequality (2.10) which is equivalent to the Clausius-Duhem inequality. We express the constitutive functions as functions of σ and its gradients, e.g.,

$$\psi = \psi(\sigma, \nabla\sigma, \dots, \nabla^{M-1}\sigma, \nabla v, \dots, \nabla^K v, \theta, \nabla\theta), \tag{4.9}$$

etc. This form of the constitutive functions together with the definitions of the elastic and viscous parts of the stress and the continuity equation in the form

$$\dot\sigma + v_{k,k} = 0 \tag{4.10}$$

give the following explicit version of the dissipation inequality:

$$\begin{aligned}
&\rho\left(-\sum_{m=0}^{M-1}\frac{\partial\psi}{\partial\sigma_{i_1\cdots i_m}}v_{l,li_1\cdots i_m} + \sum_{n=1}^{K}\frac{\partial\psi}{\partial v_{j,j_1\cdots j_n}}\dot v_{j,j_1\cdots j_n} + \frac{\partial\psi}{\partial\theta}\dot\theta + \frac{\partial\psi}{\partial\theta_{,i}}\dot\theta_{,i}\right)\\
&\quad\leq -\rho\eta\dot\theta + \sum_{k=0}^{N-1}\left(T^{(k,E)}_{jj_1\cdots j_k i} + \sum_{m=0}^{M-1}\frac{\partial T^{(k+1,E)}_{jj_1\cdots j_k ip}}{\sigma_{,i_1\cdots i_m p}}\sigma_{,i_1\cdots i_m p} + \frac{\partial}{\partial\theta}T^{(k+1,E)}_{jj_1\cdots j_k ip}\theta_{,p}\right)v_{j,j_1\cdots j_k i}\\
&\quad\quad + \sum_{k=0}^{N-1}\left(T^{(k,V)}_{jj_1\cdots j_k i} + \sum_{m=0}^{M-1}\frac{\partial T^{(k+1,V)}_{jj_1\cdots j_k ip}}{\partial\sigma_{,i_1\cdots i_m}}\sigma_{,i_1\cdots i_m p}\right.\\
&\quad\quad\quad \left. + \sum_{n=1}^{K}\frac{\partial T^{(k+1,V)}_{jj_1\cdots j_k ip}}{\partial v_{l,l_1\cdots l_n}}v_{l,l_1\cdots l_n p} + \frac{\partial}{\partial\theta}T^{(k+1,V)}_{jj_1\cdots j_k ip}\theta_{,p} + \frac{\partial T^{(k+1,V)}_{jj_1\cdots j_k ip}}{\partial\theta_{,l}}\theta_{,lp}\right)\\
&\quad\quad \cdot v_{j,j_1\cdots j_k i} - q_i\theta_{,i}/\theta.
\end{aligned} \tag{4.11}$$

The values of $\psi, \eta, q, T^{(k,V)}$ and of their partial derivatives are evaluated at

$$(\sigma, \nabla\sigma, \dots, \nabla^{M-1}\sigma, \nabla v, \dots, \nabla^K v, \theta, \nabla\theta) \tag{4.12}$$

while the values of $T^{(k,E)}$ and their derivatives are evaluated at

$$(\sigma, \nabla\sigma, \dots, \nabla^{M-1}\sigma, \theta). \tag{4.13}$$

The above inequality must be satisfied in every thermodynamic process. In view of the remarks made at the beginning of this section this means that the gradients of velocity, gradients of acceleration, time derivatives of temperature and its gradients, and the gradients of temperature occurring in (4.11) may be chosen in a completely arbitrary way. Observing that the inequality contains the variables

$$\dot{v}_{j,j_1\cdots j_n},\ \dot{\theta}_{,i},\ \dot{\theta} \tag{4.14}$$

linearly, and using the independence of the variables (4.14) of the remaining variables in (4.11), we see that (4.11) implies

$$\frac{\partial\psi}{\partial v_{j,j_1\cdots j_n}}=0,\quad \frac{\partial\psi}{\partial\theta_{,l}}=0,\quad \eta=-\frac{\partial\psi}{\partial\theta}. \tag{4.15}$$

In particular, the function ψ is independent of the gradients of velocity and of the gradient of temperature,

$$\psi=\psi(\sigma,\nabla\sigma,\dots,\nabla^{M-1}\sigma,\theta). \tag{4.16}$$

The entropy relation (4.15), also implies that η is a function of the variables (4.13), and finally by (2.11) this completes the proof of assertion (1) of Theorem 4.2 and of (4.7).

Inequality (4.11) now simplifies in the following way: on the left-hand side the terms containing the derivatives of ψ with respect to θ, and gradients of v and θ may be omitted and on the right-hand side the term $\rho\eta\dot{\theta}$ may be omitted. Using the independence of the gradients of velocity and temperature of the variables (4.13), we see that the gradients $\nabla v,\nabla^2 v,\dots,\nabla\theta,\nabla^2\theta$, can be replaced systematically by $\alpha\nabla v,\alpha\nabla^2 v,\dots,\alpha\nabla\theta,\alpha\nabla^2\theta$ at a given point, where $\alpha>0$ is an arbitrary number. The inequality so obtained is written symbolically as follows:

$$\begin{aligned}
&-\rho\sum_{m=0}^{M-1}\frac{\partial\psi}{\partial\nabla^m\sigma}\alpha\nabla^{m+1}v\\
&\quad\le\sum_{k=0}^{N-1}\left(T^{(k,E)}+\sum_{m=0}^{M-1}\frac{\partial T^{(k+1,E)}}{\partial\nabla^m\sigma}\nabla^{m+1}\sigma+\frac{\partial T^{(k+1,E)}}{\partial\theta}\alpha\nabla\theta\right)\alpha\nabla^{k+1}v\\
&\quad+\sum_{k=0}^{N-1}\left(T^{(k,V)}+\sum_{m=0}^{M-1}\frac{\partial T^{(k+1,V)}}{\partial\nabla^m\sigma}\nabla^{m+1}\sigma+\sum_{n=1}^{K}\frac{\partial T^{(k+1,V)}}{\partial\nabla^n v}\alpha\nabla^{n+1}v\right.\\
&\qquad\left.+\frac{\partial T^{(k+1,V)}}{\partial\theta}\alpha\nabla\theta+\frac{\partial T^{(k+1,V)}}{\partial\nabla\theta}\alpha\nabla^2\theta\right)\alpha\nabla^{k+1}v-q_i\alpha\theta_{,i}/\theta,
\end{aligned} \tag{4.17}$$

where the derivatives of the viscous stresses and of q are evaluated at

$$(\nabla^r\sigma,\alpha\nabla^m v,\theta,\alpha\nabla\theta), \tag{4.18}$$

and the derivatives of ψ and of the equilibrium stresses are evaluated at (4.16). The tensorial indices are omitted from (4.17); their position is analogous to that of the corresponding terms in (4.11). We now divide (4.17) by α and let α tend to 0. The arguments of the derivatives of ψ and of the equilibrium parts of the stresses remain unchanged while the arguments of the viscous parts of the stresses, their derivatives, and the arguments of q tend to $(\nabla^r\sigma, 0, \theta, 0)$. The limiting procedure thus gives

$$-\rho\sum_{m=0}^{M-1}\frac{\partial\psi}{\partial\nabla^m\sigma}\nabla^{m+1}v \leq \sum_{k=0}^{N-1}\left(T^{(k,E)}+\sum_{m=0}^{M-1}\frac{\partial T^{(k+1,E)}}{\partial\nabla^m\sigma}\nabla^{m+1}\sigma\right)\cdot\nabla^{k+1}v$$
$$+\sum_{k=0}^{N-1}\left(T^{(k,V)}+\sum_{m=0}^{M-1}\frac{\partial T^{(k+1,V)}}{\partial\nabla^m\sigma}\nabla^{m+1}\sigma\right)\cdot\nabla^{k+1}v$$
$$-q\cdot\nabla\theta/\theta, \tag{4.19}$$

where the viscous stresses, their derivatives, and q are evaluated at $(\nabla^r\sigma, 0, \theta, 0)$. But for these values of arguments we have $T^{(k,V)} = 0$ by the very definition and hence, as a consequence, $\partial T^{(k,V)}/\partial\nabla^r\sigma = 0$ also. Furthermore, since the only term containing $\nabla\theta$ is the heat conduction term, we deduce from (4.19) that

$$q(\nabla^r\sigma, 0, \theta, 0) = 0. \tag{4.20}$$

All these facts further simplify (4.19) to

$$-\rho\sum_{m=0}^{M-1}\frac{\partial\psi}{\partial\nabla^m\sigma}\nabla^{m+1}v \leq \sum_{k=0}^{N-1}\left(T^{(k,E)}+\sum_{m=0}^{M-1}\frac{\partial T^{(k+1,E)}}{\partial\nabla^m\sigma}\nabla^{m+1}\sigma\right)\nabla^{k+1}v. \tag{4.21}$$

As this inequality is linear in the gradients of velocity, relations (4.8) follow. The proof of Theorem 4.2 is complete.

Obviously, the assertions of Theorem 4.2 imply assertion (1) of Theorem 4.1, i.e., the generalized Gibbs equation. The subtraction of the generalized Gibbs relation form the full dissipation inequality (4.11) gives the residual dissipation inequality. We have thus shown that the second law of thermodynamics implies assertions (1), (2) of Theorem 4.1. Conversely, the addition of the generalized Gibbs equation with the residual dissipation inequality gives the full dissipation inequality, i.e., assertions (1), (2) imply the second law. The proof is complete.

PROPOSITION 4.1. If a multipolar viscous fluid of type (N, M, K) satisfies the second law of thermodynamics and $k \geq \max\{M, K+1\}$, we have

$$\operatorname{Sym}(T^{(k)}_{jj_1\cdots j_k i} + T^{(k+1)}_{jj_1\cdots j_k ip,p}) = 0 \tag{4.22}$$

in every process. If $T^{(k)}_{jj_1\cdots j_k i}$ is symmetric in $j_1\cdots j_k i$ for $k \geq \max\{M, K+1\}$, then

$$N \leq K+1, \qquad N \leq M. \tag{4.23}$$

(4.22) says that the multipolar stresses $T^{(k)}$, with k exceeding the number of gradients of density and the number of gradients of velocity, essentially vanish, i.e., they vanish modulo the symmetrization indicated. These stresses vanish exactly if the mentioned symmetry holds; consequently, we have the inequalities (4.23). Thus we have essentially an upper bound for the number of nonvanishing stresses N in terms of the number of gradients of density and the number of gradients of velocity. For the specific model to be treated in the next two sections, we shall also give a lower bound for N in terms of K.

Proof. We first split the working term in the dissipation inequality (4.11) as follows:

$$\begin{aligned}\sum_{k=0}^{N-1}(T^{(k)}+\operatorname{div}T^{(k+1)})\cdot\nabla^{k+1}v \\ &= \sum_{k=0}^{N_0-1}(T^{(k+1)}+\operatorname{div}T^{(k+1)})\cdot\nabla^{k+1}v \\ &+ \sum_{k=N_0}^{N-1}(T^{(k)}+\operatorname{div}T^{(k+1)})\cdot\nabla^{k+1}v\,,\end{aligned}\tag{4.24}$$

where N_0 is defined by

$$N_0=\max\{M\,,\,K+1\}.\tag{4.25}$$

We now insert the split expression in the dissipation inequality (4.11) and examine the dependence of various terms of the resulting inequality on the gradients of velocity of different orders. We first consider all the terms except for the second sum on the right-hand side of (4.24). We have the following dependences: $\dot{\psi}$ depends on

$$\nabla v\,,\,\dots\,,\,\nabla^{M}v\,,\tag{4.26}$$

$T^{(k)}$ and its divergence depend on

$$\nabla v\,,\,\dots\,,\,\nabla^{K+1}v\,,\tag{4.27}$$

and the working terms in the first sum in (4.24) depend explicitly on

$$\nabla v\,,\,\dots\,,\,\nabla^{N_0}v.\tag{4.28}$$

We have $N_0\geq M$, $N_0\geq K+1$ and hence, except for the second term in (4.24), everything else depends on the variables in (4.28). On the other hand, the second term in (4.24) depends explicitly on

$$\nabla^{N_0+1}v\,,\,\nabla^{N_0+2}v\,,\,\dots\tag{4.29}$$

and the coefficients in front of these gradients are independent of (4.29). To summarize, the dissipation inequality depends on the variables (4.29) only through the second sum in (4.24), and the dependence is linear. Equation (4.22) then follows. The rest is obvious. The proof is complete.

We concludle this section with a proposition concerning the validity of the reduced equation of balance of angular momentum (2.7). Its simple proof is omitted.

PROPOSITION 4.2. If a multipolar viscous fluid satisfies the reduced equation of balance of angular momentum in every process compatible with the constitutive equations, then

$$\varepsilon_{ijk}\left(T_{kj}^{(0,E)}+T_{kjp,p}^{(1,E)}-\frac{\partial}{\partial\theta}T_{kjp}^{(1,E)}\theta_{,p}\right)=0\tag{4.30}$$

and

$$\varepsilon_{ijk}\left(T_{kj}^{(0,V)}+T_{kjp,p}^{(1,V)}+\frac{\partial}{\partial\theta}T_{kjp}^{(1,E)}\theta_{,p}\right)=0\tag{4.31}$$

in every process.

5. Linear viscous fluids. Starting from this section we restrict ourselves to the case when only the density and not its gradients enter the constitutive equations, i.e., $M = 1$, and to the case when the viscous stresses and the heat flux vector depend linearly on the gradients of velocity and the gradient of temperature.

PROPOSITION 5.1. If a multipolar viscous fluid of type $(N, 1, K)$ satisfies the second law of thermodynamics and the principle of material frame-indifference, then

$$T_{ij}^{(0,E)} = -p\delta_{ij}, \tag{5.1}$$

where $p = p(\rho, \theta)$ is given by

$$p = \rho^2 \frac{\partial \psi}{\partial \rho}; \tag{5.2}$$

if k is odd, then

$$T^{(k,E)} = 0; \tag{5.3}$$

if k is even and $k > 0$, then

$$T^{(k,E)} = T^{(k,E)}(\theta) \tag{5.4}$$

and

$$\operatorname{Sym} T_{jj_1\cdots j_k i}^{(k,E)} = 0, \tag{5.5}$$

where Sym denotes the symmetrization with respect to the indices $j_1 \cdots j_k i$.

Proof. We first note that the transformation law (3.15) for the total multipolar stress implies analogous laws for the equilibrium and viscous parts of the stress. Hence

$$T_{j_1\cdots j_{k+1}}^{(k,E)}(\rho, \theta) = Q_{j_1 m_1} \cdots Q_{j_{k+2} m_{k+2}} T_{m_1\cdots m_{k+2}}^{(k,E)}(\rho, \theta) \tag{5.6}$$

for every orthogonal tensor Q. Consequently, $T^{(k,E)}$ is an isotropic tensor of order $k+2$. If k is odd, then the only isotropic tensor of order $k+2$ is 0 (it is enough to set $Q_{ij} = -\delta_{ij}$ in (5.6)). This proves (5.3). Formulas (5.1) and (5.2) are a combination of the generalized stress relation (4.8) for $k = 0$ with the fact that $T^{(1,E)} = 0$ by (5.3). If k is even and $k > 0$, then the generalized stress relation (4.8) with k replaced by $k-1$ gives

$$\operatorname{Sym}\left(T_{jj_1\cdots j_{k-1}i}^{(k-1,E)} + \frac{\partial}{\partial \rho} T_{jj_1\cdots j_{k-1}ip}^{(k,E)} \rho_{,p}\right) = 0. \tag{5.7}$$

But since $k-1$ is odd, the first term in (5.7) vanishes by (5.3); moreover, the symmetrization is irrelevant as $T_{jj_1\cdots j_{k-1}ip}^{(k,E)}$ is symmetric in $j_1 \cdots j_{k-1} i$. Thus

$$\frac{\partial}{\partial \rho} T_{jj_1\cdots j_{k-1}ip}^{(k,E)} \rho_{,p} = 0, \tag{5.8}$$

which implies

$$\frac{\partial}{\partial \rho} T_{jj_1\cdots j_{k-1}ip}^{(k,E)} = 0 \tag{5.9}$$

and hence also (5.4). Finally, combining the generalized stress relation (4.8) with the fact that $T^{(k+1,E)} = 0$ for k even, we obtain (5.5). The proof is complete.

The analysis of the viscous part of stresses will be given only for linear fluids. A multipolar viscous fluid of type (N, M, K) is said to be linear if $M = 1$ and if for every ρ, θ the quantities

$$T^{(k,V)}(\rho, \nabla v, \dots, \nabla^K v, \theta, \nabla\theta) \tag{5.10}$$

and

$$q(\rho, \nabla v, \dots, \nabla^K v, \theta, \nabla\theta) \tag{5.11}$$

depend linearly on $\nabla v, \dots, \nabla^K v, \nabla\theta$.

PROPOSITION 5.2. If a linear viscous fluid of type $(N, 1, K)$ satisfies the principle of material frame-indifference, then

(1) $T^{(k,V)}$ with k even depends on ρ, θ, and on the odd-order gradients of v;

$$T^{(k,V)} = T^{(k,V)}(\rho, \theta, \nabla v, \nabla^3 v, \dots, \nabla^L v); \tag{5.12}$$

(2) q and $T^{(k,V)}$ with k odd depend on ρ, θ, and on the even-order spatial gradients of v:

$$q = q(\rho, \theta, \nabla^2 v, \dots, \nabla^{L-1} v, \nabla\theta), \tag{5.13}$$

$$T^{(k,V)} = T^{(k,V)}(\rho, \theta, \nabla^2 v, \dots, \nabla^{L-1} v, \nabla\theta); \tag{5.14}$$

here L is a suitable odd number depending on K.

Proof. By linearity, the expressions for q and $T^{(k,V)}$ are sums of a number of terms depending linearly on the gradients of velocity of different orders and of a term depending linearly on $\nabla\theta$. The transformation laws for the stresses and the heat flux vector under changes of frame must be satisfied by each of these linear terms separately. Using these transformation laws with $Q_{ij} = -\delta_{ij}$ and counting the number of rotations in these transformation laws for each of the linear terms, one readily learns that only the terms with the orders of gradients of velocity indicated in assertions (1), (2) can be nonzero. The details are omitted.

THEOREM 5.1. We have the following expressions for $T^{(0,V)}$, $T^{(1,V)}$, and q in a linear viscous fluid satisfying the principle of material frame-indifference:

$$\begin{aligned} T_{ij}^{(0,V)} &= \lambda v_{k,k}\delta_{ij} + \mu(v_{i,j} + v_{j,i}) \\ &\quad + \sum_{r=0}^{P}(\alpha^{(r)}\delta_{ij}\Delta^{r+1}v_{k,k} + \beta_1^{(r)}\Delta^{r+1}v_{i,j} \\ &\quad + \beta_2^{(r)}\Delta^{r+1}v_{j,i} + \gamma^{(r)}\Delta^r v_{k,kij}), \end{aligned} \tag{5.15}$$

$$
\begin{aligned}
T_{ijk}^{(1,V)} = \sum_{r=0}^{P}(& c_1^{(r)}\delta_{ij}\Delta^{r+1}v_k + c_2^{(r)}\delta_{ij}\Delta^r v_{n,nk} \\
&+ c_3^{(r)}\delta_{ik}\Delta^{r+1}v_j + c_4^{(r)}\delta_{ik}\Delta^r v_{m,mj} \\
&+ c_5^{(r)}\delta_{jk}\Delta^{r+1}v_i + c_6^{(r)}\delta_{jk}\Delta^r v_{m,mi} \\
&+ c_7^{(r)}\Delta^r v_{i,jk} + c_8^{(r)}\Delta^r v_{k,ij} \\
&+ c_9^{(r)}\Delta^r v_{j,ki} + c_{10}^{(r)}\Delta^{r-1}v_{m,mijk}) \\
&+ c_{11}\delta_{ij}\theta_{,k} + c_{12}\delta_{ik}\theta_{,j} + c_{13}\delta_{jk}\theta_{,i},
\end{aligned}
\tag{5.16}
$$

$$
q_i = \sum_{r=0}^{P}(d_1^{(r)}\Delta^r v_{m,mi} + d_2^{(r)}\Delta^{r+1}v_i) - k\theta_{,i}. \tag{5.17}
$$

Here λ, μ, $\alpha^{(r)}$, $\beta_1^{(r)}$, $\beta_2^{(r)}$, $\gamma^{(r)}$, $c_1^{(r)}$, ..., $c_{13}^{(r)}$, $d_1^{(r)}$, $d_2^{(r)}$, and k are scalar functions of ρ, θ such that $c_{10}^{(r)}$ satisfies $c_{10}^{(0)} = 0$ and P is an appropriate integer determined by K. If the body satisfies the reduced equation of balance of angular momentum, then

$$
\beta_1^{(r)} + c_5^{(r)} + c_7^{(r)} = \beta_2^{(r)} + c_3^{(r)} + c_9^{(r)} \tag{5.18}
$$

for $r = 0, 1, \ldots, P$.

Similar but more complicated expressions can be obtained also for $T^{(k,V)}$ with $k \geq 2$.

Proof. In view of the transformation laws for $T^{(k)}$ and q under changes of frame, the coefficients in the expressions for q and $T^{(k)}$ in front of the gradients of velocity and the gradient of temperature must be isotropic tensors. Using the general forms of isotropic tensors (see Spencer [13]) and the symmetry of the gradients of v, one eventually arrives at (5.15)–(5.17). The details are omitted.

The following proposition shows that for linear viscous materials for which the scalar coefficients in the linear expressions for $T^{(k,V)}$ and q are independent of ρ, θ, the residual dissipation inequality (5.5) splits into two independent inequalities. For simplicity we make the assumption that

$$
T^{(k,E)} = 0 \tag{5.19}
$$

for $k \geq 1$, which, at least for k odd, is a consequence of the second law (cf. (5.3)–(5.5)).

Proposition 5.3. Consider a linear viscous fluid such that $T^{(k,V)}$ and q are independent of ρ, θ and (5.19) holds identically. If the fluid satisfies the principle of material frame-indifference and the second law, then the following two inequalities hold:

$$\sum_{\substack{k=0\\k\text{ even}}}^{N-1}(T^{(k,V)}_{jj_1\cdots j_ki}+T^{(k+1,V)}_{jj_1\cdots j_kip,p})v_{j,j_1\cdots j_ki}\geq 0,\tag{5.20}$$

$$\sum_{\substack{k=1\\k\text{ odd}}}^{N-1}(T^{(k,V)}_{jj_1\cdots j_ki}+T^{(k+1,V)}_{jj_1\cdots j_kip,p})v_{j,j_1\cdots j_ki}-q_i\theta_{,i}/\theta\geq 0.\tag{5.21}$$

The expressions for q, $T^{(k,V)}$, and its divergences are linear in the gradients of velocity and in the gradients of temperature. When these linear expressions are inserted in (5.20), (5.21), these inequalities express nonnegativeness of quadratic forms. Sylvester's criterion can be used to find the inequalities which the coefficients of the linear expressions for $T^{(k,V)}$ and q must satisfy in order that (5.20), (5.21) be guaranteed. We do not pursue this possibility here.

Proof. Under the hypothesis of the proposition, the residual dissipation inequality (5.5) can be written as the sum of the left-hand sides of (5.20) and (5.21). Now the left-hand side of (5.20) depends on

$$\nabla v\,,\ \nabla^3 v\,,\ \ldots\,,\ \nabla^{L+1}v\,,\ \nabla^2\theta\,,\tag{5.22}$$

while the left-hand side of (5.21) depends on

$$\nabla^2 v\,,\ \nabla^4 v\,,\ \ldots\,,\ \nabla\theta\tag{5.23}$$

(cf. Proposition 5.2). As the sets (5.22), (5.23) are disjoint and the expressions are quadratic, the splitting of the residual inequality into (5.20), (5.21) follows. The proof is complete.

We are now going to show that for linear materials the Clausius-Duhem inequality restricts the dependence of the viscous stresses $T^{(k)}$ on the gradients of order larger than $2N-1-k$ very strongly. The effect of the restriction will be shown to be so severe that the terms with the gradients of velocity of order larger than $2N-1-k$ do not contribute to the equations of balance of energy and linear momentum. (See Propositions 5.4 and 5.5 below for precise statements.)

We introduce additional notation to state the results. We shall deal exclusively with linear fluids for which the viscous stresses and the heat flux do not depend on ρ, θ. We split the viscous stresses and the heat flux into the regular and singular parts, denoted $T^{(k,R)}$, $T^{(k,S)}$, $q^{(R)}$, $q^{(S)}$ as follows:

$$\begin{aligned}T^{(k,V)}&=T^{(k,R)}(\nabla v\,,\ \nabla^2 v\,,\ \ldots\,,\ \nabla^{2N-1-k}v\,,\ \nabla\theta)\\&\quad+T^{(k,S)}(\nabla^{2N-k}v\,,\ \nabla^{2N-k+1}v\,,\ \ldots),\end{aligned}\tag{5.24}$$

$k=0\,,\,1\,,\,\ldots\,,\,N-1\,,$

$$\begin{aligned}q&=q^{(R)}(\nabla v\,,\ \nabla^2 v\,,\ \ldots\,,\ \nabla^N v\,,\ \nabla\theta)\\&\quad+q^{(S)}(\nabla^{N+1}v\,,\ \nabla^{N+2}v\,,\ \ldots),\end{aligned}\tag{5.25}$$

so that the lower the order of the stress tensor, the bigger the number of the gradients of velocity in the regular part $T^{(k,R)}$ of the stress. In particular, the highest order regular stress $T^{(N-1,R)}$ depends on the gradients of velocity up to the order N, while the lowest order regular stress $T^{(0,R)}$ depends on the gradients of velocity up to the order $2N-1$. The singular part of the stresses depends on the complementary set of variables. The singular parts of the stresses will eventually turn out to be negligible from the point of view of the balance equations. Also note that in view of Proposition 5.2 the viscous stresses and q depend either on only the even-order gradients or on only the odd-order gradients, but this is irrelevant here.

PROPOSITION 5.4. Consider a linear viscous fluid such that $T^{(k,V)}$ and q are independent of ρ, θ and (5.19) holds identically. If the fluid satisfies the second law of thermodynamics, then the following three assertions hold:

(1) The inequality

$$\sum_{k=0}^{N-1}(T^{(k,R)}_{jj_1\cdots j_k i} + T^{(k+1,R)}_{jj_1\cdots j_k ip,p})v_{j,j_1\cdots j_k i} - q_i^{(R)}\theta_{,i}/\theta \geq 0 \tag{5.26}$$

holds in every process.

(2) We have the following relations:

$$q_i^{(S)} = 0, \tag{5.27}$$

$$\mathrm{Sym}(T^{(N-1,S)}_{jj_1\cdots j_{N-1}i}) = 0, \tag{5.28}$$

$$\mathrm{Sym}(T^{(k,S)}_{jj_1\cdots j_k i} + T^{(k+1,S)}_{jj_1\cdots j_k ip,p}) = 0, \tag{5.29}$$

$k = 0, 1, \ldots, N-2$.

(3) If $T^{(k,V)}_{jj_1\cdots j_k i}$ is symmetric in $j_1\cdots j_k i$ for every k with $1 \leq k \leq N-1$, then

$$K \leq 2N-1. \tag{5.30}$$

Assertion (1) says that the regular parts of the stresses satisfy the residual dissipation inequality similar to that satisfied by the total viscous stresses. Equations (5.27) to (5.29) of assertion (2) say that the singular part of the heat flux vanishes and that also the singular stresses vanish modulo the symmetrization. The singular stresses vanish exactly if the stresses are symmetric in the indicated indices and hence the fluid satisfies inequality (5.30). Recall, on the other hand, that we also have inequality (4.23) of Proposition 4.1. We thus have a lower and an upper bound for N in terms of K for linear viscous fluids.

Proof. Consider first the residual dissipation inequality with the following choice of the arguments:

$$\nabla v = \nabla^2 v = \cdots = \nabla^N v = 0, \tag{5.31}$$

$$\nabla^{N+1}v, \nabla^{N+2}v, \ldots, \nabla\theta \quad \text{arbitrary}. \tag{5.32}$$

The inequality then reduces to

$$(q_i^{(R)} + q_i^{(S)})\theta_{,i} \leq 0, \tag{5.33}$$

where $q^{(R)}$ is evaluated at the velocity gradients indicated in (5.32). The linearity of the dependence of $q^{(S)}$ on the arguments $\nabla^{N+1}v, \dots$ then gives

$$q_i^{(S)}\theta_{,i} \le 0. \tag{5.34}$$

Since $\nabla\theta$ is arbitrary and $q^{(S)}$ is independent of it, equation (5.27) follows.

Next consider the residual dissipation inequality with the following choice of the arguments: let $k \in \{0, 1, \dots, N-1\}$ be given and set

$$\nabla^m v = 0 \quad \text{if } 1 \le m \le N,\ m \ne k+1, \tag{5.35}$$

$$\nabla^m v \text{ arbitrary if } m > N \text{ or if } m = k+1, \tag{5.36}$$

$$\nabla\theta = 0. \tag{5.37}$$

The residual inequality then reduces to

$$(T^{(k,R)} + \operatorname{div} T^{(k+1,R)} + T^{(k,S)} + \operatorname{div} T^{(k+1,S)})\nabla^{k+1}v \ge 0, \tag{5.38}$$

where $T^{(k,R)}$ and $\operatorname{div} T^{(k+1,R)}$ are evaluated at the arguments

$$(0, 0, \dots, \nabla^{k+1}v, 0, \dots, \nabla^{N+1}v, \dots, \nabla^{2N-1-k}v, \nabla\theta), \tag{5.39}$$

while $T^{(k,S)}$ and $\operatorname{div} T^{(k+1,S)}$ are evaluated at the arguments

$$(\nabla^{2N-k}v, \nabla^{2N-k+1}v, \dots). \tag{5.40}$$

These can have arbitrary values and in view of (5.24) they do not enter the regular parts of the stresses and their divergences occurring in (5.38). The linearity of the dependences on the variables (5.40) together with $T^{(N,S)} = 0$ (cf. (2.9)) gives (5.28) and (5.29). This proves assertion (2).

Assertion (1) now follows from the residual dissipation inequality and the relations of assertion (2). Also assertion (3) is an obvious consequence of assertion (2). The proof is complete.

Proposition 5.5. Consider a linear viscous fluid of type $(N, 1, K)$ such that $T^{(k,V)}$ and q are independent of ρ, θ and (5.19) holds identically. If $K > 2N-1$ and if the fluid satisfies the second law of thermodynamics, then there exists another linear viscous fluid satisfying the second law of thermodynamics such that it is of type $(N, 1, K')$, $K' \le 2N-1$, and the equations of balance of linear momentum and energy of the two fluids are identical. (This means that when the constitutive functions of the two fluids are inserted in the balance equations, then the resulting differential equations for the unknown functions $v(x, t)$ and $\theta(x, t)$ are the same for the two fluids.) The new fluid is defined as follows: its constitutive functions $\psi, \eta, e, q, T^{(k,E)}$ are identical with those of the original fluid, but its constitutive functions for the viscous stresses are the regular parts of the viscous stresses of the original fluid. Hence the singular parts of the stresses of the new fluid vanish.

If the boundary conditions of the fluid are formulated in terms of the stresses, they may be different for the two fluids.

Proof. It is clear from (5.27)–(5.29) that the new fluid has the power of the viscous stresses identical to that of the original fluid, i.e.,

$$\sum_{k=0}^{N-1}(T^{(k,R)}+\operatorname{div}T^{(k+1,R)})\cdot\nabla^{k+1}v=\sum_{k=0}^{N-1}(T^{(k,V)}+\operatorname{div}T^{(k+1,V)})\cdot\nabla^{k+1}v. \tag{5.41}$$

Since the other quantities of the two fluids are identical by the definition, it is then clear that the reduced equation of balance of energy and the dissipation inequalities of the two fluids are identical. Hence, in particular, the new fluid satisfies the second law of thermodynamics.

We shall now prove that the equations of balance of linear momentum of the two fluids coincide. These two equations can differ only in the term $T^{(0,S)}_{jp,p}$ and our task is to prove that

$$T^{(0,S)}_{jp,p}=0. \tag{5.42}$$

But using (5.29) with $k=0,1,\dots,N-2$, this term can be rewritten in the following way:

$$\begin{aligned}T^{(0,S)}_{jj_1,j_1}&=-(T^{(1,S)}_{jj_1j_2,j_2})_{,j_1}=-(T^{(1,S)}_{jj_1j_2})_{,j_1j_2}\\&=+(T^{(2,S)}_{jj_1j_2j_3,j_3})_{,j_1j_2}=+T^{(2,S)}_{jj_1j_2j_3,j_1j_2j_3}\\&=\cdots=(-1)^{N-1}T^{(N-1,S)}_{jj_1\cdots j_N,j_1\cdots j_N}.\end{aligned} \tag{5.43}$$

We have omitted the symmetrization symbols in the above computation because of the interchangeability of the partial derivatives occurring above. To complete the proof of (5.42), it is enough to note that the last term in (5.43) vanishes in view of (5.28). Hence the equations of balance of linear momentum of the two fluids coincide. As we have already proved, the reduced equations of balance of energy also coincide. Since the full equation of balance of energy is a consequence of these two, we see that the full equations of balance of energy of the two fluids also coincide. The proof is complete.

6. Dipolar fluids. Dipolar materials provide the lowest order example ($N=2$) of a genuine multipolar material. We shall restrict here to linear dipolar fluids. It is assumed that the fluid satisfies the principle of material frame-indifference and the second law. In addition to the notation introduced in (5.24) and (5.25) concerning the regular and singular parts of the stresses, we also write

$$T^{(0,R)}=T^{(0,1)}(\nabla v)+T^{(0,3)}(\nabla^3 v), \tag{6.1}$$

$$T^{(1,R)}=T^{(1,2)}(\nabla^2 v)+T^{(1,\theta)}(\nabla\theta), \tag{6.2}$$

$$q^{(R)}=q^{(\theta)}(\nabla\theta)+q^{(2)}(\nabla^2 v), \tag{6.3}$$

to split the dependences of the regular parts of the viscous stresses on the gradients of velocity of different orders.

THEOREM 6.1. If a dipolar linear fluid of type $(2,1,K)$ satisfies the principle of material frame-indifference and the second law of thermodynamics and if $T^{(0,V)}$

and $T^{(1,V)}$ are independent of ρ and θ, then

$$T_{ij}^{(0,1)} v_{i,j} \geq 0, \tag{6.4}$$

$$T_{ij}^{(0,3)} = -T_{ijp,p}^{(1,2)}, \tag{6.5}$$

$$T_{ijp,p}^{(1,\theta)} = 0, \tag{6.6}$$

$$(T_{ijk}^{(1,2)} + T_{ijk}^{(1,\theta)}) v_{i,jk} + (q_i^{(2)} + q_i^{(\theta)})\theta_{,i}/\theta \geq 0, \tag{6.7}$$

$$q_i^{(S)} = 0, \tag{6.8}$$

$$T_{ijk}^{(1,S)} = -T_{ikj}^{(1,S)}, \tag{6.9}$$

$$T_{ij}^{(0,S)} = -T_{ijp,p}^{(1,S)}. \tag{6.10}$$

The proof of this proposition is similar to that of Proposition 5.4 and is omitted.

Propositions 5.4(3) and 5.5 show that for dipolar fluids the response depends most typically on the gradients of velocity up to order 3 (i.e., $K = 3$) and we conclude this section with explicit expressions for the viscous stresses and the heat flux in materials of type $(2, 1, 3)$.

PROPOSITION 6.1. If a linear viscous fluid of type $(2, 1, 3)$ satisfies the principle of material frame-indifference and the second law of thermodynamics, then

$$\begin{aligned} T_{ij}^{(0,V)} &= \lambda v_{k,k}\delta_{ij} + \mu(v_{i,j} + v_{j,i}) \\ &\quad + \alpha \Delta v_{k,k}\delta_{ij} + \beta_1 \Delta_{i,j} + \beta_2 \Delta v_{j,i} + \gamma v_{k,kij}, \end{aligned} \tag{6.11}$$

$$\begin{aligned} T_{ijk}^{(1,V)} &= c_1\delta_{ij}\Delta v_k + c_2\delta_{ij} v_{m,mk} \\ &\quad + c_3\delta_{ik}\Delta v_j + c_4\delta_{ik} v_{m,mj} \\ &\quad + c_5\delta_{jk}\Delta v_i + c_6\delta_{jk} v_{m,mi} \\ &\quad + c_7 v_{i,jk} + c_8 v_{k,ij} + c_9 v_{j,ki} \\ &\quad + c_{10}\delta_{ij}\theta_{,k} + c_{11}\delta_{ik}\theta_{,j} + c_{12}\delta_{jk}\theta_{,i}, \end{aligned} \tag{6.12}$$

and

$$q_i = -k\theta_{,i} + d_1 v_{m,mi} + d_2 \Delta v_i, \tag{6.13}$$

where $\lambda, \mu, \alpha, \beta_1, \beta_2, \gamma, c_1, \ldots, c_{12}, k, d_1, d_2$ are real-valued functions of ρ and θ. If these functions are independent of ρ, θ, then the following relations hold:

$$\alpha + c_1 + c_2 = 0, \tag{6.14}$$

$$\beta_1 + c_5 + c_7 = 0, \tag{6.15}$$

$$\beta_2 + c_3 + c_9 = 0, \tag{6.16}$$

$$\gamma + c_4 + c_6 + c_8 = 0, \tag{6.17}$$

$$c_{10} = c_{11} + c_{12} = 0. \tag{6.18}$$

If the body satisfies the reduced equation of balance of angular momentum, then

$$\beta_1 + c_5 + c_7 = \beta_2 + c_3 + c_9. \tag{6.19}$$

Proof. Formulas (6.11)–(6.13) and (6.19) are just specializations of the corresponding general formulas of Theorem 5.1. Equations (6.14)–(6.18) follow from Eqs. (6.5), (6.6) of Theorem 6.1. The details are omitted.

Acknowledgment. The authors wish to express their thanks to M. Růžička for many valuable discussions on a previous draft of the manuscript, and to A. Novotný for pointing out an error in the draft.

References

[1] J. Nečas, A. Novotný, and M. Šilhavý, *Global solution to the ideal compressible heat conductive multipolar fluid*, Comment. Math. Univ. Carolinae **30**, 551–564 (1989)

[2] J. Nečas and M. Růžička, *Global solution to the incompressible viscous-multipolar material*, to appear

[3] A. E. Green and R. S. Rivlin, *Simple force and stress multipoles*, Arch. Rational Mech. Anal. **16**, 325–353 (1964)

[4] A. E. Green and R. S. Rivlin, *Multipolar continuum mechanics*, Arch. Rational Mech. Anal. **17**, 113–147 (1964)

[5] K. Bucháček, *Thermodynamics of monopolar continuum of grade n*, Apl. Mat. **16**, 370–383 (1971)

[6] W. Noll, *A mathematical theory of behavior of continuous media*, Arch. Rational Mech. Anal. **2**, 119–226 (1957)

[7] J. J. Cross, *Mixtures of fluids and isotropic solids*, Arch. Mech. **25**, 1025–1039 (1973)

[8] I. Samohýl, *Symmetry groups in the mass conserving, second grade materials*, Arch. Mech. **33**, 983–987 (1981)

[9] M. E. Gurtin, M. Vianello, and W. O. Williams, *On fluids of grade n*, Meccanica—J. Ital. Assoc. Theoret. Appl. Mech. **21**, 179–181 (1986)

[10] C. Truesdell and W. Noll, *The Non-linear Field Theories of Mechanics*, Handbuch der Physik III/3 (ed. by S. Flügge), Springer, Berlin-Heidelberg-New York, 1965

[11] C. Truesdell, *A First Course in Rational Continuum Mechanics*, The Johns Hopkins University, Baltimore, 1972

[12] B. D. Coleman and W. Noll, *The thermodynamics of elastic materials with heat conduction and viscosity*, Arch. Rational Mech. Anal. **13**, 167–178 (1963)

[13] A. J. M. Spencer, *Theory of invariants*, Continuum Physics, vol. I (ed. by A. C. Eringen), Academic Press, New York, 1971

COMMUN. IN PARTIAL DIFFERENTIAL EQUATIONS, 19(11&12), 1763-1803 (1994)

YOUNG MEASURE-VALUED SOLUTIONS FOR NON-NEWTONIAN INCOMPRESSIBLE FLUIDS[1]

Hamid Bellout[2] and Frederick Bloom

Department of Mathematical Sciences
Northern Illinois University
DeKalb, IL 60115

Jindřich Nečas[3]

Mathematical Institute
Sokolovska 83
Charles University
Prague, Czechoslovakia

Abstract

For the model of a nonlinear bipolar fluid, in which the highest order viscosity vanishes, and the viscous part of the stress tensor satisfies a growth condition of the form $|\tau_{ij}(\mathbf{e})| \leq C(1+|\mathbf{e}|)^{p-1}$, $C > 0$, $\mathbf{e}$ the rate of strain tensor, we demonstrate the existence of Young-measure valued solutions for $p > 1$ (in dim $n = 2$) and for $p > 6/5$ (in dim $n = 3$); these solutions are proven to be weak solutions for $3/2 < p < 2$ (in dim $n = 2$) and for $9/5 < p < 11/5$ (in dim $n = 3$) and unique regular weak solutions for $p \geq 2$

[1]Research supported, in part, by ONR Grant N00014-91-J-1002

[2]e-mail address: bellout@math.niu.edu

[3]e-mail address: necas@math.niu.edu

Copyright © 1994 by Marcel Dekker, Inc.

© Taylor & Francis 1994. Reprinted with kind permission from: Bellout, H., Bloom, F. & Nečas, J., Young Measure-Valued Solutions for non-Newtonian Incompressible Fluids, Communications in Partial Differential Equations, 19, 1763-1803.

(in dim $n = 2$) and for $p \geq 11/5$ (in dim $n = 3$). Much of the analysis deals with the associated space periodic problems.

1 Introduction

The standard Navier-Stokes equations for an incompressible viscous fluid are based on the relations expressing balance of momentum

$$\rho \frac{\partial v_i}{\partial t} + \rho v_j \frac{\partial v_i}{\partial x_j} = \frac{\partial \tau_{ij}}{\partial x_j} + \rho b_i, \tag{1.1}$$

$i = 1, 2, 3$ (summation on repeated indices being understood) and the condition

$$\frac{\partial v_i}{\partial x_i} = 0 \tag{1.2}$$

expressing the incompressible nature of the fluid. In (1.1), (1.2), $\boldsymbol{v}$ is the velocity field of the fluid, $\rho = \text{const.} > 0$ is the density, $\boldsymbol{b}$ the specific body force vector, and τ_{ij} the components of the stress tensor.

The stress tensor $\boldsymbol{\tau}$ in a viscous fluid is decomposed in the usual way as

$$\tau_{ij} = -p\,\delta_{ij} + (\tau^v)_{ij} \tag{1.3}$$

where p is the pressure, δ_{ij} the Kronecker delta, and $\boldsymbol{\tau}^{\boldsymbol{v}}$ the viscous part of the stress which must be defined by a set of constitutive relations. The Navier-Stokes relations are based on the Stokes' law which has the form

$$\tau_{ij}^v \equiv (\tau_0^v)_{ij} = 2\,\mu_0\, e_{ij} \tag{1.4}$$

with μ_0 (the viscosity) being a positive constant and

$$e_{ij} = \frac{1}{2}\left(\frac{\partial v_i}{\partial x_j} + \frac{\partial v_j}{\partial x_i}\right) \tag{1.5}$$

the components of the rate of deformation tensor; the zero subscript in $(\tau_0^v)_{ij}$ is inserted to indicate that this part of the viscous stress (which is the only part in the standard Stokes' law) depends only on first order derivatives of the velocity field and, indeed, for the usual Stokes' law this dependence is given by the linear relation (1.4).

In this paper we will consider a more general relation for $\boldsymbol{\tau}_0^v$ of the form

$$\boldsymbol{\tau}_0^v = \boldsymbol{\tau}_0^v(\boldsymbol{e}) \tag{1.6}$$

with $\boldsymbol{\tau}_0^v$ having continuous components $(\boldsymbol{\tau}_0^v)_{ij}$ that satisfy the polynomial growth condition

$$|(\tau_0^v)_{ij}(\boldsymbol{e})| \leq C(1 + |\boldsymbol{e}|)^{p-1}\,;\; C > 0 \tag{1.7}$$

with $1 < p < \infty$ in space dimension $n = 2$ and $1 < p < 6$ for $n = 3$; the Stokes' law conforms to (1.7) with $p = 2$. We will also assume the coercivity condition

$$(\tau_0^v)_{ij}(\boldsymbol{e})\, e_{ij} \geq 0 \tag{1.8}$$

but in several instances will require the stronger condition

$$(\tau_0^v)_{ij}(\boldsymbol{e})\, e_{ij} \geq c_1 |\boldsymbol{e}|^{p_1} \tag{1.9}$$

with $p \geq p_1 > p-1$, e.g., (1.9) will be assumed when we consider measure-valued solutions for (1.1), (1.2), (1.3), with $\boldsymbol{\tau}^v = \boldsymbol{\tau}_0^v$, and $\boldsymbol{\tau}_0^v$ subject to (1.7), and will also be employed in § 2.

Models of viscous fluids conforming to (1.1), (1.2), (1.3), with $\boldsymbol{\tau}^v = \boldsymbol{\tau}_0^v$ and $\boldsymbol{\tau}_0^v$ satisfying a growth condition of the form (1.7), with p not necessarily equal to 2, have appeared in several instances in the literature: Ladyzhenskaya [18], [19], Lions [21] and, more recently, Du and Gunzburger [11] have considered the nonlinear version of (1.4) in which

$$\tau_{ij}^v = (\tau_0^v)_{ij} \equiv 2(\mu_0 + \mu_0' |\nabla \boldsymbol{v}|^r)\,\partial_j v_i, \quad r > 0 \tag{1.10}$$

so that the growth condition (1.7) applies with $p = 2 + r$. The paper [11] contains a very lucid account of the rationale for considering viscous fluid models exhibiting nonlinear viscosity; other work in this direction includes the analysis of Kaniel [17]. The recently developed theory of nonlinear multipolar viscous fluids (Nečas, Novotny, and Silhavy [27], Bellout, Bloom and Nečas [3]) is based on the work of Nečas and Silhavy [28], and has its roots in earlier work of Green and Rivlin [15], [16] and Bleustein and Green [6] on multipolar continuum theories for solids and fluids; for a nonlinear bipolar viscous fluid the constitutive relations considered in [3] have the form

$$\tau_{ij} = -p\,\delta_{ij} + 2\,\mu_0\,(\varepsilon + |\boldsymbol{e}|^2)^{-\alpha/2}\,e_{ij} - 2\,\mu_1\,\Delta\,e_{ij} \qquad (1.11)$$

$$\tau_{ijk} = 2\,\mu_1\,\frac{\partial e_{ij}}{\partial x_k} \qquad (1.12)$$

where $\varepsilon > 0$, $\mu_1 > 0$ and the τ_{ijk} are the components of the first multipolar stress tensor; a semi-inverse argument based on the study of plane Poiseuille flow in [3] leads to the restriction: $0 \leq \alpha < 1$. The Stokes' relation (1.4) is that special case of (1.11), (1.12) in which $\alpha = \mu_1 = 0$.

The theory of multipolar viscous fluids, as developed in [3], [28] is compatible with the basic principles of thermodynamics, including the Clausius-Duhem inequality, as well as with the principle of material frame indifference; results obtained to date indicate that the theory may lead to a better understanding of the phenomena of turbulence [4] and may provide, in many practical situations, an alternative [3], [2] to the standard boundary-layer theory employed in the Stokes' model for laminar flows at large Reynolds numbers. The existence of unique weak solutions for initial-boundary value problems associated with the constitutive theory (1.11), (1.12) has been demonstrated, recently, in [5] and will be elaborated upon, briefly, in § 2.

From (1.11), we see that

$$(\tau_0^v)_{ij} = 2\,\mu_0\,(\varepsilon + |\boldsymbol{e}|^2)^{-\alpha/2}\,e_{ij}, \quad 0 \leq \alpha < 1, \qquad (1.13)$$

so that (1.7) is satisfied with $p = 2 - \alpha$, and (1.8) is trivially satisfied. It is worth noting here, in terms of relating the models defined by (1.10), (1.13), and the growth condition (1.7), that $p = 2 - \alpha = r + 1$, and that for $0 \leq \alpha < 1$, $-1 < r \leq 0$ and $1 < p \leq 2$. In [18] and [11] only the range $r \geq 0$ was considered, with $r = 0$ corresponding to the Stokes' model. In this paper we will prove the existence of measure-valued solutions to (1.1)–(1.3), with $\boldsymbol{\tau}^v = \boldsymbol{\tau}_0^v$ satisfying the growth condition (1.7), by studying the relevant initial-boundary value problem associated with the more general case of (1.11), (1.12) in which we retain the constitutive relation (1.12) for the higher order stress but allow, in lieu of (1.11) the more general relation

$$\tau_{ij}^v = (\tau_0^v)_{ij} - 2\,\mu_1\,\Delta\,e_{ij} \qquad (1.14)$$

with $(\tau_0^v)_{ij}(\boldsymbol{e})$ satisfying (1.7); the measure-valued solutions will be obtained by letting the higher order viscosity $\mu_1 \to 0^+$; in this case (1.9) is needed. For the problem governed by (1.11), (1.12)—or (1.14), (1.12)—the only influence of (1.12) is with respect to the boundary conditions, i.e., aside from the usual requirement that $\boldsymbol{v} = \boldsymbol{0}$ on $\partial\Omega$, $\Omega \subseteq \mathbb{R}^n$ a bounded domain, it is easily seen, as a consequence, for example, of a principle of virtual work [37], that we must also have

$$\tau_{ljk}\, \nu_j\, \nu_k\, t_l^i = 0, \quad i = 1,2 \quad \text{on } \partial\Omega \tag{1.15}$$

(the ν_i being the components of the exterior unit normal to $\partial\Omega$ and t_l^i are the components of the unit tangent vector $\boldsymbol{t}^i$ to $\partial\Omega$), whenever the moments of the tractions vanish on $\partial\Omega$. To simplify our calculations we will suppose further, in the sequel, the existence of a potential $\vartheta(\boldsymbol{e})$, for which

$$(\tau_0^v)_{ij} = \frac{\partial\vartheta}{\partial e_{ij}} \tag{1.16}$$

with $\vartheta(\cdot)$ once continuously differentiable in $\mathbb{R}^{n^2}$, $\vartheta \geq 0$, $\vartheta(0) = 0$, and

$$\vartheta(\boldsymbol{e}) \leq C(1 + |\boldsymbol{e}|)^p \tag{1.17}$$

Such a scalar potential ϑ clearly exists for the nonlinear model defined by (1.13), i.e.

$$\vartheta(\boldsymbol{e}) = \mu_0 \int_0^{e_{ij}e_{ij}} (\varepsilon + s)^{-\alpha/2}\, ds \tag{1.18}$$

In § 2, we shall sketch the proof of the existence of a unique weak solution to the initial-boundary value problem associated with the nonlinear bipolar viscous fluid defined by (1.14), (1.12); the main goal will then be to extract the limit, in an appropriate sense, of such solutions as $\mu_1 \to 0^+$. In § 3 we follow the basic ideas in the recent work of Nečas [29] and Malek, Nečas, and Novotny [24] on measure-valued solutions for non-Newtonian fluids with the theoretical background for Young measures following the presentation in the work of J. Ball [1]. For other related work on the application of the Young measure concept to partial differential equations and systems the reader may consult the following papers as well as the references therein: [7]–

[10], [25], [26], [32]-[36]. The existence of measure-valued solutions for the model governed by (1.1)-(1.3), with τ_{ij}^v given by (1.14) with $\mu_1 = 0$, and $(\tau_0^v)_{ij}$ satisfying (1.7), (1.9) will be proven in the case $n = 2$, for $p > 1$ and in the case $n = 3$ for $p > \frac{6}{5}$.

For the bipolar incompressible viscous fluid defined by the solution of (1.1)-(1.3), (1.14) with $\boldsymbol{\tau}_0^v$ subject to (1.7), (1.9), a weak solution problem is defined to be, in the case where $\mu_1 > 0$, a function $\boldsymbol{v} \in V$ which satisfies (2.1), a.e. on $(0,T)$, for all $\boldsymbol{w} \in V$, where $V = \{\boldsymbol{v} \in W^{2,2}(\Omega) | \nabla \cdot \boldsymbol{v} = 0 \text{ in } \Omega, \boldsymbol{v} = 0, \text{ on } \partial\Omega\}$. Theorem 2.1 guarantees the existence of a unique regular weak solution $\boldsymbol{v}_{\mu_1}$ to this problem, while Theorem 3.3 establishes the fact that $\boldsymbol{v} = \text{weak} \lim_{\mu_1 \to 0} \boldsymbol{v}_{\mu_1}$ is a Young measure-valued solution of our problem with $\mu_1 = 0$ for $p > 1$, when $n = 2$, and for $p > 6/5$ when $n = 3$. The remainder of the paper is confined to the space-periodic problem and it has been proven in § 4-§ 9 that for this situation the Young measured-value solutions of § 3 are unique regular weak solutions of the problem with $\mu_1 = 0$, when $n = 2$, for $p \geq 2$, weak solutions when $n = 2$, for $3/2 < p < 2$, unique regular weak solutions when $n = 3$, for $11/5 \leq p < 6$, and weak solutions when $n = 3$, for $9/5 < p < 11/5$. These results are summarized in the line graphs depicted below:

(i) $n = 2$, $\mu_1 = 0$

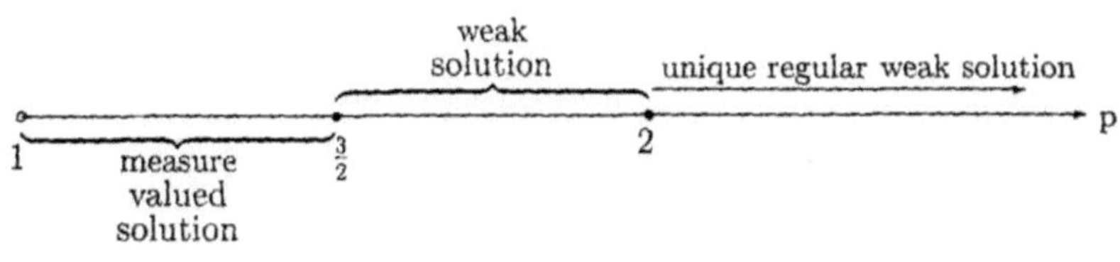

(ii) $n = 3$, $\mu_1 = 0$

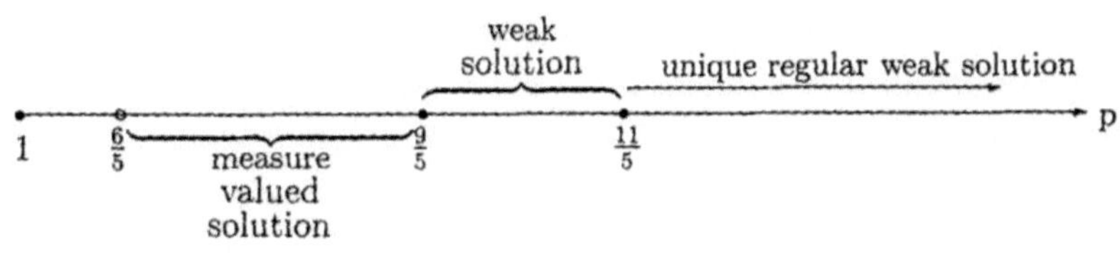

The earlier work of Ladyzhenskaya [18], [19] established existence of weak solutions for our model, with $\mu_1 = 0$, when $p \geq 1 + \frac{2n}{n+2}$ (so for $p \geq 2, n = 2$, and $p \geq 11/5, n = 3$) and uniqueness of the weak solution for $p \geq 2 (n = 2)$ and $p \geq 5/2 (n = 3)$. More recently, Du and Gunzburger [11] reported for the model considered here, with $\mu_1 = 0$, the existence of a unique weak solution for $p \geq 11/5$ when $n = 3$, although several points in their proofs are not clear to the authors. Thus, for space dimension $n = 2$ (but in the space-periodic case, of course, the work in [18] and [11] being for the case of a bounded domain) we have lowered the range for existence of a weak solution to $p > 3/2$ (in lieu of $p \geq 2$, with $p = 2$ being Navier-Stokes) while for space dimension $n = 3$ existence of a weak solution holds provided $p > 9/5$. For uniqueness of weak solutions in dim $n = 3$, taking, again, into account that our results are for the space-periodic problem, while those in [18] and [11] are for the boundary-value problem, our lower bound of $p \geq 11/5$ is better than the bound of $p \geq 5/2$ in Ladyzhenskaya [18], [19] and coincides with the bound of $p \geq 11/5$ reported in Du and Gunzburger [11]. The results on the existence of Young-measured value solutions for $p > 1 (n = 2)$ and $p > 6/5 (n = 3)$ are, of course, entirely new. For $p \leq 3/2 (n = 2)$ and $p \leq 9/5 (n = 3)$ it is entirely possible (indeed, probable) that the Young-measured value solutions are not weak solutions (i.e., the corresponding Young measure $\nu_{\boldsymbol{x},t}(\cdot)$ has a non-pointwise support) and this is, in the opinion of the authors, related to the chaotic behavior of solutions as $t \to \infty$, in other words, to the phenomena of turbulence; in some sense this idea can be thought of as the 'measure-valued' version of the original idea of Leray [20] who supposed that the origin of turbulence was connected with the development of non-regularity in solutions due to vorticity.

It is clear, at least to the authors, that the results of this paper can probably be extended to the classical situation of a bounded domain with smooth boundary (the results on existence of measure-valued solutions in § 3 already have been established for such domains); what is required are improved regularity results for the unique weak solution of the boundary-value problem for the incompressible bipolar fluid with $\mu_1 > 0$. The authors intend to come back to this problem in another paper.

2 The Incompressible Bipolar Fluid

We begin by assuming that the components b_i of the specific body force vector satisfy $b_i \in L^2(Q_T)$, $Q_T = \Omega \times (0,T)$, $0 < T < \infty$, $\Omega \subset \mathbb{R}^n$ with $n = 2$ or $n = 3$; we also suppose that Ω is a bounded domain with Lipschitz continuous boundary.

Let $V = \{ \boldsymbol{v} \in W^{2,2}(\Omega); \nabla\cdot\boldsymbol{v} = 0 \text{ in } \Omega \text{ and } \boldsymbol{v} = \boldsymbol{0} \text{ on } \partial\Omega \}$[4]; by $W^{k,p}(\Omega)$ we denote the usual Sobolev space of $L^p(\Omega)$ functions whose (generalized) derivatives up through order k are in $L^p(\Omega)$. We want to consider the initial-boundary value problem consisting of (1.1)–(1.3), (where the τ_{ij}^v are given by (1.14), with the $(\tau_0^v)_{ij}$ satisfying the growth condition (1.7)) the boundary conditions $\boldsymbol{v} = \boldsymbol{0}$, on $\partial\Omega \times (0,T)$, and (1.15), on $\partial\Omega \times (0,T)$, with the τ_{ijk} as in (1.12), and the initial condition $\boldsymbol{v}(\cdot,0) = \boldsymbol{v}_0 \in V$. We will look for solutions $\boldsymbol{v}$ of this problem which satisfy $\boldsymbol{v} \in L^\infty([0,T], V)$ and $\dot{\boldsymbol{v}} \in L^2((0,T); L^2(\Omega))$.

By a weak solution of the incompressible, bipolar initial-boundary value problem we will understand a function $\boldsymbol{v}$ such that $\forall\, \boldsymbol{w} \in V$ we have, a. e. on $(0,T)$,

$$\begin{aligned} &\int_\Omega \dot{v}_i w_i \, d\boldsymbol{x} + \int_\Omega v_j \frac{\partial v_i}{\partial x_j} w_i \, d\boldsymbol{x} \\ &+ \frac{1}{\rho}\int_\Omega (\tau_0^v)_{ij}(\boldsymbol{v})\, e_{ij}(\boldsymbol{w})\, d\boldsymbol{x} + \frac{2\mu_1}{\rho}\int_\Omega \frac{\partial e_{ij}(\boldsymbol{v})}{\partial x_k}\,\frac{\partial e_{ij}(\boldsymbol{w})}{\partial x_k}\, d\boldsymbol{x} \\ &- \int_\Omega b_i w_i \, d\boldsymbol{x} = 0. \end{aligned} \tag{2.1}$$

Now let $\{\boldsymbol{w}^k\}_{k=1}^\infty$ be eigenfunctions in V for the problem

$$\begin{cases} \displaystyle\int_\Omega \frac{\partial e_{ij}(\boldsymbol{u})}{\partial x_k}\frac{\partial e_{ij}(\boldsymbol{w})}{\partial x_k}\, d\boldsymbol{x} - \lambda \int_\Omega u_i w_i \, d\boldsymbol{x} = 0 \\ \forall\, \boldsymbol{u} \in V, \end{cases}$$

[4]We should write, more precisely, that $\boldsymbol{v} \in W^{2,2}(\Omega; \mathbb{R}^n)$ but will use, instead, the simplified notation

and consider the usual Galerkin approximations $\boldsymbol{v}^N$ defined by $v_i^N = \sum_{m=1}^{N} g_m^N w_i^m$, $i = 1, 2, 3$. Employing $\boldsymbol{v}^N$ in (2.1) and integrating the resulting equation over $(0, t), t > 0$ we obtain for $c_i > 0$, $i = 1, 2, 3$

$$
\begin{aligned}
\frac{1}{2}\int_{\Omega_t} |\boldsymbol{v}^N|^2\, d\boldsymbol{x} \;+\; c_1 \int_0^t \|\boldsymbol{v}^N\|^{p_1}_{W^{1,p_1}(\Omega)}\, d\tau \\
+\, c_2 \int_0^t \|\boldsymbol{v}^N\|^2_{W^{2,2}(\Omega)}\, d\tau \le \frac{1}{2}\int_{\Omega_0} |\boldsymbol{v}^N|^2\, d\boldsymbol{x} \\
+\, c_3 \int_0^t \|\boldsymbol{b}\|^2_{L^2(\Omega)}\, d\tau
\end{aligned} \tag{2.2}
$$

where we have employed (1.9) and the generalized form of the Korn inequality given in the Appendix to [5], namely,

$$
\int_{\Omega} \left| \frac{\partial e_{ij}(\boldsymbol{u})}{\partial x_k} \frac{\partial e_{ij}(\boldsymbol{u})}{\partial x_k} \right|^{p/2} d\boldsymbol{x} \;\ge\; c\|\boldsymbol{u}\|^p_{2,p} \tag{2.3}
$$

By $|\bullet|$ we mean the Euclidean norm on $\mathbb{R}^n$.

Now, we use $\dot{\boldsymbol{v}}^N$ in (2.1) and find that

$$
\begin{aligned}
\int_{\Omega_t} |\dot{\boldsymbol{v}}^N|^2\, d\boldsymbol{x} \;+\; \int_{\Omega_t} v_j^N \frac{\partial v_i^N}{\partial x_j} \dot{v}_i^N\, d\boldsymbol{x} \\
+\, \frac{1}{\rho}\int_{\Omega_t} (\tau_0^v)_{ij}(\boldsymbol{v}^N) e_{ij}(\dot{\boldsymbol{v}}^N)\, d\boldsymbol{x} \\
+\, \frac{\mu_1}{\rho}\frac{\partial}{\partial t}\int_{\Omega_t} \frac{\partial e_{ij}(\boldsymbol{v}^N)}{\partial x_k}\frac{\partial e_{ij}(\boldsymbol{v}^N)}{\partial x_k}\, d\boldsymbol{x} \;-\; \int_{\Omega_t} b_i \dot{v}_i^N\, d\boldsymbol{x} \;=\; 0
\end{aligned} \tag{2.4}
$$

Integrating (2.4) over $(0, t)$ we get

$$
\begin{aligned}
\int_{Q_t} |\dot{\boldsymbol{v}}^N|^2\, d\boldsymbol{x}d\tau \;+\; \frac{1}{\rho}\int_{\Omega_t} \vartheta(\boldsymbol{e}(\boldsymbol{v}^N))\, d\boldsymbol{x} \\
-\, \frac{1}{\rho}\int_{\Omega_0} \vartheta(\boldsymbol{e}(\boldsymbol{v}^N))\, d\boldsymbol{x} \;+\; \frac{\mu_1}{\rho}\int_{\Omega_t} \frac{\partial e_{ij}(\boldsymbol{v}^N)}{\partial x_k}\frac{\partial e_{ij}(\boldsymbol{v}^N)}{\partial x_k}\, d\boldsymbol{x} \\
-\, \frac{\mu_1}{\rho}\int_{\Omega_0} \frac{\partial e_{ij}(\boldsymbol{v}^N)}{\partial x_k}\frac{\partial e_{ij}(\boldsymbol{v}^N)}{\partial x_k}\, d\boldsymbol{x} \;-\; \int_{Q_t} b_i \dot{v}_i^N\, d\boldsymbol{x}d\tau
\end{aligned} \tag{2.5}
$$

$$+ \int_{Q_t} v_j^N \frac{\partial v_i^N}{\partial x_j} \dot{v}_i^N \, dx d\tau = 0,$$

where ϑ is the scalar potential given by (1.16). By virtue of the imbedding $W^{2,2}(\Omega) \hookrightarrow\hookrightarrow L^\infty(\Omega)$ and the standard (interpolation) inequality (e.g. Lions [21]):

$$\|v\|^2_{W^{1,2}(\Omega)} \le \delta \|v\|^2_{W^{2,2}(\Omega)} + \lambda(\delta)\|v\|^2_{L^2(\Omega)} \tag{2.6}$$

we have, for some $c_4 > 0$

$$\begin{aligned} &\int_{Q_T} |v^N|^2 \, |\nabla v^N|^2 \, dx dt \le \\ &c_4 \int_0^T \|v^N\|^2_{W^{2,2}(\Omega_t)} \Big[\delta \int_{\Omega_t} \|v^N\|^2_{W^{2,2}(\Omega_t)} \\ &\qquad + \lambda(\delta) \|v^N\|^2_{L^2(\Omega_t)} \Big] \, dt \end{aligned} \tag{2.7}$$

From (2.5) and (2.7) it follows that

$$\begin{aligned} &\int_{Q_t} |\dot{v}^N|^2 \, dx dt + c_5 \sup_{(0,T)} \|v^N\|^2_{W^{2,2}(\Omega_t)} \\ &\le c_6 \|v^N\|_{W^{2,2}(\Omega_0)} + c_6 \|v^N\|^p_{W^{1,p}(\Omega_0)} \\ &+ c_6 \int_{Q_T} |b|^2 \, dx dt + \delta c_4 \sup_{(0,T)} \|v^N\|^2_{W^{2,2}(\Omega_t)} \cdot \int_0^T \|v^N\|^2_{W^{2,2}(\Omega_t)} \, dt \\ &+ c_4 \lambda(\delta) \int_0^T \|v^N\|^2_{W^{2,2}(\Omega_t)} \cdot \|v^N\|^2_{L^2(\Omega_t)} \, dt, \end{aligned} \tag{2.8}$$

for some positive constant c_6, where we have again used the Korn type inequality stated in the Appendix to [5]. By virtue of (2.2), however,

$$\int_0^T \|v^N\|^2_{W^{2,2}(\Omega_t)} \, dt \le c_7 \tag{2.9}$$

for some $c_7 > 0$. Choosing δ now so small that $\delta c_4 c_7 \le \frac{1}{2} c_5$ we infer the existence of $c_8 > 0$ such that

$$\int_{Q_T} |\dot{v}^N|^2 \, dx dt + \sup_{(0,T)} \|v^N\|^2_{W^{2,2}(\Omega_t)} \le c_8 \tag{2.10}$$

A standard limit process as $N \to \infty$ now yields the following:

Theorem 2.1 *There exists a unique* $\boldsymbol{v} \in L^\infty((0,T);\, W^{2,2}(\Omega))$, $\dot{\boldsymbol{v}} \in L^2((0,T);\, L^2(\Omega))$ *which satisfies (2.1),* $\forall\, \boldsymbol{w} \in V$, *a. e. on* $(0,T)$.

Proof: We begin by choosing, as a consequence of (2.10), a subsequence $\{\boldsymbol{v}^{N_k}\}$ such that

$$\begin{aligned} \boldsymbol{v}^{N_k} &\rightharpoonup \boldsymbol{v}, \text{ in } L^2((0,T);\, W^{2,2}(\Omega)) \\ \dot{\boldsymbol{v}}^{N_k} &\rightharpoonup \dot{\boldsymbol{v}}, \text{ in } L^2((0,T);\, L^2(\Omega)) \end{aligned} \tag{2.11}$$

Then, as a consequence of Lemma 5.2 of Lions [21], (see, also, Lemma (3.1)), it follows that

$$\begin{gathered} \boldsymbol{v}^{N_k} \to \boldsymbol{v}, \text{ in } L^2((0,T);\, W^{1,\tilde{p}}(\Omega)), \\ \forall\, \tilde{p}, \text{ for } n = 2, \text{ and for } \tilde{p} < 6 \text{ for } n = 3 \end{gathered} \tag{2.12}$$

In view of (2.10) we have as $N_k \to \infty$

$$\begin{aligned} &\int_{Q_T} v_j^{N_k} \frac{\partial v_i^{N_k}}{\partial x_j} (v_i^{N_k} - v_i)\, d\boldsymbol{x}dt \to 0 \\ &\int_{Q_T} (\tau_0^v)_{ij}(\boldsymbol{e}^{N_k})\, e_{ij}\, (\boldsymbol{v}^{N_k} - \boldsymbol{v})\, d\boldsymbol{x}dt \to 0 \\ &\int_{Q_T} \dot{v}_i^{N_k}\, (v_i^{N_k} - v_i)\, d\boldsymbol{x}dt \to 0 \end{aligned} \tag{2.13}$$

It then follows from (2.1) that $\boldsymbol{v}^{N_k} \to \boldsymbol{v}$ in $L^2((0,T);\, W^{2,2}(\Omega))$; we infer, therefore, that $\boldsymbol{v}^{N_k} \to \boldsymbol{v}$ in $W^{2,2}(\Omega)$, a. e. in $(0,T)$, and the existence, as well as the uniqueness, of the weak solution follows. ∎

Remark: As a consequence of (2.2) the solution $\boldsymbol{v}$ satisfies

$$\int_0^T \|\boldsymbol{v}\|^{p_1}_{W^{1,p_1}(\Omega_t)}\, dt \;\le\; c \tag{2.14}$$

$$\sup_{(0,T)} \|\boldsymbol{v}\|^2_{L^2(\Omega_t)} \;\le\; c \tag{2.15}$$

where c is independent of μ_1.

3 Young Measure-Valued Solutions

In this section we will be concerned with the limit, as $\mu_1 \to 0$, of the unique weak solution to the initial-boundary value problem of § 2 for the incompressible bipolar fluid; we will rely strongly in this section on the condition (1.9) with $p \geq p_1 > p-1$ and $p_1 > \frac{2n}{n+2}$. We begin with the lemma 5.2 of Lions [21] that was referenced in Theorem 2.1:

Lemma 3.1 *Let B be a Banach space and B_i, $i = 0, 1$ separable, reflexive Banach spaces. Suppose that $B_0 \hookrightarrow\hookrightarrow B \hookrightarrow B_1$ where $\hookrightarrow$ denotes continuous imbedding and $\hookrightarrow\hookrightarrow$ compact imbedding.*

Let

$$W = \{ v \in L^{p_0}(I; B_0) \text{ with } \frac{dv}{dt} \in L^{p_1}(I; B_1) \} \tag{3.1}$$

for some interval $I \subset \mathbb{R}$, I finite, where p_i, $i = 0, 1$ satisfies $1 < p_i < \infty$. Then $W \hookrightarrow\hookrightarrow L^{p_0}(I; B)$.

It is now a simple matter to obtain the

Lemma 3.2 *Suppose that*

$$\begin{cases} \|v^N\|_{L^\infty(I;L^2(\Omega))} & \leq c < \infty \\ \|v^N\|_{L^{p_1}(I;W^{1,p_1}(\Omega))} & \leq c < \infty, p_1 > \dfrac{2n}{n+2} \end{cases} \tag{3.2}$$

and, for some $k > 0$

$$\begin{cases} \|\dot{v}^N\|_{L^2(I;W^{-k,2}(\Omega))} & \leq c, \quad \text{if } p_1 \leq 2, \\ \|\dot{v}^N\|_{L^{p_1'}(I;W^{-k,2}(\Omega))} & \leq c, \quad \text{if } p_1 > 2 \end{cases} \tag{3.3}$$

where p_1' is conjugate to p_1, i.e. $1/p_1 + 1/p_1' = 1$. Suppose that $v^N \rightharpoonup v$ in $L^2(Q_T)$; then $v^N \to v$ in $L^2(Q_T)$.

Proof: Inasmuch as $W^{1,p_1}(\Omega) \hookrightarrow\hookrightarrow L^2(\Omega)$, we obtain, for $p_1 \geq 2$, that $v^N \to v$ in $L^2(Q_T)$ because of Lemma 3.1. For $p_1 < 2$ we obtain, by virtue of Lemma 3.1, that $v^N \to v$ in $L^{p_1}(I; L^2(\Omega))$. However, with $a = 1 - p_1/2$ we have for scalar u

$$\int_{Q_T} u^2 \, dx dt \leq \int_0^T (\int_\Omega u^2 \, dx)^a (\int_\Omega u^2 \, dx)^{p_1/2} \, dt \tag{3.4}$$

and the required result follows. ∎

We now introduce the probability measure of L. C. Young [39] in the form proposed by J. Ball [1]. To this end, we consider the space $C_0(\mathbb{R}^{n^2})$ of continuous functions from $\mathbb{R}^{n^2} \to \mathbb{R}^1$ which satisfy $\lim_{|\boldsymbol{\lambda}|\to 0} f(\boldsymbol{\lambda}) = 0$. By the Riesz representation theorem we have that $(C_0(\mathbb{R}^{n^2}))^* = M(\mathbb{R}^{n^2})$, $M(\mathbb{R}^{n^2})$ being the space of Radon measures. Now, consider the Bochner space $L^1(Q_T;\, C_0(\mathbb{R}^{n^2}))$. It is proven in Edwards [13] that

$$(L^1(Q_T;\, C_0(\mathbb{R}^{n^2})))^* = L_w^\infty(Q_T;\, M(\mathbb{R}^{n^2})),$$

where $g \in L_w^\infty(Q_T;\, M(\mathbb{R}^{n^2}))$ means that $\forall\, f \in L^1(Q_T;\, C_0(\mathbb{R}^{n^2}))$ the function $< f(\boldsymbol{x},t), g(\boldsymbol{x},t) >$ is measurable in Q_T and essentially bounded with $< \,,\, >$ denoting the duality between $C_0(\mathbb{R}^{n^2})$ and $M(\mathbb{R}^{n^2})$. We have, in fact, the following:

Lemma 3.3 *Let $G \in (L^1(Q_T;\, C_0(\mathbb{R}^{n^2})))^*$ be a bounded, positive, linear functional. Then there exists a unique $g \in L_w^\infty(Q_T;\, M(\mathbb{R}^{n^2}))$, with $g(\boldsymbol{x},t)$ a positive Radon measure, such that $\forall f \in L^1(Q_T; C_0(\mathbb{R}^{n^2}))$*

$$< G, f > = \int_{Q_T} < f(\boldsymbol{x},t), g(\boldsymbol{x},t) > \, d\boldsymbol{x}dt$$

and

$$\|G\| = \operatorname*{ess\,sup}_{Q_T} \|g(\boldsymbol{x},t)\|_{M(\mathbb{R}^{n^2})}.$$

We now return to the weak solution of the initial-boundary value problem for the incompressible bipolar fluid, i.e., to the solution of (2.1). For our purposes here we will suppose that

$$b_i \in L^{p_1'}((0,T);\, L^2(\Omega)),\ \frac{1}{p_1} + \frac{1}{p_1'} = 1. \tag{3.5}$$

It follows from (2.1) and (1.9) that, if we denote by $\boldsymbol{v}^N$ the unique weak solution of the initial-boundary value problem corresponding to $\mu_1 = \mu_1^N$, the conditions of Lemma 3.2 are satisfied[5]; we can, therefore, suppose that $\boldsymbol{v}^N \overset{*}{\rightharpoonup} \boldsymbol{v}$ in $L^\infty(I;\, L^2(\Omega))$, as well as $\boldsymbol{v}^N \rightharpoonup \boldsymbol{v}$ in $L^{p_1}((0,T);\, W^{1,p_1}(\Omega))$.

Let $\Psi(\boldsymbol{x},t,\boldsymbol{\lambda}) \in L^1(Q_T;\, C_0(\mathbb{R}^{n^2}))$, so that a.e. for $(\boldsymbol{x},t) \in Q_T$ we have

[5] The sequence $\{\mu_1^N\}$ is chosen so that $\mu_1^N \to 0^+$ as $N \to \infty$.

that $\Psi(\boldsymbol{x},t,\bullet) \in C_0(\mathbb{R}^{n^2})$. Set

$$\int\limits_{Q_T} < \Psi(\boldsymbol{x},t,\bullet),\, \nu^N_{\boldsymbol{x},t}(\bullet) > d\boldsymbol{x}dt \;=\; \int\limits_{Q_T} \Psi(\boldsymbol{x},t,\nabla \boldsymbol{v}^N(\boldsymbol{x},t))\, d\boldsymbol{x}dt \qquad (3.6)$$

This serves to define the $\nu^N \in L^\infty_w(Q_T;\, M(\mathbb{R}^{n^2}))$. We have, a. e. in Q_T, that

$$\nu^N_{\boldsymbol{x},t} \geq 0,\ \|\nu^N_{\boldsymbol{x},t}\|_{M(\mathbb{R}^{n^2})} \;=\; 1, \quad \text{and}$$

$$|\int\limits_{Q_T} \Psi(\boldsymbol{x},t,\nabla \boldsymbol{v}^N(\boldsymbol{x},t))\, d\boldsymbol{x}dt| \;\leq\; \|\Psi\|_{L^1(Q_T;\, C_0(\mathbb{R}^{n^2}))}.$$

Therefore, we can suppose that for $\nu^N \in L^\infty_w(Q_T;\, M(\mathbb{R}^{n^2}))$ we have $\nu^N \overset{*}{\rightharpoonup} \nu$ (ν^N a subsequence of the original sequence, if necessary) as a consequence of the separability of the space $L^1(Q_T;\, C_0(\mathbb{R}^{n^2}))$. Hence, by Lemma 3.3,

$$\nu_{\boldsymbol{x},t} \;\geq\; 0, \quad \|\nu_{\boldsymbol{x},t}\|_{M(\mathbb{R}^{n^2})} \;\leq\; 1 \qquad (3.7)$$

In fact, the following is true.

Theorem 3.1 *The measure $\nu_{\boldsymbol{x},t}$ satisfies $\|\nu_{\boldsymbol{x},t}\|_{M(\mathbb{R}^{n^2})} \;=\; 1$ a. e. in Q_T.*

Proof: We sketch the proof (which may be found in Ball [1]). Set

$$\vartheta^k(\boldsymbol{\lambda}) \;=\; \begin{cases} 1, & |\boldsymbol{\lambda}| \;\leq\; k \\ 1 \;+\; k \;-\; |\boldsymbol{\lambda}|, & k \;\leq\; |\boldsymbol{\lambda}| \;\leq\; 1 \;+\; k \\ 0, & |\boldsymbol{\lambda}| \;\geq\; 1 \;+\; k \end{cases}$$

Then for any measurable $E \subset Q_T$:

$$\begin{aligned} &\lim_{N\to\infty} \frac{1}{\operatorname{meas} E} \int\!\!\int\limits_{E} \vartheta^k(\nabla \boldsymbol{v}^N(\boldsymbol{x},t))\, d\boldsymbol{x}dt \\ &= \frac{1}{\operatorname{meas} E} \int\!\!\int\limits_{E} <\vartheta^k, \nu_{\boldsymbol{x},t}>\, d\boldsymbol{x}dt \qquad (3.8) \\ &\leq \frac{1}{\operatorname{meas} E} \int\!\!\int_{E} \|\nu_{\boldsymbol{x},t}\|_{M(\mathbb{R}^{n^2})}\, d\boldsymbol{x}dt \end{aligned}$$

On the other hand

$$\begin{aligned}
&\frac{1}{\operatorname{meas} E}\int_E (1-<\vartheta^k,\nu^N_{\boldsymbol{x},t}>)\,d\boldsymbol{x}dt \\
\leq\ &\frac{1}{\operatorname{meas} E}\left[\operatorname{meas}\left\{(\boldsymbol{x},t)\in E;\ |\nabla \boldsymbol{v}^N(\boldsymbol{x},t)|\ \geq\ k\right\}\right] \\
&\leq\ \frac{c'}{\operatorname{meas} E}\cdot k^{-1}\left[\operatorname{meas} E\right]^{1/p_1'}.
\end{aligned} \tag{3.9}$$

Therefore,

$$1-k^{-1}[\operatorname{meas} E]^{\frac{1}{p_1'}-1}\cdot c'\ \leq\ \frac{1}{\operatorname{meas} E}\int_E <\vartheta^k,\nu^N_{\boldsymbol{x},t}>\ d\boldsymbol{x}dt \tag{3.10}$$

so that

$$\begin{aligned}
1-k^{-1}[\operatorname{meas} E]^{\frac{1}{p_1'}-1}\cdot c'\ &\leq\ \frac{1}{\operatorname{meas} E}\int_E <\vartheta^k,\nu_{\boldsymbol{x},t}>\ d\boldsymbol{x}dt \\
&\leq\ \frac{1}{\operatorname{meas} E}\int_E \|\nu_{\boldsymbol{x},t}\|_{M(\mathbb{R}^{n^2})}\,d\boldsymbol{x}dt
\end{aligned} \tag{3.11}$$

Letting $k\to\infty$ yields the required result. ■

We now state:

Theorem 3.2 *Let $h(\cdot)\in C(\mathbb{R}^{n^2})$ be such that $|h(\boldsymbol{\lambda})|\leq C(1+|\boldsymbol{\lambda}|)^{p-1}$ and choose $\psi\in L^q(Q_T)$ with $q\geq p_1/(p-1)$. Then the duality*

$$\int_{Q_T} <\Psi(\boldsymbol{x},t,\cdot),\nu_{\boldsymbol{x},t}(\cdot)>\ d\boldsymbol{x}dt$$

can be continuously extended to $\Psi(\boldsymbol{x},t,\boldsymbol{\lambda})=\psi(\boldsymbol{x},t)h(\boldsymbol{\lambda})$ by the limit of the functions $\psi(\boldsymbol{x},t)b_m(\boldsymbol{\lambda})$, where $b_m\in C_0(\mathbb{R}^{n^2})$ satisfies $|h_m(\boldsymbol{\lambda})|\leq C(1+|\boldsymbol{\lambda}|)^{p-1}$ and $h_m(\cdot)\to h(\cdot)$ locally in $\mathbb{R}^{n^2}$.

Proof: We have

$$\|<h_m(\cdot),\nu_{\boldsymbol{x},t}>\|_{L^{\frac{p_1}{p-1}}(Q_T)}\ \leq\ C \tag{3.12}$$

Let $1<p_2<\dfrac{p_1}{p-1}$; then $\exists\chi\in L^{p_2}(Q_T)$ such that

$$<h_m(\cdot),\nu_{\boldsymbol{x},t}(\cdot)>\ \to\ \chi(\boldsymbol{x},t)\ \text{in}\ L^{p_2}(Q_T) \tag{3.13}$$

In order to verify (3.13) we compute that

$$\int_{Q_T} | <h_{m_1}(\cdot), \nu_{\boldsymbol{x},t}(\cdot)> - <h_{m_2}(\cdot), \nu_{\boldsymbol{x},t}(\cdot)> |^{p_2}\, d\boldsymbol{x}dt$$

$$= \lim_{N\to\infty} \int_{Q_T} |h_{m_1}(\nabla \boldsymbol{v}^N(\boldsymbol{x},t)) - h_{m_2}(\nabla \boldsymbol{v}^N(\boldsymbol{x},t))|^{p_2}\, d\boldsymbol{x}dt$$

$$= \overline{\lim_{N\to\infty}} \int_{Q_{T,k,N}} |h_{m_1}(\nabla \boldsymbol{v}^N(\boldsymbol{x},t)) - h_{m_2}(\nabla \boldsymbol{v}^N(\boldsymbol{x},t))|^{p_2}\, d\boldsymbol{x}dt$$

$$+ \overline{\lim_{N\to\infty}} \int_{Q_T/Q_{T,k,N}} |h_{m_1}(\nabla \boldsymbol{v}^N(\boldsymbol{x},t)) - h_{m_2}(\nabla \boldsymbol{v}^N(\boldsymbol{x},t))|^{p_2}\, d\boldsymbol{x}dt$$

where

$$Q_{T,k,N} = \{(\boldsymbol{x},t) \in Q_T \mid |\nabla \boldsymbol{v}^N(\boldsymbol{x},t)| \geq k\} \tag{3.14}$$

However, we have

$$\operatorname{meas} Q_{T,k,N} \leq ck^{-1} \tag{3.15}$$

and by virtue of the growth assumption for $|h_m(\cdot)|$

$$\int_{Q_{T,k,N}} |h_{m_1}(\nabla \boldsymbol{v}^N(\boldsymbol{x},t)) - h_{m_2}(\nabla \boldsymbol{v}^N(\boldsymbol{x},t))|\, d\boldsymbol{x}dt$$

$$\leq C \int_{Q_{T,k,N}} (1 + |\nabla \boldsymbol{v}^N(\boldsymbol{x},t)|)^{(p-1)p_2}\, d\boldsymbol{x}dt \tag{3.16}$$

$$\leq C\Big(\int_{Q_{T,k,N}} (1 + |\nabla \boldsymbol{v}^N|)^{p_1}\, d\boldsymbol{x}dt\Big)^{\frac{(p-1)p_2}{p_1}} \bullet [\operatorname{meas} Q_{T,k,N}]^{\frac{1-(p-1)p_2}{p_1}}$$

Therefore, for any $\varepsilon > 0$ we may choose k so large that

$$\int_{Q_{T,k,N}} |h_{m_1}(\nabla \boldsymbol{v}^N(\boldsymbol{x},t)) - h_{m_2}(\nabla \boldsymbol{v}^N(\boldsymbol{x},t))|\, d\boldsymbol{x}dt \leq \frac{\varepsilon}{2} \tag{3.17}$$

Fixing k so large that (3.17) holds, we now have for m_1, m_2 sufficiently large, say, $m_1, m_2 \geq m$:

$$\overline{\lim_{N\to\infty}} \int_{Q_T/Q_{T,k,N}} |h_{m_1}(\nabla \boldsymbol{v}^N(\boldsymbol{x},t)) - h_{m_2}(\nabla \boldsymbol{v}^N(\boldsymbol{x},t))|^{p_2}\, d\boldsymbol{x}dt \leq \frac{\varepsilon}{2} \tag{3.18}$$

and the desired result follows. ∎

Once again we recall that $\boldsymbol{v}^N$ is the unique weak solution of the initial-boundary value problem for the incompressible bipolar fluid corresponding to the choice $\mu_1 = \mu_1^N$ of the higher order viscosity; as above we take $\nu_{\boldsymbol{x},t}$ to be the Young measure constructed from the sequence of solutions $\{\boldsymbol{v}^N\}$ as

$N \to \infty$ (in which case, of course, we have $\mu_1^N \to 0^+$). Then the results of this section easily imply the following theorem on the existence of measure-valued solutions:

Theorem 3.3 *Let $\boldsymbol{v}$ denote the weak limit of the sequence $\{\boldsymbol{v}^N\}$ of unique weak solutions to the initial-boundary value problem for the incompressible bipolar fluid. Then provided $p > 1$, when $n = 2$, and $p > 6/5$, when $n = 3$, $\boldsymbol{v}$ satisfies (1.1)–(1.3), and* (1.14) *with $\mu_1 = 0$, in the following sense: $\forall \boldsymbol{\phi} \in C_0^\infty(Q_T)$ such that $\operatorname{div} \boldsymbol{\phi} = 0$ (and with $\rho = 1$):*

$$\int\limits_{Q_T} \Big[-v_i \frac{\partial \phi_i}{\partial t} - v_i v_j \frac{\partial \phi_i}{\partial x_j} + e_{ij}(\boldsymbol{\phi}) \int\limits_{\mathbb{R}^{n^2}} (\tau_0^v)_{ij}(\boldsymbol{e}(\boldsymbol{\lambda}))\, d\nu_{\boldsymbol{x},t}(\boldsymbol{\lambda}) \;-\; b_i \phi_i \Big]\, d\boldsymbol{x}dt \;=\; 0 \tag{3.19}$$

4 Young Measures are Dirac and the Weak Solutions are Regular for $p \geq 2,\ n = 2$

In this section we will suppose that $(\tau_0^v)_{ij}(\boldsymbol{e})$ is once continuously differentiable in $\boldsymbol{e}$, with $(\tau_0^v)_{ij}(\boldsymbol{0}) = 0$, and that $\boldsymbol{\tau}_0^v$ satisfies a strong monotonicity condition of the form

$$\frac{\partial (\tau_0^v)_{ij}}{\partial e_{kl}} \xi_{ij}\xi_{kl} \;\geq\; \beta_v(1 + |\boldsymbol{e}|^{p-2})|\boldsymbol{\xi}|^2,\ \xi_{ij} \;=\; \xi_{ji} \tag{4.1}$$

for some $\beta_v > 0$. By virtue of (4.1), it follows that $p_1 = p$ in (1.9). For the sake of convenience we now also set $\gamma = 2\mu_1/\rho$.

The method of obtaining the results of this section depends in an essential fashion on the derivation of some new inequalities for nonlinear parabolic equations. So as not to have to confront here some very delicate questions of regularity for incompressible fluids, we will now confine our attention to the space-periodic problem; the corresponding results for bounded domains with sufficiently smooth boundaries will be addressed by the authors in another paper. For the space-periodic problem in dimensions $n = 2$ or $n = 3$ the basic existence and uniqueness results may be summarized as follows: We take $\Omega = (0, L)^n,\ L > 0$ and denote by $L^2_{\text{per}}(\Omega), W^{k,p}_{\text{per}}(\Omega)$ the obvious spaces of L-periodic functions in any spatial variable. We set $V = W^{2,2}_{\text{per}}(\Omega)$; then (2.1) is again used to define what we mean by a weak solution provided $\boldsymbol{v}, \boldsymbol{w} \in V$ and are divergence free. If we suppose that $b_i \in L^2((0,T);\, L^2_{\text{per}}(\Omega))$ then we will obtain a unique weak solution $\boldsymbol{v} \in L^\infty((0,T);\, V)$ satisfying $\dot{\boldsymbol{v}} \in L^2((0,T);\, L^2_{\text{per}}(\Omega))$.

We begin with the case in which $p \geq 2$. By virtue of the estimate (2.7) the standard method of differences yields for the weak solution $\boldsymbol{v}$:

$$\begin{aligned} \sup_{(0,T)} \|\boldsymbol{v}\|_{1,2} &\leq c(\gamma) \\ \int_0^T\!\!\int_\Omega (1+|\boldsymbol{e}|)^{p-2} e_{ij}\Big(\frac{\partial \boldsymbol{v}}{\partial x_l}\Big) e_{ij}\Big(\frac{\partial \boldsymbol{v}}{\partial x_l}\Big)\, d\boldsymbol{x}dt &\leq c_1(\gamma) \\ \int_0^T\!\!\int_\Omega \sum_{|\boldsymbol{\alpha}|=2} \Big(D^{\boldsymbol{\alpha}} \frac{\partial \boldsymbol{v}}{\partial x_l} \cdot D^{\boldsymbol{\alpha}} \frac{\partial \boldsymbol{v}}{\partial x_l}\Big)\, d\boldsymbol{x}dt &\leq c_2(\gamma). \end{aligned} \tag{4.2}$$

where $\|\cdot\|_{1,2} \equiv \|\cdot\|_{W^{1,2}(\Omega)}$ (and we will continue to employ this more compact notation in the sequel). Our goal, however, is to obtain some estimates which are independent of $\gamma \equiv 2\mu_1/\rho$.

For the sake of simplicity of notation, let us assume that the initial function as well as the forcing term $\boldsymbol{b}$ have zero average in space. From this it easily follows that the solution also has zero average in space and therefore the Poincare inquality is valid. The results can be easily extended to the general case. We will also assume that

$$\dot{b}_i \in L^2((0,T);\, L^2_{\text{per}}(\Omega)). \tag{4.3}$$

Employing, once again, the difference method, or directly from Galerkin approximations, we can then prove that

$$\begin{aligned} \sup_{(0,T)} \|\dot{\boldsymbol{v}}\|_{0,2} &\leq c(\gamma) \\ \int_0^T\!\!\int_\Omega (1+|\boldsymbol{e}|)^{p-2} e_{ij}(\dot{\boldsymbol{v}}) e_{ij}(\dot{\boldsymbol{v}})\, d\boldsymbol{x}dt &\leq c_3(\gamma) \\ \int_0^T\!\!\int_\Omega \sum_{|\boldsymbol{\alpha}|=2} |D^{\boldsymbol{\alpha}} \dot{\boldsymbol{v}}|^2\, d\boldsymbol{x}dt &\leq c_4(\gamma). \end{aligned} \tag{4.4}$$

Remark: In the above regularity result we made use of the smoothness of the decomposition result, Corollary 3.7 of [31]. In [31] the result is stated for smooth (C^2) bounded domains. However this result extends easily to the case under consideration here, i.e., the space-periodic case.

Now, let us take $0 \leq \lambda$ and denote, for any fixed subscript l, $\boldsymbol{v}' = \frac{\partial}{\partial x_l}\boldsymbol{v}$. In (2.1) we take for the test function

$$w_i = \frac{v_i''}{(1+\|\boldsymbol{v}\|_{1,2}^2)^\lambda}$$

and we obtain, for $\lambda \neq 1$ and almost all t:

$$\begin{aligned}\frac{1}{2(1-\lambda)}\frac{d}{dt}(1+\|\boldsymbol{v}\|_{1,2}^2)^{1-\lambda} &+ \int_\Omega v_j \frac{\partial v_i}{\partial x_j}\frac{v_i''}{(1+\|\boldsymbol{v}\|_{1,2}^2)^\lambda}\,d\boldsymbol{x} \\ &+ \frac{\beta_2}{\rho}\int_\Omega (1+|\boldsymbol{e}|)^{p-2} e_{ij}(\boldsymbol{v}')e_{ij}(\boldsymbol{v}')\cdot(1+\|\boldsymbol{v}\|_{1,2}^2)^{-\lambda}\,d\boldsymbol{x} \\ &+ \gamma(1+\|\boldsymbol{v}\|_{1,2}^2)^{-\lambda}\int_\Omega \frac{\partial e_{ij}(\boldsymbol{v}')}{\partial x_k}\frac{\partial e_{ij}(\boldsymbol{v}')}{\partial x_k}\,d\boldsymbol{x} \\ &- (1+\|\boldsymbol{v}\|_{1,2}^2)^{-\lambda}\int_\Omega b_i v_i''\,d\boldsymbol{x} \leq 0.\end{aligned} \tag{4.5}$$

For the case in which $\lambda = 1$, the first term in (4.5) must be replaced by $\frac{d}{dt}\log(1+\|\boldsymbol{v}\|_{1,2}^2)$. Now, via integration by parts and our assumption of spatial periodicity

$$-\int_\Omega v_j \frac{\partial v_i}{\partial x_j} v_i''\,d\boldsymbol{x} = \int_\Omega v_j' \frac{\partial v_i}{\partial x_j} v_i'\,d\boldsymbol{x} \tag{4.6}$$

thus,

$$\left|\int_\Omega v_j \frac{\partial v_i}{\partial x_j}\frac{v_i''}{(1+\|\boldsymbol{v}\|_{1,2}^2)^\lambda}\,d\boldsymbol{x}\right| \leq c\,(1+\|\boldsymbol{v}\|_{1,2}^2)^{-\lambda}\,\|\boldsymbol{v}\|_{1,3}^3. \tag{4.7}$$

Also, using that $p \geq 2$, after summing over l we have

$$\begin{aligned}(1+\|\boldsymbol{v}\|_{1,2}^2)^{-\lambda}\left|\int_\Omega b_i v_i''\,d\boldsymbol{x}\right| &\leq \\ c\left(\int_\Omega (1+|\boldsymbol{e}|)^{p-2} e_{ij}(\boldsymbol{v}')e_{ij}(\boldsymbol{v}')\cdot(1+\|\boldsymbol{v}\|_{1,2}^2)^{-\lambda}\,d\boldsymbol{x}\right)^{1/2}& \\ \times\left(\|\boldsymbol{b}\|_{0,2}^2(1+\|\boldsymbol{v}\|_{1,2}^2)^{-\lambda}\right)^{1/2}.&\end{aligned} \tag{4.8}$$

After integration in time it then follows from (4.5) that

$$\frac{1}{2(1-\lambda)}(1+\|\boldsymbol{v}(T)\|_{1,2}^2)^{1-\lambda}$$
$$+\frac{1}{2\rho}\int_0^T\int_\Omega(1+|\boldsymbol{e}|)^{p-2}e_{ij}(\boldsymbol{v}')e_{ij}(\boldsymbol{v}')\cdot(1+\|\boldsymbol{v}\|_{1,2}^2)^{-\lambda}\,d\boldsymbol{x}dt$$
$$+\gamma\textstyle\int_0^T(1+\|\boldsymbol{v}\|_{1,2}^2)^{-\lambda}\displaystyle\int_\Omega\frac{\partial e_{ij}(\boldsymbol{v}')}{\partial x_k}\frac{\partial e_{ij}(\boldsymbol{v}')}{\partial x_k}\,d\boldsymbol{x}dt$$
$$\le\frac{1}{2(1-\lambda)}(1+\|\boldsymbol{v}(0)\|_{1,2}^2)^{1-\lambda}+c\int_0^T\|\boldsymbol{b}\|_{0,2}^2dt$$
$$+c\int_0^T(1+\|\boldsymbol{v}\|_{1,2}^2)^{-\lambda}\,\|\boldsymbol{v}\|_{1,3}^3\,dt \tag{4.9}$$

Thus, we want to estimate the integral:

$$\int_0^T(1+\|\boldsymbol{v}\|_{1,2}^2)^{-\lambda}\,\|\boldsymbol{v}\|_{1,3}^3\,dt.$$

We will make repeated use of the usual interpolation estimate (see [38] p. 186, for example)

$$\|\boldsymbol{v}\|_{s,r}\ \le\ c\,\|\boldsymbol{v}\|_{s_1,r_1}^{\theta}\cdot\|\boldsymbol{v}\|_{s_2,r_2}^{1-\theta} \tag{4.10}$$

where $\dfrac{1}{r}=\dfrac{\theta}{r_1}+\dfrac{1-\theta}{r_2}$, $s=\theta s_1+(1-\theta)s_2$ and $\theta\in[0,1]$, as well as of the general embedding estimate (see [38] p. 328, for example)

$$\|\boldsymbol{v}\|_{s,r}\ \le\ c\,\|\boldsymbol{v}\|_{s_1,r_1},$$

where $\dfrac{1}{r}=\dfrac{1}{r_1}-\dfrac{s_1-s}{n}$.

Now by virtue of the imbedding $W^{4/3,2}(\Omega)\hookrightarrow W^{1,3}(\Omega)$, for $n=2$, and the interpolation estimate

$$\|\boldsymbol{v}\|_{4/3,2}\ \le\ c\,\|\boldsymbol{v}\|_{1,2}^{2/3}\cdot\|\boldsymbol{v}\|_{2,2}^{1/3}$$

we have

$$\|\boldsymbol{v}\|_{1,3}\ \le\ c\,\|\boldsymbol{v}\|_{1,2}^{2/3}\cdot\|\boldsymbol{v}\|_{2,2}^{1/3} \tag{4.11}$$

Thus, it follows that

$$\begin{aligned}
&\int_0^T \|\boldsymbol{v}\|_{1,3}^3 (1+\|\boldsymbol{v}\|_{1,2}^2)^{-\lambda}\,dt \\
&\quad \le c\int_0^T (1+\|\boldsymbol{v}\|_{1,2}^2)^{1-\lambda}\,\|\boldsymbol{v}\|_{2,2}\,dt \qquad (4.12)\\
&\quad \le c\int_0^T (1+\|\boldsymbol{v}\|_{1,2}^2)^{1-\lambda/2}\,\|\boldsymbol{v}\|_{2,2}\,(1+\|\boldsymbol{v}\|_{1,2}^2)^{-\lambda/2}\,dt \\
&\quad \le c\Big(\int_0^T (1+\|\boldsymbol{v}\|_{1,2}^2)^{2-\lambda} dt\Big)^{1/2}\Big(\int_0^T \|\boldsymbol{v}\|_{2,2}^2 (1+\|\boldsymbol{v}\|_{1,2}^2)^{-\lambda}\,dt\Big)^{1/2}
\end{aligned}$$

Employing (4.12) in (4.9), with $\lambda = 1$, we obtain

$$\log(1+\|\boldsymbol{v}(t)\|_{1,2}^2) + \tilde{c}_1\int_0^T \|\boldsymbol{v}(t)\|_{2,2}^2\,(1+\|\boldsymbol{v}(t)\|_{1,2}^2)^{-1}\,dt \le \tilde{c}_2 \qquad (4.13)$$

for some positive constants $\tilde{c}_1, \tilde{c}_2$; we have, therefore, established the following:

Lemma 4.1 *The unique weak solution for the incompressible bipolar fluid*[6] $\boldsymbol{v} \in L^\infty((0,T); V)$, $(V = W^{2,2}_{\text{per}}(\Omega), \Omega = (0,L)^2)$, *in the case where* $p \ge 2$, *satisfies, for any* $\gamma > 0$,

$$\begin{cases} \sup_{(0,T)} \|\boldsymbol{v}(t)\|_{1,2} & \le \bar{c} \\ \int_0^T \|\boldsymbol{v}(t)\|_{2,2}^2\,dt & \le \bar{c} \end{cases} \qquad (4.13)$$

where $\bar{c}$, $0 < \bar{c} < \infty$, *is independent of* γ.

We are now in a position to prove the following:

Theorem 4.1 *For the space-periodic problem with* $n = 2$ *and* $p \ge 2$, *under the assumptions (4.1), (1.7), (1.16), and* $b_i, \dot{b}_i \in L^\infty((0,T); L^2(\Omega))$, *if* $\boldsymbol{v}_{\gamma_k} \rightharpoonup \boldsymbol{v}$ *in* $L^p((0,T); W^{1,p}(\Omega))$ *then*

---

[6] We should write $\boldsymbol{v}_\gamma$ in (4.13), and similar estimates in the sequel, but will decline from doing so if the meaning is clear. It is also understood that the elements of V are divergence-free.

$$v_{\gamma_k} \to v, \; in \; L^2((0,T); W^{1,q}(\Omega)), \; q > 1 \tag{4.15}$$

$$v_{\gamma_k} \to v, \; in \; L^{\tilde{p}}((0,T); W^{1,p}(\Omega)), \; \tilde{p} < p \tag{4.16}$$

where v_{ν_k} is the solution of the space-periodic problem corresponding to $\gamma_k = 2\mu_1^k/p$ with $\mu_1^k \to 0^+$ as $k \to \infty$. Furthermore, the bounds (4.13) apply to the limiting function v.

Remark: We may conclude from (4.15), (4.16) that the Young measure $\nu_{x,t}$ generated by $\{v_{\gamma_k}\}$ is Dirac, i.e. $\nu_{x,t} = \delta(\lambda - \nabla v(x,t))$.

Proof: (Theorem 4.1): We set $I = (0,T)$. Then by (4.13) and Lemma 3.2 with $k > \frac{n+2}{2}$ and $p_1 = 2$:

$$\|v_{\gamma_j}\|_{L^p(I;W^{1,p}(\Omega))} \le c < \infty \tag{4.17}$$

$$\|\dot{v}_{\gamma_j}\|_{L^2(I;W^{-k,2}(\Omega))} \le c < \infty, \tag{4.18}$$

the conclusions of the theorem, i.e. (4.15) and (4.16) now follow as a direct consequence of (4.17), (4.18), and Lemma 3.1. ∎

5 Young Measures are Dirac for $3/2 < p < 2, n = 2$

We continue here, in some sense, the considerations of § 4 but will have to resort to more delicate estimates than those previously employed. As was the case in § 4, the estimates (4.2), (4.4) remain valid here. Also we may take $\lambda > 1$ and use the same test function $w_i = v_i''(1 + \|v\|_{1,2}^2)^{-\lambda}$ so as to obtain (4.5).

Since $p < 2$, instead of (4.8), we will use the estimate t

$$(1 + \|v\|_{1,2}^2)^{-\lambda} \Big| \int_\Omega b_i v_i'' \, dx \Big| \le c\|v\|_{2,p}\|b\|_{0,p'}(1 + \|v\|_{1,2}^2)^{-\lambda}. \tag{5.1}$$

We will also need the estimate

$$\|v\|_{2,p} \le c\Big(\int_\Omega (1 + |e|)^{p-2} e_{ij}(\frac{\partial v}{\partial x_k}) e_{ij}(\frac{\partial v}{\partial x_k}) \, dx\Big)^{1/2} \times \Big(\int_\Omega (1 + |e|)^p \, dx\Big)^{\frac{1}{p} - \frac{1}{2}}, \tag{5.2}$$

which is a direct consequence of the Holder inequality and the estimate in the Appendix of [5].

Hence,

$$(1+\|v\|_{1,2}^2)^{-\lambda}\Big|\int_\Omega b_i v_i'' \, dx\Big| \le c\|b\|_{0,p'}(1+\|v\|_{1,2}^2)^{-\lambda/2}\|v\|_{1,p}^{\frac{2-p}{2}} \times$$
$$\Big(\int_\Omega (1+|e|)^{p-2} e_{ij}\Big(\frac{\partial v}{\partial x_k}\Big) e_{ij}\Big(\frac{\partial v}{\partial x_k}\Big)(1+\|v\|_{1,2}^2)^{-\lambda}\, dx\Big)^{1/2} \qquad (5.3)$$

Therefore we again have inequality (4.9) except that the term $\int_0^T \|b\|_{0,2}^2 dt$ on the right hand side is now replaced by the term $\sup_{(0,T)}\|b\|_{0,p'}^2$.

We again seek to estimate the integral $\int_0^T (1+\|v\|_{1,2}^2)^{-\lambda}\|v\|_{1,3}^3\, dt$. However, with $\delta = 2-\frac{2}{p}$, $W^{2,p}(\Omega)\hookrightarrow W^{1+\delta,2}(\Omega)$ and, thus, as $W^{4/3,2}(\Omega)\hookrightarrow W^{1,3}(\Omega)$ we have

$$\begin{aligned}\|v\|_{1,3} &\le \bar{c}_1\|v\|_{4/3,2} \\ &\le \bar{c}_2\|v\|_{1,2}^{1-\frac{p}{6(p-1)}}\|v\|_{1+\delta,2}^{\frac{p}{6(p-1)}}\end{aligned} \qquad (5.4)$$

For (5.4) to be meaningful we must have $1-p/6(p-1)>0$, or $p\ge 6/5$. Because of the imbedding $W^{2,p}(\Omega)\hookrightarrow W^{1,q}(\Omega)$, $\frac{1}{q}=\frac{1}{p}-\frac{1}{2}$ we also have

$$\|v\|_{1,3} \le \tilde{c}\|v\|_{1,p}^{\frac{5p-6}{3p}}\|v\|_{2,p}^{\frac{6-2p}{3p}} \qquad (5.5)$$

Therefore, for any $\beta, 0<\beta<1$, we have

$$\int_0^T \|v\|_{1,3}^3 (1+\|v\|_{1,2}^2)^{-\lambda}\, dt$$
$$\le \hat{c}\int_0^T \Big(\|v\|_{1,2}^{(3-\frac{p}{2(p-1)})\beta}\|v\|_{1,p}^{\frac{5p-6}{p}(1-\beta)}\|v\|_{2,p}^{\frac{p\beta}{2(p-1)}}\|v\|_{2,p}^{\frac{6-2p}{p}(1-\beta)} \qquad (5.6)$$
$$\times(1+\|v\|_{1,2})^{-\lambda}\Big)\, dt$$

Setting $\beta=\frac{2(p-1)(3-p)}{(5p-6)}$, and $\lambda=\frac{3-p}{p-1}$ in (5.6), using the Holder inequality, and the estimate (5.2), we obtain for $3/2<p<2$, uniformly with respect to

γ and p, the estimate

$$\begin{aligned}\int_0^T \|v\|_{1,3}^3 (1+\|v\|_{1,2}^2)^{-\lambda}\, dt \\ \leq c\left(\int_0^T\Big[\int_\Omega (1+|e|)^{p-2}\frac{\partial e_{ij}}{\partial x_k}\frac{\partial e_{ij}}{\partial x_k}\,dx\Big](1+\|v\|_{1,2}^2)^{-\lambda}\,dt\right)^{\frac{3-p}{2}} \\ \times\Big(\int_0^T (1+\|v\|_{1,p})^p\,dt\Big)^{\frac{p-1}{2}}.\end{aligned} \tag{5.7}$$

Since $\frac{3-p}{2} < 1$ using (2.14) and (4.9) we also find that

$$\int_0^T\Big(\int_\Omega (1+|e|)^{p-2}\frac{\partial e_{ij}}{\partial x_k}\frac{\partial e_{ij}}{\partial x_k}\,dx\Big)(1+\|v\|_{1,2}^2)^{-\lambda}\,dt \leq c < \infty \tag{5.8}$$

and, by (5.2)

$$\int_0^T \|v\|_{2,p}^2 (1+\|v\|_{1,p})^{p-2}(1+\|v\|_{1,2}^2)^{-\lambda}\,dt \leq c < \infty \tag{5.9}$$

Also

$$\begin{aligned}\|v\|_{1,2} &\leq \bar{c}_1\|v\|_{2/p,p} \\ &\leq \bar{c}_2\|v\|_{1,p}^{2-\frac{2}{p}}\|v\|_{2,p}^{\frac{2}{p}-1}\end{aligned} \tag{5.10}$$

It then follows from (5.9) that

$$\int_0^T \|v\|_{2,p}^2 (1+\|v\|_{1,p}^{2-p+2\lambda(2-\frac{2}{p})})^{-1}(1+\|v\|_{2,p}^{2\lambda(\frac{2}{p}-1)})^{-1}\,dt \leq c < \infty. \tag{5.11}$$

Now let $M = \{t\in(0,T) \mid \|v\|_{2,p} > 1\}$, and recall that $\lambda = \frac{3-p}{p-1}$; then as a direct consequence of (5.11) we find that

$$\int_M \|v\|_{2,p}^{\frac{4(2p-3)}{p(p-1)}}(1+\|v\|_{1,p}^{\frac{12-2p-p^2}{p}})^{-1}\,dt \leq c < \infty. \tag{5.12}$$

Setting $P = \frac{12-2p}{p^2}$, $\theta(p) = \frac{4(2p-3)p}{(12-2p)(p-1)}$ and using the Holder inequality we find that

$$\int_M \|\boldsymbol{v}\|_{2,p}^{\theta(p)}\,dt \;\le\; \left(\int_M \|\boldsymbol{v}\|_{2,p}^{\frac{4(2p-3)}{p(p-1)}}\,(1+\|\boldsymbol{v}\|_{1,p}^{\frac{12-2p-p^2}{p}})^{-1}\,dt\right)^{\frac{1}{p}} \times \left(\int_M (1+\|\boldsymbol{v}\|_{1,p})^p\,dt\right)^{\frac{p-1}{p}},$$

so that for $p \in (3/2, 2)$ we have uniformly with respect to $\gamma > 0$,

$$\int_0^T \|\boldsymbol{v}\|_{2,p}^{\theta(p)}\,dt \le c < \infty. \tag{5.13}$$

Now, choose q such that $\theta(p) < q < p$. As

$$\int_0^T \|\boldsymbol{v}\|_{1+\sigma,p}^{q}\,dt \le c^* \int_0^T \|\boldsymbol{v}\|_{1,p}^{(1-\sigma)q}\,\|\boldsymbol{v}\|_{2,p}^{\sigma q}\,dt \tag{5.14}$$

with $\sigma = \theta(p)(p-q)/q(p-\theta(p))$, by (2.14) and (5.13) we get uniformly with respect to $\gamma > 0$ and p (in (3/2, 2)), the bound

$$\int_0^T \|\boldsymbol{v}\|_{1+\sigma,p}^{q}\,dt \le c < \infty, \tag{5.15}$$

and as a consequence the following theorem:

Theorem 5.1 *For the space-periodic problem, in dimension $n = 2$ with $3/2 < p < 2$ (under the same assumptions which apply in Theorem 4.1) if $\boldsymbol{v}_{\gamma_k} \rightharpoonup \boldsymbol{v}$ in $L^p((0,T); W^{1,p}(\Omega))$ then, for any $\tilde{p} < p$, $\boldsymbol{v}_{\gamma_k} \to \boldsymbol{v}$ in $L^{\tilde{p}}((0,T); W^{1,p}(\Omega))$ and $\boldsymbol{v}_{\gamma_k} \to \boldsymbol{v}$ in $L^q((0,T); W^{1+\tilde{\sigma},p}(\Omega))$, for $\tilde{\sigma} < \sigma$. Furthermore the limit function $\boldsymbol{v}$ satisfies the bound (5.15) and the Young measure $\nu_{\boldsymbol{x},t}(\boldsymbol{\lambda}) = \delta(\boldsymbol{\lambda} - \nabla\boldsymbol{v}(\boldsymbol{x},t))$ a. e. in $\Omega \times (0,T)$.*

6 Young Measures are Dirac and the Weak Solutions are Regular for $7/3 \le p < 6, n = 3$

In this section we will show that for $7/3 \le p < 6$, and space dimension $n = 3$, the sequence $\boldsymbol{v}_{\gamma_k} \to \boldsymbol{v}$ (in a sense to be made precise below) and that $\boldsymbol{v}$ is, for $7/3 \le p < 6$, the unique weak solution of (2.1), with $\mu_1 = 0$, in the space-periodic case. We indicate here that the constraint $p < 6$ is rather artificial and results from our use of the imbedding (for $n = 3$): $W^{1,2}(\Omega) \hookrightarrow L^6(\Omega)$; the constraint may, in fact, be easily removed and our efforts are directed principally at identifying the smallest value of p, in dimension $n = 3$, for which $\boldsymbol{v}$ is a unique weak solution.

We begin this time by choosing $w_i = v_i''(1 + \|\boldsymbol{v}\|_{1,2}^2)^{-\lambda}$; following the pattern of § 4, 5 we seek to estimate

$$\int_0^T \Big(\int_\Omega v_j' \frac{\partial v_i}{\partial x_j} v_i' \, d\boldsymbol{x}\Big)(1 + \|\boldsymbol{v}\|_{1,2}^2)^{-\lambda} \, dt$$

In dimension $n = 3$, the interpolation-embedding estimates

$$\|\boldsymbol{v}\|_{1,3} \le \bar{c}_1 \|\boldsymbol{v}\|_{1,2}^{1/2} \|\boldsymbol{v}\|_{2,2}^{1/2} \tag{6.1}$$

and

$$\|\boldsymbol{v}\|_{1,3} \le \bar{c}_2 \|\boldsymbol{v}\|_{1,p}^{p/2-1/2} \|\boldsymbol{v}\|_{1,3p}^{3/2-p/2} \tag{6.2}$$

hold, for some $\bar{c}_1, \bar{c}_2 > 0$. We now set

$$\mathcal{I} = \int_\Omega (1 + |e|)^{p-2} \frac{\partial e_{ij}}{\partial x_k} \frac{\partial e_{ij}}{\partial x_k} \, d\boldsymbol{x}. \tag{6.3}$$

We will need the estimates

$$\|\boldsymbol{v}\|_{2,2} \le \bar{c}_3 \mathcal{I}^{1/2} \tag{6.4}$$

and

$$\|\boldsymbol{v}\|_{1,3p} \le \bar{c}_4 \mathcal{I}^{1/p}. \tag{6.5}$$

Estimate (6.4) is proved in the appendix of [5]. Estimate (6.5) follows from the Friedrichs inequality

$$\|(e_{ij}e_{ij})^{p/2}\|_{1,2} \le c\mathcal{I}^{1/2}$$

the embedding of $W^{1,2}$ into L^6, and the usual L^r (with $r = 3p$) Korn inequality (as proven, e.g., in Nečas [30]).

For $p \ge 3$ the estimation of the key integral is trivial, i.e., $\exists\, \bar{c}_5 > 0$ such that

$$\Big|\int_0^T \Big(\int_\Omega v_j' \frac{\partial v_i}{\partial x_j} v_i'\,d\boldsymbol{x}\Big)(1+\|\boldsymbol{v}\|_{1,2}^2)^{-\lambda}\,dt\Big| \le \bar{c}_5 \int_0^T \|v\|_{1,p}^3\,dt \tag{6.6}$$

Therefore, we will turn our attention to the case in which $7/3 \le p < 3$. By virtue of (6.2), (6.1), for any β such that $0 < \beta < 1$, and some $\bar{c}_6 > 0$:

$$\|\boldsymbol{v}\|_{1,3} \le \bar{c}_6 \|\boldsymbol{v}\|_{1,2}^{\beta/2} \|\boldsymbol{v}\|_{2,2}^{\beta/2} \|\boldsymbol{v}\|_{1,p}^{(1-\beta)\frac{(p-1)}{2}} \|\boldsymbol{v}\|_{1,3p}^{(1-\beta)\frac{(3-p)}{2}} \tag{6.7}$$

Then, with $\beta = \frac{4(3-p)}{3(4-p)}$ and $\lambda = 2(\frac{3-p}{p-1})$ we have

$$\begin{aligned}
&\int_0^T \|\boldsymbol{v}\|_{1,3}^3 (1+\|\boldsymbol{v}\|_{1,2}^2)^{-\lambda}\,dt \\
&\le \bar{c}_7 \int_0^T \Big\{ \|\boldsymbol{v}\|_{1,2}^{3\beta/2} \|\boldsymbol{v}\|_{1,p}^{\frac{3(1-\beta)(p-1)}{2}} \cdot \mathcal{I}^{(\frac{3\beta}{4} + \frac{3(3-p)(1-\beta)}{2p})} \\
&\qquad \times (1+\|\boldsymbol{v}\|_{1,2}^2)^{-\lambda} \Big\}\,dt \\
&\le \bar{c}_8 \Big(\int_0^T \mathcal{I}(1+\|\boldsymbol{v}\|_{1,2}^2)^{-\lambda}\,dt\Big)^{\frac{3(3-p)}{2(4-p)}} \\
&\qquad \times \Big(\int_0^T \|\boldsymbol{v}\|_{1,p}^p\,dt\Big)^{\frac{p-1}{2(4-p)}}
\end{aligned} \tag{6.8}$$

Combining our estimates we are led to the following two results:

$$\int_0^T \mathcal{I}(1+\|\boldsymbol{v}\|_{1,2}^2)^{-\lambda}\,dt \le c \tag{6.9}$$

$$(1+\|\boldsymbol{v}\|_{1,2}^2)^{1-\lambda} \le c \tag{6.10}$$

where c is independent of γ.

The estimates (6.9), (6.10) hold for $p \ge 2$; however, λ is clearly decreasing in p and for $p = 7/3$, $\lambda = 1$. We may, therefore, state

Lemma 6.1 *For $7/3 \le p < 6$, independently of γ, the unique regular weak solution $\boldsymbol{v} \in L^\infty((0,T);V)$, $(V = W^{2,2}_{per}(\Omega), \Omega = (0,L)^3)$ for the incompressible bipolar fluid satisfies, for some $c > 0$*

$$\sup_{(0,T)} \|\boldsymbol{v}\|_{1,2} \le c < \infty \tag{6.11}$$

$$\int_0^T\!\!\int_\Omega (1+|\boldsymbol{e}|)^{p-2} \frac{\partial e_{ij}}{\partial x_k}\frac{\partial e_{ij}}{\partial x_k}\, d\boldsymbol{x} \le c < \infty \tag{6.12}$$

and

Theorem 6.1 *For $p \geq 7/3$ and $n = 3$, in the case of the space-periodic problem (and under the same assumptions as apply in Theorem 4.1) if $\boldsymbol{v}_{\gamma_k} \rightharpoonup \boldsymbol{v}$ in $L^p((0,T);W^{1,p}(\Omega))$ then $\boldsymbol{v}_{\gamma_k} \to \boldsymbol{v}$ in $L^{\tilde{p}}((0,T);W^{1,3\tilde{p}}(\Omega))$, for any $\tilde{p} < p$, and the bounds (6.11), (6.12) remain valid for the limit function $\boldsymbol{v}$. Therefore, $\nu_{\boldsymbol{x},t}(\boldsymbol{\lambda}) = \delta(\boldsymbol{\lambda} - \nabla \boldsymbol{v}(\boldsymbol{x},t))$, a.e. in Q_T, and, in addition, we have for the limit function $\boldsymbol{v}$:*

$$\int_0^T \|\boldsymbol{v}\|_{2,2}^2\, dt \le c' < \infty \tag{6.13}$$

$$\int_0^T \|\boldsymbol{v}\|_{1,3p}^p\, dt \le c' < \infty \tag{6.14}$$

for some $c' > 0$.

We now make the following definition:

<u>Definition 6.1</u>.
A regular weak solution to the equations (1.1)–(1.3), (1.14), with $\gamma = 0 (\Leftrightarrow \mu_1 = 0)$ is a function $\boldsymbol{v} \in L^\infty((0,T);W^{1,2}(\Omega)) \cap L^p((0,T);W^{1,p}(\Omega)) \cap L^2((0,T);W^{2,2}(\Omega))$, with $\operatorname{div} \boldsymbol{v} = 0$, which satisfies (2.1), with $\mu_1 = 0$, and for which $\dot{\boldsymbol{v}} \in L^p((0,T);W^{-1,p}(\Omega))$.

Theorem 6.2 *The weak solution to (1.1)–(1.3), (1.14) with $\gamma = 0$, which is given by Theorem 6.1 for the space-periodic problem in space dimension $n = 3$, is unique provided $p \geq 7/3$.*

<u>Proof</u>: We begin by noting that for $\boldsymbol{\phi} \in (C_0^\infty(\Omega))^3$

$$\int_0^T \sup_{\|\boldsymbol{\phi}\|_{1,p} \le 1} \left| \int_\Omega v_j \frac{\partial v_i}{\partial x_j} \phi_i\, d\boldsymbol{x} \right|^p dt \tag{6.15}$$

$$= \int_0^T \sup_{\|\boldsymbol{\phi}\|_{1,p}\le 1} \Big| \int_\Omega v_j v_i \frac{\partial \phi_i}{\partial x_j}\, d\boldsymbol{x} \Big|^p dt$$

$$\le c \int_0^T \Big(\int_\Omega |\boldsymbol{v}|^{2\frac{p}{p-1}}\, d\boldsymbol{x} \Big)^{p-1} dt$$

However, $\dfrac{2p}{p-1} \le 6$, and $\sup_{(0,T)} \|\boldsymbol{v}\|_{1,2} < \infty$, so for some $\tilde{c} > 0$

$$\int_0^T \sup_{\|\boldsymbol{\phi}\|_{1,p}\le 1} \Big| \int_\Omega v_j \frac{\partial v_i}{\partial x_j} \phi_i \, d\boldsymbol{x} \Big|^p dt \;\le\; \tilde{c} < \infty \tag{6.16}$$

Now, let $\boldsymbol{v}^1, \boldsymbol{v}^2$ be two solutions of (2.1), with $\operatorname{div} \boldsymbol{v}^j = 0$, $\boldsymbol{v}^j \in L^\infty((0,T); W^{1,2}(\Omega))$, $j = 1,2$. If we take the difference of (2.1) for $\boldsymbol{v} = \boldsymbol{v}^1$ and $\boldsymbol{v} = \boldsymbol{v}^2$ and integrate over $(0,T)$, it is clear that the troublesome term which must be dealt with is

$$\begin{aligned}
&\int_0^T \Big| \int_\Omega (v_j^2 - v_j^1) \frac{\partial v_i^1}{\partial x_j} (v_i^2 - v_i^1)\, d\boldsymbol{x} \Big| dt \\
&\qquad \le \int_0^T \Big(\int_\Omega |\boldsymbol{v}^2 - \boldsymbol{v}^1|^4 \, d\boldsymbol{x} \Big)^{1/2} \|\boldsymbol{v}^1\|_{1,2}\, dt \qquad (6.17)\\
&\qquad \le c_1 \int_0^T \Big(\int_\Omega |\boldsymbol{v}^2 - \boldsymbol{v}^1|^2 \, d\boldsymbol{x} \Big)^{1/4} \Big(\int_\Omega |\boldsymbol{v}^2 - \boldsymbol{v}^1|^6 \, d\boldsymbol{x} \Big)^{1/4} dt
\end{aligned}$$

where we have used the Hölder inequality and the fact that $\sup_{(0,T)} \|\boldsymbol{v}^1\|_{1,2} \le c_1 < \infty$. Therefore, by Young's inequality and the imbedding $W^{1,2}(\Omega) \hookrightarrow L^6(\Omega)$, we have for $\delta > 0$

$$\begin{aligned}
&\int_0^T \Big| \int_\Omega (v_j^2 - v_j^1) \frac{\partial v_i^1}{\partial x_j} (v_i^2 - v_i^1)\, d\boldsymbol{x} \Big| dt \\
&\qquad \le c_1 \delta \int_0^T \Big(\int_\Omega |\boldsymbol{v}^2 - \boldsymbol{v}^1|^6 \, d\boldsymbol{x} \Big)^{1/3} dt \qquad (6.18)\\
&\qquad + c_2 \lambda(\delta) \int_0^T \Big(\int_\Omega |\boldsymbol{v}^2 - \boldsymbol{v}^1|^2 \, d\boldsymbol{x} \Big) dt
\end{aligned}$$

$$\leq c_3\delta \int_0^T \|v^2 - v^1\|_{1,2}^2 \, dt + c_2\lambda(\delta) \int_0^T\int_\Omega |v^2 - v^1|^2 \, dx dt$$

Thus, if we difference (2.1) for $v = v^1, v = v^2$ and integrate over $(0, T)$, for δ sufficiently small the term

$$\delta \int_0^T \|v^2 - v^1\|_{1,2}^2 \, dt$$

in the estimate (6.18), above, can be absorbed, by virtue of (1.9), by the integral involving the nonlinear viscosity in (2.1); the uniqueness theorem then follows by a standard application of the usual Gronwall lemma. ∎

7 Young Measures are Dirac and the Weak Solutions are Regular for $11/5 \leq p < 7/3, n = 3$

In this section we will show that the $L^\infty(0, T\ W^{1,2})$ estimate for v holds also for $p \in (11/5,\ 7/3)$.

We first note that the following interpolation estimate holds

$$\|v\|_{1,3} \leq \bar{c}_2 \|v\|_{1,2}^{\frac{2(p-1)}{3p-2}} \|v\|_{1,3p}^{\frac{p}{3p-2}}. \tag{7.1}$$

By virtue of (6.2) and (7.1), for $\beta \in [0, 1]$, and some $\bar{c}_6 > 0$:

$$\|v\|_{1,3} \leq \bar{c}_6 \|v\|_{1,2}^{\beta\frac{2(p-1)}{3p-2}} \|v\|_{1,p}^{(1-\beta)\frac{(p-1)}{2}} \|v\|_{1,3p}^{(1-\beta)\frac{(3-p)}{2}+\beta\frac{p}{3p-2}} \tag{7.2}$$

Then, with $\beta = \frac{(3p-2)(3-p)}{6(p-1)}$ and

$$\lambda_1 = \frac{(3-p)}{2}(1 + \frac{3\lambda}{2}) \tag{7.3}$$

with λ is as in the previous section, i.e., $\lambda = 2(\frac{3-p}{p-1})$

$$\int_0^T \|\boldsymbol{v}\|_{1,3}^3 (1+\|\boldsymbol{v}\|_{1,2}^2)^{-\lambda_1}\,dt$$

$$\leq \bar{c}_7 \int_0^T \Big\{ \|\boldsymbol{v}\|_{1,2}^{\beta\frac{6(p-1)}{3p-2}} \|\boldsymbol{v}\|_{1,p}^{(1-\beta)\frac{3(p-1)}{2}} \|\boldsymbol{v}\|_{1,3p}^{(1-\beta)\frac{3(3-p)}{2}+\beta\frac{3p}{3p-2}}$$

$$\times\ (1+\|\boldsymbol{v}\|_{1,2}^2)^{-\lambda_1} \Big\}\,dt \tag{7.4}$$

$$\leq \bar{c}_8 \Big(\int_0^T \|\boldsymbol{v}\|_{1,3p}^p (1+\|\boldsymbol{v}\|_{1,2}^2)^{-\lambda}\,dt\Big)^{\frac{3(3-p)}{4}}$$

$$\times \Big(\int_0^T \|\boldsymbol{v}\|_{1,p}^p\,dt\Big)^{\frac{3p-5}{4}}$$

Now, by (6.5) and (6.9) the term $\Big(\int_0^T \|\boldsymbol{v}\|_{1,3p}^p (1+\|\boldsymbol{v}\|_{1,2}^2)^{-\lambda}\,dt\Big)$ is bounded independently of γ; therefore, we have that , for some $c>0$, independent of γ,

$$\int_0^T \Big(\int_\Omega (1+|\boldsymbol{e}|)^{p-2} \frac{\partial e_{ij}}{\partial x_k}\frac{\partial e_{ij}}{\partial x_k}\,d\boldsymbol{x}\Big)(1+\|\boldsymbol{v}\|_{1,2}^2)^{-\lambda_1}\,dt \leq c \tag{7.5}$$

and

$$\sup_{(0,T)} (1+\|\boldsymbol{v}(t)\|_{1,2}^2)^{1-\lambda_1} \leq c \tag{7.6}$$

Set

$$\lambda_k = \frac{(3-p)}{2}\Big(1+\frac{3\lambda_{k-1}}{2}\Big) \tag{7.7}$$

and assume that we have proved that for some c independent of γ

$$\int_0^T \Big(\int_\Omega (1+|\boldsymbol{e}|)^{p-2} \frac{\partial e_{ij}}{\partial x_k}\frac{\partial e_{ij}}{\partial x_k}\,d\boldsymbol{x}\Big)(1+\|\boldsymbol{v}\|_{1,2}^2)^{-\lambda_{k-1}}\,dt \leq c \tag{7.8}$$

By repeating the above procedure we find that

$$\int_0^T \|\boldsymbol{v}\|_{1,3}^3 (1+\|\boldsymbol{v}\|_{1,2}^2)^{-\lambda_k}\,dt$$

$$\leq \bar{c}_8 \Big(\int_0^T \|\boldsymbol{v}\|_{1,3p}^p (1+\|\boldsymbol{v}\|_{1,2}^2)^{-\lambda_{k-1}}\,dt\Big)^{\frac{3(3-p)}{4}}$$

$$\times \Big(\int_0^T \|\boldsymbol{v}\|_{1,p}^p \, dt\Big)^{\frac{3p-5}{4}} \tag{7.9}$$

from which it then follows that, for some $C_k > 0$, independent of γ, we have

$$\int_0^T \Big(\int_\Omega (1+|\boldsymbol{e}|)^{p-2} \frac{\partial e_{ij}}{\partial x_k}\frac{\partial e_{ij}}{\partial x_k}\, d\boldsymbol{x}\Big)(1+\|\boldsymbol{v}\|_{1,2}^2)^{-\lambda_k}\, dt \equiv A_k \le C_k \tag{7.10}$$

and

$$\sup_{(0,T)} (1+\|\boldsymbol{v}(t)\|_{1,2}^2)^{1-\lambda_k} \equiv B_k \le C_k. \tag{7.11}$$

The sequence λ_k is decreasing and convergent, i.e.,

$$\lim_{k\to\infty} \lambda_k = \lambda_\infty = 2\Big(\frac{3-p}{3p-5}\Big). \tag{7.12}$$

We have that $\lambda_\infty < 1$ for $p > 11/5$ therefore $\forall p > 11/5$, $\exists k(p)$ such that $\lambda_{k(p)} < 1$; thus, after finitely many iterations of the above scheme we find that Lemma (6.1) holds for $p > 11/5$.

The case $p = 11/5$ requires infinitely many iterations and we need to look at the limit of the sequence A_k. We rewrite (4.9) in the form

$$\begin{aligned} &\frac{1}{2\rho}\int_0^T\int_\Omega (1+|\boldsymbol{e}|)^{p-2} e_{ij}(\boldsymbol{v}')e_{ij}(\boldsymbol{v}')\cdot(1+\|\boldsymbol{v}\|_{1,2}^2)^{-\lambda}\, d\boldsymbol{x}dt \\ &\qquad +\gamma\int_0^T (1+\|\boldsymbol{v}\|_{1,2}^2)^{-\lambda}\int_\Omega \frac{\partial e_{ij}(\boldsymbol{v}')}{\partial x_k}\frac{\partial e_{ij}(\boldsymbol{v}')}{\partial x_k}\, d\boldsymbol{x}dt \\ &\le \frac{1}{2(1-\lambda)}\Big((1+\|\boldsymbol{v}(0)\|_{1,2}^2)^{1-\lambda} - (1+\|\boldsymbol{v}(T)\|_{1,2}^2)^{1-\lambda}\Big) \\ &\qquad + c\int_0^T \|\boldsymbol{b}\|_{0,2}^2 dt + c\int_0^T (1+\|\boldsymbol{v}\|_{1,2}^2)^{-\lambda}\, \|\boldsymbol{v}\|_{1,3}^3\, dt, \end{aligned} \tag{7.13}$$

and note that either $\dfrac{1}{2(1-\lambda)}\Big((1+\|\boldsymbol{v}(0)\|_{1,2}^2)^{1-\lambda} - (1+\|\boldsymbol{v}(T)\|_{1,2}^2)^{1-\lambda}\Big)$ is nonpositive or else

$$\frac{1}{2(1-\lambda)}\Big((1+\|\boldsymbol{v}(0)\|_{1,2}^2)^{1-\lambda} - (1+\|\boldsymbol{v}(T)\|_{1,2}^2)^{1-\lambda}\Big) \le \frac{1}{2}\log(1+\|\boldsymbol{v}(0)\|_{1,2}^2)$$

Using (7.9) and noting that $3(3-p)/4 = 0.6$, for $p = 11/5$, we then have that for some c_1, c_2 independent of both k and γ

$$A_{k+1} \le c_1 + c_2 \cdot A_k^{0.6} \tag{7.14}$$

from which it can be easily seen that the sequence A_k is bounded independently of γ. Letting k go to infinity in (7.10) we find that for some $c > 0$ independent of γ

$$\int_0^T \Big(\int_\Omega (1 + |e|)^{p-2} \frac{\partial e_{ij}}{\partial x_k} \frac{\partial e_{ij}}{\partial x_k}\, dx \Big) (1 + \|v\|_{1,2}^2)^{-1}\, dt \le c \tag{7.15}$$

from which we then deduce via (4.9) and (7.9) that

$$\sup_{(0,T)} (1 + \|v(t)\|_{1,2}^2) \le c. \tag{7.16}$$

We can, therefore, state

Theorem 7.1 *Lemma* (6.1), *Theorem* (6.1) *and Theorem* (6.2) *hold for* $11/5 \le p < 6$.

8 Young Measures are Dirac for $2 \le p < 11/5, n = 3$

In this section we again choose $w_i = v_i''(1 + \|v\|_{1,2}^2)^{-\lambda}$and proceeding as we did for the case $p = \dfrac{11}{5}$ we get

$$A_{k+1} \le c_1 + c_2 \cdot A_k^{\frac{3(3-p)}{4}} \tag{8.17}$$

instead of (7.14). Since $\frac{3(3-p)}{4} < 1$ in this range of p, it then follows that for $\lambda = \lambda_\infty = 2\dfrac{3-p}{3p-5}$ there exists $c > 0$, independent of γ, such that

$$\int_0^T \Big(\int_\Omega (1 + |e|)^{p-2} \frac{\partial e_{ij}}{\partial x_k} \frac{\partial e_{ij}}{\partial x_k}\, dx \Big) (1 + \|v\|_{1,2}^2)^{-\lambda}\, dt \le c \tag{8.18}$$

Using the Holder inequality we then have that for $\theta = \frac{p}{2\lambda+p}$, $0 < \theta < 1$, and

$$\int_0^T \mathcal{I}^\theta \, dt = \int_0^T \mathcal{I}^\theta (1 + \|\boldsymbol{v}\|_{1,2}^2)^{-\lambda\theta} (1 + \|\boldsymbol{v}\|_{1,2}^2)^{\lambda\theta} \, dt \qquad (8.19)$$
$$\leq \Big(\int_0^T \mathcal{I}(1 + \|\boldsymbol{v}\|_{1,2}^2)^{-\lambda} \, dt\Big)^\theta$$
$$\times \Big(\int_0^T (1 + \|\boldsymbol{v}\|_{1,2}^2)^{\frac{p}{2}} \, dt\Big)^{1-\theta}$$

where $\mathcal{I}$ is given by (6.3). By (8.19) and (6.5) we have

$$\int_0^T \|\boldsymbol{v}\|_{2,2}^{2\theta} \, dt \leq c < \infty, \quad \theta = \frac{p(3p-5)}{3(p^2 - 3p + 4)}. \qquad (8.20)$$

We now choose $q > 1$ so that $2\theta < q < p$. Because $W^{2,2}(\Omega) \hookrightarrow W^{1+s,p}(\Omega)$, with $s = (6-p)/2p$, we get, with $\Theta = \dfrac{qp - 2\theta p}{qp - 2\theta q}$ and $\sigma = (1 - \Theta)s$

$$\int_0^T \|\boldsymbol{v}\|_{1+\sigma,p}^q \, dt \leq c' \int_0^T \|\boldsymbol{v}\|_{1,p}^{\Theta q} \|\boldsymbol{v}\|_{1+s,p}^{(1-\Theta)q} \, dt$$
$$\leq \bar{c}\Big(\int_0^T \|\boldsymbol{v}\|_{2,2}^{2\theta} \, dt\Big)^{\frac{(1-\Theta)q}{2\theta}} \Big(\int_0^T \|\boldsymbol{v}\|_{1,p}^p \, dt\Big)^{\frac{2\theta - q + \Theta q}{2\theta}} \qquad (8.21)$$
$$\leq c^* < \infty$$

We can, therefore, state the following:

Theorem 8.1 *For $2 \leq p < 11/5$ and $n = 3$, in the case of the space-periodic problem (and under the same assumptions as apply in Theorem 4.1) the weak convergence $\boldsymbol{v}_{\gamma_k} \rightharpoonup \boldsymbol{v}$ in $L^p((0,T); W^{1,p}(\Omega))$ implies the strong convergence $\boldsymbol{v}_{\gamma_k} \to \boldsymbol{v}$, in $L^{\tilde{p}}((0,T); W^{1,p}(\Omega))$ for any $\tilde{p} < p$, and in $L^q((0,T); W^{1+\tilde{\sigma},p}(\Omega))$ for $\tilde{\sigma} < \sigma$. Thus, in this situation $\nu_{\boldsymbol{x},t}(\boldsymbol{\lambda}) = \delta(\boldsymbol{\lambda} - \nabla \boldsymbol{v}(\boldsymbol{x},t))$ and the bound represented by (8.21) applies to the limit function $\boldsymbol{v}$.*

Remark: After this work was completed it was brought to our attention that for the particular case of the standard Navier-Stokes equations $p = 2$, $(2\theta = \frac{2}{3})$, estimate (8.20) was already known. More precisely in 1981 Foias, Guillope, and Temam, [14] proved it for the space periodic case and in 1990 Duff, [12] established it for a general bounded domain.

9 Young Measures are Dirac for $9/5 < p < 2, n = 3$

For $9/5 < p < 2, n = 3$, we have $W^{2,p}(\Omega) \hookrightarrow W^{1,q}(\Omega)$ with $\frac{1}{q} = \frac{1}{p} - \frac{1}{3} = (3-p)/3p$; so, for some $c > 0$

$$\|\boldsymbol{v}\|_{1,3} \leq c\,\|\boldsymbol{v}\|_{1,p}^{\frac{2p-3}{p}}\,\|\boldsymbol{v}\|_{1,q}^{\frac{3-p}{p}} \tag{9.1}$$

which is, in fact, valid for $p > 3/2$. Furthermore, we have $W^{2,p}(\Omega) \hookrightarrow W^{1+\sigma,2}(\Omega)$, for $p > 6/5$, with $\sigma = \frac{5}{2} - \frac{3}{p}$, and as $W^{1+\delta,2}(\Omega) \hookrightarrow W^{1,3}(\Omega)$, with $\delta = 1/2$, we get

$$\|\boldsymbol{v}\|_{1,3} \leq c\,\|\boldsymbol{v}\|_{1,2}^{\frac{4p-6}{5p-6}} \cdot \|\boldsymbol{v}\|_{2,p}^{\frac{p}{5p-6}} \tag{9.2}$$

Thus, the relevant integral, generated by the convective term, may be estimated by

$$\begin{aligned}
&\int_0^T \|\boldsymbol{v}\|_{1,3}^3\,(1+\|\boldsymbol{v}\|_{1,2}^2)^{-\lambda}\,dt \\
&\leq c'\int_0^T \Big[\|\boldsymbol{v}\|_{1,2}^{3(\frac{4p-6}{5p-6})\beta}\,\|\boldsymbol{v}\|_{2,p}^{\frac{3p\beta}{5p-6}}\,dt \\
&\times \|\boldsymbol{v}\|_{1,p}^{3(\frac{2p-3}{p})(1-\beta)}\,\|\boldsymbol{v}\|_{2,p}^{3(\frac{3-p}{p})(1-\beta)}\,(1+\|\boldsymbol{v}\|_{1,2}^2)^{-\lambda}\Big]\,dt
\end{aligned} \tag{9.3}$$

for $0 < \beta < 1$. We set

$$\beta = -\frac{(5p-6)(3-p)}{6(3-2p)}, \qquad \lambda = \frac{2(3-p)}{3p-5} \tag{9.4}$$

and note that for $p > 9/5$, $\beta < 1$. By employing (5.2), we see that (9.3) yields the estimate

$$\begin{aligned}
&\int_0^T \|\boldsymbol{v}\|_{1,3}^3\,(1+\|\boldsymbol{v}\|_{1,2}^2)^{-\lambda}\,dt \\
&\leq \bar{c}\Big(\int_0^T\Big(\int_\Omega (1+|\boldsymbol{e}|)^{p-2}\,\frac{\partial e_{ij}}{\partial x_k}\,\frac{\partial e_{ij}}{\partial x_k}\,dx\Big)\cdot(1+\|\boldsymbol{v}\|_{1,2}^2)^{-\lambda}\,dt\Big)^{\frac{3(3-p)}{4}} \\
&\times \Big(\int_0^T (1+\|\boldsymbol{v}\|_{1,p})^p\,dt\Big)^{\frac{3p-5}{4}}
\end{aligned} \tag{9.5}$$

Therefore, for some $\hat{c} > 0$, $\hat{c}$ independent of γ,

$$\int_0^T \Big(\int_\Omega (1 + |\boldsymbol{e}|)^{p-2} \frac{\partial e_{ij}}{\partial x_k} \frac{\partial e_{ij}}{\partial x_k} \, d\boldsymbol{x} \Big) \cdot (1 + \|\boldsymbol{v}\|_{1,2}^2)^{-\lambda} \, dt \ \le \ \hat{c} < \infty \tag{9.6}$$

in which case, we also have for $c^* > 0$ independent of γ, that

$$\int_0^T \|\boldsymbol{v}\|_{2,p}^2 \Big(\int_\Omega (1 + |\boldsymbol{e}|^p \, d\boldsymbol{x} \Big)^{1-\frac{2}{p}} (1 + \|\boldsymbol{v}\|_{1,2}^2)^{-\lambda} \, dt \ \le c^* < \infty \tag{9.7}$$

As

$$\|\boldsymbol{v}\|_{1,2} \ \le \ \tilde{c} \|\boldsymbol{v}\|_{1,p}^{\frac{5p-6}{2p}} \cdot \|\boldsymbol{v}\|_{2,p}^{\frac{6-3p}{2p}} \tag{9.8}$$

we find that for c independent of γ:

$$\int_0^T \|\boldsymbol{v}\|_{2,p}^{\gamma_1} \Big[\|\boldsymbol{v}\|_{2,p}^{\gamma_2} + \|\boldsymbol{v}\|_{1,p}^{\gamma_3} \Big]^{-1} \cdot (1 + \|\boldsymbol{v}\|_{1,p})^{p-2} \, dt \ \le c < \infty \tag{9.9}$$

with

$$\gamma_1 \ = \ 2 - \frac{2(6-3p)(3-p)}{p(3p-5)} \tag{8.9a}$$

$$\gamma_2 \ = \ -\frac{2(6-3p)(3-p)}{p(3p-5)} \tag{8.9b}$$

and

$$\gamma_3 \ = \ \frac{2(3-p)(5p-6)}{p(3p-5)} \tag{8.9c}$$

Integrating (2.1) first over $M = \{t \in (0,T) | \, \|\boldsymbol{v}\|_{2,p} \ > \ 1\}$, we then find that for c' independent of γ

$$\int_M \|\boldsymbol{v}\|_{2,p}^{\frac{4(5p-9)}{p(3p-5)}} \cdot (1 + \|\boldsymbol{v}\|_{1,p})^{\gamma_4} \, dt \ \le c' < \infty \tag{9.10}$$

with

$$\gamma_4 \ = \ -\frac{2(3-p)(5p-6)}{p(3p-5)} \ + \ p \ - \ 2 \tag{9.11}$$

However, it is clear that with $I = (0,T)$

$$\int_{I \setminus M} \|\boldsymbol{v}\|_{2,p}^{\frac{4(5p-9)}{p(3p-5)}} \cdot (1 + \|\boldsymbol{v}\|_{1,p})^{\gamma_4} \, dt \ \le c'' < \infty$$

Setting $P = \frac{4(9-2p-p^2)}{p(3p-5)}$, $\theta = \frac{4(5p-9)}{p(3p-5)} \frac{1}{P}$, $\gamma_5 = \frac{\gamma_4}{P}$, $s = p(1 - \frac{1}{P})$ and using the

Holder inequality we find

$$\int_0^T \|\boldsymbol{v}\|_{2,p}^{\theta} \cdot (1+\|\boldsymbol{v}\|_{1,p})^{\gamma_5}(1+\|\boldsymbol{v}\|_{1,p})^{s}\,dt$$

$$\leq \left(\int_0^T \|\boldsymbol{v}\|_{2,p}^{\frac{4(5p-9)}{p(3p-5)}} \cdot (1+\|\boldsymbol{v}\|_{1,p})^{\gamma_4}\,dt\right)^{1/P}$$

$$\times \left(\int_0^T (1+\|\boldsymbol{v}\|_{1,p})^{p}\,dt\right)^{1-1/P} \leq c' < \infty \tag{9.12}$$

Noting that

$$\|\boldsymbol{v}\|_{1,p} \;\leq\; \tilde{c}\|\boldsymbol{v}\|_{0,p}^{\frac{1}{2}} \cdot \|\boldsymbol{v}\|_{2,p}^{\frac{1}{2}} \leq c\|\boldsymbol{v}\|_{2,p}^{\frac{1}{2}} \tag{9.13}$$

and letting $\eta = \min\{\theta + \frac{1}{2}(\gamma_5 + s);\ \theta\} = \min\{\theta;\ \frac{p-1}{2}\} > 0$ we obtain, for $\bar{c}$ independent of γ, the estimate

$$\int_0^T \|\boldsymbol{v}\|_{2,p}^{\eta}\,dt \;\leq \bar{c} < \infty \tag{9.14}$$

Finally, choosing q such that $\eta < q < p$ we get, for $9/5 < p < 2$, with $\sigma = \dfrac{\eta(p-q)}{q(p-\eta)}$

$$\int_0^T \|\boldsymbol{v}\|_{1+\sigma,p}^{2}\,dt \;\leq c_1 \int_0^T \|\boldsymbol{v}\|_{1,p}^{(1-\sigma)q} \cdot \|\boldsymbol{v}\|_{2,p}^{\sigma q}\,dt \;\leq c_2 < \infty \tag{9.15}$$

with c_1, c_2 both independent of γ. We may now state

Theorem 9.1 *The conclusions of Theorem 6.1 hold for $n = 3$, when $9/5 < p < 2$, including the estimate (9.15) for the limit function $\boldsymbol{v}$, provided we take $\sigma = \eta(p-q)/q(p-\eta)$, and $2\eta < q < p$.*

ACKNOWLEDGMENT

It is a pleasure for the authors to acknowledge useful exchanges with their colleagues in Prague, J. Malek and M. Ruzicka, as well as with our students W. Hao and S. Wills.

References

[1] J.M. Ball, "A Version of the Fundamental Theorem for Young Measures", preprint.

[2] H. Bellout, and F. Bloom, "Steady Plane Poiseuille Flows of Incompressible Multipolar Fluids", submitted for publication.

[3] H. Bellout, F. Bloom, and J. Nečas, "Phenomenological Behavior of Multipolar Viscous Fluids", to appear in the *Quarterly of Applied Mathematics.*

[4] H. Bellout, F. Bloom, and J. Nečas, "Bounds for the Dimensions of the Attractors of Nonlinear Bipolar Viscous Fluids", submitted for publication.

[5] H. Bellout, F. Bloom, and J. Nečas, "Existence, Uniqueness, and Stability of Solutions to the Initial Boundary Value Problem for Bipolar Viscous Fluids", submitted for publication.

[6] J.L. Bleustein, and A. E. Green, "Dipolar Fluids", *Int. J. Engrg. Sci.*, *5*, (1967), 323-340.

[7] R. DiPerna, "Compensated Compactness and General Systems of Conservation Laws", *Trans. A.M.S.*, vol. 292, (1985), 383-419.

[8] R. DiPerna, "Oscillations in Solutions to Nonlinear Differential Equations", in *Oscillation Theory, Computation, and Methods of Compensated Compactness*, Dafermos, C., Ericksen, J., and D. Kinderlehrer, eds., *IMA Volume in Math and Applications*, vol. 2, (1986), Springer-Verlag, N.Y.

[9] R. DiPerna, and A. Majda, "Oscillations and Concentrations in Weak Solutions of the Incompressible Fluid Equations", *Comm. Math. Phys.*, 108, (1987), 667-689.

[10] R. DiPerna, and A. Majda, "Concentrations in Regularizations for 2-D Incompressible Flow", *Comm. Pure Appl. Math.*, 40, (1987), 301-345.

[11] Q. Du, and M. D. Gunzburger, "Analysis of a Ladyzhenskaya Model for Incompressible Viscous Flow", *J. Math. Anal. Applic., 155*, (1991), 21-45.

[12] G.F.D. Duff, " Derivative estimates for the Navier-Stokes equations in a three dimensional region", *Acta Math.*, 164 (1990), 145-210.

[13] R.E. Edwards, *Functional Analysis*, Holt, Rinehart, and Winston (1965).

[14] C. Foias, C. Guillope, and R. Temam, " New a priori estimates for Navier-Stokes equations in dimension 3", *Comm. P.D.E.* , 6 (1981), 329-359.

[15] A.E. Green, and R.S. Rivlin, "Simple Force and Stress Multipoles", *Arch. Rat. Mech. Analy., 16*, (1964), 325-353.

[16] A.E. Green, and R.S. Rivlin, "Multipolar Continuum Mechanics", *Arch. Rat. Mech. Anal., 17*, (1964), 113-147.

[17] S. Kaniel, "On the Initial-Value Problem for an Incompressible Fluid with Nonlinear Viscosity", *J. Math. and Mech., 19*, (1970), 681-706.

[18] O. Ladyzhenskaya, *The Mathematical Theory of Viscous Incompressible Flow*, Gordon and Breach, Pub., New York 1965.

[19] O. Ladyzhenskaya, "New Equations for the Description of the Viscous Incompressible Fluids and Solvability in the Large of the Boundary Value Problems for them", in *Boundary Value Problems of Mathematical Physics V.*, Amer. Math. Soc., Providence, R.I. 1970.

[20] J. Leráy, *Étude de diverses équations intégrales non linéaires et de quelques problémes que pose l'Hydrodynamique*, J. Math. Pures. Appl., 12, (1933), 1-82.

[21] J.L. Lions, *Quelques Méthodes de Resolution des Problémes aux Limites non Linéaires*, Dunrod, Gauthier-Villars, Paris (1969).

[22] F. Murat, "Compacité par Compensation II", in *Recent Methods in Non-linear Analysis*, De Giorgi, Magenes, and Mosco, eds., (1979), Pitagora Editrice, Bologna, Italy.

[23] F. Murat, "Compacité par Compensation", *Annali Scuola Norm. Sup. Pisa*, vol. 5, (1978), 485-507.

[24] J. Malek, J. Nečas, and A. Novotny, "Measure-Valued Solutions and Asymptotic Behavior of a Multipolar Model of a Boundary Layer", preprint.

[25] F. Murat, "Compacité par Compensation II", in *Recent Methods in Nonlinear Analysis*, De Giorgi, Magenes, and Mosco, eds., (1979), Pitagora Editrice, Bologna, Italy.

[26] F. Murat, "Compacité par Compensation", *Annali Scuola Norm. Sup. Pisa*, vol. 5, (1978), 485-507.

[27] J. Nečas, A. Novotny, and M. Silhavy, " Global Solution to the Compressible Isothermal Multipolar Fluid", *J. Math. Anal. Appl.*, 162, (1991), 223-242.

[28] J. Nečas, and M. Silhavy, "Multipolar Viscous Fluids", *Quarterly of Applied Mathematics, Vol. XLIX*, no. 2, (1991), 247-266.

[29] J. Nečas, "Theory of Multipolar Viscous Fluids", in the *Proceedings of the 2nd International Conference on Finite Element Methods*, Brunel University, U.K. (1990).

[30] J. Nečas, *Equations aux Derivees Partielles*, Presse de Université de Montréal, Montreal, 1965.

[31] V. Girault, P. Raviart, P., " Finite Element Methods for Navier-Stokes Equations", Springer-Verlag, 1986.

[32] D. Serre, "La Compacité par Compensation pour les Systèmes Hyperboliques Nonlinéaires à une Dimension d'espace", *J. Math Pure et Appl.*, vol. 65, (1987), 423-468.

[33] D. Serre, "Solutions a Variation Borneé pour Certains Systemes Hyperboliques de Lois de Conservation", *J. Diff. Eqs.*, vol. 67, (1987).

[34] L. Tartar, "Compacité par Compensation: Résultats et Perspectives", in *Nonlinear P.D.E. and Their Applications: Collége de France Seminar*, vol. IV, (1983), Pitman Pub., Marshfield, Mass.

[35] L. Tartar, "The Compensated Compactness Method Applied to Systems of Conservation Laws", in *Systems of Nonlinear P.D.E.*, J. M. Ball, ed., 263-285, (1983), D. Reidel Pub. Co., Dordrecht.

[36] L. Tartar, *Cours Peccot*, Collége de France (1977).

[37] R.A. Toupin, "Theories of Elasticity with Couple Stress", *Arch. Rat. Mech. Anal.*, *17*, (1964), 85-112.

[38] H. Triebel, "Interpolation Theory, Function Spaces, Differential Operators", *North-Holland*, 1978.

[39] L.C. Young, *Lectures on the Calculus of Variations and Optimal Control Theory*, Saunders, (1969).

Received February 1992

Acta Math., 176 (1996), 283–294
© 1996 by Institut Mittag-Leffler. All rights reserved

On Leray's self-similar solutions of the Navier–Stokes equations

by

J. NEČAS

Northern Illinois University
De Kalb, IL, U.S.A.
and
Charles University
Prague, Czech Republic

M. RŮŽIČKA

University of Bonn
Bonn, Germany

and

V. ŠVERÁK

University of Minnesota
Minneapolis, MN, U.S.A.

1. Introduction

In the 1934 paper [Le] Leray raised the question of the existence of self-similar solutions of the Navier–Stokes equations

$$\left.\begin{aligned} u_t-\nu\Delta u+(u\cdot\nabla)u+\nabla p&=0\\ \operatorname{div} u&=0\end{aligned}\right\}\quad \text{in } \mathbf{R}^3\times(t_1,t_2), \tag{1.1}$$

where, as usual, $\nu>0$. These are the solutions of the form

$$u(x,t)=\frac{1}{\sqrt{2a(T-t)}}\,U\biggl(\frac{x}{\sqrt{2a(T-t)}}\biggr), \tag{1.2}$$

where $T\in\mathbf{R}$, $a>0$, and $U=(U_1,U_2,U_3)$ is defined in $\mathbf{R}^3$. (Hence u is defined in $\mathbf{R}^3\times(-\infty,T)$.) One also requires that certain natural energy norms of u are finite. If $U\not\equiv0$, then u given by (1.2) develops a singularity at time $t=T$. The Navier–Stokes equations for u give the system

$$\left.\begin{aligned} -\nu\Delta U+aU+a(y\cdot\nabla)U+(U\cdot\nabla)U+\nabla P&=0\\ \operatorname{div} U&=0\end{aligned}\right\}\quad \text{in } \mathbf{R}^3 \tag{1.3}$$

for U (where we use y to denote a generic point in $\mathbf{R}^3$). The main result of this paper is that the only solution of (1.3) belonging to $L^3(\mathbf{R}^3)$ is $U\equiv0$.

We make a few remarks regarding the integrability condition $U\in L^3(\mathbf{R}^3)$. If one requires that u defined by (1.2) has finite kinetic energy and satisfies the natural energy equality

$$\int_{\mathbf{R}^3}\tfrac12|u(x,t_1)|^2\,dx=\int_{\mathbf{R}^3}\tfrac12|u(x,t_2)|^2\,dx+\int_{t_1}^{t_2}\int_{\mathbf{R}^3}\nu|\nabla u(x,t)|^2\,dx\,dt \tag{1.4}$$

© Institut Mittag-Leffler 1996. Reprinted with kind permission from Springer Science+Business Media from: Nečas, J., Růžička, M. & Šverák, V., On Leray's self-similar solutions of the Navier-Stokes equations, Acta Mathematica, 176, 283-294.

for each $t_1 < t_2 < T$, one obtains $\int_{\mathbf{R}^3} |U|^2 \, dy < +\infty$ and $\int_{\mathbf{R}^3} |\nabla U|^2 \, dy < +\infty$, which implies $U \in L^3(\mathbf{R}^3)$ by standard imbedding theorems. (We remark that Leray suggests to seek for bounded solutions of (1.3) satisfying $\int_{\mathbf{R}^3} |U|^2 \, dy < +\infty$, see [Le, p. 225]. Such solutions obviously belong to $L^3(\mathbf{R}^3)$.)

On the other hand, if one only requires that for some ball $B = B_R(0)$ and some $t_0 < T$ we have (for u given by (1.2))

$$\operatorname*{ess\,sup}_{t_0<t<T} \int_B |u(x,t)|^2 \, dx < +\infty \quad \text{and} \quad \int_{t_0}^T \int_B |\nabla u(x,t)|^2 \, dx \, dt < +\infty, \tag{1.5}$$

one gets conditions which do not imply $U \in L^3(\mathbf{R}^3)$. Therefore our result does not exclude the possibility of self-similar singularities which satisfy the natural energy estimates only locally. Finally, we remark that our method is also applicable in higher dimensions (specifically $3 \leqslant n \leqslant 7$), but we will not pursue the case $n \geqslant 4$ in this paper.

2. Preliminaries

We shall use the following notation for balls. For $x \in \mathbf{R}^n$ we denote $B_R(x) = \{y \in \mathbf{R}^n : |y-x| < R\}$, and for $(x,t) \in \mathbf{R}^n \times \mathbf{R}$ we denote $Q_R(x,t) = B_R(x) \times (t-R^2, t)$. If there is no danger of confusion, we will skip the center of the ball from the notation and write simply B_R or Q_R. We will use the following result.

PROPOSITION 2.1. *For each $\nu > 0$ there exists $\varepsilon_0 > 0$ and $C_0, C_1, \ldots > 0$ such that the following statement is true: let $u = (u_1, u_2, u_3)$ and p be smooth functions satisfying the Navier–Stokes equations (1.1) in $Q_R = B_R \times (-R^2, 0)$, and assume that*

$$R^{-2} \int_{Q_R} (|u(x,t)|^3 + |p(x,t)|^{3/2}) \, dx \, dt < \varepsilon_0.$$

Then, for each $k = 0, 1, \ldots$,

$$\sup_{Q_{R/2}} |\nabla^k u| \leqslant C_k R^{-1-k},$$

where ∇ means, as usual, the gradient with respect to the space variables x_1, x_2, x_3.

Proof. The statement can be deduced from [CKN, Proposition 1, p. 775] (see also [Sch]), and from [Se1] (see also [Oh]) in the following way. (We remark that one could also use other results on the regularity of the Navier–Stokes equations, such as [Gi], [So], [Str1], [Ta] and [vW].) We see from the scaling properties of the Navier–Stokes equations (see, for example, [CKN, p. 774]) that it is enough to consider the case $R = 1$. Using the local energy equality

$$\int_{B_R} |u(x,t)|^2\varphi(x,t)\,dx + 2\nu\int_{-R^2}^{t}\int_{B_R} |\nabla u|^2\varphi\,dx\,ds$$
$$= \int_{-R^2}^{t}\int_{B_R} |u|^2(\varphi_t+\nu\Delta\varphi)\,dx\,ds + \int_{-R^2}^{t}\int_{B_R} (|u|^2+2p)u\cdot\nabla\varphi\,dx\,dt,$$

where φ is a suitable "cut-off" function, we can estimate $\sup_{t\in(-R_1^2,0)}\int_{B_{R_1}}|u(x,t)|^2\,dx$ and $\int_{Q_{R_1}}|\nabla u|^2\,dx\,dt$ in terms if $\int_{Q_R}(|u|^3+|p|^{3/2})\,dx\,dt$ for a suitable $R_1<R$, for example $R_1=\frac{7}{8}R$. Now we can apply [CKN, Proposition 1, p. 775], to obtain the required estimate for $k=0$ in $Q_{3R/4}$. (Strictly speaking, this proposition is proved only for $\nu=1$, but it is easy to see that it holds for any $\nu>0$. The constants in the estimates depend on ν, of course. Also, the proposition as stated in [CKN] gives the estimates on the ball whose radius is the half of the original radius, but it is easily seen that the factor $\frac{1}{2}$ can be replaced by any factor strictly less than 1. Of course, the constants in the estimates depend on the factor we choose.) To prove the estimates for $k=1,2,\ldots$, we use the estimates we have obtained for $k=0$ together with the estimate for $\int_{Q_{R_1}}|\nabla u|^2\,dx\,dt$, and follow the proof of higher regularity in Serrin's paper [Se1]. Serrin does not write explicit estimates, but we can follow his proof line-by-line and replace the statements about smoothness properties of the functions which appear in the proof by the corresponding estimates. The elliptic and parabolic estimates needed to execute this procedure are listed in the appendix. Since this works without any additional tricks, we feel that it is not necessary to reproduce this purely mechanical procedure here in detail. □

We recall some well-known facts about the equation for the pressure

$$-\Delta P = \partial_j\partial_k U_j U_k \quad \text{in } \mathbf{R}^n. \tag{2.2}$$

(Here and in what follows we use the usual convention and sum over the repeated indices.) If $U\in L^{2q}(\mathbf{R}^n)$ for some $q>1$, we can use the classical Riesz transformation to solve (2.2). Let us recall, that for $j=1,\ldots,n$ the Riesz transformation R_j is the singular integral operator given by the Fourier multiplier $-i\xi_j/|\xi|$, see [St] for details.

From the classical results regarding the operators R_j (see e.g. [CZ] or [St]) we infer that P defined by

$$P = R_jR_k(U_jU_k)$$

satisfies $\|P\|_{L^q(\mathbf{R}^n)}\leqslant c_q\|U\|^2_{L^{2q}(\mathbf{R}^n)}$ and solves (2.2) in the sense of distributions. If moreover U is smooth, then P is smooth by the classical regularity theory for the Laplace operator.

Of course, for a given U, the equation (2.2) determines P only up to a harmonic function, and we have to specify the behavior of P at infinity to get uniqueness. For example the requirement $P\in L^q(\mathbf{R}^n)$ obviously guarantees uniqueness since the only harmonic function belonging to $L^q(\mathbf{R}^n)$ is the identical zero.

LEMMA 2.2. *For $j=1,\dots,n$ let R_j denote the Riesz transform given by the Fourier multiplyer $-i\xi_j/|\xi|$. For $j,k=1,\dots,n$ let $F_{jk}\in L^1(\mathbf{R}^n)\cap L^2(\mathbf{R}^n)$. Finally, let ϕ be a smooth, compactly supported function on $\mathbf{R}^n$ depending only on $|x|$ and satisfying $\phi(0)=1$. Then*

$$\lim_{\varepsilon\to 0_+}\int_{\mathbf{R}^n} R_jR_k(F_{jk})(x)\phi(\varepsilon x)\,dx=-\frac{1}{n}\int_{\mathbf{R}^n}F_{jj}(x)\,dx.$$

Proof. Let $\phi_\varepsilon(x)=\phi(\varepsilon x)$. Denoting by $\widehat{\ }$ the Fourier transformation (with the kernel given by $e^{2\pi i x\cdot\xi}$) we have from the Parseval formula

$$\int_{\mathbf{R}^n} R_jR_k(F_{jk})\phi_\varepsilon\,dx=\int_{\mathbf{R}^n}-\frac{\xi_j\xi_k}{|\xi|^2}\widehat{F}_{jk}(\xi)\bar{\hat{\phi}}_\varepsilon(\xi)\,d\xi=\int_{\mathbf{R}^n}-\frac{\xi_j\xi_k}{|\xi|^2}\widehat{F}_{jk}(\xi)\bar{\hat{\phi}}\Big(\frac{\xi}{\varepsilon}\Big)\varepsilon^{-n}\,d\xi.$$

By using the substitution $\xi/\varepsilon\mapsto\xi$, the last integral can be rewritten as

$$\int_{\mathbf{R}^n}-\frac{\xi_j\xi_k}{|\xi|^2}\widehat{F}_{jk}(\varepsilon\xi)\bar{\hat{\phi}}(\xi)\,d\xi.$$

Since $\widehat{F}_{jk}$ is bounded and continuous, and $\hat{\phi}$ is rapidly decreasing, this integral converges to

$$\widehat{F}_{jk}(0)\int_{\mathbf{R}^n}-\frac{\xi_j\xi_k}{|\xi|^2}\bar{\hat{\phi}}(\xi)\,d\xi$$

as $\varepsilon\to 0_+$. Using the fact that $\hat{\phi}$ is radial and integrating first over the spheres and then over the radii, we see that

$$\int_{\mathbf{R}^n}-\frac{\xi_j\xi_k}{|\xi|^2}\bar{\hat{\phi}}(\xi)\,d\xi=-\frac{\delta_{jk}}{n}\int_{\mathbf{R}^n}\bar{\hat{\phi}}(\xi)\,d\xi=-\frac{\delta_{jk}}{n}\phi(0)=-\frac{\delta_{jk}}{n},$$

where δ_{jk} is the usual Kronecker symbol. Since $\delta_{jk}\widehat{F}_{jk}=\widehat{F}_{jj}$ and $\widehat{F}_{jk}(0)=\int_{\mathbf{R}^n}F_{jk}(x)\,dx$, the result follows. □

3. The main theorem

By a weak solution of (1.3) we mean a function $U=(U_1,U_2,U_3)$ on $\mathbf{R}^3$ which belongs *locally* to $W^{1,2}$, is divergence-free, and satisfies

$$\int_{\mathbf{R}^3}(\nu\nabla U\cdot\nabla\varphi+aU\cdot\varphi+a(y\cdot\nabla)U\cdot\varphi+(U\cdot\nabla)U\cdot\varphi)\,dy=0$$

for each smooth, compactly supported, divergence-free vector field $\varphi=(\varphi_1,\varphi_2,\varphi_3)$ in $\mathbf{R}^3$. Using the regularity theory for the linear Stokes operator (see e.g. [Ca]), standard imbeddings and the bootstrapping argument, we see that every weak solution U of (1.3) is smooth. (The proof is more or less the same as the proof of regularity for steady solutions of the Navier–Stokes equations in three dimensions, cf. [La] or [Ga].)

We note that the pressure does not explicitly appear in the definition of the weak solution. Therefore the following lemma will be useful.

LEMMA 3.1. *Let $U=(U_1,U_2,U_3)\in L^3(\mathbf{R}^3)$ be a weak solution of* (1.3). *Assume that P is defined by $P=R_jR_k(U_jU_k)$, where we use the notation introduced in* §2. *Then both U and P are smooth, P belongs to $L^{3/2}(\mathbf{R}^3)$, and moreover*

$$-\nu\Delta U+aU+a(y\cdot\nabla)U+(U\cdot\nabla)U+\nabla P=0 \quad \text{in } \mathbf{R}^3. \tag{3.1}$$

Proof. Above we have proved all the statements of the lemma except for the equation (3.1). To prove (3.1), we let

$$F=-\nu\Delta U+aU+a(y\cdot\nabla)U+(U\cdot\nabla)U+\nabla P.$$

We must prove that $F\equiv 0$. Our assumptions imply that $\operatorname{curl} F=0$ and $\operatorname{div} F=0$ in $\mathbf{R}^3$, hence $\Delta F=0$ in $\mathbf{R}^3$. Let ϕ be a radial (i.e. depending only on $|x|$), smooth, compactly supported function in $\mathbf{R}^3$ with $\int_{\mathbf{R}^3}\phi\,dx=1$. Since $\Delta F=0$, we have for each $\varepsilon>0$ (cf. [St, p. 275])

$$F(0)=\int_{\mathbf{R}^3}F(x)\varepsilon^3\phi(\varepsilon x)\,dx.$$

For each multiindex $\alpha=(\alpha_1,\alpha_2,\alpha_3)$ we have $\Delta(D_\alpha F)=0$, and hence, for each $\varepsilon>0$,

$$D_\alpha F(0)=\int_{\mathbf{R}^3}D_\alpha F(x)\varepsilon^3\phi(\varepsilon x)\,dx=(-1)^{|\alpha|}\int_{\mathbf{R}^3}F(x)\varepsilon^{3+|\alpha|}(D_\alpha\phi)(\varepsilon x)\,dx. \tag{3.2}$$

Since F is analytic (see e.g. [Mo, p. 166]), it is enough to prove that $D_\alpha F(0)=0$ for each α with $|\alpha|\geqslant 0$. From formula (3.2) we see that this will follow if we prove that for each smooth, compactly supported function ψ in $\mathbf{R}^3$ we have

$$\lim_{\varepsilon\to 0_+}\varepsilon^3\int_{\mathbf{R}^3}F(x)\psi(\varepsilon x)\,dx=0.$$

Using the definition of F, we see that we must prove

$$\lim_{\varepsilon\to 0_+}\varepsilon^3\int_{\mathbf{R}^3}(-\nu\Delta U+aU+a(y\cdot\nabla)U+(U\cdot\nabla)U+\nabla P)\psi(\varepsilon y)\,dy=0.$$

This can be done term-by-term, by using integration by parts and the assumption $U\in L^3(\mathbf{R}^3)$, together with the fact $P\in L^{3/2}(\mathbf{R}^3)$. For example, we have

$$\begin{aligned}\int_{\mathbf{R}^3}(y\cdot\nabla)U(y)\psi(\varepsilon y)\,dy &= -\int_{\mathbf{R}^3}3U(y)\psi(\varepsilon y)\,dy-\int_{\mathbf{R}^3}U(y)(\varepsilon y_k)(\partial_k\psi)(\varepsilon y)\,dy\\ &= -\int_{\mathbf{R}^3}3U(y)\psi(\varepsilon y)\,dy-\int_{\mathbf{R}^3}U(y)\tilde{\psi}(\varepsilon y)\,dy,\end{aligned}$$

where $\tilde{\psi}(y)=y_k\partial_k\psi(y)$. Using the Hölder inequality, we can write

$$\left|\int_{\mathbf{R}^3}U(y)\psi(\varepsilon y)\,dy\right|\leqslant\|U\|_{L^3(\mathbf{R}^3)}\left\{\int_{\mathbf{R}^3}|\psi(\varepsilon y)|^{3/2}\,dy\right\}^{2/3}=\varepsilon^{-2}\|U\|_{L^3(\mathbf{R}^3)}\|\psi\|_{L^{3/2}(\mathbf{R}^3)}.$$

Hence

$$\lim_{\varepsilon\to 0_+}\varepsilon^3\int_{\mathbf{R}^3}U(y)\psi(\varepsilon y)\,dy=0.$$

The other terms can be dealt with in the same way, and the proof is easily finished. □

Lemma 3.2. *Let $U=(U_1,U_2,U_3)$ and P be as in Lemma 3.1. Then, for each $k=0,1,2,\ldots$, we have*

$$|\nabla^k U(y)|=O(|y|^{-3-k}),\quad |y|\to\infty$$

and

$$|\nabla^k P(y)|=O(|y|^{-2-k}),\quad |y|\to\infty.$$

Proof. Let $T\in\mathbf{R}$. For $x\in\mathbf{R}^3$ and $t<T$ we let

$$u(x,t)=\frac{1}{\sqrt{2a(T-t)}}\,U\left(\frac{x}{\sqrt{2a(T-t)}}\right) \tag{3.3}$$

and

$$p(x,t)=\frac{1}{2a(T-t)}\,P\left(\frac{x}{\sqrt{2a(T-t)}}\right). \tag{3.4}$$

The equation (3.1) implies that u and p satisfy the Navier–Stokes equations (1.1) in $\mathbf{R}^3\times(-\infty,T)$. We have $U\in L^3(\mathbf{R}^3)$ and $P\in L^{3/2}(\mathbf{R}^3)$, and hence

$$\lim_{\varrho\to\infty}\int_{|y|\geqslant\varrho}(|U|^3+|P|^{3/2})\,dy=0.$$

Using this, we check easily by an elementary change of variables that for each $x_0\in\mathbf{R}^3\setminus\{0\}$ there exists $R>0$ such that

$$\sup_{T-R^2<t<T}\int_{B_R(x_0)}(|u(x,t)|^3+|p(x,t)|^{3/2})\,dx<\varepsilon_0, \tag{3.5}$$

where ε_0 has the same meaning as in Proposition 2.1. Since (3.5) obviously implies

$$R^{-2}\int_{Q_R(x_0,T)}(|u(x,t)|^3+|p(x,t)|^{3/2})\,dx\,dt<\varepsilon_0,$$

we see from Proposition 2.1 that

$$\sup_{Q_{R/2}(x_0,T)}|\nabla^k u(x,t)|\leqslant C_k R^{-1-k} \tag{3.6}$$

for each $k=0,1,2,\ldots$. Let $0<\varrho_1'<\varrho_1<\varrho_2<\varrho_2'$ and let $t_1<T$. Since u is smooth in $\mathbf{R}^3\times(-\infty,T)$ and x_0 was an arbitrary point of $\mathbf{R}^3\setminus\{0\}$, we see from (3.6) and the obvious compactness argument that the functions $\nabla^k u(x,t)$ are bounded in $\Omega_{\varrho_1',\varrho_2'}\times(t_1,T)$, where $\Omega_{\varrho_1',\varrho_2'}=\{x\in\mathbf{R}^3:\varrho_1'<|x|<\varrho_2'\}$. Let also $\Omega_{\varrho_1,\varrho_2}=\{x\in\mathbf{R}^3:\varrho_1<|x|<\varrho_2\}$. We claim that, for $k=0,1,\ldots$, the space derivatives $\nabla^k p$ are bounded in $\Omega_{\varrho_1,\varrho_2}\times(t_1,T)$. To see this, we note that for each $t<T$ we have

$$\int_{\mathbf{R}^3}|p(x,t)|^{3/2}\,dx=\int_{\mathbf{R}^3}|P(y)|^{3/2}\,dy. \tag{3.7}$$

Since for each $t<T$ we have

$$-\Delta p(x,t)=\partial_i\partial_j(u_i(x,t)u_j(x,t)) \quad \text{in } \mathbf{R}^3 \tag{3.8}$$

we see from the classical elliptic estimates (cf. [Br]) that

$$\|p(\cdot,t)\|_{C^{k,\alpha}(\Omega_{\varrho_1,\varrho_2})}\leqslant c_k\sum_{i,j}\|u_i(\cdot,t)u_j(\cdot,t)\|_{C^{k,\alpha}(\Omega_{\varrho_1',\varrho_2'})}+\tilde{c}_k\|p(\cdot,t)\|_{L^{3/2}(\mathbf{R}^3)},$$

where we use the usual notation for the Hölder norms. Since, as we have seen above, we control the terms on the right-hand side, we see that $\nabla^k p$ is bounded in $\Omega_{\varrho_1,\varrho_2}\times(t_1,T)$ for each k, as claimed.

We can now take the space derivatives of the Navier–Stokes equations to see that for each $k=0,1,2\ldots$ the derivatives $\nabla^k\partial u/\partial t=\partial(\nabla^k u)/\partial t$ are bounded in $\Omega_{\varrho_1,\varrho_2}\times(t_1,T)$. Since

$$\int_{\Omega_{\varrho_1,\varrho_2}}|u(x,t)|^3\,dx\to 0 \quad \text{as } t\to T_-$$

(which can be seen for (3.3) and the assumption $U\in L^3(\mathbf{R}^3)$), we see from the above estimates and the Arzela–Ascoli theorem that $\nabla^k u(\cdot,t)\to 0$ as $t\to T_-$ uniformly in $\Omega_{\varrho_1,\varrho_2}$ for each $k=0,1,\ldots$. Therefore we can write, for $x\in\Omega_{\varrho_1,\varrho_2}$,

$$\nabla^k u(x,t)=-\int_t^T\nabla^k\frac{\partial}{\partial t}u(x,s)\,ds,$$

and by the estimates for $\nabla^k \partial u(x,t)/\partial t$ above we have, for $k=0,1,\ldots$,

$$|\nabla^k u(x,t)| \leqslant M_k(T-t) \quad \text{in } \Omega_{\varrho_1,\varrho_2}\times(t_1,T), \tag{3.9}$$

where M_k are suitable constants. We have seen above that we also have, for $k=0,1,\ldots$,

$$|\nabla^k p(x,t)| \leqslant \widetilde{M}_k \quad \text{in } \Omega_{\varrho_1,\varrho_2}\times(t_1,T). \tag{3.10}$$

It is now a matter of an elementary calculation to verify that (3.9) together with (3.3) implies $|\nabla^k U(y)|=O(|y|^{-3-k})$ as $|y|\to\infty$, and that (3.10) together with (3.4) implies that $|\nabla^k P(y)|=O(|y|^{-2-k})$ as $|y|\to\infty$. The proof is finished. □

LEMMA 3.3. *Assume that $U=(U_1,U_2,U_3)$ and P are smooth functions satisfying* (1.3) *in $\mathbf{R}^3$. Then the quantity*

$$\Pi(y)=\tfrac{1}{2}|U(y)|^2+P(y)+ay\cdot U(y)$$

satisfies the maximum principle, i.e. for each bounded domain $\Omega\subset\mathbf{R}^3$ we have

$$\sup_{y\in\Omega}\Pi(y)\leqslant \sup_{y\in\partial\Omega}\Pi(y).$$

Proof. We let $\widetilde{U}(y)=U(y)+ay$ and $\widetilde{P}(y)=P(y)-\frac{1}{2}a^2|y|^2$. An easy calculation shows that (1.3) is equivalent to

$$\left.\begin{aligned} -\nu\Delta\widetilde{U}+(\widetilde{U}\cdot\nabla)\widetilde{U}+\nabla\widetilde{P}&=0\\ \operatorname{div}\widetilde{U}&=3a\end{aligned}\right\}\quad\text{in } \mathbf{R}^3.$$

Multiplying the first equation by $\widetilde{U}$ we get

$$-\nu\Delta\widetilde{U}\cdot\widetilde{U}+(\widetilde{U}\cdot\nabla)\big(\tfrac{1}{2}|\widetilde{U}|^2+\widetilde{P}\big)=0.$$

Hence

$$\begin{aligned}-\nu\Delta\big(\tfrac{1}{2}|\widetilde{U}|^2+\widetilde{P}\big)+(\widetilde{U}\cdot\nabla)\big(\tfrac{1}{2}|\widetilde{U}|^2+\widetilde{P}\big)&=-\nu|\nabla\widetilde{U}|^2-\nu\Delta\widetilde{P}\\ &=-\nu|\nabla U+aI|^2-\nu\Delta P+3\nu a^2\\ &=-\nu|\nabla U|^2+\nu(\partial_i U_j)(\partial_j U_i)\leqslant 0,\end{aligned}$$

where we denote by I the identity matrix. Since $\frac{1}{2}|\widetilde{U}|^2+\widetilde{P}=\Pi$, we see that

$$-\nu\Delta\Pi+(\widetilde{U}\cdot\nabla)\Pi\leqslant 0\quad\text{in } \mathbf{R}^3,$$

and the result follows from the standard maximum principle for elliptic equations (see e.g. [GT]). □

Remark. The fact that the quantity $\frac{1}{2}|u|^2+p$ satisfies a maximum principle for the steady Navier–Stokes equations is well-known (see e.g. [Se2, p. 261], or [GW]), and has played an important role in recent results ([FR1], [FR2] and [Str2]) regarding the regularity of solutions of the steady Navier–Stokes equations in higher dimensions.

THEOREM 1. *Let U be a weak solution of* (1.3) *belonging to* $L^3(\mathbf{R}^3)$. *Then* $U\equiv 0$ *in* $\mathbf{R}^3$.

Proof. Let $P=R_jR_k(U_jU_k)$, where we use the notation introduced in §2. By Lemma 3.1 the functions U and P are smooth and satisfy (1.3). Hence $\Pi(y)=\frac{1}{2}|U(y)|^2+P(y)+ay\cdot U(y)$ satisfies the maximum principle by Lemma 3.3. Since $\Pi(y)=O(|y|^{-2})$ as $|y|\to\infty$ by Lemma 3.2, the maximum principle (applied to large balls) implies that $\Pi\leqslant 0$ in $\mathbf{R}^3$. Let ϕ be as in Lemma 2.2 and assume moreover that $\phi\geqslant 0$ in $\mathbf{R}^3$. For $\varepsilon>0$ let $\phi_\varepsilon(y)=\phi(\varepsilon y)$. Since $\Pi\leqslant 0$ in $\mathbf{R}^3$, we have $\int_{\mathbf{R}^3}\Pi\phi_\varepsilon\,dy\leqslant 0$ for each $\varepsilon>0$. Since ϕ_ε is radial and $\operatorname{div} U=0$, we have $\int_{\mathbf{R}^3}a(y\cdot U)\phi_\varepsilon\,dy=0$ for each $\varepsilon>0$. Since $U_jU_k\in L^1(\mathbf{R}^3)\cap L^2(\mathbf{R}^3)$ by Lemma 3.2, we see from Lemma 2.2 that $\lim_{\varepsilon\to 0+}\int_{\mathbf{R}^3}P\phi_\varepsilon\,dy=-\frac{1}{3}\int_{\mathbf{R}^3}|U|^2\,dy$. We infer that

$$\int_{\mathbf{R}^3}\tfrac{1}{2}|U|^2\,dy-\int_{\mathbf{R}^3}\tfrac{1}{3}|U|^2\,dy\leqslant 0,$$

and hence $U\equiv 0$. The proof is finished. □

An alternative proof of Theorem 1. We feel that the following alternative way of proving Theorem 1 is of interest. We let $Q(y)=P(y)+ay\cdot U(y)$. The system (1.3) can be rewritten as

$$\begin{aligned} -\nu\Delta U_j+U_k\frac{\partial U_j}{\partial y_k}+\frac{\partial Q}{\partial y_j}+ay_k\left(\frac{\partial U_j}{\partial y_k}-\frac{\partial U_k}{\partial y_j}\right)&=0,\\ \operatorname{div} U&=0. \end{aligned}\tag{3.11}$$

After multiplying the first equation by $y_j/|y|$, we obtain

$$\frac{y_j}{|y|}\cdot\frac{\partial Q}{\partial y_j}=\nu\Delta U_j\frac{y_j}{|y|}-U_k\frac{\partial U_j}{\partial y_k}\cdot\frac{y_j}{|y|}.\tag{3.12}$$

Denoting $\varrho=|y|$ and using Lemma 3.2, we see that $\partial Q/\partial\varrho=O(\varrho^{-5})$ as $\varrho\to\infty$. Since also $Q(y)=O(|y|^{-2})$ as $|y|\to\infty$ by Lemma 3.2, we see (by integrating along the rays through the origin) that

$$Q(y)=O(|y|^{-4}),\quad |y|\to\infty.$$

Multiplying (3.11) by y_j and integrating over $\mathbf{R}^3$ (which we can because all the terms are integrable by the decay estimates for Q above and the decay estimates for U and P established in Lemma 3.2), we obtain

$$\int_{\mathbf{R}^3}\left(-\nu y_j\Delta U_j+y_jU_k\frac{\partial U_j}{\partial y_k}+y_j\frac{\partial Q}{\partial y_j}\right)dy=0.$$

Integrating by parts (which, again, we can because of the decay estimates we have for U and Q), we obtain

$$\int_{\mathbf{R}^3}(|U|^2+3Q)\,dy=0.$$

On the other hand, $|U|^2+2Q\leqslant 0$ in $\mathbf{R}^3$ by Lemma 3.2 and Lemma 3.3 (see also our first proof of the Theorem 1), and hence

$$\int_{\mathbf{R}^3}(|U|^2+2Q)\,dy\leqslant 0.$$

Subtracting $\int_{\mathbf{R}^3}(\frac{2}{3}|U|^2+2Q)\,dy=0$ from this inequality, we obtain $\int_{\mathbf{R}^3}\frac{1}{3}|U|^2\,dy\leqslant 0$ and the proof is finished. □

Appendix

Here we recall the estimates which can be used for the proof of Proposition 2.1.

PROPOSITION A.1. *Let $Q=B_1(0)\times(-1,0)\subset\mathbf{R}^n\times\mathbf{R}$ and $Q'=B_{1-\delta}\times(-(1-\delta)^2,0)$, where $0<\delta<1$ is a fixed number. (Typically we imagine that δ is small.) Let $1<p<+\infty$ and $f=(f_1,...,f_n)\in L^p(Q)$. Assume that $v\in L^2(Q)$ is a weak solution of the heat equation*

$$\frac{\partial v}{\partial t}-\Delta v=\frac{\partial f_j}{\partial x_j}\quad \textit{in } Q.$$

Then v satisfies the following estimates (in which we use c to denote the "generic constant" which can depend on p and δ):

$$\|Dv\|_{L^p(Q')}\leqslant c\|f\|_{L^p(Q)}+c\|v\|_{L^2(Q)},$$
$$\|v\|_{L^q(Q')}\leqslant c\|f\|_{L^p(Q)}+c\|v\|_{L^2(Q)}\quad \textit{if } \frac{1}{q}=\frac{1}{p}-\frac{1}{n+2}>0,$$
$$\|v\|_{C^\alpha_{\mathrm{par}}(Q')}\leqslant c\|f\|_{L^p(Q)}+c\|v\|_{L^2(Q)}\quad \textit{if } \alpha=1-\frac{n+2}{p}>0,$$

where $C^\alpha_{\mathrm{par}}(Q')$ denotes the space of functions which are α-Hölder continuous with respect to the parabolic metric in which the distance between (x,t) and (x',t') is given by $|x-x'|+|t-t'|^{1/2}$.

Proof. These estimates are well-known and can be obtained for example by combining the L^p-estimates established in [LUS, Chapter IV, §9, Theorem 9.1] with the imbeddings established in [LUS, Chapter II, §3, Lemmas 3.2 and 3.3]. □

PROPOSITION A.2. *Let $B=B_1(0)\subset\mathbf{R}^n$ and let $B'=B_{1-\delta}(0)$, where $0<\delta<1$ is a fixed number. (Typically we imagine that δ is small.) Let $u=(u_1,...,u_n)\in L^2(B)$. Then, for $k=1,2\ldots$ we have*

$$\|D^k u\|_{L^p(B')}\leqslant c\|D^{k-1}\operatorname{curl} u\|_{L^p(B)}+c\|D^{k-1}\operatorname{div} u\|_{L^p(B)}+c\|u\|_{L^2(B)}$$

and

$$\|D^k u\|_{C^\alpha(B')} \leqslant c\|D^{k-1}\operatorname{curl} u\|_{C^\alpha(B)} + c\|D^{k-1}\operatorname{div} u\|_{C^\alpha(B)} + c\|u\|_{L^2(B)},$$

where $1<p<\infty$, $0<\alpha<1$, *and c denotes the "generic constant" which can depend on* δ, p *and* α.

Proof. This is a well-known consequence of the classical L^p- and C^α-estimates for the Laplace equation. □

Acknowledgement. We thank G. P. Galdi and J. Málek for many helpful discussions and continual interest in this work. We also thank E. Fabes, R. Kohn, M. Steinhauer and M. Taylor for helpful remarks.

The research of J. Nečas was partially supported by Grant 201/93/2177 GAČR. The research of M. Růžička was supported by a DFG fellowship.

References

[Br] BROWDER, F., Regularity of solutions of elliptic equations. *Comm. Pure Appl. Math.*, 9 (1956), 351–361.

[Ca] CATTABRIGA, L., Su un problema al retorno relativo al sistema di equazioni di Stokes. *Rend. Sem. Mat. Univ. Padova*, 31 (1961), 308–340.

[CKN] CAFFARELLI, L., KOHN, R. & NIRENBERG, L., Partial regularity of suitable weak solutions of the Navier–Stokes equations. *Comm. Pure Appl. Math.*, 35 (1982), 771–831.

[CZ] CALDERÓN, A. P. & ZYGMUND, A., On existence of certain singular integrals. *Acta Math.*, 88 (1952), 85–139.

[FR1] FREHSE, J. & RŮŽIČKA, M., On the regularity of the stationary Navier–Stokes equations. *Ann. Scuola Norm. Sup. Pisa Cl. Sci. (4)*, 21 (1994), 63–95.

[FR2] — Regularity for the stationary Navier–Stokes equations in bounded domains. *Arch. Rational Mech. Anal.*, 128 (1994), 361–381.

[Ga] GALDI, G. P., *An Introduction to the Mathematical Theory of Navier–Stokes Equations, Vol. 1: Linearized stationary problems.* Springer Tracts Nat. Philos., 38. Springer-Verlag, New York, 1994.

[Gi] GIGA, Y., Solutions for semilinear parabolic equations in L^p and regularity of weak solutions of the Navier–Stokes equations with data in L^p. *J. Differential Equations*, 62 (1986), 186–212.

[GT] GILBARG, D. & TRUDINGER, N. S., *Elliptic Partial Differential Equations of Second Order*, 2nd edition. Grundlehren Math. Wiss., 224. Springer-Verlag, Berlin–New York, 1983.

[GW] GILBARG, D. & WEINBERGER, H. F., Asymptotic properties of Leray's solution of the stationary two-dimensional Navier–Stokes equations. *Russian Math. Surveys*, 29:2 (1974), 109–123.

[La] LADYZHENSKAYA, O., *The Mathematical Theory of Viscous Incompressible Flow*, 2nd English edition. Math. Appl., 2. Gordon and Breach, New York, 1969.

[Le] LERAY, J., Sur le mouvement d'un liquide visqueux emplissant l'espace. *Acta Math.*, 63 (1934), 193–248.

[LUS] Ladyzhenskaya, O., Ural'ceva, N. & Solonnikov, V., *Linear and Quasi-Linear Equations of Parabolic Type.* Transl. Math. Monographs, 23. Amer. Math. Soc., Providence, RI, 1968.

[Mo] Morrey, Ch. B., *Multiple Integrals in the Calculus of Variations.* Grundlehren Math. Wiss., 130. Springer-Verlag, New York, 1966.

[Oh] Ohyama, T., Interior regularity of weak solutions of the time-dependent Navier–Stokes equation. *Proc. Japan Acad. Ser. A Math. Sci.*, 36 (1960), 273–277.

[Sch] Scheffer, V., Partial regularity of solutions to the Navier–Stokes equations. *Pacific J. Math.*, 66 (1976), 535–552.

[Se1] Serrin, J., On the interior regularity of weak solutions of the Navier–Stokes equations. *Arch. Rational Mech. Anal.*, 9 (1962), 187–195.

[Se2] — Mathematical principles of classical fluid mechanics, in *Handbuch der Physik*, 8:1, pp. 125–263. Springer-Verlag, Berlin–Göttingen–Heidelberg, 1959.

[So] Sohr, H., Zur Regularitätstheorie der instationären Gleichungen von Navier–Stokes. *Math. Z.*, 184 (1983), 359–376.

[St] Stein, E., *Singular Integrals and the Differentiability Properties of Functions.* Princeton Math. Ser., 30. Princeton Univ. Press, Princeton, NJ, 1970.

[Str1] Struwe, M., On partial regularity results for the Navier–Stokes equations. *Comm. Pure Appl. Math.*, 41 (1988), 437–458.

[Str2] — Regular solutions of the stationary Navier–Stokes equations on $\mathbf{R}^5$. To appear in *Math. Ann.*

[Ta] Takahashi, S., On interior regularity criteria for weak solutions of the Navier–Stokes equations. *Manuscripta Math.*, 67 (1990), 237–254.

[vW] Wahl, W. von, *The Equations of Navier–Stokes and Abstract Parabolic Equations.* Aspects of Math., E8. Vieweg, Braunschweig, 1985.

J. Nečas
Department of Mathematics
Northern Illinois University
De Kalb, IL 60115
U.S.A.
and
Mathematical Institute
Charles University
Sokolovská 83
Prague
Czech Republic

M. Růžička
Institute of Applied Mathematics
University of Bonn
Beringstraße 4–6
53115 Bonn
Germany

V. Šverák
School of Mathematics
University of Minnesota
206 Church Street S.E.
Minneapolis, MN 55455
U.S.A.

Received August 8, 1995

ON WEAK SOLUTIONS TO A CLASS OF NON-NEWTONIAN INCOMPRESSIBLE FLUIDS IN BOUNDED THREE-DIMENSIONAL DOMAINS: THE CASE $p \geq 2$

J. MÁLEK AND J. NEČAS
Mathematical Institute of Charles University
Sokolovská 83, 18675 Prague 8, Czech Republic

M. RŮŽIČKA
Institute of Applied Mathematics, University of Bonn
Beringstraße 4–6, 53115 Bonn, Germany

(Submitted by: Emmanuele DiBenedetto)

Abstract. Existence and regularity properties of solutions for the evolutionary system describing unsteady flows of incompressible fluids with shear dependent viscosity are studied. The problem is considered in a bounded, smooth domain of $\mathbb{R}^3$ with Dirichlet boundary conditions. The nonlinear elliptic operator, which is related to the stress tensor, has p structure. The paper deals with the case $p \geq 2$, for which the existence of weak solutions is proved. If $p \geq \frac{9}{4}$ then a weak solution is strong and unique among all weak solutions.

1. Introduction and setting of the problem. This paper deals with unsteady flows of an incompressible fluid in a bounded domain $\Omega \subset \mathbb{R}^d$, $d > 1$ described by the system of equations

$$\begin{aligned} \operatorname{div} \mathbf{v} &= 0\,, \\ \rho \frac{\partial \mathbf{v}}{\partial t} - \operatorname{div} \mathbf{T}^E + \rho v_k \frac{\partial \mathbf{v}}{\partial x_k} &= -\nabla \pi + \rho \mathbf{f}\,, \end{aligned} \tag{1.1}$$

where $\mathbf{v} = (v_1, v_2, \ldots, v_d)$ is the velocity, π represents the pressure, ρ is a positive constant determining the density of a material, $\mathbf{f} = (f_1, f_2, \ldots, f_d)$ stands for the given external body forces and $\mathbf{T}^E$ denotes the extra stress tensor.

Accepted for publication in revised form December 1999.
AMS Subject Classifications: 35K35, 35K55, 76D99, 76D05.

© Khayyam Publishing 2001. Reprinted with kind permission from: Málek, J., Nečas, J. & Růžička, M., On the weak solutions to a class of non-Newtonian incompressible fluids in bounded three-dimensional domains: the case $p \geq 2$, Advances in Differential Equations 6, 257-302.

Let a finite $T > 0$ be given. Then all functions are evaluated at (t, x), where $t \in [0, T]$ and $x \in \Omega$. We also use the usual summation convention throughout the whole text. Let us further denote by $\mathbf{D}$ the symmetric part of the velocity gradient; i.e., $\mathbf{D} = \mathbf{D}(\mathbf{v}) \equiv \frac{1}{2}\left[(\nabla \mathbf{v}) + (\nabla \mathbf{v})^T\right]$. We consider a constitutive relation for $\mathbf{T}^E$ of the form

$$\mathbf{T}^E = \mathbf{T}(\mathbf{D})\,. \tag{1.2}$$

Before specifying precise assumptions for $\mathbf{T}$, let us first think of the following four examples: for $p > 1$ and $\nu_0 > 0$ we set

$$\begin{aligned}
&\text{(a)} \qquad \mathbf{T}^E = 2\nu_0 |\mathbf{D}|^{p-2}\mathbf{D}\,,\\
&\text{(b)} \qquad \mathbf{T}^E = 2\nu_0 (1 + |\mathbf{D}|^2)^{\frac{p-2}{2}}\mathbf{D}\,,\\
&\text{(c)} \qquad \mathbf{T}^E = 2\nu_0 (1 + |\mathbf{D}|)^{p-2}\mathbf{D}\,,\\
&\text{(d)} \qquad \mathbf{T}^E = 2\nu_0 (1 + |\mathbf{D}|^{p-2})\mathbf{D}\,,
\end{aligned} \tag{1.3}$$

where $|\mathbf{D}|$ denotes the usual Euclidean matrix-norm.

On the one hand, the models (a)–(d) have some joint properties. Firstly, for $p = 2$ all formulas reduce to the Stokes law, i.e., $\mathbf{T}^E = 2\nu_0\mathbf{D}$, and (1.1) turns into the Navier-Stokes system, which is a model for Newtonian fluids.

Secondly, we can easily construct scalar potentials to $\mathbf{T}^E$ in (1.3) (a)–(d). In fact, all examples for $\mathbf{T}^E$ can be written in the form

$$\mathbf{T}^E = 2\nu_0 \mu(|\mathbf{D}|^2)\mathbf{D}\,, \tag{1.4}$$

where $\mu : \mathbb{R}_0^+ \to \mathbb{R}_0^+$ is the generalized viscosity function. The corresponding potential $\Phi : \mathbb{R}_0^+ \to \mathbb{R}_0^+$ to $\mathbf{T}^E$ given in (1.4) is defined by

$$\Phi(|\mathbf{D}|^2) \equiv \nu_0 \int_0^{|\mathbf{D}|^2} \mu(s)\, ds\,,$$

and we have for $r, s = 1, 2, \ldots, d$

$$T_{rs}(\cdot) = \partial_{rs}\Phi(\cdot) \equiv \frac{\partial \Phi(\cdot)}{\partial D_{rs}}\,, \qquad \Phi(0) = \partial_{rs}\Phi(0) = 0\,.$$

Considering a simple shear flow, i.e., $\mathbf{v} = (v_1(x_2), 0, 0)$, the quantity $\kappa \equiv |v_1'(x_2)|$ $(= 2|\mathbf{D}|$ in the considered motion) is called shear rate. This is why the fluids constituted by (1.3) are sometimes named *fluids with shear-dependent viscosity.* Models belonging to this class of non-Newtonian fluid mechanics are frequently used in engineering practice, as discussed in Málek, Rajagopal, Růžička [13], for example.

Finally, it is worth remarking that all models (a)–(d) in (1.3) satisfy the p-coercivity condition, i.e.,

$$\mathbf{T}^E \cdot \mathbf{D} \geq 2\nu_0 |\mathbf{D}|^p \tag{1.5}$$

and they have $(p-1)$ growth, which means $|\mathbf{T}^E| \leq c(1+|\mathbf{D}|)^{p-1}$, $c > 0$. On the other hand, despite their similar structure, the graphs of $\mu(|\mathbf{D}|^2)$ differ dramatically from each other (cf. Figures 1.1 and 1.2 in [11]). The different asymptotic behaviour of $\mu(s)$ as $s \to 0^+$ or $s \to \infty$, makes the class of investigated models robust and therefore very useful.

We complement the equations (1.1) by an initial condition

$$\mathbf{v}(0,\cdot) = \mathbf{v}_0(\cdot) \qquad \text{in } \Omega\,, \tag{1.6}$$

and by Dirichlet boundary conditions

$$\mathbf{v}(t,x) = \mathbf{0} \qquad \text{for all } (t,x) \in [0,T) \times \partial\Omega\,. \tag{1.7}$$

Simpler than (1.7) are space-periodic boundary conditions. In that case, Ω is a d-dimensional cube with sides of finite length $L > 0$ and

$$\mathbf{v}, \pi \text{ are periodic with period } L \text{ in each variable } x_i\,, i = 1, 2, \ldots, d\,. \tag{1.8}$$

By the *Problem (NS-Dir)$_p$*, we will mean the problem (1.1), (1.2), (1.6), (1.7), while we will call the problem (1.1), (1.2), (1.6), (1.8) the *Problem (NS-Per)$_p$*. Since no assumptions for $\mathbf{T}$ in (1.2) have been specified yet, one should have in mind the examples (1.3) (a)–(d).

Our long-lasting aim is to study the global-in-time existence of (weak) solutions to both the *Problem (NS-Dir)$_p$* and the *Problem (NS-Per)$_p$*, and to investigate their further qualitative properties[1] in dependence on the parameter p.

Regarding the *Problem (NS-Dir)$_p$*, the first mathematical investigations go back to Ladyzhenskaya's lecture at the International Mathematical Congress in 1966, where she proposed, among others, to study the system (1.1),

[1] Here, however, we discuss the questions of uniqueness and regularity of weak solutions only.

(1.6), (1.7) and (1.3)(d) with $p = 4$. Later on these first results were extended, and presented in further contributions of Ladyzhenskaya; cf. [5], [6] and [7]. Combining monotone operator theory and compactness arguments, she proved the existence of weak solutions to all models (a)–(d) in (1.3) if $p \geq 1 + \frac{2d}{d+2}$ and their uniqueness if $p \geq \frac{d+2}{2}$. See also Lions [8] for a comparable proof of the same results. Recently, Amann [1] showed the existence of regular (classical) solutions to the *Problem (NS-Dir)$_p$* provided that the data $\mathbf{f}$ and $\mathbf{v}_0$ are small and assuming that the tensor function $\mathbf{T}$ in (1.2) satisfies

$$\partial_{rs}\mathbf{T}(\mathbf{0}) = const. > 0\,.$$

Thus, in particular, these results are related to the models (b), (c) in (1.3) if $p \geq 1$ and to (d) if $p \geq 2$.

More results are known about the *Problem (NS-Per)$_p$* due to a series of papers Bellout, Bloom, Nečas [3] (see also Bellout, Bloom, Nečas [2]), Málek, Nečas, Růžička [9], Málek, Rajagopal, Růžička [13] and finally the monograph Málek, Nečas, Rokyta, Růžička [11] which will be sometimes used as a reference for detailed explanations and proofs of some assertions.

Excluding the example (a) in (1.3) for $p > 2$, the following has been proved in these publications:

the existence of a weak solution	for $p \in (\frac{3d}{d+2}, \frac{2d}{d-2})$	if $d = 3\,,4\,;$
the existence of a strong solution	for $p \geq 1 + \frac{2d}{d+2}$	if $d \geq 3\,;$
	for $p > 1$	if $d = 2\,;$
uniqueness of the weak solution	for $p \geq 1 + \frac{2d}{d+2}$	if $d \geq 2\,.$

It is natural to ask whether the same results are valid for the *Problem (NS-Dir)$_p$*, too. Since the superquadratic case ($p > 2$) and the subquadratic case ($p < 2$) require different approaches, we investigate the former in this paper, and we devote to the latter a forthcoming paper. We also concentrate, mainly for methodological reasons, on three-dimensional situations, i.e., $\Omega \subset \mathbb{R}^3$. If $d = 2$ and $p \geq 2$, the existence and uniqueness of weak solutions follows already from Ladyzhenskaya [5], [6] and [7] and Lions [8], while the regularity (i.e., the existence of the unique strong solution) can be obtained following the lines of the present paper. Finally, we also restrict ourselves to the very interesting[2] and most complicated case $p \in [2, 3)$. The case $p \geq 3$ will require a slightly different technique which we intend to investigate later.

[2]The interest in studying the case $p \in [2, 3)$ is based on the fact that up to the present

It might be worth remarking that if the nonlinearity $\mathbf{T}$ had a potential Φ depending on the modulus of the full velocity gradient $|\nabla \mathbf{v}|$ and not only on the modulus of the symmetric velocity gradient $|\mathbf{D}(\mathbf{v})|$, then the results in the preprint [10] hold not only for a larger set of parameters p ($p \in [2,6)$), but also provide stronger information on regularity of $\mathbf{v}$. We want to emphasize, however, that although the preprint [10] is formulated for models where the potential Φ depends on $|\mathbf{D}|$, it holds only for cases when the potential Φ of the nonlinearity $\mathbf{T}$ depends on $|\nabla \mathbf{v}|$, since Lemma 7.18 in [10] is correct only in this case and wrong if Φ depends on $|\mathbf{D}|$.

Before formulating the main result, we will define some useful function spaces and notions. Let $\big(X(\Omega), \|\cdot\|_{X(\Omega)}\big)$ be a Banach space of scalar functions defined in Ω. Then $X(\Omega)^3$ (respectively $X(\Omega)^{3\times 3}$) represents the space of vector-valued (respectively tensor-valued) functions whose components belong to $X(\Omega)$. Let further p, $q > 1$ and $k > 0$. Then $\big(L^p(\Omega), \|\cdot\|_p\big)$ denotes the usual Lebesgue spaces and $\big(W^{k,p}(\Omega), \|\cdot\|_{k,p}\big)$ is used for standard Sobolev spaces. Finally, by

$$\Big(L^q((0,T); X(\Omega)), \Big(\int_0^T \|\cdot\|_{X(\Omega)}^q dt\Big)^{1/q}\Big)$$

we denote Bochner spaces. Sometimes we write I instead of $(0,T)$ and Q_T instead of $I \times \Omega$. Also as usual, $\mathcal{D}(\Omega)$ denotes the space of smooth functions with compact support in Ω. We further define

$$\begin{aligned}
\mathcal{V} &\equiv \{\boldsymbol{\psi} \in \mathcal{D}(\Omega)^3 : \operatorname{div} \boldsymbol{\psi} = 0\}\,,\\
H &\equiv \text{the closure of } \mathcal{V} \text{ with respect to the } \|\cdot\|_2\text{–norm}\,,\\
V_p &\equiv \text{the closure of } \mathcal{V} \text{ with respect to the } \|\nabla\cdot\|_p\text{–norm}\,.
\end{aligned}$$

In order to give the definition of weak and strong solutions to the *Problem (NS-Dir)$_p$*, we will specify assumptions on the tensor function $\mathbf{T}$ from (1.2).

results, the existence of weak solutions to the *Problem (NS-Dir)$_p$* was known only for $p \geq 11/5$ and for the special linear case when $p = 2$ (this means for the Navier-Stokes equations). The results presented here cover this gap, providing the existence of weak solutions to the *Problem (NS-Dir)$_p$* for $p \geq 2$. In addition, we also give information about the integrability of second derivatives if $p \geq 9/4$, which is important for the investigation of large-time behavior in the range $p \in [9/4, 5/2)$, since in this range this information is essential while if $p \in [5/2, \infty)$ the global existence of a finite-dimensional attractor can be proved using $L^\infty(0,\infty; W^{1,p})$ regularity only, which is known if $p \geq 5/2$ for a weak solution (see [12] for more details).

Let $\mathbb{R}^{3\times3}_{sym} \equiv \{\mathbf{D} \in \mathbb{R}^{3\times3} : D_{ij} = D_{ji}, \quad i,j = 1,2,3\}$. We assume the existence of a potential $\Phi : \mathbb{R}_0^+ \to \mathbb{R}_0^+$ and constants $C_1, C_2 > 0$ such that for some $p > 1$ and for all $r, s, m, n = 1, 2, 3$, $\mathbf{B}$,$\mathbf{D} \in \mathbb{R}^{3\times3}_{sym}$

$$T_{rs}(\mathbf{D}) = \partial_{rs}\Phi(|\mathbf{D}|^2)\,, \quad \Phi(0) = \partial_{rs}\Phi(0) = 0\,, \tag{1.9}$$

$$\partial_{ij}\partial_{kl}\Phi(|\mathbf{D}|^2)B_{ij}B_{kl} \geq C_1(1+|\mathbf{D}|)^{p-2}|\mathbf{B}|^2\,, \tag{1.10}$$

$$\left|\partial_{rs}\partial_{mn}\Phi(|\mathbf{D}|^2)\right| \leq C_2(1+|\mathbf{D}|)^{p-2}\,. \tag{1.11}$$

Remark 1.12. It is easy to check that examples (b)–(d) in (1.3) satisfy (1.9)–(1.11), while the analogy of the condition (1.10) for the example (a) reads

$$\partial_{ij}\partial_{kl}\Phi(|\mathbf{D}|^2)B_{ij}B_{kl} \geq C_1|\mathbf{D}|^{p-2}|\mathbf{B}|^2\,.$$

We do not consider this case in this paper.

Definition 1.13. A function $\mathbf{v}$ is said to be a weak solution to the *Problem (NS-Dir)$_p$* if and only if $\mathbf{v} \in L^\infty(I;H) \cap L^p(I;V_p)$ and the weak formulation

$$\int_{Q_T} \big[-\mathbf{v}\cdot\frac{\partial\boldsymbol{\varphi}}{\partial t} + T_{ij}(\mathbf{D}(\mathbf{v}))D_{ij}(\boldsymbol{\varphi}) - v_k\mathbf{v}\cdot\frac{\partial\boldsymbol{\varphi}}{\partial x_k}\big]\,dx\,dt = \int_{Q_T} \mathbf{f}\cdot\boldsymbol{\varphi}\,dx\,dt + \int_\Omega \mathbf{v}_0\cdot\boldsymbol{\varphi}\,dx \tag{1.14}$$

is fulfilled for all $\boldsymbol{\varphi} \in \mathcal{D}(-\infty, T; \mathcal{V})$.

Definition 1.15. A function $\mathbf{v}$ is said to be a strong solution to the *Problem (NS-Dir)$_p$* if and only if

$$\mathbf{v} \in C(I;H) \cap L^\infty(I;V_p) \cap L^{\frac{2}{p-1}}(I;W^{2,\frac{6}{p+1}}(\Omega)^3)\,, \qquad \frac{\partial\mathbf{v}}{\partial t} \in L^2(I;L^2(\Omega)^3)\,,$$

and for all $\boldsymbol{\varphi} \in V_p$ and almost every $t \in I$

$$\left(\frac{\partial\mathbf{v}(t)}{\partial t}, \boldsymbol{\varphi}\right) + \int_\Omega T_{ij}(\mathbf{D}(\mathbf{v}(t)))D_{ij}(\boldsymbol{\varphi})\,dx + \left(v_k(t)\frac{\partial\mathbf{v}(t)}{\partial x_k}, \boldsymbol{\varphi}\right) = \left(\mathbf{f}(t), \boldsymbol{\varphi}\right). \tag{1.16}$$

The brackets $(\mathbf{h}, \mathbf{g})$ stand for $\int_\Omega \mathbf{h}\cdot\mathbf{g}\,dx$, where $\mathbf{h}\cdot\mathbf{g} \in L^1(\Omega)$.

Now we formulate the main result of this paper.

Theorem 1.17. *Let $\Omega \subset \mathbb{R}^3$ be a bounded domain, $\partial\Omega \in C^3$. Let $p \in [2,3)$, $\mathbf{v}_0 \in V_p$ and $\mathbf{f} \in L^2(I;L^2(\Omega)^3)$. Assume that $\mathbf{T}$ in (1.2) satisfies (1.9)–(1.11). Then there exists a weak solution $\mathbf{v}$ to the Problem (NS-Dir)$_p$. Moreover, if*

$$p \geq \frac{9}{4} = \frac{11}{5} + 0.05 \tag{1.18}$$

then the solution is strong and unique, and for all $\Omega_0 \subset\subset \Omega$ *and* $0 \le t_1 < t_2 \le T$

$$\int_{t_1}^{t_2} \int_{\Omega_0} (1 + |\mathbf{D}(\mathbf{v})|)^{p-2} |\mathbf{D}(\nabla \mathbf{v})|^2 dx\, dt \ < \infty\,. \tag{1.19}$$

In particular, $\mathbf{v} \in L^2(I; W^{2,2}_{\mathrm{loc}}(\Omega)^3)$. *The tangential derivatives* $\frac{\partial \mathbf{v}}{\partial \tau^r}$, $r = 1, 2$, *satisfy an estimate analogous to* (1.19). *Thus,* $\nabla \frac{\partial \mathbf{v}}{\partial \tau^r} \in L^2(I; L^2(\Omega)^{3\times 3})$.

Remark 1.20. By the results in [7] and [8], the existence of weak solutions is obtained for $p \ge \frac{11}{5}$. Thus Theorem 1.17 fills the gap between 2 and $\frac{11}{5}$. Further, for p satisfying (1.18) we obtain regularity properties for $\mathbf{v}$. However, these results are slightly worse than those for the *Problem (NS-Per)$_p$*, where strong solutions exists for all $p \ge \frac{11}{5}$ and belong to the space $L^2(I; W^{2,2}(\Omega)^3)$. It might be useful to emphasize that we are able to strengthen our results significantly if the nonlinearity $\mathbf{T}$ comes from a potential depending on the modulus of the full velocity gradient of $\mathbf{v}$. More precisely, if $\mathbf{T} = \partial\Phi(|\nabla \mathbf{v}|)$, then the weak solutions exist for $p \in [2, 6)$, and the existence of strong solutions is guaranteed for $p \in (\frac{20}{9}, 6)$. In addition, the strong solutions belong to $L^{p'}(I; W^{2,p'}(\Omega)^3)$, $p' = p/(p-1)$; see [10]. We want to warn the reader before possible confusions: the preprint [10] is formulated for $\mathbf{T} = \partial\Phi(|\mathbf{D}|)$; however, due to the wrong proof of Lemma 7.18 it is valid only for $\mathbf{T}$ of the type $\mathbf{T} = \partial\Phi(|\nabla \mathbf{v}|)$.

The proof of Theorem 1.17 is split into several sections. In the next section we clarify, using some formal arguments, the main difficulty of the problem connected with low regularity of the pressure. This difficulty is overcome by constructing an appropriate twofold approximation of the original problem based on both the mollification of the convective term $v_{\varepsilon k} \frac{\partial \mathbf{v}}{\partial x_k}$ and on a quadratic approximation of the potential Φ, denoted by Φ_A. Section 3 deals with the existence and regularity of the weak solution to the approximate problem. In Section 4 we carry out the limiting process from Φ_A to Φ, while in Section 5 we let $\varepsilon \to 0$ and we finally obtain the results stated in Theorem 1.17. The appendix contains some helpful assertions.

2. Definition of the approximate problem. We start with a lemma collecting some consequences of the assumption (1.9)–(1.11). The proof can be found for example in Málek, Nečas, Rokyta, Růžička [11], Chapter 5, Lemma 1.19 and Lemma 1.35.

Lemma 2.1. *Let* $\mathbf{T}$ *and* Φ *satisfy* (1.9)–(1.11). *Assume that* $p \ge 2$. *Then there exist* $C_i, i = 3, \dots, 7$, *such that for all* $\mathbf{B} \in \mathbb{R}^{3\times 3}_{\mathrm{sym}}$

$$\mathbf{T}(\mathbf{B}) \cdot \mathbf{B} \ge C_3 (1 + |\mathbf{B}|^{p-2}) |\mathbf{B}|^2\,, \tag{2.2}$$

$$\big|\mathbf{T}(\mathbf{B})\big| \le C_4(1+|\mathbf{B}|)^{p-1}\,, \tag{2.3}$$

$$C_5(1+|\mathbf{B}|^{p-2})|\mathbf{B}|^2 \le C_3(\tfrac{1}{2}+\tfrac{1}{p}|\mathbf{B}|^{p-2})|\mathbf{B}|^2 \le \Phi(|\mathbf{B}|^2) \le C_6(1+|\mathbf{B}|)^p \tag{2.4}$$

and for all $\mathbf{B}$, $\mathbf{D} \in \mathbb{R}^{3\times 3}_{\text{sym}}$

$$\big(\mathbf{T}(\mathbf{B})-\mathbf{T}(\mathbf{D})\big)\cdot\big(\mathbf{B}-\mathbf{D}\big) \ge C_7|\mathbf{B}-\mathbf{D}|^2\,. \tag{2.5}$$

Learning from the methods used for the *Problem (NS-Per)*$_p$, we start with a mollification of the convective term. Let $\omega \in \mathcal{D}(\mathbb{R}^3)$ be a usual mollification kernel with support in $B_1(0) \equiv \{x \in \mathbb{R}^3 : |x| < 1\}$ and $\int_{\mathbb{R}^3} \omega\,dx = 1$. For every $\varepsilon > 0$ we define $\omega_\varepsilon = \frac{1}{\varepsilon^3}\omega(\frac{x}{\varepsilon})$ and we put $\mathbf{v}_\varepsilon \equiv \mathbf{v} * \omega_\varepsilon$; i.e.,

$$\mathbf{v}_\varepsilon(x) = \frac{1}{\varepsilon^3}\int_{\mathbb{R}^3} \omega\big(\frac{x-y}{\varepsilon}\big)\mathbf{v}(y)\,dy\,.$$

It is easy to observe that

$$\|\mathbf{v}_\varepsilon\|_{2,2} \le \frac{1}{\varepsilon^2}\|\mathbf{v}\|_2\,, \tag{2.6}$$

and for $\mathbf{v} \in L^p(\Omega)$, always $\|\mathbf{v}_\varepsilon\|_p \le \|\mathbf{v}\|_p$.

Now, let us consider[3] the *Problem (NS-Dir)*$^\varepsilon_p$: to find $(\mathbf{v},\pi) = (\mathbf{v}^\varepsilon,\pi^\varepsilon)$ such that

$$\begin{aligned}
&\operatorname{div}\mathbf{v} = 0 \quad \text{in } Q_T\,,\\
&\frac{\partial \mathbf{v}}{\partial t} - \operatorname{div}\mathbf{T}(\mathbf{D}(\mathbf{v})) + v_{\varepsilon k}\frac{\partial \mathbf{v}}{\partial x_k} = -\nabla\pi + \mathbf{f} \quad \text{in } Q_T\,,\\
&\mathbf{v}(0,\cdot) = \mathbf{v}_0(\cdot) \quad \text{in } \Omega \quad \text{and} \quad \mathbf{v} = \mathbf{0} \text{ at } I\times\partial\Omega\,,
\end{aligned} \tag{2.7}$$

where $\mathbf{T}$ satisfies (1.9)–(1.11).

Now we formally derive a priori estimates. First, we multiply equation $(2.7)_2$ by $\mathbf{v}$. Integrating over Q_T, using (2.2), partial integration and a p-version of Korn's inequality (see (6.2) in the appendix), we obtain

$$\|\mathbf{v}(t)\|_2^2 + C_3K_2^2\int_0^t \|\nabla\mathbf{v}\|_2^2\,d\tau + 2C_3\,K_p^p\int_0^t \|\nabla\mathbf{v}\|_p^p\,d\tau \le c(\mathbf{f},\mathbf{v}_0)\,, \tag{2.8}$$

[3]We put for simplicity $\rho \equiv 1$ in the sequel, because the size of the constant density plays no role in the whole analysis.

where $c(\mathbf{f},\mathbf{v}_0)$ denotes a generic constant depending only on the data; in case of (2.8) $c(\mathbf{f},\mathbf{v}_0)=\|\mathbf{v}_0\|_2^2+c\|\mathbf{f}\|_2^2$. The estimate (2.8) is independent of ε, and we have

$$\mathbf{v}\in L^\infty(I;H)\cap L^p(I;V_p)\,. \tag{2.9}$$

Multiplying $(2.7)_2$ by $\frac{\partial \mathbf{v}}{\partial t}$ (which is also divergence-free and vanishing at $\partial\Omega$), integrating over Q_t and using the fact that

$$\int_\Omega T_{ij}(\mathbf{D}(\mathbf{v}))D_{ij}(\frac{\partial \mathbf{v}}{\partial t})\,dx=\frac{d}{dt}\int_\Omega \Phi(|\mathbf{D}(\mathbf{v})|^2)\,dx\,,$$

along with (2.4) we get

$$\int_0^t \|\frac{\partial \mathbf{v}}{\partial t}\|_2^2\,d\tau+K_p^p\|\nabla\mathbf{v}(t)\|_p^p\le c(\mathbf{f},\mathbf{v}_0)+\int_{Q_T}|\mathbf{v}_\varepsilon|^2|\nabla\mathbf{v}|^2\,dx\,dt\equiv J\,. \tag{2.10}$$

Now, since $\mathbf{v}$ satisfies (2.9) it is possible to prove that

$$J\le c(\mathbf{f},\mathbf{v}_0)\qquad \text{if } p\ge\frac{12}{5}\,, \tag{2.11}$$

and

$$J\le c\,(\frac{1}{\varepsilon},\mathbf{f},\mathbf{v}_0)\qquad \text{if } p\in[2,\frac{12}{5})\,. \tag{2.12}$$

Indeed, using for $p\in[\frac{12}{5},3)$ the interpolation inequality

$$\|\mathbf{v}\|_{\frac{2p}{2-p}}\le\|\mathbf{v}\|_2^{\frac{5p-12}{5p-6}}\,\|\mathbf{v}\|_{\frac{3p}{3-p}}^{\frac{6}{5p-6}}$$

we have

$$\begin{aligned}\int_\Omega|\mathbf{v}_\varepsilon|^2|\nabla\mathbf{v}|^2dx&\le\|\nabla\mathbf{v}\|_p^2\|\mathbf{v}_\varepsilon\|_{\frac{2p}{p-2}}^2\le c\|\nabla\mathbf{v}\|_p^{p+\frac{p(16-5p)}{5p-6}}\,\|\mathbf{v}\|_2^{2\frac{5p-12}{5p-6}}\\&\le c\|\mathbf{v}\|_2^{2\frac{5p-12}{5p-6}}\,\|\nabla\mathbf{v}\|_p^{p\frac{16-5p}{5p-6}}\,\|\nabla\mathbf{v}\|_p^p=:h\,\|\nabla\mathbf{v}\|_p^p\,.\end{aligned} \tag{2.13}$$

Observe that $\frac{16-5p}{5p-6}\le 1$ if $p\ge\frac{11}{5}$. Thus due to (2.9) and $h\in L^1(0,T)$ we can derive (2.11) for $p\in[\frac{12}{5},3)$ from (2.10) and Gronwall's lemma. The estimate (2.12) is a consequence of (2.6) and the imbedding $W^{2,2}(\Omega)\hookrightarrow L^\infty(\Omega)$. So we have from (2.10)

$$\int_0^t\|\frac{\partial\mathbf{v}}{\partial t}\|_2^2\,d\tau+\|\nabla\mathbf{v}(t)\|_p^p\le\begin{cases}c\,(\frac{1}{\varepsilon},\mathbf{f},\mathbf{v}_0) & \text{if } p\in[2,\frac{12}{5})\,,\\ c(\mathbf{f},\mathbf{v}_0) & \text{if } p\ge\frac{12}{5}\,.\end{cases} \tag{2.14}$$

In order to pass to the limit in the nonlinear term given by $\mathbf{T}$ (at least for $p \in [2, \frac{11}{5})$) and in order to obtain higher regularity properties (for $p \geq \frac{11}{5}$) we need more information about $\nabla \mathbf{v}$. For the *Problem (NS-Per)$_p$* we used $-\Delta \mathbf{v}$ as a test function which is divergence-free, but in the case of the *Problem (NS-Dir)$_p$* the term $-\Delta \mathbf{v}$ does not vanish at $\partial\Omega$.

Let $\Omega' \subset \bar{\Omega}' \subset \Omega$ and let $\xi \in \mathcal{D}(\Omega)$ be a cut-off function such that $\xi \in [0,1]$ in Ω'. Now, we multiply $(2.7)_2$ by $-\Delta \mathbf{v}\xi^2$, a test function vanishing at the boundary, but not being divergence-free. Thus the pressure π enters the picture. Using the divergence-free constraint we have $-\int_\Omega \nabla \pi \Delta \mathbf{v} \xi^2 \, dx = \int_\Omega \pi \Delta \mathbf{v} \nabla \xi^2 \, dx$.

In order to clarify a key point that leads to the final approximation, we will consider only the main terms, "forgetting" cut-off functions. This yields

$$
\begin{aligned}
\tfrac{1}{2}\|\nabla \mathbf{v}(t)\|_2^2 &+ \int_0^t \int_\Omega \partial_{ij}\partial_{kl}\Phi(|\mathbf{D}(\mathbf{v})|^2) D_{ij}(\nabla \mathbf{v}) \cdot D_{kl}(\nabla \mathbf{v}) \, dx \, d\tau \\
&\leq \int_0^t \int_\Omega |\mathbf{f} \cdot \Delta \mathbf{v}| + |\mathbf{v}_\varepsilon| \, |\nabla \mathbf{v}| |\Delta \mathbf{v}| \, dx \, d\tau + \int_0^t \int_\Omega |\pi| |\Delta \mathbf{v}| \, dx \, d\tau \,.
\end{aligned}
\tag{2.15}
$$

By (1.10) and Young's inequality, we get

$$
\begin{aligned}
\tfrac{1}{2}\|\nabla \mathbf{v}(t)\|_2^2 + \int_0^t \|\nabla^{(2)} \mathbf{v}\|_2^2 \, d\tau &\leq c(\delta) \int_0^T \|\mathbf{f}\|_2^2 \, dt + c(\delta) \int_0^t \int_\Omega |\mathbf{v}_\varepsilon|^2 |\nabla \mathbf{v}|^2 dx \, d\tau \\
&\quad + \delta \int_0^t \|\Delta \mathbf{v}\|_2^2 \, d\tau + \int_0^t \int_\Omega |\pi| \, |\Delta \mathbf{v}| \, dx \, d\tau \\
&\underset{(2.11)}{\overset{(2.12)}{\leq}} c\big(\frac{1}{\varepsilon}, \mathbf{f}, \mathbf{v}_0\big) + \int_0^t \int_\Omega |\pi| \, |\Delta \mathbf{v}| \, dx \, d\tau + \delta \int_0^t \|\Delta \mathbf{v}\|_2^2 \, d\tau \,.
\end{aligned}
\tag{2.16}
$$

Since $\operatorname{div} \mathbf{T}(\mathbf{D}(\cdot))$ maps $L^p(I; V_p) \to L^{p'}(I; V_p^*)$, $p' = \frac{p}{p-1}$, we see that $\pi \in L^{p'}(Q_T)$, where $p' \leq 2$. This fact causes the main difficulty, because we have no information about $\Delta \mathbf{v}$ in $L^p(Q_T)$. If we estimate

$$
\int_0^t \int_\Omega |\pi| \, |\Delta \mathbf{v}| \, dx \, d\tau \leq c(\delta) \int_0^t \|\pi\|_2^2 \, d\tau + \delta \int_0^t \|\nabla^{(2)} \mathbf{v}\|_2^2 \, d\tau \,,
$$

then we need that $\pi \in L^2(Q_T)$ at least on the level of approximations. In order to obtain this, we approximate the p-potential Φ by quadratic potentials Φ_A in the following way: for $A > 1$ and $\mathbf{B} \in \mathbb{R}^{3\times 3}_{\mathrm{sym}}$ we define

$$
\Phi_A(|\mathbf{B}|^2) = \begin{cases} \Phi(|\mathbf{B}|^2) & \text{if } |\mathbf{B}| \leq A \,, \\ \alpha_2 |\mathbf{B}|^2 + \alpha_1 |\mathbf{B}| + \alpha_0 & \text{if } |\mathbf{B}| > A \,, \end{cases}
\tag{2.17}
$$

where α_0, α_1, α_2 are such that $\Phi_A \in C^2(\mathbb{R}_0^+)$. This gives

$$\begin{aligned} \alpha_0 &= \Phi(A^2) + 2\Phi''(A^2)A^4 - \Phi'(A^2)A^2, \\ \alpha_1 &= -4\Phi''(A^2)A^3, \qquad \alpha_2 = 2\Phi''(A^2)A^2 + \Phi'(A^2). \end{aligned} \tag{2.18}$$

Defining $\mathbf{T}^A$ by

$$\mathbf{T}^A(\mathbf{B}) \equiv \partial\Phi_A(|\mathbf{B}|^2), \tag{2.19}$$

we see (compare with $(2.23)_1$ below) that $\mathbf{T}^A$ has linear growth and $\operatorname{div}\mathbf{T}^A(\mathbf{D}(\cdot))$ maps $L^2(I;V_2)$ into $L^2(I;V_2^*)$. Thus we finally come to the definition of the approximate problem.

Definition 2.20. (the *Problem (NS-Dir)$_p^{\varepsilon,A}$*) For given $\varepsilon > 0$ and $A > 1$, we look for $\mathbf{v} = \mathbf{v}^{\varepsilon,A}$ and $\pi = \pi^{\varepsilon,A}$ solving the system

$$\begin{gathered} \operatorname{div}\mathbf{v} = 0 \quad \text{in } Q_T\,, \\ \frac{\partial\mathbf{v}}{\partial t} - \operatorname{div}\mathbf{T}^A(\mathbf{D}(\mathbf{v})) + v_{\varepsilon k}\frac{\partial\mathbf{v}}{\partial x_k} = -\nabla\pi + \mathbf{f} \quad \text{in } Q_T\,, \\ \mathbf{v}(0,\cdot) = \mathbf{v}_0(\cdot) \ \text{in } \Omega \quad \text{and} \quad \mathbf{v} = \mathbf{0} \quad \text{at } I\times\partial\Omega\,. \end{gathered} \tag{2.21}$$

We will need several inequalities connected with the potential Φ_A, which we summarize in the following lemma.

Lemma 2.22. *Assume that* (1.9)–(1.11) *hold. Then* Φ_A, $A > 1$, *given by* (2.17), (2.18) *and* $\mathbf{T}_A$ *given by* (2.19) *satisfy* (*for* $i, j, k, l, r, s = 1,2,3$ *and all* $\mathbf{B},\mathbf{D} \in \mathbb{R}^{3\times3}_{\mathrm{sym}}$)

$$\begin{aligned} &|\partial\Phi_A(|\mathbf{D}|^2)| \le \tilde{C}_1(A)(1+|\mathbf{D}|)\,, \quad |\partial^2\Phi_A(|\mathbf{D}|^2)| \le \tilde{C}_2(A)\,, \\ &\partial_{ij}\Phi_A(|\mathbf{D}|^2)\,D_{ij} \ge C_3|\mathbf{D}|^2\,, \quad C_8|\mathbf{D}|^2 \le \Phi_A(|\mathbf{D}|^2) \le C_9(1+|\mathbf{D}|^2)^{\frac{p}{2}}\,, \end{aligned} \tag{2.23}$$

$$\partial_{ij}\partial_{kl}\Phi_A(|\mathbf{D}|^2)B_{ij}B_{kl} \ge C_{10}\begin{cases} (1+|\mathbf{D}|)^{p-2}|\mathbf{B}|^2 & \text{if } |\mathbf{D}| \le A\,, \\ (1+A)^{p-2}|\mathbf{B}|^2 & \text{if } |\mathbf{D}| > A\,, \end{cases} \tag{2.24}$$

$$|\partial_{rs}\Phi_A| \le C_{11}(1+|\Phi_A|)^{\frac{p-1}{p}}\,, \tag{2.25}$$

$$|\partial_{kl}\partial_{rs}\Phi_A(|\mathbf{D}|^2)| \le C_{12}\begin{cases} (1+|\mathbf{D}|)^{p-2} & \text{if } |\mathbf{D}| \le A\,, \\ (1+A)^{p-2} & \text{if } |\mathbf{D}| > A\,, \end{cases} \tag{2.26}$$

$$1+|\partial\Phi_A(|\mathbf{D}|^2)| \ge C_{13}\begin{cases} (1+|\mathbf{D}|)^{p-1} & \text{if } |\mathbf{D}| \le A\,, \\ (1+A)^{p-1} & \text{if } |\mathbf{D}| > A\,. \end{cases} \tag{2.27}$$

Proof. All assertions are evident if $|\mathbf{D}| \le A$. Therefore we assume $|\mathbf{D}| > A$ in what follows. We first observe that (1.10) yields (for all $\mathbf{B},\mathbf{D} \in \mathbb{R}^{3\times3}_{\mathrm{sym}}$)

$$4\Phi''(|\mathbf{D}|^2)(\mathbf{D}\cdot\mathbf{B})^2 + 2\Phi'(|\mathbf{D}|^2)|\mathbf{B}|^2 \ge C_1(1+|\mathbf{D}|)^{p-2}|\mathbf{B}|^2\,.$$

In particular, for $\mathbf{B}=\frac{1}{\sqrt{3}}\mathbf{I}$ and $\mathbf{D}=\frac{A}{\sqrt{3}}\mathbf{I}$ we obtain

$$4\Phi''(A^2)A^2+2\Phi'(A^2)\geq C_1(1+A)^{p-2}. \tag{2.28}$$

Similarly, from (2.2), we have

$$2\Phi'(|\mathbf{D}|^2)\geq C_3(1+|\mathbf{D}|^{p-2}),$$

and C_3 is given by $C_3=\frac{C_1}{2(p-1)}$ (cf. [11], Chapter 5, Lemma 1.19). Choosing $\mathbf{D}=\frac{A}{\sqrt{3}}\mathbf{I}$ gives

$$2\Phi'(A^2)\geq C_3(1+A^{p-2}). \tag{2.29}$$

On the other hand, (1.11) and (2.3) imply

$$4\Phi''(|\mathbf{D}|^2)|\mathbf{D}|^2+2\Phi'(|\mathbf{D}|^2)\leq \tilde{C}_2(1+|\mathbf{D}|)^{p-2}. \tag{2.30}$$

$$2\Phi'(|\mathbf{D}|^2)|\mathbf{D}|\leq C_4(1+|\mathbf{D}|^{p-1}). \tag{2.31}$$

As

$$\begin{aligned}\partial_{ij}\Phi_A(|\mathbf{D}|^2)&=2\alpha_2 D_{ij}+\alpha_1\frac{D_{ij}}{|\mathbf{D}|},\\ \partial_{ij}\partial_{rs}\Phi_A(|\mathbf{D}|^2)&=2\alpha_2\delta_{ir}\delta_{js}+\alpha_1\frac{\delta_{ir}\delta_{js}}{|\mathbf{D}|}-\alpha_1\frac{D_{ij}D_{rs}}{|\mathbf{D}|^3},\end{aligned} \tag{2.32}$$

we see that $(2.23)_{1-3}$, (2.24) and (2.26) are consequences of (2.28)–(2.32). Let us verify (2.24), for example. We have ($|\mathbf{D}|>A$)

$$\begin{aligned}\partial_{ij}\partial_{rs}\Phi_A(|\mathbf{D}|^2)B_{ij}B_{kl}&=2\alpha_2|\mathbf{B}|^2+\alpha_1\frac{|\mathbf{B}|^2}{|\mathbf{D}|}-\alpha_1\frac{(\mathbf{B}\cdot\mathbf{D})^2}{|\mathbf{D}|^3}\\ &=\big(4\Phi''(A^2)A^2+2\Phi'(A^2)\big)|\mathbf{B}|^2-4\Phi''(A^2)A^3\frac{|\mathbf{B}|^2}{|\mathbf{D}|}+4\Phi''(A^2)A^3\frac{(\mathbf{B}\cdot\mathbf{D})^2}{|\mathbf{D}|^3}.\end{aligned} \tag{2.33}$$

Then either $\Phi''(A^2)\geq 0$ or $\Phi''(A^2)<0$. If $\Phi''(A^2)\geq 0$, we rewrite (2.33) as

$$\begin{aligned}&2\Phi'(A^2)|\mathbf{B}|^2+4\Phi''(A^2)\frac{A^3}{|\mathbf{D}|^3}\big(|\mathbf{B}|^2|\mathbf{D}|^2(|\mathbf{D}|-A)^2+A(\mathbf{D}\cdot\mathbf{B})^2\big)\\ &\geq 2\Phi'(A^2)|\mathbf{B}|^2\overset{(2.29)}{\geq} C_3(1+A^{p-2})|\mathbf{B}|^2.\end{aligned} \tag{2.34}$$

If $\Phi''(A^2)$ is negative then (2.33) can be written as

$$\begin{aligned}&\big(4\Phi''(A^2)A^2+2\Phi'(A^2)\big)|\mathbf{B}|^2-4\Phi''(A^2)\frac{A^3}{|\mathbf{D}|^3}\big(|\mathbf{B}|^2|\mathbf{D}|^2-(\mathbf{B}\cdot\mathbf{D})^2\big)\\&\overset{(2.28)}{\ge} C_1(1+A)^{p-2}|\mathbf{B}|^2\ge\frac{C_1}{2}(1+A^{p-2})|\mathbf{B}|^2\ge C_3(1+A^{p-2})|\mathbf{B}|^2,\end{aligned}\tag{2.35}$$

where we used the relation between C_1 and C_3 and the simple inequality $(1+A)^{p-2}\ge\frac{1}{2}(1+A^{p-2})$ valid for $p\ge 2$. Thus, we conclude that (2.24) holds independently of the sign of $\Phi''(A^2)$. Moreover, (2.33)–(2.35) with $\mathbf{B}=\mathbf{D}$ leads to

$$4\Phi''(A^2)A^2+2\Phi'(A^2)\ge C_3(1+A^{p-2})\,.\tag{2.36}$$

Next, (2.4) yields $(2.23)_4$. Indeed, we have

$$\Phi_A(|\mathbf{D}|^2)=2\Phi''(A^2)A^2(|\mathbf{D}|^2-2A|\mathbf{D}|+A^2)+\Phi'(A^2)(|\mathbf{D}|^2-A^2)+\Phi(A^2).$$

If $\Phi''(A^2)<0$ we rewrite and estimate the right-hand side as

$$\begin{aligned}&\big(2\Phi''(A^2)A^2+\Phi'(A^2)\big)\big(|\mathbf{D}|^2-A^2\big)+4\Phi''(A^2)A^3(A-|\mathbf{D}|)+\Phi(A^2)\\&\underset{(2.4)}{\overset{(2.36)}{\ge}}\frac{C_3}{2}(1+A^{p-2})(|\mathbf{D}|^2-A^2)+C_3(\frac{1}{2}+\frac{A^{p-2}}{p})A^2\\&=\frac{C_3}{2}|\mathbf{D}|^2+\frac{C_3}{2}A^{p-2}(|\mathbf{D}|^2-A^2)+\frac{C_3}{p}A^p\ge\frac{C_3}{2}|\mathbf{D}|^2\,,\end{aligned}$$

while if $\Phi''(A^2)\ge 0$ we have

$$\begin{aligned}\Phi_A(|\mathbf{D}|^2)&=2\Phi''(A^2)A^2(|\mathbf{D}|-A)^2+\Phi'(A^2)(|\mathbf{D}|^2-A^2)+\Phi(A^2)\\&\underset{(2.4)}{\overset{(2.29)}{\ge}}\frac{C_3}{2}(1+A^{p-2})(|\mathbf{D}|^2-A^2)+C_3(\frac{1}{2}+\frac{A^{p-2}}{p})A^2\,,\end{aligned}$$

which gives the same conclusion as above; i.e., the lower bound in $(2.23)_4$ is proved. The upper bound is clear. We also have

$$\partial_{ij}\Phi_A(|\mathbf{D}|^2)=\Big[\big(4\Phi''(A^2)A^2+2\Phi'(A^2)\big)\big(1-\frac{A}{|\mathbf{D}|}\big)+2\Phi'(A^2)\frac{A}{|\mathbf{D}|}\Big]D_{ij}\,.$$

Since all terms in front of D_{ij} are positive we obtain ($A > 1$)

$$|\partial\Phi_A(|\mathbf{D}|^2)| = 2\Phi'(A^2)A + \big(4\Phi''(A^2)A^2 + 2\Phi'(A^2)\big)(|\mathbf{D}| - A)$$
$$\overset{(2.29)}{\underset{(2.36)}{\geq}} C_3(1 + A^{p-2})A \geq C_{13}(1 + A)^{p-1}\,.$$

This proves (2.27). Finally, to prove (2.25) for $|\mathbf{D}| > A$, it is enough ($\Phi_A \in C^2(\mathbb{R}_0^+)$) to show that for $F(|\mathbf{D}|) \equiv \big|\partial\Phi_A(|\mathbf{D}|^2)\big|^p\big(1+\Phi_A(|\mathbf{D}|^2)\big)^{1-p}$ the limit $\lim_{|\mathbf{D}|\to\infty} F(|\mathbf{D}|)$ is finite. However,

$$F(|\mathbf{D}|) = \frac{\big|2\alpha_2\mathbf{D} + \alpha_1\frac{\mathbf{D}}{|\mathbf{D}|}\big|^p}{\big(1 + \alpha_0 + \alpha_1|\mathbf{D}| + \alpha_2|\mathbf{D}|^2\big)^{p-1}} \leq \frac{(2\alpha_2)^p|\mathbf{D}|^{2-p}\big(1 + \frac{|\alpha_1|}{2\alpha_2|\mathbf{D}|}\big)^p}{\alpha_2^{p-1}\big(\frac{1+\alpha_0}{\alpha_2|\mathbf{D}|^2} + \frac{\alpha_1}{\alpha_2|\mathbf{D}|} + 1\big)^{p-1}}.$$

Therefore,

$$\lim_{|\mathbf{D}|\to\infty} F(|\mathbf{D}|) \leq 2^p\alpha_2\frac{1}{A^{p-2}} \overset{(2.30)}{\leq} 2\tilde{C}_2\,.$$

3. On strong solutions to the *Problem (NS-Dir)*$_p^{\varepsilon,A}$**.** The goal of this section is to prove the existence of a strong solution to the *Problem (NS-Dir)*$_p^{\varepsilon,A}$ provided that $\varepsilon > 0$ and $A > 1$ are fixed. We say that a couple $(\mathbf{v},\pi) \equiv (\mathbf{v}^{\varepsilon,A},\pi^{\varepsilon,A})$ is a strong solution to the *Problem (NS-Dir)*$_p^{\varepsilon,A}$ if

$$\begin{aligned} &\mathbf{v} \in L^\infty(I;V_2) \cap L^2(I;W^{2,2}(\Omega)^3)\,,\\ &\frac{\partial\mathbf{v}}{\partial t} \in L^2(I;L^2(\Omega)^3)\,, \quad \pi \in L^2(I;L^2(\Omega))\,, \end{aligned} \tag{3.1}$$

and the weak formulation

$$\begin{aligned} &\Big(\frac{\partial\mathbf{v}(t)}{\partial t},\boldsymbol{\varphi}\Big) + \int_\Omega T_{ij}^A(\mathbf{D}(\mathbf{v}(t)))D_{ij}(\boldsymbol{\varphi})\,dx + \Big(v_{\varepsilon k}(t)\frac{\partial\mathbf{v}(t)}{\partial x_k},\boldsymbol{\varphi}\Big)\\ &\qquad = (\pi(t),\operatorname{div}\boldsymbol{\varphi}) + (\mathbf{f}(t),\boldsymbol{\varphi}) \end{aligned} \tag{3.2}$$

is fulfilled for all $\boldsymbol{\varphi} \in W_0^{1,2}(\Omega)^3$ and almost all $t \in I$. In fact, due to the linear growth of $\mathbf{T}^A$ and (3.1), a strong solution $(\mathbf{v},\pi)$ satisfies the equation $(2.20)_2$ almost everywhere in Q_T.

A couple $(\mathbf{v},\pi) = (\mathbf{v}^{\varepsilon,A},\pi^{\varepsilon,A})$ will be called a weak solution to the *Problem (NS-Dir)*$_p^{\varepsilon,A}$ if $\mathbf{v} \in L^\infty(I;V_2)$ and $(3.1)_{2-3}$ and (3.2) are valid.

Theorem 3.3. *Let $\Omega \subset \mathbb{R}^3$ be a bounded domain, $\partial\Omega \in \mathcal{C}^3$. Let $\mathbf{v}_0 \in V_p$ and $\mathbf{f} \in L^2(I;L^2(\Omega)^3)$ and $p \geq 2$. Then, for all $\varepsilon > 0$ and $A > 1$ there exists a strong solution $(\mathbf{v},\pi) = (\mathbf{v}^{\varepsilon,A},\pi^{\varepsilon,A})$ to the Problem (NS-Dir)$_p^{\varepsilon,A}$ such that $\mathbf{v}$ is unique in the class of weak solutions of the Problem (NS-Dir)$_p^{\varepsilon,A}$ and it holds that*

$$\|\mathbf{v}\|_{L^\infty(I;H)} + \|\mathbf{v}\|_{L^2(I;V_2)} \leq c(\mathbf{f},\mathbf{v}_0) \tag{3.4}$$

$$\begin{aligned}\|\frac{\partial \mathbf{v}}{\partial t}\|^2_{L^2(I;L^2(\Omega)^3)} &+ \|\Phi_A(|\mathbf{D}(\mathbf{v})|^2)\|_{L^\infty(I;L^1(\Omega))} \\ &\leq c(\mathbf{f},\mathbf{v}_0) + \int_{Q_T} |\mathbf{v}_\varepsilon|^2|\nabla\mathbf{v}|^2 dx\,dt \leq c\,(\frac{1}{\varepsilon},\mathbf{f},\mathbf{v}_0)\end{aligned} \tag{3.5}$$

$$\begin{aligned}\|\pi\|^2_{L^2(I;L^2(\Omega))} &\leq \int_0^T \|\partial\Phi_A(|\mathbf{D}(\mathbf{v})|^2)\|_2^2\,dt + \int_{Q_T} |\mathbf{v}_\varepsilon|^2|\nabla\mathbf{v}|^2 dx\,dt + c(\mathbf{f}) \\ &\leq c(\frac{1}{\varepsilon},A,\mathbf{f},\mathbf{v}_0)\,.\end{aligned} \tag{3.6}$$

Moreover,

$$\|\nabla^{(2)}\mathbf{v}\|^2_{L^2(I;L^2(\Omega)^{3\times 3})} \leq c(\frac{1}{\varepsilon},A,\mathbf{f},\mathbf{v}_0), \tag{3.7}$$

and $(2.20)_2$ holds almost everywhere in Q_T.

Proof. The proof consists of two parts. While the first one, including the existence of a unique weak solution, is standard (and we only sketch the proof), the second one, proving the regularity result (3.7), uses methods that are not so common, and we will give a detailed proof.

Using the Galerkin approach we can justify the a priori estimates (3.4) and (3.5), which are derived in the same way as (2.8) and (2.12). By (2.6) and (3.4) $v_{\varepsilon k}\frac{\partial \mathbf{v}}{\partial x_k} \in L^2(I;L^2(\Omega)^3)$. Due to (2.17), the operator

$$-\operatorname{div}\mathbf{T}^A(\mathbf{D}(\cdot)) : L^2(I;W_0^{1,2}(\Omega)^3) \to L^2(I;(W_0^{1,2}(\Omega)^3)^*)$$

is a strongly monotone operator. Thus combining the Galerkin method with the monotone operator theory we can prove the existence of a weak solution $\mathbf{v} = \mathbf{v}^{\varepsilon,A}$ satisfying (3.4) and (3.5). See Lions [8] for details, for example.

Defining $\mathbf{F} \in L^2(I;(W_0^{1,2}(\Omega)^3)^*)$ by

$$\begin{aligned}\langle \mathbf{F}(t),\boldsymbol{\varphi}\rangle \equiv (\frac{\partial \mathbf{v}(t)}{\partial t},\boldsymbol{\varphi}) &+ \int_\Omega T^A_{ij}(\mathbf{D}(\mathbf{v}(t)))D_{ij}(\boldsymbol{\varphi})\,dx \\ &+ (v_{\varepsilon k}(t)\frac{\partial \mathbf{v}(t)}{\partial x_k},\boldsymbol{\varphi}) - (\mathbf{f}(t),\boldsymbol{\varphi})\end{aligned} \tag{3.8}$$

we see that for almost all $t \in I$ and all $\boldsymbol{\varphi} \in V_2$, $\langle \mathbf{F}(t), \boldsymbol{\varphi} \rangle = 0$. By De Rham's theorem and Theorem 6.7 there exists $\pi \in L^2(I; L^2(\Omega))$, $\int_\Omega \pi \, dx = 0$ such that

$$\langle \mathbf{F}(t), \boldsymbol{\varphi} \rangle = \langle \nabla \pi(t), \boldsymbol{\varphi} \rangle \qquad \forall \boldsymbol{\varphi} \in W_0^{1,2}(\Omega)^3 .$$

From (3.8) and known a priori estimates, we obtain (3.6), where we also used $(2.23)_1$. Thus the first part of the proof is finished.

In the rest of this section we will focus on (3.7), having at our disposal (3.4)–(3.6) and the weak formulation (3.2). The proof of (3.7) consists, as usual, of interior regularity and regularity near the boundary. In the latter case we will not flatten the boundary. We prefer to use a curvilinear system in order to derive estimate (3.7) in tangential and normal directions. Because of the missing boundary conditions for the pressure we obtain the estimates in the normal direction by applying the (pressure-eliminating) curl operator to the system $(2.21)_2$.

Let $T : \Omega_0 \subset \Omega \to \Omega$ be a diffeomorphism. Using (3.2) we get the identity

$$\begin{aligned}
0 &= \Big(\frac{\partial \mathbf{v}}{\partial t}(Tx) - \frac{\partial \mathbf{v}}{\partial t}(x), \boldsymbol{\varphi}(x)\Big) \\
&+ \int_\Omega \Big[\partial_{ij}\Phi_A(|\mathbf{D}(\mathbf{v}(Tx))|^2) - \partial_{ij}\Phi_A(|\mathbf{D}(\mathbf{v}(x))|^2)\Big] D_{ij}(\boldsymbol{\varphi}(x))\, dx \\
&+ \Big(\big(v_{\varepsilon k}\frac{\partial \mathbf{v}}{\partial x_k}\big)(Tx) - \big(v_{\varepsilon k}\frac{\partial \mathbf{v}}{\partial x_k}\big)(x), \boldsymbol{\varphi}(x)\Big) \\
&+ \big(\nabla \pi(Tx) - \nabla \pi(x), \boldsymbol{\varphi}(x)\big) - \big(\mathbf{f}(Tx) - \mathbf{f}(x), \boldsymbol{\varphi}(x)\big) \\
&\equiv I_1 + I_2 + I_3 + I_4 + I_5
\end{aligned} \tag{3.9}$$

valid for "correct" $\boldsymbol{\varphi}$. In (3.9) we suppressed the dependence of $\mathbf{v}$, π, $\mathbf{f}$ and $\boldsymbol{\varphi}$ on t. If $\mathbf{g}$ denotes one of these quantities, then

$$\big(\mathbf{g}(Tx) - \mathbf{g}(x), \boldsymbol{\varphi}(x)\big) \equiv \int_\Omega \big(\mathbf{g}(t, Tx) - \mathbf{g}(t,x)\big) \cdot \boldsymbol{\varphi}(t,x)\, dx .$$

This convention is used in the sequel. In case of interior regularity, let $V' \subset\subset \Omega_0 \subset\subset \Omega$ be such that $\operatorname{dist}(\partial\Omega_0, \partial\Omega) = h_0 > 0$. Let $\mathbf{e}^r$, $r = 1, 2, 3$, be a basis of a coordinate system in $\mathbb{R}^3$. Setting for $r = 1, 2, 3$ and $h \in (0, h_0)$

$$T = T_{r,h} : x \mapsto x + h\mathbf{e}^r , \tag{3.10}$$

we then get $T : \Omega_0 \to \Omega$. Let us consider a cut-off function $\xi \in \mathcal{D}(\Omega_0)$ such that $\xi \in [0,1]$ in Ω and $\xi \equiv 1$ in V'.

In the case of regularity (near the boundary) in tangential directions, let us consider one of the maps a_k, $k = 1, 2, \dots, N$, that locally describe $\partial\Omega$ (cf. [4], page 305). We know that for a certain $\alpha > 0$, $\partial\Omega$ is covered by sets $V^k \equiv \{x = (x', x_3) \in \mathbb{R}^3 : \; x' = (x_1, x_2); |x'| \leq \alpha$ and $a_k(x') - \alpha < x_3 < a_k(x') + \alpha\}$, where $a_k \in C^3(B_\alpha(0'))$ and

$$\frac{\partial a_k(0')}{\partial x_s} = 0, \qquad (s = 1, 2). \tag{3.11}$$

Let us also define $V_+^k \equiv \{x \in \mathbb{R}^3 : |x'| \leq \alpha$ and $a_k(x') < x_3 < a_k(x') + \alpha\}$, $V_-^k \equiv \{x \in \mathbb{R}^3 : |x'| \leq \alpha$ and $a_k(x') - \alpha < x_3 < a_k(x')\}$; then $V^k = V_+^k \cup V_-^k \cup \{x \in \partial\Omega : |x'| \leq \alpha$ and $x_3 = a_k(x')\}$. We finally choose sets Ω_0^k covering $\partial\Omega$ such that $\Omega_0^k \subset V^k$, $\operatorname{dist}(\partial\Omega_0^k, \partial V^k) \geq h_0 > 0$.

Let us fix k and drop for simplicity the index k. Setting $\hat{\mathbf{e}}^1 \equiv (1, 0)$ and $\hat{\mathbf{e}}^2 \equiv (0, 1)$, we can define for $s = 1, 2$ and $h \in (0, h_0)$ the mapping $T = T_{s,h} : \Omega_0 \to V$ by

$$x \mapsto (x' + h\hat{\mathbf{e}}^s, x_3 + a(x' + h\hat{\mathbf{e}}^s) - a(x')) \equiv y\,. \tag{3.12}$$

Then the inverse mapping T^{-1} is given by $(x = T^{-1}(y))$

$$y \mapsto (y' - h\hat{\mathbf{e}}^s, x_3 + a(y' - h\hat{\mathbf{e}}^s) - a(y'))\,. \tag{3.13}$$

Put

$$\Delta^{\pm} a(x') = a(x' \pm h\hat{\mathbf{e}}^s) - a(x')\,. \tag{3.14}$$

Then

$$\left(\frac{\partial T_i}{\partial x_j}(x)\right)_{i,j=1,2,3} = \begin{pmatrix} 1 & 0 & 0 \\ 0 & 1 & 0 \\ \frac{\partial \Delta^+ a}{\partial x_1}(x') & \frac{\partial \Delta^+ a}{\partial x_2}(x') & 1 \end{pmatrix} \tag{3.15}$$

and

$$\left(\frac{\partial T_i^{-1}}{\partial y_j}(y)\right)_{i,j=1,2,3} = \begin{pmatrix} 1 & 0 & 0 \\ 0 & 1 & 0 \\ \frac{\partial \Delta^- a}{\partial y_1}(y') & \frac{\partial \Delta^- a}{\partial y_2}(y') & 1 \end{pmatrix}. \tag{3.16}$$

Both matrices in (3.15) and (3.16) have determinant equal to 1 (the same is true for T defined in (3.10)). The s-th tangential derivative $(s = 1, 2)$ of any (scalar or vector) function g, denoted $\frac{\partial g}{\partial \tau^s}$, is defined by

$$\frac{\partial g}{\partial \tau^s}(x) \equiv \lim_{h \to 0} \frac{g(Tx) - g(x)}{h},$$

and it holds that

$$\frac{\partial g}{\partial \tau^s}(x) = \frac{\partial g}{\partial x_s}(x) + \frac{\partial g}{\partial x_3}(x)\frac{\partial a}{\partial x_s}(x') . \tag{3.17}$$

For the reader's convenience let us note that if $g \in W_0^{1,p}(\Omega)$, $p > 1$, then for all $h \in (0, h_0)$

$$\int_{\Omega_0} \left|\frac{g(Tx) - g(x)}{h}\right|^p dx \le c(a)\|\nabla g\|_p^p . \tag{3.18}$$

Indeed, setting $T_\lambda(x) = \big(x' + \lambda\hat{\mathbf{e}}^s, x_3 + a(x' + \lambda\hat{\mathbf{e}}^s) - a(x')\big)$, we can write

$$\begin{aligned}\int_{\Omega_0} \left|\frac{g(Tx) - g(x)}{h}\right|^p dx &= \int_{\Omega_0} \left|\int_0^1 \frac{\partial g(T_\lambda(x))}{\partial \tau^s}\, d\lambda\right|^p dx \\ \le \int_{\Omega_0}\int_0^1 \left|\frac{\partial g(T_\lambda(x))}{\partial \tau^s}\right|^p d\lambda\, dx &\le \int_0^1\int_\Omega \left|\frac{\partial g(y)}{\partial \tau^s}\right|^p dy\, d\lambda \overset{(3.17)}{\le} c(a)\|\nabla g\|_p^p .\end{aligned}$$

On the other hand, if $g \in L^p(\Omega)$ and if for all $h \in (0, h_0)$

$$\int_\Omega \left|\frac{g(Tx) - g(x)}{h}\right|^p dx \le c_0 < \infty , \tag{3.19}$$

then $\dfrac{\partial g}{\partial \tau^s}$ exists (for $s = 1, 2$) in the sense of distributions and

$$\int_\Omega \left|\frac{\partial g}{\partial \tau^s}\right|^p dx \le c_0 . \tag{3.20}$$

If further $V' \subset\subset \Omega_0$, we consider $\xi \in [0, 1]$, $\xi \in \mathcal{D}(\Omega_0)$, $\xi \equiv 1$ in V'. Setting

$$\varphi = \frac{1}{h^2}\big(\mathbf{v}(Tx) - \mathbf{v}(x)\big)\xi^2(x) \tag{3.21}$$

we see that interior regularity and regularity in tangential directions can be treated analogously. Since the mapping T is more complicated in the latter case, we will present a detailed proof for it. In order to shorten formulas let us denote $\mathbf{w}(x) \equiv \mathbf{v}(Tx) - \mathbf{v}(x)$. Let us take (3.21) as a test function in (3.9), and let us calculate the terms I_k, $k = 1, \dots, 5$, separately. We get

$$I_1 = \frac{1}{2}\frac{d}{dt}\int_{\Omega_0} \Big|\frac{\mathbf{w}}{h}\Big|^2 \xi^2 \, dx\,, \tag{3.22}$$

$$\begin{aligned} I_2 &= \frac{1}{h^2}\int_{\Omega_0} \big[\partial_{ij}\Phi_A(|\mathbf{D}(\mathbf{v}(Tx))|^2) - \partial_{ij}\Phi_A(|\mathbf{D}(\mathbf{v}(x))|^2)\big] D_{ij}(\mathbf{w}(x))\xi^2(x)\, dx \\ &+ \frac{2}{h^2}\int_{\Omega_0} \big[\partial_{ij}\Phi_A(|\mathbf{D}(\mathbf{v}(Tx))|^2) - \partial_{ij}\Phi_A(|\mathbf{D}(\mathbf{v}(x))|^2)\big] w_i(x)\xi(x)\frac{\partial \xi(x)}{\partial x_j}\, dx \\ &= J_1 + J_2\,. \end{aligned} \tag{3.23}$$

By (2.24) and Lemma 6.5 from the appendix we have

$$\begin{aligned} J_1 &= \frac{1}{h^2}\int_{\Omega_0}\int_0^1 \partial_{ij}\partial_{kl}\Phi_A(|\mathbf{D}(\mathbf{v}) + \lambda\mathbf{D}(\mathbf{w})|^2) D_{ij}(\mathbf{w}) D_{kl}(\mathbf{w})\xi^2 \, d\lambda\, dx \\ &\ge C_{10}\int_{\Omega_0} \Big|\frac{\mathbf{D}(\mathbf{w})}{h}\Big|^2 \xi^2\, dx \ge C_{14}\int_{\Omega_0}\Big|\frac{\nabla\mathbf{w}}{h}\Big|^2\xi^2\, dx - c\int_{\Omega_0}\Big|\frac{\mathbf{w}}{h}\Big|^2 |\nabla\xi|^2\, dx\,, \end{aligned} \tag{3.24}$$

while

$$\begin{aligned} |J_2| &= \frac{2}{h^2}\Big|\int_{\Omega_0}\int_0^1 \partial_{ij}\partial_{kl}\Phi_A(|\mathbf{D}(\mathbf{v}) + \lambda\mathbf{D}(\mathbf{w})|^2) D_{kl}(\mathbf{w})\xi w_i \frac{\partial\xi}{\partial x_j}\, d\lambda\, dx\Big| \\ &\overset{(2.23)_2}{\le} \frac{2\tilde{C}_2(A)}{h^2}\int_{\Omega_0} |\nabla\mathbf{w}|\xi\,|\mathbf{w}||\nabla\xi|\, dx \\ &\le \frac{C_{14}}{8}\int_{\Omega_0}\Big|\frac{\nabla\mathbf{w}}{h}\Big|^2\xi^2\, dx + c\int_{\Omega_0}\Big|\frac{\mathbf{w}}{h}\Big|^2|\nabla\xi|^2\, dx\,. \end{aligned} \tag{3.25}$$

Note that in this section all generic constants c can depend on A. Further, the convective term gives

$$\begin{aligned} h^2 I_3 &= \int_{\Omega_0} w_{\varepsilon k}(x)\frac{\partial v_i(Tx)}{\partial x_k} w_i(x)\xi^2(x)\, dx + \int_{\Omega_0} v_{\varepsilon k}(x)\frac{\partial w_i(x)}{\partial x_k} w_i(x)\xi^2(x)\, dx \\ &= -\int_{\Omega_0}\frac{\partial v_{\varepsilon k}(Tx)}{\partial x_k} v_i(Tx) w_i(x)\xi^2(x)\, dx - \int_{\Omega_0} w_{\varepsilon k}(x) v_i(Tx)\frac{\partial w_i(x)}{\partial x_k}\xi^2(x)\, dx \\ &- 2\int_{\Omega_0} w_{\varepsilon k}(x) v_i(Tx) w_i(x)\xi(x)\frac{\partial\xi(x)}{\partial x_k}\, dx - \int_{\Omega_0} v_{\varepsilon k}(x)|\mathbf{w}(x)|^2\xi(x)\frac{\partial\xi(x)}{\partial x_k}\, dx \\ &= J_3 + J_4 + J_5 + J_6\,. \end{aligned} \tag{3.26}$$

Since $\operatorname{div} \mathbf{v}_\varepsilon = 0$, by (3.15)

$$J_3 = -\int_{\Omega_0} \frac{\partial v_{\varepsilon s}(Tx)}{\partial y_3} \frac{\partial \Delta^+ a(x')}{\partial x_s} v_i(Tx) w_i(x) \xi^2 \, dx \,, \tag{3.27}$$

where we have started to use the convention that whenever the index s appears twice in an expression we sum over s from 1 to 2. Using the regularity of the mollified function, the regularity of the boundary and (3.4), we get from (3.26) and (3.27)

$$|I_3| \leq \frac{C_{14}}{8} \int_{\Omega_0} \Big|\frac{\nabla \mathbf{w}}{h}\Big|^2 \xi^2 \, dx + c \int_{\Omega_0} \Big|\frac{\mathbf{w}}{h}\Big|^2 \big(|\nabla \xi|^2 + |\xi|^2\big) \, dx \,. \tag{3.28}$$

The pressure term I_4 requires more calculations. It is useful to start with an equivalent form

$$I_4 = \int_{\Omega_0} \frac{\partial \pi(x)}{\partial x_i} \big[\boldsymbol{\varphi}_i(Tx) - \boldsymbol{\varphi}_i(x)\big] \, dx = \int_{\Omega_0} \pi(x) \operatorname{div} \big(\boldsymbol{\varphi}(x) - \boldsymbol{\varphi}(T^{-1}x)\big) \, dx \,.$$

With $\boldsymbol{\varphi}$ as in (3.21) we obtain

$$\begin{aligned} h^2 I_4 &= \int_{\Omega_0} \pi(x) \big[\operatorname{div} \mathbf{w}(x) \, \xi^2(x) - \operatorname{div} \mathbf{w}(T^{-1}x) \, \xi^2(T^{-1}x) \big] \, dx \\ &\quad + 2 \int_{\Omega_0} \pi(x) \big[w_i(x) \xi(x) \frac{\partial \xi(x)}{\partial x_i} - w_i(T^{-1}x) \xi(T^{-1}x) \frac{\partial \xi(T^{-1}x)}{\partial x_i} \big] \, dx \\ &= \int_{\Omega_0} \pi(x) \operatorname{div} \big(\mathbf{v}(Tx) - 2\mathbf{v}(x) + \mathbf{v}(T^{-1}x)\big) \xi^2(x) \, dx \\ &\quad + \int_{\Omega_0} \pi(x) \operatorname{div} \mathbf{w}(T^{-1}x) \big(\xi^2(x) - \xi^2(T^{-1}x)\big) \, dx \\ &\quad + 2 \int_{\Omega_0} \pi(x) \big[v_i(Tx) - 2v_i(x) + v_i(T^{-1}x) \big] \xi(x) \frac{\partial \xi(x)}{\partial x_i} \, dx \\ &\quad + 2 \int_{\Omega_0} \pi(x) w_i(T^{-1}x) \Big(\xi(x) \frac{\partial \xi(x)}{\partial x_i} - \xi(T^{-1}x) \frac{\partial \xi(T^{-1}x)}{\partial x_i} \Big) \, dx \\ &= J_7 + J_8 + J_9 + J_{10} \,. \end{aligned} \tag{3.29}$$

Using further (3.15), (3.16) and the fact that $\operatorname{div} \mathbf{v} = 0$ in Q_T, we rewrite J_7

as follows:

$$
\begin{aligned}
J_7 &= \int_{\Omega_0} \pi(x)\left(\frac{\partial v_s(Tx)}{\partial y_3}\frac{\partial \Delta^+ a(x')}{\partial x_s} + \frac{\partial v_s(T^{-1}x)}{\partial y_3}\frac{\partial \Delta^- a(x')}{\partial x_s}\right)\xi^2(x)\,dx \\
&= \int_{\Omega_0} \pi(x)\left(\frac{\partial v_s(Tx)}{\partial y_3} - \frac{\partial v_s(x)}{\partial y_3}\right)\frac{\partial \Delta^+ a(x')}{\partial x_s}\xi^2(x)\,dx \\
&\quad + \int_{\Omega_0} \pi(x)\left(\frac{\partial v_s(x)}{\partial y_3} - \frac{\partial v_s(T^{-1}x)}{\partial y_3}\right)\frac{\partial \Delta^- a(x')}{\partial x_s}\xi^2(x)\,dx \\
&\quad + \int_{\Omega_0} \pi(x)\frac{\partial v_s(x)}{\partial y_3}\frac{\partial}{\partial x_s}\Big(a(x'+h\hat{\mathbf{e}}^r) - 2a(x') + a(x'-h\hat{\mathbf{e}}^r)\Big)\xi^2(x)\,dx \\
&= J_{11} + J_{12} + J_{13}\,. \qquad (3.30)
\end{aligned}
$$

The last term in (3.9) is treated similarly; hence,

$$
\begin{aligned}
h^2 I_5 &= \int_{\Omega_0} f_i(x)\big(v_i(Tx) - 2v_i(x) + v_i(T^{-1}x)\big)\xi^2(x)\,dx \qquad (3.31) \\
&\quad + \int_{\Omega_0} f_i(x)\big(v_i(x) - v_i(T^{-1}x)\big)\big(\xi^2(x) - \xi^2(T^{-1}x)\big)\,dx = J_{14} + J_{15}\,.
\end{aligned}
$$

Now, since $\operatorname{div}\mathbf{w}(Tx) = \dfrac{\partial v_s(Tx)}{\partial y_3}\dfrac{\partial \Delta^+ a(x)}{\partial x_s}$, we see that

$$
|J_8| \le c\int_{\Omega_0} |\pi(Tx)|\,|\nabla \mathbf{v}(Tx)|\,|\nabla \Delta^+ a(x)|\,|\xi(Tx) - \xi(x)|\,dx\,. \qquad (3.32)
$$

Hence

$$
\frac{1}{h^2}\big(|J_8| + |J_{10}| + |J_{13}| + |J_{15}|\big) \le c\big(\|\pi\|_2^2 + \|\mathbf{f}\|_2^2 + \|\nabla \mathbf{v}\|_2^2\big) \qquad (3.33)
$$

and

$$
\frac{1}{h^2}\big(|J_{11}| + |J_{12}|\big) \le c\|\pi\|_2^2 + \frac{C_{14}}{8}\int_{\Omega_0}\Big|\frac{\nabla \mathbf{w}}{h}\Big|^2\xi^2\,dx\,. \qquad (3.34)
$$

It remains to estimate J_9 and J_{14}. Denoting $\mathbf{g}(x) \equiv \frac{\mathbf{w}(x)}{h} = \frac{\mathbf{v}(Tx)-\mathbf{v}(x)}{h}$, we can rewrite J_9 as follows:

$$
\begin{aligned}
\frac{J_9}{h^2} &= \int_{\Omega_0} \pi(x)\frac{g_i(x) - g_i(T^{-1}x)}{h}\xi(x)\frac{\partial \xi(x)}{\partial x_i}\,dx \\
&= \int_{\Omega_0} \pi(x)\frac{1}{h}\big[g_i(x)\xi(x) - g_i(T^{-1}x)\xi(T^{-1}x)\big]\frac{\partial \xi(x)}{\partial x_i}\,dx \qquad (3.35) \\
&\quad + \int_{\Omega_0} \pi(x)\frac{1}{h}\big[\xi(T^{-1}x) - \xi(x)\big]g_i(T^{-1}x)\frac{\partial \xi(x)}{\partial x_i}\,dx\,.
\end{aligned}
$$

Since $\mathbf{g}\xi \in W_0^{1,2}(\Omega_0)$ for fixed h, we can use (3.18) to conclude that

$$\frac{|J_9|}{h^2} \le c(a)\|\pi\|_2\|\nabla\xi\|_\infty\Big(\int_{\Omega_0}\Big|\frac{\nabla\mathbf{w}}{h}\Big|^2\xi^2\,dx\Big)^{1/2} + \|\pi\|_2\|\nabla\xi\|_\infty\|\nabla\mathbf{v}\|_2\,. \tag{3.36}$$

Also

$$\frac{|J_{14}|}{h^2} \le \|\mathbf{f}\|_2\bigg(\int_{\Omega_0}\Big|\frac{\nabla\mathbf{w}}{h}\Big|^2\xi^2\,dx\bigg)^{1/2} + \|\nabla\mathbf{v}\|_2\|\mathbf{f}\|_2\|\nabla\xi\|_\infty\,. \tag{3.37}$$

Putting all calculations between (3.22) and (3.37) together, using Young's inequality and integrating over $(0,T)$ we finally obtain

$$\int_0^T\int_{\Omega_0}\Big|\frac{\nabla\mathbf{w}}{h}\Big|^2\xi^2\,dx\,dt \le c\int_0^T \|\nabla\mathbf{v}\|_2^2 + \|\pi\|_2^2 + \|\mathbf{f}\|_2^2\,dt + c(\mathbf{v}_0)\,. \tag{3.38}$$

Hence, (see (3.19) and (3.20))

$$\sum_{s=1}^{2}\sum_{i=1}^{3}\int_0^T\int_\Omega\Big|\frac{\partial^2\mathbf{v}}{\partial x_i\partial\tau^s}\Big|^2\xi^2\,dx\,dt \le c\,. \tag{3.39}$$

In case of interior regularity, we proceed in an analogous way with some simplifications due to the simpler structure of the mapping T; cf. (3.10). Thus (3.7) holds for all $\Omega' \subset\subset \Omega$. This implies that the equation $(2.21)_2$ holds almost everywhere.

In order to get (3.7) globally, we need an estimate of type (3.39) in the normal direction (which is locally x_3). We avoid the missing information about the pressure by taking the curl of $(2.21)_2$. Recall that for $\mathbf{h} = (h_1, h_2, h_3)$

$$\operatorname{curl}\mathbf{h} = \Big(\frac{\partial h_3}{\partial x_2} - \frac{\partial h_2}{\partial x_3}, \frac{\partial h_1}{\partial x_3} - \frac{\partial h_3}{\partial x_1}, \frac{\partial h_2}{\partial x_1} - \frac{\partial h_1}{\partial x_2}\Big)\,.$$

Applying the curl operator to $(2.21)_2$ we obtain three equations in $W^{-1,2}(\Omega)$; however, only the first two are useful for us. The first equation reads

$$\begin{aligned}\frac{\partial}{\partial x_2}\frac{\partial v_3}{\partial t} - \frac{\partial}{\partial x_3}\frac{\partial v_2}{\partial t} - \frac{\partial f_3}{\partial x_2} + \frac{\partial f_2}{\partial x_3} + \frac{\partial}{\partial x_2}\Big[v_{\varepsilon k}\frac{\partial v_3}{\partial x_k}\Big] - \frac{\partial}{\partial x_3}\Big[v_{\varepsilon k}\frac{\partial v_2}{\partial x_k}\Big]&\\ - \frac{\partial^2}{\partial x_1\partial x_2}\partial_{31}\Phi_A - \frac{\partial^2}{\partial x_2\partial x_2}\partial_{32}\Phi_A - \frac{\partial^2}{\partial x_2\partial x_3}\partial_{33}\Phi_A&\\ + \frac{\partial^2}{\partial x_3\partial x_1}\partial_{21}\Phi_A + \frac{\partial^2}{\partial x_3\partial x_2}\partial_{22}\Phi_A + \frac{\partial^2}{\partial x_3\partial x_3}\partial_{23}\Phi_A = 0\,,&\end{aligned} \tag{3.40}$$

while the second equation has the form

$$\begin{aligned}&\frac{\partial}{\partial x_3}\frac{\partial v_1}{\partial t}-\frac{\partial}{\partial x_1}\frac{\partial v_3}{\partial t}-\frac{\partial f_1}{\partial x_3}+\frac{\partial f_3}{\partial x_1}+\frac{\partial}{\partial x_3}\Big[v_{\varepsilon k}\frac{\partial v_1}{\partial x_k}\Big]-\frac{\partial}{\partial x_1}\Big[v_{\varepsilon k}\frac{\partial v_3}{\partial x_k}\Big]\\&+\frac{\partial^2}{\partial x_1\partial x_1}\partial_{31}\Phi_A+\frac{\partial^2}{\partial x_1\partial x_2}\partial_{32}\Phi_A+\frac{\partial^2}{\partial x_1\partial x_3}\partial_{33}\Phi_A\\&-\frac{\partial^2}{\partial x_1\partial x_3}\partial_{11}\Phi_A-\frac{\partial^2}{\partial x_2\partial x_3}\partial_{12}\Phi_A-\frac{\partial^2}{\partial x_3\partial x_3}\partial_{13}\Phi_A=0\,.\end{aligned}\tag{3.41}$$

We will get the desired estimate from the last terms in equations (3.40) and (3.41) and from the equation

$$\frac{\partial^2 v_3}{\partial x_3^2}=-\frac{\partial^2 v_1}{\partial x_1\partial x_3}-\frac{\partial^2 v_2}{\partial x_2\partial x_3}\,,\tag{3.42}$$

which follows from the divergence-free constraint, taking the derivative with respect to x_3. Let us denote

$$G_1\equiv\frac{\partial}{\partial x_3}\frac{\partial\Phi_A(|\mathbf{D}(\mathbf{v})|^2)}{\partial D_{13}}\quad\text{and}\quad G_2\equiv\frac{\partial}{\partial x_3}\frac{\partial\Phi_A(|\mathbf{D}(\mathbf{v})|^2)}{\partial D_{23}}\,.\tag{3.43}$$

Clearly, for $i=1,2$,

$$\int_0^T\|\xi G_i\|_{-1,2}^2\,dt\le c\tag{3.44}$$

due to (3.4) and (2.23). From (3.40) and (3.41), we can also observe, using (3.39) and (3.17), that

$$\int_0^T\|\frac{\partial G_i}{\partial x_3}\xi\|_{-1,2}^2\,dt\le c+c\int_0^T\int_\Omega\sum_{i=1}^3\sum_{s=1}^2\Big|\frac{\partial^2\mathbf{v}}{\partial x_i\partial x_s}\Big|^2\,dx\,dt\,.\tag{3.45}$$

Hence, by Theorem 6.7 on negative norms

$$\int_0^T\|\xi G_i\|_2^2\,dt\le c+c\int_0^T\int_\Omega\sum_{i=1}^3\sum_{s=1}^2\Big|\frac{\partial^2\mathbf{v}}{\partial x_i\partial x_s}\Big|^2\,dx\,dt\,.\tag{3.46}$$

Directly from (3.43) we obtain the system

$$\begin{aligned}&\partial_{23}\partial_{13}\Phi_A\frac{\partial D_{13}}{\partial x_3}+\partial_{23}\partial_{23}\Phi_A\frac{\partial D_{23}}{\partial x_3}\\&=\frac{G_2}{2}-\frac12\sum_{r,s=1}^2\partial_{23}\partial_{rs}\Phi_A\frac{\partial D_{rs}}{\partial x_3}-\frac12\partial_{23}\partial_{33}\Phi_A\frac{\partial^2 v_3}{\partial x_3^2}\end{aligned}\tag{3.47}$$

$$\partial_{13}\partial_{13}\Phi_A \frac{\partial D_{13}}{\partial x_3} + \partial_{13}\partial_{23}\Phi_A \frac{\partial D_{23}}{\partial x_3}$$
$$= \frac{G_1}{2} - \frac{1}{2}\sum_{r,s=1}^{2} \partial_{13}\partial_{rs}\Phi_A \frac{\partial D_{rs}}{\partial x_3} - \frac{1}{2}\partial_{13}\partial_{33}\Phi_A \frac{\partial^2 v_3}{\partial x_3^2}\,.$$

Since the matrix

$$\mathbb{A} \equiv \begin{pmatrix} \partial_{23}\partial_{13}\Phi_A & \partial_{23}\partial_{23}\Phi_A \\ \partial_{13}\partial_{13}\Phi_A & \partial_{13}\partial_{23}\Phi_A \end{pmatrix} \tag{3.48}$$

is positive definite thanks to (2.24), we can compute $\frac{\partial}{\partial x_3} D_{s3}$ $(s = 1, 2)$ from (3.47). Moreover, the coefficients of the matrix (3.48) are bounded due to $(2.23)_2$, so we get

$$\frac{\partial D_{s3}(\mathbf{v})}{\partial x_3} \approx \text{right-hand side of (3.47)}\,. \tag{3.49}$$

Because of

$$\frac{\partial D_{s3}(\mathbf{v})}{\partial x_3} = \frac{\partial^2 v_s}{\partial x_3^2} + \frac{\partial^2 v_3}{\partial x_s \partial x_3}\,,$$

we conclude from (3.49), (3.42) and (3.17) that

$$\sum_{i=1}^{3} \Big\|\frac{\partial^2 v_i}{\partial x_3^2}\xi\Big\|_2^2 \le c + \sum_{i,j=1}^{3}\sum_{s=1}^{2} \Big\|\frac{\partial^2 v_i}{\partial x_j \partial \tau^s}\xi\Big\|_2^2 + c\sup_{\Omega_0}\sum_{s=1}^{2}\Big|\frac{\partial a}{\partial x_s}\Big| \sum_{i=1}^{3}\Big\|\frac{\partial^2 v_i}{\partial x_3^2}\xi\Big\|_2^2\,. \tag{3.50}$$

Now, if we take Ω_0 small enough, we can arrange due to (3.11) that

$$c\sup_{\Omega_0}\sum_{s=1}^{2}\Big|\frac{\partial a}{\partial x_s}\Big| \le \frac{1}{2},$$

and the last term in (3.50) can be moved to the left-hand side. The proof of (3.7) is finished. Theorem 3.3 is proved. □

We will finish this section by proving a variant of inequality (3.5) which will be useful in Sections 4 and 5. For $t \in I$ we denote

$$E_A(t) \equiv 1 + \|\Phi_A(|\mathbf{D}(\mathbf{v}^{\varepsilon,A}(t))|^2)\|_1\,.$$

Lemma 3.51. *Set $\gamma_\mu(s) \equiv \frac{1}{1-\mu} s^{1-\mu}$ for $1 \neq \mu \geq 0$ and $\gamma_1(s) \equiv ln\,(s)$ (the derivative of γ_μ is denoted by γ'_μ). Let $p \in [2,6)$. Then it holds for all $t \in I$ that*

$$\begin{aligned} &\int_0^t \|\frac{\partial \mathbf{v}^{\varepsilon,A}}{\partial t}\|_2^2 \, \gamma'_\mu(E_A(\tau))\, d\tau + \sup_{\tau\in[0,t]} \gamma_\mu(E_A(\tau)) \\ &\quad \leq c(\mathbf{f}, \mathbf{v}_0) + \int_0^t \gamma'_\mu(E_A(\tau)) \int_\Omega |\mathbf{v}_\varepsilon^{\varepsilon,A}|^2 |\nabla \mathbf{v}^{\varepsilon,A}|^2 \, dx\, d\tau\,. \end{aligned} \tag{3.52}$$

Proof. Let again $\mathbf{v} = \mathbf{v}^{\varepsilon,A}$ and $\gamma(s) = \gamma_\mu(s)$. Multiplying equation $(2.21)_2$ by $\frac{\partial \mathbf{v}}{\partial t}\,\gamma'(E_A(\tau))$ and integrating over Q_t yields

$$\begin{aligned} &\int_0^t \|\frac{\partial \mathbf{v}}{\partial t}\|_2^2 \, \gamma'(E_A(\tau))\, d\tau - \int_0^t \gamma'(E_A(\tau)) \int_\Omega \frac{\partial}{\partial x_j} \partial_{ij} \Phi_A(|\mathbf{D}(\mathbf{v})|^2) \frac{\partial \mathbf{v}}{\partial \tau}\, dx\, d\tau \\ &\quad \leq c(\mathbf{f}) + \int_0^t \gamma'(E_A(\tau)) \int_\Omega |\mathbf{v}_\varepsilon|^2 |\nabla \mathbf{v}|^2 \, dx\, d\tau\,. \end{aligned} \tag{3.53}$$

Denoting the second term in (3.53) by J we need to show that $J = \gamma(E_A(t)) - \gamma(E_A(0))$ in order to obtain (3.52). Since $\mathbf{v} \in C(I; V_2) \cap L^2(I; W^{2,2}(\Omega)^3)$ and $\frac{\partial \mathbf{v}}{\partial t} \in L^2(I; L^2(\Omega)^3)$ there exists a sequence $\{\mathbf{v}^n\} \subset C^\infty(I; \mathcal{V})$ such that

$$\begin{aligned} \mathbf{v}^n &\to \mathbf{v} && \text{strongly in } L^2(I; W^{2,2}(\Omega)^3)\,, \\ \mathbf{v}^n(t) &\to \mathbf{v}(t) && \text{strongly in } V_2 \text{ for all } t \in I\,, \\ \frac{\partial \mathbf{v}^n}{\partial t} &\to \frac{\partial \mathbf{v}}{\partial t} && \text{strongly in } L^2(I; L^2(\Omega)^3)\,. \end{aligned} \tag{3.54}$$

Then we have for all $t \in I$

$$\lim_{n\to\infty} \int_\Omega \Phi_A(|\mathbf{D}(\mathbf{v}^n(t))|^2)\, dx = \int_\Omega \Phi_A(|\mathbf{D}(\mathbf{v}(t))|^2)\, dx\,. \tag{3.55}$$

To see (3.55), we use the inequality

$$|\Phi_A(|\mathbf{B}|^2) - \Phi_A(|\mathbf{A}|^2)| \leq c\big(1 + |\mathbf{A}| + |\mathbf{B}|\big)|\mathbf{A} - \mathbf{B}|\,,$$

which follows from $(2.23)_1$, and $(3.54)_2$.

Observing that $E_A(\mathbf{v}^n(t)) \geq 1$ we have $\|\gamma'(E_A(\mathbf{v}^n(t)))\|_\infty \leq 1$. Thus we obtain $\gamma'(E_A(\mathbf{v}^n(t))) \overset{*}{\rightharpoonup} \chi(t)$ in $L^\infty(I)$ at least for a subsequence. Since

$\gamma'(E_A(\mathbf{v}^n(t))) \to \gamma'(E_A(\mathbf{v}(t)))$ for all $t \in I$ due to (3.55), it is easy to conclude that

$$\gamma'(E_A(\mathbf{v}^n(t))) \stackrel{*}{\rightharpoonup} \gamma'(E_A(\mathbf{v}(t))) \qquad \text{weakly-* in } L^\infty(I)\,. \tag{3.56}$$

Now, using (3.54) together with (3.56) gives

$$J = -\lim_{n\to\infty} \int_0^t \gamma'(E_A(\mathbf{v}^n(\tau))) \int_\Omega \frac{\partial}{\partial x_j} \partial_{ij}\Phi_A(|\mathbf{D}(\mathbf{v}^n)|^2) \frac{\partial \mathbf{v}^n}{\partial \tau}\, dx\, d\tau$$

from which we deduce

$$\begin{aligned} J &= \lim_{n\to\infty} \int_0^t \frac{d}{d\tau} \|\Phi_A(|\mathbf{D}(\mathbf{v}^n)|^2)\|_1 \, \gamma'(E_A(\mathbf{v}^n(\tau)))\, d\tau \\ &= \lim_{n\to\infty} \int_0^t \frac{d}{d\tau} (\gamma(E_A(\mathbf{v}^n(\tau))))\, d\tau = \lim_{n\to\infty} \{\gamma(E_A(\mathbf{v}^n(t))) - \gamma(E_A(\mathbf{v}^n(0)))\}. \end{aligned}$$

Passing to the limit, which is allowed due to (3.55), the assertion follows.

4. Limiting process $A \to \infty$. The goal of this section is to pass to the limit as $A \to \infty$. This means to return from the quadratic approximations Φ_A to the original potential Φ. Since the convective term will still be mollified, we will come from the *Problem (NS-Dir)$_p^{\varepsilon,A}$* to the *Problem (NS-Dir)$_p^\varepsilon$* defined in (2.7). Nevertheless, in preparation for the limiting process $\varepsilon \to 0^+$, we will often indicate the dependence of the estimates on ε. We denote

$$K_{\varepsilon,A} \equiv \int_\Omega |\mathbf{v}_\varepsilon^{\varepsilon,A}|^2 |\nabla \mathbf{v}^{\varepsilon,A}|^2\, dx.$$

We already know from (3.4)–(3.6) that strong solutions $(\mathbf{v}^{\varepsilon,A}, \pi^{\varepsilon,A})$ of the *Problem (NS-Dir)$_p^{\varepsilon,A}$* satisfy the following estimates:

$$\|\mathbf{v}^{\varepsilon,A}\|^2_{L^\infty(I;H)} + \|\mathbf{v}^{\varepsilon,A}\|^2_{L^2(I;V_2)} \le c(\mathbf{f}, \mathbf{v}_0), \tag{4.1}$$

$$\begin{aligned} \Big\|\frac{\partial \mathbf{v}^{\varepsilon,A}}{\partial t}\Big\|^2_{L^2(Q_T)} + \big\|\Phi_A(|\mathbf{D}(\mathbf{v}^{\varepsilon,A})|^2)\big\|_{L^\infty(I;L^1(\Omega))} \\ \le c(\mathbf{f}, \mathbf{v}_0)\Big(1 + \int_0^T K_{\varepsilon,A}\, dt\Big)\,, \end{aligned} \tag{4.2}$$

$$\|\pi^{\varepsilon,A}\|^2_{L^2(I;L^2(\Omega))} \le c(\mathbf{f},\mathbf{v}_0)\Big(1+\int_0^T K_{\varepsilon,A}\,dt\Big) + \int_0^T \big\|\partial\Phi_A(|\mathbf{D}(\mathbf{v}^{\varepsilon,A})|^2)\big\|_2^2\,dt\,. \tag{4.3}$$

Note that due to (4.1) and (2.6)

$$\int_0^T K_{\varepsilon,A}(t)\,dt \le c\,(\frac{1}{\varepsilon},\mathbf{f},\mathbf{v}_0)\,. \tag{4.4}$$

From Lemma 4.15 below with $\mu=0$ and (4.4) it follows that

$$\int_0^T \big\|\nabla\partial\Phi_A(|\mathbf{D}(\mathbf{v}^{\varepsilon,A})|^2)\big\|_{\frac{6}{p+1}}^{\frac{2}{p-1}}\,dt \le c\,(\frac{1}{\varepsilon},\mathbf{f},\mathbf{v}_0)\Big(1+\int_0^T \big\|\partial\Phi_A(|\mathbf{D}(\mathbf{v}^{\varepsilon,A})|^2)\big\|_2^2\,dt\Big)\,. \tag{4.5}$$

Due to (2.25) and (4.2) we also see

$$\big\|\partial\Phi_A(|\mathbf{D}(\mathbf{v}^{\varepsilon,A})|^2)\big\|^{p'}_{L^\infty(I;L^{p'}(\Omega))} \le c\,(\frac{1}{\varepsilon},\mathbf{f},\mathbf{v}_0)\,. \tag{4.6}$$

Recalling an interpolation inequality (valid for all $p\in[2,4]$)

$$\|g\|_2^2 \le c\|g\|_{p'}^{2(1-\lambda)}\|g\|_{1,\frac{6}{p+1}}^{2\lambda} \qquad \text{with } \lambda=\frac{3(p-2)}{(6-p)(p-1)}$$

we obtain

$$\int_0^T \big\|\partial\Phi_A\big\|_2^2\,dt \le T\big\|\partial\Phi_A\big\|^2_{L^\infty(I;L^{p'}(\Omega))} + c\,\big\|\partial\Phi_A\big\|^{2(1-\lambda)}_{L^\infty(I;L^{p'}(\Omega))}\int_0^T \big\|\nabla\partial\Phi_A\big\|^{2\lambda}_{\frac{6}{p+1}}\,dt\,. \tag{4.7}$$

Hence from Hölder's inequality and (4.5)–(4.7) we obtain

$$\int_0^T \big\|\nabla\partial\Phi_A(|\mathbf{D}(\mathbf{v}^{\varepsilon,A})|^2)\big\|_{\frac{6}{p+1}}^{\frac{2}{p-1}}\,dt \le c\,(\frac{1}{\varepsilon},\mathbf{f},\mathbf{v}_0)\Big(1+\int_0^T \big\|\nabla\partial\Phi_A(|\mathbf{D}(\mathbf{v}^{\varepsilon,A})|^2)\big\|_{\frac{6}{p+1}}^{\frac{2}{p-1}}\,dt\Big)^{(p-1)\lambda}, \tag{4.8}$$

which implies for $p \in [2,3)$

$$\int_0^T \|\nabla \partial \Phi_A(|\mathbf{D}(\mathbf{v}^{\varepsilon,A})|^2)\|_{\frac{6}{p+1}}^{\frac{2}{p-1}} \, dt \le c\left(\frac{1}{\varepsilon}, \mathbf{f}, \mathbf{v}_0\right). \tag{4.9}$$

Inequality (4.9) gives (see the appendix, Lemma 6.12, inequality (6.14))

$$\int_0^T \|\mathbf{D}(\nabla \mathbf{v}^{\varepsilon,A})\|_{\frac{6}{p+1}}^{\frac{2}{p-1}} \, dt \le c\left(\frac{1}{\varepsilon}, \mathbf{f}, \mathbf{v}_0\right), \tag{4.10}$$

and by inequality (6.4) we finally obtain

$$\int_0^T \|\nabla^{(2)} \mathbf{v}^{\varepsilon,A}\|_{\frac{6}{p+1}}^{\frac{2}{p-1}} \, dt \le c\left(\frac{1}{\varepsilon}, \mathbf{f}, \mathbf{v}_0\right). \tag{4.11}$$

Considering also (4.1)–(4.4), $(2.23)_4$, (4.6), (4.7), (4.9) and the Aubin-Lions lemma (see Lions [8]) we get the existence of $\mathbf{v}^\varepsilon, \pi^\varepsilon$ such that for $A \to \infty$ (or at least for subsequences)

$$\begin{aligned}
\mathbf{v}^{\varepsilon,A} &\rightharpoonup \mathbf{v}^\varepsilon && \text{*-weakly in } L^\infty(I;V_2) \cap L^{\frac{2}{p-1}}(I;W^{2,\frac{6}{p+1}}(\Omega)^3)\,, \\
\frac{\partial \mathbf{v}^{\varepsilon,A}}{\partial t} &\rightharpoonup \frac{\partial \mathbf{v}^\varepsilon}{\partial t} && \text{weakly in } L^2(I;L^2(\Omega)^3)\,, \\
\pi^{\varepsilon,A} &\rightharpoonup \pi^\varepsilon && \text{weakly in } L^2(I;L^2(\Omega))\,, \\
\nabla \mathbf{v}^{\varepsilon,A} &\to \nabla \mathbf{v}^\varepsilon && \text{strongly in } L^{\frac{2}{p-1}}(I;L^{\tilde{q}}(\Omega)^{3\times 3})\,, \\
\nabla \mathbf{v}^{\varepsilon,A} &\to \nabla \mathbf{v}^\varepsilon && \text{strongly in } L^s(I;L^s(\Omega)^{3\times 3})\,,
\end{aligned} \tag{4.12}$$

where $\tilde{q} < \frac{6}{p-1}$ and $s < \frac{2(2p+1)}{3(p-1)}$.[4] Because the Φ_A's approximate Φ locally uniformly, it is not difficult to conclude from (4.12) that $\mathbf{v}^\varepsilon$ solves the *Problem (NS-Dir)*$_p^\varepsilon$.

Remark 4.13. Since $\frac{2(2p+1)}{3(p-1)} > 2$ for $p \in [2,4)$ we see from (4.12), (2.6) and Vitali's convergence lemma that for almost all $t \in I$

$$\lim_{A\to\infty} \int_0^t K_{\varepsilon,A}(\tau)\, d\tau = \int_0^t K_\varepsilon(\tau)\, d\tau\,, \tag{4.14}$$

[4] In $(4.12)_5$ we used the following parabolic imbedding: for $r = \frac{2}{p-1}$ and $q = \frac{6}{p+1}$ the space $L^\infty(I;L^2(\Omega)) \cap L^r(I;W^{1,q}(\Omega))$ is continuously imbedded into $L^\tau(Q_T)$, where $\tau = \frac{2(2p+1)}{3(p-1)}$, which follows from the interpolation inequality $\|\mathbf{u}\|_s \le \|\mathbf{u}\|_2^{1-\alpha} \|\mathbf{u}\|_{\frac{3q}{3-q}}^\alpha$ with $\alpha = \frac{3(s-2)}{s(4-p)}$.

where $K_\varepsilon = \int_\Omega |\mathbf{v}_\varepsilon^\varepsilon|^2 |\nabla \mathbf{v}^\varepsilon|^2 \, dx$. Thus the estimates (4.2), (4.3) and (4.11) for $\frac{\partial \mathbf{v}^\varepsilon}{\partial t}$, π^ε and $\nabla^{(2)}\mathbf{v}^\varepsilon$ remain valid due to the weak lower semicontinuity of norms and (4.14).

Now, we formulate and prove Lemma 4.15, which implies (4.5), by setting $\mu = 0$, $q = \frac{6}{p+1}$ and $r = \frac{2}{p-1}$ in (4.16).

Before doing so, we shall relabel the interior cut-off function by ξ_0 and the cut-off function localized near the boundary by ξ_k, $k = 1, 2, \dots, N$. Let us also use the convention that if ξ_k occurs in any expression then we sum up over k from 1 to N.

Recall that $\gamma_\mu(s) = \frac{1}{1-\mu} s^{1-\mu}$ for $1 \neq \mu \geq 0$ and $\gamma_1(s) = \ln s$. By γ_μ' we mean the derivative of γ_μ. Finally, we set $\theta_A(s) \equiv \max(1+s, 1+A)$ and $E_A(t) \equiv 1 + \|\Phi_A(|\mathbf{D}(\mathbf{v}^{\varepsilon,A}(t))|^2)\|_1$.

Lemma 4.15. *For all $\varepsilon > 0$ and all $A > 1$ let $(\mathbf{v}, \pi) = (\mathbf{v}^{\varepsilon,A}, \pi^{\varepsilon,A})$ be a solution of the Problem (NS-Dir)$_p^{\varepsilon,A}$. For $q \in [1, \frac{6}{p+1}]$, $r \in (0, \frac{2}{p-1}]$ and $\mu \geq 0$ we have*

$$\int_0^T \|\nabla \partial \Phi_A(|\mathbf{D}(\mathbf{v})|^2)\|_q^r \gamma_\mu'(E_A) \, dt + \int_0^T Y \, dt \tag{4.16}$$
$$\leq c(\mathbf{f}, \mathbf{v}_0)\Big(1 + \int_0^T (K_{\varepsilon,A} + \|\partial \Phi_A\|_2^2) \gamma_\mu'(E_A) \, dt\Big),$$

where

$$Y \equiv \gamma_\mu'(E_A)\Big(\int_\Omega \theta_A(|\mathbf{D}(\mathbf{v})|)^{p-2}\big(|\mathbf{D}(\nabla \mathbf{v})|^2 \xi_0^2 + \sum_{r=1}^{2} |\mathbf{D}(\frac{\partial \mathbf{v}}{\partial \tau^r})|^2 \xi_k^2\big) \, dx\Big)^{\frac{r(p-1)}{2}}.$$

Proof. We split the proof into four steps. Step 1 deals with estimates of the second derivatives in the interior of the domain and tangential derivatives near the boundary. Step 2 gives the estimates of the full second velocity gradient near the boundary by means of the gradient of the tangential derivatives. Inequalities for the full second velocity gradient are derived in Step 3. Step 4 provides the derivation of (4.16).

Step 1: *Second derivative estimates in the interior and in tangential directions near the boundary.*

Since $(2.21)_2$ is valid almost everywhere in Q_T, we can directly multiply by $-\Delta \mathbf{v}^{\varepsilon,A} \xi_0^2$. Let us drop the indices ε, A, so that $\mathbf{v} = \mathbf{v}^{\varepsilon,A}$, $\pi = \pi^{\varepsilon,A}$ and let $I_1, \dots, I_5$ denote the terms coming from $(2.21)_2$ after multiplying by

$-\Delta\mathbf{v}\xi_0^2$ and integrating over Ω (precise definitions of I_k are given below). We have

$$|I_1| \equiv |\int_\Omega \frac{\partial \mathbf{v}}{\partial t}\cdot \Delta\mathbf{v}\xi_0^2\,dx| \le \frac{C_{14}}{8}\big\|\nabla^{(2)}\mathbf{v}\xi_0\big\|_2^2 + c\|\frac{\partial \mathbf{v}}{\partial t}\|_2^2\,, \tag{4.17}$$

$$\begin{aligned} I_2 &\equiv \int_{\Omega_0} \frac{\partial}{\partial x_j}\frac{\partial \Phi_A(|\mathbf{D}(\mathbf{v})|^2)}{\partial D_{ij}}\Delta v_i\xi_0^2\,dx \\ &= -\int_{\Omega_0} \frac{\partial \Phi_A(|\mathbf{D}(\mathbf{v})|^2)}{\partial D_{ij}} D_{ij}(\Delta\mathbf{v})\xi_0^2 dx - 2\int_{\Omega_0} \frac{\partial \Phi_A(|\mathbf{D}(\mathbf{v})|^2)}{\partial D_{ij}}\Delta v_i\xi_0\frac{\partial \xi_0}{\partial x_j}dx \\ &= \int_{\Omega_0} \frac{\partial^2 \Phi_A(|\mathbf{D}(\mathbf{v})|^2)}{\partial D_{ij}\partial D_{kl}} D_{ij}(\nabla\mathbf{v})D_{kl}(\nabla\mathbf{v})\xi_0^2\,dx \\ &\quad + 2\int_{\Omega_0} \frac{\partial \Phi_A(|\mathbf{D}(\mathbf{v})|^2)}{\partial D_{ij}} D_{ij}(\frac{\partial \mathbf{v}}{\partial x_k})\xi_0\frac{\partial \xi_0}{\partial x_k}\,dx \\ &\quad - 2\int_{\Omega_0} \frac{\partial \Phi_A(|\mathbf{D}(\mathbf{v})|^2)}{\partial D_{ij}}\Delta v_i\xi_0\frac{\partial \xi_0}{\partial x_j}\,dx = J_1 + J_2 + J_3\,. \end{aligned} \tag{4.18}$$

Again we obtain

$$\begin{aligned} J_1 &\overset{(2.24)}{\ge} \frac{1}{2}J_1 + \frac{C_{10}}{2}\int_{\Omega_0} |\mathbf{D}(\nabla\mathbf{v})|^2\xi_0^2\,dx \\ &\overset{(6.6)}{\ge} \frac{1}{2}J_1 + C_{14}\sum_{i,j=1}^{3}\int_{\Omega_0} |\frac{\partial^2\mathbf{v}}{\partial x_i\partial x_j}|^2\xi_0^2\,dx - c\int_{\Omega_0}|\nabla\mathbf{v}|^2|\nabla\xi_0|^2\,dx\,, \end{aligned} \tag{4.19}$$

and

$$|J_2 + J_3| \le \frac{C_{14}}{8}\big\|\nabla^{(2)}\mathbf{v}\xi_0\big\|_2^2 + \|\partial\Phi_A\|_2^2\,. \tag{4.20}$$

The convective term, the pressure and the body force are estimated easily. We have

$$|I_3| \equiv | - \int_{\Omega_0} \mathbf{v}_{\varepsilon k}\frac{\partial \mathbf{v}}{\partial x_k}\Delta\mathbf{v}\xi_0^2\,dx| \le \frac{C_{14}}{8}\big\|\nabla^{(2)}\mathbf{v}\xi_0\big\|_2^2 + c\,K_{\varepsilon,A} \tag{4.21}$$

and

$$I_4 \equiv \int_{\Omega_0}\frac{\partial \pi}{\partial x_i}\Delta v_i\xi_0^2\,dx = -2\int_{\Omega_0}\pi\Delta v_i\xi_0\frac{\partial \xi_0}{\partial x_i}\,dx\,.$$

Hence

$$|I_4| \le c\|\pi\|_2\big\|\nabla^{(2)}\mathbf{v}\xi_0\big\|_2 \le c\|\pi\|_2^2 + \frac{C_{14}}{8}\big\|\nabla^{(2)}\mathbf{v}\xi_0\big\|_2^2\,. \tag{4.22}$$

Finally,

$$|I_5| \equiv \Big| \int_\Omega \mathbf{f} \cdot \Delta \mathbf{v} \xi_0^2 \, dx \Big| \le c \|\mathbf{f}\|_2^2 + \frac{C_{14}}{8} \big\| \nabla^{(2)} \mathbf{v} \xi_0 \big\|_2^2 . \tag{4.23}$$

Collecting (4.17)–(4.23) we obtain (interior estimates)

$$\begin{aligned} \int_{\Omega_0} \frac{\partial^2 \Phi_A(|\mathbf{D}(\mathbf{v})|^2)}{\partial D_{ij} \partial D_{kl}} D_{ij}(\nabla \mathbf{v}) D_{kl}(\nabla \mathbf{v}) \xi_0^2 \, dx + \int_{\Omega_0} \sum_{i,j,k=1}^{3} \Big| \frac{\partial^2 v_i}{\partial x_j \partial x_k} \Big|^2 \xi_0^2 \, dx \\ \le c \Big(\|\mathbf{f}\|_2^2 + K_{\varepsilon,A} + \big\| \partial \Phi_A \big\|_2^2 + \Big\| \frac{\partial \mathbf{v}}{\partial t} \Big\|_2^2 + \|\pi\|_2^2 \Big) \\ \le c \Big(\|\mathbf{f}\|_2^2 + K_{\varepsilon,A} + \big\| \partial \Phi_A \big\|_2^2 + \Big\| \frac{\partial \mathbf{v}}{\partial t} \Big\|_2^2 \Big) , \end{aligned} \tag{4.24}$$

where we used the inequality (cf. (4.3) and (3.8))

$$\|\pi(t)\|_2^2 \le c \Big(\|\mathbf{f}(t)\|_2^2 + K_{\varepsilon,A}(t) + \big\| \partial \Phi_A(t) \big\|_2^2 + \Big\| \frac{\partial \mathbf{v}(t)}{\partial t} \Big\|_2^2 \Big) ,$$

valid for almost all $t \in I$. The same procedure works if we estimate the tangential derivatives. We obtain

$$\begin{aligned} \int_{\Omega_0} \frac{\partial^2 \Phi_A(|\mathbf{D}(\mathbf{v})|^2)}{\partial D_{ij} \partial D_{kl}} D_{ij}\Big(\frac{\partial \mathbf{v}}{\partial \tau^s}\Big) D_{kl}\Big(\frac{\partial \mathbf{v}}{\partial \tau^s}\Big) \xi_k^2 \, dx + \sum_{i,j=1}^{3} \sum_{s=1}^{2} \int_{\Omega_0} \Big| \frac{\partial^2 v_i}{\partial x_j \partial \tau^s} \Big|^2 \xi_k^2 \, dx \\ \le c \Big(\|\mathbf{f}\|_2^2 + K_{\varepsilon,A} + \big\| \partial \Phi_A \big\|_2^2 + \Big\| \frac{\partial \mathbf{v}}{\partial t} \Big\|_2^2 \Big) . \end{aligned} \tag{4.25}$$

Because of the careful estimates of the tangential derivatives presented in the previous section, we think it is not necessary to repeat more or less the same steps once more.

Step 2: *Evaluation of all second derivatives of the velocity near the boundary in terms of tangential derivatives.*

The aim of this step is to show that for $q \in [1, 2]$

$$\begin{aligned} \|\theta_A(|\mathbf{D}(\mathbf{v})|)^{p-2} \nabla^{(2)} \mathbf{v} \xi_k\|_q \le c \Big(\|\mathbf{f}\|_2 + K_{\varepsilon,A}^{1/2} + \big\| \partial \Phi_A \big\|_2 + \Big\| \frac{\partial \mathbf{v}}{\partial t} \Big\|_2 \Big) \\ + c \sum_{r=1}^{2} \|\theta_A(|\mathbf{D}|)^{p-2} \nabla \frac{\partial \mathbf{v}}{\partial \tau^r} \xi_k \|_q . \end{aligned} \tag{4.26}$$

Set $\boldsymbol{\zeta} \equiv (D_{13}(\frac{\partial \mathbf{v}}{\partial x_3}), D_{23}(\frac{\partial \mathbf{v}}{\partial x_3}))$ and write θ_A instead of $\theta_A(|\mathbf{D}(\mathbf{v})|)$. Since

$$\sum_{i,j=1}^{3} \frac{\partial^2 \mathbf{v}}{\partial x_i \partial x_j} = \sum_{i=1}^{3}\sum_{r=1}^{2} \frac{\partial^2 \mathbf{v}}{\partial x_i \partial x_r} + \frac{\partial^2 \mathbf{v}}{\partial x_3^2} \tag{4.27}$$

and

$$\begin{aligned} \frac{\partial^2 v_s}{\partial x_3^2} &= 2D_{s3}(\frac{\partial \mathbf{v}}{\partial x_3}) - \frac{\partial^2 v_3}{\partial x_3 \partial x_s} \qquad s = 1, 2\,, \\ \frac{\partial^2 v_3}{\partial x_3^2} &= -\frac{\partial^2 v_1}{\partial x_1 \partial x_3} - \frac{\partial^2 v_2}{\partial x_2 \partial x_3}\,, \end{aligned}$$

we see that

$$|\nabla^{(2)} \mathbf{v}| \le c \Big| \sum_{i=1}^{3}\sum_{r=1}^{2} \frac{\partial^2 \mathbf{v}}{\partial x_i \partial x_r} \Big| + c\,|\boldsymbol{\zeta}|\,, \tag{4.28}$$

which leads (multiplying the last inequality by $\xi_k \theta_A^{p-2}$) to

$$\theta_A^{p-2} |\nabla^{(2)} \mathbf{v}| \xi_k \le c \sum_{i=1}^{3}\sum_{r=1}^{2} \theta_A^{p-2} \Big| \frac{\partial^2 \mathbf{v}}{\partial x_i \partial x_r} \Big| \xi_k + c\,\theta_A^{p-2} |\boldsymbol{\zeta}| \xi_k\,. \tag{4.29}$$

In order to estimate the last term in (4.29) we use (3.47), which can be re-written as (see (3.43) and (3.48) for the definition of $\mathbf{G} = (G_2, G_1)$ and $\mathbb{A}$)

$$\mathbb{A}\boldsymbol{\zeta} = \tfrac{1}{2}\mathbf{G} - \mathbb{H}\,, \tag{4.30}$$

where $\mathbb{H}$ includes the remaining terms in (3.47). Thanks to (3.42) and (2.26), $\mathbb{H}$ can be estimated by

$$|\mathbb{H}| \le c \sum_{r,s=1}^{2} \theta_A^{p-2} \Big| D_{rs}(\frac{\partial \mathbf{v}}{\partial x_3}) \Big| \le \tilde{c} \sum_{r=1}^{2} \theta_A^{p-2} \Big| \nabla \frac{\partial \mathbf{v}}{\partial x_r} \Big|\,.$$

As $(\mathbb{A}\boldsymbol{\eta}, \boldsymbol{\eta}) \ge C_0 \theta_A^{p-2} |\boldsymbol{\eta}|^2$, we see that (4.30) together with the last inequality yields

$$C_0 \theta_A^{p-2} |\boldsymbol{\zeta}| \le \tfrac{1}{2}|\mathbf{G}| + \tilde{c} \sum_{r=1}^{2} \theta_A^{p-2} \Big| \nabla \frac{\partial \mathbf{v}}{\partial x_r} \Big|\,.$$

Plugging this into (4.29) results in

$$\theta_A^{p-2}|\nabla^{(2)}\mathbf{v}|\xi_k \le c|\mathbf{G}|\xi_k + c\sum_{i=1}^{3}\sum_{r=1}^{2}\theta_A^{p-2}\Big|\frac{\partial^2\mathbf{v}}{\partial x_i\partial x_r}\Big|\xi_k\,. \tag{4.31}$$

With the help of the definition for $\frac{\partial}{\partial\tau_r}$, $r=1,2$ (cf. (3.17)), we deduce from (4.31) that

$$\begin{aligned}\theta_A^{p-2}|\nabla^{(2)}\mathbf{v}|\xi_k &\le c|\mathbf{G}|\xi_k + c\sum_{i=1}^{3}\sum_{r=1}^{2}\theta_A^{p-2}\Big|\frac{\partial^2\mathbf{v}}{\partial x_i\partial\tau^r}\Big|\xi_k\\ &\quad + c\sum_{r=1}^{2}\sup_{\Omega_k}\Big|\frac{\partial a}{\partial x_r}\Big|\,\theta_A^{p-2}\Big|\nabla^{(2)}\mathbf{v}\Big|\xi_k\,.\end{aligned}$$

Due to (3.11) we can arrange the covering of the boundary in such a way that $c\,\sup_{\Omega_k}|\frac{\partial a}{\partial x_r}|<\frac{1}{2}$, which leads to

$$\theta_A^{p-2}|\nabla^{(2)}\mathbf{v}|\xi_k \le c|\mathbf{G}|\xi_k + c\sum_{r=1}^{2}\theta_A^{p-2}\Big|\nabla\frac{\partial\mathbf{v}}{\partial\tau^r}\Big|\xi_k\,. \tag{4.32}$$

The inequality (4.26) will follow from (4.32) by taking the L^q-norm of all terms provided that we can estimate $\|\mathbf{G}\xi_k\|_q$ in a suitable way. For this purpose we proceed similarly to Section 3 using the curl-operator.

By the negative norm theorem (cf. the appendix, Theorem 6.7) and equations (3.40) and (3.41) we have for $s=1,2$ (cf. (3.43))

$$\begin{aligned}\big\|G_s\xi_k\big\|_q &\le c\Big\|\frac{\partial}{\partial x_3}\frac{\partial\Phi_A}{\partial D_{s3}}\xi_k\Big\|_{-1,q} + \sum_{r=1}^{2}\sum_{i,k,l=1}^{3}c\Big\|\frac{\partial^2}{\partial x_i\partial x_r}\frac{\partial\Phi_A}{\partial D_{kl}}\xi_k\Big\|_{-1,q}\\ &\quad + c\Big\|\operatorname{curl}\Big(\frac{\partial\mathbf{v}}{\partial t}+\mathbf{v}_\varepsilon\cdot\nabla\mathbf{v}+\mathbf{f}\Big)\xi_k\Big\|_{-1,q} + c\Big\|\frac{\partial}{\partial x_3}\frac{\partial\Phi_A}{\partial D_{s3}}\nabla\xi_k\Big\|_{-1,q}\\ &= I_6+\cdots+I_9\,.\end{aligned} \tag{4.33}$$

We easily see that

$$I_6+I_9 \le c\|\partial\Phi_A\|_2 \tag{4.34}$$

and

$$I_8 \le c(\mathbf{f},\mathbf{v}_0)\big(1+K_{\varepsilon,A}^{1/2}+\|\frac{\partial\mathbf{v}}{\partial t}\|_2\big)\,. \tag{4.35}$$

Further, we get using (2.26) ($q' = q/(q-1)$)

$$I_7 \le \sup_{\substack{\phi\in W_0^{1,q'}(\Omega)\\ \|\phi\|_{1,q'}\le 1}} \int_{\Omega_0} \Big|\frac{\partial}{\partial x_r}\frac{\partial \Phi_A}{\partial D_{kl}}\xi_k\frac{\partial\phi}{\partial x_i}\Big| + \Big|\frac{\partial \Phi_A}{\partial D_{kl}}\Big(\frac{\partial \xi_k}{\partial x_i}\frac{\partial\phi}{\partial x_r} + \frac{\partial^2\xi_k}{\partial x_i\partial x_r}\phi\Big)\Big|\,dx$$

$$\le c\Big(\int_{\Omega_0}\theta_A^{(p-2)q}\Big|\nabla\frac{\partial \mathbf{v}}{\partial x_r}\Big|^q\xi_k^q\,dx\Big)^{1/q} + c\big\|\partial\Phi_A\big\|_2\,. \tag{4.36}$$

Now, using (4.33)–(4.36), (4.32) integrated over Ω and again the definition of $\frac{\partial}{\partial\tau^r}$, we finally come to (4.26).

Step 3: *Inequalities for the full second velocity gradient.*

We are going to estimate the second gradient of the velocity putting (4.24), (4.25) and (4.26) together. We first observe that by the Hölder's inequality we have

$$\begin{aligned}\|\theta_A^{p-2}\nabla^{(2)}\mathbf{v}\xi_0\|_q &+ \|\theta_A^{p-2}\nabla\frac{\partial\mathbf{v}}{\partial\tau^r}\xi_k\|_q \\ &\le c\Big(\|\nabla^{(2)}\mathbf{v}\xi_0\|_2 + \|\nabla\frac{\partial\mathbf{v}}{\partial\tau^r}\xi_k\|_2\Big)\|\theta_A\|^{p-2}_{\frac{(p-2)2q}{2-q}} \\ &\overset{(2.27)}{\le} c\Big(\|\nabla^{(2)}\mathbf{v}\xi_0\|_2 + \|\nabla\frac{\partial\mathbf{v}}{\partial\tau^r}\xi_k\|_2\Big)\|1+\partial\Phi_A\|^{\frac{p-2}{p-1}}_{\frac{(p-2)2q}{(p-1)(2-q)}}\,.\end{aligned} \tag{4.37}$$

Requiring $W^{1,q}(\Omega)\hookrightarrow L^{\frac{(p-2)2q}{(p-1)(2-q)}}(\Omega)$ we obtain the condition $q\in[1,\frac{6}{p+1}]$ and

$$\|1+\partial\Phi_A\|^{\frac{p-2}{p-1}}_{\frac{(p-2)2q}{(p-1)(2-q)}} \le c\Big(\|1+\partial\Phi_A\|_1^{\frac{p-2}{p-1}} + \|\nabla\partial\Phi_A\|_q^{\frac{p-2}{p-1}}\Big)\,.$$

Using this, (4.37), Lemma 6.3, (4.24)–(4.26) and Lemma 6.12 we come to the conclusion that

$$\begin{aligned}\|\nabla\partial\Phi_A\|_q + \|\nabla^{(2)}\mathbf{v}\|_q &\le c\Big(\|\mathbf{f}\|_2 + K_{\varepsilon,A}^{1/2} + \big\|\partial\Phi_A\big\|_2 + \|\frac{\partial\mathbf{v}}{\partial t}\|_2\Big) \\ &+c\Big(\|\nabla\partial\Phi_A\|_q^{\frac{p-2}{p-1}} + \|1+\partial\Phi_A\|_1^{\frac{p-2}{p-1}}\Big)\Big(\|\mathbf{f}\|_2 + K_{\varepsilon,A}^{1/2} + \big\|\partial\Phi_A\big\|_2 + \|\frac{\partial\mathbf{v}}{\partial t}\|_2\Big),\end{aligned}$$

which yields with the help of Young's inequality

$$\begin{aligned}\|\nabla\partial\Phi_A\|_q + \|\nabla^{(2)}\mathbf{v}\|_q &\le c\Big(1 + \|\mathbf{f}\|_2 + K_{\varepsilon,A}^{1/2} + \big\|\partial\Phi_A\big\|_2 + \|\frac{\partial\mathbf{v}}{\partial t}\|_2\Big) \\ &+ c\Big(1 + \|\mathbf{f}\|_2 + K_{\varepsilon,A}^{1/2} + \big\|\partial\Phi_A\big\|_2 + \|\frac{\partial\mathbf{v}}{\partial t}\|_2\Big)^{p-1} \\ &+ c\Big(1 + \Big(\|\mathbf{f}\|_2 + K_{\varepsilon,A}^{1/2} + \|\frac{\partial\mathbf{v}}{\partial t}\|_2\Big)\|\partial\Phi_A\|_2^{\frac{p-2}{p-1}} + \|\partial\Phi_A\|_2^{\frac{2p-3}{p-1}}\Big)\,.\end{aligned} \tag{4.38}$$

Since $\frac{2p-3}{p-1} \leq p-1$ for $p \geq 2$ and $\frac{p-1}{p-2}$ is the dual exponent to $p-1$ we finally obtain by considering the r-th power of the inequality ($r > 0$)

$$\|\nabla\partial\Phi_A\|_q^r + \|\nabla^{(2)}\mathbf{v}\|_q^r \leq c\Big(1 + \|\mathbf{f}\|_2^2 + K_{\varepsilon,A} + \big\|\partial\Phi_A\big\|_2^2 + \big\|\frac{\partial\mathbf{v}}{\partial t}\big\|_2^2\Big)^{\frac{r(p-1)}{2}}, \quad (4.39)$$

which we wanted to prove.

Step 4: *Derivation of* (4.16).

Taking the $\frac{r(p-1)}{2}$-th power of (4.24) and (4.25) and adding them to (4.39) we obtain the inequality with the right-hand side bounded by the right-hand side of (4.39) (up to a multiplicative constant). Then we multiply the so-obtained inequality by $\gamma_\mu'(E_A)$ and integrate it with respect to time over $(0, T)$. Inequality (4.16) then easily follows by requiring $\frac{r(p-1)}{2} \leq 1$ and also taking (3.52) into account. □

We want to finish this section with four lemmas, which are consequences of (3.52) and (4.16) if we pass to the limit as $A \to \infty$.

Lemma 4.40. *Let* $p \in [2, 6)$. *Then we have for almost all* $t \in I$

$$\|\nabla\mathbf{v}^\varepsilon(t)\|_2^2 \leq c(\mathbf{f}, \mathbf{v}_0) + \int_0^t K_\varepsilon(\tau)\, d\tau. \quad (4.41)$$

Proof. From (3.52) for $\mu = 0$ and $(2.23)_4$ we obtain

$$\|\nabla\mathbf{v}^{\varepsilon,A}(t)\|_2^2 \leq c(\mathbf{f}, \mathbf{v}_0) + \int_0^t K_{\varepsilon,A}(\tau)\, d\tau\,. \quad (4.42)$$

From (4.14) and the lower semicontinuity of norms we immediately get (4.41). □

Lemma 4.43. *Let* $p \in [2, 3)$ *and let* γ_μ *be defined as in Lemma* 3.51, $\mu \geq 0$. *Then we have for almost all* $t \in I$

$$\operatorname*{ess\,sup}_{\tau\in(0,t)} \gamma_\mu(E(\tau)) \leq c(\mathbf{f}, \mathbf{v}_0) + \int_0^t K_\varepsilon(\tau)\, \gamma_\mu'(E(\tau))\, d\tau\,, \quad (4.44)$$

where $E(t) \equiv 1 + \|\Phi(|\mathbf{D}(\mathbf{v}^\varepsilon(t))|^2)\|_1$.

Proof. First we will show that for almost all $t \in I$

$$\lim_{A\to\infty} \int_\Omega \Phi_A(|\mathbf{D}(\mathbf{v}^{\varepsilon,A}(t))|^2)\, dx = \int_\Omega \Phi(|\mathbf{D}(\mathbf{v}^\varepsilon(t))|^2)\, dx\,. \quad (4.45)$$

We have

$$
\begin{aligned}
&\Big|\int_\Omega \Phi_A(|\mathbf{D}(\mathbf{v}^{\varepsilon,A}(t))|^2)\,dx - \int_\Omega \Phi(|\mathbf{D}(\mathbf{v}^\varepsilon(t))|^2)\,dx\Big| \\
&\le \Big|\int_\Omega \Phi_A(|\mathbf{D}(\mathbf{v}^{\varepsilon,A}(t))|^2)\,dx - \int_\Omega \Phi_A(|\mathbf{D}(\mathbf{v}^\varepsilon(t))|^2)\,dx\Big| \\
&\quad + \Big|\int_\Omega \Phi_A(|\mathbf{D}(\mathbf{v}^\varepsilon(t))|^2)\,dx - \int_\Omega \Phi(|\mathbf{D}(\mathbf{v}^\varepsilon(t))|^2)\,dx\Big| \equiv I_1 + I_2\,.
\end{aligned}
\tag{4.46}
$$

From (2.27) and $(2.23)_4$ follows

$$
I_1 \le C\big(1 + \|\nabla\mathbf{v}^{\varepsilon,A}(t)\|_p^{p-1} + \|\nabla\mathbf{v}^\varepsilon(t)\|_p^{p-1}\big)\|\nabla\mathbf{v}^{\varepsilon,A}(t) - \nabla\mathbf{v}^\varepsilon(t)\|_p\,,
$$

which together with $(4.12)_4$ gives for almost all $t \in I$

$$
\lim_{A\to\infty} I_1 = 0\,, \tag{4.47}
$$

since $\frac{6}{p-1} > p$ for $p \in [2,3)$. Further, from the definition of Φ_A (cf. (2.17)) we see that

$$
I_2 = \Big|\int_{\{x:|\mathbf{D}(\mathbf{v}^\varepsilon(t,x))|\ge A\}} \Phi_A(|\mathbf{D}(\mathbf{v}^\varepsilon(t))|^2)\,dx - \Phi(|\mathbf{D}(\mathbf{v}^\varepsilon(t))|^2)\,dx\Big|\,. \tag{4.48}
$$

But from $(4.12)_5$ we get for almost all $t \in I$

$$
\lim_{A\to\infty} |\{x : |\mathbf{D}(\mathbf{v}^\varepsilon(t,x))| \ge A\}| = 0\,. \tag{4.49}
$$

This and the bound $|\Phi_A(|\mathbf{D}(\mathbf{v}^\varepsilon)|^2) - \Phi(|\mathbf{D}(\mathbf{v}^\varepsilon)|^2)| \le c(1 + |\mathbf{D}(\mathbf{v}^\varepsilon)|^p)$ gives $\lim\limits_{A\to\infty} I_2 = 0$, which together with (4.47) proves (4.45). This in turn implies that

$$
\begin{aligned}
&\gamma_\mu(E_A(t)) \to \gamma_\mu(E(t)) &&\text{for almost every } t \in I\,,\\
&\gamma'_\mu(E_A(t)) \overset{*}{\rightharpoonup} \gamma'_\mu(E(t)) &&\text{weakly-}^* \text{ in } L^\infty(I)\,.
\end{aligned}
\tag{4.50}
$$

Furthermore, we have (cf. (4.14))

$$
\int_\Omega |\mathbf{v}_\varepsilon^{\varepsilon,A}|^2|\nabla\mathbf{v}^{\varepsilon,A}|^2\,dx \to \int_\Omega |\mathbf{v}_\varepsilon^\varepsilon|^2|\nabla\mathbf{v}^\varepsilon|^2\,dx \qquad \text{in } L^1(I)\,. \tag{4.51}
$$

Since the first term in (3.52) is nonnegative we can pass to the limit in (3.52) as $A \to \infty$ due to (4.50) and (4.51). This gives (4.44). □

Lemma 4.52. *Let $p \in [2,3)$ and let γ_μ be as in Lemma* 3.51, $\mu \geq 0$. *Then we have for all $q \in [1, \frac{6}{p+1}]$ and all $r \in [1, \frac{2}{p-1}]$*

$$\int_0^T \|\nabla \partial \Phi(|\mathbf{D}(\mathbf{v}^\varepsilon(t))|^2)\|_q^r \gamma_\mu'(E(t))\,dt + \int_0^T Z(\mathbf{v}^\varepsilon(t))\gamma_\mu'(E(t))\,dt \tag{4.53}$$
$$\leq c(\mathbf{f},\mathbf{v}_0)\Big(1 + \int_0^T K_\varepsilon(t)\,\gamma_\mu'(E(t))\,dt + \int_0^T \|\partial\Phi(|\mathbf{D}(\mathbf{v}^\varepsilon(t))|^2)\|_2^2\,\gamma_\mu'(E(t))\,dt\Big),$$

where

$$Z(\mathbf{v}^\varepsilon(t)) = \Big(\|\nabla^{(2)}\mathbf{v}^\varepsilon(t)\xi_0\|_2^2 + \sum_{r=1}^2 \|\mathbf{D}(\frac{\partial \mathbf{v}^\varepsilon(t)}{\partial \tau^r})\,\xi_k\|_2^2\Big)^{\frac{r(p-1)}{2}}.$$

Proof. Let us for simplicity denote $\gamma(s) = \gamma_\mu(s)$. In Lemma 4.15 we proved for $r \leq \frac{2}{p-1}$

$$\int_0^T \|\nabla\partial\Phi_A(|\mathbf{D}(\mathbf{v}^{\varepsilon,A})|^2)\|_q^r\,\gamma'(E_A(t))\,dt + \int_0^T Z(\mathbf{v}^{\varepsilon,A}(t))\gamma'(E_A(t))\,dt \tag{4.54}$$
$$\leq c(\mathbf{f},\mathbf{v}_0)\Big(1+\int_0^T K_{\varepsilon,A}\,\gamma'(E_A(t))\,dt + \int_0^T \|\partial\Phi_A(|\mathbf{D}(\mathbf{v}^{\varepsilon,A})|^2)\|_2^2\,\gamma'(E_A(t))\,dt\Big).$$

Since Φ_A approximates Φ locally uniformly we get from $(4.12)_5$ that

$$\partial\Phi_A(|\mathbf{D}(\mathbf{v}^{\varepsilon,A})|^2) \to \partial\Phi(|\mathbf{D}(\mathbf{v}^\varepsilon)|^2) \qquad \text{almost everywhere in } Q_T. \tag{4.55}$$

From (4.6) and (4.9) we obtain[5]

$$\partial\Phi_A(|\mathbf{D}(\mathbf{v}^{\varepsilon,A})|^2) \quad \text{is bounded in } L^{\frac{2(p+3)}{3(p-1)}}(Q_T), \tag{4.56}$$

which together with (cf. (2.25), $(2.23)_4$)

$$|\partial\Phi_A(|\mathbf{A}|^2)| \leq c(1+|\mathbf{A}|)^{p-1}$$

[5]Here we use the following parabolic imbedding: the space $L^\infty(I; L^{p'}(\Omega)) \cap L^r(I; W^{1,q}(\Omega))$ is continuously imbedded into $L^\tau(Q_T)$, where $\tau = \frac{2(3+p)}{3(p-1)}$, which follows from the interpolation inequality $\|\mathbf{u}\|_s \leq \|\mathbf{u}\|_{p'}^{1-\alpha}\|\mathbf{u}\|_{\frac{3q}{3-q}}^\alpha$ with $\alpha = \frac{6(r(p-1)-p)}{r(p-1)(6-p)}$.

and Vitali's convergence lemma gives (for $p \in [2,3)$)

$$\partial\Phi_A(|\mathbf{D}(\mathbf{v}^{\varepsilon,A})|^2) \to \partial\Phi(|\mathbf{D}(\mathbf{v}^{\varepsilon})|^2) \qquad \text{strongly in } L^2(Q_T)\,. \tag{4.57}$$

Thus, (4.57) and $(4.50)_2$ enable the limiting process in the second term on the right-hand side of (4.54). Similarly, using (4.14) and $(4.50)_2$, we can justify the limiting process in the first term on the right-hand side in (4.54). It remains to pass to the limit in the left-hand side of (4.54). For all $q < \infty$ we have

$$\gamma'(E_A(t)) \to \gamma'(E(t)) \qquad \text{strongly in } L^q(I)\,.$$

Therefore, by the Egorov theorem we know that there is a set I_δ, $|I \setminus I_\delta| \le \delta$, and a subsequence $A_\delta \to \infty$ such that

$$\gamma'(E_{A_\delta}(t)) \rightrightarrows \gamma'(E(t)) \qquad \text{uniformly in } I_\delta\,.$$

From this and boundedness of $\|\nabla\partial\Phi_A(|\mathbf{D}(\mathbf{v}^{\varepsilon,A})|^2)\|_q^r$ in $L^1(I)$ we conclude that

$$\liminf_{A_\delta\to\infty} \int_{I_\delta} \|\nabla\partial\Phi_{A_\delta}(|\mathbf{D}(\mathbf{v}^{\varepsilon,A_\delta})|^2)\|_q^r \Big(\gamma'(E_{A_\delta}(t)) - \gamma'(E(t))\Big)\,dt = 0\,.$$

Thus we have

$$\begin{aligned}\liminf_{A_\delta\to\infty} &\int_{I_\delta} \|\nabla\partial\Phi_{A_\delta}(|\mathbf{D}(\mathbf{v}^{\varepsilon,A_\delta})|^2)\|_q^r\,\gamma'(E_{A_\delta}(t))\,dt \\ &= \liminf_{A_\delta\to\infty} \int_{I_\delta} \|\nabla\partial\Phi_{A_\delta}(|\mathbf{D}(\mathbf{v}^{\varepsilon,A_\delta})|^2)\|_q^r\,\gamma'(E(t))\,dt\,. \end{aligned} \tag{4.58}$$

Since

$$0 < \gamma'(E(t)) \le 1 \tag{4.59}$$

we can define by $\nu \equiv \gamma'(E(t))\,dt$ a new measure, which is absolutely continuous with respect to dt. From (4.57) and (4.9) we can identify the weak limit of $\nabla\partial\Phi_A$ in $L^r(I;L^q(\Omega))$ as $\nabla\partial\Phi$. From this, (4.59), the boundedness of the right-hand side of (4.54) and the uniqueness of the weak limit we obtain that $\nabla\partial\Phi_A(|\mathbf{D}\mathbf{v}^{\varepsilon,A}|^2) \rightharpoonup \nabla\partial\Phi(|\mathbf{D}\mathbf{v}^{\varepsilon}|^2)$ also in $L^r(I,\nu;L^q(\Omega))$, and therefore

$$\begin{aligned}\liminf_{A_\delta\to\infty} &\int_{I_\delta} \|\nabla\partial\Phi_{A_\delta}(|\mathbf{D}(\mathbf{v}^{\varepsilon,A_\delta})|^2)\|_q^r\,\gamma'(E(t))\,dt \\ &\ge \int_{I_\delta} \|\nabla\partial\Phi(|\mathbf{D}(\mathbf{v}^{\varepsilon})|^2)\|_q^r\,\gamma'(E(t))\,dt\,. \end{aligned}$$

Since $\delta > 0$ was arbitrary and the terms on the right-hand side of (4.54) are bounded independent of δ we can conclude the proof for the first quantity on the left-hand side in (4.53) by a diagonal argument. The proof for the second term on the left-hand side of (4.53) follows along the same lines. □

Lemma 4.60. *For $p \in [2,3)$, we have*

$$\int_0^T \int_\Omega (1+|\mathbf{D}(\mathbf{v}^\varepsilon)|)^{p-2}\Big(\|\nabla^{(2)}\mathbf{v}^\varepsilon\xi_0\|_2^2 + \sum_{r=1}^2 \|\mathbf{D}(\frac{\partial \mathbf{v}^\varepsilon}{\partial \tau^r})\,\xi_k\|_2^2\Big)\,dx\,dt$$
$$\leq c(\mathbf{f},\mathbf{v}_0)\Big(1+\int_0^T K_\varepsilon(t)\,dt + \int_0^T \|\partial\Phi(|\mathbf{D}(\mathbf{v}^\varepsilon(t))|^2)\|_2^2\,dt\Big). \tag{4.61}$$

Proof. In Lemma 4.15 we proved

$$\int_0^T \int_\Omega \theta_A(|\mathbf{D}(\mathbf{v}^{\varepsilon,A})|)^{p-2}\Big(\|\nabla^{(2)}\mathbf{v}^{\varepsilon,A}\xi_0\|_2^2 + \sum_{r=1}^2 \|\mathbf{D}(\frac{\partial \mathbf{v}^{\varepsilon,A}}{\partial \tau^r})\,\xi_k\|_2^2\Big)\,dx\,dt$$
$$\leq c(\mathbf{f},\mathbf{v}_0)\left(1+\int_0^T K_{\varepsilon,A}\,dt + \int_0^T \|\partial\Phi_A(|\mathbf{D}(\mathbf{v}^{\varepsilon,A})|^2)\|_2^2\,dt\right). \tag{4.62}$$

We have already shown in the proof of the previous lemma that we can pass to the limit as $A \to \infty$ in the right-hand side of (4.62). For the left-hand side we proceed similarly and use (4.12) and the Egorov theorem to conclude that for all $\delta > 0$ there is a set Q_δ, $|Q \setminus Q_\delta| \leq \delta$, and a subsequence A_δ such that

$$\theta_{A_\delta}(|\mathbf{D}(\mathbf{v}^{\varepsilon,A_\delta})|) \rightrightarrows 1+|\mathbf{D}(\mathbf{v}^\varepsilon)| \qquad \text{uniformly in } Q_\delta\,.$$

Now we finish the proof along the lines of Lemma 4.52.

5. Limiting process $\varepsilon \to 0$. In the previous section we obtained solutions $(\mathbf{v},\pi) = (\mathbf{v}^\varepsilon,\pi^\varepsilon)$ of the *Problem (NS-Dir)$_p^\varepsilon$*, that were based on estimates independent of A. These estimates, however, were dependent on ε through $K_\varepsilon = \int_\Omega |\mathbf{v}_\varepsilon^\varepsilon|^2|\nabla\mathbf{v}^\varepsilon|^2\,dx$, or precisely through $\int_0^T K_\varepsilon\,dt$.

From the energy inequality (2.8) for the *Problem (NS-Dir)$_p^\varepsilon$*, Lemma 4.40 and Remark 4.13 we get for all $p \in [2,3)$ and almost all $t \in I$

$$\operatorname*{ess\,sup}_{t\in I} \|\mathbf{v}^\varepsilon(t)\|_2^2 + \int_0^T \|\nabla\mathbf{v}^\varepsilon\|_2^2\,dt + \int_0^T \|\nabla\mathbf{v}^\varepsilon\|_p^p\,dt \leq c(\mathbf{f},\mathbf{v}_0)\,, \tag{5.1}$$

$$\|\nabla\mathbf{v}^\varepsilon(t)\|_2^2 \leq c(\mathbf{f},\mathbf{v}_0) + \int_0^t K_\varepsilon(\tau)\,d\tau\,, \tag{5.2}$$

$$\int_0^T \|\frac{\partial\mathbf{v}^\varepsilon}{\partial t}\|_2^2\,dt \leq c(\mathbf{f},\mathbf{v}_0)\Big(1+\int_0^T K_\varepsilon(t)\,dt\Big). \tag{5.3}$$

It is worth noticing that in addition to $\int_0^T \|\nabla\mathbf{v}^\varepsilon\|_2^2\,dt$, which has already occurred in (4.1) or (3.4), we now also obtain the estimate of $\int_0^T \|\nabla\mathbf{v}^\varepsilon\|_p^p\,dt$

in (5.1), as follows from (2.2). Moreover, we proved in Lemma 4.43 and Lemma 4.52 that for $\mu \geq 0$

$$\operatorname*{ess\,sup}_{\tau\in(0,t)} \gamma_\mu(E(\tau)) \leq c(\mathbf{f},\mathbf{v}_0) + \int_0^t K_\varepsilon(\tau)\, \gamma'_\mu(E(\tau))\, d\tau\,, \tag{5.4}$$

$$\begin{aligned} &\int_0^T \|\nabla\partial\Phi(|\mathbf{D}(\mathbf{v}^\varepsilon(t))|^2)\|_{\frac{6}{p+1}}^{\frac{2}{p-1}} \gamma'_\mu(E(t))\, dt + \int_0^T Z(\mathbf{v}^\varepsilon(t))\, \gamma'_\mu(E(t))\, dt \\ &\qquad \leq c(\mathbf{f},\mathbf{v}_0)\Big(1 + \int_0^T K_\varepsilon(t)\, \gamma'_\mu(E(t))\, dt + \int_0^T \|\partial\Phi\|_2^2\, \gamma'_\mu(E(t))\, dt\Big)\,, \end{aligned} \tag{5.5}$$

where $\gamma_\mu(s) = \frac{1}{1-\mu}s^{1-\mu}$, for $\mu \geq 0$ and $\mu \neq 1$, and $\gamma_1(s) = \ln(s)$ and $Z(\mathbf{v}^\varepsilon(t))$ is defined in Lemma 4.52. Moreover, in the case $\mu = 0$ we also have from Lemma 4.60

$$\begin{aligned} &\int_0^T \int_\Omega (1 + |\mathbf{D}(\mathbf{v}^\varepsilon)|)^{p-2}\Big(\|\nabla^{(2)}\mathbf{v}^\varepsilon\xi_0\|_2^2 + \sum_{r=1}^2 \|\mathbf{D}(\frac{\partial\mathbf{v}^\varepsilon}{\partial\tau^r})\,\xi_k\|_2^2\Big)\, dx\, dt \\ &\qquad \leq c(\mathbf{f},\mathbf{v}_0)\Big(1 + \int_0^T K_\varepsilon(t)\, dt + \int_0^T \|\partial\Phi(|\mathbf{D}(\mathbf{v}^\varepsilon(t))|^2)\|_2^2\, dt\Big)\,. \end{aligned} \tag{5.6}$$

Let us make two observations. Firstly, if $\mu \in [0,1]$ in (5.4) and the right-hand side is bounded, then $\{\mathbf{v}^\varepsilon\}$ lies in a ball of $L^\infty(I;V_p)$. If $\mu > 1$, then inequality (5.4) gives us no information. Secondly, denoting

$$I_{p,\frac{6}{p+1}}(t) \equiv \int_\Omega (1 + |\mathbf{D}(\mathbf{v}(t))|)^{\frac{6(p-2)}{p+1}}\, |\mathbf{D}(\nabla\mathbf{v}(t))|^{\frac{6}{p+1}}\, dx \tag{5.7}$$

and using Lemma 6.9, we obtain from (5.5) and (5.4) that

$$\begin{aligned} &\gamma_\mu(E(t)) + \int_0^T \gamma'_\mu(E(t)) I_{p,\frac{6}{p+1}}^{\frac{p+1}{3(p-1)}}(t)\, dt + \int_0^T Z(\mathbf{v}(t))\, \gamma'_\mu(E(t))\, dt \\ &\qquad \leq c(\mathbf{f},\mathbf{v}_0)\Big(1 + \int_0^T \gamma'_\mu(E(t)) K_\varepsilon(t)\, dt + \int_0^T \gamma'_\mu(E(t)) \|\nabla\mathbf{v}\|_{2(p-1)}^{2(p-1)}\, dt\Big)\,, \end{aligned} \tag{5.8}$$

in which the first term is omitted if $\mu > 1$.

Our last task is to estimate (uniformly with respect to ε) the right-hand sides of (5.2)–(5.4), (5.6) and (5.8).

(i) The case $p \in [\frac{12}{5}, 3)$. Let us first recall that by (2.4) the norm $\|\Phi(|\mathbf{D}(\mathbf{v})|^2)\|_1$ is equivalent to $\|\mathbf{D}(\mathbf{v})\|_p^p$. Thus, we can estimate K_ε in (5.4), with $\mu = 0$, in the same way as in (2.13), and we obtain

$$\|\mathbf{v}(t)\|_p^p \le c(\mathbf{f}, \mathbf{v}_0) + c \int_0^t \|\mathbf{v}\|_2^{2\frac{5p-12}{5p-6}} \|\nabla \mathbf{v}\|_p^{p\frac{16-5p}{5p-6}} \|\nabla \mathbf{v}\|_p^p \, d\tau \,,$$

which together with (5.1) and Gronwall's lemma gives the fact that

$$\mathbf{v}^\varepsilon \text{ is bounded in } L^\infty(I; V_p) \tag{5.9}$$

uniformly with respect to ε. Thus $\int_0^T K_\varepsilon \, dt$ is bounded independent of ε and it remains to bound the second term on the right-hand sides of (5.6) and (5.8) for $\mu = 0$. From (6.11) we deduce that

$$\|\nabla \mathbf{v}\|_6^{\frac{6(p-1)}{p+1}} \le C_{18}\big(1 + \|\mathbf{v}\|_p^{\frac{6(p-1)}{p+1}} + I_{p,\frac{6}{p+1}}(t)\big) \,, \tag{5.10}$$

which together with the interpolation inequality

$$\|\nabla \mathbf{v}\|_{2(p-1)} \le \|\nabla \mathbf{v}\|_6^{\frac{3(p-2)}{(6-p)(p-1)}} \|\nabla \mathbf{v}\|_p^{\frac{p(4-p)}{(6-p)(p-1)}}$$

implies that

$$\begin{aligned} \|\nabla \mathbf{v}\|_{2(p-1)}^{2(p-1)} &\overset{(5.10)}{\le} c\|\nabla \mathbf{v}\|_p^{\frac{2p(4-p)}{6-p}} \big(1 + \|\nabla \mathbf{v}\|_p^{\frac{6(p-2)}{6-p}} + I_{p,\frac{6}{p+1}}^{\frac{(p-2)(p+1)}{(6-p)(p-1)}}\big) \\ &\overset{(2.4)}{\le} c\|\Phi(|\mathbf{D}(\mathbf{v})|^2)\|_1^{\frac{2(4-p)}{6-p}} \big(1 + \|\Phi(|\mathbf{D}(\mathbf{v})|^2)\|_1^{\frac{6(p-2)}{p(6-p)}} + I_{p,\frac{6}{p+1}}^{\frac{(p-2)(p+1)}{(6-p)(p-1)}}\big) \,. \end{aligned} \tag{5.11}$$

We get from (5.8) ($\mu = 0$), (5.9) and the first inequality in (5.11)

$$\int_0^T I_{p,\frac{6}{p+1}}^{\frac{p+1}{3(p-1)}}(t) \, dt \le c(\mathbf{f}, \mathbf{v}_0)\Big(1 + \int_0^T I_{p,\frac{6}{p+1}}^{\frac{(p-2)(p+1)}{(6-p)(p-1)}}(t) \, dt\Big) \,. \tag{5.12}$$

However for $p \in [2, 3)$ we see that the exponent on the right-hand side of (5.12) is strictly less than the exponent on the left-hand side of (5.12). Thus the right-hand sides of (5.6) and (5.8) with $\mu = 0$ are bounded independent of ε, and we proved that

$$\int_0^T \|(1+|\mathbf{D}(\mathbf{v}^\varepsilon)|)^{\frac{p-2}{2}}\mathbf{D}(\nabla\mathbf{v}^\varepsilon)\|_{2,\mathrm{loc}}^2\,dt + \int_0^T \|(1+|\mathbf{D}(\mathbf{v}^\varepsilon)|)^{\frac{p-2}{2}}\mathbf{D}(\frac{\partial \mathbf{v}^\varepsilon}{\partial \tau^r})\|_2^2\,dt$$

$$\int_0^T \|\frac{\partial \mathbf{v}^\varepsilon}{\partial t}\|_2^2\,dt + \operatorname*{ess\,sup}_{t\in I}\|\nabla\mathbf{v}^\varepsilon(t)\|_p^p + \int_0^T \|\nabla^{(2)}\mathbf{v}^\varepsilon\|_{\frac{6}{p+1}}^{\frac{2}{p-1}}\,dt \le c(\mathbf{f},\mathbf{v}_0)\,. \tag{5.13}$$

From (5.13) and the Aubin-Lions lemma we obtain that for $\tilde{q} \in [1, \frac{6}{p-1})$

$$\nabla\mathbf{v}^\varepsilon \to \nabla\mathbf{v} \qquad \text{in } L^{\frac{2}{p-1}}(I; L^{\tilde{q}}(\Omega)^{3\times 3})\,,\ \text{almost everywhere in } Q_T\,. \tag{5.14}$$

Using (5.1), (5.13), (5.14), (2.3) and Vitali's convergence lemma we pass to the limit as $\varepsilon \to 0$ in the weak formulation of $(2.7)_2$. Concerning the uniqueness we argue as in [11], Theorem 5.4.37.

(*ii*) The case $p \in [2, 12/5)$. We are going to estimate the terms on the right-hand side of (5.8). For the convective term we have

$$\begin{aligned} K_\varepsilon &\le c\|\nabla\mathbf{v}\|_p^2\|\nabla\mathbf{v}\|_{\frac{6p}{5p-6}}^2 \le c\|\nabla\mathbf{v}\|_p^{2(2-\lambda)}\|\nabla\mathbf{v}\|_6^{2\lambda} \\ &\overset{(2.4)}{\le} c\|\Phi(|\mathbf{D}(\mathbf{v})|^2)\|_1^{\frac{2(2-\lambda)}{p}}\big(1+\|\Phi(|\mathbf{D}(\mathbf{v})|^2)\|_1^{\frac{2\lambda}{p}} + I_{p,\frac{6}{p+1}}^{\frac{2\lambda(p+1)}{6(p-1)}}\big)\,, \end{aligned}$$

where we have used the interpolation inequality $\|\mathbf{z}\|_{\frac{6p}{5p-6}} \le \|\mathbf{z}\|_p^{1-\lambda}\|\mathbf{z}\|_6^{\lambda}$ with $\lambda = \frac{12-5p}{6-p}$ and $1-\lambda = \frac{2(2p-3)}{6-p}$ and (5.10). From this, (5.11) and (5.8) we obtain that (recall $\gamma_\mu'(s) = s^{-\mu}$)

$$\begin{aligned} &\gamma_\mu(E(t)) + \int_0^T E(t)^{-\mu} I_{p,\frac{6}{p+1}}^{\frac{p+1}{3(p-1)}}(t)\,dt \\ &\le c(\mathbf{f},\mathbf{v}_0)\Big(1 + \int_0^T E(t)^{\frac{4}{p}-\mu}\,dt + \int_0^T E(t)^{\frac{2(p-1)}{p}-\mu}\,dt \\ &+ \int_0^T \big\{E(t)^{-\mu} I_{p,\frac{6}{p+1}}^{\frac{p+1}{3(p-1)}}(t)\big\}^{\frac{3(p-2)}{6-p}} E(t)^{\frac{p+2}{6-p}+\mu(\frac{3(p-2)}{6-p}-1)}\,dt \\ &+ \int_0^T \big\{E(t)^{-\mu} I_{p,\frac{6}{p+1}}^{\frac{p+1}{3(p-1)}}(t)\big\}^{\lambda} E(t)^{\lambda+\frac{2(2-\lambda)}{p}+\mu(\lambda-1)}\,dt\Big). \end{aligned} \tag{5.15}$$

Requiring now for the first two integrals on the right-hand side that the exponent is less than 1 yields

$$\mu_1 = \frac{4-p}{p}\,, \qquad \mu_2 = \frac{p-2}{p}\,. \tag{5.16}$$

Note that for $p \in [2,3)$ $\mu_2 \le \mu_1$. Furthermore, we require for the third integral on the right-hand side of (5.15)

$$\frac{3(p-2)}{6-p}\gamma = 1\,, \qquad [\frac{p+2}{6-p} + \mu(\frac{3(p-2)}{6-p} - 1)]\gamma' \le 1\,,$$

where $\frac{1}{\gamma} + \frac{1}{\gamma'} = 1$, which implies

$$\mu_3 = \frac{p-2}{2(3-p)}\,, \tag{5.17}$$

and for the fourth integral on the right-hand side

$$\lambda\delta = 1\,, \qquad [\lambda + \frac{2(2-\lambda)}{p} + \mu(\lambda - 1)]\delta' \le 1\,,$$

where $\frac{1}{\delta} + \frac{1}{\delta'} = 1$, which gives

$$\mu_4 = 2\frac{3-p}{2p-3}\,. \tag{5.18}$$

Note that for $p \in [2, \frac{12}{5})$, we have $\mu_3 \le \mu_4$ and $\mu_1 \le \mu_4$, and therefore we choose $\mu = \mu_4$, since $E(t)^{-1} \le 1$. By Young's inequality and (5.1) we obtain

$$\gamma_\mu(E(t)) + \int_0^T E(t)^{-\mu} I_{p,\frac{6}{p+1}}^{\frac{p+1}{3(p-1)}}(t)\,dt \le c(\mathbf{f}, \mathbf{v}_0)\,, \tag{5.19}$$

where the first term is omitted if $\mu > 1$. Note that

$$\mu \le 1 \quad \text{if and only if} \quad p \ge \frac{9}{4}\,,$$

and therefore we obtain (5.9) for these p's. Now we go again into (5.15) and set $\mu = 0$. Using (5.9) and the fact that the exponents of the terms in the squiggly brackets in (5.15) are strictly less than 1 we easily obtain that the right-hand side of (5.15) is finite independent of ε. This together with (5.6) yields (5.12), and we can conclude the proof for these p's as in case (i). Due to the definition of $I_{p,\frac{6}{p+1}}$ we get from (5.19) for all $p \in [2, 9/4)$

$$\int_0^T \left(1 + \|\nabla \mathbf{v}\|_p^p\right)^{-\mu} \|\nabla^2 \mathbf{v}\|_{\frac{6}{p+1}}^{\frac{2}{p-1}}\,dt \le c(\mathbf{f}, \mathbf{v}_0)\,. \tag{5.20}$$

Now we proceed analogously as in the *Problem (NS-Per)$_p$* (cf. [11], Chapter 5, pp. 237–238, or [9]) to obtain from (5.20) $\nabla \mathbf{v}^\varepsilon \to \nabla \mathbf{v}$ strongly in $L^1(Q_T)$. The proof of Theorem 1.17 is complete.

6. Appendix. Here we first collect some general auxiliary definitions and assertions used in the previous text, and then we prove more technical assertions concerning the potentials Φ and Φ_A.

Proposition 6.1 (Korn's inequality). *Let $1 < p < \infty$ and let $\Omega \subset \mathbb{R}^d$ be of class $\mathcal{C}^1$. Then there exists a constant $K_p = K_p(\Omega)$ such that the inequality*

$$K_p \|\mathbf{v}\|_{1,p} \leq \|\mathbf{D}(\mathbf{v})\|_p \tag{6.2}$$

is fulfilled for all $\mathbf{v} \in W_0^{1,p}(\Omega)^d$.

Proof.
See e.g. [14].

Lemma 6.3. *Let $1 < p < \infty$ and let $\Omega \subset \mathbb{R}^d$ be a domain of class $\mathcal{C}^1$. Then there exists a constant $c_p = c_p(\Omega)$ such that for all $\mathbf{v} \in W_0^{1,p}(\Omega)^d \cap W^{2,p}(\Omega)^d$*

$$c_p \|\nabla^{(2)}\mathbf{v}\|_p \leq \|\mathbf{D}(\nabla \mathbf{v})\|_p \,. \tag{6.4}$$

Proof. The assertion follows immediately from the algebraic identity

$$\frac{\partial^2 v_i}{\partial x_j \partial x_k} = \frac{\partial D_{ik}(\mathbf{v})}{\partial x_j} + \frac{\partial D_{ij}(\mathbf{v})}{\partial x_k} - \frac{\partial D_{jk}(\mathbf{v})}{\partial x_i} \,.$$

Lemma 6.5. *Let $\Omega \subset \mathbb{R}^d$ be a domain, $\partial\Omega \in \mathcal{C}^1$ and let $\mathbf{v} \in W^{1,2}(\Omega)^d$, $\xi \in \mathcal{D}(\Omega)$. Then*

$$\int_\Omega |\mathbf{D}(\mathbf{v})|^2 \xi^2 \, dx \geq C_{15} \int_\Omega |\nabla \mathbf{v}|^2 \xi^2 \, dx - C_{16} \int_\Omega |\mathbf{v}|^2 |\nabla \xi|^2 \, dx \,. \tag{6.6}$$

Proof. We have

$$\int_\Omega \mathbf{D}(\mathbf{v}) \cdot \mathbf{D}(\mathbf{v}) \xi^2 \, dx = \frac{1}{2} \int_\Omega \frac{\partial v_i}{\partial x_j} \frac{\partial v_i}{\partial x_j} \xi^2 \, dx + \frac{1}{2} \int_\Omega \frac{\partial v_i}{\partial x_j} \frac{\partial v_j}{\partial x_i} \xi^2 \, dx = I_1 + I_2 \,.$$

The integration by parts yields

$$I_2 = \frac{1}{2} \int_\Omega |\operatorname{div} \mathbf{v}|^2 \xi^2 \, dx + \int_\Omega v_i \frac{\partial v_j}{\partial x_j} \xi \frac{\partial \xi}{\partial x_i} \, dx - \int_\Omega v_i \frac{\partial v_j}{\partial x_i} \xi \frac{\partial \xi}{\partial x_j} \, dx \,.$$

Thus we obtain

$$\begin{aligned}
&\frac{1}{2} \int_\Omega |\nabla \mathbf{v}|^2 \xi^2 \, dx + \frac{1}{2} \int_\Omega |\operatorname{div} \mathbf{v}|^2 \xi^2 \, dx \\
&\quad = \int_\Omega |\mathbf{D}(\mathbf{v})|^2 \xi^2 \, dx - \int_\Omega v_i \frac{\partial v_j}{\partial x_j} \xi \frac{\partial \xi}{\partial x_i} \, dx + \int_\Omega v_i \frac{\partial v_j}{\partial x_i} \xi \frac{\partial \xi}{\partial x_j} \, dx \\
&\quad \leq \int_\Omega |\mathbf{D}(\mathbf{v})|^2 \xi^2 \, dx + \frac{1}{4} \int_\Omega |\operatorname{div} \mathbf{v}|^2 \xi^2 \, dx + \frac{1}{4} \int_\Omega |\nabla \mathbf{v}|^2 \xi^2 \, dx + c \int_\Omega |\mathbf{v}|^2 |\nabla \xi|^2 \, dx,
\end{aligned}$$

and the assertion follows.

Theorem 6.7 (On negative norms). *Let $1 < p < \infty$ and let $\mathbf{v} \in W_0^{1,p}(\Omega)^d$. Then there exists a constant such that*

$$c\|\mathbf{v}\|_p \leq \|\mathbf{v}\|_{-1,p} + \|\nabla \mathbf{v}\|_{-1,p} \,. \tag{6.8}$$

Proof. See e.g. [12].

In the rest of this section we assume that the potential Φ satisfies (1.9)–(1.11). Then, by Lemma 2.22, the potential Φ_A satisfies (2.23)–(2.26).

We denote $I_{p,q}(\mathbf{v}) \equiv \int_\Omega (1 + |\mathbf{D}(\mathbf{v})|)^{q(p-2)} |\mathbf{D}(\nabla \mathbf{v})|^q \, dx$.

Lemma 6.9. *Let $p \geq 2$ and $q \in [1,\infty)$. Then there exist constants C_{17}, C_{18} and C_{19} such that*

$$C_{17} I_{p,q}(\mathbf{v}) \leq \int_\Omega |\nabla \partial \Phi(|\mathbf{D}(\mathbf{v})|^2)|^q \, dx \leq C_{18} I_{p,q}(\mathbf{v}) \tag{6.10}$$

$$\|\nabla \mathbf{v}\|^{q(p-1)}_{(p-1)\frac{3q}{3-q}} \leq C_{19}\big(1 + \|\nabla \mathbf{v}\|_p^{q(p-1)} + I_{p,q}(\mathbf{v})\big)\,. \tag{6.11}$$

Proof. We have

$$\begin{aligned} I_{p,q}(\mathbf{v}) &= \int_\Omega \big[(1+|\mathbf{D}(\mathbf{v})|)^{p-2}|\mathbf{D}(\nabla \mathbf{v})|^2\big]^{\frac{q}{2}} (1+|\mathbf{D}(\mathbf{v})|)^{(p-2)\frac{q}{2}} \, dx \\ &\overset{(1.10)}{\leq} \int_\Omega \big[\partial_{ij}\partial_{kl}\Phi(|\mathbf{D}(\mathbf{v})|^2) D_{ij}(\nabla \mathbf{v}) \cdot D_{kl}(\nabla \mathbf{v})\big]^{\frac{q}{2}} (1+|\mathbf{D}(\mathbf{v})|)^{(p-2)\frac{q}{2}} \, dx \\ &\leq \int_\Omega |\nabla \partial \Phi(|\mathbf{D}(\mathbf{v})|^2)|^{\frac{q}{2}} |\mathbf{D}(\nabla \mathbf{v})|^{\frac{q}{2}} (1+|\mathbf{D}(\mathbf{v})|)^{(p-2)\frac{q}{2}} \, dx \\ &\overset{\text{Young}}{\leq} \frac{1}{2} I_{p,q}(\mathbf{v}) + c \int_\Omega |\nabla \partial \Phi(|\mathbf{D}(\mathbf{v})|^2)|^q \, dx\,, \end{aligned}$$

which immediately gives the first inequality in (6.10). The second one follows easily from the chain rule and (1.11). Further, we have

$$\begin{aligned} I_{p,q}(\mathbf{v}) &\geq \int_\Omega |\nabla(1+|\mathbf{D}|)|^q (1+|\mathbf{D}|)^{q(p-2)} \, dx = \frac{1}{(p-1)^q} \int_\Omega |\nabla(1+|\mathbf{D}|)^{p-1}|^q \, dx \\ &\geq c\Big(\int_\Omega (1+|\mathbf{D}|)^{(p-1)\frac{3q}{3-q}} \, dx\Big)^{\frac{3-q}{3}} - \|(1+|\mathbf{D}|)^{p-1}\|^q_{p'} \\ &\overset{(6.2)}{\geq} c\|\nabla \mathbf{v}\|^{q(p-1)}_{(p-1)\frac{3q}{3-q}} - c\|\nabla \mathbf{v}\|_p^{q(p-1)} - c\,, \end{aligned}$$

which is (6.11).

Lemma 6.12. *Let $p \geq 2$ and $q \in [1,\infty)$ and let χ_A be the characteristic function of the set $\{x \in \Omega : |\mathbf{D}(\mathbf{v}(x))| \leq A\}$. Then there exist constants C_{20}, C_{21}, C_{22} such that*

$$\begin{aligned} C_{20} \int_\Omega &\big((1+|\mathbf{D}(\mathbf{v})|)^{q(p-2)})\chi_A + (1+A)^{q(p-2)}(1-\chi_A)\big)|\mathbf{D}(\nabla \mathbf{v})|^q \, dx \\ &\leq \int_\Omega |\nabla \partial \Phi_A(|\mathbf{D}(\mathbf{v})|^2)|^q \, dx \qquad (6.13) \\ \leq C_{21} \int_\Omega &\big((1+|\mathbf{D}(\mathbf{v})|)^{q(p-2)}\chi_A + (1+A)^{q(p-2)}(1-\chi_A)\big)|\mathbf{D}(\nabla \mathbf{v})|^q \, dx, \end{aligned}$$

and therefore

$$\int_\Omega |\mathbf{D}(\nabla \mathbf{v})|^q \, dx \leq C_{22} \int_\Omega |\nabla \partial \Phi_A(|\mathbf{D}(\mathbf{v})|^2)|^q \, dx\,. \tag{6.14}$$

Proof. We proceed analogously to the proof of Lemma 6.9 just using (2.24) instead of (1.10) and (2.26) instead of (1.11).

Acknowledgment. The authors have been partially supported by grants of SFB256 at Institute of Applied Mathematics of the University of Bonn. J. Nečas and J. Málek were partially supported by the grant 201/96/0228 of the Grant Agency of the Czech Republic, and by CEZ:J13/98113200007. M. Růžička was partially supported by fellowships from DFG and CNR. Finally the authors thank the referees for valuable remarks and suggestions.

REFERENCES

[1] H. Amann, *Stability of the rest state of a viscous incompressible fluid*, Arch. Rat. Mech. Anal. Vol. 126 (1994), 231–242.

[2] H. Bellout, F. Bloom, and J. Nečas, *Solutions for incompressible non–Newtonian fluids*, C. R. Acad. Sci. Paris Vol. 317 Série I (1993), 795–800.

[3] H. Bellout, F. Bloom, and J. Nečas, *Young measure-valued solutions for non-Newtonian incompressible fluids*, Comm. in PDE Vol. 19 (11 & 12) (1994), 1763–1803.

[4] A. Kufner, O. John, and S. Fučík, "Function Spaces", Academia, Praha, 1977.

[5] O.A. Ladyzhenskaya, *On some new equations describing dynamics of incompressible fluids and on global solvability of boundary value problems to these equations*, Trudy Steklov's Math. Institute Vol. 102 (1967), 85–104.

[6] O.A. Ladyzhenskaya, *On some modifications of the Navier-Stokes equations for large gradients of velocity*, Zapiski Naukhnych Seminarov LOMI Vol. 7 (1968), 126–154.

[7] O.A. Ladyzhenskaya, "The Mathematical Theory of Viscous Incompressible Flow", Gordon and Breach, New York, 1969.

[8] J.L. Lions, "Quelques Méthodes de Résolution des Problèmes aux Limites Non Linéaires", Dunod, Paris, 1969.

[9] J. Málek, J. Nečas, and M. Růžička, *On the non–Newtonian incompressible fluids*, Math. Models Methods Appl. Sci. Vol. 3 (1993), 35–63.

[10] J. Málek, J. Nečas, and M. Růžička, *On weak solutions to a class of non-Newtonian incompressible fluids in bounded bomains. The case* $p \geq 2$., Preprint no. 481 SFB 256 of the University of Bonn, October 1996.

[11] J. Málek, J. Nečas, M. Rokyta, and M. Růžička, "Weak and Measure–Valued Solutions to Evolutionary Partial Differential Equations", Applied Mathematics and Mathematical Computation, vol. 13, Chapman and Hall, 1996.

[12] J. Málek, and D. Pražák, *Finite fractal dimension of the global attractor for a class of non-Newtonian fluids*, (to appear), Applied Math. Letters **13** (2000), 105–110.

[13] J. Málek, K.R. Rajagopal, and M. Růžička, *Existence and regularity of solutions and stability of the rest state for fluids with shear dependent viscosity*, Math. Models Methods Appl. Sci. Vol. 5 (6) (1995), 789–812.

[14] J. Nečas, *Sur le normes équivalentes dans* $W_p^k(\Omega)$ *et sur la coercivité des formes formellement positives*, in Séminaire Equations aux Dérivées Partielles, Montreal (1966), 102–128.

Journal of Differential Equations – DE3057

JOURNAL OF DIFFERENTIAL EQUATIONS **127**, 498–518 (1996)
ARTICLE NO. 0080

A Finite-Dimensional Attractor for Three-Dimensional Flow of Incompressible Fluids

JOSEF MÁLEK* AND JINDŘICH NEČAS

Mathematical Institute of Charles University, Sokolovská 83, 18600 Prague 8, Czech Republic

Received September 20, 1993; revised January 17, 1995

The asymptotic behavior of the flow for a system of the Navier–Stokes type is investigated. In the considered model, the viscous part of the stress tensor is generally a nonlinear function of the symmetric part of the velocity gradient. Provided that the function describing this dependence satisfies the polynomial $(p-1)$ growth condition, a unique weak solution exists if either $p \geqslant (2+n)/2$ and $\mathbf{u}_0 \in H$ or $p \geqslant 1+2n/(n+2)$ and $\mathbf{u}_0 \in W^{1,2}(C)^n \cap H$. In the first case, the existence of a global attractor in H is proved. In order to indicate the finite-dimensional behavior of the flow at infinity, the fractal dimension of the new invariant set, composed from all short δ-trajectories with initial value in the attractor, is estimated in the $L^2(0, \delta; H)$ topology. Having uniqueness only for more regular data in the second case, many trajectories can start from the initial value $\mathbf{u}_0 \in H$. This does not allow one to define a semigroup on the space H. Therefore, the set of short trajectories X^s_δ, closed in $L^2(0, \delta; H)$, is introduced along with a semigroup working on this set. The existence of a global attractor with a finite fractal dimension is then demonstrated. © 1996 Academic Press, Inc.

1. INTRODUCTION AND PROBLEM FORMULATION

Let C be a bounded set of $\mathbb{R}^n$, $n=2$ or 3, with a sufficiently smooth boundary ∂C. For $T>0$, we will denote by Q_T the time cylinder $C \times I$, where $I \equiv (0, T)$ is a time interval.

The motion of incompressible, viscous fluids in C, characterized by the velocity field $\mathbf{u} = (u_1, u_2, ..., u_n)$ and the pressure π, is governed by the system of $n+1$ equations

$$\operatorname{div} \mathbf{u} = 0, \tag{1.1}$$

$$\rho_0 \frac{\partial u_i}{\partial t} + \rho_0 u_j \frac{\partial u_i}{\partial x_j} = -\frac{\partial \pi}{\partial x_i} + \frac{\partial \tau^V}{\partial x_j} + f_i, \qquad i = 1, 2, ..., n, \tag{1.2}$$

* E-mail: malek@karlin.mff.cuni.cz.

0022-0396/96 $18.00

Copyright © 1996 by Academic Press, Inc.
All rights of reproduction in any form reserved.

© Elsevier 1996. Reprinted with kind permission from: Málek, J. & Nečas, J., A Finite-Dimensional Attractor for Three-Dimensional Flow of Incompressible Fluids, the Journal of Differential Equations, 127, 498-518.

expressing the balance of mass and balance of momentum. Here $\mathbf{f} = (f_1, f_2, ..., f_n)$ is the field of body forces and τ^{V} is the viscous part of the stress tensor for which the constitutive relation has to be prescribed. If the linear Stokes law

$$\tau^{\mathrm{V}} = 2\mu\mathbf{e} \tag{1.3}$$

is used then the system (1.1)–(1.2) turns to the Navier–Stokes one. As usual, $\mathbf{e}$ denotes the symmetric part of the velocity gradient with components

$$e_{ij} = e_{ij}(\mathbf{u}) = \frac{1}{2}\left(\frac{\partial u_i}{\partial x_j} + \frac{\partial u_j}{\partial x_i}\right)$$

and μ is a positive viscosity coefficient. All quantities are evaluated at $(\mathbf{x}, t)$, where $\mathbf{x}$ states the actual position and t represents time. Hereafter, we put a constant density ρ_0 equal to 1.

Let us supplement the equations (1.1) and (1.2) with the initial condition

$$\mathbf{u}(\mathbf{x}, 0) = \mathbf{u}_0(\mathbf{x}) \qquad \forall \mathbf{x} \in C \tag{1.4}$$

and let us investigate the Dirichlet problem that reads as

$$\mathbf{u}(\mathbf{x}, t) = \mathbf{0} \qquad \forall \mathbf{x} \in \partial C \times I. \tag{1.5}$$

It is known that the uniqueness of a (weak) solution of the three-dimensional Navier–Stokes equations is a challenging open problem. It consequently is not possible to prove the existence of an absorbing set for this dissipative system and this motivates us to consider a nonlinear perturbation of the Stokes law (1.3) of the type

$$\tau^{\mathrm{V}} = 2(\mu_0 + \mu_1\,|\mathbf{e}|^r)\,\mathbf{e} \tag{1.6}$$

with μ_0, μ_1 being positive constants and r being a nonnegative parameter characterizing the quality of the perturbation. We can then ask for what r the system (1.1)–(1.2), (1.4)–(1.6) is well posed and possesses an absorbing set, a finite-dimensional global attractor. Since a function $\vartheta_1 : \mathbb{R}^{n^2} \mapsto \mathbb{R}$ given by

$$\vartheta_1(\mathbf{e}) = \int_0^{|\mathbf{e}|^2} (\mu_0 + \mu_1 s^{r/2})\, ds \tag{1.7}$$

is a scalar potential to the tensor τ^{V} from (1.6), i.e., $\tau^{\mathrm{V}} = \partial\vartheta_1/\partial e_{ij}$ for all $i, j = 1, 2, ..., n$, we include this system in a model which is characterized by a more general tensor function τ of $\mathbf{e}$ represented through a nonnegative potential ϑ, that is

$$\tau^{\mathrm{V}} = \tau(\mathbf{e}) \qquad \text{and} \qquad \tau_{ij}(\mathbf{e}) = \frac{\partial \vartheta(\mathbf{e})}{\partial e_{ij}} \quad \forall i, j = 1, 2, ..., n. \tag{1.8}$$

We assume that

$$\vartheta(\mathbf{o}) = 0, \qquad \frac{\partial \vartheta(\mathbf{o})}{\partial e_{ij}} = 0 \quad \forall i, j = 1, 2, ..., n,$$

$$c_1(1 + |\mathbf{e}|)^{p-2} |\xi|^2 \leqslant \frac{\partial^2 \vartheta(\mathbf{e})}{\partial e_{kl}\, \partial e_{ij}} \xi_k^l \xi_i^j$$
$$\leqslant c_2(1 + |\mathbf{e}|)^{p-2} |\xi|^2 \quad \forall \xi \in \mathbb{R}^{n^2}, \tag{1.9}$$

for some $p > 1$. Here, c_i, $i = 1, 2$ are positive constants. Notice that $p = 2$ for the Stokes law (1.3) and $p = r + 2$ for the relation (1.6).

We wish to emphasize that the constitutive law (1.8) covers many models used in chemical and process engineering to explain phenomena like shear thinning or shear thickening which appear in the flow of non-Newtonian fluids.

1.10. Definition. Let $\mathbf{u}_0$, $\mathbf{f}$ be given. Let $p > 1$. The problem $(\mathrm{NS})_p$ denotes the initial boundary value problems (1.1)–(1.2), (1.4)–(1.5), (1.8)–(1.9).

The objective of this paper is to prove the existence of an attracting set with finite fractal dimension for the problem $(\mathrm{NS})_p$ with minimal p starting from the recent existence, uniqueness and regularity results proved in [1] and [13] that essentially extend previous ones proved in [6] or [9].

The paper is organized as follows: In Section 2, we give an overview of the existence and uniqueness results and sketch proofs of some of them.

Section 3 is devoted to the asymptotic study of solutions of $(\mathrm{NS})_p$ if $p \geqslant (n+2)/2$. In this case the existence of a unique weak solution $\mathbf{u} \in C(I; H)$ requires the initial value $\mathbf{u}_0$ to belong into the space H only. We then define a family of operators $S_t: H \mapsto H$,

$$S_t : \mathbf{u}_0 \mapsto \mathbf{u}(t) \qquad \forall t \geqslant 0,$$

that has all the semigroup properties and the general theory giving the existence of the global attractor $\mathscr{A}$ is applied. Let us recall that the global attractor $\mathscr{A}$ is defined as a nonempty compact subset of H which is invariant with respect to S_t and attracts all bounded sets of H. In order to indicate finite-dimensional behavior of the flow at infinity we need to estimate the fractal dimension of the attractor $\mathscr{A}$.

Usually, the theory of Lyapunov exponents based on regularity of weak solutions for a linearized system is applied (for instance see [2–4, 16]). In our case, however, this required regularity of the solution is not known. Consequently, we are not able to control the quantity

$$\|S_t \mathbf{u}_0 - S_t \mathbf{v}_0 - S_t'(\mathbf{v}_0)(\mathbf{u}_0 - \mathbf{v}_0)\|_2$$

where $S_t'(\mathbf{v}_0)(\mathbf{u}_0 - \mathbf{v}_0)$ denotes the solution of the linearized system to (1.1)–(1.2) with initial value $\mathbf{u}_0 - \mathbf{v}_0$.[1]

The other method of estimating the dimension of invariant sets was developed in [7] and it is based on the theorem which says that an invariant set $\mathscr{B}$ has a finite fractal (Hausdorff) dimension if for all $\mathbf{u}_0$, $\mathbf{v}_0 \in \mathscr{A}$ there exist positive constants L and $\delta < 1$ such that

$$\|S_t \mathbf{u}_0 - S_t \mathbf{v}_0\|_2 \leqslant L \, \|\mathbf{u}_0 - \mathbf{v}_0\|_2, \tag{i}$$

$$\|Q_N S_t \mathbf{u}_0 - Q_N S_t \mathbf{v}_0\|_2 \leqslant \delta \|\mathbf{u}_0 - \mathbf{v}_0\|_2. \tag{ii}$$

By Q_N we denote $I - P_N$, where I is the identity map and P_N denotes an orthogonal projection to the space generated by the first N eigenvectors of the Stokes operator. In contrast to [11], where we applied this criterion on the system where the leading dissipative part is linear, we are not able to verify condition (ii) above for the elements of attractor $\mathscr{A}$ associated to the problem $(\mathrm{NS})_p$. These are the reasons why we proceed here in a different way.

First we add to the attractor $\mathscr{A}$ the set $\mathscr{A}_\delta$ composed from all short trajectories having the length δ and starting in $\mathscr{A}$ and then, we estimate the fractal dimension of this new invariant set $\mathscr{A}_\delta$ in topology $L^2(0, \delta; H)$. Let us note that due to uniqueness, the elements of $\mathscr{A}$ are in one–to–one correspondence to the elements of $\mathscr{A}_\delta$.

In Section 4, we concentrate on the investigation of the asymptotic behavior of solutions to $(\mathrm{NS})_p$ if $p \geqslant 1 + (2n/(n+2))$. Here we can dispose again by a unique weak solution, but higher regularity of the initial values is needed. Thus, a semigroup on the space H cannot be defined. Therefore, the set of short trajectories X_δ, closed in $L^2(0, \delta; H)$, along with a semigroup L_t working on this set is introduced. Using new regularity results, the existence of a global attractor with finite fractal dimension is then demonstrated.

As can be easily checked, the lower bound of p for both cases studied separately in Sections 3 and 4 coincides for $n = 2$ and is equal to 2, which corresponds to the two-dimensional Navier–Stokes equations. For this system the existence of a global attractor with finite fractal dimension is

[1] See Lemma 14.3 in [4] for comparison and detailed explanation of the notation.

well–known (see [7, 8, 3]). Nevertheless, even in such a case the presented results are valuable since our approach admits also nonlinear relations between stress τ^V and tensor $\mathbf{e}$.[2]

Before ending this section some notation of function spaces and helpful assertions are recalled.

Let $q > 1$ and q' be the dual index to q, i.e. $1/q + 1/q' = 1$. The Lebesgue spaces $L^q(C)$, $L^q(C)^n$ are the sets of scalar and $\mathbb{R}^n$-valued measurable functions defined on C or on Q_T with norm $\|\mathbf{u}(t)\|_q = (\int_C |\mathbf{u}(\mathbf{x}, t)|^q \, dx)^{1/q}$. Mostly, the dependence of the norm on time will not be explicitly noted. The norm of Sobolev spaces $W^{m,q}(C)^n$ is denoted by $\|\cdot\|_{m,q}$. The set $W_0^{1,q}(C)^n$ represents a subspace of functions of $W^{1,q}(C)^n$ with zero traces. Then the L^q-norm of the gradient of φ, denoted by $|\varphi|_{1,q}$, is the equivalent norm on $W_0^{1,q}(C)^n$.

The following Korn inequality proved in [15] will be often used: let $\varphi \in W_0^{1,q}(C)^n$. Then

$$\left(\int_C |\mathbf{e}(\varphi)|^q \, dx\right)^{1/q} \geqslant K_q \, |\varphi|_{1,q}, \qquad K_q > 0. \tag{1.11}$$

Let us denote

$$\mathscr{V} = \{\phi \in C^\infty(C)^n; \operatorname{div} \phi = 0\}; \tag{1.12}$$

$$H = \text{closure of } \mathscr{V} \text{ in the } L^2(C)^n\text{-norm}; \tag{1.13}$$

$$V_q = \text{closure of } \mathscr{V} \text{ in the } W^{1,q}(C)^n\text{-norm}. \tag{1.14}$$

The dual space of V_q is denoted by V_q^* and $\langle \cdot, \cdot \rangle$ denotes the duality between V_q and V_q^*. The scalar product in H is marked by $(\cdot, \cdot)$.

Let X be a Banach space with the norm $\|\cdot\|_X$. The space $C(I; X)$ consists of all continuous functions $\mathbf{u}\colon \bar{I} \equiv [0, T] \mapsto X$. The Bochner spaces $L^q(I; X)$ consist of all measurable functions $\mathbf{u}\colon I \mapsto X$ the norm of which $\|u\|^q_{L^q(I;X)} := \int_0^T \|u(t)\|^q_X \, dt$ is finite. For $q = \infty$, the norm is $\|u\|_{L^\infty(I;X)} \equiv \operatorname{ess\,sup}_{t \in I} \|u(t)\|_X$.

Due to the continuous imbeddings $V_q \hookrightarrow H \hookrightarrow V_q^*$ it holds that

$$\left\{\mathbf{u} \in L^q(I; V_q); \frac{d\mathbf{u}}{dt} \in L^{q'}(I; V_q^*)\right\} \hookrightarrow C(I; H). \tag{1.15}$$

For the proof, see [5, Theorem 1.17, Chap. 4]. In (1.15) $d\mathbf{u}/dt$ denotes the generalized derivative of the function $\mathbf{u}$ on I.

[2] Nonlinear constitutive relations, especially power laws, also play an important role in the dynamics of in the Earth's mantle. In such case the Boussinesq approximation is used to govern the motions in planetary body. The asymptotic study of solutions is a subject of the paper [14].

1.16. LEMMA (Aubin–Lions). *Let X be a Banach space, and X_0, X_1 separable, reflexive Banach spaces. Provided that $X_0 \hookrightarrow\hookrightarrow X \hookrightarrow X_1$, then*

$$\left\{v \in L^\alpha(I; X_0) \text{ with } \frac{dv}{dt} \in L^\beta(I; X_1)\right\} \hookrightarrow\hookrightarrow L^\alpha(I; X),$$

where $1 < \alpha, \beta < +\infty$.

Proof. See for instance [9]. ∎

2. OVERVIEW OF THE EXISTENCE AND UNIQUENESS RESULTS

Let

$$p \geqslant 1 + \frac{2n}{n+2} \qquad \text{with} \quad p' = \frac{p}{p-1} \tag{2.1}$$

and let

$$\mathbf{u}_0 \in H \qquad \text{and} \qquad \mathbf{f} \in L^{p'}(I; V_p^*). \tag{2.2}$$

2.3. DEFINITION. Let (2.1)–(2.2) be satisfied. The function

$$\mathbf{u} \in C(I; H) \cap L^p(I; V_p) \tag{2.4}$$

with

$$\frac{d\mathbf{u}}{dt} \in L^{p'}(I; V_p^*) \tag{2.5}$$

is called the weak solution of the problem $(\mathrm{NS})_p$ if

$$\left\langle \frac{d\mathbf{u}(t)}{dt}, \varphi \right\rangle + \int_C u_j(t) \frac{\partial u_i(t)}{\partial x_j} \varphi_i \, dx + \int_C \tau_{ij}(\mathbf{e}(\mathbf{u}(t)))\, e_{ij}(\varphi)\, dx = \langle \mathbf{f}(t), \varphi \rangle \tag{2.6}$$

is fulfilled almost everywhere in I and for every $\varphi \in V_p$.

2.7. DEFINITION. Let us define

(i) the operator T by

$$\int_0^T \langle T(\mathbf{u})(t), \mathbf{v}(t) \rangle \, dt \equiv \int_0^T \int_C \tau_{ij}(\mathbf{e}(\mathbf{u}(\mathbf{x}, t)))\, e_{ij}(\mathbf{v}(\mathbf{x}, t))\, dx\, dt;$$

(ii) the operator B by

$$\int_0^T \langle B(\mathbf{u})(t), \mathbf{v}(t)\rangle \, dt \equiv \int_0^T \int_C u_j(\mathbf{x}, t) \frac{\partial u_i(\mathbf{x}, t)}{\partial x_j} v_i(\mathbf{x}, t) \, dx \, dt.$$

2.8. LEMMA. (i) *Let* $p > 1$. *Then*

$$T: L^p(I; V_p) \mapsto L^{p'}(I; V_p^*) \quad \textit{continuously}.$$

(ii) *Let* $p \geqslant 1 + 2n/(n+2)$. *Then*

$$B: L^p(I; V_p) \cap L^\infty(I; H) \mapsto L^{p'}(I; V_p^*) \quad \textit{continuously}.$$

Proof. (i). It follows from (1.9)

$$|\tau_{ij}(\mathbf{e})| \leqslant c_3(1 + |\mathbf{e}|)^{p-1}, \qquad c_3 > 0. \tag{2.9}$$

This growth condition along with the Hölder inequality finishes the proof of (i).

In order to prove (ii), let us start with the interpolation inequality[3]

$$\|\mathbf{u}\|_\sigma \leqslant c \, \|\mathbf{u}\|_2^{1-\alpha} \, \|\mathbf{u}\|_{np/(n-p)}^\alpha \qquad \text{with} \quad \alpha = \frac{(\sigma - 2)\, np}{\sigma(np - 2(n-p))}, \tag{2.10}$$

provided that $2 \leqslant \sigma \leqslant np/(n-p)$. We verify now that if $\mathbf{u} \in L^p(I; V_p) \cap L^\infty(I; H)$ then

$$\mathbf{u} \in L^\sigma(Q_T) \qquad \text{with} \quad \sigma = \frac{2+n}{n} p. \tag{2.11}$$

Indeed,

$$\begin{aligned} \int_0^T \|\mathbf{u}\|_\sigma^\sigma \, dt &\leqslant c \int_0^T \|\mathbf{u}\|_2^{(1-\alpha)\sigma} \, \|\mathbf{u}\|_{np/(n-p)}^{\alpha\sigma} \, dt \\ &\leqslant \tilde{c} \int_0^T \|\mathbf{u}\|_{1,p}^{p(\sigma-2)\, n/(np - 2(n-p))} \, dt. \end{aligned} \tag{2.12}$$

The right side of (2.12) is finite provided that

$$\frac{(\sigma - 2)\, n}{np - 2(n-p)} = 1.$$

Thus (2.11) is proved.

[3] This inequality is valid if $p < n$ due to imbedding $W^{1,p}(C)$ into $L^{np/(n-p)}(C)$. The case for $p \geqslant n$ is easier.

Now

$$\int_0^T \left| \int_C \left(u_j \frac{\partial u_i}{\partial x_j} v_i \right) (\mathbf{x}, t)\, dx \right| dt \leqslant \|\nabla \mathbf{v}\|_{L^p(Q_T)} \|\mathbf{u}\|^2_{L^{2p'}(Q_T)}. \tag{2.13}$$

The right side of (2.13) is bounded if $2p/(p-1) \leqslant (2+n)\, p/n$ or in any case if $p \geqslant 1 + 2n/(n+2)$. The proof of the lemma is complete. ▮

The previous lemma allows us to rewrite the weak formulation in the operator equation

$$\frac{d\mathbf{u}}{dt} + T(\mathbf{u}) + B(\mathbf{u}) = \mathbf{f}$$

with

$$\frac{d\mathbf{u}}{dt} \in L^{p'}(I;\, V_p^*).$$

Due to (1.15) the initial condition (1.4) is meaningful under assumption (2.2).

2.14. THEOREM. *Let the assumptions* (2.1), (2.2) *be satisfied. Then*

(i) *there exists a weak solution of the problem* $(\mathrm{NS})_p$ *and*

$$\mathbf{u} \in L^2_{\mathrm{loc}}(I;\, W^{2,2}(C)^3) \cap L^\infty_{\mathrm{loc}}(I;\, W^{1,p}(C)^3); \tag{2.15}$$

$$\mathbf{u} \in L^p_{\mathrm{loc}}(I;\, W^{1,3p}(C)^3), \qquad \text{if} \quad n = 3; \tag{2.16}$$

$$\frac{d\mathbf{u}}{dt} \in L^2_{\mathrm{loc}}(I;\, H). \tag{2.17}$$

Moreover, if $\mathbf{u}_0 \in W^{1,p}(C)^n \cap H$ *then the regularity results* (2.15)–(2.17) *hold globally on* I *and the weak solution is unique.*

(ii) *If* $\mathbf{u}_0 \in H$ *and* $p \geqslant (n+2)/2$ *then the weak solution is also unique.*

Proof. The proof of the existence is omitted. It can be found in [6, 9] or together with regularity results (2.15)–(2.17) in [1 or 13] for the space periodic problem or in [10] for the Dirichlet problem. We only sketch the proof of uniqueness if $n = 3$.

First, let $\mathbf{u}_0 \in W^{1,2}(C)^n \cap H$ and p satisfy (2.1). Let us suppose the existence of two weak solutions $\mathbf{u}$, $\mathbf{v}$ to the problem $(\mathrm{NS})_p$ with the same initial value $\mathbf{u}_0$. Put $\mathbf{w} \equiv \mathbf{u} - \mathbf{v}$. Subtracting the weak formulation for $\mathbf{v}$ from one for $\mathbf{u}$ and taking $\mathbf{w}$ as the test function in the resulting equation, we get

$$\frac{1}{2}\frac{d}{dt}\|\mathbf{w}\|_2^2 + \int_C [\tau_{ij}(\mathbf{e}(\mathbf{u})) - \tau_{ij}(\mathbf{e}(\mathbf{v}))]\, e_{ij}(\mathbf{w})\, dx = -\int_C w_j \frac{\partial u_i}{\partial x_j} w_i\, dx. \tag{2.18}$$

From (1.9)

$$\int_C [\tau_{ij}(\mathbf{e}(\mathbf{u})) - \tau_{ij}(\mathbf{e}(\mathbf{v}))]\, e_{ij}(\mathbf{w})\, dx \geqslant c_1 \sum_{i,j=1}^{3} \int_C |e_{ij}(\mathbf{w})|^2\, dx, \qquad c_1 > 0. \tag{2.19}$$

Using (1.11) and the Hölder inequality on the convective term, we obtain from (2.18)

$$\frac{1}{2}\frac{d}{dt}\|\mathbf{w}\|_2^2 + \hat{c}\,|\mathbf{w}|_{1,2}^2 \leqslant |\mathbf{u}|_{1,3p}\,\|\mathbf{w}\|_{6p/(3p-1)}^2, \qquad \hat{c} = c_1 K_2^2. \tag{2.20}$$

The interpolation inequality $\|\mathbf{w}\|_{6p/(3p-1)} \leqslant d\,\|\mathbf{w}\|_2^{(2p-1)/2p}\,|\mathbf{w}|_{1,2}^{1/2p}$ and the Young inequality then imply

$$\frac{d}{dt}\|\mathbf{w}\|_2^2 + \hat{c}\,|\mathbf{w}|_{1,2}^2 \leqslant c_4\,|\mathbf{u}|_{1,3p}^{2p/(2p-1)}\,\|\mathbf{w}\|_2^2, \qquad c_4 = \frac{2p-1}{p}\left(\frac{d^2}{(\hat{c}p)^{1/(2p)}}\right)^{2p/(2p-1)}. \tag{2.21}$$

Since the function $g(t) \equiv |\mathbf{u}|_{1,3p}^{2p/(2p-1)}$ is integrable on I thanks to (2.16) and $\mathbf{w}(0) = \mathbf{0}$, the classical Gronwall lemma gives

$$\|\mathbf{w}(t)\|_2^2 = 0 \qquad \text{for all} \quad t \geqslant 0.$$

(ii). Now $p \geqslant (n+2)/2$ and $\mathbf{u}_0 \in H$. As in the previous proof we have

$$\frac{1}{2}\frac{d}{dt}\|\mathbf{w}\|_2^2 + \hat{c}\,|\mathbf{w}|_{1,2}^2 \leqslant \int_C |\mathbf{w}|^2\,|\nabla \mathbf{u}|\, dx \leqslant |\mathbf{u}|_{1,p}\,\|\mathbf{w}\|_{2p'}^2. \tag{2.22}$$

Using the fact that $\|\mathbf{w}\|_{2p'} \leqslant d\,\|\mathbf{w}\|_2^{(2p-3)/2p}\,|\mathbf{w}|_{1,2}^{3/2p}$ and the Young inequality, we obtain

$$\frac{d}{dt}\|\mathbf{w}\|_2^2 + \hat{c}\,|\mathbf{w}|_{1,2}^2 \leqslant c_5\,|\mathbf{u}|_{1,p}^{2p/(2p-3)}\,\|\mathbf{w}\|_2^2 \tag{2.23}$$

with c_5 given by the similar relation as c_4 in (2.21). All assumptions of the Gronwall lemma are again satisfied, namely, the function

$$g(t) = |\mathbf{u}(t)|_{1,p}^{2p/(2p-3)} \text{ is integrable on } (0, T) \qquad \text{if and only if } p \geqslant \tfrac{5}{2}.$$

Thus, the proofs of uniqueness are complete. ▮

2.24. *Remark.* In Theorem 2.14 we state results that are needed in the following sections. Let us remark, however, that weak solutions of the problem $(\mathrm{NS})_p$ exist already for $p > 3n/(n+2)$. This is proved in [1, 12, 13]. If $p > 2n/(n+2)$ (and less than $3n/(n+2)$) then the existence of measure-valued solutions can be proved. See for instance [11] or [13].

3. Asymptotic Study of Solutions if $p \geqslant (n+2)/2$

Let $p \geqslant (n+2)/2$, $\mathbf{u}_0 \in H$ and $\mathbf{f} \in L^2(C)^n$. According to the part (ii) of Theorem 2.14, just one weak solution $\mathbf{u} \in C(I; H)$ of the problem $(\mathrm{NS})_p$ exists. For such p, we can define a family of operators $(S_t)_{t \geqslant 0}$: $H \mapsto H$ by

$$S_t \colon \mathbf{u}_0 \mapsto \mathbf{u}(t). \tag{3.1}$$

3.2. Lemma. (i) *The operators $(S_t)_{t \geqslant 0}$ form a semigroup, i.e.,*

$$S_{t+s}\mathbf{u}_0 = S_t(S_s\mathbf{u}_0) \qquad \forall \mathbf{u}_0 \in H\ \forall s, t > 0$$
$$S_0 = I \text{ the identity map}$$
$$\lim_{t \to 0+} S_t\mathbf{u}_0 = \mathbf{u}_0.$$

(ii) *$S_t \colon H \mapsto H$ is a continuous mapping for all $t \geqslant 0$.*

Proof. (i) is a consequence of the uniqueness and (2.4). (ii) follows from (2.23) after using the Gronwall lemma. ∎

3.3. Definition. A subset $\mathscr{A} \subset H$ is called a global attractor for the semigroup $(S_t)_{t \geqslant 0}$ if the following properties are satisfied:

(α) $S_t\mathscr{A} = \mathscr{A} \quad \forall t \geqslant 0$;

(β) $\mathscr{A}$ is the compact set;

(γ) $\operatorname{dist}(S_t\mathscr{B}_0, \mathscr{A}) \xrightarrow[t \to \infty]{} 0$ for every bounded set $\mathscr{B}_0 \subset H$.

3.4. *Remark.* It is known that such an attractor is an unique, connected set which is maximal among all invariant sets. Recall that $\operatorname{dist}(\mathscr{A}, \mathscr{B}) \equiv \sup_{\mathbf{u} \in \mathscr{A}} \inf_{\mathbf{v} \in \mathscr{B}}(\|\mathbf{u} - \mathbf{v}\|_2)$. For details see [16].

3.5. Theorem. *Let B_ρ be a ball in a topology which is compactly embedded in the topology of H and let B_ρ be an absorbing set in H; i.e., for every bounded $\mathscr{B} \subset H$ there exists $t_0 = t_0(\mathscr{B})$ such that $S_t\mathscr{B} \subset B_\rho$ for $t \geqslant t_0$. Then*

$$\mathscr{A} = \bigcap_{t \geqslant 0} \bigcup_{s \geqslant t} \overline{S_s B_\rho}$$

is the global attractor.

Proof. See for instance [16, p. 23]. ∎

3.6. THEOREM. *Let $p \geqslant (n+2)/2$. Then there exists a ball*

$$B_\rho = \{\mathbf{u} \in V_p;\ |\mathbf{u}|_{1,p} \leqslant \rho\}$$

which absorbs all bounded sets in H.

Proof. Let $n=3$. It follows from (1.9) that

$$\tau_{ij} e_{ij} \geqslant c_1 (1 + |\mathbf{e}|^{p-2}) |\mathbf{e}|^2.$$

From the first energy estimate, we have

$$\frac{d}{dt} \|\mathbf{u}\|_2^2 + c_1 K_2^2 |\mathbf{u}|_{1,2}^2 + 2c_1 K_p^p |\mathbf{u}|_{1,p}^p \leqslant \frac{\|\mathbf{f}\|_2^2}{c_1 \lambda_1 K_2^2}. \tag{3.7}$$

Since $\lambda_1 \|\mathbf{u}\|_2^2 \leqslant |\mathbf{u}|_{1,2}^2$, where λ_1 is the first eigenvalue of the Stokes operator, one obtains

$$\frac{d}{dt} (\|\mathbf{u}(s)\|_2^2 e^{\lambda_1 c_1 K_2^2 t}) \leqslant \frac{\|\mathbf{f}\|_2^2}{\lambda_1 c_1 K_2^2} e^{\lambda_1 c_1 K_2^2 t}. \tag{3.8}$$

By integrating (3.8) over $(0, t)$ we get

$$\|\mathbf{u}(t)\|_2^2 \leqslant \|\mathbf{u}_0\|_2^2 e^{-\lambda_1 c_1 K_2^2 t} + \frac{\|\mathbf{f}\|_2^2}{\lambda_1^2 c_1^2 K_2^4} (1 - e^{-\lambda_1 c_1 K_2^2 t}). \tag{3.9}$$

Denote $\rho_0 = \|\mathbf{f}\|_2 / \lambda_1 c_1 K_2^2$. Clearly, balls in H with radius $\hat{\rho}$, $\hat{\rho} > \rho_0$ absorb bounded sets in H-topology. More precisely, for every $\hat{\rho} > \rho_0$ and for every $\mathscr{B} \subset H$ bounded there exists a $t_0 \equiv t_0(\mathscr{B})$ such that $\|\mathbf{u}(t)\|_2 \leqslant \hat{\rho}$ for $t \geqslant t_0$.

Let us take such $t \geqslant t_0$ and integrate (3.7) between t and $t+1$; then

$$2c_1 K_p^p \int_t^{t+1} |\mathbf{u}|_{1,p}^p \leqslant \frac{\|\mathbf{f}\|_2^2}{c_1 \lambda_1 K_2^2} + \|\mathbf{u}(t)\|_2^2 \leqslant \frac{\|\mathbf{f}\|_2^2}{c_1 \lambda_1 K_2^2} + \hat{\rho}^2 \equiv \hat{C}(\hat{\rho}). \tag{3.10}$$

Put $C_{(\hat{\rho})} \equiv \hat{C}(\hat{\rho}) / 2c_1 K_p^p$.

Now, let us use $d\mathbf{u}/dt$ as a test function in (2.6). We obtain

$$\begin{aligned} \frac{1}{2} \left\| \frac{d\mathbf{u}}{dt} \right\|_2^2 + \frac{d}{dt} \|\vartheta(\mathbf{e})\|_1 &\leqslant \|\mathbf{f}\|_2^2 + \int_C |\mathbf{u}|^2 |\nabla \mathbf{u}|^2 \, dx \\ &\leqslant \|\mathbf{f}\|_2^2 + |\mathbf{u}|_{1,p}^2 \|\mathbf{u}\|_{2p/(p-2)}^2. \end{aligned} \tag{3.11}$$

Let $p<3$. Due to the interpolation inequality[4] $\|\mathbf{u}\|_{2p(p-2)} \leqslant d\,|\mathbf{u}|_{1,p}^{6/(5p-6)}\,\|\mathbf{u}\|_2^{(5p-12)/(5p-6)}$ and since

$$c_7\,|\mathbf{u}|_{1,p}^p \leqslant \|\vartheta(\mathbf{e})\|_1 \leqslant c_8(1+|\mathbf{u}|_{1,p}^p) \tag{3.12}$$

the right side of (3.11) can be estimated as

$$\begin{aligned}\|\mathbf{u}\|_{2p/p-2}^2\,|\mathbf{u}|_{1,p}^2 &\leqslant d\,|\mathbf{u}|_{1,p}^p\,|\mathbf{u}|_{1,p}^{p(16-5p)/(5p-6)}\,\|\mathbf{u}\|_2^{2(5p-12)/(5p-6)}\\ &\leqslant d\,\|\vartheta(\mathbf{e})\|_1\,|\mathbf{u}|_{1,p}^{p(16-5p)/(5p-6)}\,\|\mathbf{u}\|_2^{2(5p-12)/(5p-6)}.\end{aligned} \tag{3.13}$$

From (3.11) and (3.13)

$$\frac{d}{dt}\|\vartheta(\mathbf{e})\|_1 \leqslant \|\mathbf{f}\|_2^2 + c\,\|\vartheta(\mathbf{e})\|_1\,|\mathbf{u}|_{1,p}^{p(16-5p)/(5p-6)}\,\|\mathbf{u}\|_2^{2(5p-12)/(5p-6)}. \tag{3.14}$$

Let us put $y=\|\vartheta(\mathbf{e})\|_1$, $h=\|\mathbf{f}\|_2^2$, and $g=|\mathbf{u}|_{1,p}^{p(16-5p)/(5p-6)}\,\|\mathbf{u}\|_2^{2(5p-12)/(5p-6)}$ in the Uniform Gronwall Lemma (3.18) below. Then for $t\geqslant t_0$

$$\int_t^{t+1} y(\tau)\,d\tau \overset{(3.10),\,(3.12)}{\leqslant} c_8(1+C(\hat{\rho}))\equiv\alpha_3, \tag{3.15}$$

$$\begin{aligned}\int_t^{t+1} g(\tau)\,d\tau &\leqslant \hat{\rho}^{2(5p-12)/(5p-6)}\int_t^{t+1}\|u\|_{1,p}^{p(16-5p)/(5p-6)}\,d\tau\\ &\overset{(3.10)}{\leqslant} \hat{\rho}^{2(5p-12)/(5p-6)}\,C(\hat{\rho})\equiv\alpha_1.\end{aligned} \tag{3.16}$$

Finally,

$$|\mathbf{u}(t)|_{1,p}^p \leqslant (\alpha_3+\alpha_2)\,e^{\alpha_1}\equiv\alpha^p \qquad \forall t\geqslant t_0+1 \tag{3.17}$$

with $\alpha_2=\|\mathbf{f}\|_2^2$. Since $\alpha=\alpha(\hat{\rho})$ one can easily see that the ball $B_\rho=\{\mathbf{u}(t)\in V_p;\ \|\mathbf{u}\|_{1,p}\leqslant\rho\}$ absorbs all bounded sets in H as far as $\rho\geqslant\alpha$. ∎

3.18. Lemma (Uniform Gronwall). *Let g, h, y be three positive locally integrable functions for $t_0\leqslant t<\infty$ which satisfy*

$$\frac{dy}{dt}\leqslant gy+h \qquad \textit{for all}\quad t\geqslant t_0$$

[4] If $p\geqslant 3$ the consideration is simple.

and

$$\int_t^{t+1} g(s)\,ds \leqslant \alpha_1, \qquad \int_t^{t+1} h(s)\,ds \leqslant \alpha_2,$$

$$\int_t^{t+1} y(s)\,ds \leqslant \alpha_3 \qquad \textit{for all} \quad t \geqslant t_0,$$

where $\alpha_1, \alpha_2, \alpha_3$ *are positive constants. Then*

$$y(t+1) \leqslant (\alpha_3 + \alpha_2) \exp \alpha_1 \qquad \textit{for all} \quad t \geqslant t_0.$$

Proof. see [16, p. 89]. ∎

We close the first part of this section by summarizing the proved result:

3.19. THEOREM. *Let* $p \geqslant (2+n)/2$. *The dynamical system given by the semigroup* $(S_t)_{t \geqslant 0}$ *to the problem* $(\mathrm{NS})_p$ *possesses the global attractor* $\mathscr{A}$.

Proof. The theorem follows from Theorems 3.5 and 3.6. ∎

In order to indicate that the chaotic behavior of the solution of the problem $(\mathrm{NS})_p$ is characterized by a finite number of degrees of freedom, it is usual to show the finiteness of the Hausdorff or fractal dimension of the attractor $\mathscr{A}$. This question is usually connected with the regularity of the linearized problem and the introducing of Lyapunov exponents (see for example [2, 3]).

Because regularity of the linearized system is lacking we proceed here in a different way. Having the attractor $\mathscr{A}$ we define the set of short trajectories $\mathscr{A}_\delta$ of the attractor $\mathscr{A}$.

3.20. DEFINITION. Let δ be positive and fixed. An element $[\chi(\tau, \mathbf{u}_0)]_{\tau \in [0, \delta]}$ belongs to $\mathscr{A}_\delta$ if and only if $\mathbf{u}_0 \in \mathscr{A}$ and $\chi(\tau, \mathbf{u}_0) = S_\tau \mathbf{u}_0$ for all $\tau \in [0, \delta]$. We will call an element $[\chi(\tau, \mathbf{u}_0)]_{\tau \in [0, \delta]}$ a short trajectory and abbreviate by $\chi(\cdot, \mathbf{u}_0)$.

3.21. LEMMA. *The set* $\mathscr{A}_\delta$ *is an invariant set with respect to the semigroup* $(S_t)_{t \geqslant 0}$, *i.e.,*

$$S_t \mathscr{A}_\delta = \mathscr{A}_\delta \qquad \forall t \geqslant 0.$$

Proof. (1) Let $\chi \in (\cdot, \mathbf{u}_0) \in \mathscr{A}_\delta$ and let $t > 0$ be arbitrary but fixed. Because $\mathbf{u}_0 \in \mathscr{A}$ and $\mathscr{A}$ is invariant there exists a $\mathbf{u} \in \mathscr{A}$ such that $S_t \mathbf{u} = \mathbf{u}_0$. But due to the semigroup property $\chi(\tau, \mathbf{u}_0) = S_t S_\tau \mathbf{u}$ for all $\tau \in [0, \delta]$. Let us denote the short trajectory starting from $\mathbf{u}$ by $\chi^1(\cdot, \mathbf{u})$. Then $\chi(\cdot, \mathbf{u}_0) = S_t \chi^1(\cdot, \mathbf{u})$ and $\chi^1(\cdot, \mathbf{u}) \in \mathscr{A}_\delta$. Since t is chosen arbitrarily, $\mathscr{A}_\delta \subset S_t \mathscr{A}_\delta$ for all $t \geqslant 0$.

(2) The opposite inclusion is proved in the same way. ∎

3.22. Definition. Let X be a Banach space and let $\mathscr{A} \subset X$ be an invariant set with respect to the semigroup $S_t : X \mapsto X$. Then the fractal dimension of $\mathscr{A}$ is defined by

$$d_f(\mathscr{A}) = \limsup_{\varepsilon \to 0} \frac{\log n_\varepsilon(\mathscr{A})}{\log 1/\varepsilon}, \tag{3.23}$$

where $n_\varepsilon(\mathscr{A})$ denotes the minimum number of balls of radius ε in X which is necessary to cover $\mathscr{A}$.

Let us denote by Y the space $W^{3,2}(C)^n \cap V_2$. In the sequel we will mainly work with the spaces $L^2(0, \delta; H)$ and $W \equiv \{\mathbf{u} \in L^2(0, \delta; V_2)$; $d\mathbf{u}/dt \in L^2(0, \delta; Y^*)\}$, respectively. Let us therefore denote

$$\|\mathbf{u}\|_0^2 \equiv \int_0^\delta \|\mathbf{u}\|_2^2 \, d\tau \qquad \text{and} \qquad \|\mathbf{u}\|_1^2 \equiv \int_0^\delta |\mathbf{u}|_{1,2}^2 \, d\tau + \int_0^\delta \left| \frac{d\mathbf{u}}{dt} \right|_{Y^*}^2 d\tau. \tag{3.24}$$

Similarly, B_ρ^0 denotes a ball of radius ρ with respect to the norm $\|\cdot\|_0$ in contrast to B_ρ^1, which denotes the ball in the $\|\cdot\|_1$-norm of radius ρ. Thus

$$B_\varepsilon^0(\mathbf{u}_0) \equiv \{\mathbf{u};\ \|\mathbf{u} - \mathbf{u}_0\|_0 \leqslant \varepsilon\},$$

$$B_\varepsilon^1(\mathbf{u}_0) \equiv \{\mathbf{u};\ \|\mathbf{u} - \mathbf{u}_0\|_1 \leqslant \varepsilon\}.$$

3.25. Lemma. *Let $p \in [(2+n)/2, 4]$. Let $\chi^1(\cdot, \mathbf{u}_0)$, resp. $\chi^2(\cdot, \mathbf{v}_0)$, be two short trajectories starting from $\mathbf{u}_0 \in \mathscr{A}$, resp. $\mathbf{v}_0 \in \mathscr{A}$. If δ satisfies* (3.30) *then there exists a $\kappa > 0$ such that for arbitrary $t \geqslant \delta$*

$$\|S_t \chi^1(\cdot, \mathbf{u}_0) - S_t \chi^2(\cdot, \mathbf{v}_0)\|_1^2 \leqslant \kappa \, \|\chi^1(\cdot, \mathbf{u}_0) - \chi^2(\cdot, \mathbf{v}_0)\|_0^2. \tag{3.26}$$

Proof. Let $\mathbf{u}(t) \equiv S_t \mathbf{u}_0$, $\mathbf{v}(t) \equiv S_t \mathbf{v}_0$. Then the difference $\mathbf{w} = \mathbf{u} - \mathbf{v}$ satisfies the inequality (2.23). If we integrate this inequality between $t - s$ and $t + r$, $s \in [0, \delta/2]$, $r \in [0, \delta]$ with $t \geqslant \delta$, we have

$$\|\mathbf{w}(t+r)\|_2^2 - \|\mathbf{w}(t-s)\|_2^2 + \hat{c} \int_{-s}^{r} |\mathbf{w}(t+\tau)|_{1,2}^2 \, d\tau$$

$$\leqslant c_5 \int_{-s}^{r} \|\mathbf{w}(t+\tau)\|_2^2 \, |\mathbf{u}(t+\tau)|_{1,p}^{2p/(2p-3)} \, d\tau. \tag{3.27}$$

Denote

$$\gamma(t) \equiv \sup_{r \in [-\delta/2, \delta]} \|\mathbf{w}(t+r)\|_2^2.$$

Then we get from (3.27)

$$\gamma(t) + \hat{c} \int_0^\delta |\mathbf{w}(t+\tau)|_{1,2}^2 \, d\tau$$

$$\leqslant 2\gamma(t) \int_{-\delta/2}^{\delta} |\mathbf{u}(t+\tau)|_{1,p}^{2p/(2p-3)} \, d\tau + 2 \, \|\mathbf{w}(t-s)\|_2^2. \tag{3.28}$$

Further, from (3.17) and from the fact that $\mathscr{A}$ is invariant we know that

$$|\mathbf{u}(t+\tau)|_{1,p} \leqslant \alpha. \tag{3.29}$$

Now, if

$$(1 - 3(\alpha)^{2p/(2p-3)} \, \delta) \geqslant 0. \tag{3.30}$$

then it follows from (3.28)

$$\hat{c} \int_0^\delta |\mathbf{w}(t+\tau)|_{1,2}^2 \, d\tau \leqslant \|\mathbf{w}(t-s)\|_2^2. \tag{3.31}$$

After integrating (3.31) with respect to s between 0 and $\delta/2$ and using continuity of S_t, we obtain the estimate of $L^2(0, \delta; V_2)$-norm by $L^2(0, \delta; H)$-norm.

Now we need the estimate of $L^2(0, \delta; Y^*)$-norm. Let $\varphi \in L^2(t, t+\delta; Y)$ and $\|\varphi\|_{L^2(t, t+\delta; Y)} \leqslant 1$.Then

$$\int_0^\delta \left(\frac{d\mathbf{w}}{dt}(t+\tau), \varphi(t+\tau) \right) d\tau$$

$$= -\int_0^\delta \int_C [\tau_{ij}(\mathbf{e}(\mathbf{u}(t+\tau))) - \tau_{ij}(\mathbf{e}(\mathbf{v}(t+\tau)))] \, e_{ij}(\varphi(t+\tau)) \, dx \, d\tau$$

$$-\int_0^\delta \int_C \left(u_j \frac{\partial u_i}{\partial x_j} - v_j \frac{\partial v_i}{\partial x_j} \right) (t+\tau) \, \varphi_i(t+\tau) \, dx \, d\tau \equiv I_1 + I_2.$$

But

$$|I_1| \leqslant \int_0^\delta \int_C (1 + |\nabla \mathbf{u}(t+\tau)| + |\nabla \mathbf{v}(t+\tau)|)^{p-2}$$

$$\times |\mathbf{e}(\mathbf{w}(t+\tau))| |\mathbf{e}(\varphi(t+\tau))| \, dx \, d\tau$$

$$\leqslant \int_0^\delta \|\varphi(t+\tau)\|_{3,2}\, |\mathbf{w}(t+\tau)|_{1,2} \left(\int_C (1+|\nabla \mathbf{u}(t+\tau)| + |\nabla \mathbf{v}(t+\tau)|)^{2(p-2)}\, dx \right)^{1/2} d\tau.$$

Since $p \leqslant 4$ then $2(p-2) \leqslant p$. Using (3.29) we get

$$|I_1| \leqslant \kappa(\alpha) \left(\int_0^\delta |\mathbf{w}(t+\tau)|_{1,2}^2\, d\tau \right)^{1/2} \leqslant \left(\int_0^\delta \|\mathbf{w}(\tau)\|_2^2\, d\tau \right)^{1/2}.$$

The last inequality follows from the estimates above. Similarly we bound the convective term I_2. Thus the lemma is proved. ▮

3.32. Theorem. *Let δ satisfy* (3.30) *and $p \in [(2+n)/2, 4]$. Then the set $\mathscr{A}_\delta$ introduced in Definition* 3.20 *has a finite fractal dimension.*

Proof. Since $\mathscr{A}_\delta$ is bounded there exists an $R > 0$ such that

$$\mathscr{A}_\delta \subset B_R^0.$$

By using Lemma 3.21 and Lemma 3.25 we have

$$\mathscr{A}_\delta = S_t \mathscr{A}_\delta \subset S_t B_R^0 \overset{(3.26)}{\subset} B_{\kappa R}^1, \qquad \text{for} \quad t \geqslant \delta.$$

But due to Lemma 1.16, this last set is relatively compact in $L^2(0, \delta; H)$ and consequently can be covered by a finite number N of balls of radius $R/2$. It is easy to verify that the number N does not depend on the centrum of a ball and the dilatation R. Thus

$$\mathscr{A}_\delta \subset \bigcup_{i=1}^N B_{R/2}^0(\xi_i).$$

Now, we proceed iteratively:

$$\mathscr{A}_\delta = S_t \mathscr{A}_\delta \subset \bigcup_{i=1}^N S_t B_{R/2}^0 \overset{(3.26)}{\subset} \bigcup_{i,j=1}^N B_{\kappa R/2}^1(\xi_{ij}).$$

Again, with the help of 1.16, we get

$$\mathscr{A}_\delta \subset \bigcup_{i,j=1}^N B_{R/4}^0(\xi_{ij})$$

and we continue the iteration process with $\varepsilon_j = R/2^j$. Thus

$$\frac{\log n_{\varepsilon_j}(\mathscr{A}_\delta)}{\log 1/\varepsilon_j} \leqslant \frac{j \log N}{j \log 2 - \log R}. \tag{3.33}$$

It it easy to check that the estimate of the fraction in (3.33) is finite not only for the sequence ε_j (as $j \to \infty$) but also for an arbitrary sequence $\hat{\varepsilon}_k$ tending to zero as $k \to \infty$. ∎

4. Asymptotic Study of Solutions if $p \geqslant 1 + 2n/(n+2)$

Let $p \geqslant 1 + 2n/(n+2)$ and $\mathbf{u}_0 \in H$, $\mathbf{f} \in L^2(C)^n$. Notice that this case coincides with the previous one (Section 3) in two dimensions.

Since we have uniqueness only for smoother data, see Theorem 2.14(ii), many trajectories can start from the initial value $\mathbf{u}_0 \in H$. Let us denote these trajectories by $[\chi^\beta(\tau, \mathbf{u}_0)]_{\tau \in [0,\delta]}$, for short $\chi^\beta(\cdot, \mathbf{u}_0)$, $\beta \in \Gamma_{\mathbf{u}_0}$, where $\Gamma_{\mathbf{u}_0}$ is the set of indices marking trajectories starting from $\mathbf{u}_0$.

4.1. Definition (Set of Short Trajectories). Let $\delta > 0$. We define

$$\begin{aligned} X_\delta &\equiv \bigcup_{\mathbf{u}_0 \in H} \bigcup_{\beta \in \Gamma_{\mathbf{u}_0}} \chi^\beta(\cdot, \mathbf{u}_0), \\ X_\delta^s &\equiv \overline{X_\delta}, \end{aligned} \tag{4.2}$$

where the closure is with respect to the norm of $L^2(0, \delta; H)$.

Then the space X_δ^s equipped by the topology of $L^2(0, \delta; H)$ is a metric space. Let us define the operators $L_t\colon X_\delta^s \mapsto X_\delta^s$ by the relation

$$L_t(\chi^\beta(\cdot, \mathbf{u}_0)) = \chi(\cdot, \chi^\beta(t, \mathbf{u}_0))\colon L^2(0, \delta; H) \mapsto L^2(0, \delta; H) \tag{4.3}$$

if $\chi^\beta(\cdot, \mathbf{u}_0) \in X_\delta$ and by the natural extension (as a limit of a Cauchy sequence) if $\chi^\beta(\cdot) \in X_\delta^s \setminus X_\delta$. Due to Lipschitz continuity of L_t we will work only with elements of X_δ.

4.4. Lemma. (i) *The operators $(L_t)_{t \geqslant 0}$ form a semigroup on X_δ^s.*

(ii) *The mapping $L_t\colon X_\delta^s \mapsto X_\delta^s$ is continuous for all $t \geqslant 0$.*

Proof. (i) follows from the fact that $\chi(\delta, \mathbf{u}_0) \in W^{1,2}(C)^3$ and the operators $S_t\colon W^{1,2}(C)^3 \mapsto W^{1,2}(C)^3$ defined by $S_t(\mathbf{u}_0) = \mathbf{u}(t)$ have the semigroup property. (But it is not clear whether these operators are continuous for fixed $t \geqslant 0$ with respect to the $W^{1,2}$ topology.)

(ii) Let $t>0$ be arbitrary, but fixed. Let $\mathbf{u}(t)$ respectively $\mathbf{v}(t)$, be two solutions with the initial values $\mathbf{u}_0$, respectively, $\mathbf{v}_0$. As in the proof of the uniqueness, we have the inequality (2.21) for the difference $\mathbf{w}=\mathbf{u}-\mathbf{v}$.

Take $s\in(0,\delta)$, $\tau\in(\delta/2,\delta)$ and integrate (2.21) between τ and $t+s$. We obtain

$$\|\mathbf{w}(t+s)\|_2^2 \leqslant \|\mathbf{w}(\tau)\|_2^2 + c_4 \int_\tau^{t+s} \|\mathbf{w}\|_2^2 \, |\mathbf{u}|_{1,3p}^{2p(2p-1)} \, d\xi. \tag{4.5}$$

The classical Gronwall Lemma then implies

$$\begin{aligned}\|\mathbf{w}(t+s)\|_2^2 &\leqslant \|\mathbf{w}(\tau)\|_2^2 \exp\left(\int_\tau^{t+s} |\mathbf{u}|_{1,3p}^{2p/(2p-1)} \, d\xi\right) \\ &\leqslant \|\mathbf{w}(\tau)\|_2^2 \, M_\delta,\end{aligned} \tag{4.6}$$

where $M_\delta \equiv \exp(\int_{\delta/2}^{t+\delta} |\mathbf{u}|_{1,3p}^{2p/(2p-1)} \, d\xi)$ is a finite number due to (2.16). Now, we integrate (4.7) with respect to s between 0 and δ and then the resulting inequality with respect to τ over $(\delta/2, \delta)$. We get

$$\int_0^\delta \|\mathbf{w}(t+s)\|_2^2 \, ds \leqslant 2M_\delta \int_{\delta/2}^{2} \|\mathbf{w}(\tau)\|_2^2 \, d\tau \leqslant 2M_\delta \int_0^\delta \|\mathbf{w}(\tau)\|_2^2 \, d\tau. \tag{4.7}$$

The assertion (ii) is proved. ∎

We want to prove the existence of the global attractor of the semigroup $(L_t)_{t\geqslant 0}$ according to Theorem 3.5. Let us look for an absorbing set B_W in the space

$$W \equiv \left\{\mathbf{u} \in L^p(0,\delta;V_p); \frac{d\mathbf{u}}{dt} \in L^{p'}(0,\delta;V_p^*)\right\}$$

which is compactly embedded in $L^2(0,\delta;H)$ due to Lemma 1.16.

Let us start with an easy observation:

4.8. Lemma (Absorbing Set in X_δ^s). *Put $\rho_0 = \delta \, \|\mathbf{f}\|_2/\lambda_1 c_1 K_2^2$. If $\rho' > \rho_0$ then a ball $B_{\rho'}^0$ is absorbing in X_δ^s.*

Proof. After integrating (3.8) between τ and $t+\tau$ one obtains

$$\|\mathbf{u}(t+\tau)\|_2^2 \leqslant \|\mathbf{u}(\tau)\|_2^2 \, e^{-\lambda_1 c_1 K_2^2 t} + \frac{\|\mathbf{f}\|_2^2}{\lambda_1^2 c_1^2 K_2^4} (1 - e^{-\lambda_1 c_1 K_2^2 t}). \tag{4.9}$$

The proof is finished by integrating (4.9) with respect to τ over $(0,\delta)$. ∎

4.10. LEMMA. *There exists a ball B_W in W which absorbs all bounded sets of X^s_δ.*

Proof. Let $\mathscr{B} \subset X^s_\delta$ be a bounded set. Let $\rho' > \rho_0$. Then there exists (according to Lemma 4.8) a $t_0 = t_0(\mathscr{B})$ such that

$$L_t \mathscr{B} \subset \mathscr{B}^0_{\rho'} \qquad \forall t \geqslant t_0.$$

Let us integrate (3.7) between $t-s$ and $t+\delta$ with $t \geqslant t_0 + \delta/2$, $s \in (0, \delta/2)$. Then

$$\|\mathbf{u}(t+\delta)\|_2^2 + \tilde{c} \int_0^\delta |\mathbf{u}(t+\tau)|_{1,p}^p \, d\tau$$
$$\leqslant \frac{3\delta}{2} \frac{\|\mathbf{f}\|_2^2}{c_1 \lambda_1 K_2^2} + \|\mathbf{u}(t-s)\|_2^2 \qquad \text{with} \quad \tilde{c} = 2c_1 K_p^p. \tag{4.11}$$

After integrating (4.11) over $(0, \delta/2)$ with respect to s, we get

$$\int_0^\delta |\mathbf{u}(t+\tau)|_{1,p}^p \, d\tau \leqslant \frac{3\delta}{2\tilde{c}} \frac{\|\mathbf{f}\|_2^2}{c_1 \lambda_1 K_2^2} + \frac{2}{\delta \tilde{c}} \int_0^{\delta/2} \left\| \mathbf{u}\left(t - \frac{\delta}{2} + \tau\right)\right\|_2^2 d\tau$$
$$\leqslant \frac{3\delta}{2\tilde{c}} \frac{\|\mathbf{f}\|_2^2}{c_1 \lambda_1 K_2^2} + \frac{2}{\delta \tilde{c}} (\rho')^2 \equiv \hat{\rho}. \tag{4.12}$$

We see that the ball $B_{\hat{\rho}}$ in $L^p(0, \delta; V_p)$ absorbs all bounded sets of X^s_δ.

It remains to prove the existence of a constant $C_1 = C_1(\rho')$ such that

$$\int_0^\delta \left\| \frac{d\mathbf{u}(t+\tau)}{dt} \right\|_{V_p^*}^{p'} \leqslant C_1 \qquad \text{for all} \quad t \geqslant t_0 + \frac{\delta}{2}. \tag{4.13}$$

But this is a consequence of Lemma 2.8, where we use (4.12) and also (3.9). ▮

4.14. THEOREM. *The dynamical system given by the semigroup $(L_t)_{t \geqslant 0}$ to the problem $(NS)_p$, $p \geqslant 1 + 2n/(n+2)$, possesses the global attractor, denoted by $\mathscr{A}_{X^s_\delta} \subset X^s_\delta$. Moreover, if δ satisfies* (4.19) *below, then*

$$d_f(\mathscr{A}_{X^s_\delta}) < \infty. \tag{4.15}$$

Proof. The existence of the global attractor $\mathscr{A}_{X^s_\delta}$ follows from Theorem 3.5 and from Lemma 4.10.

The proof of (4.15) is the analogue of the proof of Theorem 3.32 with S_t replaced by L_t and Lemma 4.16 below used instead of Lemma 3.25. ▮

4.16. LEMMA. *Let δ satisfy* (4.19) *and* $p \in [1+2n/(n+2), 4]$. *Let* $\chi^1(\cdot, \mathbf{u}_0)$ *and* $\chi^2(\cdot, \mathbf{v}_0)$ *be two short trajectories belonging to* $\mathscr{A}_{X^s_\delta}$. *Then there exists a* $\kappa > 0$ *such that for arbitrary* $t \geqslant \delta$

$$\|L_t\chi^1(\cdot, \mathbf{u}_0) - L_t\chi^2(\cdot, \mathbf{v}_0)\|_1^2 \leqslant \kappa \, \|\chi^1(\cdot, \mathbf{u}_0) - \chi^2(\cdot, \mathbf{v}_0)\|_0^2. \tag{4.17}$$

Proof. The proof is similar to that of Lemma 3.25. But now we start from the inequality (2.21).

Let $t \geqslant \delta$. After integrating (2.21) between $t-s$ and $t+r$, $s \in [0, \delta/2]$, $r \in [0, \delta]$, we get

$$\|\mathbf{w}(t+r)\|_2^2 - \|\mathbf{w}(t-s)\|_2^2 + \hat{c} \int_{-s}^{r} |\mathbf{w}(t+\tau)|_{1,2}^2 \, d\tau$$

$$\leqslant c_4 \int_{-s}^{r} \|\mathbf{w}(t+\tau)\|_2^2 \, |\mathbf{u}(t+\tau)|_{1,3p}^{2p/(2p-1)} \, d\tau. \tag{4.18}$$

Denote

$$\gamma(t) = \sup_{r \in [-\delta/2, \delta]} \|\mathbf{w}(t+r)\|_2^2.$$

Then we get from (4.18)

$$\gamma(t) + \hat{c} \int_0^\delta |\mathbf{w}(t+\tau)|_{1,2}^2 \, d\tau \leqslant 2\delta^\alpha \gamma(t) \int_{-\delta/2}^{\delta} |\mathbf{u}(t+\tau)|_{1,3p}^p \, d\tau + \|\mathbf{w}(t-s)\|_2^2,$$

where $\alpha = (2p-3)/(2p-1) > 0$. But

$$\int_{-\delta/2}^{\delta} |\mathbf{u}(t+\tau)|_{1,3p}^p \, d\tau \leqslant K_T$$

according to (2.16), where K_T depends only on some “large” T but not on $t \in (\delta/2, T+\delta)$. Provided that

$$(1 - K_T\delta^\alpha) \geqslant 0 \qquad \alpha = \frac{2p-3}{2p-1} \tag{4.19}$$

we get (4.17) in the same way as in the proof of Lemma 3.25.

The estimate of the differences of time derivatives in $L^2(0, \delta; (W^{3,2}(C)^n \cap V_2))^*)$ is completely the same as in Lemma 3.25. The proof is finished. ∎

Acknowledgments

This paper was written during a stay of the first author at the Friedrich-Wilhelms-University of Bonn, supported by grants of the Sonderforschungsbereich 256.

References

1. H. Bellout, F. Bloom, and J. Nečas, Young measure-valued solutions for non-Newtonian incompressible fluids, *Comm. Partial Differential Equations* **19**, Nos. 11–12 (1994), 1763–1803.
2. P. Constantin and C. Foias, Global Lyapunov exponents, Kaplan–Yorke formulas and the dimension of the attractors for twodimensional Navier–Stokes equations, *Comm. Pure Appl. Math.* **38** (1985), 1–27.
3. P. Constantin and C. Foias, "Navier–Stokes Equations," Chicago Lectures in Mathematics, Univ. of Chicago Press, Chicago, 1989.
4. P. Constantin, C. Foias, and R. Temam, On the dimension of the attractors in two-dimensional turbulence, *Physica D* (1992).
5. H. Gajewski, K. Gröger, and K. Zacharias, "Nichtlineare Operatorgleichungen und Operatordifferentialgleichungen," Akademie-Verlag, Berlin, 1974.
6. O. A. Ladyzhenskaya, "The Mathematical Theory of Viscous Incompressible Flow," Gordon & Breach, New York, 1969.
7. O. A. Ladyzhenskaya, On finite dimensionality of bounded invariant sets for the Navier–Stokes equations and some other dissipative systems, *Zapi. Nauchn. Semi. LOMI* **115** (1982), 3–15.
8. O. A. Ladyzhenskaya, On the dynamical system generated by the Navier–Stokes equations, (English translation in *J. Sov. Math.* **3** (1975), 458–479), *Zap. Nauchn. Semi. LOMI* **27** (1972), 91–114.
9. J. L. Lions, "Quelques méthodes de résolution des problèmes aux limites non linéaires," Dunod, Paris, 1969.
10. J. Málek, J. Nečas, and M. Růžička, Weak solutions to a class of non-Newtonian incompressible fluids in bounded domains, in preparation.
11. J. Málek, J. Nečas, and A. Novotný, Measure-valued solutions and asymptotic behavior of a multipolar model of a boundary layer, *Czech. Math. J.* **42**, No. 117 (1992), 549–576.
12. J. Málek, J. Nečas, M. Rokyta, and M. Růžička, "Weak and Measure-Valued Solutions to Evolutionary PDEs," Chapman & Hall, London, 1996, to appear.
13. J. Málek, J. Nečas, and M. Růžička, On the non-Newtonian incompressible fluids, *Math. Meth. Models Appl. Science* (M^3AS) **3**, No. 1 (1993), 35–63.
14. J. Málek, M. Růžička, and G. Thäter, Fractal dimension, attractors and Boussinesq approximation in three dimensions, *Acta Appl. Math.* **37** (1994), 83–97.
15. J. Nečas, "Sur les normes équivalentes dans $W_p^k(\Omega)$ et sur la coercivité des formes formellement positives," Presses Univ. Montréal, Montréal, 1966.
16. R. Temam, "Infinite-Dimensional Dynamical Systems in Mechanics and Physics," Springer-Verlag, New York, 1988.

Arch. Rational Mech. Anal. 165 (2002) 243–269
Digital Object Identifier (DOI) 10.1007/s00205-002-0219-4

Global Analysis of the Flows of Fluids with Pressure-Dependent Viscosities

JOSEF MÁLEK, JINDŘICH NEČAS & K. R. RAJAGOPAL

Communicated by S. S. ANTMAN

Abstract

To describe the flows of fluids over a wide range of pressures, it is necessary to take into account the fact that the viscosity of the fluid depends on the pressure. That the viscosity depends on the pressure has been verified by numerous careful experiments. While the existence of solutions local-in-time to the equations governing the flows of such fluids are available for small, special data and rather unrealistic dependence of the viscosity on the pressure, no global existence results are in place. Our interest here is to establish the existence of weak solutions for spatially periodic three-dimensional flows that are global in time, for a large class of physically meaningful viscosity-pressure relationships.

1. Introduction

In classical fluid mechanics, it is customary to assume that the Cauchy stress tensor $\mathbf{S}$ and the velocity gradient $\nabla \boldsymbol{v}$, $\boldsymbol{v}$ being the velocity, and the density ρ are related through

$$\mathbf{S} = \mathbf{T}_0(\rho, \nabla \boldsymbol{v}). \tag{1}$$

While models of the class (1) can describe shear thinning and shear thickening, they are incapable of describing the elastic response or memory effects exhibited by certain fluids. This inadequacy notwithstanding, models belonging to the class (1) are the most widely used to date. The classical Navier-Stokes model and the power-law fluid model are two examples of models that belong to the class (1).

The principle of material frame indifference reduces (1) to the form

$$\mathbf{S} = \mathbf{T}_1(\rho, \mathbf{D}), \tag{2}$$

where $\mathbf{D} := \mathbf{D}(\boldsymbol{v}) \equiv \frac{1}{2}(\nabla \boldsymbol{v} + (\nabla \boldsymbol{v})^T)$ is the symmetric part of the velocity gradient.

© Springer-Verlag 2002. Reprinted with kind permission from: Málek, J., Nečas, J. & Rajagopal, K.R., Global Analysis of the Flows of Fluids with Pressure-Dependent Viscosities, Archive of Rational Mechanics and Analysis 165, 243-269.

For many fluids the variations in the fluid density is small, say of the order of a few percent, while the variations in its viscosity can differ by many orders of magnitude, due to significant changes in the pressure (see BRIDGMAN [7]). Such fluids can be effectively modeled as incompressible fluids with pressure-dependent viscosity. In this work, we shall consider such a class of fluids.

When a body is constrained to move in a specific manner, a part of the stress in the body, which we shall call the constraint stress, goes towards ensuring that the constraint is maintained. The stress in the material is then the sum of this constraint stress for which a constitutive relation cannot be prescribed and another part that is constitutively determined. The constraint stress is orthogonal to the constraint manifold and does no work in deformations tangent to the constraint manifold. ANTMAN [1] recognized that the constitutively specified part of the stress could depend on the constraint stress though in certain materials (e.g., hyperelastic materials) it cannot. Later, working with Marlow (see [2]), he pointed out that it could be difficult meeting the compatibility equations if the classical notions are not appropriately generalized. ANTMAN & MARLOW [2] consider a variety of constraints to illustrate their ideas and we refer the reader to their interesting treatment of the subject. Of particular relevance to our work are the implications of their analysis for incompressible materials. As the constitutively determinate part of the stress can depend on the Lagrange multiplier P, it immediately follows that in general the Cauchy stress $\mathbf{S}$ is defined through an implicit relationship. We shall discuss this issue later.

If the fluid is incompressible, it follows from standard representation theorems that the most general form for $\mathbf{S}$ is

$$\mathbf{S} = -P\mathbf{I} + \alpha_1(P, \mathrm{II}_{\mathbf{D}}, \mathrm{III}_{\mathbf{D}})\mathbf{D} + \alpha_2(P, \mathrm{II}_{\mathbf{D}}, \mathrm{III}_{\mathbf{D}})\mathbf{D}^2, \tag{3}$$

where P is usually called the pressure (in general P may not be the mean normal stress as $\operatorname{tr}\mathbf{D}^2$ is not zero), and

$$\mathrm{II}_{\mathbf{D}} := \operatorname{tr}\mathbf{D}^2 = |\mathbf{D}|^2, \quad \mathrm{III}_{\mathbf{D}} := \det\mathbf{D}.$$

The fact that material constants α_1, α_2 can depend on the Lagrange multiplier P is a consequence of the results in [1, 2].

In this exposition we consider a subclass of (3), fluids whose stresses are given by the relation

$$\mathbf{S} = -P\mathbf{I} + \nu(P, \mathrm{II}_{\mathbf{D}})\mathbf{D} = -P\mathbf{I} + \nu(P, |\mathbf{D}|^2)\mathbf{D}. \tag{4}$$

Since $\operatorname{tr}\mathbf{D} = 0$, it follows that $-P = \frac{1}{3}\operatorname{tr}\mathbf{S}$ and thus the viscosity is a function of the mean normal stress. This in turn implies that (4) provides an implicit relation for the stress $\mathbf{T}$ and the symmetric part of the velocity gradient $\mathbf{D}$.

Let us consider a special subclass of fluids defined by (3), namely those whose stress is given by

$$\mathbf{S} = -P\mathbf{I} + \alpha_1(\mathrm{II}_{\mathbf{D}}, \mathrm{III}_{\mathbf{D}})\mathbf{D} + \alpha_2(\mathrm{II}_{\mathbf{D}}, \mathrm{III}_{\mathbf{D}})\mathbf{D}^2. \tag{5}$$

The most widely used models are a subclass of (5) that are special cases of

$$\mathbf{S} = -P\mathbf{I} + \alpha_1(\mathbb{II}_{\mathbf{D}})\mathbf{D}. \tag{6}$$

These include models due to Ostwald, de Waele, Spriggs, Eyring, Carreau, Ellis, Bingham and many others.

The model (6) includes the power-law model as a special subclass. Power-law models are by far the most commonly used models for describing the response of a variety of fluids. Many colloids and suspensions are very well described by such models (see [8, 11, 38], and the books by SCHOWALTER [36] and BIRD *et al.* [6] and the many references cited therein). It is also a well accepted fact that blood, in a homogenized sense, can be modeled as a power-law fluid while flowing in large blood vessels (see for example [10, 9, 12, 14, 16, 29, 42]).

The use of power-law fluids to describe the flow of lubricants can be traced back to the work of NG & SAIBEL [30]. JOHNSON & MANGKOESOEBROTO [23] have developed a lubrication approximation, in much the same manner as REYNOLDS [34] for the Navier-Stokes fluid, for power-law fluids. The importance of the role of pressure in the response of a lubricant and the manner in which the viscosity is assumed to depend on the pressure is discussed at length in the recent book on lubrication by SZERI [39]. There is also experimental evidence (see BATES *et al.* [4] and HUTTON *et al.* [21]) that bears witness to the viscosity depending on the shear rate (based on a power law) and exponentially on the pressure. A computational study of the flow of such a fluid in a journal bearing has been carried out in [19]. Such power-law models are also quite popular in the unlikeliest of areas – in ice-mechanics, glaciology (see KJARTANSON *et al.* [24], VAN DER VEEN & WHILLANS [41]) and geology – and are the models of choice in food rheology.

The model (5) is now referred to as the Reiner-Rivlin model. Reiner introduced a model of the form (5), where α_1 and α_2 are polynomials in the invariants. REINER [31] developed the model to describe the phenomenon of "dilatancy" (that property of a material by which it increases in volume when stirred), a term coined by REYNOLDS [33]. It is best to provide the definition for "dilatancy" in Reynolds' own words: "I have called this unique property of granular masses "dilatancy", because the property consists in a definite change of bulk, consequent on a definite change of shape or distortional strain, any disturbance whatever causing a change of volume and generally dilation". While Reynolds' work pertained to the behavior of granular materials and granular materials infused with liquids, Reiner saw more general relevance to his work. He remarks: "In the following a mathematical theory of dilatancy is expounded from the phenomenological point of view, i.e., no account will be taken of the internal structure to which the phenomenon of dilatancy may be due. It will be shown on purely mathematical grounds that dilatancy should be present in a generalized Newtonian liquid". Nowhere in his work does he claim that his model is to be applied to any specific polymeric liquid.

The Reiner-Rivlin model (5) does not predict the normal stress differences observed in the simple shear flow of many dilute polymeric liquids that exhibit elastic response. However, it should be borne in mind that there are many non-Newtonian fluids which do not exhibit significant normal stress differences, the viscosity however depending strongly on the rate of shear, and such fluids can be modeled by a subclass of Reiner-Rivlin fluids, namely the power-law fluids. The remark by TRUESDELL & NOLL [40] with regard to the Reiner-Rivlin fluid that "Data was collected; they seem to indicate that for polyisobutylene solutions, $T^{\langle 11\rangle}$ and $T^{\langle 22\rangle}$ are not equal when κ becomes appreciably large. This experimental evidence was the greatest motive for the rejection of the Reiner-Rivlin equation (119.4) as an adequate basis of a physical theory and for the search of greater generality, in the last few years....", and those of COLEMAN, MARKOVITZ & NOLL [13] that "Note that the two normal stress function (30.4) resulting from the Reiner-Rivlin equation (30.2) are the same, which implies for example that the normal stresses in simple shearing flow must coincide (see E11). A relation of this kind can be compared with experiment. Data reported by Padden and DeWitt show that $T\langle xx\rangle$ and $T\langle yy\rangle$ are not the same in some polymer solutions. This observation was a motive for the rejection of the Reiner-Rivlin theory. There is a very large literature on the Reiner-Rivlin theory, much of which must be regarded as obsolete today", are too sweeping to be acceptable. The last two sentences in the remark by Coleman *et al.* indicting the whole class of Reiner-Rivlin fluids is an unacceptable leap from a result that they themselves recognized as being true for "some" viscoelastic fluids. After all, the Navier-Stokes model is a special case of the Reiner-Rivlin model. Do we then reject this model on the basis of experiments on polyisobutylene and the supposed quest for greater generality? The comments of Truesdell & Noll, and Coleman *et al.*, ought to have been tempered to reflect the fact that the Reiner-Rivlin model is unsuitable for certain classes of polymeric liquids. In their preoccupation with the study of viscoelastic polymeric liquids which exhibit different normal stress differences, TRUESDELL & NOLL [40] and COLEMAN *et al.* [13] seem to have lost sight of the possibility that there could be other fluids that are not viscoelastic that have little or no memory and for which certain subclasses of Reiner-Rivlin fluids are suitable. Many fluids seem to be approximated reasonably well by shear-thinning power-law fluids for a sufficiently large class of flows and we have discussed some of them in the paragraphs above.

For Navier-Stokes (Newtonian) fluids the viscosity function ν is assumed to be constant. STOKES [37] recognized that the viscosity of water would depend on the pressure but he also recognized that its dependence on the pressure can be neglected for a large class of flows, and based on the experimental work of Du Buat offered one such example: flows in canals at normal pressures. The work of Andrade and Bridgman (see BRIDGMAN [7]) and the recent experimental studies related to lubricants (see the lists of references quoted in [39] or [20]) however, confirm that the viscosity has to increase with the increasing pressure in high-pressure regimes. The importance of including the effect of the viscosity on the pressure cannot be overemphasized. For instance, in the theory of elastohydrodynamic lubrication (i.e., hydrodynamic lubrication applied to solid surfaces capable of elastic deformation), the existence of continuous films under severe loading cannot be explained unless the dependence of the fluid viscosity on the pressure and elastic deformation of the solid are taken into account. See [39] for a detailed discussion of this issue.

Experiments (see [22, 21] and [19] for example) also show that for many fluids the viscosity ν should decrease with respect to the shear rate and thus with respect to $|\mathbf{D}|^2$.

We observe that we can rewrite (4) in the form

$$\mathbf{S} = -P\mathbf{I} + \partial_{\mathbf{D}} U(P, |\mathbf{D}|^2), \tag{7}$$

where U is the potential of the viscous part of $\mathbf{S}$ defined through

$$U(P, |\mathbf{D}|^2) = \tfrac{1}{2} \int_0^{|\mathbf{D}|^2} \nu(P, s)\, ds. \tag{8}$$

Here we use the following notation:

$$\begin{aligned}
\partial_{\mathbf{D}} U(P, |\mathbf{D}|^2) &= 2\frac{\partial U(P, |\mathbf{D}|^2)}{\partial |\mathbf{D}|^2}\mathbf{D} \equiv \nu(P, |\mathbf{D}|^2)\mathbf{D} =: \mathbf{V}(P, \mathbf{D}),\\
\partial_{\mathbf{D}}^{(2)} U(P, |\mathbf{D}|^2) &= \frac{\partial \mathbf{V}(P, \mathbf{D})}{\partial \mathbf{D}},\\
\partial_P \nu(P, |\mathbf{D}|^2) &= \frac{\partial \nu(P, |\mathbf{D}|^2)}{\partial P}.
\end{aligned}$$

We shall require that the viscosity ν satisfy certain conditions, and these are presented in the next section. We give an example for the viscosity that satisfies all of them

$$\begin{aligned}
\nu(P, |\mathbf{D}|^2) &= \left(A + (1 + \exp(\alpha P))^{-q} + |\mathbf{D}|^2\right)^{\frac{p-2}{2}}\\
&\quad \text{with } p \in \left(\tfrac{9}{5}, 2\right), \alpha > 0, A > 0 \text{ and } q \in \left(0, \frac{1}{2\alpha}\frac{p-1}{2-p}A^{\frac{2-p}{2}}\right).
\end{aligned} \tag{9}$$

This form for the viscosity reduces to that for a Navier-Stokes fluid (i.e., the viscosity is a constant) when $p = 2$, and to that for a fluid with shear-dependent viscosity when $q = 0$ and $p \neq 2$.

In this paper we provide the first global-in-time existence result for the class of fluids considered, without restricting the data to be small. We prove the existence of a weak solution for the spatially periodic case. We also deal only with the three-dimensional case; in two dimensions the results can be strengthened due to simplification of the estimates and better embedding properties of function spaces.

The question of uniqueness of the solution remains open as in the case of the three-dimensional Navier-Stokes equations. On the other hand, our method of proof also provides the existence of a strong solution for a short time interval or a global strong solution for small data that are unique in the class of weak solutions.

It is worth remarking that the pressure is not determined uniquely from the velocity field once the velocity field is determined; it is determined up to a function of time. While in the theory of the Navier-Stokes equations this indeterminacy does not play a significant role (as only pressure gradients occur in the equations), it is an important feature for model (4). If the solutions are sufficiently smooth, this indeterminacy can be fixed by assigning an appropriate value at some instant of time and at some point in space, and this is indeed the solution adopted in engineering applications. However, as we are dealing with integrable functions here, we cannot assign the value of the pressure at just one point. One way out of this quandary is to prescribe the pressure on a set of nonzero Lebesgue measure, say a small region in the flow or on the boundary. In the case of spatially periodic flow, we can specify the mean value of the pressure over the representative cell.

An interesting issue that arises concerning the pressure is whether it needs to be bounded from below. The answer to this is not clear. Example 2 below shows, however, that this is true in some special cases.

Earlier papers dealing with mathematical analysis of the equations of the type (10) below have been carried out by RENARDY [32] and GAZZOLA [18]. They established some results concerning the local-in-time existence of smooth solution for the model where $\nu = \nu(P)$ under certain, rather restrictive and unrealistic (nonphysical) assumptions concerning the manner in which the viscosity depends on the pressure. For example, contrary to experimental evidence (see BRIDGMAN [7], BARUS [3], BATES *et al.* [4] and all the experimental papers pertinent to the variation of the viscosity on the pressure, at sufficiently high pressures), RENARDY [32] requires that

$$\frac{\nu(P)}{P} \to 0 \quad \text{as } P \to \infty.$$

The plan of the paper is as follows. In Section 2 we state the main theorem after documenting the governing equation.

Section 3 deals with approximations: we use a perturbation by a fourth order operator and the solvability of the approximate problem, its uniqueness and regularity are studied. The significant difference in comparison with the mathematical theory of the Navier-Stokes equations is in treating the pressure P. While in the Navier-Stokes theory (as also for fluids with shear-dependent viscosity) the pressure can be eliminated from the analysis by using divergenceless test functions, here pressure has to be retained in the analysis as the viscosity function ν depends on P.

In Section 4 we derive the crucial estimates that are sufficient to establish the compactness of the velocity gradient and the pressure for the approximations. The limiting process for the approximations is also outlined; the reader may consult [27] for details since the method of the proof has some features in common with the "regularity method" explained in [27], Section 5. (See also [5] and [28].) Since the techniques and results discussed at length in the book [27] will be frequently referred to here, we refer the reader to it for relevant details (derivation of several inequalities, definition of weak solution, the limiting processes, etc.).

For fluids with shear-dependent viscosities independent of pressure, the existence of a weak solution to (10) has been proved if $p > \frac{9}{5}$ in [5, 27, 28]. This result has been improved in [17] for $p > \frac{8}{5}$, via a different approach.

2. Equations, assumptions and main theorem

For $T \in (0, \infty)$ and $\Omega = (0, 2\pi)^3$ we denote $(0, T) \times \Omega$ by Q. We consider the system of equations for $\boldsymbol{v} : (0, T) \times \mathbb{R}^3 \mapsto \mathbb{R}^3$ and $P : (0, T) \times \mathbb{R}^3 \mapsto \mathbb{R}$ expressing the balance of momentum and the fact that an incompressible fluid can undergo only isochoric motions:

$$\begin{aligned} \partial_t \boldsymbol{v} + \operatorname{div}(\boldsymbol{v} \otimes \boldsymbol{v}) - \operatorname{div}(\nu(P, |\mathbf{D}(\boldsymbol{v})|^2)\mathbf{D}(\boldsymbol{v})) &= -\nabla P, \\ \operatorname{div} \boldsymbol{v} &= 0. \end{aligned} \tag{10}$$

For the sake of simplicity, we set the external body forces to be zero. The ν and P that appear in (10) are not to be confused with the ν and P that appear in (4), they stand for the quantities in (4) divided by ρ. Further, we suppose that

$$\begin{aligned} &\boldsymbol{v} \text{ and } P \text{ are spatially } 2\pi\text{-periodic functions,} \\ &\int_\Omega \boldsymbol{v}(t, \cdot) = \mathbf{0}, \quad \int_\Omega P(t, \cdot) = g(t) \quad \text{for a.e. } t \in (0, T), \end{aligned} \tag{11}$$

where $g : (0, T) \to \mathbb{R}$ is given. We also require that, for a given $\boldsymbol{v}_0 : (0, T) \times \mathbb{R}^3 \mapsto \mathbb{R}^3$, the initial condition

$$\boldsymbol{v}(0, \cdot) = \boldsymbol{v}_0 \qquad (\operatorname{div} \boldsymbol{v}_0 = 0, \boldsymbol{v}_0 \in L^2) \tag{12}$$

is satisfied at least in a weak sense.

Recalling from (7) that $\nu(P, |\mathbf{D}|^2)\mathbf{D} = \partial_{\mathbf{D}} U(P, |\mathbf{D}|^2)$, we make the following assumptions concerning the viscosity function ν.

(A1) For a given $p \in (1, 2)$ there are positive constants C_1 and C_2 such that, for all symmetric matrices $\mathbf{B}$, $\mathbf{D}$ and all $P \in \mathbb{R}$,

$$\begin{aligned} C_1 \left(1 + |\mathbf{D}|^2\right)^{\frac{p-2}{2}} |\mathbf{B}|^2 &\leqq \left(\partial_{\mathbf{D}}^{(2)} U(P, |\mathbf{D}|^2)\mathbf{B}, \mathbf{B}\right) \equiv \partial_{\mathbf{D}}[\nu(P, |\mathbf{D}|^2)\mathbf{D}] \cdot (\mathbf{B} \otimes \mathbf{B}) \\ &\leqq C_2 (1 + |\mathbf{D}|^2)^{\frac{p-2}{2}} |\mathbf{B}|^2; \end{aligned} \tag{13}$$

(A2) For all symmetric matrices $\mathbf{D}$ and all $P \in \mathbb{R}$,

$$\left|\partial_P \nu(P, |\mathbf{D}|^2)\right| |\mathbf{D}| \leqq \gamma_0 (1 + |\mathbf{D}|^2)^{\frac{p-2}{4}} \leqq \gamma_0 \ \text{ with } \gamma_0 = \min\left\{\frac{1}{2}, \frac{C_1}{4C_2}\right\}. \tag{14}$$

Viscosity forms presented below in Examples 1 and 2 show that there is a realistic class of viscosities satisfying (14). Note that the viscosities given in Example 1 also fulfil (13), while the viscosity in Example 2 does not.

Example 1. Let $A \in (0, 1]$ be arbitrary (typically small). Consider the model (4) with

$$\nu_i(P, |\mathbf{D}|^2) = \left(A + \gamma_i(P) + |\mathbf{D}|^2\right)^{\frac{p-2}{2}}, \quad i = 1, 2, 3, \tag{15}$$

where $\gamma_i(P)$ takes one of the following forms for $\alpha, q > 0$:

$$\gamma_1(P) = (1 + \alpha^2 P^2)^{-\frac{q}{2}}, \tag{16}$$

$$\gamma_2(P) = (1 + \exp(\alpha P))^{-q}, \tag{17}$$

$$\gamma_3(P) = \begin{cases} \exp(-\alpha q P) & \text{if } P > 0, \\ 1 & \text{if } P \leqq 0. \end{cases} \tag{18}$$

We see that for all $i = 1, 2, 3$

$$0 < \gamma_i(P) \leqq 1 \qquad \text{for all } P \in \mathbb{R}, \tag{19}$$

$$\gamma_i(\cdot) \text{ is decreasing} \qquad \text{as } |P| \to \infty. \tag{20}$$

Since $p < 2$ we observe that ν_i takes values in $(0, A^{\frac{p-2}{2}})$ and (20) implies that

$$\begin{aligned} &\nu_i(\cdot, |\mathbf{D}|^2) \text{ is increasing,} \\ &\nu_i(P, \cdot) \quad \text{ is decreasing.} \end{aligned} \tag{21}$$

Thus fluids (4) defined through (15)–(18) are pressure thickening and shear thinning, which is in agreement with experimental observations.

Let $W_i = A + \gamma_i(P) + |\mathbf{D}|^2$. Then $\partial_{\mathbf{D}} U_i\left(P, |\mathbf{D}|^2\right) = W_i^{\frac{p-2}{2}} \mathbf{D}$. Thus

$$\begin{aligned} \left(\partial_D^{(2)} U_i(P, |\mathbf{D}|^2)\mathbf{B}, \mathbf{B}\right) &= W_i^{\frac{p-2}{2}} |\mathbf{B}|^2 + (p-2) W_i^{\frac{p-4}{2}} (\mathbf{D}\cdot\mathbf{B})^2 \\ &= W_i^{\frac{p-4}{2}} \left(W_i |\mathbf{B}|^2 + (p-2)(\mathbf{D}\cdot\mathbf{B})^2\right) \\ &= W_i^{\frac{p-4}{2}} \left(\gamma_i(P)|\mathbf{B}|^2 + |\mathbf{D}|^2\,|\mathbf{B}|^2 + (p-2)(\mathbf{D}\cdot\mathbf{B})^2\right). \end{aligned} \tag{22}$$

Since $p < 2$ we obtain

$$(p-1) W_i^{\frac{p-2}{2}} |\mathbf{B}|^2 \leqq \partial_D^{(2)} U_i((P, |\mathbf{D}|^2)\mathbf{D}, \mathbf{D}) \leqq W_i^{\frac{p-2}{2}} |\mathbf{B}|^2. \tag{23}$$

Then (13) follows from (23) with $C_1 = p - 1$ and $C_2 = A^{\frac{p-2}{2}}$.

Next,

$$\partial_P \nu_i(P, |\mathbf{D}|^2) = \frac{p-2}{2} W_i^{\frac{p-4}{2}} \gamma_i'(P). \tag{24}$$

We are going to verify that for all forms of $\gamma_i(P)$, $i = 1, 2, 3$, the following inequality holds:

$$\left|\partial_P \nu_i(P, |\mathbf{D}|^2)\,\mathbf{D}\right| \leqq \alpha\, q \frac{2-p}{2} (1 + |\mathbf{D}|^2)^{\frac{p-3}{2}}. \tag{25}$$

Indeed,

$$\begin{aligned} &\left|\partial_P \nu_1(P, |\mathbf{D}|^2)\,\mathbf{D}\right| \\ &\leqq \alpha q \frac{2-p}{2} \left(\frac{1}{(1+\alpha^2 P^2)^{\frac{q}{2}}} + |\mathbf{D}|^2\right)^{\frac{p-4}{2}} \frac{\alpha P |\mathbf{D}|}{(1+\alpha^2 P^2)^{\frac{q+2}{2}}} \\ &= \alpha q \frac{2-p}{2} (1+\alpha^2 P^2)^{\frac{q}{2}\left(\frac{4-p}{2} - \frac{1}{2} - 1\right) - 1} \frac{\alpha P (|\mathbf{D}|^2 (1+\alpha^2 P^2)^{\frac{q}{2}})^{\frac{1}{2}}}{(1 + (1+\alpha^2 P^2)^{\frac{q}{2}} |\mathbf{D}|^2)^{\frac{4-p}{2}}} \\ &\leqq \alpha q \frac{2-p}{2} \frac{1}{(1+\alpha^2 P^2)^{\frac{q}{4}(p-1)+1}} \frac{1}{(1 + (1+\alpha^2 P^2)^{\frac{q}{2}} |\mathbf{D}|^2)^{\frac{3-p}{2}}} \\ &\leqq \alpha q \frac{2-p}{2} \frac{1}{(1+|\mathbf{D}|^2)^{\frac{3-p}{2}}}; \end{aligned}$$

$$
\begin{aligned}
&\left|\partial_P \nu_2(P,|\mathbf{D}|^2)\,\mathbf{D}\right| \\
&\quad = \alpha q \frac{2-p}{2} \frac{(1+(1+\exp(\alpha P))^q|\mathbf{D}|^2)^{\frac{p-4}{2}}}{(1+\exp(\alpha P))^{q\frac{p-4}{2}}} \frac{\exp(\alpha P)|\mathbf{D}|}{(1+\exp(\alpha P))^{q+1}} \\
&\quad \leqq \alpha q \frac{2-p}{2} (1+\exp(\alpha P))^{q\left(\frac{4-p}{2}-\frac{1}{2}-1\right)} \frac{(|\mathbf{D}|^2(1+\exp(\alpha P))^q)^{\frac{1}{2}}}{(1+(1+\exp(\alpha P))^q|\mathbf{D}|^2)^{\frac{4-p}{2}}} \\
&\quad \leqq \alpha q \frac{2-p}{2} \frac{1}{(1+\exp(\alpha P))^{\frac{q}{2}(p-1)}} \frac{1}{(1+(1+\exp(\alpha P))^q|\mathbf{D}|^2)^{\frac{3-p}{2}}} \\
&\quad \leqq \alpha q \frac{2-p}{2} \frac{1}{(1+|\mathbf{D}|^2)^{\frac{3-p}{2}}}.
\end{aligned}
$$

The estimate of $|\partial_P \nu_3(P,|\mathbf{D}|^2)\mathbf{D}|$ is carried out in a completely analogous way.

Since $\frac{3-p}{2} > \frac{2-p}{4}$ we see that (14) holds provided that we fulfil the condition on smallness of γ_0, which is tantamount to

$$
\alpha q(2-p) \leqq \frac{p-1}{2} A^{\frac{2-p}{2}}. \tag{26}
$$

This is fulfilled by taking one of the parameters α, q sufficiently small or p close enough to 2.

To summarize, the model (4) with the viscosity given by (15) where γ_i is any one of functions (16)–(18), with α, A, q and p satisfying (26), fulfils the assumptions (A1), (A2).

On the other hand the model proposed in [15] and numerically tested in [19] does not satisfy the assumptions (A1), (A2). The viscosity in these articles is given by

$$
\nu(P,|\mathbf{D}|^2) = \left(\eta_\infty + \frac{\eta_0-\eta_\infty}{(1+\beta|\mathbf{D}|^{2-p})}\right)\exp(\alpha P) \qquad (\text{with } p = 1.46), \tag{27}
$$

which does not fall into the class studied here. We can however "approximate" (27) by (15) with γ given by (17) or (18). The linear growth in $\exp(\alpha P)$ is then achieved by setting $q = \frac{2}{2-p}$, which takes values in $[2,\infty)$ if p runs between 1 and 2. Then we can write ν_2 as

$$
\nu_2(P,|\mathbf{D}|^2) = \frac{1+\exp(\alpha P)}{\left(1+(A+|\mathbf{D}|^2)(1+\exp(\alpha P))^{\frac{2}{2-p}}\right)^{\frac{2-p}{2}}}. \tag{28}
$$

In view of the previous work, it would be interesting to compare the numerical solutions for model (10) with the viscosity given by (27) versus (10) with the viscosity given by (28) or its variant

$$
\tilde{\nu}_2(P,|\mathbf{D}|^2) = \frac{\varepsilon_0+\eta_\infty\exp(\alpha P)}{\left(1+|\mathbf{D}|^2(\varepsilon_0+\eta_\infty\exp(\alpha P))^{\frac{2}{2-p}}\right)^{\frac{2-p}{2}}}, \qquad \varepsilon_0 \geqq 0. \tag{29}
$$

Example 2. Another interesting class of fluids are those in which the dependence of the viscosity on pressure is linear, i.e.,

$$\nu(P, |\mathbf{D}|^2) = \alpha\, P\, \tilde{\nu}(|\mathbf{D}|^2). \tag{30}$$

For simplicity we consider the case

$$\nu(P, |\mathbf{D}|^2) = \alpha\, P\, |\mathbf{D}|^{p-2}. \tag{31}$$

Our interest in such models stems from the fact that such fluids admit interesting characteristics, both from the engineering and mathematical standpoint, as clarified below.

In a certain regime, the viscosity-pressure relationship can be approximated by a linear function and it has been shown in [20] that steady unidirectional flows for models of the class (31) are possible. More precisely, considering steady flows between two parallel plates located at $x_2 = \pm 1$ (in the Cartesian coordinate system (x_1, x_2, x_3)) of the form

$$\boldsymbol{v} = (v_1(x_2), 0, 0), \quad P = P(x_1, x_2, x_3), \quad v_1(\pm 1) = 0, \tag{32}$$

we obtain the derivative of v_1 (see [20] for details) as

$$v_1'(x_2) = \left[\frac{1}{\alpha}\frac{\sinh(C_0 x_2)}{\cosh(C_0 x_2)}\right]^{\frac{1}{p-1}}, \quad C_0 < 0, \tag{33}$$

which can be explicitly integrated if $p = \frac{n}{n-1}$, $n \in \mathbb{N}$. In particular, if $p = 2$ (i.e., $n = 2$) we have (see Fig. 1)

$$\begin{aligned} v_1(x_2) &= \frac{1}{\alpha C_0} \ln\left(\frac{\cosh(C_0 x_2)}{\cosh(C_0)}\right), \\ P(x_1, x_2) &= L \exp(C_0 x_1) \cosh(C_0 x_2), \quad L > 0. \end{aligned} \tag{34}$$

However, flows of the type (32) are not possible if the viscosity depends on the pressure exponentially, as also explained in [20].

In addition, for $p = \frac{n}{n-1}$, $n \in \mathbb{N}$, $n > 2$, it was shown that unidirectional flows (32) between plates moving with constant speed V subjected thus to $v_1(-1) = 0$ and $v_1(1) = V$ *allow for a multiplicity of solutions. This type of non-uniqueness does not occur in either Navier-Stokes fluids or power-law fluids (with viscosity independent of pressure).* For example, if $n = 3$ (i.e, $p = \frac{3}{2}$) then there are two solutions $(\boldsymbol{v}^1, P^1)$, $(\boldsymbol{v}^2, P^2)$ of the form (32) satisfying (10) with (31) and $\partial_t \boldsymbol{v} = 0$. Their analytic forms for $i = 1, 2$ are

$$\begin{aligned} v_1^i(x_2) &= \frac{1}{\alpha^2 C_0}\left(\frac{2}{M_i e^{2C_0 x_2} + 1} + C_0(x_2 + 1) - \frac{2}{M_i e^{-2C_0} + 1}\right), \\ P^i(x_1, x_2) &= L(M_i e^{2C_0 x_2} + 1) e^{C_0(x_1 - x_2)}, \quad L > 0, \end{aligned}$$

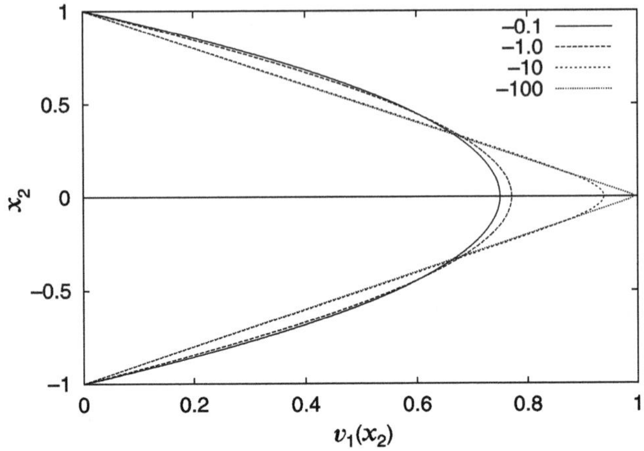

Fig. 1. Poiseuille flow. Velocity profiles for different values of $C_0 = -0.1, -1, -10, -100$ and $p = 2$.

with

$$M_i = \frac{-C_0(\alpha V - 2)\big(e^{4C_0} + 1\big) - 2\big(e^{4C_0} - 1\big)}{2C_0(\alpha^2 V - 2)e^{2C_0}}$$
$$+(-1)^i \frac{\Big[\big(C_0(\alpha V - 2)\big(e^{4C_0} + 1\big) + 2\big(e^{4C_0} - 1\big)\big)^2 - 4\alpha^2 C_0^2(\alpha^2 V - 2)e^{4C_0}\Big]^{\frac{1}{2}}}{2C_0(\alpha^2 V - 2)e^{2C_0}}.$$

These solutions are depicted for $C_0 = -1.9$ in Fig. 2.

Furthermore, *the explicitly computed pressure and consequently also the viscosity are positive although this was not required a priori*. This naturally leads to the question as to whether in a general three-dimensional setting the pressure can be shown to be bounded from below, or even more generally, whether solutions of the equations with a linear pressure-viscosity relation have better properties than (suitable) weak solutions of the Navier-Stokes equations in three dimensions.

We also observe that models given by (31) satisfy (14) only when $p = 1$. This particular model was introduced by SCHAEFFER (see [35]) in order to describe certain flows of granular materials. The preliminary numerical tests presented in [20] also confirmed that fluids with viscosities depending linearly on the pressure exhibit not only flows that differ from Navier-Stokes fluids in the distribution of the pressure, but also they may exhibit significant departures in the structure of the velocity field.

Although there is not much known theoretically concerning models with $p = 1$ even when ν does not vary with the pressure, there is some hope for proving the existence of a global-in-time strong solution in the two-dimensional spatially periodic setting in view of certain recent results [26].

Before formulating the main theorem we recall the definition of some function spaces. The standard Lebesgue and Sobolev spaces (of spatially 2π-periodic functions) are denoted by L^r and $W^{1,r}$ respectively. We do not distinguish notationally

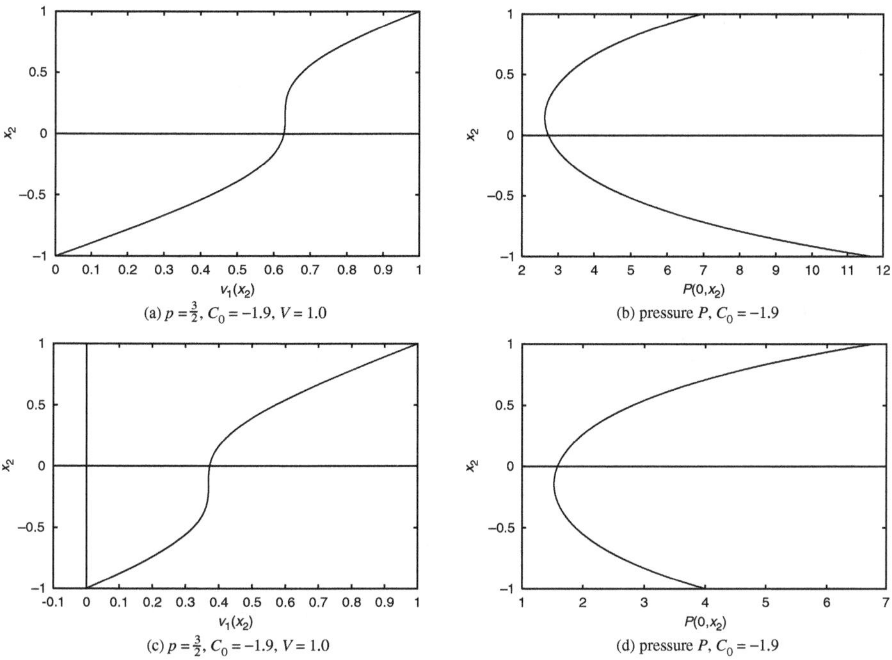

Fig. 2. Couette flow. Velocity profiles (a, c) and pressure across the channel (b, d) for $C_0 = -1.9$ and $p = \frac{3}{2}$; there are two possible solutions.

spaces of scalar- , vector- or tensor-valued functions. A subspace of any function space X consisting of functions belonging to X and having mean value zero is denoted by $\dot{X}$. A subspace of X consisting of vector-valued functions belonging to X and being divergenceless is denoted by X_{div}. The notation of Bochner spaces is mostly also standard. The space $C(0, T; X_{\text{weak}})$ consists of all $v \in L^\infty(0, T; X)$ such that for all $h \in X$ the mapping $t \to \langle v(t), h\rangle$ belongs to $C(0, T)$.

We formulate the main result.

Theorem 1. *Let* (A1) *and* (A2) *hold, let p in* (13) *satisfy*

$$p \in \left(\tfrac{9}{5}, 2\right), \tag{35}$$

and let

$$\boldsymbol{v}_0 \in \dot{W}^{1,2}_{\text{div}}, \quad g \in L^2(0, T). \tag{36}$$

Then there is a weak solution $(\boldsymbol{v}, P)$ to (10)–(12) *such that*

$$\boldsymbol{v} \in C\left(0, T; \dot{L}^2_{\text{weak}}\right) \cap L^p\left(0, T; \dot{W}^{1,p}_{\text{div}}\right), \tag{37}$$

$$P \in L^{\frac{5p}{6}}\left(0, T; L^{\frac{5p}{6}}\right) \tag{38}$$

and (10) *holds in the sense of distributions.*

Moreover, if $p \in (\frac{5}{3}, 2)$, then there exists a strong solution $(\boldsymbol{v}, P)$ to (10)–(12) *satisfying*

$$\boldsymbol{v} \in L^\infty\left(0, T^*; \dot{W}^{1,2}_{\text{div}}\right) \cap L^p\left(0, T^*; \dot{W}^{2,p}_{\text{div}}\right), \tag{39}$$

$$P \in L^2\left(0, T^*; W^{1,2}\right). \tag{40}$$

Here, $T^* > 0$ is arbitrary if $\boldsymbol{v}_0$ is sufficiently small, or T^* is small enough if $\boldsymbol{v}_0$ is arbitrary.

The proof of Theorem is presented in the remaining two sections.

3. Approximations

For an $\varepsilon > 0$ we introduce the following approximation to the problem (10):

$$\begin{aligned} \partial_t \boldsymbol{v} + \operatorname{div}(\boldsymbol{v} \otimes \boldsymbol{v}) - \operatorname{div}(\nu(P, |\mathbf{D}(\boldsymbol{v})|^2)\mathbf{D}(\boldsymbol{v})) + \varepsilon \operatorname{div}(\Delta \mathbf{D}(\boldsymbol{v})) &= -\nabla P, \\ \operatorname{div} \boldsymbol{v} &= 0, \end{aligned} \tag{41}$$

with the requirement that the approximations satisfy the spatial periodicity condition (11) and the initial condition (12). In (41), Δ denotes the Laplace operator. Note that since $\operatorname{div} \boldsymbol{v} = 0$, we have $\operatorname{div}(\Delta(\mathbf{D}(\boldsymbol{v}))) = \Delta^2 \boldsymbol{v}$.

Let $\mathbb{P}$ project an vector field onto its divergenceless part. Formally applying the projector $\mathbb{P}$ to (41), and then taking the divergence of (41), we have

$$\partial_t \boldsymbol{v} + \mathbb{P} \operatorname{div}(\boldsymbol{v} \otimes \boldsymbol{v}) - \mathbb{P} \operatorname{div}(\nu(P, |\mathbf{D}(\boldsymbol{v})|^2)\mathbf{D}(\boldsymbol{v})) + \varepsilon \operatorname{div}(\Delta \mathbf{D}(\boldsymbol{v})) = \mathbf{0}, \tag{42}$$

$$P = (-\Delta)^{-1} \operatorname{div} \operatorname{div}(\boldsymbol{v} \otimes \boldsymbol{v} - \nu(P, |\mathbf{D}(\boldsymbol{v})|^2)\mathbf{D}(\boldsymbol{v})). \tag{43}$$

Such a procedure for the Navier-Stokes model provides separate equations for the velocity and pressure. However, in our case, the system (42), (43) is still coupled. Nevertheless, the above procedure leads to a "pseudo-separation" as evidenced by results presented below.

First of all, by virtue of (13) and Lemma 5.1.19 in [27], equation (42) leads to the energy inequality

$$\tfrac{1}{2}\|\boldsymbol{v}(t)\|_2^2 + C_4 \int_0^T \|\mathbf{D}(\boldsymbol{v})\|_p^p \, dt \leqq \tfrac{1}{2}\|\boldsymbol{v}_0\|_2^2 + C_4 \,, \tag{44}$$

and we see that P does not appear in (44).

The second observation is that by virtue of (14) the pressure is determined in a unique way by $\boldsymbol{v}$ and, roughly speaking, the strong convergence of the velocity gradients implies the strong convergence of the pressures. To see this, let us consider for $\varepsilon_n \to 0$ the sequence $(\boldsymbol{v}^n, P^n) := (\boldsymbol{v}^{\varepsilon_n}, P^{\varepsilon_n})$ of solutions to (41). Thus, in particular, we compute from (43) considered for P^n and P^m (for $\theta_1, \theta_2 \in (0, 1)$),

$$\begin{aligned} P^n - P^m &= \mathcal{F}(\boldsymbol{v}^n \otimes \boldsymbol{v}^n - \boldsymbol{v}^m \otimes \boldsymbol{v}^m) \\ &\quad + \mathcal{F}(\nu(P^m, |\mathbf{D}(\boldsymbol{v}^m)|^2)\mathbf{D}(\boldsymbol{v}^m) - \nu(P^m, |\mathbf{D}(\boldsymbol{v}^n)|^2)\mathbf{D}(\boldsymbol{v}^n)) \\ &= \mathcal{F}((\boldsymbol{v}^n - \boldsymbol{v}^m) \otimes \boldsymbol{v}^n + \boldsymbol{v}^m \otimes (\boldsymbol{v}^n - \boldsymbol{v}^m)) \\ &\quad + \mathcal{F}([\nu(P^m, |\mathbf{D}(\boldsymbol{v}^m)|^2) - \nu(P^m, |\mathbf{D}(\boldsymbol{v}^m)|^2)]\mathbf{D}(\boldsymbol{v}^m)) \\ &\quad + \mathcal{F}(\nu(P^n, |\mathbf{D}(\boldsymbol{v}^m)|^2)\mathbf{D}(\boldsymbol{v}^m) - \nu(P^m, |\mathbf{D}(\boldsymbol{v}^n)|^2)\mathbf{D}(\boldsymbol{v}^n)) \\ &= \mathcal{F}((\boldsymbol{v}^n - \boldsymbol{v}^m) \otimes \boldsymbol{v}^n + \boldsymbol{v}^m \otimes (\boldsymbol{v}^n - \boldsymbol{v}^m)) \\ &\quad + \mathcal{F}(\partial_P \nu(P^n + \theta_2(P^m - P^n), |\mathbf{D}(\boldsymbol{v}^m)|^2)(P^m - P^n)\mathbf{D}(\boldsymbol{v}^m)) \\ &\quad + \mathcal{F}(\partial^2_{\mathbf{D}} U(P^n, |\mathbf{D}(\boldsymbol{v}^n) + \theta_1 \mathbf{D}(\boldsymbol{v}^m - \boldsymbol{v}^n)|^2)\mathbf{D}(\boldsymbol{v}^m - \boldsymbol{v}^n)) \,. \end{aligned} \tag{45}$$

Since $\mathcal{F} = (-\Delta)^{-1} \operatorname{div}\operatorname{div}$ is a Fourier multiplier, we take the L^q norm of (45) and use (A1), (A2) to conclude that

$$\begin{aligned} \|P^n - P^m\|_q &\leqq \| \, |\boldsymbol{v}^n + \boldsymbol{v}^m| \, |\boldsymbol{v}^n - \boldsymbol{v}^m| \, \|_q + \gamma_0 \|P^n - P^m\|_q \\ &\quad + C_2 \left\| \frac{\mathbf{D}(\boldsymbol{v}^m - \boldsymbol{v}^n)}{(1 + |\mathbf{D}(\boldsymbol{v}^n) + \theta_1 \mathbf{D}(\boldsymbol{v}^m - \boldsymbol{v}^n))|^2)^{\frac{2-p}{2}}} \right\|_q \\ &\leqq \|\boldsymbol{v}^n + \boldsymbol{v}^m\|_{2q} \, \|\boldsymbol{v}^n - \boldsymbol{v}^m\|_{2q} + \gamma_0 \|P^n - P^m\|_q \\ &\quad + C_2 \|\mathbf{D}(\boldsymbol{v}^m - \boldsymbol{v}^n)\|_q \,. \end{aligned} \tag{46}$$

Therefore

$$\|P^n - P^m\|_q \leqq \frac{1}{1-\gamma_0} \|\boldsymbol{v}^n + \boldsymbol{v}^m\|_{2q} \|\boldsymbol{v}^n - \boldsymbol{v}^m\|_{2q} + \frac{C_2}{1-\gamma_0} \|\mathbf{D}(\boldsymbol{v}^m - \boldsymbol{v}^n)\|_q \,. \tag{47}$$

From (47) we immediately see that corresponding to any $\boldsymbol{v}^n$ there is just one pressure P^n. (Recall that we fixed the function g through (11), which is the reason why the pressure is unique.) Furthermore, taking the q-th power of (47), integrating the result with respect to time from 0 to T and then using Hölder's inequality leads to

$$\begin{aligned} \int_0^T \|P^n - P^m\|_q^q \, dt &\leqq C \left(\int_0^T \|\boldsymbol{v}^n + \boldsymbol{v}^m\|_{2q}^{2q} \, dt \right)^{1/2} \left(\int_0^T \|\boldsymbol{v}^n - \boldsymbol{v}^m\|_{2q}^{2q} \, dt \right)^{1/2} \\ &\quad + C \int_0^T \|\nabla \boldsymbol{v}^n - \nabla \boldsymbol{v}^m\|_q^q \, dt. \end{aligned} \tag{48}$$

Thus, if $\{\boldsymbol{v}^n\}_{n=1}^\infty$ is a Cauchy sequence in $L^{2q}(0, T; L^{2q})$ and $\{\nabla \boldsymbol{v}^n\}_{n=1}^\infty$ is a Cauchy sequence in $L^q(0, T; L^q)$ for some $q > 1$, then (48) implies that $\{P^n\}_{n=1}^\infty$ is a Cauchy sequence in $L^q(0, T; L^q)$.

3.1. Solvability of the approximating problem

Lemma 1. *Let* $\varepsilon > 0$ *be fixed. Assume that* (A1), (A2) *and* (36) *hold. Then there is a weak solution to* (41), (11), (12) *such that*

$$\boldsymbol{v} \in C\left(0, T; L^2_{\text{weak}}\right) \cap L^p\left(0, T; \dot{W}^{1,p}_{\text{div}}\right) \cap L^2\left(0, T; W^{2,2}\right), \tag{49}$$

$$P \in L^2(0, T; L^2), \tag{50}$$

and (41) *holds in the sense of distributions.*

Proof. Let $\{\boldsymbol{\omega}^r\}_{r=1}^{\infty}$ be a basis of 2π-periodic eigenvectors of the Stokes operator orthonormal in L^2 and orthogonal in $W^{1,2}$; i.e., $\boldsymbol{\omega}^r$ satisfies $-\Delta\boldsymbol{\omega}^r + \nabla q = \lambda_r \boldsymbol{\omega}^r$, $\operatorname{div} \boldsymbol{\omega}^r = 0$.

Let $\boldsymbol{v}^N(t, x)$ be defined by

$$\boldsymbol{v}^N(t, x) = \sum_{r=1}^{N} c_r^N(t) \boldsymbol{\omega}^r(x),$$

where $c_r^N(t)$, $r = 1, \ldots, N$, are found by solving the Galerkin system

$$\begin{aligned}(\partial_t \boldsymbol{v}^N, \boldsymbol{\omega}^r) - (\boldsymbol{v}^N \otimes \boldsymbol{v}^N, \nabla\boldsymbol{\omega}^r) + (\nu(P^N, |\mathbf{D}(\boldsymbol{v}^N)|^2)\mathbf{D}(\boldsymbol{v}^N), \mathbf{D}(\boldsymbol{\omega}^r))\\ + \varepsilon(\nabla^2 \boldsymbol{v}^N, \nabla^2 \boldsymbol{\omega}^r) = 0, \qquad r = 1, \ldots, N.\end{aligned} \tag{51}$$

The approximations of the pressure P^N are given via the Fourier series by the formula

$$P^N(t, x) = \sum_{\boldsymbol{k} \in \mathbb{Z}^3, |\boldsymbol{k}| \leqq N} P_{\boldsymbol{k}}^N(t) \mathrm{e}^{i(x, \boldsymbol{k})}, \qquad \boldsymbol{k} = (k_1, k_2, k_3), \tag{52}$$

where $P_{\mathbf{0}}^N(\cdot) = g(\cdot)$ and $P_{\boldsymbol{k}}^N(\cdot)$, $\boldsymbol{k} \neq \mathbf{0}$, are determined from a discrete analogue of (43), namely,

$$P^N = (-\Delta)^{-1} \operatorname{div} \operatorname{div} \left(\Pi^N(\boldsymbol{v}^N \otimes \boldsymbol{v}^N - \partial_{\mathbf{D}} U(P^N, |\mathbf{D}(\boldsymbol{v}^N)|^2))\right). \tag{53}$$

Here Π^N projects any Fourier series $\sum_{\boldsymbol{k} \in \mathbb{Z}^3} g_{\boldsymbol{k}} \mathrm{e}^{i(x,\boldsymbol{k})}$ onto its N-dimensional part, i.e., $\sum_{\boldsymbol{k} \in \mathbb{Z}^3, |\boldsymbol{k}| \leqq N} g_{\boldsymbol{k}} \mathrm{e}^{i(x,\boldsymbol{k})}$.[1] We recall that P^N is uniquely determined by $\boldsymbol{v}^N$ (see remark that immediately follows (47)).

---

[1] To be even more specific, if

$$\left(v_r^N v_s^N - \partial_{D_{rs}} U(P^N, |\mathbf{D}(\boldsymbol{v}^N)|^2)\right)(t, x) = \sum_{\boldsymbol{k} \in \mathbb{Z}^3} g_{\boldsymbol{k}}^{rs}(t) \mathrm{e}^{i(x,\boldsymbol{k})},$$

then $P_{\boldsymbol{k}}^N(\cdot)$, $\boldsymbol{k} \neq \mathbf{0}$, introduced in (52) are given by

$$P_{\boldsymbol{k}}^N(t) = \frac{k_r k_s}{|\boldsymbol{k}|^2} g_{\boldsymbol{k}}^{rs}(t).$$

The *a priori* estimates (obtained rigorously by multiplying the r-th equation in (51) by $c_r^N(t)$, summing over $r = 1, \ldots, N$ and integrating over time from 0 to t)

$$\tfrac{1}{2}\|\boldsymbol{v}^N(t)\|_2^2 + C_4 \int_0^t \|\mathbf{D}(\boldsymbol{v}^N)\|_p^p\, ds + \varepsilon \int_0^t \|\nabla^2 \boldsymbol{v}^N\|_2^2\, ds \leqq \tfrac{1}{2}\|\boldsymbol{v}_0\|_2^2 + C_4\,, \quad (54)$$

and the imbedding and interpolation inequalities

$$\|z\|_4 \leqq c\|z\|_{\frac{3}{4},2} \leqq c\|z\|_2^{\frac{5}{8}}\|z\|_{2,2}^{\frac{3}{8}} \quad (55)$$

imply that

$$\boldsymbol{v}^N \otimes \boldsymbol{v}^N - \partial_{\mathbf{D}} U(P^N, |\mathbf{D}(\boldsymbol{v}^N)|^2) \quad \text{is uniformly bounded in } L^2(0,T;L^2). \quad (56)$$

Consequently,

$$P^N \quad \text{is uniformly bounded in } L^2(0,T;L^2). \quad (57)$$

Thus the Fourier series are convergent in L^2.

The proof of existence of a solution to (51), (53) for fixed N can be carried out in a standard way via the Banach fixed-point theorem, for example. The uniquely defined solution is constructed on a short time interval first and then extended onto $(0, T)$ by virtue of the uniform energy estimates.

In order to pass to the limit we also need information concerning $\partial_t \boldsymbol{v}^N$. It can be concluded from (51) and (54) that

$$\partial_t \boldsymbol{v}^N \quad \text{is uniformly bounded in } L^2(0,T;(W^{2,2})^*). \quad (58)$$

Because of (54), (57), (58) and the Aubin-Lions compactness lemma (see [25], Section 1.5) we can find $(\boldsymbol{v}, P)$ and a subsequence of $(\boldsymbol{v}^N, P^N)$ (not relabeled) such that

$$\boldsymbol{v}^N \rightharpoonup \boldsymbol{v} \qquad \text{weakly in } L^2(0,T;W^{2,2}), \quad (59)$$

$$\boldsymbol{v}^N \rightharpoonup \boldsymbol{v} \qquad \text{*weakly in } L^\infty(0,T;L^2), \quad (60)$$

$$\boldsymbol{v}^N \to \boldsymbol{v} \qquad \text{strongly in } L^2(0,T;W^{1,2}), \quad (61)$$

$$\partial_t \boldsymbol{v}^N \rightharpoonup \partial_t \boldsymbol{v} \qquad \text{weakly in} L^2(0,T;(W^{2,2})^*), \quad (62)$$

$$P^N \rightharpoonup P \qquad \text{weakly in } L^2(0,T;L^2). \quad (63)$$

Due to (54), (61) and the use of interpolation, it also follows that

$$\boldsymbol{v}^N \to \boldsymbol{v} \quad \text{strongly in } L^4(0,T;L^4). \quad (64)$$

Choosing $q = 2$ in (48) and using (61) and (64) we see that $\{P^N\}$ is a Cauchy sequence in $L^2(0,T;L^2)$. Therefore

$$P^N \to P \quad \text{strongly in } L^2(0,T;L^2). \quad (65)$$

The rest of the proof is standard.

3.2. Uniqueness of the solution of the approximating problem

Lemma 2. *Let all the assumptions of Lemma 1 hold. Then the weak solution constructed above is unique.*

Proof. Define $\boldsymbol{w} = \boldsymbol{v}^1 - \boldsymbol{v}^2$ and $Q = P^1 - P^2$ where $(\boldsymbol{v}^1, P^1)$ and $(\boldsymbol{v}^2, P^2)$ are two solutions of (41). Testing the weak form of (42) by $\boldsymbol{w}$ we obtain

$$\frac{1}{2}\frac{d}{dt}\|\boldsymbol{w}\|_2^2 + J + \varepsilon\|\nabla^2\boldsymbol{w}\|_2^2 \leqq \int_\Omega |\boldsymbol{w}|^2|\nabla\boldsymbol{v}^1|\,dx \leqq C\|\boldsymbol{w}\|_{\frac{12}{5}}^2\|\boldsymbol{v}^1\|_{2,2}\,, \tag{66}$$

where J is given by

$$J := \Big(\partial_{\mathbf{D}}U(P^1, |\mathbf{D}(\boldsymbol{v}^1)|^2) - \partial_{\mathbf{D}}U(P^2, |\mathbf{D}(\boldsymbol{v}^2)|^2), \mathbf{D}(\boldsymbol{w})\Big), \tag{67}$$

and where we use Hölder's inequality and the imbedding of $W^{2,2}$ into $W^{1,6}$.

Adding to and subtracting from J the term $(\partial_{\mathbf{D}}U(P^1, |\mathbf{D}(\boldsymbol{v}^2)|^2), \mathbf{D}(\boldsymbol{w}))$ we can split J into two terms J_1 and J_2 defined by

$$J_1 := \Big(\partial^2_{\mathbf{D}}U(P^1, |\mathbf{D}(\boldsymbol{v}^2) + \theta_1\mathbf{D}(\boldsymbol{w})|^2)\mathbf{D}(\boldsymbol{w}), \mathbf{D}(\boldsymbol{w})\Big), \tag{68}$$

$$J_2 := \Big(\partial_P\nu(P^2 + \theta_2 Q, |\mathbf{D}(\boldsymbol{v}^2)|^2)\, Q\,\mathbf{D}(\boldsymbol{v}^2), \mathbf{D}(\boldsymbol{w})\Big). \tag{69}$$

By (13) (and Lemma 5.1.19 in [27])

$$J_1 \geqq 0. \tag{70}$$

Further, by (14) and the interpolation inequality

$$\|\nabla\boldsymbol{w}\|_2 \leqq \|\boldsymbol{w}\|_2^{\frac{1}{2}}\|\nabla^2\boldsymbol{w}\|_2^{\frac{1}{2}}, \tag{71}$$

we estimate J_2 as follows:

$$|J_2| \leqq \gamma_0\|Q\|_2\|\mathbf{D}(\boldsymbol{w})\|_2 \leqq \frac{\gamma_0}{2}\|Q\|_2^2 + \frac{\varepsilon}{4}\|\nabla^2\boldsymbol{w}\|_2^2 + C(\varepsilon)\|\boldsymbol{w}\|_2^2. \tag{72}$$

Since $W^{\frac{1}{4},2}$ is imbedded in $L^{\frac{12}{5}}$, we obtain

$$\|f\|_{\frac{12}{5}}^2 \leqq c\|f\|_{\frac{1}{4},2}^2 \leqq c\|f\|_2^{\frac{7}{4}}\|\nabla^2 f\|_{2,2}^{\frac{1}{4}}. \tag{73}$$

We observe that (66)–(73) yields

$$\frac{1}{2}\frac{d}{dt}\|\boldsymbol{w}\|_2^2 + \frac{1}{2}\varepsilon\|\nabla^2\boldsymbol{w}\|_2^2 \leqq \frac{\gamma_0}{2}\|Q\|_2^2 + C(\varepsilon)\Big(1 + \|\boldsymbol{v}^1\|_{2,2}^{\frac{8}{7}}\Big)\|\boldsymbol{w}\|_2^2. \tag{74}$$

Next, starting with (43) for the pressure, we similarly obtain

$$\begin{aligned}\|Q\|_2^2 \leqq &\int_\Omega |\boldsymbol{v}^1\otimes\boldsymbol{v}^1 - \boldsymbol{v}^2\otimes\boldsymbol{v}^2|^2\,dx\\ &+ \int_\Omega |\partial^2_{\mathbf{D}}U(P^1, |\mathbf{D}(\boldsymbol{v}^1) + \theta_1\mathbf{D}(\boldsymbol{w})|^2)\mathbf{D}(\boldsymbol{w})|^2\,dx\\ &+ \int_\Omega |\partial_P\nu(P^1 + \theta_2 Q, |\mathbf{D}(\boldsymbol{v}^2)|^2)|\,|\mathbf{D}(\boldsymbol{v}^2)|\,|Q|\,dx =: J_4 + J_5 + J_6\,.\end{aligned} \tag{75}$$

By virtue of (55) we have

$$
\begin{aligned}
|J_4| &\leqq \int_\Omega |\boldsymbol{v}^1 \otimes \boldsymbol{w}|^2 + |\boldsymbol{w} \otimes \boldsymbol{v}^2|^2 \, dx \leqq C \int_\Omega \left(|\boldsymbol{v}^1|^2 + |\boldsymbol{v}^2|^2\right) |\boldsymbol{w}|^2 dx \\
&\leqq C\|\boldsymbol{w}\|_4^2 \left(\|\boldsymbol{v}^1\|_4^2 + \|\boldsymbol{v}^2\|_4^2\right) \leqq C\left\|\boldsymbol{v}^1 + \boldsymbol{v}^2\right\|_{2,2}^{3/4} \left\|\boldsymbol{w}\right\|_2^{5/4} \left\|\nabla^2 \boldsymbol{w}\right\|_2^{3/4} \\
&\leqq \frac{\varepsilon}{8}\left\|\nabla^2 \boldsymbol{w}\right\|_2^2 + C(\varepsilon)\left\|\boldsymbol{v}^1 + \boldsymbol{v}^2\right\|_{2,2}^{6/5} \left\|\boldsymbol{w}\right\|_2^2 .
\end{aligned}
\tag{76}
$$

It follows from (13) (since $p < 2$) that

$$
|J_5| \leqq c\|\mathbf{D}(\boldsymbol{w})\|_2^2 \leqq \frac{\varepsilon}{8}\left\|\nabla^2 \boldsymbol{w}\right\|_2^2 + C(\varepsilon)\|\boldsymbol{w}\|_2^2 . \tag{77}
$$

Finally, by (14)

$$
|J_6| \leqq \gamma_0 \|Q\|_2^2 . \tag{78}
$$

Adding (74) and (75), and using (76)–(78) we obtain (since $1 - \gamma_0 \geqq \frac{1}{2}$)

$$
\frac{d}{dt}\|\boldsymbol{w}\|_2^2 + \varepsilon\|\nabla^2(\boldsymbol{w})\|_2^2 + \|Q\|_2^2 \leqq C(\varepsilon)\left(1 + \|\boldsymbol{v}^1 + \boldsymbol{v}^2\|_{2,2}^{6/5} + \|\boldsymbol{v}^1\|_{2,2}^{8/7}\right)\|\boldsymbol{w}\|_2^2 .
$$

Application of the Gronwall lemma completes the proof.

3.3. Regularity of the solution of the approximating problem

Lemma 3. *Approximations $(\boldsymbol{v}^N, P^N)$ satisfy*

$$
\begin{aligned}
&\frac{1}{2}\frac{d}{dt}\|\nabla \boldsymbol{v}^N\|_2^2 + C_1 I_p(\boldsymbol{v}^N) + \varepsilon\|\nabla^{(3)}\boldsymbol{v}^N\|_2^2 \\
&\qquad \leqq C\left(\|\nabla \boldsymbol{v}^N\|_3^3 + \int_\Omega |\boldsymbol{v}^N|^2 |\nabla \boldsymbol{v}^N|^2 \, dx\right),
\end{aligned}
\tag{79}
$$

where

$$
I_p(\boldsymbol{v}^N) := \int_\Omega \left(1 + |\mathbf{D}(\boldsymbol{v}^N)|^2\right)^{\frac{p-2}{2}} |\mathbf{D}(\boldsymbol{v}^N)|^2 \, dx.
$$

Also,

$$
\left\|\partial_t \boldsymbol{v}^N\right\|_2^2 + 2\frac{d}{dt}\int_\Omega U(P^N, |\mathbf{D}(\boldsymbol{v}^N)|^2) \, dx \leqq \int_\Omega \left|\boldsymbol{v}^N\right|^2 \left|\nabla \boldsymbol{v}^N\right|^2 dx. \tag{80}
$$

Proof. First, we take the L^2 norm of the gradient of (53). With the help of (14) and (13) we obtain

$$
\begin{aligned}
\left\|\nabla P^N\right\|_2 &\leqq \left(\int_\Omega |\boldsymbol{v}^N|^2 |\nabla \boldsymbol{v}^N|^2\right)^{1/2} + \left\|\partial_P \nu(P^N, |\mathbf{D}(\boldsymbol{v}^N)|^2)\mathbf{D}(\boldsymbol{v}^N)\nabla P^N\right\|_2 \\
&\quad + \left\|\partial_{\mathbf{D}}^{(2)} U(P^N, |\mathbf{D}(\boldsymbol{v}^N)|^2)\mathbf{D}(\nabla \boldsymbol{v}^N)\right\|_2 \\
&\leqq \left(\int_\Omega |\boldsymbol{v}^N|^2 |\nabla \boldsymbol{v}^N|^2\right)^{1/2} + \gamma_0\|\nabla P^N\|_2 + C_2\left(I_p(\boldsymbol{v}^N)\right)^{1/2} .
\end{aligned}
$$

Therefore

$$\left\|\nabla P^N\right\|_2 \leqq 2C_2\left(I_p(\boldsymbol{v}^N)\right)^{1/2} + 2\left(\int_\Omega |\boldsymbol{v}^N|^2|\nabla\boldsymbol{v}^N|^2\right)^{1/2}. \tag{81}$$

Next, we multiply (51) by $\lambda_r c_r^N(t)$, sum over $r = 1, \dots, N$, and use integration by parts. (In short, we use $-\Delta\boldsymbol{v}^N$ as a test function.) This leads to

$$\begin{aligned}
&\frac{1}{2}\frac{d}{dt}\left\|\nabla\boldsymbol{v}^N\right\|_2^2 + \left(\partial_{\mathbf{D}}^{(2)}U(P^N, |\mathbf{D}(\boldsymbol{v}^N)|^2)\mathbf{D}(\nabla\boldsymbol{v}^N), \mathbf{D}(\nabla\boldsymbol{v}^N)\right) + \varepsilon\left\|\nabla^{(3)}(\boldsymbol{v}^N)\right\|_2^2 \\
&\quad\leqq \left\|\nabla\boldsymbol{v}^N\right\|_3^3 + \left|\left(\partial_P\nu(P^N, |\mathbf{D}(\boldsymbol{v}^N)|^2)\nabla P^N\mathbf{D}(\boldsymbol{v}^N), \mathbf{D}(\nabla\boldsymbol{v}^N)\right)\right|.
\end{aligned} \tag{82}$$

Assumptions (13) and (14) then imply that

$$\begin{aligned}
&\frac{1}{2}\frac{d}{dt}\left\|\nabla\boldsymbol{v}^N\right\|_2^2 + C_1 I_p(\boldsymbol{v}^N) + \varepsilon\left\|\nabla^{(3)}\boldsymbol{v}^N\right\|_2^2 \\
&\quad\leqq \left\|\nabla\boldsymbol{v}^N\right\|_3^3 + \gamma_0\int_\Omega (1 + |\mathbf{D}(\boldsymbol{v}^N)|^2)^{\frac{p-2}{4}}|\mathbf{D}(\nabla\boldsymbol{v}^N)|\,|\nabla P^N|\,dx \\
&\quad\leqq \left\|\nabla\boldsymbol{v}^N\right\|_3^3 + \gamma_0\left\|\nabla P^N\right\|_2\left(I_p(\boldsymbol{v}^N)\right)^{1/2}.
\end{aligned} \tag{83}$$

Substituting (81) into the right-hand side of (83) and using the Young inequality and the assumptions for γ_0 (see (14)) we obtain (79).

To derive (80) we multiply (51) by $\dot{c}_r^N(t)$ and sum over $r = 1, \dots, N$. (In short, we test by $\partial_t\boldsymbol{v}^N$.) The inequality (80) is then obtained by using the Schwarz inequality to evaluate the right-hand term. The proof of the lemma is complete.

Lemma 4. *Let the assumptions of Lemma 1 be satisfied. Then for every fixed $\varepsilon > 0$, a solution $(\boldsymbol{v}^\varepsilon, P^\varepsilon)$ to* (41), (11), (12) *satisfies*

$$\boldsymbol{v}^\varepsilon \in L^\infty(0, T; \dot{W}^{1,2}_{\mathrm{div}}) \cap L^2(0, T; \dot{W}^{4,2}), \tag{84}$$

$$\partial_t\boldsymbol{v}^\varepsilon \in L^2(0, T; \dot{L}^2), \tag{85}$$

$$\nabla P^\varepsilon \in L^2(0, T; L^2). \tag{86}$$

Proof. First of all we observe that the imbedding and interpolation inequalities lead to

$$\begin{aligned}
\left\|\nabla\boldsymbol{v}^N\right\|_3^3 &\leqq C\left\|\boldsymbol{v}^N\right\|_{\frac{3}{2},2}^3 \leqq C\left\|\boldsymbol{v}^N\right\|_{1,2}^{9/4}\left\|\boldsymbol{v}^N\right\|_{3,2}^{3/4} \leqq \frac{\varepsilon}{4}\left\|\nabla^{(3)}\boldsymbol{v}^N\right\|_2^2 + C(\varepsilon)\left\|\boldsymbol{v}^N\right\|_{1,2}^{18/5} \\
&\leqq \frac{\varepsilon}{4}\left\|\nabla^{(3)}\boldsymbol{v}^N\right\|_2^2 + C(\varepsilon)\left\|\boldsymbol{v}^N\right\|_{2,2}^{8/5}\left\|\nabla\boldsymbol{v}^N\right\|_2^2,
\end{aligned}$$

and

$$\begin{aligned}
\int_\Omega |\boldsymbol{v}^N|^2|\nabla\boldsymbol{v}^N|^2\,dx &\leqq C\|\boldsymbol{v}^N\|_\infty^2\|\nabla\boldsymbol{v}^N\|_2^2 \leqq C\|\boldsymbol{v}^N\|_2\|\nabla^{(3)}\boldsymbol{v}^N\|_2\|\nabla\boldsymbol{v}^N\|_2^2 \\
&\leqq \frac{\varepsilon}{4}\|\nabla^{(3)}\boldsymbol{v}^N\|_2^2 + C(\varepsilon)\|\boldsymbol{v}^N\|_2^2\|\nabla\boldsymbol{v}^N\|_2^2.
\end{aligned}$$

Using these estimates, we observe that the inequality (79) yields

$$\frac{d}{dt}\|\nabla \boldsymbol{v}^N\|_2^2 + 2C_1 I_p(\boldsymbol{v}^N) + \varepsilon\|\nabla^{(3)}\boldsymbol{v}^N\|_2^2 \leqq C(\varepsilon)\left(\|\boldsymbol{v}^N\|_{2,2}^{\frac{8}{3}} + 1\right)\|\nabla \boldsymbol{v}^N\|_2^2. \tag{87}$$

In view of (54) we can apply the Gronwall lemma to (87) to conclude that

$$\sup_{t\in(0,T)} \|\nabla \boldsymbol{v}^N(t)\|_2^2 \leqq C(\varepsilon). \tag{88}$$

These computations, (80) and (87) lead us to conclude that

$$\int_0^T \|\partial_t \boldsymbol{v}^N(t)\|_2^2 \leqq C(\varepsilon). \tag{89}$$

Consequently, the limit $\boldsymbol{v}^\varepsilon$ of any weakly or *weakly convergent subsequence of $\{\boldsymbol{v}^\varepsilon\}$ satisfies $(84)_1$ and (85). The regularity of the pressure follows from (53). To see that $\boldsymbol{v}^\varepsilon$ belongs to $L^2(0, T; W^{4,2})$ we need only to appeal to (41).

4. Proof of the theorem

Let us recall some known estimates that are uniform with respect to ε and which will be helpful in our proof. From the energy inequality (54) we have

$$\|\boldsymbol{v}^\varepsilon(t)\|_2^2 + \int_0^T \|\nabla \boldsymbol{v}^\varepsilon\|_p^p\, ds + \varepsilon \int_0^T \|\nabla^2 \boldsymbol{v}^\varepsilon\|_2^2\, ds \leqq K, \tag{90}$$

from which we can conclude, on using standard interpolation techniques, that

$$\int_0^T \|\boldsymbol{v}^\varepsilon\|_{\frac{5p}{3}}^{\frac{5p}{3}}\, ds \leqq K. \tag{91}$$

From equation (53) for P^ε and (90), (91) we observe that

$$\int_0^T \|P^\varepsilon\|_{\frac{5p}{6}}^{\frac{5p}{6}}\, ds \leqq K. \tag{92}$$

We also have the following bound on $\partial_t \boldsymbol{v}^\varepsilon$ (see (5.2.25) in [27], p 207):

$$\|\partial_t \boldsymbol{v}^\varepsilon\|_{L^{p'}(0,T;(W^{s,2})^*)} \leqq K, \qquad s > \frac{5}{2},\ p' = \frac{p}{p-1}. \tag{93}$$

These uniform estimates and the Aubin-Lions compactness lemma then imply the following convergence of the various quantities.

Lemma 5. *There is a sequence* $\varepsilon^n \to 0$ *and* $(\boldsymbol{v}, P)$ *such that*

$$\boldsymbol{v}^{\varepsilon_n} \rightharpoonup \boldsymbol{v} \qquad \text{weakly in } L^p(0,T;W^{1,p}), \tag{94}$$

$$\boldsymbol{v}^{\varepsilon_n} \rightharpoonup \boldsymbol{v} \qquad \text{*weakly in } L^\infty(0,T;L^2), \tag{95}$$

$$\partial_t\boldsymbol{v}^{\varepsilon_n} \rightharpoonup \partial_t\boldsymbol{v} \qquad \text{weakly in } L^{p'}\left(0,T;(W^{s,2})^*\right), \qquad s > \frac{5}{2}, \tag{96}$$

$$P^{\varepsilon_n} \rightharpoonup P \qquad \text{weakly in } L^{\frac{5p}{6}}\left(0,T;L^{\frac{5p}{6}}\right), \tag{97}$$

$$\boldsymbol{v}^{\varepsilon_n} \to \boldsymbol{v} \qquad \text{strongly in } L^q\left(0,T;L^q\right) \cap L^r\left(0,T;L^2\right) \quad \text{for } q \in \left\langle 1, \frac{5p}{3}\right), r \in \langle 1, \infty). \tag{98}$$

In the process of going from the weak formulations for $(\boldsymbol{v}^\varepsilon, P^\varepsilon)$ to a weak form for $(\boldsymbol{v}, P)$ the linear terms, namely those involving the pressure and time derivative, and the convective term can be handled easily. The limiting process for the terms with pressure and time derivative follows from (94) and (97), while the limit for the convective term follows from (98). The key point is to show that for any smooth 2π-periodic function φ we have

$$\int_0^T\int_\Omega \nu\left(P^{\varepsilon_n}, |\mathbf{D}(\boldsymbol{v}^{\varepsilon_n})|^2\right) D_{ij}(\boldsymbol{v}^{\varepsilon_n})D_{ij}(\varphi)\,dx\,dt \to \int_0^T\int_\Omega \nu\left(P, |\mathbf{D}(\boldsymbol{v})|^2\right) D_{ij}(\boldsymbol{v})D_{ij}(\varphi)\,dx\,dt. \tag{99}$$

At the first glance, it seems that estimates (90)–(93) are not sufficient. However, we will show later that (90)–(93) leads to

$$\int_0^T \|\nabla^2\boldsymbol{v}^{\varepsilon_n}\|_p^{2\beta}\,dt \leqq K \qquad \text{with } \beta \in (0, \tfrac{1}{3}) \text{ if } p \in \left(\tfrac{9}{5}, 2\right). \tag{100}$$

All these estimates together with an interpolation technique for certain $\sigma > 0$ and $p_0 \in (1, p)$ lead then to

$$\left.\begin{aligned}\int_0^T \|\boldsymbol{v}^{\varepsilon_n}\|_{1,p}^p\,dt \leqq K\\ \int_0^T \|\boldsymbol{v}^{\varepsilon_n}\|_{2,p}^{2\beta}\,dt \leqq K\end{aligned}\right. \implies \int_0^T \|\boldsymbol{v}^{\varepsilon_n}\|_{1+\sigma,p}^{p_0}\,dt \leqq K. \tag{101}$$

It is worth noting that although $2\beta < 1$, we can find $p_0 > 1$ and $\sigma > 0$ such that (101) holds.

Having (101) and (93) at hand, we apply the Aubin-Lions compactness lemma again and conclude that

$$\boldsymbol{v}^{\varepsilon_n} \to \boldsymbol{v} \qquad \text{strongly in } L^{p_0}(0,T;W^{1,p}). \tag{102}$$

On the basis of (90), we can strengthen (102) to obtain

$$\boldsymbol{v}^{\varepsilon_n} \to \boldsymbol{v} \qquad \text{strongly in } L^{\tilde{p}}(0, T; W^{1,p}) \qquad \text{for any } \tilde{p} \in [1, p). \tag{103}$$

Then, taking $q = \frac{5\tilde{p}}{6}$ in (48) and using (91) and (103) we come to the conclusion that

$$P^{\varepsilon_n} \to P \qquad \text{strongly in } L^s(0, T; L^s) \qquad \text{for all } s \in \left[1, \frac{5p}{6}\right). \tag{104}$$

Finally, (103) and (104) imply almost everywhere convergence of a subsequence of $(\boldsymbol{v}^{\varepsilon_n}, P^{\varepsilon_n})$ to $(\boldsymbol{v}, P)$. This together with Vitali's theorem yields (99).

In order to complete the proof of global-in-time existence of a weak solution we need to derive the estimate (100). We now outline the procedure.

Since all terms in (41) for $\boldsymbol{v}^{\varepsilon_n}$ are L^2-integrable, we can take the scalar product of (41) with $-\Delta \boldsymbol{v}^{\varepsilon_n}$. Following the same procedure as in the proof of Lemma 3 we obtain the inequality (79) provided we replace $\boldsymbol{v}^N$ by $\boldsymbol{v}^{\varepsilon_n}$ and P^N by P^{ε_n}.

In a manner similar to the procedure adopted in the proof of Lemma 4, we have to estimate both the terms on the right-hand side of (79), this time uniformly with respect to ε.

First of all, let us recall that (cf. Lemma 5.3.24 in [27])

$$\|\nabla \boldsymbol{v}^{\varepsilon_n}\|_{3p}^p \leqq c\, I_p(\boldsymbol{v}^{\varepsilon_n}). \tag{105}$$

Also, we have the following interpolation inequalities:

$$\|z\|_3 \leqq \|z\|_2^{\frac{2(p-1)}{3p-2}} \|z\|_{3p}^{\frac{p}{3p-2}}, \tag{106}$$

$$\|z\|_3 \leqq \|z\|_p^{\frac{p-1}{2}} \|z\|_{3p}^{\frac{3-p}{2}}. \tag{107}$$

Using these inequalities we compute (for $\alpha \in [0, 1]$)

$$\begin{aligned}
\|\nabla \boldsymbol{v}^{\varepsilon_n}\|_3^3 &= \|\nabla \boldsymbol{v}^{\varepsilon_n}\|_3^{3\alpha} \|\nabla \boldsymbol{v}^{\varepsilon_n}\|_3^{3(1-\alpha)} \\
&\leqq \|\nabla \boldsymbol{v}^{\varepsilon_n}\|_p^{3(1-\alpha)\frac{p-1}{2}} \|\nabla \boldsymbol{v}^{\varepsilon_n}\|_2^{\frac{6\alpha(p-1)}{3p-2}} \|\nabla \boldsymbol{v}^{\varepsilon_n}\|_{3p}^{\frac{3\alpha p}{3p-2}+3(1-\alpha)\frac{3-p}{2}} \\
&\leqq \left(\|\nabla \boldsymbol{v}^{\varepsilon_n}\|_p^p\right)^{3(1-\alpha)\frac{p-1}{2p}} \left(\|\nabla \boldsymbol{v}^{\varepsilon_n}\|_2^2\right)^{\frac{3\alpha(p-1)}{3p-2}} \left(\|\nabla \boldsymbol{v}^{\varepsilon_n}\|_{3p}^p\right)^{\frac{3\alpha}{3p-2}+3(1-\alpha)\frac{3-p}{2p}} \\
&\leqq \frac{C_1}{4} I_p(\boldsymbol{v}^{\varepsilon_n}) + C\left(\|\nabla \boldsymbol{v}^{\varepsilon_n}\|_p^p\right)^{\mu} (1 + \|\nabla \boldsymbol{v}^{\varepsilon_n}\|_2^2)^{\lambda},
\end{aligned} \tag{108}$$

where

$$\mu := \frac{3(1-\alpha)(p-1)(3p-2)}{(3p-2)(2p-3(1-\alpha)(3-p)) - 6\alpha p},$$
$$\lambda := \frac{6p\alpha(p-1)}{(3p-2)(2p-3(1-\alpha)(3-p)) - 6\alpha p}.$$

Requiring that $\mu = 1$, we obtain α which we substitute into the definition of λ and obtain

$$\lambda = \frac{2(3-p)}{3p-5}. \tag{109}$$

The estimate of the second term on the right-hand side of (79) is carried out in a similar manner. With the help of (107) we have

$$\begin{aligned}
\int_\Omega |\boldsymbol{v}^{\varepsilon_n}|^2 |\nabla \boldsymbol{v}^{\varepsilon_n}|^2 \, dx &\leqq \|\nabla \boldsymbol{v}^{\varepsilon_n}\|_2^2 \|\nabla \boldsymbol{v}^{\varepsilon_n}\|_3^2 \\
&\leqq C \|\nabla \boldsymbol{v}^{\varepsilon_n}\|_2^2 \|\nabla \boldsymbol{v}^{\varepsilon_n}\|_p^{p-1} \|\nabla \boldsymbol{v}^{\varepsilon_n}\|_{3p}^{3-p} \\
&\leqq C \|\nabla \boldsymbol{v}^{\varepsilon_n}\|_2^2 \|\nabla \boldsymbol{v}^{\varepsilon_n}\|_p^{p-1} \left(I_p(\boldsymbol{v}^{\varepsilon_n})\right)^{\frac{3-p}{p}} \\
&\leqq \frac{C_1}{4} I_p(\boldsymbol{v}^{\varepsilon_n}) + C \left\|\nabla \boldsymbol{v}^{\varepsilon_n}\right\|_p^{p\frac{p-1}{2p-3}} \left\|\nabla \boldsymbol{v}^{\varepsilon_n}\right\|_2^{\frac{2p}{2p-3}} \\
&\leqq \frac{C_1}{4} I_p(\boldsymbol{v}^{\varepsilon_n}) + C \left\|\nabla \boldsymbol{v}^{\varepsilon_n}\right\|_p^{p} \left\|\nabla \boldsymbol{v}^{\varepsilon_n}\right\|_p^{\frac{2-p}{2p-3}} \left\|\nabla \boldsymbol{v}^{\varepsilon_n}\right\|_2^{\frac{2p}{2p-3}} \\
&\leqq \frac{C_1}{4} I_p(\boldsymbol{v}^{\varepsilon_n}) + C \|\nabla \boldsymbol{v}^{\varepsilon_n}\|_p^{p} \Big(\|\nabla \boldsymbol{v}^{\varepsilon_n}\|_2^2 \Big)^{\frac{p}{2p-3} + \frac{p}{2}\frac{2-p}{2p-3}} \\
&\leqq \frac{C_1}{4} I_p(\boldsymbol{v}^{\varepsilon_n}) + C \|\nabla \boldsymbol{v}^{\varepsilon_n}\|_p^{p} \left(1 + \|\nabla \boldsymbol{v}^{\varepsilon_n}\|_2^2 \right)^{\frac{p(4-p)}{2(2p-3)}} .
\end{aligned} \tag{110}$$

Since $\lambda > \frac{p(4-p)}{2(2p-3)}$ for $p \in (\frac{3}{2}, 2)$ we can conclude from (79), (108) and (110) that

$$\frac{d}{dt} \|\nabla \boldsymbol{v}^{\varepsilon_n}\|_2^2 + C_1 I_p(\boldsymbol{v}^{\varepsilon_n}) \leqq C \|\nabla \boldsymbol{v}^{\varepsilon_n}\|_p^p (1 + \|\nabla \boldsymbol{v}^{\varepsilon_n}\|_2^2)^\lambda . \tag{111}$$

This implies that

$$\frac{1}{1-\lambda} \frac{d}{dt} \left(1 + \|\nabla \boldsymbol{v}^{\varepsilon_n}\|_2^2\right)^{1-\lambda} + C_1 \frac{I_p(\boldsymbol{v}^{\varepsilon_n})}{\left(1 + \|\nabla \boldsymbol{v}^{\varepsilon_n}\|_2^2\right)^\lambda} \leqq C \|\nabla \boldsymbol{v}^{\varepsilon_n}\|_p^p. \tag{112}$$

As $\lambda > 1$ for $p < 2$, integrating (112) over time from 0 to T and then using (90) yields

$$\int_0^T \frac{I_p(\boldsymbol{v}^{\varepsilon_n})}{\left(1 + \|\nabla \boldsymbol{v}^{\varepsilon_n}\|_2^2\right)^\lambda} \, ds \leqq C \int_0^T \|\nabla \boldsymbol{v}^{\varepsilon_n}\|_p^p \, ds + \frac{1}{\lambda - 1} \leqq K. \tag{113}$$

From this we can derive (100) easily with help of Hölder's inequality and (90) (see [27], p. 239 for details).

As regards the existence of a strong solution for small data, we start by integrating (112) with respect to time from 0 to t. Omitting the second term, this yields

$$-\frac{1}{\left(1+\|\nabla \boldsymbol{v}^{\varepsilon_n}(t)\|_2^2\right)^{\lambda-1}}+\frac{1}{\left(1+\|\nabla \boldsymbol{v}_0\|_2^2\right)^{\lambda-1}} \leqq C(\lambda-1)\int_0^t \|\nabla \boldsymbol{v}^{\varepsilon_n}\|_p^p\,dt. \tag{114}$$

Since

$$\int_0^\infty \|\nabla \boldsymbol{v}^{\varepsilon_n}\|_p^p\,dt \leqq \|\boldsymbol{v}_0\|_2^2, \tag{115}$$

we observe that for each $\eta > 0$ there is T^* such that

$$\int_0^{T^*} \|\nabla \boldsymbol{v}^{\varepsilon_n}\|_p^p\,dt \leqq \eta = \eta(T^*). \tag{116}$$

Considering either (114) with (115) or (114) with (116) we have

$$\|\nabla \boldsymbol{v}^{\varepsilon}(t)\|_2^2 \leqq \begin{cases} \dfrac{1+\|\nabla \boldsymbol{v}_0\|_2^2}{\left(1-C(\lambda-1)\|\boldsymbol{v}_0\|_2^2(1+\|\nabla \boldsymbol{v}\|_2^2)^{\lambda-1}\right)^{\frac{1}{\lambda-1}}}, \\[2ex] \dfrac{1+\|\nabla \boldsymbol{v}_0\|_2^2}{\left(1-C(\lambda-1)\eta(T^*)(1+\|\nabla \boldsymbol{v}\|_2^2)^{\lambda-1}\right)^{\frac{1}{\lambda-1}}}. \end{cases}$$

Thus, choosing either $\boldsymbol{v}_0 \in \dot{W}^{1,2}_{\mathrm{div}}$ so small that

$$C(\lambda-1)\|\boldsymbol{v}_0\|_2^2(1+\|\nabla \boldsymbol{v}_0\|_2^2)^{\lambda-1} < \tfrac{1}{2},$$

or taking, for an arbitrarily given $\boldsymbol{v}_0 \in \dot{W}^{1,2}_{\mathrm{div}}$, a time T^* so small that

$$\eta(T^*) < \frac{1}{2C(\lambda-1)\left(1+\|\nabla \boldsymbol{v}_0\|_2^2\right)^{\lambda-1}},$$

we obtain

$$\sup_{t\in(0,T^*)} \|\nabla \boldsymbol{v}^{\varepsilon_n}(t)\|_2^2 \leqq 2^{\frac{1}{\lambda-1}}\left(1+\|\nabla \boldsymbol{v}_0\|_2^2\right). \tag{117}$$

(In the first case, T^* is arbitrary.)

Clearly, from (111) and (117) we have

$$\int_0^{T^*} I_p(\boldsymbol{v}^{\varepsilon_n})(s)\,ds < \infty. \tag{118}$$

From (117) and (118) we conclude the existence of a strong solution, defined on $(0, T^*)$ and satifying (39), (40). We refer the reader to [27], pp. 257–260 for details.

Thus, the proof of the theorem is complete.

Note added in proof. The issue of the boundedness of the pressure from below for the Navier-Stokes equations is discussed in the recent paper by Seregin and Sverák – SEREGIN, G., ŠVERÁK, V.: Navier-Stokes equations with lower bounds on the pressure. *Arch. Rational Mech. Anal.* **163**, 65–86 (2002).

Acknowledgements. This research is supported by the project Kontakt ME012 of the National Science Foundation and the Ministry of Education of the Czech Republic. J. MÁLEK was also supported by MSM 113200007 and by the project of the Grant Agency of the Czech

Republic No. 201/00/0768. K. R. RAJAGOPAL thanks the National Science Foundation for its support.

References

1. ANTMAN, S.S.: Material constraints in continuum mechanics. *Atti della Accademia Nazionale dei Lincei, Rendiconti, Classe di Scienza Fisiche, Matematiche e Naturali* **70**, 256–264 (1982)
2. ANTMAN, S.S., MARLOW, R.S.: Material constraints Lagrange multipliers and compatibility. Applications to rod and shell theories. *Arch. Rational Mech. Anal.* **116**, 257–299 (1991)
3. BARUS, C.: Isothermals, isopiestics, and isometrics relative to viscosity. *Am. J. Sci.* **45**, 87–96 (1893)
4. BATES, T.W., WILLIAMSON, J.A., SPEAROT, J.A., MURPHY, C.K.: A correlation between engine oil rheology and oil film thickness in engine journal bearing Society of Automotive Engineers, 1986. Paper No. 860376
5. BELLOUT, H., BLOOM, F., NEČAS, J.: Young measure-valued solutions for non-Newtonian incompressible fluids. *Comm. Partial Differential Equations* **19**, 1763–1803 (1994)
6. BIRD, R.B., AMSTRONG, R.C., HASSAGER, O.: *Dynamics of Polymer Liguids*. John Wiley & Sons, New York, 2nd edition, 1993
7. BRIDGMAN, P.W.: *The physics of high pressure*. The MacMillan Company, New York, 1931
8. CARREAU, P.J.: *Rheological equations of state from molecular network theories*. PhD thesis, University of Wisconsin, Madison, 1968
9. CHO, Y.I., KENSEY, K.R.: Effects of the non-Newtonian viscosity of blood on hemodynamics of diseased arterial flows. *Adv. Bioengineering* **15**, 147–148 (1989)
10. CHO, Y.I., KENSEY, K.R.: Effects of the non-Newtonian viscosity of blood on flows in a diseased arterial vessel. Part 1: Steady flows. *Biorheology* **28**, 241–262 (1991)
11. CHRISTIANSEN, E.B., KELSEY, S.J.: Isothermal and nonisothermal, laminar, inelastic, non-Newtonian tube-entrance flow. *Chem. Engrg. Sci.* **28**, 1099–1133, (1973)
12. COKELET, G.R.: The rheology of human blood In Y.C. FUNG, N. PERRONE & M. ALIKER, editors, *Biomechanics: Its function and Objectives*, pages 63–103. Prentice-Hall, 1972
13. COLEMAN, B.D., MARKOVITZ, H., NOLL, W.: *Viscometric Flows of Non-Newtonian Fluids*. Springer-Verlag, 1966
14. CROSS, M.M.: Rheology of non-Newtonian fluids: a new flow equation for pseudoplastic systems. *J. Colloid Sci.* **20**, 417–437 (1965)
15. DAVIES, A.R., LI, X.K.: Numerical modeling of pressure and temperature effects in viscoelastic flow between eccentrically rotating cylinders. *J. Non-Newt. Fluid Mech.* **54**, 331–350 (1994)
16. DAVIES, P.H., MAZHER, H.K., GIDDENS, D.P., ZARINS, C.K., GLAGOV, S.: Effects of non-Newtonian fluid behavior on wall shear in a separated flow region. In *Proc. 1st World Conf. of Biomech.*, volume 1, page 301, University of California, San Diego, 1990
17. FREHSE, J., MÁLEK, J., STEINHAUER, M.: On existence results for fluids with shear dependent viscosity – unsteady flows. In W. JÄGER, O. JOHN, K. NAJZAR, J. NEČAS & J. STARÁ, editors, *Partial Differential Equations, Theory and Numerical Solution*, volume 406 of *CRC Research Notes in Mathematics series*, pages 121–129, Boca Raton - London - New York - Washington, D.C., 2000. CRC Press UK
18. GAZZOLA, F.: A note on the evolution of Navier-Stokes equations with a pressure-dependent viscosity. *Z. Angew. Math. Phys.* **48**, 760–773 (1997)
19. GWYNLLYW, D.Rh., DAVIES, A.R., PHILLIPS, T.N.: On the effects of a piezoviscous lubricant on the dynamics of a journal bearing. *J. Rheol.* **40**, 1239–1266 (1996)
20. HRON, J., MÁLEK, J., RAJAGOPAL, K.R.: Simple flows of Fluids with pressure dependent viscosities. *Proceedings of The Royal Society London A: Mathematical, Physical and Engineering Sciences* **457**, 1603–1622 (2001)

21. HUTTON, J.F., JONES, B., BATES, T.W.: Effects of isotropic pressure on the high temperature high shear viscosity of motor oils. Technical report, Society of Automotive Engineers, 1983
22. JOHNSON, K.L., TEVAARWERK, J.L.: Shear behaviour of elastohydrodynamic oil films. *Proc. R. Soc. Lond. A* **356**, 215–236 (1977)
23. JOHNSON, M.W., MANGKOESOEBROTO, S.: Analysis of lubrication theory for the power-law fluid. *ASME Journal of Tribology* **115**, 71–77 (1993)
24. KJARTANSON, B.H., SHIELDS, D.H., DOMASCHUK, L., MAN, C.S.: The creeep of ice measured with the pressuremeter. *Can. Geotech. J.* **25**, 250–261 (1988)
25. LIONS, J.L.: *Quelques Méthodes de Résolution des Problémes aux Limites Non Linéares.* Dunod, 1969
26. LUCKHAUS, S., MÁLEK, J.: On an evolutionary nonlinear fluid model in the limiting case. *Mathematica Bohemica* **126**, 421–428 (2001)
27. MÁLEK, J., NEČAS, J., ROKYTA, M., RŮŽIČKA, M.: *Weak and measure-valued solutions to evolutionary PDEs.* Chapman & Hall, London 1996
28. MÁLEK, J., NEČAS, J., RŮŽIČKA, M.: On the non-Newtonian incompressible fluids. *Math. Models Methods Appl. Sci.* **3**, 35–63 (1993)
29. NAKAMURA, M., SAWADA, T.: Numerical study of the flow of non-Newtonian fluid through an axisymmetric stenosis. *J. Biomech. Engrg.* **110**, 137–143 (1988)
30. NG, C.W., SAIBEL, E.: Non-linear viscosity effects in slider bearing lubrication. *Journal of Lubrication Technology* **7**, 192–196 (1962)
31. REINER, M.: A mathematical theory of dilatancy. *American Journal of Mathematics* **67**, 350–362 (1945)
32. RENARDY, M.: Some remarks on the Navier-Stokes equations with a pressure-dependent viscosity. *Comm. Partial Differential Equations* **11**, 779–793 (1986)
33. REYNOLDS, O.: On the dilatancy of media composed of rigid particles in contact. With experimental illustrations. *Philosophical Magazine* **20**, 469–481 (1885)
34. REYNOLDS, O.: On the theory of lubrication and its application to Mr. Beauchamp Tower's experiments, including an experimental determination of the viscosity of olive oil. *Philos. Trans. R. Soc.* **177**, 157–234 (1886)
35. SCHAEFFER, D.G.: Instability in the evolution equations describing incompressible granular flow. *J. Differential Equations* **66**, 19–50 (1987)
36. SCHOWALTER, W.R.: *Mechanics of non-Newtonian fluids.* Pergamon press, 1978
37. STOKES, G.G.: On the theories of the internal friction of fluids in motion and of the equilibrium and motion of elastic solids. *Trans. Cambridge Phil. Soc.* **8**, 287–305 (1845)
38. SUTTERBY, J.L.: Laminar convergent flow of dilute polymer solutions in conical sections ii. *Trans. Soc. Rheology* **9**, 227–241 (1965)
39. SZERI, A.Z.: *Fluid Film Lubrication: Theory and Design.* Cambridge University Press, 1998
40. TRUESDELL, C., NOLL, W.: The nonlinear field theories of mechanics. In S. FLÜGGE, editor, *Handbuch der Physik*, volume III/3. Springer-Verlag, 1969
41. VAN DER VEEN, C.J., WHILLANS, I.M.: New and improved determinations of the velocity of ice stream-B and ice stream-C. *West Antartica J. Glaciology* **36**, 324–339 (1990)
42. WALAWENDER, W.P., CHEN, T.Y., CALA, D.F.: An approximate Casson fluid model for the flow of blood. *Biorheology* **12**, 111–119 (1975)

Mathematical Institute of Charles University,
Sokolovská 83, 186 75 Prague 8,
Czech Republic

and

Department of Mathematical Sciences,
Northern Illinois University, DeKalb, IL 60115,
USA

and

Department of Mechanical Engineering,
Texas A&M University,
College Station,
TX 77843, USA

(Accepted May 1, 2002)
Published online November 15, 2002 – © Springer-Verlag (2002)

Mathematical Models and Methods in Applied Sciences
Vol. 3, No. 1 (1993) 35–63
© World Scientific Publishing Company

ON THE NON-NEWTONIAN INCOMPRESSIBLE FLUIDS

JOSEF MÁLEK and JINDŘICH NEČAS
Mathematical Institute of Charles University, Sokolovská 83, 18600 Prague 8, Czechoslovakia

and

MICHAEL RŮŽIČKA
Institute of Applied Mathematics, Beringstraße 4–6, D-5300 Bonn 1, Germany

Communicated by G. P. Galdi
Received 31 March 1992

The Navier–Stokes equations can be included as a special case into the class of non-Newtonian incompressible fluids with the nonlinear stress tensor $\tau = \tau(\mathbf{e})$, the components of which satisfy the p-growth condition. Measure-valued solutions already exist for $p > 2n/(n+2)$. For the space periodic problem, the existence of the weak solution is then obtained for $p > 3n/(n+2)$. These solutions are regular and unique for $p \geq 1 + 2n/(n+2)$.

1. Introduction

1.1. *Equations*

Let Ω be a fixed domain in $\mathbb{R}^n$, $n = 2$ or 3, with boundary $\partial\Omega$. For $T > 0$, we will denote by Q_T the time cylinder $\Omega \times I$, where $I \equiv (0, T)$ is a time interval.

Viscous, incompressible[a] fluid motions in Ω seem to be quite accurately described by the momentum equations

$$\frac{\partial u_i}{\partial t} + u_j \frac{\partial u_i}{\partial x_j} = -\frac{\partial p}{\partial x_i} + \frac{\partial \tau_{ij}^V}{\partial x_j} + f_i\,, \quad i = 1, 2, \ldots, n\,; \tag{1.1}$$

and the continuity equation

$$\frac{\partial u_i}{\partial x_i} = \operatorname{div} \mathbf{u} = 0\,, \tag{1.2}$$

where $\mathbf{f} = (f_1, f_2, \ldots, f_n)$ is a given forcing term and the unknowns are the velocity field $\mathbf{u} = (u_1, u_2, \ldots, u_n)$ and the pressure p. All functions depend on position $\mathbf{x} = (x_1, x_2, \ldots, x_n)$ and time t.

[a]For simplicity, we put the constant density equal to unity.

© World Scientific Publishing Co. Pte. Ltd 1993. Reprinted with kind permission from: Málek, J., Nečas, J. & Růžička, M., On the non-Newtonian incompressible fluids, Mathematical Models and Methods in Applied Sciences 3(1), 35-63.

The viscous part of the stress tensor, τ_{ij}^V, is expressed by the constitutive law. For the Navier-Stokes equations, it takes the known form

$$\tau_{ij}^V = 2\,\nu\, e_{ij}(\mathbf{u}) = \nu \left(\frac{\partial u_i}{\partial x_j} + \frac{\partial u_j}{\partial x_i} \right) , \tag{1.3}$$

where $\mathbf{e} = (e_{ij})_{i,j=1}^n$ is the symmetric part of the velocity gradient and $\nu > 0$ is a viscosity coefficient.

Here we will take into account a more general relation of τ_{ij}^V of the form

$$\tau^V = \tau(\mathbf{e}) \quad \text{or} \quad \tau_{ij}^V = \tau_{ij}(\mathbf{e}) \tag{1.4}$$

provided that τ has continuous components $\tau_{ij}(\mathbf{e})$, i.e., $\tau_{ij} \in C(\mathbb{R}^{n^2})$, which are symmetric and satisfy the polynomial growth condition

$$|\tau_{ij}(\mathbf{e})| \le c_1(1 + |\mathbf{e}|)^{p-1} , \quad c_1 > 0 \tag{1.5}$$

with $1 < p < \infty$, as well as the strong coercivity condition

$$\tau_{ij}(\mathbf{e})e_{ij} \ge c_2|\mathbf{e}|^{p_1} , \quad c_2 > 0 \tag{1.6}$$

with

$$p \ge p_1 > p - 1 , \quad p_1 > 1 . \tag{1.7}$$

Here $|\mathbf{e}| = \left(\sum_{i,j=1}^n e_{ij}^2(\mathbf{u}) \right)^{1/2}$.

The system (1.1), (1.2) is completed by the no-slip boundary condition

$$\mathbf{u}(\mathbf{x}, t) = 0 \quad \forall\, (\mathbf{x}, t) \in \partial\Omega \times I \tag{1.8}$$

and the initial condition

$$\mathbf{u}(\mathbf{x}, 0) = \mathbf{u}_0(\mathbf{x}) \quad \forall\, \mathbf{x} \in \Omega . \tag{1.9}$$

The initial boundary value problem (1.1), (1.2), (1.4)–(1.9) is denoted by (NonNIF), which abbreviates Non-Newtonian Incompressible Fluids.

1.2. *Motivation*

There are many physical situations when the Stokes law (1.3) provides a sufficiently accurate model for the motion of the incompressible fluids. On the other hand, there are phenomena, for example in the study of dynamics in the Earth's mantle, see Ref. 17, when the linear model fails and when scientists use the nonlinear version of the constitutive law.

Notice that in the case of the standard Navier-Stokes equations $p = 2$ in the condition (1.5). From this point of view it is natural to consider the Navier-Stokes equations as a special case of nonlinear models.

Our investigation of nonlinear constitutive relations is motivated by the recent work of Bellout, Bloom, Nečas (see Ref. 4) and their detailed study of the behavior of the flow in several special cases (Poisseuille flow, Coutte flow, flow in the pipe,...). Let us consider one example from there.

It is well known (provided that $\mathbf{u} = (u_1, 0)$, $u_1 = u_1(x_2)$) that the steady plane flow between two parallel plates has the known parabolic profile

$$u_1(x_2) = \frac{3}{4}Q(1 - |x_2|^2) \quad \text{with} \quad Q = \int_{-1}^{1} u_1(x_2)\, dx_2 \tag{1.10}$$

as far as we reflect the system (1.1), (1.2) with the Stokes' law (1.3).

Suppose for a moment that the viscosity coefficient ν is not a constant in (1.3) but depends on $|\mathbf{e}|$. It means in our special case that

$$\nu = \nu(|u_1'|) \,. \tag{1.11}$$

Let us solve the inverse problem: to find the form of $\nu = \nu(|u_1'|)$, if the profile of the flow, $u_1(x_2)$, is given.

From (1.1), (1.2), we have

$$\begin{aligned} &\frac{\partial p}{\partial x_1} - \frac{\partial}{\partial x_2}(\nu(|u_1'|)u_1') = 0 \,, \\ &\frac{\partial p}{\partial x_2} = 0 \end{aligned} \tag{1.12}$$

together with the boundary condition $u_1(\pm 1) = 0$.

It is easy to compute that

$$p_1 x_2 - \nu(|u_1'|)u_1' = 0 \,, \quad p_1 \text{ is a constant.} \tag{1.13}$$

Provided that u_1' is positive in $\langle -1, 0\rangle$, odd in $\langle -1, 1\rangle$ and denoting by $(u_1')^{-1}$ the inverse function on $\langle -1, 0\rangle$, we obtain

$$\nu(|u_1'|) = \frac{p_1 (u_1')^{-1}(u_1')}{u_1'} \,. \tag{1.14}$$

If, for example,

$$u_1(x_2) = \frac{Q}{2}\frac{(1+\delta)}{\delta}(1 - |x_2|^\delta) \,, \quad \delta \geq 2 \,, \tag{1.15}$$

then

$$\nu(|u_1'|) = \beta |u_1'|^{(2-\delta)/(\delta-1)} \quad \text{with} \quad \beta = |p_1| \left(\frac{2}{Q(1+\delta)}\right)^{1/(\delta-1)} \,. \tag{1.16}$$

For $\delta \to \infty$, we get in the limit the straightline shape, which corresponds very well to experiments (see Ref. 12). The function ν in the limit $\delta \to \infty$ is $\nu = \tilde{\beta}|u_1'|^{-1}$.

On the basis of the previous considerations (and similarly for other special cases) Bellout, Bloom and Nečas have postulated the constitutive law in the form

$$\tau_{ij}(\mathbf{e}) = \nu_0 \left(\varepsilon + |\mathbf{e}|^2\right)^{-\alpha/2} e_{ij} \ , \quad 0 < \alpha \leq 1 \ . \tag{1.17}$$

Notice that now $p = 2 - \alpha$ (for p see (1.5)). Consequently, if α is close to 1_-, then p is close to 1_+. Therefore, our effort in this paper will be concentrated on finding solutions which solve the problem (NonNIF) in some weak sense provided that the nonlinear constitutive function τ satisfies so-called bad growth conditions, i.e., $p < 2$.

1.3. *Historical remarks*

As long as we assume that the viscous part of the stress tensor depends on the velocity gradient

$$\tau_{ij}^V = \tau_{ij}(\nabla \mathbf{u}) \ , \tag{1.18}$$

it can be shown in consideration of the principle of material frame indifference that τ_{ij} can depend on the velocity gradient only through its symmetric part $\mathbf{e}$, i.e., $\tau_{ij}^V = \tau_{ij}(\mathbf{e})$ (see for example Ref. 14).

Moreover, τ_{ij} is an isotropic symmetric tensor function of second order which has the usual decomposition

$$\tau_{ij}(\mathbf{e}) = \alpha(I_1, I_2, I_3)\delta_{ij} + \beta(I_1, I_2, I_3)e_{ij} + \gamma(I_1, I_2, I_3)e_{is}e_{sj} \ , \tag{1.19}$$

where $I_1 = \operatorname{div} \mathbf{u}$ (trace of $\mathbf{e}$), $I_2 = e_{ij}e_{ij}$, $I_3 = \det(\mathbf{e})$, are the invariants of the tensor $\mathbf{e}$. Materials satisfying (1.19) are called Reiner-Rivlin fluids. For the incompressible fluids, I_1 is identically equal to zero. The fluids where $\tau_{ij}(\mathbf{e})$ depends on the spatial derivatives of the velocity field linearly, are called Newtonian ones, the rest are non-Newtonian ones.

The mathematical theory of the non-Newtonian incompressible fluids was first introduced by Ladyzenskaya in 1966–1970. She has considered the relation (1.3) supposing the monotonicity, coercivity and growth conditions similar to (1.5), (1.6) and has proved the existence of the weak solutions for $p \geq 11/5$ [see again (1.5), $p_1 = p$] which is unique for $p \geq 5/2$ in three space dimensions. For details see Ref. 13.

The same results were proved in Ref. 15 for p-Laplacian; the existence result for $p \geq 1 + 2n/(n+2)$ and the uniqueness for $p \geq (n+2)/n$, $n \leq 4$.

In 1989, Du and Gunzburger[7] have improved the three-dimensional unity result up to $p \geq 11/5$ considering

$$\tau_{ij}^V = \left(\nu_0 + \nu_1 |\nabla \mathbf{u}|^{p-2}\right) \frac{\partial u_i}{\partial x_j} \tag{1.20}$$

with ν_0, ν_1 positive constants.

1.4. *Survey of the new results*

Starting from the work of Ball,[2] the existence of Young measure-valued solutions for the model (NonNIF) for $p > 1$ has been proved, if $n = 2$ and for $p > 6/5$, if $n = 3$. See Refs. 16 and 18.

In 1991, for the space periodic problem $(\text{NonNIF})_{\text{per}}$ Bloom, Bellout, Nečas[3] have proved, on the basis of the new estimates for parabolic systems, that this limiting measure is a Dirac one for $p > 3/2$ (if $n = 2$) and for $p > 9/5$ (if $n = 3$) (notice that the case $p = 2$ is included), which immediately implies the existence of the weak solution for these p. These solutions are unique and regular for $p \geq 2$ ($n = 2$) and for $p \geq 11/5$ ($n = 3$).

All these results are written in the context of bipolar incompressible fluids which can be shortly characterized by the dependence of the τ_{ij} not only on the first velocity gradient but also on the (higher) third derivatives of the velocity. The above-mentioned results are then obtained when we follow the behavior of the unique bipolar weak solutions with the vanishing higher viscosity (i.e., the viscosity coefficient standing at terms with the third derivatives tends to zero).

To prove similar results for smooth bounded domains is the objective of Bloom, Bellout, Málek, Nečas, Růžička at this time.

In this paper, we will repeat completely the new results without introducing the notion of the bipolar or multipolar fluids.

2. Galerkin Approximations

2.1. *Function spaces*

For scalar and vector-valued function spaces, we will use the standard notation ($p > 1$):

$$L^p(\Omega) = \left\{ v : \Omega \times I \to \mathbb{R} \text{ measurable; } \|v\|_p = \left(\int_\Omega |v(\mathbf{x},t)|^p dx \right)^{1/p} < \infty \right\};$$

$$L^p(\Omega)^n = \left\{ \mathbf{v} : \Omega \times I \to \mathbb{R}^n \text{ measurable; } \|\mathbf{v}\|_p = \left(\int_\Omega \sum_{i=1}^n |v_i(\mathbf{x},t)|^p dx \right)^{1/p} < \infty \right\}.$$

The dependence of the norm on time will not be explicitly noted. Similarly, $W^{m,p}(\Omega)$ and $W^{m,p}(\Omega)^n$ consist of all scalar and $\mathbb{R}^n$-valued functions which belong to $L^p(\Omega)$ and $L^p(\Omega)^n$, respectively, and all the same holds for their generalized derivatives up to the order m. For m fractional, the definition of $W^{m,p}$-spaces is introduced, for example, in Ref. 1. The norm in $W^{m,p}$ will be denoted by $\|\cdot\|_{m,p}$ both for scalar and for vector valued functions.

We will need some inequalities which are formulated, for a sufficiently smooth domain, in the following lemmas.

Lemma 2.1 (interpolation in p). *Let $\mathbf{u} \in W^{k,p_1}(\Omega)^n$ and $p_1 \geq p \geq p_2$. Then there exists a constant c such that*

$$||\mathbf{u}||_{k,p} \leq c||\mathbf{u}||_{k,p_1}^{\alpha} \, ||\mathbf{u}||_{k,p_2}^{1-\alpha} \, , \tag{2.1}$$

where $1/p = \alpha/p_1 + (1-\alpha)/p_2$, $\alpha \in (0,1)$.

Proof. see Ref. 5. □

Lemma 2.2 (interpolation in k). *Let $\mathbf{u} \in W^{k_1,p}(\Omega)^n$ and $k_1 \geq k_2 \geq 0$. Then there exists a constant c such that*

$$||\mathbf{u}||_{k_2,p} \leq c||\mathbf{u}||_{k_1,p}^{k_2/k_1} \, ||\mathbf{u}||_{p}^{1-k_2/k_1} \, . \tag{2.2}$$

Proof. see for example Ref. 1. □

Lemma 2.3. *The space $W^{m,p}(\Omega)$ is imbedded into $W^{k,q}(\Omega)$ if $1/q \geq 1/p-(m-k)/n$, where $0 \leq k \leq m$, $n > mp-k$. If $1/q > 1/p-(m-k)/n$ and $m > k$ then this imbedding is compact.*

Proof. see for example Ref. 1. □

Let B be a Banach space with the norm $||\cdot||_B$, $p_1 > 1$. The Bochner spaces $L^{p_1}(I;B)$ are defined as the set of all functions $u = u(\mathbf{x},t)$ the norm of which $||u||_{L^{p_1}(I;B)} = \int_0^T ||u(t)||_B^{p_1} \, dt = \int_0^T ||u||_B^{p_1} \, dt$ is finite. For $p_1 = \infty$, the norm is $||u||_{L^\infty(I;B)} = \operatorname{ess\,sup}_{t\in I} ||u(t)||_B$.

We denote

$$\mathcal{V} = \{\phi \in C_0^\infty(\Omega)^n \, ; \, \operatorname{div}\phi = 0\} \; ; \tag{2.3}$$

$$H = \text{closure of } \mathcal{V} \text{in the} L^2(\Omega)^n\text{-norm} \; ; \tag{2.4}$$

$$V = \text{closure of } \mathcal{V} \text{in the} W^{1,2}(\Omega)^n\text{-norm} \; . \tag{2.5}$$

Remark 2.4. The spaces H, V can be characterized as follows:

$$H = \{\mathbf{u} \in L^2(\Omega)^n; \, \operatorname{div}\mathbf{u} = 0; \, \hat{\gamma}(\mathbf{u}) = 0\} \; ;$$
$$V = \{\mathbf{u} \in W_0^{1,2}(\Omega)^n; \, \operatorname{div}\mathbf{u} = 0\} \; ,$$

where $\hat{\gamma}$ is the trace operator from $E(\Omega) = \{\mathbf{u} \in L^2(\Omega)^n; \operatorname{div}\mathbf{u} \in L^2(\Omega)\}$ onto $H^{-1/2}(\partial\Omega)$(dual of $H^{1/2}(\partial\Omega)$). For proofs and details from the next section see Ref. 6.

2.2. *The choice of basis*

The space $L^2(\Omega)^n$ has the decomposition $L^2(\Omega)^n = H + H^\perp$, where $H^\perp = \{\mathbf{u} \in L^2(\Omega)^n; \, \exists \, p \in W^{1,2}(\Omega)$ such that $\mathbf{u} = \nabla p\}$; it means that for every $\mathbf{u} \in L^2(\Omega)^n$ there exist $\mathbf{u}_1$, $\mathbf{u}_2 \in L^2(\Omega)^n$ such that

$$\mathbf{u} = \mathbf{u}_1 + \mathbf{u}_2 \; , \quad \mathbf{u}_1 \in H \quad \text{and} \quad \mathbf{u}_2 \in H^\perp \; , \tag{2.6}$$

or anyway, for every $\mathbf{u} \in L^2(\Omega)^n$ there exist $\mathbf{u}_1 \in L^2(\Omega)^n$ and $p \in W^{1,2}(\Omega)$ such that the vector field $\mathbf{u}_1$ is solenoidal ($\operatorname{div} \mathbf{u}_1 = 0$) and parallel with the boundary ($\mathbf{u}_1 \cdot \mathbf{n} = 0$ on $\partial\Omega$) and

$$\mathbf{u} = \mathbf{u}_1 + \nabla p \,, \tag{2.7}$$

where p is defined as the solution of

$$\begin{aligned} \Delta p &= 0 \quad \text{in} \quad \Omega \,, \\ \frac{\partial p}{\partial n} &= \mathbf{u} \cdot \mathbf{n} \quad \text{on} \quad \partial\Omega \,. \end{aligned} \tag{2.8}$$

We define the mapping $\mathbb{P} : L^2(\Omega)^n \to H$ by $\mathbb{P}\,\mathbf{u} = \mathbf{u}_1$.

Definition 2.5. The operator $A := -\mathbb{P}\Delta\mathbf{u} : H \to H$; $D(A) = W^{2,2}(\Omega)^n \cap V$ is called *Stokes operator*.

As the operator A is positive, self-adjoint and its inverse is consequently compact, we know that there exists a basis $\{\omega^r\}_{r=1}^{\infty}$ having the following properties:

$$\begin{aligned} &\{\omega^r\} \text{ is orthonormal in } V; \\ &\{\omega^r\} \text{ is orthogonal in } H; \\ &A\omega^r = \lambda_r \omega^r \text{ a.e. in } \Omega; \\ &\omega^r \in D(A) \text{ (regularity of the Stokes system, } \Omega \text{ of class } C^2); \\ &\omega^r \in D(A) \cap W^{3,2}(\Omega)^n \text{ as far as } \Omega \text{ is of class } C^3 \,. \end{aligned} \tag{2.9}$$

If $\mathbf{u} \in D(A)$, then $A\mathbf{u} \in H$. Put $f \equiv A\mathbf{u}$. From (2.7), we see that $-\Delta\mathbf{u} = A\mathbf{u} - \nabla p$ with $p \in H^1(\Omega)$.

If Ω is a domain of class C^2, then from the regularity of the unique solution $\mathbf{u}$ of the Stokes system

$$\begin{aligned} -\Delta\mathbf{u} + \nabla p &= f \quad \text{in } \Omega \,, \\ \operatorname{div} \mathbf{u} &= 0 \quad \text{in } \Omega \,, \\ \mathbf{u} &= 0 \quad \text{in } \partial\Omega \,, \end{aligned} \tag{2.10}$$

it follows

$$\|\mathbf{u}\|_{2,2} + \|p\|_{1,2} \le c\|f\|_2 = c\|A\mathbf{u}\| \,. \tag{2.11}$$

On the other hand, for $\mathbf{u} \in D(A)$

$$\|A\mathbf{u}\|_2 = \| - \mathbb{P}\Delta\mathbf{u}\|_2 \le c\|\Delta\mathbf{u}\|_2 \le \tilde{c}\|\mathbf{u}\|_{2,2} \,. \tag{2.12}$$

The comparison of (2.11) with (2.12) gives us the equivalence of $W^{2,2}$-norm of $\mathbf{u}$ and L^2-norm of $A\mathbf{u}$ on $D(A)$, i.e., there exist c_1, $c_2 > 0$ such that

$$c_1\|\mathbf{u}\|_{2,2} \le \|A\mathbf{u}\|_2 \le c_2\|\mathbf{u}\|_{2,2} \quad \forall\, \mathbf{u} \in D(A) \,. \tag{2.13}$$

Recall that for all $\mathbf{u} \in D(A)$, $\mathbf{v} \in V$

$$(A\mathbf{u}, \mathbf{v}) = ((\mathbf{u}, \mathbf{v})) \,; \tag{2.14}$$

especially,

$$\lambda_r(\omega^r, \mathbf{v}) = (A\omega^r, \mathbf{v}) = ((\omega^r, \mathbf{v})) \ . \tag{2.15}$$

We denote, as usual,

$$H^m = \overline{\langle \omega^1, \dots, \omega^m \rangle}^{L^2(\Omega)^n}$$
$$V^m = \overline{\langle \omega^1, \dots, \omega^m \rangle}^{W^{1,2}(\Omega)^n} \ .$$

Lemma 2.6. *For* $\mathbf{u} \in H$*, define* $P_m(\mathbf{u}) = \sum_{r=1}^m \lambda_r(\omega^r, \mathbf{u})\omega^r$*. Then* P_m *is the projector both from* H *into* H^m *and from* V *into* V^m*.*
Moreover,

$$AP_m\mathbf{u} = P_m A\mathbf{u} \quad \forall\, \mathbf{u} \in D(A) \ , \tag{2.16}$$

and

$$\|P_m\mathbf{u}\|_{2,2} \le c\|\mathbf{u}\|_{2,2} \quad \forall\, \mathbf{u} \in D(A) \ . \tag{2.17}$$

Proof. It follows from definition of P_m and from (2.15) that

$$P_mP_m = P_m \quad \text{and} \quad P_m = P_m^* \tag{2.18}$$

both in H and in V.

Further,

$$\|P_m\mathbf{u}\|_2^2 = (P_m\mathbf{u}, P_m\mathbf{u}) = (P_m\mathbf{u}, \mathbf{u}) \le \|P_m\mathbf{u}\|_2\|\mathbf{u}\|_2$$

and

$$\|P_m\mathbf{u}\|_{1,2}^2 = ((P_m\mathbf{u}, P_m\mathbf{u})) \le \|P_m\mathbf{u}\|_{1,2}\|\mathbf{u}\|_{1,2} \ .$$

So, the continuity of the projector P_m in both norms is proved.

Let $\mathbf{u} \in D(A)$. It holds (see (2.14))

$$(AP_m\mathbf{u}, \omega) = ((P_m\mathbf{u}, \omega)) \quad \forall\, \omega \in V \ . \tag{2.19}$$

But

$$\begin{aligned}(P_mA\mathbf{u}, \omega) &= \sum_{r=1}^m \lambda_r(\omega^r, A\mathbf{u})(\omega^r, \omega) \\ &\overset{(2.15)}{=} \sum_{r=1}^m (A\mathbf{u}, \omega^r)((\omega^r, \omega)) \\ &\overset{(2.14)}{=} \sum_{r=1}^m ((\mathbf{u}, \omega^r))((\omega^r, \omega)) = ((P_m\mathbf{u}, \omega)) \ . \end{aligned} \tag{2.20}$$

The relation (2.16) follows from (2.19), (2.20). The estimate (2.17) is a direct consequence of (2.13). Indeed,

$$\begin{aligned}\|P_m\mathbf{u}\|_{2,2} &\le c_1^{-1}\|AP_m\mathbf{u}\|_2 = \frac{1}{c_1}\|P_mA\mathbf{u}\|_2 \le \text{(from the continuity of } P_m \text{ in } L^2)\\ &\le \tilde{c}\|A\mathbf{u}\|_2\\ &\overset{(2.13)}{\le} \tilde{c}\|\mathbf{u}\|_{2,2}\,. \quad \square\end{aligned}$$

Remark 2.7. Let $\mathbf{u} \in W_0^{3,2} \cap V$. Then $\mathbf{u} \in D(A)$ and

$$\|P_m\mathbf{u}\|_{3,2} \le \|\mathbf{u}\|_{3,2}\,. \tag{2.21}$$

Proof. It is similar as the proof of (2.17):

$$\begin{aligned}\|P_m\mathbf{u}\|_{3,2} &= \|P_m\mathbf{u}\|_2 + \|\nabla(P_m\mathbf{u})\|_{2,2} \le \|\mathbf{u}\|_2 + \|A(\nabla P_m\mathbf{u})\|_2\\ &= \|\mathbf{u}\|_2 + \|AP_m(\nabla\mathbf{u})\|_2 = \|\mathbf{u}\|_2 + \|P_mA(\nabla\mathbf{u})\|_2\\ &\le \|\mathbf{u}\|_2 + \|A(\nabla\mathbf{u})\|_2 \le \|\mathbf{u}\|_2 + \|\nabla\mathbf{u}\|_{2,2} \le \|\mathbf{u}\|_{3,2}\,. \quad \square\end{aligned}$$

2.3. *Galerkin system and a priori estimates*

Definition 2.8. Let $\mathbf{f} \in L^2(I; L^2(\Omega)^n)$ and $\mathbf{u}_0 \in L^2(\Omega)^n$ be given. Put

$$\mathbf{u}^N(\mathbf{x}, t) = \sum_{i=1}^{N} c_i^N(t)\omega^i(\mathbf{x})\,. \tag{2.22}$$

We say that $\mathbf{u}^N$ *solves the Galerkin system*, if the coefficients $c_i^N(t)$ satisfy the system

$$\begin{aligned}&\left(\frac{d}{dt}\mathbf{u}^N, \omega^r\right) + \int_\Omega \tau_{ij}(\mathbf{e}(\mathbf{u}^N))e_{ij}(\omega^r)\,dx + \int_\Omega u_j^N \frac{\partial u_i^N}{\partial x_j}\omega_i^r\,dx\\ &\quad = (\mathbf{f}, \omega^r)\,, \quad 1 \le r \le N\,.\end{aligned} \tag{2.23}$$

As usual, the system of the ordinary differential equations is solvable at least locally in time. The global solvability then follows from the following estimates. See Lemma 2.9.

By using L^q-version of the Korn's inequality[19] which has the form ($q > 1$)

$$\left(\int_\Omega |\mathbf{e}(\varphi)|^q dx\right)^{1/q} \ge d\,\|\varphi\|_{1,q} \quad \forall\, \varphi \in W_0^{1,q}(\Omega) \tag{2.24}$$

with $d > 0$, it is easy to prove:

Lemma 2.9. *For* $\mathbf{u}_0$, $\mathbf{f}$ *given above, the sequence* $(\mathbf{u}^N)$ *satisfies the following estimates*

$$\|\mathbf{u}^N\|_{L^\infty(I;L^2(\Omega)^n)} \le \text{ const.} \tag{2.25}$$

and

$$\|\mathbf{u}^N\|_{L^{p_1}(I;W^{1,p_1}(\Omega)^n)} \leq \text{const.} \tag{2.26}$$

Moreover,

$$\begin{cases} \|\dot{\mathbf{u}}^N\|_{L^2(I;W^{-3,2}(\Omega)^n)} \leq \text{const.}\ , & \text{if } q_1 \equiv p_1/(p-1) \geq 2\ , \\ \|\dot{\mathbf{u}}^N\|_{L^{q_1}(I;W^{-3,2}(\Omega)^n)} \leq \text{const.} & \text{if } q_1 < 2\ . \end{cases} \tag{2.27}$$

Proof. (i) Multiply the rth equation of the Galerkin system by $c_r^N(t)$ and then sum up the equations. We have

$$\frac{1}{2}\frac{d}{dt}\int_\Omega |\mathbf{u}^N|^2 dx + \int_\Omega \tau_{ij}(\mathbf{e}(\mathbf{u}^N))e_{ij}(\mathbf{u}^N)\, dx = (\mathbf{f}, \mathbf{u}^N) \tag{2.28}$$

because of $\int_\Omega u_j^N \frac{\partial u_i^N}{\partial x_j} u_j^N\, dx \equiv 0$. After integrating (2.28) over $(0,t)$ and using (2.24) we obtain

$$\frac{1}{2}\int_{\Omega_t} |\mathbf{u}^N|^2 dx + c_2\int_0^t \|\mathbf{u}^N\|_{1,p_1}^{p_1} ds \leq \int_0^t\int_\Omega \mathbf{f}\,\mathbf{u}^N\, dx ds + \frac{1}{2}\int_\Omega |\mathbf{u}_0^N|^2\, dx\ , \tag{2.29}$$

which demonstrates (2.25) and (2.26) after employing the Hölder and Gronwall inequality (on the forcing term).

(ii) Let $\varphi \in L^{q_2}(I;W_0^{3,2}(\Omega)^n)$, where $1/q_2 + 1/q_1 = 1$, $q_2 = p_1/(p_1 - p + 1)$ and $\|\varphi\|_{L^{q_2}(I;W_0^{3,2}(\Omega)^n)} \leq 1$. Then

$$\begin{aligned}\left(\frac{d\mathbf{u}^N}{dt}, \varphi\right) = \left(\frac{d}{dt}\mathbf{u}^N, P_N\varphi\right) &= -\int_\Omega u_j^N \frac{\partial u_i^N}{\partial x_j} P_N\varphi_i\, dx \\ &\quad - \int_\Omega \tau_{ij}(\mathbf{e}(\mathbf{u}^N))P_N e_{ij}(\varphi) dx + (\mathbf{f}, P_N\varphi)\ .\end{aligned}$$

Recall that $\|P_N\varphi\|_{W_0^{3,2}(\Omega)^n} \leq c\|\varphi\|_{3,2}$ (cf. Remark 2.7). Thus, the proof is done by the following two estimates:

(a)
$$\begin{aligned}\left|\int_{Q_T} u_j^N \frac{\partial u_i^N}{\partial x_j} P^N\varphi_i\, dxdt\right| &\leq \int_{Q_T} \left|u_j^N u_i^N \frac{\partial}{\partial x_j}(P^N\varphi_i)\right|\, dxdt \\ &\leq \int_{Q_T} |\mathbf{u}^N|^2\, |\nabla(P^N\varphi)| dxdt \leq [W^{2,2}(\Omega) \underset{n=2,3}{\hookrightarrow} L^\infty(\Omega)] \\ &\leq \int_0^T \|\mathbf{u}^N\|_2^2 \|P^N\varphi\|_{3,2} \overset{(2.21)}{\leq} \|\mathbf{u}^N\|^2_{L^\infty(I;L^2(\Omega)^n)} \int_0^T \|\varphi\|_{3,2} dt \\ &\overset{(2.25)}{\leq} \text{const.}\end{aligned}$$

(b)
$$\int_{Q_T} |\tau_{ij}(\mathbf{e}(\mathbf{u}^N))|\,|e_{ij}(P_N\varphi)|dxdt$$
$$\overset{(1.5)}{\leq} c_1 \int_{Q_T} (1+|\mathbf{e}(\mathbf{u}^N)|)^{p-1}|\nabla(P_N\varphi)|dxdt$$
$$\leq c\left(\int_0^T (1+\|\mathbf{u}^N\|_{1,p_1})^{p_1}\right)^{(p-1)/p_1}\left(\int_0^T \|P_N\varphi\|_{1,q_2}^{q_2}\right)^{1/q_2}$$
(for the definition q_2 see above)
$$\leq c\left(\int_0^T \|\varphi\|_{3,2}^{q_2}\right)^{1/q_2} \leq \text{const.}$$

The first estimate in (2.27) follows from the imbedding $L^2(I) \hookrightarrow L^{q_2}(I)$ for $q_2 \leq 2$ which holds just for $\frac{p_1}{p-1} \geq 2$. □

3. Measure-valued Solutions

Let $\varphi \in \mathcal{C}^1(I,\mathcal{V})$ with $\varphi(T)=0$ and $\varphi^m \equiv P_m\varphi$.

Our aim is to study the limiting process, for $N\to\infty$, in the system

$$\int_{Q_T}[-u_i^N\frac{\partial\varphi_i^m}{\partial t} - u_j^N u_i^N\frac{\partial\varphi_i^m}{\partial x_j} + \tau_{ij}(\mathbf{e}(\mathbf{u}^N))e_{ij}(\varphi^m) - f_i\varphi_i^m]dxdt$$
$$= -\int_\Omega u_{0i}^N\varphi_i^m dx\ . \tag{3.1}$$

First, we concentrate on the convective term starting from Aubin-Lions lemma (see Refs. 15 and 20).

Lemma 3.1. *Let X be a Banach space, and X_0, X_1 separable, reflexive Banach spaces. Provided that $X_0 \hookrightarrow\hookrightarrow X \hookrightarrow X_1$, then*

$$\left\{v\in L^\alpha(I;X_0) \text{ with } \frac{dv}{dt}\in L^\beta(I;X_1)\right\} \hookrightarrow\hookrightarrow L^\alpha(I;X)\ ,$$

where $1<\alpha,\ \beta<+\infty$.

Lemma 3.2. *Let the sequence $(\mathbf{u}^N)$ satisfy the uniform estimates (2.25)–(2.27) and let $p_1 > \frac{2n}{n+2}$. If*

$$\mathbf{u}^N \rightharpoonup \mathbf{u} \quad \textit{weakly in } L^2(Q_T)\ , \tag{3.2}$$

then

$$\mathbf{u}^N \to \mathbf{u} \quad \textit{strongly in } L^2(Q_T)\ . \tag{3.3}$$

Proof. Because of $W^{1,p_1}(\Omega)^n \hookrightarrow\hookrightarrow L^2(\Omega)^n \hookrightarrow W^{-3,2}(\Omega)^n$ we can use Lemma 3.1 and the assertion (3.3) is its direct consequence, if $p_1 \geq 2$. For $p_1 < 2$, Lemma 3.1 gives the strong convergence $\mathbf{u}^N \to \mathbf{u}$ in $L^{p_1}(I; L^2(\Omega))$. But

$$\begin{aligned}\|\mathbf{u}^N - \mathbf{u}\|^2_{L^2(Q_T)} &= \int_0^T \|\mathbf{u}^N - \mathbf{u}\|_2^{2-p_1} \|\mathbf{u}^N - \mathbf{u}\|_2^{p_1} dt \\ &\leq \|\mathbf{u}^N - \mathbf{u}\|^{2-p_1}_{L^\infty(I;L^2(\Omega)^n)} \cdot \|\mathbf{u}^N - \mathbf{u}\|_{L^{p_1}(I;L^2(\Omega)^n)} ,\end{aligned}$$

which demonstrates (3.3). □

Remark 3.3. The limiting process in the convective term has restricted our consideration in the space dimension $n = 3$ on p_1 greater than $6/5$ in contrast to the two-dimensional case, where we have no restriction ($p_1 > 1$).

In order to find the limiting element in the nonlinear term, we establish the notion of the probability measure introduced by Young. For our purpose the Ball's results are the most convenient.[2] As the function τ_{ij} works in the space of $\nabla \mathbf{u}$, i.e., in the space of $\mathbb{R}^{n^2}$, we consider the Banach space $C_0(\mathbb{R}^{n^2})$ consisting of the continuous functions $f : \mathbb{R}^{n^2} \to \mathbb{R}$ satisfying $\lim_{|\lambda|\to\infty} f(\lambda) = 0$. A well known form of the Riesz representation theorem states that the dual of $C_0(\mathbb{R}^{n^2})$ can be identified with $M(\mathbb{R}^{n^2})$, the space of finite Radon measures.

If we consider the Bochner space $L^1(Q_T; C_0(\mathbb{R}^{n^2}))$, then its dual space can be described and we have, see Ref. 10,

$$\left[L^1(Q_T; C_0(\mathbb{R}^{n^2}))\right]^* = L^\infty_\omega(Q_T; M(\mathbb{R}^{n^2})) . \tag{3.4}$$

Definition 3.4. A function $g \in L^\infty_\omega(Q_T; M(\mathbb{R}^{n^2}))$ if and only if the function $(\mathbf{x}, t) \mapsto \langle h(\mathbf{x}, t, \cdot), g_{\mathbf{x},t}\rangle$ is measurable for every $h \in L^1(Q_T; C_0(\mathbb{R}^{n^2}))$ in the Lebesgue sense and

$$\|g\|_{L^\infty_\omega(Q_T;M(\mathbb{R}^{n^2}))} \equiv \operatorname{ess\,sup}_{(\mathbf{x},t)\in Q_T} \|g_{\mathbf{x},t}\|_{M(\mathbb{R}^{n^2})} < +\infty . \tag{3.5}$$

Here, the brackets $\langle\cdot,\cdot\rangle$ denote the duality between $C_0(\mathbb{R}^{n^2})$ and $M(\mathbb{R}^{n^2})$, i.e., $\langle h(\mathbf{x}, t, \cdot), g_{\mathbf{x},t}\rangle = \int_{\mathbb{R}^{n^2}} h(\mathbf{x}, t, \lambda) dg_{\mathbf{x},t}(\lambda)$, and

$$\|g_{\mathbf{x},t}\|_{M(\mathbb{R}^{n^2})} = \sup_{\substack{\Phi \in C_0(\mathbb{R}^{n^2}) \\ \|\Phi\|_{C_0(\mathbb{R}^{n^2})} \leq 1}} \int_{\mathbb{R}^{n^2}} \Phi(\lambda) dg_{\mathbf{x},t}(\lambda) .$$

Let $N \to \infty$. Due to (2.25), (2.26), we can suppose that

$$\mathbf{u}^N \rightharpoonup \mathbf{u} \quad \text{weakly in } L^2(Q_T) \tag{3.6}$$

$$\mathbf{u}^N \rightharpoonup \mathbf{u} \quad \text{weakly in } L^{p_1}(I; W^{1,p_1}(\Omega)^n) \tag{3.7}$$

at least for some subsequence.

Now, define

$$\nu^N_{\mathbf{x},t} = \delta_{\nabla \mathbf{u}^N(\mathbf{x},t)} \quad \text{for a.e.}(\mathbf{x},t) \in Q_T \tag{3.8}$$

or equivalently by the relation

$$\int_{Q_T} \langle \psi(\mathbf{x},t,\cdot), \nu^N_{\mathbf{x},t} \rangle dxdt = \int_{Q_T} \psi(\mathbf{x},t,\nabla \mathbf{u}^N(\mathbf{x},t))dxdt$$

$$\text{for all } \psi \in L^1(Q_T; C_0(\mathbb{R}^{n^2})) \,. \tag{3.9}$$

Clearly $\nu^N_{\mathbf{x},t} \ge 0$ and $\|\nu^N_{\mathbf{x},t}\|_{M(\mathbb{R}^{n^2})} \le 1$ a.e. in Q_T.

From the separability of $L^1(Q_T; C_0(\mathbb{R}^{n^2}))$ and from (3.4) immediately follows the existence of an element $\nu \in L^\infty_\omega(Q_T; M(\mathbb{R}^{n^2}))$ and a subsequence of ν^N (still denoted by ν^N) such that

$$\nu^N \rightharpoonup \nu \quad * -\text{weakly in} L^\infty_\omega(Q_T; M(\mathbb{R}^{n^2})) \,. \tag{3.10}$$

The limiting measure ν is positive a.e. in Q_T and $\|\nu_{\mathbf{x},t}\|_{M(\mathbb{R}^{n^2})} \le 1$. Moreover, the following lemma asserts that ν is a.e. probability measure.

Lemma 3.5. $\|\nu_{\mathbf{x},t}\|_{M(\mathbb{R}^{n^2})} = 1$ *a.e. in* Q_T.

Proof. Choose

$$\vartheta^k(\lambda) = \begin{cases} 1 & |\lambda| \le k \\ 1 + k - |\lambda| & k \le |\lambda| \le 1 + k \\ 0 & |\lambda| \ge k+1 \,. \end{cases}$$

Let $E \subset Q_T$ be an arbitrary measurable set and let $|E|$ denote its Lebesgue measure in $\mathbb{R}^{n+1}$. Then

$$\begin{cases} \lim\limits_{N\to\infty} \frac{1}{|E|} \int_E \vartheta^k(\nabla \mathbf{u}^N(\mathbf{x},t))dxdt = \lim\limits_{N\to\infty} \frac{1}{|E|} \int_E \langle \vartheta^k, \nu^N_{\mathbf{x},t}\rangle dxdt \\ \quad = \frac{1}{|E|} \int_E \langle \vartheta^k, \nu_{\mathbf{x},t}\rangle dxdt \le \frac{1}{|E|} \int_E \|\nu_{\mathbf{x},t}\|_{M(\mathbb{R}^{n^2})} dxdt \le 1 \,. \end{cases} \tag{3.11}$$

Now, denote

$$A^N = \{(\mathbf{x},t) \in E; |\nabla \mathbf{u}^N(\mathbf{x},t)| \ge k\} \tag{3.12}$$

and notice that from the estimate of $|\nabla \mathbf{u}^N|$ in $L^{p_1}(Q_T)$ we get $(1/p_1' = 1 - 1/p_1)$

$$\begin{cases} k|A^N| \le \int_{A^N} |\nabla \mathbf{u}^N(\mathbf{x},t)|dxdt \le \int_E |\nabla \mathbf{u}^N(\mathbf{x},t)|dxdt \\ \quad \le \|\nabla \mathbf{u}^N\|_{L^{p_1}(Q_T)} |E|^{1/p_1'} \le c|E|^{1/p_1'} \,, \end{cases} \tag{3.13}$$

which gives $\sup\limits_{N=1,2,\ldots} \frac{|A^N|}{|E|} \le \frac{c|E|^{\frac{1}{p_1'}-1}}{k} \equiv \frac{\tilde{c}}{k}$. From there

$$\frac{1}{|E|} \int_E (1 - \vartheta^k(\nabla \mathbf{u}^N(\mathbf{x},t))dxdt \le \sup_N \frac{|A^N|}{|E|} \le \frac{\tilde{c}}{k} \,. \tag{3.14}$$

Therefore, letting $N \to \infty$ in (3.14),

$$1 - \frac{\tilde{c}}{k} \le \frac{1}{|E|} \int_E \langle \vartheta^k, \nu_{\mathbf{x},t} \rangle dx dt \le \frac{1}{|E|} \int_E \|\nu_{\mathbf{x},t}\|_{M(\mathbb{R}^{n^2})} dx dt \tag{3.15}$$

The comparison of (3.11) and (3.15) finishes the proof. □

We do a short recapitulation:

For the function $\psi \in L^1(Q_T; C_0(\mathbb{R}^{n^2}))$, in particular, for $\psi(\mathbf{x}, t, \lambda) = g(\mathbf{x}, t)h(\lambda)$, where $g \in L^1(Q_T)$ and $h \in C_0(\mathbb{R}^{n^2})$ we have

$$\lim_{N \to \infty} \int_{Q_T} g(\mathbf{x}, t) h(\nabla \mathbf{u}^N(\mathbf{x}, t)) dx dt = \int_{Q_T} g(\mathbf{x}, t) \langle h, \nu_{\mathbf{x},t} \rangle dx dt . \tag{3.16}$$

Unfortunately, τ_{ij} does not vanish at the infinity. But we can extend the relation (3.16) for ψ of the form $\psi(\mathbf{x}, t, \lambda) = e_{ij}(\varphi(\mathbf{x}, t))\tau_{ij}(e(\lambda))$, where $i, j = 1, \dots, n$ are fixed and $e(\lambda) = \frac{1}{2}(\lambda_i^j + \lambda_j^i)$, as τ_{ij} satisfies the further assumptions (1.5), (1.6) and φ is a smooth (test) function.

More precisely,

Lemma 3.6. *For $\tau_{ij} \in C(\mathbb{R}^{n^2})$ satisfying $|\tau_{ij}(e(\lambda))| \le c(1 + |\lambda|)^{p-1}$, where $p - 1 < p_1 \le p$, and for $g \in L^q(Q_T)$ with $q = p_1/(1 + p_1 - p)$, the relation (3.16) holds (put τ_{ij} instead of h and $e_{ij}(\varphi)$ instead of g).*

Proof. is omitted. It was proved, in Ref. 18, that if τ is some continuous function which satisfies

$$\sup_{N=1,2,\dots} \int_{Q_T} \Psi(|\tau(|\nabla \mathbf{u}^N|)|) dx dt < \infty$$

($\Psi : (0, \infty) \to \mathbb{R}$, any Young function with $\lim_{\zeta \to \infty} \frac{\Psi(\zeta)}{\zeta} = \infty$), then

$$\tau(|\nabla \mathbf{u}^N|) \rightharpoonup \langle \tau, \nu_{\mathbf{x},t} \rangle \quad * -\text{weakly in } L_\Psi(Q) .$$

So, it is sufficient to take $\Psi(\zeta) = \zeta^{p_1/(p-1)}$. Indeed, from (2.26)

$$\int_{Q_T} |\tau_{ij}(|\nabla \mathbf{u}^N(\mathbf{x}, t)|)|^{\frac{p_1}{p-1}} \overset{(1.5)}{\le} \int_{Q_T} (1 + |\mathbf{e}(\mathbf{u}^N)|)^{p_1} \le \text{const.}$$

and $p_1/(p-1) > 1$ according to the assumption. □

At the same time we have proved the following:

Theorem 3.7. *Let $(\mathbf{u}^N)$ satisfy the uniform estimates (2.25)–(2.27). Then there exists a subsequence $\mathbf{u}^{N'}$ and there exist*

$$\mathbf{u} \in L^{p_1}(I; W_0^{1,p_1}(\Omega)^n) \cap L^2(I; L^2(\Omega)^n)$$

and

$$\nu \in L_\omega^\infty(Q_T; M(\mathbb{R}^{n^2}))$$

such that

$$\mathbf{u}^{N'} \to \mathbf{u} \quad in\ L^2(Q_T) \quad for\ p_1 > \frac{2n}{n+2} \tag{3.17}$$

and the couple $(\mathbf{u}, \nu)$ *solves the problem (NonNIF) in the following sense:*

$$\int_{Q_T} [-u_i \frac{\partial \varphi_i}{\partial t} - u_j u_i \frac{\partial \varphi_i}{\partial x_j} + e_{ij}(\varphi) \int_{\mathbb{R}^{n^2}} \tau_{ij}(\mathbf{e}(\lambda)) d\nu_{\mathbf{x},t}(\lambda) - f_i \varphi_i] dx dt$$
$$= \int_\Omega u_{0i} \varphi_i dx \quad \forall\, \varphi \in C^1(I; \mathcal{V}),\ \varphi(T) = 0\ . \tag{3.18}$$

Moreover,

$$\nabla u(x,t) = \int_{\mathbb{R}^{n^2}} \lambda d\nu_{x,t}(\lambda) \quad a.e. in\ \ Q_T\ . \tag{3.19}$$

Definition 3.8. The couple $(\mathbf{u}, \nu)$ is called *Young measure-valued solution.*

4. New Estimates Force the Young Measure to Become Dirac

4.1. *The other assumptions and main theorem*

Let us assume

$$\begin{aligned} &\tau_{ij} \in C^1(\mathbb{R}^{n^2})\ ; \\ &\mathbf{u}_0 \in W^{1,2}(\Omega)^n\ ; \\ &\mathbf{f} \in \begin{cases} L^2(I; L^2(\Omega)^n) \text{ for } p \geq 2 \\ L^{p'}(I; L^{p'}(\Omega)^n) \text{ for } p < 2\ , \quad p' = \dfrac{p}{p-1}\ . \end{cases} \end{aligned} \tag{4.1}$$

Let L be a positive real number, $\Gamma_j = \partial\Omega \cap \{x_j = 0\}$ and $\Gamma_{j+n} = \partial\Omega \cap \{x_j = L\}$. In the rest of this paper, we restrict our considerations to problem $(\text{NonNIF})_{\text{per}}$ which can be formulated as follows:

We find $\mathbf{u}, p$ satisfying (1.1), (1.2) in $\Omega = (0, L)^n$ with the nonlinear stress tensor (1.4). Instead of (1.8) we consider

$$\begin{cases} \mathbf{u}|_{\Gamma_j} = \mathbf{u}|_{\Gamma_{j+n}}\ ; \quad p|_{\Gamma_j} = p|_{\Gamma_{j+n}}\ ; \\ \dfrac{\partial \mathbf{u}}{\partial x_k}|_{\Gamma_j} = \dfrac{\partial \mathbf{u}}{\partial x_k}|_{\Gamma_{j+n}} \quad \forall\, j, k = 1, 2, \dots, n\ . \end{cases} \tag{4.2}$$

Together with (1.5), (1.6), we will assume the existence of a scalar potential ϑ to τ_{ij}, i.e.,

$$\tau_{ij}(\mathbf{e}) = \frac{\partial \vartheta(\mathbf{e})}{\partial e_{ij}} \quad \forall\, i, j = 1, 2, \dots, n\ ; \tag{4.3}$$

$$\vartheta(0) = 0\ , \quad \vartheta \geq 0\ ; \tag{4.4}$$

$$\frac{\partial^2 \vartheta(\mathbf{e})}{\partial e_{kl} \partial e_{ij}} \xi_k^l \xi_i^j \geq c_3\, (1 + |\mathbf{e}|)^{p-2}\, |\xi|^2 \quad \text{with } c_3 > 0\ . \tag{4.5}$$

It follows from the strong monotonicity condition (4.5) that $p_1 = p$ (compare with (1.6)). Of course, the condition (4.2) is accompanied by the changes in the definitions of the function spaces ($L^p_{per}(\Omega)$, $W^{1,p}_{per}(\Omega), \ldots$ are considered instead of $L^p(\Omega)$, $W^{1,p}(\Omega), \ldots$),[b] but all previous results hold henceforth. For the exact definition and notion, see, for example, Temam.[21]

Put $V_p = \{\varphi \in W^{1,p}(\Omega)^n = (W^{1,p}_{per}(\Omega)^n)\,;\, \operatorname{div}\varphi = 0\}$.

Definition 4.1. Let $\mathbf{u}_0$ and $\mathbf{f}$ satisfy (4.1). The function

$$\mathbf{u} \in L^p(I; V_p) \cap L^\infty(I; H) \tag{4.6}$$

is called *the weak solution* to the problem $(\text{NonNIF})_{\text{per}}$ if $\mathbf{u}$ satisfies

$$\left(\frac{d\mathbf{u}}{dt}, \varphi\right) + \int_\Omega u_j \frac{\partial u_i}{\partial x_j} \varphi_i \, dx + \int_\Omega \tau_{ij}(\mathbf{e}) e_{ij}(\varphi)\, dx = (\mathbf{f}, \varphi) \tag{4.7}$$

a.e. in I for every $\varphi \in V_p$.

Theorem 4.2 (Main). *Let $\Omega = (0, L)^n$ be a cube in $\mathbb{R}^n$, $n = 2$ or 3. Let $\mathbf{u}_0$ and $\mathbf{f}$ satisfy (4.1). It holds:*

(i) if $p > 3n/(n+2)$, then the Young measure-valued solution $(\mathbf{u}, \nu)$ is the weak solution of problem $(\text{NonNIF})_{\text{per}}$, it means
  (a) measure $\nu_{\mathbf{x},t}$ is a Dirac one which lives on $\nabla\mathbf{u}(\mathbf{x}, t)$ for a.e. $(\mathbf{x}, t) \in Q_T$;
  (b) the equality (4.7) is fulfilled.

(ii) if $p \geq 1 + 2n/(n+2)$, then the weak solution is
  (a) unique;
  (b) regular, i.e., $\mathbf{u} \in L^\infty(I; W^{1,2}(\Omega)^n)$ and $L^2(I; W^{2,2}(\Omega)^n)$.

Remark 4.3. Notice that

$$\text{“}\mathbf{x} \in \Omega \mapsto \left(u_j \frac{\partial u_i}{\partial x_j} v_i\right)(\mathbf{x})\text{”} \in L^1(\Omega) \text{ if and only if } p \geq \frac{3n}{n+2}$$

for $\mathbf{u}$, $\mathbf{v} \in V_p$ and

$$\text{“}(\mathbf{x}, t) \in Q_T \mapsto \left(u_j \frac{\partial u_i}{\partial x_j} v_i\right)(\mathbf{x}, t)\text{”} \in L^1(Q_T) \text{ if and only if } p \geq 1 + \frac{2n}{n+2}$$

for $\mathbf{u}$, $\mathbf{v} \in L^p(I; V_p)$. These bounds correspond to that ones mentioned in the above theorem.

Before proving Theorem 4.2, it is convenient to introduce some notations and helpful assertion. Let us denote

$$D^{(2)}\mathbf{u} = \sum_{|\alpha|=2} D^\alpha \mathbf{u}\,; \tag{4.8}$$

$$|\mathbf{u}|_{1,r} = \|\nabla\mathbf{u}\|_r \quad \forall\, r > 1\,; \tag{4.9}$$

$$\mathcal{I} = \int_\Omega (1 + |\mathbf{e}|)^{p-2} \frac{\partial e_{ij}(\mathbf{u})}{\partial x_k} \frac{\partial e_{ij}(\mathbf{u})}{\partial x_k}\, dx\,;\, p > 1\,. \tag{4.10}$$

[b]However, we will decline from the explicit writing of “per.”

The following lemma provides the estimates of some norms by means of $\mathcal{I}$.

Lemma 4.4. *It holds (for* $\mathbf{u}$ *sufficiently regular)*

$$\|D^{(2)}\mathbf{u}\|_2 \le \mathcal{I}^{1/2} \text{ if } p \ge 2 \; ; \tag{4.11}$$

$$\|D^{(2)}\mathbf{u}\|_p \le k_1 \mathcal{I}^{1/2} \left(1 + |\mathbf{u}|_{1,p}\right)^{(2-p)/2p} \text{ if } 1 < p < 2 \; . \tag{4.12}$$

Moreover, for $p \ge 2$

$$|\mathbf{u}|_{1,3p} \le k_2 \left(1 + |\mathbf{u}|_{1,p}^p + \mathcal{I}\right)^{1/p} \text{ if } n = 3 \; ; \tag{4.13}$$

and

$$\|\mathbf{u}\|_{2,2} \le k_3 \left(1 + |\mathbf{u}|_{1,p}^p + \mathcal{I}\right)^{1/p} \text{ (independently of } n) \; . \tag{4.14}$$

Proof. The inequality (4.11) follows immediately from the definition of $\mathcal{I}$ and the Korn inequality (2.24), (4.12):

$$\begin{aligned}
\|D^{(2)}\mathbf{u}\|_p^p &\le \text{according to Korn inequality (2.24)} \\
&\le c \int_\Omega \left(\frac{\partial}{\partial x_k} e_{ij}(\mathbf{u}) \frac{\partial}{\partial x_k} e_{ij}(\mathbf{u}) \right)^{p/2} dx \\
&= c \int_\Omega \left\{ (1 + |\mathbf{e}|)^{p-2} \frac{\partial e_{ij}}{\partial x_k} \frac{\partial e_{ij}}{\partial x_k} \right\}^{p/2} (1 + |\mathbf{e}|)^{-\frac{p}{2}(p-2)} \, dx \\
&\le \text{ with Hölder inequality} \\
&\le c \mathcal{I}^{p/2} \left(\int_\Omega (1 + |\mathbf{e}|)^p \, dx \right)^{(2-p)/2} .
\end{aligned}$$

It can be checked that

$$\|(1 + |e|)^{p/2}\|_{1,2} \le c(1 + |\mathbf{u}|_{1,p}^p + \mathcal{I})^{1/2} \; . \tag{4.15}$$

(4.13) then follows from the imbedding $W^{1,2}(\Omega) \hookrightarrow L^6(\Omega)$ ($n = 3$) and the Korn inequality. (4.14) is a direct consequence of

$$\|\mathbf{u}\|_{2,2} \le c \| \, (1 + |\mathbf{e}|)^{p/2} \, \|_{1,2} \; . \tag{4.16}$$

The proof of Lemma 4.14 is complete. ☐

4.2. *Idea of the proof*

Our objective is to find a chosen subsequence $\mathbf{u}^{N'}$ of $\mathbf{u}^N$ (Galerkin approximations) such that

$$\int_{Q_T} \tau_{ij}(\mathbf{e}(\mathbf{u}^{N'})) e_{ij}(\varphi) \, dxdt \underset{N' \to \infty}{\longrightarrow} \int_{Q_T} \tau_{ij}(\mathbf{e}(\mathbf{u})) e_{ij}(\varphi) \, dxdt \tag{4.17}$$

for every smooth function φ which is divergence free.

Thanks to our choice of the basis we can multiply the rth equation of the Galerkin system (2.23) by $c_r^N(t)\lambda_r$. After using (2.15) and summing up over r, $r = 1, 2, \dots, N$, we get

$$\frac{1}{2}\frac{d}{dt}|\mathbf{u}^N|_{1,2}^2 + \int_\Omega u_j^N \frac{\partial u_i^N}{\partial x_j}(A\mathbf{u}^N)_i\,dx + \int_\Omega \tau_{ij}(\mathbf{e}(\mathbf{u}^N))e_{ij}(A\mathbf{u}^N)\,dx = (\mathbf{f}, A\mathbf{u}^N)\,. \tag{4.18}$$

Because of

(i) $A\mathbf{u} = -\Delta\mathbf{u} \quad \forall\ \mathbf{u} \in D(A)$ for space periodic problem;

(ii) $\displaystyle\int_\Omega u_j^N \frac{\partial u_i^N}{\partial x_j}(-\Delta u_i^N)\,dx = \int_\Omega \frac{\partial u_j^N}{\partial x_k}\frac{\partial u_i^N}{\partial x_j}\frac{\partial u_i^N}{\partial x_k}\,dx$

(iii) $\displaystyle\int_\Omega \tau_{ij}(\mathbf{e}(\mathbf{u}^N))e_{ij}(A\mathbf{u}^N)\,dx = \int_\Omega \frac{\partial^2\vartheta(\mathbf{e}(\mathbf{u}^N))}{\partial e_{kl}\partial e_{ij}}\frac{\partial e_{kl}(\mathbf{u}^N)}{\partial x_s}\frac{\partial e_{ij}(\mathbf{u}^N)}{\partial x_s}\,dx$

we have

$$\begin{aligned}\frac{1}{2}\frac{d}{dt}(1 + |\mathbf{u}|_{1,2}^2) &+ \int_\Omega \frac{\partial^2\vartheta}{\partial e_{kl}\partial e_{ij}}\frac{\partial e_{kl}}{\partial x_s}\frac{\partial e_{ij}}{\partial x_s}\,dx \\ &= -\int_\Omega \frac{\partial u_j}{\partial x_k}\frac{\partial u_i}{\partial x_j}\frac{\partial u_i}{\partial x_k}\,dx - (\mathbf{f}, \Delta\mathbf{u})\,.\end{aligned} \tag{4.19}$$

For simplicity, we have started to omit the index N.

Multiplying (4.19) by $\left(1 + |\mathbf{u}|_{1,2}^2\right)^{-\lambda}$, where λ is a positive parameter, $\lambda \neq 1$, using (4.5) and integrating over $(0, T)$, we finally get

$$\begin{aligned}&\frac{1}{2(1-\lambda)}\left(1 + |\mathbf{u}(T)|_{1,2}^2\right)^{1-\lambda} + \int_0^T \left(1 + |\mathbf{u}|_{1,2}^2\right)^{-\lambda}\mathcal{I}\,dt \\ &\quad \le c_{\mathbf{u}_0} + \int_0^T \left(1 + |\mathbf{u}|_{1,2}^2\right)^{-\lambda}|\mathbf{u}|_{1,3}^3\,dt + \int_0^T \left(1 + |\mathbf{u}|_{1,2}^2\right)^{-\lambda}(\mathbf{f}, \Delta\mathbf{u})\,dt\,.\end{aligned} \tag{4.20}$$

Here $c_{\mathbf{u}_0} = \frac{1}{2(1-\lambda)}\left(1 + |\mathbf{u}_0|_{1,2}^2\right)^{1-\lambda}$. The forcing term can be estimated by means of (4.11) – if $p \ge 2$ – or of (4.12) – if $p < 2$. Let us demonstrate it only in the second case. Thus

$$\begin{aligned}\int_0^T &\left(1 + |\mathbf{u}|_{1,2}^2\right)^{-\lambda}(\mathbf{f}, \Delta\mathbf{u})dt \le \int_0^T \left(1 + |\mathbf{u}|_{1,2}^2\right)^{-\lambda}\|\mathbf{f}\|_{p'}\|D^{(2)}\mathbf{u}\|_p \\ &\le (4.12) \\ &\le k_1\int_0^T \left(1 + |\mathbf{u}|_{1,2}^2\right)^{-\lambda}\|\mathbf{f}\|_{p'}\mathcal{I}^{1/2}\left(1 + |\mathbf{u}|_{1,p}\right)^{(2-p)/2p}dt \\ &\le \text{ with Young inequality} \\ &\le k_1\varepsilon\int_0^T \mathcal{I}\left(1 + |\mathbf{u}|_{1,2}^2\right)^{-\lambda}dt + c(\varepsilon)\int_0^T \left(1 + |\mathbf{u}|_{1,2}^2\right)^{-\lambda}\|\mathbf{f}\|_{p'}^2\left(1 + |\mathbf{u}|_{1,p}\right)^{(2-p)/p}.\end{aligned}$$

The first term is transported to the left-hand side with a small coefficient and the second term is estimated by means of Hölder inequality:

$$\begin{aligned}
&\int_0^T \|\mathbf{f}\|_{p'}^2 \left(1+|\mathbf{u}|_{1,p}\right)^{\frac{2-p}{p}}\, dt \\
&\quad \le \left(\int_0^T (1+|\mathbf{u}|_{1,p})\, dt\right)^{\frac{2-p}{p}} \left(\int_0^T \|\mathbf{f}\|_{p'}^{\frac{p}{p-1}}\right)^{\frac{2(p-1)}{p}} \\
&\quad \le \text{ according to (2.26), (4.1)} \le c_f\,.
\end{aligned}$$

Let us denote by $\mathcal{K} = \left(1+|\mathbf{u}|_{1,2}^2\right)^{-\lambda}\mathcal{I}$. The key inequality then has the form

$$\frac{1}{2(1-\lambda)}\left(1+|\mathbf{u}(T)|_{1,2}^2\right)^{1-\lambda} + \int_0^T \mathcal{K}\, dt \le c(\mathbf{u}_0,\mathbf{f}) + \int_0^T \left(1+|\mathbf{u}|_{1,2}^2\right)^{-\lambda} |\mathbf{u}|_{1,3}^3\, dt\,. \tag{4.21}$$

Let us note that as far as $\lambda = 1$, the first term of (4.21) is then replaced by $\log\left(1+|\mathbf{u}|_{1,2}^2\right)$. In those cases when we are able to subdue the convective term

$$\int_0^T \left(1+|\mathbf{u}|_{1,2}^2\right)^{-\lambda} |\mathbf{u}|_{1,3}^3\, dt\,, \tag{4.22}$$

we can dispose in the least of the estimate

$$\int_0^T \mathcal{K}\, dt \le \text{const.} \tag{4.23}$$

Notice that the time term provides some information only when $\lambda \le 1$. As it will be shown later, it is possible to obtain from (4.23) the strong convergence of $\nabla\mathbf{u}^N \to \nabla\mathbf{u}$ in $L^{\tilde{p}}(Q_T)$ for certain $\tilde{p} \in (1,p)$, which gives the a.e. convergence of $\nabla\mathbf{u}^N$ at least for some subsequence. The following classical lemma then provides the convergence in (4.17):

Lemma 4.5. *Let $f_N : Q_T \to \mathbb{R}$ be integrable for every N and let*

(i) $\lim f_N(y)$ exists for a.e. $y \in Q_T$ and is finite
(ii) $\forall\, \varepsilon > 0\ \exists\, \delta > 0$ such that

$$\sup \int_H |f_N(y)|\, dy < \varepsilon \quad \forall\, H \subset Q_T; |H| < \delta\,.$$

Then

$$\lim_{N\to\infty} \int_{Q_T} f_N(y)\, dy = \int_{Q_T} \lim_{N\to\infty} f_N(y)\, dy\,.$$

5. Proof in Three Dimensions

In this chapter, n is always equal to 3.

Step 1. *The estimate of the convective term (4.22).*

Let us consider three possibilities: (i) $p \ge 3$, (ii) $p \in \langle 2,3)$, (iii) $p \in (1,2)$.

(i) We have by means of Hölder inequality and the imbedding $L^p(\Omega)^n \hookrightarrow L^3(\Omega)^n$

$$\int_0^T \left(1+|\mathbf{u}|_{1,2}^2\right)^{-\lambda} |\mathbf{u}|_{1,3}^3 \, dt \le \tilde{c}(T,\Omega) \left(\int_0^T |\mathbf{u}|_{1,p}^p \, dt\right)^{3/p} .$$

The right-hand side is bounded according to (2.26). In this trivial case, put $\lambda = 0$ and clearly $\mathbf{u}^N$ is uniformly estimated in $L^\infty(I; W^{1,2}(\Omega)^n) \cap L^2(I; W^{2,2}(\Omega)^n)$.

(ii) Because of $2 < 3 < 3p$ resp. $p < 3 < 3p$, we can use

$$|\mathbf{u}|_{1,3} \le c|\mathbf{u}|_{1,2}^{2\frac{p-1}{3p-2}} |\mathbf{u}|_{1,3p}^{\frac{p}{3p-2}} , \tag{5.1}$$

$$|\mathbf{u}|_{1,3} \le c|\mathbf{u}|_{1,p}^{\frac{p-1}{2}} |\mathbf{u}|_{1,3p}^{\frac{3-p}{2}} . \tag{5.2}$$

Along with (4.13) we get

$$\begin{aligned} |\mathbf{u}|_{1,3}^3 \left(1+|\mathbf{u}|_{1,2}^2\right)^{-\lambda} &= |\mathbf{u}|_{1,3}^{3(\alpha+1-\alpha)} \left(1+|\mathbf{u}|_{1,2}^2\right)^{-\lambda} \\ &\le c\left(1+|\mathbf{u}|_{1,p}^p + \mathcal{I}\right)^{Q_1} |\mathbf{u}|_{1,p}^{Q_2} \left(1+|\mathbf{u}|_{1,2}^2\right)^{-\lambda+\frac{3(1-\alpha)(p-1)}{3p-2}} \equiv Y , \end{aligned} \tag{5.3}$$

where $Q_1 := \frac{3(1-\alpha)}{3p-2} + 3\alpha\frac{3-p}{2p}$ and $Q_2 := 3\alpha\frac{(p-1)}{2}$. Let us split Y into two parts, i.e., $Y = Y_1 + Y_2$, where

$$Y_1 = c\mathcal{I}^{Q_1} |\mathbf{u}|_{1,p}^{Q_2} \left(1+|\mathbf{u}|_{1,2}^2\right)^{-\lambda+\frac{3(1-\alpha)(p-1)}{3p-2}} \tag{5.4}$$

and

$$Y_2 = c\left(1+|\mathbf{u}|_{1,p}^p\right)^{Q_1+\frac{Q_2}{p}} \left(1+|\mathbf{u}|_{1,2}^2\right)^{-\lambda+\frac{3(1-\alpha)(p-1)}{3p-2}} . \tag{5.5}$$

If we claim

$$\frac{3(1-\alpha)(p-1)}{3p-2} - \lambda + \lambda Q_1 = 0 \quad \text{(the equation for } \lambda) \tag{5.6}$$

then

$$\begin{aligned} \int_0^T Y_1 \, dt &\le \int_0^T \mathcal{K}^{Q_1} |\mathbf{u}|_{1,p}^{Q_2} \, dt \\ &\le \left(\int_0^T \mathcal{K}^{Q_1\delta} \, dt\right)^{1/\delta} \left(\int_0^T |\mathbf{u}|_{1,p}^{Q_2\delta'} \, dt\right)^{1/\delta'} \end{aligned} \tag{5.7}$$

where $\frac{1}{\delta} + \frac{1}{\delta'} = 1$, $Q_1\delta = 1$ and $Q_2\delta' = p$. From this

$$\alpha = \frac{p}{6}\frac{(3p-5)}{p-1} , \tag{5.8}$$

which enables us[c] to compute λ from (5.6):

$$\lambda = 2\frac{3-p}{3p-5} \, . \tag{5.9}$$

Finally, with Young inequality we get

$$\int_0^T Y_1\,dt \le \varepsilon \int_0^T \mathcal{K}\,dt + K(\varepsilon)\int_0^T |\mathbf{u}|_{1,p}^p\,dt \, . \tag{5.10}$$

Again, the first term is transported to the left-hand side of (4.21) and the second term is bounded according to (2.26) independently of N. It is elementary to verify that $\int_0^T Y_2\,dt$ is also uniformly bounded; indeed, $\frac{3(1-\alpha)(p-1)}{3p-2} - \lambda \le 0$ and $Q_1 + \frac{Q_2}{p} = 1$ and with (2.26) the proof is finished.

(iii) Let us recall that (see Lemma 2.3)

$$W^{1,p}(\Omega) \hookrightarrow L^{\frac{3p}{3-p}}(\Omega) \quad (n=3)$$

which allows us to rewrite (4.12) in the form

$$|\mathbf{u}|_{1,\frac{3p}{3-p}} \le |\mathbf{u}|_{1,p} + k_1 \mathcal{I}^{1/2}\left(1+|\mathbf{u}|_{1,p}^p\right)^{(2-p)/2p} \, . \tag{5.11}$$

Let $p \ge 3/2$. Then $2 < 3 \le 3p/(3-p)$ and $p < 3 \le 3p/(3-p)$ and we use the following interpolation inequalities:

$$|\mathbf{u}|_{1,3} \le c|\mathbf{u}|_{1,2}^{2\frac{2p-3}{5p-6}}|\mathbf{u}|_{1,\frac{3p}{3-p}}^{\frac{p}{5p-6}} \, , \tag{5.12}$$

$$|\mathbf{u}|_{1,3} \le c|\mathbf{u}|_{1,p}^{\frac{2p-3}{p}}|\mathbf{u}|_{1,\frac{3p}{3-p}}^{\frac{3-p}{p}} \, . \tag{5.13}$$

Now, we proceed similarly as in the previous case using (5.12), (5.13) and (5.11) instead of (5.1), (5.2) and (4.13). Denoting

$$Q_1 = \frac{1}{2}\frac{3p\alpha}{5p-6} + \frac{3}{2}\frac{3-p}{p}(1-\alpha) \, ,$$

$$Q_2 = 3\frac{2p-3}{p^2}(1-\alpha) + \frac{2-p}{p}Q_1$$

and claiming

$$-\lambda + 3\frac{2p-3}{5p-6}\alpha + \lambda Q_1 = 0 \, , \tag{5.14}$$

we get

$$\int_0^T |\mathbf{u}|_{1,3}^{3(\alpha+1-\alpha)}\left(1+|\mathbf{u}|_{1,2}^2\right)^{-\lambda}$$
$$\le \int_0^T \mathcal{K}^{Q_1}\left(1+|\mathbf{u}|_{1,p}^p\right)^{Q_2}dt + \int_0^T Y_3\,dt \, . \tag{5.15}$$

[c]Notice that $\alpha \in (0,1)$, if $p \in \langle 2,3)$.

The right-hand side of (5.15) can be subdued by means of Hölder and Young inequalities as far as

$$Q_1 = 1 - Q_2 \,, \tag{5.16}$$

which gives

$$\alpha = \frac{(3-p)(5p-6)}{6(2p-3)} \geq 0 \quad \text{if and only if} \quad p \in \langle \frac{3}{2}, 2) \tag{5.17}$$

and

$$1 - \alpha = \frac{p(5p-9)}{6(2p-3)} \geq 0 \quad \text{if and only if} \quad p \in \langle \frac{9}{5}, 2) \,. \tag{5.18}$$

So, we get the lower bound for p, $p \geq 9/5$. From (5.14) we compute

$$\lambda = 2\frac{3-p}{3p-5} \,. \tag{5.19}$$

(Compare with (5.9).) It remains to check that $\int_0^T Y_3 \, dt \leq$ const., where Y_3 mentioned in (5.15) is defined as

$$Y_3 \equiv \left(1 + |\mathbf{u}|_{1,2}^2\right)^{-\lambda + \frac{2p-3}{5p-6}3\alpha} \left(1 + |\mathbf{u}|_{1,p}^p\right)^{\frac{2Q_1}{p} + \frac{2p-3}{p^2}3(1-\alpha)} \,.$$

But $-\lambda + \frac{2p-3}{5p-6}3\alpha \leq 0$ and the exponent standing at the term $\left(1 + |\mathbf{u}|_{1,p}^p\right)$ is equal to 1.

Step 2. $\lambda \leq 1$ *implies uniqueness and regularity.*

From (5.9) resp. (5.19), we get

$$\lambda \leq 1 \quad \text{if and only if} \quad p \geq \frac{11}{5} \,. \tag{5.20}$$

In that case, we have from (4.25)

$$\left(1 + |\mathbf{u}^N(t)|_{1,2}^2\right) \leq \text{const.} \,, \tag{5.21}$$

$$\int_0^T \left(1 + |\mathbf{u}^N|_{1,2}^2\right)^{-\lambda} \int_\Omega \left(1 + |\mathbf{e}(\mathbf{u}^N)|\right)^{p-2} \frac{\partial e_{ij}(\mathbf{u}^N)}{\partial x_k} \frac{\partial e_{ij}(\mathbf{u}^N)}{\partial x_k} \, dx \, dt \leq \text{const.} \tag{5.22}$$

From (5.21), it follows not only that

$$\mathbf{u}^N \text{ is uniformly bounded in } L^\infty(I; W^{1,2}(\Omega)^n) \,, \tag{5.23}$$

but also

$$\int_0^T \int_\Omega \left(1 + |\mathbf{e}(\mathbf{u}^N)|\right)^{p-2} \frac{\partial e_{ij}(\mathbf{u}^N)}{\partial x_k} \frac{\partial e_{ij}(\mathbf{u}^N)}{\partial x_k} \, dx \, dt \leq \text{const.} \tag{5.24}$$

which implies with (4.11), (4.13)

$$\mathbf{u}^N \text{ is uniformly bounded in } L^2(I;W^{2,2}(\Omega)^n) \tag{5.25}$$

$$\mathbf{u}^N \text{ is uniformly bounded in } L^p(I;W^{1,3p}(\Omega)^n)\ . \tag{5.26}$$

Now, we can choose a subsequence which we still denote by $\mathbf{u}^N$ and find $\mathbf{u} \in L^p(I;W^{1,3p}(\Omega)^n) \cap L^2(I;W^{2,2}(\Omega)^n)$ such that due to Lemma 2.3

$$\nabla \mathbf{u}^N \to \nabla \mathbf{u} \text{ in } L^2(Q_T)\ .$$

According to Lemma 4.5 and the explanation before this, we are able to do the limiting process in (4.17), which is the key point in proving the existence of the weak solution to the problem $(\text{NonNIF})_{\text{per}}$. Moreover, we have that

$$\nu_{\mathbf{x},t}(\lambda) = \delta(\lambda - \nabla \mathbf{u}(\mathbf{x},t)) \quad \text{for a.e.} \quad (\mathbf{x},t) \in Q_T\ .$$

(Compare (4.17) with (3.18).)

Uniqueness. Let $\mathbf{u}$ and $\mathbf{v}$ be two weak solutions to the problem $(\text{NonNIF})_{\text{per}}$ and let $\mathbf{w} = \mathbf{u} - \mathbf{v}$. Taking $\mathbf{w}$ as the test function in (4.7) we get

$$\begin{aligned}\frac{1}{2}\frac{d}{dt}\|\mathbf{w}\|_2^2 &+ \int_\Omega [\tau_{ij}(\mathbf{e}(\mathbf{u})) - \tau_{ij}(\mathbf{e}(\mathbf{v}))]\, e_{ij}(\mathbf{u}-\mathbf{v})\, dx \\ &= -\int_\Omega w_j \frac{\partial u_i}{\partial x_j} w_i\, dx\ .\end{aligned} \tag{5.27}$$

From (4.5)

$$\int_\Omega [\tau_{ij}(\mathbf{e}(\mathbf{u})) - \tau_{ij}(\mathbf{e}(\mathbf{v}))]\, e_{ij}(\mathbf{u}-\mathbf{v})\, dx \geq c_3 \sum_{i,j=1}^n \int_\Omega |e_{ij}(\mathbf{u}-\mathbf{v})|^2\, dx\ . \tag{5.28}$$

If we use Korn inequality, we finally obtain from (5.27)

$$\frac{1}{2}\frac{d}{dt}\|\mathbf{w}\|_2^2 + c|\mathbf{w}|_{1,2}^2 \leq \|\mathbf{w}\|_4^2 |\mathbf{u}|_{1,2}\ . \tag{5.29}$$

Because of $\mathbf{w}(0) = 0$ and $\|\mathbf{w}\|_4 \leq c\|\mathbf{w}\|_2^{1/4}\|\mathbf{w}\|_6^{3/4}$, we get

$$\frac{1}{2}\|\mathbf{w}(t)\|_2^2 + c\int_0^t |\mathbf{w}|_{1,2}^2\, d\tau \leq c\int_0^t \|\mathbf{w}\|_2^{1/2}\|\mathbf{w}\|_{1,2}^{3/2}|\mathbf{u}|_{1,2}\, d\tau\ .$$

According to

(i) the fact that $\sup_I |\mathbf{u}|_{1,2} < \infty$;

(ii) $\|\mathbf{w}\|_2^{1/2}\|\mathbf{w}\|_{1,2}^{3/2} \leq \frac{\varepsilon}{k_5}\|\mathbf{w}\|_{1,2}^2 + c(\varepsilon)\|\mathbf{w}\|_2^2$;

(iii) Poincaré inequality $\|\mathbf{w}\|_{1,2}^2 \leq k_5|\mathbf{w}|_{1,2}^2$;

(iv) Gronwall inequality

we finally have

$$\|\mathbf{w}(t)\|_2 = 0 \quad \forall\, t \in I\ ,$$

which proves the uniqueness.

Step 3. *The existence of the weak solution for* $p \in \langle\frac{9}{5}, \frac{11}{5})$.

We will consider two situations: (i) $p \in \langle 2, \frac{11}{5})$, (ii) $p \in \langle\frac{9}{5}, 2)$.

In both cases, however, we start from

$$\int_0^T \left(1+|\mathbf{u}|_{1,2}^2\right)^{-2\frac{3-p}{3p-5}} \mathcal{I}\,dt \le \text{const.} \tag{5.30}$$

Let us emphasize that we permanently write down $\mathbf{u}$ instead of $\mathbf{u}^N$.

(i) From (5.30) and (4.11), it is easy to show that

$$\int_0^T \|D^{(2)}\mathbf{u}\|_2^{2\beta}\,dt \le \text{const.} \quad \text{with } \beta = \frac{3p-5}{p+1}\,. \tag{5.31}$$

Notice that $\beta \in \langle\frac{1}{3}, \frac{1}{2})$ on $\langle 2, \frac{11}{5})$. Choose $q \in (1,p)$. The imbedding $W^{2,2}(\Omega) \hookrightarrow W^{1+s,p}(\Omega)$ with $s = (6-p)/2p$, the interpolation inequality

$$\|\mathbf{u}\|_{1+\sigma,p} \le c\|\mathbf{u}\|_{1,p}^{1-\frac{\sigma}{s}} \|\mathbf{u}\|_{1+s,p}^{\frac{\sigma}{s}}$$

with $0<\sigma<s$, (5.31) and the Hölder inequality enable us to find out that

$$\int_0^T \|\mathbf{u}\|_{1+\sigma,p}^q\,dt \le \text{const.} \quad \text{with } \sigma = s\frac{2\beta(p-q)}{q(p-2\beta)}\,. \tag{5.32}$$

Now, let us assume that

$$\mathbf{u}^N \rightharpoonup \mathbf{u} \quad \text{weakly in } L^p(I; W^{1,p}(\Omega)^n)\,. \tag{5.33}$$

According to Lemma 3.2, (2.27) and (5.32), we also have

$$\mathbf{u}^N \to \mathbf{u} \quad \text{strongly in } L^q(I; W^{1,p}(\Omega)^n) \tag{5.34}$$

at least for a chosen subsequence. As a consequence, we obtain

$$\nabla\mathbf{u}^N \to \nabla\mathbf{u} \quad \text{strongly in } L^q(Q_T)\,. \tag{5.35}$$

(ii) From (5.30) and (4.12) it follows

$$\int_0^T |\nabla\mathbf{u}|_{1,p}^2 (1+|\mathbf{u}|_{1,p})^{p-2} \left(1+|\mathbf{u}|_{1,2}^2\right)^{-2\frac{3-p}{3p-5}} dt \le \text{const.} \tag{5.36}$$

Because of $|\mathbf{u}|_{1,2} \le |\mathbf{u}|_{1,p}^{\frac{5p-6}{2p}} |\mathbf{u}|_{1,\frac{3p}{3-p}}^{\frac{3(2-p)}{2p}}$ and $W^{1,p}(\Omega)^n \hookrightarrow L^{\frac{3p}{3-p}}(\Omega)^n$

$$\left(1+|\mathbf{u}|_{1,2}^2\right)^{-2\frac{3-p}{3p-5}} \ge c\left(1+|\mathbf{u}|_{1,p}^{Q_1}|\nabla\mathbf{u}|_{1,p}^{Q_2}\right)^{-1}\,, \tag{5.37}$$

where

$$Q_1 = \frac{5p-6}{p}\frac{2(3-p)}{3p-5}$$

$$Q_2 = \frac{6(2-p)(3-p)}{p(3p-5)} .$$

Finally

$$\int_0^T |\nabla \mathbf{u}|_{1,p}^{\frac{4(5p-9)}{p(3p-5)}} (1+|\mathbf{u}|_{1,p})^{p-2} \left[|\nabla \mathbf{u}|_{1,p}^{-Q_2} + |\mathbf{u}|_{1,p}^{Q_1}\right]^{-1} dt \le \text{const.} \tag{5.38}$$

Denoting $M = \{t \in I; |\nabla \mathbf{u}|_{1,p} > 1\}$, then on M

$$|\nabla \mathbf{u}|_{1,p}^{-Q_2} < 1$$

which provides

$$\int_M |\nabla \mathbf{u}|_{1,p}^{\frac{4(5p-9)}{p(3p-5)}} (1+|\mathbf{u}|_{1,p})^{p-2-Q_1} \le \text{const.} \tag{5.39}$$

But the inequality (5.39) is also true on $I \setminus M$ as $p-2-Q_1 \le 0$ and $|\nabla \mathbf{u}|_{1,p} \le 1$. So, (5.39) is satisfied on the whole interval I which enables us to derive by means of the Hölder inequality that

$$\int_0^T \|\mathbf{u}\|_{2,p}^{2\beta}\, dt \le \text{const.} \quad \text{with } \beta = \frac{p(5p-9)}{2(-p^2+8p-9)} . \tag{5.40}$$

Let us note that $\beta \in (0, \frac{1}{3})$ on $(\frac{9}{5}, 2)$. If $p > 9/5$, then we proceed similarly as in the previous case and we also get (5.35). The case $p = 9/5$ cannot be included.

6. Sketch of the Proof in Two Dimensions

Let us assume in this section that $n = 2$. As the case $p \ge 3$ is trivial we pay our attention to the following two cases: (i) $p \in (2,3)$, (ii) $p \in (1,2)$.

(i) Instead of (5.1), (5.2) and (4.13), we consider

$$|\mathbf{u}|_{1,3} \le c|\mathbf{u}|_{1,2}^{\frac{2}{3}} \|\mathbf{u}\|_{2,2}^{\frac{1}{3}} , \tag{6.1}$$

$$|\mathbf{u}|_{1,3} \le c|\mathbf{u}|_{1,p}^{\frac{p}{3}} \|\mathbf{u}\|_{2,2}^{\frac{3-p}{3}} \tag{6.2}$$

and (4.14). In order to verify the validity of (6.1) and (6.2), respectively, we use:

- the imbeddings $W^{1+\frac{1}{3},2}(\Omega)^n \hookrightarrow W^{1,3}(\Omega)^n$, $W^{1+2\frac{3-p}{3p},p}(\Omega)^n \hookrightarrow W^{1,3}(\Omega)^n$,
- $|\mathbf{u}|_{1+\frac{1}{3},2} \le c|\mathbf{u}|_{1,2}^{2/3} |\mathbf{u}|_{2,2}^{1/3}$, $|\mathbf{u}|_{1+2\frac{3-p}{3p},p} \le c|\mathbf{u}|_{1,p}^{p/3} |\mathbf{u}|_{1+\frac{2}{p},p}^{1-p/3}$,
- the imbedding , $W^{2,2}(\Omega)^n \hookrightarrow W^{1+\frac{2}{p},p}(\Omega)^n$.

The rest is the same as in Sec. 5, step 1(ii). Easily, we find out that

$$\lambda = \frac{3-p}{p-1}\,, \tag{6.3}$$

and as $\lambda \le 1$ if and only if $p \ge 2$, we have not only the existence of the weak solution but also its uniqueness[d] and regularity. Compare with step 2 in Sec. 5. Now, we have already proved Theorem 4.9, (ii).

(ii) Let us replace (5.11) by

$$|\mathbf{u}|_{1,\frac{2p}{2-p}} \le |\mathbf{u}|_{1,p} + k_1 \mathcal{I}^{\frac{1}{2}} \left(1 + |\mathbf{u}|_{1,p}^p\right)^{(2-p)/2p} \tag{6.4}$$

and consequently (5.12), (5.13), by $(p > 6/5)$

$$|\mathbf{u}|_{1,3} \le c|\mathbf{u}|_{1,2}^{2\frac{5p-6}{6(p-1)}} |\mathbf{u}|_{1,\frac{2p}{2-p}}^{\frac{p}{6(p-1)}}\,, \tag{6.5}$$

$$|\mathbf{u}|_{1,3} \le c|\mathbf{u}|_{1,p}^{\frac{5p-6}{3p}} |\mathbf{u}|_{1,\frac{2p}{2-p}}^{\frac{2(3-p)}{3p}}\,. \tag{6.6}$$

Again, it is possible to gain control of the convective term using (6.5), (6.6) and (6.4) with λ given by (6.3) and α expressed by

$$\alpha = \frac{2(p-1)(3-p)}{5p-6}\,.$$

Notice that $\alpha \in (0, 1\rangle$ if and only if $p \in \langle \frac{3}{2}, 2)$. Compare this with (5.17) and (5.18). Thus

$$\int_0^T \left(1 + |\mathbf{u}|_{1,2}^2\right)^{-\frac{3-p}{p-1}} \mathcal{I}\,dt \le \text{const.} \tag{6.7}$$

Similarly, like in Sec. 5, step 3, (ii) we can derive

$$\int_0^T |\nabla \mathbf{u}|_{1,p}^2 \left(1 + |\mathbf{u}|_{1,p}\right)^{p-2} \left(1 + |\mathbf{u}|_{1,2}^2\right)^{-\frac{3-p}{p-1}} dt \le \text{const.} \tag{6.8}$$

The imbedding $W^{1,p}(\Omega) \hookrightarrow L^{\frac{2p}{2-p}}(\Omega)$ along with

$$|\mathbf{u}|_{1,2} \le |\mathbf{u}|_{1,p}^{\frac{2(p-1)}{p}} |\mathbf{u}|_{1,\frac{2p}{2-p}}^{\frac{(2-p)}{p}} \tag{6.9}$$

then provides

$$\int_0^T |\nabla \mathbf{u}|_{1,p}^{\frac{4(2p-3)}{p(p-1)}} \left(1 + |\mathbf{u}|_{1,p}\right)^{p-2} \left[|\nabla \mathbf{u}|_{1,p}^{-\frac{2(2-p)(3-p)}{p(p-1)}} + |\mathbf{u}|_{1,p}^{\frac{4(3-p)}{p}} \right]^{-1} dt \le \text{const.} \tag{6.10}$$

[d] In the proof of unit use $\|\mathbf{w}\|_4 \le c\|\mathbf{w}\|_2^{1/2} \|\mathbf{w}\|_{1,2}^{1/2}$ instead of $\|\mathbf{w}\|_4 \le c\|\mathbf{w}\|_2^{1/4} \|\mathbf{w}\|_6^{3/4}$.

Let $M = \{t \in I; |\nabla \mathbf{u}|_{1,p} > 1\}$. Then

$$\int_M |\nabla \mathbf{u}|_{1,p}^{\frac{4(2p-3)}{p(p-1)}} (1 + |\mathbf{u}|_{1,p})^{p-2-4\frac{3-p}{p}} \leq \text{const.} \tag{6.11}$$

On $I \setminus M$, however, the same is true. Finally,

$$\int_0^T |\nabla \mathbf{u}|_{1,p}^{\frac{4(2p-3)}{p(p-1)}} (1 + |\mathbf{u}|_{1,p})^{p-2-4\frac{3-p}{p}} \leq \text{const.} \tag{6.12}$$

From (6.12) and with the Hölder inequality, it can be computed that

$$\int_0^T \|\mathbf{u}\|_{2,p}^{2\beta} \, dt \leq \text{const.} \quad \text{with } \beta = \frac{p(2p-3)}{(p-1)(6-p)} \,. \tag{6.13}$$

The case $p = 3/2$ must be excluded. The rest is an analog of (5.32)–(5.35).

7. The New Estimates of Time Derivatives of $\mathbf{u}^N$

In this section we complete the results by the *a priori* estimates of the time derivatives of $\mathbf{u}^N$ for the case $n = 3$, $p \in \langle 2, \frac{11}{5})$. If we multiply the rth equation of the Galerkin system by $\dot{c}_r^N(t)$ and sum up, then, we obtain (without the explicit notation of N)

$$\begin{aligned} \|\dot{\mathbf{u}}\|_2^2 &+ \int_\Omega \frac{\partial \vartheta(\mathbf{e})}{\partial e_{ij}} \frac{d e_{ij}(\mathbf{u})}{dt} \, dx \\ &- \int_\Omega f_i \dot{u}_i \, dx + \int_\Omega u_j \frac{\partial u_i}{\partial x_j} \dot{u}_i \, dx = 0 \end{aligned} \tag{7.1}$$

or

$$\begin{aligned} \frac{1}{2} \|\dot{\mathbf{u}}\|_2^2 &+ \frac{d}{dt} (1 + \|\vartheta(\mathbf{e})\|_1) \\ &\leq \|\mathbf{f}\|_2^2 + \int_\Omega |\mathbf{u}|^2 |\nabla \mathbf{u}|^2 \, dx \,. \end{aligned} \tag{7.2}$$

Let $\hat{\lambda} \geq 0$. Multiplying (7.2) by $(1 + \|\vartheta(\mathbf{e})\|_1)^{-\hat{\lambda}}$ and integrating over I, we get

$$\begin{aligned} \int_0^T (1 + \|\vartheta(\mathbf{e})\|_1)^{-\hat{\lambda}} \|\dot{\mathbf{u}}\|_2^2 \, dt &+ \frac{1}{1-\hat{\lambda}} (1 + \|\vartheta(\mathbf{e}(T))\|_1)^{1-\hat{\lambda}} \\ &\leq c(\mathbf{f}, \mathbf{u}_0) + c \int_0^T (1 + \|\vartheta(\mathbf{e})\|_1)^{-\hat{\lambda}} \|\mathbf{u}\|_{\frac{3p}{3-p}}^2 |\mathbf{u}|_{1,\frac{6p}{5p-6}}^2 \, dt \,. \end{aligned} \tag{7.3}$$

As $\|\vartheta(\mathbf{e})\|_1$ is the equivalent norm with $|\mathbf{u}|_{1,p}^p$ our attention in the sequel will be concentrated on the estimate of

$$\int_0^T \left(1 + |\mathbf{u}|_{1,p}^p\right)^{-\hat{\lambda}} \|\mathbf{u}\|_{\frac{3p}{3-p}}^2 |\mathbf{u}|_{1,\frac{6p}{5p-6}}^2 \, dt \,. \tag{7.4}$$

Theorem 7.1. *Let $\mathbf{u}_0 \in W^{1,p}(\Omega)^n$ and $\mathbf{f}$ satisfy (4.1). Let $n = 3$ and $p \in \langle 2, \frac{11}{5})$. Then*

$$\int_0^T \frac{\|\dot{\mathbf{u}}\|_2^2}{\left(1+|\mathbf{u}|_{1,p}^p\right)^{\hat{\lambda}}}\,dt \le \text{const.} \quad \textit{with } \hat{\lambda} = \frac{2(p^2-5p+8)}{(3p-5)p}\,, \tag{7.5}$$

which implies

$$\int_0^T \|\dot{\mathbf{u}}\|_2^{2\beta}\,dt \le \text{const.}\,, \tag{7.6}$$

where $\beta = \frac{(3p-5)p}{5p^2-15p+16}$.

Proof. Let us start with

$$|\mathbf{u}|_{1,\frac{6p}{5p-6}} \le c\,|\mathbf{u}|_{1,p}^{\frac{5p-8}{4}}\,|\mathbf{u}|_{1,3p}^{\frac{12-5p}{4}}\,, \tag{7.7}$$

which implies along with (4.13)

$$(7.4) \le \int_0^T \left(1+|\mathbf{u}|_{1,p}^p\right)^{\frac{2}{p}-\hat{\lambda}+\frac{5p-8}{2p}}\,\mathcal{I}^{\frac{12-5p}{2p}}\,dt \equiv Y\,. \tag{7.8}$$

Considering (5.30) and $\left(1+|\mathbf{u}|_{1,2}^2\right) \le c\left(1+|\mathbf{u}|_{1,p}^p\right)$, we have

$$Y \le \int_0^T \mathcal{K}^{\frac{12-5p}{2p}}\left(1+|\mathbf{u}|_{1,p}^p\right)^{\frac{7p-12}{2p}}\,dt \tag{7.9}$$

provided that

$$\hat{\lambda} = \frac{2(p^2-5p+8)}{(3p-5)p}\,. \tag{7.10}$$

Using the Hölder inequality on the right-hand side of (7.9) together with (2.26) and (5.30), we get the estimate (7.5). Having it, (7.6) is the direct consequence of (7.5). □

Remark 7.2. Let us state as the example the case $p = 2$. In this situation, it has been derived (see (5.31) and (7.6))

$$\int_0^T \|\mathbf{u}\|_{2,2}^{2/3}\,dt \le \text{const.} \tag{7.11}$$

and

$$\int_0^T \|\dot{\mathbf{u}}\|_2^{2/3}\,dt \le \text{const.} \tag{7.12}$$

For the standard Navier-Stokes equations, the estimate (7.11) have been proved in Ref. 11 for the space periodic problem. For "smooth" bounded domains, the estimates (7.11), (7.12) were proved by Duff.[8,9]

Acknowledgments

We began to write this paper during a study stay of the first author at the University Bonn which was supported by the SFB 256.

References

1. R. A. Adams, **Sobolev Spaces** (Academic Press, 1975).
2. J. Ball, *A version of the fundamental theorem for Young measures*, in **PDE's and Continuum Models of Phase Transitions**, *Lecture Notes in Physics, Vol. 344* (1989) 241–259.
3. H. Bellout, F. Bloom and J. Nečas, *Young measure-valued solutions for non-newtonian Incompressible fluids*, preprint 1991.
4. H. Bellout, F. Bloom and J. Nečas, *Phenomenological behavior multipolar viscous fluids*, to appear in *Quart. Appl. Math.*
5. J. Bergh and J. Löfström, **Interpolation Spaces** (Springer, 1976).
6. P. Constantin and C. Foias, **Navier-Stokes Equations** (Univ. Chicago Press, 1988).
7. Q. Du and M. D. Gunzburger, *Analysis of a Ladyzhenskaya model for incompressible viscous flow*, *J. Math. Anal. Appl.* **155** (1991) 21–45.
8. G. F. D. Duff, *On derivative estimates for the Navier-Stokes equations in dimension* $\mathbb{R}^3$, C. R. Math. Rep. Acad. Sci. Canada **VII** (1990).
9. G. F. D. Duff, *Derivative estimates for the Navier-Stokes equations in a three dimensional region*, *Acta. Math.* **164** (1990) 145–210.
10. R. E. Edwards, **Functional Analysis** (Rinehart and Winston, 1965).
11. C. Foias, C. Guiloppe and R. Temam, *New a priori estimates for Navier-Stokes equations in dimension 3*, *Comm. Partial Differential Equations* **6** (1981) 329–359.
12. D. D. Joseph, **Stability of Fluid Motion** (Springer, 1976).
13. O. A. Ladyženskaya, **The Mathematical Theory of Viscous Incompressible Flow** (Gordon and Beach, 1969).
14. D. C. Leigh, **Nonlinear Continuum Mechanics** (McCraw-Hill, 1968).
15. J. L. Lions, **Quelques méthodes de résolution des problèmes aux limites non linéaires** (Dunod, 1969).
16. J. Málek, J. Nečas and A. Novotný, *Measure-valued solutions and asymptotic behavior of a multipolar model of a boundary layer*, to appear in *Czechoslovak. Math. J.*
17. A. V. Malevsky and D. A. Yuen, *Strongly chaotic non-Newtonian mantle convection* AHPCRC Preprint 91-67 (1991).
18. J. Nečas, *Theory of multipolar viscous fluids*, in **The Mathematics of Finite Elements and Applications** VII, MAFELAP 1990, ed. J. R. Whiteman (Academic Press, 1991) 233–244.
19. J. Nečas, *Sur les normes équivalentes dans* $W_p^k(\Omega)$ *et sur la coercivité des formes formellement positives*, Les presses de l'Univesité de Montréal (1966).
20. J. Simon, *Compact sets in the space* $L^p(0,T;B)$, *Ann. Mat. Pura Appl.* **146** (1987) 65–96.
21. R. Temam, **Infinite-Dimensional Dynamical Systems in Mechanics and Physics** (Springer, 1988).

Chapter 14
Commentary

14.1 Comments on Papers on Fluid Mechanics

Josef Málek, Gregory Seregin, and Vladimír Šverák

Abstract We discuss contributions of Jindřich Nečas to the analysis of the equations of Fluid Mechanics. As in other areas, Jindřich Nečas's interests in this field were broad, ranging from the analysis of the classical Navier–Stokes equations to the design and study of new models.

The partial differential equations of Fluid Mechanics represent a vast area of research, from pragmatic engineering computations, to the theoretical analysis of the Navier–Stokes equations based on advanced PDE and Harmonic Analysis techniques. In addition, the task of finding the right equations for modeling various fluid mechanics phenomena can also be considered as belonging to the area. The ten papers selected for this volume will hopefully give a good idea about the breadth and depth of Jindřich Nečas's contributions to different aspects of the subject.

The main themes of these works can perhaps be considered as follows:

1. Classical Navier–Stokes equations (paper [JN-C176]),
2. Modeling issues and PDE analysis of non-Newtonian and multipolar fluid models (papers [JN-C151, JN-C153, JN-C155, JN-C167, JN-C184, JN-C178, JN-C187, JN-C165]),
3. Attractors and long-time behavior (paper [JN-C167]),
4. Transonic Flows (paper [JN-C136]),
5. Young measure solutions (paper [JN-C167]).

In addition, results from paper [JN-C24] in the section on Linear Elliptic Operators have found important applications in the context of Stokes and Navier–Stokes equations.

We now comment on these themes in more detail.

© Springer Basel 2015

Š. Nečasová et al. (eds.), *Selected Works of Jindřich Nečas*, Advances in Mathematical Fluid Mechanics, DOI 10.1007/978-3-0348-0231-4_14

1. In the classical paper [20] J. Leray proved the local-in-time existence for classical solutions of the Cauchy problem for the 3d Navier–Stokes equation[1]

$$\left.\begin{aligned} u_t + u\nabla u + \nabla p - \Delta u &= 0 \\ \operatorname{div} u &= 0 \end{aligned}\right\} \quad \text{in } \mathbb{R}^3 \times (0,\infty)\,, \tag{14.1}$$

$$u|_{t=0} \;=\; u_0 \qquad \text{in } \mathbb{R}^3\,, \tag{14.2}$$

and also the global-in-time existence of *weak solutions*. The problem of global existence of classical solutions is still open at the time of this writing. Leray suggested that singular solutions starting with smooth initial data might perhaps be constructed by looking at solutions of the form

$$u(x,t) = \frac{1}{\sqrt{T-t}}\, U\left(\frac{x}{\sqrt{T-t}}\right), \tag{14.3}$$

where U is a suitable divergence-free vector field in $\mathbb{R}^3$. Such solutions are perhaps the simplest and most natural candidates for singularities. If we assume that a solution $u(x,t)$ (not necessarily given by (14.3)) of the Navier–Stokes equation is defined in some time interval $(t_1, 0)$ (with $t_1 < 0$), and is smooth for $t < 0$ but develops a singularity at $t = 0$ which is located in space at the point $x = 0$, one can use the invariance of the equations under the scaling

$$\begin{aligned} u(x,t) \to u_\lambda(x,t) &= \lambda u(\lambda x, \lambda^2 t)\,, \\ p(x,t) \to p_\lambda(x,t) &= \lambda^2 p(\lambda x, \lambda^2 t)\,, \end{aligned} \tag{14.4}$$

where $\lambda > 0$, and "zoom in" on the singularity by considering the scaled functions u_λ, p_λ for $\lambda \to 0_+$. It is easy to check that, *assuming* the functions u_λ converge (in a suitable sense) to a limit $\bar{u}$ as $\lambda \to 0_+$, the function $\bar{u}$ has to be a (suitably defined) Navier–Stokes solution of the form (14.3) for a suitable U. In general it is not known that the functions u_λ approach some limiting function as $\lambda \to 0_+$. Also, the question in which sense the limit should be approached is subtle. Leray's suggestion was to search for a solution of the form (14.3) with $U \in L^2(\mathbb{R}^3)$ and $\nabla U \in L^2(\mathbb{R}^3)$. These conditions imply that $U \in L^3(\mathbb{R}^3)$. The Navier–Stokes equation gives the following equation in $\mathbb{R}^3$ for U:

$$-\Delta U + \frac{1}{2}U + \frac{1}{2}x\nabla U + U\nabla U + \nabla P = 0, \qquad \operatorname{div} U = 0\,. \tag{14.5}$$

The main result of [JN-C176] is that the system (14.5) has no non-trivial solutions with $U \in L^3(\mathbb{R}^3)$. This result answers the original question of Leray, but it is

[1] We will use the notation $u\nabla u$ for the vector $\sum_{j=1}^{n} u_k \frac{\partial u}{\partial x_k}$, as is standard in fluid mechanics.

natural to ask if one can have solutions of the form (14.3) for which the energy is locally finite, in the sense that

$$u \in L_t^\infty L_x^2(B_1 \times (t_1, 0)) \cap L_t^2 \dot{H}_x^1(B_1 \times (t_1, 0)) \,, \tag{14.6}$$

where $B_1 = \{x \in \mathbb{R}^3, \ |x| < 1\}$. This question was answered in the negative in the paper [32], which uses a hidden maximum principle discovered in [JN-C176], together with additional ideas. Later work [13] showed that the result for $U \in L^3(\mathbb{R}^3)$ can be in fact proved without the use of the maximum principle. (The main tools in [13] are partial regularity and backward uniqueness for parabolic equations.) This is important when we try to generalize the result to a half-space. In this case we ask the same questions, except our equation is considered in the half-space $\mathbb{R}^3_+$ with the boundary condition $u|_{\partial\mathbb{R}^3_+} = 0$. The analogue of Leray's question under the condition (14.6) is still open because the presence of the boundary and the lack of control of pressure near the boundary prevents the technique of [32] being applied. On the other hand, the result under the assumption $U \in L^3(\mathbb{R}^3_+)$ can be recovered by applying the techniques in [13], as pointed out in [29].

2. In dimension $n = 3$ it is not known whether the Navier–Stokes equations have any predictive power for large data and times scales of order one, due to the fact that the uniqueness of the weak solutions remains unknown. The reason for this unsatisfactory state of affairs may be just our technical inability to prove uniqueness, but it is also conceivable that the problem is more fundamental: perhaps the equations do exhibit non-uniqueness of weak solutions in certain situations and uniqueness simply fails. At the time of this writing we do not know which is the case. Several researchers have proposed various ways to modify the equations so that the uniqueness proofs can be achieved by modifications of existing techniques. These proposals usually involve a modification of the classical linear law for the deviatoric part S of the Cauchy stress tensor,[2]

$$S = 2\mu D, \tag{14.7}$$

where[3] $D = \{D_{ij}\}$ with $D_{ij} = \frac{1}{2}(u_{i,j} + u_{j,i})$. (For simplicity we limit the discussion only to the incompressible case.) There are various ways in which (14.7) can be modified.[4] For example

---

[2] S is often called the viscous stress and denoted by σ_{visc}.

[3] The tensor D is often called "the deformation tensor", although this terminology may not be quite precise. The term "rate of deformation tensor" is also used by some authors.

[4] We note that the need for such modifications comes also from experimental observations of behavior of certain non-Newtonian fluids.

(a) one can assume nonlinear dependences such as $S = 2\mu(|D|^2)D$;
(b) one can consider the dependence of S on higher-order derivatives of u. In this case one usually considers linear dependence, although nonlinear models also make sense, of course;
(c) one can bring other quantities into the equation, for example $S = 2\mu(|D|^2, p)\, D$, where p is pressure.

Similar approaches have a long history, see [24] for a more detailed account.[5]

Concerning (a), J. Nečas (with co-authors—see [JN-C155, JN-C165, JN-C187]) mostly considered proposals related to suggestions of Ladyzhenskaya [19] for the nonlinear dependence, and studied the consequences of such assumptions for the resulting equations. (We note that some of the models also appear in the classical book [21].) Under suitable assumptions one can restore uniqueness (in some cases) and the equations can be viewed as a usual dynamical system. A host of open and difficult questions arises about optimal regularity of solutions, their long-time behavior, etc. The main conclusions involve various critical exponents in the nonlinear viscosity law at which the type of results which can be proved by developed techniques change. For example, there are critical exponents for the existence of weak solutions, and different critical exponents for uniqueness of weak solutions.[6] If we can neglect the convective term $u\nabla u$ in the Newtonian fluids, we obtain the linear Stokes system $u_t + \nabla p - \Delta u = 0, \quad \operatorname{div} u = 0$. In the non-Newtonian case the analogue of this equation will already be non-linear, which brings a host of technical issues which make the problems subtle in many respects. In general, many questions for the non-Newtonian fluid remain open, even at the level of the system without convective terms. For example, the optimal regularity of the solutions remains open.

Concerning (b), J. Nečas and co-authors addressed two fundamental issues:

(i) What is a general way to introduce higher order derivatives into the equations which is still consistent with the principle of frame-indifference (and possibly other basic principles, such as those coming from Thermodynamics)? This is addressed in [JN-C155]. It is not the first work where the dependence on higher derivatives is considered, but it appears to be the first work where a systematic study of the consistency of the dependence on higher spatial derivatives with other basic laws is carried out.

[5]It is worth noting that in turbulence models the relation $S_{\text{turb}} = 2\mu_*|D|D$ proposed in the 1960s by Smagorinskii is often used. Related ideas for turbulence modeling were also proposed by Kolmogorov.

[6]Today, the results can be characterized in terms of two critical parameters. In subcritical case $r > 1 + 2d/(d+2)$ the existence and uniqueness for L^2 data is known and leads via the method of trajectories to the existence of a finite-dimensional global attractor and to the existence of an exponential attractor as well, see for example [5, 23]. The large data existence of a weak solution in the sense of Leray is known for $r > 2d/(d+2)$ using advanced tools, such as the improved properties of Lipschitz truncations of Sobolev and Bochner–Sobolev functions, see [12] or [6].

(ii) What are the mathematical conclusions one can obtain from the new models? Examples include [JN-C153, JN-C151]. For the linear multipolar fluids the main message Nečas wished to convey to the mathematical community was that within such a framework one can develop robust long time and large data theory for relevant problems, see [JN-C153, JN-C151]. This applied not only to fluids, but also to dynamical problems in nonlinear elasticity, see also the paper [26]. It is worth noting that these results pre-date the important developments in P.-L. Lions' monograph [22] that were later on developed further by Feireisl and his co-workers, see [14, 15].

Concerning (c), the possibility that the viscosity may depend on the pressure was already discussed by Stokes. Barus [3], and later on Andrade [2] (see also the book by Bridgman [4]) experimentally showed that the viscosity grows with increasing pressure exponentially. The variation in the density is often negligible in which case one can model these liquids as being incompressible. Flows of such fluids seem to be first analyzed by Renardy [27] (see also Gazzola [17] and Gazzola and Secchi [18]), but their theory is not complete. To date there is no global existence for either steady or unsteady flows of fluids whose viscosity depends purely on the pressure (except for special flows in channels for special boundary conditions). In [JN-C187], the authors show that a suitable additional dependence of the viscosity on the shear rate may help in establishing large data existence of weak solutions to relevant initial and boundary value problems. The result is proved for the spatially periodic problem. The extension to a different boundary value problem has been given in several subsequent studies, see [8] for the most general extension including thermal effects in a bounded domain and [7] for the extension to flows in unbounded domains.

3. For the classical 3d Navier–Stokes equations we do not know whether the dynamics is well-defined, so one cannot really treat the equations as a usual dynamical system. In non-Newtonian situations where uniqueness can be established, it is natural to study the usual notions of Dynamical Systems Theory, such as attractors. In [JN-C178], the authors came with the idea to study long time dynamics of evolutionary systems differently, by analyzing the long time dynamics of (short or ℓ-) trajectories instead of dynamics of points in the phase space. This framework is useful for example if the regularity of the evolutionary system is not available, which is very frequently the case if the leading operator of the problem is nonlinear. This idea resulted in [JN-C178] in the proof of the existence of a finite-dimensional global attractor in the sense of trajectories. Later on, Málek and Pražák [23] observed that once such an attractor is established one can define the set consisting of the endpoint of the trajectories belonging to the attractor and prove that this set is the finite-dimensional attractor of the original problem. Nowadays, the dynamics in the sense of trajectories provides very useful standard tools in the theory of infinite-dimensional dynamical systems, see the monograph [16] for details.

4. The study of transonic flows is an area mathematically quite different from the topics above. In this case the flow is compressible, and one often considers the inviscid limit and potential flows, which are already difficult enough, in contrast with the incompressible case. In the incompressible inviscid case the potential flows are simply the flows given by harmonic functions

$$\Delta h = 0 \tag{14.8}$$

which satisfies the condition $\frac{\partial h}{\partial n} = g$ at the boundaries, where g is a given function (representing flux through the boundary). In the compressible case this equation changes to a nonlinear equation

$$-\operatorname{div} a(|\nabla h|^2)\nabla h = 0 \tag{14.9}$$

for a suitable scalar function a. A characteristic feature of the problem is that the equation loses ellipticity once the speed $|\nabla h|$ becomes large and shock waves can develop. We refer the reader to the work [25] for background. In [JN-C136] the authors obtain results about various passages to the limit for sequences of solutions of transonic flow with certain assumed properties. In general, the area of transonic flows is very difficult and many fundamental questions remain open. Significant breakthroughs have been achieved recently by Chen-Feldman, see for example [9].

5. If existence of a classical solution for a PDE cannot be proved,[7] one can sometimes still prove existence of a *weak solution*. This is the case of the 3d Navier–Stokes equation. The price one often has to pay is that one can say much less about the weak solutions. In particular, in many cases (including the 3d Navier–Stokes equations), one is not able to establish their uniqueness. As usual, there can be two reasons for this: either we know there is no uniqueness, or we simply are unable to decide at present. On the other hand, if uniqueness can be established (as is the case for example with certain linear equations or convex problems), one is clearly working with the right mathematical object. One can take this trend even further, and consider an additional weakening of the notion of the solution in cases for which even the existence of weak solutions is not known (or known to fail). This includes for example the notion of a measure-valued solution. Solutions of this type were introduced for example by L.C. Young [33], and they also appeared in the theory of minimal surfaces under the notion of *varifolds*, see for example [1]. In the 1980s these notions started to be used also in Fluid Mechanics and paper [JN-C167] gives a good example of how one can work with these solutions. In addition, the paper makes important contributions in other directions, in particular by establishing compactness for the velocity gradient in certain cases. Striking examples of non-uniqueness in the context of weak or measure valued for incompressible Euler's solutions were given in the works [28, 30] and, more recently in [10, 11, 31].

[7] The reasons can be for example that we know that classical solutions do not exist, or—somewhat less appealingly—that we are unable to decide whether they exist.

6. In paper [JN-C28] (which is discussed in the context of the linear equation) Nečas proves the estimate

$$||p||_{L^2(\Omega)} \leq C \left(||p||_{H^{-1}(\Omega)} + ||\nabla p||_{H^{-1}(\Omega)}\right) \tag{14.10}$$

for bounded Lipschitz domains Ω. This estimate turned out to be very useful in connection with the Navier–Stokes and Stokes equations (as well as other problems).

Nečas' Works Quoted Above

The references [Cxx] refer to the respective papers listed in Chapter 2 "Complete Bibliography of Jindřich Neças". The references [JN-Cyy] refer to papers [Cyy] that are reproduced either in Chapter 13 "Selected Papers on Fluid Mechanics" or in Chapter 5 "Selected Papers on Linear Elliptic Operators".

[JN-C28] J. Nečas, *Sur les normes equivalentes dans* W_p^k *et sur la coercivité des formes formellement positives*, Sem. math. sup., Université Montréal, 1966, pp. 102–128.

[JN-C136] M. Feistauer and J. Nečas, *Remarks on the solvability of transonic flow problems*, Manuscripta Math., 1988, vol. 61, pp. 417–428.

[JN-C151] J. Nečas, A. Novotný, and M. Šilhavý, *Global solution to the compressible isothermal multipolar fluid*, J. Math. Anal. Appl., 1991, vol. 162, pp. 223–241.

[JN-C153] J. Nečas and A. Novotný, *Some qualitative properties of the viscous compressible heat conductive multipolar fluid*, Comm. Partial Differential Equations, 1991, vol. 16, pp. 197–220.

[JN-C155] J. Nečas and M. Šilhavý, *Multipolar viscous fluids*, Quart. Appl. Math., 1991, vol. 49, pp. 247–265.

[JN-C165] J. Málek, J. Nečas, and M. Růžička, *On the non-Newtonian incompressible fluids*, Math. Models Methods Appl. Sci., 1993, vol. 3, pp. 35–63.

[JN-C167] H. Bellout, F. Bloom, and J. Nečas, *Young measure-valued solutions for non-Newtonian incompressible fluids*, Comm. Partial Differential Equations, 1994, vol. 19, pp. 1763–1803.

[JN-C176] J. Nečas, M. Růžička, and V. Šverák, *On Leray's self-similar solutions of the Navier-Stokes equations*, Acta Math., 1996, vol. 176, pp. 283–294.

[JN-C178] J. Málek and J. Nečas, *A finite-dimensional attractor for three-dimensional flow of incompressible fluids*, J. Differential Equations, 1996, vol. 127, pp. 498–518.

[JN-C184] J. Málek, J. Nečas, and M. Růžička, *On the weak solutions to a class of non-Newtonian incompressible fluids in bounded three-dimensional domains: the case* $p \geq 2$, Adv. Differential Equations, 2001, vol. 6, pp. 257–302.

[JN-C187] J. Málek, J. Nečas, and K.R. Rajagopal, *Global analysis of the flows of fluids with pressure-dependent viscosities*, Arch. Rational. Mech. Anal., 2002, vol. 165, pp. 243–269.

References

[1] W. K. ALLARD, *On the first variation of a varifold*, The Annals of Mathematics, 95 (1972), pp. 417–491.

[2] C. ANDRADE, *Viscosity of liquids*, Nature, 125 (1930), pp. 309–310.

[3] C. BARUS, *Isothermals, isopiestics and isometrics relative to viscosity*, American Jour. Sci., 45 (1893), pp. 87–96.

[4] P. W. BRIDGMAN, *The physics of high pressure*, the MacMillan Company, New York, 1931.

[5] M. BULÍČEK, F. ETTWEIN, P. KAPLICKÝ, AND D. PRAŽÁK, *On uniqueness and time regularity of flows of power-law like non-Newtonian fluids*, Mathematical Methods in the Applied Sciences, 33 (2010), pp. 1995–2010.

[6] M. BULÍČEK, P. GWIAZDA, J. MÁLEK, AND A. ŚWIERCZEWSKA-GWIAZDA, *On Unsteady Flows of Implicitly Constituted Incompressible Fluids*, SIAM J. Math. Anal., 44 (2012), pp. 2756–2801.

[7] M. BULÍČEK, M. MAJDOUB, AND J. MÁLEK, *Unsteady flows of fluids with pressure dependent viscosity in unbounded domains*, Nonlinear Anal. Real World Appl., 11 (2010), pp. 3968–3983.

[8] M. BULÍČEK, J. MÁLEK, AND K. R. RAJAGOPAL, *Mathematical analysis of unsteady flows of fluids with pressure, shear-rate, and temperature dependent material moduli that slip at solid boundaries*, SIAM J. Math. Anal., 41 (2009), pp. 665–707.

[9] G.-Q. CHEN AND M. FELDMAN, *Global solutions of shock reflection by large-angle wedges for potential flow*, Annals of mathematics, 171 (2010), pp. 1067–1182.

[10] C. DE LELLIS AND L. SZÉKELYHIDI JR., *On admissibility criteria for weak solutions of the euler equations*, Archive for rational mechanics and analysis, 195 (2010), pp. 225–260.

[11] ———, *The h-principle and the equations of fluid dynamics*, Bulletin of the American Mathematical Society, 49 (2012), pp. 347–375.

[12] L. DIENING, M. RŮŽIČKA, AND J. WOLF, *Existence of weak solutions for unsteady motions of generalized Newtonian fluids*, Ann. Sc. Norm. Super. Pisa Cl. Sci., IX (2010), pp. 1–46.

[13] L. ESCAURIAZA, G. A. SEREGIN, AND V. SVERAK, *$L^{3,\infty}$-solutions of Navier-Stokes equations and backward uniqueness*, Uspekhi Mat. Nauk, 58 (2003), pp. 3–44. Translation in Russian Math. Surveys 58 (2003), no. 2, 211–250.

[14] E. FEIREISL, *Dynamics of viscous compressible fluids*, vol. 26 of Oxford Lecture Series in Mathematics and its Applications, Oxford University Press, Oxford, 2004.

[15] E. FEIREISL AND A. NOVOTNÝ, *Singular limits in thermodynamics of viscous fluids*, Advances in Mathematical Fluid Mechanics, Birkhäuser Verlag, Basel, 2009.

[16] E. FEIREISL AND D. PRAŽÁK, *Asymptotic behavior of dynamical systems in fluid mechanics*, vol. 4 of AIMS Series on Applied Mathematics, American Institute of Mathematical Sciences (AIMS), Springfield, MO, 2010.

[17] F. GAZZOLA, *A note on the evolution of Navier-Stokes equations with a pressure-dependent viscosity*, Z. Angew. Math. Phys., 48 (1997), pp. 760–773.

[18] F. GAZZOLA AND P. SECCHI, *Some results about stationary Navier-Stokes equations with a pressure-dependent viscosity*, in Navier-Stokes equations: theory and numerical methods, R. Salvi, ed., vol. 388 of Pitman Res. Notes Math. Ser., Harlow, 1998, Longman, pp. 31–37.

[19] O. A. LADYZHENSKAYA, *The mathematical theory of viscous incompressible flow*. Second English edition, revised and enlarged. Translated from the Russian by Richard A. Silverman and John Chu. Mathematics and its Applications, Vol. 2 Gordon and Breach, 1969.

[20] J. LERAY, *Sur le mouvement d'un liquide visqueux emplissant l'espace*, (French) Acta mathematica, 63 (1934), pp. 193–248.

[21] J.-L. LIONS, *Quelques méthodes de résolution des problèmes aux limites non linéaires*, Dunod, 1969.

[22] P.-L. LIONS, *Mathematical topics in fluid mechanics. Vol. 2*, vol. 10 of Oxford Lecture Series in Mathematics and its Applications, The Clarendon Press Oxford University Press, New York, 1998. Compressible models, Oxford Science Publications.

[23] J. MÁLEK AND D. PRAŽÁK, *Large time behavior via the method of ℓ-trajectories*, J. Differential Equations, 181 (2002), pp. 243–279.
[24] J. MÁLEK AND K. R. RAJAGOPAL, *Mathematical issues concerning the Navier-Stokes equations and some of its generalizations*, in Evolutionary equations. Vol. II, Handb. Differ. Equ., Elsevier/North-Holland, Amsterdam, 2005, pp. 371–459.
[25] C. S. MORAWETZ, *Non-existence of transonic flow past a profile*, Communications on Pure and Applied Mathematics, 17 (1964), pp. 357–367.
[26] J. NEČAS AND M. RŮŽIČKA, *Global solution to the incompressible viscous-multipolar material problem*, J. Elasticity, 29 (1992), pp. 175–202.
[27] M. RENARDY, *Some remarks on the Navier-Stokes equations with a pressure-dependent viscosity*, Comm. Partial Differential Equations, 11 (1986), pp. 779–793.
[28] V. SCHEFFER, *An inviscid flow with compact support in space-time*, Journal of Geometric Analysis, 3 (1993), pp. 343–401.
[29] G. SEREGIN, *On smoothness of $L^{3,\infty}$-solutions to the Navier-Stokes equations up to boundary*, Mathematische Annalen, 332 (2005), pp. 219–238.
[30] A. SHNIRELMAN, *On the nonuniqueness of weak solution of the Euler equation*, Communications on Pure and Applied Mathematics, 50 (1997), pp. 1261–1286.
[31] L. SZÉKELYHIDI JR. AND E. WIEDEMANN, *Young measures generated by ideal incompressible fluid flows*, Archive for Rational Mechanics and Analysis, 206 (2012), pp. 333–366.
[32] T.-P. TSAI, *On Leray's self-similar solutions of the Navier-Stokes equations satisfying local energy estimates*, Archive for Rational Mechanics and Analysis, 143 (1998), pp. 29–51.
[33] L. C. YOUNG, *Lectures on the calculus of variations and optimal control theory, Foreword by Wendell H. Fleming*, Saunders, Philadelphia, (1969).

Josef Málek
Faculty of Mathematics and Physics, Charles University in Prague, Sokolovská 83, 186 75 Prague 8, Czech Republic
email: malek@karlin.mff.cuni.cz

Gregory Seregin
Mathematical Institute, University of Oxford, 24-29 St. Giles',
Oxford OX1 3LB, UK
email: Gregory.Seregin@maths.ox.ac.uk

Vladimír Šverák
School of Mathematics, University of Minnesota, 206 Church St. SE,
Minneapolis, MN 55455, USA
email: sverak@math.umn.edu

Part VII
Recollections, Correspondence and Photographs

Chapter 15
Comments and Recollections by Collaborators and Students

15.1 My Memories of My Friend Jindřich Nečas

Philippe G. Ciarlet

I began to know Jindřich Nečas by his name around 1968 when I came across his profoundly influential book "Les Méthodes Directes en Théorie des Equations Elliptiques" (published by Masson in 1967), which, following a very well established tradition among famous Eastern European mathematicians (at least until the 1970s), he had the elegance to write in French. Incidentally, another, more "mathematical", of the reasons I so much like this book is that it was the first to consider domains with only a Lipschitz-continuous boundary, instead as with a "sufficiently smooth" boundary.

I first met Jindřich in person in Plzen in 1978, where we both attended a conference during the week of 04–08 September (that was my first trip to Eastern Europe; crossing the border, i.e., the no-man's land between Czechoslovakia and West Germany, by car was quite an experience!). We immediately sympathised from the very beginning of the conference, and held many wildly interesting conversations about just about everything during that week.

I then invited him twice to the University Pierre et Marie Curie and to the Ecole Normale Supérieure, for 3 months each time, during which he greatly enjoyed giving series of lectures, as part of our doctoral programme, to brilliant students coming not only from the University, but also from the Ecole Polytechnique and the Ecole Normale Supérieure. These lectures were published (again in French!) in 1989 under the title "Ecoulements de Fluide - Compacité par entropie", thanks in particular to the lasting efforts of Sylvie Mas-Gallic.

I vividly remember that, on a very cold morning (about minus ten degrees, which is quite unusual in Paris), I waited for him at Gare de l'Est, where he arrived at 7:20 am by a direct train from Prague via Stuttgart. When I greeted him, he only said: "I am coming from another world"... We then took a taxi to his home in Paris,

© Springer Basel 2015

Š. Nečasová et al. (eds.), *Selected Works of Jindřich Nečas*, Advances in Mathematical Fluid Mechanics, DOI 10.1007/978-3-0348-0231-4_15

the "Hotel Parisiana", 4 rue Tournefort, ideally located in the heart of the Latin Quarter.

During his stays in Paris, we became very close friends. Once I invited him to my home for dinner. I played "Die Moldau" by Smetana in his honor, and tears came to his eyes... Besides, I had the great pleasure to collaborate with him and, as a result, to co-author four papers with him. It was a real joy to work with him. I remember how hard-working he was (this really impressed me!) and also that he would always write drafts of lectures or papers with a not well-sharpened pencil... It was very pleasant and lively to work together, with moments of pure laughter together. For instance, at some point we proposed that our paper "unilateral problems in nonlinear, three-dimensional elasticity" have as a subtitle "Or how to properly carry a bottle of Slivovitz" (but this proposal was eventually turned down by the publisher...).

Jindřich knew how to convey his passion for Mathematics, a passion that I fully share! He was a great mathematician, an inspiring colleague, always extraordinarily generous with his ideas, and above all a warm and caring friend. It is an honor to have known, collaborated with, and become a friend of, such an eminent and noble mind.

Philippe G. Ciarlet
City University of Hong Kong, 83 Tat Chee Avenue, Kowloon Tong, Hong Kong
email: mapgc@cityu.edu.hk

15.2 My Memories of My Supervisor Professor Jindřich Nečas

Pavol Quittner

Jindřich Nečas supervised both my diploma and Ph.D. theses at the Charles University in Prague. Due to the tradition and rules at the Charles University, I knew already at the beginning of the third year of my (5-year-long) undergraduate study that he would supervise my diploma thesis. Although I did not have any knowledge in partial differential equations at that time, I was asked to follow his advanced course on nonlinear elliptic problems. The course was followed not only by his undergraduate and Ph.D. students but also by many of his colleagues and collaborators and the level of the course was definitely not adapted to undergraduate students. In spite of this fact, the course was very interesting and attractive for me. This was mainly due to the charisma of Jindřich Nečas and the style of his lectures. Before he provided us with technical details of the proofs, he always motivated the problem under study, explained a lot of related phenomena and interlaced his talk with many funny stories and comments whose connections to the topic of his talk were often quite unexpected.

I can still remember the first of his lectures in October 1979. This was a period of planned economy and we had to fulfil the so-called 5-year plans which often had to be corrected. Jindřich Nečas started his talk with an outline of the whole course (consisting of many points) and then he immediately mentioned that the outline would probably be like a plan which means that it would not be fulfilled .... Having said so he started to develop an entire theory of ill-posedness of planning and he continued smoothly with interesting observations and facts about the theory of cognition, Mao Tse Tung, Stalin, philosophy, philosophers and their relation to mathematicians ("philosophers consider notions which are not well defined; they often discuss tautologies but they are not aware of it ..."). After this entrée, Jindřich Nečas started to present beautiful mathematics, soon moving the audience to enthusiasm.

The supervision of my diploma thesis did not consist just of interviews, courses and participation in several schools and conferences. Once Jindřich Nečas had decided about the topic of my diploma thesis, he provided me with several very recent papers and my first task was to give a talk on them in his scientific seminar. The seminar was held regularly on Tuesday mornings and the talks were up to 3 h long. In spite of this length, my talk had to be continued for 3 weeks and, due to many questions and comments, it helped me a lot to gain more insight and also understand connections to other problems.

Once Jindřich Nečas was sure that I understand the topic, he posed me several questions and let me work on them. He only read the final version of my diploma thesis and this was also the case of my Ph.D. thesis. Due to the absence of computers and typesetting software, such practice was common at that time (theses and papers had been typeset just once) and now, as a supervisor of various theses, I really regret that those times are over . . . .

During the first interview of my Ph.D. study Jindřich Nečas created a list of 12 thick monographs that I was supposed to study (and pass the corresponding exam within the subsequent 2 years). He also asked me which foreign languages I speak. After he learned that I spoke Russian, English and German, he shook his head and told me: "Well, you have to learn also French and Italian, at the least . . . ". Of course, soup is never eaten as hot as it is cooked. After some time, when I visited my supervisor and complained about the difficulty with one section in one of those 12 books, he calmed me saying: "Don't worry. Nobody in Prague understands these issues so who could examine you on that . . . ". I also have to say that I have not learned Italian yet and my French is also far from being satisfactory.

Jindřich Nečas was interested in many topics. He was able to become enthusiastic about some problem very fast but sometimes his enthusiasm faded out also quite rapidly. Unfortunately, this was also the case of the topic of my Ph.D. thesis. In spite of this fact and my leave from Prague (caused by family reasons), he continued to support me in various ways until the end of his fruitful life.

Nowadays I have been spending much time with Liouville-type theorems for elliptic and parabolic problems. Doing so, I always have to recollect Jindřich Nečas and his interest in Liouville theorems. In fact, precisely at the time when he became my supervisor, he had a big hope that Liouville theorems could be very helpful in proving the regularity of weak solutions of elliptic systems. In addition, his most famous result on the regularity of solutions of the Navier–Stokes equations can also be considered as a Liouville theorem.

The abilities of Jindřich Nečas to forget about real life problems and to enjoy the beauty of mathematics were very infectious and I am thankful for this infection.

Pavol Quittner
Comenius University, Mlynska dolina, SK-84248 Bratislava, Slovakia
email: quittner@fmph.uniba.sk

15.3 My Memories of Professor Jindřich Nečas

Miloslav Feistauer

At the International Mathematical Congress that took place in 1900 in Paris, German mathematician David Hilbert presented probably one of the most important mathematical lectures of all time. He formulated 23 mathematical problems, which were in his opinion fundamental for the development of mathematics in the twentieth century.

One of the mathematicians who contributed substantially to the solution of Hilbert's problems was the Czech mathematician Prof. Jindřich Nečas. Namely, his results play an important role in the solution of the 19th problem, which can be formulated as the question: "Are the solutions of regular variational problems always necessarily analytical?" and the 20th problem: "General problem of boundary values". Both problems are closely related to each other and can be simply formulated as the question of the existence of solutions of boundary value problems and their regularity.

Prof. Nečas was born in Prague on 14th December 1929, but since 1932 lived in the town of Mělník, where he attended a high school, which he finished in 1948. Then he studied mathematics at the Charles University in Prague, where he graduated after 4 years. After 1 year of assistantship at the Czech Technical University in Prague, he started Ph.D. study under the supervision of Prof. Ivo Babuška. The influence of Babuška on J. Nečas was very deep, causing his student to always be oriented to the solution of difficult fundamental problems.

In the year 1956, J. Nečas finished his Ph.D. study and started to work in the Mathematical Institute of the Czechoslovak Academy of Sciences. In 1960 he became the head of the department of partial differential equations. Moreover, he simultaneously worked at the Faculty of Mathematics and Physics of the Charles University, where he was the head of the department of mathematical analysis in the years 1965–1972.

Very soon J. Nečas became a distinguished specialist recognized in Czechoslovakia as well as abroad. However, his political orientation (against the communist normalization regime after 1968) prevented him from traveling abroad and was therefore not fully appreciated even in his native country until after 1990. He spent the last years of his life partially at the Charles University and partially in DeKalb in the USA at Northern Illinois State University, where he passed away on 5th December 2002 after a long and debilitating disease.

It is impossible to speak in detail about his vast scientific work, which consists of more than 200 papers and 11 monographs. Allow me only to characterize several periods of his most important scientific activity.

The first period is represented by the theory of linear elliptic PDE's crowned by his famous monograph "Les Methodes Directes..." published in 1967. J. Nečas started to write this book in 1964 during his visit in Nancy, where he was invited by J.L. Lions for a 6 months stay.

Professor Nečas told me that one day Prof. Lions came to his office and asked him what he was writing. When he heard that J. Nečas was writing a book, he told him "Yes, yes, it is an excellent idea to write a book. It will help you to order in your knowledge, . . ., but be assured that nobody will read your book . . . there are already too many books on PDE's." In this case the opinion of Prof. Lions was absolutely wrong, because this book became soon very famous and was considered as a "Bible" for specialists in partial differential equations in spite having been written in French. On the other hand, this fact motivated a number of mathematicians and students of all nations to learn French in order to be able to study this book. It is possible to say that, thanks to his distinguished results and his book, J. Nečas became "overnight" famous.

The second period of J. Nečas's career can be called nonlinear. It contains the theory of the Fredholm alternative for nonlinear equations, new results in monotone operators, and the Ljusternik-Schnirlemann theory. His results obtained during his elliptic linear and nonlinear period strongly contributed to the solution of Hilbert's 20th problem

The third period represents his contribution to the solution of Hilbert's 19th problem. Namely, the problem of the regularity of solutions to PDE's.

It could seem that Prof. Nečas was a pure mathematician, dealing with abstract theories only. However, it would be a great mistake. He was deeply interested in applied mathematics, where he obtained a number of important results. I can mention, for example, the theory of elastic and elasto-plastic bodies, contact problems, theory of von Karman equations, contact problems with friction, nonlinear thermoelasticity, material optimization, entropy compactification of transonic flow, theory of multipolar fluids.

Prof. Nečas obtained high international evaluation of his work and was awarded a number of prizes. He was an invited plenary speaker at many prestigious conferences. He obtained the title of Honorary Doctor of the Technical University of Dresden and Professor Emeritus of the Charles University. On 28 October 1998 the Czech President Václav Havel awarded him the Medal for Merits of the Czech Republic. At Northern Illinois University he obtained the title of Presidential Research Professor.

I had the luck to get to know Professor Nečas very closely. I studied applied mathematics at the Charles University. I first came in contact with him in the year 1963, when I was recommended to him as a student scientific fellow-worker. At this time he did not work at the Faculty of Mathematics and Physics of the Charles University, but he was employed in the Mathematical Institute of the Academy of Sciences in Prague. After some time, after having finally caught him on the telephone, he deluged me with a list of approximately five books and some papers which I should study and after 3 weeks refer to at his seminar. It frightened me so much that I gave up the possibility to cooperate with him. I did not suspect that after more than 10 years our paths would meet again. In the meantime I graduated in applied mathematics from the Charles University and obtained a Ph.D. in mathematical fluid dynamics under the supervision of Prof. J. Polášek. In 1975 I became a member of the department governed by Prof. Nečas. First I was concerned with numerical analysis and computational fluid dynamics, but the personality of Prof. Nečas attracted me nearly hypnotically and it was quite natural that I started to cooperate with him on problems of partial differential equations, particularly on the problem of transonic flow. During 12 years of our cooperation he had a great influence on me. Everything that I learned from him helped me in the solution of problems of fluid dynamics as well as in numerical mathematics, where I returned later.

During this time we became good friends and I had the opportunity to get to know Prof. Nečas not only as a mathematician. It is possible to say that Prof. Nečas was a real renaissance personality. Besides mathematics and physics he was interested in philosophy, history and particularly in music. (His daughters continued in his footsteps. Šárka is a successful mathematician and Jindřiška is a piano virtuoso and composer.) Prof. Nečas liked to sing and also played violin. Once I had the opportunity to practice music with him in the Mathematical Institute in Oberwolfach, where we played duos for violin and viola. He had a very nice tone, but he had problems with tempo, i.e., with counting. Simply said, he was rather temperamental. He told me that it was much better to play with his daughter Jindřiška—a professional musician. He could play with an arbitrary speed and she accompanied him all the time without problems.

Prof. Nečas did everything with a great enthusiasm and he had the ability to transfer his enthusiasm to his students and colleagues. He was also interested in numerical mathematics and despite the fact that he did not work in this area, he influenced a number of numerical analysts. His seemingly inexhaustible energy and ability to inspire other colleagues played an important role in the creation of the modern school of applied mathematics. His results form a golden treasure of the theory of PDE's and will influence the development of this field for a long time.

Miloslav Feistauer
Faculty of Mathematics and Physics, Charles University in Prague, Sokolovská 83, 18675 Prague 8, Czech Republic
email: feist@karlin.mff.cuni.cz

15.4 Jindřich Nečas and Industrial Mathematics

Tomáš Roubíček*

Although the main focus of Professor Nečas' work was mathematical research and education, he was invariably open to real-world applications and always interested in working with engineers. Besides his participation in the calculations on the Orlík's dam project (as a member of the team of Prof. Ivo Babuška, cf. his contribution in this volume), from the mid of 1970s until the Fall in 1989 he conducted several joint projects with the Central Research Institute of the Škoda company in Plzen. The members of the team in Plzen were engineers and mathematicians led by Jiří Hasman, and involved Petr Martinec, Josef Miler, Josef Voldřich, and Štěpán Zámečník, while in Prague J. Nečas' team involved Jaroslav Haslinger, Ivan Hlaváček, Jan Mandel (before he moved to the United States), and—since about 1983—myself, too.

This collaboration took place on the background of the grey reality of the so-called "Real Socialism" (= an official name, used colloquially rather ironically, for the malfunctioning regime to distinguish it from a theoretically well-functioning but only virtual one) with industry often spending an assigned portion of its budget for academic research of problematic value, just to satisfy some bureaucratic requirements. Yet, rather exceptionally, these particular projects were—as far as I can tell—quite useful for the team in Plzen. And they certainly served as an excellent source of nontrivial and sensible problems for our mathematical team in Prague, which was otherwise quite isolated because chances to travel abroad (especially in the direction West) were obviously quite restricted by the regime.

I participated in the identification of boundary conditions for a steady-state heat-transfer problem. This led to a linear-quadratic but ill-posed optimization problem, and we used essentially the Tikhonov regularization to stabilize it. Of course, we had then only relatively inefficient computers at our disposal. Thus only 2D problems were amenable, even if we used quite coarse discretisation. However, effects of the Tikhonov regularization had surprisingly been seen even on quite coarse meshes of a few dozen triangles in the domain and less than ten boundary segments involved in the boundary condition to be identified. The ultimate requirement of Prof. Nečas was that I had to forget the finite differences and focus on the finite element method. This was initially quite demanding, as I had to code P1-elements directly, without having any FEM-package at my disposal. Of course, it opened an elegant way for merging analysis of the continuous problem with numerics and I am benefiting from this work even today, when occasionally writing articles focused purely on numerical analysis.

*This contribution reflects also memories of RNDr. Petr Martinec and doc.RNDr. Josef Voldřich, CSc., Center of New Technologies, Univ. of West Bohemia.

Another task we worked on was an optimal shape design problem related to a positioning of cooling channels in a casting mould with the goal of ensuring good cooling effects without causing too big temperature gradients (and thus presumably prevent too excessive thermoelastic stresses). This was a nonconvex optimization problem and we used a gradient optimization algorithm with analytical evaluation of the cost-function gradient. It still gives me a headache today to recall the coding of the related adjoint system (without automatic differentiation), used to calculate sensitivity on variation of geometry via varying all node coordinates entering the coefficients of the conductivity matrix.

A piece of my personal memory concerns a very minor mistake I found in some weak formulation devised by Prof. Nečas when I tried to code it. I was fascinated by Prof. Nečas' reaction: although he was obviously slightly surprised that a young engineer without a profound mathematical education would find something not entirely correct in his mathematical notes, he did not react negatively (which itself is not surprising in mathematics) but just the opposite, he made it public and appreciated it more than I would have had expected. And after all, this indeed minor episode was the beginning of his constant support of my professional career.

The so-called Velvet Revolution in 1989 terminated not only the mentioned real socialism but also the collaboration with the Czech industry. In turn, a possibility to join (that time rather tiny) J. Nečas' mathematical modeling group at Faculty of Mathematics and Physics in Prague became more realistic. Eventually, in 1995, after returning from my nearly 2-year stay abroad I got an offer which was hard to refuse and moved from the Academy of Sciences to Charles University. I started regularly teaching and enjoying the new experience at the University. Most of my classes were intimately connected with the main interests of Professor Nečas which, in a general sense, can be perhaps called "Nonlinear partial differential equations with applications". To a great extent, it reflected my long lasting (though occasionally not very intensive, but always extremely influencing) collaboration with him in Prague and later in De Kalb, too. Eventually I used the same title for one of my books that resulted from these teaching and research activities, published in 2005 and, in a second enlarged edition, in 2013 by Birkhäuser in Basel, being dedicated to the memory of Professor Nečas.

Decades passed and I was forgetting these old episodes concerning the industrial research under Prof. Nečas' guidance. Yet, surprisingly, echoes of them have reached me even recently while getting involved in a hydrogen fuel-cell project in the Center of New Technologies at the University of West Bohemia in Plzen where I was happy to meet some members of the above mentioned old team in distinguished positions.

Tomáš Roubíček
Faculty of Mathematics and Physics, Charles University in Prague, Sokolovská 83, 18675 Prague 8, Czech Republic
email: roubicek@karlin.mff.cuni.cz

Chapter 16
Some Notable Personal Correspondence

Letter of Ronald J. Di Perna to J. Nečas

J N

:ditorial Board:
. ALINHAC, 'niversity of Paris 11, 'rsay, France
/. ALLARD, 'uke University, 'urham, NC
I. S. BAOUENDI, 'urdue University, Vest Lafayette, IN
.. BEALS, 'ale University, 'ox 2155 Yale Station, 'ew Haven, CT
. M. BONY, 'niversity of Paris 11, 'rsay, France
. BOUTET DE MONVEL, 'cole Normale Superieure, 'ue d'Ulm, Paris
I. BREZIS, 'niversity of Paris 6, 'lace Jussieu, Paris, France
BROWDER, 'niversity of Chicago, 'hicago, IL
. CAFFARELLI, 'nstitute for Advanced Study, 'rinceton, NJ
:. FEFFERMAN, 'rinceton University, 'rinceton, NJ
:. FOIAS, 'ndiana University, 'loomington, IN
.. FRIEDMAN, 'urdue University, Vest Lafayette, IN
: GREINER, Iniversity of Toronto, 'oronto, Canada M5S 1A4
I. HELFFER, Iniversity of Nantes, 'lantes, France
. HÖRMANDER, Iniversity of Lund, 'und, Sweden
'. KANNAI, Veizmann Institute, 'ehovot, Israel
'. KATO, Iniversity of California, 'erkeley, CA
'. L. LIONS, Iniversity of Paris IL-Dauphine, 'aris, France
\. MAJDA, 'rinceton University, 'rinceton, NJ
I. MELROSE, 'assachusetts Institute of Technology, 'ambridge, MA
'. MIZOHATA, 'yoto University, 'yoto, Japan
:. MORAWETZ, 'ourant Institute, 'ew York University, N.Y., NY
.. NIRENBERG, 'ourant Institute, 'ew York University, N.Y., NY
). OLEINIK, 'oscow University, 'oscow, U.S.S.R.
'. RABINOWITZ, Iniversity of Wisconsin, 'adison, WI
.. ROTHSCHILD, Iniversity of California San Diego, .a Jolla, CA
A. SCHECHTER, Iniversity of California Irvine, rvine, CA
R. SCHOEN, Iniversity of California San Diego, .a Jolla, CA
I. SERRIN, Iniversity of Minnesota, 'inneapolis, MN
I. SJÖSTRAND, Iniversity of Lund, 'und, Sweden
E. STEIN, 'rinceton University, 'rinceton, NJ
A. TAYLOR, SUNY Center at Stony Brook, Stony Brook, NY
R. TEMAM, Iniversity of Paris 11, Orsay, France
F. TREVES, 'utgers University, New Brunswick, NJ
K. UHLENBECK, Iniversity of Chicago, Chicago, IL

Communications in Partial Differential Equations

Editors:
R. Di PERNA
Department of Mathematics
University of California, Berkeley
Berkeley, CA 94720

J. RALSTON
Department of Mathematics
University of California, Los Angeles
Los Angeles, CA 90024

Oct 28, 1987

Dear Professor Nečas,

I am writing with regard your paper "Viscosity method in transonic flow" with M. Feistauer. I have had a great deal of trouble getting it reviewed. One of the referees' took several months before making a report. I apologize for the delay. We usually try to get papers reviewed much faster. In any case your paper is accepted for publication subject to some revisions in the introduction. The referee remarked that the paper would benefit from an additional paragraph or two in the introduction explaining to the reader

published by MARCEL DEKKER, INC. • New York • Basel

© R.J. DiPerna 1987. Reproduced with kind permission from his wife Maria Schonbek.

the meaning and/or interpretation of the hypotheses of your theorems. For example, the hypotheses rule out discontinuities ie shock waves. One is lead to ask: Are the hypotheses physically or mathematically motivated? I agree with the referee's main remark that it should be made clear what is proved and what is not proved by your theorems in the introduction.

At present the introduction is contained in papers 2 and 3 of your manuscript. Please make the revisions which you feel ~~and~~ are consistent with the requests of the referee and send me new copies of page 2 and 3. Please feel free to expand the introduction by several pages to give the reader an idea of what you are going to do in the paper. Pages 4–41 are fine so we can use them for photoreproduction. Thank you for submitting your work.

Sincerely,

Ronald J. DiPerna

Transcription of hand written letter of Jean Leray by editors to J. Nečas

Destinataires :
Prof. J. Nečas, M. Růžička, V. Šverák

15 mars 1996

Chers Collègues,

Vous m'avez écrit le 26 septembre 1995 une lettre dont j'ai aussitôt compris la grande importance. Mais c'est seulement aujourd'hui que j'y réponds ; car j'ai été souffrant et notre médecin nous a ordonné à moi-même et à ma Femme de nous retirer dans une maison de retraite, où mon activité scientifique est minime :

Quietus, app. 39
B.P 191
24 avenue Victor Duruy
44 505 La Baule Cedex France
Tél. 40.62.74.39

Ce déménagement a été lent et pénible. Je n'ai pu sauvegarder ma documentation scientifique. Cela m'empêche de comprendre votre lettre du 26 septembre 1995, malgré tout le temps que j'y ai passé. C'est donc avec un retard monstrueux que j'adresse aujourd'hui à l'Académie, sans pouvoir garantir l'exactitude de son contenu, votre lettre. C'est contraire aux habitudes que j'avais.

Je garantis à l'Académie mon intérêt pour vos documents :
– un projet de Note à l'Académie, en double exemplaire ;
– un texte justifiant ce projet, à paraître ailleurs.

Ma Femme se joint à moi pour adresser nos meilleurs souvenirs au Professeur Nečas. Très cordialement,

J. Leray

P.S. : En 1934, j'avais prouvé qu'en dimension spatiale 3, pour Navier-Stokes, le problème de Cauchy sans bord possède toujours au moins une solution, régulière ou non ; (si elle est régulière, elle est unique ; elle est régulière en dimension spatiale 2).

Je croyais avoir trouvé, en dimension spatiale 3, un moyen de construire, peut-être, une solution non régulière "self-similar".

Votre Note et votre texte plus détaillé réussissent à prouver très ingénieusement qu'une telle solution "self-similar" n'existe pas. Je vous en félicite vivement.

© J. Leray 1996. Reproduced with kind permission from his daughter Françoise Pecker.

T.S.V.P.

J'ai longuement réfléchi, après votre lettre, à ma question de 1934 : pour Navier-Stokes la solution du problème de Cauchy peut-elle être non-régulière ?

A mon grand regret il est sage que je cesse d'y penser. Henri Lebesgue me l'avait conseillé dès 1935 !

Bien à vous,

J. Leray

Letter of Philippe G. Ciarlet to J. Nečas

CENTRE NATIONAL DE LA RECHERCHE SCIENTIFIQUE
ET
UNIVERSITÉ PIERRE et MARIE CURIE (PARIS VI)

LABORATOIRE ASSOCIÉ 189 – ANALYSE NUMÉRIQUE

ADRESSE POSTALE
POSTAL ADDRESS

ANALYSE NUMÉRIQUE, TOUR 55-65, 5e étage
UNIVERSITÉ PIERRE et MARIE CURIE
4, place Jussieu
75230 PARIS CEDEX 05

TÉL. : 336.25.25, Poste 42.98

Philippe G. CIARLET

Le 18 Mai 1982

Le Directeur,

Cher Professeur Nečas,

Merci beaucoup pour votre lettre du 14 Avril. J'ai commencé à lire votre projet d'article en commun, qui me semble effectivement très intéressant ! Alors je vous propose de travailler de la façon suivante : Je vais écrire une nouvelle version, vous l'envoyer (dans environ un mois), etc...

À cet égard, je vous ai envoyé récemment un article "Quelques remarques sur les problèmes d'existence en élasticité non linéaire", où vous trouverez un exemple qui (je crois, je n'ai pas encore vérifié) ressemble beaucoup au vôtre ! (votre exemple 2.9).

En tout cas, je me permets de vous dire que je suis très heureux de cette collaboration ; je vous en remercie très vivement, et je me réjouis à l'idée

© P.G. Ciarlet 1982. Reproduced with kind permission.

de travailler avec vous!

Merci également pour votre invitation pour aller à Prague (et Brno) 1 semaine en 1983. En principe, je crois que c'est possible; je vous confirmerai dans un mois environ.

Par ailleurs, il y a une possibilité de vous inviter 3 mois à l'Ecole Normale Supérieure, en 1983-1984 (pendant l'année scolaire). Seriez-vous intéressé?

Enfin, j'ai récemment démontré la convergence des méthodes incrémentales en élasticité non linéaire (mais avec la restriction $u = 0$ sur tout $\partial\Omega$!). Je vous enverrai l'article.

Ma femme se joint à moi pour vous adresser ~~l'expression de notre~~ meilleur souvenir ~~et~~ avec l'espoir de vous revoir bientôt,

Philippe G. Ciarlet

Letter of Stefan Hildebrandt to J. Nečas

SK

Prof. Dr. S. Hildebrandt
Mathematisches Institut
der Universität Bonn

53 BONN, den January 8th, 1974
Wegelerstraße 10

Prof. Dr. J. Nečas
Mathematisches Institut
der Universität Prag

P r a g , CSSR

Dear Professor Nečas!

At the end of July of this year there will be a meeting on "Calculus of Variations" in Oberwolfach organized by E. Heinz, W. Jäger and myself. We should like to invite you to this conference as well as some of your collaborators you want to recommend to us. Also, I want to ask you whether you are interested to have some contact between your research group in Prague and the group at our Institute in Bonn. Frehse mentioned to me that this might be so. In this case, we should like to invite you and (or) some of your coworkers to come for a longer or shorter period to Bonn. Our group in p.d.e. consists of Frehse, Leis, Reeken, v.Wahl, myself, and a number of younger people. Also, there are always some guests invited, for example we had Giusti, Allard, Osserman and Finn as visitors.
Please, let me know if you are interested.

Sincerely yours,

Stefan Hildebrandt

© S. Hildebrandt 1974. Reproduced with kind permission.

Letter of Mariano Guiquinta to J. Nečas

S/IV

UNIVERSITÀ DEGLI STUDI DI FIRENZE - FACOLTÀ DI INGEGNERIA
ISTITUTO DI MATEMATICA APPLICATA

December 18, 1982

Dear Jindřich

I'm writing you to illustrate an idea I have discussed a little with Enrico Giusti.
We would be interested in a program of collaboration and exchange with your group in Prague (I mean you, Oldřich John, Jana Stará and so on). Maybe there are some possibilities of achieving that, if you are interested. So I thought I'd write you just to ask for your opinion and suggestions.
If you agree in principle, we may inquire with more details on the real possibilities we have in Italy. Maybe you could see what can be done from the side of your institutions.
Many greetings to you and your family. Laura, Cecilia and I wish you also a Merry Christmas and Happy New Year.

Yours

Mariano.

50139 FIRENZE - Via di S. Marta, 3 - tel. 49.91.32 - 47.20.01

© M. Giaquinta 1982. Reproduced with kind permission.

Letter of Karel Rektorys to J. Nečas

JR

Prof. RNDr. Karel Rektorys, DrSc.

v Praze dne 12. května 1998.

Milý Jindřichu,

srdečně zdravím a posílám Ti jednu malou knížku. Povstala při příležitosti oslav mých pětasedmdesátin, kterých jsi se nemohl zúčastnit, neboť jsi byl daleko za mořem. Nevím, do jaké míry Tě bude zajímat, jsou tam většinou drobnosti. Napsal jsem tam krátký článeček – pár slov o sobě – kde pravím, že Ty a Ivo Babuška jste dvě osobnosti, které nejvíce ovlivnily mé odborné zaměření. Nebyl jsem nikdy náboženský založený, ale přece jen trochu věřím na Ducha svatého. A ten mě před mnoha lety osvítil, když jsem louskal Tvoje Les méthodes directes. Je to kniha, která mi dala zabrat, nepřečte se za pět minut. Ale nebýt jí, neuměl bych do dneška nic. Píši v článečku, že bez Tvé knihy bych byl sotva napsal své variační

© K. Rektorys 1998. Reproduced with kind permission from his son Jiří Rektorys.

metody a svou časovou diskretizaci. Neměl jsem napsat „sotva", ale „určitě ne". Takže, Jindříšku, díky!

Je pravděpodobné, že se uvidíme v létě na velké konferenci o parciálních diferenciálních rovnicích, jejíž jsi hlavou. Kdyby ne, buď tak hodný a omluv mě, slíbil jsem našim, že ohlídám vnoučata až oni pojedou ven, přičemž stále se neví, kdy to bude.

Buď tak hodný, pozdravuj rodinku, a kdybych nepřišel, také naše společné přátele.

Moc zdraví

Karol

Letter of William J. LeVeque to J. Nečas

JK

American Mathematical Society
P. O. Box 6248, Providence, Rhode Island 02940
Location: 201 Charles Street at Randall Square
Telephone (401) 272-9500

William J. LeVeque
Executive Director

May 19, 1983

Professor J. Necas
Matematicko-fyzikalini Fakulta
Univerzity karlovy
Katedra Aplikovane Matematiky
Malostranski nam.25
118 00 Praha 1, Czechoslavakia

Dear Professor Necas:

Each year the American Mathematical Society sponsors a summer institute for the purpose of bringing together a group of mathematicians interested in a particular field of mathematics. In 1983 the institute will be devoted to Nonlinear Functional Analysis and Applications and will take place on the campus at the University of California, Berkeley, from July 11-29. Members of the Organizing Committee are Haim Brezis, Felix Browder (chairman), Tosio Kato, J. L. Lions, Louis Nirenberg, and Paul Rabinowitz.

Professor Browder has now asked me to send you this formal invitation and inform you that you will receive a subsistence allowance of $35 for each day of your residence at the institute (for a maximum of 19 days). This amount will cover the room and board costs in the residence halls at UCB. In addition, travel expenses not exceeding the cost of round trip economy air fare from your home base to San Francisco wll be reimbursed, up to a maximum of $1,000. Many special fares are presently available, and if the cost of your transportation will be less than this amount, please write to Mrs. Dorothy Smith in the Meetings Department at the above address and inform her how much is actually required for your travel. As the Conference Coordinator, she will be maintaining all financial records and will have a check for you at the institute to cover the above expenses. Airline travel must be by U.S. flag carrier, as outlined in the enclosed NSF regulations. In the event you are able to obtain funds from other sources to help support your attendance at the institute that information would also be appreciated. Any funds thus saved could be used toward the support of some additional participants.

© W.J. LeVeque 1983. Reproduced with kind permission.

Letter of Roland Glowinski to J. Nečas

INRIA

INSTITUT NATIONAL DE RECHERCHE EN INFORMATIQUE ET EN AUTOMATIQUE

Domaine de Voluceau - Rocquencourt
B.P. 105 - 78153 LE CHESNAY CEDEX (France)
Tél.: (3) 954 90 20 Télex: 697 033 F
R. GLOWINSKI

September 14, 1982

Professor Dr. J. NECAS
Univerzita Karlova
Fakulta Matematicko Fyzikalni
Katedra aplikované matematiky
Malostranské nam. 2/25
110 00 PRAHA 1
Tchécoslovaquie

Cher Professeur Necas,

Je suis très honoré par votre invitation de passer une semaine au Département de Mathématiques de la Charles University à Prague.

Pourriez-vous m'indiquer la période de l'année scolaire 82-83 qui vous parait la plus satisfaisante. En ce qui concerne les frais de transport, je vais voir si le CNRS ou le Ministère des Relations Extérieures peut financer une telle mission.

Amicalement,

R. GLOWINSKI

RG/FW

CENTRE DE SOPHIA ANTIPOLIS - 06560 VALBONNE - Tél.: (93) 74 80 80
CENTRE DE RENNES - IRISA - Campus Universitaire de Beaulieu - 35042 RENNES CEDEX - Tél.: (99) 36 20 00

Établissement Public National - Décret du 27 décembre 1979

© R. Glowinski 1982. Reproduced with kind permission.

Letter of Olga A. Oleinik to J. Nečas

JN

15. XI. 1981 г.

Дорогой Jindrich,

Получила Ваше письмо от 4. XI. Мы очень надеемся, что Вы приедете к нам в феврале. Сейчас я жду соответствующее распоряжение нашего Министерства и тогда Вам будет послано официальное приглашение быть гостем нашей кафедры.

Мне очень понравилось Ваше усовершенствование доказательства теоремы Лиувилля, которое было в нашей заметке в ДАН СССР. Некоторые из Ваших работ мы будем изучать на нашем семинаре.

Я часто с удовольствием вспоминаю конференцию в Братиславе. Жаль только, что было очень мало времени на математические разговоры. Надеюсь, что это можно будет восполнить в Москве. Я сейчас должна заниматься написанием учебника (сроки уже подходят для его сдачи в печать), но времени на это остается очень мало.

Что у Вас нового? Чем занимаетесь? Привет Вашим коллегам.

До скорой встречи

С наилучшими пожеланиями

Ваша О. Олейник

© O.A. Oleinik 1981. Reproduced with kind permission from her nephew Vladimir Ya.Shevchenko.

Letter of John Ball to J. Nečas

JL

Heriot-Watt University

Department of Mathematics

Riccarton, Edinburgh EH14 4AS
Telephone 031-449 5111

Professor R J Knops, BSc, PhD, FRSE
(Head of Department)

Professor J H Williamson, MA, PhD

your ref

our ref

date 20 November 1981

Dear Professor Necas,

Thank you very much for the invitation to visit Prague. I would like to do this if suitable dates can be found. I plan to combine the visit with a week in Pavia, Italy, where I have also been invited. Could you let me know the weeks in the period April 11 – May 22* that are convenient for you. I can then write again in January when the Italians have also replied giving precise dates.

With best wishes,

Yours sincerely,

John Ball

* If there are no possibilities in this period, please suggest other dates.

© J. Ball 1981. Reproduced with kind permission.

Letter of Sergei L. Sobolev to J. Nečas

СССР, 630090
Новосибирск,90,
Институт математики
Сибирского отделения АН СССР

Глубокоуважаемый доктор J.Necas,

Организационный Комитет приглашает Вас принять участие в работе СОВЕТСКО-ЧЕХОСЛОВАЦКОГО Совещания, которое проводится с 28 сентября по 3 октября 1981 года в г. Ереване. Работа Совещания будет проходить по трем основным проблемам:

1. Теория аппроксимаций и кубатурные формулы.
2. Дифференциальные уравнения.
3. Вычислительная математика.

Программа Совещания состоит из пленарных и секционных докладов.

Мы надеемся, что планируемое Совещание будет представлять для Вас интерес.

О Вашем согласии участвовать в Советско-Чехословацком Совещании и название Вашего доклада просим сообщить по адресу:

630090, Новосибирск, 90,
Университетский проспект, 4,
Институт математики СО АН СССР,
Оргкомитет Советско-Чехословацкого Совещания.
Тел.: 65-44-50, 65-44-64

Председатель Оргкомитета
Советско-Чехословацкого Совещания
академик С.Л.Соболев

ИМ СО АН СССР Зак. 122 Тир. 80

© S.L. Sobolev 1981. We tried to trace the rights holder. If someone has legitimate claims, please contact the publisher.

Letter of John A. Nohel to J. Nečas

Alice Peters, Springer Verlag, 69, Heidelberg 1,
Postfach 105280, Spolh. [illegible]!
Od [illegible]

UNIVERSITY OF WISCONSIN–MADISON
MADISON 53706
MATHEMATICS RESEARCH CENTER

610 WALNUT STREET
TELEPHONE (608) 263-2661

April 4, 1978

Dear Professor Nečas:

Thank you very much for your letter (and enclosures) of Feb. 2, 1978. I am delighted that you are interested in spending 4 1/2 months (Jan 15 – May 31, 1980) in Madison. Paul Rabinowitz, Mike Crandall and I, as well as several other colleagues, look forward very much to your stay and we sincerely hope that your visit will be arranged. (I have all the documents from you which I need and) For the moment there is little that can be done until the summer. At that time I will apply for the necessary funding, and I should know by the end of 1978 about these funds, and I will let you know immediately so that you can then proceed with visa formalities and other necessary "red tape". I am confident of success at this end.

© J.A. Nohel 1978. Reproduced with kind permission from his wife Liselotte Karrer Nohel.

If it can still be done, I would like to make three small changes in the manuscript which I sent for the Proceedings of EQUADIFF 4. All the changes occur on page 4 of my manuscript, and I enclose a retyped page 4 which, if possible, I would like to be used to replace the old page 4. Thank you very much for your assistance in this matter.

By the way, I never received any ~~acknowledgement that the manuscript~~ entitled "A forced quasilinear wave equation with dissipation" which I sent for the Proceedings of Equadiff IV on about Oct. 15, 1977, was actually received in Prague! I assume yes, since I didn't hear.

I trust that you are well. Please convey my warm regards and greetings to my many friends. Na shledanou — Yours sincerely, John Nohel

Answer of J. Nečas to John A. Nohel

Prague, April 27, 1978

1978-Mü

Professor John A. Nohel
University of Wisconsin-Madison
Mathematics Research Center
610 Walnut Street
MADISON 53706 - U S A

Dear Professor Nohel:

Thank you very much for your letter of April 4, 1978. I have been really delighted to learn you had suggested my stay in Madison for 4 1/2 months. I firmly hope that our authorities will give me the permission to visit you. It is needless to say that my work during my stay could be useful.

My coworker Dr. M. Kučera has just finished his work dealing with a non-trivial application of P. Rabinowitz's results on the solution of spectral theory to variational inequalities.

As concerns your manuscript to EQUADIFF 4 it was sent to the publishing house. After receiving your correction of the page 4 I arranged its sending as well. In case you would like to contact the publishers yourself, be so kind as to write to the address mentioned below:

Springer Verlag
Mrs. Alice Peters
69 Heidelberg 1
Postfach 105280 - BRD.

I am looking forward to meet you, P. Rabinowitz and M. Crandall at your University in 1980.

Yours sincerely,

J. Nečas

© J. Nečas 1978. Reproduced with kind permission from his daughter Šárka Nečasová.

Letter of Guido Stampacchia to J. Nečas

SK

SCUOLA NORMALE SUPERIORE

PIAZZA DEI CAVALIERI

PISA

CLASSE DI SCIENZE

Pisa, li September 7, 1977

Prof. J. Nečas
Mathematics Institute
Žitná 25
115 67 PRAHA 1
Czechoslovakia

Re: Czechoslovakia-Italy Cultural Cooperation Program.
University Teachers Exchange Academic Year 1976/77.

Dear Professor Necas,

we have received the attached letter from the Italian Department of Education saying:

"... this Department comunicates not being able to consider valid the invitation addressed to Prof. Jindřich NEČAS, of the University of Prague, because the protocol requires that the Czechoslovakians' candidatures be presented by the Czechoslovakians themselves."

I therefore suggest you to forward your candidature in accordance with the above mentioned agreement on cultural cooperation between the two Countries, in order to make our invitation operating.

Yours sincerely,

Guido Stampacchia
Guido Stampacchia

© G. Stampacchia 1977. Reproduced with kind permission from his wife Sara Naldini.

Answer of J. Nečas to Guido Stampacchia

Prof. Dr. J. Nečas, DrSc.

1977-Mü Prague, November 3, 1977

Professor Guido Stampacchia
Scuola Normale Superiore - Classe di Scienze
Piazza dei Cavalieri
P i s a - I T A L Y

Dear Professor Stampacchia:

Very many thanks for your letter of September 7, 1977. I am sorry to inform you that our Ministry of Education adopted a successful practice to use the by you mentioned cultural co-operation between our countries for younger people only. For that reason it is not adequate to forward my proposition.

I hope that I shall be able to come to Pisa in some other time and on another financial basis.

I would be very glad to see you in Prague for a couple of weeks sometimes in autumn 1979. If you think that you could eventually come, please, let me know and I shall send you the official invitation.

Looking forward to your answer I remain,

Yours sincerely,

J. Nečas

© J. Nečas 1977. Reproduced with kind permission from his daughter Šárka Nečasová.

Letter of Jacques-Louis Lions to J. Nečas

COLLÈGE
DE
FRANCE

CHAIRE D'ANALYSE
MATHÉMATIQUE DES SYSTÈMES
ET DE LEUR CONTROLE

J.L. LIONS

Paris, le 5 Avril 1976.

Monsieur le Professeur J. NEČAS
Matematicky Ústav
PRAHA 1
Zitna Ulice 25 TCHECOSLOVAQUIE

Cher Professeur Nečas,

Je vous remercie vivement de votre invitation à Prague, du 22 au 26 Août 1977. Je suis heureux d'accepter.

Je parlerai, ou bien d'Inéquations Quasi Variationnelles ou bien de Problèmes d'Homogénéisation.

Avec encore tous les remerciements, et mes meilleurs souvenirs, veuillez croire, cher Professeur Nečas, à l'expression de mes sentiments les meilleurs.

J.L. LIONS

© J.-L. Lions 1976. Reproduced with kind permission from his son Pierre-Louis Lions.

Answer of J. Nečas to Jacques-Louis Lions

Prof. J. Nečas, DrSc.

sine - 1977-Mü

Prague, April 28, 1977

Monsieur le Professeur
J. L. L i o n s
Collège de France
Chaire d'Analyse Mathématique
des Systèmes et de leur Controle
75231 P a r i s - F R A N C E
Cedex 05 - 11, Place Marcelin-Berthelot

Cher Professeur Lions,

Je vous remercie de votre lettre du 5 Avril 1977 et vous prie de bien choisir selon votre désir un de vos thèmes proposés dans votre lettre.

Actuellement, nous précisons le programme de l'EQUADIFF IV et il faut savoir le thème do votre conférence que vous présenterez à cette occasion.

Avec l'expression de mes sentiments les plus distingués

J. Nečas

© J. Nečas 1977. Reproduced with kind permission from his daughter Šárka Nečasová.

Letter of David R. Owen to J. Nečas

August 16, 1988

Professor J. Necas
Mathematical Institute of the Czechoslovakian Akademy of Sciences
Zitna 25
11567 Praha 1
Czechoslovakia

Dear Professor Necas,

It has been almost 15 years since we met in Prague when I was visiting Jan Kratochvil and Mirek Silhavy. In the meantime I have continued my interests both in plasticity and thermodynamics, and I have tried to follow your research, particularly in the area of plasticity. I have a student who is interested in dynamic problems in plasticity and who came across a reference to the thesis of L. Travnicek on boundary value problems in plasticity. We would be very grateful if you could arrange to have sent to us a copy of the thesis and/or copies of any papers related to the thesis. (We are assuming that Dr. Travnicek worked under your supervision. If not, or if you cannot for any reason carry out our request, please let us know where we can contact Dr. Travnicek ourselves or send us the journal listings of any publications of yours and his that might be related.)

With best regards,

David R. Owen
Professor of Mathematics
Department of Mathematics
Carnegie Mellon University
Pittsburgh, PA 15213

© D.R. Owen 1988. Reproduced with kind permission.

Letter of Klaus Kirchgässner to J. Nečas

PROF. DR. K. KIRCHGÄSSNER
MATHEMATISCHES INSTITUT A

UNIVERSITÄT
STUTTGART

Math. Institut A, 1. Lehrstuhl, Postfach 80 11 40, D-7000 Stuttgart 80

Pfaffenwaldring 57
7000 Stuttgart 80
Telefon (0711) 685-5545
Telefax (0711) 685-5338
kirchgaessner@
rus.uni-stuttgart.de

Professor
Jindřich Nečas
Mathematical Institute
of the Charles University
Sokolovská 83
186 00 Praha 8
Czechoslovakia

21.10.92
kk/si

Dear Jindřich:

this morning, when I discovered that you had already left, I scolded myself not to have stopped in your office a day before. But this was hardly possible, since, after my class until 11.15 h I had a meeting which lasted until 5 p. m. and then the colloquium at 5.15 h p. m. So, I want to thank you for your visit by writing this letter. I enjoyed very much your lectures which I was able to attend, and in particular, our exchange of thought during our memorable lunch.

Have you had any ideas about invited speakers from Eastern Europe for ICIAM 95. Of course, I would be eager to know them.

Looking forward to meet you again in the near future,

(Klaus Kirchgässner)

© K. Kirchgässner 1992. Reproduced with kind permission from Maria Kirchgässner.

Letter of Jens Frehse to J. Nečas

⌐ Jens Frehse, until. Oct. 28, 1976
UNIVERSIDADE DE BRASÍLIA
INSTITUTO DE CIÊNCIAS EXATAS
DEPARTAMENTO DE MATEMÁTICA

22/9/1976

Dear Prof. Nečas,

I need the exact reference of your counterexample $\frac{x_i x_k}{|x|}$ concerning non-linear elliptic systems*. (I've produced a paper on essential selfadjointness and included an appendix on Liouville-theorems for elliptic equations which can be proved by related techniques. It turns out that there is a connection between regularity and Liouville-theorems – your counter-example is also a counterexample for Liouville-type theorems for elliptic systems in high dimensions). Do you know Brazil? It's a mixture between USA and Italy and I fell quite well here. I hope I'll see you next year.

♡-diali saluti, also to your colleagues
Yours Jens Frehse

* Could you send it to Bonn, please?

70.000 – BRASÍLIA – DISTRITO FEDERAL – BRASIL – FONE: 72-0000 – R.: 2-441

© J. Frehse 1976. Reproduced with kind permission.

Letter of Hugo Beiro da Veiga to J. Nečas

INSTITUTO DE FÍSICA E MATEMÁTICA

Lisbon, 24.11.76

Dear Professor Nečas:

I have received your demand of a reprint of my paper "Existence of solutions ...". However the reprints of this paper have not yet arrived, but as soon as I have them I will send one to you.

I am taking the opurtunity of enclosing a note of mine.

Sincerely yours

Hugo da Veiga

(HUGO BEIRÃO DA VEIGA)

© H.B. da Veiga 1976. Reproduced with kind permission.

Letter of Thomas I. Seidmann to J. Nečas

T. Seidman

University of Maryland Baltimore County
umbc 5401 Wilkens Avenue Baltimore, Maryland 21228
Division of Mathematics and Physics Mathematics 301-455-2401 Physics 301-455-2513

3 Jan., '84

Dear Prof. Nečas,

I enclose a copy of the paper which I discussed with you in Nice and Paris a few years ago together with another paper written at about that time and two more recent preprints which I hope you will find of interest. My apologies for not having sent much sooner a copy of the paper, "Boundary inequalities..."

Do send me copies of some of your recent work – and may I hope that you will be better at such correspondence than I have been!

Best wishes for 1984.

Sincerely,

[signature]

© T.I. Seidmann 1984. Reproduced with kind permission.

Letter of Denis Serre to J. Nečas

Université de Saint-Etienne

EQUIPE D'ANALYSE NUMERIQUE

(membre de l' Unité Associée au C.N.R.S. n° 740)

23, rue du Docteur Paul Michelon
42023 Saint-Etienne cedex 2

Saint-Etienne, le 20 Avril 1987

Tél. 77.25.38.97
Telex: UNISTET 300816
Pr D. Serre

V. ref.
N. ref.

Dear Professor J. Nečas

It is a great honnour for me that you want to know my work about Compensated Compactness: I join to the main article some other related papers.

Let me inform you that I study also

- BV solutions for some convenient hyperbolic 2x2 systems.
- compressible Navier - Stokes equations.
- stationnary incompressible N-S equations.
- dynamical systems arising in combustion theory.
- phase transition, by means of the Glimm scheme.

Sincerely yours

Serre

D. Serre.

© D. Serre 1987. Reproduced with kind permission.

Letter of Karel Drbohlav to J. Nečas

JN

DĚKAN
MATEMATICKO-FYZIKÁLNÍ FAKULTY
UNIVERSITY KARLOVY V PRAZE

V Praze dne 11. 4. 1990

Vážený pane docente,

období, které právě prožíváme, je pro mnohé z nás vyvrcholením našeho života a naplněním našich nejvnitřnějších tužeb. Cítíme upřímnou radost z nově nabyté svobody, bojujeme o demokratický zítřek, ale také, naplněni hrůznou a úžasem, se ohlížíme zpět na obludné rozměry zla, které po sobě minulý politický režim zanechal. Škody, jež způsobil, není možno vyčíslit, křivdy, jichž se dopustil, není možno změřit.

Nové vedení MFF UK se snaží, pokud to vůbec lze, alespoň trochu zmírnit dopad všech těchto neblahých skutečností minulých časů. Jsme si vědomi toho, že ve Vašem případě Vám nejenom bylo zabráněno v tom, abyste zcela oprávněně získal vědecko-pedagogickou hodnost univerzitního profesora, ale že jste byl zároveň s tím vystaven i lidskou důstojnost urážejícímu zacházení, ponižujícímu přezírání, nátlaku.

Jménem fakulty se Vám za všechny tyto neblahé události upřímně omlouvám.

Podnikli jsme samozřejmě i všechny potřebné kroky k Vaší úplné rehabilitaci.

V dokonalé úctě

Karel

Prof. RNDr. K. Drbohlav, DrSc.

Vážený pan
doc. RNDr. Jindřich **Nečas**, DrSc.
Matematický ústav Univerzity Karlovy
Karlín

SČt 14-0215-79

© K. Drbohlav 1990. We tried to trace the rights holder. If someone has legitimate claims, please contact the publisher.

Letter of Pierre Grisvard to J. Nečas

SW GRISVARD-
UNIVERSITÉ D'ALGER
FACULTÉ DES SCIENCES
Mathématiques
~~Tél. : 64-67-00 et 64-83-15~~ Alžír

ALGER, le 23/11/72

un peux

Cher Professeur Nečas

Je me permets d'ajouter un mot personnel à la lettre officielle d'invitation de l'Université. Je suis détaché auprès de l'Université d'Alger pour Toute l'année. Il y a ici un groupe assez important d'analystes, certains travaillant sur des problèmes non linéaires, d'autres sur les problèmes elliptiques dans des domaines non réguliers. Nous serions interessés par quelques conférences sur le sujet que vous avez développé au Colloque de Brno cet été (valeurs propres

© P. Grisvard 1972. Reproduced with kind permission from his children Olivier, Béatrice and Etienne Grisvard.

d'opérateurs non linéaires) entre autres.

S'il vous était impossible de venir, nous vous demandons de nous indiquer le nom de quelque mathématicien tchécoslovaque qui pourrait nous parler sur les sujets que vous travaillez.

Recevez s'il vous plaît mes meilleures salutations.

GRISVARD. Pierre.

Letter of Gaetano Fichera to J. Nečas

PROF. GAETANO FICHERA
ORDINARIO NELL'UNIVERSITÀ DI ROMA
VIA PIETRO MASCAGNI 7   TEL. 8310362
ROMA

Rome, December 17,1974

Prof.J.Nečas
Katedra matematické analýzy na matematicko-fyzikálni fakultě
University Kerlovy v Praze
8,Karlín, Sokolovská 83

Dear Professor Necås,

Thank you for your letter of 3.12.74 with your paper dedicated to Picone. Your paper is very interesting indeed. However I would allow myself to submit you two remarks.

1) The interior regularity as well as the boundary regularity for unilateral problems connected with linear elliptic 2nd order systems has been investigated in my paper "Boundary value problem of elasticity with unilateral constraints" published on Vol.VIa/2 of Handbuch der Physik (1972) (see theorems 8.1 & 9.1). The results seem to be exactly like yours. I enclose herewith a copy of my paper, which you could quote - if you like - by adding a note in your paper when correcting the galley proofs.

2) The non-linearity in your problem seems to be given by the circumstance that you assume a non-linear Hooke's law for the linearized strain components. This is something that - in spite of the fact that it occurs in many papers of applied mathematicians - has, all the times, puzzled me very much. In fact, assuming linearized strain components means that the non-linear part of the strain components is negligible. In other words it is assumed that the functions $\frac{\partial u_i}{\partial x_h}\frac{\partial u_j}{\partial x_k}$ may be assumed to be so small that they can be considered as zero. However when you write the Hooke law

$$\tau_{ij} = \frac{\partial W}{\partial e_{ij}}$$

and W is not quadratic, the terms which you have assumed previously as negligible are reconsidered. For instance, let us consider a one--dimensional problem where the linearized strain function is $e = \frac{du}{dx}$ and $\left(\frac{du}{dx}\right)^2 = \left(\frac{du}{dx}\right)^3 = \cdots \simeq 0$. The Hooke law, by using Taylor development of W should be

$$\tau = W'(e) = a_1 e + a_2 e^2 + \cdots$$

where the non-linear terms should be neglected unless you assume that the derivatives of $W(e)$ of order ≥ 3 become infinite for $e = 0$. That is inadmissible for obvious physical considerations.

Best thanks for your Season wishes which I cordially reciprocate.

Yours

© G. Fichera 1974. Reproduced with kind permission from his wife Matelda Colautti Fichera.

Letter of Gerard Tronel to J. Nečas

CENTRE NATIONAL DE LA RECHERCHE SCIENTIFIQUE
ET
UNIVERSITÉ PIERRE et MARIE CURIE (PARIS VI)

LABORATOIRE ASSOCIÉ 189 – ANALYSE NUMÉRIQUE

ADRESSE POSTALE / POSTAL ADDRESS
ANALYSE NUMÉRIQUE, TOUR 55-65, 5e étage
UNIVERSITÉ PIERRE et MARIE CURIE
4, place Jussieu
75230 PARIS CEDEX 05

TÉL. : 336.25.25, Poste 42.98

G. TRONEL

Paris le 18 Novembre 1982,

Monsieur le Professeur NEČAS,

Tout d'abord je vous prie de bien vouloir excuser le retard avec lequel je réponds à vos lettres et je vous remercie d'avoir fait le nécessaire pour que je puisse répondre à votre invitation. J'ai attendu, jusqu'au dernier moment, un avis favorable de notre administration pour avancer d'une semaine mon voyage, je n'ai pas obtenu satisfaction car la semaine du 13 au 20 décembre est une semaine d'examens, je vous propose donc la semaine du 20 au 26 décembre je pense que les premiers jours de cette semaine seront actifs et que je pourrai participer à votre séminaire du mardi. Par ailleurs en ce qui concerne mon voyage le problème du financement n'est pas encore réglé j'ai tout de même acheté mon billet, si vous êtes toujours d'accord j'arriverai à Prague par le vol Air-France le dimanche 19 décembre à 22h30, c'est une heure

© G. Tronel 1982. Reproduced with kind permission.

bien tardive et un jour peu commode mais je n'ai pas pu faire autrement, mon retour est prévu le dimanche 25.

En ce qui concerne les exposés que je pourrais faire figurent des résultats de régularité des solutions de problèmes paraboliques non-linéaires. Peut-être serez vous intéressé par ce qui se fait à Paris dans le cadre de notre Laboratoire : actuellement, sont à la mode, les problèmes hyperboliques linéaires ou non-linéaires avec, bien entendu l'application aux problèmes de contrôle optimal et aux problèmes de perturbations singulières. Il y a aussi des résultats intéressants obtenus par H. BRÉZIS et son équipe je pourrai faire un exposé d'ensemble. Maintenant que les problèmes materiels de ce voyage semblent réglés je vais préparer activement les sujets que je vous propose et je vous écrirai prochainement pour vous donner des informations plus précises sur le contenu des exposés.

Par ailleurs, permettez moi de vous préciser que, par nature, mes goûts en ce qui concerne les conditions matérielles sont assez modestes et je vous serais infiniment reconnaissant de ne prévoir que le strict minimum compatible avec le budget de votre Laboratoire.

En espérant de vous voir et d'avoir le plaisir de parler "mathématiques" avec Vous et Vos collaborateurs je vous prie d'accepter mes remerciements anticipés et de croire à mon amitié.

Tonel.

Letter of Pekka Neittaanmäki to J. Nečas

Jyväskylä 16.2.-85

Dear Professor Nečas

It really nice to meet you in Praha and have an interesting discussion with you.

Enclosed you find some papers you borrowed for me. There are also two preprints on superconvergence.

You suggested a project on bifurcation problems where we could apply your method (for even function etc.). Next summer I'll get a student to test your method in computer. I'll give a report immediately when we have results. Theoretically I have not yet problems.

With best regards
P. Neittaanmäki

© P. Neittaanmäki 1985. Reproduced with kind permission.

Letter of Felix E. Browder to J. Nečas

THE UNIVERSITY OF CHICAGO
DEPARTMENT OF MATHEMATICS
5734 UNIVERSITY AVENUE
CHICAGO · ILLINOIS 60637

December 3, 1982

Professor Jidrich Necas

The American Mathematical Society is sponsoring a Summer Institute on Nonlinear Functional Analysis and its Applications in Berkeley, California on July 11-29, 1983. This is the thirty-first of the series of mathematics research institutes sponsored by the AMS during the summer. On behalf of the organizing committee, Haim Brezis, Felix Browder, Tosio Kato, J. L. Lions, Louis Nirenberg, and Paul Rabinowitz, I write to invite you to participate in the 1983 Institute. Financial support for the Institute has been requested from both the National Science Foundation and the Army Research Office. We hope to pay the living expenses of as many of the participants as possible, and traveling expenses for some. Particularly because of financial restrictions on foreign travel, we urge as many of the participants as possible to obtain travel support from other sources.

The basic format of the Institute will consist of several short series of expository lectures centering primarily upon crucial areas of development in nonlinear problems in partial differential equations (particularly in differential geometry and mathematical physics) and a number of more extensive research seminars on more detailed developments in these directions and developments in nonlinear functional analysis proper.

I attach the prospectus of the Summer Institute as it was approved by the AMS, which serves to give some further indications of the thrust of this meeting. We hope very much that you will be able to participate.

Sincerely yours,

Felix E. Browder

enclosure

© F.E. Browder 1982. Reproduced with kind permission.

Letter of Joseph J. Kohn to J. Nečas

JK

Princeton University DEPARTMENT OF MATHEMATICS
FINE HALL - BOX 37
PRINCETON, NEW JERSEY 08540

October 5, 1973

Professor Jindřich Nečas
Professor Alois Švec

Dear Professors Nečas and Švec:

As you know, throughout the academic year 1972-73 I was in Europe and was very anxious to visit Prague at that time. The American National Academy of Sciences had proposed my name to the Czechoslovakian Academy under the Exchange Program. Only after many inquiries from our Academy, the Czechoslovakian Academy answered by telegram in April 1973 saying that it was impossible to receive me at that time.

When I returned to Princeton in August, I obtained permission from the University to take a short leave during the Spring of 1974 and the U. S. Academy again proposed my visit to Czechoslovakia under the Exchange Program. I am enclosing a copy of the response they received to this request.

Of course I am very sorry that I will not be able to visit you. However, I am pleased that you have so many foreign scientific workers. I sincerely hope that it will be possible for me to visit during the academic year 1974-75.

Sincerely yours,

J. J. Kohn

J. J. Kohn

JJK:mt
Enc.

© J.J. Kohn 1973. Reproduced with kind permission.

Chapter 17
Photographs

17.1 Childhood

J. Nečas at music school, class of violin
© J. Nečas' family. Reprinted with kind permission from his daughter Šárka Nečasová.

© Springer Basel 2015

Š. Nečasová et al. (eds.), *Selected Works of Jindřich Nečas*, Advances in Mathematical Fluid Mechanics, DOI 10.1007/978-3-0348-0231-4_17

J. Nečas with his mother
© J. Nečas' family. Reprinted with kind permission from his daughter Šárka Nečasová.

J. Nečas - abitur
© J. Nečas' family. Reprinted with kind permission from his daughter Šárka Nečasová.

17.2 Jindřich Nečas in the 1960s and 1970s

J. Nečas and S. L. Sobolev - Novosibirsk

© unknown. If someone has legitimate claims, please contact the publisher.

J. Nečas and S. L. Sobolev - Novosibirsk
© unknown. If someone has legitimate claims, please contact the publisher.

J. Nečas and S. L. Sobolev - Novosibirsk
© unknown. If someone has legitimate claims, please contact the publisher.

J. Nečas, S. L. Sobolev and V. N. Maslenikova - Novosibirsk
© unknown. If someone has legitimate claims, please contact the publisher.

J. Nečas and S. L. Sobolev - Novosibirsk
© unknown. If someone has legitimate claims, please contact the publisher.

I. Marek, A. Kufner, J. Nečas and S. L. Sobolev - Novosibirsk
© unknown. If someone has legitimate claims, please contact the publisher.

J. Nečas surrounded by friends
© unknown. If someone has legitimate claims, please contact the publisher.

J. Nečas giving a lecture
© unknown. If someone has legitimate claims, please contact the publisher.

J. Nečas and K. Gröger
© unknown. If someone has legitimate claims, please contact the publisher.

J. Nečas
© unknown. If someone has legitimate claims, please contact the publisher.

J. Nečas in Pisa
© unknown. If someone has legitimate claims, please contact the publisher.

J. Nečas and J. Souček
© unknown. If someone has legitimate claims, please contact the publisher.

17.3 Jindřich Nečas in the 1980s and 1990s

J. Nečas, J. Kurzweil, K. Rektorys and I. Netuka
© Faculty of Mathematics and Physics, Charles University. Reprinted with kind permission.

J. Nečas
© Faculty of Mathematics and Physics, Charles University. Reprinted with kind permission.

J. Nečas and I. Hlaváček
© Faculty of Mathematics and Physics, Charles University. Reprinted with kind permission.

Meeting at the home of Prof. Vejvoda (head of Dept. Evolutionary Differential Equations)
© unknown. If someone has legitimate claims, please contact the publisher.

J. Nečas and J. Polášek
© J. Franců. Reprinted with kind permission.

J. Nečas and W. Jäger
© Faculty of Mathematics and Physics, Charles University. Reprinted with kind permission.

Christmas party
© unknown. If someone has legitimate claims, please contact the publisher.

K. Segeth, J. Kurzweil, I. Netuka and J. Nečas
© Faculty of Mathematics and Physics, Charles University. Reprinted with kind permission.

K. Gröger, J. Nečas, J. Lovíšek
© unknown. If someone has legitimate claims, please contact the publisher.

J. Nečas and J. Málek
© M. Rokyta, Charles University. Reprinted with kind permission.

J. Nečas with collaborator
© unknown. If someone has legitimate claims, please contact the publisher.

J. Nečas
© Faculty of Mathematics and Physics, Charles University. Reprinted with kind permission.

J. Nečas and K. Drbohlav
© Faculty of Mathematics and Physics, Charles University. Reprinted with kind permission.

J. Nečas and J. Neustupa
© Faculty of Mathematics and Physics, Charles University. Reprinted with kind permission.

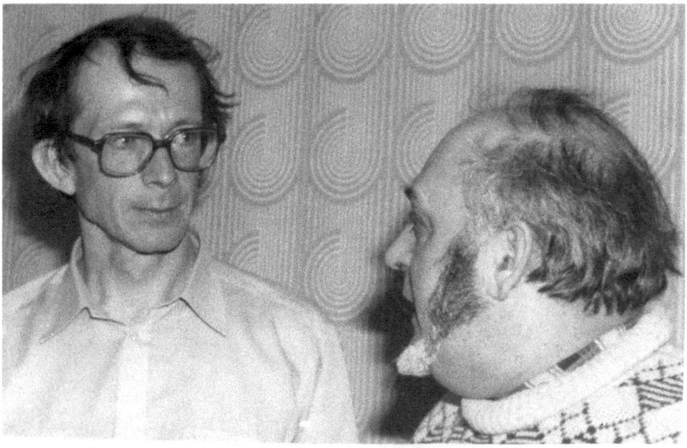

J. Nečas and O. John
© Faculty of Mathematics and Physics, Charles University. Reprinted with kind permission.

J. Nečas and J. Kurzweil
© Faculty of Mathematics and Physics, Charles University. Reprinted with kind permission.

J. Nečas
© Faculty of Mathematics and Physics, Charles University. Reprinted with kind permission.

J. Nečas
© Faculty of Mathematics and Physics, Charles University. Reprinted with kind permission.

Celebration
© unknown. If someone has legitimate claims, please contact the publisher.

J. Nečas
© Faculty of Mathematics and Physics, Charles University. Reprinted with kind permission.

J. Nečas
© unknown. If someone has legitimate claims, please contact the publisher.

President Václav Havel awarding J. Nečas the Order of Merit of the Czech Republic on October 28, 1998
© CTK Fotobanka 1998. Reprinted with kind permission.

J. Nečas and K. Segeth
© Faculty of Mathematics and Physics, Charles University. Reprinted with kind permission.

J. Nečas and P. Drábek - Olomouc 1999
© J. Franců. Reprinted with kind permission.

I. Hlaváček, H. Wu, J. Nedelec, J. Nečas and J. Kačur - Olomouc 1999
© J. Franců 1999. Reprinted with kind permission.

J. Nečas and J. Kačur - Olomouc 1999
© J.Franců 1999. Reprinted with kind permission.

A. Kufner, J. Kačur, I. Hlaváček, K. Kozel and J. Nečas - Olomouc 1999
© J. Franců 1999. Reprinted with kind permission.

V. Mazya, E. Meister, J. Nečas - Olomouc 1999
© J. Franců 1999. Reprinted with kind permission.

17.4 Italy 2000 and USA 2001

A. Sequeira, J. Nečas and H. Morimoto
© A. Robertson 2000. Reprinted with kind permission.

A. Sequeira, H. Beirão da Veiga, J. Nečas and H. Morimoto
© A. Robertson 2000. Reprinted with kind permission.

J. Nečas, G.P. Galdi, A. Robertson and Š. Nečasová
© A. Sequeira 2000. Reprinted with kind permission.

J. Nečas and A. Novotný
© A. Robertson 2000. Reprinted with kind permission.

Nečas becoming Presidential Research professor at Northern Illinois University (NIU), DeKalb
© unknown. If someone has legitimate claims, please contact the publisher.

J. Nečas' office at the Northern Illinois University (NIU), DeKalb
© Šárka Nečasová 2001. Reprinted with kind permission.

MIX
Papier aus verantwortungsvollen Quellen
Paper from responsible sources
FSC® C105338
FSC
www.fsc.org

If you have any concerns about our products,
you can contact us on
ProductSafety@springernature.com

In case Publisher is established outside the EU,
the EU authorized representative is:
Springer Nature Customer Service Center GmbH
Europaplatz 3, 69115 Heidelberg, Germany

Printed by Libri Plureos GmbH
in Hamburg, Germany